Remote Sensing Applications in Ocean Observation

Remote Sensing Applications in Ocean Observation

Editors

Chung-Ru Ho
Antony K. Liu
Xiaofeng Li

MDPI • Basel • Beijing • Wuhan • Barcelona • Belgrade • Manchester • Tokyo • Cluj • Tianjin

Editors

Chung-Ru Ho
National Taiwan Ocean University
Taiwan

Antony K. Liu
National Aeronautics and Space Administration
USA

Xiaofeng Li
Chinese Academy of Sciences
China

Editorial Office
MDPI
St. Alban-Anlage 66
4052 Basel, Switzerland

This is a reprint of articles from the Special Issue published online in the open access journal *Remote Sensing* (ISSN 2072-4292) (available at: https://www.mdpi.com/journal/remotesensing/special_issues/RS_applications_ocean_observation).

For citation purposes, cite each article independently as indicated on the article page online and as indicated below:

LastName, A.A.; LastName, B.B.; LastName, C.C. Article Title. *Journal Name* **Year**, *Volume Number*, Page Range.

ISBN 978-3-0365-6438-8 (Hbk)
ISBN 978-3-0365-6439-5 (PDF)

Contents

Chung-Ru Ho and Antony K. Liu
Preface: Remote Sensing Applications in Ocean Observation
Reprinted from: *Remote Sens.* **2023**, *15*, 415, doi:10.3390/rs15020415 . . . . **1**

Tianxia Jia, Yonglin Zhang and Rencai Dong
A Universal Fuzzy Logic Optical Water Type Scheme for the Global Oceans
Reprinted from: *Remote Sens.* **2021**, *13*, 4018, doi:10.3390/rs13194018 . . . . **13**

Emna Amri, Pierre Dardouillet, Alexandre Benoit, Hermann Courteille, Philippe Bolon, Dominique Dubucq and Anthony Credoz
Offshore Oil Slick Detection: From Photo-Interpreter to Explainable Multi-Modal Deep Learning Models Using SAR Images and Contextual Data
Reprinted from: *Remote Sens.* **2022**, *14*, 3565, doi:10.3390/rs14153565 . . . . **43**

Xiaomeng Geng, Lei Shi, Jie Yang, Pingxiang Li, Lingli Zhao, Weidong Sun and Jinqi Zhao
Ship Detection and Feature Visualization Analysis Based on Lightweight CNN in VH and VV Polarization Images
Reprinted from: *Remote Sens.* **2021**, *13*, 1184, doi:10.3390/rs13061184 . . . . **79**

Tonghan Fei, Binghu Huang, Xiang Wang, Junxing Zhu, Yan Chen, Huizan Wang and Weimin Zhang
A Hybrid Deep Learning Model for the Bias Correction of SST Numerical Forecast Products Using Satellite Data
Reprinted from: *Remote Sens.* **2022**, *14*, 1339, doi:10.3390/rs14061339 . . . . **101**

Lin Dong, Jifeng Qi, Baoshu Yin, Hai Zhi, Delei Li, Shuguo Yang, et al.
Reconstruction of Subsurface Salinity Structure in the South China Sea Using Satellite Observations: A LightGBM-Based Deep Forest Method
Reprinted from: *Remote Sens.* **2022**, *14*, 3494, doi:10.3390/rs14143494 . . . . **121**

Léa Schamberger, Audrey Minghelli, Malik Chami and François Steinmetz
Improvement of Atmospheric Correction of Satellite Sentinel-3/OLCI Data for Oceanic Waters in Presence of *Sargassum*
Reprinted from: *Remote Sens.* **2022**, *14*, 386, doi:10.3390/rs14020386 . . . . **141**

Mingjun He, Shuangyan He, Xiaodong Zhang, Feng Zhou and Peiliang Li
Assessment of Normalized Water-Leaving Radiance Derived from GOCI Using AERONET-OC Data
Reprinted from: *Remote Sens.* **2021**, *13*, 1640, doi:10.3390/rs13091640 . . . . **159**

Shiming Lu, Mingjun He, Shuangyan He, Shuo He, Yunhe Pan, Wenbin Yin and Peiliang Li
An Improved Cloud Masking Method for GOCI Data over Turbid Coastal Waters
Reprinted from: *Remote Sens.* **2021**, *13*, 2722, doi:10.3390/rs13142722 . . . . **175**

Hongwei Zhang, Yuanshuai Zhang, Ziwang Li, Bingyi Liu, Bin Yin and Songhua Wu
Small Angle Scattering Intensity Measurement by an Improved Ocean Scheimpflug Lidar System
Reprinted from: *Remote Sens.* **2021**, *13*, 2390, doi:10.3390/rs13122390 . . . . **195**

Junyi Li, Huiyuan Zheng, Lingling Xie, Quanan Zheng, Zheng Ling and Min Li
Response of Total Suspended Sediment and Chlorophyll-a Concentration to Late Autumn Typhoon Events in The Northwestern South China Sea
Reprinted from: *Remote Sens.* **2021**, *13*, 2863, doi:10.3390/rs13152863 . . . . **215**

Khalid A. Hussein, Khameis Al Abdouli, Dawit T. Ghebreyesus, Pakorn Petchprayoon, Naeema Al Hosani and Hatim O. Sharif
Spatiotemporal Variability of Chlorophyll-a and Sea Surface Temperature, and Their Relationship with Bathymetry over the Coasts of UAE
Reprinted from: *Remote Sens.* **2021**, *13*, 2447, doi:10.3390/rs13132447 . . . . **237**

Shaochun Xu, Shuai Xu, Yi Zhou, Shidong Yue, Xiaomei Zhang, Ruiting Gu, et al.
Long-Term Changes in the Unique and Largest Seagrass Meadows in the Bohai Sea (China) Using Satellite (1974–2019) and Sonar Data: Implication for Conservation and Restoration
Reprinted from: *Remote Sens.* **2021**, *13*, 856, doi:10.3390/rs13050856 . . . . **263**

Hiroshi Kuroda and Takashi Setou
Extensive Marine Heatwaves at the Sea Surface in the Northwestern Pacific Ocean in Summer 2021
Reprinted from: *Remote Sens.* **2021**, *13*, 3989, doi:10.3390/rs13193989 . . . . **279**

Zhihua Zhang, Difeng Wang, Yinhe Cheng and Fang Gong
Long-Term Changes and Factors That Influence Changes in Thermal Discharge from Nuclear Power Plants in Daya Bay, China
Reprinted from: *Remote Sens.* **2022**, *14*, 763, doi:10.3390/rs14030763 . . . . **293**

Mengtian Qiao, Anzhou Cao, Jinbao Song, Yunhe Pan and Hailun He
Enhanced Turbulent Mixing in the Upper Ocean Induced by Super Typhoon Goni (2015)
Reprinted from: *Remote Sens.* **2022**, *14*, 2300, doi:10.3390/rs14102300 . . . . **313**

Zhen Li, Anton Verhoef and Ad Stoffelen
Bayesian Sea Ice Detection Algorithm for CFOSAT
Reprinted from: *Remote Sens.* **2022**, *14*, 3569, doi:10.3390/rs14153569 . . . . **329**

Wenxiang Ding, Caiyun Zhang, Jianyu Hu and Shaoping Shang
Unusual Fish Assemblages Associated with Environmental Changes in the East China Sea in February and March 2017
Reprinted from: *Remote Sens.* **2021**, *13*, 1768, doi:10.3390/rs13091768 . . . . **345**

Heping Liu, Wei Yang, Hao Wei, Chengfei Jiang, Changgen Liu and Liang Zhao
On Characteristics and Mixing Effects of Internal Solitary Waves in the Northern Yellow Sea as Revealed by Satellite and In Situ Observations
Reprinted from: *Remote Sens.* **2022**, *14*, 3660, doi:10.3390/rs14153660 . . . . **361**

Xiaoyu Zhao, Zhenhua Xu, Ming Feng, Qun Li, Peiwen Zhang, Jia You, et al.
Satellite Investigation of Semidiurnal Internal Tides in the Sulu-Sulawesi Seas
Reprinted from: *Remote Sens.* **2021**, *13*, 2530, doi:10.3390/rs13132530 . . . . **379**

Weidong Wang, Robin Robertson, Yang Wang, Chen Zhao, Zhanjiu Hao, Baoshu Yin and Zhenhua Xu
Distinct Variability between Semidiurnal and Diurnal Internal Tides at the East China Sea Shelf
Reprinted from: *Remote Sens.* **2022**, *14*, 2570, doi:10.3390/rs14112570 . . . . **399**

Thomas Meunier, Paula Pérez-Brunius and Amy Bower
Reconstructing the Three-Dimensional Structure of Loop Current Rings from Satellite Altimetry and In Situ Data Using the Gravest Empirical Modes Method
Reprinted from: *Remote Sens.* **2022**, *14*, 4174, doi:10.3390/rs14174174 . . . . **419**

Zhanjiu Hao, Zhenhua Xu, Ming Feng, Qun Li and Baoshu Yin
Spatiotemporal Variability of Mesoscale Eddies in the Indonesian Seas
Reprinted from: *Remote Sens.* **2021**, *13*, 1017, doi:10.3390/rs13051017 . . . . **443**

Shipeng Ying, Hongsong Qu, Shuping Tao, Liangliang Zheng and Xiaobin Wu
Radiation Sensitivity Analysis of Ocean Wake Information Detection System Based on Visible Light Remote Sensing
Reprinted from: *Remote Sens.* **2022**, *14*, 4054, doi:10.3390/rs14164054 . . . . **471**

Zhi Huang, Jianyu Hu and Weian Shi
Mapping the Coastal Upwelling East of Taiwan Using Geostationary Satellite Data
Reprinted from: *Remote Sens.* **2021**, *13*, 170, doi:10.3390/rs13020170 . . . . **487**

Yuhui Li, Yun Qiu, Jianyu Hu, Cherry Aung, Xinyu Lin and Yue Dong
Springtime Upwelling and Its Formation Mechanism in Coastal Waters of Manaung Island, Myanmar
Reprinted from: *Remote Sens.* **2020**, *12*, 3777, doi:10.3390/rs12223777 . . . . **507**

Senyang Xie, Zhi Huang and Xiao Hua Wang
Remotely Sensed Seasonal Shoreward Intrusion of the East Australian Current: Implications for Coastal Ocean Dynamics
Reprinted from: *Remote Sens.* **2021**, *13*, 854, doi:10.3390/rs13050854 . . . . **523**

Chun-Hoe Chow, Yi-Chen Lin, Wee Cheah and Jen-Hua Tai
Injection of High Chlorophyll-a Waters by a Branch of Kuroshio Current into the Nutrient-Poor North Pacific Subtropical Gyre
Reprinted from: *Remote Sens.* **2022**, *14*, 1531, doi:10.3390/rs14071531 . . . . **535**

Zhenyu Sun, Jianyu Hu, Zhaozhang Chen, Jia Zhu, Longqi Yang, Xirong Chen and Xuewen Wu
A Strong Kuroshio Intrusion into the South China Sea and Its Accompanying Cold-Core Anticyclonic Eddy in Winter 2020–2021
Reprinted from: *Remote Sens.* **2021**, *13*, 2645, doi:10.3390/rs13142645 . . . . **549**

Ching-Yuan Lu, Po-Chun Hsu, Quanan Zheng and Chung-Ru Ho
Variations in Flow Patterns in the Northern Taiwan Strait Observed by Satellite-Tracked Drifters
Reprinted from: *Remote Sens.* **2022**, *14*, 2154, doi:10.3390/rs14092154 . . . . **567**

Rodrigo N. Vasconcelos, André T. Cunha Lima, Carlos A. D. Lentini, Garcia V. Miranda, Luís F. Mendonça, Marcus A. Silva, et al.
Oil Spill Detection and Mapping: A 50-Year Bibliometric Analysis
Reprinted from: *Remote Sens.* **2020**, *12*, 3647, doi:10.3390/rs12213647 . . . . **583**

remote sensing

MDPI

Editorial

Preface: Remote Sensing Applications in Ocean Observation

Chung-Ru Ho [1,*] and Antony K. Liu [2]

1 Department of Marine Environmental Informatics, National Taiwan Ocean University, 2 Pei-Ning Road, Keelung 202301, Taiwan
2 NASA Goddard Space Flight Center, Greenbelt, MD 20771, USA
* Correspondence: b0211@mail.ntou.edu.tw

Citation: Ho, C.-R.; Liu, A.K. Preface: Remote Sensing Applications in Ocean Observation. *Remote Sens.* **2023**, *15*, 415. https://doi.org/10.3390/rs15020415

Received: 3 January 2023
Accepted: 9 January 2023
Published: 10 January 2023

The launch of Seasat, TIROS-N and Nimbus-7 satellites equipped with ocean observation sensors in 1978 opened the way for remote sensing applications in ocean observation. After more than 40 years of development, the application of satellite remote sensing in the ocean has expanded from marine environmental observation to the extraction of ocean dynamic information, and from traditional analysis methods to the use of artificial intelligence technology. With remote sensing technology, the parameters of the ocean surface, such as ocean chlorophyll-a (Chl-a), sea surface temperature (SST), sea surface height (SSH), sea surface wind (SSW), sea surface salinity (SSS), sea ice and sea surface current, can be observed.

SST was one of the first ocean variables studied by earth observation satellites. The first satellite instrument to sense SST was an infrared sensor that used channel combining to correct for atmospheric attenuation of infrared signals from the ocean surface [1]. Besides the infrared SST, imaging microwave radiometers with SST capabilities can also be used [2]. With improvements in the performance of satellite radiometers and SST retrieval algorithms, accurate, global, high-resolution, frequently sampled SST fields have become the basis for many research and operational activities. In 1982, the first SST product was made available for operational use, was the Multi-Channel SST (MCSST), which was derived from the advanced high-resolution radiometer (AVHRR) onboard the NOAA series of satellites [1,3–5]. A nonlinear algorithm was later used to generate another SST product, named the Nonlinear SST (NLSST), or the Pathfinder SST [6]. These algorithms have also been applied to other satellite infrared sensors, such as the MODIS onboard the Aqua and Terra satellites [7]. Data from these datasets have been widely used in various ocean and atmospheric studies, such as those on ocean fronts [8–10], ocean–atmosphere interactions [11], variability and long-term changes [12–16] and ocean modeling [17,18]. This Special Issue contains several articles on SST applications, which will be introduced in the later sections.

Following the successful demonstration of quantitative estimations of Chl-a derived from the Coastal Zone Color Scanner (CZCS) onboard the Nimbus-7 satellite, the Chl-a products derived from the Sea-Viewing Wide-Field-of-View Sensor (SeaWiFS) onboard the Orbview-2 satellite were made available for public use in September 1997. The two Moderate Resolution Imaging Spectroradiometers (MODISs) onboard the Terra (since 200) and Aqua (since 2002) satellites provide the same products. The Chl-a product was derived using bio-optical algorithms such as the chlorophyll 2 algorithm (OC2) and chlorophyll 4 algorithm (OC4) for SeaWiFSs [19] and the chlorophyll 3 algorithm (OC3) for MODISs [20]. These ocean color sensors provide not only Chl-a products, but also several other ocean-water-quality products and colored dissolved organic matter, turbidity, dissolved organic carbon and suspended sediment concentration [21]. These missions provided data of exceptional quality and continuity, allowing scientific investigation of a variety oceanographic research topics [22]. For a description of remote sensing methods and statistical techniques for evaluating ocean color, refer to [23]. The good-quality SeaWiFS data were also compared with data from other ocean color sensors [24–28]. As there are several ocean color sensors, new algorithms for ocean Chl-a products have been proposed [29]. In addition to the band ratio algorithm, artificial intelligence models have been applied to retrieve Chl-a data from ocean color sensors, including models based on neural networks [30–32], machine learning [33–37] and deep learning [38,39]. This Special Issue also presents several articles detailing applied research using ocean color sensors. These will be introduced in later sections.

Satellite altimeters measure sea surface height from space with an accuracy of approximately 2 cm [40,41], making altimeter data useful for various ocean studies [42,43]. The TOPEX/Poseidon altimeter provided sea surface height data from 1992 to 2005, and its follow-up satellites, the Jason series, have since provided continuous data. These high-quality data have allowed oceanographers to conduct many studies, such as those on ocean circulation [44], ocean currents [45,46], oceanic eddies [47–49], ocean tides [50,51], sea surface waves [52,53] and sea level change [54–57]. To reflect the practicability of satellite altimeter data in the study of ocean dynamics, this Special Issue has gathered several studies using satellite altimetry, which will be introduced in later sections.

Similar to satellite altimeters using radar signal, satellite scatterometers provide wind field data over the oceans [58,59]. Satellite scatterometers are not only applied to study weather and ocean–atmosphere interactions, but also land and ice [60]. This Special Issue includes an article that introduces the application of a scatterometer onboard a China–French satellite to identify sea ice.

Besides the above-mentioned applications in ocean observation, this Special Issue presents various advanced ocean remote sensing technologies and their applications, including the use of artificial intelligence (AI) technology to explore ocean information [61–64] and reconstruct missing values [65,66]. The applications of ocean remote sensing detailed in this Special Issue include methods for the observation of changes in the ocean environment [67–77] and fishing ground [78], as well as the dynamics of the ocean, such as internal tides [79], internal waves [80,81], eddies and wakes [82,83], upwelling [84,85], ocean current [86–89] and even bibliometric analysis applied to oil detection and mapping [90]. A brief overview of the articles collected in this Special Issue is given below.

AI has been widely used for image classification. Fuzzy logic is a method of reasoning that resembles human reasoning. This approach is similar to how humans perform decision making. Thus, fuzzy logic can be treated as a kind of AI technology. The classification of seawater is very important for the study of ocean water color, because the various substances contained in seawater cause differences in seawater color. Therefore, the classification of seawater color can help us determine and systematically understand the substances contained in seawater. However, currently there is no commonly recognized template for the classification of water color. Therefore, Jia et al. [61] used a synthetic hyperspectral dataset of plankton, aerosols, clouds and marine ecosystems for unsupervised classification to categorize global ocean waters into 15 classes, resulting in a set of fuzzy logic optical water pattern schemes. These schemes were applied to several satellite multispectral sensors, including the Sea-Viewing Wide-Field-of-View Sensor (SeaWiFS), Medium-Resolution Imaging Spectrometer (MERIS), Moderate-Resolution Imaging Spectroradiometer (MODIS), Operational Land Imager (OLI), Visible Infrared Imaging Radiometer Suite (VIIRS), Multispectral Instrument (MSI) and Ocean and Land Colour Instrument (OLCI), and are considered more appropriate than existing optical water-type classification methods for global oceans.

Oil spills have always been an important issue in environmental protection efforts. However, it is difficult to automatically distinguish between man-made (spill) and natural (seep) oil slicks from synthetic aperture radar (SAR) images using limited datasets. Amri et al. [62] introduced the application of deep learning for automated offshore oil slick detection in SAR images. The data used were derived from a large database of real and recent oil slick monitoring for both types of oil slicks. Compared with the monomodal model, the proposed method yields a detection performance of up to 94% and reduces the false-positive rate by 14% to 34%. These results provide new solutions to improving the detection of natural and man-made oil slicks by providing tools that allow image interpreters to more efficiently monitor global ocean surfaces. Such a tool would speed up oil slick detection tasks, helping researchers to keep up with the continuous acquisition of sensors.

SAR images can be used not only for oil spill detection, but also for ship detection. With the development of artificial intelligence and big data technology, the data-driven convolutional neural network (CNN) has been widely used in ship detection. However, the accuracy of ship detection, feature visualization and analysis when using CNN methods need to be further improved. Geng et al. [63] proposed a two-stage ship detection method for land-contained sea area without the traditional sea–land segmentation process. Based on Sentinel-1 SAR images, the proposed method is suitable for ships smaller than 32 × 32 in size and can achieve a very high accuracy.

SST plays an important role in air–sea interactions, which have a significant impact on global change. Satellite-remote-sensing-derived SST data are often used as input for numerical forecasting models, but the results produced by numerical forecasting models often deviate from the observation data. In this era of big data, artificial intelligent technology can also be used to correct the bias of numerical forecast products. Fei et al. [64] presented an SST correction method with a convolutional long short-term memory network with multiple attention mechanisms. The method has been tested in the South China Sea and can effectively reduce errors.

Developing a system for accurately estimating changes below the sea surface from data on the sea surface is a challenge in ocean remote sensing. Dong et al. [65] applied the machine learning method of artificial intelligent technology to subsurface salinity reconstruction. The input parameters were SSS, SST, SSH and SSW, all of which can be obtained from satellite remote sensing. The parameters also included marine geographic information, that is, longitude and latitude. The Argo data were used to train and validate the machine learning model. The results show that reconstructed subsurface salinity is mainly affected by wind speed and its location, and better estimation accuracy is obtained in winter and autumn due to weaker stratification. This study offers new insight into oceanic observations.

Sargasso is abundant in the Caribbean Sea. To assess the presence and abundance of Sargasso algae from satellite ocean color data, atmospheric corrections are first required. However, atmospheric correction procedures commonly used in ocean waters need to be adjusted when dealing with Sargassum's emergence, because the non-zero water reflectance in the near-infrared band caused by Sargassum's optical signature may lead its misidentification as aerosols. Schamberger et al. [66] relied on the local uniformity of aerosol reflectance between Sargassum and Sargassum-free regions, overcoming this difficulty by interpolating aerosol and sunlight reflectance between nearby Sargassum-free pixels. The proposed method was shown to generate more reasonable aerosol and sunlight reflectance estimations.

The Geostationary Ocean Color Imager (GOCI) is the first operational geostationary ocean color sensor designed to monitor short-term and small-scale changes in the Northwest Pacific. Atmospheric corrections are especially important for small-scale short-term changes. To examine the uncertainty of GOCI-derived normalized water-leaving radiance (nLw) products, He et al. [67] applied the nLw data provided by Aerosol Robotic Network Ocean Color (AERONET-OC) to analyze the results of two GOCI algorithms. The results show that the nLw data generated by the GOCI Data Processing System were slightly better than those of the Sea-Viewing Wide-Field-of-View Sensor Data Analysis System (SeaDAS) in the visible band; however, the average relative error percentage in the blue band was over 30% for both algorithms. Both algorithms perform better at noon, and worse in the early morning and early evening. It is speculated that the uncertainty in the nLw measurement comes from the aerosol model, the near-infrared water-leaving radiometric correction method, and the bidirectional reflectance distribution function correction method in the corresponding atmospheric correction procedure.

In addition to atmospheric corrections, correcting detected clouds is critical for visible-light remote sensing because they severely impede the radiative transmission of visible light. However, cloud occlusion on turbid waters is prone to misjudgment, resulting in the loss of non-cloud pixel data. Lu et al. [68] proposed an improved GOCI cloud-masking method for turbid water. Compared with other existing cloud-masking methods, this improved method can more realistically identify the spatial location and shape of clouds and thus preserves more accurate turbid water pixels.

Unlike traditional passive ocean color remote sensing technology to detect phytoplankton and suspended particles on the sea surface, lidar technology is also used to simulate the biogeochemical processes of the upper ocean, providing data on the vertical distribution of suspended particles and the optical properties of the ocean. Zhang et al. [69] presented a new optical method to distinguish between water with different concentrations of algae through data generation (the initial width of the laser beam and the width decay rate) using the marine Scheimpflug lidar system. The applications of backscattered intensity and laser beam width measurements are explored with spatial resolution with millimeter accuracy over distances of up to several meters.

The application of satellite observation in studying typhoons has become a research hotspot in recent years. Strong typhoons enhance turbulent mixing, causing sediment re-suspension and promoting Chl-a blooms. Li et al. [70] found that the three late-autumn typhoons in the northwest of the South China Sea had limited responses to Chl-a, with only a slight increase of 23%, but a 280% increase in total suspended sediment (TSS). However, in the southern region, approximately 100 km away from the typhoon track, after the typhoon passed, the concentrations of TSS and Chl-a increased by 160% and 150%, respectively, showing different mechanisms for the increase in Chl-a concentration. This study contributes to a further detailed evaluation of the biological responses induced by typhoons.

The Chl-a and SST products from MODIS Aqua were used to study the events of harmful algae bloom (HAB) in the Arabian Gulf [71]. The results of the study show that the highest Chl-a concentration was in the Strait of Hormuz, with an average of 2.8 mg m^{-3}, which was 1.1 mg m^{-3} higher than the average of the entire study area. While the shallow-water region showed a strong positive correlation between Chl-a and SST, the deep-water region showed the opposite, with a negative correlation.

Seagrass meadows play a key role in supporting high levels of biodiversity, but are constantly threatened by human activity. To understand changes in the distribution of seagrass meadow, Xu et al. [72] used Landsat-8 OLI imagery (1974–2019) to assess large-scale changes in seagrass (*Zostera marina* L.) in the Caofeidian shoal port in the northern Bohai Sea from reclamation activities. By mapping changes in the distribution of seagrass meadows, it was shown that changes in seagrass meadows increased rapidly as the amount of land reclaimed increased. Storm surges caused by typhoons were shown to be causing habitat degradation. Fortunately, a land reclamation project created an artificial "coastal dam" that buffered seagrass meadows from wave action and provided relative shelter, which has resulted in a substantial increase in habitat since 2012.

In July–August 2021, a severe marine heat wave (MHW) occurred over a wide area of the Pacific Northwest, including the entire Sea of Japan and parts of the Sea of Okhotsk [73]. These MHWs are the largest observed since satellite measurements of global SST began in 1982. The MHWs in summer 2021 were observed at the sea surface and occurred with a stable shallow oceanic surface boundary layer. The distribution of MHWs is closely related to the sea surface heat flux, indicating that MHWs are mainly generated by atmospheric forcing. The atmospheric westerly jet moved extremely northward, and MHWs developed rapidly under the atmospheric high-pressure system close to the sea surface, which is related to the northwestward expansion of the North Pacific Subtropical High. Subsequently, as the westerly jet shifted southward, the MHWs weakened and then contracted abruptly, synchronizing with the rapid deepening of the ocean surface boundary layer.

Thermal discharge from power plants is a form of ocean pollution. It needs to be monitored. Zhang et al. [74] analyzed the thermal discharge of the Daya Bay Nuclear Power Plant (NPP) in China. To determine temporal and spatial patterns and factors affecting heat emissions, Landsat imagery acquired for the period 1993–2020 was used. SST data were retrieved from Landsat imagery using the radiative transfer equation and split window algorithm. The retrieved SST data were then used to analyze seasonal and interannual variations in areas affected by thermal discharge from the NPP, as well as the effects of the installed capacity of the NPP, tides and wind on the spread of thermal discharge. The analysis showed that with the increase in the installed capacity of the NPP, the SST increases. There is a significant linear correlation between SST and the installed capacity of an NPP. Tides affect areas of the warming zone, and the low tide has a greater impact. Regardless of whether the wind is favorable or adverse, the tides affect the warming zone more than the wind.

Based on satellite-observed SST data, Argo observations and model reanalysis results, Qiao et al. [75] investigated the upper ocean response to Super Typhoon Goni (2015) in the western North Pacific. The results show that the maximum SST cooling caused by Goni was larger than that caused by most typhoons, which is related to the enhancement of turbulent mixing caused by Goni. On the right side of the typhoon's path, the Goni-induced diapycnal diffusivity in the upper ocean increased by three orders of magnitude and persisted for at least 9 days after Goni's passage. In contrast, the diapycnal diffusivity on the left side of the typhoon path did not show a significant change. The enhancement of turbulent mixing was consistent with the Goni-induced near-inertial kinetic energy, which suggests that the enhanced turbulent mixing was caused by Goni-induced near-inertial waves.

Sea ice plays an important role in global climate change issues. Therefore, methods of detecting sea ice are also valued in ocean telemetry. Li et al. [76] introduced an improvement of the Bayesian Sea ice detection algorithm for the rotating fan-beam scatterometer CSCAT on the China–France Ocean Satellite (CFOSAT). This also serves as a guide for the recently launched dual-frequency rotating fan-beam scatterometer WindRAD.

Fishing grounds have a significant relationship with the marine environment. The use of satellites to observe changes in the marine environment has often been used for fishery management. Ding et al. [77] used satellite-derived SST, SSW, Chl-a and reanalysis data to explore the relationship between the observed aggregation of large fish and environmental factors. In the winter of 2017, the bottom water of the fishing grounds in the East China Sea was abnormally warm, and there was a significant cooling caused by the eastward movement of the Yellow Sea Cold Current with intensified northwesterly winds. Unusually warm fisheries may have provided a suitable environment for warm fish, resulting in the observation of large fish assemblages. This abnormal temperature change may be related to changes in local ocean circulation.

Although the ocean internal solitary wave (ISW) is a phenomenon that occurs under the sea surface, it modulates the convergence and divergence of the sea surface. Therefore, by observing the roughness of the sea surface, the characteristics of ISWs can be detected. Combining SAR images and mooring stations located between offshore islands with rough topographic features, Liu et al. [78] presented the characteristics of ISWs observed in the northern Yellow Sea during the summers of 2018 and 2019. ISWs with vertical displacements of up to 10 m induced prevailing high-frequency temperature variations. SAR images show that ISW intensity exhibits a clear spring–neap cycle corresponding to the local tidal forcing. The propagation speed derived from SAR images is consistent with the Korteweg–de Vries (KdV) model. The prevalence of ISWs in the study area is believed to play a crucial role in the regulation of vertical heat and nutrient transport and thus the biogeochemistry cycle.

Multi-satellite altimeter data from 1993 to 2020 were used to study the model-1 semidiurnal diurnal tides in the Sulu and Sulawesi Seas [79]. To extract multiple coherent internal tides separately, a practical plane wave analysis method was used. The complex radiation paths and interference patterns of internal tides were revealed, showing the spatial contrast between the Sulu and Sulawesi Seas. Mode-1 semidiurnal tides in the Sulawesi Sea are efficiently generated by the Sulu and Sangchi island chains, creating a spatially inhomogeneous disturbance pattern in the deep basin. These high-energy semidiurnal internal tidal beams contribute to the frequent occurrence of solitary waves (ISWs) in the study area.

Following similar techniques to [79], Wang et al. [80] used satellite altimeter data to study internal tides and explore the difference between the semidiurnal and diurnal internal tides on the East China Sea Shelf. The semidiurnal and diurnal internal tides exhibited distinct temporal trends. The semidiurnal internal tides increased by an order of magnitude, while the diurnal internal tides followed quasi-spring–neap cycles with a generally stable intensity. These internal tides probably originated from the shelf–slope area in northeastern Taiwan. Time-varying stratification was the most important factor for the internal tidal magnitude. Although both semidiurnal and diurnal internal tides were mode-1-dominated, the semidiurnal internal tide intensified at the sea surface and the diurnal internal tide intensified at the bottom.

Satellite altimeter data can be used not only to observe internal tides but also to observe core rings. Meunier et al. [81] reconstructed the 3D structure of the Loop Current Rings (LCRs) in the Gulf of Mexico using satellite altimeter data and a large set of ARGO float profiles. Between 1993 and 2021, 40 LCRs were detected in altimetry, and their 3D thermohaline structures were determined. The dynamically correlated variables and their cumulative effects on the heat, salt and energy balances in the Gulf of Mexico were discussed. The results show that LCRs have a significant effect on these balances.

Satellite altimeter data can also be used to detect mesoscale eddies. Mesoscale eddies can be found in the global ocean and have been studied on global and regional scales. Hao et al. [82] investigated the spatiotemporal variation and generation mechanism of mesoscale eddies across Indonesian seas. Eddies were detected from altimetry sea-level anomalies. The Sulu Sea, Sulawesi Sea, Maluku Sea and Banda Sea were the main eddy-generating areas. More than 80% of eddies are short-lived, with a lifespan of less than 30 days. The eddies exhibit high spatial inhomogeneity, with typical amplitudes and radii of 2–6 cm and 50–160 km, respectively. Approximately 48% of the eddies in the Sulawesi Sea are highly nonlinear, compared to less than 30% in the Sulu and Banda Seas. In the Sulu and Sulawesi Seas, barotropic instability of the mean flow dominates the eddy generation, while in the Maluku and Banda Seas, baroclinic instability is slightly greater.

As the visible trajectory information left by moving objects on the sea surface, wake has developed into one of the key detection targets of ocean visible-light remote sensing. In the case of slow ship speed, deep draft and the existence of clouds and fog, due to the low reflectivity of the sea surface and interfering objects, the wake target signal is weak and the signal-to-noise ratio is low. To solve the above problems, Ying et al. [83] calculated the difference in noise equivalent reflectance of eight bands commonly used in oceanographic remote sensing and found that the index is generally in the order of 10^{-4} and stabilizes within a certain range of values. This research has helped to improve the ability of imaging systems to detect weak wake signals.

Coastal upwelling is important for coastal ecosystems because it increases nutrients and supports plankton growth in the upper ocean. Huang et al. [84] used the SST data of the Himawari-8 geostationary satellite to map the upwelling area on the east coast of Taiwan in summer during the southwest monsoon season, which provides favorable conditions for upwelling. The results show that the range of upwelling was larger from June to August, but the upwelling duration was longest in the north center from May to September.

In addition to using satellite remote sensing SST data, data from other sources can also be applied for the study of upwelling. Li et al. [85] analyzed the spatial distribution, variability and possible forcing mechanisms of the upwelling off Manaung Island, Myanmar, using multisource satellite remote sensing data and the World Ocean Atlas 2018 (WOA18) temperature and salinity dataset. The results show that upwelling occurs in February, peaks in March and decays in May, and this upwelling is caused by the rise of seawater at depths below 100 m. The evolution of upwelling was mainly induced by remote forcing from the equator, while local wind forcing also contributed to enhancing the strength of the upwelling.

Whether the East Australian Current (EAC) has a seasonal intrusion has been debated. Xie et al. [86] applied Topographic Position Index (TPI)-based image processing techniques and a 26-year satellite SST dataset to quantitatively map the EAC in northern New South Wales (NSW). The results show that EAC intrusion exhibited seasonal cycles, being closer to the coast in austral summer than winter. The spatial analysis also showed that the EAC had a seasonal shift upstream of 29°40′S latitude and a seasonal expansion downstream. This study confirmed the seasonality of EAC intrusion observed in long-term remote sensing data. The findings provide new information on seasonal upwelling and shelf circulation off the NSW coast.

During June and July 2010, an anomalous branch of the Kuroshio Current near Taiwan in the western North Pacific was observed meandering eastward around 21°N [87]. This branch carries high-Chl-a waters into the nutrient-poor North Pacific Subtropical Gyre from 125°E. The thermohaline characteristics of this branch are similar to those of the Kuroshio. This branch has an average surface speed of 0.5 m s^{-1}, as shown by satellite altimeter data, Lagrangian drifters and Japan Meteorological Agency meridian cruise transects at 137°E. The branch appears to be associated with a surface cyclonic wind anomaly to the north at approximately 22–24°N.

The change that occurs in the Kuroshio Current as it passes through the Luzon Strait is a frequently discussed topic. Sun et al. [88] used multiple remote sensing datasets, combined with in situ drift observations, to analyze the Kuroshio intrusion into the South China Sea through the Luzon Strait. The results show the presence of a strong Kuroshio branch and accompanying anticyclonic eddy (ACE) in the winter of 2020–2021. Both the orographic negative wind stress curl southwest of Taiwan and the westward Ekman transmission through the Luzon Strait had higher values than the historical maximum. Hence, wind forcing is considered to be the main mechanism of this event.

Using coastal radar to observe the oceans is another form of ocean remote sensing technology. Lu et al. [89] used coastal high-frequency radar observations, satellite tracking drifters and numerical models to explore the ocean current variations in the northern Taiwan Strait in summer. The results show an obvious interaction between the intra-diurnal tides and ocean currents northwest of Taiwan. As the tide changed f high tide and low tide, the change in direction of the nearshore current occurred before the change in the offshore current. The drifter trajectories showed that there were three different drifting paths in the Taiwan Strait in summer. The regional ocean modeling system model was applied to clarify the factors influencing the three pathways. Simulation results and high-frequency radar data show that the difference in the drift path is caused by the transition of tidal ebb and flood and the difference in the speed of nearshore and offshore ocean currents.

In addition to analyzing observed data or using numerical models, marine researchers can also apply bibliometric and network analysis to analyze research trends. Oil spill detection and mapping (OSPM) is an extremely relevant problem due to spills' environmental impact on coastal and marine ecosystems. Vasconcelos et al. [90] evaluated the scientific literature from the last 50 years from a scientific point of view. The authors conducted a literature review on OSPM applications to perform bibliometric and network analysis to assess research and trends in this scientific field. Data were taken from the Scopus database, and then bibliometric tools were used to obtain information and reveal quantitative patterns in the literature. The findings indicate that the detection of oil in the ocean has undergone tremendous development over the past few decades and that there is a close relationship between technological developments aimed at detection and improvements in remote sensing data acquisition methods.

Author Contributions: Writing—original draft preparation, C.-R.H. and A.K.L.; writing—review and editing, A.K.L. and C.-R.H. All authors have read and agreed to the published version of the manuscript.

Acknowledgments: We thank the staff of the "Remote Sensing" Editorial Office for their support in helping to make the publication of this Special Issue a success.

Conflicts of Interest: The authors declare no conflict of interest.

References

1. McClain, E.P.; Pichel, W.G.; Walton, C.C. Comparative performance of AVHRR-based multichannel sea surface temperatures. *J. Geophys. Res. Oceans* **1985**, *90*, 11587–11601. [CrossRef]
2. Minnett, P.J.; Alvera-Azcárate, A.; Chin, T.M.; Corlett, G.K.; Gentemann, C.L.; Karagali, I.; Li, X.; Marsouin, A.; Maturi, E.; Santoleri, R.; et al. Half a century of satellite remote sensing of sea-surface temperature. *Remote Sens. Environ.* **2019**, *233*, 111366. [CrossRef]
3. McMillin, L.M. Estimation of sea surface temperatures from two infrared window measurements with different absorption. *J. Geophys. Res.* **1975**, *80*, 5113–5117. [CrossRef]
4. Barton, I.J. Dual channel satellite measurements of sea surface temperature. *Q. J. R. Meteorol. Soc.* **1983**, *109*, 365–378. [CrossRef]
5. Llewellyn-Jones, D.T.; Minnett, P.J.; Saunders, R.W.; Zavody, A.M. Satellite multichannel infrared measurements of sea surface temperature of the NE Atlantic Ocean using AVHRR/2. *Q. J. R. Meteorol. Soc.* **1984**, *110*, 613–631. [CrossRef]
6. Walton, C.C. Nonlinear multichannel algorithms for estimating sea surface temperature with AVHRR satellite data. *J. Appl. Meteorol. Climatol.* **1988**, *27*, 115–124. [CrossRef]
7. Brown, O.B.; Minnett, P.J.; Evans, R.; Kearns, E.; Kilpatrick, K.; Kumar, A.; Sikorski, R.; Závody, A. *MODIS Infrared Sea Surface Temperature Algorithm-Algorithm Theoretical Basis Document Version 2.0*; University of Miami: Miami, FL, USA, 1999; Volume 31, 098-33.
8. Castelao, R.M.; Mavor, T.P.; Barth, J.A.; Breaker, L.C. Sea surface temperature fronts in the California Current System from geostationary satellite observations. *J. Geophys. Res. Oceans* **2006**, *111*, C09026. [CrossRef]
9. Belkin, I.M.; Cornillon, P.C.; Sherman, K. Fronts in large marine ecosystems. *Prog. Oceanogr.* **2009**, *81*, 223–236. [CrossRef]
10. Kahru, M.; Di Lorenzo, E.; Manzano-Sarabia, M.; Mitchell, B.G. Spatial and temporal statistics of sea surface temperature and chlorophyll fronts in the California Current. *J. Plankton Res.* **2012**, *34*, 749–760. [CrossRef]
11. Small, R.J.; de Szoeke, S.P.; Xie, S.P.; O'neill, L.; Seo, H.; Song, Q.; Cornullon, P.; Spall, M.; Minobe, S. Air–sea interaction over ocean fronts and eddies. *Dyn. Atmos. Oceans* **2008**, *45*, 274–319. [CrossRef]
12. Yan, X.-H.; Ho, C.-R.; Zheng, Q.; Klemas, V. Temperature and size variabilities of the Western Pacific Warm Pool. *Science* **1992**, *258*, 1643–1645. [CrossRef]
13. Ho, C.-R.; Yan, X.-H.; Zheng, Q. Satellite observations of upper-layer variabilities in the western Pacific warm pool. *Bull. Am. Meteorol. Soc.* **1995**, *76*, 669–679. [CrossRef]
14. Lin, C.-Y.; Ho, C.-R.; Zheng, Q.; Kuo, N.-J.; Chang, P. Warm pool variability and heat flux change in the global oceans. *Glob. Planet. Chang.* **2011**, *77*, 26–33. [CrossRef]
15. Bouali, M.; Sato, O.T.; Polito, P.S. Temporal trends in sea surface temperature gradients in the South Atlantic Ocean. *Remote Sens. Environ.* **2017**, *194*, 100–114. [CrossRef]
16. Bouali, M.; Polito, P.S.; Sato, O.T.; Vazquez-Cuervo, J. On the use of NLSST and MCSST for the study of spatio-temporal trends in SST gradients. *Remote Sens. Lett.* **2019**, *10*, 1163–1171. [CrossRef]
17. Thomas, L.; Ferrari, R. Friction, frontogenesis, and the stratification of the surface mixed layer. *J. Phys. Oceanogr.* **2008**, *38*, 2501–2518. [CrossRef]
18. Ferrari, R. A frontal challenge for climate models. *Science* **2011**, *332*, 316–317. [CrossRef] [PubMed]
19. O'Reilly, J.E.; Maritorena, S.; Mitchell, B.G.; Siegel, D.A.; Carder, K.L.; Garver, S.A.; Kahru, M.; McClain, C. Ocean color chlorophyll algorithms for SeaWiFS. *J. Geophys. Res. Oceans* **1998**, *103*, 24937–24953. [CrossRef]
20. Campbell, J.W.; Feng, H. The empirical chlorophyll algorithm for MODIS: Testing the OC3M algorithm using NOMAD data. In Proceedings of the Ocean Color Bio-optical Algorithm Mini-workshop, Durham, NH, USA, 27–29 September 2005; pp. 27–29.
21. Mohseni, F.; Saba, F.; Mirmazloumi, S.M.; Amani, M.; Mokhtarzade, M.; Jamali, S.; Mahdavi, S. Ocean water quality monitoring using remote sensing techniques: A review. *Mar. Environ. Res.* **2022**, *180*, 105701. [CrossRef]
22. McClain, C.R. A decade of satellite ocean color observations. *Annu. Rev. Mar. Sci.* **2009**, *1*, 19–42. [CrossRef]
23. Blondeau-Patissier, D.; Gower, J.F.; Dekker, A.G.; Phinn, S.R.; Brando, V.E. A review of ocean color remote sensing methods and statistical techniques for the detection, mapping and analysis of phytoplankton blooms in coastal and open oceans. *Prog. Oceanogr.* **2014**, *123*, 123–144. [CrossRef]
24. Wang, M.; Franz, B.A. Comparing the ocean color measurements between MOS and SeaWiFS: A vicarious intercalibration approach for MOS. *IEEE Trans. Geosci. Remote Sens.* **2000**, *38*, 184–197. [CrossRef]
25. Ho, C.-R.; Lee, L.-S.; Kuo, N.-J.; Li, H.-W.; Chen, C.-T. Intercomparison of spaceborne ocean color measurements between OCI and SeaWiFS. *Geophys. Res. Lett.* **2001**, *28*, 1255–1258. [CrossRef]

26. Barbini, R.; Colao, F.; Fantoni, R.; Fiorani, L.; Okladnikov, I.G.; Palucci, A. Comparison of SeaWiFS, MODIS-Terra and MODIS-Aqua in the Southern Ocean. *Int. J. Remote Sens.* **2005**, *26*, 2471–2478. [CrossRef]
27. Zibordi, G.; Mélin, F.; Berthon, J.F. Comparison of SeaWiFS, MODIS and MERIS radiometric products at a coastal site. *Geophys. Res. Lett.* **2006**, *33*, L06617. [CrossRef]
28. Barnes, B.B.; Hu, C. Dependence of satellite ocean color data products on viewing angles: A comparison between SeaWiFS, MODIS, and VIIRS. *Remote Sens. Environ.* **2016**, *175*, 120–129. [CrossRef]
29. O'Reilly, J.E.; Werdell, P.J. Chlorophyll algorithms for ocean color sensors-OC4, OC5 & OC6. *Remote Sens. Environ.* **2019**, *229*, 32–47.
30. Keiner, L.E. Estimating oceanic chlorophyll concentrations with neural networks. *Int. J. Remote Sens.* **1999**, *20*, 189–194. [CrossRef]
31. Tanaka, A.; Kishino, M.; Doerffer, R.; Schiller, H.; Oishi, T.; Kubota, T. Development of a neural network algorithm for retrieving concentrations of chlorophyll, suspended matter and yellow substance from radiance data of the ocean color and temperature scanner. *J. Oceanogr.* **2004**, *60*, 519–530. [CrossRef]
32. Su, F.-C.; Ho, C.-R.; Zheng, Q.; Kuo, N.-J.; Chen, C.-T. Satellite chlorophyll retrievals with a bipartite artificial neural network model. *Int. J. Remote Sens.* **2006**, *27*, 1563–1579. [CrossRef]
33. Blix, K.; Eltoft, T. Machine learning automatic model selection algorithm for oceanic chlorophyll-a content retrieval. *Remote Sens.* **2018**, *10*, 775. [CrossRef]
34. Hu, C.; Feng, L.; Guan, Q. A machine learning approach to estimate surface chlorophyll a concentrations in global oceans from satellite measurements. *IEEE Trans. Geosci. Remote Sens.* **2020**, *59*, 4590–4607. [CrossRef]
35. Pahlevan, N.; Smith, B.; Schalles, J.; Binding, C.; Cao, Z.; Ma, R.; Alikas, K.; Kangro, K.; Gurlin, D.; Ha, N.; et al. Seamless retrievals of chlorophyll-a from Sentinel-2 (MSI) and Sentinel-3 (OLCI) in inland and coastal waters: A machine-learning approach. *Remote Sens. Environ.* **2020**, *240*, 111604. [CrossRef]
36. Su, H.; Lu, X.; Chen, Z.; Zhang, H.; Lu, W.; Wu, W. Estimating coastal chlorophyll-a concentration from time-series OLCI data based on machine learning. *Remote Sens.* **2021**, *13*, 576. [CrossRef]
37. Kolluru, S.; Tiwari, S.P. Modeling ocean surface chlorophyll-a concentration from ocean color remote sensing reflectance in global waters using machine learning. *Sci. Total Environ.* **2022**, *844*, 157191. [CrossRef]
38. Jin, D.; Lee, E.; Kwon, K.; Kim, T. A deep learning model using satellite ocean color and hydrodynamic model to estimate chlorophyll-a concentration. *Remote Sens.* **2021**, *13*, 2003. [CrossRef]
39. Chen, H.; Jiang, J.; Han, G.; Lin, X.; Liu, Y.; Jia, X.; Ji, Q.; Li, B. Applying deep learning in the prediction of chlorophyll-a in the East China Sea. *Remote Sens.* **2022**, *14*, 5461.
40. Cheney, R.; Miller, L.; Agreen, R.; Doyle, N.; Lillibridge, J. TOPEX/POSEIDON: The 2-cm solution. *J. Geophys. Res. Oceans* **1994**, *99*, 24555–24563. [CrossRef]
41. Nouël, F.; Berthias, J.P.; Deleuze, M.; Guitart, A.; Laudet, P.; Piuzzi, A.; Pradines, D.; Valorge, C.; Dejoie, C.L.; Susini, M.F.; et al. Precise Centre National d'Etudes Spatiales orbits for TOPEX/POSEIDON: Is reaching 2 cm still a challenge? *J. Geophys. Res. Oceans* **1994**, *99*, 24405–24419. [CrossRef]
42. Fu, L.-L.; Cazenave, A. *Satellite Altimetry and Earth Sciences: A Handbook of Techniques and Applications*; Academic Press: San Diego, CA, USA, 2001; pp. 1–463.
43. Fu, L.-L.; Le Traon, P.Y. Satellite altimetry and ocean dynamics. *Comptes Rendus Geosci.* **2006**, *338*, 1063–1076. [CrossRef]
44. Fu, L.-L.; Smith, R.D. Global ocean circulation from satellite altimetry and high-resolution computer simulation. *Bull. Am. Meteorol. Soc.* **1996**, *77*, 2625–2636. [CrossRef]
45. Fu, L.-L.; Chelton, D.B. Observing large-scale temporal variability of ocean currents by satellite altimetry: With application to the Antarctic Circumpolar Current. *J. Geophys. Res. Oceans* **1985**, *90*, 4721–4740. [CrossRef]
46. Ho, C.-R.; Zheng, Q.; Soong, Y.-S.; Kuo, N.-J.; Hu, J.-H. Seasonal variability of sea surface height in the South China Sea observed with TOPEX/Poseidon altimeter data. *J. Geophys. Res. Oceans* **2000**, *105*, 13981–13990. [CrossRef]
47. Chelton, D.B.; Schlax, M.G.; Samelson, R.M.; de Szoeke, R.A. Global observations of large oceanic eddies. *Geophys. Res. Lett.* **2007**, *34*, L15606. [CrossRef]
48. Fu, L.-L.; Chelton, D.B.; Le Traon, P.Y.; Morrow, R. Eddy dynamics from satellite altimetry. *Oceanography* **2010**, *23*, 14–25. [CrossRef]
49. Cheng, Y.-H.; Ho, C.-R.; Zheng, Q.; Qiu, B.; Hu, J.; Kuo, N.J. Statistical features of eddies approaching the Kuroshio east of Taiwan Island and Luzon Island. *J. Oceanogr.* **2017**, *73*, 427–438. [CrossRef]
50. Schrama, E.J.O.; Ray, R.D. A Preliminary tidal analysis of TOPEX/POSEIDON altimetry. *J. Geophys. Res. Oceans* **1994**, *99*, 24799–24808. [CrossRef]
51. Ray, R.D. *A Global Ocean Tide Model from TOPEX/POSEIDON Altimetry: GOT99. 2*; National Aeronautics and Space Administration, Goddard Space Flight Center: Greenbelt, MD, USA, 1999.
52. Lefevre, J.-M.; Cotton, P.D. Ocean Surface Waves. In *Satellite Altimetry and Earth Sciences: A Handbook of Techniques and Applications*; Fu, L.-L., Cazenave, A., Eds.; Academic Press: San Diego, CA, USA, 2001; Volume 69, pp. 305–328.
53. Young, I.R.; Ribal, A. Multiplatform evaluation of global trends in wind speed and wave height. *Science* **2019**, *364*, 548–552. [CrossRef]
54. Cazenave, A.; Llovel, W. Contemporary Sea Level Rise. *Ann. Rev. Mar. Sci.* **2010**, *2*, 145–173. [CrossRef]

55. Nerem, R.S.; Chambers, D.P.; Choe, C.; Mitchum, G.T. Estimating mean sea level change from the TOPEX and Jason altimeter missions. *Mar. Geod.* **2010**, *33*, 435–446. [CrossRef]
56. Willis, J.K.; Chambers, D.P.; Kuo, C.Y.; Shum, C.K. Global Sea Level Rise: Recent progress and challenges for the decade to come. *Oceanography* **2010**, *23*, 26–35. [CrossRef]
57. Hamlington, B.D.; Frederikse, T.; Thompson, P.R.; Willis, J.K.; Nerem, R.S.; Fasullo, J.T. Past, present, and future Pacific sea-level change. *Earth's Future* **2021**, *9*, e2020EF001839. [CrossRef]
58. Freilich, M.H.; Chelton, D.B. Wavenumber spectra of Pacific winds measured by the Seasat scatterometer. *J. Phys. Oceanogr.* **1986**, *16*, 741–757. [CrossRef]
59. Chelton, D.B.; Schlax, M.G.; Freilich, M.H.; Milliff, R.F. Satellite measurements reveal persistent small-scale features in ocean winds. *Science* **2004**, *303*, 978–983. [CrossRef] [PubMed]
60. Liu, W.T. Progress in scatterometer application. *J. Oceanogr.* **2002**, *58*, 121–136. [CrossRef]
61. Jia, T.; Zhang, Y.; Dong, R. A universal fuzzy logic optical water type scheme for the global oceans. *Remote Sens.* **2021**, *13*, 4018. [CrossRef]
62. Amri, E.; Dardouillet, P.; Benoit, A.; Courteille, H.; Bolon, P.; Dubucq, D.; Credoz, A. Offshore oil slick detection: From photo-interpreter to explainable multi-modal deep learning models using SAR images and contextual data. *Remote Sens.* **2022**, *14*, 3565. [CrossRef]
63. Geng, X.; Shi, L.; Yang, J.; Li, P.; Zhao, L.; Sun, W.; Zhao, J. Ship detection and feature visualization analysis based on lightweight CNN in VH and VV polarization images. *Remote Sens.* **2021**, *13*, 1184. [CrossRef]
64. Fei, T.; Huang, B.; Wang, X.; Zhu, J.; Chen, Y.; Wang, H.; Zhang, W. A Hybrid Deep Learning Model for the Bias Correction of SST Numerical Forecast Products Using Satellite Data. *Remote Sens.* **2022**, *14*, 1339. [CrossRef]
65. Dong, L.; Qi, J.; Yin, B.; Zhi, H.; Li, D.; Yang, S.; Wang, W.; Cai, H.; Xie, B. Reconstruction of Subsurface Salinity Structure in the South China Sea Using Satellite Observations: A LightGBM-Based Deep Forest Method. *Remote Sens.* **2022**, *14*, 3494. [CrossRef]
66. Schamberger, L.; Minghelli, A.; Chami, M.; Steinmetz, F. Improvement of Atmospheric Correction of Satellite Sentinel-3/OLCI Data for Oceanic Waters in Presence of Sargassum. *Remote Sens.* **2022**, *14*, 386. [CrossRef]
67. He, M.; He, S.; Zhang, X.; Zhou, F.; Li, P. Assessment of normalized water-leaving radiance derived from GOCI using AERONET-OC data. *Remote Sens.* **2021**, *13*, 1640. [CrossRef]
68. Lu, S.; He, M.; He, S.; He, S.; Pan, Y.; Yin, W.; Li, P. An improved cloud masking method for GOCI data over turbid coastal waters. *Remote Sens.* **2021**, *13*, 2722. [CrossRef]
69. Zhang, H.; Zhang, Y.; Li, Z.; Liu, B.; Yin, B.; Wu, S. Small angle scattering intensity measurement by an improved ocean scheimpflug lidar system. *Remote Sens.* **2021**, *13*, 2390. [CrossRef]
70. Li, J.; Zheng, H.; Xie, L.; Zheng, Q.; Ling, Z.; Li, M. Response of total suspended sediment and chlorophyll-a concentration to late autumn typhoon events in the northwestern South China Sea. *Remote Sens.* **2021**, *13*, 2863. [CrossRef]
71. Hussein, K.A.; Al Abdouli, K.; Ghebreyesus, D.T.; Petchprayoon, P.; Al Hosani, N.; Sharif, H.O. Spatiotemporal variability of chlorophyll-a and sea surface temperature, and their relationship with bathymetry over the coasts of UAE. *Remote Sens.* **2021**, *13*, 2447. [CrossRef]
72. Xu, S.; Xu, S.; Zhou, Y.; Yue, S.; Zhang, X.; Gu, R.; Zhang, Y.; Qiao, Y.; Liu, M. Long-term changes in the unique and largest seagrass meadows in the Bohai Sea (China) using satellite (1974–2019) and sonar data: Implication for conservation and restoration. *Remote Sens.* **2021**, *13*, 856. [CrossRef]
73. Kuroda, H.; Setou, T. Extensive marine heatwaves at the sea surface in the northwestern Pacific Ocean in summer 2021. *Remote Sens.* **2021**, *13*, 3989. [CrossRef]
74. Zhang, Z.; Wang, D.; Cheng, Y.; Gong, F. Long-term changes and factors that influence changes in thermal discharge from nuclear power plants in Daya Bay, China. *Remote Sens.* **2022**, *14*, 763. [CrossRef]
75. Qiao, M.; Cao, A.; Song, J.; Pan, Y.; He, H. Enhanced turbulent mixing in the upper ocean induced by super Typhoon Goni (2015). *Remote Sens.* **2022**, *14*, 2300. [CrossRef]
76. Li, Z.; Verhoef, A.; Stoffelen, A. Bayesian sea ice detection algorithm for CFOSAT. *Remote Sens.* **2022**, *14*, 3569. [CrossRef]
77. Ding, W.; Zhang, C.; Hu, J.; Shang, S. Unusual fish assemblages associated with environmental changes in the East China Sea in February and March 2017. *Remote Sens.* **2021**, *13*, 1768. [CrossRef]
78. Liu, H.; Yang, W.; Wei, H.; Jiang, C.; Liu, C.; Zhao, L. On characteristics and mixing effects of internal solitary waves in the northern Yellow Sea as revealed by satellite and in situ observations. *Remote Sens.* **2022**, *14*, 3660. [CrossRef]
79. Zhao, X.; Xu, Z.; Feng, M.; Li, Q.; Zhang, P.; You, J.; Gao, S.; Yin, B. Satellite investigation of semidiurnal internal tides in the Sulu-Sulawesi Seas. *Remote Sens.* **2021**, *13*, 2530. [CrossRef]
80. Wang, W.; Robertson, R.; Wang, Y.; Zhao, C.; Hao, Z.; Yin, B.; Xu, Z. Distinct variability between semidiurnal and diurnal internal tides at the East China Sea shelf. *Remote Sens.* **2022**, *14*, 2570. [CrossRef]
81. Meunier, T.; Pérez-Brunius, P.; Bower, A. Reconstructing the three-dimensional structure of loop current rings from satellite altimetry and in situ data using the Gravest empirical modes method. *Remote Sens.* **2022**, *14*, 4174. [CrossRef]
82. Hao, Z.; Xu, Z.; Feng, M.; Li, Q.; Yin, B. Spatiotemporal variability of mesoscale eddies in the Indonesian Seas. *Remote Sens.* **2021**, *13*, 1017. [CrossRef]
83. Ying, S.; Qu, H.; Tao, S.; Zheng, L.; Wu, X. Radiation sensitivity analysis of ocean wake information detection system based on visible light remote sensing. *Remote Sens.* **2022**, *14*, 4054. [CrossRef]

84. Huang, Z.; Hu, J.; Shi, W. Mapping the coastal upwelling east of Taiwan using geostationary satellite data. *Remote Sens.* **2021**, *13*, 170. [CrossRef]
85. Li, Y.; Qiu, Y.; Hu, J.; Aung, C.; Lin, X.; Dong, Y. Springtime upwelling and its formation mechanism in coastal waters of Manaung Island, Myanmar. *Remote Sens.* **2020**, *12*, 3777. [CrossRef]
86. Xie, S.; Huang, Z.; Wang, X.H. Remotely sensed seasonal shoreward intrusion of the East Australian Current: Implications for coastal ocean dynamics. *Remote Sens.* **2021**, *13*, 854. [CrossRef]
87. Chow, C.-H.; Lin, Y.-C.; Cheah, W.; Tai, J.-H. Injection of high chlorophyll-a waters by a branch of Kuroshio Current into the nutrient-poor north Pacific Subtropical Gyre. *Remote Sens.* **2022**, *14*, 1531. [CrossRef]
88. Sun, Z.; Hu, J.; Chen, Z.; Zhu, J.; Yang, L.; Chen, X.; Wu, X. A strong Kuroshio intrusion into the South China Sea and its accompanying cold-core anticyclonic eddy in winter 2020–2021. *Remote Sens.* **2021**, *13*, 2645. [CrossRef]
89. Lu, C.-Y.; Hsu, P.-C.; Zheng, Q.; Ho, C.-R. Variations in flow patterns in the northern Taiwan Strait observed by satellite-tracked drifters. *Remote Sens.* **2022**, *14*, 2154. [CrossRef]
90. Vasconcelos, R.N.; Lima, A.T.C.; Lentini, C.A.; Miranda, G.V.; Mendonça, L.F.; Silva, M.A.; Cambuí, E.C.B.; Lopes, J.M.; Porsani, M.J. Oil spill detection and mapping: A 50-year bibliometric analysis. *Remote Sens.* **2020**, *12*, 3647. [CrossRef]

remote sensing MDPI

Article

A Universal Fuzzy Logic Optical Water Type Scheme for the Global Oceans

Tianxia Jia [1,2], Yonglin Zhang [1] and Rencai Dong [1,*]

1 State Key Laboratory of Urban and Regional Ecology, Research Center for Eco-Environmental Sciences, Chinese Academy of Sciences, Beijing 100085, China; jiatianxia17@mails.ucas.edu.cn (T.J.); ylzhang_rcees@foxmail.com (Y.Z.)
2 University of Chinese Academy of Sciences, Beijing 100049, China
* Correspondence: dongrencai@rcees.ac.cn

Abstract: The classification of natural waters is a way to generalize and systematize ocean color science. However, there is no consensus on an optimal water classification template in many contexts. In this study, we conducted an unsupervised classification of the PACE (Plankton, Aerosols, Cloud, and Ocean Ecosystem) synthetic hyperspectral data set, divided the global ocean waters into 15 classes, then obtained a set of fuzzy logic optical water type schemes (abbreviated as the U-OWT in this study) that were tailored for several multispectral satellite sensors, including SeaWiFS, MERIS, MODIS, OLI, VIIRS, MSI, and OLCI. The consistency analysis showed that the performance of U-OWT on different satellite sensors was comparable, and the sensitivity analysis demonstrated the U-OWT could resist a certain degree of input disturbance on remote sensing reflectance. Compared to existing ocean-aimed optical water type schemes, the U-OWT can distinguish more mesotrophic and eutrophic water classes. Furthermore, the U-OWT was highly compatible with other water classification taxonomies, including the trophic state index, the multivariate absorption combinations, and the Forel-Ule Scale, which indirectly demonstrated the potential for global applicability of the U-OWT. This finding was also helpful for the further conversion and unification of different water type taxonomies. As the fundamental basis, the U-OWT can be applied to many oceanic fields that need to be explored in the future. To promote the reproducibility of this study, an IDL®-based standalone U-OWT calculation tool is freely distributed.

Keywords: ocean color; water type taxonomies; trophic state; inherent optical properties; Forel-Ule Scale

Citation: Jia, T.; Zhang, Y.; Dong, R. A Universal Fuzzy Logic Optical Water Type Scheme for the Global Oceans. *Remote Sens.* **2021**, *13*, 4018. https://doi.org/10.3390/rs13194018

Academic Editors: Chung-Ru Ho, Antony K. Liu and Xiaofeng Li

Received: 20 August 2021
Accepted: 7 October 2021
Published: 8 October 2021

Publisher's Note: MDPI stays neutral with regard to jurisdictional claims in published maps and institutional affiliations.

1. Introduction

Since the first ocean color-aimed spaceborne instrument, the Coastal Zone Color Scanner Experiment (CZCS), was launched in 1978, together with subsequent ocean color satellite programs SeaWiFS, MODIS, MERIS, VIIRS, OLCI, etc., the synoptic and accurate measurement of ocean ecosystems on a global scale has become a reality [1]. Ocean color remote sensing has been widely applied to and revolutionized many fields of oceanography, such as oceanic modeling, ocean physics, biogeochemical cycles, fisheries, water quality, and natural and man-made hazards [2,3]. The classification of natural waters using different water type taxonomies is a way to generalize and systematize the science of ocean color [3]. Because of the desire to move toward water classification based directly on satellite data [4], some water classification schemes have relied on the apparent optical properties (AOPs), namely reflectances (various ratios of upwelling to downwelling intensity) and diffuse attenuation functions [5]. Jerlov introduced an ocean water classification based on spectral optical attenuation depth (the inverse of the diffuse attenuation coefficient), and he divided the observations into five open oceanic and nine coastal water types [6,7], which laid the foundation for the AOP-based optical water type (OWT) classification. Moore et al. proposed a fuzzy logic-based OWT scheme that was used for blending class-specific

Chlorophyll (Chl) inverse algorithms by the membership weight [8]. Over the past 20 years, much effort has been devoted to AOP-based OWT schemes, which have been developed based on huge spectral reflectance datasets from in situ measurements or satellite data (see Section 2.3.1 for more details) [9–30].

Although the existing AOP-based OWT schemes were effective and elegant in their respective contexts, from the authors' point of view, there is still the possibility of making some improvements based on the following aspects. First, some existing AOP-based OWT schemes that were developed from in situ or satellite data may include uncertainties. Specifically, in situ data may contain uncertainties caused by experimental and environmental factors, such as calibration, dark signal, data processing, deployment strategies, and sea and sky states. The uncertainties of satellite data can be introduced by a variety of factors, such as pre-launch characterization of the sensor, atmospheric and bi-directional corrections, geo-location, and contamination by adjacent pixels [31]. Furthermore, the spatial mismatches arise when matching up in situ data and satellite data: horizontally, in situ data often cover an area of 1−10 m while ocean color satellite pixels are often more than 100 m; vertically, in situ measurements are usually conducted on discrete depths while satellite measurements represent a water column weighted average [31]. Thus, there are some potential uncertainties when the in situ data-based OWT schemes are applied to satellite data. Second, the spatial and temporal distributions of the in situ or the pixel spectra affect the representativeness of water classifications [30,32]. For instance, the in situ data were mostly collected in the Northern Hemisphere, with few data points representing very oligotrophic gyre areas [33]. On the contrary, when developing OWT schemes based on satellite data, the pixels corresponding to oligotrophic waters cover a large proportion of the overall spectrum. As a result, the satellite data-based-OWT schemes performed better in open oceans, while the in situ data-based OWT schemes often worked better in coastal waters [19]. Third, with the development of hyperspectral remote sensing technology, some of the existing OWT schemes that initially were developed within limited multispectral bands may not encompass all characteristic bands reflecting the bio-optical features of water. Fourth, the existing OWT schemes were mostly designed for one specific ocean color instrument, such as SeaWiFS [11,13,14,17,22], MODIS [10,15], MERIS [15,16,19,20,30], or OLCI [18]; thus it is difficult to migrate these OWT schemes to other satellite sensors. Therefore, it is necessary to develop an OWT scheme that is measurement error-free, suitable for the global ocean waters, and multi-satellite sensors.

In addition to the AOP-based OWT schemes, there are other water classification taxonomies that are suitable for quantitative water quality evaluation, such as the trophic state index (TSI) [34], the absorption coefficient ternary diagram [4,5,34], and the Forel-Ule Scale [23,35–37]. Further introduction to these water classification taxonomies is in Section 2.4. Although the abovementioned water classification taxonomies are commonly used, to the best of our knowledge, scant effort has been applied to exploring the relationships and correspondences between the AOP-based OWT schemes and other water type taxonomies which will hinder their comparison and unification in many cases.

Against the above background, the aims of this study were to: (1) develop a fuzzy logic AOP-based OWT scheme (abbreviated as the U-OWT hereafter) for global oceans and multi-satellite sensors; (2) test the robustness and reliability of the U-OWT; (3) explore the relationships between the U-OWT and other water classification taxonomies; (4) apply the U-OWT to the global oceans preliminarily.

2. Materials and Methods

2.1. Materials

2.1.1. Synthetic Data

The Hydrolight simulated hyperspectral data set (350−800 nm, 5 nm resolution) that was created by the first NASA PACE (Plankton, Aerosol, Cloud, and Ocean Ecosystem) Science Team (NNH13ZDA001N-PACEST) was downloaded from PANGAEA [38]. This data set contains 714 measurement error-free hyperspectral synthetic sea surface remote sensing reflectance $R_{rs}(\lambda)$ (sr^{-1}), total absorption $a_{tot}(\lambda)$ (m^{-1}), phytoplankton absorption $a_{ph}(\lambda)$ (m^{-1}), CDOM absorption $a_g(\lambda)$ (m^{-1}), and detrital absorption $a_d(\lambda)$ (m^{-1}) with a solar zenith angle of 30°. The Hydrolight model run was constrained with a set of in situ $a_{ph}(\lambda)$) spectra collected from NASA's SeaBASS repository, using the principles outlined in IOCCG Report No. 5 [31]. Because the above in situ $a_{ph}(\lambda)$ spectra were selected from more than 4000 SeaBASS spectra, and together with the random combination of optically active constituents, this synthetic data set was expected to cover all possible natural ocean water IOP-AOP scenarios [30,39], theoretically without duplicate data. This data set was used for the U-OWT scheme development and the IOPs analysis of each U-OWT class.

The Hydrolight synthesized $R_{rs}(\lambda)$ and Chla data set with the sun zenith angle of 30° from IOCCG Report No. 5 [31] was used for exploring the robustness and the water quality properties of the U-OWT. The data set contains 500 $R_{rs}(\lambda)$ and IOPs spectra pairs for testing and comparing ocean color algorithms. It covers a wide range of natural waters with discrete Chla concentrations from 0.03 to 30 mg/m^3. The data set was used for the consistency analysis of the U-OWT among different multispectral satellite sensors, and the water quality parameter analysis of each U-OWT class.

2.1.2. In Situ Data

The NOMAD (NASA bio-Optical Marine Algorithm Data set) version 2 data set is a publicly available, global, high quality in situ bio-optical data set for use in ocean color algorithm development and satellite data product validation activities [40]. The data set includes surface water-leaving radiance L_w, surface downward irradiance E_s, and Chla concentration data. After the data filtering, 1052 qualified SeaWiFS-band $R_{rs}(\lambda)$ ($R_{rs} = L_w/E_s$) and Chla data pairs were retrieved from the NOMAD data set.

The CCRR (Coast Colour Round Robin) in situ data set was collected to test algorithms and to assess their accuracy for retrieving water quality parameters [41]. A total number of 336 MERIS level 2 $R_{rs}(\lambda)$ and Chla match-ups were extracted from this data set. The $R_{rs}(\lambda)$ and Chla data pairs from the NOMAD and the CCRR data sets were used for exploring the water quality properties of each U-OWT class.

Two data sets containing in situ $R_{rs}(\lambda)$ and FUI data pairs were used for exploring the relationship between the U-OWT and the Forel-Ule Scale [39]: the first data set was collected from 612 sites covering coastal and oceanic waters around the world [42], the second data set was collected from 195 sites in coastal and oceanic waters off China [43]. Specifically, the hyperspectral $R_{rs}(\lambda)$ spectra of these data sets were interpolated to OLCI bands to calculate their dominant OWT (the water class with maximum membership) of the U-OWT.

2.1.3. Satellite Images

Two Landsat-8 OLI and Sentinel-3A OLCI images overpassing Pearl River Estuary on 23 October 2017 were downloaded from the USGS EarthExplore and the EUMETSAT Earth Observation Portal, respectively. The scene center times of the OLI image and the OLCI image were UTC 2:53 and UTC 2:34, respectively; thus, the two images can be regarded as approximately concurrent. The water in Pearl River Estuary is highly turbid and optically complicated [44]; thus, this area was suitable for testing the performance of the U-OWT.

The ESA OC-CCI climatology monthly composite R_{rs} data set (version 4.2) [45], with 4 km nominal spatial resolution at the equator, was downloaded from the ESA OC-CCI ftp server. As a level 3 binning product, the OC-CCI was composited from multiple sensors (SeaWiFS, MODIS, MERIS, and VIIRS) to SeaWiFS R_{rs} bands and values. The data set was used to study the global seasonal variability of the relative U-OWT indicators from 1998 to 2019 [46].

2.1.4. Other Data

The relative spectral response functions $RSR(\lambda)$ of several multispectral satellite sensors were obtained to convolute the hyperspectral $R_{rs}(\lambda)$ to the multispectral $R_{rs}(\lambda)$. Specifically, the $RSR(\lambda)$ of SeaWiFS, MERIS, MODIS-Aqua, MODIS-Terra, OLI, VIIRS-NPP, OLCI-S3A, and OLCI-S3B were downloaded from the NASA Ocean Biology Processing Group. The $RSR(\lambda)$ of MSI-S2A and MSI-S2B were downloaded from the ESA Sentinel website (see Data Availability Statement for more information).

2.2. Development of the U-OWT

(1) The 714 PACE synthetic R_{rs} spectra were normalized by their Root-Sum-Squares (RSS) [15]:

$$nR_{rs}(\lambda) = \frac{R_{rs}(\lambda)}{\sqrt{\sum_1^i R_{rs}(\lambda_i)^2}} \tag{1}$$

where $nR_{rs}(\lambda)$ is normalized remote sensing reflectance, index i represents the total number of wavelengths, ranging from 1 to 91, and λ represents the wavelength varying from 350 to 800 nm, with 5 nm intervals. The $R_{rs}(\lambda)$ spectral shape characteristics are highlighted after the normalized transformation and are more related to the absorption than to the backscattering [11,14].

(2) The optimal clustering number of $nR_{rs}(\lambda)$ was estimated using the gap statistic method [47]. This step was accomplished using the "clusGap" function in the R® platform. Repeated experiments with the gap method determined that the optimal clustering number was 15.

(3) The spherical k-means clustering (skmeans) method [48] was used to cluster 714 $nR_{rs}(\lambda)$ spectra into 15 groups. The skmeans is an unsupervised clustering method that employs cosine dissimilarity:

$$d(x,p) = 1 - \cos(x,p) = 1 - \frac{\langle x,p \rangle}{|x||p|} \tag{2}$$

where $d(x, p)$ is the cosine dissimilarity between the feature vectors x (here, the nR_{rs} spectra) and centroids p (here, the nR_{rs} mean of each group). The skmeans partitions data into a given number k of groups via minimizing $d(x, p)$ over all samples x to cluster centroids p. Compared to the ordinary k-means method, the skmeans is more suitable for clustering $nR_{rs}(\lambda)$ spectra, because the cosine dissimilarity places emphasis more on the spectral shape rather than on the spectral amplitude. In addition, the skmeans method was ran 10 times to reduce the effects of random initialization of unsupervised clustering [48]. Finally, the 714 $nR_{rs}(\lambda)$ spectra (and their original $R_{rs}(\lambda)$ spectra) fell into 15 groups (OWT1–OWT15) according to their most frequent clustering results. This step was accomplished using the R® package "skmeans".

(4) After the unsupervised clustering of each $R_{rs}(\lambda)/nR_{rs}(\lambda)$ spectrum that had a 5 nm interval, the cubic spline interpolation was used to create the hyperspectral $R_{rs}(\lambda)$ with 1 nm increments (350−800 nm, 451 bands).

(5) In order to obtain the different multispectral satellite sensors' $R_{rs}(\lambda)$ spectra for each OWT group, spectral bandpass integration was performed on each hyperspectral R_{rs} spectrum:

$$R_{rs\text{-}multi}(\lambda) = \frac{\int_{\lambda_1}^{\lambda_2} R_{rs\text{-}hyper} * RSR(\lambda)d\lambda}{\int_{\lambda_1}^{\lambda_2} RSR(\lambda)d\lambda} \tag{3}$$

where $R_{rs\text{-}multi}(\lambda)$ is the band-averaged multispectral remote sensing reflectance, $R_{rs\text{-}hyper}$ is the hyperspectral remote sensing reflectance spectrum. $RSR(\lambda)$ represents the relative spectral response function of the multispectral satellite sensor. λ is the specific multispectral wavelength between 400−800 nm, i.e., the visible to near-infrared wavelength. λ_1 and λ_2 are the lower (400 nm) and upper (800 nm) limits of integration, respectively. In this paper, several commonly used multispectral satellite sensors, including SeaWiFS-SEASTAR, MODIS-Aqua, MODIS-Terra, VIIRS-NPP, OLI-Landsat 8, MERIS-ENVISAT, MSI-Sentinel 2A, MSI-Sentinel 2B, OLCI- Sentinel 3A, and OLCI- Sentinel 3B, were selected to develop their corresponding U-OWT schemes.

(6) For each multispectral sensor, the $R_{rs\text{-}multi}(\lambda)$ spectra of each OWT group were normalized to determine the $nR_{rs\text{-}multi}(\lambda)$ spectra.

(7) Finally, the $nR_{rs\text{-}multi}(\lambda)$ mean and covariance of each OWT were calculated, which are two key statistics for the OWT membership function calculation [13]. Then, the squared Mahalanobis distances of each OWT were acquired [8]:

$$Z_i^2 = (nR_{rs} - M_i)^t Cov_i^{-1}(nR_{rs} - M_i) \tag{4}$$

where Z_i^2 is the squared Mahalanobis distance, nR_{rs} is the multispectral normalized remote sensing reflectance of a target spectrum or a target pixel, M_i is the nR_{rs} mean of the *i*th OWT, t represents the transpose of the vector (nR_{rs}–M_i), and Cov_i^{-1} is the inverse covariance matrix of the *i*th OWT. In this study, a common weighted covariance matrix was used for the membership calculation of all 15 OWT classes [8].

Then, the membership function was obtained:

$$f_i = 1 - F_n\left(Z_i^2\right) \tag{5}$$

where f_i is the members belonging to the *i*th OWT, $F_n\left(Z_i^2\right)$ is the cumulative Chi-square distribution function with n degrees of freedom. Here, for each satellite sensor, n equals their multispectral band numbers. After the membership calculation, three other OWT indicators were also obtained: first, the dominant OWT was defined as the OWT of maximum membership [14]; second, the total membership was defined as the sum of OWT membership; third, the normalized membership was defined as the *i*th OWT membership divided by the total membership [8]; thus, the normalized membership is constrained between 0 to 1.

In addition, a 3-sigma denoising mechanism was used in the OWT membership calculation, i.e., a too-small membership (membership less than 0.01, approximately equal to the 3-sigma threshold) was regarded as a small probability event, and this membership was assigned the value 0. This denoising processing removed many outliers and simplified the calculation. The OWT membership, dominant OWT, total membership, and normalized membership calculation procedures were all implemented in the IDL® development environment. The above steps are depicted in Figure 1.

2.3. *Reliability Analysis of the U-OWT*

2.3.1. Existing AOP-Based OWT Schemes

In order to compare the U-OWT with other AOP-based OWT schemes and to indirectly prove the rationality of the U-OWT, 9 OWT schemes from the previous studies were selected. They were acquired from Jerlov (1968) [6] (JL68), Moore et al. (2009) [10] (MO09), Moore et al. (2014) [13] (MO14), Wei et al. (2016) [15] (WE16), Jackson et al. (2017) [19] (JK17), Pitarch et al. (2019) [23] (PT19), Coastal [49] (CST), GLaSS 5C [49] (G5C), and GLaSS 6C [49] (G6C). Brief introductions to these OWT schemes appear in Table 1, and the mean reflectance vectors of each OWT class from other OWT schemes are shown in Figure S1. Different OWT schemes had their own reflectance shape and point spread; specifically, the reflectance mean vectors determined the spectral shape, and the covariance matrixes determined the spectral point spread [13]. Considering that it is not easy to compare different point cloud distributions, we calculated the memberships of the reflectance mean vectors from other OWT schemes in the U-OWT context. This method illustrated the relationships between other OWT schemes and the U-OWT scheme to a large extent.

The Jerlov OWT scheme was initially suggested by the form of irradiance transmittance through 1 m of seawater; therefore, the Jerlov water types cannot be directly compared to the U-OWT. Solonenko and Mobley derived the total absorption $a(\lambda)$ and the total scattering $b(\lambda)$ for each Jerlov water type in the wavelength range of 300–700 nm [50].

The backscattering ratio B is defined as the backscattering coefficient b_b divided by the total scattering coefficient b:

$$B = b_b / b \tag{6}$$

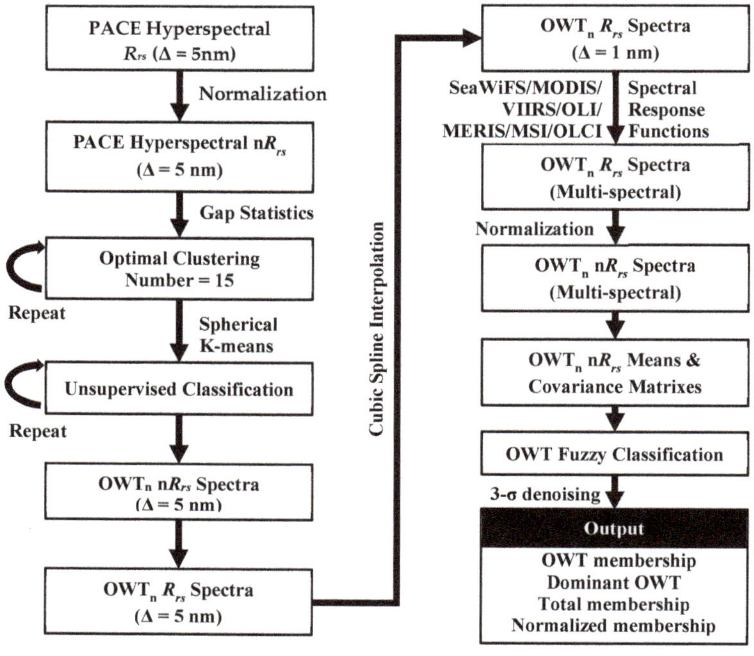

Figure 1. Workflow of the U-OWT development.

Table 1. Introduction to different AOP-based water classification schemes.

Source	Abbreviation	AOP Type	Sampling Area	Band Setting	OWT Class Number
Jerlov (1968) [6]	JL68	Irradiance transmittance for surface water/%	Global oceans	310–700 nm, Δ = 25 nm	10
Moore et al. (2009) [10]	MO09	Subsurface remote sensing reflectance/sr^{-1}	Global oceans	SeaWiFS 6 bands	8
Moore et al. (2014) [13]	MO14	Subsurface remote sensing reflectance/sr^{-1}	Coastal and inland waters	MERIS 10 bands	7
Wei et al. (2016) [15]	WE16	Normalized remote sensing reflectance	Global oceans	MODIS 9 bands	23
Jackson et al. (2017) [19]	JK17	Remote sensing reflectance/sr^{-1}	Global oceans	SeaWiFS 6 bands	14
Pitarch et al. (2019) [23]	PT19	Remote sensing reflectance/sr^{-1}	Global oceans	SeaWiFS 6 bands	21
Coastal [49]	CST	Subsurface remote sensing reflectance/sr^{-1}	Coastal waters	SeaWiFS 5 bands	16
GLaSS 5C [49]	G5C	Subsurface remote sensing reflectance/sr^{-1}	Inland waters	OLCI 13 bands	5
GLaSS 6C [17,49]	G6C	Subsurface remote sensing reflectance/sr^{-1}	Inland waters	OLCI 13 bands	6

However, the exact B values for the Jerlov water types were unknown. In the natural world most oceanic particles have backscatter ratios between 0.001 (e.g., very large phytoplankton) and 0.1 (e.g., very small mineral particles) [3]. To derive the b_b, we assumed B = 0.001, 0.01, and 0.1 to represent the different scenarios as much as possible. Gordon et al. proposed the relationship between the subsurface remote sensing reflectance $r_{rs}(\lambda)$ and the inherent optical properties $a(\lambda)$ and $b_b(\lambda)$ [51],

$$r_{rs}(\lambda) = 0.0949\frac{b_b(\lambda)}{a(\lambda)+b_b(\lambda)} + 0.0794\left(\frac{b_b(\lambda)}{a(\lambda)+b_b(\lambda)}\right)^2 \quad (7)$$

The semianalytical relationship between the subsurface remote sensing reflectance $r_{rs}(\lambda)$ and the surface remote sensing reflectance $R_{rs}(\lambda)$ proposed by Lee et al. (2002) [52] is as follows:

$$r_{rs}(\lambda) = \frac{R_{rs}(\lambda)}{0.52+1.7R_{rs}(\lambda)} \quad (8)$$

Following Equations (6)–(8), the $R_{rs}(\lambda)$ of all Jerlov water types were obtained, then we interpolated the Jerlov OWT $R_{rs}(\lambda)$ to the SeaWiFS multispectral bands and calculated their memberships in the U-OWT context.

For the other OWT schemes (MO09, MO14, CST, G5C, and G6C) displaying with the form of $r_{rs}(\lambda)$, Equation (8) was applied to convert their water type means $r_{rs}(\lambda)$ corresponding to different band settings to $R_{rs}(\lambda)$. Then, each water type of MO09, MO14, CST, G5C, and G6C was compared with the SeaWiFS 6-band, MERIS 10-band, SeaWiFS 5-band, OLCI 13-band, and OLCI 13-band U-OWT schemes, respectively. Likewise, for the OWT schemes (WE16, JK17, PT19) with the form of $R_{rs}(\lambda)$ or $nR_{rs}(\lambda)$, each water type of WE16, JK17, and PT19 was directly compared with the MODIS-A 8-band, SeaWiFS 6-band, and SeaWiFS 6-band U-OWT scheme, respectively.

2.3.2. Consistency Evaluation of U-OWT Performance on Different Sensors

The consistency performance of the U-OWT on different multispectral sensors was still unknown. To this end, the IOCCG Hydrolight Synthetic Chla and $R_{rs}(\lambda)$ data pairs (sun zenith angle = 30°) were used as a consistent set of evaluation benchmarks for all sensors. Specifically, 75 hyperspectral $R_{rs}(\lambda)$ corresponding to Chla = 0.3, 3 and 30 mg/m^3 were selected. The above hyperspectral $R_{rs}(\lambda)$ were convoluted to different multispectral band settings, then the dominant OWT and average memberships of the U-OWT were calculated to explore the consistency between different sensors. The trophic state indexes (TSIs) corresponding to Chla = 0.3, 3, and 30 mg/m^3 were equal to 18.7, 41.4, and 63.9, which represent the oligotrophic, mesotrophic, and eutrophic water states, respectively (see Section 2.4.1).

In addition, two concurrent OLI and OLCI images, which were taken overpassing Pearl River Estuary on 23 October 2017, were selected to evaluate the consistency between the two sensors with the biggest difference in band settings (4 bands for OLI and 13 bands for OLCI) from a spatial distribution perspective. Before the membership and dominant OWT calculations of the above two level 1 images, they were atmospherically corrected using the C2RCC module embedded in the SNAP® platform to transform the top of atmosphere reflectance to the surface remote sensing reflectance (Figure S2).

2.3.3. Sensitivity Analysis of the U-OWT

As a local sensitivity analysis method, the one factor at a time (OAT) method [53] was used in this study to evaluate the robustness of the U-OWT scheme by varying the input $R_{rs}(\lambda)$. The $R_{rs}(\lambda)$ values in the wavelength ranges 400–500 nm (blue bands), 500–600 nm (green bands), 600–700 nm (red bands), and 700–800 nm (near-infrared bands) were regarded as factors [25]. For the 15 U-OWT $R_{rs}(\lambda)$ mean vectors of the multispectral sensors, the perturbation range changed from −100 to 100%, with 10% increments. In this process, the dominant OWT and the memberships of that 15 $R_{rs}(\lambda)$ were recalculated to evaluate the robustness of the U-OWT. Three sensors, OLI, MODIS-A and OLCI, were selected to conduct the above sensitivity analysis, which equipped the minimum (4 bands), the medium (8 bands) and the maximum (13 bands) band numbers.

2.4. Other Water Classification Taxonomies

2.4.1. Chla-Based TSI

In this section, the synthetic (SeaWiFS 6 bands for IOCCG) and in situ (SeaWiFS 6 bands for NOMAD and MERIS 9 bands for CCRR) Chla samples, together with their multispectral $R_{rs}(\lambda)$ data pairs, were used to determine the relationship between the U-OWT and the water quality parameter (Chla concentration). Meanwhile, the trophic state index (TSI) [34] was used to evaluate the trophic state of the dominant OWT of the U-OWT:

$$\mathrm{TSI}(Chla) = 10\left(6 - \frac{2.04 - 0.68 \ln Chla}{\ln 2}\right) \tag{9}$$

where Chla is the concentration of Chla in micrograms per liter. The water trophic state is regarded as oligotrophic, mesotrophic and eutrophic when TSI < 30, 30 ≤ TSI < 50, and TSI ≥ 50, respectively [37].

2.4.2. IOP-Based Classification

Water color classification schemes are primarily based on absorption rather than scattering, as scattering contributes more to brightness but less to color [3]. The seawater absorption coefficient can be subdivided into four principal parts [54]:

$$a_{tot}(\lambda) = a_w(\lambda) + a_{ph}(\lambda) + a_g(\lambda) + a_d(\lambda) \tag{10}$$

where $a_{tot}(\lambda)$, $a_w(\lambda)$, $a_{ph}(\lambda)$, $a_g(\lambda)$, $a_d(\lambda)$ are the absorption coefficients of total, pure seawater, phytoplankton, colored dissolved organic matter (CDOM), and non-algal particles (NAPs), respectively. An optical classification based on $a_{ph}(\lambda)$, $a_g(\lambda)$, and $a_d(\lambda)$ was proposed by Prieur and Sathyendranath [54]. For 440 nm, the contributions of phytoplankton, CDOM, and NAP to the total absorption after subtracting the water contribution were defined:

$$\eta\left(a_{ph}\right) = \frac{a_{ph}(440)}{a_{tot}(440) - a_w(440)} \times 100\% \tag{11}$$

$$\eta(a_g) = \frac{a_g(440)}{a_{tot}(440) - a_w(440)} \times 100\% \tag{12}$$

$$\eta(a_d) = \frac{a_d(440)}{a_{tot}(440) - a_w(440)} \times 100\% \tag{13}$$

At a given wavelength, the ternary plot is a way to illustrate the relative proportions of each absorption subdivision to the total value [55]; it shows how different subdivision constituents dominate absorption [3]. In this study, the absorption spectra corresponding to the 714 raw $R_{rs}(\lambda)$ of the PACE synthetic data set were retrieved to explore the relationship between the U-OWT and the absorption properties.

2.4.3. Forel-Ule Scale

In order to explore the relationship between the U-OWT and Forel-Ule Scale, a hue angle and Forel-Ule Index (FUI) retrieval workflow was used for the visible bands of OLCI; the steps are summarized below:

(1) CIE tristimulus calculation:

$$X = \sum_{i=1}^{11} x_i * R_{rs}(\lambda_i) \tag{14}$$

$$Y = \sum_{i=1}^{11} y_i * R_{rs}(\lambda_i) \tag{15}$$

$$Z = \sum_{i=1}^{11} z_i * R_{rs}(\lambda_i) \quad (16)$$

where X, Y, Z are CIE tristimulus variables, $R_{rs}(\lambda_i)$ denotes the OLCI multispectral remote sensing reflectance spectra, and x_i, y_i, z_i are linear conversion coefficients to calculate the chromaticity values based on OLCI bands (Table A1) [36].

(2) CIE chromaticity coordinates calculation:

$$Chr_x = \frac{X}{X + Y + X} \quad (17)$$

$$Chr_y = \frac{Y}{X + Y + X} \quad (18)$$

where Chr_x and Chr_y are chromaticity coordinates.

(3) Hue angle calculation. The hue angle under the Woerd and Wernand (2015) definition (namely the first hue definition) [36] is obtained:

$$\alpha_{OLCI} = 90 - \mathrm{ATAN}\left(Chr_x - 1/3,\ Chr_y - 1/3\right) * 180/\pi \quad (19)$$

where α_{OLCI} is the hue angle under the first hue definition, ATAN denotes the arctangent function. To reduce the color difference between the hyperspectral true color and the multispectral sensor band setting, a systematic deviation (Δ) is defined as the hyperspectral hue angle α_{hyper} minus the multispectral hue angle, and Δ is the function of the multispectral hue angle [36]. For OLCI sensor:

$$\Delta = \alpha_{hyper} - \alpha_{OLCI} \backsim f(\alpha_{OLCI}) \quad (20)$$

If $b = \alpha_{OLCI}\ /\ 100$, Δ can be approximated by [36]:

$$\Delta = -12.508\, b^5 + 91.635\, b^4 - 249.848\, b^3 + 308.656\, b^2 - 165.482\, b + 28.561 \quad (21)$$

Therefore, the corrected hue angle can be obtained:

$$\alpha_{hyper} = \alpha_{OLCI} + \Delta \quad (22)$$

Then, the transformed hue angle under the Wang et at. (2018) definition (namely, the second hue definition) is calculated [37,56]:

$$\alpha = 270 - \alpha_{hyper} \quad (23)$$

The only difference between the first and the second hue definition is that as the hue angle increases, the FUI also increases under the second hue definition, whereas the FUI decreases under the first hue definition (Figure 2). The second hue definition is more coordinated between the hue angle and the FUI. Therefore, the following analysis in this study is under the second hue definition.

(4) FUI determination. Based on the transformed hue angle, the FUI was calculated using the FUI look-up table (LUT, Table A2): the transformed hue angle was pointed toward the nearest LUT 21 class standard hue angle, and the FUI corresponding to that standard hue angle was what we wanted [57].

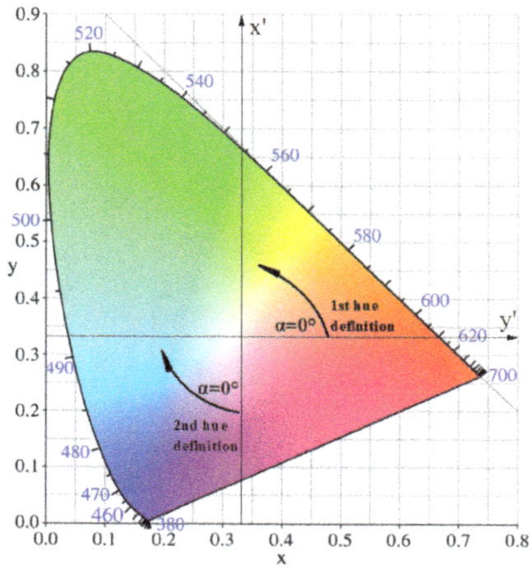

Figure 2. The CIE 1931 chromaticity diagram. The white point located at (x, y) = (1/3, 1/3). The first hue definition is the angle between the vector to a point and the positive y′-axis (at y = 1/3), giving higher angles in an anti-clockwise direction [36]; the second hue definition is the angle between the vector to a point and the negative x′-axis (at x = 1/3), giving higher angles in a clockwise direction [37].

2.5. Global Ocean Applications of the U-OWT

The U-OWT scheme was applied to the ESA-OC-CCI v4.2 R_{rs} Monthly Climatology (1998–2019) data sets to explore the global seasonal variability of ocean optical classes. Specifically, the dominant OWT and memberships of each U-OWT class for each month were calculated. For brevity, the memberships of January, April, June, and September are shown in this paper, to represent the seasonal variability. What is more, the monthly Shannon index of the U-OWT normalized memberships was calculated to evaluate the ocean optical diversity [14,20]. For a given bin:

$$H = -\sum_{i=1}^{N} P_i * \ln(P_i) \tag{24}$$

where H is the Shannon diversity index, P_i is the normalized membership of each optical water type of the U-OWT, and N denotes the number of optical classes; here, N = 15. If the 15 optical water types have the same membership, i.e., 1/15, H reaches a maximum value of ln(15) = 2.7, which means the maximum optical diversity; in contrast, if one bin is composed of only one optical class, H reaches a minimum value of 0, which indicates the minimum optical diversity.

3. Results

3.1. U-OWT Cluster Analysis

The mean normalized remote sensing reflectances n$R_{rs}(\lambda)$ of 15 U-OWT classes for the PACE hyperspectral sensor and different multispectral sensors are depicted in Figure 3. For the original hyperspectral U-OWT scheme, the n$R_{rs}(\lambda)$ spectra of different classes were well separated from others. The peaks of OWT1-15 increased from blue wavelength to near-infrared wavelength: the n$R_{rs}(\lambda)$ peaks of OWT1-3 were located near 410nm; the n$R_{rs}(\lambda)$ peaks of OWT4-6 were located near 490 nm; the n$R_{rs}(\lambda)$ peaks of OWT7-12 were located near 560 nm; the n$R_{rs}(\lambda)$ peaks of OWT13-15 were 700–800 nm. In addition, another spectral peaks for OWT1-12 classes was observed near 685 nm, and the value of this peak increased gradually from OWT1 to OWT12. The 685 nm mark is often used in the calculation of fluorescence line height (FLH) and maximum Chlorophyll index (MCI), which are the indicators of the biological activity of phytoplankton [58]. Therefore, the increasing 685 nm peak values may imply gradually stronger phytoplankton signals from OWT1 to OWT12. OWT13-15 did not show obvious 685 nm peaks, and we deduced that they were turbid water types and were more influenced by suspended sediments and organic particles [10], causing the light availability to decrease for the primary producers, thus limiting Chla concentration [23].

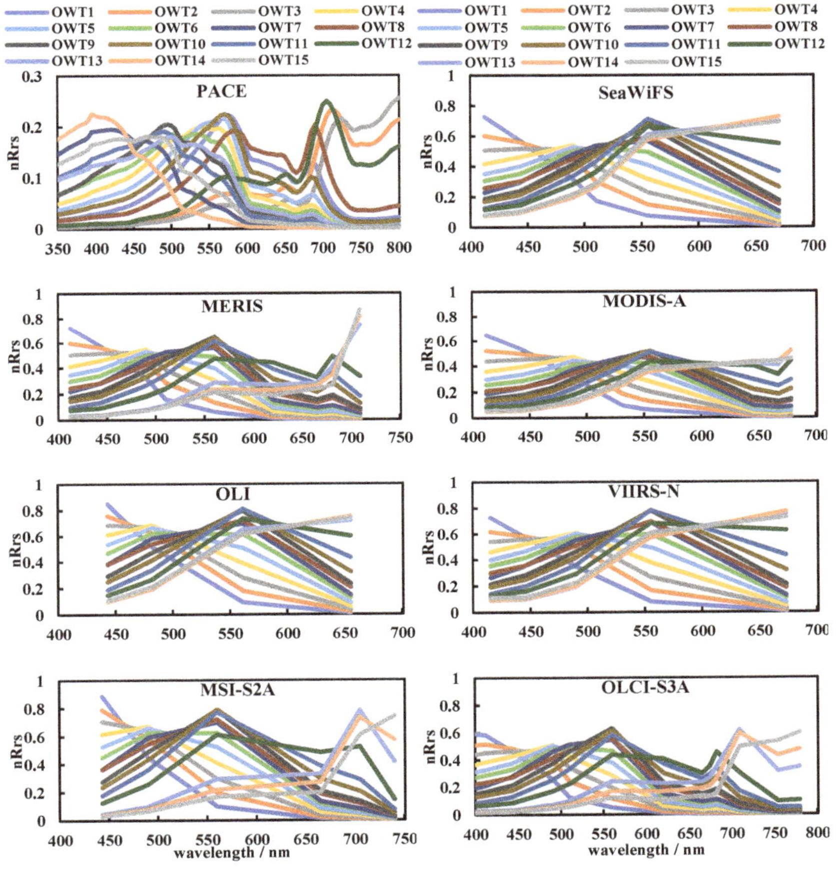

Figure 3. Mean $nR_{rs}(\lambda)$ spectra of U-OWT classes 1–15 corresponding to the hyperspectral (PACE) and different multispectral (SeaWiFS, MERIS, MODIS-A, OLI, VIIRS-N, MSI-S2A, and OLCI-S3A) satellite sensors.

As the set of U-OWT schemes of multispectral sensors were converted from the PACE hyperspectral scheme, the shape and magnitude of the multispectral U-OWT $nR_{rs}(\lambda)$ were similar to the original hyperspectral scheme, especially for the multispectral sensors with relatively more band numbers, e.g., OLCI and MERIS. On the contrary, the U-OWT schemes of the multispectral sensors with fewer band numbers lost some spectral characteristics of the original hyperspectral scheme, e.g., OLI, which had only four visible bands.

3.2. Reliability of the U-OWT

3.2.1. Inter-Comparison with Other AOP-based OWT Schemes

Regarding OWT schemes aiming at global ocean classification, including JL68, MO09, WE16, JK17, and PT19, they showed good correspondence with the U-OWT. Each class corresponded to one or more specific U-OWT classes, and the correlations were relatively high (membership > 0.5). The OWT classes of JL68, MO09, and JK17 approximately corresponded to OWT1-10 of the U-OWT successively; however, they had no corresponding classes with OWT11-15 of the U-OWT. Specifically, JL68, MO09, and JK17 appeared to classify oligotrophic water classes redundantly, i.e., they had several OWT classes corresponding to OWT1-4 of the U-OWT; while for the mesotrophic and eutrophic waters, one of their OWT classes often corresponded to several OWT5-10 classes of the U-OWT. It is worth noting that when the backscattering ratio B varied, the above phenomena were still valid for the JL68 scheme. As for WE16, its OWT classes well corresponded to OWT1-15 of the U-OWT successively, though there was only WE16.OWT19 corresponded to OWT13-15 of the U-OWT. OWT1-17 of PT19 were well related to OWT1-12 of the U-OWT; however, OWT18-21 of PT19, i.e., the CDOM-dominated classes [23], had no correspondence with the U-OWT (Figure 4). In general, the U-OWT showed good comparability with other global ocean-aimed OWT schemes, which indirectly proved the representativeness and reliability of the U-OWT.

Though the U-OWT had good correlation with other ocean-aimed OWT schemes, the correlation with the OWT schemes for the inland and coastal waters (including MO14, CST, G5C, and G6C) was not significant. Many classes of these inland and coastal water-aimed OWT schemes did not have classes corresponding to the U-OWT (Figure 4).

3.2.2. Consistency Analysis between Different Multispectral Sensors

For the 75 IOCCG oligotrophic, mesotrophic and eutrophic synthetic spectra (each trophic state had 25 spectra), their dominant OWT (Figure 5) and average OWT memberships (Figure 6) under the U-OWT context, which were calculated between different multispectral sensors, showed relatively good consistency, although the dominant OWT calculated under a few sensors' U-OWT schemes was not equal to the majority counterparts, the deviations were fewer than three dominant OWT classes.

For the two concurrent OLI and OLCI images, their spatial distribution patterns of the OWT1-12 memberships were similar, although the membership values around the rims of OWT1-9 were higher in the OLI image (Figure 7a), and the membership values around the rims of OWT10-12 were higher in the OLCI image (Figure 7b). From the perspective of the dominant OWT, both the OLI and OLCI images had similar spatial distributions and values for the oligotrophic (OWT1-5) and eutrophic (OWT9-12) water types, however, there was significant inconsistency in the spatial distributions in the mesotrophic water types (OWT6-8) (Figure 8). The band settings of different multispectral sensors, specifically the band numbers and the band locations, may have affected the U-OWT performance between sensors.

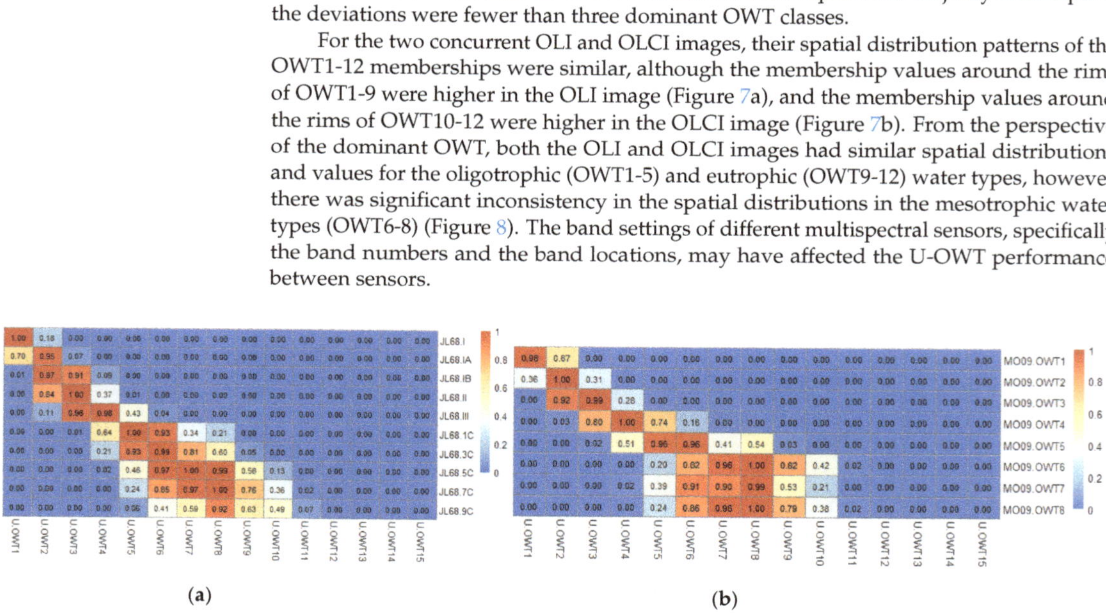

Figure 4. *Cont.*

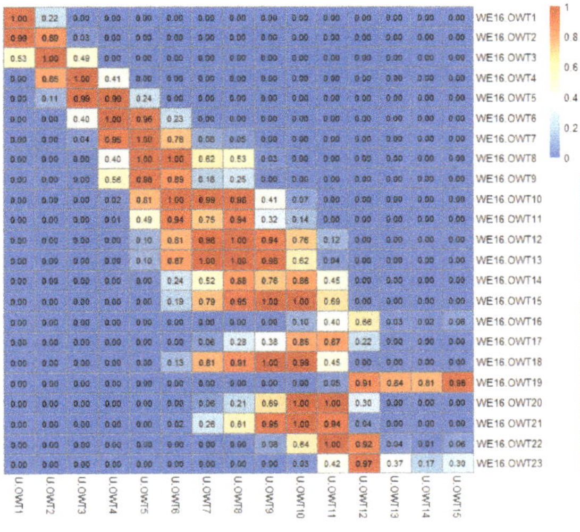

(c)

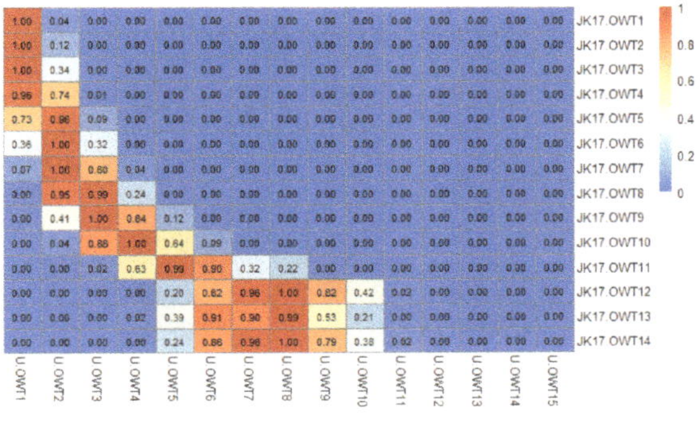

(d)

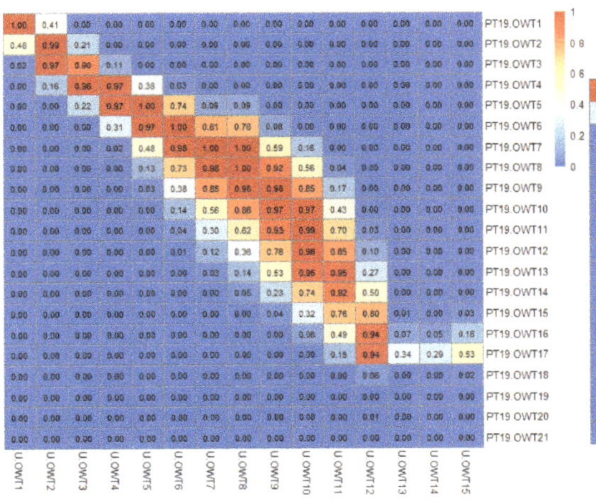

(e)

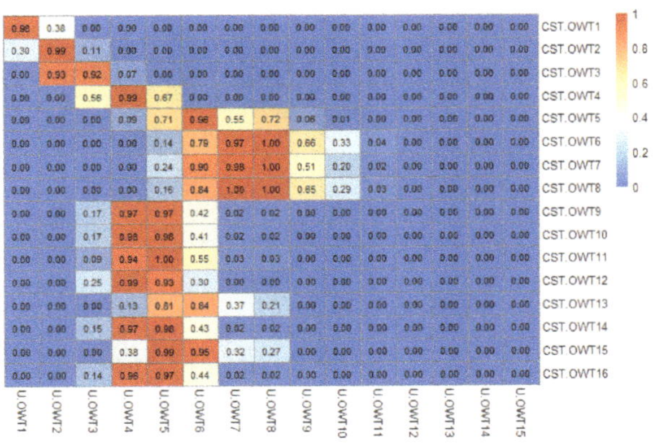

(f)

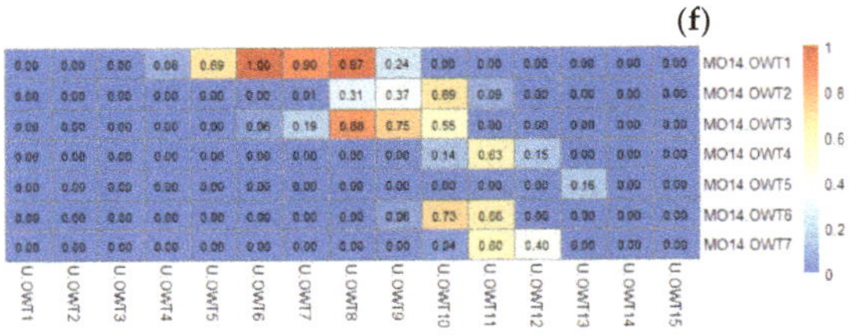

(g)

Figure 4. *Cont.*

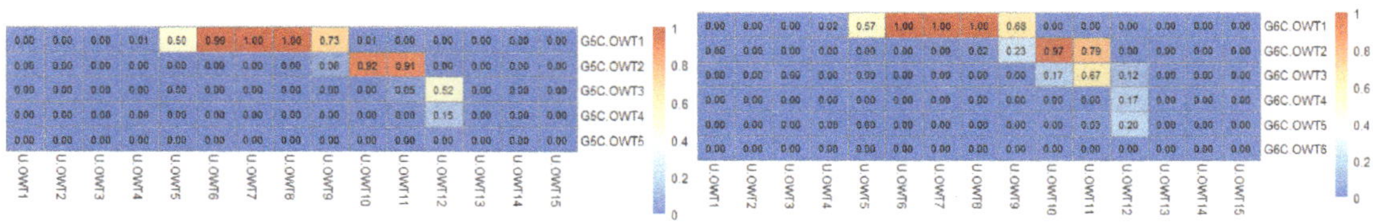

(h) (i)

Figure 4. The U-OWT memberships of each water class (represented by mean $nR_{rs}(\lambda)$ vector) corresponding to different AOP-based OWT schemes: (**a**) JL68 (B = 0.1), (**b**) MO09, (**c**) WE16, (**d**) JK17, (**e**) PT19, (**f**) CST, (**g**) MO14, (**h**) G5C, and (**i**) G6C. The relationships of the U-OWT with JL68 (B = 0.01) and with JL68 (B = 0.001) are not shown here, because of their high similarity to the JL68 (B = 0.1).

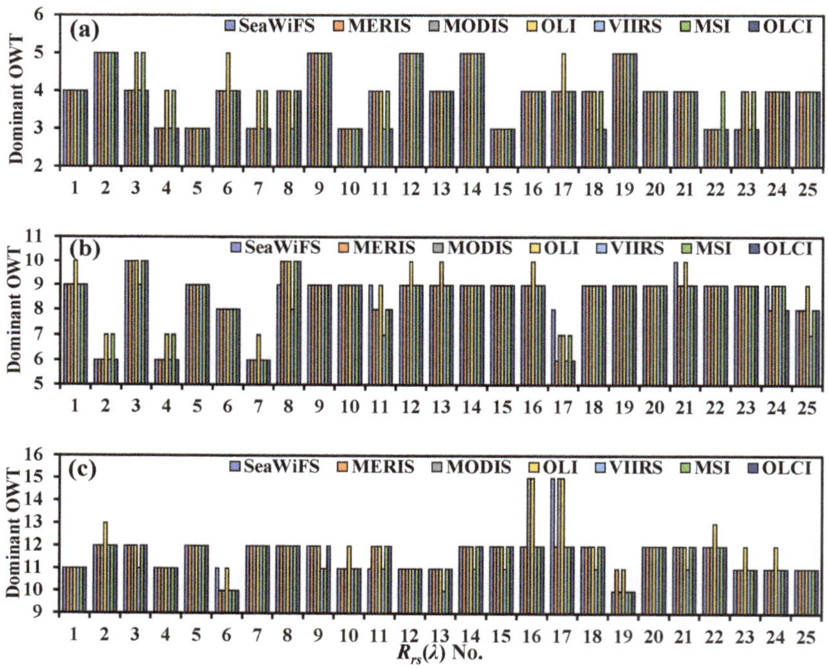

Figure 5. The U-OWT dominant OWT of the IOCCG synthetic $R_{rs}(\lambda)$ for different water trophic states: (**a**) 25 $R_{rs}(\lambda)$ spectra with Chla = 0.3 μg/L (oligotrophic), (**b**) 25 $R_{rs}(\lambda)$ spectra with Chla = 3 μg/L (mesotrophic), (**c**) 25 $R_{rs}(\lambda)$ spectra with Chla = 30 μg/L (eutrophic).

3.2.3. Sensitivity Analysis

The local sensitivity of the U-OWT schemes corresponding to OLI, MODIS, and OLCI are seen in Figures S3 and S4, and Figure 9, respectively. Taking OLCI as an example, the perturbation of R_{rs} values in the wavelength range of blue bands (400–500 nm) mainly affected the oligotrophic and mesotrophic water classes (OWT1-8), in terms of the dominant OWT and the membership values. The increase in blue band R_{rs} values could have decreased the OWT class numbers, while the decrease in blue band R_{rs} values could have increased the OWT class numbers. Additionally, the perturbations of the green bands (500–600 nm), red bands (600–700 nm), and near-infrared bands (700–800 nm) R_{rs} values mainly affected the mesotrophic (approximately OWT5-10), eutrophic (approximately OWT9-12), and over-eutrophic (approximately OWT13-15) U-OWT classes, respectively. In addition, according to the membership sensitivity analysis, the OLCI U-OWT scheme can endure nearly ±30% $R_{rs}(\lambda)$ perturbations on different wavelength ranges.

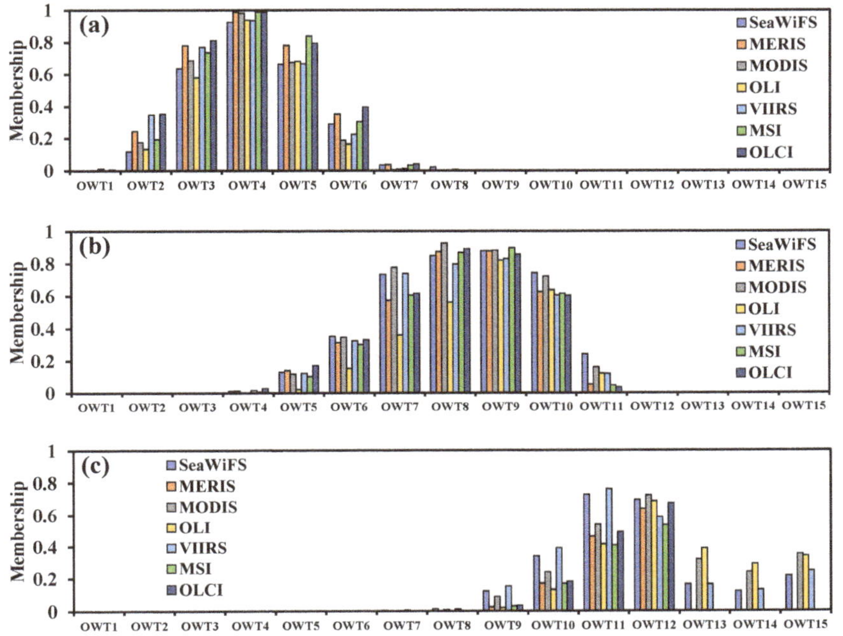

Figure 6. The U-OWT average memberships of the IOCCG synthetic $R_{rs}(\lambda)$ for different water trophic states: (**a**) 25 $R_{rs}(\lambda)$ spectra with Chla = 0.3 μg/L (oligotrophic), (**b**) 25 $R_{rs}(\lambda)$ spectra with Chla = 3 μg/L (mesotrophic), and (**c**) 25 $R_{rs}(\lambda)$ spectra with Chla = 30 μg/L (eutrophic).

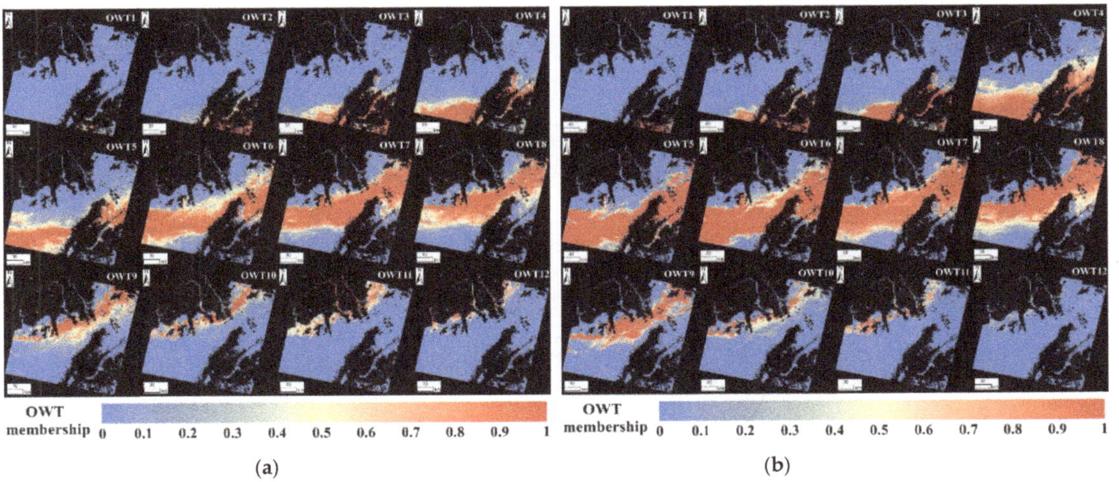

(a) (b)

Figure 7. The U-OWT memberships of the Pearl River Estuary's (**a**) Landsat-8 OLI image on 23 October 2017, and (**b**) Sentinel-3A OLCI image on 23 October 2017. The memberships of U-OWT13/14/15 are not shown here because of their negligible values. The land and cloud pixels are masked.

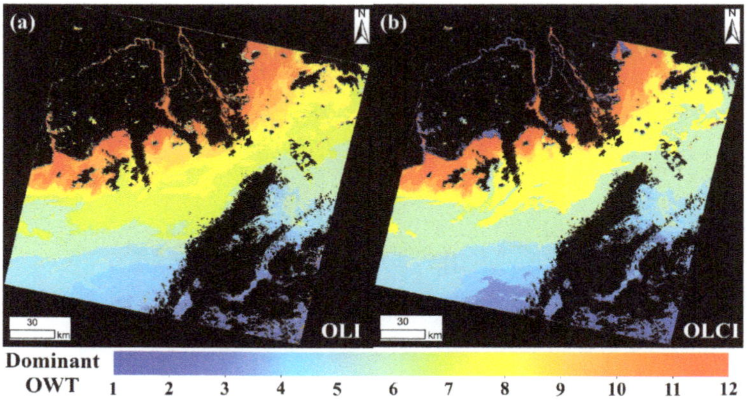

Figure 8. The U-OWT dominant OWT of (**a**) the Pearl River Estuary's Landsat-8 OLI image on 23 October 2017, and (**b**) the Pearl River Estuary's Sentinel-3A OLCI image on 23 October 2017. The land and cloud pixels are masked.

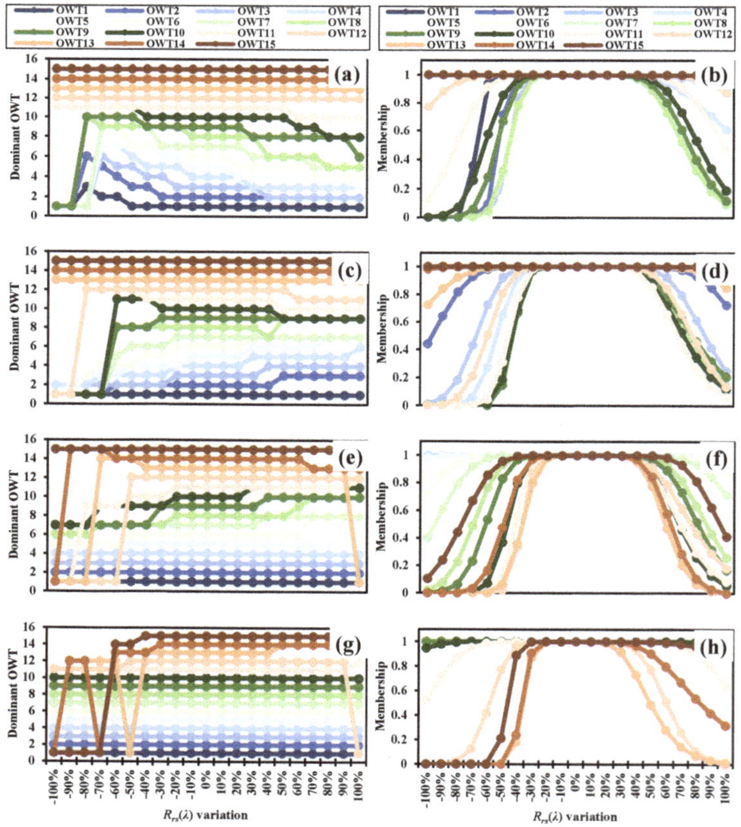

Figure 9. The U-OWT dominant OWT and membership changes in 15 OLCI mean $R_{rs}(\lambda)$ vectors under the variations in (**a**,**b**) the blue bands (400–500 nm), (**c**,**d**) the green bands (500–600 nm), (**e**,**f**) the red bands (600–700 nm), and (**g**,**h**) the near-infrared bands (700–800 nm).

The above phenomena were also detected in the OLI and MODIS U-OWT schemes. However, the OLI and MODIS U-OWT schemes could only endure nearly ±10 and ±20% $R_{rs}(\lambda)$ perturbations on different wavelength ranges, respectively. That may be due to the fewer band numbers of these two sensors. Generally, the sensitivity of the U-OWT was influenced differently by the OWT classes, sensor types, and wavelength ranges [25].

3.3. Relationships to Other Water Type Taxonomies

3.3.1. Relationship to the Chla-Based TSI

The synthetic and in situ data showed that each U-OWT class corresponded to a different Chla concentration range, whose values had an exponential increment (Figure S5). The in situ measurements were globally located, and the high-numbered OWT samples were mainly in the coastal regions, while the low-numbered OWT samples were mainly in the open ocean waters (Figure S6). The Chla concentrations of oligotrophic water classes (OWT1-4), mesotrophic water classes (OWT5-9), and eutrophic water classes (OWT10-12), were mainly in the ranges of 0.01–1, 1–10, and >10 mg/m^3, respectively. The synthetic and in situ data did not contain enough OWT13-15 spectra samples; thus, the Chla concentrations of OWT13-15 are not shown here. However, it was expected that the Chla concentrations of OWT13-15 were greater or roughly equal to OWT10-12. In addition, the Chla concentrations of the adjacent OWT classes had overlapping intervals, which indicated that the Chla inversion algorithms corresponding to the adjacent OWT classes can be weighted blended based on memberships to improve the Chla inversion accuracy in the full concentration range.

We regarded the water bodies with TSI < 30, 30 < TSI < 50, and TSI > 50 as oligotrophic, mesotrophic, and eutrophic states, respectively [37]. From the analysis of the synthetic data set, OWT1-4 of the U-OWT represented oligotrophic waters, OWT5-6 were transitional between oligotrophic and mesotrophic states, OWT7-8 were mesotrophic waters, OWT9-10 were transitional between mesotrophic and eutrophic waters, and OWT11-12 were eutrophic waters (Figure 10a). The above phenomena were also confirmed by the in situ observations (Figure 10b). Although in the synthetic and in situ data sets, there were not enough $R_{rs}(\lambda)$ spectra and Chla pairs corresponding to OWT13-15 classes, these water classes were expected to belong to eutrophic waters.

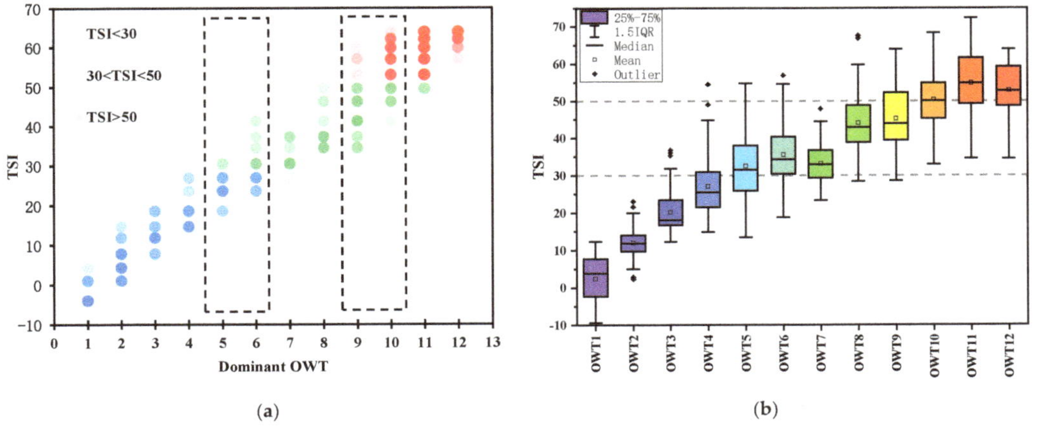

Figure 10. The relationships between the U-OWT and Chla-based TSI. (**a**) Scatterplot of data pairs of Chla-based TSI and U-OWT dominant OWT from the IOCCG synthetic SeaWiFS convolutional $R_{rs}(\lambda)$ (N = 500). Two boxes with black dashes mark the dominant OWT5-6 (transition between oligotrophic and mesotrophic waters) and the dominant OWT9-10 (transition between mesotrophic and eutrophic waters). The points are plotted with 85% transparency to show the data density. (**b**) Chla-based TSI distributions of NOMAD and CCRR in situ measurements corresponding to different U-OWT dominant OWT (OWT1-OWT12). OWT13/14/15 are not shown here because of the lack of observations.

3.3.2. Relationship to the Absorption Properties

The mean $a_{tot}(\lambda)$, $a_{ph}(\lambda)$, $a_g(\lambda)$, and $a_d(\lambda)$ spectra of the PACE synthetic data for each U-OWT class appear in Figure 11, and they were highly spectrally dependent. The magnitude of absorption varies linearly with the concentration of the absorbing material [3]; therefore, the exponential increase in the absorption spectra may illustrate the increasing absorption materials for the U-OWT classes. Because of the presence of Chla, two peaks of the $a_{ph}(\lambda)$ spectra were in the blue and red wavelength regions [3], one was near 440 nm, and another was near 680 nm. The $a_g(\lambda)$ values were high in the blue wavelength and decreased exponentially to the red regions. The $a_d(\lambda)$ spectra were very similar to the $a_g(\lambda)$, but with a more gently decreasing slope. The total absorption was composed of pure water and three individual absorption components; thus, it had an exponential decreasing slope and two peaks around the blue and red regions.

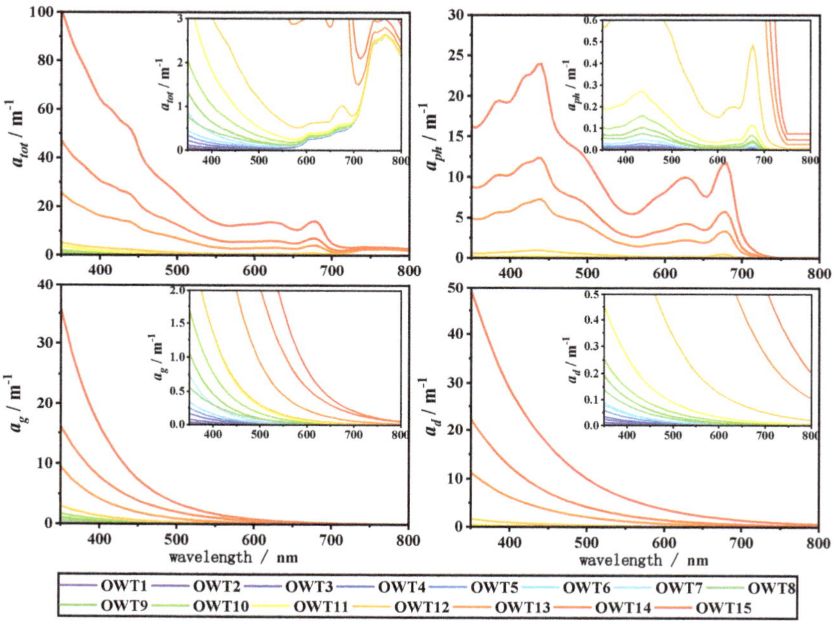

Figure 11. Average total absorption coefficients $a_{tot}(\lambda)$, average phytoplankton absorption coefficients $a_{ph}(\lambda)$, average gelbstoff coefficients $a_g(\lambda)$, and average non-algal particles absorption coefficients $a_d(\lambda)$ of each U-OWT class.

Similar to a ternary plot of the individual absorption contributions, a line chart depicting the contributions of $a_{ph}(440)$, $a_g(440)$, and $a_d(440)$ to total absorption after subtracting water contribution is in Figure 12. From OWT1 to OWT11, $\eta(a_{ph})$ and $\eta(a_d)$ decreased gradually, while $\eta(a_g)$ increased gradually and exceeded $\eta(a_{ph})$ and $\eta(a_d)$. The transition was at OWT12; in this water class, $\eta(a_{ph})$ and $\eta(a_d)$ increased while $\eta(a_g)$ decreased. At OWT13-15, the comparison between the individual contributions was $\eta(a_{ph}) > \eta(a_d) > \eta(a_g)$.

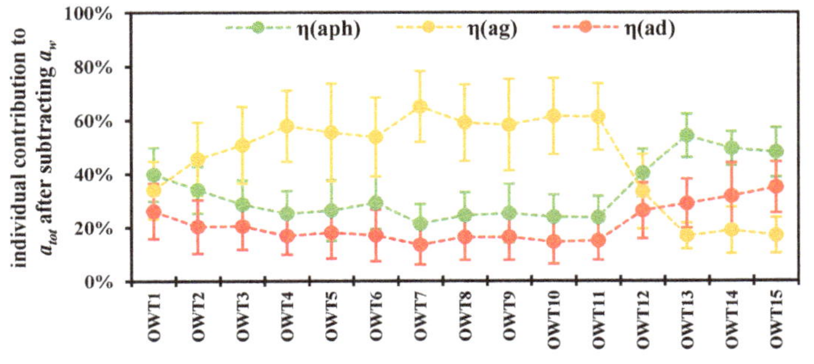

Figure 12. The contributions of the phytoplankton absorption coefficient at 440 nm $a_{ph}(440)$, CDOM absorption coefficient at 440 nm $a_g(440)$, and non-algal particle absorption coefficient at 440 nm $a_d(440)$ to total absorption $a_{tot}(440)$ after subtracting water contribution $a_w(440)$ for each dominant OWT class of the U-OWT. The error bars are the standard deviations of corresponding individual contributions.

3.3.3. Relationship to the Forel–Ule Scale

Analysis of the IOCCG synthetic data showed that in the oligotrophic water types, i.e., OWT1-6, one U-OWT class roughly corresponded to one FUI class; in the mesotrophic water types, i.e., OWT7-10, one U-OWT class often corresponded to several FUI classes; in the eutrophic water types, i.e., OWT11-15, one FUI class corresponded to several U-OWT classes (Figure 13a). In other words, for the purpose of water quality assessment, the distinguishable abilities of the U-OWT and FUI were equal in oligotrophic waters, FUI was better in mesotrophic waters, while the U-OWT was better in eutrophic waters. The in situ data from the global oceans and offshore China also confirmed the case (Figure 13b).

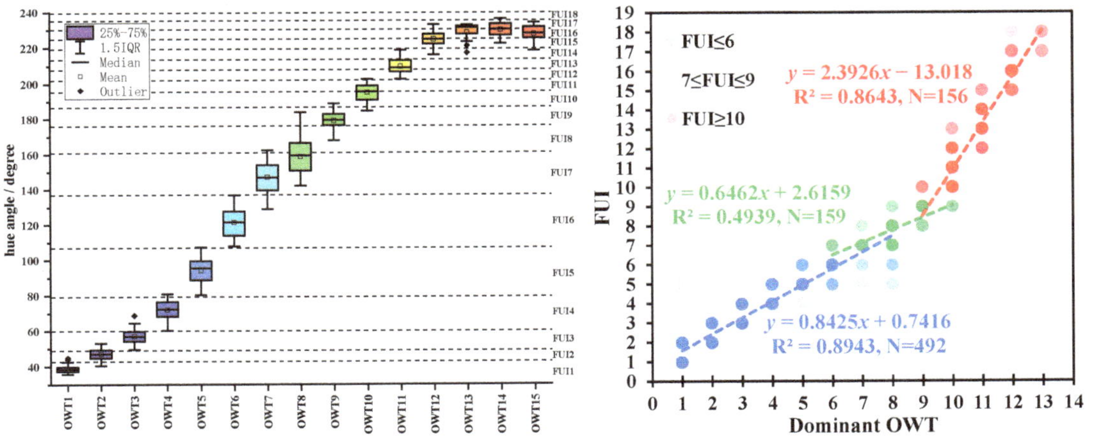

Figure 13. (**a**) Hue angle distribution of each U-OWT class from the PACE synthetic data set ($N = 714$). The separations of adjacent FUI are shown with dashed lines. (**b**) Scatterplot of data pairs of FUI and U-OWT dominant OWT from the in situ measurements collected from 612 sites covering coastal and oceanic waters around the world ($N = 612$) and 195 sites in coastal and oceanic waters off China ($N = 195$). The points are plotted with 85% transparency to show the data density. FUI ≤ 6, $7 \leq$ FUI ≤ 9, and FUI ≥ 10 represent oligotrophic, mesotrophic, and eutrophic waters, respectively [37].

In addition to the spectral analysis of comparison between the U-OWT and the FUI, one OLCI image overpassing the Pearl River Estuary was processed for the spatial consistency analysis of the U-OWT and FUI. Figure 14 shows that the dominant U-OWT class had a spatial pattern that was high consistent with the FUI, especially in terms of the boundary distribution of the corresponding OWT and FUI classes. Similar to the spectral consistency analysis, the FUI had a better ability to distinguish in mesotrophic waters than the U-OWT. For instance, in the waters off Hong Kong, the U-OWT only classified waters into one class, i.e., OWT8; however, the FUI classified the same areas into three classes, i.e., FUI7-9. The U-OWT has its limitations; even so, its dominant OWT classes can potentially be regarded as water quality indicators that are as useful as the FUI.

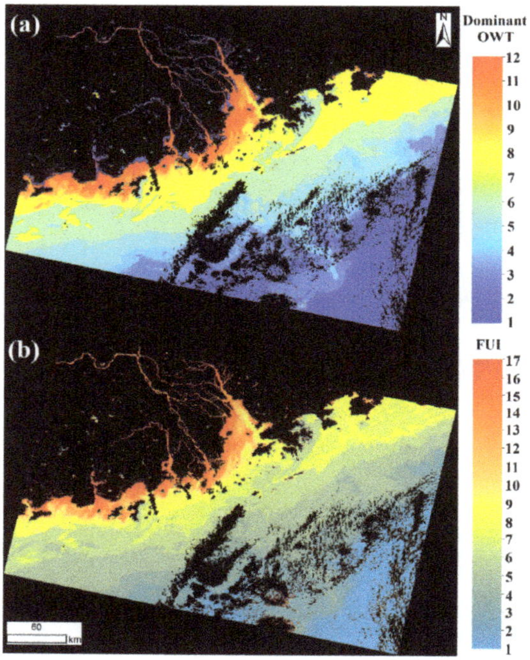

Figure 14. The (**a**) U-OWT dominant OWT and the (**b**) FUI spatial distributions of the Sentinel-3A OLCI image taken overpassing the Pearl River Estuary on 23 October 2017. The land and cloud pixels are masked.

3.4. Global Applications of the U-OWT

The U-OWT scheme was applied to the level-3 ESA-OC-CCI climatology monthly R_{rs} data to obtain the global oceanic dominant OWT (Figure 15), monthly OWT memberships (Figure 15), and Shannon indexes of normalized memberships (Figure 16). The dominant OWT and memberships showed that most ocean waters were occupied by the two clearest water classes, i.e., OWT1 and OWT2. In addition, the OWT1 ocean regions corresponded to the subtropical gyres surrounded by the OWT2 waters. The monthly variations in subtropical gyres over the past 20 years were obtained: the subtropical gyres in the Northern Hemisphere had a smaller area in January (Northern Hemisphere winter) than in July (Northern Hemisphere summer), while the subtropical gyres in the Southern Hemisphere had a larger area in January (Southern Hemisphere summer) than in July (Southern Hemisphere winter). The OWT classes belonging to mesotrophic and eutrophic waters were mainly located off the shores, including China coastal waters, Red Sea–Persian Gulf, West Africa coastal waters, North Sea–Baltic Sea, East Coastal waters of the United States, Caspian Sea, Black Sea, Argentine Sea, Bering Strait, etc. More details of these coastal areas must be studies in the future. Generally speaking, the closer to the coast, the higher the eutrophic degree of the water bodies. The optical properties of some offshore areas were the result of land–sea interaction. The coastal waters were affected both by runoffs from land sources and by marine factors such as currents and tides.

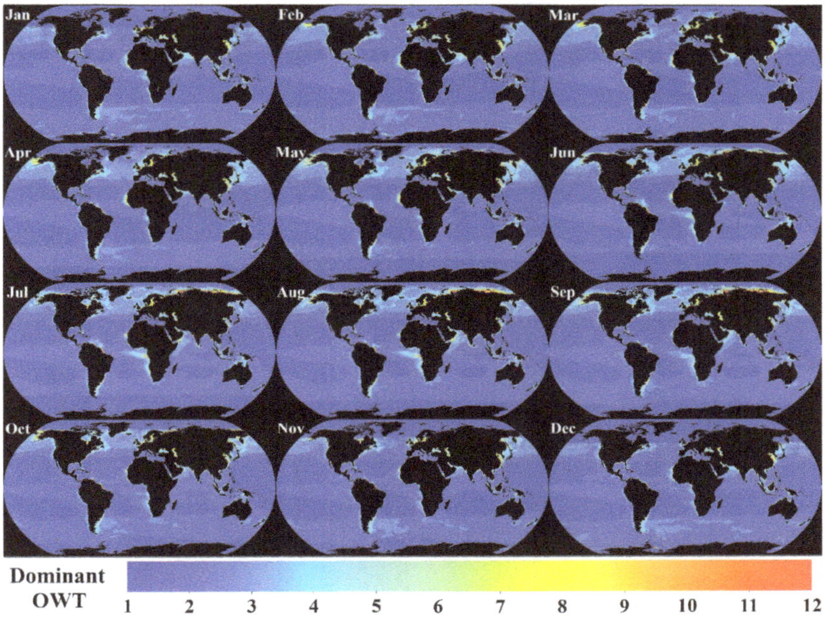

Figure 15. Global U-OWT dominant OWT monthly variability based on the ESA-OC-CCI v4.2 climatology monthly R_{rs} (1998–2019).

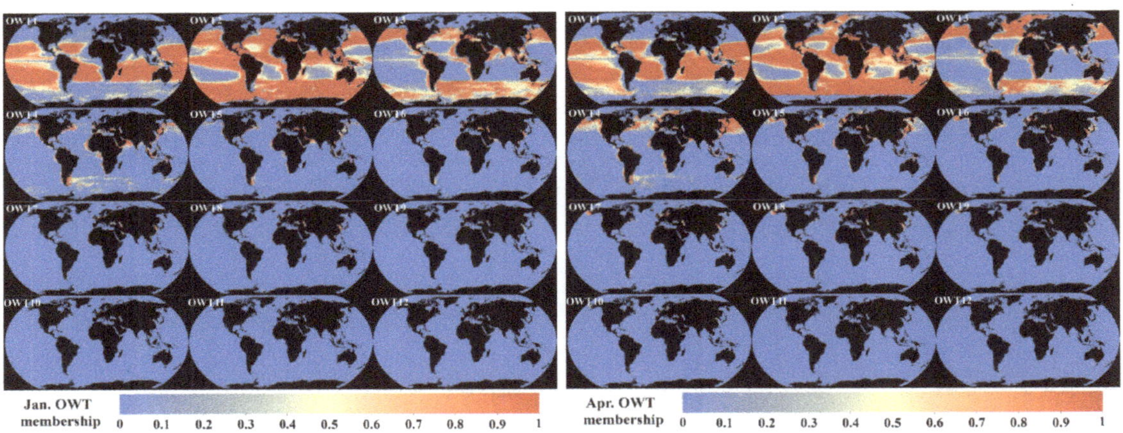

(a) (b)

Figure 15. *Cont.*

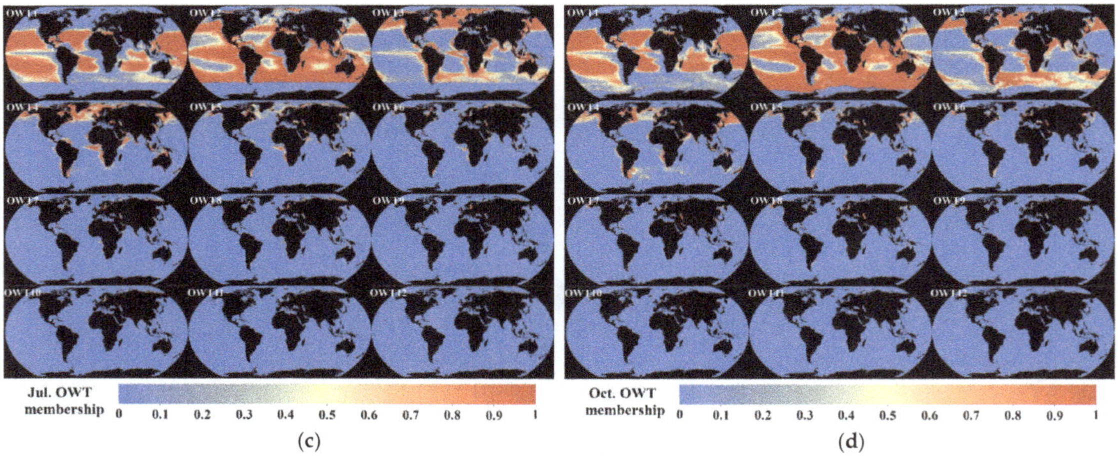

Figure 15. Global U-OWT memberships of (**a**) January, (**b**) April, (**c**) July, and (**d**) October, based on the ESA-OC-CCI v4.2 climatology monthly R_{rs} (1998–2019). The memberships of OWT13/14/15 are not shown here because of their negligible low values.

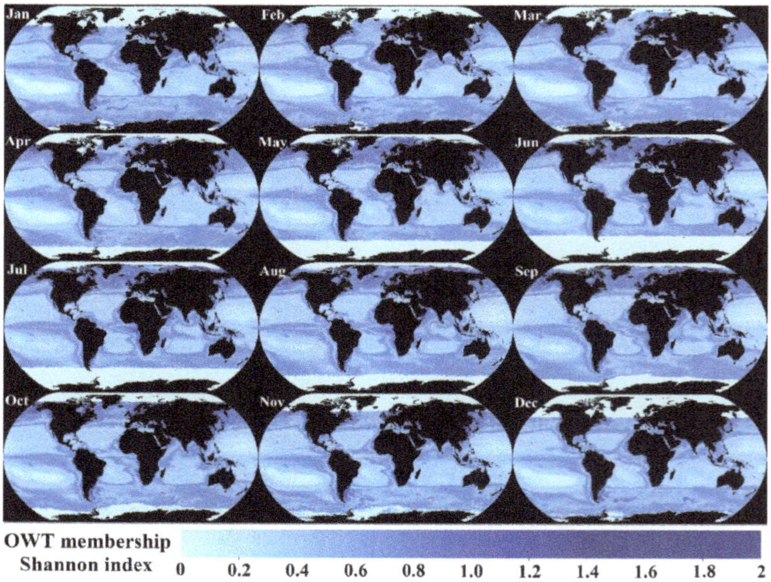

Figure 16. Global monthly variability in the U-OWT normalized membership Shannon index based on the ESA-OC-CCI v4.2 climatology monthly R_{rs} (1998–2019).

The ocean optical diversity was observed from the Shannon index; the higher Shannon index indicated higher diversity in the OWT classes, and vice versa. The Shannon index appeared to be relatively high in the transition areas between different dominant OWT classes, and this phenomenon was especially obvious in the edge areas of gyres. As the optical diversity was related to the biological diversity [14,20], further studies are needed to explore the bio-optical properties of the U-OWT Shannon index.

4. Discussion

4.1. How Many Optical Water Types in the World?

According to the Section 3.2.1, the AOP-based OWT schemes which were designed for inland and coastal waters, including MO14, CST, G5C and G6C, did not show good consistency and compatibility with the U-OWT. We speculated that these optically complex water-aimed OWT schemes may contain some unique water types, which the U-OWT does not yet include. At the same time, several ocean-aimed OWT schemes, including JL68, MO09, WE16, JK17, and PT19, were quite compatible with the U-OWT, and most of their water classes could be explained by OWT1-12 of the U-OWT. However, compared to the U-OWT scheme, other ocean-aimed OWT schemes had relatively more oligotrophic water classes and fewer mesotrophic and eutrophic water classes. From the authors' point of view, one of the most important applications of the optical water classification is in the class-specific algorithms blending for the in-water constituents, such as Chla and total suspended substance (TSS). The coastal and inland waters are more optically complex than the oceanic waters, and some conventional inversion algorithms developed for the clear oceanic waters do not perform well in these turbid waters; thus, class-specific algorithms are urgently needed for these mesotrophic and eutrophic waters. Previous ocean- aimed OWT schemes largely relied on huge in situ and remote sensing spectra samples, and the clear water spectra occupied large proportions of their clustering samples, which affected the water class clustering processes and led to more clear water types in the final OWT schemes. The U-OWT was based on the synthetic spectra, which did not contain repeated spectra samples. Therefore, the U-OWT was not affected by the occurrence frequency of spectra, and more turbid water types could be distinguished, and the optical properties of the optically complex waters could be more comprehensively represented by the U-OWT.

In general, the optical water type numbers of most AOP-based oceanic schemes were between 8 and 23, while the optical water type numbers of most inland and coastal aimed schemes were between 3 and 15. Although the water type numbers of most OWT schemes were determined by the mathematical methods, some trade-offs must be considered: on the one hand, too few classifications may not fully reflect the spectral characteristics of different water bodies; on the other hand, too many classifications will reduce the efficiency of OWT calculation, especially when involving many spectra or satellite data. According to the sensitivity analysis of the U-OWT, OWT1-12 classes were mainly distinguished by the visible wavelength, while the OWT13-15 classes were mainly distinguished by the near-infrared region. The OWT13-15 classes had relatively high $nR_{rs}(\lambda)$ magnitude in 700–800 nm, which indicated that these water types were highly turbid. Because of the insufficient in situ samples corresponding to the OWT13-15 classes, their bio-optical properties were not fully explored in this study. In highly turbid coastal and estuarine waters, optical variability can be determined for wavelengths longer than 670 nm and cannot be captured by a sensor like SeaWiFS [14]. However, for some satellite instruments that are equipped with near-infrared bands, such as MERIS, MSI and OLCI, they can well detect the highly turbid water types; thus, the OWT13-15 classes of the U-OWT scheme may be more distinguishable using these sensors.

Some previous AOP-based OWT schemes were compared with the U-OWT. However, because of the difficulty in obtaining the raw OWT spectra data, some OWT schemes proposed by previous literatures were not included in such comparisons, such as Vantrepotte et al. [11], Shi et al. [12], Mélin and Vantrepotte [14], Ye et al. [16], Hieronymi et al. [18], Monolisha et al. [20], Spyrakos et al. [21], Uudeberg et al. [25], Xue et al. [26], Zhang et al. [27], Balasubramanian et al. [28], and da Silva et al. [29] The adaptability of the U-OWT in inland waters also deserves to be explored, and some unique OWT classes for inland waters may be added to the present U-OWT scheme in the future.

4.2. Unification of Different Water Type Taxonomies

One of the most intuitive water type taxonomies is the multivariate combination of in-water constituents, such as phytoplankton, detritus, CDOM, and sediment. These in-water constituents can be further subdivided into diverse groups, according to the species, sizes, shapes, builds, etc. [3]. However, accurate and synoptic in situ measurement of the in-water constituents is difficult. Thus, from a remote sensing perspective, IOPs and AOPs can be regarded as the indirect indicators of water properties. The absorption and scattering (backscattering) are the most important IOPs, and they are dependent on the in-water constituents. Taking the radiative transfer model as a bridge, different kinds of AOPs, such as ocean color, radiances, and reflectances can be expressed as the functions of IOPs, and vice versa [59]. Likewise, from in situ or remote sensing observations, in-water constituents can be inversely modeled by AOPs [60]. In general, the interrelationships between AOPs, IOPs, and in-water constituents are explicit. To date, many commonly used water type taxonomies are based on AOPs, IOPs, or in-water constituents, such as the fuzzy logic classification of AOPs spectra, the Forel–Ule Scale, the absorption ternary plot, and the Chla-based TSI. However, the interrelationships between these water classification systems are still ambiguous. To the best of our knowledge, there is no consensus on a universal fuzzy logic AOP-based OWT scheme, and this may hinder the further comparison and integration of different kinds of water type taxonomies. Though different water classification systems have their own characteristics and application scopes, the compatibility of these water type taxonomies is helpful for their inter-comparison and mutual-conversion.

In addition to the Chla-based TSI and the individual absorption proportions, the U-OWT showed high coordination with the Forel–Ule Scale. As the Forel–Ule Scale is known for its applicability to global ocean and inland waters, thus, the U-OWT was indirectly proven to have the potential to be applied globally. In essence, this indicated that the core components of the U-OWT fuzzy logic classification scheme—OWT class mean nR_{rs} vectors and covariance matrixes, were well suitable. At the same time, the U-OWT also showed its difference from and superiority to the Forel–Ule Scale: the FU scale mainly relies on the visible wavelength; thus, it cannot distinguish the spectral characteristics on near-infrared bands, where optically complex waters have non-negligible signals [60]. However, the eutrophic water classes (especially OWT13-15) of the U-OWT can distinguish those optically complex waters. Meanwhile, the fuzzy logic OWT scheme is a bridge and intermediary with many application fields (see Section 4.3); thus, it is expected the U-OWT will have a wider applicative perspective.

4.3. Future Prospects of the U-OWT

Broader applications of the U-OWT can be explored based on the direct U-OWT calculation indicators—OWT membership, OWT normalized membership, total membership, and dominant OWT. One obstacle to the promotion of the fuzzy logic OWT method is that there are few ready-made tools; thus, the authors of this study developed a freely distributed U-OWT calculation tool, and all the above direct OWT indicators can be calculated by the readers. Some indirect OWT parameters derived from the direct OWT indicators are also valuable, such as the Shannon index of the normalized membership.

The fuzzy logic OWT framework was initially introduced by Moore et al. [8], and it has been widely used in bio-optical algorithms blending, such as Chla [11,15,16,22,30] and TSS [28]. In optically complex waters, it is difficult to inverse in-water constituents over entire concentration ranges, and it may be more realistic to use a series of branching algorithms to deal with this problem, rather than a single, all-purpose algorithm [60]. The fuzzy logic OWT schemes can seamlessly integrate such a series of branching inversion algorithms of different water classes, the membership-weighted blending retrievals from the class-specific algorithms showed smooth and continuous patterns [8], which were superior to some conventional means. In the field of IOPs inversion, the spectral slope values of the absorption or backscattering spectra were usually regarded as a constant for all water states. However, the spectral slope of IOPs are related to different water types [31]; thus, the U-OWT scheme may play an active role in the IOPs inversion. In addition, the OWT total membership can be used in the uncertainty assessment of in situ spectra or remote sensing images [6,18,22]. This study also showed that the U-OWT was highly coordinated and compatible with the TSI and Forel–Ule Scale, therefore, the dominant OWT of the U-OWT can be regarded as an indicator of water quality assessment [20,23,25,28,29]. What is more, the OWT schemes were helpful in the biogeochemical province partition of the oceans [61–63], the marine mesoscale features (water mass, eddy, plume, etc.) extraction [13,23,28], and the ocean biological diversity analysis [14,20]. All the areas where previous OWT schemes have been applied are worth applying the U-OWT, and more application scenarios are also expected in the future.

5. Conclusions

In this study, we conducted an unsupervised classification of the PACE synthetic hyperspectral data set, then obtained a fuzzy logic optical water type scheme (abbreviate as the U-OWT). The U-OWT divided the oceanic and coastal waters into 15 classes, from OWT1 to OWT15. The low-numbered classes represented clear oligotrophic water types, and the high-numbered classes represented turbid eutrophic water types. Specifically, the OWT13/14/15 classes were sensitive to the over-turbid waters with high near-infrared reflectance signals. Compared to OWT schemes from previous studies, the U-OWT was not affected by the frequency of clustering spectral samples, thus, it could distinguish more mesotrophic and eutrophic state water types. The U-OWT classified water bodies according to the spectral shapes; thus, it could resist a certain degree of disturbance from input R_{rs} values.

The U-OWT was tailored for several commonly used multispectral satellite sensors, including SeaWiFS, MERIS, MODIS, OLI, VIIRS, MSI, and OLCI. The reliability analysis showed that the performance of U-OWT were consistent between different satellite sensors. If necessary, the U-OWT can also be extended to more multispectral ocean color instruments.

The relationships of the U-OWT with other water classification taxonomies, including Chla-based TSI, multivariate absorption combinations, and the Forel–Ule Scale, were also analyzed. The results showed that the U-OWT was well compatible with other water type taxonomies: OWT1-4 of the U-OWT represented oligotrophic state waters, OWT5-6 were transitional between oligotrophic and mesotrophic states, OWT7-8 were mesotrophic state waters, OWT9-10 were transitional between mesotrophic and eutrophic waters, and OWT11-15 were eutrophic state waters; OWT1-15 had the exponentially increasing trend of the absorption coefficients pairs, and there was a significant magnitude gap between OWT13-15 and OWT1-12, which may illustrate the distinctive properties of the highly turbid OWT13-15 classes; one U-OWT class roughly corresponded to one FUI class in the oligotrophic water types, one U-OWT class corresponded to several FUI classes in the mesotrophic water types, and one FUI class corresponded to several U-OWT classes in the eutrophic water types. Our analysis demonstrated that the U-OWT had the potential to be universally applied to global ocean waters, and it was also helpful for the inter-comparison and unification of the U-OWT and other water classification taxonomies.

In addition to the class-specific algorithm blending, more application scenarios of the U-OWT deserve to be mined in the future. Although the U-OWT was suitable for ocean water classification, its applicability to inland waters needs to be further explored. Additionally, an IDL®-based U-OWT calculation tool was made freely available, so the relative indicators of the U-OWT can be calculated by the readers.

Supplementary Materials: The following are available online at http://www.mdpi.com/s1, Figure S1: Mean vectors of the AOP-based OWT schemes from previous studies, Figure S2: True color composites of two concurrent (23 October 2017) Landsat-8 OLI and Sentinel-3A OLCI images, Figure S3: The U-OWT sensitivity analysis under OLI context, Figure S4: The U-OWT sensitivity analysis under MODIS-A context, Figure S5: The relationships between the U-OWT and Chla concentrations, and Figure S6: Locations of NOMAD and CCRR in situ measurements and their dominant OWT. A standalone U-OWT calculation tool designed for different multispectral sensors' R_{rs} spectrum file (with text format) and image file (with tiff format) is freely distributed as a supplementary material. This tool can calculate dominant OWT, OWT membership, normalized membership, and total membership on each multispectral Rrs spectrum or pixel. The IDL® version 8.5 was used in the U-OWT calculation tool development.r

Author Contributions: Conceptualization, T.J.; methodology, T.J.; software, T.J. and Y.Z.; validation, T.J.; formal analysis, T.J.; investigation, T.J.; resources, T.J.; data curation, T.J.; writing—original draft preparation, T.J.; writing—review and editing, T.J., Y.Z., and R.D.; visualization, T.J.; supervision, R.D.; project administration, R.D.; funding acquisition, R.D. All authors have read and agreed to the published version of the manuscript.

Funding: This research was supported by the Strategic Priority Research Program of the Chinese Academy of Sciences (XDA23030403) and State Key Laboratory of Urban and Regional Ecology (SKLURE2013-1-05).

Institutional Review Board Statement: Not applicable.

Informed Consent Statement: Not applicable.

Data Availability Statement: The PACE hyperspectral synthetic data set was downloaded from https://doi.pangaea.de/10.1594/PANGAEA.915747 (accessed on 20 October 2020). The Hydrolight synthesized $R_{rs}(\lambda)$ and Chla data set with the sun zenith angle of 30° from IOCCG Report No. 5 was downloaded from https://www.ioccg.org/groups/OCAG_data.html (accessed on 20 October 2020). The NOMAD version 2 data set was downloaded from https://seabass.gsfc.nasa.gov/wiki/NOMAD (accessed on 10 February 2021). The CCRR in situ data set was collected from https://doi.pangaea.de/10.1594/PANGAEA.841950 (accessed on 10 February 2021). Two in situ Rrs(λ) and FUI data pairs were downloaded from https://osapublishing.figshare.com/articles/dataset/Data_File_1_csv/7355903 (accessed on 15 February 2021) and https://osapublishing.figshare.com/articles/dataset/Data_File_2_csv/7355906 (accessed on 15 February 2021), respectively. The Landsat-8 OLI image was downloaded from https://earthexplorer.usgs.gov/ (accessed on 10 May 2021). The Sentinel-3A OLCI image was downloaded from https://codarep.eumetsat.int/#/home (accessed on 10 May 2021). The ESA OC-CCI climatology monthly composite data set (version 4.2) was downloaded from ftp://oc-cci-data:ELaiWai8ae@oceancolour.org (accessed on 22 December 2020). The relative spectral response functions of several multispectral satellite sensors were obtained from https://oceancolor.gsfc.nasa.gov/docs/rsr/rsr_tables/ (accessed on 26 October 2020) and https://sentinels.copernicus.eu/web/sentinel/user-guides/sentinel-2-msi/document-library/-/asset_publisher/Wk0TKajiISaR/content/sentinel-2a-spectral-responses (accessed on 26 October 2020). All the above data were open access.

Acknowledgments: The authors would thank the reviewers and the editors for their valuable comments and suggestions.

Conflicts of Interest: The authors declare no conflict of interest.

Appendix A

Table A1. Linear coefficients to calculate the chromaticity values based on OLCI bands [36].

*i*th Band	1	2	3	4	5	6	7	8	9	10	11
λ_i(nm)	400	413	443	490	510	560	620	665	673.5	681.25	708.75
x_i	0.154	2.957	10.861	3.744	3.750	34.687	41.853	7.323	0.591	0.549	0.189
y_i	0.004	0.112	1.711	5.672	23.263	48.791	23.949	2.836	0.216	0.199	0.068
z_i	0.731	14.354	58.356	28.227	4.022	0.618	0.026	0.000	0.000	0.000	0.000

Table A2. The FUI look-up table under the second hue definition [57].

FUI	*a*	FUI	*a*	FUI	*a*
1	40.467	8	170.463	15	222.115
2	45.196	9	181.498	16	227.629
3	52.852	10	191.835	17	232.830
4	67.169	11	199.038	18	237.352
5	91.298	12	205.062	19	241.759
6	122.585	13	210.577	20	245.551
7	151.479	14	216.557	21	248.953

References

1. Werdell, P.J.; Behrenfeld, M.J.; Bontempi, P.S.; Boss, E.; Cairns, B.; Davis, G.T.; Franz, B.A.; Gliese, U.B.; Gorman, E.T.; Hasekamp, O.; et al. The Plankton, Aerosol, Cloud, Ocean Ecosystem Mission: Status, Science, Advances. *Bull. Am. Meteorol. Soc.* **2019**, *100*, 1775–1794. [CrossRef]
2. IOCC. *Why Ocean Colour? The Societal Benefits of Ocean-Colour Technology*; Reports of the International Ocean-Colour Coordinating Group, No. 7; Platt, T., Hoepffner, N., Stuart, V., Brown, C., Eds.; IOCCG: Dartmouth, NS, Canada, 2008.
3. Mobley, C.; Boss, E.; Roesler, C. Ocean Optics Web Book. Available online: https://www.oceanopticsbook.info/ (accessed on 11 August 2021).
4. Uudeberg, K.; Aavaste, A.; Koks, K.L.; Ansper, A.; Uusoue, M.; Kangro, K.; Ansko, I.; Ligi, M.; Toming, K.; Reinart, A. Optical water type guided approach to estimate optical water quality parameters. *Remote Sens.* **2020**, *12*, 391. [CrossRef]
5. Mobley, C.D.; Stramski, D.; Bissett, W.P.; Boss, E. Optical Modeling of Ocean Waters: Is the Case 1–Case 2 Classification Still Useful? *Oceanography* **2004**, *17*, 60–67. [CrossRef]
6. Jerlov, N.G. *Optical Oceanography*; Elsevier Oceanography Series 5; Elsevier: Amsterdam, The Netherlands, 1968.
7. Jerlov, N.G. *Marine Optics*; Elsevier Oceanography Series 14; Elsevier: Amsterdam, The Netherlands, 1976.
8. Moore, T.S.; Campbell, J.W.; Feng, H. A fuzzy logic classification scheme for selecting and blending satellite ocean color algorithms. *IEEE Trans. Geosci. Remote Sens.* **2001**, *39*, 1764–1776. [CrossRef]
9. Martin Traykovski, L.V. Feature-based classification of optical water types in the Northwest Atlantic based on satellite ocean color data. *J. Geophys. Res.* **2003**, *108*. [CrossRef]
10. Moore, T.S.; Campbell, J.W.; Dowell, M.D. A class-based approach to characterizing and mapping the uncertainty of the MODIS ocean chlorophyll product. *Remote Sens. Environ.* **2009**, *113*, 2424–2430. [CrossRef]
11. Vantrepotte, V.; Loisel, H.; Dessailly, D.; Mériaux, X. Optical classification of contrasted coastal waters. *Remote Sens. Environ.* **2012**, *123*, 306–323. [CrossRef]
12. Shi, K.; Li, Y.; Li, L.; Lu, H.; Song, K.; Liu, Z.; Xu, Y.; Li, Z. Remote chlorophyll-a estimates for inland waters based on a cluster-based classification. *Sci. Total Environ.* **2013**, *444*, 1–15. [CrossRef]
13. Moore, T.S.; Dowell, M.D.; Bradt, S.; Verdu, A.R. An optical water type framework for selecting and blending retrievals from bio-optical algorithms in lakes and coastal waters. *Remote Sens. Environ.* **2014**, *143*, 97–111. [CrossRef]
14. Mélin, F.; Vantrepotte, V. How optically diverse is the coastal ocean? *Remote Sens. Environ.* **2015**, *160*, 235–251. [CrossRef]
15. Wei, J.; Lee, Z.; Shang, S. A system to measure the data quality of spectral remote sensing reflectance of aquatic environments. *J. Geophys. Res. Ocean.* **2016**. [CrossRef]
16. Ye, H.; Li, J.; Li, T.; Shen, Q.; Zhu, J.; Wang, X.; Zhang, F.; Zhang, J.; Zhang, B. Spectral Classification of the Yellow Sea and Implications for Coastal Ocean Color Remote Sensing. *Remote Sens.* **2016**, *8*, 321. [CrossRef]
17. Eleveld, M.; Ruescas, A.; Hommersom, A.; Moore, T.; Peters, S.; Brockmann, C. An Optical Classification Tool for Global Lake Waters. *Remote Sens.* **2017**, *9*, 420. [CrossRef]
18. Hieronymi, M.; Müller, D.; Doerffer, R. The OLCI Neural Network Swarm (ONNS): A Bio-Geo-Optical Algorithm for Open Ocean and Coastal Waters. *Front. Mar. Sci.* **2017**, *4*. [CrossRef]
19. Jackson, T.; Sathyendranath, S.; Mélin, F. An improved optical classification scheme for the Ocean Colour Essential Climate Variable and its applications. *Remote Sens. Environ.* **2017**, *203*, 152–161. [CrossRef]
20. Monolisha, S.; Platt, T.; Sathyendranath, S.; Jayasankar, J.; George, G.; Jackson, T. Optical Classification of the Coastal Waters of the Northern Indian Ocean. *Front. Mar. Sci.* **2018**, *5*. [CrossRef]
21. Spyrakos, E.; O'Donnell, R.; Hunter, P.D.; Miller, C.; Scott, M.; Simis, S.G.H.; Neil, C.; Barbosa, C.C.F.; Binding, C.E.; Bradt, S.; et al. Optical types of inland and coastal waters. *Limnol. Oceanogr.* **2018**, *63*, 846–870. [CrossRef]
22. Hieronymi, M. Spectral band adaptation of ocean color sensors for applicability of the multi-water biogeo-optical algorithm ONNS. *Opt. Express* **2019**, *27*, A707–A724. [CrossRef]
23. Pitarch, J.; van der Woerd, H.J.; Brewin, R.J.W.; Zielinski, O. Optical properties of Forel-Ule water types deduced from 15 years of global satellite ocean color observations. *Remote Sens. Environ.* **2019**, *231*, 111249. [CrossRef]
24. Soomets; Uudeberg; Jakovels; Zagars; Reinart; Brauns; Kutser. Comparison of Lake Optical Water Types Derived from Sentinel-2 and Sentinel-3. *Remote Sens.* **2019**, *11*, 2883. [CrossRef]
25. Uudeberg, K.; Ansko, I.; Põru, G.; Ansper, A.; Reinart, A. Using Optical Water Types to Monitor Changes in Optically Complex Inland and Coastal Waters. *Remote Sens.* **2019**, *11*, 2297. [CrossRef]
26. Xue, K.; Ma, R.; Wang, D.; Shen, M. Optical Classification of the Remote Sensing Reflectance and Its Application in Deriving the Specific Phytoplankton Absorption in Optically Complex Lakes. *Remote Sens.* **2019**, *11*, 184. [CrossRef]
27. Zhang, F.; Li, J.; Shen, Q.; Zhang, B.; Tian, L.; Ye, H.; Wang, S.; Lu, Z. A soft-classification-based chlorophyll-a estimation method using MERIS data in the highly turbid and eutrophic Taihu Lake. *Int. J. Appl. Earth Obs. Geoinf.* **2019**, *74*, 138–149. [CrossRef]
28. Balasubramanian, S.V.; Pahlevan, N.; Smith, B.; Binding, C.; Schalles, J.; Loisel, H.; Gurlin, D.; Greb, S.; Alikas, K.; Randla, M.; et al. Robust algorithm for estimating total suspended solids (TSS) in inland and nearshore coastal waters. *Remote Sens. Environ.* **2020**, *246*, 111768. [CrossRef]
29. da Silva, E.F.F.; Novo, E.M.L.d.M.; Lobo, F.d.L.; Barbosa, C.C.F.; Noernberg, M.A.; Rotta, L.H.d.S.; Cairo, C.T.; Maciel, D.A.; Flores Júnior, R. Optical water types found in Brazilian waters. *Limnology* **2020**, 1–12. [CrossRef]

30. Vandermeulen, R.A.; Mannino, A.; Craig, S.E.; Werdell, P.J. 150 shades of green: Using the full spectrum of remote sensing reflectance to elucidate color shifts in the ocean. *Remote Sens. Environ.* **2020**, *247*, 111900. [CrossRef]
31. IOCCG. *Remote Sensing of Inherent Optical Properties: Fundamentals, Tests of Algorithms, and Applications*; Reports of the International Ocean-Colour Coordinating Group, No. 5; Lee, Z.P., Ed.; IOCCG: Dartmouth, NS, Canada, 2006.
32. Arnone, R.; Wood, M.; Gould, R. The Evolution of Optical Water Mass Classification. *Oceanography* **2004**, *17*, 14–15. [CrossRef]
33. Claustre, H.; Maritorena, S. The Many Shades of Ocean Blue. *Science* **2003**, *302*, 1514–1515. [CrossRef] [PubMed]
34. Carlson, R.E. A tophic state index for lakes. *Limnol. Oceanogr.* **1977**, *22*, 361–369. [CrossRef]
35. Novoa, S.; Wernand, M.R.; Woerd, H.J.v.d. The Forel-Ule scale revisited spectrally_preparation, protocol, transmission messurements and chromaticity. *J. Eur. Opt. Soc. Rapic Publ.* **2013**, *8*, 1–8. [CrossRef]
36. Woerd, H.J.; Wernand, M.R. True colour classification of natural waters with medium-spectral resolution satellites: SeaWiFS, MODIS, MERIS and OLCI. *Sensors* **2015**, *15*, 25663–25680. [CrossRef] [PubMed]
37. Wang, S.; Li, J.; Zhang, B.; Spyrakos, E.; Tyler, A.N.; Shen, Q.; Zhang, F.; Kuster, T.; Lehmann, M.K.; Wu, Y.; et al. Trophic state assessment of global inland waters using a MODIS-derived Forel-Ule index. *Remote Sens. Environ.* **2018**, *217*, 444–460. [CrossRef]
38. Craig, S.E.; Lee, Z.; Du, K. Top of Atmosphere, Hyperspectral Synthetic Dataset for PACE (Phytoplankton, Aerosol, and ocean Ecosystem) Ocean Color Algorithm Development. Available online: https://doi.org/10.1594/PANGAEA.915747 (accessed on 20 October 2020).
39. Wang, S.; Lee, Z.; Shang, S.; Li, J.; Zhang, B.; Lin, G. Deriving inherent optical properties from classical water color measurements: Forel-Ule index and Secchi disk depth. *Opt. Express* **2019**, *27*, 7642–7655. [CrossRef] [PubMed]
40. Werdell, P.J.; Bailey, S.W. An improved in situ data set for bio-optical algorithm development and ocean color satellite validation. *Remote Sens. Environ.* **2005**, *98*, 122–140. [CrossRef]
41. Nechad, B.; Ruddick, K.; Schroeder, T.; Blondeau-Patissier, D.; Cherukuru, N.; Brando, V.E.; Dekker, A.G.; Clementson, L.; Banks, A.; Maritorena, S.; et al. CoastColour Round Robin datasets, Version 1. Available online: https://doi.org/10.1594/PANGAEA.841950 (accessed on 10 February 2021).
42. Wang, S.; Lee, Z.; Shang, S.; Li, J.; Zhang, B.; Lin, G. Data File 1.csv. Available online: https://doi.org/10.6084/m9.figshare.7355903.v1 (accessed on 15 February 2021).
43. Wang, S.; Lee, Z.; Shang, S.; Li, J.; Zhang, B.; Lin, G. Data File 2.csv. Available online: https://doi.org/10.6084/m9.figshare.7355906.v1 (accessed on 15 February 2021).
44. Wang, J.; Tong, Y.; Feng, L.; Zhao, D.; Zheng, C.; Tang, J. Satellite-Observed Decreases in Water Turbidity in the Pearl River Estuary: Potential Linkage With Sea-Level Rise. *J. Geophys. Res. Ocean.* **2021**, *126*. [CrossRef]
45. Sathyendranath, S.; Jackson, T.; Brockmann, C.; Brotas, V.; Calton, B.; Chuprin, A.; Clements, O.; Cipollini, P.; Danne, O.; Dingle, J.; et al. ESA Ocean Colour Climate Change Initiative (Ocean_Colour_cci): Global chlorophyll-a data products gridded on a sinusoidal projection, Version 4.2. Available online: https://catalogue.ceda.ac.uk/uuid/99348189bd33459cbd597a58c30d8d10 (accessed on 22 December 2020).
46. Sathyendranath, S.; Brewin, R.J.W.; Brockmann, C.; Brotas, V.; Calton, B.; Chuprin, A.; Cipollini, P.; Couto, A.B.; Dingle, J.; Doerffer, R.; et al. An Ocean-Colour Time Series for Use in Climate Studies: The Experience of the Ocean-Colour Climate Change Initiative (OC-CCI). *Sensors* **2019**, *19*, 4285. [CrossRef]
47. Tibshirani, R.; Walther, G.; Hastie, T. Estimating the number of clusters in a data set via the gap statistic. *J. R. Stat. Soc. Ser. B* **2001**, *63*, 411–423. [CrossRef]
48. Hornik, K.; Feinerer, I.; Kober, M.; Buchta, C. Spherical k-means clustering. *J. Stat. Softw.* **2012**, *50*, 1–22. [CrossRef]
49. Zuhlke, M.; Fomferra, N.; Brockmann, C.; Peters, M.; Veci, L.; Malik, J.; Regner, P. SNAP (Sentinel Application Platform) and the ESA Sentinel 3 Toolbox. *Sentinel-3 for Science Workshop* **2015**, *734*, 21.
50. Solonenko, M.G.; Mobley, C.D. Inherent optical properties of Jerlov water types. *Appl. Opt.* **2015**, *54*, 5392–5401. [CrossRef] [PubMed]
51. Gordon, H.R.; Brown, O.B.; Evans, R.H.; Brown, J.W.; Smith, R.C.; Baker, K.S.; Clark, D.K. A semianalytic radiance model of ocean color. *J. Geophys. Res.* **1988**, *93*, 10909. [CrossRef]
52. Lee, Z.; Carder, K.L.; Arnone, R.A. Deriving inherent optical properties from water color: A multiband quasi-analytical algorithm for optically deep waters. *Appl. Opt.* **2002**, *41*, 5755–5772. [CrossRef] [PubMed]
53. Morio, J. Global and local sensitivity analysis methods for a physical system. *Eur. J. Phys.* **2011**, *32*, 1577–1583. [CrossRef]
54. Prieur, L.; Sathyendranath, S. An optical classification of coastal and oceanic waters based on the specific spectral absorption of phytoplankton pigments, dissolved organic matter, and other particulate materials. *Limnol. Oceanogr* **1981**, *26*, 671–689. [CrossRef]
55. Babin, M. Variations in the light absorption coefficients of phytoplankton, nonalgal particles, and dissolved organic matter in coastal waters around Europe. *J. Geophys. Res.* **2003**, *108*. [CrossRef]
56. Wang, S.; Li, J.; Shen, Q.; Zhang, B.; Zhang, F.; Lu, Z. MODIS-Based Radiometric Color Extraction and Classification of Inland Water With the Forel-Ule Scale: A Case Study of Lake Taihu. *IEEE J. Sel. Top. Appl. Earth Obs. Remote Sens.* **2015**, *8*, 907–918. [CrossRef]
57. Wang, S. Large-scale and Long-term Water Quality Remote Sensing Monitoring over Lakes Based on Water Color Index. Ph.D. Thesis, University of Chinese Academy of Sciences, Beijing, China, 2018.
58. Gower, J.F.R.; Doerffer, R.; Borstad, G.A. Interpretation of the 685nm peak in water-leaving radiance spectra in terms of fluorescence, absorption and scattering, and its observation by MERIS. *Int. J. Remote Sens.* **1999**, *20*, 1771–1786. [CrossRef]

59. Fan, Y.; Li, W.; Calzado, V.S.; Trees, C.; Stamnes, S.; Fournier, G.; McKee, D.; Stamnes, K. Inferring inherent optical properties and water constituent profiles from apparent optical properties. *Opt. Express* **2015**, *23*, A987–A1009. [CrossRef] [PubMed]
60. IOCCG. *Remote Sensing of Ocean Colour in Coastal, and Other Optical-Complex, Waters;* Reports of the International Ocean-Colour Coordinating Group, No. 3; Sathyendranath, S., Ed.; IOCCG: Dartmouth, NS, Canada, 2000.
61. Longhurst, A.; Sathyendranath, S.; Platt, T.; Caverhill, C. An estimate of global primary production in the ocean from satellite radiometer data. *J. Plankton Res.* **1995**, *17*, 1245–1271. [CrossRef]
62. IOCCG. *Partition of the Ocean into Ecological Provinces: Role of Ocean-Colour Radiometry;* Reports of the International Ocean-Colour Coordinating Group; No. 9; Dowell, M., Platt, T., Eds.; IOCCG: Dartmouth, NS, Canada, 2009.
63. Devred, E.; Sathyendranath, S.; Platt, T. Delineation of ecological provinces using ocean colour radiometry. *Mar. Ecol. Prog. Ser.* **2007**, *346*, 1–13. [CrossRef]

remote sensing

MDPI

Article

Offshore Oil Slick Detection: From Photo-Interpreter to Explainable Multi-Modal Deep Learning Models Using SAR Images and Contextual Data

Emna Amri [1,2,*], Pierre Dardouillet [1], Alexandre Benoit [1], Hermann Courteille [1], Philippe Bolon [1], Dominique Dubucq [2] and Anthony Credoz [2]

1 LISTIC Laboratory, Polytech Annecy-Chambery, University of Savoie Mont Blanc, F-74944 Annecy le Vieux, France; pierre.dardouillet@univ-smb.fr (P.D.); alexandre.benoit@univ-smb.fr (A.B.); hermann.courteille@univ-smb.fr (H.C.); philippe.bolon@univ-smb.fr (P.B.)

2 TotalEnergies S.E., Avenue Larribau, F-64018 Pau, France; dominique.dubucq@totalenergies.com (D.D.); anthony.credoz@totalenergies.com (A.C.)

* Correspondence: emna.amri@univ-smb.fr; Tel.: +33-0751-394-533

Abstract: Ocean surface monitoring, emphasizing oil slick detection, has become essential due to its importance for oil exploration and ecosystem risk prevention. Automation is now mandatory since the manual annotation process of oil by photo-interpreters is time-consuming and cannot process the data collected continuously by the available spaceborne sensors. Studies on automatic detection methods mainly focus on Synthetic Aperture Radar (SAR) data exclusively to detect anthropogenic (spills) or natural (seeps) oil slicks, all using limited datasets. The main goal is to maximize the detection of oil slicks of both natures while being robust to other phenomena that generate false alarms, called "lookalikes". To this end, this paper presents the automation of offshore oil slick detection on an extensive database of real and recent oil slick monitoring scenarios, including both types of slicks. It relies on slick annotations performed by expert photo-interpreters on Sentinel-1 SAR data over four years and three areas worldwide. In addition, contextual data such as wind estimates and infrastructure positions are included in the database as they are relevant data for oil detection. The contributions of this paper are: (i) A comparative study of deep learning approaches using SAR data. A semantic and instance segmentation analysis via FC-DenseNet and Mask R-CNN, respectively. (ii) A proposal for Fuse-FC-DenseNet, an extension of FC-DenseNet that fuses heterogeneous SAR and wind speed data for enhanced oil slick segmentation. (iii) An improved set of evaluation metrics dedicated to the task that considers contextual information. (iv) A visual explanation of deep learning predictions based on the SHapley Additive exPlanation (SHAP) method adapted to semantic segmentation. The proposed approach yields a detection performance of up to 94% of good detection with a false alarm reduction ranging from 14% to 34% compared to mono-modal models. These results provide new solutions to improve the detection of natural and anthropogenic oil slicks by providing tools that allow photo-interpreters to work more efficiently on a wide range of marine surfaces to be monitored worldwide. Such a tool will accelerate the oil slick detection task to keep up with the continuous sensor acquisition. This upstream work will allow us to study its possible integration into an industrial production pipeline. In addition, a prediction explanation is proposed, which can be integrated as a step to identify the appropriate methodology for presenting the predictions to the experts and understanding the obtained predictions and their sensitivity to contextual information. Thus it helps them to optimize their way of working.

Keywords: oil slicks; data fusion; offshore detection; SAR images; meteorological data; deep learning; AI explanation

Citation: Amri, E.; Dardouillet, P.; Benoit, A.; Courteille, H.; Bolon, P.; Dubucq, D.; Credoz, A. Offshore Oil Slick Detection: From Photo-Interpreter to Explainable Multi-Modal Deep Learning Models Using SAR Images and Contextual Data. *Remote Sens.* **2022**, *14*, 3565. https://doi.org/10.3390/rs14153565

Academic Editor: Merv Fingas

Received: 12 June 2022
Accepted: 19 July 2022
Published: 25 July 2022

1. Introduction

Throughout the era of offshore data, the detection of oil slicks originating from anthropogenic (spills) or natural (seeps) sources has always been a long-standing challenge. Offshore oil slick monitoring is a relevant topic for a broad audience, including scientists, environmentalists, and local authorities. The devastating effects of marine pollution, including the deterioration of aquatic ecosystems, make oil slick detection a mandatory task [1].

Previous studies [2,3] have highlighted the usefulness of Remote Sensing (RS) technology for offshore monitoring. In particular, Synthetic Aperture Radar (SAR) technology has been identified as an effective technology for detecting marine pollution [4,5]. The main advantage of SAR technology is that it is independent of sunlight, weather, and clouds and allows for global coverage. These results provide new solutions to improve the detection of natural and anthropogenic oil slicks by providing tools that allow photo-interpreters to work more efficiently on a wide range of marine surfaces to be monitored worldwide. Such a tool will accelerate the oil slick detection task to keep up with the continuous sensor acquisition. This upstream work will allow us to study its possible integration into an industrial production pipeline. Nevertheless, slick detection has remained a challenge due to the high variability of their nature, shape, and extent. Oil slicks have no typical characteristics and vary according to the environment and surrounding conditions, which makes their detection very complex. As oil slicks are present on the sea surface, their shape depends on weather conditions and their source (e.g., an elongated dark patch if the origin is a moving ship or a random pattern if it comes from a platform or natural seepage). In addition, the acquisition time of the SAR image containing the oil is delayed relative to the time of its appearance. Thus, the shape of the oil slick then has time to evolve significantly with the help of weather conditions that influence the physio-chemical properties of the oil slick (e.g., fragmentation of the slick into droplets, dissolution in seawater, etc.). Another challenge is the potential confusion with similar patterns such as algae, low wind areas, and up-welling [2].

Indeed, multiple studies such as Brekke et al. [4], Alpers et al. [2], Solberg et al. [6], and Espedal [7] point out that improving slick detection using SAR requires the inclusion of more ancillary contextual information such as meteorological information. Finally, from an application point of view, an additional challenge relates to the fast processing of large quantities of data to assist human experts in real-time monitoring. Such an aim is actually no more feasible by sole human experts when considering the high spatial resolution and high revisit frequency of the sensors required for the task.

The state-of-the-art on this topic involves manual inspection, pattern detection, and thresholding methods based on various feature categories [8,9]. However, the latter exhibit poor generalization behavior and lack robustness against false detection due to lookalikes.

Neural networks and, more particularly, deep neural networks (DNNs) [10] have recently shown an increasing interest in improving over classical approaches both in terms of detection accuracy and generalization capability [11]. However, those works rely on the sole use of SAR images.

In this paper, we study offshore oil slick detection using deep neural network approaches in a supervised manner, taking full advantage of massive annotated datasets of real recent slicks monitoring scenarios manually annotated by human experts (photo-interpreter). The objective is to provide new solutions to improve the detection of natural and anthropogenic oil slicks by providing tools that allow photo-interpreters to work more efficiently on a wide range of marine surfaces to be monitored worldwide. This tool will speed up their detection task to keep up with continuous sensor acquisition. As this is a difficult task with many challenges, including high target variability and the potential for false alarms, providing rapid predictions should be valuable.

To this end, we consider multi-modal deep learning approaches, allowing heterogeneous data fusion taking into account SAR images and wind information.

To the best of our knowledge, this is the first study combining meteorological information with SAR data and evaluating the impact of wind speed on slick detection on a broad collection of data spanning different regions in the world with a wide diversity of real slick cases. This upstream work will allow us to study its possible integration into an industrial production pipeline.

This paper presents the following contributions: first, a slick detection performance analysis of structurally different deep neural networks is conducted. Second, a new deep neural network model structure considers the fusion of SAR information with wind speed information. Third, a refined performance analysis method is proposed, taking into account contextual factors such as wind speed and human infrastructure position. Finally, model prediction explanations are proposed. It relies on adapting the SHAP [12] method to the semantic segmentation problem and allows the input features contributing to the local decision to be highlighted.

1.1. Offshore Oil Slick Detection and the Related Literature

1.1.1. Oil Slick Observation on the Sea Surface

Oil slicks observed on the sea surface are commonly of two types: spills and seeps. The causes of oil spills can be discharges of crude oil from tankers, offshore platforms, ships, drilling rigs, and spills of refined petroleum products or used oil. On the other hand, seeps are naturally occurring oil flows that escape from the ground through soil fractures and sediments to the sea surface.

These differences in nature lead to variations in oil slick characteristics, such as viscosity and thickness, resulting in different behaviors and increasing the variability of oil slick observations.

The observation of offshore oil slicks is conducted mainly by RS, specifically by active sensors such as SAR. SAR technology relies on ElectroMagnetic (EM) signals sensitive to the sea surface roughness. The intensity of SAR images is related to the strength of the backscattered radar signal. In more detail, the energy transmitted by the SAR sensor is backscattered with characteristics that highlight the properties of the areas involved. In the case of a calm sea, most of the transmitted energy is reflected away from the radar, resulting in minimal backscatter to the sensor and a darker area on the resulting image. Conversely, in the case of a rough surface due to wind, a more significant part of the EM energy is backscattered from the surface [2] and thus yields a brighter area with speckle noise.

Further, if an oil slick appears, it dampens the waves on the sea surface, reducing the surface roughness and the corresponding radar backscatter. This is due to the viscous damping of short gravity/capillary waves (wavelengths of a few centimeters) by the oil slick or oil/water mixture, whose viscosity is much higher than water [2]. As a result, oil slicks appear on SAR images as dark patches compared to the surrounding clean sea, as illustrated in Figure 1. Slick characteristics are widely variable, such as the contrast value, which depends on the local sea state, the slick type, the image resolution, the SAR frequency, and the incidence angle.

Besides the oil slicks on the sea surface, several phenomena referred to as lookalikes can generate similar radar signatures (low backscatter areas) that can yield false alarms (FA). This generally originates from algal blooms, sargassum, and upwelling [4]. More generally, any patch that is darker than the surrounding area could be an oil slick. Illustrations of spills, seeps, and lookalikes are shown in Figure 2.

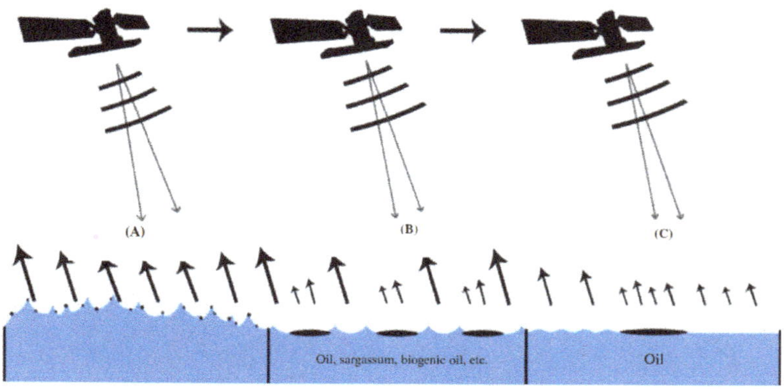

Figure 1. Illustration of SAR backscatter with different sea roughness. The wind conditions at the time of the data collection constrain the backscattered energy properties [13]. (A) represent strong winds (winds > 10 m/s), (B) represent ideal winds (7 m/s> Winds > 3 m/s) and (C) represent weak winds (winds < 3 m/s).

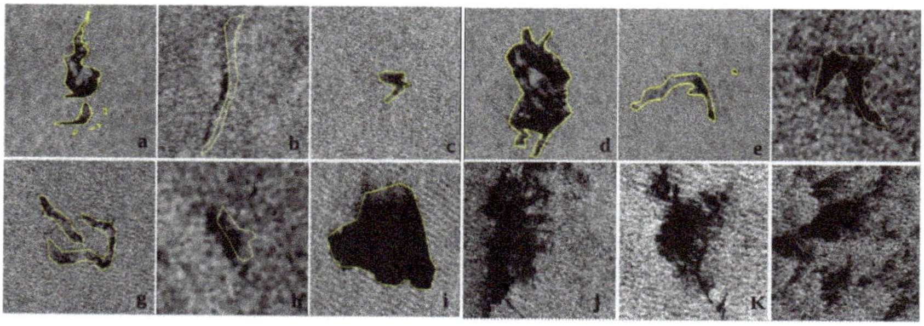

Figure 2. Examples of oil slicks marked in yellow polygons (**a**–**i**) and lookalike phenomena (**j**–**l**) extracted from Sentinel-1 data.

To summarize, the main factors involved in oil slick detection from SAR images are related to contextual information such as wind conditions, sensor characteristics, and the presence of lookalikes.

1.1.2. Contextual Data: Impact on Oil Slick Detection

In oil slick detection, numerous studies have emphasized the importance of contextual information as the main factor impacting the monitoring of offshore oil slicks using SAR imagery. The study proposed by Brekke et al. [4] highlights the interest in weather conditions, distance from ships, and infrastructure position (platform, pipeline, etc.). Meteorological weather conditions are emphasized, namely wind speed, which affects the oil slick profile (oxidation, biodegradation, dispersion, sedimentation, etc.) and changes its characteristics (size, shape, etc.) [14]. In addition, wind speed impacts the backscatter contrast between the sea and slick areas. On the one hand, oil slicks tend to scatter gradually into smaller parts at moderate wind speeds and disappear as wind speed increases. On the other hand, local low wind speeds can generate areas of low backscatter and, therefore, lookalikes [15]. Observing the effect of wind on slick characterization reveals the apparent importance of this information in the slick detection and characterization process.

Moreover, during oil slick manual detection by the photo-interpreters, the evaluation can be adjusted by taking the instantaneous wind speed into account. According to Fingas et al. [3], the wind speed range for oil detectability is (1.5 m/s, 6–10 m/s). Beyond this range, the oil signature will be further removed. The most accepted limits are 1.5 to 10 m/s. La et al. [16] and Brekke et al. [4] have further refined this range, a summary of which is provided in Table 1. However, it may remain specific to small-scale, local studies that do not permit generalization. As a general rule, these experiments report the requirement for moderate wind speeds. This paper considers a trade-off between the proposed wind speed ranges that seems relevant to our large-scale study.

Table 1. Range of wind speeds for oil slick detection according to the literature.

Wind Speed m/s	Reference	Year
1.5 to 6	[3]	2014
2 to 7	[16]	2018
2.09 to 8.33	[17]	2017
3 to 7–10	[4]	2005

Further, in the process of oil spill detection, the photo-interpreters are informed about the positions of the infrastructure since pollution can originate from human activities. The deballasting of the ship hold is easily recognizable thanks to the strong backscattering point created by the ship at the end of the oil slick when the SAR image is captured. The geometry of the oil slick is also generally straight along the ship's path due to the speed effect. Oil spills can also originate from underwater infrastructures and conduits (pipes) designed to carry oil. For the situations mentioned above, ship and platform positions can be spatially detected based on the diffraction points observed in the SAR images. This can help distinguish anthropogenic oil from natural oil based on the distance of the slicks from the infrastructure in the area [2].

1.1.3. Classical Methods for Oil Slick Detection

The state-of-the-art of offshore oil slick monitoring is extensive. A brief classification of the main approaches is presented below based on the surveys proposed by Alpers et al. [2], Brekke et al. [4], and Al-Ruzouq et al. [18].

Entirely Manual Inspection: Oil slick detection on SAR images is essentially manual. Operators (photo-interpreters) are trained to analyze images to detect oil slicks versus lookalikes and differentiate between natural and anthropogenic oil. The class assignment (spill, seep, sea including lookalikes) is based on the following features: the contrast level with the surroundings, the homogeneity of the surroundings, the wind speed, the oil platforms, ships and natural slicks in the proximity, as well as the shape and edge of the patch. This detection method is tedious, time-consuming, and costly regarding resources.

Conventional Approaches: This category of approaches focuses mainly on three steps: the first is to detect the dark patterns, the second is to extract their features, and the third is to classify them [3]. Conventional features can belong to several categories, such as geometric, statistical, and polarimetric features [4,19,20]. Such approaches, however, exhibit poor generalization behavior and lack robustness against lookalikes [18,21].

Semi-Automatic Approaches: Some processing stages of the conventional approaches, such as dark pattern detection, can rely on machine learning. For instance, the integration of Neural Networks improves the traditional process. Nevertheless, such a general approach still has limited generalization behaviors and keeps high false alarm rates [2].

1.1.4. Deep Learning Methods for Oil Slick Detection

Various facts have directed the search for the automation of oil detection toward end-to-end approaches driven by deep Convolutional Neural Networks (CNNs). Among these is overcoming the shortcomings of conventional approaches reported by state-of-the-art; lack of studies on both oil types, lack of relevant features to distinguish oil slicks from lookalike phenomena and the limited generalization capability. The remarkable results of CNN-based approaches have led to dramatic advances in the state-of-the-art for fundamental computer vision problems such as object detection, object localization, and semantic and instance segmentation [22–24]. Several studies have compared deep learning techniques to classical classifiers and indicated better performance with deep learning techniques [20,25]. The studies report the ability of CNNs to perform both feature extraction and classification, allowing the exploration of relevant features for better discrimination between the oil slick and background patterns. The ability of the CNN network to leverage the extensive existing data can ensure a certain level of generalization capability [18].

Table 2 reports examples of recent NN models applied to oil slick detection and segmentation from Sentinel-1 images, indicating the type of oil spill targeted by the study along with the number and size of the used images. Various models of NNs are proposed; some perform semantic segmentation, such as Unet [26], and others perform object detection and instance segmentation, such as Mask R-CNN. It should be noted that few annotated SAR data are available, which limits supervised learning of large models and explains the use of Transfer Learning (TL) strategies, i.e., representations of data preliminary learned in other tasks and domains.

Table 2. Commonly used CNN models for oil slick detection.

Architecture	Crop Number	Image Size	Spill & Seep	TL
OFCN/UNet [26]	713	160 × 160	spill	-
Fully CNNs [27]	-	128 × 128, 2048 × 2048	spill	-
Mask R-CNN [28,29]	9302	512 × 512	spill and seep	✓
DeepLab [30]	677	1252 × 609	spill	-
DeepLabv3+ [11]	1002	321 × 321	spill	✓
AutoEncoders [25]	-	256 × 256, 384 × 384	spill	-
GANs [31]	-	256 × 256	spill	-

2. Materials and Methods

2.1. Oil Slicks Segmentation Methods

In this study, we consider both transfer learning and full model learning (learning from scratch). For the first method, the model is pre-trained on a different dataset and task and is next transferred and fine-tuned for oil slick detection on dedicated datasets. As for the second method, we focus on model optimization exclusively performed on the data of interest without any pre-training.

Since the goal is to find the precise location of oil slicks in SAR images, standard object detection methods are not appropriate because they commonly provide object bounding boxes that may be too large compared to the thin oil slicks to be detected. In this case, the segmentation of oil slick instances is more relevant. Such a strategy opens wide doors to a variety of approaches. In this work, a comparison is made between a convolutional neural network, for instance, segmentation: Mask Region-Based Convolutional Neural Networks (Mask R-CNN [32]) with object segmentation capability and a semantic segmentation neural network Fully Convolutional DenseNet (FC-DenseNet [33]). Instance and semantic segmentation methods are comparable in this case study since we do not face occlusion issues of oil slicks on the sea surface. Consequently, applying connected components to the semantic segmentation predictions yields instance segmentation.

As for our experimental strategy, we first compare the performance of models using only SAR data in order to identify the best parameter configuration (number of layers, learning rate, etc.). Then, we investigate heterogeneous data fusion strategies to fuse wind speed information with SAR information.

2.1.1. Instance Segmentation: Mask R-CNN

Mask R-CNN is a multi-task model that generates bounding boxes of the target object regions, as well as masks for object classification and segmentation. The first step is to extract features r from the input x using the parameter set θ_b, such that $f_b(x, \theta_b) = r$. Then, the features are passed to several heads $f_{t_i}(r, \theta_{t_i}) = y_{t_i}$, with t_i referring to the $i-th$ task.

The model is shown in Figure 3. Nevertheless, this complex deep architecture must be trained by transfer learning when targets are scarce. As described in a previous study [28], the parameters of a pre-trained model are transferred from the COCO dataset [34]. Subsequently, we gradually adjust the transferred weights on our data, starting with the heads and moving towards the backbone. This whole process is achieved while reducing the learning rate as described in the following.

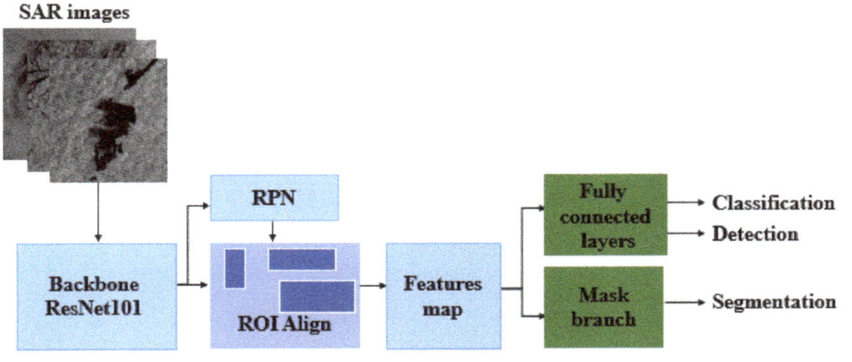

Figure 3. Mask R-CNN architecture. RPN stands for region Proposal Network, ROI stands for Region Of Interest.

2.1.2. Model Parameters Configuration

The implementation from Abdulla [35] was adopted to accommodate the Mask R-CNN model. Several experiments were performed to select the parameters and adapt the problem properly; the main ones are described below.

Backbone selection: The backbone network could be any CNN designed for image classification, such as ResNet-50 or ResNet-101.

Loss selection: Since oil slicks are scarce and diverse, one needs to down-weight easy examples and focus the training on hard ones. Thus, we use the focal loss [36] instead of the cross-entropy (CE) for mask loss computation. This loss is also more adapted than CE for highly imbalanced classes. It is formulated as follows: $L(p_t) = -\alpha_t(1-p_t)^{\lambda} log(p_t)$ where p_t is the model's slick detection probability. The role of the α_t and λ parameters is to down-weight easy examples (error loss) and thus focus training on hard negatives [36].

Learning strategies: as reported in [28], relevant optimization of Mask R-CNN can be achieved with multiple training phases. This decomposition can give us training flexibility. The primary approach is to train all the networks in a single stage. A second approach consists of two steps: training the model heads first while using transferred weights for the rest of the network, then fine-tuning all networks. Finally, the three-step learning strategy trains the model heads over a few epochs first while also using transferred weights. Then, all but the first four ResNet layers are trained over additional epochs. Finally, the entire network is fine-tuned while reducing the learning rate. For this work, the approach used is to train the model based on the three-step learning strategy.

2.1.3. Semantic Segmentation: FC-DenseNet

We propose a refined version of the FC-DenseNet model [33], a well-known extension of densely connected convolutional networks (DenseNets) [37]. The DenseNets architecture has been proposed to maximize feature reuse and limit the model depth and computational costs compared to the classical U-net structure [38]. The FC-DenseNet extends the DenseNets classification architecture to perform semantic segmentation by adding an upsampling path to perform pixel-level classification. This architecture belongs to the category of encoder-decoders, as shown in Figure 4. It is built from dense blocks and sampling operations.

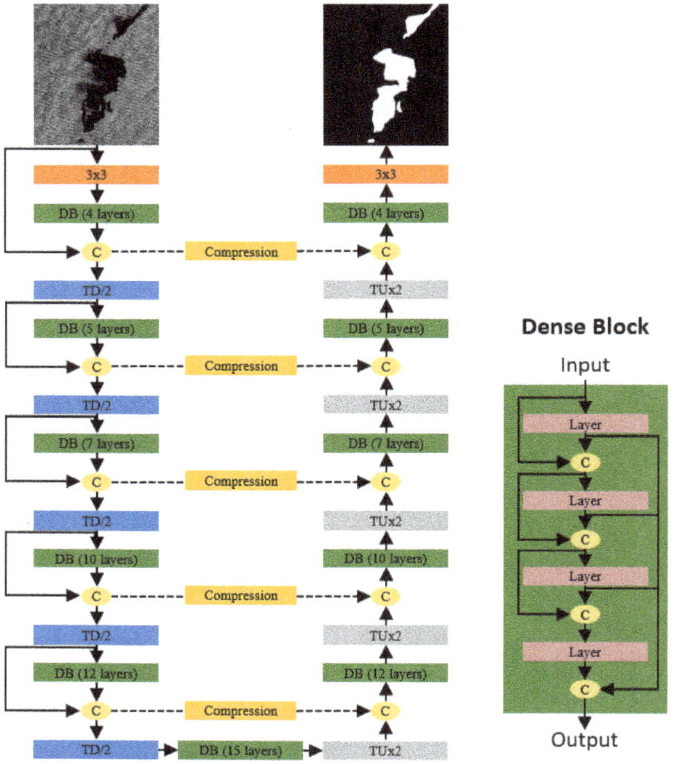

Figure 4. Architecture of the enhanced FC-DenseNet version. Orange blocks are convolutions, DB: dense blocks, TD: transitions down (2 times spatial downscale), TU: transitions up (2 times spatial upscale), C corresponds to the concatenation operation, and the dashed lines are skip connections.

This architecture uses the so-called skip connections, allowing the transmission of low-level feature maps from the encoder to the decoder. The decoder thus performs a concatenation of the low-level abstract feature provided by the encoder with the high-level semantic feature available in the decoder. This results in a more refined and more accurate prediction map that fully exploits the details of the high-resolution first model features. Moreover, skip connections facilitate the model learning by transmitting the gradient error directly to the first layers of the model and thus reducing gradient fading. Similarly, at the dense block level, skip connections are used. All previous layer feature maps are concatenated and used as inputs for each layer, and their feature maps are used as inputs for all subsequent layers. Each layer generates an activation map designed according to a connectivity pattern that iteratively concatenates all feature outputs in a feed-forward approach, according to Equation (1) [33]:

$$f_l = H_l(\{f_{l-1}, f_{l-2}, \ldots, f_0\}, \theta_l) \tag{1}$$

where f_l indicates the output feature map of the lth layer. It is computed by applying a non-linear transformation H_l on the concatenation of the previous layers' feature maps $f_{l-1}, f_{l-2}, \ldots, f_0$. H_l is a non-linear transformation defined as a convolution with trained parameters θ_l, followed by a *ReLU* and dropout. It then maximizes feature reuse and facilitates the training of deep structures.

2.1.4. Model Parameters Configuration

Intensive experiments were conducted to search for the best model and to select the appropriate parameters. The main ones are described below.

- *Optimizer:* Different gradient-based optimization algorithms of objective functions have been introduced so far, such as Stochastic Gradient Descent (SGD) [39], Adam [40], and RMSprop [40]. Driven by the work of Kingma et al. [40], we use the Adam algorithm in our work. It is computationally efficient, requires little memory, is invariant to the diagonal scaling of gradients, and is well suited to large problems in terms of data and/or parameters. We consider an initial learning rate of 0.00005 to begin the learning. Subsequently, a learning rate decay policy is applied: it is reduced by a factor of ten if the validation loss has not improved in the last 150 epochs. This latency time does not result in an over-fitting, thanks to the dropout existing in all the model layers.
- *Loss function:* An extreme imbalance is observed between foreground (slicks) and background (sea) classes on the considered data collections. A set of losses has been experimented with, such as focal loss [36], dice loss [41], and cross-entropy. The main objective is to select a loss taking into account the class imbalance, focusing the training on difficult cases. The loss chosen based on the experiment is the sum of the dice loss and the cross-entropy.
- *Batch size:* Batch size, rather than optimizing the network from one sample at a time, leading to non-optimal solutions, averaging the errors over a set of samples has proven to be more efficient. We use a batch of 4 samples due to hardware limitations.
- *Model depth and width:* We experiment FC-DenseNet with various hyper-parameter combinations controlling model depth (number of dense blocks) and width (number of neurons per layer, also referred to as growth rate). Experiments show that a higher depth gives the best results. We chose to set the number of dense blocks to 5 and the number of feature maps per layer to 16. Thus, the total number of parameters of the corresponding model is about 8 M.

2.1.5. FC-DenseNet Model Enhancements

We propose an improved version of the FC-DenseNet model based on several optimizations listed below.

- *Layer initialization:* The choice of initial parameter values for gradient-based optimization is very crucial. Following [42], we chose random orthogonal initial weights to start with complementary operators and to accelerate the convergence compared to a Gaussian initialization. Such an approach indeed leads to faithful gradient propagation, even in deep non-linear networks, by combating exploding and vanishing gradients. Further, regarding the initialization of the last linear classification layer, it is common to use a bias b = 0. However, Lin et al. [36] point out that this could cause instability during training for obtaining class probabilities. Therefore, for training, we initialize the bias of the last layer as $b = -log((1-cf)/cf)$, where $cf = 1/C$ where C is the number of classes.
- *Non-linearity:* In the original FC-DenseNet model, $ReLU(x) = max(0, x)$ is considered as an activation function. Its main advantage is the non-saturation of its gradient, which leads to faster convergence of the training process [43]. Improved versions of ReLU activation have been proposed, such as Leaky-ReLU [44]. Such activation enables the transformation of the negative input signal instead of canceling it as for ReLU. This activation can interest classical model structures with no or few skip connections to avoid losing important information. However, the dense blocks of the model allow all the data to be shared across layers with dense connections, thus maximizing feature reuse before applying activation functions and justifying the use of RELU.

- *Model regularization:* Original FC-DenseNet relies on batch normalization with large batch size. However, small batch size is required to handle large images while working with limited GPU memory. In this situation, batch normalization is discarded. A chosen alternative is the use of Spectrum Restricted Isometry (SRIP) weights regularization [45] for all but the last layer. This regularization technique does not induce additional computation at test time and provides notable advantages: it ensures feature normalization while maintaining the orthogonality of neuron parameters throughout the training, which is complementary to the initialization strategy [45]. This approach then ensures that neural kernels act in a complementary way, which is relevant when dealing with few feature maps per layer as for the FC-DenseNet structure.
- *Skip connection compression:* It aims at reducing the size of the features passed from the encoding part to the decoding one. Actually, for the vanilla FC-DenseNet, the dimensions of the low-level features outing from the encoder to be fused with the decoder features are higher than their counterpart. A compression then allows for a balance of feature dimensions and reduces model complexity on the decoder side. In this work, compression consists of 1×1 convolutions. The compression ratio is adjusted so that the feature maps of the skip connections and the previous dense block have the same size before they are concatenated.
- *Upsampling:* Transposed convolutions are generally considered but tend to introduce checkerboard artifacts on the outputs [46]. An alternative is to separate upsampling to a higher resolution from convolutions to compute features. We then first resize the image using nearest-neighbor interpolation and then apply 2D convolution layers as proposed in [47].

2.1.6. Detecting from Heterogeneous Sources, Fuse-FC-DenseNet

Since the radar backscatter at the sea surface is deeply affected by the wind speed, as detailed in Section 1.1.2, the fusion of wind speed with SAR information is an exciting approach for oil detection. In this work, the fusion of wind and SAR data is performed in two different ways, as illustrated in Figure 5 (1 and 2).

(1) Early fusion: Wind and SAR modalities are considered at the same level. Thus, the network input consists of the SAR data channel and the corresponding wind speed channel, resampled during the data processing step to match the SAR image resolution.

(2) Late fusion: The SAR and wind modalities are considered as two specific input channels with their separate layers before their fusion. In this approach, while keeping the enhanced FC-DenseNet structure for the SAR data, the wind modality is introduced in subsequent dense blocks along the model encoding path. The merging process is performed after the second or third block so that the wind features at their original scale are fused with the SAR features when the latter reach a similar resolution after a few downscaling steps along the encoding section. As for the general DenseNet approach, this fusion operation consists of concatenating the features of each channel. Regarding the processing of wind channels, several strategies can be investigated. In this paper, we consider two approaches: (i) applying a simple average pooling to adjust feature scales and (ii) applying mean clustering and then extracting higher-level features with a dense block.

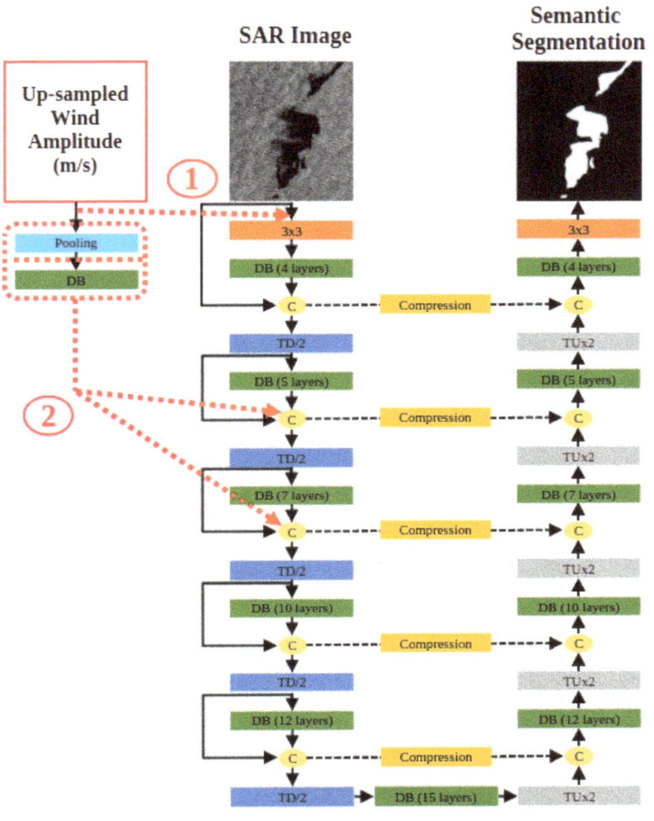

Figure 5. Fuse-FC-DenseNet architectures that fuse heterogeneous data, wind, and SAR information. The left branch is the wind branch, and the operations surrounded by red dotted lines are the different ways to fuse the wind speed information with the SAR information. (1) represents the early fusion, (2) represents the late fusion. The operations surrounded by red dotted lines are selected depending on the experiment.

2.2. Experimental Data

The experimental data considered in this work are acquired over four years (2015–2019) in three separate study areas shown in Figure 6. The training, validation, and test datasets were acquired in these three areas but differed in acquisition date and coverage. This location diversity enriches our database with a larger area to explore and more variety in terms of statistics (weather, infrastructure, etc.). Below is a description of all the data considered in our study.

- *Sentinel-1 SAR data*: Acquired by the European Space Agency (ESA) organization in Interferometric Wide-swath Mode, C-band (5.40 GHz) and with a 10 m resolution per pixel. The level of signal backscattered by the sea surface is higher for vertically polarized waves (V) than for horizontally polarized waves (H) [48]. Hence, vertical polarizations for transmission and reception (VV channel) are selected as they are generally preferred to the HH channel for ocean studies [49]. A set of 1428 images is considered in this study.
- *Slick annotation (Ground Truth)*: Human experts performed manual annotation of natural and anthropogenic oil slicks. The considered classes are: sea, spill, and seep. Importantly, this study is based on real-world monitoring scenarios. Under these conditions,

the photo-interpreters cannot provide accurate annotations on object boundaries due to the fuzzy contours of the slicks and the annotation rhythm. Therefore, the annotations show sharp transitions that do not reflect the actual slick shape. Further, slick annotation is performed by five photo-interpreters, revealing a diversity of annotation criteria. For example, some slick annotations do not match the black slick perfectly and are shifted by a few pixels. Others account for small slick patches separated from the main slick and displaced. The difference is shown in Figure 2a–j. Figure 7 shows an example of image annotation. One can then consider that annotation is noisy but satisfies operational monitoring requirements.

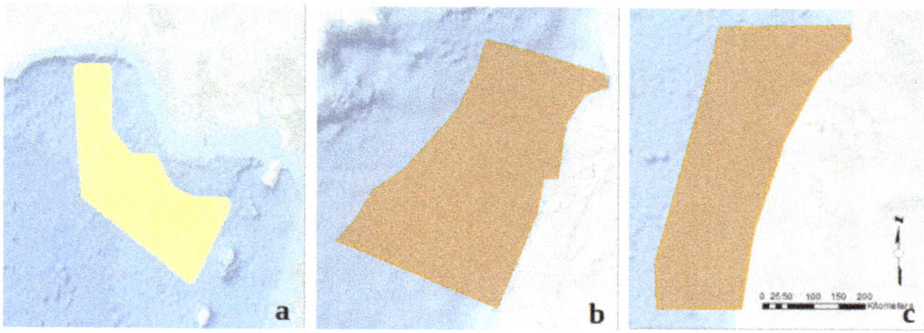

Figure 6. Location maps of the considered Areas Of Interest (AOI) captured from the Sentinel-1 sensor: (**a**,**b**) two areas of the Atlantic Ocean coast are located in Southern Africa (Nigeria and Namibia), (**c**) the Mediterranean Sea in Western Asia (Lebanon). The orange areas are used for training and validation, and the yellow area is used for testing.

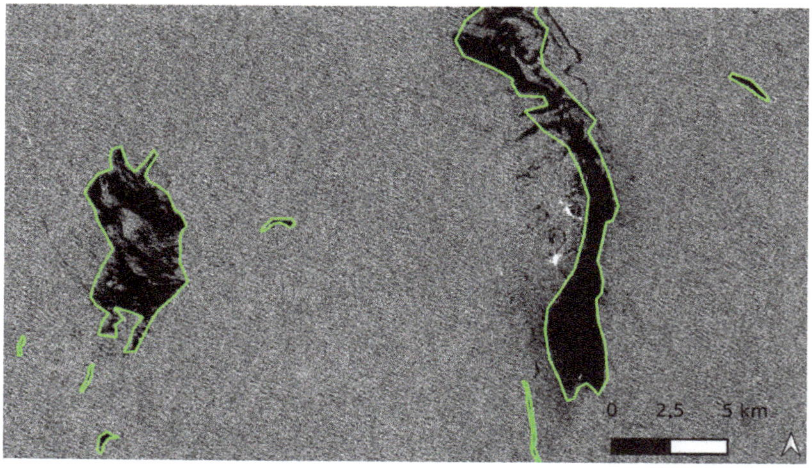

Figure 7. Example of slick annotation on a Sentinel-1 image on 21 January 2020 in the Nigeria area, where annotations are highlighted in green.

- *Wind speed information*: Wind speed estimates for each SAR image are also provided. The estimation is performed based on empirical Geophysical Model Functions (GMFs), relying on the relationship between the backscatter and the wind speed. For C-band and VV polarization, several variants of GMFs named CMODs have been developed [50]. The CMOD5 variant is a version that has been used successfully as an improvement of the above variants. It is formulated as Equation (2):

$$\sigma^0_{CMOD5}(V, \phi, \theta) = b_0(1 + b_1 \cos\phi + b_2 \cos\phi)^{1.6} \quad (2)$$

where the variables are:

σ^0_{CMOD5}: backscatter value of the model CMOD5,
V: wind speed (m/s),
ϕ: relative direction between the radar look direction and the wind direction,
θ: angle of incidence,
b_0, b_1, b_2: functions of wind speed V and incidence angle θ.

CMOD5 estimates the radar backscatter in a scene as a function of the surface wind speed (V) and the angle that the wind makes with respect to the direction of the pulse (ϕ) and the incidence angle (θ) [51]. However, for surfaces such as slick areas, the wind speed is underestimated due to the wave damping effect. This discrepancy has been considered at the model performance assessment step.
In our context, SAR and wind resolutions are 10 and 100 m per pixel, respectively, such that wind is upsampled by a factor of 10 using linear interpolation when required. An example of a Sentinel-1 SAR image before and after adaptation to CMOD5 is shown in Figure 8.

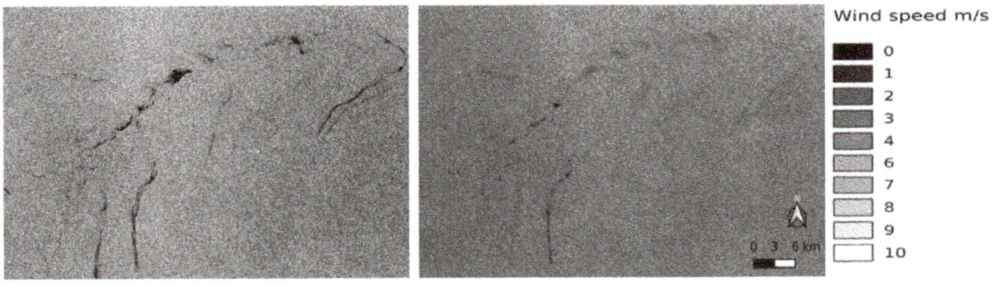

Figure 8. Illustration of the wind information associated with the SAR image on 21 January 2020 in the Nigeria area. The SAR image with slicks and lookalike phenomena on the left, on the right, the associated wind speed data.

- *Infrastructure Position*: This information is known using global referencing, which represents support for photo-interpreter analysis [4]. Through the use of infrastructure position, we further improve the evaluation of model performance in terms of detecting anthropogenic oil. The types of infrastructure considered are pipelines, wells, ports, platforms and ships. For each of them, there is a possibility of leakage accorded to the experts. The location of the infrastructures is associated with the SAR images, as shown in Figure 9.

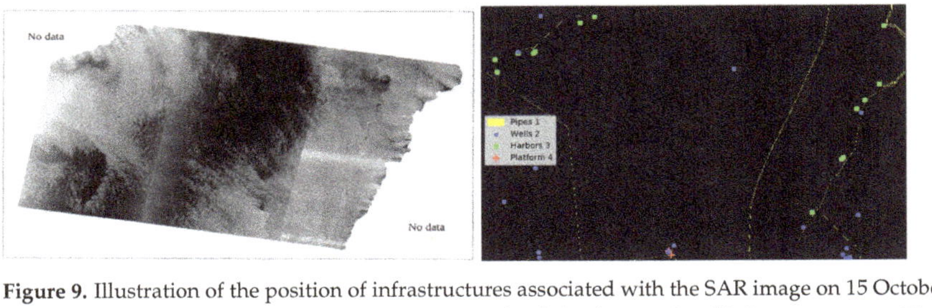

Figure 9. Illustration of the position of infrastructures associated with the SAR image on 15 October 2018 in the Lebanon area. On the **left**, SAR image, on the **right**, the associated infrastructures position.

To summarize, for each SAR image available in the dataset, the associated wind speed, infrastructure position and slick annotation are provided.

2.2.1. Data Preprocessing

To enable heterogeneous data processing, we describe below the preprocessing of each data and its purpose.

- SAR Data: The preprocessing of initial SAR data consists of low-level transformations to improve the qualitative and quantitative interpretation of image components, thus facilitating the visibility of slicks for photo-interpreters. SAR preprocessing can be grouped into four processes: radiometric calibration, geo-referencing, filtering, and masking [18].
  (1) First, radiometric correction and calibration: Its purpose is to remove or minimize radiometric distortions and to ensure the correlation of pixel values with the backscatter coefficient of the reflecting surface [52]. Thus, quantitative measurements (backscattered microwave energy) restored from image pixel values can be compared with object characteristics in multi-temporal SAR images acquired with different sensors and SAR modes [53].
  (2) A *geo-referencing* step of the SAR data is then performed to correct eventual geometric distortions and to locate each pixel of the image on the Earth [54]. This step is also applied to all the considered data sources to ensure their alignment.
  (3) A speckle noise filtering step is then performed. As reported by [3,55], such noise must be reduced in order to facilitate the analysis and interpretation of the data. An optimal *speckle filtering* technique should preserve useful radiometric information and avoid the loss of features, such as the local mean of backscatter, texture, edges, and point targets [56]. Several filter types have been used in previous studies to reduce speckles and enhance SAR images for oil slicks, such as Lee, Frost, Kuan, median, and Lopez [57]. The considered preprocessing pipeline relies on such filters, but its detailed implementation remains confidential.
  (4) The final preprocessing step consists of *masking the land and shorelines* from the SAR images. This process restricts sea surface analysis and prevents land from interfering with oil slick detection [58].
- Wind data: For each 10 m resolution SAR image, the associated wind intensity map is provided with the same geographic coverage but at 100 m resolution. Therefore, to align the two modalities, bi-linear interpolation is applied to the wind map. However, the wind speed is underestimated over the slick areas due to the wave damping effect. Then, specifically for the test set and evaluation process but not for the training set, the wind speed, in the vicinity of 50 m around the annotated slicks, is merged within the slick area, relying on iterative median filtering.
- Infrastructure data: It is based on the infrastructure position; a map that represents the proximity of existing infrastructure in the neighborhood for each pixel is realized. These distance maps will be considered only in the evaluation process.

2.2.2. Preparation of the Training/Validation Datasets

After processing the different data modalities separately, an alignment of each SAR image with the corresponding slick annotations, wind speed map, and infrastructure distance map is performed. Then, datasets of heterogeneous data are built for training, validation, and testing sets. The strategy is described below.

(1) Dataset Splitting: Following our previous study [28], both training and validation sets rely on the same geographical areas. However, they do not share the same images; each has a different capture date and sea coverage. A third area (Nigeria) is chosen for testing to validate the generalization capability of the model.
(2) Image Crop Selection: We built a collection of smaller image crops of 512 × 512 pixels from the large images of the training and validation datasets. This resolution is a

compromise between the size of the layer, the field of view of the model, and the memory constraints of the GPUs (NVidia V100 16 Gb).
The crop selection strategy takes into account the slick annotations and ensures the presence of slicks within crops following the logical function presented in Equation (3):

$$C_{slick}(X,Y) = \neg B(X) \wedge (E(Y) > T) \tag{3}$$

where $\wedge$ and $\neg$ are the logical AND and NOT operators, respectively, X is a random crop in a large SAR image, Y is the corresponding annotation, $B(X)$ a Boolean function that checks if there is a border (no data area) inside the crop, $E(Y)$ is the entropy function used to check the sea and slick classes statistics within the crop, and T is a threshold fixed heuristically to 0.3.
Further, to make the model robust against slick lookalikes, an additional crop selection of such potential patterns is built from the slick-free image areas. Since no specific annotation reports them, we rely on a heuristic reported in Equation (4), the objective of which is to choose crops with contrasting patterns:

$$C_{lookalikes}(X) = \neg B(X) \wedge V(X) \wedge \neg E(Y))) \tag{4}$$

where $\wedge$ is the logical OR operator, $B(X)$ and $E(Y)$ correspond to the ones described in Equation (3), and $V(X)$ is a Boolean function that randomly selects the crop that probably contains lookalikes based on the normalized variance of the pixel values. A minimal variance threshold fixed on the basis of experiments must be reached to highlight a contrasted area.

(3) Data Augmentation: It is applied to increase the variability in the dataset artificially. It is comprised of random horizontal and vertical flipping and ±90° random rotation in both training and validation datasets [59].
Table 3 outlines the details of the training and validation datasets. A total of 85% of the crops in the dataset belong to the training set. We notice that the oil slicks are small to medium in size compared to the large sea area, covering less than 11% of the total area. This emphasizes the strong imbalance in the number of slicks pixels and the clean sea pixels that include lookalikes.

Table 3. Statistics of the training and validation dataset consisting of crops with a size of 512 × 512 pixels.

	Spill	**Seep**	**Sea**	**Total**
Image crops number	1.964	244	860	3.068
Studied area surface (hm^2)	251,200	21,900	7,769,500	8,042,600
Surface rate	3.12%	0.27%	96.6%	100%

2.2.3. Test Dataset

Many images are captured in an additional region close to Africa that has not been used for training/validation. This region provides the opportunity to test the generalization ability of our models through a variety of meteorological conditions. Table 4 indicates the characteristics of the test set. It contains 214 oil slicks that give a broad representation of the shape and variation in slick type (spill, seep). One must highlight that this diversity is originally compared to state-of-the-art, which is mainly limited to a single type of oil slicks. This data collection also contains several lookalikes phenomena, such as windless areas. The cost of preparing this data limits the number of samples available, but it represents relevant real-world monitoring and annotation scenarios.

Table 4. Statistics of the test dataset.

	Spill	Seep	Sea	Total
Slick instances number	150	64	-	214
Studied area surface (hm²)	49,771	25,622	267,112,507	267,187,900
Surface rate	3.12%	0.27%	96.6%	100%

Figure 10 shows the distribution of oil slick annotations as a function of wind speed, slick size, and proximity to infrastructure. Regarding wind speed, most slicks have been annotated at medium wind speeds (3 m/s), few were annotated at very low wind speeds, and none at wind speeds above 6 m/s. As for the size of the slicks, it varies from a few hm² to 10.000 hm². The proximity to infrastructure ranges from being very close to being 70 km away from one. This figure illustrates the diversity of the targets and the strong imbalance of their behaviors, which explains the difficulty of their detection.

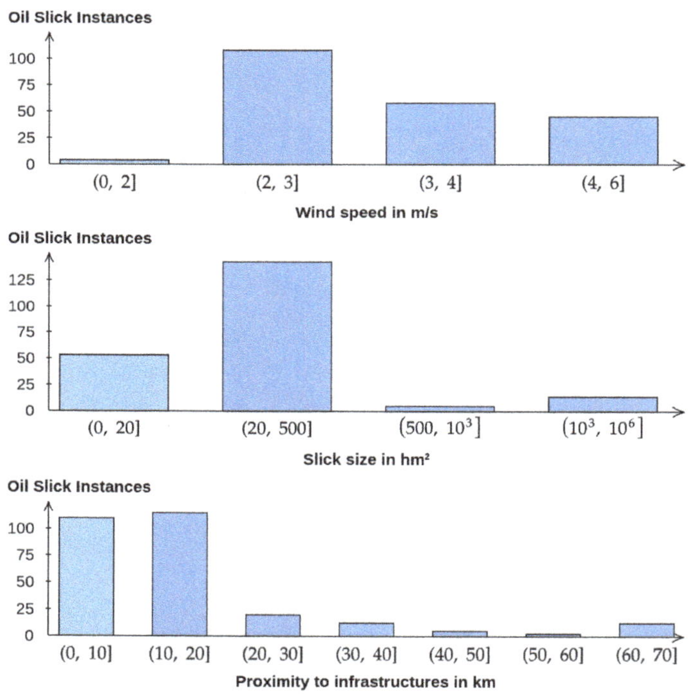

Figure 10. Distribution of slick instances in the test dataset as a function of (from **top** to **bottom**) wind speed (m/s), slick size (hm²), and infrastructure proximity (km).

2.3. *Performance Assessments*

A set of metrics adapted to the task and operational context is selected. The first category corresponds to standard measures considered for semantic segmentation and object detection. These metrics are reported by taking into account contextual information such as the size of the slick, the local wind speed, and the position of infrastructures. We also rely on ROC curves to visualize the trade-off between detector hit (true positive) rates and false alarm (false positive) rates. The second category of metrics relies on a visual explanation of the model predictions.

2.3.1. Standard Metrics

Semantic Segmentation Quality Metric (pixel-level): A pixel-level classification metric such as the Intersection over Union (IoU) is the most used metric to evaluate models in image segmentation tasks [60]. However, its relevance is moderate in our case since the annotation is noisy.

Standard Object Detection Metrics (instance level): Such metrics describe the detection potential of the model better rather than the segmentation quality. Detection rates and the associated false positives and negative rates are thus computed. Further, in the proposed context, the fragmentation of a single slick in multiple detection instances is not an issue. Therefore, if more than one prediction intersects the same annotated slick, detection is considered valid for this slick. The number of well-detected instances is calculated according to Equation (5).

$$Detected_{Instances} = \sum_{i=1}^{N} \mathbb{1}_{Gt_i \cap Pred \neq \emptyset} \quad (5)$$

where N is the number of annotated slick instances, $Pred$ is the predicted instances, and $\mathbb{1}$ is the indicator function.

Receiver Operating Characteristic curves (ROC): These curves are used to characterize a model segmentation quality in a more detailed way. Such ROC curves are obtained by plotting the True Positive Rate (TPR) as a function of the False Positive Rate (FPR), thus quantifying the performance of a detector as its discrimination threshold varies. In other words, ROC curves describe the trade-off between detector hit rates and false alarm rates [61].

2.3.2. Prediction Explanation Methods

Several model explainability techniques have been proposed to facilitate the understanding of complex model predictions. The best known, according to the study of Linardatos et al. [62], are listed below.

- Class Activation Maps (CAM)-based methods, e.g., Grad-CAM [63,64], are designed to generate heat maps of the input, indicating which areas most influence the network decision. It relies on a linear combination of activation maps of a given layer, weighted with the gradient of the class score, with regard to the feature map activation. CAM-based methods have some drawbacks: first, explanation precision is limited given that the produced heatmap is computed based on low-resolution activation maps and further upsampled to match the original image size. Second, it does not provide information on negative contributions (inhibition effects).
- Local Interpretable Model-agnostic Explanations (LIME) [65] is a model-agnostic method that aims to locally (e.g., for one set of inputs) approximate the complex model to a more easily understandable one. This method aims to produce visual artifacts that provide a good understanding of the model choice. However, the LIME method can be criticized for its lack of stability and the discrepancy of its results with human intuition.
- Layer-wise Relevance Propagation (LRP) [66] uses calculation rules to backpropagate the score of a specified output of the network until the first layer, thus showing areas that affected the network decision for the specific output. This method is specific to neural networks and may not provide a trustworthy comparison when applied to different network architectures, as the score backpropagation will proceed differently. Moreover, the backpropagation through FC-DenseNet architecture can lead to conflicts caused by skip connections.
- SHapley Additive exPlanation (SHAP) [12] is a game-theory-inspired method that attempts to enhance interpretability by computing the importance values of each input feature on individual predictions. By definition, the Shapley value calculated by SHAP is the average marginal contribution of an input feature to a model output across all possible coalitions. Different methods are proposed for estimating Shapley

values, such as KernelSHAP or DeepSHAP [12]. They provide results demonstrating the expressiveness of SHAP values in terms of discrimination ability between different output classes and better alignment with human intuition compared to many other existing methods [62]. Several works have adopted the SHAP method for image classification or object detection [67,68].

In this work, explanations are based on the SHAP method, mainly because it is one of the most comprehensive and relevantly used methods in the literature for visualizing interactions and feature significance. SHAP is not only model agnostic but also applies to any data.

However, SHAP must be adapted to the semantic segmentation problem. To do so, we use an algorithm based on the KernelSHAP function [12], computing model-agnostic explanations linked to a specific input image. The proposed workflow is illustrated in Figure 11. It can be summarized by the following steps:

(1) The first step aims at creating groups of pixels, or super-pixels, from the input image. The method result shows the contribution of each super-pixel in the model's decision. Experiments have shown that relying on super-pixels of equal size and shape generates clearer explanations. Moreover, hexagonal super-pixels allow for a more natural explanation than square super-pixels, mainly due to the higher number of direct neighbors of each super-pixels. Thus, explanations presented in this paper rely on a grid of super-pixels shaped as a regular hexagon.

(2) Then, masking is applied to the input image super-pixels in order to generate several masked samples, as shown in Figure 11. Note that masking super-pixels with the zero value is often considered in this step but is avoided in our case, as it would introduce ambiguity with the target (dark oil slicks). Thus, the explanations presented in this paper are based on a grid of super-pixels shaped like a regular hexagon.

(3) The third step is to feed the resulting masked samples through a semantic black-box image segmentation model, which yields prediction probability maps for each sample.

(4) The fourth step is to select the pixel and class of interest (e.g., the pixel highlighted in red in Figure 11 and the class "slick"), for which the explanation of the model decision is to be conducted.

(5) After that, explanation computation based on Shapley values computation rules is applied [69], considering the mask and the pixel-level decision for every sample. In more detail, the Kernel SHAP algorithm estimates the impact of each super-pixel on the probability that the selected pixel belongs to the selected class (e.g., slick).

(6) Finally, a heat map explanation is generated, highlighting the areas (super-pixel) of the input image that contributed positively (excited) and negatively (inhibited) to the model decision.

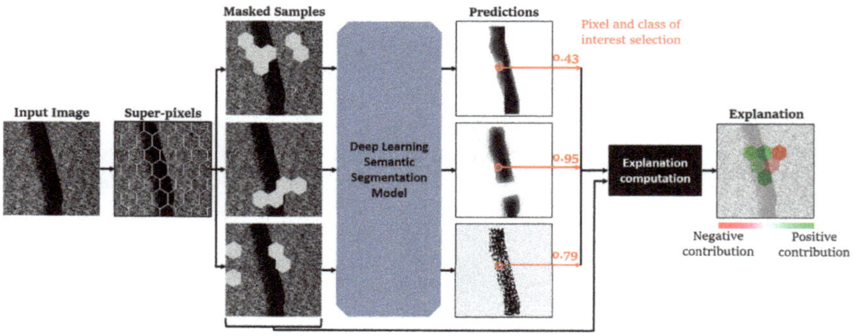

Figure 11. Overview of SHAP adaptation to image semantic segmentation.

The result is obtained as an image, presenting the interpretation of the model decision for a given local (pixel level) classification. The interpretation (SHAP values) regarding a selected pixel (outlined in red on prediction matrices) is shown on the right in Figure 11. A SHAP value is assigned to each super-pixel, and these values are assigned to a color range. Red colors are attributed to a super-pixel decreasing the prediction value for the given class (inhibition, negative contribution to the class probability), and green colors are attributed to a super-pixel increasing this value (excitation, positive contribution to the class probability). The intensity of the color is directly linked to the amplitude of the SHAP value.

3. Results

In this section, the first part concerns the experiments conducted exclusively on SAR images using the selected deep neural network architectures. A presentation and comparison of the results are established. The second part discusses the fusion of SAR and wind information by the proposed Fuse-FC-DenseNet model. The identification of the most suitable approach is conducted with respect to the baseline model that relies on only the SAR information. The third part focuses on interpreting the results of both models (trained on SAR data alone and trained on SAR and wind speed data) by the adapted SHAP explanation method.

All comparisons and assessments of the results presented are based on the expertly selected test set presented in Section 2.2.3. It consists of SAR monitoring images of a real monitoring case with 214 slick instances of both types (spills and seeps) with various sizes, shapes, and slick characteristics.

3.1. *Evaluation of SAR Data Experiments*

3.1.1. Evaluation of Models Based on SAR Data

The evaluation of the proposed approaches is established on a test set of 214 instances of spill and seep, shown in Figure 4. The test set contains large SAR images (10,000 × 10,000 pixels) that are never seen during the training and validation of the models. It can also be noted that the test set belongs to an entirely new area (which was not included in the training/validation sets), implying a variation in the context. Thus, the performance of the models includes their generalization capabilities. The performance presented in Table 5 shows the instance and pixel level metrics of the enhanced FC-DenseNet and Mask-RCNN models. It shows that the improved FC-DenseNet model has a detection rate of 0.93%, while the Mask-RCNN model has a rate of 0.83%. However, the Mask-RCNN model has a slightly lower number of false alarms in the test set. In terms of pixel-level metrics, the improved FC-DenseNet significantly outperforms the Mask-RCNN in terms of IoU, precision, and recall.

Table 5. Results of FC-DenseNet versus Mask R-CNN on a test set of 214 slicks.

	Metrics	FC-DenseNet	Mask R-CNN
Instance level	Good detection number	198	177
	Miss-detection number	16	37
	False-detection number	1658	1103
Pixel level	IoU Slick	0.3	0.06
	Precision	0.42	0.33
	Recall	0.53	0.06

Figure 12 shows an illustration of the result of a test image containing different slicks. The predictions of both models are shown together with the ground truth. Predictions in red are placed under the yellow ones for a more trustworthy visualization. As we observe, Mask R-CNN fails to detect more annotated oil slicks and overlaps less with expert annotations (IoU 0.06). On the other hand, the improved version of FC-DenseNet covers the surface of slicks better (IoU 0.3). In terms of miss-detection rates, Mask R-CNN misses 2.3 times more slick instances than FC-DenseNet. More in-depth analysis shows that the spill miss-detection rate is around 40% lower than the seep miss-detection rate for both models. This can be explained by the fact that the spill instance in our training set is higher than the seep instance. Regarding false alarm rates, the enhanced FC-DenseNet is higher and can be observed in the bottom part of Figure 12.

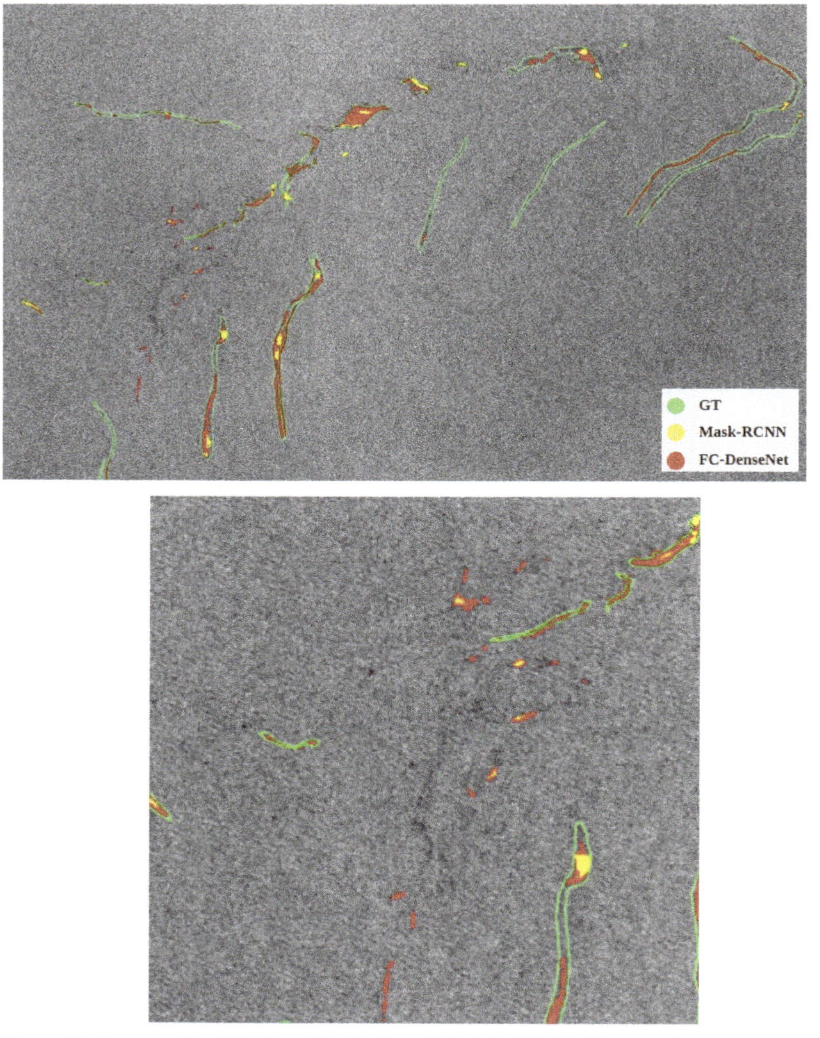

Figure 12. An example of FC-DenseNet (red) and Mask R-CNN (yellow) prediction on Sentinel-1 images where green polygons represent manual slick annotations. A zoom in on an area of interest is placed at the bottom. Note that since Mask-RCNN is systematically within FC-DenseNet ones, Mask-RCNN masks are applied above FC-DenseNet to facilitate analysis.

We observe that in the case of huge oil slicks, as seen in Figure 13, the improved version of the FC-DenseNet detection is fragmented. One general reason for this is the lack of large slicks in the training database. In addition, the detection is performed on image crops of 512 × 512 pixels for training and applied in a sliding window fashion on large test images. The models, therefore, only have a partial view of large slicks. In the situations where a slick extends over the entire crop area, its detection will be inhibited by the lack of contextual information. This could be combated by training on larger image crops but requires more memory on the GPUs.

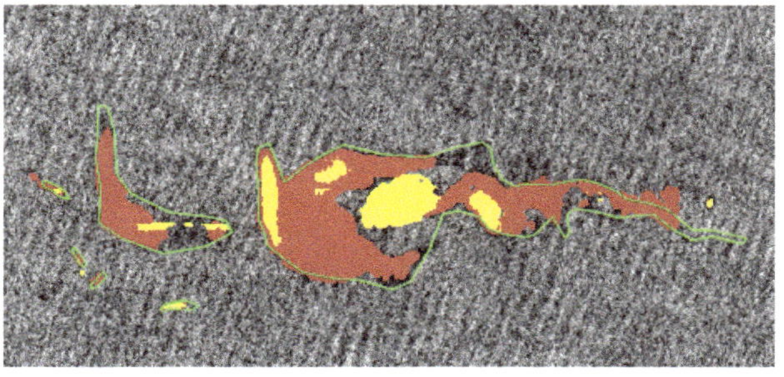

Figure 13. Sample of large slick and the corresponding prediction of FC-DenseNet and Mask R-CNN. Same legend as Figure 12.

3.1.2. Evaluation Based on Contextual Data

Figures 14 and 15 show the results of the two approaches as a function of wind speed level and slick size. Similar general behaviors can be noted for both models, detecting slicks over the full range of wind speed and slick size. For both models, false detection areas are greater than 20 hm^2 and mostly correspond to windless areas, but their rate decreases as wind speed increases.

Figure 16 shows the distribution of the predictions of the two models regarding the proximity of the slicks (spill and seep) to infrastructure (ship, platform, pipeline, etc.). The information on proximity to infrastructure mainly relates to spill type, representing 150 instances in the test database. The improved version of FC-DenseNet model detects 149/150 spill instances, missing only one instance.

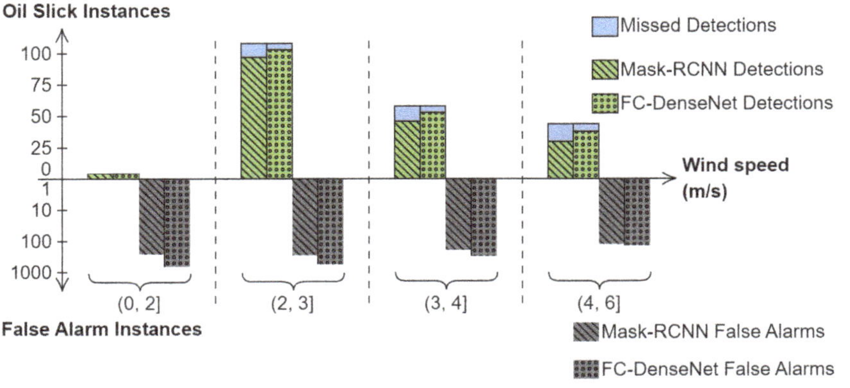

Figure 14. Detection performance of the improved version of FC-DenseNet and Mask R-CNN as a function of wind speed (m/s). Green bars represent good detection; blue shows missed detection. False alarms are represented by gray bars on a logarithmic scale.

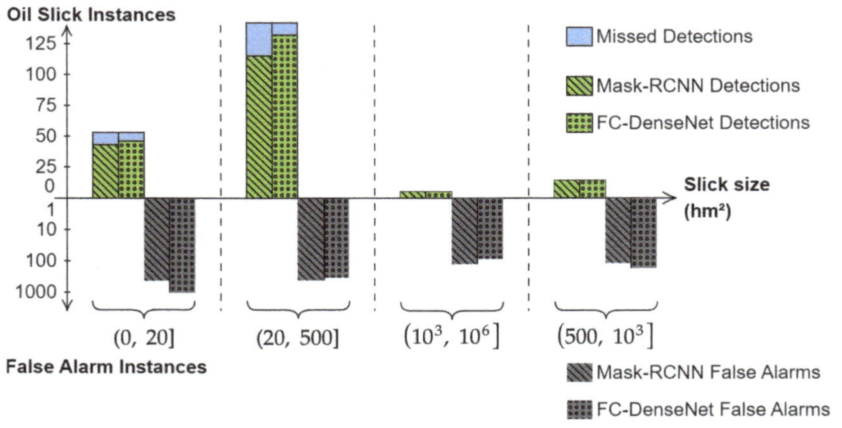

Figure 15. Detection performance of FC-DenseNet and Mask R-CNN as a function of ground truth size (hm²), same legend as Figure 14.

The first two bars ([0, 20] km) in Figure 16 shows both spill and seep instances, indicating that few slicks near the infrastructure are missed (blue). The R-CNN mask behaves similarly, detecting fewer slicks and missing some slicks farther from the infrastructure ([30, 40] km). The information on the proximity of the infrastructure represents relevant information for identifying the oil slick, notably the oil spill. This information serves as a reassurance to experts about the existence and type of slick.

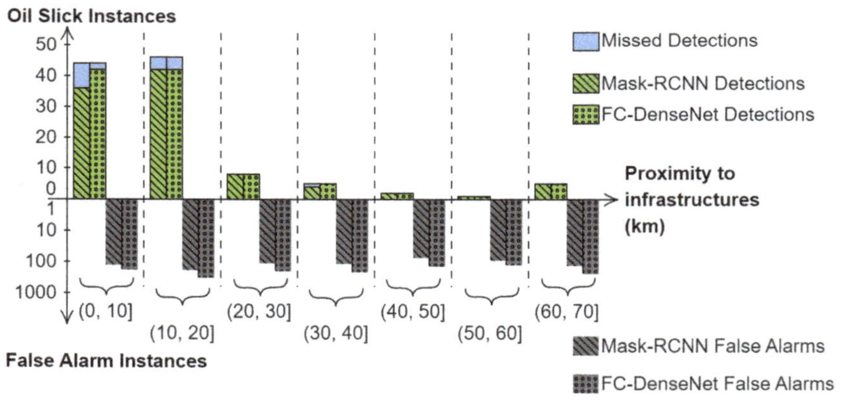

Figure 16. Detection performance of FC-DenseNet and Mask R-CNN as a function of infrastructure proximity (km), same legend as Figure 14.

As enhanced FC-DenseNet shows a better IoU value, reasonable detection rate, and miss-detection rate, this model is chosen as the baseline for the following experiments. An additional argument for this choice is also based on the fact that it is fully trained on the target data, unlike Mask R-CNN, which is pre-trained on other data.

3.2. *Evaluation of Data Fusion Models*

In this section, models conducting heterogeneous data fusion (SAR and wind speed information) are evaluated and compared against the SAR-based baseline model referred to as T_{SAR}. Building upon the fusion strategies illustrated in Figure 5, several experiments have been performed. The most pertinent ones are reported in the following. T_0 corresponds to early fusion (the first arrow of Figure 5, only an upsampling of the wind maps is applied). Late fusion experiments correspond to the second arrow on the same figure, and we derive experiments T_1, T_2, T_3, which systematically apply a subsampling (pooling) of the wind data to match the feature size of the SAR feature maps in later blocks. Those experiments differ in terms of feature extraction strategies on the wind information and fusion step positioning with respect to the SAR branch. Relevant experiment summaries are reported in Table 6, and the global results are presented in Table 7 and discussed in the following.

Table 6. Experiment descriptions.

Name	Description
T_{SAR}	Enhanced FC-DenseNet baseline model relying exclusively on SAR data
T_0	Early fusion of upsampled wind data at the same level as the SAR data
T_1	Average pooling of the wind data and fusion with SAR data after dense block 2 (late fusion)
T_2	Average pooling and Denseblock applied to the wind data prior fusion with the SAR data after dense block 2 (late fusion)
T_3	Average pooling and Denseblock applied to the wind data prior fusion with the SAR data after dense block 3 (late fusion)

Table 7. Comparison of the FC-DenseNet results on a test set of 214 slicks using only SAR data with different data fusion models.

	Metrics	Only SAR	Early Fusion	Late Fusion		
		T_{SAR}	T_0	T_1	T_2	T_3
Instance level	Good detection number	198	198	200	201	192
	Miss-detection number	16	16	14	13	22
	False-detection number	1658	1357	1376	1430	1094
Pixel level	IoU slick	0.30	0.22	0.31	0.28	0.21
	Precision	0.42	0.35	0.42	0.38	0.24
	Recall	0.53	0.38	0.55	0.49	0.60

3.2.1. Early Fusion Experiment

When directly fusing SAR information with the upsampled wind data and providing this as a unified input to the model, the mitigated results are reported in Table 7. The number of good detection remains the same, and an 18% reduction of false detections is observed. This is consistent with the process of photo-interpreters using wind data to identify suspicious slicks and distinguish them from lookalikes. However, such an early fusion model reduces the IoU metric by 0.1 and also lowers precision and recall. The distribution of model detection as a function of wind speed and slick size is represented by the detection plots in Figure 17. Interpretations are similar to the baseline model.

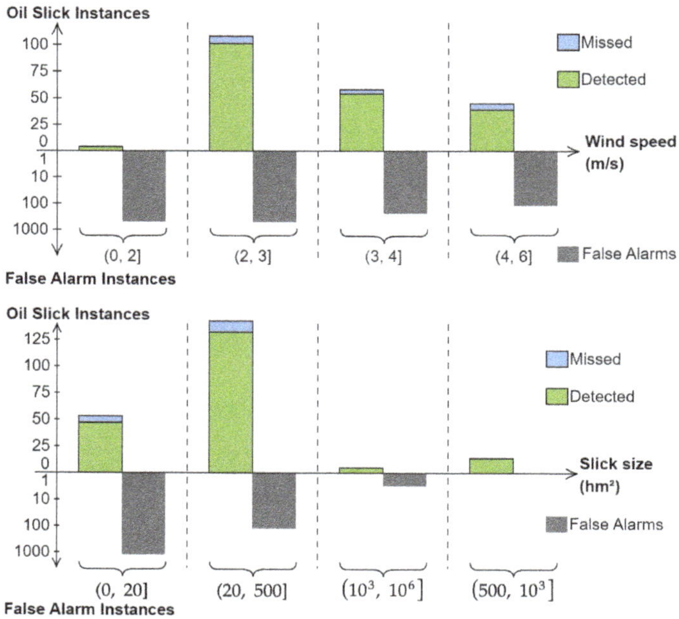

Figure 17. Detection performance of Fuse-FC-DenseNet as a function of wind speed (m/s) and ground truth size (hm^2), early fusion case.

Figure 18 shows a local comparison of the baseline SAR-based model and the early fusion strategy (displayed on top). The number of false detections of the baseline model is more noticeable, especially outside of the fine manual annotations.

Figure 18. An example of the baseline SAR-based enhanced FC-DenseNet (red) and the early fusion version (blue) predictions to be compared with the green polygons representing the photo-interpretation slick annotation.

Compared to the baseline model, the early fusion indicates a decrease in the IoU value. It may be noted that this limitation may be due to the manipulation of two data channels at the same scale. Meanwhile, the wind modality is highly interpolated. Extracting the fine resolution patterns do not seem appropriate. The fusion of wind data at a more appropriate scale should yield more insights.

3.2.2. Late Fusion Experiments

As shown in Table 7, in the case of late fusion at the block 2 levels (experiments T_1 and T_2), the detection rate (93.4%), IoU (0.31), precision (0.42), and recall (0.55) improve over the early fusion experiment T_0 and can reach or outperform the baseline. Further, by comparing the late fusion experiments, we notice that the later we fuse wind information with SAR, the lower the false detections at the price of a slight decrease in the number of good detection and a limited increase in the number of false detections. The slick IoU also decreases. However, the results are based only on selecting the most likely class between sea and slick and, therefore, they do not focus solely on slick probability levels, and further analysis is required.

Regarding the number of false detections, we notice that they are often either around a large slick or grouped in the no-wind areas and fringed in small fragments such as Figure 19, showing an example of false detections of an improved version of the FC-DenseNet. A post-processing step of the dilation-erosion operation can be applied to minimize the number of false alarm instances.

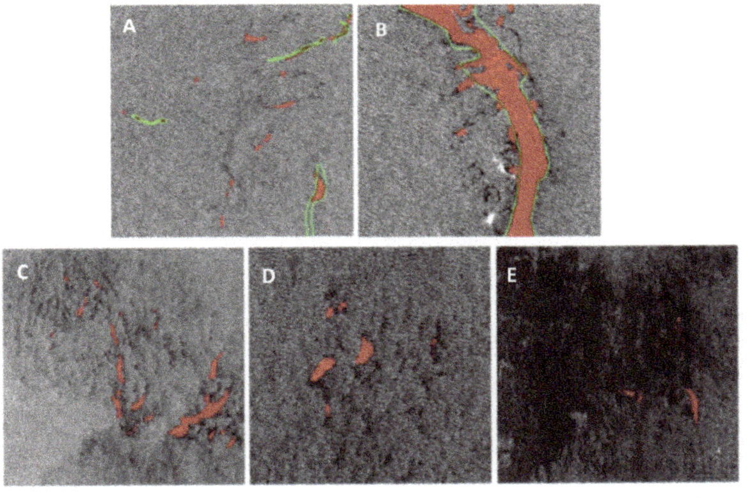

Figure 19. False alarm samples for the improved baseline version of the FC-DenseNet model (T_{SAR}). False alarms are marked in red and slick annotations are marked with green polygons. (**A**,**B**) represent false alarms around a slick, and (**C**–**E**) represent other false alarms.

Finally, for all the fusion experiments, a decrease in the number of false detections is observed. This then confirms the importance of wind speed for slick detection, especially for distinguishing slicks from lookalikes.

3.2.3. ROC Curves Analysis

To better understand the prediction of the considered improved version of FC-DenseNet models, particularly their false detections, the slick probability maps outing from the model are analyzed. Figure 20 shows an example with a heat map representing the slick probability levels. One can visually observe that the slick inner surfaces generally have a high probability (red) and that the low probability values are always on the boundaries of the slicks (yellow). Given the quality of the annotation reported in Section 2.2, the IoU level remains strongly proportional to the quality of the ground truth boundary and cannot reach very high values.

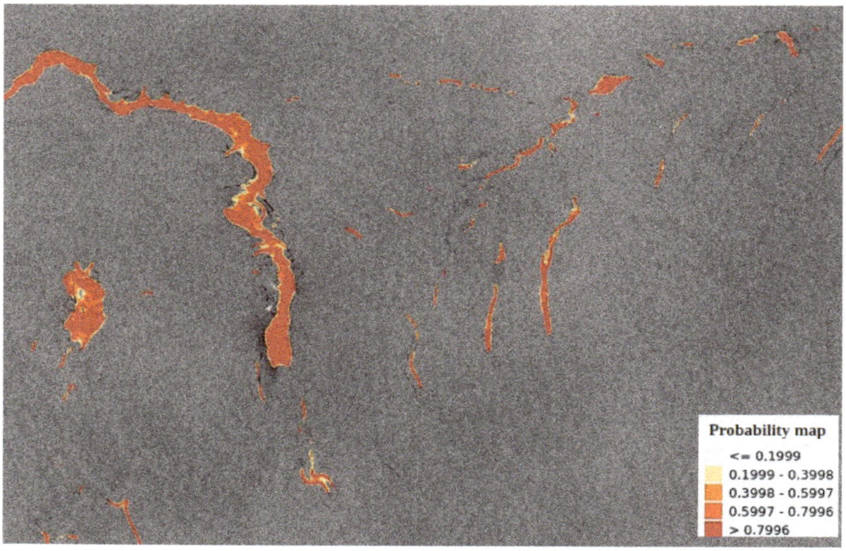

Figure 20. Example of a prediction probability map of the improved version of FC-DenseNet model.

For a larger-scale investigation, these slick probability maps are considered to plot the ROC curves aggregating all the predictions on the test database for the different late fusion models. ROC curves for the fusion experiments are shown in Figure 21.

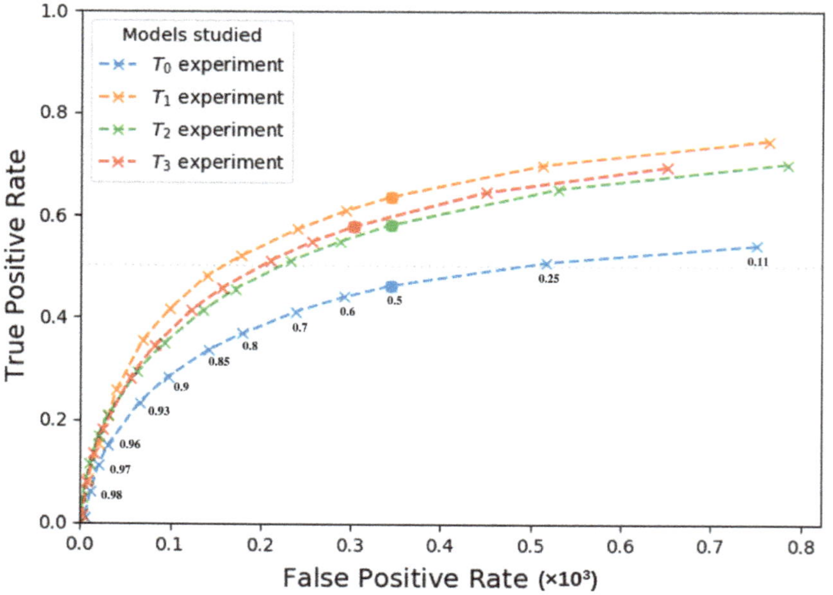

Figure 21. ROC curves of fusion experiments (T_0, T_1, T_2, T_3). The threshold measurements marked on the blue curves are identical for all curves. The solid point represents the 0.5 threshold.

Experiment T_1 shows high values for each threshold measure compared to the other fusion experiments curves, which have a similar shape. It also shows a lower FPR for almost the entire range of TPR, we can assume that this fusion model provides the best segmentation quality.

Moreover, there is a noticeable gap between the late fusion experiments and the early fusion ones. This gap shows that late fusion is more appropriate in our context, as both input channels present too different characteristics to be considered at the same level in the architecture.

One can note that slick detection improvement can be made by selecting an adjusted threshold on the slick probability map instead of choosing the most probable class for every pixel. The threshold involves a different trade-off between TPR and FPR and thus can be decided by the domain experts.

3.3. SHAP Explanation

Explanation of the network decision processes can be made using the SHAP adaptation proposed in Section 2.3.2. This method is used to compare two models of interest: one corresponds to the T_{SAR} experiment, and the other corresponds to the T_1 late fusion experiment. Then, based on the explanations obtained from the two model predictions on a varied set of the same pixels from the same input images, one can receive insights into the model behaviors and their differences in terms of the effective field of view and sensitivity to neighboring patterns. The following will focus on the three main observed behaviors illustrated in Figures 22–24. These figures show the input images, a set of explained pixels (red circles), the applied features delimitation, and the explanation maps:

- For a pixel classified as sea, as for explained pixels in Figures 22a,c and 23a: the surface affecting the decision of the networks is large and depends on the presence of slicks (or lookalikes) in the entire image. Specifically, a general observation reveals that the prediction is mainly influenced by the presence of oil slicks in the vicinity of the explained pixel: a positive impact (reinforcing the classification of the pixel as slick) in the case of a relatively close slick and a negative one in the case of a distant slick. However, their contribution to the decision is extremely low and always countered by the considered pixel area. As a result, the prediction associated with the slick class is always close to 0%.
- For a pixel classified as slick, as for the explained pixel in Figures 22b, 23c, and 24b: the prediction is based on a limited area centered around the considered pixel. The maximal SHAP value in these cases is the highest observed (about 0.5). Typically, the network prediction for the pixel is impacted by one input feature containing mainly black pixels, which is enough to classify the pixel as a slick, with a probability above 80%.
- For a pixel located on a slick edge or within a narrow slick, as for the explained pixel in Figure 23b,d: networks tend to detect and base their decision on the slick edge or the narrow slick length. This shows that networks can detect the slick edges around the selected pixel in all observed images, which significantly influences their decision. The impact of each area containing a slick edge tends to decrease as the area moves away from the pixel.

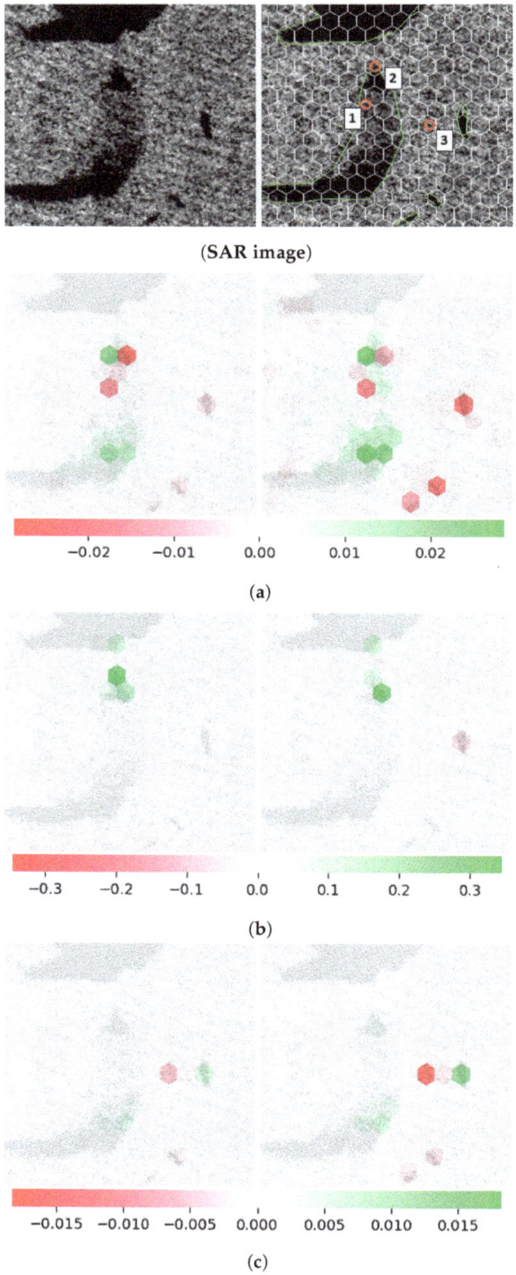

Figure 22. Results of SHAP explanation function. (**a**) SHAP images for pixel number 1, classified as sea; (**b**) SHAP images for pixel number 2, classified as slick; (**c**) SHAP images for pixel number 3, classified as sea. (**SAR image**) **right**: SAR image, **left**: SAR image with a grid of white super-pixels representing the SHAP input. The ground truth is outlined in green and the pixels considered are numbered and circled in red. For SHAP image rows, right corresponds to the information obtained through the network T_{SAR}, and left corresponds to information obtained through the T_1 experiment.

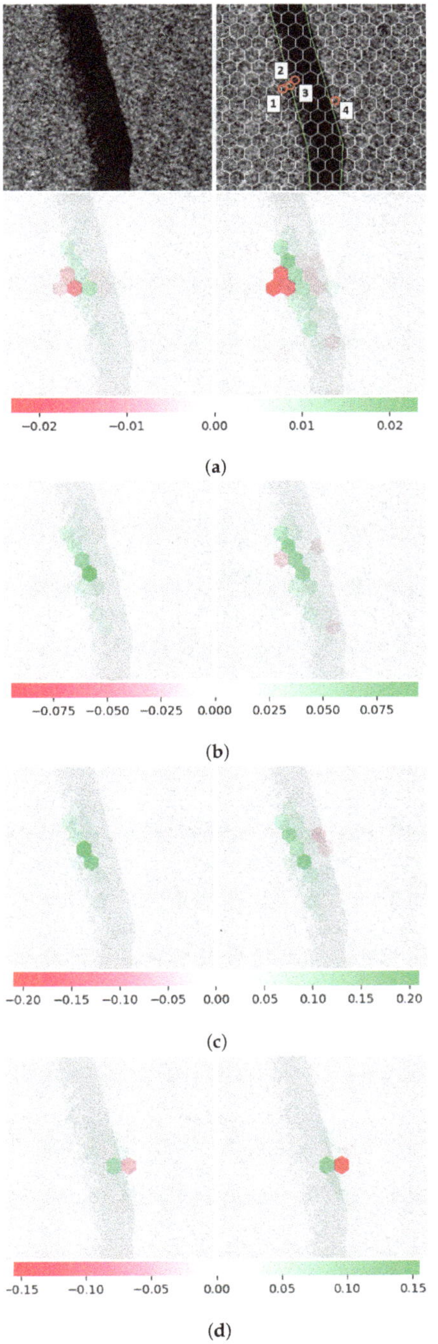

Figure 23. Results of SHAP explanation function, showing effects of slick edges on networks. (**a**) SHAP images for pixel number 1, classified as sea; (**b**) SHAP images for pixel number 2, between slick and sea; (**c**) SHAP images for pixel number 3, classified as slick; (**d**) SHAP images for pixel number 4, between slick and sea. Similar legend to Figure 22.

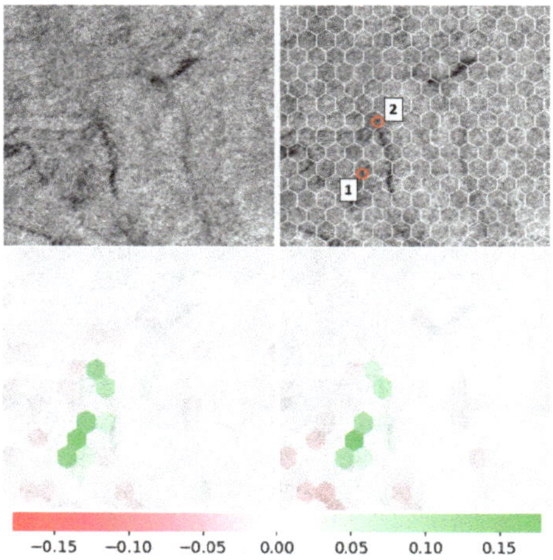

(a)

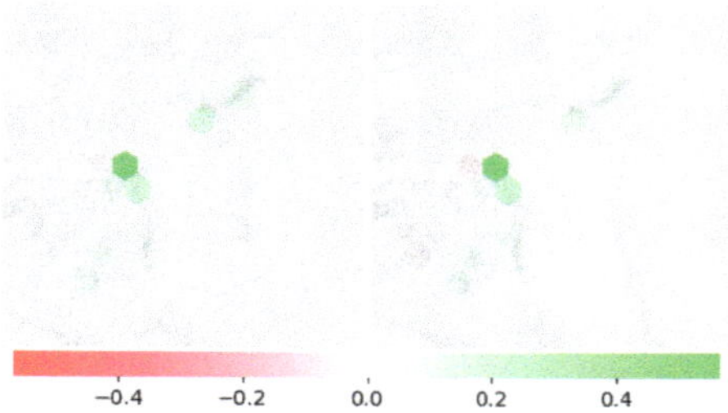

(b)

Figure 24. Results of SHAP explanation function, showing the difference between both studied experiments. (**a**) SHAP images for pixel number 1, classified as a slick by T_{SAR} model and as sea by T_1; (**b**) SHAP images for pixel number 2, classified as slick. Same legend as Figure 22.

4. Discussion

Based on the results obtained, we can confirm the ability of deep learning to detect both types of oil slicks (natural and anthropogenic) in a large-scale and real-world application, which contrasts with previous work dedicated to single types of oil slicks.

The result evaluation based solely on the SAR data shows a good detection rate but notable false alarms. The false alarms obtained can be divided into two categories: the first represents false alarms related to several lookalikes, such as low wind areas, algal blooms, and upwelling, as shown in Figure 19C–E, and the second represents false alarms related to small spots around an oil slick annotation, as shown in Figure 19A,B. These small oil slick patches, confirmed by photo-interpreters, are related to environmental effects, particularly meteorological conditions that influence the physicochemical properties of the oil slick, inducing its dispersion (fragmentation into droplets) and spreading process. These small oil slicks are also noticeable in Figure 15, where we can see the presence of false alarms ranging in size from 0 to 500 hm^2.

One observation is that the ground truth annotation made by the photo-interpreters is not perfect and suffer from imprecise boundaries, as shown in Figure 2, revealing a diversity of annotations that position our detection task in a noisy reference context [70], as mentioned in Section 2.2. Hence the validation process requires a second pass of expertise on the large images to check the false alarms and eventually correct the annotations to integrate some false alarms as real slicks. We are thus, in a real context, allowing a continuous progression of the models and improvement of the annotations. This also becomes apparent in the prediction probability maps obtained by the models, as shown in Figure 20, where the lowest probabilities are always found on the slick boundaries. The imprecision of the annotation boundaries is apparent in the slick IoUs limited to 0.31 in Table 5 and also in the prediction probability maps obtained by the models, where the lowest probabilities are always found on the slick boundaries, as shown in Figure 20. Despite this, the semantic and instance segmentation models are quite capable of detecting slicks but suffer from false alarms.

To further reduce false alarms, we proposed Fuse-FC-DenseNet models that allow the fusion of wind speed and SAR data. Comparing the early and late fusion strategies using ROC curves shows that fusing wind data later in the models provides a satisfactory trade-off between good detection and false alarm rates. It can be noted that fusing wind speed data later in the model, approaching its resolution, provides better results than fusing too early, which causes the layers to learn inconsistent features, leading to undesired results. The obtained results with Fuse-FC-DenseNet have a better detection rate, and its detection better matches the photo interpretations (up to 93% of the dataset targets are detected).

Late fusion experiments improve the performance while reducing false detections in the range of 14% to 34%. Regarding the late fusion experiments (T_1 and T_3), the comparison is made based on the trade-off between good detection and false detection. T_3 shows a lower false alarm rate and detection rate than the single-modality (SAR) model results. This comparison is based on oil distribution in the dataset (214 oil instances). In the end, the choice of the best model is to be decided by the end-users to find the best compromise, taking into account the global costs that may include on-site verification flights.

Explainable Results

Going further, based on the comparison of the explanations presented in Section 3.3, it is possible to understand the effect of wind speed data on the model predictions. Figure 24a shows a lookalike that corresponds to waves. More precisely, one focuses on the explanation of the pixel of interest. This lookalike is detected as a slick only by the network that does not manage wind speed information (T_{SAR}). The explanatory maps for both models show the impact of near-pixel areas on their decisions, increasing the probability of the pixel classification as a slick. However, the T_1 model that performs SAR and wind fusion gives more importance to contextual information: other dark patches in the image harm the classification of the considered pixel as a slick. The interpretation of this phenomenon is linked to the presence of wind speed data in the network decision process: the T_1 experiment can better differentiate slicks and lookalikes in this configuration, which makes contextual information more reliable. For all interpretations, the T_1 model shows a higher intensity of the SHAP value than the T_{SAR} (red and green intensity in the SHAP images).

The proposed explanation method already provides valuable information to understand better and compare models. Its application to oil slick detection is interesting and aims to provide monitoring teams with additional information to help in the diagnosis of automated alarms. Several case studies and extensions of this approach are under consideration but are beyond the scope of this paper and will be discussed in future publications.

5. Conclusions

Deep learning for health, safety, and the environment is an active research trend, and oil spill monitoring is a relevant case study. To address this task, we demonstrate the interest in optimized semantic segmentation models based on the improved version of FC-Densenet and an instance segmentation approach transferred from another domain, Mask R-CNN. We evaluate and compare them relying on Sentinel-1 SAR imagery depending on domain expert photo-interpreters and taking into account contextual information regarding wind speed and proximity to human activities. First, we show the relevance of these approaches and propose a set of dedicated metrics that allow refined comparison of the models in various contextual situations in worldwide real-monitoring scenarios. Both models can detect slicks, but the improved version of FC-DenseNet has a better detection rate, and its detections better match the photo interpretations. Further, this paper shows the models applicability to detect oil slicks (both natural and anthropogenic), thus contrasting with previous work dedicated to a single slick type. In addition, we propose Fuse-FC-DenseNet model to fuse SAR and wind speed data to improve performances and diminish false alarm rates.

Compared to the baseline model, a decrease in false alarms is observed for all the fusion experiments. This then confirms the importance of wind speed for slick detection, especially for distinguishing slicks from lookalikes. The proposed Fuse-FC-DenseNet networks are capable of improving performance while reducing false detections in the range of 14% to 34%. As for human experts, such data fusion helps the models select relevant information to provide enhanced predictions and better distinguish oil slicks from lookalikes. The model predictions would be integrated into the industrial production pipeline to provide ready predictions to photo-interpreters. This will speed up the oil slick detection task and keep up with continuous sensor acquisition.

Finally, to offer ready-made predictions that are understandable by the photo-interpreters and enable human validation, we propose an extension of the SHAP explanation method that allows semantic segmentation predictions to be explained. It also allows a refined comparison of model behaviors on local decisions and their sensitivity to neighboring patterns in the data.

This work yields several insights, including the possibility of further studying explainability by extending analyses and explanations using the SHAP technique. In addition, different spatial organizations of the super-pixels can be tested, and analysis based on variability in the size of the contextual information can be established.

Another perspective may be to adapt the data fusion model further by introducing infrastructure position data as additional information during the learning process. The aim is to train the network to detect oil slicks and discriminate between spills and seeps.

Author Contributions: Investigation, E.A. and P.D.; Methodology; E.A., data curation; E.A. and H.C., Project administration; A.B., P.B., D.D. and A.C., Resources: D.D. and A.C., Software: E.A., Validation; P.B.; Supervision: P.B., A.B., A.C. and D.D., Writing—original draft: E.A. and P.D., Writing—review and editing; A.B. and D.D., Funding acquisition; A.C. and D.D. All authors have read and agreed to the published version of the manuscript.

Funding: This research was funded by TotalEnergies grant number FR00064703.

Data Availability Statement: Not applicable.

Acknowledgments: This work was carried out thanks to the ENVISAT and Sentinel-1 data provided by the European Space Agency.

Conflicts of Interest: Author E.A. received research grants from the company TotalEnergies. The funders have a role in the collection, partial preprocessing of the data, and the decision to publish the results.

Abbreviations

The following abbreviations are used in this manuscript:

AI	Artificial Intelligence
EM	ElectroMagnetic
FA	False Alarms
FC-DenseNet	Fully Convolutional DenseNet
GMF	Geophysical Model Functions
Mask R-CNN	Mask Region Based Convolutional Neural Network
SAR	Synthetic Aperture Radar
SHAP	SHapley Additive exPlanation
ROC	Receiver Operating Characteristic
RS	Remote Sensing

References

1. Girard-Ardhuin, F.; Mercier, G.; Garello, R. Oil slick detection by SAR imagery: Potential and limitation. In Proceedings of the Oceans 2003. Celebrating the Past... Teaming Toward the Future (IEEE Cat. No. 03CH37492), San Diego, CA, USA, 22–26 September 2003; Volume 1, pp. 164–169.
2. Alpers, W.; Holt, B.; Zeng, K. Oil spill detection by imaging radars: Challenges and pitfalls. *Remote Sens. Environ.* **2017**, *201*, 133–147. [CrossRef]
3. Fingas, M.; Brown, C. Review of oil spill remote sensing. *Mar. Pollut. Bull.* **2014**, *83*, 9–23. [CrossRef] [PubMed]
4. Brekke, C.; Solberg, A.H. Oil spill detection by satellite remote sensing. *Remote Sens. Environ.* **2005**, *95*, 1–13. [CrossRef]
5. Angelliaume, S.; Dubois-Fernandez, P.C.; Jones, C.E.; Holt, B.; Minchew, B.; Amri, E.; Miegebielle, V. SAR imagery for detecting sea surface slicks: Performance assessment of polarization-dependent parameters. *IEEE Trans. Geosci. Remote Sens.* **2018**, *56*, 4237–4257. [CrossRef]
6. Solberg, A.S.; Storvik, G.; Solberg, R.; Volden, E. Automatic detection of oil spills in ERS SAR images. *IEEE Trans. Geosci. Remote Sens.* **1999**, *37*, 1916–1924. [CrossRef]
7. Espedal, H. Satellite SAR oil spill detection using wind history information. *Int. J. Remote Sens.* **1999**, *20*, 49–65. [CrossRef]
8. Karathanassi, V.; Topouzelis, K.; Pavlakis, P.; Rokos, D. An object-oriented methodology to detect oil spills. *Int. J. Remote Sens.* **2006**, *27*, 5235–5251. [CrossRef]
9. Nirchio, F.; Sorgente, M.; Giancaspro, A.; Biamino, W.; Parisato, E.; Ravera, R.; Trivero, P. Automatic detection of oil spills from SAR images. *Int. J. Remote Sens.* **2005**, *26*, 1157–1174. [CrossRef]
10. Benoit, A.; Ghattas, B.; Amri, E.; Fournel, J.; Lambert, P. Deep learning for semantic segmentation. In *Multi-Faceted Deep Learning*; Springer: Berlin/Heidelberg, Germany, 2021.
11. Krestenitis, M.; Orfanidis, G.; Ioannidis, K.; Avgerinakis, K.; Vrochidis, S.; Kompatsiaris, I. Oil spill identification from satellite images using deep neural networks. *Remote Sens.* **2019**, *11*, 1762. [CrossRef]
12. Lundberg, S.; Lee, S.I. A unified approach to interpreting model predictions. *arXiv* **2017**, arXiv:1705.07874.
13. Li, X.; Nunziata, F.; Garcia, O. Oil spill detection from single-and multipolarization SAR imagery. In *Reference Module in Earth Systems and Environmental Sciences*; Elsevier: Amsterdam, The Netherlands, 2018.
14. Espedal, H.; Hamre, T.; Wahl, T.; Sandven, S. *Oil Spill Detection Using Satellite Based SAR, Pre-Operational Phase A*; Technical Report; Nansen Environmental and Remote Sensing Center: Bergen, Norway, 1995.
15. Wang, P.; Zhang, H.; Patel, V.M. SAR image despeckling using a convolutional neural network. *IEEE Signal Process. Lett.* **2017**, *24*, 1763–1767. [CrossRef]
16. La, T.V.; Messager, C.; Honnorat, M.; Channelliere, C. Detection of convective systems through surface wind gust estimation based on Sentinel-1 images: A new approach. *Atmos. Sci. Lett.* **2018**, *19*, e863. [CrossRef]
17. Najoui, Z.; Deffontaines, B.; Xavier, J.P.; Riazanoff, S.; Aurel, G. Wind Speed and instrument modes influence on the detectability of oil slicks using SAR images: A stochastic approach. *Remote Sens. Environ.* **2017**. Available online: www-igm.univ-mlv.fr/~riazano/publications/NAJOUI_Zhour_thesis_paper1_Oil_slicks_detectability_from_SAR_images_draft31.pdf (accessed on 2 February 2022).
18. Al-Ruzouq, R.; Gibril, M.B.A.; Shanableh, A.; Kais, A.; Hamed, O.; Al-Mansoori, S.; Khalil, M.A. Sensors, Features, and Machine Learning for Oil Spill Detection and Monitoring: A Review. *Remote Sens.* **2020**, *12*, 3338. [CrossRef]
19. Chehresa, S.; Amirkhani, A.; Rezairad, G.A.; Mosavi, M.R. Optimum features selection for oil spill detection in SAR image. *J. Indian Soc. Remote Sens.* **2016**, *44*, 775–787. [CrossRef]
20. Topouzelis, K.; Karathanassi, V.; Pavlakis, P.; Rokos, D. Detection and discrimination between oil spills and look-alike phenomena through neural networks. *ISPRS J. Photogramm. Remote Sens.* **2007**, *62*, 264–270. [CrossRef]
21. Hamedianfar, A.; Barakat A. Gibril, M. Large-scale urban mapping using integrated geographic object-based image analysis and artificial bee colony optimization from worldview-3 data. *Int. J. Remote Sens.* **2019**, *40*, 6796–6821. [CrossRef]
22. Guo, H.; Wu, D.; An, J. Discrimination of oil slicks and lookalikes in polarimetric SAR images using CNN. *Sensors* **2017**, *17*, 1837. [CrossRef]
23. Yaohua, X.; Xudong, M. A sar oil spill image recognition method based on densenet convolutional neural network. In Proceedings of the 2019 International Conference on Robots & Intelligent System (ICRIS), Haikou, China, 15–16 June 2019; pp. 78–81.
24. Chen, Y.; Li, Y.; Wang, J. An end-to-end oil-spill monitoring method for multisensory satellite images based on deep semantic segmentation. *Sensors* **2020**, *20*, 725. [CrossRef]
25. Gallego, A.J.; Gil, P.; Pertusa, A.; Fisher, R. Segmentation of oil spills on side-looking airborne radar imagery with autoencoders. *Sensors* **2018**, *18*, 797. [CrossRef]
26. Bianchi, F.M.; Espeseth, M.M.; Borch, N. Large-scale detection and categorization of oil spills from SAR images with deep learning. *Remote Sens.* **2020**, *12*, 2260. [CrossRef]
27. Cantorna, D.; Dafonte, C.; Iglesias, A.; Arcay, B. Oil spill segmentation in SAR images using convolutional neural networks. A comparative analysis with clustering and logistic regression algorithms. *Appl. Soft Comput.* **2019**, *84*, 105716. [CrossRef]
28. Emna, A.; Alexandre, B.; Bolon, P.; Véronique, M.; Bruno, C.; Georges, O. Offshore Oil Slicks Detection From SAR Images Through The Mask-RCNN Deep Learning Model. In Proceedings of the 2020 International Joint Conference on Neural Networks (IJCNN), Glasgow, UK, 19–24 July 2020; pp. 1–8.

29. Amri, E.; Courteille, H.; Benoit, A.; Bolon, P.; Dubucq, D.; Poulain, G.; Credoz, A. Deep learning based automatic detection of offshore oil slicks using SAR data and contextual information. In Proceedings of the Remote Sensing of the Ocean, Sea Ice, Coastal Waters, and Large Water Regions 2021, Online, 13–17 September 2021; Volume 11857, pp. 35–42.
30. Orfanidis, G.; Ioannidis, K.; Avgerinakis, K.; Vrochidis, S.; Kompatsiaris, I. A deep neural network for oil spill semantic segmentation in Sar images. In Proceedings of the 2018 25th IEEE International Conference on Image Processing (ICIP), Athens, Greece, 7–10 October 2018; pp. 3773–3777.
31. Yu, X.; Zhang, H.; Luo, C.; Qi, H.; Ren, P. Oil spill segmentation via adversarial f-divergence learning. *IEEE Trans. Geosci. Remote Sens.* **2018**, *56*, 4973–4988. [CrossRef]
32. He, K.; Gkioxari, G.; Dollár, P.; Girshick, R. Mask r-cnn. In Proceedings of the International Conference on Computer Vision (ICCV), Venice, Italy, 22–29 October 2017; pp. 2961–2969.
33. Jégou, S.; Drozdzal, M.; Vazquez, D.; Romero, A.; Bengio, Y. The one hundred layers tiramisu: Fully convolutional densenets for semantic segmentation. In Proceedings of the IEEE Conference on Computer Vision and Pattern Recognition Workshops, Honolulu, HI, USA, 21–26 July 2017; pp. 11–19.
34. Lin, T.; Maire, M.; Belongie, S.J.; Bourdev, L.D.; Girshick, R.B.; Hays, J.; Perona, P.; Ramanan, D.; Dollár, P.; Zitnick, C.L. Microsoft COCO: Common Objects in Context. In *European Conference on Computer Vision*; Springer: Berlin/Heidelberg, Germany, 2014.
35. Abdulla, W. Mask R-CNN for Object Detection and Instance Segmentation on Keras and TensorFlow. 2017. Available online: https://github.com/matterport/Mask_RCNN (accessed on 2 February 2022).
36. Lin, T.Y.; Goyal, P.; Girshick, R.; He, K.; Dollár, P. Focal loss for dense object detection. In Proceedings of the IEEE International Conference on Computer Vision, Venice, Italy, 22–29 October 2017; pp. 2980–2988.
37. Huang, G.; Liu, Z.; Van Der Maaten, L.; Weinberger, K.Q. Densely connected convolutional networks. In Proceedings of the IEEE Conference on Computer Vision and Pattern Recognition, Honolulu, HI, USA, 21–26 July 2017; pp. 4700–4708.
38. Ronneberger, O.; Fischer, P.; Brox, T. U-net: Convolutional networks for biomedical image segmentation. In *Proceedings of the International Conference on Medical Image Computing and Computer-Assisted Intervention*; Springer: Berlin/Heidelberg, Germany, 2015; pp. 234–241.
39. Bottou, L. Stochastic gradient descent tricks. In *Neural Networks: Tricks of the Trade*; Springer: Berlin/Heidelberg, Germany, 2012; pp. 421–436.
40. Kingma, D.P.; Ba, J. Adam: A method for stochastic optimization. *arXiv* **2014**, arXiv:1412.6980.
41. Sudre, C.H.; Li, W.; Vercauteren, T.; Ourselin, S.; Cardoso, M.J. Generalised dice overlap as a deep learning loss function for highly unbalanced segmentations. In *Deep Learning in Medical Image Analysis and Multimodal Learning for Clinical Decision Support*; Springer: Berlin/Heidelberg, Germany, 2017; pp. 240–248.
42. Saxe, A.M.; McClelland, J.L.; Ganguli, S. Exact solutions to the nonlinear dynamics of learning in deep linear neural networks. *arXiv* **2013**, arXiv:1312.6120.
43. Krizhevsky, A.; Sutskever, I.; Hinton, G.E. Imagenet classification with deep convolutional neural networks. *Adv. Neural Inf. Process. Syst.* **2012**, *25*, 1097–1105. [CrossRef]
44. Nair, V.; Hinton, G.E. Rectified linear units improve restricted boltzmann machines. In Proceedings of the International Conference on Machine Learning, Haifa, Israel, 21–24 June 2010.
45. Bansal, N.; Chen, X.; Wang, Z. Can we gain more from orthogonality regularizations in training deep cnns? *arXiv* **2018**, arXiv:1810.09102.
46. Hénaff, O.J.; Simoncelli, E.P. Geodesics of learned representations. *arXiv* **2015**, arXiv:1511.06394.
47. Dong, C.; Loy, C.C.; He, K.; Tang, X. Image super-resolution using deep convolutional networks. *IEEE Trans. Pattern Anal. Mach. Intell.* **2015**, *38*, 295–307. [CrossRef]
48. Valenzuela, G.R. Theories for the interaction of electromagnetic and oceanic waves—A review. *Bound.-Layer Meteorol.* **1978**, *13*, 61–85. [CrossRef]
49. Goodman, R. Overview and future trends in oil spill remote sensing. *Spill Sci. Technol. Bull.* **1994**, *1*, 11–21. [CrossRef]
50. Attema, E. An experimental campaign for the determination of the radar signature of the ocean at C-band. In Proceedings of the Third International Colloquium on Spectral Signatures of Objects in Remote Sensing, Les Arcs, France, 16–20 December 1986; pp. 791–799.
51. Mouche, A. Sentinel-1 Ocean Wind Fields (OWI) Algorithm Definition; Sentinel-1 IPF Reference:(S1-TN-CLS-52-9049) Report; CLS: Brest, France, 2010; pp. 1–75.
52. Freeman, A.; Curlander, J.C. Radiometric correction and calibration of SAR images. *Photogramm. Eng. Remote Sens.* **1989**, *55*, 1295–1301.
53. Lihai, Y.; Jialong, G.; Kai, J.; Yang, W. Research on efficient calibration techniques for airborne SAR systems. In Proceedings of the 2009 2nd Asian-Pacific Conference on Synthetic Aperture Radar, Shanxi, China, 26–30 October 2009; pp. 266–269.
54. Moreira, A.; Prats-Iraola, P.; Younis, M.; Krieger, G.; Hajnsek, I.; Papathanassiou, K.P. A tutorial on synthetic aperture radar. *IEEE Geosci. Remote Sens. Mag.* **2013**, *1*, 6–43. [CrossRef]
55. Lillesand, T.; Kiefer, R.W.; Chipman, J. *Remote Sensing and Image Interpretation*; John Wiley & Sons: Hoboken, NJ, USA, 2015.
56. Gao, F.; Xue, X.; Sun, J.; Wang, J.; Zhang, Y. A SAR image despeckling method based on two-dimensional S transform shrinkage. *IEEE Trans. Geosci. Remote Sens.* **2016**, *54*, 3025–3034. [CrossRef]

57. Tong, S.; Liu, X.; Chen, Q.; Zhang, Z.; Xie, G. Multi-feature based ocean oil spill detection for polarimetric SAR data using random forest and the self-similarity parameter. *Remote Sens.* **2019**, *11*, 451. [CrossRef]
58. Singha, S.; Vespe, M.; Trieschmann, O. Automatic Synthetic Aperture Radar based oil spill detection and performance estimation via a semi-automatic operational service benchmark. *Mar. Pollut. Bull.* **2013**, *73*, 199–209. [CrossRef]
59. Wang, J.; Perez, L. The effectiveness of data augmentation in image classification using deep learning. *Convolutional Neural Netw. Vis. Recognit.* **2017**, *11*, 1–8.
60. Powers, D.M. Evaluation: From precision, recall and F-measure to ROC, informedness, markedness and correlation. *arXiv* **2020**, arXiv:2010.16061.
61. Fawcett, T. An introduction to ROC analysis. *Pattern Recognit. Lett.* **2006**, *27*, 861–874. [CrossRef]
62. Linardatos, P.; Papastefanopoulos, V.; Kotsiantis, S. Explainable ai: A review of machine learning interpretability methods. *Entropy* **2021**, *23*, 18. [CrossRef]
63. Selvaraju, R.R.; Cogswell, M.; Das, A.; Vedantam, R.; Parikh, D.; Batra, D. Grad-cam: Visual explanations from deep networks via gradient-based localization. In Proceedings of the IEEE International Conference on Computer Vision, Venice, Italy, 22–29 October 2017; pp. 618–626.
64. Vinogradova, K.; Dibrov, A.; Myers, G. Towards interpretable semantic segmentation via gradient-weighted class activation mapping (student abstract). In Proceedings of the AAAI Conference on Artificial Intelligence, New York, NY, USA, 7–12 February 2020; Volume 34, pp. 13943–13944.
65. Ribeiro, M.T.; Singh, S.; Guestrin, C. "Why should i trust you?" Explaining the predictions of any classifier. In Proceedings of the 22nd ACM SIGKDD International Conference on Knowledge Discovery and Data Mining, San Francisco, CA, USA, 13–17 August 2016; pp. 1135–1144.
66. Montavon, G.; Samek, W.; Müller, K.R. Methods for interpreting and understanding deep neural networks. *Digit. Signal Process.* **2018**, *73*, 1–15. [CrossRef]
67. Van der Velden, B.H.; Janse, M.H.; Ragusi, M.A.; Loo, C.E.; Gilhuijs, K.G. Volumetric breast density estimation on MRI using explainable deep learning regression. *Sci. Rep.* **2020**, *10*, 18095. [CrossRef]
68. Knapič, S.; Malhi, A.; Salujaa, R.; Främling, K. Explainable Artificial Intelligence for Human Decision-Support System in Medical Domain. *arXiv* **2021**, arXiv:2105.02357.
69. Shapley, L.S. *17. A Value for n-Person Games*; Princeton University Press: Princeton, NJ, USA, 2016.
70. Yu, S.; Chen, M.; Zhang, E.; Wu, J.; Yu, H.; Yang, Z.; Ma, L.; Gu, X.; Lu, W. Robustness study of noisy annotation in deep learning based medical image segmentation. *Phys. Med. Biol.* **2020**, *65*, 175007. [CrossRef]

remote sensing

MDPI

Article

Ship Detection and Feature Visualization Analysis Based on Lightweight CNN in VH and VV Polarization Images

Xiaomeng Geng [1], Lei Shi [1], Jie Yang [1,*], Pingxiang Li [1], Lingli Zhao [2], Weidong Sun [1] and Jinqi Zhao [1]

[1] State Key Laboratory of Information Engineering in Surveying, Mapping and Remote Sensing, Wuhan University, Wuhan 430079, China; gengxm@whu.edu.cn (X.G.); shi.lei@whu.edu.cn (L.S.); pxli@whu.edu.cn (P.L.); widensun2012@whu.edu.cn (W.S.); masurq@whu.edu.cn (J.Z.)

[2] School of Remote Sensing and Information Engineering, Wuhan University, Wuhan 430079, China; zhaolingli@whu.edu.cn

* Correspondence: yangj@whu.edu.cn

Abstract: Synthetic aperture radar (SAR) is a significant application in maritime monitoring, which can provide SAR data throughout the day and in all weather conditions. With the development of artificial intelligence and big data technologies, the data-driven convolutional neural network (CNN) has become widely used in ship detection. However, the accuracy, feature visualization, and analysis of ship detection need to be improved further, when the CNN method is used. In this letter, we propose a two-stage ship detection for land-contained sea area without a traditional sea-land segmentation process. First, to decrease the possibly existing false alarms from the island, an island filter is used as the first step, and then threshold segmentation is used to quickly perform candidate detection. Second, a two-layer lightweight CNN model-based classifier is built to separate false alarms from the ship object. Finally, we discuss the CNN interpretation and visualize in detail when the ship is predicted in vertical–horizontal (VH) and vertical–vertical (VV) polarization. Experiments demonstrate that the proposed method can reach an accuracy of 99.4% and an F1 score of 0.99 based on the Sentinel-1 images for a ship with a size of less than 32 × 32.

Keywords: SAR; CNN; Sentinel-1; ship detection

Citation: Geng, X.; Shi, L.; Yang, J.; Li, P.; Zhao, L.; Sun, W.; Zhao, J. Ship Detection and Feature Visualization Analysis Based on Lightweight CNN in VH and VV Polarization Images. *Remote Sens.* **2021**, *13*, 1184. https://doi.org/10.3390/rs13061184

Academic Editor: Chung-Ru Ho

Received: 5 February 2021
Accepted: 17 March 2021
Published: 19 March 2021

1. Introduction

Ship detection plays a crucial role in maritime transportation, maritime surveillance applications in fishing, and maritime rights maintenance. Synthetic aperture radar (SAR), as active remote sensing, is most suitable for ship detection because it is sensitive to hard targets. Furthermore, SAR works throughout the day and in all weather conditions. In recent years, many SAR satellites, such as Radarsat1/2, TerraSAR-X, Sentinel-1, COSMO-SkyMed, and GF-3, have been providing a wide variety of SAR images with different resolutions, modes, and polarizations for maritime application, thereby enabling ship detection.

According to previous research, ship detection usually involves land-ocean segmentation, preprocessing, prescreening, and discrimination. Constant false alarm rate (CFAR) [1–4], as a traditional method, is typically used in ship detection. Furthermore, these methods are dependent on the statistical distribution of sea clutter, which is difficult to accurately estimate because of sea waves and ocean currents. Besides, the window size of protection and background influences the detection effectiveness. Land-ocean segmentation is also unavoidable, thereby causing poor robustness for SAR imagery in those methods. These traditional ship-detection methods require extensive calculations to address the parameters of statistical distribution, which is not sufficiently flexible and intelligent, and the detection speed does not meet actual needs.

At present, with the development of big data and deep learning technologies, convolutional neural networks (CNN) are widely used in mapping ice-wedge polygon (IWP) [5,6], identifying damaged buildings [7], classifying sea ice cover and land type [8–10], and so on.

Those CNN models successfully developed an automatic extraction framework for high spatial resolution remote sensing applications in a large-scale application. However, those CNN models need the input data and ground truth annotation one-to-one correspondence. In some research fields, the ground truth data are not easy to obtain due to lack of expert knowledge and time consumption. Besides, a growing number of researchers are beginning to study object detection based on convolutional neural network (CNN) methods. Single-stage methods, such as a proposed region-based convolutional network (R-CNN) [11], Fast R-CNN [12], and Faster R-CNN [13], and two-stage methods such as SSD [14], YOLO V1/V2/V3/V4 [14–18], and RetinaNet [19], have exhibited impressive results on various object detection benchmarks based on PASCAL VOC [20] and MS COCO [21] datasets. However, the natural images differ from the SAR images, which are produced through a coherent imaging process that leads to foreshortening, layover, and shadowing. Apart from the image mechanisms, targets in SAR images vary, such as ghosts, islands, artificial objects, island, or a harbor that displays similar backscattering mechanisms to ships, which lead to a high rate of false alarms. Therefore, to apply the deep learning algorithm to the SAR data, researchers have constructed SAR Ship Detection Dataset [22], SAR-Ship-Dataset [23], OpenSAR [24], and high-resolution SAR image dataset [25] containing Sentinel-1, Radarsat-2, TerraSAR-X, COSMO-SkyMe, and GaoFen-3 images. These datasets vary in polarization (HH, HV, VH, and VV), resolution (0.5, 1, 3, 5, 8, and 10 m), incidence angle, imaging mode, and background.

Compared with the PASCAL and COCO datasets, the SAR datasets have a low volume. When training the object detectors for ship detection in SAR images, finetuning or transfer learning is widely used. These CNN methods have been used for target detection in SAR images, ship detection [26], and land target detection [27], and have performed better than the traditional methods.

The deficiency of the method is that average precision is low because the models fail to consider the SAR image mechanisms [22]. However, the pretraining time and detection speed of classical object detectors usually do not meet the requirements of real-time ship detection, maritime rescue, and emergency military decision-making. In recent years, many researchers have paid attention to ship detection using CNN objectors. A grid CNN was proposed and proved to improve the accuracy and speed of ship detection [28]. Receptive pyramid network extraction strategies and attention mechanism technology are proved to improve the accuracy of ship detection [29]. These methods have relatively deep convolutional layers, hundreds of millions of parameters, and involve a long training time. Besides, in the data-driven CNN model, it is not easy and time-consuming to obtain the true value of the target bounding box corresponding to the input image. Therefore, these methods do not meet the requirements of fast processing, real-time response, and large-scale detection.

To achieve low complexity and high reliability through a CNN, some researchers have begun to split the images into small patches in the pre-screening stage and then use a relatively lightweight CNN model to classify the patches. Thereafter, the classification results are mapped onto the original images. A two-stage framework involves pre-screening and a relatively simple CNN architecture have been proposed [30,31], but in the pre-screening stage where a simple constant false alarm rate detector is used. As mentioned, the CFAR detector falls into a large number of calculations to solve the parameters of the statistical distribution and ignores small targets. Six convolutional layers, three max-pooling layers, and two full-connection layers are proposed to ship classification based on GF3-SAR images [26]. In these methods, CFAR and Ostu are typically used to obtain candidate targets in the pre-screening stage, and then a simple CNN model is used to reduce false alarms and recognize the ship. Unfortunately, time consumption is increased when sea-land segmentation and CFAR detector are applied in the pre-screening stage. Although the Ostu improves the speed of the pre-screening stage, the threshold may not work effectively and may cause an excessive number of false alarms. After the SAR image preprocessing, the CNN model can perform ship detection from all patches, but the accuracy of ship detection needs to be improved for the small-level ship. Besides, scholars had analyzed and discussed the ship detection in the CNN method, but the feature visualization and analysis of ship detection in both VH and VV polarization were less discussed, which is important in understanding ship detection through the CNN method. Thus, in this letter, we are mainly concerned with ship detection accuracy and feature visualization and analysis by using the VH and VV polarization.

Considering these difficulties, we propose a two-stage ship-detection method. In the first stage, Lee and island filters are used to reduce the noise and false alarms. Then, an exponential inverse cumulative distribution function (EICDF) [32,33] is applied to quickly estimate the segmentation threshold and obtain candidate detection results with relatively few false alarms. Then, all candidates are put in a lightweight CNN to accurately recognize the ships. Finally, the feature visualization and analysis of ship detection are carried out by the Grad-class activation mapping (Grad-CAM). The main contributions of the work are as follows:

1. The first ship detection method for SAR images is proposed. To quickly obtain candidate detection results, this study presents a fast threshold segmentation for candidate detection, which has been proved to reduce false alarms, obtain all candidate ships with different scales, and save time in the offshore area.
2. Most detectors consist of deep architecture and millions of parameters, thereby resulting in complex extraction features and lengthy pretraining time. In this study, a simple lightweight CNN architecture, which is fast and effective, was proposed to detect the ship.
3. The Grad-CAM was introduced to explain and visualize the CNN model, and then analyze the great attention pixel when the ship and false alarm were predicted.

The rest of this paper is organized as follows. In Section 2, we present the details of the dataset, data pre-processing, and the proposed method. Section 3 reports the experiment results. Sections 4 and 5 present the discussion and conclusions, respectively. Finally, a summary of this paper is provided.

2. Dataset and Proposed Methodology

In this section, first, the Sentinel-1 SAR images are introduced in detail. Second, the data progress, candidate targets, and dataset conduction are described. Third, the lightweight CNN model is presented.

2.1. Dataset

In this section, three Sentinel-1 SAR images located in the East Sea of China and one Sentinel-1SAR image located in the Huanghai Sea were used in the experiment as shown in Figure 1. The SAR images contain VH and VV polarizations with a pixel resolution of 10 m × 10 m, and the real resolution is 22 m × 20 m in azimuth and range. The information of SAR images includes the acquisition time of the image, the swath width, and the image mode is presented in Table 1.

2.2. Data Pre-Processing

In the section, the process of Sentinel-1 SAR images is described in detail. Figure 2 shows the complete workflow of ship detection. The workflow consists of four steps: preprocessing, candidate target and dataset construction, CNN model building, and training and ship detection.

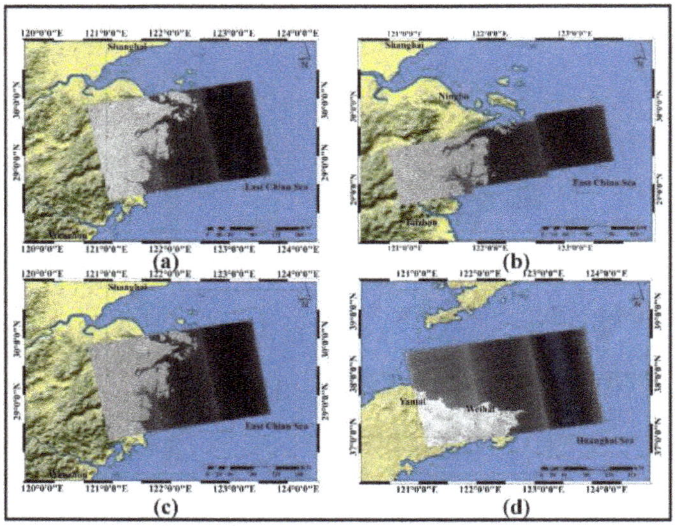

Figure 1. Image coverage of three Sentinel-1 images: (**a**) 23 June 2020; (**b**) 11 July 2020; (**c**) 17 July 2020; (**d**) 13 February 2021.

Table 1. Detailed information of Sentinel-1.

No.	Time	Image Mode	Polarization	Resolution (Azimuth × Range)	Swath (km)	Position
1.	23 June 2020	IW	VH, VV	22 × 20	250	East China Sea
2.	11 July 2020	IW	VH, VV	22 × 20	250	East China Sea
3.	17 July 2020	IW	VH, VV	22 × 20	250	East China Sea
4.	13 February 2021	IW	VH, VV	22 × 20	250	Huanghai Sea

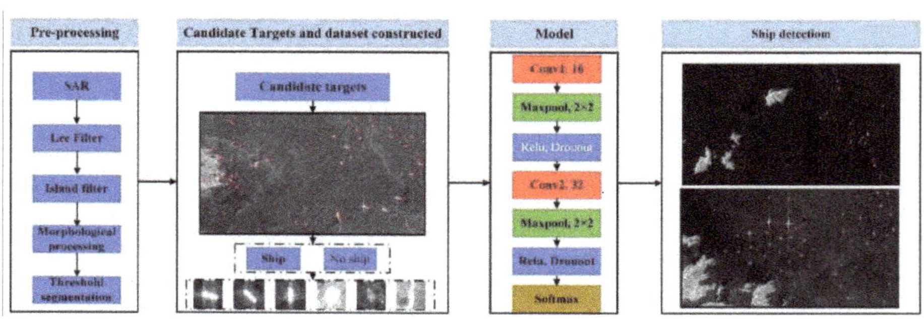

Figure 2. Workflow of ship detection.

The presence of speckle noise in SAR images causes difficulty in interpretation, thereby degrading the image quality. Therefore, the refined Lee filter [34] was used to improve the quality of the image and eliminate the coherence noise before the SAR image input (Figure 3a). In previous research, land-ocean segmentation was unavoidable in the pre-process to reduce the false alarms from land, harbor, and island. In the SAR images, the false alarms are mainly from stones, rocks, artificial targets, and island can usually provide similar backscatter coefficients of ships. Therefore, the island filter is applied to reduce the false alarms from the island, which is proved effective [30,35], as shown in Figure 3b. The ship candidates on the dark sea surface are reserved, and similar changes are visible on the edge of the island and reef (see Figure 3c). In summary, the false alarm of the island and reef is reduced more than that of the ship, leading to an increase in the ship-island contrast. To achieve ship target enhancement, we applied the morphological process consisting of erosion and dilation to improve contrast (see Figure 3d). Generally, the ship for detection is assumed to fit a permanent distribution, and then the threshold is calculated through probability density function. Similar to the CFAR method, after the morphological process, the image is similar to exponent distribution, and then the threshold is estimated by EICDF method with the input image mean value and a priori value of 0.999. The segmentation result is shown in Figure 3e. Finally, eight-connected domain processing was applied in the segmentation, and the preliminary result is shown in Figure 3f. After eight-connected domain processing, we can get the minimum bounding rectangle of the candidate target. Then, we can obtain all candidate slices according to the minimum bounding rectangle. Noteworthy, taken the centroid of the target as the origin, the slices are extended to 32 × 32 for the minimum bounding rectangle less than 32 × 32 and the slices are resized 32 × 32 for the minimum bounding rectangle of more than 32 × 32.

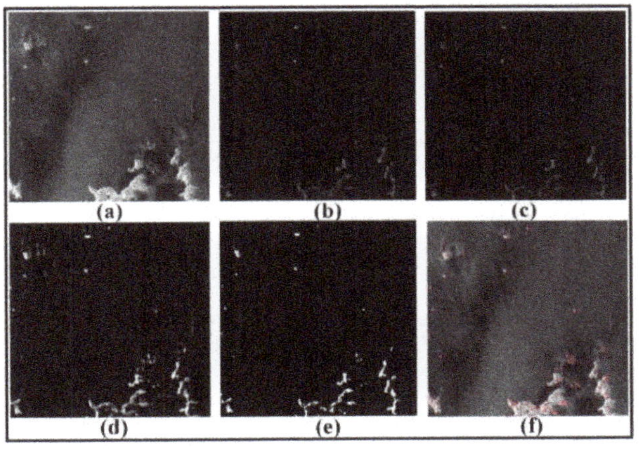

Figure 3. Pre-processing of SAR image: (**a**) SAR image, (**b**) image of island filter, (**c**) image of Gaussian filter, (**d**) image of the morphological process, (**e**) binary image, and (**f**) candidate result.

2.3. Candidate Detection

In this section, the candidate detection is discussed in detail based on the threshold calculation. In the candidate detection stage, the SAR images acquired on 11 July 2020 (No.2) and 17 July 2020 (No.3) were used. The two SAR images were cropped into 1000 × 1000 sub-images, with 50% overlap, then all sub-images were preprocessed according to the data-process method described above. Figure 4a shows four SAR image background scenes with the size of 1000 × 1000 in the flow of candidate detection. Scene 1, screen 2, and screen 3 include different land and islands and different scale ships, and scene 4 includes ships in inhomogeneous conditions. The contrast between the candidate target and sea background of the SAR data is more obvious after data preprocessing. Thus, we had an opportunity to detect the targets using the traditional method. The CFAR was proved available when it was used to detect the candidates in [30]. However, the CFAR was usually slow due to the parameter calculation of the sliding window in the entire image. In addition, the parameter estimation may not work well under the inhomogeneous conditions of the sea background. Thus, the threshold segmentation was considered to save time and avoid excessive calculation. In previous studies, the Otsu method was one of the most successful technologies used in image segmentation. However, the method was not effective in cases where the difference of the variance in object and background was significant [36]. In our experiment, the image presented exponential distribution (Figure 4c) after the morphological process (Figure 4b), and the EICDF could effectively estimate the threshold. To quantitatively compare the effectiveness of the method, we further discuss the aforementioned methods. Figure 4d–f shows the candidate results of the Ostu, CFAR, and EICDF methods. The results indicated that the ships and false alarms can be detected in both methods. The remarkable difference was the EICDF estimation with less time cost and false alarms compared with Otsu and CFAR. Table 2 lists the number of candidates and time cost. Our goal was to detect all ships with the least time cost and false alarms. In the four different screens, although all of the ships were detected in both methods, the number of false alarms and time cost were different. CFAR had a large number of false alarms and time cost than Otsu and ELCDF. The number of candidates was closer in Otsu and ELCDF, but ELCDF took less time than Ostu. Hence, it is proved that the ELCDF is effective with the least time cost and candidates.

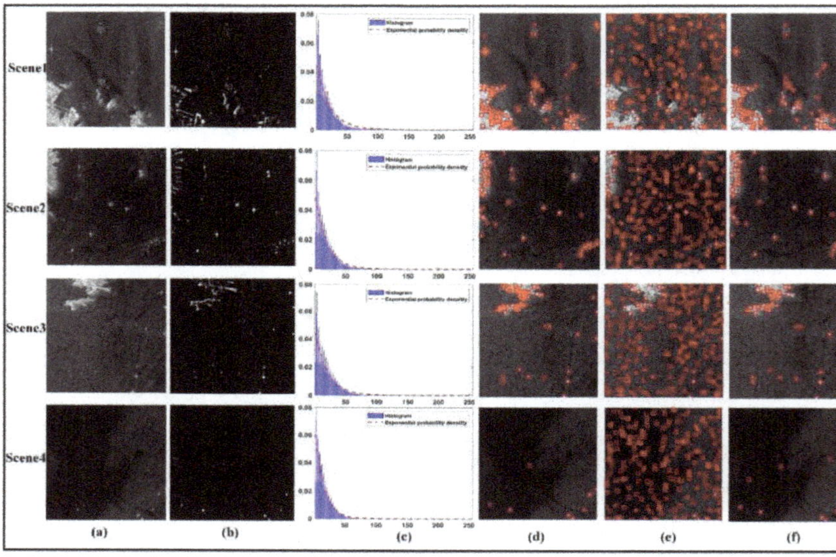

Figure 4. Progress of candidate detection: (**a**) original image, (**b**) morphological process, (**c**) histogram and probability density function, (**d**) candidate detection by Otsu method, (**e**) candidate detection by constant false alarm rate (CFAR) (applying Gaussian distribution and the probability of false alarm is 0.0001), and (**f**) candidate detection by exponential inverse cumulative distribution function (EICDF).

Table 2. Calculation of candidates and time.

	Method	Number of Candidates	Time (s)
Scene 1	Otsu	112	0.153
	CFAR	236	46.137
	ELCDF	92	0.051
Scene 2	Otsu	26	0.034
	CFAR	232	46.567
	ELCDF	29	0.009
Scene 3	Otsu	63	0.008
	CFAR	206	49.428
	ELCDF	49	0.002
Scene 4	Otsu	8	0.034
	CFAR	207	46.027
	ELCDF	10	0.011

2.4. Policies for Construction of Ship Detection Dataset

As a result of data pre-processing in Section 2.2, all targets contain false alarms and ships can be detected by ELCDF from the sub-images of No. 2 and No. 3, as shown in Figure 4f. Figure 5 presents the details of the false alarm and ship slices. For most of the ship slices, the backscatter intensity is relatively larger than the false alarms, and the rest are dark sea surface pixels distributed in the edges and corners. By contrast, the false alarm slices vary widely, some targets have strong backscatter intensity and the other has relatively weak backscatter intensity.

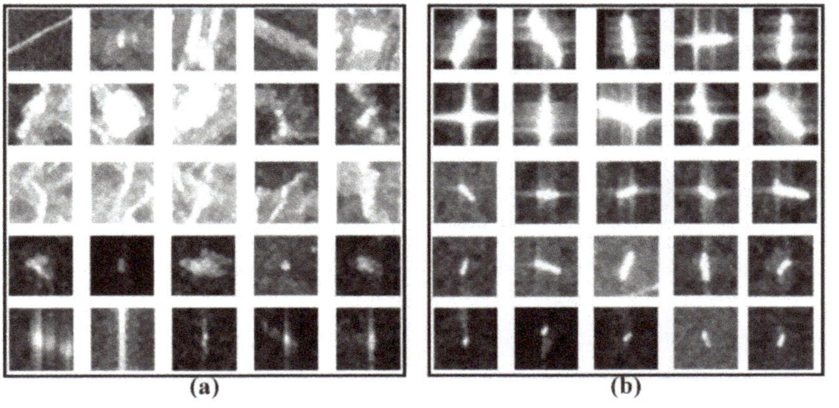

Figure 5. Candidate target: (**a**) false alarm and (**b**) ships.

A dataset including both false alarm slices and ship slices is constructed to make the CNN model more robust in the training stage. The policies of the dataset construction are considering both ships of different sizes and non-ship objects with a very similar shape and structure to that of a ship. Hence, we divide both the ship and false alarm slices into different categories, and the detailed categories are listed as follow:

The false alarms are divided into four categories:

False alarm #1: The characteristics of this type of false alarm mainly come from artificial targets such as cross-sea bridges, tall buildings, lighthouses and others, which are similar to the ship slices, as shown in the first two rows in Figure 5a. Most of these targets have strong scattering intensity across the center.

False alarm #2: The characteristics of this type of false alarm mainly come from the land targets, where has a bright ridge line as shown in the third row of Figure 5a.

False alarm #3: The characteristics of this type of false alarm mainly come from natural targets such as small islands, reefs, and rocks, as shown in the fourth row of Figure 5a. Most of those targets are similar to the medium-level ship in the third row of Figure 5b, which has low scattering intensity across the center and is surrounded by a dark sea surface.

False alarm #4: The characteristics of this type of false alarm mainly come from azimuth ambiguity as explained in [37], which usually brings a great challenge in ship detection through the traditional method. This type of false alarm is distributed at the center of the slice, close to the backscattering intensity of the ships, as shown in the fifth row of Figure 5a.

The ships are divided into three categories:

Ship #1: The characteristics of these ships have a strong scattering intensity and a large-level size, as shown in the first two rows in Figure 5b.

Ship #2: The characteristics of these ships have a strong scattering intensity and a medium-level size, as shown in the third and fourth row in Figure 5b.

Ship #3: The characteristics of these ships have a low scattering intensity and a small-level size, which is similar to the ghost, as shown in the fifth row of Figure 5a,b.

After the pre-process and candidate detection by the VH and VV polarizations of No. 2 and No. 3 SAR images, then, labeled the false alarm and ship by comparing manually the SAR image and the Google Earth high-resolution optical image, and by considering the scattering characteristics and context information of targets. The results of a dataset of the ship and false alarms are listed in Table 3. VH and VH polarization have a total of 4198 false alarm slices and 3132 ship slices.

Table 3. Pre-process detection result.

Name	Number of Slices	Polarization
False alarms	4198	VH, VV
Ships	3132	

2.5. CNN Model

As mentioned, the classic object detectors tend to have deep convolutional layers and more training parameters, which often take a long time to train. In this letter, we introduce a two-layer lightweight CNN model similar to the classic LeNet-5 model [38], called the modified LeNet-5 (M-LeNet). Detailed information on the proposed CNN model is listed in Table 4. The CNN model contains convolution, MaxPool, rectified linear units (ReLU), dropout, and full-connection layers.

$$y = \phi\left(\sum_i w_i x_i - \theta\right). \tag{1}$$

Table 4. Details of M-LeNet model.

Name	Layer Type	Input Size	Kernel Size	Output Size
Input	Input	-	-	1 × 32 × 32
C1	Convolution + ReLU	1 × 32 × 32	3 × 3	16 × 32 × 32
P1	MaxPooling + Dropout	16 × 32 × 32	2 × 2	16 × 16 × 16
C3	Convolution + ReLU	16 × 16 × 16	3 × 3	32 × 16 × 16
P2	MaxPooling + Dropout	32 × 16 × 16	2 × 2	32 × 8 × 8
FC1	Fully connected + ReLU + Dropout	32 × 8 × 8	-	512
FC2	Fully connected + ReLU + Dropout	512	-	128
FC3	Fully connected	128	-	2

In general, the convolution operation computes its output as a nonlinear function ϕ of the weighted sum of its inputs and of a bias term θ, as shown in Equation (1).

In the previous studies, the input size was set to 60 × 60, 64 × 64, and 128 × 128, respectively [26,30,31,39]. In this letter, the input size was set to 32 × 32 to reduce the calculation and simplify the model. As the length and width of some marine objects in this study were larger than 32 pixels, the resize process was applied in the pre-process stage to ensure the same input size. To limit the number of weights to learn, all the filter kernels were set to 3 × 3. In the beginning, 16 convolutional kernels of size work on the input images to extract features, after which the outputs are downsampled by max-pooling kernels with a size of 2 × 2. Then, the second convolutional layer filters the outputs of the first pooling layer with 32 filter kernels. Thereafter, the convolutional layers are downsampled by the second pooling layer to shrink the feature maps. Finally, three fully connected layers (FC1 with 512 output neurons, FC2 with 128 output neurons, and FC3 with 2 output neurons) take the outputs of the dropout layers as input, and then the softmax function is used to predict the labels of the targets after the final output vector. The strides of all the convolutional layers and all the pooling layers are set to 1 and 2, respectively. Furthermore, overfitting may occur easily when a neural network is trained on a small dataset. The dropout layer [27] is used for every max-pooling and fully connected layer to prevent overfitting and improve the performance of the neural network. Furthermore, rectified Linear Unit (ReLU) is used for every convolutional layer and fully connected layer to prevent a vanishing or exploding gradient. The mathematical derivation of the forward and backpropagation algorithm was proved and discussed in [40].

The cross-entropy loss function is used to minimize the error between the ground truth and the CNN prediction output, which can be written as follows:

$$L(w) = \frac{1}{m}\sum_{i=1}^{m} P\left(y^{(i)}\middle|x^{(i)};w\right), \quad (2)$$

where w is the trainable weight parameter, m represents the total number of training samples, and y^i, x^i refer to the true label and predicted label of the ith example, respectively.

3. Experimental Results and Analysis

For the comprehensive evaluation of the ship detection result using the proposed method, two sub-images of the No. 1 SAR image acquired on 23 June 2020 located in the East China Sea area were clipped, as shown in Figure 6a–d. The sizes of the two sub-images were 3791 × 2847 and 2589 × 1565. Besides, a sub-image of 4339 × 3258 was also clipped from the No.4 SAR image acquired on 13 February 2021, located in the Huanghai Sea area, as shown in Figure 7a,b. Figure 6 shows that the land and island contain sea areas in both regions. The azimuth ambiguities are often caused by the sampling of the Doppler spectrum at finite intervals of the pulse repetition frequency (PRF) due to the acquisition mode of two channels [37]. Thus, in the SAR images, a small amount of "ghost" appears around the ship in high-speed movement, but is not negligible in ship detection. Figure 6e,f shows the azimuth ambiguities caused by ships moving at high speed. In general, the scattering intensity of co-polarization (see Figure 6b,d) is higher than that of cross-polarization (see Figure 6a,b). Thus, the same targets may present different scattering intensity in VH and VV polarization. The characteristic of the target in VH polarization is less than that in VV polarization, especially for the small targets. In previous studies, the co-polarization data were also selected for ship detection. However, the VH polarization is less influenced by azimuth ambiguities. Thus, in PolSAR images, the azimuth ambiguity was usually suppressed by two cross-polarization channels [37]. However, in previous studies, the performance of ship detection by VH and VV polarization was less discussed. Thus, considering the characteristics of dual-polarization SAR in marine imaging, we utilized VH and VV to detect the ship using the CNN method. Figure 8 shows the candidate results of ship detection based on the method described in Section 2.3. The sub-image with complex background presents that all ships can be detected, and false alarm caused by land, island and azimuth ambiguity also can be detected. There are 122 true ships and 244 false alarm targets in the sub-images of No. 1 SAR image and 17 true ship and 137 false alarm targets in the sub-image of No. 4 SAR image. The ground truth can be obtained by using SAR expert knowledge interpretation and Google Earth in order to evaluate the performance of the proposed method in the next section. It should be noted that the interpretation of those ground truth is to identify false alarms by comparing the SAR image with the high-resolution optical image on Google Earth and then identifying the ship based on the scattering characteristics and context of the ship on the SAR image.

3.1. Training Details

In this section, the implementation of the hardware and platform is introduced in our experiments. We perform the experiments on the Ubuntu 14.04 operating system with an 11.9 GB memory NVIDIA TITAN Xp GPU. Inspired by the hyperparameters set of the literature [23,41,42], the learning rate, batch size, max epoch, moment, and momentum were set at 0.01, 32, 0.9, 1000, and 0.0005, respectively. Considering the SAR characteristics, we discarded the data augmentation in our experiment [43]. A set of optimal hyperparameters for a learning algorithm list in Table 5.

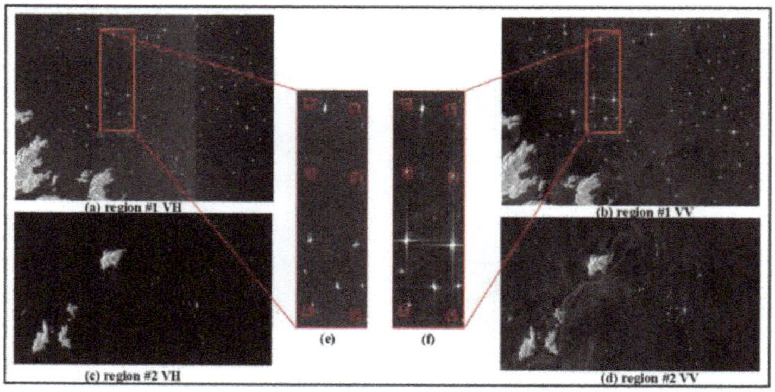

Figure 6. Sub-images of No. 1 image vertical–horizontal (VH) and vertical–vertical (VV) polarizations in the East China Sea area. The red rectangle in (**e**,**f**) shows the azimuth ambiguities caused by ships in VH and VV polarization.

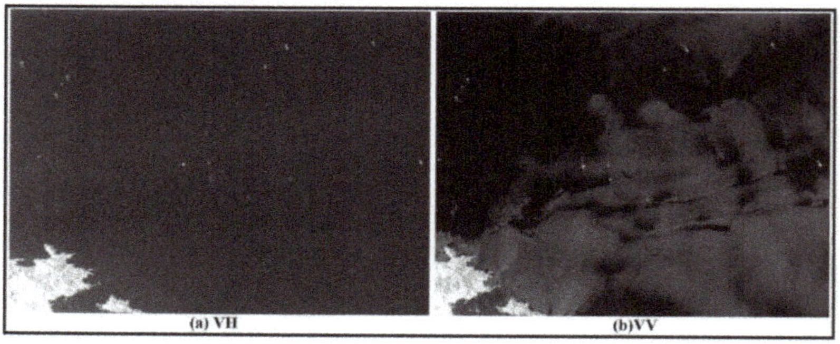

Figure 7. Sub-images of No. 4 image VH and VV polarizations in the Huanghai Sea area.

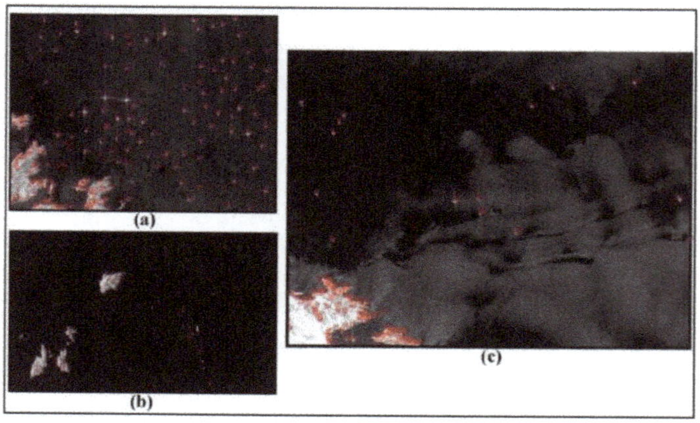

Figure 8. Results of candidate detection. (**a**,**b**) The sub-image of the No. 1 SAR image located in the East China Sea area. (**c**) The sub-image of the No. 4 SAR image location in the Huanghai Sea area.

Table 5. The hyperparameters settings.

Hyperparameters	Value
Learning rate	0.01
Momentum	0.9
Weight decay	0.005
Epochs	1000
Batch size	32
Learning rate scheduler	StepLR (step size = 200, gamma = 0.1)
Optimizer	SGD
Loss function	cross-entropy

To compare with our method, we also introduced machine-learning methods such as KNN, SVM, RF, and the classic CNN LeNet-5 method, which was commonly used and showed good performance in the classification task. In this letter, KNN, SVM, and random forest (RF) were implemented on the Ubuntu 14.04 operating system and Scikit-learn in Python. The parameters of KNN, SVM, and RF can be set with the default parameters. Besides, the classic CNN LeNet-5 method was also used. The hyperparameters of LeNet-5 were set as the M-LeNet. To ensure similarity in input data, these data were normalized to 0 and 1, with the values of mean and variance set to 0.5.

The training and validation samples are listed in Table 6. In all methods, the training and validation sample comes from 11 July 2020 (No. 2) and 23 July 2020 (No. 3), and the ratio is set at 8:2 in the training stage. In the testing stage, the test sample comes from the SAR data acquired on 23 June 2020 (No. 1) and 13 February 2021 (No. 4).

Table 6. The information of training, validation, and test data.

Samples	Location	SAR Image	Width × Heigh	False Alarm	Ship
Training and validation	East China Sea	11 July 2020 17 July 2020	25,138 × 8667 25,493 × 16,718	4198	3132
Testing	East China Sea	23 June 2020	3791 × 2847, 2589 × 1565	244	122
	Huanghai Sea	13 February 2021	4339 × 3258	137	17

3.2. VH Polarization Results

In this section, we first conducted the experiments on VH polarization by KNN, SVM, RF, LeNet-5, and our method. Figure 8 shows the candidate results of the ship and false alarm. Apart from the true ship, many false alarm targets are detected in the land and island areas. As mentioned above, in order to detect ships more accurately, a lightweight CNN method is proposed. Meanwhile, the KNN, SVM, RF, and classic LeNet-5 methods were introduced to indicate the effectiveness of our methods. Figure 9 shows the results of different methods through which all ships could be detected and false alarms were reduced further. The KNN method presents more false alarms and fewer true ships than the other methods. The performance of different machine learning methods was discussed [30,44]. Noi and Kappas [44] confirmed that when the number of training samples increases from 1267 pixels to 2619 pixels (each class has 135 polygons) in land cover classification experiments, the accuracy of SVM and RF is significantly better than that of KNN. Wang et al. [30] also demonstrated that the performance of KNN is less than that of RF and SVM in ship detection. Thus, the performance of RF and SVM is reasonably better than that of KNN. In the CNN method, the performance of M-LeNet is better than that of the LeNet-5 method.

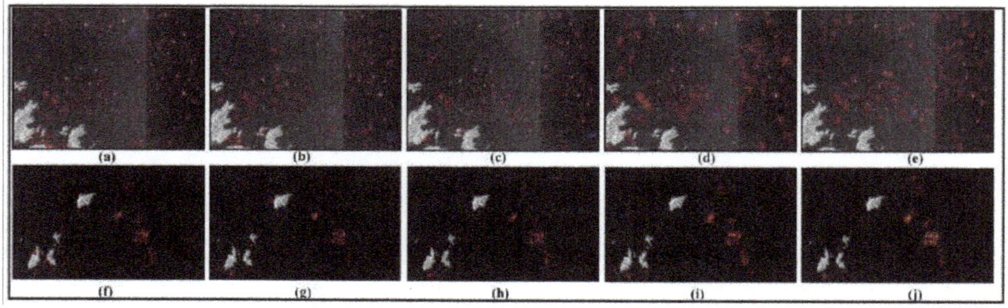

Figure 9. Sub-image 1 (**up**) and sub-image 2 (**down**) detection results of VH polarization in the East China Sea area. (**a–e**) KNN, SVM, RF, LeNet-5, and M-LeNet, respectively. (**f–j**) KNN, SVM, RF, LeNet-5, and M-LeNet, respectively. (Red rectangle: ship, red rectangular box with arrow: false alarm, and blue rectangle: missed ship).

To quantitatively evaluate the performance of KNN, SVM, RF, LeNet-5, and M-LeNet, we introduced the evaluation indicator, such as accuracy, precision, recall, and F1 score. In these evaluation indicators, the F1 score is the weighted average of precision and recall, and is usually more useful than accuracy. The equations are as follows:

$$\text{Accuracy} = \frac{\text{TP} + \text{TN}}{\text{TP} + \text{FP} + \text{FN} + \text{TN}}, \tag{3}$$

$$\text{Precision} = \frac{\text{TP}}{\text{TP} + \text{FP}}, \tag{4}$$

$$\text{Recall} = \frac{\text{TP}}{\text{TP} + \text{FN}}, \tag{5}$$

$$\text{F1 Score} = 2 \times \frac{\text{Recall} \times \text{Precision}}{\text{Recall} + \text{Precision}}, \tag{6}$$

where true positive (TP) means that the ships are correctly predicted, true negative (TN) means that the ships are predicted to be false alarms, false positive (FP) means that the actual class is a false alarm and the predicted class is the ship, and false negative (FN) means that the actual class is the ship but the predicted class is a false alarm. In the CNN, the input data are the slices, the output is the probability of ships and false alarms. Hence, the evaluation performance is based on the number of ships and false alarms. Then, the accuracy, precision, recall, and F1 score were evaluated based on ground truth and the number of predictions of the ship and false alarm slices.

In addition, the number of missed ships and the number of false alarms were also calculated. In the sub-images, 122 true ships were obtained through expert knowledge interpretation using the SAR scattering mechanism. Table 7 presents the detailed evaluation indicators. RF provides the best evaluation indicators compared with KNN and SVM for the machine learning method. M-LeNet presents the best evaluation indicators for the CNN method. The number of the least missed ship is one in the RF method, and the number of the most missed ship is five in the KNN method. The number of the least false alarms is zero in M-LeNet, and the number of the most false alarms is eleven. The false alarm mainly occurs in the land areas in the lower-left corner of the image, which shows a structure similar to a ship, with low surrounding background. Furthermore, the false alarms caused by azimuth ambiguity are also incorrectly detected. Compared with VV polarization, VH polarization has lower backscattering, especially for small targets. Thus, the poor performance of this type of ship fails to be detected in the CNN methods, as indicated by the blue rectangles in Figure 9. The CNN method generally exhibits better performance than the machine learning method. Although several ships are missed, the overall performance of M-LeNet is better than that of RF. The reason is that the RF classifier based on the statistical model is sensitive to the image pixels, while the convolution and pooling kernel operations lead to the small targets miss detailed texture information and rich semantic information in the CNN method. Thus, the performance of small target detection in RF is better than that of M-LeNet and the performance of false alarm detection in M-LeNet is better than that of RF. Although the number of correct ship detections by M-LeNet is not as much as that of RF, the false number of ship detections is less than that of RF and LeNet-5. The comprehensive evaluation indicators such as the F1 score, accuracy, and recall show better performance than RF. M-LeNet showed the best performance with an F1 score of 0.99 and an accuracy of 99.40%. Besides, in order to show our CNN model more transferability, a SAR image located in the Huanghai Sea area was used to test the performance of ship detection. In order to quickly evaluate the accuracy, a sub-image with the size of 4339 × 3258 was clipped. In the sub-images, 17 true ships were obtained through expert knowledge interpretation and Google Earth. Although the number of the ship is less than the sub-images in No. 1, the VH and VV polarization shows different sea background. Figure 10 shows the detection results and Table 8 presents the detailed evaluation indicators. The LeNet-5 presents better performance than KNN, SVM, and RF with an F1 score of 0.90 and an accuracy of 98.05%. The M-LeNet shows the best performance in those methods with an F1 score of 0.97 and an accuracy of 99.35%.

Table 7. Detailed evaluation index of VH polarization in the East China Sea area.

Polarization	Method	Accuracy (%)	F1	Recall	Precision	Correct	False	Miss
VH	KNN	95.20	0.94	0.96	0.91	117	11	5
	SVM	96.98	0.96	0.98	0.94	119	7	3
	RF	98.81	0.98	**0.99**	**0.98**	**121**	3	**1**
	LeNet-5	97.60	0.97	0.95	0.98	116	2	6
	M-LeNet	**99.40**	**0.99**	**0.98**	**1.0**	120	**0**	2

The best evaluation criteria are highlighted in bold and underlined in each column.

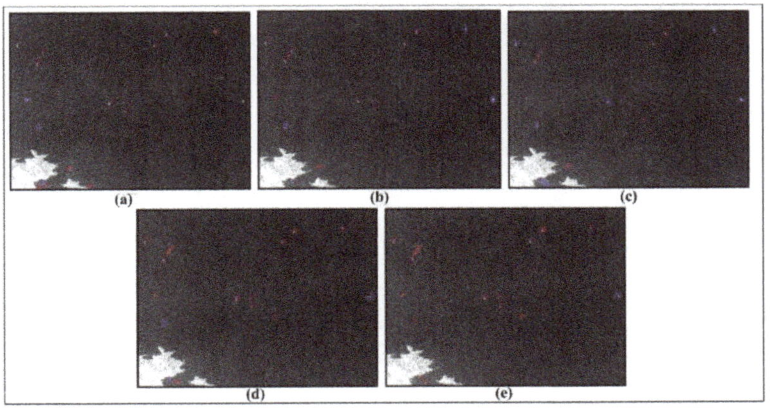

Figure 10. Sub-image detection results of VH polarization in the Huanghai Sea. (**a**–**e**) represent KNN, SVM, RF, LeNet-5, and M-LeNet, respectively. (Red rectangle: ship, red rectangular box with arrow: false alarm, and blue rectangle: missed ship).

Table 8. Detailed evaluation index of VH polarization in the Huanghai Sea area.

Polarization	Method	Accuracy (%)	F1	Recall	Precision	Correct	False	Miss
VH	KNN	96.10	0.82	0.82	0.82	14	3	3
	SVM	97.75	0.84	0.76	0.92	13	**1**	4
	RF	94.16	0.67	0.53	0.90	9	**1**	8
	LeNet-5	**98.05**	0.90	0.82	**1.0**	14	0	3
	M-LeNet	**99.35**	**0.97**	**0.94**	**1.0**	**16**	**0**	**1**

The best evaluation criteria are highlighted in bold and underlined in each column.

3.3. VV Polarization Results

Figure 11 shows the detection results of VV polarization. Similar to VH polarization, the more false alarms were reduced, the more ships were retained. In sub-image 1, the more false alarms mainly appeared in Figure 11a,b,d. In sub-image 2, the false alarms mainly existed in Figure 11f,g. Figure 11 shows that RF performs best in machine learning and M-LeNet performs best in deep learning. To quantitatively compare the performance of different methods, we calculated the accuracy, precision, recall, and F1 score. Table 9 presents the results of the evaluation indicators. The number of the least missed ship is zero in the RF and SVM method, and the number of the most missed ship is ten in the LeNet-5 method. The number of the least false alarms is three in M-LeNet, and the number of the most false alarms is twenty-one in the KNN method. RF and SVM could detect all the true ships, but a few false alarms were retained compared with VH polarization. Similar to the performance of VH polarization, KNN had more missed ships and false alarms. LeNet-5 performed worst with more missed ships. Although the number of missed ships in the M-LeNet method was more than that of RF and SVM, the comprehensive evaluation indicators showed the best performance with an F1 score of 0.98 and an accuracy of 98.2%. The characteristics of false alarms caused by azimuth ambiguity are similar to those of the true ship, so distinguishing the false alarms is difficult. Although the M-LeNet method could reduce false alarms caused by azimuth ambiguity more effectively than other methods, the false alarms still existed. In [31], 680 ships and 170 ghosts were selected for training; the experiments on the Sentinel-1 images showed encouraging results, but further improvement is needed. In our experiment, the number of ghosts was under 0.2%, which indicated a great imbalance for ship and ghost training samples. Thus, the predicted performance for the ghost is poor. Figure 12 shows the detection result and Table 10 presents the evaluation index of VV polarization in the No.4 sub-image of the Huanghai Sea area. Different from the VH polarization in Figure 10, the VV polarization image shows an inhomogeneous pattern in the SAR scene due to other marine phenomena that may exist in the images, e.g., moderate-to-high wind, upwelling, and eddies [45,46]. In those methods, the RF shows the better performance with an F1 score of 0.97 and an accuracy of 99.35% than other methods. Although the M-LeNet achieves an F1 score of 0.92 and an accuracy of 98.05%, the M-LeNet enables all ships detected in the inhomogeneous.

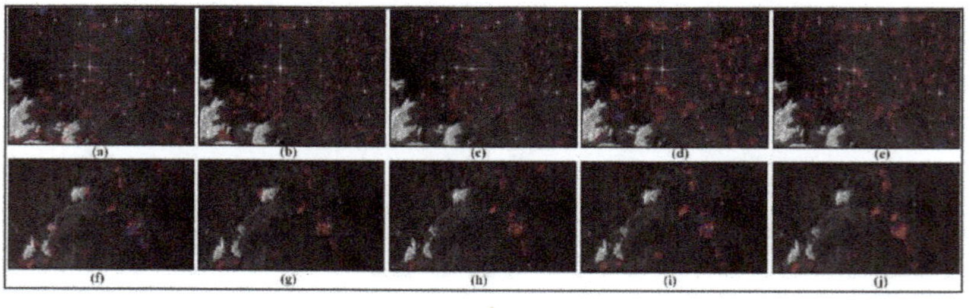

Figure 11. Sub-image 1 (**up**) and sub-image 2 (**down**) detection results of VV polarization in the East China Sea area: (**a–e**) show KNN, SVM, RF, LeNet-5, and M-LeNet, respectively. (**f–j**) represent KNN, SVM, RF, LeNet-5, and M-LeNet, respectively. (Red rectangle: ship, red rectangular box with arrow: false alarm, and blue rectangle: missed ship).

Table 9. Detailed evaluation index of VV polarization in the East China Sea area.

Polarization	Method	Accuracy (%)	F1	Recall	Precision	Correct	False	Miss
VV	KNN	91.59	0.89	0.94	0.85	115	21	7
	SVM	95.80	0.95	1.0	**0.90**	**<u>122</u>**	14	**<u>0</u>**
	RF	97.00	0.96	1.0	**0.92**	**<u>122</u>**	10	**<u>0</u>**
	LeNet-5	94.89	0.93	0.92	0.94	112	7	10
	M-LeNet	**<u>98.20</u>**	**<u>0.98</u>**	**<u>0.98</u>**	**<u>0.98</u>**	119	**<u>3</u>**	3

The best evaluation criteria are highlighted in bold and underlined in each column.

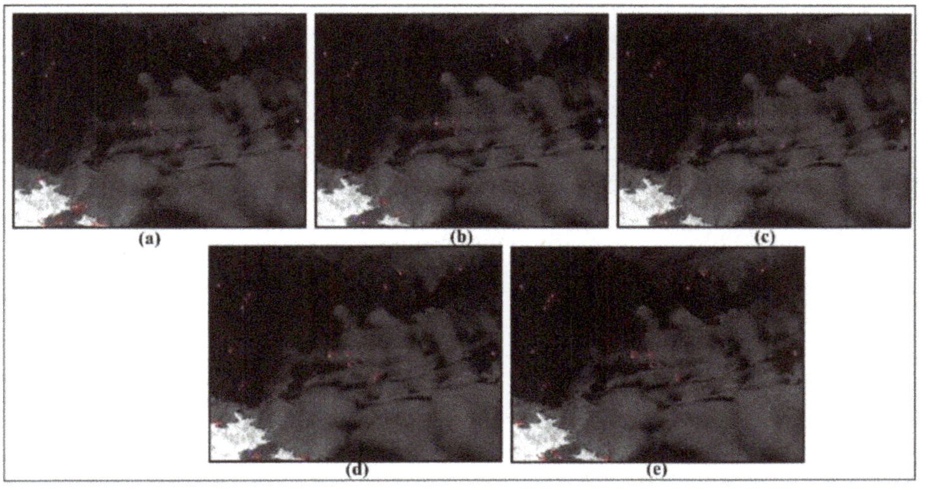

Figure 12. Sub-image detection results of VV polarization in the Huanghai Sea area. (**a**–**e**) represent KNN, SVM, RF, LeNet-5, and M-LeNet, respectively. (Red rectangle: ship, red rectangular box with arrow: false alarm, and blue rectangle: missed ship).

Table 10. Detailed evaluation index of VV polarization in the Huanghai Sea area.

Polarization	Method	Accuracy (%)	F1	Recall	Precision	Correct	False	Miss
VV	KNN	95.20	0.81	**1.0**	0.68	**17**	8	**0**
	SVM	96.98	0.85	0.82	0.88	14	2	3
	RF	**99.35**	**0.97**	0.94	**1.0**	16	**0**	1
	LeNet-5	98.05	0.92	0.94	0.89	16	2	1
	M-LeNet	98.05	0.91	**1.0**	0.85	**17**	3	**0**

The best evaluation criteria are highlighted in bold and underlined in each column.

3.4. CNN Feature Visualization Analysis

Deep neural networks have enabled unprecedented breakthroughs in classification, semantic segmentation, and object detection task. Although those CNN networks enable superior performance, interpreting and visualizing them are difficult due to the lack of decomposability into intuitive and understandable components [47]. CAM was proposed to identify discriminative regions by a restricted class of image classifications and to gain a better understanding of a model. However, any fully connected layer of the model was removed, and instead of global average pooling (GAP) to obtain the localization of a class [48]. Thus, altering the model architecture was unavoidable, training is needed again, and the available staffing scenarios are restricted. Grad-CAM improved the CAM by using the gradient information flowing into the last convolutional layer of CNN to understand the importance of each neuron for a classification decision [49]. Similar to CAM, Grad-CAM uses the feature maps produced by the last convolutional layer of a CNN. In CAM, we weigh these feature maps using weights taken out of the last fully connected layer of the network. In Grad-CAM, we obtained neuron importance weight using a_k^c (Equation (5)) calculated based on the global average pool, with the gradients over the height dimension (indexed by i) and the width dimension (indexed by j). Therefore, Grad-CAM obtained the class discriminative localization map $L^c_{\text{Grad-CAM}}$ without a particular model architecture because we can calculate gradients through any kind of neural network layer we want.

$L^c_{\text{Grad-CAM}}$ performs a weighted combination of forward activation maps, and follows it by ReLU to obtain the final class discriminative saliency map, as shown in Equation (6).

$$a^c_k = \frac{1}{Z}\sum_i \sum_j \frac{\partial y^c}{\partial A^k_{ij}}, \tag{7}$$

$$L^c_{\text{Grad-CAM}} = ReLU\left(\sum_k a^c_k A^k\right), \tag{8}$$

where weight a^c_k is the feature map k of a target class. A^k_{ij} represents feature map k. A^k is the feature map of a convolutional layer, $L^c_{\text{Grad-CAM}} \in \mathbb{R}^{u \times v}$ of height v, and width u for any class c, y^c is the feature map A^k of a convolutional layer, i.e., $\frac{\partial y^c}{\partial A^k_{ij}}$. Detailed information can be found in [49].

The output of Grad-CAM is a "class-discriminative localization map," i.e., a heatmap where the hot part corresponds to a particular class. Figures 13 and 14 show the Grad-CAM visualization heatmap for "false alarm" and "ship" of VH and VV polarization, respectively. The heatmap represents the image region with the greatest attention from CNN for the correct prediction of images belonging to a particular class. Figures 13a and 14a show great attention through the CNN prediction of images belonging to false alarms. These image slices belong to the same area of the VH and VV polarization, which contain buildings near the sea-land, small island, reef, and azimuth ambiguity. The heatmap of false alarms shows that the surrounding background was conducive to the false alarm recognition. The azimuth ambiguity presented different characteristics in VH and VV polarization; a similar phenomenon has been discussed in Section 3. Fortunately, the azimuth ambiguity could be observed in the first row in VH and VV polarization. The azimuth ambiguity scattering intensity in VV polarization was more obvious than that in VH polarization. Furthermore, the false alarm in VV polarization presented different characteristics. One focused on the surrounding background from the heatmap, and another focused on the azimuth ambiguity itself, which was why the azimuth ambiguity of false alarm could not predict better in polarization. Figures 13b and 14b show great attention through the CNN prediction of images belonging to the ship. The different scale ships with high scattering intensity had an important contribution to ship recognition than the surrounding sea surface, which was different from the false alarm in the VH and VV polarization.

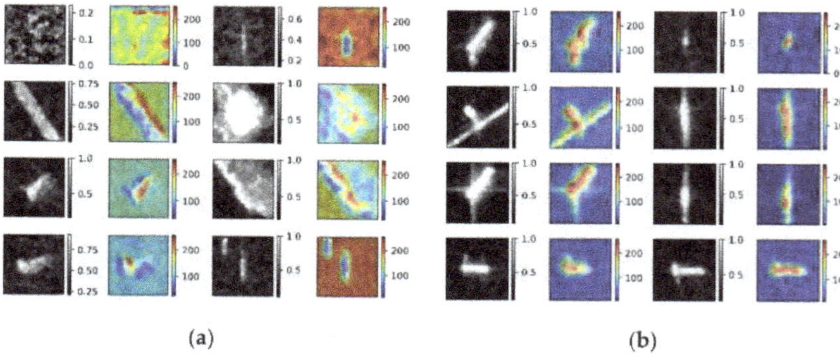

Figure 13. Visualization of VH polarization: (**a**) heatmap of false alarm and (**b**) heatmap of ship.

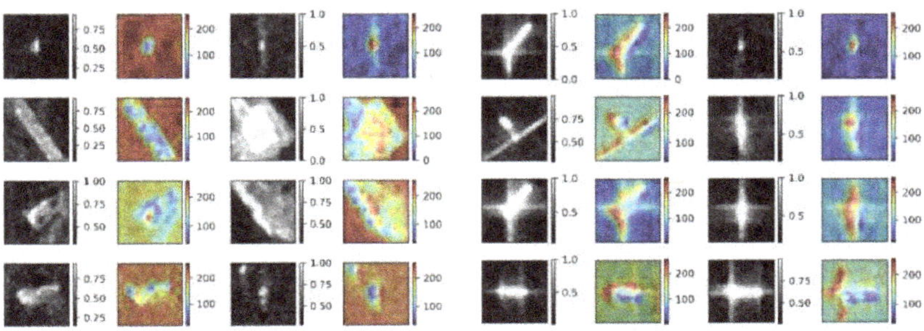

(a) (b)

Figure 14. Visualization of VV polarization: (**a**) heatmap of false alarm and (**b**) heatmap of ship.

4. Discussion

The performance of ship detection in CNN methods proves its great potential in different backgrounds such as incidence angles, wind speeds, sea states, and ocean dynamic parameters that mainly influence the backscattering coefficient between the ocean surface and the ship [23,50,51]. Besides, the scattering characteristics of ghosts caused by azimuth ambiguity when the ship is moving at high speed is similar to the characteristics of the ship, thereby causing difficulty in distinguishing between the ship and ghost in a single-polarization image. The CNN method also shows great potential. In this study, the performance of lightweight CNN does not completely suppress the ghost due to the lack of adequate training samples in VV polarization. Fortunately, the ghost in VH polarization is less affected, and thus, the performance of lightweight CNN shows the best result in VH polarization. Future work will be conducted to add the training samples of the ghost.

In the object detectors, the size of small targets is less than 32 × 32 for nature images [21]. However, the SAR images are different from the nature images, and the size of the ship is usually much less than 32 × 32, especially for ships operating offshore. Figure 15 shows the size of the ship in the test SAR image. Almost all ships have an area of less than 32 × 32, and most ships have an area of less than 24 × 25. The SVM and RF methods based on statistical characteristics show good performance with the fewest ships missed, especially the small ships in VV polarization; however, some false alarms cannot be avoided. The PFN module and feature fusion strategy are often used to improve the detection accuracy and reduce the false alarms of the small target [26,42,52]. Furthermore, those modules always integrate into the VGG16 and ResNet-50 networks [42,53]; the CNN models are complex and have many parameters to train. The PFN module and feature fusion strategy show effectiveness for small goals in object detectors, but may show poor effectiveness for much less than 32 × 32. Thus, in this study, we provide a dataset and two-stage method for ship detection with the SAR image, where even extremely small ships can be completely recognized in the first stage. In the second stage, the different scale candidates in the test SAR images can be accurately detected by considering context background information. The best and stable performance of ship detection is demonstrated by M-LeNet, which can reduce the false alarms and missed ships, and obtain higher precision in VH and VV polarization than other methods in different ocean areas and scenarios.

In the previous studies, the ship detection using sentinel-1 SAR images was carried out by Wang et al. [54]. The performance of ship detection can reach an accuracy of 98.07% and an F1 score of 0.90 by Faster RCNN, thus, the number of false alarms was detected to be relatively large [54]. The accuracy could reach 90.05% based on YOLOv2 for imagery [55]. The test precision and F1 score were 91.3% and 0.92 for detecting multiscale ships and small ships, when using the GF-3 dataset, respectively [42]. In [29], the attention module was used to improve the performance of ship detection, the recall, precision, and F1 score could reach 0.96, 96.4%, and 0.96, respectively. Although the performance of ship detection was improved, the model complexity had increased. To reduce model complexity, a simple CNN was used to detect the ship, and the accurate rate of ship detection was 97.2% when using the spaceborne image [30]. The lightweight CNN was proposed to improve the accuracy and F1 score in our experiments. The performance of lightweight CNN shows that the best result can reach an accuracy of 99.4% and an F1 score of 0.99 based on Sentinel-1 images. Figure 15 shows the most ship has an area of less than 24 × 25 pixel. The test accuracy and F1 score also demonstrate the proposed method can detect the small-level ship. To sum up, the proposed method can detect the ship effectively in contrast to that with the detector above. Unfortunately, it was rarely analyzed and visualized the feature to gain a better understanding of a model in the previous studies. In order to understand and visualize the model, the Grad-CAM was used, and the result demonstrated it could help us understand the mechanism of how the ship and false alarm was predicted by the lightweight CNN model work. Hence, based on the visualization and analysis of the Grad-CAM, it can be used to help to detect the ship with the weakly unsupervised method in future work.

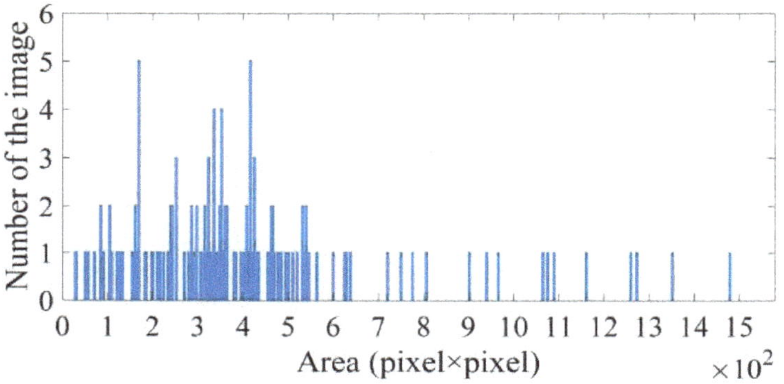

Figure 15. Area of the ship in test data.

From the above discussion, the lightweight CNN we proposed can show good performance in different ocean areas and scenarios. The difference with those detectors [29,54,55] does not need the input data and ground truth bounding box one-to-one correspondence, and only labeled in ship and no-ship. Besides, the CNN model we proposed is simplified as a shallow convolution neural network and improves efficiency in comparison with Faster RCNN, SSD, and Yolo, etc. However, comparing with those detectors, the CNN model we proposed is not end-to-end. To obtain the detect result, the data preprocess first needs to be applied to SAR images, then the lightweight CNN is used to accurately detect the ship. Although, the proposed method shows an accuracy of 99.4% and an F1 score of 0.99, how to simplify the data preprocess and integrate it into the CNN model to achieve end-to-end training is worth considering in future work. Besides, the ocean surface is modulated and complexed by ocean dynamics processes such as wind, waves, upwelling, and eddies, as well as sea state. Due to the limited data for training, it cannot cover all sea state conditions. The CNN model was not truly explored with comparably limited training data by Zhang et al. [5]. Hence, in order to make the CNN model to have more generalization capability, more data should be added in future work.

5. Summary and Conclusions

In this paper, the two-stage ship detection method is proposed in a complex background, i.e., in the offshore area. First, the SAR data pre-process contains the image filter, island filter, and threshold segmentation. The island filter is proposed to improve the ship contrast using a convolutional kernel, and threshold segmentation is proposed for slice production and candidate detection for time-saving. Second, the CNN model is proposed for slice fine classification and recognition. The experiment demonstrates that compared with the KNN, SVM, RF, and LeNet-5 methods, the proposed method can obtain stable accuracy in VH and VV polarization. Furthermore, although the proposed method cannot eliminate the false alarm caused by azimuth ambiguity in VV polarization because the few ghosts of false alarms sample in the training stage are insufficient to maintain the balance between the ship and false alarm, the false alarm caused by azimuth ambiguity in VH polarization give a little contribution. The detection performance shows better results in both VH and VV polarization for the ship size of much less than 32 × 32. Fortunately, the CNN interpretation and visualization of the ship and false alarm are accurate predictions through Grad-CAM visualized analysis. The experiments demonstrate that the high scattering intensity of the ship itself provides an important contribution to ship recognition rather than the surrounding sea surface in VH and VV polarization. However, the surrounding sea surface is useful for false alarm recognition in VH and VV polarization.

Author Contributions: Conceptualization, J.Z. and W.S.; data curation, L.Z.; methodology, X.G. and L.S.; supervision, P.L. and J.Y.; validation, X.G.; writing—original draft, X.G.; writing—review and editing, X.G. All authors have read and agreed to the published version of the manuscript.

Funding: This study was funded by the National Natural Science Foundation of China under Grant Nos. 42071295, 41771377 and 41901286. The Key Laboratory of Surveying and Mapping Science and Geospatial Information Technology of Ministry of Natural Resources under Grant No. 201906, and the Open Research Fund of Jiangsu Key Laboratory of Resources and Environmental Information Engineering, CUMT under Grant No. JS201909.

Acknowledgments: The authors are grateful to the Hubei Province Postdoctoral Science and Technology Preferred Project for funding this study. They thank the European Space Agency and Alaska Satellite Facility for providing free Sentinel-1 data online. The authors would also like to thank the anonymous reviewers for their comments to improve this article.

Conflicts of Interest: The authors declare no conflict of interest.

References

1. Ai, J.; Qi, X.; Yu, W.; Deng, Y.; Liu, F.; Shi, L. A New CFAR Ship Detection Algorithm Based on 2-D Joint Log-Normal Distribution in SAR Images. *IEEE Geosci. Remote Sens. Lett.* **2010**, *7*, 806–810. [CrossRef]
2. Ai, J.-Q.; Qi, X.-Y.; Yu, W.-D. Improved Two Parameter CFAR Ship Detection Algorithm in SAR Images. *J. Electron. Inf. Technol.* **2009**, *31*, 2881–2885.
3. Dai, H.; Du, L.; Wang, Y.; Wang, Z. A Modified CFAR Algorithm Based on Object Proposals for Ship Target Detection in SAR Images. *IEEE Geosci. Remote Sens. Lett.* **2016**, *13*, 1925–1929. [CrossRef]
4. Wang, C.; Bi, F.; Zhang, W.; Chen, L. An Intensity-Space Domain CFAR Method for Ship Detection in HR SAR Images. *IEEE Geosci. Remote Sens. Lett.* **2017**, *14*, 529–533. [CrossRef]
5. Zhang, W.; Liljedahl, A.K.; Kanevskiy, M.; Epstein, H.E.; Jones, B.M.; Jorgenson, M.T.; Kent, K. Transferability of the deep learning mask R-CNN model for automated mapping of ice-wedge polygons in high-resolution satellite and UAV images. *Remote Sens.* **2020**, *12*, 1085. [CrossRef]
6. Bhuiyan, M.A.E.; Witharana, C.; Liljedahl, A.K. Use of Very High Spatial Resolution Commercial Satellite Imagery and Deep Learning to Automatically Map Ice-Wedge Polygons across Tundra Vegetation Types. *J. Imaging* **2020**, *6*, 137. [CrossRef]
7. Yang, W.; Zhang, X.; Luo, P. Transferability of Convolutional Neural Network Models for Identifying Damaged Buildings Due to Earthquake. *Remote Sens.* **2021**, *13*, 504. [CrossRef]
8. Zhang, C.; Wei, S.; Ji, S.; Lu, M. Detecting Large-Scale Urban Land Cover Changes from Very High Resolution Remote Sensing Images Using CNN-Based Classification. *ISPRS Int. J. Geo-Inf.* **2019**, *8*, 189. [CrossRef]
9. Wang, Y.-R.; Li, X.-M. Arctic sea ice cover data from spaceborne SAR by deep learning. *Earth Syst. Sci. Data Discuss.* **2020**, 1–30. [CrossRef]
10. Shao, Z.; Zhou, W.; Deng, X.; Zhang, M.; Cheng, Q. Multilabel Remote Sensing Image Retrieval Based on Fully Convolutional Network. *IEEE J. Sel. Top. Appl. Earth Obs. Remote Sens.* **2020**, *13*, 318–328. [CrossRef]
11. Girshick, R.; Donahue, J.; Darrell, T.; Malik, J. Region-based convolutional networks for accurate object detection and segmentation. *IEEE Trans. Pattern Anal. Mach. Intell.* **2015**, *38*, 142–158. [CrossRef] [PubMed]
12. Girshick, R. Fast r-cnn. In Proceedings of the IEEE International Conference on Computer Vision, Santiago, Chile, 7–13 December 2015; pp. 1440–1448.
13. Ren, S.; He, K.; Girshick, R.; Sun, J. Faster r-cnn: Towards real-time object detection with region proposal networks. In Proceedings of the Advances in Neural Information Processing Systems, Barcelona, Spain, 5–10 December 2016; pp. 91–99.

14. Liu, W.; Anguelov, D.; Erhan, D.; Szegedy, C.; Reed, S.; Fu, C.-Y.; Berg, A.C. Ssd: Single shot multibox detector. In Proceedings of the European Conference on Computer Vision, Amsterdam, The Netherlands, 8–16 October 2016; pp. 21–37.
15. Redmon, J.; Farhadi, A. Yolov3: An incremental improvement. *arXiv* **2018**, arXiv:1804.02767.
16. Redmon, J.; Farhadi, A. YOLO9000: Better, faster, stronger. In Proceedings of the IEEE Conference on Computer Vision and Pattern Recognition, Honolulu, HI, USA, 21–26 July 2017; pp. 7263–7271.
17. Bochkovskiy, A.; Wang, C.-Y.; Liao, H.-Y.M. YOLOv4: Optimal Speed and Accuracy of Object Detection. *arXiv* **2020**, arXiv:2004.10934.
18. Redmon, J.; Divvala, S.; Girshick, R.; Farhadi, A. You only look once: Unified, real-time object detection. In Proceedings of the IEEE Conference on Computer Vision and Pattern Recognition, Las Vegas, NV, USA, 27–30 June 2016; pp. 779–788.
19. Lin, T.-Y.; Goyal, P.; Girshick, R.; He, K.; Dollár, P. Focal loss for dense object detection. In Proceedings of the IEEE International Conference on Computer Vision, Venice, Italy, 22–29 October 2017; pp. 2980–2988.
20. Everingham, M.; Van Gool, L.; Williams, C.K.; Winn, J.; Zisserman, A. The pascal visual object classes (voc) challenge. *Int. J. Comput. Vis.* **2010**, *88*, 303–338. [CrossRef]
21. Lin, T.-Y.; Maire, M.; Belongie, S.; Hays, J.; Perona, P.; Ramanan, D.; Dollár, P.; Zitnick, C.L. Microsoft coco: Common objects in context. In Proceedings of the European Conference on Computer Vision, Zurich, Switzerland, 6–12 September 2014; pp. 740–755.
22. Li, J.; Qu, C.; Shao, J. Ship detection in SAR images based on an improved faster R-CNN. In Proceedings of the Sar in Big Data Era: Models, Methods & Applications, Beijing, China, 13–14 November 2017.
23. Wang, Y.; Wang, C.; Zhang, H.; Dong, Y.; Wei, S. A SAR dataset of ship detection for deep learning under complex backgrounds. *Remote Sens.* **2019**, *11*, 765. [CrossRef]
24. Li, B.; Liu, B.; Huang, L.; Guo, W.; Zhang, Z.; Yu, W. OpenSARShip 2.0: A large-volume dataset for deeper interpretation of ship targets in Sentinel-1 imagery. In Proceedings of the 2017 SAR in Big Data Era: Models, Methods and Applications (BIGSARDATA), Beijing, China, 13–14 November 2017; pp. 1–5.
25. Wei, S.; Zeng, X.; Qu, Q.; Wang, M.; Su, H.; Shi, J. HRSID: A high-resolution SAR images dataset for ship detection and instance segmentation. *IEEE Access* **2020**, *8*, 120234–120254. [CrossRef]
26. Ma, M.; Chen, J.; Liu, W.; Yang, W. Ship Classification and Detection Based on CNN Using GF-3 SAR Images. *Remote Sens.* **2018**, *10*, 2043. [CrossRef]
27. Krizhevsky, A.; Sutskever, I.; Hinton, G.E. Imagenet classification with deep convolutional neural networks. *Commun. ACM* **2017**, *60*, 84–90. [CrossRef]
28. Zhang, T.; Zhang, X. High-Speed Ship Detection in SAR Images Based on a Grid Convolutional Neural Network. *Remote Sens.* **2019**, *11*, 1206. [CrossRef]
29. Zhao, Y.; Zhao, L.; Xiong, B.; Kuang, G. Attention receptive pyramid network for ship detection in SAR images. *IEEE J. Sel. Top. Appl. Earth Obs. Remote Sens.* **2020**, *13*, 2738–2756. [CrossRef]
30. Wang, Z.; Yang, T.; Zhang, H. Land contained sea area ship detection using spaceborne image. *Pattern Recognit. Lett.* **2020**, *130*, 125–131. [CrossRef]
31. Cozzolino, D.; Di Martino, G.; Poggi, G.; Verdoliva, L. A fully convolutional neural network for low-complexity single-stage ship detection in Sentinel-1 SAR images. In Proceedings of the 2017 IEEE International Geoscience and Remote Sensing Symposium (IGARSS), Fort Worth, TX, USA, 23–28 July 2017; pp. 886–889.
32. Martinez, W.L.; Martinez, A.R. *Computational Statistics Handbook with MATLAB*; CRC Press: Boca Raton, FL, USA, 2015; Volume 22.
33. Davis, T.A. *MATLAB Primer*; CRC Press: Boca Raton, FL, USA, 2010.
34. Lee, J.S.; Grunes, M.R.; De Grandi, G. Polarimetric SAR speckle filtering and its implication for classification. *IEEE Trans. Geosci. Remote Sens.* **2002**, *37*, 2363–2373.
35. Wang, Z.; Wang, C.; Zhang, H.; Wang, F.; Jin, F.; Xie, L. SAR-based ship detection in sea areas containing small islands. In Proceedings of the 2015 IEEE 5th Asia-Pacific Conference on Synthetic Aperture Radar (APSAR), Singapore, 1–4 September 2015; pp. 591–595.
36. Xu, X.; Xu, S.; Jin, L.; Song, E. Characteristic analysis of Otsu threshold and its applications. *Pattern Recognit. Lett.* **2011**, *32*, 956–961. [CrossRef]
37. Velotto, D.; Soccorsi, M.; Lehner, S. Azimuth ambiguities removal for ship detection using full polarimetric X-band SAR data. *IEEE Trans. Geosci. Remote Sens.* **2013**, *52*, 76–88. [CrossRef]
38. El-Sawy, A.; Hazem, E.-B.; Loey, M. CNN for handwritten arabic digits recognition based on LeNet-5. In Proceedings of the International Conference on Advanced Intelligent Systems and Informatics, Cairo, Egypt, 24–26 October 2016; pp. 566–575.
39. Sharifzadeh, F.; Akbarizadeh, G.; Kavian, Y.S. Ship Classification in SAR Images Using a New Hybrid CNN-MLP Classifier. *J. Indian Soc. Remote Sens.* **2019**, *47*, 551–562. [CrossRef]
40. Wu, J. Introduction to convolutional neural networks. *Natl. Key Lab Nov. Softw. Technol. Nanjing Univ. China* **2017**, *5*, 23.
41. Kim, K.; Hong, S.; Choi, B.; Kim, E. Probabilistic Ship Detection and Classification Using Deep Learning. *Appl. Sci.* **2018**, *8*, 936. [CrossRef]
42. Dai, W.; Mao, Y.; Yuan, R.; Liu, Y.; Pu, X.; Li, C. A Novel Detector Based on Convolution Neural Networks for Multiscale SAR Ship Detection in Complex Background. *Sensors* **2020**, *20*, 2547. [CrossRef]
43. Goodfellow, I.; Bengio, Y.; Courville, A. *Deep Learning*; Mit Press: Cambridge, MA, USA, 2016.

44. Thanh Noi, P.; Kappas, M. Comparison of Random Forest, k-Nearest Neighbor, and Support Vector Machine Classifiers for Land Cover Classification Using Sentinel-2 Imagery. *Sensors* **2018**, *18*, 18. [CrossRef]
45. Zhu, S.; Shao, W.; Armando, M.; Shi, J.; Sun, J.; Yuan, X.; Hu, J.; Yang, D.; Zuo, J. Evaluation of Chinese quad-polarization Gaofen-3 SAR wave mode data for significant wave height retrieval. *Can. J. Remote Sens.* **2018**, *44*, 588–600. [CrossRef]
46. Corcione, V.; Grieco, G.; Portabella, M.; Nunziata, F.; Migliaccio, M. A novel azimuth cutoff implementation to retrieve sea surface wind speed from SAR imagery. *IEEE Trans. Geosci. Remote Sens.* **2018**, *57*, 3331–3340. [CrossRef]
47. Lipton, Z.C. The mythos of model interpretability. *Queue* **2018**, *16*, 31–57. [CrossRef]
48. Lin, M.; Chen, Q.; Yan, S. Network in network. *arXiv* **2013**, arXiv:1312.4400.
49. Selvaraju, R.R.; Cogswell, M.; Das, A.; Vedantam, R.; Parikh, D.; Batra, D. Grad-cam: Visual explanations from deep networks via gradient-based localization. In Proceedings of the IEEE International Conference on Computer Vision, Venice, Italy, 22–29 October 2017; pp. 618–626.
50. Tings, B.R.; Bentes, C.; Velotto, D.; Voinov, S. Modelling ship detectability depending on TerraSAR-X-derived metocean parameters. *Ceas Space J.* **2019**, *11*, 81–94. [CrossRef]
51. Wang, Y.; Wang, C.; Zhang, H.; Dong, Y.; Wei, S. Automatic Ship Detection Based on RetinaNet Using Multi-Resolution Gaofen-3 Imagery. *Remote Sens.* **2019**, *11*, 531. [CrossRef]
52. Zhang, G.; Li, Z.; Li, X.; Yin, C.; Shi, Z. A Novel Salient Feature Fusion Method for Ship Detection in Synthetic Aperture Radar Images. *IEEE Access* **2020**, *8*, 215904–215914. [CrossRef]
53. He, J.; Guo, Y.; Yuan, H. Ship Target Automatic Detection Based on Hypercomplex Flourier Transform Saliency Model in High Spatial Resolution Remote-Sensing Images. *Sensors* **2020**, *20*, 2536. [CrossRef]
54. Wang, Y.; Wang, C.; Zhang, H. Combining a single shot multibox detector with transfer learning for ship detection using sentinel-1 SAR images. *Remote Sens. Lett.* **2018**, *9*, 780–788. [CrossRef]
55. Chang, Y.-L.; Anagaw, A.; Chang, L.; Wang, Y.C.; Hsiao, C.-Y.; Lee, W.-H. Ship Detection Based on YOLOv2 for SAR Imagery. *Remote Sens.* **2019**, *11*, 786. [CrossRef]

remote sensing

MDPI

Article

A Hybrid Deep Learning Model for the Bias Correction of SST Numerical Forecast Products Using Satellite Data

Tonghan Fei [1], Binghu Huang [1], Xiang Wang [2,*], Junxing Zhu [2], Yan Chen [2], Huizan Wang [2] and Weimin Zhang [2]

[1] College of Oceanography and Space Informatics, China University of Petroleum, Qingdao 266580, China; s19160016@s.upc.edu.cn (T.F.); huangbhu@upc.edu.cn (B.H.)

[2] College of Meteorology and Oceanography, National University of Defense Technology, Changsha 410073, China; junxingzhu@nudt.edu.cn (J.Z.); chenyan16@nudt.edu.cn (Y.C.); wanghuizan17@nudt.edu.cn (H.W.); wmzhang@nudt.edu.cn (W.Z.)

* Correspondence: xiangwangcn@nudt.edu.cn

Abstract: Sea surface temperature (SST) has important practical value in ocean related fields. Numerical prediction is a common method for forecasting SST at present. However, the forecast results produced by the numerical forecast models often deviate from the actual observation data, so it is necessary to correct the bias of the numerical forecast products. In this paper, an SST correction approach based on the Convolutional Long Short-Term Memory (ConvLSTM) network with multiple attention mechanisms is proposed, which considers the spatio-temporal relations in SST data. The proposed model is appropriate for correcting SST numerical forecast products by using satellite remote sensing data. The approach is tested in the region of the South China Sea and reduces the root mean squared error (RMSE) to 0.35 °C. Experimental results reveal that the proposed approach is significantly better than existing models, including traditional statistical methods, machine learning based methods, and deep learning methods.

Keywords: SST; bias correction; deep learning; ConvLSTM; 3D-C BAM

Citation: Fei, T.; Huang, B.; Wang, X.; Zhu, J.; Chen, Y.; Wang, H.; Zhang, W. A Hybrid Deep Learning Model for the Bias Correction of SST Numerical Forecast Products Using Satellite Data. *Remote Sens.* **2022**, *14*, 1339. https://doi.org/10.3390/rs14061339

Academic Editors: Chung-Ru Ho, Antony K. Liu and Xiaofeng Li

Received: 15 January 2022
Accepted: 17 February 2022
Published: 10 March 2022

1. Introduction

Oceans take up almost 71% of the entire surface of the globe, and are closely related to human activities. Sea surface temperature (SST) is the water temperature near the surface of the ocean. SST is an important physical quantity for global climate studies [1], marine ecosystem studies, and related applications. The forecast accuracy of SST is essential for marine disaster prevention, navigation, ocean fishery [2], and other ocean-related cases. SST prediction methods can be classified into two major categories [3]. One category is the numerical model, which is based on physics [4]. The other category is data-driven models, based on data analysis. With the improvement and development of the numerical model, the accuracy of numerical prediction has been improved. However, the numerical model cannot completely describe various physical processes in the ocean [5,6]—an uncertainty of the initial field [7,8]—and calculation errors exist in the numerical solution process of the model. Therefore, the prediction results of numerical forecast products need to be further corrected.

Currently, there are mainly three kinds of methods for numerical forecast products correction: traditional statistical methods, machine learning based methods, and deep learning methods. Statistical post-processing [9] is the typical one, such as model output statistics (MOS) approach [10,11], Kalman filtering [12,13], and Bayesian probability decision [14], all of which have achieved some results. As the machine learning, deep learning development [15], and computing performance improved, data-driven approaches were introduced into numerical forecast product correction, such as SVM [16], BP neural network [17], and CNN [18]. However, current methods for numerical forecast products correction have weaknesses, as they do not include the spatio-temporal relationships among the datasets.

Meanwhile, observation data of buoys in the ocean have been lacking for a long time. Since the launch of satellites equipped with ocean observation sensors, ocean remote sensing data observed from satellites have been widely used in in coastal erosion calculation [19], offshore oil spill [20], disaster warning [21], and other related research. Therefore, we consider combining deep learning methods with numerical models [22] and applying satellite data into numerical prediction models for SST numerical forecast products correction.

In this paper, we propose a new hybrid SST correction model, which not only takes into account the influence of spatial distribution of the dataset, but also takes into account the importance of temporal information. This approach is inspired by the outstanding performance of the ConvLSTM in capturing the spatio-temporal relationships and the attention mechanism in improving feature utilization. Combining these novel methodologies together will create a more effective model to correct SST as it will create a greater synergy than the individual models on their own.

In recent years, with the rapid development of machine learning, deep learning methods have been widely used in many fields, such as natural language processing [23], audio classification [24], community detection [25], and image restoration [26]. Some researchers have already used these methods in areas related to our research. For example, Shi et al. [27] proposed the ConvLSTM method for precipitation prediction. D. Liu et al. [28] proposed a combination of empirical mode decomposition (EMD) algorithm and encoder decoder long short-term memory (EN-DE-LSTM) architecture for water flow prediction. Z.I. Petrou et al. [29] proposed an encoder–decoder network with a convolutional long short-term memory unit for sea ice prediction. Chen.R et al. [30] proposed the hybrid CNN-LSTM model for typhoon forecasting, which improved the accuracy of typhoon forecasting. A. Y. Winona et al. [31] use the so-called LSTM method to forecast the sea level and X. Kun et al. [32] proposed LSTM-Attention temperature prediction model I by combining LSTM with Attention mechanism in order to make full use of historical data and improve the accuracy of temperature prediction.

Our new hybrid SST correction model can be used to correct SST numerical forecast products more accurately. 3DCNN is used to determine the spatial relations of various marine variables. Simultaneously, 3D-CBAM model is used to improve the utilization of spatial features and marine environmental features. ConvLSTM is used to determine the spatio-temporal relationships of the data. The attention model is used to assign the weight of historical information. Our proposed model can effectively determine the spatio-temporal dependencies between SST field data, and at the same time introduce an attentional mechanism to correct the ConvLSTM output by learning the appropriate weights at each step, thus achieving high-precision SST correction. A series of experimental results show that the proposed method can achieve better accuracy in SST correction.

The contributions of this paper include:

1. We propose a new hybrid model for SST correction, which uses satellite remote sensing observation data and spatio-temporal data of sea surface variables. The performance of our model is then evaluated;
2. The attention mechanism is used to assign weights to the information in the dataset, which reflect the influence of spatio-temporal information on the SST correction, so that the key information is highlighted and thus we obtain better correction results;
3. Taking the South China Sea area (10°N–15°N, 125°E–130°E) as an example, the accuracy rate was improved by 41.9% after the correction. We analyze the influence of input sequence with different time steps, different model parameters and other variables on the correction effect through the experiments. Experiments on the dataset of the South China Sea show that our new hybrid model is more effective than existing methods, including some classical machine learning methods.

The paper is structured in the following manner: Section 2 takes a look at the current state of correction methods for numerical forecast products and deep learning in the discipline; Section 3 elucidates the central problems of this paper; Section 4 introduces the new hybrid SST correction model; Section 5 introduces the evaluation scheme, the experimental set-up, and presents the experimental results; and, finally, Section 6 concludes this work and deals with recommendations for future work.

2. Related Work

Our research focuses on bias correction of SST numerical forecast products. In order to improve the accuracy of numerical forecast products, many scholars have proposed several methods to correct numerical prediction results. For example, Vannitsem et al. [10] and Tian et al. [11] used the mode output statistics (MOS) approach to establish a linear statistical relationship between model predictions and actual observations to improve SST forecasting accuracy, respectively. Krishnamurti et al. [33] used multiple regression to determine coefficients from multi-model forecasts and observations to improve weather and seasonal climate forecasts. Xu et al. [34] used the classical moving average method to analyze and correct the temperature forecast of the model, which improved the forecast accuracy to a certain extent. Libonati et al. [12] and Pelosi et al. [13] used Kalman filter to improve the quality of ensemble forecast in view of the existing deviation of ensemble forecast. X. Zhang et al. [35] proposed a method for correcting wave height prediction results of SWAN model based on Gaussian process regression (GPR).

With the continuous development of technology, the theory of machine learning has shown its extraordinary ability and great potential in the field of ocean and weather prediction and correction [36]. The correction model based on machine learning can capture the nonlinear variation [37] between the numerical model simulation results and the observation, so as to obtain more accurate model correction results. For example, J. Zeng et al. [16] used SVM to correct the weather forecast model, and the accuracy was effectively improved. Wang A et al. [38] designed a Random Forests-based adjusting method to correct the output of the WRF model and the RMSE of wind achieved an average decrease of 40% compared with the WRF model.

In addition, deep learning has injected fresh blood into artificial intelligence and machine learning. Deep learning is used to extract potential features and learn complex relationships in meteorological and oceanographic data, which provides a new idea for ocean and weather forecast and correction [39]. For example, Makarynskyy [40] improved wave parameter short-term forecasts based on artificial neural networks. Xu X. et al. [41] presented an ordinal Distribution Autoencoder (ODA) model, which can effectively correct numerical precipitation prediction based on ECMWF and SMS-WARMS model meteorological data. T. Wang et al. [42] proposed a residual single-hidden layer feedforward neural network, which is able to obtain effective corrections of numerical models. Rasp S et al. [43] proposed a flexible alternative based on neural networks to correct 2 m temperature. A. Sayeed et al. [18] used convolutional neural network (CNN) as post-processing technology to improve mesoscale weather research and prediction (WRF) daily simulated output. A. N. Deshmukh et al. [44] applied a wavelet neural network in improving numerical ocean wave predictions of significant wave height and peak wave period. It can be seen that deep learning has shown some potential in the temperature correction of model prediction, but it is still in the initial research and application stage. The above research also indicates the potential of machine learning in the correction of numerical model results.

However, there is little work to correct the forecast of SST. Zhang R used artificial neural network BP model to correct SST [17]. Yang X Q et al. [45] applied the prognostic trend (PT) correction method to reduce systematic errors in coupled GCM seasonal forecasts. Han Y.K. [46] proposed a new error-correction model based on the AR(p). Zhang P.J. [47] tried to correct numerical prediction SST product using GHRSST, and established a correction method for SST model prediction in the South China Sea—the effect of SST forecast correction was quite significant. The above methods for SST correction do not consider the temporal and spatial correlation between SST data. Therefore, we consider combining deep learning with the numerical model, using the deep learning method to mine the temporal and spatial correlation of SST data and correcting forecast products to improve forecast accuracy.

We attempt to use the deep learning method to mine the spatio-temporal relationship between SST forecast data and carry out the correction of daily mean SST in the study area. On the one hand, it is helpful to obtain more accurate prediction results, and on the other hand, it is also an exploratory application of the deep learning correction model in oceanography.

3. Problem Definitions

Our goal is to use historical ocean data and reanalysis data as truth values to modify model forecast data and establish an SST correction model. SST data is a time series data without considering spatial information. In order to analyze and obtain the time sequence relationship between the data, historical ocean data at multiple times should be used for correction. However, SST and other marine environmental variables are spatial fields at any time, so SST correction can be defined as a spatial-temporal series correction problem. Different from previous methods that take the SST of a single site as the model input data, this paper corrects the SST within the region as a whole, that is, a matrix, to facilitate the model to extract the temporal and spatial correlation of SST.

The input data with multiple elements can be represented as a matrix $W \times H \times C \times T$, where W and H represent longitude and latitude, C represents the number of elements, and T represents the length of time series. SST and sequence of marine environmental variables $T = T_1, T_2, \cdots, T$, where $|T|$ is the SST sequence length of time, $T_i (1 \leq i \leq |T|,\ i \in Z)$ is the marine environmental variables matrix of all the record points of day i in the region, which is a $W \times H \times C$ matrix. The sequence of these matrices is the input to the model. The SST correction problem can be defined as a series of historical marine environmental variables data of the previous N days $X_{t-n}(n = 1, 2, 3 \cdots)$, used to correct the SST at time t, where $X_{t-n}(n = 1, 2, 3 \cdots)$ is a sequential matrix, which is $W \times H \times C \times N$. Define the current moment to be corrected as t, and the SST and marine environment variable of the current moment to be corrected as X_t. Y_t is the corrected SST value and n is the previous days before the current time, each time step is 1 day. X_{t-n} is the grid data set of each variable at the predicted time and the previous n days.

The model can also be expressed as:

$$Y_t^{W \times H} = f\left(X_{t-n}^{W \times H \times C \times N}, X_t^{W \times H \times C}\right), n = 1,\ 2,\ 3 \cdots. \quad (1)$$

This is our target function, where f is the final model learnt by the historical data. On this basis, we design and train the deep learning model. During the training, the data is divided into two parts. First, we train our model with the training set, where the "truth value" is known, and use it to adjust the parameters in the model. Finally, we use the test set to evaluate the correction effect of the training model. Figure 1 shows the data structure of SST and the related variables.

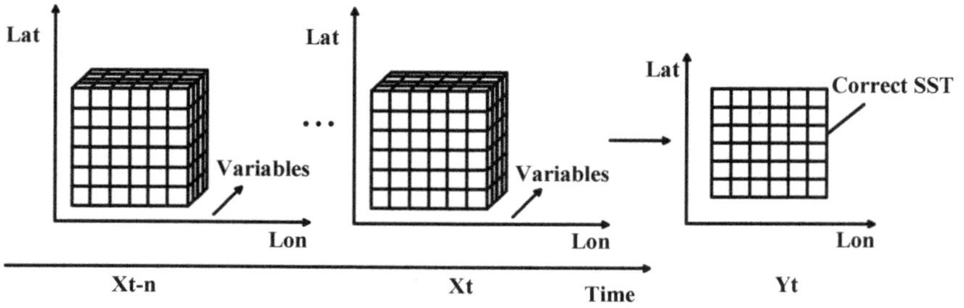

Figure 1. The structure of spatio-temporal variables sequence.

4. Method

In order to solve the problem of SST numerical prediction correction, we propose a new hybrid model for SST correction, which is based on ConvLSTM and the 3DCBAM model with attention mechanisms. It makes full use of the spatio-temporal information and marine environmental variables information.

4.1. The Framework of the New Hybrid SST Correction Model

The framework of the new hybrid SST correction model is shown in Figure 2, which is mainly divided into five stages: spatial feature extraction, spatial and channel attention mechanism, time-dependent learning, time attention mechanism, and output results. The main idea is to use convolution operation to extract and integrate spatial features of multiple variables, and use CBAM mechanism to improve the utilization rate of the spatial features of a 3D convolution network and show the importance of different environmental variables to the results. At the same time, ConvLSTM is used to learn the spatiotemporal relationship in the process of SST change, and attention mechanism is used to adjust the importance of information at different historical moments in variables. It not only considers the spatial correlation of SST field data, but also the time dependence between SST field data at different time and the interaction between marine environmental variables. Therefore, it can correct SST more accurately.

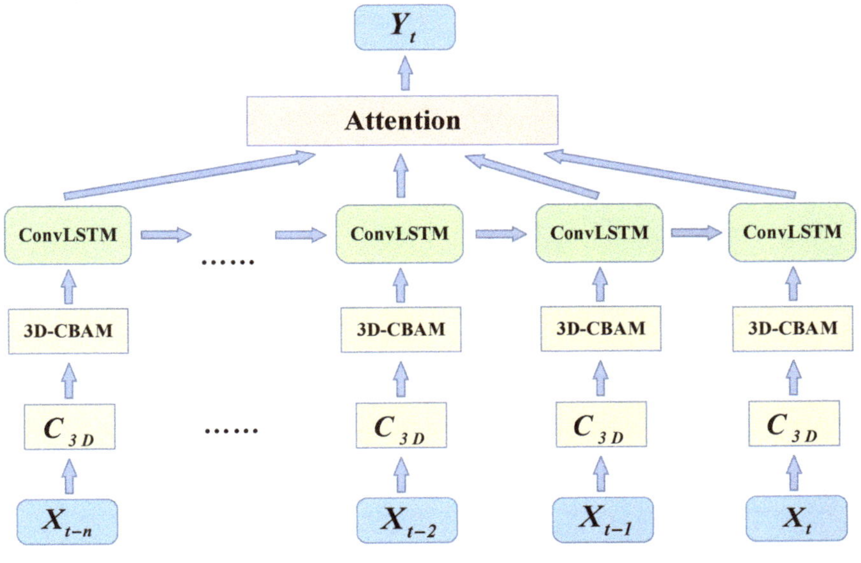

Figure 2. The framework of 3DCBAM-ConvLSTM method.

Therefore, the whole SST forecast revision model can be expressed as follows:

$$Y_t^{W\times H} = AT(\text{ConvLSTM}(\text{CBAM}(C_{3D}(X_t, X_{t-n})))), n = 1,2,3\cdots. \quad (2)$$

For historical series data X, the X_t composed of SST data and other marine environmental variables at any time of t is grid data with $W \times H \times C$ specifications. Therefore, the input of the whole model is a five-dimensional tensor, which is expressed as $B \times T \times C \times W \times H$. Here B is the number of a batch of training samples, and T is the length of sequence data. W and H are the width and length of the SST field, and C is the number of marine environmental variables. In our experiments, length H and width W are longitude and latitude. The length of the time step can be obtained through the sliding window. For example, if the historical data of the past three days are used to correct the SST of the day, then the length of the time step is 4, that is, the value of T. In the experiments, in addition to SST, salinity and water velocity u and water velocity v are added, so C is here 4. The five-dimensional tensor serves as input to the model.

As the correction of SST is a regression problem, this paper chooses MSE as the loss function. The calculation formula is shown in Equation (3), where n represents the number of points in grid data, $\hat{y}_i$ is the truth value of the point i, and y_i is the revised value of the point i. The training set is input into the model, and N iterations are carried out until the model converges.

$$\text{LOSS} = \sum_{i=1}^{n} \frac{(y_i - \hat{y}_i)^2}{n}. \tag{3}$$

4.2. Spatial Feature Extraction with 3D-CBAM

In the spatial feature extraction part, we use 3D convolution to extract spatial features from the input training data. 3D convolution is developed on the basis of 2D convolution [48]. 3D convolution is achieved by convolving a three-dimensional kernel with a cube formed by stacking multiple continuous matrices. Through this construction, the feature map of the convolution layer is connected with the previous layer to capture spatial information. The input of 3D convolution is sample X, $X \in R^{B \times T \times H \times W \times C}$. The 3D convolution operation C_{3D} mainly completes the spatial feature extraction and it can be computed as:

$$C_{3D}(X) = \sum_{p=0}^{P-1} \sum_{q=0}^{Q-1} \sum_{r=0}^{R-1} \omega(p,q,r) * X, \tag{4}$$

where $*$ and $\omega(p,q,r)$ represent the convolution operation and kernel, and P, Q, R represent width, height, and temporal length of the data.

Then, we use 3D-CBAM attention mechanism to improve the utilization rate of the spatial features of the 3D convolution network and to show the importance of different environmental variables to the results. Convolutional Block Attention Module (CBAM) is a simple and effective attentional module that can be directly applied to a feedforward convolutional neural network, consisting of a channel attentional module and a spatial attentional module [49]. Figure 3 shows the structure of the 3D CBAM attention module.

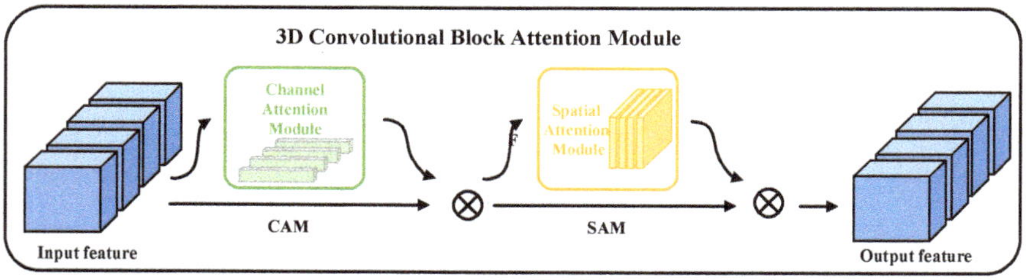

Figure 3. 3D convolutional block attention module.

The input of CBAM is $F = C_{3D}(X), X \in R^{B \times T \times H \times W \times C}$, the feature map from a previous 3D convolution layer. The 3D CBAM will apply channel attention module (CAM) and spatial attention module (SAM) in sequence to the input F. As shown in Figure 3, the CBAM can be designed as:

$$\text{CBAM}(F) = \text{SAM}(\text{CAM}(F)). \tag{5}$$

The channel attention module of 3D-CBAM pays attention to which feature plays a role in the final correction result. Firstly, we apply the global max pooling and global average pooling based on width, height, and time to the input feature matrix F, respectively, and we get F_{avg} and F_{max}. Both F_{avg} and F_{max} are one dimensional feature maps: $F_{avg} \in R^{B \times 1 \times 1 \times 1 \times C}$ and $F_{max} \in R^{B \times 1 \times 1 \times 1 \times C}$. Then, multilayer perceptron (MLP), a fully connected layer is used to efficiently combine the channel statistical information F_{avg} and F_{max}. To reduce the parameter resources, the hidden size of MLP is set to $R^{C/r}$, where r is defined as the reduction rate, and the formula is shown below:

$$F_{\text{mlp\_}avg} = \text{MLP}(F_{avg}) = W_2(relu(W_1(F_{avg}))), \tag{6}$$

$$F_{\text{mlp\_}max} = \text{MLP}(F_{max}) = W_2(relu(W_1(F_{max}))), \tag{7}$$

where $W_1 \in R^{C/r \times C}{}_1$, $W_2 \in R^{C \times C/r}$ stands for the MLP weights and $relu$ represents the active function ReLU, respectively. W_1 and W_2 are shared by both F_{avg} and F_{max}.

After obtaining the statistical information $F_{\text{mlp\_}avg}$ and $F_{\text{mlp\_}max}$ by MLP, the probability prediction matrix, which is the importance of each channel, can be obtained by element-wise summing and passing through the sigmoid function. Finally, the matrix generated by a sigmoid function is element-wise multiplied with the input matrix F to obtain the output, which is calculated by equation:

$$\text{CAM}(F) = F \times \sigma\left(F_{\text{mlp\_}avg} + F_{\text{mlp\_}max}\right), \tag{8}$$

where σ is the sigmoid function. Figure 4 shows the flowchart of CAM.

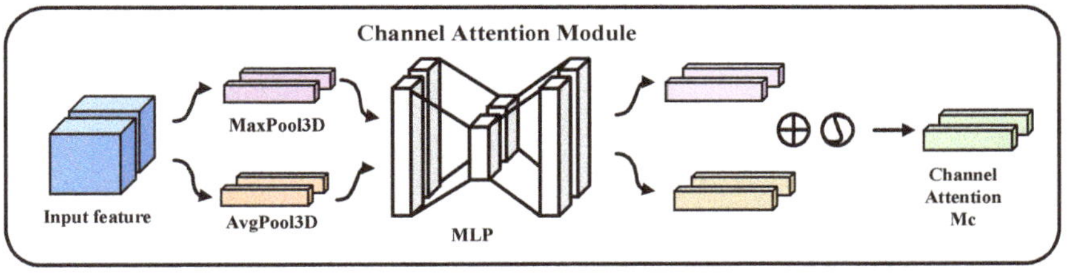

Figure 4. Channel attention module.

The feature matrix $F_c = \text{CAM}(F)$, which is output by the channel attention module, is taken as the input feature matrix of a spatial attention module. Firstly, we use global max pooling and global average pooling based on the channel to get two feature maps: $F_{c\_avg} \in R^{B \times T \times H \times W \times 1}$ and $F_{c\_max} \in R^{B \times T \times H \times W \times 1}$. Then, they are concatenated at the channel dimension and passed through a $3 \times 3 \times 3$ convolution to generate a feature descriptor. The spatial attention feature is generated through sigmoid activation function. Then, we multiply the spatial attention matrix with the input matrix F_c to obtain the output result, which is calculated by equation:

$$\text{SAM}(F_c) = F_c \times \sigma\left(f_{conv}^{3 \times 3 \times 3}([F_{c\_avg}; F_{c\_max}])\right), \tag{9}$$

where σ is the sigmoid function. Figure 5 shows the flowchart of SAM.

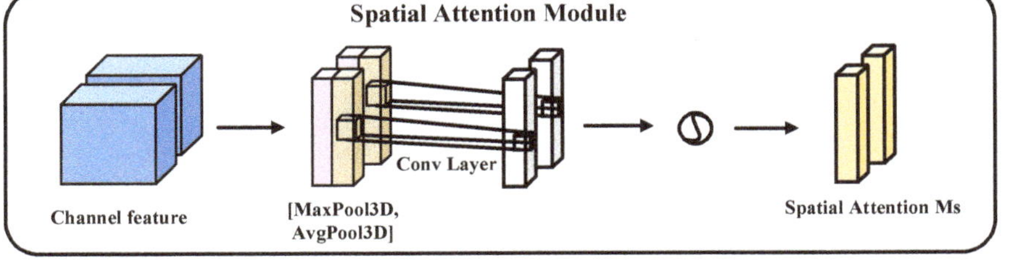

Figure 5. Spatial attention module.

4.3. Time Feature Extraction with Attention Mechanism

SST forecast correction is actually a spatio-temporal series problem with historical information as the input and revised SST as the output. LSTM has a strong ability to modeling time series data. ConvLSTM [27] inherits the merits of convolution operator and retains the advantages of LSTM to capture long-term memory, and can also reduce the redundancy of the fully connected structure. So that, ConvLSTM is used to model the temporal and spatial correlation of SST data. The input of the ConvLSTM in correction model is $X = \text{CBAM}(F),\ X \in R^{B\times T\times H\times W\times C}$. The formula is shown in Equation (10):

$$H = \text{ConvLSTM}(X), \tag{10}$$

where H consists of the results h_t computed by ConvLSTM for each sample x of input data X, $x \in R^{T\times H\times W\times C}$. At each moment, since the interval time of data is one day, the ConvLSTM unit accepts the input $x_t,\ t = 1,2,\cdots T$ at the moment of t, the state of the hidden layer at the last time h_{t-1}, and the state of the memory cell at the last time c_{t-1} as inputs, and outputs the hidden state h_t and the cell state c_t. The calculation process is as follows:

$$h_t = o_t \cdot tanh(c_t)\ , \tag{11}$$

$$c_t = f_t \cdot c_{t-1} + i_t \cdot tanh(w_{xc} * x_t + w_{hc} * h_{t-1} + b_c). \tag{12}$$

As shown in Equations (10) and (11), $*$ and $\cdot$ denote the convolution operator and Hadamard product. w is the weight matrix, b_c is the offset, and $tanh$ represents the activation function.

The ConvLSTM forgets and remembers the input information through four gates. The forgetting gate determines what information should be discarded from the c_{t-1} of the previous moment, the input gate determines what new information should be stored in the memory of the ConvLSTM, and the output gate determines what information should be selected from the c_t to be passed as output to the next ConvLSTM unit. The involved computation is given as follows in Equations (12)–(14):

$$i_t = \sigma(w_{xi} * x_t + w_{hi} * h_{t-1} + w_{ci} \cdot c_{t-1} + b_i)\ , \tag{13}$$

$$f_t = \sigma\Big(w_{xf} * x_t + w_{hf} * h_{t-1} + w_{cf} \cdot c_{t-1} + b_f\Big), \tag{14}$$

$$o_t = \sigma(w_{xo} * x_t + w_{ho} * c_{t-1} + w_{co} \cdot c_t + b_o)\ , \tag{15}$$

where i_t indicates the input gates, f_t indicates the forgotten gates, o_t indicates the output gates, and c_t indicates the cell state. In the above formulas, σ is the active function sigmoid, x_t represents the moment's input, ht−1 represents last time's hidden state, w is the weight matrix, and b is the offset from the input gate to the output gate, which are the characteristics that the ConvLSTM model must learn during training.

In order to improve the quality of the model by giving different weights to different parts of the model and make the model more focused on the parts that are more relevant to the task, a temporal attention layer is added after the ConvLSTM layer. To make full use of the hidden layer state of each step of the ConvLSTM model, we allocate the temporal attention weight to the hidden state of each time step, and adjust the final ConvLSTM output and thus obtain better correction results.

The attention [50] module assigns weight coefficients to the outputs of the ConvLSTM layer. It pays more attention to the features that contribute more to the important information and ignores useless information to reduce the calculations of the network and save storage space. The attention mechanism shown in Figure 6 provides an efficient way to aggregate the output sequence of ConvLSTM layer and it implements the following equation:

$$AT(h_t) = \frac{exp(W \cdot h_t)}{\sum_{t=1}^{T} exp(W \cdot h_t)}. \tag{16}$$

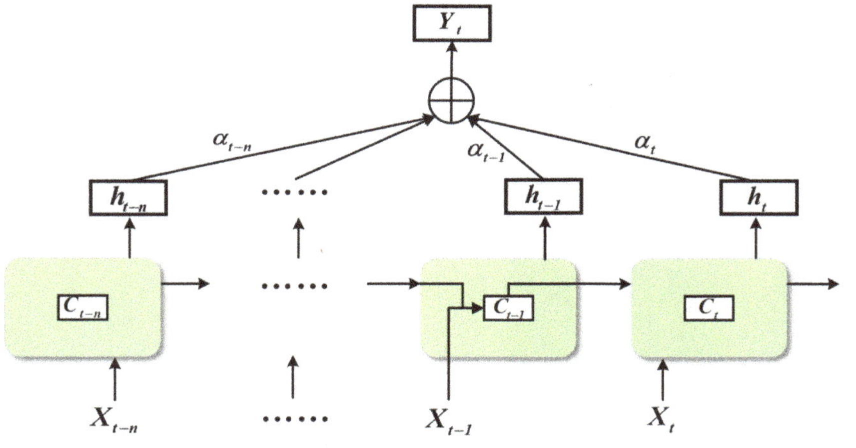

Figure 6. The ConvLSTM layer and attention layer.

The attention layer takes the output h_t of each iteration of ConvLSTM as input. At time t, normalized weights $AT(h_t)$ are computed by the softmax function through the weight W and the output h_t of the ConvLSTM, the calculation formula is shown in (16).

$$Y_t = \sum_{t=1}^{T} h_t AT(h_t). \tag{17}$$

Finally, the output Y_t can be obtained by multiplying the attention weight AT with the hidden layer state h_t. The calculation formula is shown in (17).

5. Experiments and Results

5.1. Data Preparation and Evaluation Metrics

The HYbrid Coordinate Ocean Model (HYCOM) [51] is a data-assimilative hybrid isopycnal-sigma-pressure (generalized) coordinate ocean model. The US Navy Operational Global Ocean Prediction System based on the HYCOM model is a relatively advanced and widely used ocean prediction system [52]. In the experiments, we use HYCOM model forecast product from National Oceanic and Atmospheric Administration (NOAA) as the prediction data to be corrected. The HYCOM model prediction product used in our experiments is a prediction product of 24 h in the future, which is reported every 3 h and includes ocean temperature, salinity, and current structure. Its horizontal resolution is $1/12^\circ$ and the temporal resolution is 3 h.

There is a lack of ocean observation data to support our experiments, SST products data is a relatively good choice as the truth value. We use the NOAA OI SST [53] Analysis version 2(v2), which is acquired from the NOAA's National Climatic Data Center (NCDC) with a high spatial resolution of $0.25^\circ \times 0.25^\circ$ as truth values to evaluate correction accuracy. Liu et al. [54] showed that NOAA OI SST is the best one among the SST products when they were compared with in situ SST data This dataset was generated from several data sources including SST data from the Advanced Very High Resolution Radiometer (AVHRR), sea-ice data, and in situ data from ships and buoys. In order to unify the spatial and temporal resolution of the forecast data and remote sensing observation data, we average the HYCOM data daily and the daily average HYCOM model forecast data is interpolated to OI SST data grid points by using bilinear interpolation method. We select a dataset from January 2019 to December 2019, that covers the area from 8°N to 12°N in latitude and 110° E to 114° E in longitude. Figure 7 shows the location of test area.

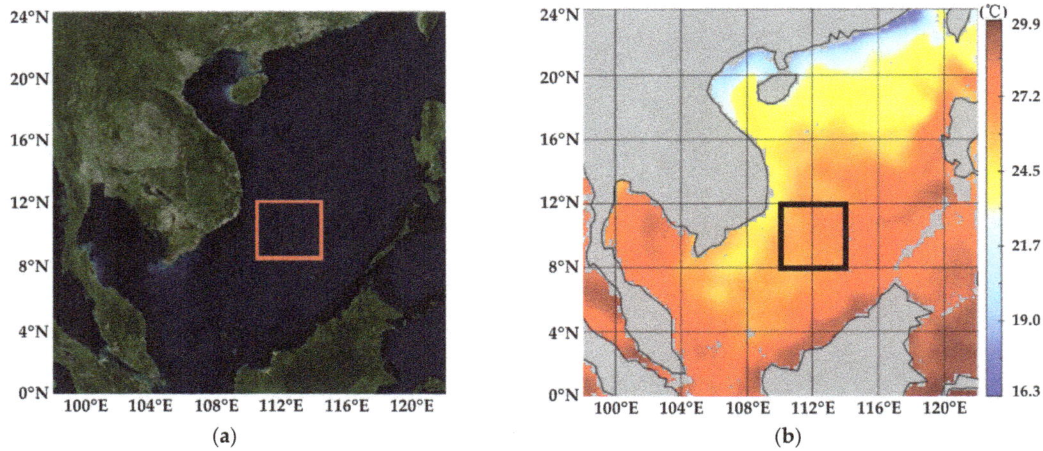

Figure 7. The location of test area: (**a**) Satellite image of the test area location, the box is the test area; (**b**) SST map of the test area location, the box is the test area.

Due to the obvious discrepancy of data value, we apply a normalization to each input sequence of data before inputting it into our model. The normalization operation can not only improve the convergence speed of the model, but also improve the accuracy of the model and prevent the gradient explosion of the model. The normalization function is shown in Equation (19):

$$X = \frac{X - X_{min}}{X_{max} - X_{min}}. \tag{18}$$

After correction, the output data and the truth would go through a de-normalization. The parameters of the de-normalization are based on the temperature span of the original input sequence of data.

In order to verify the validity of the new hybrid SST correction model, this study evaluated the model with four indexes, namely mean square error (MSE), root mean square error (RMSE), mean absolute error (MAE), and mean absolute percentage error (MAPE). MSE, RMSE, MAE, and MAPE can be defined as:

$$\text{MSE}(y, \hat{y}) = \frac{1}{n}\sum_{i=1}^{n}(y_i - \hat{y}_i)^2, \tag{19}$$

$$\text{RMSE}(y, \hat{y}) = \sqrt{\frac{1}{n}\sum_{i=1}^{n}(y_i - \hat{y}_i)^2}, \tag{20}$$

$$\text{MAE}(y, \hat{y}) = \frac{1}{n}\sum_{i=1}^{n}|y_i - \hat{y}_i|, \tag{21}$$

$$\text{MAPE}(y, \hat{y}) = \frac{1}{n}\sum_{i=1}^{n}\left|\frac{y_i - \hat{y}_i}{y_i}\right|, \tag{22}$$

where y_i represents the actual observed value, $\hat{y}$ represents the average of the actual observed values, and $\hat{y}_i$ represents the correction value.

The model is built by Pytorch. In order to prove the effectiveness of the model proposed in this paper, all SST data were divided into two parts. The dataset from January 2019 to September 2019 is used as the training set to train the parameters of the new hybrid SST correction model, and the remaining dataset from October 2019 to December 2019 is used as the verification set to verify the learning effect of the model. We adjusted the shape of the training data and the input data to the required Tensor format in the Pytorch framework. Then the parameters of the new hybrid SST correction model were defined, including the input step length, the length of input sequence, the hidden layers, the length of output sequence, and the number of neurons in each layer. In our experiments, the convolution part of the model includes a Conv3D layer and a Batch Normalization layer. The main function of the Batch Normalization layer is to make the distribution of the input data of each layer in the network relatively stable, to accelerate the model learning speed, alleviate the problem of gradient disappearance, and have a certain regularization effect. When the network is set up, the size of the convolution kernel in Conv3D layer is $3 \times 3 \times 3$ and the size of the convolution kernel in the convolution attention in the 3D-CBAM part of the model is $3 \times 3 \times 3$. The activation function of all layers is 'relu', which can keep the convergence speed of the model in a steady state. In the ConvLSTM part of the model, due to the short sequence selected in the experiment, only a single layer ConvLSTM was selected, the number of neurons in the hidden layer was 32, and the number of neurons in the output layer was 1. After the model parameters are defined, we select MSE and Adam as the loss function and optimizer. Then, the appropriate training times are defined to start training the model. After the model training is completed, the test data is input into the model for testing, and the output results of the model are reversely normalized to obtain the deviation of SST. The effect of correction is tested by comparing the evaluation metrics before and after deviation correction.

5.2. Comparison of Correction Methods

In order to prove the validity of the proposed new hybrid SST correction model, the experimental results will be compared with two traditional machine learning methods for SST correction. They are Linear Regression (LR) and Support Vector Regression (SVR), respectively. The LR model has the advantages of strong anti-interference ability and fast training speed, but it cannot simulate nonlinear relations, as the accuracy is not very high, and it is easy to lack fitting. The SVR model has good generalization performance, it is not easy to overfit, and it can achieve good performance with less data. However, SVR is sensitive to missing data, parameters, and kernel function.

The process of realizing the two methods in this paper is to expand all samples into the form that the algorithm can handle, and use the machine learning package, sklearn, for correction analysis. In order to match the input form of these methods, SST and ocean variables were generally regarded as independent features, thus the spatial-temporal relationship between variables could not be considered in these methods. For both LR and SVR, we combined vector of SST, SSS, and water current u and v of HYCOM forecasts for days n, n-1, n-2, n-3 as 4×4 features for one-day correction. The performances of these two methods are shown in Table 1. For SVR, we use the Radial Basis Function (RBF) kernel for correction, which can realize nonlinear mapping with few parameters.

In addition, we set up the comparison experiment, which only considers the temporal relationship without considering the spatial relationship, namely, compared with the traditional sequence model LSTM to enhance the contrast. For the LSTM network, we set the learning rate = 0.01, epochs = 300, timestep = 3, and use SST, SSS, and water current u and v for one-day correction. Our new hybrid correction model also compares the traditional sequence model LSTM and its improved model ConvLSTM, which considers temporal relations and spatial relations.

Furthermore, we develop and compare a series of models with the new hybrid SST correction model (3DCNN-CBAM-CONVLSTM-AT), because there were fewer methods previously used to correct SST. These include an improved ConvLSTM model that combines 3D-convolution, a ConvLSTM model that only adds temporal attention mechanism (AT), and a ConvLSTM hybrid model where both 3D-convolution and AT are added. Here, we set the experimental parameters learning rate = 0.01 and epochs = 300, the size of convolution kernel in Conv3D layer is $3 \times 3 \times 3$, and used three days of historical data for SST correction. The input data form of these models is consistent with our new model input data.

Table 1. The experimental results of SST correction. Bold entries show the best results.

	MAPE	MAE	MSE	RMSE	Improve
Forecast	1.6118	0.4587	0.3600	0.6000	
Linear Regression (LR)	1.4592	0.4075	0.3005	0.5482	8.67%
Support Vector Regression (SVR)	1.3767	0.3832	0.2536	0.5036	16.17%
LSTM	1.2781	0.3553	0.2115	0.4599	23.35%
CONVLSTM	1.1679	0.3312	0.1842	0.4292	28.47%
CONVLSTM-AT	1.1071	0.3139	0.1623	0.4028	32.92%
3DCNN-CONVLSTM-AT	1.0033	0.2839	0.3600	0.3690	38.5%
3DCNN-CBAM-CONVLSTM-AT	**0.9546**	**0.2641**	**0.1239**	**0.3520**	**41.33%**

Table 1 shows the experimental results of different correction methods for SST correction. When we use the traditional machine learning method to correct SST, the accuracy of SVR is higher than Linear Regression; the RMSE value is 0.5036 and the MAE value is 0.3832. The accuracy is not greatly improved after the correction.

Among other deep learning models, 3DCNN-ConvLSTM-AT has the best results; the RMSE value is 0.3690 and the MAE value is 0.2839. However, our new hybrid correction model can achieve a level where the MSE value is 0.3520 and the MAE value is 0.2641 in the correction experiment, which is better than the other models.

It can be seen from Table 1 that the effects of LSTM are better than traditional machine learning methods, which illustrates the importance of time correlation in SST data. However, the original LSTM does not consider the spatial relations in data. The result of ConvLSTM, an improved method, is better than the result of LSTM, which verifies the importance of spatial correlation to SST correction. Experimental results show that ConvLSTM-AT model, which adds attention mechanism, has better performance than ConvLSTM. Attention mechanism can assign different weights to historical data, allowing the model to focus more on the parts that are more important, thus improving the quality of the model. We compared the results of 3DCNN-CONVLSTM-AT with CONVLSTM-AT, in which 3DCNN-CONVLSTM-AT added a convolution layer. With the same parameters and the same input, the RMSE of ConvLSTM-AT and 3DCNN-CONVLSTM-AT were 0.4028 and 0.3690, respectively. 3DCNN-ConvLSTM-AT has a higher correction accuracy than ConvLSTM-AT. The experimental result shows that the addition of the convolution layer can improve the accuracy of SST correction to a certain extent. The main reason for this is that the local features extracted from input data through ConvLSTM's own convolution operation is not obvious enough. A convolution layer is added into the model, which improves the feature extraction ability of the model and makes the spatial features of the data more obvious in the ConvLSTM model, which is beneficial to improve the accuracy of SST correction.

After adding 3D-CBAM attention mechanism on the basis of the 3DCNN-CONVLSTM-AT model, the RMSE index is 0.3520, and the correction effect is the best in our experiments. 3DCBAM mechanism and AT mechanism were used based on ConvLSTM in our new hybrid correction model to improve the utilization rate of spatial features, environmental variables, and historical time series information.

To further prove the effectiveness of our new hybrid SST correction model, we visualize the correction results, forecast results, and the truth in Figure 8, which shows the comparison of the revised SST of several models. To put things into places in the overall view, there is high similarity between the correction that is shown in Figure 8i and the truth that is shown in Figure 8a. Combined with Figures 8 and 9 and Table 1, it can be seen intuitively from the figure that the result of the new hybrid SST correction model is closest to the truth value. The new hybrid correction model further extracts spatial features and adds weights to environmental information and spatial features to improve information utilization, making the model closer to reality and containing more comprehensive information, and finally improving the accuracy of SST prediction. In conclusion, compared with LR, SVR, and other traditional machine learning correction methods, as well as deep learning methods LSTM, ConvLSTM, and ConvLSTM-AT, the new hybrid correction model has the best performance in SST correction, which verifies the effectiveness of this method.

For SST correction, Zhang et al. [47] proposed a new bias correction model for sea surface temperature in 2020, which used satellite remote sensing data for correction of the numerical forecast model on SST in the South China Sea as well. After being corrected, the RMSE of the SST forecast results was dropped from 0.8 °C to 0.5 °C, reducing by 37.5%, whereas the RMSE of our model is approximately 0.35 °C after being corrected, reducing by 41.33%. The SST correction by our new hybrid SST correction offers higher accuracy.

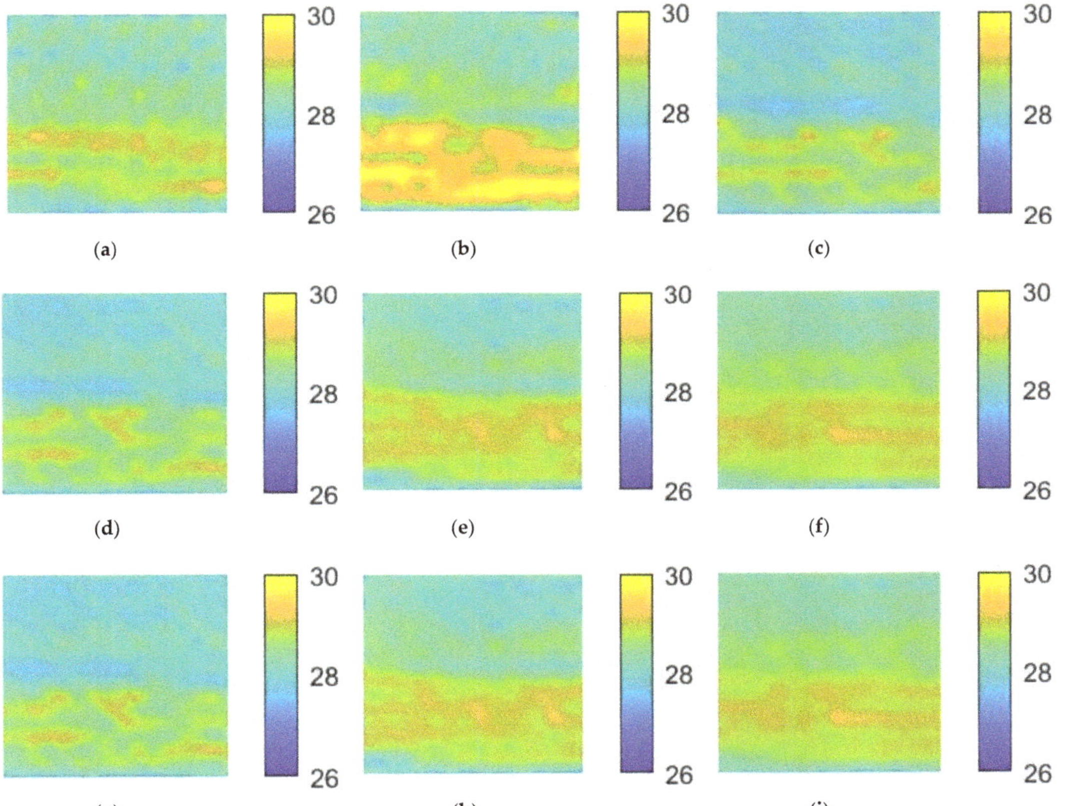

Figure 8. The experimental results of different methods for SST correction. (**a**) Truth; (**b**) forecast; (**c**) linear regression; (**d**) SVR; (**e**) LSTM; (**f**) CONVLSTM; (**g**) CONVLSTM-AT; (**h**) 3DCNN-CONVLSTM-AT; (**i**) 3DCNN-CBAM-CONVLSTM-AT.

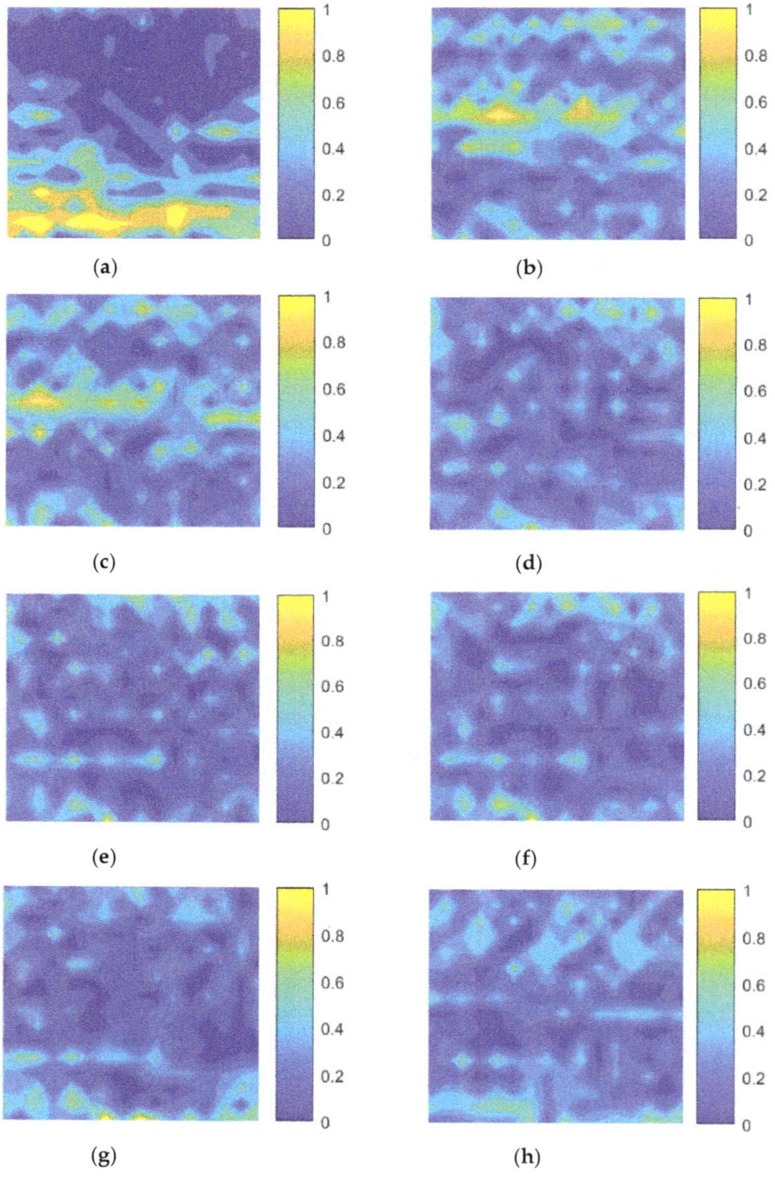

Figure 9. The comparisons of difference between the truth and the correction output. (**a**) Difference between the truth and the forecast; (**b**) difference between the truth and the linear regression result; (**c**) difference between the truth and the SVR result; (**d**) difference between the truth and the LSTM result; (**e**) difference between the truth and the CONVLSTM result; (**f**) difference between the truth and the CONVLSTM-AT result; (**g**) difference between the truth and the 3DCNN-CONVLSTM-AT result; (**h**) difference between the truth and the 3DCNN-CBAM-CONVLSTM-AT result.

5.3. Complexity and Training Time Analysis

The experimental environment is Windows10, Intel Core i5 11, 2.4 GHz, 16G RAM, with algorithm implementation using python3.

Table 2 lists the training time and the parameters of models used in the experiment. It can be found that the training parameters of the new hybrid SST correction model are about three times less than those of ConvLSTM, which makes the training much faster and more suitable for practical application. Our proposed new hybrid SST correction model consumes the least time and has fewer parameters. The parameter of 3DCNN-ConvLSTM-AT model is close to that of the new hybrid SST correction model, indicating that the 3D-CBAM module is very small and the training time of the model is reduced. Our proposed new hybrid SST correction model consumes the least time and has fewer parameters, and it has good performance.

Table 2. The number of network parameters and training time for each model.

	Parameters	Train(s)	Test(s)
LSTM	13,601	271.52	0.55
CONVLSTM	44,993	236.98	0.46
CONVLSTM-AT	46,079	437.67	0.94
3DCNN-CONVLSTM-AT	13,197	272.57	0.55
3DCNN-CBAM-CONVLSTM-AT	13,560	223.15	0.43

5.4. Parameters Analysis

5.4.1. Time Step Analysis

In the previous experiments to determine the model structure, the previous three days of data is used to correct the SST according to expert empirical knowledge. Time step is an important parameter for the model to learn time series character. Considering that the size of timestep has an impact on the accuracy of SST correction, timestep = 1, 3, 5, 7, 10, 15 is used to correct SST in our experiments to determine the appropriate timestep for SST correction.

Timestep represents the information of the time dimension, which has an impact on the performance of the model. Figure 10 shows the variation of the model of several evaluation indicators with the timestep size. When timestep = 3, RMSE is 0.35, which is better than others when timestep = 1, 5, 7, 10, and 15. It is obviously seen from the figure that timestep = 3 works best to revise SST. When the timestep is greater than 10, the results of correction tend to be stable, and the time information has less influence on the revised results. When correcting SST, the information of temporal dimensions should be moderate, as too much or too little will affect the performance of the model. To sum up, timestep = 3 is used in this paper to correct SST.

5.4.2. Learning Rate Analysis

Learning rate is an important hyperparameter, which determines whether and when the objective function converges to the local minimum. The proper learning rate can make the objective function converge to the local minimum in the proper time. Then we adjust the learning rate and other hyperparameters within the fixed model frame. The first step is to drop from 0.1 to 0.001, at a speed of 10. Then, when the learning rate is at the level between 0.01 and 0.001, the training and validation loss of the model will be in a steady state. The experiment is conducted by adjusting the learning rate, and the experimental results are shown in Figure 11. The figure shows that RMSE, MAPE, and other indicators change with the learning rate. According to RMSE, the optimal learning rate is 0.01. According to MAPE, the optimal LR is 0.004. Thus, the best learning rate in our data set is at the level between 10^{-2} and 10^{-3}.

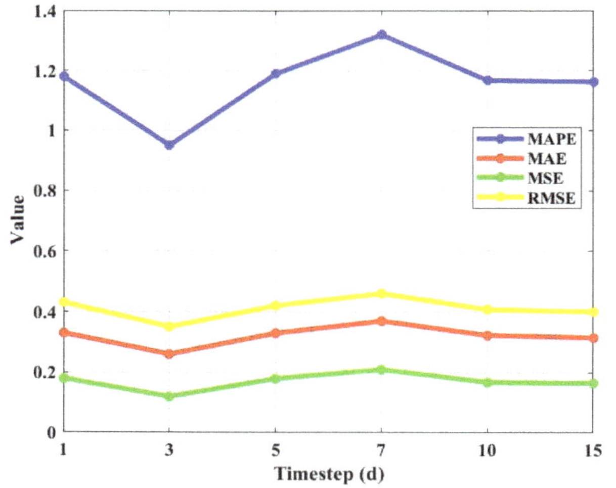

Figure 10. The experimental results of the new hybrid SST correction model in different timesteps. The units of RMSE, MAE, and MSE are °C, the unit of MAPE is %, and the unit of timestep is day.

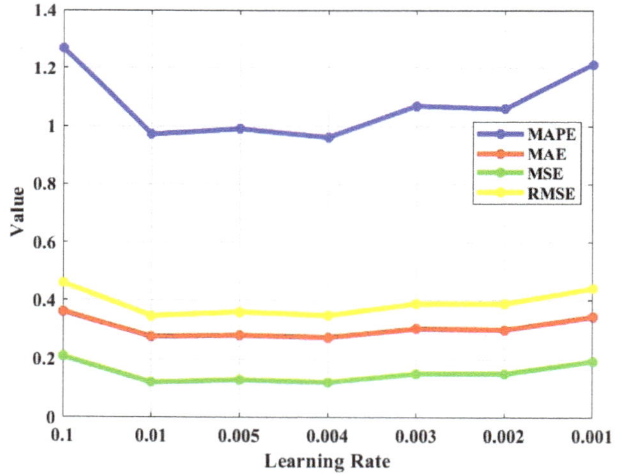

Figure 11. The experimental results of the new hybrid SST correction model in different learning rate. The units of RMSE, MAE, and MSE are °C and the unit of MAPE is %.

5.4.3. Epochs Analysis

In order to determine the best epochs for the dataset, different epochs were set for the experiments. The experimental results are shown in Figure 12. The figure shows that RMSE reaches a stable state at 300 epochs. Therefore, 300 epochs are suitable for our experiments with consideration of model accuracy and performance.

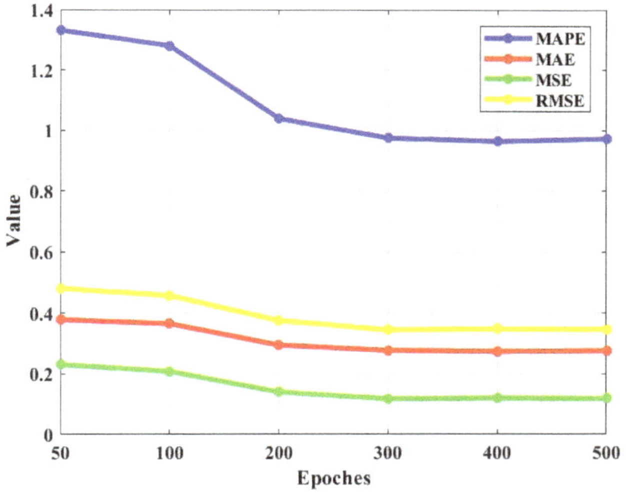

Figure 12. The experimental results of the new hybrid SST correction model in different epochs. The units of RMSE, MAE, and MSE are °C and the unit of MAPE is %.

6. Conclusions

In this paper, the new hybrid SST correction model is applied to correct the HYCOM forecasts and it is evaluated for its performance. Our proposed model combines spatio-temporal information and marine environmental variables information to correct the SST forecast and improve the accuracy of the SST forecast. The model defines the SST correction problem as the spatio-temporal series regression problem, which mainly consists of three parts: first, 3D convolution and 3D-CBAM are used to improve the utilization rate of spatial features and marine environmental variables. Secondly, time and space characteristics of SST were extracted by ConvLSTM. Thirdly, the attention mechanism is used to enhance the historical temporal information. What is more, the new hybrid SST correction has a better correction effect than the other models we compared in this paper, and it can reduce the RMSE of the HYCOM forecast results by 41.33%.

As for future development, further refinements to the new hybrid SST correction model will be undertaken. Our study only corrects the temperature of the sea surface, but the subsurface temperature in the inner ocean is much more important. Therefore, in the next step, we consider extending the model to three-dimensional space to realize the forecast correction of ocean internal temperature. Meanwhile, in this paper, we only revised the forecast data for the next day due to the limitation of forecast data. For future development, our correction model can be considered to improve and apply to correct the forecast of three days, five days, or one month into the future.

Author Contributions: Conceptualization, X.W. and B.H.; methodology, T.F.; validation, J.Z. and T.F.; formal analysis, T.F., X.W., B.H., J.Z., H.W., Y.C. and W.Z.; supervision, X.W.; data curation, H.W. and T.F.; writing—original draft preparation, T.F.; writing—review and editing, X.W., B.H. and Y.C.; visualization, T.F.; project administration, W.Z.; funding acquisition, W.Z. All authors have read and agreed to the published version of the manuscript.

Funding: This research is partially supported by the National Key Research and Development Program of China (No. 2018YFC1406206) and National Natural Science Foundation of China (Grant No. 61802424).

Institutional Review Board Statement: Not applicable.

Informed Consent Statement: Not applicable.

Data Availability Statement: SST OI data sets were obtained from https://psl.noaa.gov/data/gridded/data.noaa.oisst.v2.highres.html (accessed on 10 July 2021); HYCOM forecast data sets were obtained from https://www.ncei.noaa.gov/thredds-coastal/catalog/hycom_sfc/catalog.html (accessed on 3 July 2021).

Conflicts of Interest: The authors declare no conflict of interest.

References

1. Funk, C.C.; Hoell, A. The leading mode of observed and cmip5 enso-residual sea surface temperatures and associated changes in indo-pacific climate. *J. Clim.* **2015**, *28*, 150202132719008. [CrossRef]
2. Solanki, H.U.; Bhatpuria, D.; Chauhan, P. Integrative analysis of altika-ssha, modis-sst, and ocm-chlorophyll signatures for fisheries applications. *Mar. Geod.* **2015**, *38* (Suppl. 1), 672–683. [CrossRef]
3. Yang, Y.; Dong, J.; Sun, X.; Lima, E.; Mu, Q.; Wang, X. A cfcc-lstm model for sea surface temperature prediction. *IEEE Geosci. Remote Sens. Lett.* **2017**, *15*, 207–211. [CrossRef]
4. Stockdale, T.N.; Balmaseda, M.A.; Vidard, A. Tropical atlantic sst prediction with coupled ocean-atmosphere gcms. *J. Clim.* **2006**, *19*, 6047. [CrossRef]
5. Song, Z.; Qiao, F.; Yang, Y.; Yuan, Y. An improvement of the too cold tongue in the tropical pacific with the development of an ocean-wave-atmosphere coupled numerical model. *Prog. Nat. Sci.* **2007**, *17*, 576–583.
6. Xu, Z.; Li, M.; Patricola, C.M.; Ping, C. Oceanic origin of southeast tropical atlantic biases. *Clim. Dyn.* **2014**, *43*, 2915–2930. [CrossRef]
7. Peng, S.Q.; Xie, L. Effect of determining initial conditions by four-dimensional variational data assimilation on storm surge forecasting. *Ocean Model.* **2006**, *14*, 1–18. [CrossRef]
8. Li, X.; Wang, Q.; Mu, M. Optimal initial error growth in the prediction of the kuroshio large meander based on a high-resolution regional ocean model. *Adv. Atmos. Sci.* **2018**, *35*, 1362–1371. [CrossRef]
9. Hemri, S.; Scheuerer, M.; Pappenberger, F.; Bogner, K.; Haiden, T. Trends in the predictive performance of raw ensemble weather forecasts. *Geophys. Res. Lett.* **2014**, *41*, 9197–9205. [CrossRef]
10. Vannitsem, S. Dynamical properties of mos forecasts: Analysis of the ecmwf operational forecasting system. *Weather Forecast.* **2010**, *23*, 1032–1043. [CrossRef]
11. Tian, D.; Martinez, C.J.; Graham, W.D.; Hwang, S. Statistical downscaling multimodel forecasts for seasonal precipitation and surface temperature over the southeastern united states. *J. Clim.* **2014**, *27*, 8384–8411. [CrossRef]
12. Libonati, R.; Trigo, I.; Dacamara, C.C. Correction of 2m-temperature forecasts using kalman filtering technique. *Atmos. Res.* **2008**, *87*, 183–197. [CrossRef]
13. Pelosi, A.; Medina, H.; Bergh, J.V.D.; Vannitsem, S.; Chirico, G.B. Adaptive kalman filtering for postprocessing ensemble numerical weather predictions. *Mon. Weather Rev.* **2017**, *145*, 4837–4854. [CrossRef]
14. Wang, J.; Chen, C.; Long, K.; Feng, L. Temporal and spatial distribution of short-time heavy rain of Sichuan Basin in summer. *Plateau Mt. Meteorol. Res.* **2015**, *35*, 16–20.
15. Zhang, Q.; Yu, Y.; Zhang, W.; Luo, T.; Wang, X. Cloud detection from fy-4a's geostationary interferometric infrared sounder using machine learning approaches. *Remote Sens.* **2019**, *11*, 3035. [CrossRef]
16. Zeng, J.; Zhang, C.; Wang, H.; Chu, H. Correction model for the temperature of numerical weather prediction by SVM. *Second Target Recognit. Artif. Intell. Summit Forum* **2020**, *11427*, 114270Z.
17. Zhang, R.; Yu, Z.H.; Jiang, Q.R. Neural network bp model approximation and prediction of complicated weather systems. *Acta Meteorol. Sin.* **2001**, *15*, 105–115.
18. Sayeed, A.; Choi, Y.; Jung, J.; Lops, Y.; Eslami, E.; Salman, A.K. A deep convolutional neural network model for improving WRF forecasts. *Atmos. Environ.* **2020**, *253*, 118376. [CrossRef]
19. Kupilik, M.; Witmer, F.D.W.; MacLeod, E.-A.; Wang, C.; Ravens, T. Gaussian Process Regression for Arctic Coastal Erosion Forecasting. *IEEE Trans. Geosci. Remote Sens.* **2019**, *57*, 1256–1264. [CrossRef]
20. Brekke, C.; Solberg, A. Oil spill detection by satellite remote sensing. *Remote Sens. Environ.* **2005**, *95*, 1–13. [CrossRef]
21. Yu, Y.; Yang, X.; Zhang, W.; Duan, B.; Cao, X.; Leng, H. Assimilation of sentinel-1 derived sea surface winds for typhoon forecasting. *Remote Sens.* **2017**, *9*, 845. [CrossRef]
22. Chen, R.; Zhang, W.; Wang, X. Machine learning in tropical cyclone forecast modeling: A review. *Atmosphere* **2020**, *11*, 676. [CrossRef]
23. Xi, X.F.; Zhou, G.D. A survey on deep learning for natural language processing. *Acta Autom. Sin.* **2016**, *42*, 1445–1465.
24. Lee, H.; Pham, P.T.; Largman, Y.; Ng, A.Y. Unsupervised feature learning for audio classification using convolutional deep belief networks. *Adv. Neural Inf. Process. Syst.* **2009**, *22*, 1096–1104.
25. Sattar, N.S.; Arifuzzaman, S. Community Detection using Semi-supervised Learning with Graph Convolutional Network on GPUs. In Proceedings of the 2020 IEEE International Conference on Big Data (Big Data), Atlanta, GA, USA, 10–13 December 2020; pp. 5237–5246.

26. Jain, V.; Murray, J.F.; Roth, F.; Turaga, S.; Zhigulin, V.; Briggman, K.L.; Helmstaedter, M.N.; Denk, W.; Seung, H.S. Supervised Learning of Image Restoration with Convolutional Networks. In Proceedings of the 2007 IEEE 11th International Conference on Computer Vision, Rio de Janeiro, Brazil, 14–21 October 2007; pp. 1–8.
27. Shi, X.; Chen, Z.; Wang, H.; Yeung, D.Y.; Wong, W.K.; Woo, W.C. *Convolutional Lstm Network: A Machine Learning Approach for Precipitation Nowcasting*; MIT Press: Cambridge, MA, USA, 2015.
28. Liu, D.; Jiang, W.; Mu, L.; Wang, S. Streamflow Prediction Using Deep Learning Neural Network: Case Study of Yangtze River. *IEEE Access* **2020**, *8*, 90069–90086. [CrossRef]
29. Petrou, Z.I.; Tian, Y. Prediction of sea ice motion with convolutional long short-term memory networks. *IEEE Trans. Geosci. Remote Sens.* **2019**, *99*, 1–12. [CrossRef]
30. Chen, R.; Wang, X.; Zhang, W.; Zhu, X.; Li, A.; Yang, C. A hybrid cnn-lstm model for typhoon formation forecasting. *GeoInformatica* **2019**, *23*, 375–396. [CrossRef]
31. Winona, A.Y.; Adytia, D. Short Term Forecasting of Sea Level by Using LSTM with Limited Historical Data. In Proceedings of the 2020 International Conference on Data Science and Its Applications (ICoDSA), Bandung, Indonesia, 5–6 August 2020; pp. 1–5.
32. Kun, X.; Shan, T.; Yi, T.; Chao, C. Attention-based long short-term memory network temperature prediction model. In Proceedings of the 2021 7th International Conference on Condition Monitoring of Machinery in Non-Stationary Operations (CMMNO), Guangzhou, China, 11–13 June 2021.
33. Krishnamurti, T.N.; Kishtawal, C.M.; LaRow, T. Improved weather and seasonal climate forecasts from multimodel superensemble. *Science* **1999**, *285*, 1548–1550. [CrossRef]
34. Xu, Z.; Wang, Y.; Fan, G. A two-stage quality control method for 2-m temperature observations using biweight means and a progressive eof analysis. *Mon. Weather Rev.* **2013**, *141*, 798–808. [CrossRef]
35. Zhang, X.; Gao, S.; Wang, T.; Li, Y.; Ren, P. Correcting Predictions from Simulating Wave Nearshore Model via Gaussian Process Regression. In Proceedings of the Global Oceans 2020: Singapore—U.S. Gulf Coast, Biloxi, MS, USA, 5–30 October 2020; pp. 1–4.
36. Doroshenko, A.; Shpyg, V.; Kushnirenko, R. Machine Learning to Improve Numerical Weather Forecasting. In Proceedings of the 2020 IEEE 2nd International Conference on Advanced Trends in Information Theory (ATIT), Kyiv, Ukraine, 25–27 November 2020.
37. Wang, X.; Li, X.; Zhu, J.; Xu, Z.; Yu, K. A local similarity-preserving framework for nonlinear dimensionality reduction with neural networks. In Proceedings of the The 26th International Conference on Database Systems for Advanced Applications (Dasfaa 2021), Tai Pei, China, 11–14 April 2021.
38. Wang, A.; Xu, L.; Li, Y.; Xing, J.; Zhou, Z. Random-forest based adjusting method for wind forecast of WRF model. *Comput. Geosci.* **2021**, *55*, 104842. [CrossRef]
39. Zheng, G.; Li, X.; Zhang, R.H.; Liu, B. Purely satellite data–driven deep learning forecast of complicated tropical instability waves. *Sci. Adv.* **2020**, *6*, eaba1482. [CrossRef] [PubMed]
40. Makarynskyy, O. Improving wave predictions with artificial neural networks. *Ocean Eng.* **2004**, *31*, 709–724. [CrossRef]
41. Xu, X.; Liu, Y.; Chao, H.; Luo, Y.; Chu, H.; Chen, L. Towards a precipitation bias corrector against noise and maldistribution. *arXiv* **2019**, arXiv:1910.07633.
42. Wang, T.; Gao, S.; Xu, J.; Li, Y.; Li, P.; Ren, P. Correcting Predictions from Oceanic Maritime Numerical Models via Residual Learning. In Proceedings of the 2018 OCEANS—MTS/IEEE Kobe Techno-Ocean. (OTO), Kobe, Japan, 28–31 May 2018; pp. 1–4.
43. Rasp, S.; Lerch, S. Neural networks for post-processing ensemble weather forecasts. *Mon. Weather Rev.* **2018**, *146*, 3885–3900. [CrossRef]
44. Deshmukh, A.N.; Deo, M.C.; Bhaskaran, P.K.; Nair, T.; Sandhya, K.G. Neural-network-based data assimilation to improve numerical ocean wave forecast. *IEEE J. Ocean. Eng.* **2016**, *4*, 944–953. [CrossRef]
45. Yang, X.Q.; Anderson, J.L. Correction of systematic errors in coupled gcm forecasts. *J. Clim.* **2000**, *13*, 2072–2085. [CrossRef]
46. Han, Y.K.; Dan, Y.U.; Shen, X.Y.; Zhou, Y.Y. Study on the correction of SST prediction of HYCOM. *Mar. Forecast.* **2018**, *35*, 5. (In Chinese)
47. Zhang, P.J.; Zhou, S.H.; Liang, C.X. Study on the correction of SST prediction in South China Sea using remotely sensed SST. *J. Trop. Oceanogr.* **2020**, *39*, 59–67. (In Chinese)
48. Ji, S.; Xu, W.; Yang, M.; Yu, K. 3d convolutional neural networks for human action recognition. *IEEE Trans. Pattern Anal. Mach. Intell.* **2013**, *35*, 221–231. [CrossRef]
49. Woo, S.; Park, J.; Lee, J.Y.; Kweon, I.S. Cbam: Convolutional block attention module. In Proceedings of the European Conference on Computer Vision (ECCV), Munich, Germany, 8–14 September 2018.
50. Mnih, V.; Heess, N.; Graves, A.; Kavukcuoglu, K. Recurrent models of visual attention. *Adv. Neural Inf. Processing Syst.* **2014**, *2*, 2204–2212.
51. Bleck, R. An oceanic general circulation model framed in hybrid isopycnic-cartesian coordinates. *Ocean Modeling* **2002**, *4*, 88. [CrossRef]
52. Metzger, E.J.; Smedstad, O.M.; Thoppil, P.G.; Hurlburt, H.E.; Cummings, J.A. US Navy Operational Global Ocean and Arctic Ice Prediction Systems. *Oceanography* **2014**, *27*, 32–43. [CrossRef]
53. Reynolds, R.W.; Smith, T.M.; Liu, C.; Chelton, D.B.; Casey, K.S.; Schlax, M.G. Daily High-Resolution-Blended Analyses for Sea Surface Temperature. *J. Clim.* **2007**, *20*, 5473–5496. [CrossRef]
54. Liu, Y.; Weisberg, R.H.; Law, J.; Huang, B. Evaluation of Satellite-Derived SST Products in Identifying the Rapid Temperature Drop on the West Florida Shelf Associated With Hurricane Irma. *Mar. Technol. Soc. J.* **2018**, *52*, 43. [CrossRef]

remote sensing

MDPI

Article

Reconstruction of Subsurface Salinity Structure in the South China Sea Using Satellite Observations: A LightGBM-Based Deep Forest Method

Lin Dong [1], Jifeng Qi [2,3,4,*], Baoshu Yin [2,3,4], Hai Zhi [5], Delei Li [2,3,4], Shuguo Yang [1], Wenwu Wang [6], Hong Cai [1] and Bowen Xie [1]

1 School of Mathematics and Physics, Qingdao University of Science and Technology, Qingdao 266061, China; 2020090004@mails.qust.edu.cn (L.D.); ysg_qust@qust.edu.cn (S.Y.); caihong@qust.edu.cn (H.C.); 2021090032@mails.qust.edu.cn (B.X.)

2 CAS Key Laboratory of Ocean Circulation and Waves, Institute of Oceanology, Chinese Academy of Sciences, Qingdao 266071, China; bsyin@qdio.ac.cn (B.Y.); deleili@qdio.ac.cn (D.L.)

3 Pilot National Laboratory for Marine Science and Technology (Qingdao), Qingdao 266237, China

4 University of Chinese Academy of Sciences, Beijing 100049, China

5 College of Atmospheric Sciences, Nanjing University of Information Science and Technology, Nanjing 210044, China; zhihai@nuist.edu.cn

6 Department of Electrical and Electronic Engineering, University of Surrey, Guildford GU2 7XH, UK; w.wang@surrey.ac.uk

* Correspondence: jfqi@qdio.ac.cn

Citation: Dong, L.; Qi, J.; Yin, B.; Zhi, H.; Li, D.; Yang, S.; Wang, W.; Cai, H.; Xie, B. Reconstruction of Subsurface Salinity Structure in the South China Sea Using Satellite Observations: A LightGBM-Based Deep Forest Method. *Remote Sens.* **2022**, *14*, 3494. https://doi.org/10.3390/rs14143494

Academic Editors: Xiaofeng Li, Chung-Ru Ho and Antony K. Liu

Received: 17 June 2022
Accepted: 20 July 2022
Published: 21 July 2022

Abstract: Accurately estimating the ocean's interior structures using sea surface data is of vital importance for understanding the complexities of dynamic ocean processes. In this study, we proposed an advanced machine-learning method, the Light Gradient Boosting Machine (LightGBM)-based Deep Forest (LGB-DF) method, to estimate the ocean subsurface salinity structure (OSSS) in the South China Sea (SCS) by using sea surface data from multiple satellite observations. We selected sea surface salinity (SSS), sea surface temperature (SST), sea surface height (SSH), sea surface wind (SSW, decomposed into eastward wind speed (USSW) and northward wind speed (VSSW) components), and the geographical information (including longitude and latitude) as input data to estimate OSSS in the SCS. Argo data were used to train and validate the LGB-DF model. The model performance was evaluated using root mean square error (RMSE), normalized root mean square error (NRMSE), and determination coefficient (R^2). The results showed that the LGB-DF model had a good performance and outperformed the traditional LightGBM model in the estimation of OSSS. The proposed LGB-DF model using sea surface data by SSS/SST/SSH and SSS/SST/SSH/SSW performed less satisfactorily than when considering the contribution of the wind speed and geographical information, indicating that these are important parameters for accurately estimating OSSS. The performance of the LGB-DF model was found to vary with season and water depth. Better estimation accuracy was obtained in winter and autumn, which was due to weaker stratification. This method provided important technical support for estimating the OSSS from satellite-derived sea surface data, which offers a novel insight into oceanic observations.

Keywords: machine learning; ocean subsurface salinity structure; South China Sea; satellite remote sensing data

1. Introduction

Ocean salinity, a vital parameter of seawater, plays a significant role in understanding marine ecosystems, ocean dynamics, and climate changes [1–5]. For example, ocean salinity can be used as an indicator for the hydrologic cycle, which provides valuable insights into the understanding of global water cycle features [6–8]. Changes in ocean salinity may also play a role in the formation of water masses [9–11]. To better understand the role of ocean salinity in dynamic ocean processes and climate changes, it is necessary to clarify the vertical structure of ocean salinity.

As the largest marginal sea of the Western Pacific, the South China Sea (SCS) has several straits along its border that connect to the Sulu Sea, the Java Sea, and the Indian Ocean (Figure 1a). The deepest water (around 5000 m) is found in the Eastern part of the SCS, while extended continental shelves (less than 200 m) have been found in the Western and Southern regions [12]. In the climatological mean, the sea surface salinity (SSS) in the SCS is north-south oriented: the SSS decreases from 34.0 psu in the north to 32.7 psu in the south (Figure 1b). The maximum SSS is in the Northern part of the SCS, which is related to the intrusion of the Kuroshio water through the Luzon Strait from the Pacific [13–15]. A low salinity tongue extends from the Southern part of the SCS, reaching as far as 10° N, which is closely related to the freshwater discharge from the Mekong and Rajang Rivers. Due to its special geographical location, the spatial distribution of the salinity in the SCS has significant features which are closely related to El Niño–Southern Oscillation (ENSO) [16–19], Asian monsoons, and the Pacific Western boundary current system [15,20]. Previous studies have suggested that the variability of the salinity in the SCS has a significant influence on the regional circulation and climate changes [21–23]. However, due to the lack of observations, little is known about the spatial and temporal variability of the salinity in the SCS. This has greatly limited the research on the thermohaline structures in the SCS. Therefore, it is of great importance to accurately retrieve the ocean subsurface salinity structure (OSSS), which remains a challenging problem for researchers.

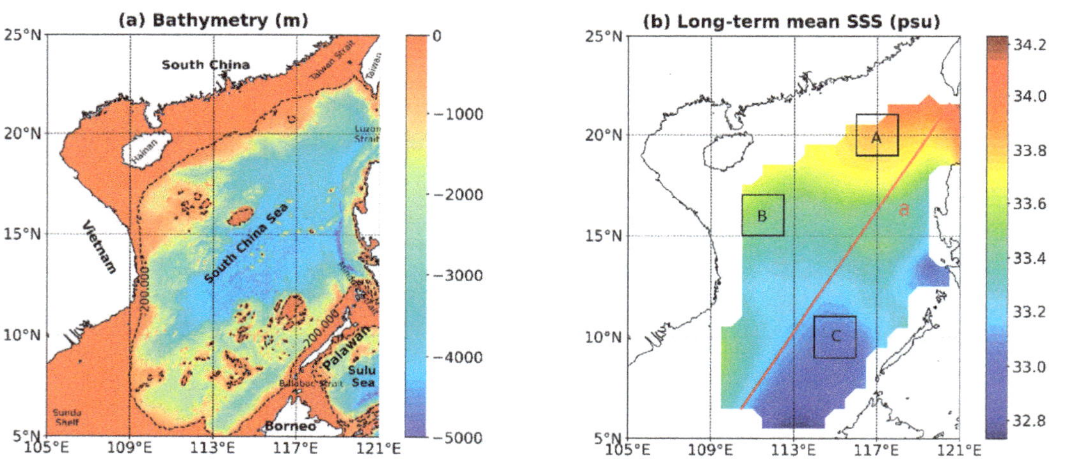

Figure 1. (**a**) Bathymetry (m) and geography of the SCS and (**b**) spatial distribution of long-term mean SSS (from January 2010 to December 2019) from Argo in the SCS. The three black boxes denote the study regions used in this study. The red line represents the location of transect used in this study.

Early studies on the estimation of ocean thermohaline structures in the SCS were usually based on numerical modeling and data assimilation [24–28]. For example, Chao et al. [24] modeled the interannual variations of thermal structure in the SCS by a three-dimensional primitive equation and found warming of the upper ocean during El Niño in the 1980s. Chu et al. [25] used the Princeton Ocean Model (POM) to investigate the seasonal variation of the thermal structure in the SCS. As researchers realized that the decreasing of dissolved oxygen was likely associated with the slowdown of thermohaline circulation, Li and Qu [26] analyzed the thermohaline circulation in the SCS on the basis of the available historical oxygen data. In order to provide better initial and boundary conditions for numerical simulations, assimilation methods have been used. Xiao et al. [27] performed an assimilation experiment for the Southern SCS, and the altimeter data were assimilated into POM. Shu et al. [28] focused on correcting temperature in the mixed layer by projecting sea surface temperature (SST) onto subsurface observations based on the optimal interpolation in the SCS using the POM. Although numerical ocean models offer important tools for estimating ocean thermohaline structures, these dynamical models are computationally expensive, as simulating physical governing equations demands intensive computational resources.

In recent decades, remote sensing technology has experienced a remarkable and rapid advancement that has provided large amounts of useful satellite-derived sea surface data, such as SSS, SST, and sea surface height (SSH). These well-sampled surface observations have significantly improved our understanding of upper ocean dynamic processes. Although satellite observations have been confined to the surface, they can be used to infer information about the vertical structures of the ocean, such as temperature and salinity structures [29–34]. Previous studies suggested that many oceanic subsurface phenomena have surface manifestations [35–40]. For example, the SSH was determined by the seawater density field, and the overall integrated effect of thermohaline is constrained by the SSH, according to observations [36]. There was a high correlation between temperature and salinity variables in the ocean; that is, the vertical distribution of the salinity could be deduced from the SST [37]. The thermocline was associated with the warming or cooling of surface ocean water through seasonal warming and the surface stratification or upwelling in deeper waters caused by offshore seawater transport [38]. Vernieres et al. [39] and Lu et al. [40] have demonstrated that there is a close link between the SSS and subsurface salinity structures. A number of methods, such as linear regression of variables, and statistical and dynamic methods, have been used to estimate vertical ocean temperature and salinity structures using satellite-derived sea surface data [41–44]. For example, Carnes et al. [41] inferred the global subsurface thermohaline structure using SSH and SST through a least-squares regression method. Based on the empirical orthogonal function (EOF) method, Maes and Behringer [45] estimated the vertical salinity structure in the Western Pacific Ocean by using sea level anomaly (SLA) and SST. Chu et al. [46] proposed a parametric model based on a layered structure that successfully reproduced the subsurface thermal structure in the SCS using SST. A coupled pattern reconstruction (CPR) method was proposed for estimating the subsurface temperature profiles from SSH and SST, which was shown to provide a substantial improvement [47]. Guinehut et al. [44] successfully reconstructed global temperature and salinity fields at a high resolution based on sea surface data and in situ measurements through a linear regression method. Yang et al. [48] developed a new method based on a transfer function and a neural network to estimate vertical profiles of the salinity in the global ocean from the SSS observed by the Soil Moisture and Ocean Salinity (SMOS) satellite, which was reasonable in contrast with climatology. Considering the spatial non-stationarity feature, a satellite-based geographically weighted regression model was proposed to estimate the subsurface temperature anomaly (STA) of the Indian Ocean by combining satellite-derived sea surface data and Argo in situ data, which has a significant improvement over the linear regression model [34]. Although the estimation accuracy of subsurface thermohaline structures based on satellite-derived sea surface data was much better than that of the numerical model-based data assimilation, further improvements are possible.

With the rapid development of machine-learning technology, it has been extensively employed in the fields of ocean and atmosphere [49–52]. A number of machine-learning approaches, such as the artificial neural networks (ANN) [53–55], self-organization mapping (SOM) [56,57], support vector machine (SVM) [58,59], random forests (RF) [34,60,61], and extreme gradient boosting (XGBoost) [62], have been widely used to retrieve vertical thermohaline structures of the ocean. Ali et al. [53] used an ANN method to estimate the vertical thermal structure from SST, SSH, wind stress, net radiation, and net heat flux data. This model could successfully reconstruct the ocean subsurface thermal structure. The SOM neural network has been applied to SST, SSH, and SSS data to estimate the STA [56]. Considering the data space correlation, Chen et al. [57] combined the SOM method with an EOF analysis to reconstruct the subsurface thermal structure by using the SST, the SSH, the longitude (LON), the latitude (LAT), and the month in the North-Western Pacific Ocean. Furthermore, machine-learning algorithms such as SVM, RF, and XGBoost were used to estimate the STA from surface remote sensing observations, which proved that the SSS and sea surface wind (SSW) were helpful in improving the accuracy of the estimations [34,58,60,62]. K-means clustering and feed-forward neural network were combined to estimate the subsurface temperature and achieve promising results in the deep ocean by taking the distribution of the ocean fields into consideration [63]. Based on a stacked long short-term memory (LSTM) neural network method, Buongiorno Nardelli [64] developed a model to estimate the ocean hydrographic profiles in the North Atlantic Ocean using surface remote sensing observations. Recently, Jiang et al. [65] proposed a bidirectional long short-term memory (Bi-LSTM) framework to estimate and analyze the subsurface temperature and salinity in the global ocean. A back-propagation neural network (BPNN) method was used to estimate the thermal structure in the North Pacific Ocean from sea surface data, such as SSH, SST, SSS, SSW, and sea surface velocity (SSV) [66].

As compared to temperature, relatively few attempts have been made to estimate OSSS from satellite-derived sea surface data using machine-learning methods [62,67,68]. For example, Gueye et al. [67] proposed a neural network model-based SOM for reconstructing salinity profiles of the tropical Atlantic Ocean from satellite-derived sea surface data. Salinity profiles in the Pacific Ocean can be estimated from satellite-derived sea surface data using a generalized regression neural network with the fruit-fly-optimization algorithm (FOAGRNN) [68]. Su et al. [62] proposed XGBoost for retrieving subsurface thermohaline anomalies of the global ocean, including the STA and the subsurface salinity anomaly (SSA). These existing studies focused on large-scale ocean regions or the global ocean.

To the best of our knowledge, in the SCS, there are no related studies conducted to estimate the OSSS from satellite-derived sea surface data using machine-learning methods. In this study, we proposed a Light Gradient Boosting Machine (LightGBM)-based Deep Forest (LGB-DF) method to estimate OSSS in the SCS from satellite-derived sea surface data, including SSS, SST, SSH, SSW (decomposed into eastward wind speed (USSW)) and northward wind speed (VSSW) components), and the geographical information (LON and LAT). To evaluate the performance of the LGB-DF model, another popular machine-learning model, LightGBM, was also used to estimate the OSSS in the SCS.

The rest of the paper is organized as follows. The data and methods are presented in Section 2. The evaluation of the model performance in estimating OSSS in the SCS is presented in Section 3. Finally, the discussion and conclusions are provided in Section 4.

2. Data and Method

2.1. Data

As an important part of the Indian–Western Pacific Ocean warm pool, salinity changes in the SCS play an important role in regulating the regional and global climate system [69]. Therefore, we selected the SCS (105°E–121°E and 5°N–23°N) as our study area.

In this study, we used two sources of ocean observational data: the sea surface data from satellite observations, such as SSS, SST, SSH, and SSW, combined with geographical information (LON and LAT); and gridded Argo data. The SSS data were obtained from the SMOS with a spatial resolution of 0.25° latitude × 0.25° longitude [70]. The SST data were obtained from the National Oceanic and Atmospheric Administration (NOAA), which consisted of optimal interpolated data observed by the satellite radiometer with a spatial resolution of 1° latitude × 1° longitude [71]. The SSH data were obtained from the Archiving, Validation, and Interpretation of Satellite Oceanographic data (AVISO) project with a spatial resolution of 0.25° latitude × 0.25° longitude [72]. The SSW data were obtained from Cross-Calibrated Multi-Platform (CCMP) gridded data, which are combined with multi-source data using a variational analysis method (VAM) to produce high-resolution (0.25° latitude× 0.25° longitude) gridded analyses [73]. The subsurface salinity data were obtained from the new version of the Roemmich–Gilson Argo Climatology (RG-Argo) data with a spatial resolution of 1° latitude × 1° longitude [74], which includes 58 vertical levels, but only 44 levels were used as training labels as well as to evaluate the model performance on the estimation of the OSSS.

Considering the differences in data resolution and time period between the input and output data available in the SCS, all data used in this study were processed into monthly averaged data and interpolated to a resolution of 0.5° latitude × 0.5° longitude with the same coverage of the SCS and the time period from January 2010 to December 2019. It should be noted that, in order to ensure uniformity, data points were deleted if any variable was null at the same point. All the data used in this paper are shown in Table 1.

Table 1. Summary of the data used in this study.

Index	Input Variable	Data Source	Output Variable	Data Source	Time Range	Time/Spatial Resolution
Data	SSS SST SSH SSW	SMOS NOAA AVISO CCMP	Salinity (2.5–1000 m)	Argo	2010–2019	Monthly 0.5° × 0.5°

2.2. Method

2.2.1. The LGB-DF Model

Deep Forest (DF) is an advanced decision-tree ensemble algorithm based on random forest, proposed by Zhou and Feng in 2017 [75,76]. Recently, the DF model has been widely used in many fields to prove its robustness in classification and prediction tasks [77–82]. A DF model would have great potential if it could go deeper. The LightGBM method has been shown to have the ability to estimate the ocean's subsurface information [83–87]. This inspired us to propose an improved DF model based on LightGBM (LGB-DF) method to estimate the OSSS in the SCS using satellite-derived sea surface data. In this study, the estimators (random forest and completely-random tree forests) of the DF model were replaced with the LightGBM to increase the accuracy of the model. The model code was based on the code of the open-source DF. The LGB-DF model was implemented and tested for all cases using Python programming on an Intel(R) Core(TM) I9-9940X CPU.

The flowchart for the proposed LGB-DF model is shown in Figure 2. The LGB-DF model had an important procedure: cascade structure, which could enhance the representational learning ability. In this study, the cascade structure of the LGB-DF model was constructed using two LightGBM (Figure 2). The number of trees in each forest was set to 150, and the maximum depth of each tree was set to 6. Having almost no adjustable hyperparameters was also one of the advantages of the LGB-DF model. As shown in Figure 2, the LGB-DF model processed the variables, layer by layer, in the cascade structure. In detail, the variables were input into the first layer, and each subsequent layer of input was spliced from the output of the preceding layer and the initial variable until the last layer to estimate the OSSS. To reduce the risk of overfitting, the vector produced by each estimator was generated by k-fold cross validation. Subsequently, the information in the last layer would be averaged as the estimation result. As compared to exiting machine-learning algorithms, the LGB-DF algorithm has the following advantages: fast speed, high accuracy, strong robustness, and simple implementation.

2.2.2. Experimental Setup

The flowchart of applying the LGB-DF model to estimate the OSSS is shown in Figure 3. The model setup was divided into three steps. The first step was the building of the training datasets. The satellite-derived sea surface data, such as SSS, SST, SSH, SSW (USSW and VSSW), and the geographical information (LON and LAT) were selected as input data for the LGB-DF model. The Argo data were used as training and testing labels. The second step was to train the model. The training data (from January 2010 to December 2018) were input into the LGB-DF model to obtain the output. Here, we used the grid search method to determine the optimal parameter combination for the LGB-DF model. Finally, with the optimal parameter combination of the LGB-DF model, we estimated the OSSS in the SCS using sea surface data from the testing set (from January 2019 to December 2019).

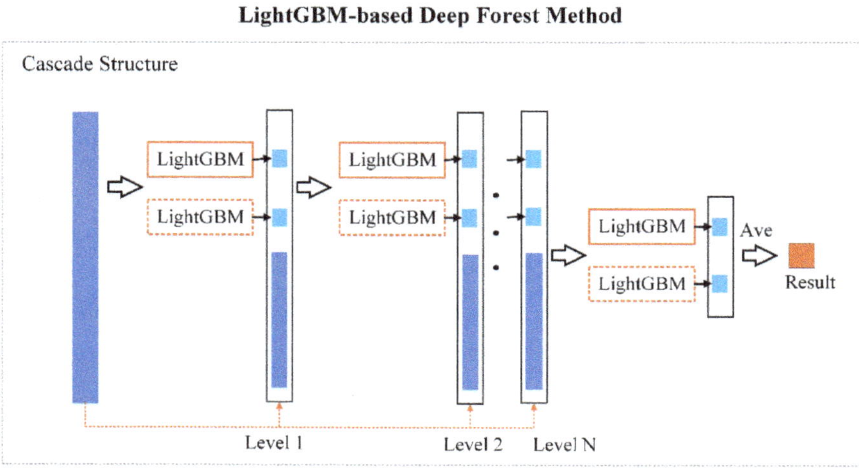

Figure 2. Flowchart for the LGB-DF model.

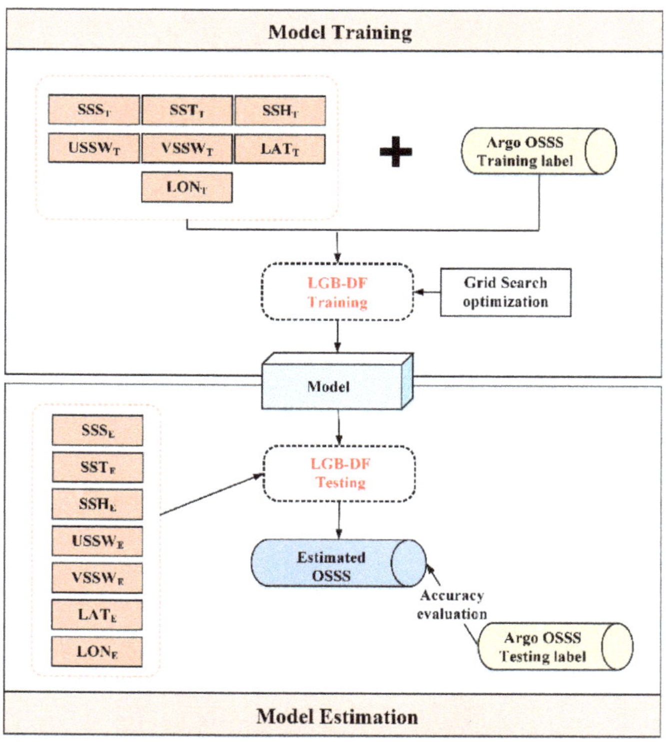

Figure 3. The flowchart of the OSSS estimation in the SCS using the LGB-DF model.

In addition to the LGB-DF model, we also set up a traditional machine-learning model (LightGBM) as a comparison to the LGB-DF model. Since the selection of input variables has an important impact on the performance of the model, three different combinations of sea surface parameters (three, five, and seven parameters) were used as LGB-DF model inputs to estimate the OSSS in the SCS. In this study, we evaluated the performance of the LGB-DF model through statistical metrics, such as root mean square error (RMSE), normalized root mean square error (NRMSE), and determination coefficient (R^2).

3. Results

3.1. Validation of Satellite-Derived SSS and SST

The accuracy of a machine-learning model is sensitive to the original input data [65]. Before utilizing the LGB-DF model to estimate OSSS in the SCS, the satellite-derived SSS and SST data were briefly validated by comparing them with the Argo data. As shown in Figure 4a, the seasonal variation of the satellite-derived SSS averaged over the SCS had good agreement with the Argo-derived SSS. For example, both of them showed that the maximum SSS value (>33.5 psu) occurred in April, and the minimum SSS value (<33.1 psu) occurred in November. The difference between the Argo SSS and satellite SSS varied from −0.02 psu to 0.14 psu. As for SST, the satellite-derived SST also showed good agreement with the Argo SST data on a seasonal scale (as shown in Figure 4b). In the SCS, the maximum SST value (>29.8 °C) occurred in May, whereas the minimum SST value (<26.3 °C) occurred in February. The difference between the Argo SST and satellite SST varied from −0.2 °C to 0.2 °C. Although the satellite data showed good agreement with the Argo observed data, some discrepancies were still observed, which may be due to different depths of measurement.

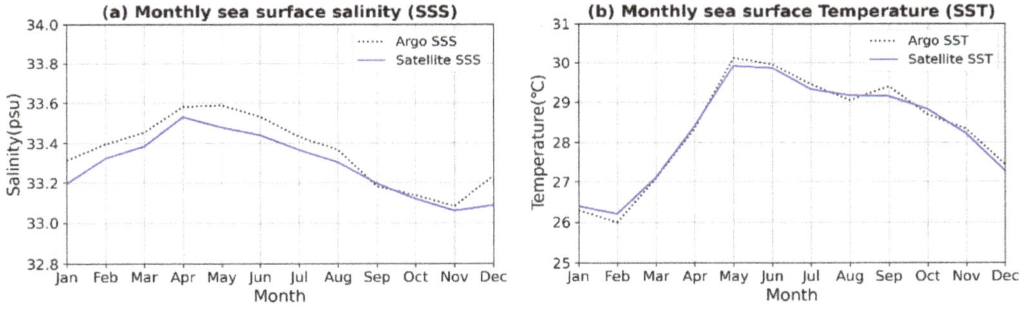

Figure 4. Comparison of the Argo (dashed black line) and satellite (solid blue line) for (**a**) the monthly mean SSS and (**b**) SST in the SCS from January 2010 to December 2019.

3.2. Identification of Input Variables

Previous studies had suggested that sea surface data could be used to infer ocean subsurface information with surface manifestations [37,55,62,86]. To determine the optimal combination of input variables for the LGB-DF model, a correlation analysis was conducted. Here, we only considered the absolute value of Pearson's correlation coefficients and focused on the magnitude of the correlation coefficients. The OSSS has a correlation with the sea surface variables at 50m, 100m, 500m, and 1000m of depth (Figure 5). The correlation coefficient between the OSSS and the SSS was relatively high at each depth, up to approximately 0.6. The correlation coefficients between the OSSS and SST/SSH/USSW (individually) were relatively small, approximately 0.2, while the VSSW was the lowest. As shown in Figure 5, the correlation coefficients between the OSSS and SSS/SSH/VSSW (individually) gradually decreased with depth, suggesting that SSS, SSH, and VSSW could play more important roles in the upper ocean. SST played a greater role in shallow and deeper layers, while the USSW performed better in the mid-ocean layers. The correlation analysis between the OSSS and sea surface parameters at different depths elucidated the impact of SSS, SST, SSH, and SSW on OSSS and explained the reasons for the selected variables.

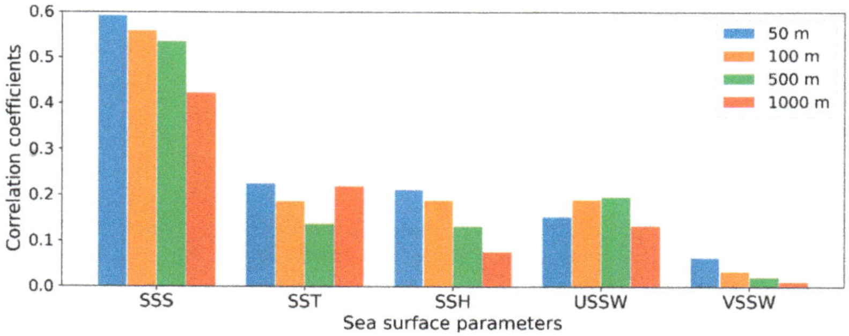

Figure 5. Correlation coefficients between the sea surface parameters (SSS, SST, SSH, USSW, and VSSW) and the Argo-observed OSSS at 50 m (blue), 100 m (orange), 500 m (green), and 1000 m (red) from January 2010 to December 2019.

As mentioned above, the unique geographical location and sparse observational data complicated the estimate of the OSSS in the SCS. Satellite-derived sea surface data captured most of the important features observed by the Argo surface data, providing an unprecedented opportunity to estimate OSSS in the SCS. Moreover, previous studies had suggested that geographical information could improve the estimation accuracy of the ocean subsurface information [67,86]. Therefore, we selected SSS, SST, SSH, SSW, and geographical information (LON and LAT) as the input variables to estimate the OSSS in the SCS.

3.3. Accuracy Comparison between the LGB-DF Model and LightGBM Model

To illustrate the improved performance of the LGB-DF model, we compared the LGB-DF model to the LightGBM model in terms of RMSE and R^2. For the LGB-DF model and LightGBM model, the average RMSE and R^2 at all depth levels were 0.0320/0.9398 and 0.0398/0.9150, respectively. The OSSS estimated by the LGB-DF model had relatively lower RMSE and higher R^2 values not only on average but also at each depth level (Figure 6), indicating that the LGB-DF model was more accurate than the LightGBM model for the estimation of the OSSS in the SCS.

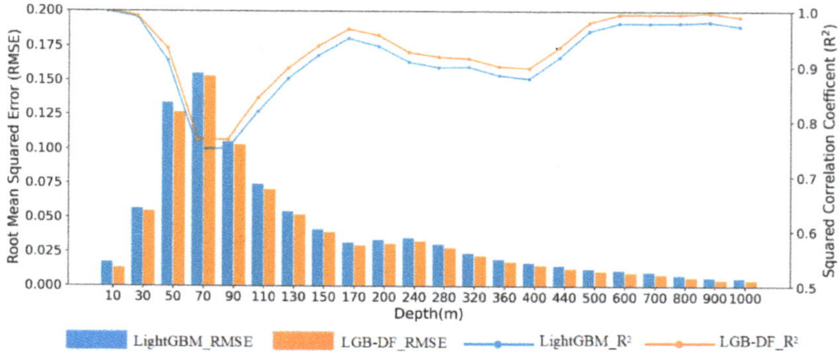

Figure 6. The average RMSE (psu) and R^2 of OSSS estimated using the LGB-DF model and LightGBM model at different depth levels in 2019 (the bars indicate the RMSE (psu) of the OSSS and the lines indicate the R^2 of the OSSS).

Another important issue in the estimation of the OSSS is the selection of input variables for models. Previous studies have suggested that the SSW and the geographical information could improve the accuracy of subsurface thermohaline estimates [62,67,85]. To further examine the influences of the SSW and the geographical information on the OSSS estimation in the SCS, we designed three sets of experiments with different input parameter combinations (Case 1, Case 2, and Case 3). In Case 1, we selected SSS, SST, and SSH as input parameters. In addition to the above parameters, we also selected SSW as an input parameter for Case 2. In Case 3, the geographical information (LON, LAT) was added as well as SSW.

The comparisons showed that both the SSW and the geographical information improved the estimation accuracy of the LGB-DF model in the SCS. The vertical mean RMSE and R^2 of the 7-parameter model in Case 3 were 0.0320 and 0.9398, respectively. For the 5-parameter model in Case 2, the vertical mean RMSE and R^2 were 0.0520 and 0.7569, respectively. For the 3-parameter model in Case 1, the vertical mean RMSE and R^2 were 0.0615 and 0.7150, respectively. The LGB-DF model in Case 3 (SSS, SST, SSH, USSW, VSSW, LON, and LAT) produced significantly lower RMSE values than the LGB-DF models in Case 1 (SSS, SST, and SSH) and Case 2 (SSS, SST, SSH, USSW, and VSSW) at all depths, while the R^2 values were higher than other cases (Figure 7). All these indicated that adding SSW and the geographical information significantly improved the estimation accuracy of the OSSS in the SCS using the LGB-DF model.

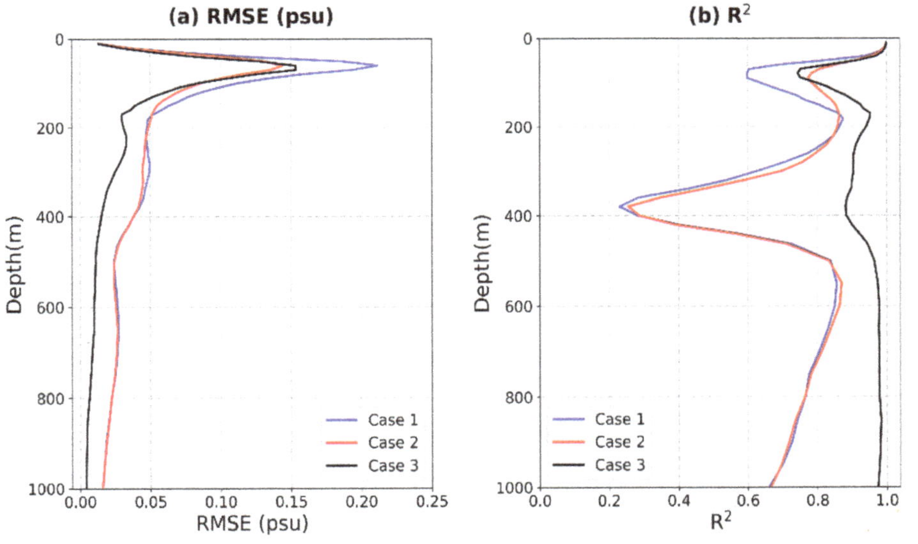

Figure 7. The estimation accuracy of the OSSS at different depths by the LGB-DF model based on (**a**) RMSE (psu) and (**b**) R^2 in different cases in 2019.

3.4. Evaluation of the LGB-DF Model

Based on the optimal parameter combination, the LGB-DF model was employed to estimate OSSS in the SCS. Next, we evaluated the performance and stability of the LGB-DF model from different aspects. Figure 8 shows the comparison of the LGB-DF-estimated OSSS and Argo-observed OSSS at depths of 50, 100, 500, and 1000 m in 2019; there were no significant differences between them. The LGB-DF model estimated OSSS showed good agreement with the Argo-observed OSSS at all depths. Most salinity features could be effectively reconstructed via sea surface data using the LGB-DF model. For example, at 50 m depth, both showed that there was a relatively high salinity tongue (>34.2 psu) in the northeast SCS. Relatively low salinity (<33.5 psu) was observed in the southeast

SCS (Figure 8a,e). The spatial distributions of the salinity at 100 m depth were similar to those at 50 m depth (Figure 8b,f). With increased depth, salinity tended to be stable. Below 500 m depth, the salinity varied from 34.4 psu to 34.6 psu. These spatial distribution features were well reconstructed by the LGB-DF model. From a horizontal point of view, the LGB-DF model had good performance in the estimation of OSSS in the SCS. More detailed descriptions of the LGB-DF model performance are discussed in the next section.

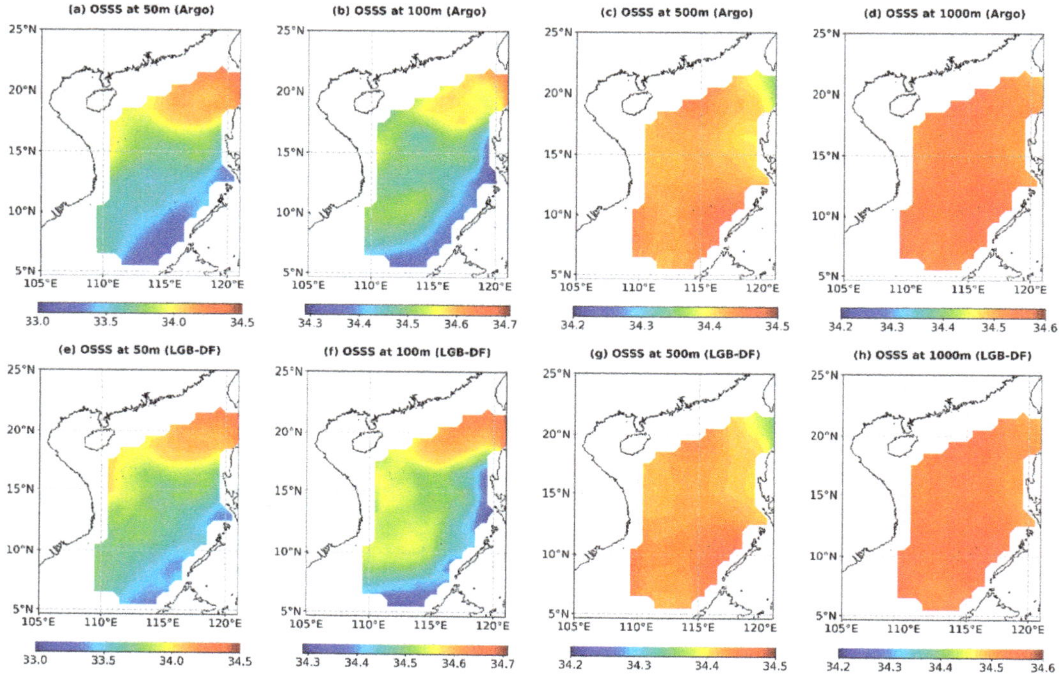

Figure 8. Argo-observed (**a–d**) and LGB-DF-estimated (**e–h**) yearly mean salinity at different depths (50, 100, 500, and 1000 m) in 2019.

To further evaluate the validity of the LGB-DF model, the accuracy of the OSSS estimation was quantitatively evaluated using the performance measures of RMSE and R^2 at different depths (Table 2). The RMSE could visually reflect the true errors at different depths. We employed the Argo-observed salinity at the same depth levels to validate the estimation results. As shown in Table 2, the RMSE value of the LGB-DF model exhibited differences at different depths; for example, RMSE = 0.1269 psu and R^2 = 0.9181 at 50 m depth, RMSE = 0.0841 psu and R^2 = 0.7919 at 100 m depth, RMSE = 0.0112 psu and R^2 = 0.9645 at 500 m depth, and RMSE = 0.0044 psu and R^2 = 0.9744 at 1000 m depth. The RMSE of the LGB-DF model decreased with depth due to the decreased range and standard deviation of the OSSS at deeper depths.

To improve the comparability of the model accuracy at different depths, we normalized the RMSE values to the relative error, i.e., NRMSE, dividing RMSE by the standard deviation of the Argo salinity at that depth. As shown in Figure 9, the NRMSE values increased from the surface to approximately 70 m, and then decreased from 70 m to 150 m, and then increased from 150 m to approximately 350 m, finally decreased from 350 m to 500 m, and stabilized from 500 m to 1000 m depth; whereas an opposite trend was observed in R^2. At approximately 70 m depth, the NRMSE value was the highest, while the R^2 was the lowest. This indicated that the estimation accuracy of the LGB-DF model was lowest at approximately 70 m depth. This was likely due to 70 m being approximately the depth of the thermocline layer in the SCS [64], where the temperature and salinity changed more drastically with depth than in the layers above or below. This led to the difficulty of reconstructing the OSSS in the SCS. Although the estimation accuracy at approximately 70 m was relatively low, the LGB-DF model was generally satisfactory. This also suggested that the LGB-DF model could accurately estimate the OSSS of the SCS using satellite-derived sea surface data with satisfactory performance.

Table 2. Vertical distributions of RMSE (psu) and R^2 for the LGB-DF model at different depths in 2019.

Depth (m)	RMSE	R^2
30	0.0547	0.9893
50	0.1269	0.9181
70	0.1533	0.7526
100	0.0841	0.7919
200	0.0310	0.9418
300	0.0249	0.9043
400	0.0153	0.8829
500	0.0112	0.9645
600	0.0100	0.9788
700	0.0087	0.9789
800	0.0066	0.9792
900	0.0047	0.9818
1000	0.0044	0.9744

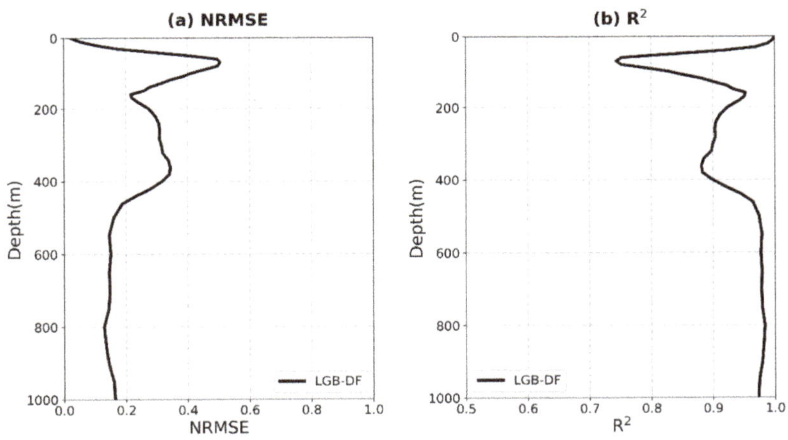

Figure 9. The estimation accuracy of OSSS in terms of NRMSE and R^2 at different depths by the LGB-DF model in 2019.

Next, to further evaluate the vertical performance of the LGB-DF model, we also compared the model estimated vertical salinity profiles with the Argo-observed salinity profiles in typical regions. Based on the characteristics of bathymetry and salinity distributions, we selected three typical boxes with a size of 2° × 2°, namely, Boxes A, B, and C (Figure 1). Box A (116°E~118°E and 19°N~21°N) was located along the continental slope south of China. Box B (110.5°E~112.5°E and 15°N~17°N) was situated in the region of the East Vietnam eddy. Box C (114°E~116°E and 9°N~11°N) was located in the Southern SCS. The vertical salinity profiles estimated by the LGB-DF model generally coincided with the Argo-observed profiles (Figure 10a–c). The vertically averaged RMSE and R^2 values between the LGB-DF estimation and the Argo observation were 0.0131 psu and 0.9950 for Box A,

0.0228 psu and 0.9942 for Box B, and 0.0594 psu and 0.9820 for Box C, respectively. Our comparison showed that the salinity difference between the LGB-DF estimation and the Argo observation for Box C was larger than those for Box A and Box B, and the maximum difference reached as high as 0.2 psu at approximately 70 m depth (Figure 10d). Although there were some differences, the LGB-DF model estimated salinity profiles were in good agreement with the Argo-observed salinity profiles. This result also demonstrated that the LGB-DF model was reliable and performed well in the estimation of OSSS in the SCS.

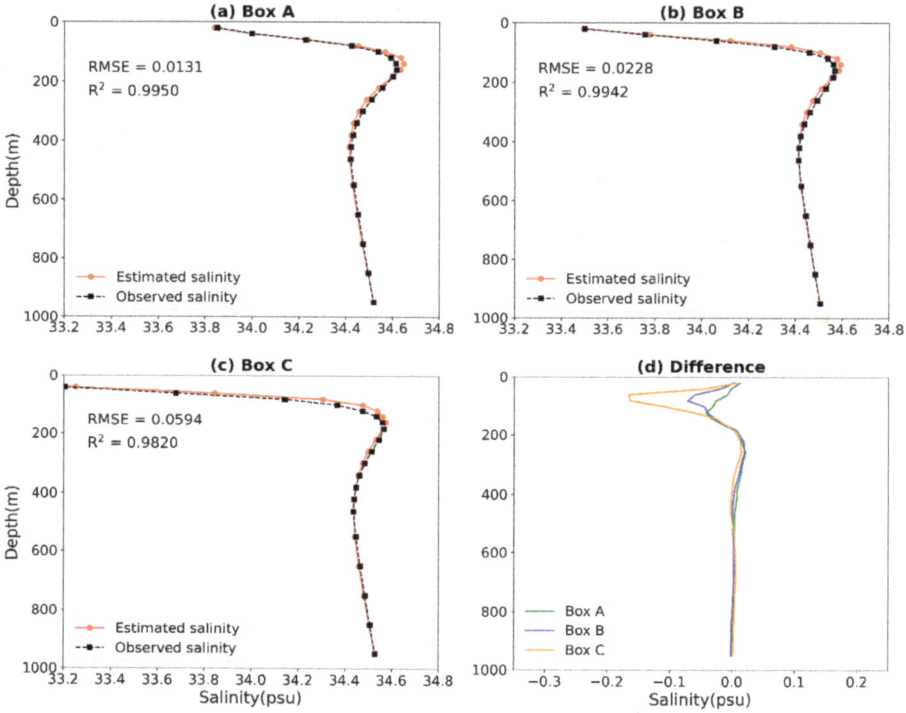

Figure 10. Comparison of the LGB-DF-estimated and Argo-observed salinity profiles averaged at different depths in Boxes (A–C) in 2019 (**a–c**) and their differences (**d**); Box A (116°E~118°E, 19°N~21°N), Box B (110.5°E~112.5°E, 15°N~17°N), and Box C (114°E~116°E, 9°N~11°N).

In addition, we selected a transect passing through the SCS from the southwest to the northeast to further evaluate the performance of the LGB-DF model. Figure 11 shows the comparison of the Argo-derived OSSS and LGB-DF model estimated OSSS in this transect. The results showed that the spatial distribution of OSSS from the LGB-DF model estimation was in good agreement with the Argo observations. Most of the observed significant features of the OSSS in this transect could be accurately reconstructed by the LGB-DF model. For example, in the upper 100 m, both of them showed that the salinity changed dramatically with depth, ranging from 33.1 psu at the surface to 34.5 psu at 100 m. The maximum salinity occurred between 100 m and 150 m in depth. Below 150 m, the salinity changed slightly but tended to be stable, ranging from 34.4 psu at 300 m depth to 34.6 psu at 1000 m depth. Figure 11c shows the salinity differences between Argo-observed and LGB-DF model estimated data (namely, Argo observation minus LGB-DF estimation). The results showed that the major differences (exceeding 0.25 psu) were present at a depth from 40 m to 150 m, between 9°N and 14°N, with Argo values less than the estimated salinity value; whereas Argo values more than the estimated salinity value were present at a depth from 40 m to 150 m, between 16°N and 19°N. Overall, the spatial distribution of the salinity from the LGB-DF model estimation had a very similar pattern as compared to the Argo observations, further indicating that the LGB-DF model had good performance in the estimation of the OSSS in the SCS.

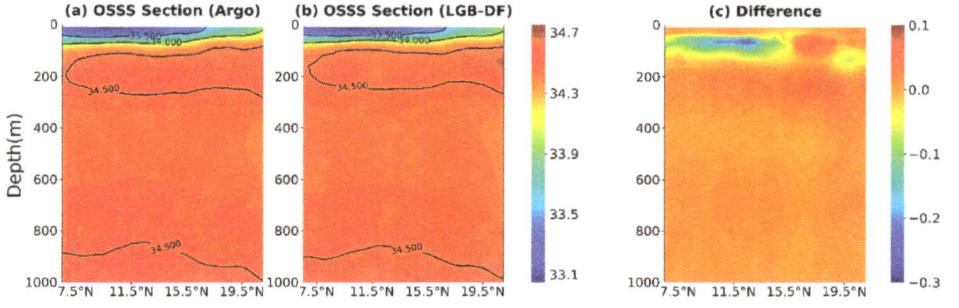

Figure 11. Argo-observed OSSS, LGB-DF-estimated OSSS and the differences between them, along a transect passing through the SCS from southwest to northeast. The black contours are, from top to bottom, 33.5 psu, 34.0 psu, and 34.5 psu, respectively.

The accuracy of the estimation by the model could also be evaluated directly using a density scatter plot. Therefore, we also calculated density scatter plots of the salinity from the Argo observations and LGB-DF model estimations to evaluate the performance of the LGB-DF model. The scatter distribution of the salinity from the LGB-DF estimation and the Argo observations at different depths for all geographical locations in 2019 is shown in Figure 12. Most of the scatter points were distributed evenly and densely along the line near 1:1 with a low RMSE. The RMSE values between the Argo-observed salinity and LGB-DF model estimated salinity were 0.0809 psu at 50 m depth, 0.0449 psu at 100 m depth, 0.0023 psu at 500 m depth, and 0.0012 psu at 1000 m depth. These also indicated that the estimated results by the LGB-DF model were reliable.

As previously discussed, the LGB-DF model had good performance in the yearly mean OSSS estimation in the SCS. However, the question of how it would perform in different seasons remained. In this study, we selected February, May, August, and November, all in 2019, to represent the winter, spring, summer, and autumn seasons of the year, respectively. Our quantitative evaluation of OSSS estimation for different seasons at the different depth levels (30, 50, 70, 100, 200, 300, 500, 600, 700, 800, 900, and 1000 m) in terms of the NRMSE and R^2 results are shown in Figure 13.

Generally, the NRMSE values in different seasons showed first an uptrend and then a downtrend, with a turning point appearing at 70 m. The highest NRMSE values occurred at 100 m in February and May at 0.3864 and 0.4085, respectively, and at 70 m for August (0.4603) and November (0.4587). The trend features of R^2 were unstable and fluctuated. They first fluctuated in the upper 500 m layer and then showed an uptrend from 500 m to 1000 m. The estimation accuracy of the LGB-DF model varied with the seasons. The average NRMSE (R^2) in February and November were 0.2052 and 0.2204 (0.9505 and 0.9377), respectively, which was lower (greater) than those in May and August. This indicated that the estimation accuracy in winter and autumn was better. The average NRMSE in May was 0.2676, and the average R^2 was 0.9112, which was the largest (smallest) value in four seasons. The average NRMSE and R^2 in August were 0.2646 and 0.9147, respectively. In general, the lower accuracy occurred in May and August, and the higher accuracy occurred in November and February, which could have been related to the different performances of the salinity at seasonal scales due to changes in the monsoonal circulation system. Specifically, the monsoon system dominated the summer pattern, and the winter pattern determined the climate of the SCS. The warm and humid southwest monsoon from the equator produced heavy precipitation and associated river runoff from mid-May to mid-

September, resulting in a double circulation pattern [14]. The dynamic ocean process was significant in May and August, resulting in poor model estimation. The results showed that there were low NRMSE and high R^2 values in all four seasons, indicating that the LGB-DF model had good seasonal applicability to estimate the OSSS in the SCS.

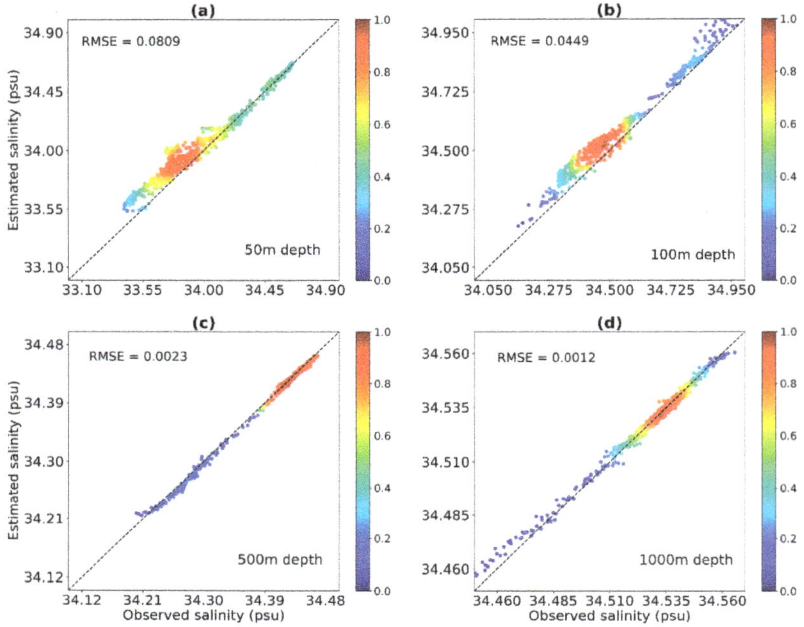

Figure 12. Density scatter plots of the salinity from LGB-DF estimations and Argo observations at (**a**) 50 m, (**b**) 100 m, (**c**) 500 m, and (**d**) 1000 m in 2019. The color bar represents the density of the scatter plots, with values closer to 1 indicating more scatter plots in the salinity range.

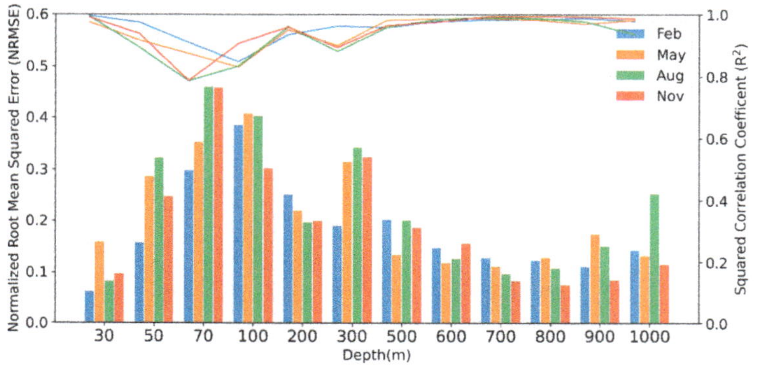

Figure 13. Performance measures by NRMSE and R^2 values of LGB-DF model for OSSS estimation at different depths in the SCS in 2019. Blue indicates February (winter), orange indicates May (spring), green indicates August (summer), red indicates November (autumn), the histograms display the NRMSE, and the lines display R^2.

4. Conclusions

Accurately estimating the vertical structure of ocean salinity in the SCS is of great importance for understanding oceanic processes due to its significant role in marine ecosystems, ocean dynamics, and climate changes. However, there is still a great lack of observational salinity data in the SCS due to the in situ observation being challenging and expensive. In this study, we proposed an LGB-DF model to estimate OSSS in the SCS. The developed LGB-DF model was used to reconstruct the OSSS in the SCS using satellite-derived sea surface data (SSS, SST, SSH, and SSW) and the geographical information (LON and LAT) as input data and in situ Argo data as label data. The LGB-DF-estimated results were measured for accuracy and reliability by RMSE, NRMSE, and R^2 using the Argo observational data.

Comparisons showed that the OSSS estimated by the LGB-DF model had relatively lower RMSE and higher R^2 values, not only on average but also at each depth level, as compared to the LightGBM model, indicating that the LGB-DF model accurately estimated the subsurface salinity of the SCS and outperformed the LightGBM model. This was attributed to the LGB-DF model combining the characteristics of deep learning and ensemble models to solve complex problems. In addition to SSH, SST, and SST, SSW and geographical information were two necessary parameters for accurately estimating the OSSS in the SCS and significantly improved the estimation accuracy of the LGB-DF model.

The results showed that the LGB-DF model had good performance in the estimation of the OSSS in the SCS with an area-averaged RMSE value of 0.0320 psu and an area-averaged R^2 value of 0.9398. The estimated salinity by the LGB-DF model and the Argo observed salinity both showed consistent spatial distribution at various depths in 2019. The performance measures showed that the performance of the LGB-DF model also varied with depth, with better performance in shallow layers due to the physical state relative to the surface being easily described. The performance of the LGB-DF model also varied with seasons: the average NRMSE (R^2) values in winter and autumn were lower (greater) than those in other seasons, indicating a better estimation accuracy was obtained in winter (NRMSE = 0.2052, R^2 = 0.9505) and autumn (NRMSE = 02204, R^2 = 0.9377). Although complex dynamic processes and the strong monsoon climate increase the difficulty of local OSSS estimation, our LGB-DF model had good performance in estimating the OSSS in the SCS according to satellite-derived sea surface data. This study demonstrated that the reconstruction of the subsurface salinity structure in the SCS using satellite observations based on the LGB-DF model was reliable and accurate.

Although the LGB-DF model has good applicability to estimate the vertical structure of the ocean salinity from the satellite-derived sea surface data, some discrepancies were observed in primarily two aspects. Data errors existed between the observed values and the true values due to objective factors such as the observation equipment itself and the environment. In the data processing, we interpolated the remote sensing data and Argo data to unify the resolution, which also caused errors. The estimation model error was also noted. The relationship between the sea surface data and the subsurface salinity could vary due to dynamic processes, such as subsidence and upwelling. Furthermore, as a data-driven method, the LGB-DF model was highly dependent on training data, which could underestimate or overlook the signal of some large anomalous events.

In future studies, we will further improve the estimation accuracy by using more accurate data and more advanced deep-learning methods combined with oceanic dynamic mechanisms to provide more explanatory results.

Author Contributions: Conceptualization, J.Q.; methodology, L.D.; validation, D.L., formal analysis, J.Q. and H.Z.; data curation, L.D.; writing—original draft preparation, L.D. and J.Q.; writing—review and editing, J.Q., S.Y. and B.Y.; visualization, B.X.; supervision, J.Q., W.W. and H.C.; All authors have read and agreed to the published version of the manuscript.

Funding: This research was funded by the Marine S&T Fund of Shandong Province for Pilot National Laboratory for Marine Science and Technology (Qingdao), grant number 2022QNLM010301-3,

National Natural Science Foundation of China, grant number 42176010, Natural Science Foundation of Shandong Province, China, grant number ZR2021MD022, Strategic Priority Research Program of the Chinese Academy of Sciences, grant number XDB42000000, National Natural Science Foundation of China, grant number 42076022.

Data Availability Statement: The datasets presented in this study are publicly available.

Acknowledgments: The SSS data were obtained from https://www.catds.fr/Products (accessed on 8 June 2021). The SST data were obtained from http://apdrc.soest.hawaii.edu/data/data.php (accessed on 6 June 2021). The SSH data were obtained from https://www.aviso.altimetry.fr/en/data/products (accessed on 12 June 2021). The SSW product was obtained from http://www.remss.com/measurements/ccmp/ (accessed on 16 June 2021). The gridded Argo data were obtained from http://sio-argo.ucsd.edu/RG_Climatology.html (accessed on 17 June 2021).

Conflicts of Interest: The authors declare no conflict of interest. The funders had no role in the design of the study, in the collection, analysis, or interpretation of the data, in the writing of the manuscript, or in the decision to publish the results.

References

1. Williams, P.D.; Guilyardi, E.; Madec, G.; Gualdi, S.; Scoccimarro, E. The role of mean ocean salinity in climate. *Dyn. Atmos. Ocean.* **2010**, *49*, 108–123. [CrossRef]
2. Felton, C.S.; Subrahmanyam, B.; Murty, V.; Shriver, J.F. Estimation of the barrier layer thickness in the Indian Ocean using Aquarius Salinity. *J. Geophys. Res. Ocean.* **2014**, *119*, 4200–4213. [CrossRef]
3. Zeng, L.; Chassignet, E.P.; Schmitt, R.W.; Xu, X.; Wang, D. Salinification in the South China Sea since late 2012: A reversal of the freshening since the 1990s. *Geophys. Res. Lett.* **2018**, *45*, 2744–2751. [CrossRef]
4. Berger, E.; Frör, O.; Schäfer, R.B. Salinity impacts on river ecosystem processes: A critical mini-review. *Philos. Trans. R. Soc. B* **2019**, *374*, 20180010. [CrossRef]
5. Qi, J.; Zhang, L.; Qu, T.; Yin, B.; Xu, Z.; Yang, D.; Li, D.; Qin, Y. Salinity variability in the tropical Pacific during the Central-Pacific and Eastern-Pacific El Niño events. *J. Mar. Syst.* **2019**, *199*, 103225. [CrossRef]
6. Schmitt, R.W. The ocean component of the global water cycle. *Rev. Geophys.* **1995**, *33*, 1395–1409. [CrossRef]
7. Barreiro, M.; Fedorov, A.; Pacanowski, R.; Philander, S.G. Abrupt climate changes: How freshening of the northern Atlantic affects the thermohaline and wind-driven oceanic circulations. *Annu. Rev. Earth Planet. Sci.* **2008**, *36*, 33–58. [CrossRef]
8. Durack, P.J.; Wijffels, S.E.; Matear, R.J. Ocean salinities reveal strong global water cycle intensification during 1950 to 2000. *Science* **2012**, *336*, 455–458. [CrossRef]
9. Curry, R.; Mauritzen, C. Dilution of the northern North Atlantic Ocean in recent decades. *Science* **2005**, *308*, 1772–1774. [CrossRef]
10. Stark, S.; Wood, R.A.; Banks, H.T. Reevaluating the causes of observed changes in Indian Ocean water masses. *J. Clim.* **2006**, *19*, 4075–4086. [CrossRef]
11. Helber, R.; Richman, J.; Barron, C. The influence of temperature and salinity variability on the upper ocean density and mixed layer. *Ocean Sci. Discuss.* **2010**, *7*, 1469–1495. [CrossRef]
12. Chu, P.C.; Tseng, H.C.; Chang, C.; Chen, J. South China Sea warm pool detected in spring from the Navy's master oceanographic observational data set (MOODS). *J. Geophys. Res. Ocean.* **1997**, *102*, 15761–15771. [CrossRef]
13. Qu, T.; Mitsudera, H.; Yamagata, T. Intrusion of the north Pacific waters into the South China Sea. *J. Geophys. Res. Ocean.* **2000**, *105*, 6415–6424. [CrossRef]
14. Hu, J.; Kawamura, H.; Hong, H.; Qi, Y. A review on the currents in the South China Sea: Seasonal circulation, South China Sea warm current and Kuroshio intrusion. *J. Oceanogr.* **2000**, *56*, 607–624. [CrossRef]
15. Yi, D.L.; Melnichenko, O.; Hacker, P.; Potemra, J. Remote sensing of sea surface salinity variability in the South China Sea. *J. Geophys. Res. Ocean.* **2020**, *125*, e2020JC016827. [CrossRef]
16. Ballabrera-Poy, J.; Murtugudde, R.; Busalacchi, A. On the potential impact of sea surface salinity observations on ENSO predictions. *J. Geophys. Res. Ocean.* **2002**, *107*, SRF 8-1–SRF 8-11. [CrossRef]
17. Qu, T.; Yu, J.-Y. ENSO indices from sea surface salinity observed by Aquarius and Argo. *J. Oceanogr.* **2014**, *70*, 367–375. [CrossRef]
18. Zhu, J.; Huang, B.; Zhang, R.-H.; Hu, Z.-Z.; Kumar, A.; Balmaseda, M.A.; Marx, L.; Kinter III, J.L. Salinity anomaly as a trigger for ENSO events. *Sci. Rep.* **2014**, *4*, 6821. [CrossRef]
19. Qi, J.; Du, Y.; Chi, J.; Yi, D.L.; Li, D.; Yin, B. Impacts of El Niño on the South China Sea surface salinity as seen from satellites. *Environ. Res. Lett.* **2022**, *17*, 054040. [CrossRef]
20. Singh, A.; Delcroix, T. Estimating the effects of ENSO upon the observed freshening trends of the western tropical Pacific Ocean. *Geophys. Res. Lett.* **2011**, *38*, L21607. [CrossRef]
21. Qu, T.; Du, Y.; Sasaki, H. South China Sea throughflow: A heat and freshwater conveyor. *Geophys. Res. Lett.* **2006**, *33*, L23617. [CrossRef]
22. Zeng, L.; Du, Y.; Xie, S.-P.; Wang, D. Barrier layer in the South China Sea during summer 2000. *Dyn. Atmos. Ocean.* **2009**, *47*, 38–54. [CrossRef]

23. Wang, G.; Xie, S.P.; Qu, T.; Huang, R.X. Deep South China Sea circulation. *Geophys. Res. Lett.* **2011**, *38*, L05601. [CrossRef]
24. Chao, S.-Y.; Shaw, P.-T.; Wu, S.Y. El Niño modulation of the South China sea circulation. *Prog. Oceanogr.* **1996**, *38*, 51–93. [CrossRef]
25. Chu, P.C.; Edmons, N.L.; Fan, C. Dynamical mechanisms for the South China Sea seasonal circulation and thermohaline variabilities. *J. Phys. Oceanogr.* **1999**, *29*, 2971–2989. [CrossRef]
26. Li, L.; Qu, T. Thermohaline circulation in the deep South China Sea basin inferred from oxygen distributions. *J. Geophys. Res. Ocean.* **2006**, *111*, C05017. [CrossRef]
27. Xiao, X.; Wang, D.; Xu, J. The assimilation experiment in the southwestern South China Sea in summer 2000. *Chin. Sci. Bull.* **2006**, *51*, 31–37. [CrossRef]
28. Shu, Y.; Zhu, J.; Wang, D.; Yan, C.; Xiao, X. Performance of four sea surface temperature assimilation schemes in the South China Sea. *Cont. Shelf Res.* **2009**, *29*, 1489–1501. [CrossRef]
29. Fox, D.N. The modular ocean data assimilation system. *Oceanography* **2002**, *15*, 22–28. [CrossRef]
30. Nardelli, B.B.; Santoleri, R. Methods for the reconstruction of vertical profiles from surface data: Multivariate analyses, residual GEM, and variable temporal signals in the North Pacific Ocean. *J. Atmos. Ocean. Technol.* **2005**, *22*, 1762–1781. [CrossRef]
31. Wang, J.; Flierl, G.R.; LaCasce, J.H.; McClean, J.L.; Mahadevan, A. Reconstructing the ocean's interior from surface data. *J. Phys. Oceanogr.* **2013**, *43*, 1611–1626. [CrossRef]
32. Akbari, E.; Alavipanah, S.K.; Jeihouni, M.; Hajeb, M.; Haase, D.; Alavipanah, S. A review of ocean/sea subsurface water temperature studies from remote sensing and non-remote sensing methods. *Water* **2017**, *9*, 936. [CrossRef]
33. Chen, J.; You, X.; Xiao, Y.; Zhang, R.; Wang, G.; Bao, S. A performance evaluation of remotely sensed sea surface salinity products in combination with other surface measurements in reconstructing three-dimensional salinity fields. *Acta Oceanol. Sin.* **2017**, *36*, 15–31. [CrossRef]
34. Su, H.; Li, W.; Yan, X.H. Retrieving temperature anomaly in the global subsurface and deeper ocean from satellite observations. *J. Geophys. Res. Ocean.* **2018**, *123*, 399–410. [CrossRef]
35. Cornillon, P.; Stramma, L.; Price, J.F. Satellite measurements of sea surface cooling during hurricane Gloria. *Nature* **1987**, *326*, 373–375. [CrossRef]
36. Cooper, M.; Haines, K. Altimetric assimilation with water property conservation. *J. Geophys. Res. Ocean.* **1996**, *101*, 1059–1077. [CrossRef]
37. Stommel, H. Note on the use of the TS correlation for dynamic height anomaly computations. *J. Mar. Res.* **1947**, *6*, 85–92.
38. Fiedler, P.C. Surface manifestations of subsurface thermal structure in the California Current. *J. Geophys. Res. Ocean.* **1988**, *93*, 4975–4983. [CrossRef]
39. Vernieres, G.; Kovach, R.; Keppenne, C.; Akella, S.; Brucker, L.; Dinnat, E. The impact of the assimilation of Aquarius sea surface salinity data in the GEOS ocean data assimilation system. *J. Geophys. Res. Ocean.* **2014**, *119*, 6974–6987. [CrossRef]
40. Lu, Z.; Cheng, L.; Zhu, J.; Lin, R. The complementary role of SMOS sea surface salinity observations for estimating global ocean salinity state. *J. Geophys. Res. Ocean.* **2016**, *121*, 3672–3691. [CrossRef]
41. Carnes, M.R.; Teague, W.J.; Mitchell, J.L. Inference of subsurface thermohaline structure from fields measurable by satellite. *J. Atmos. Ocean. Technol.* **1994**, *11*, 551–566. [CrossRef]
42. Vossepoel, F.C.; Reynolds, R.W.; Miller, L. Use of sea level observations to estimate salinity variability in the tropical Pacific. *J. Atmos. Ocean. Technol.* **1999**, *16*, 1401–1415. [CrossRef]
43. Agarwal, N.; Sharma, R.; Basu, S.; Agarwal, V.K. Derivation of salinity profiles in the Indian Ocean from satellite surface observations. *IEEE Geosci. Remote Sens. Lett.* **2007**, *4*, 322–325. [CrossRef]
44. Guinehut, S.; Dhomps, A.-L.; Larnicol, G.; Le Traon, P.-Y. High resolution 3-D temperature and salinity fields derived from in situ and satellite observations. *Ocean Sci.* **2012**, *8*, 845–857. [CrossRef]
45. Maes, C.; Behringer, D. Using satellite-derived sea level and temperature profiles for determining the salinity variability: A new approach. *J. Geophys. Res. Ocean.* **2000**, *105*, 8537–8547. [CrossRef]
46. Chu, P.C.; Fan, C.; Liu, W.T. Determination of vertical thermal structure from sea surface temperature. *J. Atmos. Ocean. Technol.* **2000**, *17*, 971–979. [CrossRef]
47. Nardelli, B.B.; Santoleri, R. Reconstructing synthetic profiles from surface data. *J. Atmos. Ocean. Technol.* **2004**, *21*, 693–703. [CrossRef]
48. Yang, T.; Chen, Z.; He, Y. A new method to retrieve salinity profiles from sea surface salinity observed by SMOS satellite. *Acta Oceanol. Sin.* **2015**, *34*, 85–93. [CrossRef]
49. Zheng, G.; Li, X.; Zhang, R.-H.; Liu, B. Purely satellite data–driven deep learning forecast of complicated tropical instability waves. *Sci. Adv.* **2020**, *6*, eaba1482. [CrossRef]
50. Zhang, X.; Li, X. Combination of satellite observations and machine learning method for internal wave forecast in the Sulu and Celebes seas. *IEEE Trans. Geosci. Remote Sens.* **2020**, *59*, 2822–2832. [CrossRef]
51. Zhang, X.; Li, X.; Zheng, Q. A Machine-Learning Model for Forecasting Internal Wave Propagation in the Andaman Sea. *IEEE J. Sel. Top. Appl. Earth Obs. Remote Sens.* **2021**, *14*, 3095–3106. [CrossRef]
52. Wang, Y.; Li, X.; Song, J.; Li, X.; Zhong, G.; Zhang, B. Carbon Sinks and Variations of pCO 2 in the Southern Ocean From 1998 to 2018 Based on a Deep Learning Approach. *IEEE J. Sel. Top. Appl. Earth Obs. Remote Sens.* **2021**, *14*, 3495–3503. [CrossRef]
53. Ali, M.; Swain, D.; Weller, R. Estimation of ocean subsurface thermal structure from surface parameters: A neural network approach. *Geophys. Res. Lett.* **2004**, *31*, L20308. [CrossRef]

54. Su, H.; Zhang, H.; Geng, X.; Qin, T.; Lu, W.; Yan, X.-H. OPEN: A new estimation of global ocean heat content for upper 2000 meters from remote sensing data. *Remote Sens.* **2020**, *12*, 2294. [CrossRef]
55. Wang, H.; Song, T.; Zhu, S.; Yang, S.; Feng, L. Subsurface temperature estimation from sea surface data using neural network models in the western pacific ocean. *Mathematics* **2021**, *9*, 852. [CrossRef]
56. Wu, X.; Yan, X.-H.; Jo, Y.-H.; Liu, W.T. Estimation of subsurface temperature anomaly in the North Atlantic using a self-organizing map neural network. *J. Atmos. Ocean. Technol.* **2012**, *29*, 1675–1688. [CrossRef]
57. Chen, C.; Yang, K.; Ma, Y.; Wang, Y. Reconstructing the subsurface temperature field by using sea surface data through self-organizing map method. *IEEE Geosci. Remote Sens. Lett.* **2018**, *15*, 1812–1816. [CrossRef]
58. Su, H.; Wu, X.; Yan, X.-H.; Kidwell, A. Estimation of subsurface temperature anomaly in the Indian Ocean during recent global surface warming hiatus from satellite measurements: A support vector machine approach. *Remote Sens. Environ.* **2015**, *160*, 63–71. [CrossRef]
59. Li, W.e.; Su, H.; Wang, X.; Yan, X. Estimation of global subsurface temperature anomaly based on multisource satellite observations. J. *Remote Sens* **2017**, *21*, 881–891.
60. Su, H.; Yang, X.; Yan, X.-H. Estimating Ocean Subsurface Salinity from Remote Sensing Data by Machine Learning. In Proceedings of the IGARSS 2019-2019 IEEE International Geoscience and Remote Sensing Symposium, Yokohama, Japan, 28 July–2 August 2019; pp. 8139–8142. [CrossRef]
61. Rajabi-Kiasari, S.; Hasanlou, M. An efficient model for the prediction of SMAP sea surface salinity using machine learning approaches in the Persian Gulf. *Int. J. Remote Sens.* **2020**, *41*, 3221–3242. [CrossRef]
62. Su, H.; Yang, X.; Lu, W.; Yan, X.-H. Estimating subsurface thermohaline structure of the global ocean using surface remote sensing observations. *Remote Sens.* **2019**, *11*, 1598. [CrossRef]
63. Lu, W.; Su, H.; Yang, X.; Yan, X.-H. Subsurface temperature estimation from remote sensing data using a clustering-neural network method. *Remote Sens. Environ.* **2019**, *229*, 213–222. [CrossRef]
64. Buongiorno Nardelli, B. A deep learning network to retrieve ocean hydrographic profiles from combined satellite and in situ measurements. *Remote Sens.* **2020**, *12*, 3151. [CrossRef]
65. Jiang, F.; Ma, J.; Wang, B.; Shen, F.; Yuan, L. Ocean Observation Data Prediction for Argo Data Quality Control Using Deep Bidirectional LSTM Network. *Secur. Commun. Netw.* **2021**, *2021*, 5665386. [CrossRef]
66. Cheng, H.; Sun, L.; Li, J. Neural network approach to retrieving ocean subsurface temperatures from surface parameters observed by satellites. *Water* **2021**, *13*, 388. [CrossRef]
67. Gueye, M.B.; Niang, A.; Arnault, S.; Thiria, S.; Crépon, M. Neural approach to inverting complex system: Application to ocean salinity profile estimation from surface parameters. *Comput. Geosci.* **2014**, *72*, 201–209. [CrossRef]
68. Bao, S.; Zhang, R.; Wang, H.; Yan, H.; Yu, Y.; Chen, J. Salinity profile estimation in the Pacific Ocean from satellite surface salinity observations. *J. Atmos. Ocean. Technol.* **2019**, *36*, 53–68. [CrossRef]
69. Chen, X.; Liu, Z.; Wang, H.; Xu, D.; Wang, L. Significant salinity increase in subsurface waters of the South China Sea during 2016–2017. *Acta Oceanol. Sin.* **2019**, *38*, 51–61. [CrossRef]
70. Boutin, J.; Vergely, J.-L.; Marchand, S.; d'Amico, F.; Hasson, A.; Kolodziejczyk, N.; Reul, N.; Reverdin, G.; Vialard, J. New SMOS Sea Surface Salinity with reduced systematic errors and improved variability. *Remote Sens. Environ.* **2018**, *214*, 115–134. [CrossRef]
71. Banzon, V.; Smith, T.M.; Chin, T.M.; Liu, C.; Hankins, W. A long-term record of blended satellite and in situ sea-surface temperature for climate monitoring, modeling and environmental studies. *Earth Syst. Sci. Data* **2016**, *8*, 165–176. [CrossRef]
72. Hauser, D.; Tourain, C.; Hermozo, L.; Alraddawi, D.; Aouf, L.; Chapron, B.; Dalphinet, A.; Delaye, L.; Dalila, M.; Dormy, E. New observations from the SWIM radar on-board CFOSAT: Instrument validation and ocean wave measurement assessment. *IEEE Trans. Geosci. Remote Sens.* **2020**, *59*, 5–26. [CrossRef]
73. Atlas, R.; Hoffman, R.N.; Ardizzone, J.; Leidner, S.M.; Jusem, J.C.; Smith, D.K.; Gombos, D. A cross-calibrated, multiplatform ocean surface wind velocity product for meteorological and oceanographic applications. *Bull. Am. Meteorol. Soc.* **2011**, *92*, 157–174. [CrossRef]
74. Roemmich, D.; Gilson, J. The 2004–2008 mean and annual cycle of temperature, salinity, and steric height in the global ocean from the Argo Program. *Prog. Oceanogr.* **2009**, *82*, 81–100. [CrossRef]
75. Zhou, Z.-H.; Feng, J. Deep forest. *Natl. Sci. Rev.* **2019**, *6*, 74–86. [CrossRef] [PubMed]
76. Zhou, Z.-H.; Feng, J. Deep Forest: Towards an Alternative to Deep Neural Networks. In Proceedings of the IJCAI, Melbourne, Australia, 19–25 August 2017; pp. 3553–3559. [CrossRef]
77. AlJame, M.; Imtiaz, A.; Ahmad, I.; Mohammed, A. Deep forest model for diagnosing COVID-19 from routine blood tests. *Sci. Rep.* **2021**, *11*, 16682. [CrossRef]
78. Chu, Y.; Kaushik, A.C.; Wang, X.; Wang, W.; Zhang, Y.; Shan, X.; Salahub, D.R.; Xiong, Y.; Wei, D.-Q. DTI-CDF: A cascade deep forest model towards the prediction of drug-target interactions based on hybrid features. *Brief. Bioinform.* **2021**, *22*, 451–462. [CrossRef]
79. Su, R.; Liu, X.; Wei, L.; Zou, Q. Deep-Resp-Forest: A deep forest model to predict anti-cancer drug response. *Methods* **2019**, *166*, 91–102. [CrossRef]
80. Wang, L.; Zhang, Z.; Zhang, X.; Zhou, X.; Wang, P.; Zheng, Y. A Deep-Forest Based Approach for Detecting Fraudulent Online Transaction. In *Advances in Computers*; Elsevier: Amsterdam, The Netherlands, 2021; Volume 120, pp. 1–38. [CrossRef]

81. Yin, L.; Sun, Z.; Gao, F.; Liu, H. Deep forest regression for short-term load forecasting of power systems. *IEEE Access* **2020**, *8*, 49090–49099. [CrossRef]
82. Fu, Q.; Li, K.; Chen, J.; Wang, J.; Lu, Y.; Wang, Y. Building energy consumption prediction using a deep-forest-based DQN method. *Buildings* **2022**, *12*, 131. [CrossRef]
83. Ke, G.; Meng, Q.; Finley, T.; Wang, T.; Chen, W.; Ma, W.; Ye, Q.; Liu, T.-Y. Lightgbm: A highly efficient gradient boosting decision tree. *Adv. Neural Inf. Processing Syst.* **2017**, *30*, 3147–3155.
84. Wang, N.; Zhang, G.; Pang, W.; Ren, L.; Wang, Y. Novel monitoring method for material removal rate considering quantitative wear of abrasive belts based on LightGBM learning algorithm. *Int. J. Adv. Manuf. Technol.* **2021**, *114*, 3241–3253. [CrossRef]
85. Su, H.; Lu, X.; Chen, Z.; Zhang, H.; Lu, W.; Wu, W. Estimating coastal chlorophyll-a concentration from time-series OLCI data based on machine learning. *Remote Sens.* **2021**, *13*, 576. [CrossRef]
86. Su, H.; Wang, A.; Zhang, T.; Qin, T.; Du, X.; Yan, X.-H. Super-resolution of subsurface temperature field from remote sensing observations based on machine learning. *Int. J. Appl. Earth Obs. Geoinf.* **2021**, *102*, 102440. [CrossRef]
87. Shi, H. Best-First Decision Tree Learning. Ph.D. Thesis, The University of Waikato, Hamilton, New Zealand, 2007.

remote sensing

MDPI

Article

Improvement of Atmospheric Correction of Satellite Sentinel-3/OLCI Data for Oceanic Waters in Presence of *Sargassum*

Léa Schamberger [1,2], Audrey Minghelli [1,2,*], Malik Chami [3] and François Steinmetz [4]

1 Laboratoire d'Informatique et Système (LIS), Université de Toulon, CNRS UMR 7020, F-83041 Toulon, France; lea-schamberger@etud.univ-tln.fr
2 Laboratoire d'Informatique et Système (LIS), Aix Marseille Université, F-13288 Marseille, France
3 Laboratoire Atmosphères Milieux Observations Spatiales (LATMOS), Sorbonne Université, CNRS-INSU, F-06304 Nice, France; malik.chami@upmc.fr
4 HYGEOS, Euratechnologies, 165 av. de Bretagne, F-59000 Lille, France; fs@hygeos.com
* Correspondence: audrey.minghelli@univ-tln.fr

Abstract: The invasive species of brown algae *Sargassum* gathers in large aggregations in the Caribbean Sea, and has done so especially over the last decade. These aggregations wash up on shores and decompose, leading to many socio-economic issues for the population and the coastal ecosystem. Satellite ocean color data sensors such as Sentinel-3/OLCI can be used to detect the presence of *Sargassum* and estimate its fractional coverage and biomass. The derivation of *Sargassum* presence and abundance from satellite ocean color data first requires atmospheric correction; however, the atmospheric correction procedure that is commonly used for oceanic waters needs to be adapted when dealing with the occurrence of *Sargassum* because the non-zero water reflectance in the near infrared band induced by *Sargassum* optical signature could lead to *Sargassum* being wrongly identified as aerosols. In this study, this difficulty is overcome by interpolating aerosol and sunglint reflectance between nearby *Sargassum*-free pixels. The proposed method relies on the local homogeneity of the aerosol reflectance between *Sargassum* and *Sargassum*-free areas. The performance of the adapted atmospheric correction algorithm over *Sargassum* areas is evaluated. The proposed method is demonstrated to result in more plausible aerosol and sunglint reflectances. A reduction of between 75% and 88% of pixels showing a negative water reflectance above 600 nm were noticed after the correction of the several images.

Keywords: *Sargassum*; atmospheric correction; aerosols; OLCI

Citation: Schamberger, L.; Minghelli, A.; Chami, M.; Steinmetz, F. Improvement of Atmospheric Correction of Satellite Sentinel-3/OLCI Data for Oceanic Waters in Presence of *Sargassum*. *Remote Sens.* **2022**, *14*, 386. https://doi.org/10.3390/rs14020386

Academic Editors: Chung-Ru Ho, Antony K. Liu and Xiaofeng Li

Received: 15 December 2021
Accepted: 12 January 2022
Published: 14 January 2022

1. Introduction

Sargassum is an algal invasive species originally found in the Sargasso Sea, which is located in the northwest of the Atlantic Ocean facing the State of Florida, USA between 20° N and 35° N. Since 2010, *Sargassum* has been found in large quantities elsewhere, typically in the south of the Caribbean Sea as far as Brazil, but also as far as West Africa. The spatial extent of *Sargassum* presence can possibly be explained by global climate changes, especially by the increase of oceanic water temperatures, by the modifications of hydrodynamical conditions and by the farming of deforested zones [1,2]. Once washed up on beaches, *Sargassum* decomposes, producing a gas impacting inhabitants' health, tourism and the coastal environment. Scientific researches are conducted to gain an understanding of the evolution of the *Sargassum* [3,4]. Remote sensing techniques can provide interesting information regarding standing stock forecasts in terms of spatial location, period of occurrence and abundance [5].

The presence of floating *Sargassum* can be detected using satellite imagery due to the increased water reflectance caused by *Sargassum* in the red and near infrared (NIR).

Hu [6] has proposed the Floating Algae Index (FAI) applied to MODIS data to detect *Sargassum* aggregation using the red edge signal. However, the interpretation of such an index can be challenging due to the fact that *Sargassum* and cloud optical signature could both lead to high FAI values. Wang and Hu [7] modified the FAI to provide the Alternative Floating Algae Index (AFAI), which is less sensitive to the presence of cloud edges or thin clouds. Based on the AFAI, these authors were able to infer the fractional coverage of *Sargassum* in the Gulf of Mexico. Wang and Hu's method was adapted to ENVISAT/MERIS data by Gower et al. [8,9] through the Maximum Chlorophyll Index (MCI). The MCI was then adapted for the data acquired by the Ocean Land Color Imager (OLCI) sensor onboard the Sentinel-3 satellite platform. The current *Sargassum* reflectance indices use the top of atmosphere reflectance corrected for the Rayleigh scattering as inputs. This is because there is currently no reliable atmospheric correction that can be applied to derive the above water reflectance for the specific case of the presence of floating algae [9]. Although the aerosol correction may have a weak influence on the Algal Indices (AFAI and MCI) [6], which are based on spectral reflectance differences, such a correction remains necessary for applications that deal with the above water reflectance values, such as the derivation of *Sargassum* fractional coverage and the consideration of the immersion depth of the *Sargassum* aggregations. The performance of the atmospheric correction procedure is therefore of critical importance for providing relevant products related to *Sargassum* properties.

While the Rayleigh (i.e., molecular) atmospheric scattering component is well known in theory, the correction of top-of-atmosphere reflectance for the aerosol reflectance remains a challenging task because their optical properties cannot locally be predicted. Standard atmospheric correction methods over open ocean waters assume the water reflectance to be small in the near infrared (NIR) domain [10–12]. If the water reflectance is close to zero for such oceanic water, the top-of-atmosphere reflectance is mainly ascribed to the aerosol reflectance in addition to the Rayleigh scattering component. The derivation of the aerosol reflectance is carried out in the NIR domain based on the comparison between satellite measurements and reflectances simulated using various models of aerosol optical properties. Then, the aerosol models are used to estimate the aerosol reflectance in the visible domain. In the presence of *Sargassum*, such a common procedure cannot apply because *Sargassum* induce a strong water reflectance in the near infrared bands due to their optical signature in the red-edge part of the spectrum. Therefore, the NIR aerosol reflectance could be overestimated regarding pixels containing *Sargassum* (called "*Sargassum* pixels"), as compared with *Sargassum*-free areas, when using standard atmospheric correction procedures [13]. Other methods such as ACOLITE [14] can be considered to correct aerosol reflectance in the presence of algal blooms. In this paper, an extension of a given standard atmospheric correction method (POLYMER [15]) is proposed to derive the water reflectance from satellite data over oceanic waters containing *Sargassum*. The POLYMER method has the advantage of taking into account sunglint as well as absorbing inhomogeneous aerosols. The standard POLYMER algorithm has been successfully applied in the past to various satellite sensors, including ENVISAT/MERIS, Sentinel-2/MSI and Sentinel-3/OLCI data for *Sargassum*-free waters [15,16].

This paper is organized as follows: the study area, the Sentinel-3/OLCI satellite data and the methodology used to adapt the standard atmospheric correction in the presence of *Sargassum* are outlined in Section 2. The evaluation of the performance of the *Sargassum*-dominated area detection and the benefits of the proposed atmospheric correction procedure are presented in Section 3. Finally, the consistency of the proposed atmospheric correction methodology is discussed in Section 4.

2. Materials and Methods

2.1. Study Area

The study area is located in the Lesser Antilles Islands (Figure 1), off which an increasing amount of *Sargassum* aggregation has been observed in the Caribbean Sea since 2010.

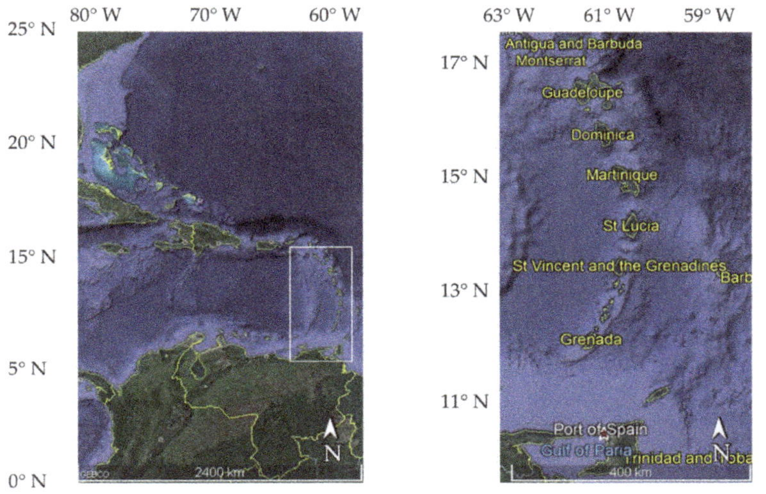

(a) (b)

Figure 1. (**a**) Area of interest for the study: the Caribbean Sea; (**b**) Focus on the lesser Antilles Islands.

2.2. Satellite Data

The Ocean Land Color Instrument's (OLCI) satellite sensor, which is onboard the Sentinel-3 satellite platform, provides images at a spatial resolution of 300 m for 21 spectral bands from the visible to near infrared, namely from 400 to 1020 nm. OLCI is highly appropriate for providing reliable ocean color observations thanks to its high radiometric sensitivity. Two satellite platforms were launched in 2016 and 2018, respectively, providing a 1-day temporal resolution [17].

Five Sentinel-3/OLCI images were analyzed for this study; they were acquired on 8 July 2017 13:00, 27 June 2018 14:24, 9 May 2020 14:00, 14 September 2020 14:20 and 28 December 2020 13:55. These images were selected because they exhibit numerous occurrences of *Sargassum* and were captured during different years and seasons. The Level-1 data product, namely the Top Of Atmosphere radiance, was downloaded from the Copernicus website [18]. Figure 2 shows the Red-Green-Blue (RGB) composite of the processed Sentinel-3/OLCI image acquired on 8 July 2017 13:00.

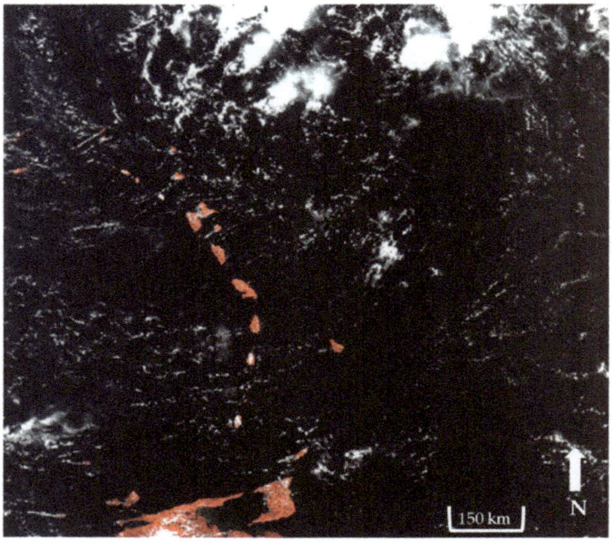

Figure 2. NIR-G-B composite image of the Sentinel-3/OLCI satellite data acquired on 8 July 2017 at 13:00.

2.3. *In Situ Data*

In situ measurements are important to correctly interpret satellite observations, especially to validate the proposed adjustment of the atmospheric correction procedure in the presence of *Sargassum*. Although in situ data were not collected for the current study, Ody et al. [18] were able to gather in situ data of *Sargassum* aggregations through two cruise campaigns in the western Atlantic Ocean between Brazil and the Caribbean Sea in the summer and autumn of 2017, which is the same period as the satellite image processed here. The above water reflectances were collected by Ody et al. using deck observations of *Sargassum* aggregations. Water reflectances were recorded for many *Sargassum* aggregations associated with their abundance (fractional coverage at the metric scale [12]). Although Ody et al.'s data cannot be rigorously used as a validation dataset in the current study, they will be used in Section 4 to discuss the benefits and the consistency of the proposed atmospheric correction method to correctly provide water reflectance in the presence of *Sargassum*.

2.4. *Methodology*

2.4.1. Standard Procedure for Atmospheric Correction over Oceanic Waters

The main steps of the atmospheric correction procedure that can typically be used above oceanic waters are reiterated here. The Level-1 satellite data consist of top-of-atmosphere radiance (L_{TOA}) that first need to be normalized by the cosine of the solar zenithal angle (θs) and the extraterrestrial solar irradiance (Es) to provide top-of-atmosphere reflectance (Equation (1)).

$$\rho_{\text{TOA}}(\lambda) = \frac{\pi L_{\text{TOA}}(\lambda)}{cos(\theta s) Es(\lambda)} \tag{1}$$

ρ_{TOA} can be decomposed as a sum of various terms (Equation (2)):

$$\rho_{\text{TOA}} = t_{\text{oz}} t_{\text{NO2}} \left(\rho_{\text{mol}} + T\rho_{\text{gli}} + \rho_{\text{aer}} + \rho_{\text{coupl}} + t\rho_{\text{w}}^{+} \right) \tag{2}$$

where t_{oz} is the transmittance of the ozone and t_{NO2} the transmittance of the nitrogen dioxide, ρ_{mol} is the Rayleigh molecular scattering, ρ_{gli} is the sunglint reflectance weighted by the direct transmittance T, ρ_{aer} is the reflectance of the aerosols, ρ_{coupl} accounts for the various coupling terms between the sunglint, the molecules and the aerosols, t is the total (i.e., direct and diffuse) transmittance for atmospheric scattering and ρ_w^+ is the reflectance just above the air-water interface (i.e., water reflectance).

Some of the atmospheric contributions can be accurately predicted using either ancillary data or theoretical calculations. Those contributions consist of the ozone transmittance $t_{oz}(\lambda)$, which is determined based on the total ozone concentration as provided by ECMWF data (European Centre for Medium-Range Weather Forecasts), and of the Rayleigh molecular scattering reflectance based on the atmospheric pressure at sea level. These inputs are stored in look-up tables to save computational time. The correction for the sunglint is performed using the ECMWF wind speed data that is derived from the sea surface roughness model by Cox and Munk (1954, [19]). The top-of-atmosphere reflectance can then be corrected by the ozone transmittance, the Rayleigh scattering effects and the sunglint reflectance. The resulting reflectance is hereafter referred to as ρ'. The aerosol reflectance ρ_{aer}, the coupling term ρ_{coupl}, and the residue of sunglint reflectance that has not been totally corrected by Cox and Munk models are gathered under the notation ρ_{ag}. Thus, the reflectance ρ' can be written based on Equation (2) as follows (Equation (3)):

$$\rho'(\lambda) = \rho_{ag}(\lambda) + t(\lambda)\rho_w^+(\lambda) \tag{3}$$

If the water reflectance spectrum is assumed to be known, then ρ_{ag} could be derived through the term $\rho' - t\rho_w^+$. The basic principle of the commonly used POLYMER algorithm [15] relies on the modeling of the atmosphere reflectance and the residual sunglint reflectances, namely the term ρ_{ag}, as a polynomial function of λ and ρ_{mol} to best represent the atmospheric signal even at high latitude (Equation (4)):

$$\rho_{ag}(\lambda) \approx t_0\Big(\lambda\Big)c_0 + c_1\lambda^{-1} + c_2\rho_{mol} \tag{4}$$

where the coefficients c_0, c_1 and c_2 are estimated by the least square fitting of the satellite measurements and t_0 is the total transmittance due to the Rayleigh scattering. The water reflectance ρ_w^+ in the visible domain can then be derived as (Equation (5)).

$$\rho_w^+(\lambda) = \frac{\rho'(\lambda) - \rho_{ag}(\lambda)}{t(\lambda)} \tag{5}$$

2.4.2. Extension of the POLYMER Algorithm to Sargassum-Dominated Waters

The spectral reflectance of pure *Sargassum* (Figure 3) is highly similar to land vegetation reflectance. The main feature is observed in the NIR bands where the reflectance strongly increases. A lower increase of the reflectance is observed in the green. Absorption properties of *Sargassum* can be observed through the decrease of the reflectance in the blue and the red bands. Although the proportion of *Sargassum* content in a given pixel is weak in the open ocean, typically lower than 5% as revealed from MODIS satellite data analysis using the monthly mean of the AFAI index [7], water reflectance remains high, especially in the NIR domain. As highlighted in Section 2.4.1, the aerosol reflectance is commonly derived from the reflectance measured in the NIR domain where the water reflectance is supposed to be negligible. Therefore, the influence of *Sargassum* on the standard atmospheric correction procedure could lead to overestimating the term ρ_{ag} in the expression of ρ' (Equation (3)) because *Sargassum* optical signature is wrongly ascribed to an aerosol signal rather than an oceanic signal in the NIR part of the spectrum.

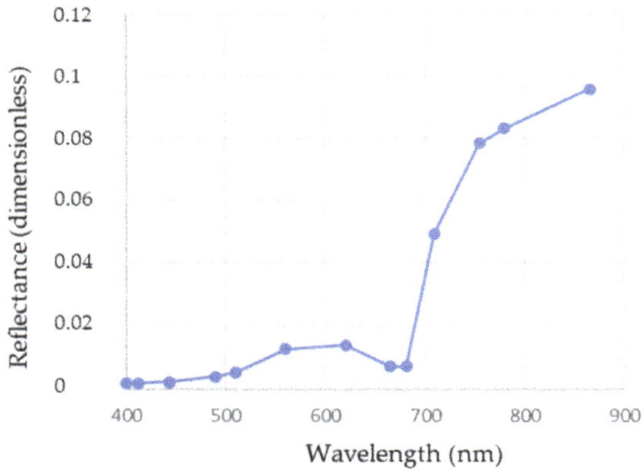

Figure 3. Spectral reflectance of pure *Sargassum* [20].

The spectral variation of ρ', ρ_{ag} and ρ_w^+ are shown in Figure 4a for a pixel containing only water (i.e., *Sargassum*-free pixel). The spectral decrease of ρ_{ag} is consistent with the typical spectral decrease of aerosol reflectance with a wavelength in the red and near infrared spectral bands [10,21]. In the case of *Sargassum*-dominated water (Figure 4b), ρ_{ag} exhibits a minimum value at 500 nm and increases from 500 nm to 800 nm, which is neither consistent with the usual spectral variation of aerosol reflectance, which typically varies as a decreasing power law, nor with the sunglint reflectance, which is fairly spectrally flat. Only a few types of aerosols, such as desert dust absorbing aerosols, might exhibit a steep increase in the reflectance with a wavelength from the blue to red/NIR range [22]. In any case, a minimum value of the aerosol spectral reflectance (here around 500 nm), as observed in Figure 4b (orange line), is not realistic from a physical point of view. The aerosol and sunglint effects are then theoretically overestimated for the case of *Sargassum*-dominated waters. The derived water reflectance ρ_w^+ thus shows negative values (Figure 4b), which is not realistic from a physical point of view. Consequently, *Sargassum*-dominated waters require a specifically dedicated atmospheric correction procedure.

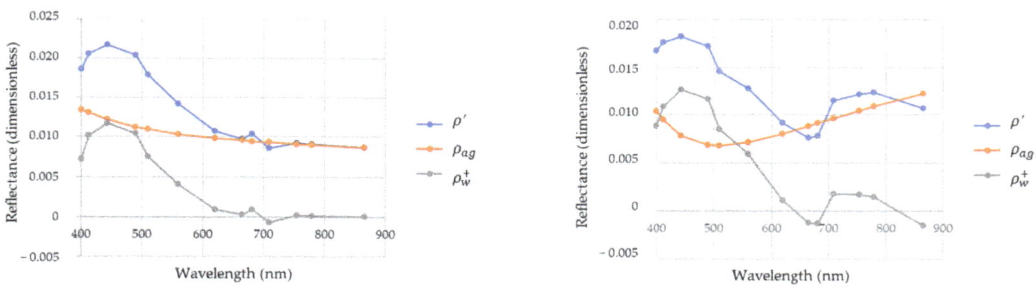

(**a**) (**b**)

Figure 4. Reflectance corrected for Rayleigh molecular scattering (ρ'), reflectance of aerosol and sunglint (ρ_{ag}) and above water reflectance: (**a**) for a *Sargassum*-free area; (**b**) for a *Sargassum*-dominated area.

Since the determination of aerosol and sunglint reflectance (ρ_{ag}) is challenging over *Sargassum*-dominated areas, we propose an extension of the POLYMER atmospheric correction procedure, which will hereafter be named POLYMER$_{ext}$. This method relies on the idea of exploiting the value of ρ_{ag}, which can be estimated over *Sargassum*-free waters using the standard method. The ρ_{ag} reflectance over *Sargassum*-dominated areas is estimated by interpolating a ρ_{ag} that has been derived using *Sargassum*-free waters located in the vicinity of the *Sargassum* waters. Then, the interpolated value of the aerosol and sunglint reflectance (ρ_{ag}) is subtracted from the top-of-atmosphere reflectance measured over the *Sargassum*-dominated areas. The underlying assumption of such a method is the consideration of a homogeneous spatial distribution of aerosols over the entire area (i.e., *Sargassum*-dominated area and its *Sargassum*-free neighborhood). To proceed in such a way, the *Sargassum*-dominated areas need to be first identified (i.e., flagged) and their inconsistent ρ_{ag} values need to be removed. The method used to identify *Sargassum* dominated areas (or pixels), which is a necessary preliminary step prior to interpolating the aerosol-glint reflectance ρ_{ag}, is described in the next section (Section 2.4.3). The method used to interpolate the *Sargassum*-free ρ_{ag} for *Sargassum*-dominated areas is the MatLab function called *"fillmissing"*, which uses a linear interpolation between the closest ρ_{ag} values that have been estimated for *Sargassum*-free areas. It should be noted that, even though the satellite data used in this study does not contain significant desert dust contamination, the applicability of the proposed method is mainly driven by the validity of the POLYMER method over *Sargassum*-free waters and thus is applicable to the case of the occurrence of moderate desert dust aerosols, especially since the red to NIR bands are not significantly affected by the aerosol absorption.

2.4.3. Implementation of a Flag Dedicated to the Identification of Sargassum Pixels

Clouds could lead to overestimation of the number of pixels containing *Sargassum* if they are wrongly identified, so cloud-contaminated pixels should be discarded from the analysis. The first step is then to flag pixels contaminated by clouds. The clouds are filtered using the method developed by Nordkvist et al. [23] for MODIS data because this method was found to be performant, including for the case of thin clouds. They demonstrated that pixels free of cloud, including thin cloud, can be identified using the Rayleigh-corrected top-of-atmosphere reflectance acquired at bands 748 nm and 869 nm (respectively, λ_i and λ_j). For the purpose of the current study, which is based on Sentinel-3/OLCI data, ρ' (Equation (3)) is used to identify the cloud-free pixels, including for thin cloud conditions. It should be noted that the proposed method does not detect the cloud shadows, but it should be highlighted that cloud shadows could be eliminated using an infrared sensor such as SLSTR following the method described by Fernandez-Moran et al. [24]. The criterion and threshold values used by Nordvist et al. to distinguish cloud-free pixels were revisited here to match with OLCI sensor spectral specifications as follows (Equation (6)):

$$\left[\frac{\rho'(865)}{t(865)} < 0.0045\ OR\ \frac{\frac{\rho'(865)}{t(865)}}{\frac{\rho'(754)}{t(754)}} < 1.01\right]\ AND\ \left[\frac{\rho'(865)}{t(865)} < 0.06\right] \quad (6)$$

Two methods are proposed to identify and detect the occurrence of *Sargassum* in this study. The first method, called "reflectance thresholding", is based on the fact that pixels containing *Sargassum* show an increase of the Rayleigh-corrected reflectance ρ' between the red and the NIR band (Figure 3). The criterion that has been established is based on the reflectance ρ' at bands 665 nm, 681 nm, 754 nm and 779 nm (Equation (7)).

$$\max\left[\rho'(665), \rho'(681)\right] < \max\left[\rho'(754), \rho'(779)\right] \quad (7)$$

The occurrence of *Sargassum* can also be detected using algal index thresholding. The MCI [8] is here used for the Rayleigh-corrected reflectance as follows (Equation (8)):

$$MCI = \rho'(709) - \left[\rho'(681) + \left(\rho'(754) - \rho'(681)\right) \times \frac{709 - 681}{754 - 681}\right] \quad (8)$$

The index value corresponding to *Sargassum* free waters, which is further called MCI background, is subtracted from the MCI to derive the MCI deviation $\delta MCI = MCI -$

MCI_{bg}. The MCI background is determined using a median filter whose purpose is to reduce noise on the MCI. It should be noted that such a median filter is relevant when the large majority of pixels defining the filter are *Sargassum*-free pixels. Our tests showed that a filtering window of 167 × 167 pixels (50 km × 50 km) is satisfactory for fulfilling such a condition. The histogram of δMCI shows two modes (see Section 3.1.2): one can be identified as the *Sargassum*-free pixels and another one corresponding to higher δMCI consists of the *Sargassum*-dominated pixels. A threshold value of δMCI can be derived in between the two modes to detect a *Sargassum*-dominated areas from *Sargassum*-free areas.

2.4.4. Flowchart of the Atmospheric Correction Algorithm Proposed for Sargassum Dominated Waters

The flowchart of the methodology proposed for adjusting the POLYMER standard atmospheric correction for *Sargassum*-dominated areas ($POLYMER_{ext}$) is shown in Figure 5. The Rayleigh molecular scattering correction of the top-of-atmosphere Level-1 reflectance is performed similarly over *Sargassum*-free and *Sargassum*-dominated areas. Thin clouds pixels are identified and masked. Then, the pixels containing *Sargassum* are identified and flagged. The aerosol and sunglint reflectances for the flagged *Sargassum*-dominated pixels are estimated through the interpolation of the derived aerosol and sunglint reflectances, as described in Section 2.4.2.

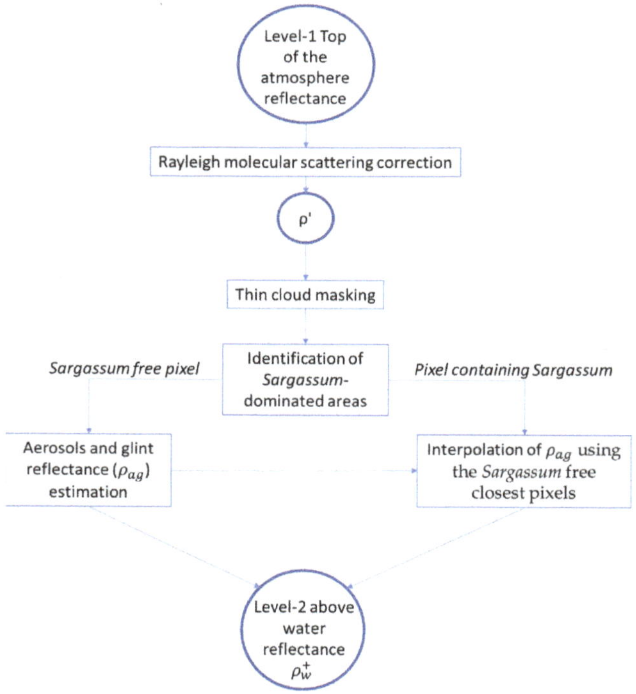

Figure 5. Flowchart of the $POLYMER_{ext}$ procedure to correct the top-of-atmosphere reflectance over *Sargassum*-dominated areas.

3. Results

3.1. Evaluation of the Performance of the Thin Cloud Masking and Sargassum Identification

3.1.1. Cloud Masking

Lands pixels are provided by the POLYMER algorithm. The thin cloud mask obtained using the method described in Section 2.4.3 is then used to eliminate cloud-contaminated pixels. The thin cloud mask applied to the OLCI reflectance data can be seen in Figure 6b and can be compared with the NIR-G-B composite data acquired by Sentinel-3/OLCI on 8 July 2017 at 13:00 (Figure 6a). Pixels containing land, water (including *Sargassum*) and cloud are colored in green, blue and white, respectively.

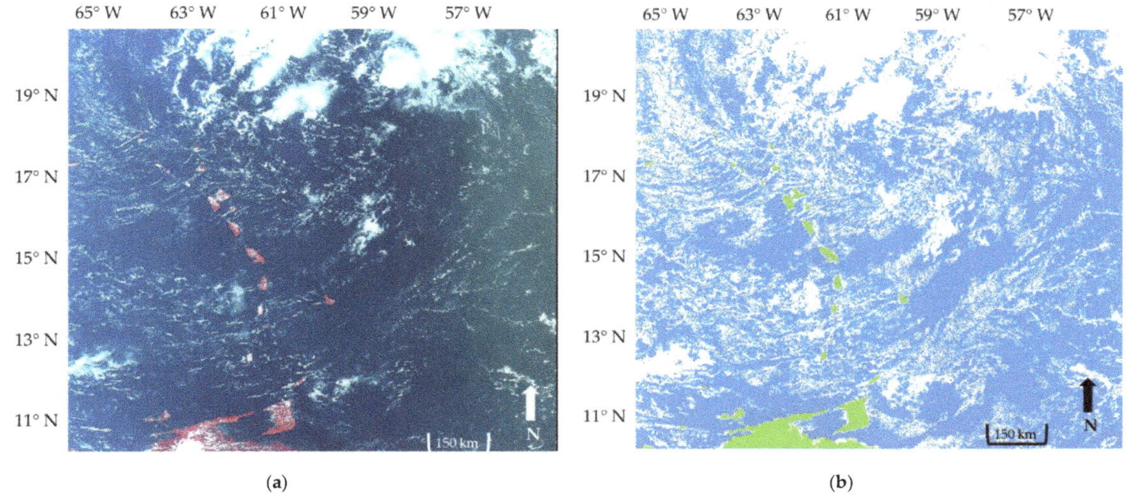

Figure 6. Data acquired by Sentinel-3/OLCI on 8 July 2017 at 13:00 (**a**) Level-1 data in NIR-G-B composite; (**b**) Water-land-cloud mask. Pixels containing land are in green, cloud-free pixels are in blue and cloud pixels are in white.

Figure 6 shows that the contribution of thin clouds, which are not observable on the NIR-G-B image (Figure 6a), is significant over the study area (Figure 6b).

3.1.2. Identification of Sargassum Areas

The identification of *Sargassum*-dominated areas is performed using a threshold value on the δMCI. For that purpose, the histograms of δMCI values over entire Sentinel-3/OLCI scenes were analyzed. For each scene, its histogram is represented in Figure 7. Those histograms have the same first significant peak. Three modes can be observed. The highest peak that is observed for the lowest values of δMCI is mostly related to *Sargassum*-free pixels, which represent a large majority of pixels across the entire scene. The second and third modes that are observed for the highest values of δMCI are mostly linked to the occurrence of *Sargassum* within the pixels. The difference between the two peaks could be due to the aggregation density. A threshold value could thus be determined between those two modes to distinguish pixels containing *Sargassum* from *Sargassum*-free pixels. The threshold-adopted value is the value of the threshold observed between the two modes, which has been decreased by 30% to make sure not to omit any *Sargassum*-dominated pixels (Figure 7). The threshold value of 0.002 is then determined (Figure 7). Such a threshold value has been estimated based on the histogram obtained for the image acquired on 8 July 2017 at 13:00 (in black in Figure 7). However, it was found to be a relevant value for several other acquisitions, namely Sentinel-3/OLCI data of 27 May 2018 14:24, 9 May 2020 14:00, 14 September 2020 14:20 and 28 December 2020 13:55.

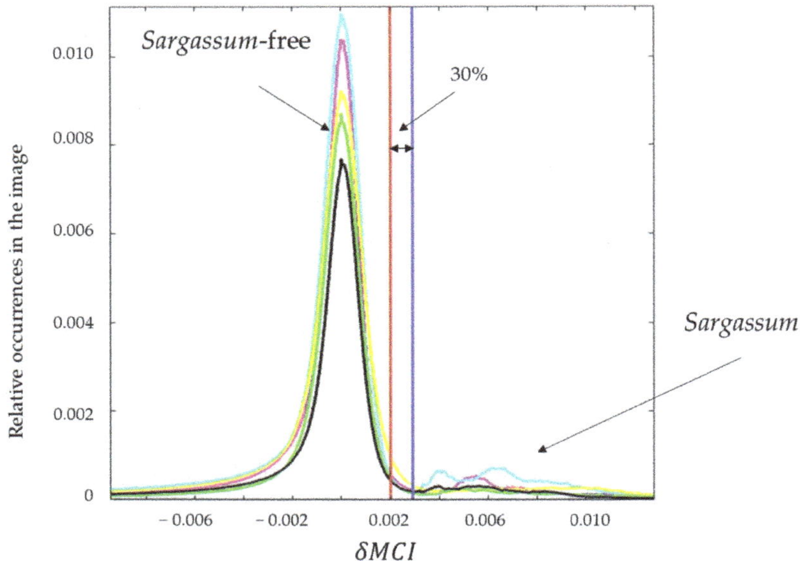

Figure 7. Histograms (number of occurrences) of δMCI for each of the Sentinel-3/OLCI scenes. The histograms are represented for 8 July 2017 13:00 (black line), 27 May 2018 14:24 (yellow line), 9 May 2020 14:00 (green line), 14 September 2020 14:20 (magenta line) and 28 December 2020 13:55 (cyan line). The red vertical line shows the threshold value of 0.002 that has been adopted for this study. Such a value is 30% lower than the minimum value of δMCI observed between the two modes (blue vertical line).

The comparison between the two approaches that have been proposed in Section 2.4.3 to detect the presence of *Sargassum* within the pixels, namely the reflectance thresholding approach and the δMCI thresholding approach, is shown in Figure 8.

Sargassum aggregation edges are better detected in the whole scene when using the reflectance thresholding approach (red color in Figure 8b). However, some aggregations are largely underestimated on the right side of the image relative to the δMCI thresholding approach, as observed in Figure 8c. The method using the δMCI thresholding approach detects the main aggregations but leaves out many pixels on the edge of the *Sargassum* aggregation. In addition, a threshold value is likely to change depending on the image processed, thus leading to an overestimation or an underestimation of the *Sargassum*-dominated pixels. On the one hand, omitting *Sargassum*-dominated pixels can be an issue for the interpolation of the aerosol and sunglint reflectance performed during the atmospheric correction procedure. On the other hand, an overestimation of *Sargassum*-dominated pixels does not prevent the proposed atmospheric correction method. Then, a hybrid approach has been considered using both approaches to increase the performance of *Sargassum* detection. The *Sargassum* flags obtained using both the reflectance and δMCI thresholding methods are merged to optimize the number of *Sargassum* aggregations detected over the entire scene.

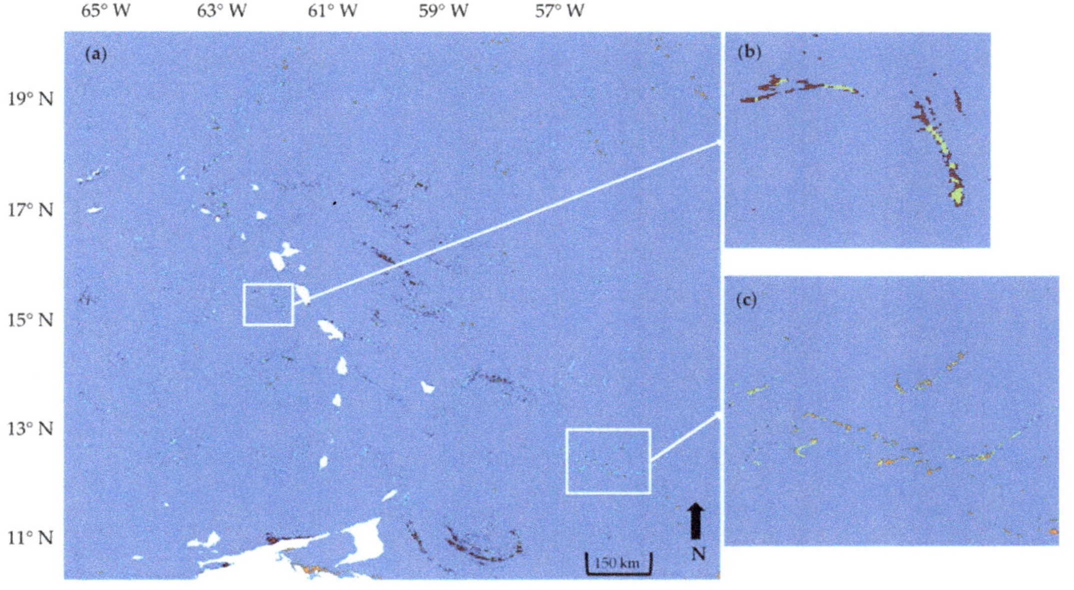

Figure 8. *Sargassum* identification using the two approaches proposed in this study (i.e., reflectance thresholding approach and δMCI thresholding approach) for the 8 July 2017 13:00 scene: (**a**) for the OLCI entire scene, (**b**) for a sub-area located westward of the Dominica Island and (**c**) for a sub-area located in the southeast of the scene. The results for each approach are presented using a different color: red color is used for the reflectance thresholding method and orange color is used for the δMCI thresholding approach. If both approaches agree, pixels are colored in green. Land pixels are colored in white while *Sargassum*-free pixels (i.e., water only) are colored in blue. Clouds are also colored in blue to enhance the visibility of *Sargassum* areas.

3.2. Benefit of the Proposed Atmospheric Correction Procedure Relative to a Standard Procedure

The standard atmospheric correction procedure (POLYMER) is compared to the proposed method ($POLYMER_{ext}$) to evaluate the improvement made for investigating the detection of *Sargassum* in oceanic waters (Figure 9).

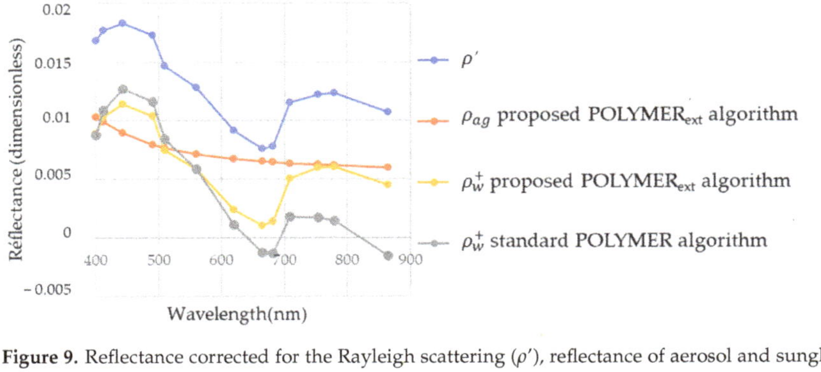

Figure 9. Reflectance corrected for the Rayleigh scattering (ρ'), reflectance of aerosol and sunglint (ρ_{ag}) and reflectance of above surface water for a pixel containing *Sargassum* derived for the case where (i) the standard POLYMER atmospheric correction procedure is used and (ii) the proposed $POLYMER_{ext}$ is used.

The above water reflectances (ρ_w^+) derived from both the standard and adapted POLYMER atmospheric correction algorithm are fairly similar in the spectral range 412 nm–620 nm. Beyond 620 nm, significant discrepancies are observed between the two methods. The standard method provides negative values of ρ_w^+ (i.e., <-0.001) from 650 nm that are likely to be due to the overestimation of the aerosol and sunglint reflectance (ρ_{ag}) by the standard POLYMER algorithm as a result of the occurrence of *Sargassum*. The interpolation of ρ_{ag} derived for *Sargassum*-free pixels and applied to *Sargassum* pixels, as proposed by the POLYMER$_{ext}$ method, leads to a consistent spectral decrease of ρ_{ag} over *Sargassum* areas (Figure 9). In addition, ρ_w^+ values remain positive and consistent with the spectral signature of *Sargassum* (i.e., increase of ρ_w^+ in the red-edge part of the spectrum, Figure 3).

The analysis of the entire scene reveals that approximately 85% of the negative values of ρ_w^+ between 620 and 681 nm derived by the standard approach have been corrected using the POLYMER$_{ext}$ algorithm. Similarly, the values of ρ_{ag} that showed a spectral increase with wavelength in the red/NIR bands when using the standard method have been replaced by spectrally consistent ρ_{ag} values derived by the adapted approach. Figure 10 compares $\rho_{ag}(754)/\rho_{ag}(681)$ for *Sargassum*-dominated areas using POLYMER (Figure 10a) and POLYMER$_{ext}$ algorithms (Figure 10b). The values are mostly higher than 1 when using POLYMER, which means an increase of ρ_{ag} in the NIR domain (Figure 10a). The proposed algorithm POLYMER$_{ext}$ is able to correct for those values, which become lower than 1 (Figure 10b). Several artefacts remain visible on the far east of the scene; they can be attributed to the strong impact of the glint in this area.

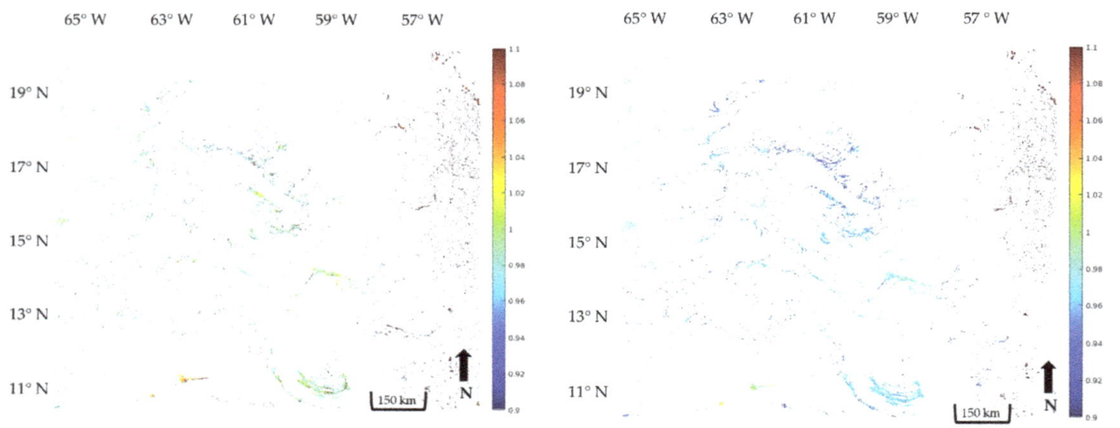

(a) **(b)**

Figure 10. Spectral ratio of the aerosol and sunglint reflectance $\rho_{ag}(754)/\rho_{ag}(681)$ over *Sargassum*-dominated areas in the scene acquired on 8 July 2017 13:00: (**a**) using POLYMER algorithm; (**b**) using POLYMER$_{ext}$ algorithm.

Therefore, the above water Level-2 reflectance derived for the adapted method is now optically relevant for improving the assessment of the *Sargassum* biomass using algal index.

Table 1 shows the percentage of negative reflectance values that has been corrected using POLYMER$_{ext}$ on *Sargassum*-dominated areas for the five scenes presented in Section 2.2. It is observed that more than 75% of the negative reflectance (between 620 and 681 nm) were corrected using POLYMER$_{ext}$ for *Sargassum*-dominated areas. It should be noted that the cloud coverage is more significant (about 50%) on 27 May 2018 14:24 and on 14 September 2020 14:20, which could explain the lower efficiency of the algorithm for these two scenes.

Table 1. Percentage of negative reflectance values corrected between (620 and 681 nm) by using $POLYMER_{ext}$ on *Sargassum*-dominated areas.

Date of the Scene	Percentage of Negative Reflectance Values Corrected between (620 and 681 nm)
8 July 2017 13:00	83%
27 May 2018 14:24	76%
9 May 2020 14:00	88%
14 September 2020 14:20	75%
28 December 2020 13:55	79%

4. Discussion

The extension of the POLYMER standard algorithm proposed in this study to perform atmospheric correction over *Sargassum*-dominated areas is important to enable the satellite data processing and monitoring of *Sargassum* variation with time and space from remote sensing techniques, especially when the above water reflectances are used rather than spectral reflectance differences. The $POLYMER_{ext}$ procedure relies on the assumption that the aerosol and sunglint reflectance ρ_{ag} is spatially homogeneous over the *Sargassum* areas relative to adjacent *Sargassum*-free areas. The relevance of such an assumption has been investigated. For that purpose, the standard atmospheric correction algorithm has been applied over a *Sargassum*-free area to provide ρ_{ag} values (Figure 11a). Then, a rectangular sub-area within this area was considered as a test area. Since the area is *Sargassum*-free waters, ρ_{ag} inside such a test area can be considered as the reference ρ_{ag}, hereafter referred to as ρ_{ag_ref}. Then, in a second step, ρ_{ag} over the test area was estimated as it could not be correctly retrieved using the standard POLYMER algorithm. Thus, the $POLYMER_{ext}$ method was used and ρ_{ag} from the surrounds of the test area was linearly interpolated in the test area, hereafter referred to as ρ_{ag_test}. The difference between ρ_{ag_test} values and ρ_{ag_ref} values are calculated to provide the Relative Root Mean Square Error (noted RRMSE). The relevance of the assumption made on the spatial homogeneity of ρ_{ag} using these errors could be determined. Practically, an area was selected within the Sentinel-3/OLCI image acquired on 8 July 2017 at 13:00 without pixels containing clouds, without *Sargassum* and without anomalous spectral variation of ρ_{ag}. Then, a 300 × 80 pixel rectangle (90 km × 24 km) was defined inside this area. It should be noted that the size of a 300 × 80 pixel rectangle is larger than the typical size of *Sargassum* aggregation [20]. The reflectance ρ_{ag} is interpolated for such 300 × 80 pixel area. A transect of the area for the 200th line is represented in Figure 11b, which shows ρ_{ag}, ρ_{ag_test} and ρ_{ag_ref} at 560 nm for the study area. The mean relative RRMSE error between the interpolated ρ_{ag} values and the reference ρ_{ag} values (i.e., *Sargassum*-free areas) varies between 0.09% and 11% inside the rectangle area. Such a low discrepancy confirms that the assumption made on the spatial local homogeneity of the aerosol is reliable.

The fill missing data method used could be improved because the interpolation is only made along lines and uses only two available values for the interpolation that can amplify uncertainty. An improved version of our approach might be to use the mean x and y of the first few pixels before the gap and the first few pixels after the gap for the interpolation, to do this both horizontally and vertically. Then, the shortest or nearest or the mean of the two interpolations weighted by gap length or distance is selected.

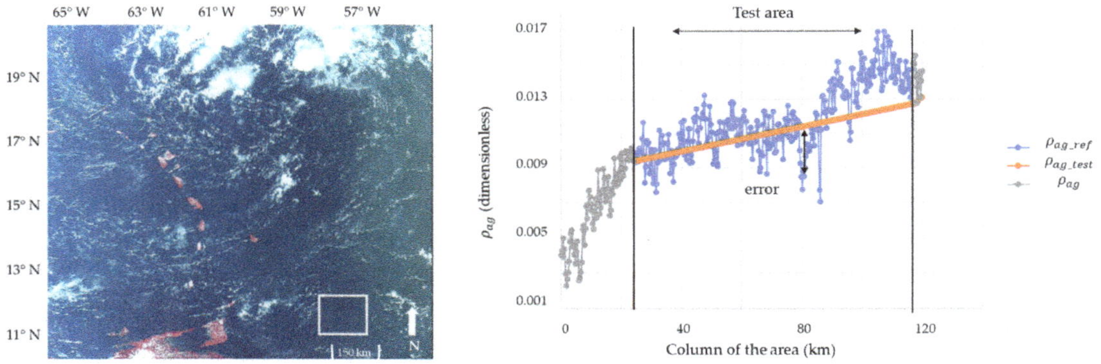

(**a**) (**b**)

Figure 11. (**a**) Location of the test area (white rectangle); (**b**) Transect (200th line) of the *Sargassum*-free area for $\rho_{ag\_test}(560\text{ nm})$ values and $\rho_{ag\_ref}(560\text{ nm})$ values.

The measurement of in situ data within *Sargassum*-dominated waters at the same date and location as a satellite data, called "match-up" data, is a challenging task because of the various scales of observation, typically a 300 × 300 m^2 area covered by the satellite observation versus a few meters covered by ship observations. However, the above water reflectance derived from the adapted atmospheric correction method was compared with the reflectance spectra measured during previous field experiments that occurred in 2017 in the western Atlantic Ocean (i.e., same study areas and time period as the current study) [18]. The typical water reflectances measured over *Sargassum* aggregation areas are shown in Figure 12a. Each reflectance spectrum corresponds to an abundance of *Sargassum* (in %) determined based on water sample analysis. Even though the in situ data cannot be used as a rigorous validation of the proposed atmospheric correction method, they are helpful for analyzing the consistency of the estimated above water reflectance. Figure 12b shows (i) two in situ measurements, (ii) the proposed POLYMER$_{ext}$ algorithm and (iii) the standard POLYMER algorithm for a pixel containing *Sargassum*. Such a figure highlights the inconsistency of the standard atmospheric correction procedure because of the negative reflectance values between 681 nm and 709 nm. The water reflectance derived from the POLYMER$_{ext}$ atmospheric correction approach shows more similarities with the in situ reflectance measurements than the water reflectance derived from the standard algorithm, especially for the case of 5% *Sargassum* abundance in the spectral range 681 nm–754 nm. Such a comparison with the in situ data thus corroborates the relevance of the adjustments proposed in this study for the atmospheric correction procedure. It should also be highlighted that in situ sampling carried out across the Atlantic Ocean in previous studies did not reveal evidence of the occurrence of any other invasive species competitive with *Sargassum* [25–27]. Thus, although it might be theoretically possible, it is unlikely that other species could lead to the same red/NIR spectral features as *Sargassum* over our study area based on Ody et al.'s [20] in situ observations and other studies [25–27].

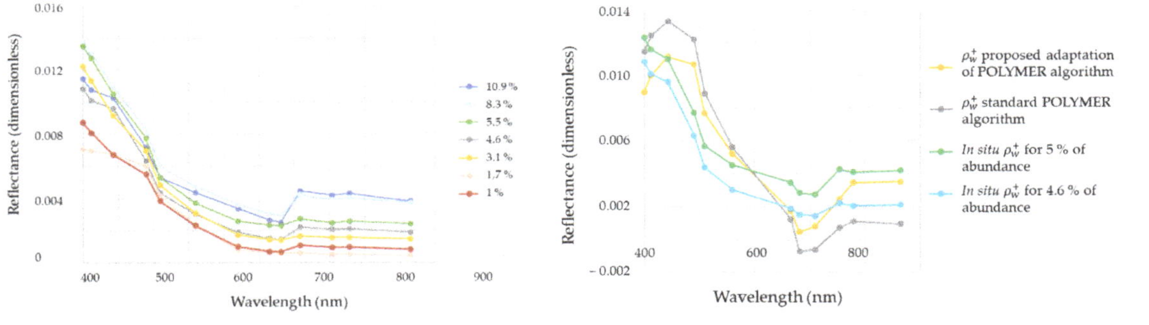

(**a**) (**b**)

Figure 12. (**a**) Above water reflectance (ρ_w^+) measured in situ in *Sargassum*-dominated areas in the Atlantic Ocean [20]. The different spectra correspond to different *Sargassum* abundances (%); (**b**) comparison of the spectrum water reflectance measured in situ when the abundance is 4.6, 4.8 and 5% with the reflectance derived from the standard and adapted atmospheric correction procedure.

5. Conclusions

In this study, an adaptation of a standard atmospheric correction procedure, namely POLYMER, is proposed to correct the surface reflectance of pixels containing *Sargassum*. This method requires, first, a mask of thin clouds over the entire scene; the POLYMER standard procedure is only applied to the *Sargassum*-free area. Then, the aerosol reflectance over *Sargassum*-dominated areas is interpolated based on its value determined over the *Sargassum*-free area located in the vicinity of *Sargassum* waters. The proposed method assumes the spatial homogeneity of the aerosol reflectance, which has been shown as a relevant assumption. The proposed $\text{POLYMER}_{\text{ext}}$ procedure was successfully applied to Sentinel-3/OLCI data, for which the inconsistencies in the spectral variation of the aerosol reflectance (i.e., increase from visible to NIR bands) vanish. A reduction of between 75% and 88% of the pixels showing a negative water reflectance above 600 nm were noticed after the correction of the five images. The efficiency of the correction is fairly related to the cloud coverage. The more clouds, the less effective is the method. The comparison between the in situ data and the water reflectance derived over *Sargassum* areas based on the improved atmospheric correction corroborates the relevance of the proposed method. Therefore, relevant estimates of *Sargassum* variables such as biomass and fractional coverage could be further derived based on the methodology developed in this study. This method can also be applied to other satellite sensors, such as Aqua/MODIS, Terra/MODIS, SUOMI-NPP/VIIRS and Sentinel-2/MSI.

Author Contributions: Conceptualization, L.S., A.M. and M.C.; data curation, L.S.; formal analysis, M.C.; funding acquisition, A.M.; methodology, L.S., A.M. and M.C.; software, L.S.; supervision, A.M.; writing—original draft, L.S.; writing—review & editing, A.M., M.C. and F.S. All authors have read and agreed to the published version of the manuscript.

Funding: This research was funded by l'Agence Nationale de la Recherche (France), grant number ANR-19-SARG-0007-07.

Institutional Review Board Statement: Not applicable.

Informed Consent Statement: Not applicable.

Data Availability Statement: Not applicable.

Conflicts of Interest: The authors declare no conflict of interest.

References

1. Louime, C.; Fortune, J.; Gervais, G. Sargassum Invasion of Coastal Environments: A Growing Concern. *Am. J. Environ. Sci.* **2017**, *13*, 58–64. [CrossRef]
2. Jouanno, J.; Moquet, J.-S.; Berline, L.; Radenac, M.-H.; Santini, W.; Changeux, T.; Thibaut, T.; Podlejski, W.; Ménard, F.; Martinez, J.-M.; et al. Evolution of the Riverine Nutrient Export to the Tropical Atlantic over the Last 15 Years: Is There a Link with Sargassum Proliferation? *Environ. Res. Lett.* **2021**, *16*, 034042. [CrossRef]
3. Djakouré, S.; Araujo, M.; Hounsou-Gbo, A.; Noriega, C.; Bourlès, B. On the Potential Causes of the Recent Pelagic Sargassum Blooms Events in the Tropical North Atlantic Ocean. *Biogeosci. Discuss.* **2017**, 1–20. [CrossRef]
4. Marx, U.C.; Roles, J.; Hankamer, B. Sargassum Blooms in the Atlantic Ocean–From a Burden to an Asset. *Algal Res.* **2021**, *54*, 102188. [CrossRef]
5. Maréchal, J.-P.; Hellio, C.; Hu, C. A Simple, Fast, and Reliable Method to Predict Sargassum Washing Ashore in the Lesser Antilles. *Remote Sens. Appl. Soc. Environ.* **2017**, *5*, 54–63. [CrossRef]
6. Hu, C. A Novel Ocean Color Index to Detect Floating Algae in the Global Oceans. *Remote Sens. Environ.* **2009**, *113*, 2118–2129. [CrossRef]
7. Wang, M.; Hu, C. Mapping and Quantifying Sargassum Distribution and Coverage in the Central West Atlantic Using MODIS Observations. *Remote Sens. Environ.* **2016**, *183*, 350–367. [CrossRef]
8. Gower, J.; Hu, C.; Borstad, G.; King, S. Ocean Color Satellites Show Extensive Lines of Floating Sargassum in the Gulf of Mexico. *IEEE Trans. Geosci. Remote Sens.* **2006**, *44*, 3619–3625. [CrossRef]
9. Hu, C.; Feng, L.; Hardy, R.F.; Hochberg, E.J. Spectral and Spatial Requirements of Remote Measurements of Pelagic Sargassum Macroalgae. *Remote Sens. Environ.* **2015**, *167*, 229–246. [CrossRef]
10. Gordon, H.R.; Wang, M. Retrieval of Water-Leaving Radiance and Aerosol Optical Thickness over the Oceans with SeaWiFS: A Preliminary Algorithm. *Appl. Opt.* **1994**, *33*, 443–452. [CrossRef] [PubMed]
11. Mobley, C.D.; Werdell, J.; Franz, B.; Ahmad, Z.; Bailey, S. Atmospheric Correction for Satellite Ocean Color Radiometry. 2016. Available online: https://oceancolor.gsfc.nasa.gov/docs/technical/NASA-TM-2016-217551.pdf (accessed on 14 December 2021). [CrossRef]
12. Siegel, D.A.; Wang, M.; Maritorena, S.; Robinson, W. Atmospheric Correction of Satellite Ocean Color Imagery: The Black Pixel Assumption. *Appl. Opt.* **2000**, *39*, 3582–3591. [CrossRef]
13. Qi, L.; Hu, C.; Mikelsons, K.; Wang, M.; Lance, V.; Sun, S.; Barnes, B.B.; Zhao, J.; Van der Zande, D. In Search of Floating Algae and Other Organisms in Global Oceans and Lakes. *Remote Sens. Environ.* **2020**, *239*, 111659. [CrossRef]
14. Vanhellemont, Q.; Ruddick, K. Atmospheric Correction of Sentinel-3/OLCI Data for Mapping of Suspended Particulate Matter and Chlorophyll-a Concentration in Belgian Turbid Coastal Waters. *Remote Sens. Environ.* **2021**, *256*, 112284. [CrossRef]
15. Steinmetz, F.; Deschamps, P.-Y.; Ramon, D. Atmospheric Correction in Presence of Sun Glint: Application to MERIS. *Opt. Express* **2011**, *19*, 9783–9800. [CrossRef] [PubMed]
16. Steinmetz, F.; Ramon, D. Sentinel-2 MSI and Sentinel-3 OLCI Consistent Ocean Colour Products Using POLYMER. In Proceedings of the Remote Sensing of the Open and Coastal Ocean and Inland Waters, Honolulu, HI, USA, 30 October 2018; International Society for Optics and Photonics: Bellingham, WA, USA, 2018; Volume 10778, p. 107780.
17. Donlon, C.; Berruti, B.; Buongiorno, A.; Ferreira, M.-H.; Féménias, P.; Frerick, J.; Goryl, P.; Klein, U.; Laur, H.; Mavrocordatos, C.; et al. The Global Monitoring for Environment and Security (GMES) Sentinel-3 Mission. *Remote Sens. Environ.* **2012**, *120*, 37–57. [CrossRef]
18. Copernicus Open Access Hub Website. Available online: https://scihub.copernicus.eu/ (accessed on 5 October 2021).
19. Cox, C.; Munk, W. Measurement of the Roughness of the Sea Surface from Photographs of the Sun's Glitter. *Josa* **1954**, *44*, 838–850. [CrossRef]
20. Ody, A.; Thibaut, T.; Berline, L.; Changeux, T.; André, J.-M.; Chevalier, C.; Blanfuné, A.; Blanchot, J.; Ruitton, S.; Stiger-Pouvreau, V.; et al. From In Situ to Satellite Observations of Pelagic Sargassum Distribution and Aggregation in the Tropical North Atlantic Ocean. *PLoS ONE* **2019**, *14*, e0222584. [CrossRef]
21. Scheirer, R.; Dybbroe, A.; Raspaud, M. A General Approach to Enhance Short Wave Satellite Imagery by Removing Background Atmospheric Effects. *Remote Sens.* **2018**, *10*, 560. [CrossRef]
22. Tanré, D.; Kaufman, Y.; Holben, B.N.; Chatenet, B.; Karnieli, A.; Lavenu, F.; Blarel, L.; Dubovik, O.; Remer, L.; Smirnov, A. Climatology of Dust Aerosol Size Distribution and Optical Properties Derived from Remotely Sensed Data in the Solar Spectrum. *J. Geophys. Res. Atmos.* **2001**, *106*, 18205–18217. [CrossRef]
23. Nordkvist, K.; Loisel, H.; Gaurier, L.D. Cloud Masking of SeaWiFS Images over Coastal Waters Using Spectral Variability. *Opt. Express* **2009**, *17*, 12246–12258. [CrossRef]
24. Fernandez-Moran, R.; Gómez-Chova, L.; Alonso, L.; Mateo-García, G.; López-Puigdollers, D. Towards a Novel Approach for Sentinel-3 Synergistic OLCI/SLSTR Cloud and Cloud Shadow Detection Based on Stereo Cloud-Top Height Estimation. *ISPRS J. Photogramm. Remote Sens.* **2021**, *181*, 238–253. [CrossRef]
25. Schell, J.M.; Goodwin, D.S.; Siuda, A.N. Recent Sargassum Inundation Events in the Caribbean: Shipboard Observations Reveal Dominance of a Previously Rare Form. *Oceanography* **2015**, *28*, 8–11. [CrossRef]

26. Dibner, S.; Martin, L.; Thibaut, T.; Aurelle, D.; Blanfuné, A.; Whittaker, K.; Cooney, L.; Schell, J.M.; Goodwin, D.S.; Siuda, A.N. Consistent Genetic Divergence Observed among Pelagic Sargassum Morphotypes in the Western North Atlantic. *Mar. Ecol.* **2021**, e12691. [CrossRef]
27. Martin, L.M.; Taylor, M.; Huston, G.; Goodwin, D.S.; Schell, J.M.; Siuda, A.N. Pelagic Sargassum Morphotypes Support Different Rafting Motile Epifauna Communities. *Mar. Biol.* **2021**, *168*, 115. [CrossRef]

remote sensing

MDPI

Article

Assessment of Normalized Water-Leaving Radiance Derived from GOCI Using AERONET-OC Data

Mingjun He [1], Shuangyan He [1,2,3,*], Xiaodong Zhang [4], Feng Zhou [1,2,5] and Peiliang Li [1,3]

1 Ocean College, Zhejiang University, Zhoushan 316021, China; 21734002@zju.edu.cn (M.H.); zhoufeng@sio.org.cn (F.Z.); lipeiliang@zju.edu.cn (P.L.)
2 State Key Laboratory of Satellite Ocean Environment Dynamics, Second Institute of Oceanography, Ministry of Natural Resources, Hangzhou 310012, China
3 Hainan Institute of Zhejiang University, Sanya 572025, China
4 Division of Marine Science, School of Ocean Science and Engineering, The University of Southern Mississippi, Stennis Space Center, MS 39529, USA; Xiaodong.zhang@usm.edu
5 Observation and Research Station of Yangtze River Delta Marine Ecosystems, Ministry of Natural Resources, Zhoushan 316022, China
* Correspondence: hesy@zju.edu.cn or hesy103@163.com

Abstract: The geostationary ocean color imager (GOCI), as the world's first operational geostationary ocean color sensor, is aiming at monitoring short-term and small-scale changes of waters over the northwestern Pacific Ocean. Before assessing its capability of detecting subdiurnal changes of seawater properties, a fundamental understanding of the uncertainties of normalized water-leaving radiance (nLw) products introduced by atmospheric correction algorithms is necessarily required. This paper presents the uncertainties by accessing GOCI-derived nLw products generated by two commonly used operational atmospheric algorithms, the Korea Ocean Satellite Center (KOSC) standard atmospheric algorithm adopted in GOCI Data Processing System (GDPS) and the NASA standard atmospheric algorithm implemented in Sea-Viewing Wide Field-of-View Sensor Data Analysis System (SeaDAS/l2gen package), with Aerosol Robotic Network Ocean Color (AERONET-OC) provided nLw data. The nLw data acquired from the GOCI sensor based on two algorithms and four AERONET-OC sites of Ariake, Ieodo, Socheongcho, and Gageocho from October 2011 to March 2019 were obtained, matched, and analyzed. The GDPS-generated nLw data are slightly better than that with SeaDAS at visible bands; however, the mean percentage relative errors for both algorithms at blue bands are over 30%. The nLw data derived by GDPS is of better quality both in clear and turbid water, although underestimation is observed at near-infrared (NIR) band (865 nm) in turbid water. The nLw data derived by SeaDAS are underestimated in both clear and turbid water, and the underestimation worsens toward short visible bands. Moreover, both algorithms perform better at noon (02 and 03 Universal Time Coordinated (UTC)), and worse in the early morning and late afternoon. It is speculated that the uncertainties in nLw measurements arose from aerosol models, NIR water-leaving radiance correction method, and bidirectional reflectance distribution function (BRDF) correction method in corresponding atmospheric correction procedure.

Keywords: geostationary ocean color imager (GOCI); GDPS; SeaDAS; normalized water-leaving radiance; atmospheric correction

Citation: He, M.; He, S.; Zhang, X.; Zhou, F.; Li, P. Assessment of Normalized Water-Leaving Radiance Derived from GOCI Using AERONET-OC Data. *Remote Sens.* **2021**, *13*, 1640. https://doi.org/10.3390/rs13091640

Academic Editor: Alexander Kokhanovsky

Received: 2 April 2021
Accepted: 19 April 2021
Published: 22 April 2021

1. Introduction

Remote sensing of ocean color has proven to be an efficient and irreplaceable technique in monitoring the ocean environment in the last two decades, contributing to biogeochemistry, physical oceanography, ocean-system modeling, fisheries, and coastal management [1]. Although polar-orbiting ocean color sensors (e.g., MODIS, SeaWiFS, MERIS) are well suited for observing seasonal or annual variations of ocean phenomena on a global scale, their once-per-day-time coverage cannot resolve diel variability. With greater temporal resolution than traditional polar-orbiting ocean color sensors, geostationary ocean color sensors may provide an alternative when observing the ocean environments that vary on short temporal scales [2–4]. Geostationary ocean color imager (GOCI), the first geostationary ocean color mission, was designed to focus on an area of 2500 × 2500 km^2 centered around the Korean Peninsula. It acquires multispectral images with a 500 m ground resolution eight times per day [4–6]. The GOCI data have been successfully used in monitoring and characterizing short-term and small-scale ocean phenomena, such as red tides, green algae, particulate organic carbon flux (POC), total suspended matter (TSM), diurnal variation of water turbidity or aerosol optical depth and $PM_{2.5}$, surface circulation, surface tidal currents and plume [7–15]. As with any ocean color sensors, the successful application of GOCI data depends on the quality of its data products, especially the normalized water-leaving radiance (nLw) or the remote sensing reflectance (Rrs) [16], which is mainly up to atmospheric correction.

Atmospheric correction is a fundamental step to remove the atmospheric influence and extract the water information. The atmospheric correction for GOCI data has been mainly performed using two operational algorithms: the Korea Ocean Satellite Center standard atmospheric correction algorithm, which can be achieved in the GOCI Data Processing System (GDPS) (hereafter, GDPS algorithm) [6], and the NASA standard atmospheric correction algorithm, which can be realized in Sea-Viewing Wide Field-of-View Sensor Data Analysis System (SeaDAS/l2gen package) (hereafter, SeaDAS algorithm) [17]. Huang et al. [18] found that the GDPS algorithm shows better performance in retrieving Rrs and aerosol optical information over the Yellow Sea region than the SeaDAS algorithm, although low accuracies were discovered at blue and near-infrared (NIR) bands, which is consistent with the research results of Concha et al. [19,20]. Kim et al. [19,21] evaluated the chlorophyll concentration derived from GOCI radiometric data acquired from the GDPS algorithm using 130 matchups between GOCI data and field data and concluded that the surface radiometric outcome needs to be improved primarily for clear waters and for the blue bands (412, 443, and 490 nm). Lamquin et al. [22] proved relative agreement between GOCI-derived seawater reflectance products based on GDPS algorithm and medium resolution imaging spectrometer (MERIS) and moderate-resolution imaging spectrometer (MODIS) results. Li et al. [23] also demonstrated that remote sensing reflectance, the concentration of standard chlorophyll-a, and suspended particulate matter products produced by the GDPS algorithm have the best agreement with in situ data in Liaodong Bay. Qi et al. [24] determined typical types of diurnal changing patterns in cyanobacteria blooms of Taihu Lake using SeaDAS algorithm-based GOCI products. However, Wu et al. [25] declared that either the GDPS algorithm or SeaDAS algorithm-based Rrs could not be retrieved successfully in the highly turbid coastal waters. Moreover, many studies have used the GDPS algorithm or SeaDAS algorithm, as well as improved algorithms, to retrieve other products, such as colored dissolved organic matter (CDOM) absorption coefficient, the diffuse attenuation coefficient, and chlorophyll-a concentration [6,16,21,22,26,27].

At present, most studies validated the feasibility and applicability of atmospheric correction algorithms on account of limited in situ data or time-limited ship-measured data. In this study, we will assess the uncertainties of GOCI nLw(λ) produced by GDPS algorithm and SeaDAS algorithm through comparing with field measurements from four AERONET-OC sites, namely, ARIAKE_TOWER, Ieodo_Station, Socheongcho, and Gageocho_Station, located within the GOCI observation range ranging in a rather long observation period covering both clear and turbid water (Figure 1; more details about these four sites are presented in Section 2.2), and aim to give data users an impression of how much uncertainties could exist in the surface radiometric data derived from these two operational atmospheric correction algorithms in different waters, and at different observation time (GOCI can provide eight every-hour observations daily). The paper is constituted as follows: Section 1 introduces the background of GOCI data processing; the data resources and methods used in the research are described in Section 2; in Section 3, the comparison results between GOCI-derived nLw(λ) based on different algorithms and nLw(λ) at four AERONET-OC sites are presented with corresponding figures and tables; Sections 4 and 5 discuss the performance of the two algorithms and propose viewpoints and suggestions.

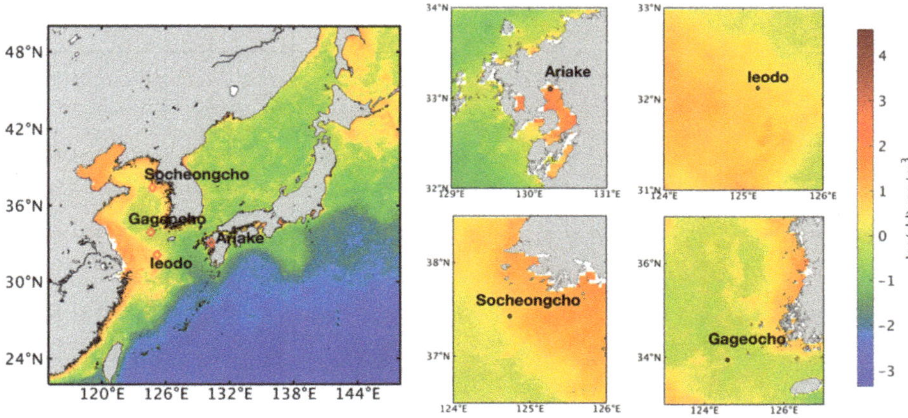

Figure 1. The distribution of chlorophyll-a concentration in the GOCI observation area (data from MODIS-Aqua) and the location of four AERONET-OC sites (Ariake, Ieodo, Socheongcho, Gageocho). The zoom-in figures at four sites are exhibited on the right side together with the color bar.

2. Materials and Methods

2.1. GOCI Data

The GOCI, launched on 26 June 2010 by the Republic of Korea, is the first geostationary spaceborne ocean color sensor. It provides images with 500 m ground resolution at hourly intervals up to eight times per day (from 00:16 UTC to 07:16 UTC) at eight spectral bands (six visible bands: 412, 443, 490, 555, 660, 680 nm, and two near-infrared bands: 745, 865 nm) around the Korean Peninsula [13,16,21,26]. The GOCI Level 1B (L1B) data are available for downloading from the KOSC website (http://kosc.kiost.ac.kr/). Data from 17 October 2011 to 5 March 2019 are collected in this study.

2.2. AERONET-OC Data

The AERONET is an assembly of ground-based remote sensing aerosol networks, which also supports marine applications through a new component called AERONET-OC through providing nLw data measured by the SeaPRISM autonomous radiometer systems deployed on offshore platforms. The SeaPRISM system acquires the radiance from the sea and the sky at viewing zenith angles of 40° and 140° with a relative azimuth angle of 90°, and then the nLw can be further calculated [28]. As shown in Figure 1, there are four AERONET-OC sites located within the GOCI's footprint: ARIAKE_TOWER (Ariake, 33.104° N, 130.272° E, PIs: Joji Ishizaka and Kohei Arai, time span: 19 April 2018–5 March 2019, number of data: 503), Ieodo_Station (Ieodo, 32.123° N, 125.182° E, PIs: Young-Je Park and Hak-Yeol You, time span: 1 December 2013–27 March 2018, number of data: 42), Socheongcho (37.423° N, 124.738° E, PI: Young-Je Park, time span: 13 October 2015–30 November 2018, number of data: 165) and Gageocho_Station (Gageocho, 33.942° N, 124.593° E, PIs: Jae-Seol Shim and Joo-Hyung Ryu, time span: 17 October 2011–16 May 2012, number of data: 31). The Level 2.0 nLw data (obtained based on the f/Q correction [29]), which is cloud screened and quality assured, at the above four sites were downloaded from the AERONET-OC website (https://aeronet.gsfc.nasa.gov/, accessed on 21 December 2020). Note that the mean nLw (660) values with standard deviations at Ariake, Ieodo, Socheongcho, and Gageocho sites are 7.7876 ± 3.8878, 4.8363 ± 3.2161, 1.1993 ± 1.2842, and 1.0931 ± 0.6589 $W/m^2 \cdot sr \cdot \mu m$, respectively, which implies the descending water turbidity. In addition, with the purpose to reveal the turbidity of four sites synoptically, annual average chlorophyll-a concentration based on MODIS-Aqua data were obtained from Ocean Color Website (https://oceancolor.gsfc.nasa.gov/, accessed on 21 December 2020) and visualized after logarithmic calculation in Figure 1. Both statistical calculation results of nLw at 660 nm wavelength and chlorophyll-a concentration distribution shown in Figure 1 illustrate that seawater at Ariake is most turbid, and become increasingly clear from Ieodo, to Socheongcho, to Gageocho sites.

Although the spectral band of AERONET-OC and GOCI data are not exactly the same (Table 1), comparable wavelengths of 412, 443, 490, 660, and 865 nm for both data were selected for comparison purpose in this study. The main difference is at 667 nm for AERONET-OC and 660 nm for GOCI. However, no spectral interpolation was performed here since we believe that the spectral difference is acceptable considering that the GOCI bandwidth is 20 nm. Hereafter, we use the wavelengths of GOCI in the following text for the sake of simplicity.

Table 1. GOCI spectral bands and corresponding wavebands of AERONET-OC data.

GOCI			AERONET-OC		
Band#	**λ (nm)**	**$\Delta\lambda$(nm)**	**Band#**	**λ (nm)**	**$\Delta\lambda$ (nm)**
B1	412	20	B4	412	10
B2	443	20	B6	443	10
B3	490	20	B7	490	10
B4	555	20	-	-	-
B5	660	20	B16	667	10
B6	680	10	-	-	-
B7	745	20	-	-	-
B8	865	40	B21	865	10

2.3. Methods

The GOCI L1B products (top-of-atmosphere radiance) were first processed into L2 products (surface water radiometric data, nLw(λ)) by both GDPS and SeaDAS/l2gen, and then we matched and assessed these two kinds of GOCI nLw(λ) data (nLw_xxx) with AERONET-OC nLw(λ) data (nLw_AERONET-OC).

2.3.1. Conversion to L2 Products

GOCI L1B data were processed both with the GDPS (Version 1.4.1, South Korea) and the SeaDAS/l2gen package (Version 7.5.3, United States). GDPS is a software for GOCI data processing dedicated to GOCI distributed by KOSC (http://kosc.kiost.ac.kr/eng/p30/kosc_p31.html, accessed on 1 October 2020) through which L2 data, including nLw, colored dissolved organic matter (CDOM), the diffused attenuation coefficient of down-welling irradiance (Kd), as well as ancillary information such as solar and sensor zenith angles and azimuth angles, can be obtained from L1B data [6]. The GDPS employed KOSC standard atmospheric algorithm to achieve data conversion. It is theoretically based on the SeaWiFS (NASA) standard atmospheric correction method developed by Gordon and Wang in 1994 (hereafter GW94) [30], with partial modification in aerosol models and additionally uses an iterative process to correct NIR reflectance alternatively in the case of case-2 waters [31]. As for aerosol models in GDPS, the GDPS algorithm adopts only 3 from 12 aerosol models used in GW94, namely, M99, M50, and C50, to reduce the processing time and avoid the image discontinuity problem. For the NIR correction model in turbid case-2 waters, it adapts an empirical relationship between normalized water-leaving reflectance (ρ_w) at 660 nm, 745 nm, and 865 nm [32].

SeaDAS distributed by NASA is a comprehensive software package for the processing, display, analysis, and quality control of ocean color data (https://seadas.gsfc.nasa.gov/). It can apply the standard NASA atmospheric correction algorithm [33] to the GOCI data through its multisensor L1 to L2 generator (l2gen). SeaDAS algorithm is also initially based on GW94 and has undergone a series of improvements [17,34,35]. When treating nonblack pixels, it first retrieves $\rho_w(443)$ and $\rho_w(555)$ based on the "black-pixel" assumption. Next, it obtains an initial estimate of the chlorophyll concentration and total absorption coefficient at 660 nm, a(660), via a bio-optical model and empirical relationship [17]. Then, particulate backscattering at 660 nm, b_{bp} (660), can be derived from a(660) and ρ_w (660), which can make a further step to calculate the total backscattering of NIR bands. Finally, ρ_w values of at NIR bands are recalculated on the basis of the above steps, and this procedure will cycle until convergence is reached.

Both GDPS and SeaDAS output L2 processing flags reflecting warnings or errors generated concerning the data quality [20,36]; we applied these flags in the following data matchup procedure.

2.3.2. Matchup Procedure

Through trial, a space–time window with 5 × 5 pixels and ±0.5 h was utilized to match GOCI data and AERONET-OC data. Specifically, the proposed space–time window is a compromise between minimizing geophysical variability and navigation error and ensuring a sufficient amount of data for statistical analysis. To process the matchup data, we followed the satellite validation protocol described in Bailey and Werdell [36] (Figure 2). We discarded GOCI pixels flagged as land, cloud edge in GDPS, and failure in products in SeaDAS. To minimize the influence of outliers, we also discarded those GOCI pixels whose values are outside ± 1.5 standard deviations of their respective median values. For the remaining data within the matchup window, we calculated arithmetic means for comparison. As GOCI observation times are every hour from 00:16 UTC to 07:16 UTC, whereas the time of AERONET-OC data is irregular; there may be a situation in which several AERONET-OC data matched simultaneously with one GOCI acquisition data. For each GOCI data, only temporally closest AERONET-OC data were first considered for availability. A total of 307 pairs of matchups were found between GDPS-derived data (nLw_GDPS, stands for the nLw results of the GDPS algorithm) and nLw_AERONET-OC, and 294 pairs between SeaDAS-derived data (nLw_SeaDAS, stands for the nLw results of SeaDAS algorithm) and nLw_AERONET-OC. For the GDPS algorithm, 221, 22, 55, and 9 matchup data were obtained at Ariake, Ieodo, Socheongcho, and Gageocho, Ieodo, respectively; for the SeaDAS algorithm, 193, 24, 68, and 9 matchup data were obtained at the four sites.

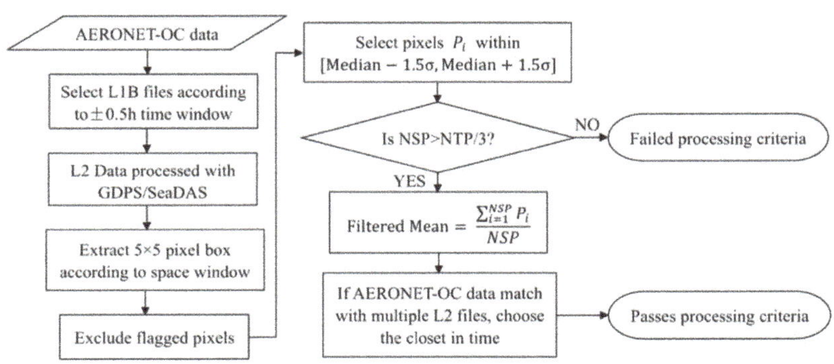

Figure 2. Flowchart of exclusion criterion and matchup procedure. Median, median of valid pixel values; σ, standard deviation of valid pixel values; NSP, number of selected pixels; NTP, number of total pixels.

2.3.3. Accuracy Indicator

In order to quantify comparison results, we use mean absolute error (MAE) and mean percentage relative error (MPRE) to evaluate the accuracy of nLw_xxx (either nLw_GDPS or nLw_SeaDAS) in contrast to in situ observations (nLw_AERONET-OC) as follows:

$$MAE = \frac{\sum_{i=1}^{N} |nLw\_\text{AERONET} - \text{OC}(i) - nLw\_\text{xxx}(i)|}{N} \quad (1)$$

$$MPRE = \frac{100}{N} \sum_{i=1}^{N} \frac{|nLw\_AERONET - OC(i) - nLw\_\text{xxx}(i)|}{nLw\_AERONET - OC(i)} \quad (2)$$

where, nLw_AERONET-OC(i) and nLw_xxx(i) denote the nLw_AERONET-OC and nLw_xxx of the i-th matchup, respectively, and N denotes the total number of matchup pairs.

3. Results

3.1. Comparison between nLw_xxx and nLw_AERONET-OC

The performance of the GDPS algorithm and SeaDAS algorithm varies from site to site (Figures 3 and 4). Figure 3 shows that most of the matchup points between nLw_SeaDAS and nLw_AERONET-OC are located below the 1:1 line, while the majority of the matchups between nLw_GDPS and nLw_AERONET-OC falls approximately evenly on both sides of the 1:1 line, although several matchups apparently lied below 1:1 line at Ariake site and with larger variance at Socheongcho site. Comparison results between nLw_SeaDAS and nLw_AERONET-OC indicate that relatively larger variance exists at Ariake and Gageocho sites than that between nLw_GDPS and nLw_AERONET-OC; besides, nLw_SeaDAS contains evident negative values at shorter wavelengths (e.g., 412-490 nm) at Ariake, Socheongcho, and Gageocho sites, while nLw_GDPS embodies negative values yet in less quantity at Ariake and Socheongcho sites. Therefore, to some extent, the GDPS algorithm is capable of higher stability and accuracy than the SeaDAS algorithm, particularly in turbid water (e.g., Ariake site).

Table 2. The MAE (W/m^2·sr·μm) and MPRE (%) of each site between nLw_AERONET-OC and nLw_xxx for all matchup pairs

Site		Ariake	Ieodo	Socheongcho	Gageocho
AERONET-OC vs. GDPS	MAE	2.0713	2.8250	2.0816	1.7582
	MPRE	28.9892	29.7820	67.6869	56.0459
	N	1098	87	210	35
AERONET-OC vs. SeaDAS	MAE	3.2454	2.9215	2.4304	2.1949
	MPRE	39.2502	17.9286	74.1094	57.0899
	N	943	96	270	36

Figure 4 reveals the density distribution of relative error between nLw_xxx and nLw_AERONET-OC. At each site, 75% of matching pairs between nLw_SeaDAS and nLw_AERONET-OC have relative errors less than zero, while the median values of relative error between nLw_GDPS and nLw_AERONET-OC are approximately zero at Ariake, Ieodo, and Socheongcho sites and slightly higher at Gageocho site. Both matching results exhibit long tail at Socheongcho and Gageocho sites, indicating that both algorithms entail large uncertainty in clean water. The two figures suggest that compared with the GDPS algorithm, the SeaDAS algorithm underestimates nLw, in general, especially at Ariake and Socheongcho sites. The related MAE and MPRE values of four sites between nLw_xxx and nLw_AERONET-OC summarized in Table 2 also support this result and further illustrate the degree of deviation. MAE between nLw_GDPS and nLw_AERONET-OC is smaller than MAE between nLw_SeaDAS and nLw_AERONET-OC at all four sites. Moreover, MPRE between nLw_GDPS and nLw_AERONET-OC is also less than that between nLw_GDPS and nLw_AERONET-OC at Ariake, Socheongcho, and Gageocho sites, however, conversely at Ieodo site, due to overestimated nLw_GDPS at 660 nm wavelength.

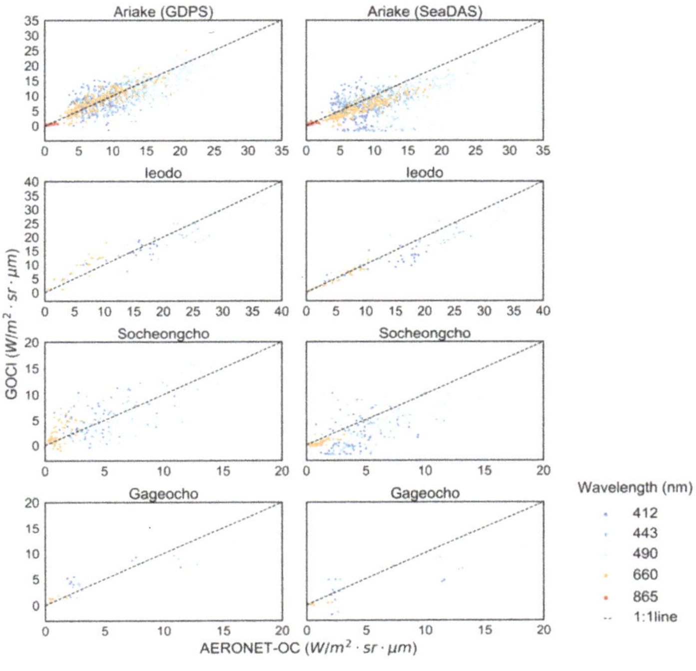

Figure 3. Scatterplots between nLw_GDPS and nLw_AERONET-OC (left column), and nLw_SeaDAS and nLw_AERONET-OC (right column) at Ariake, Ieodo, Gageocho, and Socheongcho sites, respectively.

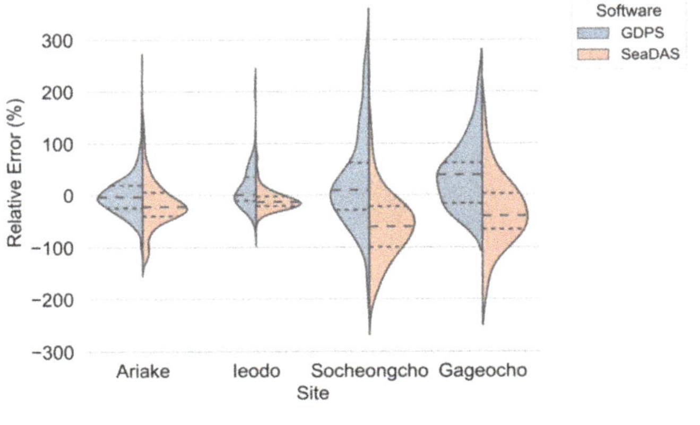

Figure 4. Violin plots of relative error distribution of nLw_GDPS (shown in blue) and nLw_SeaDAS (shown in pink), compared to nLw_AERONET-OC at Ariake, Ieodo, Socheongcho, and Gageocho sites. The dotted lines represent the upper quartile, median, and lower quartile. The statistical information between nLw_GDPS (nLw_SeaDAS) and nLw_AERONET-OC at each site was summarized in Table 2.

In conclusion, nLw_GDPS shows better agreement with nLw_AERONET-OC than nLw_SeaDAS. Noteworthy, two issues are discovered through the above analysis—one is that some blue band values of nLw_GDPS and nLw_SeaDAS are negative; the other is that the relative errors of nLw_GDPS and nLw_SeaDAS are even higher than 100% at Socheongcho and Gageocho sites, which are sites with clear waters. Overall, the applicability and serviceability of the GDPS algorithm are better than the SeaDAS algorithm regardless of turbidity in our research area; nevertheless, the stability of both the GDPS algorithm and SeaDAS algorithm in clean water is worthy of attention.

The performance of the GDPS algorithm and SeaDAS algorithm also varies with wavelengths, as shown in scatterplots of Figure 5 and violin plots of Figure 6, as well as statistical results listed in Table 3. As shown in Figure 5, for nLw_GDPS, the highest accuracy was found at 660 nm band among all bands, while at shorter wavelengths, nLw matching results are influenced by several underestimated valuations. More explicitly, when nLw_AERONET-OC is greater than 15, 20, 28 $W/m^2 \cdot sr \cdot \mu m$ at 412, 443, 490 nm wavelengths, respectively, nLw_GDPS is underestimated and slightly affects the overall results. Given this consideration, nLw_GDPS is well matched with nLw_AERONET-OC at all wavelengths except 865 nm. Similarly, the nLw_SeaDAS at 660 nm shows the best agreement with the nLw_AERONET-OC, although there still exists a linear underestimation trend; however, it can be observed that the underestimation issue is becoming severe with decreasing visible bands, and the performance is worse than nLw_GDPS at shorter bands. On the contrary, better performance at 865 nm was found with nLw_SeaDAS than that with nLw_GDPS.

Table 3. The MAE ($W/m^2 \cdot sr \cdot \mu m$) and MPRE (%) of nLw(λ) between nLw_AERONET-OC and nLw_xxx for all matchup pairs.

λ (nm)		412 nm	443 nm	490 nm	660 nm	865 nm
AERONET-OC vs. GDPS	MAE	2.6553	2.8279	2.8016	1.6127	0.1550
	MPRE	44.1857	31.7500	22.1521	41.9414	37.7339
	N	304	305	305	298	218
AERONET-OC vs. SeaDAS	MAE	3.9288	3.9614	4.3641	1.7531	0.1452
	MPRE	63.3289	47.5962	32.6843	42.8915	36.9046
	N	287	287	292	291	188

In Figure 6, the mean value of relative error distribution between nLw_GDPS and nLw_AERONET-OC is close to 0 from 412 to 660 nm and a bit lower than 0 at 865 nm, while that of nLw_SeaDAS and nLw_AERONET-OC is close to 0 only at 865 nm band and approaches −20% at visible bands. Moreover, as Table 3 shows quantitatively, for all matchup pairs, the MAE between nLw_GDPS and nLw_AERONET-OC are 2.66, 2.82, 2.80, 1.61, and 0.16 (W/m2·sr·μm) at 412, 443, 490, 660, and 865 nm, respectively; meanwhile, the MAE between nLw_SeaDAS and nLw_AERONET-OC are 3.93, 3.96, 4.36, 1.75, and 0.15 (W/m^2·sr·μm) at the corresponding wavelengths. The MPRE between nLw_GDPS and nLw_AERONET-OC are 44.19%, 31.75%, 22.15%, 41.94%, and 37.73% at 412, 443, 490, 660, and 865 nm, respectively; and the MPRE between nLw_SeaDAS and nLw_AERONET-OC are 63.33%, 47.60%, 32.68%, 42.89%, and 36.90% at the corresponding wavelengths. Clearly, MAE between nLw_xxx and nLw_AERONET-OC shows larger mean values at shorter bands, whereas MPRE shows larger mean values at both shorter and NIR bands.

Overall, both nLw_GDPS and nLw_SeaDAS show better agreement with nLw_AERONET-OC at 490 nm and 660 nm bands, followed by 865 nm and 443 nm bands, and poorest performance at 412 nm. The nLw data generated from the GDPS algorithm perform better than that from the SeaDAS algorithm at visible bands in this study area, while the SeaDAS algorithm is slightly better than the GDPS algorithm at the near-infrared band of 865 nm.

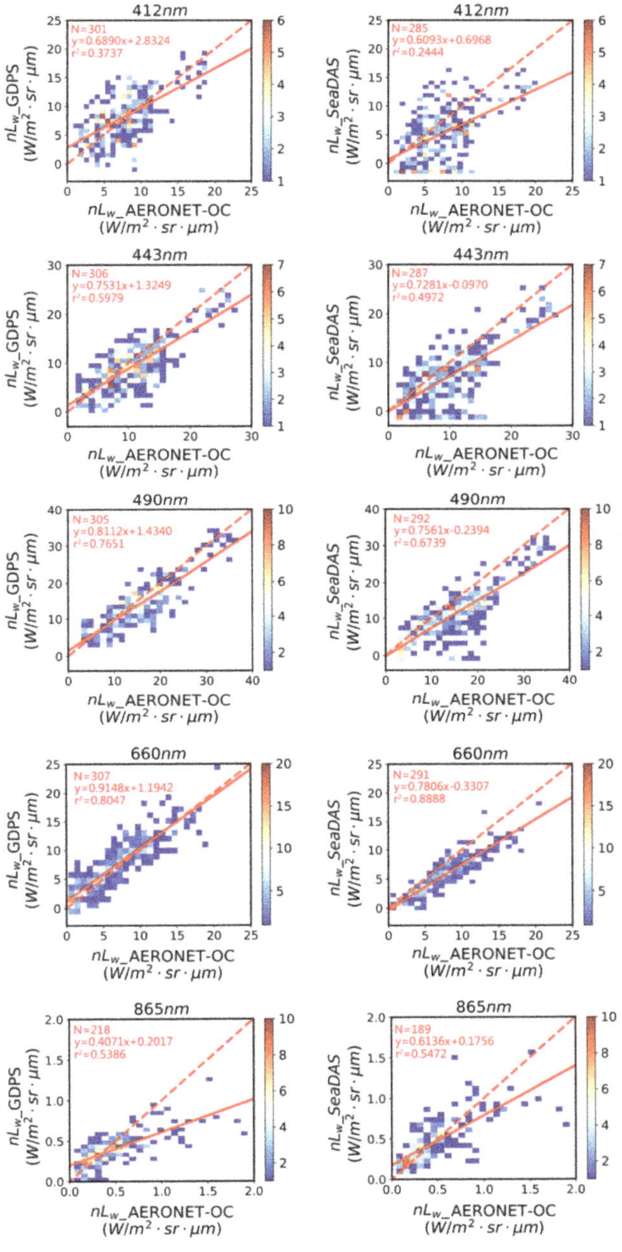

Figure 5. Scatterplots between nLw_GDPS and nLw_AERONET-OC (left column), and nLw_SeaDAS and nLw_AERONET-OC (right column) at 412, 443, 490, 660, and 865 nm, respectively. The color bar represents the number of matchup pairs. Note that the data at 865 nm are only from the Ariake site.

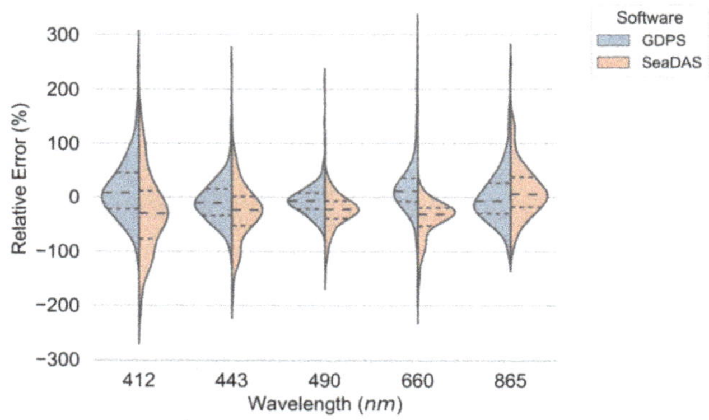

Figure 6. Violin plots of relative error distribution of nLw_GDPS (shown in blue) and nLw_SeaDAS (shown in pink), compared to nLw_AERONET-OC, at 412, 443, 490, 660, and 865 nm (note that the data at 865 nm are only from the Ariake site). The dotted lines represent the upper quartile, median, and lower quartile. The statistical information between nLw_GDPS (nLw_SeaDAS) and nLw_AERONET-OC at each wavelength is summarized in Table 3.

The GOCI data were acquired eight times per day; therefore, we also analyzed the relative errors between nLw_GDPS and nLw_AERONET-OC at different times from 00 to 07 UTC at one-hour intervals in Figure 7 in order to investigate whether the performance of the GDPS algorithm and SeaDAS algorithm changes with time. The statistical information of nLw_GDPS (nLw_SeaDAS) and nLw_AERONET-OC at each hour is also summarized in Table 4. From the 25, 50, and 75 percent percentile lines, nLw_GDPS performs better at 02, 03, and 04 hours, followed by 01 and 05 hours, and worst at 00, 06, and 07 hours. As for nLw_SeaDAS, it performs better at 02 and 03 hours, followed by 01 and 04 hours, and worsens at 05, 06, 07, and 00 hours. This means both algorithms behave better at noon and worse in the early morning and late afternoon.

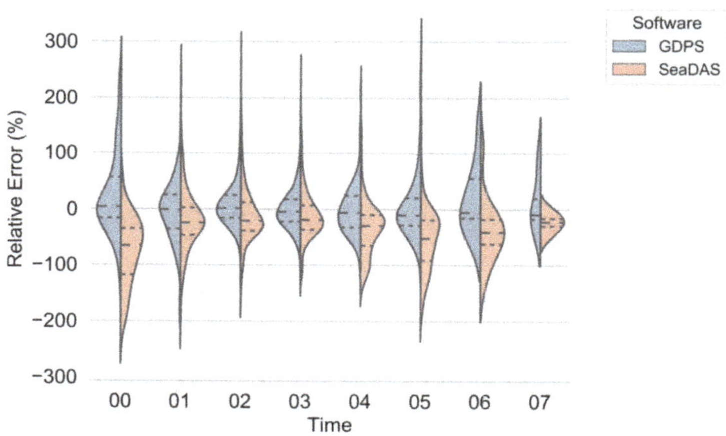

Figure 7. Violin plots of relative error distribution of nLw_GDPS (shown in blue) and nLw_SeaDAS (shown in pink), compared to nLw_AERONET-OC from 00 to 07 (UTC) at one-hour intervals. The dotted lines represent the upper quartile, median, and lower quartile. The statistical information of nLw_GDPS (nLw_SeaDAS) and nLw_AERONET-OC at each hour is summarized in Table 4.

Table 4. The MAE (W/m^2·sr·μm) and MPRE (%) at different times between nLw_AERONET-OC and nLw_xxx for all matchup pairs.

Time		00	01	02	03	04	05	06	07
AERONET-OC vs. GDPS	MAE	1.7705	2.2657	1.9321	2.0722	2.2548	2.5361	2.0484	3.7610
	MPRE	48.7573	40.8977	31.9189	26.9986	34.9606	42.5066	18.8008	32.2694
	N	96	311	425	289	155	146	4	4
AERONET-OC vs. SeaDAS	MAE	2.2521	2.5500	2.6899	2.9287	3.7351	4.3946	4.7275	5.0621
	MPRE	78.5825	44.7390	39.0737	36.4317	47.2525	62.7846	55.5676	22.6875
	N	73	259	351	341	165	139	13	4

Overall, the nLw_GDPS turns out slightly better than the nLw_SeaDAS with lower MAE and MPRE at each time (Table 4). However, for both algorithms, the relative errors at 00, 01, 05, 06, and 07 can even exceed 100% and approach 200%, which cannot be ignored in data applications.

3.2. Comparison between nLw_GDPS and nLw_SeaDAS

The nLw(λ) values derived from the GDPS algorithm and SeaDAS algorithm are also compared, and the result is presented by the density scatterplots in Figure 8. The nLw_GDPS and nLw_SeaDAS agree with each other well, with r^2 of 0.6465, 0.7340, 0.7988, 0.8971, and 0.7760 at 412, 443, 490, 660, and 865 nm, respectively. Notably, nLw_GDPS and nLw_SeaDAS show better correlation at 660 nm and degrade as the wavelength moved to the blue end of the spectral range, which is consistent with the information reflected in Figure 5. From the linear regression results (red solid lines) and 1:1 line (red dashed lines) between nLw_GDPS and nLw_SeaDAS in Figure 8, it can be observed that the nLw_GDPS values are primarily greater than the nLw_SeaDAS at visible bands shorter than 660 nm, while smaller at NIR band of 865 nm.

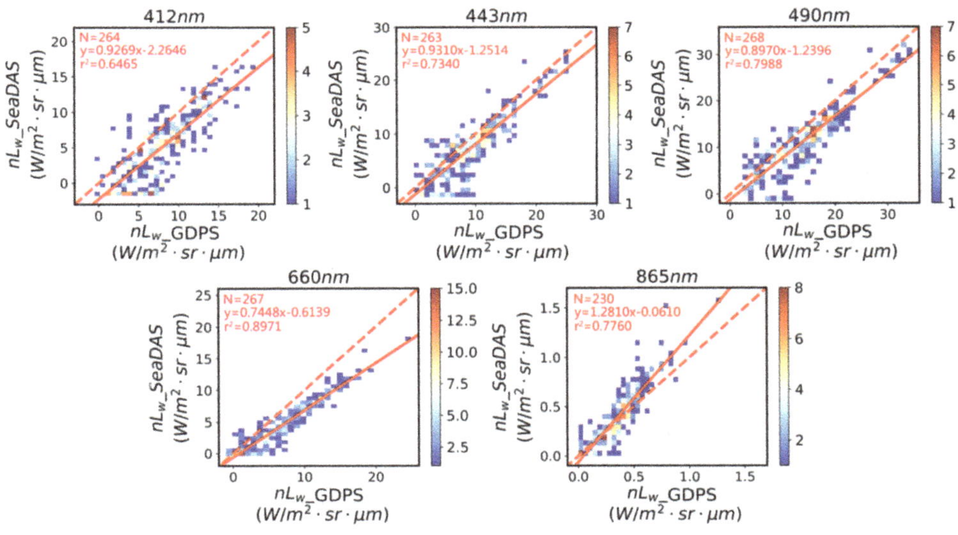

Figure 8. Density scatterplots between the GDPS-derived (abscissa) and SeaDAS-derived (ordinate) nLw values at 412, 443, 490, 660, and 865 nm bands. The color bar stands for the number of matchup pairs. The red solid lines are derived by linear regression between nLw_GDPS and nLw_SeaDAS. The red dashed lines are 1:1 lines.

4. Discussion

Although the GDPS algorithm and SeaDAS algorithm are both based on the atmospheric correction scheme developed by Gordon and Wang [30], their actual atmospheric correction processors are with different aerosol models, different near-infrared (NIR) water-leaving radiance corrections, and different vicarious calibration gains [18–20,31–33,37]; thus, as Section 3.2 indicated, the nLw_GDPS and nLw_SeaDAS show some deviations. When comparing with nLw(λ) data of four AERONET-OC sites in terms of clear and turbid water, nLw_GDPS shows better accuracy than nLw_SeaDAS. The nLw_GDPS values are underestimated at 865 nm. However, when nLw_AERONET-OC at 865 is larger than 0.5 $W/m^2{\cdot}sr{\cdot}\mu m$, a rather good agreement, albeit with a larger standard deviation, was observed at shorter bands. On the other hand, the nLw_SeaDAS values show better agreement at 865 nm but performing progressively worse toward short bands. The reason that the nLw_GDPS shows an overall better accuracy than nLw_SeaDAS is that the aerosol models in GDPS algorithm consider more coarse particles than that in SeaDAS algorithm. As aerosol reflectance of coarse aerosol particles has a smaller spectral slope, the underestimated nLw at 865 nm and a smaller spectral slope of aerosol reflectance could just lead to reasonable nLw values at green and blue bands for GDPS algorithms. Thus, even though SeaDAS algorithm uses 80 aerosol models based on AERONET observations [35], and the GDPS algorithm only uses three aerosol models (M99, M50, and C50) in atmospheric correction [31], our results show that GDPS algorithm performs better in correcting atmospheric signal and hence extracting nLw signal at the research sites than SeaDAS algorithm. The GDPS strategy is appropriate for estimating a comparative accurate surface contribution (the MPRE of all bands is ranging from 22% to 44%), and for a long-term further improvement goal, a group of more exact aerosol models is still needed. As mentioned before, the nLw_SeaDAS values at 865 nm agree well with AERONET-OC data. Because the AERONET-OC does not have the measurements at 745 nm, we speculate that if the nLw_SeaDAS values at 745 nm are also accurate, the degraded accuracy of nLw_SeaDAS with decreasing wavelength may arise from the incorrect extrapolation of aerosol reflectance approximation from NIR to short visible bands by the exponential spectral function (this was used by both GDPS and SeaDAS algorithms). However, if the nLw_SeaDAS values at 745 nm are not accurate, then the aerosol model selected using aerosol contribution at two NIR bands of 865 nm and 745 nm would not be appropriate. The aerosol properties in this study area can be further studied since aerosol properties are of high spatiotemporal variations [38], and it influences the accuracy of nLw retrieval and affects the vicarious calibration process of ocean color sensors. It is worth mentioning that the validation at 865 nm is particularly challenging since water-leaving radiance is relatively small as compared to the reflected sky radiance at the air–sea interface. A better validation at 865 nm will be helpful in better understanding the sources of nLw retrieval errors over turbid water.

As indicated in Figure 6 in Section 3.1, the mean relative error of nLw_SeaDAS at 865 nm is close to zero, while an apparent underestimation of nLw_GDPS can be observed. This indicates that the NIR correction model in the SeaDAS algorithm works better, and that of the GDPS algorithm can be further improved. In the GDPS algorithm, an empirical relationship between ρ_w at 660 nm, 745 nm, and 865 nm is adopted, and its empirical coefficients were derived using in situ data in turbid waters around Korean Peninsula [21,39]. Since the correlation between red and NIR bands is not immutable but varies with concentrations of TSM, CDOM, and chlorophyll-a [40], a dataset covering different seasons and various water properties will be needed for further improving the NIR correction method of the GDPS algorithm. It is also worth noting that the optical saturation issue of water reflectance appears earlier at short bands with increasing turbidity, and water-leaving reflectance at 660 nm and NIR band can alter incompatibly in waters with different optical properties since 660 nm is close to the second peak of the chlorophyll absorption spectrum [41,42]. Concerning this problem, the GOCI-II mission added a 709 nm band and used it as a substitution for 660 nm in a later NIR correction model [42].

Based on comparison with nLw data from four AERONET-OC sites, we observe that the nLw_GDPS performs better with the MPRE around 30% in turbid water at Ariake and Ieodo, while worse with the MPRE larger than 55% in clearer water at Gageocho and Socheongcho; the nLw_SeaDAS is underestimated with the MPRE among 18–39% in clear water and 57–74% in turbid water. Combing with results in Figure 6 and Table 3, it is discovered that except for the 865 nm band, the errors mostly come from the 412 nm band. The farther the band is from NIR bands, the larger the error is since the estimation accuracy degrades with a longer distance extrapolation. Comparing with turbid water, it seems a better accuracy of nLw at 412 nm is more difficult to achieve in clear waters. Although from scatterplots (Figure 5), the nLw_GDPS agree well with AERONET-OC data in an overall trend, the MPRE is still among 18–55%, and it is far from the ideal goal of 10%. There is still a way to go in improving the atmospheric correction in this study area.

It can be clearly observed that the relative errors of nLw_GDPS and nLw_SeaDAS vary with hours in Figure 7. The better performance of nLw_GDPS and nLw_SeaDAS are at 02 and 03 UTC (11:00 and 12:00 for Korean local time) of one day, and the farther away from noon, the greater relative errors are. This error might be partly from incorrect aerosol estimations. The changing humidity due to solar illumination can affect the aerosol optical properties, low solar altitudes in the early morning and late afternoon can also lead to increasing aerosol contributions, and these will both bring more uncertainties in aerosol estimations in geostationary ocean color remote sensing. This error might be also partly caused by the bidirectional reflectance distribution function (BRDF) correction method, which is used to remove the solar-viewing dependence and nonisotropic distribution of the in-water light field [29,43], and the measured nLw values are more affected by low solar altitudes in the early morning or late afternoon. Both the GDPS algorithm and SeaDAS algorithm use the BRDF correction method of Morel and Gentilli [19,31,33,37], based on the chlorophyll concentration estimation. Results in Figure 7 indicate that the atmospheric correction algorithm in GDPS and SeaDAS algorithms is necessary to be improved when applied to GOCI data with a large solar zenith angle [44], especially when the observed time is far away from noon, since the relative errors at 00, 06, and 07 hours are even over 100% and approach 200%.

5. Conclusions

In this study, we assessed the nLw(λ) products generated from the GDPS algorithm and SeaDAS algorithm with AERONET-OC data from October 2011 to March 2019. The nLw_GDPS and nLw_SeaDAS are, respectively, matched and compared with nLw_AERONET-OC at four AERONET-OC sites of Ariake, Ieodo, Socheongcho, and Gageocho. Results show that the performance of the GDPS algorithm and SeaDAS algorithm varies with measure time, seawater turbidity, and spectral bands.

The nLw_GDPS is of better accuracy both in turbid and clear water, although with underestimations at the 865 nm band. The nLw_SeaDAS data are underestimated in both clear and turbid water, and the underestimation worsens toward short visible bands. Additionally, the negative nLw_SeaDAS values at blue bands are evident. The nLw data generated from the GDPS algorithm perform better than that from the SeaDAS algorithm at visible bands in this study area, while the SeaDAS algorithm performs better than the GDPS algorithm at the NIR band of 865 nm. A better near-infrared water-leaving radiance correction is needed for the GDPS algorithm, while aerosol optical properties and aerosol model selections can be further investigated for the SeaDAS algorithm in this study area.

The GDPS and SeaDAS algorithms perform better at noon (02 and 03 UTC) and worse in the early morning and late afternoon. The relative errors for both processors at 00, 06, and 07 even approach 200%. Thus, a better atmospheric correction performance is required for GOCI data acquired in the early morning and late afternoon.

The GDPS-generated nLw data are slightly better than that with SeaDAS at visible bands; however, the mean percentage relative errors for both algorithms at blue bands are over 30%. The nLw data derived by GDPS is better both in clear and turbid water, although underestimation is observed at the NIR band (865 nm) in turbid water. The nLw data derived by SeaDAS are underestimated in both clear and turbid water, and the underestimation worsens toward short visible bands. Generally, based on the combination of the NIR correction model and aerosol model setting, the GDPS algorithm performs better than the SeaDAS algorithm in this study area. The comparison results lead us to recommend GDPS as a first choice when processing GOCI data. However, its aerosol models and near-infrared water-leaving radiance correction method in atmospheric correction procedure can be further improved to ensure higher quality data in the further improvement in GDPS. Note that we applied the GDPS 1.4.1 version to process GOCI L1b data in this research. The latest GDPS 2.0 version with improvement for red wavelengths is available at the time of publication, but its corresponding algorithm theoretical basis document (ATBD) in English is not currently available. The performance of the improved algorithm adopted in GDPS 2.0 will be further explored and discussed in our future work.

The analysis performed in this study so far is based on a limited dataset from finite sites (four available sites), and the analysis results can only represent the current study area. The water and aerosol conditions in different regions are diverse, and therefore, further analysis might be needed for a performance assessment of the normalized water-leaving radiance in other oceanic regions with various seawater properties.

Author Contributions: Conceptualization, S.H. and M.H.; methodology, M.H. and S.H.; software, M.H.; validation, M.H.; data curation, M.H. and S.H.; writing—original draft preparation, M.H.; writing—review and editing, M.H., S.H., X.Z., and F.Z.; visualization, M.H. and S.H.; supervision, S.H.; funding acquisition, S.H., F.Z., and P.L. All authors have read and agreed to the published version of the manuscript.

Funding: This work was supported by the National Key Research and Development Plan of China (Grant No. 2016YFC1401603), the National Natural Science Foundation of China (Grant No. 41876031), the Major Science and Technology Project of Sanya (Grant No. SKJC-KJ-2019KY03), the Key Research and Development Plan of Zhejiang Province (Grant No. 2020C03012), and the High-level Personnel of Special Support Program of Zhejiang Province (Grant No. 2019R52045).

Institutional Review Board Statement: Not applicable.

Informed Consent Statement: Not applicable.

Acknowledgments: We thank Joji Ishizaka, Kohei Arai, Young-Je Park, Hak-Yeol You, Jae-Seol Shim, and Joo-Hyung Ryu for their efforts in establishing and maintaining in situ data at four AERONET-OC sites of ARIAKE_TOWER, Ieodo_Station, Socheongcho, and Gageocho_Station. We thank Korea Ocean Satellite Center for providing the GOCI L1b data and GDPS software. We thank NASA SeaDAS software development team.

Conflicts of Interest: The authors declare no conflict of interest.

References

1. Platt, T.; Hoepffner, N.; Stuart, V.; Brown, C. *Why Ocean Colour? The Societal Benefits of Ocean-Colour Technology*; Reports of the International Ocean-Colour Coordinating Group, NO.7: Dartmouth, NS, Canada, 2008.
2. McClain, C.R. A Decade of Satellite Ocean Color Observations. *Annu. Rev. Mar. Sci.* **2009**, *1*, 19–42. [CrossRef]
3. Esaias, W.E.; Abbott, M.R.; Barton, I.; Brown, O.B.; Campbell, J.W.; Carder, K.L.; Clark, D.K.; Evans, R.H.; Hoge, F.E.; Gordon, H.R.; et al. An Overview of MODIS Capabilities for Ocean Science Observations. *IEEE Trans. Geosci. Remote Sens.* **1998**, *36*, 1250–1265. [CrossRef]
4. Ruddick, K.; Neukermans, G.; Vanhellemont, Q.; Jolivet, D. Challenges and Opportunities for Geostationary Ocean Colour Remote Sensing of Regional Seas: A Review of Recent Results. *Remote Sens. Environ.* **2014**, *146*, 63–76. [CrossRef]
5. Antoine, D. *Ocean Color Observation from the Geostationary Orbit*; Reports of the International Ocean-Colour Coordinating Group, NO.12: Dartmouth, NS, Canada, 2012.
6. Ryu, J.-H.; Han, H.-J.; Cho, S.; Park, Y.-J.; Ahn, Y.-H. Overview of Geostationary Ocean Color Imager (GOCI) and GOCI Data Processing System (GDPS). *Ocean Sci. J.* **2012**, *47*, 223–233. [CrossRef]
7. Yang, H.; Kim, M.; Park, Y.-J.; Bae, S.-S.; Han, H.-J. Marine Disaster Detection Using the Geostationary Ocean Color Imager (GOCI). *Int. J. U-E-Serv. Sci. Technol.* **2016**, *9*, 129–138. [CrossRef]
8. Liu, X.; Wang, M. Analysis of Ocean Diurnal Variations from the Korean Geostationary Ocean Color Imager Measurements Using the DINEOF Method. *Estuar. Coast. Shelf Sci.* **2016**, *180*, 230–241. [CrossRef]

9. Lennartson, E.M.; Wang, J.; Gu, J.; Castro Garcia, L.; Ge, C.; Gao, M.; Choi, M.; Saide, P.E.; Carmichael, G.R.; Kim, J.; et al. Diurnal Variation of Aerosol Optical Depth and PM2.5 in South Korea: A Synthesis from AERONET, Satellite (GOCI), KORUS-AQ Observation, and the WRF-Chem Model. *Atmos. Chem. Phys.* **2018**, *18*, 15125–15144. [CrossRef]
10. Choi, J.-K.; Park, Y.J.; Lee, B.R.; Eom, J.; Moon, J.-E.; Ryu, J.-H. Application of the Geostationary Ocean Color Imager (GOCI) to Mapping the Temporal Dynamics of Coastal Water Turbidity. *Remote Sens. Environ.* **2014**, *146*, 24–35. [CrossRef]
11. Hu, Z.; Qi, Y.; He, X.; Wang, Y.-H.; Wang, D.-P.; Cheng, X.; Liu, X.; Wang, T. Characterizing Surface Circulation in the Taiwan Strait during NE Monsoon from Geostationary Ocean Color Imager. *Remote Sens. Environ.* **2019**, *221*, 687–694. [CrossRef]
12. Lou, X.; Hu, C. Diurnal Changes of a Harmful Algal Bloom in the East China Sea: Observations from GOCI. *Remote Sens. Environ.* **2014**, *140*, 562–572. [CrossRef]
13. Son, Y.B.; Choi, B.-J.; Kim, Y.H.; Park, Y.-G. Tracing Floating Green Algae Blooms in the Yellow Sea and the East China Sea Using GOCI Satellite Data and Lagrangian Transport Simulations. *Remote Sens. Environ.* **2015**, *156*, 21–33. [CrossRef]
14. Hu, Z.; Wang, D.-P.; Pan, D.; He, X.; Miyazawa, Y.; Bai, Y.; Wang, D.; Gong, F. Mapping Surface Tidal Currents and Changjiang Plume in the East China Sea from Geostationary Ocean Color Imager. *J. Geophys. Res. Oceans* **2016**, *121*, 1563–1572. [CrossRef]
15. Son, Y.B.; Min, J.-E.; Ryu, J.-H. Detecting Massive Green Algae (Ulva Prolifera) Blooms in the Yellow Sea and East China Sea Using Geostationary Ocean Color Imager (GOCI) Data. *Ocean Sci. J.* **2012**, *47*, 359–375. [CrossRef]
16. Hu, C.; Feng, L.; Lee, Z. Evaluation of GOCI Sensitivity for At-Sensor Radiance and GDPS-Retrieved Chlorophyll-a Products. *Ocean Sci. J.* **2012**, *47*, 279–285. [CrossRef]
17. Bailey, S.W.; Franz, B.A.; Werdell, P.J. Estimation of Near-Infrared Water-Leaving Reflectance for Satellite Ocean Color Data Processing. *Opt. Express* **2010**, *18*, 7521–7527. [CrossRef]
18. Huang, X.; Zhu, J.; Han, B.; Jamet, C.; Tian, Z.; Zhao, Y.; Li, J.; Li, T. Evaluation of Four Atmospheric Correction Algorithms for GOCI Images over the Yellow Sea. *Remote Sens.* **2019**, *11*, 1631. [CrossRef]
19. Concha, J.; Mannino, A.; Franz, B.; Kim, W. Uncertainties in the Geostationary Ocean Color Imager (GOCI) Remote Sensing Reflectance for Assessing Diurnal Variability of Biogeochemical Processes. *Remote Sens.* **2019**, *11*, 295. [CrossRef]
20. Concha, J.; Mannino, A.; Franz, B.; Bailey, S.; Kim, W. Vicarious Calibration of GOCI for the SeaDAS Ocean Color Retrieval. *Int. J. Remote Sens.* **2019**, *40*, 3984–4001. [CrossRef]
21. Kim, W.; Moon, J.-E.; Park, Y.-J.; Ishizaka, J. Evaluation of Chlorophyll Retrievals from Geostationary Ocean Color Imager (GOCI) for the North-East Asian Region. *Remote Sens. Environ.* **2016**, *184*, 482–495. [CrossRef]
22. Lamquin, N.; Mazeran, C.; Doxaran, D.; Ryu, J.-H.; Park, Y.-J. Assessment of GOCI Radiometric Products Using MERIS, MODIS and Field Measurements. *Ocean Sci. J.* **2012**, *47*, 287–311. [CrossRef]
23. Li, H.; He, X.; Ding, J.; Hu, Z.; Cui, W.; Li, S.; Zhang, L. Validation of the Remote Sensing Products Retrieved by Geostationary Ocean Color Imager in Liaodong Bay in Spring. *Acta Opt. Sin.* **2016**, *36*, 17–28.
24. Qi, L.; Hu, C.; Visser, P.M.; Ma, R. Diurnal Changes of Cyanobacteria Blooms in Taihu Lake as Derived from GOCI Observations. *Limnol. Oceanogr.* **2018**, *63*, 1711–1726. [CrossRef]
25. Wu, J.; Chen, C.; Nukapothula, S. Atmospheric Correction of GOCI Using Quasi-Synchronous VIIRS Data in Highly Turbid Coastal Waters. *Remote Sens.* **2020**, *12*, 89. [CrossRef]
26. Yongchao, W.; Shen, F.; Sokoletsky, L.; Sun, X. Validation and Calibration of QAA Algorithm for CDOM Absorption Retrieval in the Changjiang (Yangtze) Estuarine and Coastal Waters. *Remote Sens.* **2017**, *9*, 1192.
27. Huang, C.; Yao, L. Semi-Analytical Retrieval of the Diffuse Attenuation Coefficient in Large and Shallow Lakes from GOCI, a High Temporal-Resolution Satellite. *Remote Sens.* **2017**, *9*, 825. [CrossRef]
28. Zibordi, G.; Holben, B.; Slutsker, I.; Giles, D.; D'Alimonte, D.; Mélin, F.; Berthon, J.-F.; Vandemark, D.; Feng, H.; Schuster, G.; et al. AERONET-OC: A Network for the Validation of Ocean Color Primary Products. *J. Atmos. Ocean. Technol.* **2009**, *26*. [CrossRef]
29. Morel, A.; Antoine, D.; Gentili, B. Bidirectional Reflectance of Oceanic Waters: Accounting for Raman Emission and Varying Particle Scattering Phase Function. *Appl. Opt.* **2002**, *41*, 6289–6306. [CrossRef]
30. Gordon, H.; Wang, M. Retrieval of Water-Leaving Radiance and Aerosol Optical-Thickness Over the Oceans with Seawifs - a Preliminary Algorithm. *Appl. Opt.* **1994**, *33*, 443–452. [CrossRef]
31. Park, Y.J.; Ahn, Y.H.; Han, H.J.; Yang, H.; Moon, J.E.; Ahn, J.H.; Lee, B.R.; Min, J.E.; Lee, S.J.; Kim, K.S. *GOCI Level 2 Ocean Color Products (GDPS 1.3) Brief Algorithm Description*; Korea Ocean Satellite Center: Busan Metropolitan City, Korea, 2014; Available online: http://kosc.kiost.ac.kr/index.nm?menuCd=55 (accessed on 21 December 2020).
32. Ahn, J.-H.; Park, Y.-J.; Ryu, J.-H.; Lee, B.; Oh, I.S. Development of Atmospheric Correction Algorithm for Geostationary Ocean Color Imager (GOCI). *Ocean Sci. J.* **2012**, *47*, 247–259. [CrossRef]
33. Mobley, C.D.; Werdell, J.; Franz, B.; Ahmad, Z.; Bailey, S. *Atmospheric Correction for Satellite Ocean Color Radiometry*; National Aeronautics and Space Administration: Greenbelt, MD, USA, 2016. Available online: https://ntrs.nasa.gov/citations/20160011399 (accessed on 21 December 2020).
34. Stumpf, R.; Arnone, R.; Gould, R.; Martinolich, P.; Ransibrahmanakul, V. A Partially Coupled Ocean-Atmosphere Model for Retrieval of Water-Leaving Radiance from SeaWiFS in Coastal Waters. *NASA Tech. Memo* **2003**, *206892*, 51–59.
35. Ahmad, Z.; Franz, B.A.; McClain, C.R.; Kwiatkowska, E.J.; Werdell, J.; Shettle, E.P.; Holben, B.N. New Aerosol Models for the Retrieval of Aerosol Optical Thickness and Normalized Water-Leaving Radiances from the SeaWiFS and MODIS Sensors over Coastal Regions and Open Oceans. *Appl. Opt.* **2010**, *49*, 5545–5560. [CrossRef]

36. Bailey, S.W.; Werdell, P.J. A Multi-Sensor Approach for the on-Orbit Validation of Ocean Color Satellite Data Products. *Remote Sens. Environ.* **2006**, *102*, 12–23. [CrossRef]
37. Ahn, J.-H.; Park, Y.-J.; Kim, W.; Lee, B.; Oh, I.S. Vicarious Calibration of the Geostationary Ocean Color Imager. *Opt. Express* **2015**, *23*, 23236–23258. [CrossRef] [PubMed]
38. Eck, T.F.; Holben, B.N.; Reid, J.S.; Dubovik, O.; Smirnov, A.; O'Neill, N.T.; Slutsker, I.; Kinne, S. Wavelength Dependence of the Optical Depth of Biomass Burning, Urban, and Desert Dust Aerosols. *J. Geophys. Res. Atmos.* **1999**, *104*, 31333–31349. [CrossRef]
39. Choi, J.-K.; Park, Y.J.; Ahn, J.H.; Lim, H.-S.; Eom, J.; Ryu, J.-H. GOCI, the World's First Geostationary Ocean Color Observation Satellite, for the Monitoring of Temporal Variability in Coastal Water Turbidity. *J. Geophys. Res. Oceans* **2012**, *117*. [CrossRef]
40. Goyens, C.; Jamet, C.; Ruddick, K.G. Spectral Relationships for Atmospheric Correction. I. Validation of Red and near Infra-Red Marine Reflectance Relationships. *Opt. Express* **2013**, *21*. [CrossRef] [PubMed]
41. Wang, M.; Shi, W.; Jiang, L. Atmospheric Correction Using Near-Infrared Bands for Satellite Ocean Color Data Processing in the Turbid Western Pacific Region. *Opt. Express* **2012**, *20*, 741–753. [CrossRef]
42. Ahn, J.-H.; Park, Y.-J. Estimating Water Reflectance at Near-Infrared Wavelengths for Turbid Water Atmospheric Correction: A Preliminary Study for GOCI-II. *Remote Sens.* **2020**, *12*, 3791. [CrossRef]
43. Morel, A.; Gentili, B. Diffuse Reflectance of Oceanic Waters. III. Implication of Bidirectionality for the Remote-Sensing Problem. *Appl. Opt.* **1996**, *35*, 4850–4862. [CrossRef]
44. Li, H.; He, X.; Bai, Y.; Shanmugam, P.; Park, Y.-J.; Liu, J.; Zhu, Q.; Gong, F.; Wang, D.; Huang, H. Atmospheric Correction of Geostationary Satellite Ocean Color Data under High Solar Zenith Angles in Open Oceans. *Remote Sens. Environ.* **2020**, *249*, 112022. [CrossRef]

remote sensing

MDPI

Article

An Improved Cloud Masking Method for GOCI Data over Turbid Coastal Waters

Shiming Lu [1], Mingjun He [1,2], Shuangyan He [1,2,*], Shuo He [1], Yunhe Pan [1], Wenbin Yin [3] and Peiliang Li [1,2]

[1] Ocean College, Zhejiang University, Zhoushan 316021, China; 21934015@zju.edu.cn (S.L.); 21734002@zju.edu.cn (M.H.); shhe@zju.edu.cn (S.H.); 21934020@zju.edu.cn (Y.P.); lipeiliang@zju.edu.cn (P.L.)

[2] Hainan Institute of Zhejiang University, Sanya 572025, China

[3] College of Marine Science and Technology, Zhejiang Ocean University, Zhoushan 316022, China; yinwenbin@zjou.edu.cn

* Correspondence: hesy@zju.edu.cn

Abstract: Clouds severely hinder the radiative transmission of visible light; thus, correctly masking cloudy and non-cloudy pixels is a preliminary step in processing ocean color remote sensing data. However, cloud masking over turbid waters is prone to misjudgment, leading to loss of non-cloudy pixel data. This research proposes an improved cloud masking method over turbid water to classify cloudy and non-cloudy pixels based on spectral variability of Rayleigh-corrected reflectance acquired by the Geostationary Ocean Color Imager (GOCI). Compared with other existing cloud masking methods, we demonstrated that this improved method can identify the spatial positions and shapes of clouds more realistically, and more accurate pixels of turbid waters were retained. This improved method can be effectively applied in typical turbid coastal waters. It has potential to be used in cloud masking procedures of spaceborne ocean color sensors without short-wave infrared bands.

Keywords: cloud masking; turbid water; remote sensing; spectral variability

Citation: Lu, S.; He, M.; He, S.; He, S.; Pan, Y.; Yin, W.; Li, P. An Improved Cloud Masking Method for GOCI Data over Turbid Coastal Waters. *Remote Sens.* **2021**, *13*, 2722. https://doi.org/10.3390/rs13142722

Academic Editor: Chung-Ru Ho

Received: 4 June 2021
Accepted: 9 July 2021
Published: 10 July 2021

1. Introduction

Spaceborne sensors observe the Earth from above the top of atmosphere (TOA); hence, the presence of clouds is often inevitable in optical remote sensing images [1]. Effective ocean color and other surface information can only be extracted from cloudless pixels of satellite remote sensing images, as clouds can block the visible light emerging from the ocean to the sensor. Therefore, the detection and masking of cloud pixels is an essential and important step before further processing in various optical remote sensing applications [2].

Current statistical threshold cloud masking algorithms are mostly based on the analysis of optical and physical characteristics of cloud pixels, such as brightness temperature and reflectance variation, at different bands [3–7]. In infrared remote sensing, the algorithm package called APOLLO (AVHRR Processing scheme Over cLouds, Land and Ocean) has been used since the late 1980s, and its physics is the backbone of a series of cloud detection schemes for AVHRR (Advanced Very High Resolution Radiometer) heritage sensors. In APOLLO, the spatial coherence and dynamic visible threshold tests were conducted after a gross cloud check involving five daytime or nighttime tests. If the reflectance test ratio at near-infrared and visible bands is lower than 1.6 over land or larger than 0.75 over sea or the thin cirrus test using brightness temperatures at 11 μm and 12 μm fail, pixels are flagged as clear [8]. Subsequent studies were conducted to improve and document the APOLLO [9–11]. Thereinto, pixel properties, such as spectral and spatial variability [12,13], and multi-spectral polarization properties [14–17] are further considered. In operational processing of ocean color remote sensing, a threshold of Rayleigh-corrected reflectance (ρ_{rc}) at near-infrared bands is commonly used in cloud masking. For example, a threshold method of $\rho_{rc}(865\ \text{nm}) \geq 0.028$ is used to automatically mask cloud pixels in standard atmospheric correction for the Geostationary Ocean Color Imager (GOCI) sensor. Considering band registration errors or cloud movements, it also masks out their neighboring (up, down, left, and right) pixels [18]. A threshold method of $\rho_{rc}(869\text{ nm}) \geq 0.027$ is used by the Moderate-resolution Imaging Spectroradiometer (MODIS) onboard NASA's Earth Observing System (EOS) Aqua and Terra satellites [19–21]. For simplicity, hereinafter, we call these cloud masking procedure the NIR threshold method. Generally, such methods perform well over open ocean water, but clear pixels with optical characteristics of complex seawater are often mistaken as clouds over coastal water areas [22,23].

Complex water optical conditions affected by high phytoplankton biomass, intense blue-green algal (cyanobacteria) blooms, high suspended matter concentrations, or some anthropogenic factors [24–27] can lead to the failure of the standard threshold method in cloud masking. This often occurs in populated coastal waters, inland lakes, and estuaries. These waters usually have high reflectance at red and near-infrared bands well beyond standard thresholds due to significant contributions from high suspended matter concentrations and eutrophication. The mistaken discrimination of water and cloud reflectance is more likely to lead to the loss of cloudless pixels.

Existing cloud masking methods over coastal waters with complex optical properties can be divided primarily into two categories: threshold methods and machine learning algorithms. For sensors with infrared bands, considering that water absorbs strongly in the short-wave infrared (SWIR) spectral range, progresses have been made in cloud detection methods over turbid coastal waters. For the MODIS sensor, Wang and Shi suggested using the threshold reflectance of 0.0235 and 0.0215 at 1240 nm and 1640 nm [28]. For sediment-laden water in Greenland fjords, Hudson et al. used a difference threshold of 0.05 between TOA reflectance at 645 nm and that estimated at 865 nm, using an empirical relationship to classify turbid water from cloud pixels [29].

For some sensors without SWIR bands, such as the GOCI and Sea-viewing Wide Field-of-view Sensor (SeaWiFS), the credibility of standard cloud identification methods is further diminished [30,31]. Wang and Shi proposed the use of Rayleigh-corrected reflectance ratios of two NIR bands and a Rayleigh-corrected reflectance threshold at 865 nm [28]. Firstly, thick cloudy pixels are masked out with $\rho_{rc}(865\text{ nm}) > 0.06$. Then, for pixels with $\rho_{rc}(865\text{ nm}) \leq 0.06$ and $\rho_{rc}(865\text{ nm}) \geq 0.027$, pixels with $\rho_{rc}(745)/\rho_{rc}(865\text{ nm}) \leq 1.15$ are also identified as clouds. For simplicity, hereinafter, we call this the Wang and Shi method. Nordkvist et al. also proposed a threshold method for cloud recognition based on spectral variability of Rayleigh-corrected reflectance over coastal waters [32]. This algorithm is based on standard ocean color wavelengths. It makes use of the lower spectral variability of clouds compared to that of water. Firstly, the Rayleigh-corrected reflectance of pixels at the four bands of 412 nm, 660 nm, 680 nm, and 865 nm was derived, and then spectral variability ε_{max} was calculated as the maximum value of the reflectance at these four bands divided by the minimum value following

$$\varepsilon_{max} = \frac{\text{MAX}[\rho_{rc}(412\text{ nm}),\ \rho_{rc}(660\text{ nm}),\ \rho_{rc}(680),\ \rho_{rc}(865\text{ nm})]}{\text{MIN}[\rho_{rc}(412\text{ nm}),\ \rho_{rc}(660\text{ nm}),\ ]\rho_{rc}(680),\ \rho_{rc}(865\text{ nm})]}. \quad (1)$$

The combination of ε_{max} and $\rho_{rc}(865\text{ nm})$ is used to identify whether the pixel is cloudless or not. When ε_{max} exceeds 2.5 and $\rho_{rc}(865\text{ nm}) \geq 0.027$, pixels are classified as clouds. For simplicity, hereinafter, we call this the Nordkvist et al. method.

In addition, machine learning algorithms, such as support vector machine (SVM) [33], artificial neural networks [34], image segmentation, and deep convolutional neural network (CNN) [35–38], were also used in many studies in different waters. Machine learning methods usually bring good performance through a large amount of training data; however, their performance is also limited by the time and space range of their training samples.

The first geostationary ocean color sensor GOCI launched in June 2010 is equipped with eight spectral bands ranging from visible to near infrared (412 nm–865 nm) [22], and a second mission GOCI-II with 13 bands (380 nm–865 nm) was launched in February 2020. They can acquire 8/10 images daily with a spatial resolution of 500/250 m [39]. They offer good opportunities for researchers to study diurnal variabilities of coastal environment parameters. However, standard GOCI atmospheric correction processing systematically masks out data over very turbid waters and requires further corrections [40]. We also observed that in sediment-dominated particularly turbid waters, such as the Hangzhou Bay of China, existing cloud masking methods designed for coastal waters also often mistake turbid water pixels as clouds. Thus, in this study, to acquire more water surface pixels in GOCI processing over turbid coastal water, an improved threshold-based cloud masking algorithm is proposed based on the spectral variability of the Rayleigh-corrected reflectance of GOCI. Its performance was further compared with other existing methods. In this paper, the improved cloud mask method is first described. Then, its feasibility is demonstrated. Finally, the performance of the algorithm is evaluated and compared based on image interpretation in different GOCI scenarios.

2. Materials and Methods

2.1. GOCI Data

The GOCI Level 1B data products used in this study were provided by Korean Ocean Satellite Center (KOSC) at http://kosc.kiost.ac.kr/eng/ (accessed on 4 June 2021). The Rayleigh-corrected reflectance at eight spectral bands (412 nm, 443 nm, 490 nm, 555 nm, 660 nm, 680 nm, 745 nm, 865 nm) was processed using the GOCI Data Processing System (GDPS version 1.4.1). GDPS is an officially recognized data processing system for GOCI data by KOSC.

In ocean color remote sensing, the optical properties of water constituents can be retrieved under the premise of an accurate atmospheric correction. The GOCI standard atmospheric correction algorithm [41] was developed based on the theoretical basis of the SeaWiFS standard atmospheric correction algorithm [42], though partially different in the turbid water near-infrared (NIR) correction method and the aerosol models. In the GOCI standard atmospheric correction procedure, the multiply scattered Rayleigh (molecular) reflectance is first removed from the TOA reflectance of $\rho_{TOA}(\lambda)$ following

$$\rho_{TOA}(\lambda) = \rho_r(\lambda) + \rho_a(\lambda) + \rho_{ra}(\lambda) + \{td_r^v(\lambda) \times td_a^v(\lambda) \times td_{ra}^v(\lambda)\}(\lambda)\, \rho_w(\lambda), \tag{2}$$

$$\begin{aligned} \rho_{rc}(\lambda) &= \rho_{TOA}(\lambda) - \rho_r(\lambda) \\ &= \rho_a(\lambda) + \rho_{ra}(\lambda) + \{td_r^v(\lambda) \times td_a^v(\lambda) \times td_{ra}^v(\lambda)\}(\lambda)\rho_w(\lambda), \end{aligned} \tag{3}$$

where $\rho_r(\lambda)$ means multiply scattered Rayleigh reflectance, $\rho_a(\lambda)$ means multiply scattered aerosol reflectance, and $\rho_{ra}(\lambda)$ means reflectance of interactively scattered between aerosols and molecules. In addition, $td_r^v(\lambda)$ means diffuse Rayleigh transmittance from the sea surface to the sensor, $td_a^v(\lambda)$ means diffuse transmittance of aerosols from the sea surface to the sensor, similarly, $td_{ra}^v(\lambda)$ represents diffuse transmittance of Rayleigh and aerosol interaction from the sea surface to the sensor, and $\rho_w(\lambda)$ represents water-leaving reflectance [18]. Then, cloud masking is performed using Rayleigh-corrected reflectance ($\rho_{rc}(\lambda)$) to retain cloud-free pixels. After that, aerosol contribution is estimated and removed, and, finally, the surface reflectance of the ocean is obtained. Thus, correctly masking cloudy and non-cloudy pixels is a preliminary step in processing ocean color remote sensing data.

2.2. An Improved Cloud Masking Method

As a preliminary check of existing cloud masking methods, we applied the NIR threshold method and the Nordkvist et al. method over the turbid water around the Hangzhou Bay and Yangtze River (120–124°E, 29.5–33°N) of China. Figure 1a shows the RGB image composited from the GOCI Rayleigh-corrected reflectance at 680 nm, 555 nm, and 443 nm at 05:16 (Coordinated Universal Time, UTC) on 10 February 2020. Figure 1b shows the pseudo-color images of $\rho_{rc}(865\ \text{nm})$. The red isolines represent the pixels with the threshold of $\rho_{rc}(865\ \text{nm}) = 0.027$ used in the NIR threshold method, and the white pixels were masked with $\varepsilon_{max} \leq 2.5$ used by the Nordkvist et al. method.

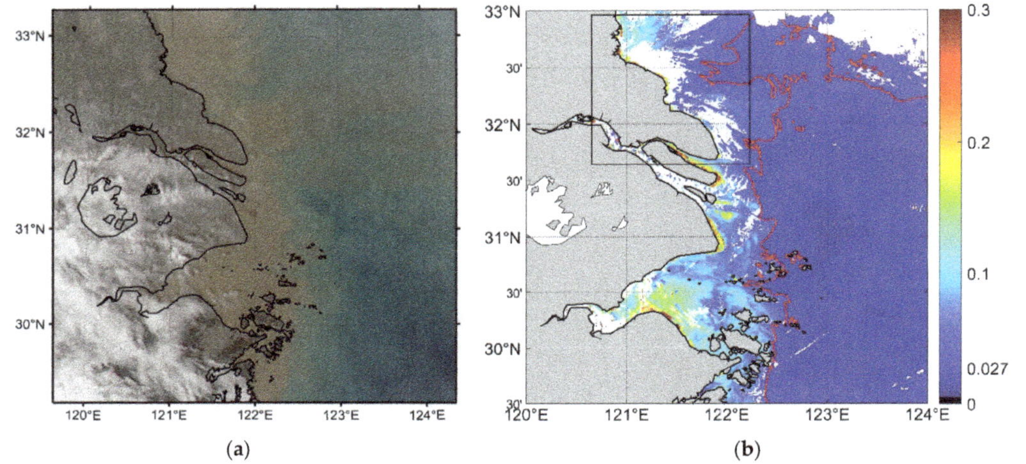

Figure 1. (**a**) The RGB image composited from GOCI Rayleigh-corrected reflectance at 680 nm, 555 nm, and 443 nm at 05:16 (UTC) on 10 February 2020; (**b**) the pseudo-color image of $\rho_{rc}(865\,\text{nm})$. The red isolines in Figure 1b represent $\rho_{rc}(865\,\text{nm}) = 0.027$, and the white pixels represent the pixels with $\varepsilon_{max} \leq 2.5$.

In standard ocean color data products, pixels with $\rho_{rc}(865\,\text{nm}) \geq 0.027$ are often masked as clouds. The red isolines in Figure 1b represent $\rho_{rc}(865\,\text{nm}) = 0.027$. We can see that a large number of cloudless coastal turbid water pixels with $\rho_{rc}(865\,\text{nm})$ values higher than the red isolines can be mistaken as cloudy ones in standard ocean color data products [43]. By contrast, the Nordkvist et al. method performs better in most coastal turbid waters. Most cloudless coastal turbid water pixels are correctly recognized, but cloudless pixels with very turbid water or silt coast are still misjudged as clouds. Therefore, it is necessary to further improve existing cloud masking methods over very turbid water. In this paper, we propose an improved cloud masking scheme by combining additional threshold methods with the Nordkvist et al. method to address this problem.

Clear water pixels generally have low Rayleigh-corrected reflectance at the 865 nm band and high spectral variations and can be easily distinguished from cloud pixels by the NIR threshold method. However, for coastal turbid water pixels with complex optical characteristics, their Rayleigh-corrected reflectance values are generally higher than that of clear water at red and near-infrared bands and lower than that of clouds at short visible bands. These turbid water pixels are easily mistaken as clouds by the standard NIR threshold method.

To demonstrate the difference of spectral variability between turbid water and cloud pixels, eight samples of Rayleigh-corrected spectral reflectance were selected around sediment-dominated Hangzhou Bay from the GOCI data acquired at 07:16 (UTC) on 10 May 2018. Their spectral Rayleigh-corrected reflectance at 8 bands are shown in Figure 2a, and the sample locations are shown in Figure 2b. The arithmetic mean values of 3 × 3 pixels centered at each sample were used in order to reduce the random noise effects of the satellite data.

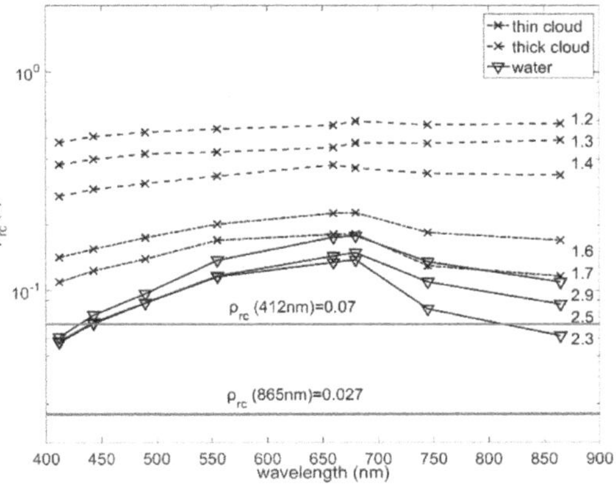

(a)

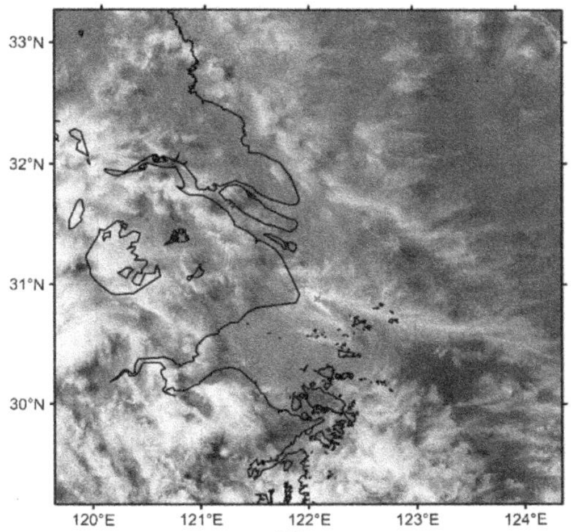

(b)

Figure 2. (a) Spectral variation of Rayleigh-corrected reflectance at 8 bands from eight typical GOCI samples in turbid waters around Hangzhou Bay of China. The ε_{max} values calculated according to Equation (1) in Nordkvist et al. method are also labeled. (b) Locations of eight typical samples indicated in red crosses and blue circles on the GOCI RGB image at 07:16 UTC on 10 May 2018. The circles represent water samples visually selected, and the crosses represent clouds.

As seen from Figure 2a, the sampled Rayleigh-corrected reflectance of thick cloud pixels has higher magnitudes and smaller spectral variations. The spectral reflectance of thin clouds and over cloudless turbid water is close, especially at red bands, and the primary difference can be observed in the blue spectral range, such as at 412 nm. We also observed that the $\rho_{rc}(412\text{ nm})$ values of most turbid waters are generally over 0.07. In terms of correctly distinguishing cloudy and non-cloudy pixels with turbid water, we tested whether a threshold of $\rho_{rc}(412\text{ nm}) > 0.07$ in combination with $\varepsilon_{max} < 2.5$ and $\rho_{rc}(865\text{ nm}) \geq 0.027$ can effectively retain water pixels.

It is known that the spectral characteristics of clear water and turbid water are different. Samples in clearer water generally have larger $\rho_{rc}(412\text{ nm})$ and smaller $\rho_{rc}(660\text{ nm})$, and it is just the reverse in turbid water. The $\rho_{rc}(412\text{ nm})$ values in clear water can be larger than 0.07. This could cause some of the turbid water pixels to be mistaken as clouds. Based on sampled reflectance data over clear and turbid water, we observed that the ratio of $\frac{\rho_{rc}(412\text{ nm})}{\rho_{rc}(660\text{ nm})}$ can be used to discriminate clear and turbid water in our cloud masking procedure. Therefore, with the purpose of distinguishing the clear water pixels from cloudy ones correctly, a second threshold of

$$\frac{\rho_{rc}(412\text{ nm})}{\rho_{rc}(660\text{ nm})} > 1 \tag{4}$$

in combination of $\varepsilon_{max} < 2.5$ and $\rho_{rc}(865\text{ nm}) \geq 0.027$ is performed.

In brief, our improved cloud masking procedure is concluded as the following:

Step 1. calculate the Rayleigh-corrected reflectance at the four bands of 412 nm, 660 nm, 680 nm, and 865 nm;
Step 2. calculate the ε_{max} according to Equation (1), which was proposed by Nordkvist et al.;
Step 3. pixels satisfying either of the following two conditions (1 or 2) are masked as clouds:

(1) $\rho_{rc}(412\text{ nm}) > 0.07$ and $\varepsilon_{max} < 2.5$ and $\rho_{rc}(865\text{ nm}) \geq 0.027$;
(2) $\frac{\rho_{rc}(412\text{ nm})}{\rho_{rc}(660\text{ nm})} > 1$ and $\varepsilon_{max} < 2.5$ and $\rho_{rc}(865\text{ nm}) \geq 0.027$.

For simplicity, hereinafter, we call this the improved method in this study.

3. Results

3.1. Comparison of Four Cloud Masking Methods Using Selected Samples

The cloud masking effects of the NIR threshold method, Wang and Shi method, Nordkvist et al. method, and the improved method in this study were compared. These four methods were, respectively, applied and tested using visually selected samples in the inland lake water and turbid coastal water off the mouth of Yangtze River, which are typical turbid waters in China.

A total of 136 samples were visually selected from the GOCI data acquired at 02:16 (UTC) on 19 October 2014, 01:16 (UTC) on 19 August 2017, 06:16 (UTC) on 11 May 2018, and 04:16 (UTC) on 19 November 2019 around Lake Tai, Hangzhou Bay, and the coastal area of Jiangsu Province of China. The sample locations are indicated in Figure 3. These samples include cloudy ones covered by thick clouds and cloud-free ones over turbid water, as well as lakes in different weather conditions and near narrow straits. The circles represent visually selected cloudless samples, and the crosses represent cloudy ones. As a performance comparison demonstration, we applied four cloud masking methods to these visually selected 136 samples.

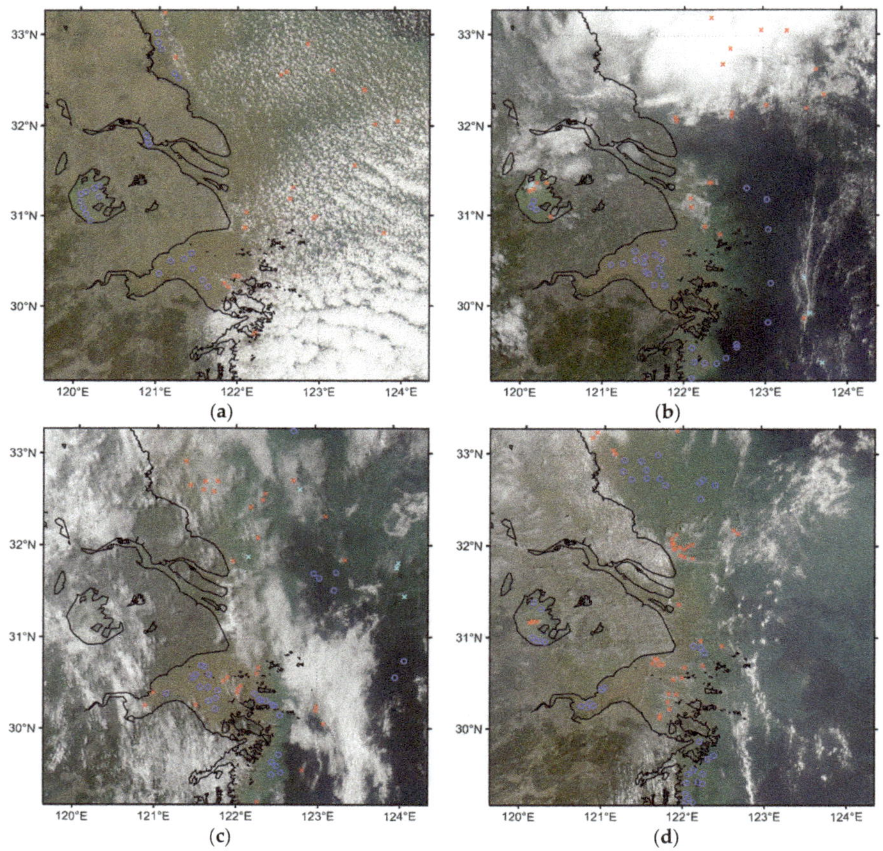

Figure 3. Positions of (**a**) 45 samples selected on 19 October 2014, (**b**) 24 samples selected on 19 November 2019, (**c**) 36 samples selected on 11 May 2018, and (**d**) 31 samples selected on 19 August 2017. The background images are RGB ones synthesized using Rayleigh-corrected reflectance at three bands of 443 nm, 555 nm, and 680 nm. The circles represent water samples visually selected, and the crosses represent clouds. Red and bright cyan represent the clouds identified by conditions 1 and 2 in the improved cloud masking method in this study, respectively.

The scatter plot of $\rho_{rc}(865\ \text{nm})$ versus ε_{max} for these 136 samples is shown in Figure 4. The vertical dashed line is $\rho_{rc}(865\ \text{nm}) = 0.027$, which represents the standard cloud masking threshold method instandard atmospheric correction procedure. The horizontal dashed line represents a threshold of 2.5 for ε_{max} used in the Nordkvist et al. method. It is clear that the standard NIR method using $\rho_{rc}(865\ \text{nm}) \geq 0.027$ is not enough in coastal waters, as many cloudless samples (circles) are with $\rho_{rc}(865\ \text{nm}) \geq 0.027$. The Nordkvist et al. method using $\varepsilon_{max} < 2.5$ and $\rho_{rc}(865\ \text{nm}) \geq 0.027$ can mask all the cloudy samples, while some cloudless samples indicated by blue circles in the fourth quadrant of Figure 4 will be incorrectly masked. These usually correspond to samples in lake water, near narrow straits, and near land.

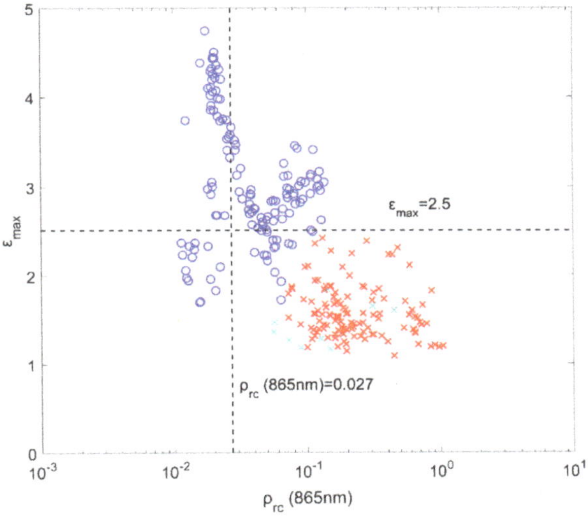

Figure 4. Scatter plot of $\rho_{rc}(865\ \text{nm})$ versus ε_{max} for 136 visually selected samples as shown in Figure 3. The circles represent water samples visually selected, and the crosses represent clouds. Samples with $\rho_{rc}(412\ \text{nm})$ exceeding a threshold of 0.07 are indicated in red, and those with $\frac{\rho_{rc}(412\ \text{nm})}{\rho_{rc}(660\ \text{nm})}$ exceeding a threshold of 1 are marked in bright cyan.

In Figure 4, by applying the improved cloud masking method in this study, samples with $\rho_{rc}(412\ \text{nm}) > 0.07$ are indicated in red, and those with $\frac{\rho_{rc}(412\ \text{nm})}{\rho_{rc}(660\ \text{nm})} > 1$ are marked in bright cyan. If the visually selected cloud pixels do not pass these two threshold tests, they are masked in black, and the visually selected water pixels are in blue. It can be seen that all crosses visually selected as clouds are correctly masked by the improved method in this study. Combining with the sample positions in Figure 3, we can see that $\rho_{rc}(412\ \text{nm}) > 0.07$ works primarily in very turbid yellow colored water, and the $\frac{\rho_{rc}(412\ \text{nm})}{\rho_{rc}(660\ \text{nm})} > 1$ takes effect primarily in clearer green colored water.

In Figure 5, by applying the Wang and Shi method, samples with $\rho_{rc}(865\ \text{nm}) > 0.06$ are indicated in bright cyan, and those with $\frac{\rho_{rc}(412\ \text{nm})}{\rho_{rc}(660\ \text{nm})} < 1.15$ in combination with $\rho_{rc}(865\ \text{nm}) \leq 0.06$ and $\rho_{rc}(865\ \text{nm}) \geq 0.027$ are marked in red. We can see that all crosses visually selected as clouds are correctly masked; however, one red circle and many bright cyan circles, which are supposed to be water pixels, are mistaken as clouds.

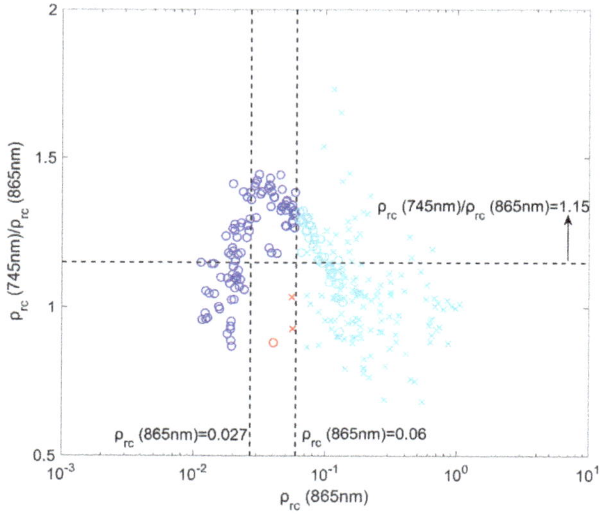

Figure 5. Scatter plot of $\rho_{rc}(865\text{ nm})$ versus $\rho_{rc}(745\text{ nm})/\rho_{rc}(865\text{ nm})$ for 136 visually selected samples as shown in Figure 3. The circles represent water samples visually selected, and the crosses represent clouds. Samples with $\frac{\rho_{rc}(745\text{ nm})}{\rho_{rc}(865\text{ nm})} < 1.15$, $\rho_{rc}(865\text{ nm}) \leq 0.06$, and $\rho_{rc}(865\text{ nm}) \geq 0.027$ are indicated in red, and those with $\rho_{rc}(865\text{ nm})$ exceeding a threshold of 0.06 are marked in bright cyan.

Therefore, based on our 136 visually selected samples, it is clear that the improved method in this study performs better than the other three methods. To further check the general application of the four cloud masking methods, we then applied them in different GOCI scenes.

3.2. Performance Comparison Over Typical Turbid Waters

Figure 6a shows the RGB image composited by Rayleigh-corrected reflectance at the three bands of 443 nm, 555 nm, and 680 nm at 05:16 (UTC) on 10 February 2020. White pixels in the RGB image are clouds. Figure 6b–e shows the results of the four cloud masking methods. Pixels identified by cloud masking methods are also indicated as white. Compared with the RGB image in Figure 6a, we can see that the NIR threshold method as well as the Wang and Shi method masked both clouds and land as clouds. The NIR threshold method masked almost all turbid coastal water (yellow colored water) as clouds in Figure 6b. The Wang and Shi method masked most of the turbid coastal water as clouds in Figure 6c. The Nordkvist et al. method is better in retaining turbid coastal water pixels, but it still misjudged some very turbid water pixels as clouds in the coastal water of Jiangsu Province of China as indicated by the red rectangle in Figure 6d. The improved method in this study performed better than the other three methods. It can mask almost all cloud pixels with fewer misjudgments over almost all turbid water pixels. It is worthy to note that there is thin fog in the upper right corner of the image in Figure 6a. The water here is clearer, which is indicated by the RGB color in Figure 6a. The Wang and Shi method can retain more pixels over this foggy clear water area. The NIR threshold method masked more pixels. The Nordkvist et al. method and the improved one in this study masked less than the NIR threshold method and more than the Wang and Shi method.

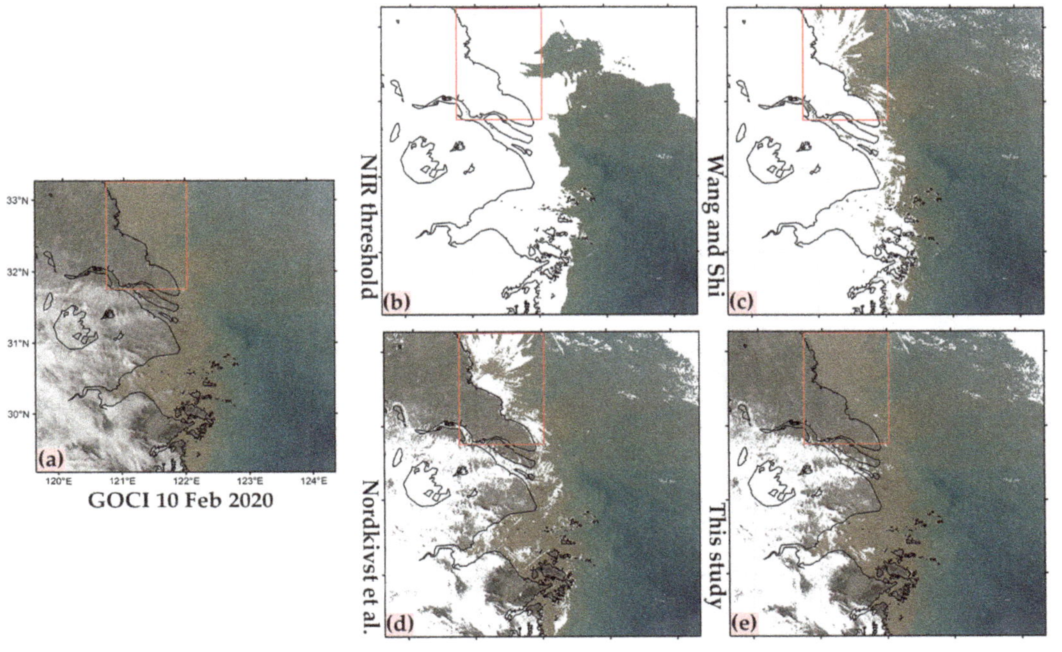

Figure 6. Comparison of four cloud masking methods in turbid coastal water off the mouth of Yangtze River at 05:16 (UTC) on 10 February 2020. The images are (**a**) the RGB image composited by GOCI-derived Rayleigh-corrected reflectance at 443 nm, 555 nm, and 680 nm; the cloud masking results using the NIR threshold method (**b**); Wang and Shi method (**c**); the Nordkvist et al. method (**d**); and the improved method in this study (**e**). Clouds are indicated as white. The red rectangle marks the turbid coastal water of Jiangsu Province of China.

Figure 7 shows cloud recognition results of the four different cloud masking methods applied to the GOCI data at 01:16 (UTC) on 19 August 2017. All four methods performed well in clear water off the coast. For the inland lake water indicated by the left red rectangle, by comparing it with the RGB image in Figure 7a, it can be observed that the NIR method retained the fewest water pixels and misjudged most water pixels as clouds, followed by the Wang and Shi method and then by the Nordkvist et al. method; the best was the improved method in this study. In addition, it is not difficult to see that the Nordkvist et al. method is prone to misjudge water near the land edge and in narrow water channels as clouds. This can also be seen in the mouth of the Yangtze River and the Hangzhou Bay of China indicated by the right red rectangle in Figure 7d.

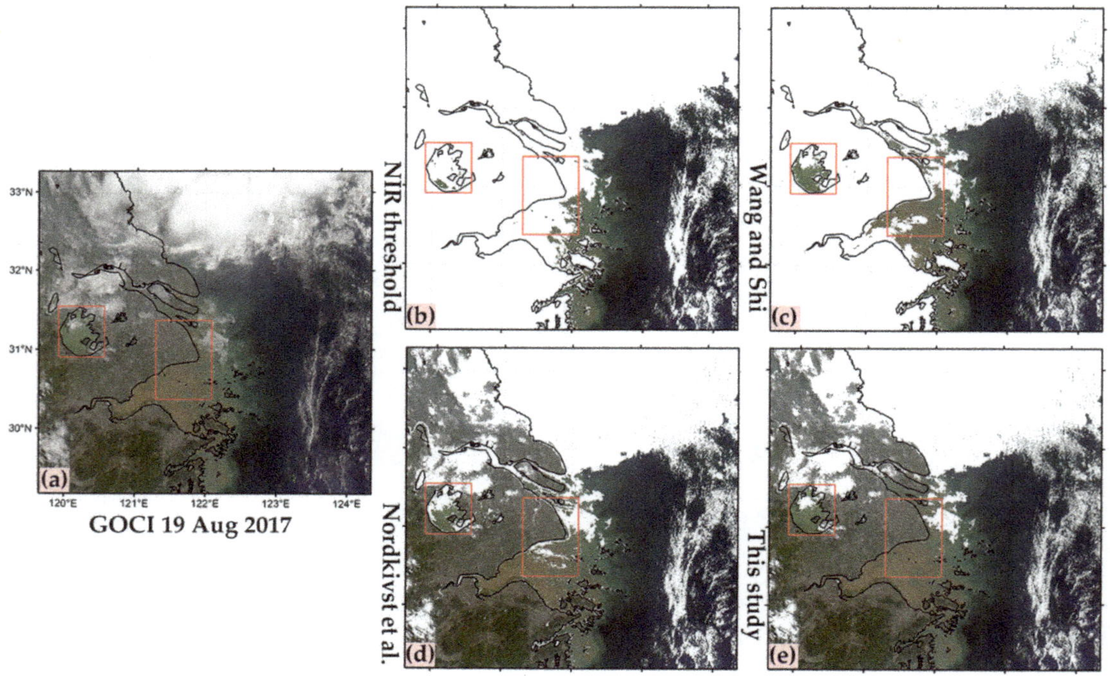

Figure 7. Comparison of four cloud masking methods over Lake Tai and coastal water around Hangzhou Bay of China at 01:16 (UTC) on 19 August 2017. The images of (**a–e**) are the same as Figure 6. Clouds are indicated as white. The smaller red rectangle (left one) marks the location of Lake Tai. The bigger red rectangle (right one) marks turbid coastal water around the mouth of Yangtze River and Hangzhou Bay of China.

To further verify the adaptability of the improved cloud masking method in hazy weather conditions, Figure 8 compares the results of the four cloud masking methods on 6 January 2020. Comparing the RGB image in Figure 8a with the water area around the mouth of Yangtze River, indicated by the right smaller red rectangle, and the west part of Hangzhou Bay, indicated by the left bigger red rectangle in Figure 8b–e, the results of the Nordkvist et al. method show serious misjudgments in water near land edges and in narrow water channels. The improved method in this study performed better than the others.

To quantitatively evaluate the performance of the four cloud masking methods, the clear pixel numbers recognized by the different methods over the selected turbid water regions as indicated in Figures 6–8 are listed in Table 1. The clear pixel percentage is the ratio of the non-cloudy pixels to the total pixels in the selected region. Based on these five selected regions, the improved method in this study can retain a higher number of clear pixels with an average clear pixel percentage of 92.77%, which is 19%, 55%, and 89% higher than that of the Nordkvist et al., the Wang and Shi, and the NIR threshold methods, respectively.

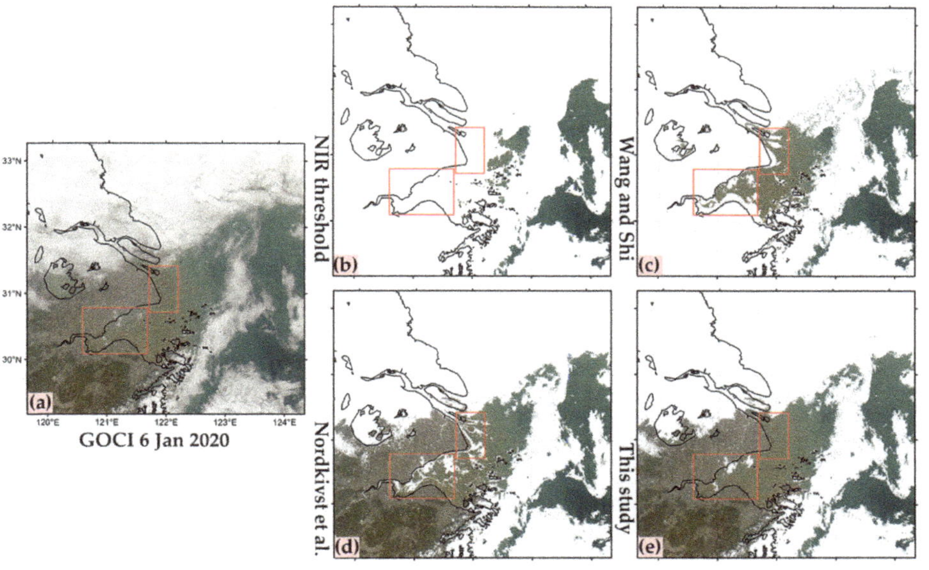

Figure 8. Comparison of four cloud masking methods over coastal water around the mouth of Yangtze River and Hangzhou Bay of China on 6 January 2020. The images of (**a**–**e**) are the same as Figure 6. Clouds are indicated as white. The smaller red rectangle (right one) marks the location around the mouth of Yangtze River, and the bigger red rectangle (left one) marks the west part of Hangzhou Bay of China.

Table 1. Comparison of clear pixel numbers recognized by four cloud masking methods over selected turbid water regions as indicated in Figures 6–8.

Region	Date	Total Pixel Number	Clear Pixel Percentage			
			NIR Threshold	Wang and Shi	Nordkvist et al.	This Study
Lake Tai	19 August 2017	17,952	4.23%	28.67%	65.61%	79.77%
Hangzhou Bay		35,321	9.65%	40.25%	82.59%	94.63%
Subei coastal	10 February 2020	83,659	2.88%	33.59%	70.65%	99.37%
Hangzhou Bay	6 January 2020	30,056	0.00%	29.12%	77.65%	92.12%
Yangtze River		15,836	0.01%	55.41%	71.47%	97.95%
Averaged clear pixel percentage			3.35%	37.41%	73.59%	92.77%

3.3. *Performance Comparison Over Other Areas*

As the improved cloud masking method in this study is designed for turbid waters, in this section, illustrations are given to demonstrate whether it performs as well as other methods in water areas other than the coastal waters off the mouth of Yangtze River, such as waters with high chlorophyll concentrations or clear water. For the above purpose, we selected GOCI scenarios acquired in Bohai of China and the water area off the south of the Korean Peninsula.

Figure 9a shows the RGB image composited by GOCI Rayleigh-corrected reflectance in Bohai at 03:16 (UTC) on 1 April 2019. As seen from the RGB color, the water in Bohai is dominated by high chlorophyll concentrations. It is surrounded by silt and tidal flats along the coast of Bohai, as shown in the red rectangle.

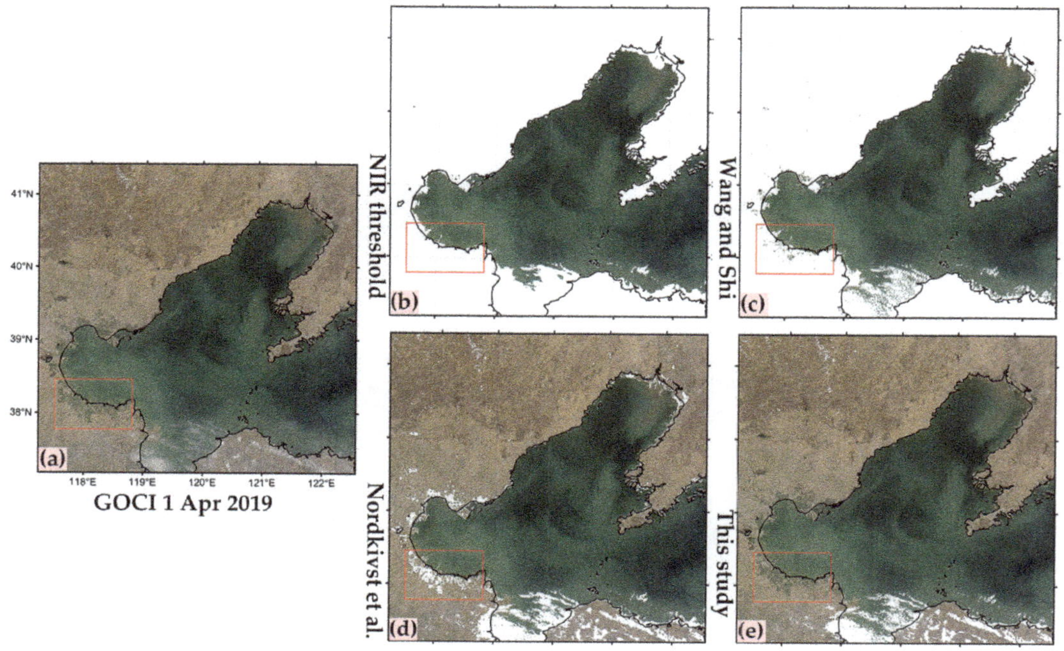

Figure 9. Comparison of four cloud masking methods over Bohai of China at 03:16 (UTC) on 1 April 2019. The images of (**a–e**) are the same as Figure 6. Clouds are indicated as white. The red rectangle marks a typical region of silt and tidal flats along the coast of Bohai.

Comparing the cloud recognition results of four cloud masking methods in Figure 9b–e, it can be seen that all the four cloud masking methods perform well in waters with high chlorophyll concentrations in Bohai. The NIR threshold method as shown in Figure 9b misidentified the sediment-dominated turbid water and shallow water over the tidal flats along the shorelines as clouds. The Wang and Shi method (Figure 9c) and Nordkvist et al. method (Figure 9d) can correctly recognize most of the turbid water pixels, while the cloudless shallow water pixels over the tidal flats along the shorelines were still misidentified as clouds. By comparison, the improved cloud masking method in this study can delineate the spatial location and morphology of clouds more accurately and retain more turbid water pixels (Figure 9e).

Figure 10a shows the RGB image composited by GOCI Rayleigh-corrected reflectance off and around the south of the Korean Peninsula at 03:16 (UTC) on 18 April 2019. The water along the southern coast of the Korean peninsula as indicated by a red rectangle in Figure 10 is turbid with sediments composed of fine sand and silt [44]. The water over the southeast of the Korean Peninsula is clear open water, and the southwest water is affected by turbid water from Yangtze River.

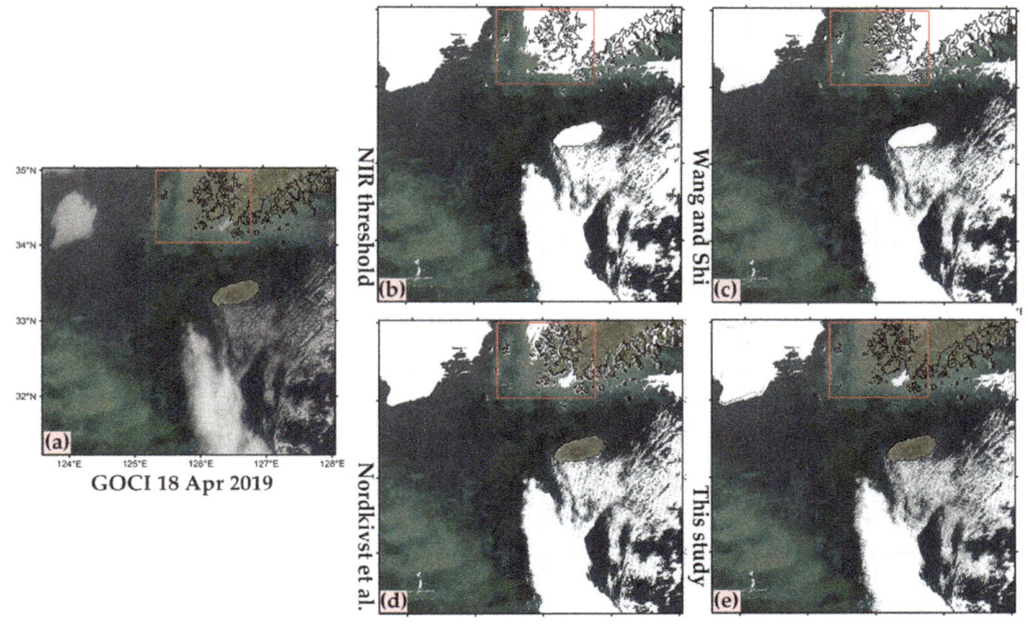

Figure 10. Comparison of four cloud masking methods over water area off the south of the Korean Peninsula at 03:16 (UTC) on 18 April 2019. The images of (**a–e**) are the same as Figure 6. Clouds are indicated as white. The red rectangle marks the coastal turbid water area around the Korean Peninsula.

In comparison to the RGB image in Figure 10a, the four cloud masking methods in Figure 10b–e all perform well over clear water toward the southeast of the Korean Peninsula, and this is also the same case in medium turbid water toward the southwest of the Korean Peninsula. As for the water with high sediment concentrations in the red box, the NIR threshold, Wang and Shi method, and Nordkvist et al. method all have misjudgment at different levels. The results of the improved method in this study (Figure 10e) agree well with the cloud distribution shown in the RGB composite image (Figure 10a). Therefore, our improved cloud masking method can be successfully applied over not only turbid waters but also clear waters.

It is worthy to note that there is thick cloud and thin fog in the upper left corner of the image in Figure 10a. Over this area, the four cloud masking methods show similar results in Figure 10b–e, while they show little difference along the edge between thick cloud and thin fog. It is difficult to judge visually whether it is cloudy or not along the edge. To reduce the misjudgment error, it is recommended to discard these edge pixels by masking out the pixels neighboring the cloud pixels (up, down, left, and right) according to the GOCI algorithm theoretical basis document [18].

The clear pixel numbers recognized by the different methods over the selected water regions as indicated in Figures 9 and 10 are listed in Table 2. Based on these two selected regions, the improved method in this study can retain a higher number of clear pixels with an average clear pixel percentage of 98.66%, which is 14%, 51%, and 61% higher than that of the Nordkvist et al., the Wang and Shi, and the NIR threshold methods, respectively.

Table 2. Comparison of clear pixel numbers recognized by four cloud masking methods over selected water regions as indicated in Figures 9 and 10.

Region	Date	Total Pixel Number	Clear Pixel Percentage			
			NIR Threshold	Wang and Shi	Nordkvist et al.	This Study
Bohai Bay	1 April 2019	31,104	31.77%	36.97%	85.58%	99.88%
Korean Peninsula	18 April 2019	57,449	41.85%	56.99%	81.85%	97.43%
Averaged clear pixel percentage			36.81%	46.98%	83.72%	98.66%

4. Discussion

The improved cloud masking method is achieved by combining the spectral variability threshold of ε_{max} by Nordkvist et al. with two other thresholds of $\rho_{rc}(412\text{ nm})$ and $\frac{\rho_{rc}(412\text{ nm})}{\rho_{rc}(660\text{ nm})}$ proposed in this study. In order to further discuss the necessity of the new threshold scheme and consider whether the same effect can be achieved by simply relaxing or tightening the ε_{max} threshold in the Nordkvist et al. method, we made a simple attempt as shown in Figure 11.

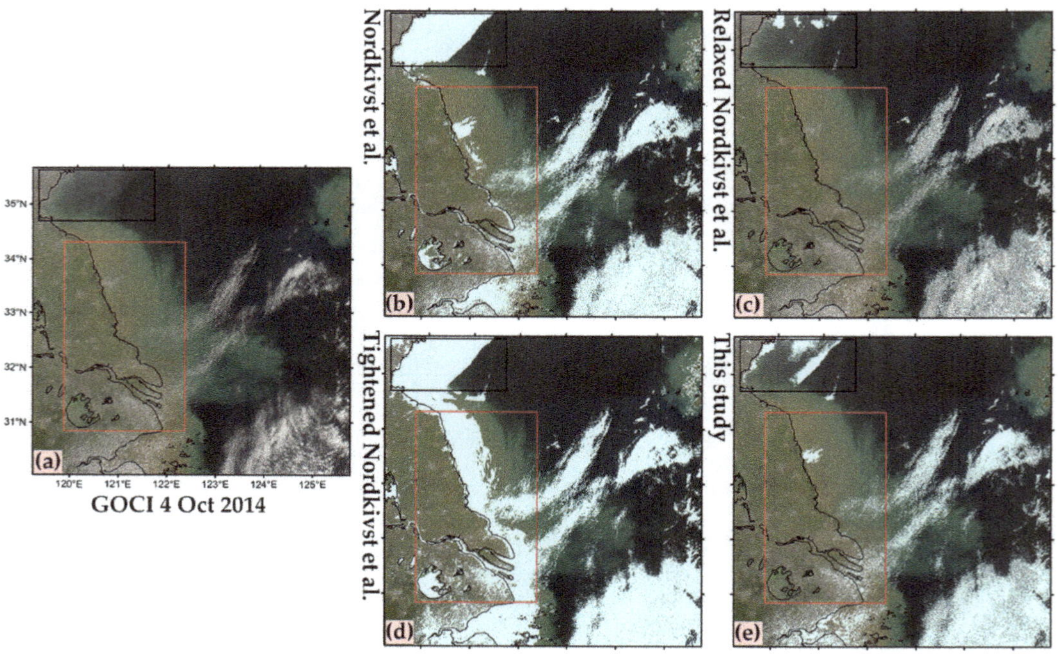

Figure 11. (**a**) GOCI RGB images (680 nm, 555 nm, 443 nm) acquired at 02:16 (UTC) on 4 October 2014; (**b**) cloud masking results using Nordkvist et al. method; (**c**) cloud masking results after adjusting the ε_{max} threshold value in Nordkvist et al. method to 1.5; (**d**) cloud masking results after adjusting the ε_{max} threshold value in Nordkvist et al. method to 3.5; (**e**) cloud masking results using the improved method in this study. Clouds are indicated as light green. The red rectangle marks the turbid coastal water of Jiangsu Province of China. The black rectangle indicates the foggy or thin cloudy water area off the south of Shandong Peninsula.

It can be seen from Figure 11 that if the ε_{max} threshold value in the Nordkvist et al. method is set to be larger than 2.5, many cloudless coastal turbid water pixels are misjudged as clouds (Figure 11d). If the ε_{max} threshold value is set to be smaller than 2.5, many cloudless turbid water pixels, such as the water pixels in lakes, narrow rivers and coastal channels, and near land, are retained (Figure 11c). However, thin cloud pixels over turbid water are also mistaken as water pixels as indicated in the red rectangle. The improved method in this study (Figure 11e) performs well in both lakes and coastal turbid waters. Thus, by simply changing the ε_{max} threshold in the Nordkvist et al. method, it is difficult to achieve the same effect over turbid water as well as the improved method in this study. This also proves the practicability and effectiveness of the improved cloud masking method in this study.

The improved method in this study can perform well over thin cloudy yellow colored turbid water pixels. It is also worthy to note that, as indicated by the black rectangle, the improved method in this study seems to retain some foggy and thin cloudy pixels over clearer waters. By comparison, the Nordkvist et al. method performs better here. Note that the performance evaluation in this study is primarily based on manual visual judgements. It is difficult to discriminate the thin clouds and foggy pixels from clear pixels. This may bring some errors in performance evaluations and comparisons. If users want to apply the improved method in this study primarily onto clearer water, we recommend adjusting the threshold value of $\frac{\rho_{rc}(412\ \text{nm})}{\rho_{rc}(660\ \text{nm})}$ to smaller than 1 to remove thin cloud pixels. The specific value needs to be adjusted by regional water samples.

The improved cloud masking method in this study is designed for GOCI, and typical operational ocean color sensors have similar spectral bands in the visible and near-infrared range. Therefore, the improved method in this study is supposed to be applicable to almost all ocean color sensors. Here, we test the improved method in this study using MODIS data.

Figure 12a shows MODIS (onboard Aqua) RGB images (645 nm, 555 nm, 469 nm) acquired on 10 February 2020, almost synchronized with the GOCI data in Figure 6. By using the Nordkvist et al. method and the improved method in this study (Figure 12b,c), the improved method in this study performs better over the Hangzhou Bay and Yangtze River coastal area than the Nordkvist et al. method. However, it seems to mask out all pixels in the Subei coastal area. If we modify the $\rho_{rc}(412\ \text{nm})$ threshold value from 0.07 to 0.09, most water pixels in the Subei coastal area are retained. This implies that if one applies the improved cloud masking method in this study to ocean color sensors other than GOCI, the $\rho_{rc}(412\ \text{nm})$ threshold value may need to be slightly adjusted. Different sensors observe the ocean with different viewing angles, and this may lead to the deviation of Rayleigh-corrected reflectance at 412 nm.

Although this improved cloud masking method generally performs better compared to other threshold methods, misjudgments may still occur in the presence of large amounts of aerosols in the atmosphere. The cloud masking methods easily mistook the aerosol-laden pixels as turbid water ones. This improved cloud masking method could be further refined based on the spectral variability analysis of a number of samples with aerosols over the coastal region. Due to huge changes in clouds, their shape aspect is not currently considered. Cloud edge and continuity aspects may need to be further improved in the future. In addition, combined with an accurate cloud masking method, feasible techniques to reconstruct missing data, such as the DINEOF (Data Interpolating Empirical Orthogonal Functions), are promising to be used to restore more water pixels obscured by clouds [45,46].

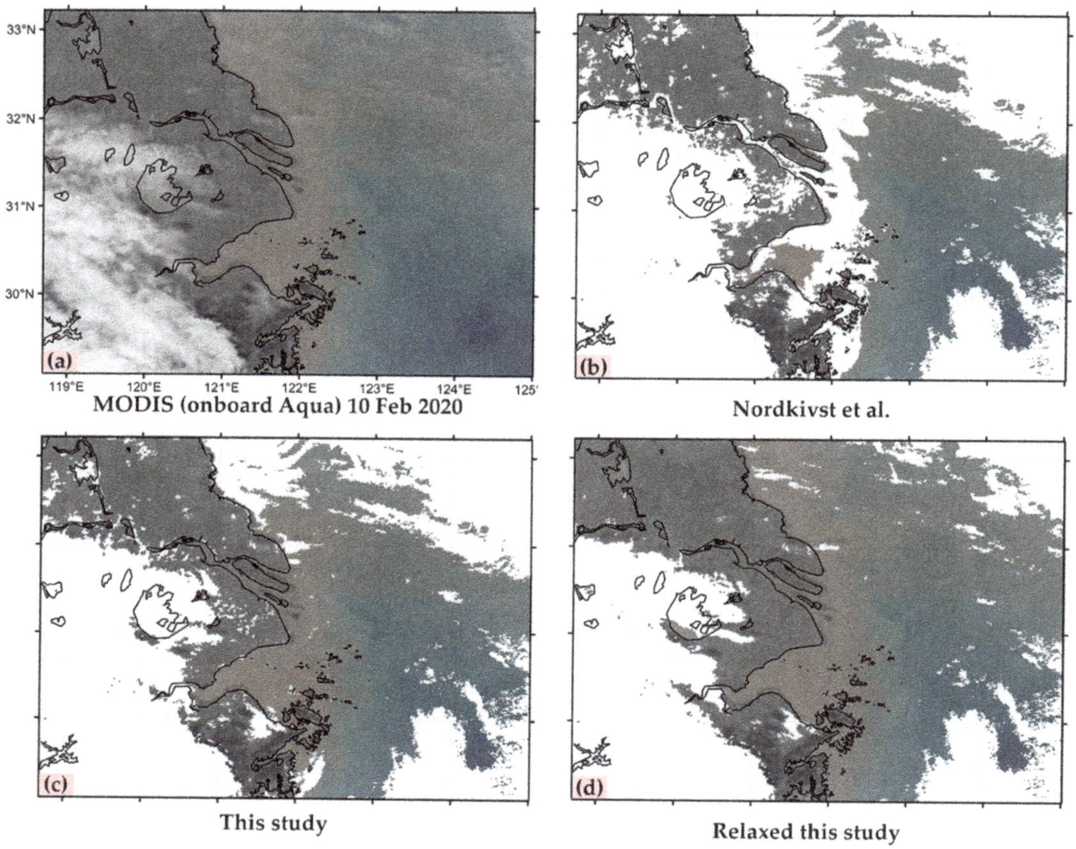

Figure 12. (**a**) MODIS (onboard Aqua) RGB images (645 nm, 555 nm, 469 nm) acquired on 10 February 2020; (**b**) cloud masking results using Nordkvist et al. method; (**c**) cloud masking results using the improved method in this study; (**d**) cloud masking results using the improved method in this study after adjusting the $\rho_{rc}(412\ \text{nm})$ threshold value from 0.07 to 0.09. Clouds are indicated as white.

5. Conclusions

The geostationary ocean color sensor GOCI offers good opportunities to study diurnal variabilities of coastal environment dynamics. However, existing cloud masking methods often mistake turbid water pixels as clouds in sediment-dominated particularly turbid waters, such as the coastal area of Jiangsu Province and the Hangzhou Bay of China. In this study, based on 136 samples selected from the GOCI data on four individual days, the spectral variability of Rayleigh-corrected reflectance over turbid water and cloud pixels was analyzed. According to the characteristics of the Rayleigh-corrected reflectance at 412 nm and 660 nm bands, an improved cloud masking method combining two threshold tests of $\rho_{rc}(412\ \text{nm}) > 0.07$ and $\frac{\rho_{rc}(412\ \text{nm})}{\rho_{rc}(660\ \text{nm})} > 1$ with the Nordkvist et al. method was proposed in this study.

The cloud masking effects of the NIR threshold method, Wang and Shi method, Nordkvist et al. method, and the improved method in this study were further compared and evaluated. These four methods were, respectively, applied and tested in inland lake water, sediment-dominated and phytoplankton-dominated turbid coastal waters, and clearer waters. Results show that the improved method in this study performs better than the others over typical turbid water and can effectively retain more turbid water pixels. However, the improved method in this study sometimes may not be able to mask out thin foggy pixels over clearer water.

The improved cloud masking method in this study is designed for GOCI based on typical ocean color spectral bands in the visible and near-infrared range. Thus, it can be applied to almost all ocean color sensors, especially for those without SWIR bands. It can retain more effective coastal water information in data products of ocean color remote sensing processing in support of various optical remote sensing studies and applications.

Author Contributions: Conceptualization, S.H. (Shuangyan He); methodology, S.L., S.H. (Shuangyan He), and S.H. (Shuo He); software, S.L., M.H., Y.P., and W.Y.; validation, S.L. and S.H. (Shuo He); writing—original draft preparation, S.L.; writing—review and editing, S.H. (Shuangyan He), M.H., and W.Y.; funding acquisition, S.H. (Shuangyan He) and P.L. All authors have read and agreed to the published version of the manuscript.

Funding: This research was funded by the National Key Research and Development Plan of China, grant number 2018YFD0900901; the National Natural Science Foundation of China, grant number 41876031; the Hainan Provincial Natural Science Foundation of China, grant number 420QN289; the Major Science and Technology Project of Sanya, grant number SKJC-KJ-2019KY03; the Key Research and Development Plan of Zhejiang Province, grant number 2020C03012; and the High-level Personnel of Special Support Program of Zhejiang Province, grant number 2019R52045.

Institutional Review Board Statement: Not applicable.

Informed Consent Statement: Not applicable.

Acknowledgments: We thank Korea Ocean Satellite Center for providing the GOCI L1b data and the GDPS software. We thank Pei Sun Loh for help with English.

Conflicts of Interest: The authors declare no conflict of interest.

References

1. Tsai, Y.-L.S.; Dietz, A.; Oppelt, N.; Kuenzer, C. Remote Sensing of Snow Cover Using Spaceborne SAR: A Review. *Remote Sens.* **2019**, *11*, 1456. [CrossRef]
2. Jeppesen, J.H.; Jacobsen, R.H.; Inceoglu, F.; Toftegaard, T.S. A cloud detection algorithm for satellite imagery based on deep learning. *Remote Sens. Environ.* **2019**, *229*, 247–259. [CrossRef]
3. Derrien, M.; Le Gleau, H. MSG/SEVIRI cloud mask and type from SAFNWC. *Int. J. Remote Sens.* **2005**, *26*, 4707–4732. [CrossRef]
4. Lyapustin, A.; Wang, Y.; Frey, R. An automatic cloud mask algorithm based on time series of MODIS measurements. *J. Geophys. Res. Atmos.* **2008**, *113*. [CrossRef]
5. Purbantoro, B.; Aminuddin, J.; Manago, N.; Toyoshima, K.; Lagrosas, N.; Sumantyo, J.T.S.; Kuze, H. Comparison of Aqua/Terra MODIS and Himawari-8 Satellite Data on Cloud Mask and Cloud Type Classification Using Split Window Algorithm. *Remote Sens.* **2019**, *11*, 2944. [CrossRef]
6. Wilson, M.J.; Oreopoulos, L. Enhancing a Simple MODIS Cloud Mask Algorithm for the Landsat Data Continuity Mission. *IEEE Trans. Geosci. Remote Sens.* **2013**, *51*, 723–731. [CrossRef]
7. Zhuge, X.; Zou, X.; Wang, Y. A Fast Cloud Detection Algorithm Applicable to Monitoring and Nowcasting of Daytime Cloud Systems. *IEEE Trans. Geosci. Remote Sens.* **2017**, *55*, 6111–6119. [CrossRef]
8. Saunders, R.W.; Kriebel, K.T. An improved method for detecting clear sky and cloudy radiances from AVHRR data. *Int. J. Remote Sens.* **1988**, *9*, 123–150. [CrossRef]
9. Kriebel, K.; Gesell, G.; Kaestner, M.; Mannstein, H. Cloud detection in AVHRR and ATSR data with APOLLO. In Proceedings of the Satellite Remote Sensing, Rome, Italy, 23 December 1994.
10. Kriebel, K.T.; Gesell, G.; Kaestner, M.; Mannstein, H. The cloud analysis tool APOLLO: Improvements and validations. *Int. J. Remote Sens.* **2003**, *24*, 2389–2408. [CrossRef]
11. Karlsson, K.G. Development of an operational cloud classification model. *Int. J. Remote Sens.* **1989**, *10*, 687–693. [CrossRef]
12. Martins, J.V.; Tanré, D.; Remer, L.; Kaufman, Y.; Mattoo, S.; Levy, R. MODIS Cloud screening for remote sensing of aerosols over oceans using spatial variability. *Geophys. Res. Lett.* **2002**, *29*, MOD4-1–MOD4-4. [CrossRef]
13. Zhu, Z.; Woodcock, C.E. Object-based cloud and cloud shadow detection in Landsat imagery. *Remote Sens. Environ.* **2012**, *118*, 83–94. [CrossRef]
14. Breon, F.M.; Colzy, S. Cloud detection from the spaceborne POLDER instrument and validation against surface synoptic observations. *J. Appl. Meteorol.* **1999**, *38*, 777–785. [CrossRef]
15. Benas, N.; Finkensieper, S.; Stengel, M.; van Zadelhoff, G.J.; Hanschmann, T.; Hollmann, R.; Meirink, J.F. The MSG-SEVIRI-based cloud property data record CLAAS-2. *Earth Syst. Sci. Data* **2017**, *9*, 415–434. [CrossRef]

16. Li, C.; Ma, J.J.; Yang, P.; Li, Z.Q. Detection of cloud cover using dynamic thresholds and radiative transfer models from the polarization satellite image. *J. Quant. Spectrosc. Radiat. Transf.* **2019**, *222–223*, 196–214. [CrossRef]
17. Spangenberg, D.A.; Doelling, D.R.; Chakrapani, V.; Minnis, P.; Uttal, T. Nighttime cloud detection over the Arctic using AVHRR data. In Proceedings of the Twelfth ARM Science Team Meeting Proceedings, St. Petersburg, FL, USA, 8–12 April 2002.
18. Park, Y.-J.; Ahn, Y.; Han, H.; Yang, H.; Moon, J.; Ahn, J.; Lee, B.; Min, J.; Lee, S.; Kim, K.; et al. *GOCI Level 2 Ocean Color Products (GDPS 1.3) Brief Algorithm Description*; Korea Ocean Satellite Center: Busan Metropolitan City, Korea, 2014.
19. Ackerman, S.A.; Strabala, K.I.; Menzel, W.P.; Frey, R.A.; Moeller, C.C.; Gumley, L.E. Discriminating clear sky from clouds with MODIS. *J. Geophys. Res. Atmos.* **1998**, *103*, 32141–32157. [CrossRef]
20. Platnick, S.; King, M.D.; Ackerman, S.A.; Menzel, W.P.; Baum, B.A.; Riedi, J.C.; Frey, R.A. The MODIS cloud products: Algorithms and examples from Terra. *IEEE Trans. Geosci. Remote Sens.* **2003**, *41*, 459–473. [CrossRef]
21. Remer, L.A.; Kaufman, Y.J.; Tanre, D.; Mattoo, S.; Chu, D.A.; Martins, J.V.; Li, R.R.; Ichoku, C.; Levy, R.C.; Kleidman, R.G.; et al. The MODIS aerosol algorithm, products, and validation. *J. Atmos. Sci.* **2005**, *62*, 947–973. [CrossRef]
22. Choi, J.-K.; Park, Y.J.; Ahn, J.H.; Lim, H.-S.; Eom, J.; Ryu, J.-H. GOCI, the world's first geostationary ocean color observation satellite, for the monitoring of temporal variability in coastal water turbidity. *J. Geophys. Res. Ocean.* **2012**, *117*. [CrossRef]
23. Ruddick, K.G.; Ovidio, F.; Rijkeboer, M. Atmospheric correction of SeaWiFS imagery for turbid coastal and inland waters. *Appl. Opt.* **2000**, *39*, 897–912. [CrossRef]
24. Arrigo, K.R.; McClain, C.R. Spring phytoplankton production in the western ross sea. *Science* **1994**, *266*, 261–263. [CrossRef]
25. Banks, A.C.; Melin, F. An assessment of cloud masking schemes for satellite ocean colour data of marine optical extremes. *Int. J. Remote Sens.* **2015**, *36*, 797–821. [CrossRef]
26. Banks, A.C.; Prunet, P.; Chimot, J.; Pina, P.; Donnadille, J.; Jeansou, E.; Lux, M.; Petihakis, G.; Korres, G.; Triantafyllou, G.; et al. A satellite ocean color observation operator system for eutrophication assessment in coastal waters. *J. Mar. Syst.* **2012**, *94*, S2–S15. [CrossRef]
27. Hallegraeff, G.M. Ocean climate change, phytoplankton community responses, and harmful algal blooms: A formidable predictive challenge. *J. Phycol.* **2010**, *46*, 220–235. [CrossRef]
28. Wang, M.; Shi, W. Cloud masking for ocean color data processing in the coastal regions. *IEEE Trans. Geosci. Remote Sens.* **2006**, *44*, 3196–3205. [CrossRef]
29. Hudson, B.D.; Overeem, I.; Syvitski, J.P.M. A novel technique to detect turbid water and mask clouds in Greenland fjords. *Int. J. Remote Sens.* **2016**, *37*, 1730–1746. [CrossRef]
30. Hu, C.; Carder, K.L.; Muller-Karger, F.E. Atmospheric correction of SeaWiFS imagery over turbid coastal waters: A practical method. *Remote Sens. Environ.* **2000**, *74*, 195–206. [CrossRef]
31. Wang, M.; Shi, W.; Jiang, L. Atmospheric correction using near-infrared bands for satellite ocean color data processing in the turbid western Pacific region. *Opt. Express* **2012**, *20*, 741–753. [CrossRef] [PubMed]
32. Nordkvist, K.; Loisel, H.; Gaurier, L.D. Cloud masking of SeaWiFS images over coastal waters using spectral variability. *Opt. Express* **2009**, *17*, 12246–12258. [CrossRef]
33. Ishida, H.; Oishi, Y.; Morita, K.; Moriwaki, K.; Nakajima, T.Y. Development of a support vector machine based cloud detection method for MODIS with the adjustability to various conditions. *Remote Sens. Environ.* **2018**, *205*, 390–407. [CrossRef]
34. Mahajan, S.; Fataniya, B. Cloud detection methodologies: Variants and development-a review. *Complex Intell. Syst.* **2020**, *6*, 251–261. [CrossRef]
35. Chai, D.; Newsam, S.; Zhang, H.K.; Qiu, Y.; Huang, J. Cloud and cloud shadow detection in Landsat imagery based on deep convolutional neural networks. *Remote Sens. Environ.* **2019**, *225*, 307–316. [CrossRef]
36. Qiu, S.; He, B.; Zhu, Z.; Liao, Z.; Quan, X. Improving Fmask cloud and cloud shadow detection in mountainous area for Landsats 4–8 images. *Remote Sens. Environ.* **2017**, *199*, 107–119. [CrossRef]
37. Hughes, M.J.; Kennedy, R. High-Quality Cloud Masking of Landsat 8 Imagery Using Convolutional Neural Networks. *Remote Sens.* **2019**, *11*, 2591. [CrossRef]
38. Park, J.H.; Inamori, T.; Hamaguchi, R.; Otsuki, K.; Kim, J.E.; Yamaoka, K. RGB Image Prioritization Using Convolutional Neural Network on a Microprocessor for Nanosatellites. *Remote Sens.* **2020**, *12*, 3941. [CrossRef]
39. Yang, H.; Han, H.; Heo, J.; Jeong, J.; Lee, T.; Hu, W.; Kwak, S. Ocean Color Algorithm Development Environment for High-Speed Data Processing of GOCI-II. In Proceedings of the IGARSS 2018–2018 IEEE International Geoscience and Remote Sensing Symposium, Valencia, Spain, 22–27 July 2018; pp. 7968–7971.
40. Lamquin, N.; Mazeran, C.; Doxaran, D.; Ryu, J.-H.; Park, Y.-J. Assessment of GOCI Radiometric Products using MERIS, MODIS and Field Measurements. *Ocean Sci. J.* **2012**, *47*, 287–311. [CrossRef]
41. Ahn, J.-H.; Park, Y.-J.; Ryu, J.-H.; Lee, B.; Oh, I.S. Development of Atmospheric Correction Algorithm for Geostationary Ocean Color Imager (GOCI). *Ocean Sci. J.* **2012**, *47*, 247–259. [CrossRef]
42. Gordon, H.R.; Wang, M.H. Retrieval of water-leaving radiance and aerosal optical-thickness over the oceans with SeaWiFS-a preliminary algorithm. *Appl. Opt.* **1994**, *33*, 443–452. [CrossRef]
43. Ruddick, K.; Vanhellemont, Q.; Yan, J.; Neukermans, G.; Wei, G.; Shang, S. Variability of Suspended Particulate Matter in the Bohai Sea from the Geostationary Ocean Color Imager (GOCI). *Ocean Sci. J.* **2012**, *47*, 331–345. [CrossRef]
44. Kim, J.B.; Park, J.-I.; Jung, C.-S.; Lee, P.-Y.; Lee, K.-S. Distributional range extension of the seagrass Halophila nipponica into coastal waters off the Korean peninsula. *Aquat. Bot.* **2009**, *90*, 269–272. [CrossRef]

45. Alvera-Azcarate, A.; Vanhellemont, Q.; Ruddick, K.; Barth, A.; Beckers, J.-M. Analysis of high frequency geostationary ocean colour data using DINEOF. *Estuar. Coast. Shelf Sci.* **2015**, *159*, 28–36. [CrossRef]
46. Liu, X.; Wang, M. Filling the Gaps of Missing Data in the Merged VIIRS SNPP/NOAA-20 Ocean Color Product Using the DINEOF Method. *Remote Sens.* **2019**, *11*, 178. [CrossRef]

remote sensing

MDPI

Article

Small Angle Scattering Intensity Measurement by an Improved Ocean Scheimpflug Lidar System

Hongwei Zhang [1], Yuanshuai Zhang [1], Ziwang Li [1], Bingyi Liu [1,2], Bin Yin [1] and Songhua Wu [1,2,3,*]

[1] Department of Marine Technology, College of Information Science and Engineering, Ocean University of China, Qingdao 266100, China; zhanghongwei8944@ouc.edu.cn (H.Z.); zhangys01@aircas.ac.cn (Y.Z.); liziwang@stu.ouc.edu.cn (Z.L.); liubingyi@ouc.edu.cn (B.L.); binyin@ouc.edu.cn (B.Y.)

[2] Laboratory for Regional Oceanography and Numerical Modeling, Pilot National Laboratory for Marine Science and Technology (Qingdao), Qingdao 266200, China

[3] Institute for Advanced Ocean Study, Ocean University of China, Qingdao 266100, China

* Correspondence: wush@ouc.edu.cn; Tel.: +86-532-66782573

Abstract: Quantification of the horizontal patterns of phytoplankton and the distribution of suspended particles across the sea's surface has been greatly improved by traditional passive oceanic color remote sensing technology. Lidar technology has already been proven to be effective positive remote sensing technology to construct high-resolution bathymetry models. Lidar technology significantly improves our ability to model biogeochemical processes in the upper ocean and provides advanced concepts regarding the vertical distribution of suspended particles and oceanic optical properties. In this paper, we present a novel optical approach to measuring the scattering intensity and characteristics of suspended particles within small angles backwards and distinguish water medium with different attenuation coefficients by a laboratory demonstration of the ocean Scheimpflug lidar system. The approach allows the direct determination of the scattering intensity over a small angle at the backward direction (175.8~178.8°) with an angular resolution of 0.38. Corrections for the effects of refraction at the air-glass-water interface were demonstrated. The data production (initial width and width attenuation rate of the laser beam) of the ocean Scheimpflug lidar system were utilized to distinguish water with different algae concentrations. Application for the measurement of backward scattering intensity and laser beam width were explored in distances up to several meters with spatial resolutions of millimeter precision.

Keywords: lidar; remote sensing sensors; backward scattering intensity; ocean Scheimpflug lidar; volume scattering function

Citation: Zhang, H.; Zhang, Y.; Li, Z.; Liu, B.; Yin, B.; Wu, S. Small Angle Scattering Intensity Measurement by an Improved Ocean Scheimpflug Lidar System. *Remote Sens.* **2021**, *13*, 2390. https://doi.org/10.3390/rs13122390

Academic Editors: Chung-Ru Ho, Antony K. Liu and Xiaofeng Li

Received: 13 April 2021
Accepted: 17 June 2021
Published: 18 June 2021

1. Introduction

Researchers are able to study large-scale horizontal patterns of phytoplankton and the distribution of suspended particles across the sea's surface and no longer want to just observe the ocean for short periods in small places, spurred by advances in spaced platforms, air–ground telecommunications, and marine detection technology [1–3]. In terms of ocean color remote sensing, optical properties and the distribution of particles are key parameters for estimating upper ocean primary production and for quantifying the spatial distributions of specific constituent concentrations from satellites [3,4]. Inherent optical properties (IOPs, including the volume scattering function (VSF) and the absorption coefficient) of aquatic medium play vital roles in optical oceanography. Remote sensing of water constituents based on traditional ocean color remote sensing technology are determined by the light field underwater [5]. Improvements in ocean color observations, the distribution of particles, and the visibility of water are benefit from the study of these optical properties, especially the measurement of the backward scattering coefficient b_b [6,7].

The dynamic range of VSF variability and the true value of the absorption coefficient in the ocean are little known, largely due to the practical difficulty in carrying out the direct measurement of these parameters [8]. Researchers must apply radiative transfer models to study the property of suspended particles. Current radiative transfer models, which were created more than 40 years ago, are based on a limited set with coarsely resolved angular distribution of scattering intensity [9,10]. Furthermore, the nature properties of the suspended particulate matter in the oceans can be retrieved from the study of the ocean water VSF [11].

Commercial instruments have been widely used for the in situ measurement of water's optical properties and the validation of the remote sensing data, e.g., Laser In Situ Scattering and Transmissometer (LISST), HydroScat Optical Backscattering Sensor (HydroScat-6), and Environmental Characterization Optics (ECO) [12–14]. All these devices are able to measure scattering at a few fixed angles or measure scattering intensity over a narrow range in the forward direction under manual operation. In terms of these point measurement devices, the shipborne platform is the only reliable platform to carry out the observation of oceanic water properties with a large demand for labor.

Lidar technology has already proven to be effective positive remote sensing technology that produces high-resolution models of bathymetry [15]. Lidar technology significantly improves our ability to model biogeochemical processes in the upper ocean and provides advanced conceptions about the vertical distribution of suspended particles and oceanic optical properties [16,17].

The predominant positive remote sensing technology for atmospheric and aquatic applications are based on the time-of-flight method by utilizing pulsed lasers, which have been widely used in the observation of atmospheric constituents such as aerosols, molecules, and meteorological parameters including temperature and wind [18–21]. In the manufacture of a Scheimpflug lidar, a high-power continuous-wave (CW) laser rather than a pulsed laser is utilized. In terms of traditional pulsed lidar systems, the spatial resolution is determined by the width of laser pulses, calculated by the time-of-flight method. The range resolution of a Scheimpflug lidar system is achieved through a triangulation method [22–24]. With further research and exploration of the Scheimpflug lidar, the application fields of the Scheimpflug lidar cover spectroscopy techniques for entomological applications based on elastic and inelastic lidar [25–28], remote sensing of atmospheric aerosol and particles by utilizing multi-channel Raman lidar [29], gas monitors by performing differential absorption lidar [30,31], and the detection of aquatic and vegetation by operating the hyperspectral lidar [26,32].

Gordon studied the influence of multiple scattering on the optical characteristics of sea water detected by marine lidar [33]. Walker formulated entirely analytical lidar equations for a turbid water medium that included beam spreading and pulse stretching. The lidar equations were reduced to simple algorithms which served as useful engineering models for systems study by accompany the lidar equations with simplified approximations [34]. Roddewig derived the diffuse attenuation coefficient of downwelling irradiance (K_d) in Yellow Lake by analyzing two sets of airborne lidar data. Reddewig compared the calculated lidar K_d values with the eight-day-averaged K_d measurement retrieved by the moderate resolution imaging spectrometer (MODIS) and converted the K_d measurements to the Secchi disk depth [35].

Spurred by the old Scheimpflug principle, a novel and compact 2D Ocean Scheimpflug lidar system was developed for the profile measurements of water medium based on the Scheimpflug lidar system to distinguish water medium with a different attenuation coefficient. The Scheimpflug lidar presented in this paper is a laboratory demonstration system. The effective measurement range was 1.31–4.54 m with a spatial resolution of 0.36–4.8 mm. The backward scattering angles varied from 175.5 to 178.8°, the angular resolution was from 72 mrad to 80.8 mrad. The future Scheimpflug lidar system will be packaged in compact housing and configured to the research ship (Dongfanghong-3, which is affiliated with the Ocean University of China) to carry out long-term observation of water properties over a large sea area (300 sailing days/year; coverage area: northwest Pacific Ocean, Yellow Sea, Indian Ocean). When applied to the field ocean observation, the measurement range of the Scheimpflug lidar system can be modified to profile the optical parameters of ocean water from 2 to 30 m in depth.

The general principle and detailed specification of the ocean Scheimpflug lidar system are introduced in Section 2. The methodology for data processing and the Monte Carlo simulation are described in Section 3 to provide a detailed explanation of the processing of the data production and the validation data. Measurement data and validation data are described in detail in Section 4. Section 5 presents the discussion and conclusion of the laboratory experiment and data analysis.

2. The Ocean Scheimpflug Lidar System

2.1. General Principle

The Scheimpflug lidar system is a novel lidar system based on the Scheimpflug principle. The Scheimpflug principle states that in a scenario with infinite focal depth, to satisfy the Scheimpflug principle, the object plane, the lens plane, and image plane should be interested in the same point. In this scenario, there is no relationship between the focal depth and the optical aperture of the optical system. Figure 1 illustrates a typical Scheimpflug lidar system that has a lens (lens plane) and object planes that are not parallel, and three points (i.e., the image plane, object plane, and the lens plane) intersect at the Scheimpflug intersection point. The backscattering echo signal can be collected by the lens and detected by the tilted 2D CMOS sensors when a laser beam transmits into the water media. In a Scheimpflug lidar system, CMOS pixels correspond with the illuminated volume. Continuous-wave laser sources are common light sources that have been widely employed for range-resolved measurements using the Scheimpflug principle. All the configurations of the Scheimpflug lidar system significantly reduce the cost and the system complexity and differ from conventional lidar systems by utilizing the pulsed light.

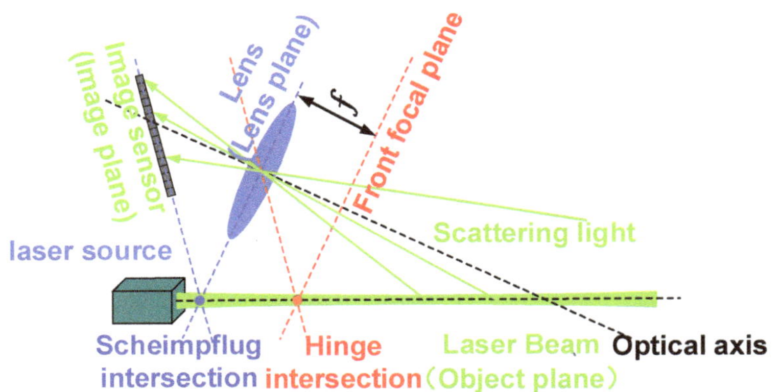

Figure 1. Scheme diagram of the Scheimpflug imaging principle.

2.2. Specifications of the Scheimpflug Lidar System

A typical lidar system includes a laser transmitting system, a receiving system, and a photoelectric detection system. In this section, detailed specifications of the Scheimpflug lidar system are introduced.

2.2.1. Transmitter

The integration time of the CMOS camera is short, since the output signals of the CMOS camera are easily affected by the radiant power fluctuations of a laser over a short period of time. The light source of the ocean Scheimpflug lidar system was a dual wavelength DPSS laser operating at 491 and 532 nm with an average energy of 50 mW at each wavelength (Cobolt Calypso 04-01, Cobolt Inc., Solna, Sweden). The laser had a very low intensity noise, achieved by running a single frequency to alleviate competing modes and subsequent mode beating. The coherence length of the laser utilized in the Scheimpflug lidar was typically over 100 m.

2.2.2. Receiver

According to the Scheimpflug principle, it is difficult to determine the focal length of the lens. To obtain the focus length of the lens, the Hinge rule was introduced into the design of the lidar system. According to the Hinge rule, there should be another intersection in which the image plane is displaced to the effective center of the lens, and the front focal plane of the lens and the object plane will coincide. In terms of a typical Scheimpflug lidar system, the laser beam transfer in the water medium is the "object plane", the telescope is the "imaging lens" and the "image plane" is typically a square CMOS array detector. Details of the geometrical relations between the Scheimpflug principle and Hinge rule is illustrated in Figure 1.

The backscattering light was collected by a 25.4 mm plane convex glass with a focus length of 10 cm. The DPSS laser operated at 532 nm with a narrow bandwidth of less than 1 MHz. To suppress the background radiation, such as fluorescence excited by transmitted laser and possible multi-scattering sunlight, a 532 nm interference filter (FL532-3, Thorlabs Inc., Newton, NJ, USA) with a 3 nm full-width at half maximum (FWHM) was utilized in the Scheimpflug lidar system. The CMOS sensor was mounted at 45°, tilted to the glass plane to satisfy the Scheimpflug principle.

2.2.3. Detector

The imaging detector utilized in this system is a 5 Megapixel Monochrome CMOS sensor (CS505MU, Thorlabs Inc., Newton, NJ, USA) with extremely low noise (<2.5 e^- RMS Read Noise) and high sensitivity. The CMOS was packaged by compact housing to provide passive thermal management. The global shutter scans the entire field of view simultaneously, allowing for imaging of fast-moving objects. These features make it ideal for low-light imaging applications. The size of the monochrome CMOS sensor was 2448 × 2048 pixels (approximately 5.0 megapixel, pixel size: 3.54 μm × 3.54 μm). The quantum efficiency was approximately 72%, over 525–580 nm. A polarizer was configured in front of the lens making it only receive the co-polarized echo signal. The FOV of the Scheimpflug lidar system was 35.4 μrad.

3. Experiment and Methodology

3.1. Experimental Setup

The configuration of the ocean Scheimpflug lidar system during the laboratory measurements of the water attenuation coefficient is illustrated in Figure 2. The imaging detector and telescope were configured by fulfilling the Scheimpflug principle and the Hinge rule. A black plastic tank with a 90 mm diameter glass window on its side was filled with water. The distance from the ocean Scheimpflug lidar system to the water-filled tank was approximately 1.3 m. The ocean Scheimpflug lidar system was mounted to make sure the laser beam was perpendicular to the surface of the tank. When the laser illuminated a cross-section of an object in the water, a photo of the echo signal could be obtained on the CMOS camera.

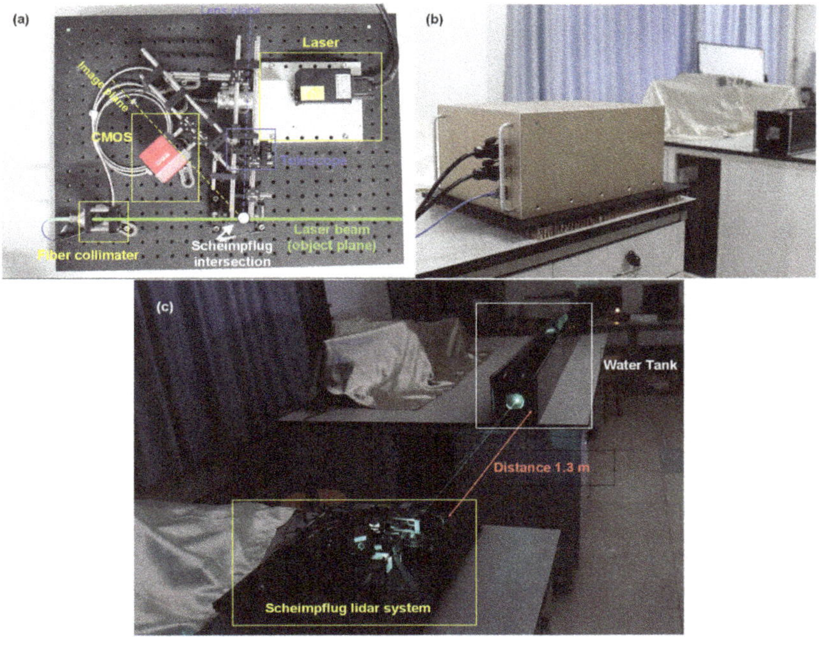

Figure 2. Configuration of the ocean Scheimpflug lidar system during the laboratory measurements of the water attenuation coefficient: (**a**) top view of the Scheimpflug lidar system; (**b**) overall appearance of the Scheimpflug lidar system; (c) configuration of the laboratory measurement campaign.

The laser beam was collimated using an achromatic fiber collimator (PAF2-7A, Thorlabs Inc., Newton, NJ, USA) before it was transmitted into the air and a 1.3 m long tank filled with water medium (the collimator was unable to collimate the laser beam to a parallel light strictly. The waist of the laser beam as focused at approximately 5 m away from the collimator. Thus, the laser beam of the Scheimpflug lidar system converged within the measurement range). The telescope in this system is an imaging lens with a diameter of 25.4 mm and a focus length of 100 mm (LB1676-A, Thorlabs Inc., Newton, NJ, USA). The distance between the center of the imaging lens and the "object plane" was 0.105 m as shown in the Figure 3. The angle α is the tilt angle between the lens plane and the object plane, and the tilt angle was set to 87° in our lidar system. The tilt angle β of the image plane to the lens plane was set to 45°, while the image plane was 132° off the optical axis. Thus, the laser beam can image onto the detector surface with a 48° tilt angle off the optical axis. The range calibration and the range correction were performed using Equation (1) as illustrated in Figure 3.

$$Z = \frac{L[D(\sin\beta - \cos\beta\tan\theta) + L']}{D(\cos\beta + \sin\beta\tan\theta) + L'\tan\theta} \tag{1}$$

where $L' = L\tan\beta/\cos\theta$. f is the focal length of the receiving telescope, L is the distance of the center of the image plane to the lens plane, β is the tilt angle of the image plane to the lens plane, and θ is the intersection angle between the scattering light and the laser beam. The focal length of the receiving telescope can be obtained by Equation (2) which is fulfilled with the Hinge rule.

$$f = \frac{L\tan\beta}{\tan\beta\sin\theta + \cos\theta} \tag{2}$$

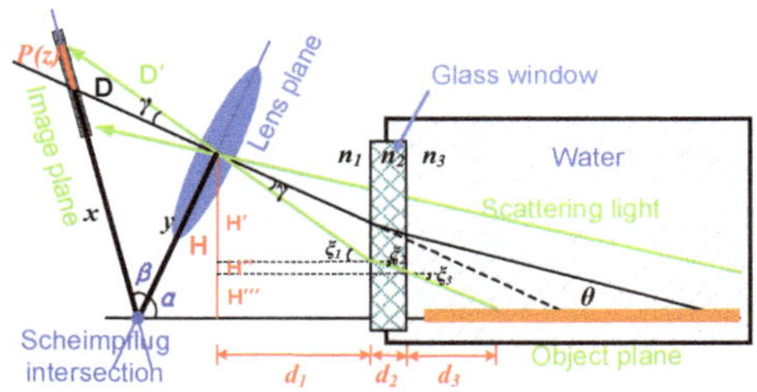

Figure 3. The scheme diagram of the measurement range correction.

In terms of the ocean Scheimpflug lidar system during the experimental measurements, the emitted laser beam and scattering light travelled through at least two interfaces (the interface between the water and glass window and the interface between the glass window and air). The actual distance was no longer consistent with the theoretical distance as Figure 3 illustrated.

For the convenience of system adjustment, a black board was placed in the tank to terminate the beam and to increase the intensity of backscattering echo. In the calibration campaign, fifteen distinct distances were chosen to validate the distance accuracy of the Scheimpflug lidar system. The backscattering echo from the hard target was calibrated to measure the relationship between pixels' position on the CMOS camera and distances, and it is possible to deduce the other distances from the corresponding pixel number (details in the Appendix A). The curves illustrated in Figure 4 depict a theoretical relationship between pixel number and distance. The red curve is the data without taking the glass window into consideration, and the blue curve is the calibrated data. The black dots are experimental test data, which coincide well with the theoretical curve. It can be concluded from the Figure 4 that the distance range of the Scheimpflug lidar system is approximately 1.40–2.80 m. The average range resolution is approximately 1 mm at 1–3 m. The range resolution is determined by the focal length and the tilt angles. Thus, the parameters of the Scheimpflug lidar system should be considered according to specific applications.

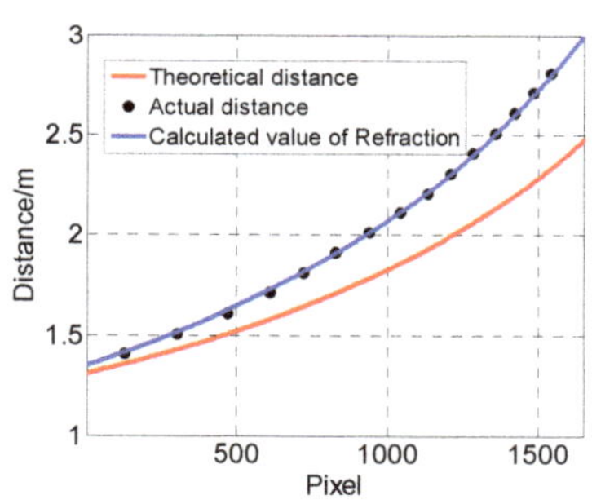

Figure 4. Relationship of the pixel and measurement distance.

3.2. Imaging Processing

During the configuration of a Scheimpflug lidar system, the divergences of the laser beam should be minimized to promise the range resolution and effective range distance. Mei L et al. discussed the importance of divergences of the transmitted laser beam to the atmospheric Scheimpflug lidar [25]. In our setup, the measurement range of the ocean Scheimpflug lidar system is limited to 4.54 m, and it is possible to achieve minimized divergence of the transmitted laser beam by aligning an achromatic fiber collimator (PAF2-7A, Thorlabs Inc., Newton, NJ, USA).

The image width in different distances (pixel) can be obtained by finding the 1/e maximum along vertical pixels as illustrated by the white, solid lines in Figure 5a. Part of the data of the CMOS imaging sensor (each square represents a CMOS pixel with 12 bits ADC resolution) are selected for fitting with a Gaussian curve. The laser beam width mentioned in this manuscript is different from the whole halo in a scattering medium due to the multiple scattering process. The traditional definition of the laser beam width is the whole width at transversal direction including the laser beam defined in this manuscript and the halo around the laser beam (the halo width is affected by the sensitivity of detector). The beam width at a related distance can be obtained by finding the 1/e maximum intensity value of the Gaussian fitting curve, while the intensity information can be retrieved by finding the maximum intensity value of the fitting curve. By fitting the curves along the laser beam traveling direction, the intensity profile and the beam width profile can be constructed as illustrated in Figure 5c,d.

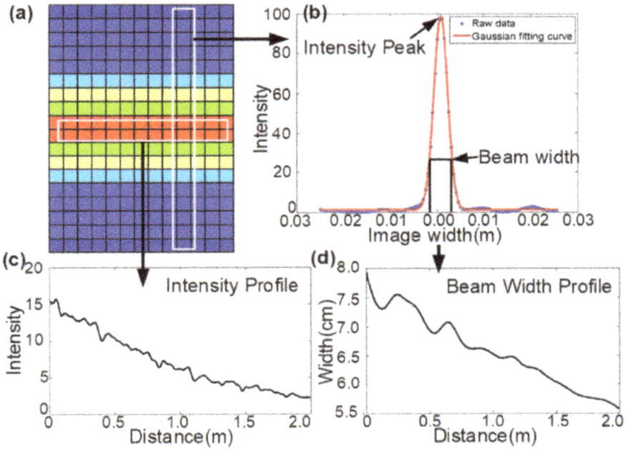

Figure 5. The scheme of the method to retrieve the width of the laser beam from the raw data: (**a**) part of the data of the CMOS imaging sensor (each square represents a CMOS pixel with 12 bits ADC resolution); (**b**) the transversal data relative to the laser beam travel direction were extracted to calculate the laser beam width by Gaussian curve fitting; (**c**) the profile of laser beam intensity which was constructed by collecting all the intensity peaks of the Gaussian curve fitting; (**d**) the profile of laser's beam width which was constructed by collecting all the beams widths of the Gaussian curve fitting at 1/e of intensity peak.

3.3. Monte Carlo Simulation

Monte Carlo simulations have been proven by previous studies to model the probability of different outcomes in a process that cannot easily be predicted due to the intervention of random variables and the accuracy has been verified experimentally. It is a technique that is used to understand the impact of risk and uncertainty in prediction and forecasting models, and it has been widely used to simulate the influence of different lidar measurement modes, different water parameters, and different laser polarization characteristics on photon transmission in seawater [36,37]. The Monte Carlo simulation method can guide the measurement and analysis of field experiments and can well show the distribution characteristics of laser light under water under different water optical characteristics. In this paper, the Monte Carlo simulation method was used to simulate the detection of water bodies to verify the measurement results of this lidar.

The laser beam was set perpendicular to the water's surface, and the photons scattered after colliding with the water medium. During simulation, the position of the photon should be confirmed first to see whether it is involved in the receiving field. If the photon is not involved in the receiving field, it will not track the path of the photon sequentially. Otherwise, the photon scattering path and the scattering points to the detector are determined according to the law of refraction, and then the energy directly scattered into the receiver will be calculated according to the analytical probability. The remaining photons within the FOV will continue to scatter. The analytical probability of the Monte Carlo method can be expressed as:

$$P(z) = \beta(\theta, z)\frac{A}{n^2(h + d/n)^2}\exp[-c(z)d]T_{atm}T_{water}w(z) \tag{3}$$

where A represents the area of the FOV at the water's surface, h represents the distance from detector's surface to the water's surface, d is the depth of the scattering photon within the water medium, $\beta(\theta, z)$ is the volume scattering function at the water depth z, $c(z)$ and $w(z)$ are the attenuation coefficient and weight coefficient of the seawater at depth z, respectively. T_{atm} and T_{water} are the Fresnel transmittance of the atmosphere and water, respectively. $w(0) = 1$, and $w_{n+1}(z) = w_n(z)w_0(z)$ where n and $n+1$ are the n times and $n+1$ times scattering [38].

Due to the differences in the attenuation and scattering properties, the multiple scattering processes vary sharply when the laser beam transmits in different water medium, which affects the lidar echo signal. Based on Walker and McLean models, Howard Gordon model, and ocean lidar equations, semi-analytical Monte Carlo simulation methods were used to analyze the effects of lidar observation modes on lidar echo signals, and the relationship between the lidar extinction coefficient α and seawater optical parameters was established [38,39].

Stokes, Mueller, and Meridian methods are typical methods to establish a polarized Monte Carlo simulation model for the radiative transmission of polarized lasers in water, and the model method was used to simulate the process of polarized laser radiation transmission in the sea under typical water conditions to analyze the polarization state of light.

When the polarized light collides with particles, the scattering angle and azimuth angle were obtained through a combination of the Stokes parameter and Mueller matrix. The Stokes parameter was used to characterize the polarization characteristics, while the Mueller matrix was used to characterize the change in polarization state and form a 4×4 phase function matrix. The Stokes parameter $S = \begin{bmatrix} S_0 & S_1 & S_2 & S_3 \end{bmatrix}^T$ was used to describe the polarization state of the laser, S_0 was the light intensity information, S_1 and S_2 were the linearly polarized light components in the x-axis and 45° directions, and S_3 was the circularly polarized light component. The meridional plane of the incident Stokes parameter was defined as the x–z plane, that is, the azimuth was 0. Assuming that the incident light was linearly polarized, the Stokes parameter was $S = \begin{bmatrix} 1 & 1 & 0 & 0 \end{bmatrix}^T$. The polarization characteristics were simulated by the Stokes parameter. The polarization scattering phase function (probability density function) composed of the azimuth angle θ and φ can be expressed as:

$$f(\theta, \varphi) = M_{11}(\theta) + M_{12}(\theta)[S_1\cos(2\varphi) + S_2\sin(2\varphi)]/S_0 \tag{4}$$

$$M(\theta) = \begin{bmatrix} M_{11}(\theta) & M_{12}(\theta) & 0 & 0 \\ M_{12}(\theta) & M_{11}(\theta) & 0 & 0 \\ 0 & 0 & M_{33}(\theta) & M_{34}(\theta) \\ 0 & 0 & -M_{34}(\theta) & M_{33}(\theta) \end{bmatrix} \tag{5}$$

$M(\theta)$ is the Mueller matrix of the spherical particle, when the radius of the scattering spheres in the scattering medium is small (in the case of Rayleigh scattering). In the case of Rayleigh scattering, the elements in $M(\theta)$ can be expressed as:

$$\begin{aligned} M_{11}(\theta) &= \tfrac{3}{16\pi}(\cos^2\theta + 1) \\ M_{12}(\theta) &= \tfrac{3}{16\pi}(\cos^2\theta - 1) \\ M_{33}(\theta) &= \tfrac{3}{8\pi}\cos\theta \\ M_{34}(\theta) &= 0 \end{aligned} \tag{6}$$

The cumulative distribution function of the scattering angle θ can be expressed as:

$$P(0 \le \vartheta \le \theta) = 2\pi \int_0^{\theta} M_{11}(\vartheta)\sin\vartheta \mathrm{d}\vartheta = \xi \tag{7}$$

The scattering angle θ can be solved by the above formula, where ξ is a random number uniformly distributed from 0 to 1. The cumulative distribution function of azimuth φ can be expressed by the following formula:

$$\begin{aligned} P(0 \le \phi \le \varphi) &= \frac{\int_0^{\varphi}\left[1 + \frac{M_{12}(\theta)}{M_{11}(\theta)}\frac{S_1\cos(2\phi)+S_2\sin(2\phi)}{S_0}\right]d\phi}{\int_0^{2\pi}\left[1 + \frac{M_{12}(\theta)}{M_{11}(\theta)}\frac{S_1\cos(2\phi)+S_2\sin(2\phi)}{S_0}\right]d\phi} \\ &= \frac{1}{2\pi}\left[\varphi + \frac{M_{12}(\theta)}{M_{11}(\theta)}\frac{S_1\cos(2\phi)+S_2\sin(2\phi)}{S_0}\right] = \xi \end{aligned} \tag{8}$$

In the simulation presented in this manuscript, both the scattering coefficients of pure water and suspended particles were considered. When the laser transmitted into the water medium, the light will be scattered by the water molecules and suspended particles. The scattering coefficient $b(\lambda)$ is defined as the ratio of the scattering energy within per unit length in the medium to the incident energy. The scattering coefficient included the scattering coefficient of pure water $b_{water}(\lambda)$ and the scattering coefficient of particles $b_{particles}(\lambda)$. The scattering coefficient of pure water was based on the model established by Morel in 1997 [40], while the scattering coefficient of plankton was based on the model mentioned by Gordon in 1983 [41]. Petzold established scattering coefficient of three typical sea waters based on in situ measurement data [42].

4. Results

The key feature of the results obtained by the ocean Scheimpflug lidar system was high-range resolve. Objects separated in space can be measured simultaneously with this technique, especially the measurement of water optical parameter profiles with high resolution. To alleviate the influence of laser energy fluctuations in the raw data of ocean Scheimpflug lidar systems, long-term observation campaigns based on three typical water mediums were carried out. The backscattering signal time series profiles of the ocean Scheimpflug lidar system with 10 s exposure time during validation experiments are illustrated in Figure 6. The color in Figure 6 represents the intensity of the echo signal detected by CMOS. The maximum value of the color bar illustrated in Figure 6a was 10, while the maximum value of the color bar depicted in Figure 6b,c was 35. We can conclude that there were no large variations in the output signal of the CMOS camera over long time periods (16 min), and it is not essential to configure a channel for energy monitor when there was no significant impact upon the magnitude of the output signal from the CMOS camera, especially when the integration time of the CMOS camera was short (10 s) during the entire laboratory experiment campaign.

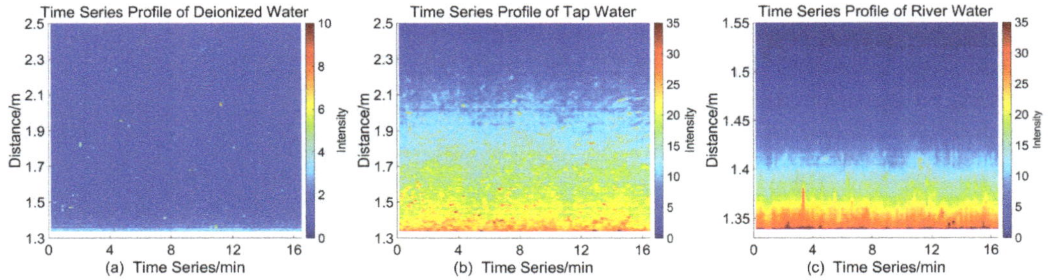

Figure 6. The backscattering signal time series profiles of the ocean Scheimpflug lidar system with 10 s exposure time during validation experiments: (**a**) the measurement medium was deionized water; (**b**) the measurement medium was tap water; (**c**) the measurement medium was river water (the river water samples were taken from the river inside the campus of the Ocean University of China as Figure A2 shows).

Water backscattering images were measured with a 10 s exposure time during nighttime to characterize the profile of the transmitted laser beam under three typical water mediums as shown in Figure 7. The color in Figure 7 represents the intensity of the echo signal detected by CMOS. The maximum value of the color bar illustrated in Figure 7a was 6, while the maximum value of the color bar depicted in Figure 7b,c were 30 and 45, respectively. The echo signal intensity of deionized water was the weakest for the main signal resource, which was the Rayleigh scattering signal of the water molecules as Figure 7a shows. The backscattering images obtained by the ocean Scheimpflug lidar system indicate the intensity of the multi-scattering processing in water medium during the small angle at backward direction. The laser beam could transmit through the whole tank when the measurement medium was tap water as Figure 7b shows. The laser beam could only transmit a short distance when the measurement medium was river water due to the high attenuation coefficient of suspended particles in the river water. From the illustration of Figure 7, we can conclude that the initial width of the laser beam had a maximum value when the medium was river water due to the strong multi-scattering processing. The width of the laser beam obtained by the Scheimpflug lidar system (retrieved by the method mentioned in Section 3.2) decreased slowly, and it transmitted through the whole tank filled with the deionized water. The lidar system performed perfectly when the tank was filled with tap water. The upward curvature in the laser beam width obtained by the Scheimpflug lidar system was the cause of a slight displacement of the CMOS camera, and a detailed analysis can be found in Appendix B.

Width and intensity information of a laser beam can be obtained by utilizing the method mentioned in Section 3.1. Then, the width variation profiles can be constructed by calculating the transversal raw data step by step. Figure 8a,b depict the laser beam width obtained by the Scheimpflug lidar system (retrieved by the method mentioned in Section 3.2) and intensity as a function of distance, respectively. The red lines are the data obtained by the Scheimpflug lidar system when the measurement medium is tap water. The black lines represent the echo signal of river water detected by CMOS. The blue lines depict the changes in the laser beam width obtained by the Scheimpflug lidar system and intensity with distance when the measurement medium is deionized water. The spatial resolution of the Scheimpflug lidar system changed with distance. In Figure 8, data points of the curves were calculated by an averaged 50 raw data points. Figure 8 depicts the laser beam width obtained by the Scheimpflug lidar system (retrieved by the method mentioned in Section 3.2) and intensity as a function of distance. The laser beams could transmit through the whole tank and the laser beam width obtained by Scheimpflug lidar system decreased relatively moderately when the measurement mediums were tap water and deionized water. The laser beam could only transmit a short distance and the laser beam attenuated sharply when the measurement medium was river water due to the abundance of suspended particles and colored dissolved organic matter (CDOM) in the river water. The width variation profiles of laser beam showed different characteristics under three typical water mediums as Figure 8 shows. The initial width and attenuation rate of laser beams could be effective tools to distinguish water medium with different attenuation coefficient.

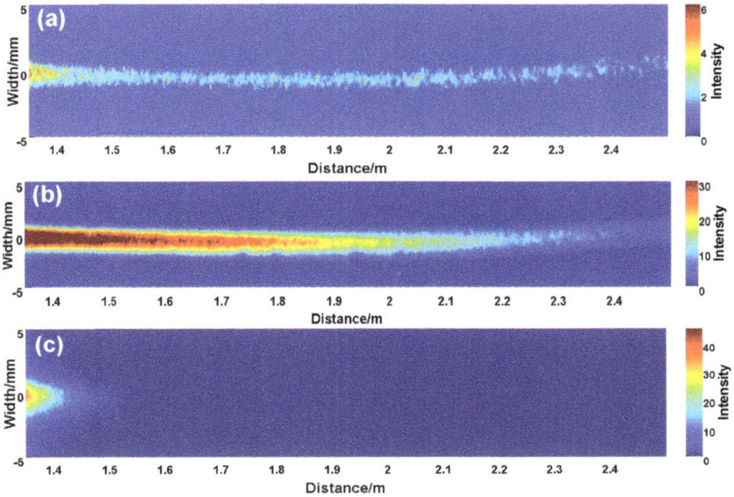

Figure 7. Intensity-range maps under three kinds of water mediums over a 10 s time window with a range interval of approximately 1.32–2.50 m: (**a**) the intensity-range maps under the measurement medium was deionized water; (**b**) the intensity-range maps under the measurement medium was tap water; (**c**) the intensity-range maps under the measurement medium was river water.

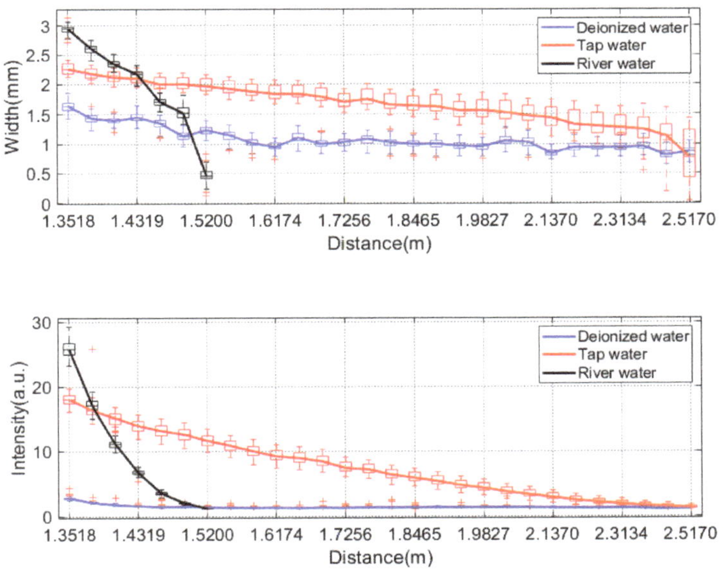

Figure 8. (**a**) Relationship between the laser beam width and the measurement range. (**b**) Relationship between the laser beam intensity and the measurement range.

To verify the ability of ocean Scheimpflug lidar system to distinguish algae at different concentrations by utilizing its data products including the initial width and width attenuation rate of laser beam, three typical algae in China's coastal area (including Isochrysis galbana, Platymonas subcordiformis, Nitzschia closterium) were used during the laboratory observation experiments. The size distributions of the three algae are listed in Table 1. Isochrysis galbana are a species of Haptophyta, and are nearly spherical in shape. The size of Platymonas subcordiformis and Nitzschia closterium are close to long sticks.

Table 1. Size distribution of three typical algae at China coastal area.

Name	Size Distribution (μm)
Isochrysis galbana	Length: 4.4~7.1 Width: 2.7~4.4 Thickness: 2.4~3.0
Platymonas subcordiformis	Length: 11.0~14.0 Width: 7.0~9.0 Thickness: 3.5~5.0
Nitzschia closterium	Length: 12.0~23.0 Width: 2.0~3.0 Thickness: 2.4~3.0

Figure 9 depicts the relationship between initial beam width and beam width attenuation rate obtained by the Scheimpflug lidar system (retrieved by the method mentioned in Section 3.2) under different mediums. The blue dots represent the data when the measurement medium is Isochrysis galbana. The data production of the Platymonas subcordiformis was illustrated by the red dots. The initial width information of Platymonas subcordiformis are larger than other algae. It is unreliable to distinguish the Isochrysis galbana and Nitzschia closterium according to the initial width of laser beam. However, it is reliable to distinguish the species of algae by taking both the initial width and attenuation rate into consideration.

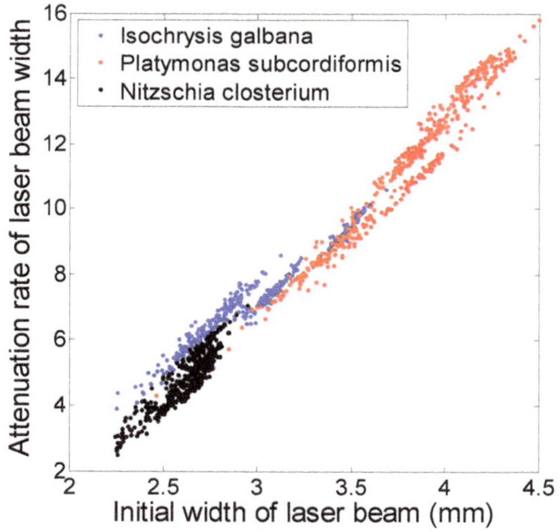

Figure 9. Relationship between the initial beam width and beam width attenuation rate under different medium. The black dots represent the water with Nitzschia closterium at different concentrations; the blue dots represent the water with Isochrysis galbana at different concentrations; the red dots represent the water with Platymonas subcordiformis at different concentrations.

Validation

Monte Carlo simulations have been widely used to simulate the influence of different lidar measurement modes, different water parameters, and different laser polarization characteristics on photon transmission in seawater. The Monte Carlo simulation method utilized in this paper is depicted in Section 3.3 in detail. In this section, the simulated data are illustrated for the validation of Scheimpflug lidar system.

The intensity-range maps simulated by the Monte Carlo method under three water mediums with different attenuation coefficients are illustrated in Figure 10. The color in Figure 10 represents the intensity of the echo signal detected by the CMOS. The maximum value of the color bar illustrated in Figure 10a was 15, while the maximum value of the color bar depicted in Figure 10b,c were 20 and 225, respectively. The intensity-range maps under the measurement mediums with an attenuation coefficient was 0.2 m^{-1}, 0.4 m^{-1}, 2 m^{-1}. From Figure 10, we can find a similar tendency between the simulated data by the Monte Carlo method and the measured data by the Scheimpflug lidar under different water mediums. The laser beam transferred through long distances (up to 4 m) when the attenuation coefficient was 0.2 m^{-1}. The backscattering echo signal within a 2 m measurement range could be detected with an attenuation coefficient that was 0.4 m^{-1}. When the attenuation coefficient was 2 m^{-1}, the maximum distance of the echo signal decreased to 0.375 m. The initial widths of the laser beam illustrated in Figure 10c were broader than the simulation data shown in Figure 10a,b due to the multiple scattering processes varying sharply in different water mediums, which were closely related to the attenuation coefficient.

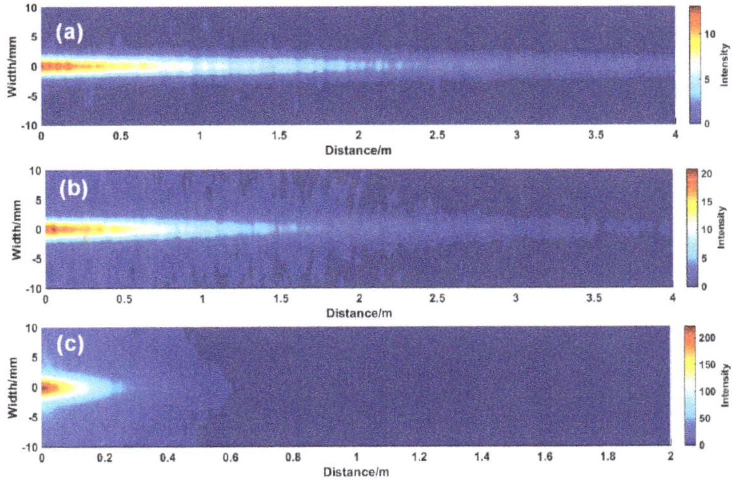

Figure 10. The intensity-range maps simulated by the Monte Carlo method under three different water mediums with different attenuation coefficients: (**a**) the medium with attenuation coefficient was 0.2 m^{-1}; (**b**) the medium with attenuation coefficient was 0.4 m^{-1}; (**c**) the medium with attenuation coefficient was 2 m^{-1}.

To quantify the difference of simulated data and measured data, laser beam information (e.g., width and intensity information) were retrieved by the utilized methods mentioned above as Figure 11 shows (the initial width of the measured data were normalized to the simulated data). Figure 11a depicts the changes in laser beams under water medium with different attenuation coefficients. The solid lines (red line represents river water; blue line means tap water; black line depicts deionized water) represent the width information obtained by the Scheimpflug lidar, and the dotted lines (red line represents the attenuation coefficient is 2 m^{-1}; blue line means the attenuation coefficient is 0.4 m^{-1};

black line depicts the attenuation coefficient is 0.2 m^{-1}) mean the data were simulated by the Monte Carlo method. The original absorption coefficient, scattering coefficient, and the attenuation coefficient utilized in this manuscript were obtained from the references as shown in Table 2 [34]. By comparing the simulated data and the measured data, we can conclude that the measured data changed faster with distance than the simulated data, which was mainly due to the attenuation coefficients of the measured medium being larger than the attenuation coefficients during the simulation work. However, there was a singular situation of the width information when the measured medium was tap water. The multi-scattering processing was dominant in the attenuation process rather than the absorption process in the tap water medium (the width of the laser beam obtained by the Scheimpflug lidar system (retrieved by the method mentioned in Section 3.2) indicated the backscattering intensity of the water medium).

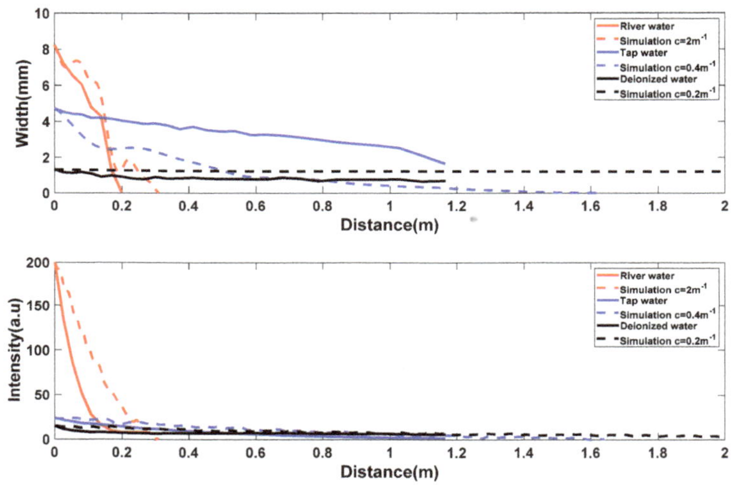

Figure 11. Comparison of measured data obtained by the Scheimpflug lidar and data simulated by the Monte Carlo method: (**a**) laser beam width (note that the scale is too large to see the small changes in the beam width simulated by the Monte Carlo method when c = 0.2 m^{-1}); (**b**) intensity of the laser beam.

Table 2. Water optical properties.

Water Types	a (m^{-1})	b (m^{-1})	c (m^{-1})
Pure sea water	0.0405	0.0025	0.043
Clear sea water	0.114	0.037	0.151
Coastal sea water	0.179	0.219	0.398
Turbid sea water	0.366	1.824	2.190

It is practically difficult to quantify the optical parameters (including the attenuation coefficient, backscattering coefficient, and absorption coefficient) of water medium. The simulated data can only be used for qualitative analysis of the trends in changes of the data. The measured data had good consistency with the simulated data in most cases.

A spectrophotometer has been proven to be good choice to satisfy the demand of precision measurements in the research of organic chemistry, biochemistry, environmental protection, water testing industry, etc. The long optical system inside the spectrophotometer ensures high accuracy and good stability of the instrument. A high-resolution Spectrophotometer (U-3900H, Hitachi Inc., Tokyo, Japan) was operated during the laboratory observation experiments to obtain the absorbance that indicates the absorbability of the measured medium to the light, then the absorption coefficient α could be calculated. The GF/FTM filter membrane with a 0.7 μm pore diameter was used to filter the 100 mL measurement medium as shown in Figure 12.

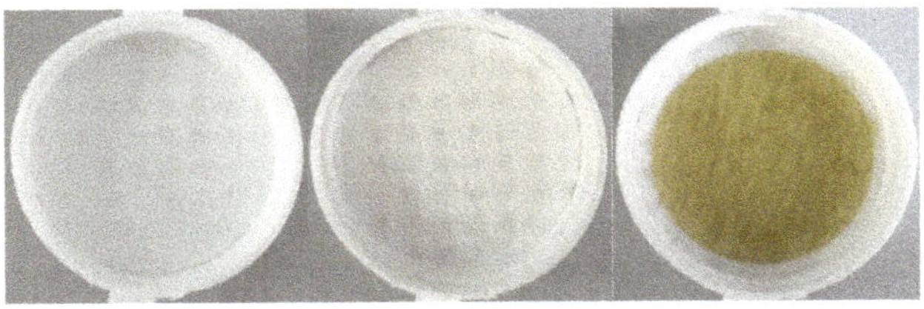

Figure 12. Filter membrane obtained by filtrating three kinds of water bodies.

Results from the data obtained by the spectrophotometer (including the absorbance (blue lines) and absorption coefficient α (red lines)) and the ocean Scheimpflug lidar system (including the attenuation rate of laser beam width (green lines) and attenuation rate of laser beam intensity (black lines)) were illustrated into one figure to study the correlation among the various results as shown in Figure 13. Figure 13a illustrates the variation in the spectrophotometer and ocean Scheimpflug lidar system retrieval results when the water was mixed with different concentration Nitzschia closterium. Figure 13b depicts the profiles when the water was mixed with different concentrations of Isochrysis galbana. Figure 13c shows the trends in the measurement results with different concentrations of Platymonas subcordiformis. As Figure 13 illustrates, the data production of the Scheimpflug lidar system were consistent with the results obtained by the spectrophotometer. By establishing the correspondence between lidar results and spectrophotometer results, the lidar data product can be utilized to obtain the absorption coefficient of the measured medium.

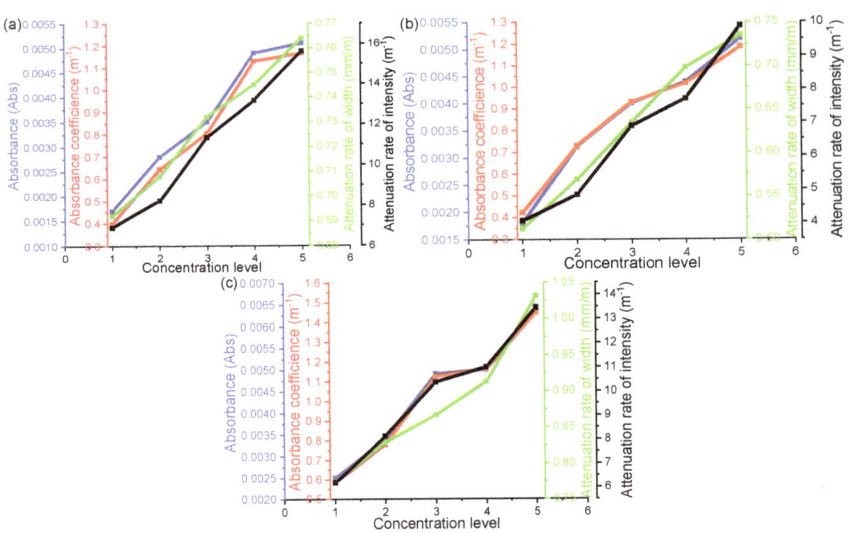

Figure 13. Correlation between the results of spectrophotometer and the data production of the ocean Scheimpflug lidar system: (**a**) the water with Nitzschia closterium at different concentrations; (**b**) the water with Isochrysis galbana at different concentrations; (**c**) the water with Platymonas subcordiformis at different concentrations. Note that the concentration level is defined by adding 100 mL of high concentration algae culture solution into the water tank step by step. The concentrations of algae are relative only.

5. Discussions and Conclusions

In this paper, a laboratory demonstration of the ocean Scheimpflug lidar system was developed with a particular focus on the optical parameters measurement of the aquatic environment. The range resolution of our Scheimpflug lidar system was achieved by utilizing the triangulation method rather than the time-of-flight method commonly used in conventional lidar systems. By positioning the detector, telescope, and the transmitted laser beam in a configuration that fulfills the Scheimpflug principle (in this scene that infinite focal depth can be obtained) and the laser beam can thus be imaged clearly onto the detector.

The conclusions may be summarized as follows:

- A novel optical approach was developed to measure the scattering intensity and to quantify the characteristics of the suspended particles within small angles at backwards and distinguish water medium with different attenuation coefficients.
- The work aimed to verify the capability of the Scheimpflug system to distinguish different water mediums with different optical parameters.
- Intensity-range maps simulated by the Monte Carlo methods under three different water mediums with different attenuation coefficients were developed.

From the carried out experimental verification, we can conclude that the results of improved Scheimpflug lidar system showed good agreement with the theoretical simulation. In all cases, the spatial distribution of the laser beam intensity deduced from the Monte Carlo simulations was found to be very close to the one measured experimentally. The field validation of the Scheimpflug lidar system is the essential work for the future.

In terms of high-precision ocean color remote sensing, IOPs of water medium play a key role in modulating aquatic light field. Unlike the traditional approaches that use a special periscope prism for the measurement of the scattering intensity, the Scheimpflug lidar system allows the direct determination of the VSF over a small angle in the backward direction with high angular resolution. However, the influence of intensity due to the backscattering angles have been ignored in this paper. When it comes to the measurement of VSF, the influence of backscattering angles cannot be ignored.

Polarization lidar systems have been widely utilized in the detection of the distribution of atmospheric aerosols and clouds [21,43]. In a typical polarization-sensitive lidar system, echo signals are easy affected by the scattering particles along the laser beam. In this paper, a polarizer was utilized in the ocean Scheimpflug lidar system to make an echo signal with a single polarized direction. Multi-channels with polarization identification and the size measurement of suspended particles will continue to be researched and developed in subsequent work.

The intensity-range maps simulated by the Monte Carlo method under three different water mediums were slightly different from the data measured experimentally due to the inaccurate settings of the attenuation coefficients.

Author Contributions: Conceptualization, H.Z. and S.W.; Methodology, H.Z., Y.Z. and Z.L.; Validation, H.Z., Y.Z. and B.L.; Formal analysis, H.Z., S.W., Y.Z. and B.L.; Investigation, H.Z.; Data curation, Y.Z. and H.Z.; Writing—original draft preparation, H.Z.; Writing—review and editing, H.Z., S.W., B.L. and B.Y.; Visualization, H.Z., Y.Z. and B.Y.; Supervision, S.W.; Project administration, S.W.; Funding acquisition, S.W. All authors have read and agreed to the published version of the manuscript.

Funding: This research was funded by the National Key Research and Development Program of China under Grant No. 2019YFC1408002 and 2016YFC1400904, and it was supported by the National Natural Science Foundation of China (NSFC) under Grant No. 61975191, U2006217, and 41905022.

Institutional Review Board Statement: Not applicable.

Informed Consent Statement: Not applicable.

Data Availability Statement: The data presented in this study are available on request from the corresponding author.

Acknowledgments: We thank our colleagues for their kind support during the laboratory experiments including Kailin Zhang and Xiangcheng Chen from Ocean University of China for performing the ocean Scheimpflug lidar system and Ke Li and Xiaojuan Kong for the Monte Carlo simulation work.

Conflicts of Interest: The authors declare no conflict of interest.

Appendix A

In a typical ocean Scheimpflug system, the echo signal travels through at least two interfaces (the interface between water and a glass window and the interface between a glass window and air). The specifications of the water Scheimpflug lidar system are fixed including the distance between the center of the CMOS camera and the Scheimpflug intersection x, the distance between the telescope center and the Scheimpflug intersection y, the perpendicular distance between the laser beam and the center of the lens H, the angle between the lens plane and the object plane α, the angle between the lens plane and the plane of CMOS camera β, the distance between the center of the lens and the center of CMOS camera D, the horizontal distance between the center of the lens and the window glass d_1, and the thickness of the window glass d_2. The corrected distance d can be derived from the following equation:

$$D' = \sqrt{\left[p(z) + \frac{y}{\cos\beta}\right]^2 + y^2 - 2\cdot\left[p(z) + \frac{y}{\cos\beta}\right]\cdot y}$$

$$\cos\gamma = \frac{D^2 + D'^2 - p^2(z)}{2\cdot D\cdot D'}$$

$$\xi_1 = 90^\circ - (\alpha - \gamma)$$

$$H' = d_1\cdot\tan\xi_1$$

$$\sin\xi_2 = \frac{n_2}{n_1}\sin\xi_1 = \frac{n_2}{n_1}\cos(\alpha - \gamma)$$

$$\sin\xi_3 = \frac{n_3}{n_2}\sin\xi_2 = \frac{n_3}{n_1}\cos(\alpha - \gamma)$$

$$H'' = d_2\cdot\tan\xi_2$$

$$H''' = H - H' - H''$$

$$d_3 = \frac{H'''}{\tan\xi_3}$$

$$d = d_1 + d_2 + d_3$$

Appendix B

The size of the monochrome CMOS sensor utilized in the Scheimpflug lidar system was a 2448 × 2048 pixel array (approximately 5.0 megapixels, pixel size: 3.54 μm × 3.54 μm). A typical Scheimpflug lidar system has a lens (lens plane) and object planes that are not parallel and three planes that intersect at the Scheimpflug intersection line as Figure A1 shows. MN is a line on the object plane in the real-world coordinate system which represents the laser beam width at distance L, and M′N′ is the image of the line in the image coordinate system. O″ is the intersect point of the Scheimpflug lidar system. O″M′ is the theoretical centerline of the CMOS. O″G′ is the actual centerline of the CMOS. Thus, the G′M′ represents the distance between the centerline of the CMOS sensor and the laser beam. In addition, the relationship between MN and M′N′ is related to the distance and the pixel number. The relationship is illustrated in Figure 4. It is difficult to avoid the CMOS camera without a little displacement during the configuration. In our opinion, the slight displacement of the CMOS was the cause of the upward curvature in the laser beam width. Data gridding and data interpolation were utilized during the data processing. Hence, the upward curvature in the laser beam width would not affect the results of laser beam width in the manuscript.

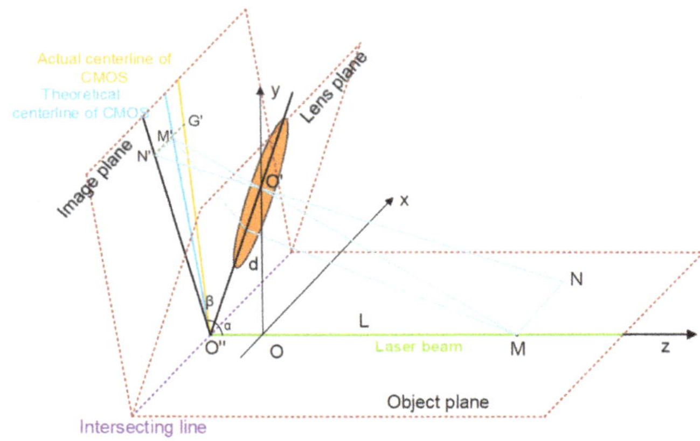

Figure A1. The relationship between the image and the real-world coordinate system. MN is a line on the object plane in the real-world coordinate system, and M′N′ is the image of the line in the image coordinate system. O″ is the intersect point of the Scheimpflug lidar system. O″M′ is the centerline of the CMOS in theory. O″G′ is the actual centerline of the CMOS.

Appendix C

Figure A2. Photograph of the river inside the campus of the Ocean University of China where we sampled the river water.

References

1. Soomets, T.; Uudeberg, K.; Jakovels, D.; Brauns, A.; Zagars, M.; Kutser, T. Validation and Comparison of Water Quality Products in Baltic Lakes Using Sentinel-2 MSI and Sentinel-3 OLCI Data. *Sensors* **2020**, *20*, 742. [CrossRef] [PubMed]
2. Aguilar-Maldonado, J.A.; Santamaría-del-Ángel, E.; Gonzalez-Silvera, A.; Sebastiá-Frasquet, M.T. Detection of Phytoplankton Temporal Anomalies Based on Satellite Inherent Optical Properties: A Tool for Monitoring Phytoplankton Blooms. *Sensors* **2019**, *19*, 3339. [CrossRef]

3. Milton, K.; Joao, A.L.; Cristina, M.B. Simultaneous measurements of Chlorophyll concentration by lidar, fluorometry, above-water radiometry, and ocean color MODIS images in the Southwestern Atlantic. *Sensors* **2009**, *9*, 528–541. [CrossRef]
4. Michael, E.L.; Marlon, R.L. A new method for the measurement of the optical volume scattering function in the upper ocean. *J. Atmos. Ocean. Technol.* **2003**, *20*, 563–571. [CrossRef]
5. Eric, R.; Norman, J.M. Inherent optical property estimation in deep waters. *Opt. Express* **2011**, *19*, 24986–25005. [CrossRef]
6. David, R.L. Remote sensing of bottom reflectance and water attenuation parameters in shallow water using aircraft and Landsat data. *Int. J. Remote Sens.* **1981**, *2*, 71–82. [CrossRef]
7. Ina, L.; Fethi, B.; Charles, T.; Rüdiger, R.; David, B.; Alex, N.; Jill, S. Optical closure in marine waters from in situ inherent optical property measurements. *Opt. Express* **2016**, *24*, 14036–14052. [CrossRef]
8. Li, C.; Cao, W. An instrument for in-situ measuring the volume scattering function of water: Design, Calibration and Primary Experiments. *Sensors* **2012**, *12*, 4514–4533. [CrossRef]
9. Malik, C.; Richard, S.; Eric, D. Radiative transfer model for the computation of radiance and polarization in an ocean–atmosphere system: Polarization properties of suspended matter for remote sensing. *Appl. Opt.* **2001**, *40*, 2398–2416. [CrossRef]
10. Jin, Z.; Stamnes, K. Radiative transfer in nonuniformly refracting layered media: Atmosphere–ocean system. *Appl. Opt.* **1994**, *33*, 431–442. [CrossRef]
11. Gordon, H.R.; Wang, M. Retrieval of water-leaving radiance and aerosol optical thickness over the oceans with SeaWiFS: A preliminary algorithm. *Appl. Opt.* **1994**, *33*, 443–452. [CrossRef]
12. Massey, G.M.; Friedrichs, C.T. *Laser In-Situ Scattering and Transmissometer (LISST) Observations in Support of the Sensor Insertion System Duck, NC October 1997*; Data Report (Virginia Institute of Marine Science) no. 57; Virginia Institute of Marine Science, College of William and Mary, Commonwealth of Virginia: Gloucester Point, VA, USA, 1997. [CrossRef]
13. Doxaran, D.; Leymarie, E.; Nechad, B.; Dogliotti, A.; Ruddick, K.; Gernez, P.; Knaeps, E. Improved correction methods for field measurements of particulate light backscattering in turbid waters. *Opt. Express* **2016**, *24*, 3615–3637. [CrossRef]
14. Roesler, C.; Uitz, J.; Claustre, H.; Boss, E.; Xing, X.; Organelli, E.; Barbieux, M. Recommendations for obtaining unbiased chlorophyll estimates from in situ chlorophyll fluorometers: A global analysis of WET Labs ECO sensors. *Limnol. Oceanogr. Methods* **2017**, *15*, 572–585. [CrossRef]
15. Kim, H.; Lee, S.B.; Min, K.S. Shoreline change analysis using airborne LiDAR bathymetry for coastal monitoring. *J. Coast. Res.* **2017**, *79*, 269–273. [CrossRef]
16. Collister, B.L.; Zimmerman, R.C.; Sukenik, C.I.; Hill, V.J.; Balch, W.M. Remote sensing of optical characteristics and particle distributions of the upper ocean using shipboard lidar. *Remote Sens. Environ.* **2018**, *215*, 85–96. [CrossRef]
17. Behrenfeld, M.J.; Hu, Y.; O'Malley, R.T.; Boss, E.S.; Hostetler, C.A.; Siegel, D.A.; Sarmiento, J.L.; Schulien, J.; Hair, J.W.; Lu, X.; et al. Annual boom–bust cycles of polar phytoplankton biomass revealed by space-based lidar. *Nat. Geosci.* **2017**, *10*, 118–122. [CrossRef]
18. Wu, S.; Liu, B.; Liu, J.; Zhai, X.; Feng, C.; Wang, G.; Gallacher, D. Wind turbine wake visualization and characteristics analysis by Doppler lidar. *Opt. Express* **2016**, *24*, A762–A780. [CrossRef]
19. Zhai, X.; Wu, S.; Liu, B. Doppler lidar investigation of wind turbine wake characteristics and atmospheric turbulence under different surface roughness. *Opt. Express* **2017**, *25*, A515–A529. [CrossRef]
20. Zhang, H.; Wu, S.; Wang, Q.; Liu, B.; Yin, B.; Zhai, X. Airport low-level wind shear lidar observation at Beijing Capital International Airport. *Infrared Phys. Technol.* **2019**, *96*, 113–122. [CrossRef]
21. Dai, G.; Wu, S.; Song, X. Depolarization ratio profiles calibration and observations of aerosol and cloud in the Tibetan Plateau based on polarization Raman lidar. *Remote Sens.* **2018**, *10*, 378. [CrossRef]
22. Mei, L.; Brydegard, M. Atmospheric aerosol monitoring by an elastic Scheimpflug lidar system. *Opt. Express* **2015**, *23*, A1613–A1628. [CrossRef]
23. Liu, Z.; Li, L.; Li, H. Preliminary Studies on Atmospheric Monitoring by Employing a Portable Unmanned Mie-Scattering Scheimpflug Lidar System. *Remote Sens.* **2019**, *11*, 837. [CrossRef]
24. Sun, G.; Qin, L.; Hou, Z.; Jing, X.; He, F. Small-scale Scheimpflug lidar for aerosol extinction coefficient and vertical atmospheric transmittance detection. *Opt. Express* **2018**, *26*, 7423–7436. [CrossRef]
25. Shaw, J.A.; Seldomridge, N.L.; Dunkle, D.L.; Nugent, P.W.; Spangler, L.H.; Bromenshenk, J.J.; Henderson, C.B.; Churnside, J.H.; Wilson, J.J. Polarization lidar measurements of honey bees in flight for locating land mines. *Opt. Express* **2005**, *13*, 5853–5863. [CrossRef]
26. Kirkeby, C.; Wellenreuther, M.; Brydegaard, M. Observations of movement dynamics of flying insects using high resolution lidar. *Sci. Rep.* **2016**, *6*, 29083. [CrossRef]
27. Tauc, M.J.; Fristrup, K.M.; Repasky, K.S.; Shaw, J.A. Field demonstration of a wing-beat modulation lidar for the 3D mapping of flying insects. *OSA Contin.* **2019**, *2*, 332–348. [CrossRef]
28. Li, Y.; Wang, K.; Quintero-Torres, R.; Brick, R.; Sokolov, A.V.; Scully, M.O. Insect flight velocity measurement with a CW near-IR Scheimpflug lidar system. *Opt. Express* **2020**, *28*, 21891–21902. [CrossRef]
29. Mei, L.; Kong, Z.; Guan, P. Implementation of a violet Scheimpflug lidar system for atmospheric aerosol studies. *Opt. Express* **2018**, *26*, A260–A274. [CrossRef]
30. Mei, L.; Brydegaard, M. Continuous-wave differential absorption lidar. *Laser Photonics Rev.* **2015**, *9*, 629–636. [CrossRef]

31. Mei, L.; Guan, P.; Kong, Z. Remote sensing of atmospheric NO 2 by employing the continuous-wave differential absorption lidar technique. *Opt. Express* **2017**, *25*, A953–A962. [CrossRef]
32. Lin, H.; Zhang, Y.; Mei, L. Fluorescence Scheimpflug LiDAR developed for the three-dimension profiling of plants. *Opt. Express* **2020**, *28*, 9269–9279. [CrossRef] [PubMed]
33. Gordon, H.R. Interpretation of airborne oceanic lidar: Effects of multiple scattering. *Appl. Opt.* **1982**, *21*, 2996–3001. [CrossRef] [PubMed]
34. Walker, R.E.; McLean, J.W. Lidar equations for turbid media with pulse stretching. *Appl. Opt.* **1999**, *38*, 2384–2397. [CrossRef] [PubMed]
35. Roddewig, M.R.; Churnside, J.H.; Shaw, J.A. Lidar measurements of the diffuse attenuation coefficient in Yellow Lake. *Appl. Opt.* **2020**, *59*, 3097–3101. [CrossRef]
36. Bogucki, D.J.; Piskozub, J.; Carr, M.E.; Spiers, G.D. Monte Carlo simulation of propagation of a short light beam through turbulent oceanic flow. *Opt. Express* **2007**, *15*, 13988–13996. [CrossRef]
37. Liu, D.; Xu, P.; Zhou, Y.; Chen, W.; Han, B.; Zhu, X.; Chen, S. Lidar remote sensing of seawater optical properties: Experiment and Monte Carlo simulation. *IEEE Trans. Geosci. Remote Sens.* **2019**, *57*, 9489–9498. [CrossRef]
38. Poole, L.R.; Venable, D.D.; Campbell, J.W. Semianalytic Monte Carlo radiative transfer model for oceanographic lidar systems. *Appl. Opt.* **1981**, *20*, 3653–3656. [CrossRef]
39. Churnside, J.H. Review of profiling oceanographic lidar. *Opt. Eng.* **2014**, *53*, 051405. [CrossRef]
40. Morel, A.; Prieur, P. Analysis of variations in ocean color. *Limnol. Oceanogr.* **1997**, *22*, 709–722. [CrossRef]
41. Gordon, H.R.; Morel, A. Remote assessment of ocean color for interpretation of satellite visible imagery: A review. In *Lecture Notes on Coastal and Estuarine Studies*; Springer: New York, NY, USA, 1983.
42. Petzold, T.J. *Volume Scattering Functions for Selected Ocean Waters*; Naval Air Development Center: Warminster, PA, USA, 1972.
43. Sassen, K.; Zhu, J.; Webley, P.; Dean, K.; Cobb, P. Volcanic ash plume identification using polarization lidar: Augustine eruption, Alaska. *Geophys. Res. Lett.* **2007**, *34*. [CrossRef]

remote sensing

Article

Response of Total Suspended Sediment and Chlorophyll-a Concentration to Late Autumn Typhoon Events in The Northwestern South China Sea

Junyi Li [1,2], Huiyuan Zheng [1,3], Lingling Xie [2], Quanan Zheng [1,4], Zheng Ling [1,2] and Min Li [2,*]

1 Laboratory of Coastal Ocean Variation and Disaster Prediction, College of Ocean and Meteorology, Guangdong Ocean University, Zhanjiang 524088, China; jyl@gdou.edu.cn (J.L.); zhenghuiyuan20@mails.ucas.ac.cn (H.Z.); qzheng2@umd.edu (Q.Z.); lingz@gdou.edu.cn (Z.L.)
2 Key Laboratory of Climate, Sources and Environments in Continent Shelf Sea and Deep Ocean, Zhanjiang 524088, China; xiell@gdou.edu.cn
3 Institute of Oceanology, Chinese Academy of Sciences, Qingdao 266071, China
4 Department of Atmospheric and Oceanic Science, University of Maryland, College Park, MD 20742, USA
* Correspondence: min_li@gdou.edu.cn

Abstract: Strong typhoon winds enhance turbulent mixing, which induces sediment to resuspend and to promote chlorophyll-a (Chl-a) blooms in the continental shelf areas. In this study, we find limited Chl-a responses to three late autumn typhoons (typhoon Nesat, Mujigae and Khanun) in the northwestern South China Sea (NWSCS) using satellite observations. In climatology, the Chl-a and total suspended sediment (TSS) concentrations are high all year round with higher value in autumn in the offshore area of the NWSCS. After the typhoon passage, the Chl-a concentration increases slightly (23%), while even TSS enhances by 280% on the wide continental shelf of the NWSCS. However, in the southern area, located approximately 100 km from the typhoon tracks, both TSS and Chl-a concentrations increase 160% and 150% after typhoon passage, respectively. In the deeper area, the increased TSS concentration is responsible for the considerable increase of the Chl-a. An empirical analysis is applied to the data, which reveals the TSS and Chl-a processes during typhoon events. The results of this study suggest a different mechanism for Chl-a concentration increase and thus contribute toward further evaluation of typhoon-induced biological responses.

Keywords: total suspended sediment; chlorophyll-a bloom; typhoon; South China Sea; alongshore current

Citation: Li, J.; Zheng, H.; Xie, L.; Zheng, Q.; Ling, Z.; Li, M. Response of Total Suspended Sediment and Chlorophyll-a Concentration to Late Autumn Typhoon Events in the Northwestern South China Sea. *Remote Sens.* **2021**, *13*, 2863. https://doi.org/10.3390/rs13152863

Academic Editor: Jorge Vazquez

Received: 28 May 2021
Accepted: 17 July 2021
Published: 21 July 2021

1. Introduction

Typhoons inject substantial amounts of energy into the ocean and initiate various ocean processes, e.g., mixing and near-inertial oscillations (NIOs). During typhoon passage, the main response of the upper ocean is related to wide-scale cooling. Hu and Kawmura [1] found that the cold core with a sea surface temperature (SST) difference greater than 2 °C compared to the surrounding areas may be observed using an Advanced Very High Resolution Radiometer (AVHRR) sensor. Since 2010, typhoon-forced near inertial waves have been a research focus for many projects [2]. Guan, et al. [3] examined the upper ocean dynamic response to typhoon Megi (2010) with the presence of a strong internal tide. Sun, et al. [4] found that the waveguide effect of the background shear flow redistributed the NIOs energy after the typhoon passage and trapped energy in the area of the typhoon's negative vorticity. Typhoons induce the oceanic geostrophic response, which perturbs the underlying ocean eddy field [5]. Typhoon intensity can then be strengthened by the warm ocean mesoscale eddy [6].

In addition to the physical typhoon-induced interactions, biogeochemical interactions between the typhoon and ocean occur under the enhanced mixing and transport of water and nutrients. In 2004, enhanced chlorophyll-a (Chl-a) concentrations were revealed in the central South China Sea (SCS) in November 2001 after the passage of typhoon Lingling [7]. The NIOs induces uplift of nutrients and Chl-a into the mixed layer from below, leading to a surface Chl-a increase [8,9]. High winds can induce upwelling, and cold eddies often promote the considerable growth of phytoplankton after a typhoon [10]. Chl-a concentrations increase after typhoon passage and remain at a high concentration for around 5 days before beginning to return to initial conditions [11,12]. Moreover, Wang [13] showed that increases of Chl-a concentration occur after 70% of the typhoons in the SCS based on 16 years of data.

The SCS is one of the areas most affected by typhoons anda local typhoon birthplace [14]. Typhoons generated in October and November are known as late autumn typhoons in the SCS, which may have a stronger impact on their local environment because of their high intensity. Chl-a concentrations during this period increase over the entire SCS, but they remain at a high level (~0.3 mg m^{-3}) in the coastal region [15]. Previous studies have described Chl-a blooms induced by late autumn typhoons. For example, the Chl-a concentration increased by 226% on the 7th day after the passage of typhoon Nesat (2011), associated with the near-inertial baroclinic shear instability [16]. The maximum increase of chlorophyll-a (Chl-a) concentration induced by typhoon Mujigae (2015) occurred in the shelf sea of the NWSCS, with features of unusual surface cooling [17,18].

Moreover, the SCS sits in the East Asian monsoon region, with northeasterly winds prevailing in winter and southwesterly wind in summer. Previous studies have indicated that a southwestward coastal current is present west of Guangdong throughout the year [19,20]. In summer, the occurrence of southwest monsoons is an important factor in the formation of the southwestward coastal current on the shelf [21], while the northeasterly wind-induced southwestward flow prevails over the continental shelf of the northern SCS in winter [22]. Sufficient nutrition and Chl-a are transported by the southwestward current from the Pearl River Estuary in the autumn and winter. Thus, the Chl-a concentration is very high on the continental shelf of the northwestern South China Sea (NWSCS).

Furthermore, the monsoon winds could mix the offshore waters well to the bottom in winter [23]. As the maximum Chl-a concentration is observed in the subsurface water (50–70 m), the vertical mixing effect would transport the Chl-a in the subsurface layer up to the surface layer and cause the Chl-a concentration to increase [24–26]. The prevailing winds produce onshore Ekman transport, which consists of freshwater discharge from the Pearl River. Thus, the coastal current also plays an important role in transporting the Pearl River water and sediment. In the NWSCS, there is a seasonal coastal upwelling system, which occurs from April to September with the strongest upwelling period in June and July [27].

Strong typhoon winds would induces sediment to resuspend and to promote Chl-a blooms in continental shelf areas. However, Lü, et al. [28] found that a high total suspended sediment (TSS) near the shore and a lack of nutrients in the bottom resulted in no Chl-a bloom in the shelf area. This study aims to investigate the mechanisms for Chl-a increase during three late autumn typhoons (Nesat (2011), Mujigae (2015) and Khanun (2017)) landed in the NWSCS.

This paper is organized as follows. Section 2 describes an introduction to the data and methods including algorithms for retrieval of the colored dissolved organic matter (CDOM) and TSS concentrations from satellite observations. Sections 3 and 4 present the analysis of the monthly variation in TSS and Chl-a concentrations and the distribution of the TSS and Chl-a concentrations during typhoon passage in the NWSCS. Section 5 discusses the Pearl River water transport, mixing, and upwelling effects to the Chl-a blooms. Section 6 contains a summary.

2. Materials and Methods

2.1. Study Area

The SCS is one of the largest semi-closed marginal seas of the northwestern Pacific Ocean as shown in Figure 1. It connects the Pacific Ocean, the East China Sea, and the

Indonesian Seas via the Luzon Strait, the Taiwan Strait, and the Karimata Strait, respectively. A deep basin with a maximum depth of about 5500 m sits in the center of the SCS. The wide continental shelf with a depth less than 200 m is distributed around the basin, which occupies about 48% of the total area [29,30].

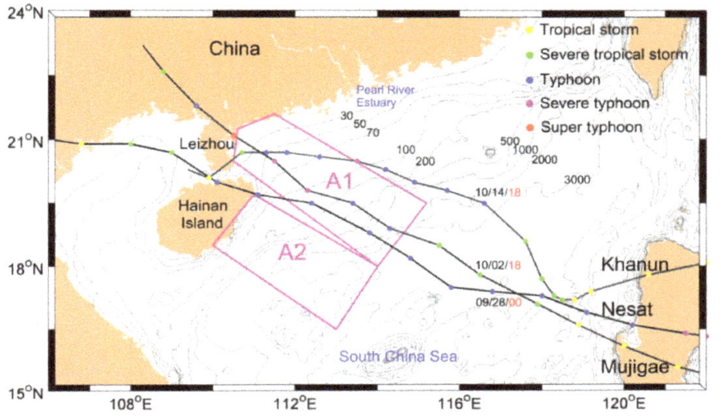

Figure 1. Study area and typhoon tracks in the NWSCS. Black curves with color dots (every 6 h) represent the typhoon tracks of typhoon Nesat (2011) Mujigae (2015), and Khanun (2017). Color dots on the black curves represent the typhoon intensity. The typhoon passing times are coded by month (first two digits), day (the middle two digits), and hour (the last two digits in red). Pink polygons represent the study area consisting of two subareas: Yuexi (A1) and Qiongdong (A2). Numerals on the isobaths are in m.

The study area sits in the NWSCS. We divide it into two sub-areas: Yuexi (A1) and Qiongdong (A2) as shown in Figure 1. A1 is featured by a broad continental shelf. In the north of A1, the largest river system in South China, the Pearl River Estuary is located. The previous studies have shown that a southwestward coastal current dominates the shelf circulation of the NWSCS in autumn and winter [29]. Qiongdong coastal upwelling occupies the area of about 20–50 km from the coastline of A2 [31–33], and it is a seasonal coastal upwelling system occurring from April to September with the strongest upwelling period in June and July [23].

2.2. Late Autumn Typhoons

There were three late autumn typhoons—Nesat (2011), Mujigae (2015), and Khanun (2017)—which made landfall on the southwest coast of China during 2010–2019. The typhoons Nesat (2011), Mujigae (2015), and Khanun (2017) originated from the Pacific Ocean (Table 1), moved northwestward, crossed the continental shelf of the NWSCS, and made landfall on the southwest coast of China. The tracks of the typhoons, as shown in Figure 1, are downloaded from the Tropical Cyclone Data Center of the China Meteorological Administration (CMA) (http://tcdata.typhoon.org.cn, accessed on 4 March 2021) [34,35]. The typhoon center locations, minimum pressure, and two-minute mean maximum sustained wind near the typhoon center are collected at a temporal resolution of 6 h.

Table 1. Summary of typhoon cases passing over the study area.

Typhoon	Date	Category	Origin
Nesat	24–30 September 2011	Typhoon	Pacific Ocean
Mujigae	02–05 October 2015	Super typhoon	Pacific Ocean
Khanun	11–16 October 2017	Typhoon	Pacific Ocean

2.3. Satellite Ocean Color Data

The daily Chl-a data for the typhoon periods and the monthly data from 2004 to 2019 are downloaded from http://oceandata.sci.gsfc.nasa.gov/ (accessed on 7 February 2021). The dataset is a level-3 product with a spatial resolution of 4 km from the moderate resolution imaging spectroradiometer (MODIS) onboard satellites Terra and Aqua (Table 2). The data from the two platforms are merged for improving the coverage of the Chl-a data [36]. The average Chl-a concentration from the Terra and Aqua satellites is estimated and used in this study. The monthly mean Chl-a concentration from 2004 to 2019, with a spatial resolution of 4 km, are derived from http://oceandata.sci.gsfc.nasa.gov/ (accessed on 5 October 2020).

Table 2. Details of the ocean color data used in this study.

Time Period	Data	Temporal Resolution	Spatial Resolution	Satellite/Sensor
19–31 September 2011	Chl-a, Rrs645	daily	4 km	Terra, Aqua/MODIS
20 September–31 October 2015	Chl-a, Rrs645	daily	4 km	Terra, Aqua/MODIS
01–31 October 2017	Chl-a, Rrs645	daily	4 km	Terra, Aqua/MODIS
2004–2019	Chl-a, Rrs412, Rrs555, Rrs645	monthly	4 km	Terra Aqua/MODIS

The Chl-a concentration varied with isobaths (z) in the study area was estimated using the following equation:

$$C_{Chl-a_{Mean}}(z) = \int\int C_{Chl-a}(x, y, z)\, dxdy. \quad (1)$$

where $C_{Chl\text{-}a}(x,y,z)$ is the Chl-a concentration obtained from daily Chl-a data.

In order to estimate the Chl-a concentration variation during the typhoon events, the Chl-a concentration anomaly is estimated with respect to a climatological mean:

$$Ac_{Chl-a_{Mean}}(z) = C_{Chl-a_{Mean}}(z) - Cc(z). \quad (2)$$

where $Cc(z)$ is the climatological mean (2004–2019) for Chl-a concentration varied with isobaths. The climatological mean for Chl-a concentration ($Cc(z)$), e.g., for October, was estimated by

$$Cc(z) = \frac{1}{16}\sum_{i=2004}^{2019} Cc(z,\ t)|_{t=10} \quad (3)$$

where $Cc(z,t)$ is the monthly mean Chl-a concentration varied with isobaths (z) from 2004 to 2019. $Cc(z,t)$ is estimated by using Equation (1) from monthly Chl-a concentration.

The MODIS reflectance (*Rrs*) data at 412, 555, and 645 nm with a spatial resolution of 4 km are obtained from http://oceandata.sci.gsfc.nasa.gov/ (accessed on 7 February 2021) and used to calculate the TSS concentration (Section 2.4) and CDOM (Section 2.5).

2.4. TSS Retrieval

Remote sensing techniques have been applied to quantify the TSS concentration. Empirical relationships between TSS concentration and equivalent reflectance values were established with the linear, polynomial models, and even artificial neural networks [37,38]. Various algorithms for estimating the TSS have been developed using *Rrs* (443), *Rrs* (488), *Rrs* (555), and *Rrs* (645) data [39–41]. As the suspended concentration is high in the continental shelf area, especially near the Pearl River estuary, the TSS concentrations (C_{TSS}) are estimated with retrieval algorithms using *Rrs* (645) data [41]:

$$C_{TSS} = 0.6455 + 1455.7 \times Rrs645. \tag{4}$$

TSS concentration varied with isobaths (z) and anomaly could be estimated from Equations (1) and (2).

2.5. CDOM Retrieval

Satellite ocean color sensors (e.g., MODIS) offers global coverage of CDOM. The amount of CDOM is expressed by the absorption coefficient at 400 nm. The CDOM absorption coefficient at 400 nm is estimated by using monthly *Rrs* (412) and *Rrs* (555) data [42,43]:

$$a_{CDOM(400)} = 0.2355 \times R^{-1.3423}, \tag{5}$$

where $R = Rrs(412)/Rrs(555)$.

2.6. Sea Level Anomaly and Geostrophic Current

Satellite altimeter sea level anomaly (SLA) data are downloaded as a gridded product from the Copernicus Marine Environment Monitoring Service (CMEMS). The temporal resolution is daily and monthly, and the spatial resolution is 0.25° in latitude and longitude with global ocean coverage. The data merged from multiple altimetry sensors are computed with respect to a twenty-year mean.

The daily geostrophic current is provided by the Copernicus Marine Environment Monitoring Service (CMEMS). The monthly mean geostrophic current velocities are calculated using monthly mean SLA data:

$$u = -\frac{g}{f}\frac{\partial \eta}{\partial y}, \tag{6}$$

$$v = \frac{g}{f}\frac{\partial \eta}{\partial x}, \tag{7}$$

where g is the acceleration of gravity, f is the Coriolis parameter and η is monthly mean SLA, respectively.

2.7. Sea Surface Wind and Ekman Pumping

The sea surface wind data at 10 m above the sea surface are obtained from the CMEMS at http://marine.copernicus.eu/ (accessed on 24 April 2021). The data are a level-2 product with a spatial resolution of 25 km, measured by the Advanced Scatterometer (ASCAT) instrument on the EUMETSAT Metop-A satellite. The monthly sea surface wind data from 2010 to 2019 with a spatial resolution of 0.25° are obtained from CMEMS. The climatological sea surface wind is calculated from monthly sea surface wind data.

The velocity of Ekman pumping is calculated by [44]:

$$We = \frac{1}{\rho f}(\nabla \times \tau), \tag{8}$$

where f, ρ and τ are the Coriolis parameter, the seawater density and the wind stress, respectively.

A finite difference scheme is applied to calculate wind stress curl [45]:

$$\nabla \times \tau_{i,j} = \frac{1}{R\cos\varphi_{i,j}}\left(\frac{(\tau_y)_{i+1,j} - (\tau_y)_{i-1,j}}{2\Delta\lambda} - \frac{(\tau_x\cos\varphi)_{i,j+1} - (\tau_x\cos\varphi)_{i,j-1}}{2\Delta\varphi}\right), \tag{9}$$

The wind stress curl is determined as [46]:

$$\tau = \rho_a C_D U|U|, \tag{10}$$

where ρ_a, C_D, and U are the air density, the drag coefficient and the 10 m wind, respectively. The drag coefficient is calculated as [47]:

$$C_D = \begin{cases} (0.75 + 0.067U) \times 10^{-3} & 0 < U \leq 26 \text{ m s}^{-1} \\ 2.5 \times 10^{-3} & U > 26 \text{ m s}^{-1} \end{cases}. \quad (11)$$

3. Climatological and Time Series Analyses

3.1. Monthly Variations of TSS and Chl-a Concentrations

Figure 2 shows the monthly TSS concentration averaged for 2004–2019 varied with isobaths (z) in A1 and A2. The depth-averaged TSS values decrease with the depth increase in A1 throughout the year. The averaged TSS concentration in the summer half year is higher than 3 mg L^{-1} in the offshore area. In the winter half year, the high TSS concentrations extend to the shelf area with a depth of ~100 m.

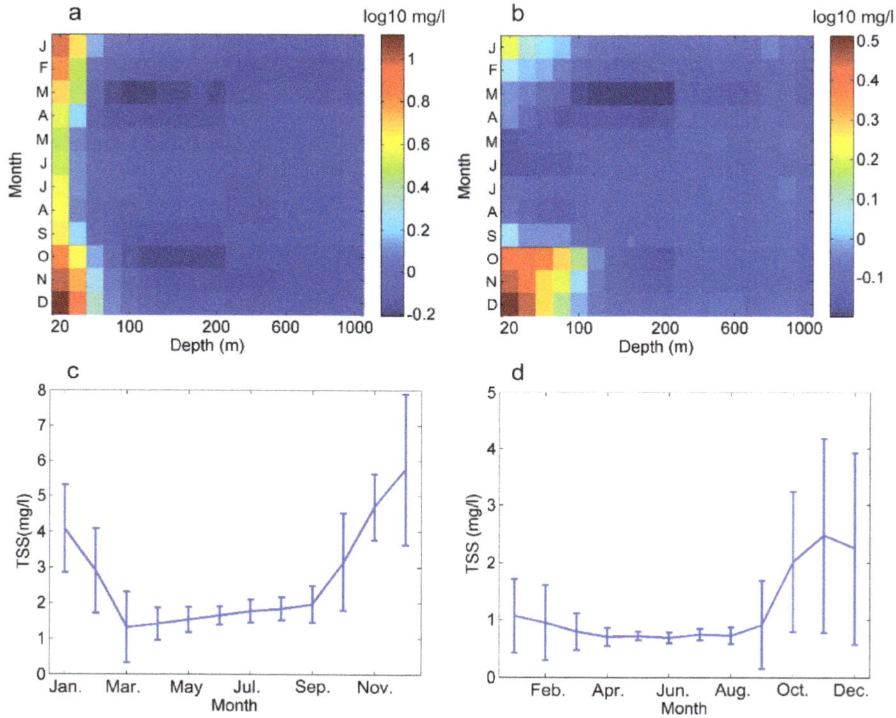

Figure 2. Climatological monthly mean TSS concentration. Time series of ocean surface TSS concentration variation with isobath depth in A1 (**a**) and A2 (**b**). Mean climatological monthly mean TSS concentration with a depth less than 80 m in A1 (**c**) and A2 (**d**). Error bars in (**c**,**d**) are standard deviations of TSS. Color codes for TSS concentration (logarithmic scale) are in mg L^{-1}.

In A2 (Figure 2b), low TSS concentration of <0.1 mg L^{-1} occurred in the offshore and shelf areas in the summer half year. However, a very high TSS concentration of >1.0 mg L^{-1} is also observed, of which the range extends to the areas as deep as 100 m.

Figure 2c,d show that the monthly TSS concentrations in both A1 and A2 in the winter half year are much higher than that in the summer half year. The TSS concentration in A1 is about 3.0 mg L^{-1}, almost twice as that in A2. Moreover, a relatively high TSS concentration (>2 mg L^{-1}) sustains during late autumn period in A2, and is twice as that of the remainder of the year.

From Figure 3, one can see that the monthly mean Chl-a concentrations in A1 and A2 decrease as the depth increases throughout the year. In A1 (Figure 3a), higher Chl-a concentrations (>1 mg m^{-3}) appear in the offshore area with depths less than 40 m. Meanwhile, the Chl-a concentrations in the shelf area (40–200 m) show clear annual variations. In the summer half of the year from April to August, Chl-a concentrations higher than 0.5 mg m^{-3} are distributed in the areas with depths less than 40 m. In contrast, in the autumn, the main areas of high Chl-a concentrations have depths < 100 m. The Chl-a concentrations in areas with depths of 100–200 m in the autumn (~0.4 mg m^{-3}) are more than twice as that in the summer half year. The concentrations in the deep-sea areas (>200 m) are much lower.

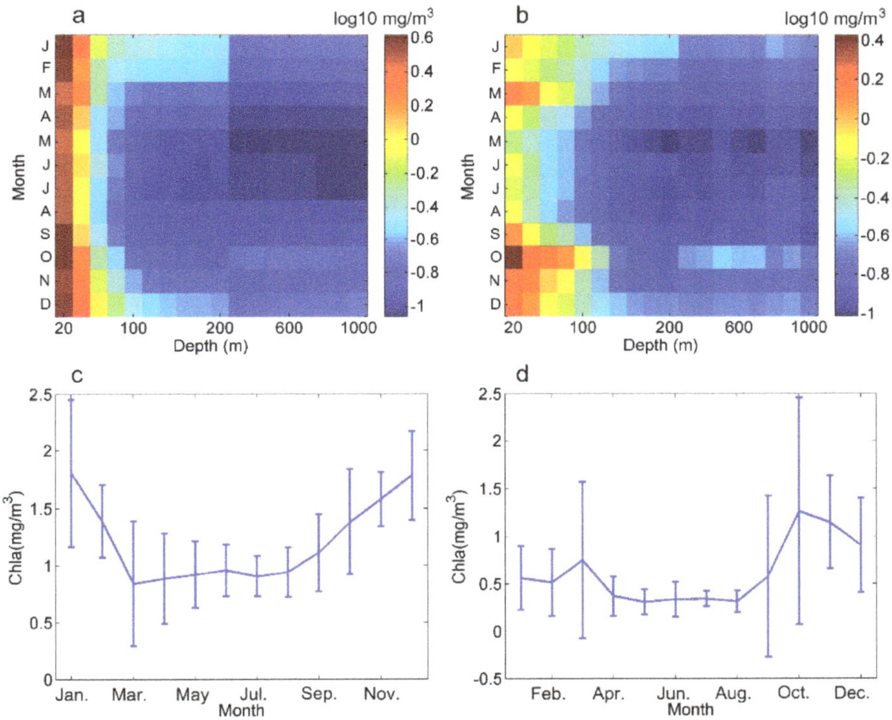

Figure 3. Climatological monthly mean Chl-a concentration variation with isobath depth in A1 (**a**) and A2 (**b**) and at depths less than 80 m (**c**,**d**). Error bars in (**c**,**d**) are the STD of Chl-a. Color codes for Chl-a concentration (logarithmic scale) are in mg m^{-3}.

From Figure 3b, one can see that in A2, the Chl-a values in the offshore areas are much lower than that in A1, and higher Chl-a concentrations are observed in March and October. However, the Chl-a concentrations in the shelf area (40–100 m) are slightly higher (~0.6 mg m^{-3}) than that in A1. Moreover, annual variations in Chl-a concentrations are observed not only in the shelf area but also in the offshore areas.

The high Chl-a concentrations are distributed in the areas with depths less than 80 m in A1 and A2 as shown in Figure 3c,d. The Chl-a concentrations in both A1 and A2 in the autumn are much higher than that in the summer half year. In the summer half year, the mean Chl-a concentrations in A1 and A2 are 0.9 and 0.4 mg m^{-3}, respectively, with 1.4 and 0.9 mg m^{-3} in the winter half year, respectively. Furthermore, two Chl-a concentration in A2 peaks occur in March and October.

In general, the distribution of TSS and Chl-a concentrations in A1 are comparable, i.e., high TSS and Chl-a concentrations are mainly concentrated in the offshore area throughout the year. In the autumn, relatively high TSS and Chl-a concentrations are found in the shelf area due to a well-mixed water column. The TSS and Chl-a concentrations concurrently increase after September. The mean TSS and Chl-a concentrations in October reach 3.1 mg m^{-3} and 1.4 mg L^{-1}, respectively.

In A2, the distributions of TSS and Chl-a concentrations are similar from April to December. The lower TSS concentrations (<1.0 mg L^{-1}) occur in March, whereas the Chl-a concentration peaks at 0.8 mg m^{-3}. Moreover, in contrast to the higher Chl-a, remarkably lower TSS concentrations (<0.1 mg L^{-1}) in the offshore area are observed from April to August. The high TSS concentrations (>1.5 mg L^{-1}) are accompanied by high Chl-a concentrations (>1 mg m^{-3}), of which the distribution extends to the depth as 100 m as observed in October.

3.2. TSS and Chl-a Concentrations in A1 during the Typhoon Period

The time series of TSS concentrations varied with isobaths (z) on the continental shelf in A1 during typhoons Nesat (2011), Mujigae (2015) and Khanun (2017) are shown in Figure 4. The TSS concentration in the offshore area (<40 m) was 6.3 mg L^{-1} before Nesat (2011), Mujigae (2015), and Khanun (2017) arrived. This phenomenon is also shown in Figure 2a, since the monthly mean TSS was high in the offshore area, especially in the winter. The TSS concentration in the offshore area increased significantly to as high as 20.0 mg L^{-1}, after typhoons' landfall (red curves in Figure 4b,d,f). The significant increase of TSS was also observed at a depth of 80 m. In the basin water, the TSS concentration increased slightly. Ten days later, the TSS concentration for three cases reduced gradually to the previous level.

Figure 5 shows time series of Chl-a concentration during the typhoon periods. The Chl-a concentration increased slightly in the offshore areas during the typhoon period, e.g., ~1.0 mg m^{-3} for typhoon Mujigae (2011). The Chl-a concentrations increased about 1.5 mg m^{-3} in the shelf area. In the basin area, the Chl-a concentrations showed only slight fluctuation. Ten days after the typhoon passage, the Chl-a concentrations increased significantly.

Interestingly, the mean Chl-a values only increased 23% in the offshore areas after the typhoon passage. However, that of TSS concurrently increased by 280% (13.6 mg L^{-1}). In contrast, the Chl-a concentration increased by 200% (1.3 mg m^{-3}) at the depth of 60 m, while TSS concentration increased by 450%. Thus, there is no Chl-a bloom even with enhanced TSS, which is different from the feature in the shelf region and many previous literatures.

3.3. TSS and Chl-a Concentrations in A2 during the Typhoon Period

Figure 6 shows how the time series of TSS concentration responded to typhoons Nesat (2011), Mujigae (2015), and Khanun (2017) in sub-region A2. There was a remarkable response of TSS concentration to typhoons at the depth of 100 m. After Mujigae (2015) passed, high TSS concentrations were observed at 200 m depth (Figure 6c,d). Moreover, the maximum increase of the TSS concentration was shown in the shelf area (i.e., ~100 m); it increased by 5.7, 3.7, and 0.6 mg L^{-1} for Nesat (2011), Mujigae (2015), and Khanun (2017), respectively. The TSS concentration only increased by about 0.3 mg L^{-1} in the offshore areas. Ten days after the typhoon passage, the TSS concentration decreased to a background level.

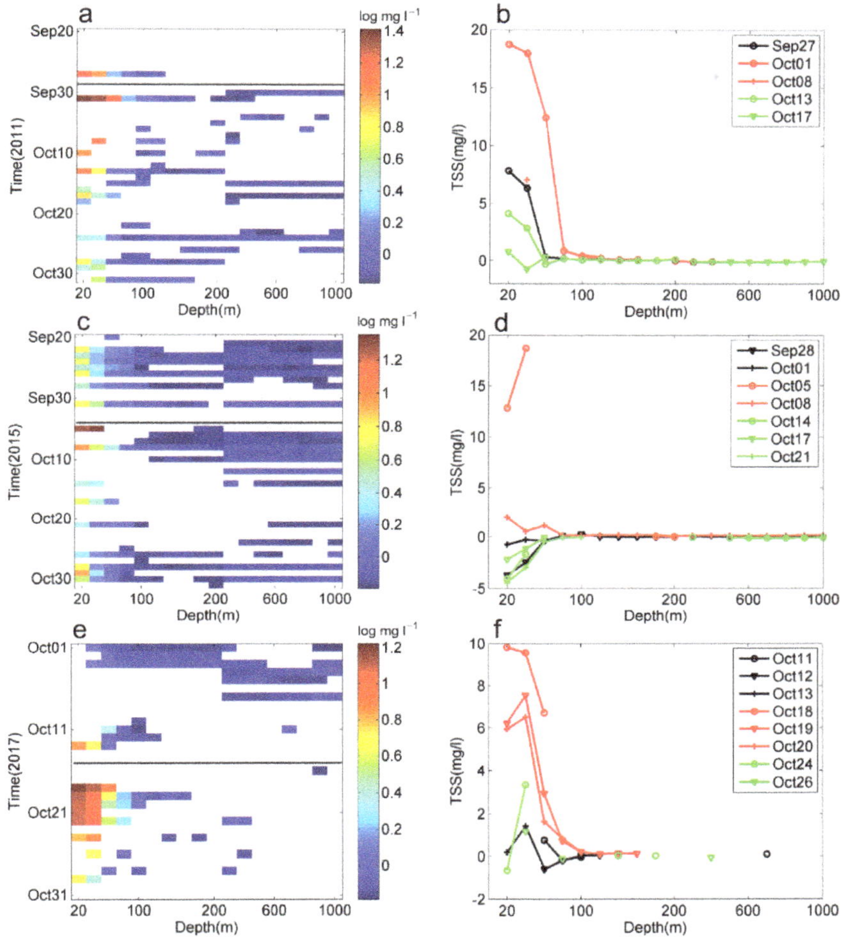

Figure 4. Time series of TSS varied with isobaths (z) on the continental shelf of A1 (**a**,**c**,**e**). TSS concentration anomaly along the continental shelf (**b**,**d**,**f**). (**a**,**c**,**e**) are from Equation (1). X-axis indicates the concentration varied with isobaths (z). Y-axis indicates the time during the typhoon event. (**b**,**d**,**f**) are from Equation (2). Data are limited to the depth of 1000 m. Black lines in (**a**,**c**,**e**) represent passing times of typhoons Nesat (2011), Mujigae (2015) and Khanun (2017), respectively. Color codes for TSS concentration (logarithmic scale) are in mg L^{-1}. In (**b**,**d**,**f**), black, red and green curves represent the Chl-a concentrations before, and after typhoon passage, respectively. The legend indicates the observation time of satellite.

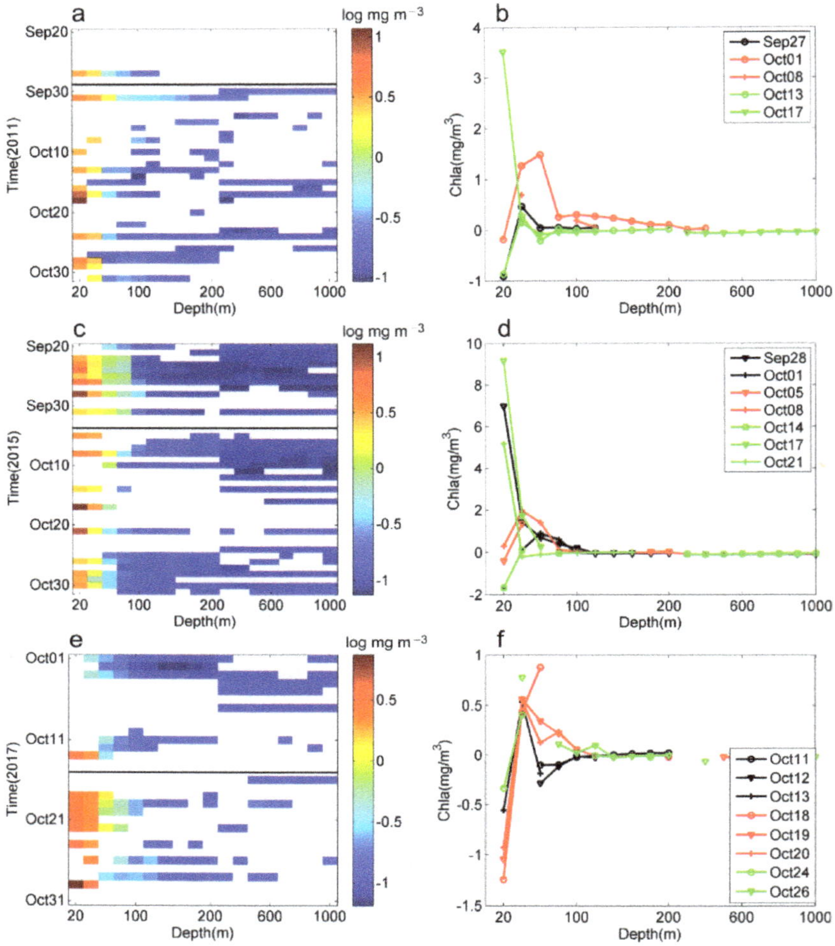

Figure 5. Time series of Chl-a concentration varied with isobaths (z) on the continental shelf of A1 (**a**,**c**,**e**). The Chl-a concentration anomaly along the continental shelf is shown in (**b**,**d**,**f**). (**a**,**c**,**e**) are from Equation (1). X-axis indicates the concentration varied with isobaths (z). Y-axis indicates the time during the typhoon event. (**b**,**d**,**f**) are from Equation (2). Data are limited to the depth of 1000 m. Black lines in (**a**,**c**,**e**) represent passing times of typhoons Nesat (2011), Mujigae (2015) and Khanun (2017). Color codes for Chl-a concentration (logarithmic scale) are in mg m^{-3}. In (**b**,**d**,**f**), black, red and green curves represent Chl-a concentration before and after typhoon passage, respectively. The legend indicates the observation time of satellite.

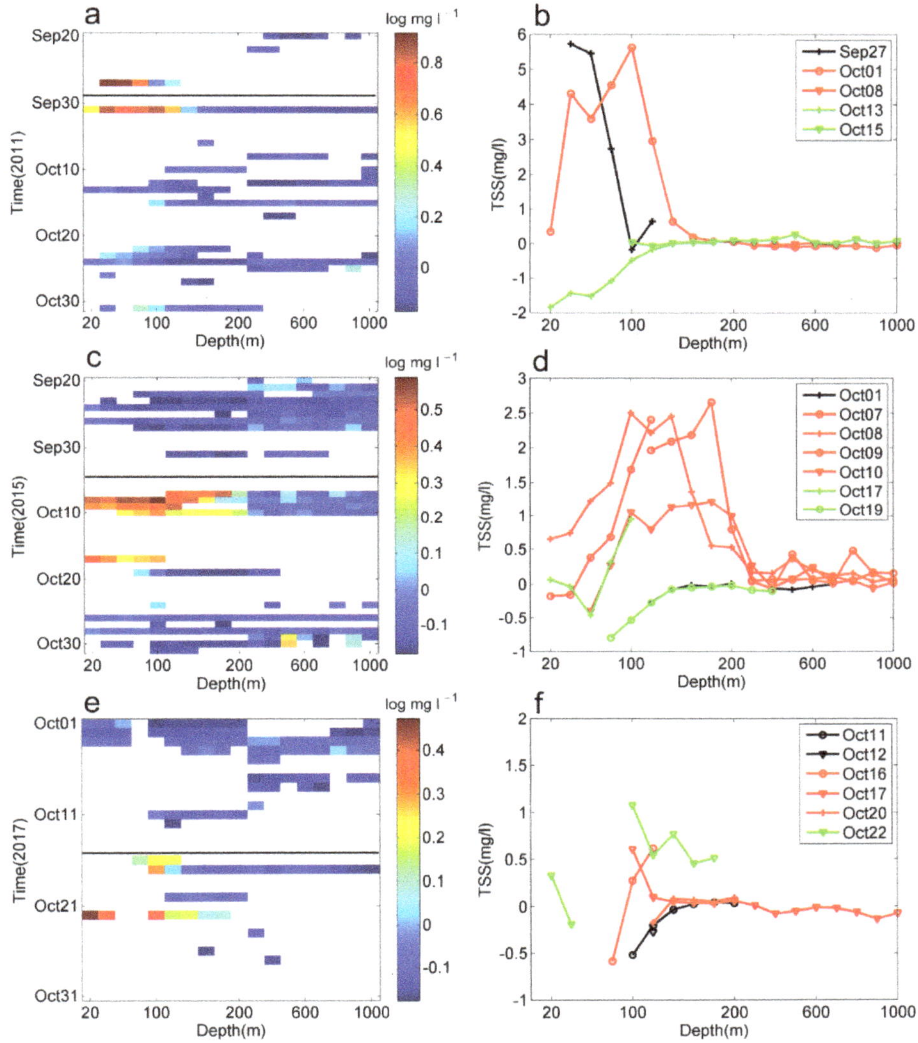

Figure 6. Time series of TSS concentration varied with isobaths (z) on the continental shelf of A2 (**a**,**c**,**e**). TSS concentrations anomaly along the continental shelf (**b**,**d**,**f**). (**a**,**c**,**e**) are from Equation (1). X-axis indicates the concentration varied with isobaths (z). Y-axis indicates the time during the typhoon event. (**b**,**d**,**f**) are from Equation (2). Data are limited to the depth of 1000 m. Black lines in (**a**,**c**,**e**) represent passing times of typhoons Nesat (2011), Mujigae (2015) and Khanun (2017). Color codes for TSS concentration (logarithmic scale) are in mg L^{-1}. In (**b**,**d**,**f**), black, red, and green curves are Chl-a concentration before and after typhoon passage, respectively. The legend indicates the observation time of satellite.

Figure 7 shows the time series of the Chl-a concentration during the three typhoon periods. The maximum increase of Chl-a concentration occurs concurrently with that of TSS in the shelf area. The concurrent phenomenon indicates that the high level concentration of Chl-a is the result of the high level concentration of TSS. The Chl-a concentrations increase by about 1.0, 3.0, and 0.5 mg m^{-3}, respectively. In the basin area, however, the Chl-a concentrations show only slight fluctuations.

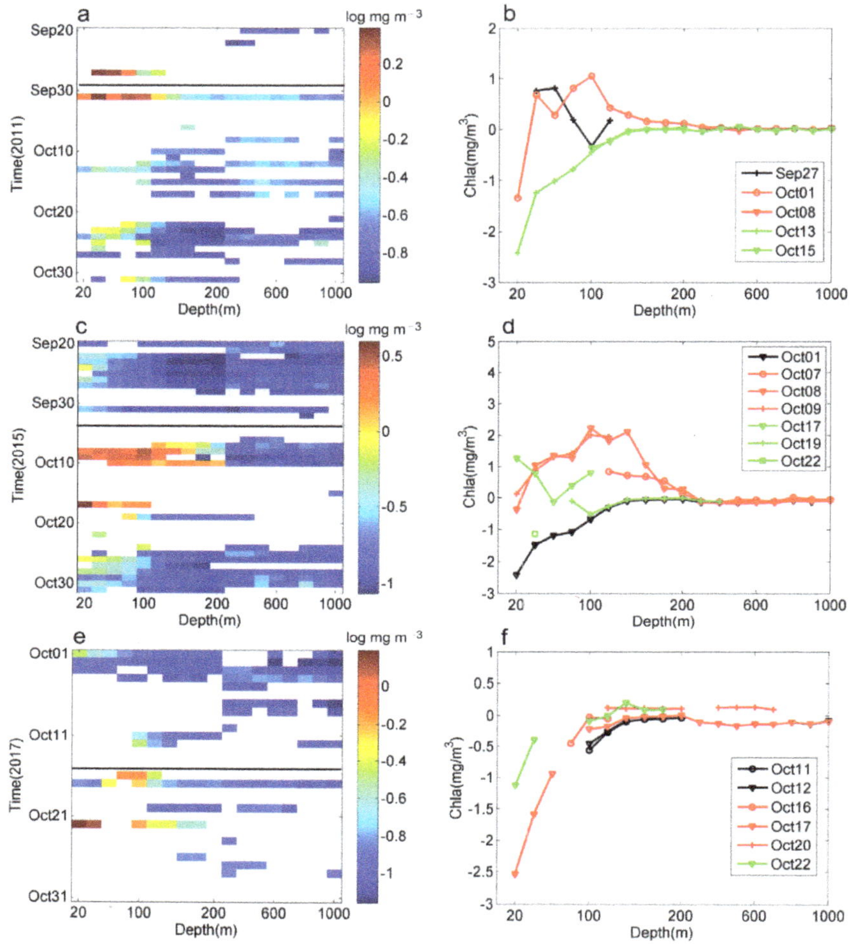

Figure 7. Time series of Chl-a concentration varied with isobaths (z) on the continental shelf of A2 (**a**,**c**,**e**). Chl-a concentrations anomaly along the continental shelf (**b**,**d**,**f**). (**a**,**c**,**e**) are from Equation (1). X-axis indicates the concentration varied with isobaths (z). Y-axis indicates the time during the typhoon event. (**b**,**d**,**f**) are from Equation (2). Data are limited to the depth of 1000 m. Black lines in (**a**,**c**,**e**) represent passing times of typhoons Nesat (2011), Mujigae (2015) and Khanun (2017). Color codes for Chl-a concentration (logarithmic scale) are in mg m^{-3}. In (**b**,**d**,**f**), black, red and green curves are Chl-a concentrations before and after typhoon passage, respectively. The legend indicates the observation time of satellite.

In October, the background TSS and Chl-a concentrations (Figures 2b and 3b) were high in the shelf area, e.g., ~2.0 mg L^{-1} and 1.0 mg m^{-3}, respectively. In all three typhoon cases, typhoons enhanced Chl-a concentrations in the offshore and shelf areas by 160% and 150%, respectively.

4. Empirical Analysis of Temporal Variations

We divide the study area into two sub-areas: the continental shelf and the basin area and calculate the mean TSS and Chl-a concentration of shelf area with a depth less than

100 m in A1, and with a depth less than 200 m in A2. Here we use a Rayleigh function to fit the time series of TSS and Chl-a concentration:

$$y = A\frac{(t+T)}{\sigma^2}\exp\left(-\frac{(t+T)^2}{2\sigma^2}\right) + \Delta \tag{12}$$

where A is the coefficient of the TSS and Chl-a concentration; t is the time (d); T is the time lag (left-ward shift with a positive value); σ^2 is the variance of TSS and Chl-a concentration. Δ is a background concentration of TSS and Chl-a.

Figure 8 shows the TSS concentrations 10 days before and after typhoons Nesat (2011), Mujigae (2015), and Khanun (2017). The TSS concentration in the shelf area of A1 was about 2–7 mg L^{-1} before typhoon landfall (Figure 8a). After typhoon landfall, the maximum of TSS concentration was as high as 22 mg L^{-1}, then decreased to 3 mg L^{-1} within 10 days. Figure 8b shows that in the basin area of A1, the mean TSS concentration increased from 0.7 to 0.8 mg L^{-1} during typhoon passage.

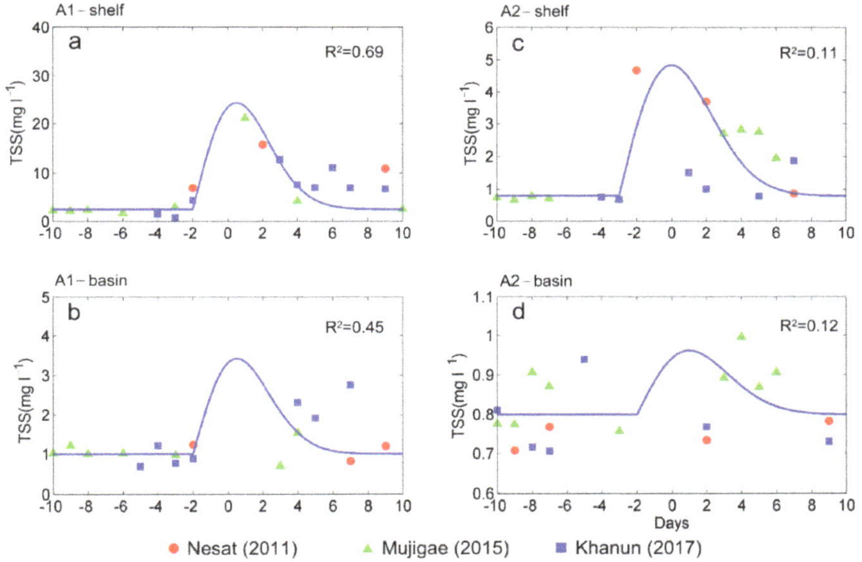

Figure 8. Rayleigh function fits of TSS time series increase after typhoon passage in A1 (**a**,**b**) and A2 (**c**,**d**). The colored symbols show the observations on the continental shelf (**a**,**c**) and basin area (**b**,**d**). The solid curves are the Rayleigh function. Averaged depths are upper 100 m and 200 m for (**a**) and (**c**), respectively.

In the shelf area of A2, the mean TSS concentration remained lower than 1 mg L^{-1} before typhoon landfall (Figure 8c), peaked to 5 mg L^{-1} and then reduced to 1 mg L^{-1} after typhoon passing. In the basin area, the mean TSS concentration (~0.7 mg L^{-1}) only showed slight fluctuations (Figure 8d).

The empirical function explains the TSS variation process during typhoon events very well, especially for the shelf of A1. The coefficient of TSS concentration in A1 is as high as 90 mg L^{-1} with R^2 value of 0.69 (Table 3), which means that TSS concentration could reach about 25 mg L^{-1} during typhoon events. In the basin, the coefficient of TSS concentration decreases to 10 mg L^{-1}.

Table 3. Summary of TSS and Chl-a decrease in A1 and A2.

Area		Δ_{TSS} (mg L^{-1})	A_{TSS} (mg L^{-1})	T (d)	σ^2	R^2	$\Delta_{Chl\text{-}a}$ (mg m^{-3})	$A_{Chl\text{-}a}$ (mg L^{-1})	T (d)	σ^2	R^2
A1	Offshore	2.5	90	2	2.5	0.69	2.6	6	2	2	0.44
	Shelf						0.3	6	2	3	0.39
	Basin	1	10	2	2.5	0.45	0.1	0.3	2	3	0.37
A2	Shelf	0.8	20	3	3	0.11	0.2	6	3	2	0.23
	Basin	0.8	0.8	3	3	0.12	0.12	1	3	2	0.37

Fore Chl-a, we divided A1 into three sub-areas: the offshore (<40 m), shelf (<100 m), and the basin area. One can see that the time series Chl-a concentrations in the offshore area (Figure 9a) changed slightly. The Chl-a concentrations after typhoon landfall were even lower than those six days before typhoon passage in the case of typhoon Mujigae (2015). Moreover, in the case of typhoon Khanun (2017), Chl-a concentrations decreased from 2.9 to 2.4 mg m^{-3} during typhoon passage. Figure 9b shows that the Chl-a concentrations increased from ~0.5 to ~1.0 mg m^{-3} in the shelf area. In the basin area (Figure 9c), the Chl-a concentrations increased from ~0.1 to ~0.15 mg m^{-3}.

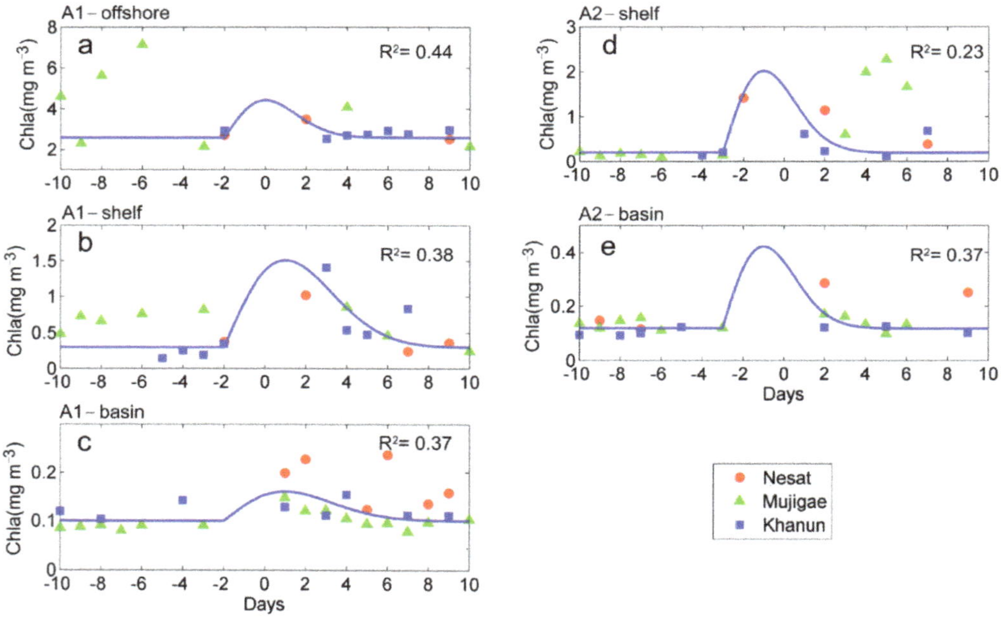

Figure 9. Rayleigh function fit of time series Chl-a concentration in A1 (**a–c**) and A2 (**d**,**e**). The colored symbols show the observations on the continental shelf (**a**,**b**,**d**) and open sea (**c**,**e**). Blue lines are the Rayleigh function.

Time series Chl-a concentrations in the offshore and shelf area of A2 are shown in Figure 9d. The Chl-a concentrations increased significantly after typhoon passage, from 0.2 to about 1.0 mg m^{-3}. However, the Chl-a concentrations increased slightly in the basin area (Figure 9e).

Rayleigh function fits revealed the Chl-a variation process during typhoon events. The coefficient of the Chl-a concentration is as high as 6 mg m^{-3} with R^2 value of 0.44 (Table 3). The R^2 value is not so high, and the main reason is the data missing during the passage of typhoon.

5. Discussion

5.1. Pearl River Water Transport

Previous studies have revealed that typhoons result in substantial increases of TSS and Chl-a concentrations [7,10,48]. An offshore bloom exhibited a Chl-a peak of 0.7 mg m^{-3} relative to a background of 0.1 mg m^{-3} before landfall of typhoon Damrey [49]. The maximum Chl-a concentration was enhanced to approximately 7.5 times the climatological value after the typhoon Nuri (2008) [50,51]. Li, et al. [19] found that the Chl-a concentration in the offshore area increased three times in five days after landfall of typhoon Mangkhut (2018). The maximum Chl-a concentration in the offshore area resulted from the suspended particulate matter from runoff or mixing [52–54]. Hu, et al. [55] found that the photochemical reactions would transfer sedimentary organic matter into dissolved organic matter and affect ultimate fate in aquatic ecosystems. The resuspended sediments on the continental shelf would release inorganic nutrients under the effect of sunlight within a few hours [56–58].

However, in the current study, the TSS concentration in the offshore area of A1 (Figure 4) was very high due to the strong wind during autumn typhoon events, whereas the Chl-a concentration only increased by 23% (about 0.3 mg m^{-3}) compared with the monthly mean Chl-a value (Figure 2). In the shelf area, the TSS concentration increased by four times, whereas the Chl-a concentration increased twice with respect to the climatological value. It remains unclear, therefore, why the increment of Chl-a concentration was so different between the offshore and shelf area during these late autumn typhoon events.

The CDOM, considered as the powerful indicators represents the degree riverine plumes to be affected by terrestrial inputs [43,59]. Figure 10a shows the climatological CDOM distribution of the study area in July. One can see that CDOM (a_{CDOM}) is mainly distributed in the offshore area. Huang, et al. [60] observed high nutrient contents and TSS concentrations in the offshore area in summer and the terrestrial organic matter contributed 38% of TSS in the Pearl River Estuary. Lao, et al. [61] found that the nitrate content along the coast of A1 was strongly influenced by discharges from local urban areas and the westward flow of diluted Pearl River water. Therefore, it is reasonable to attribute summer CDOM to influences of local urban area discharges and Pearl River water.

Figure 10b shows that in October the CDOM is much higher than that in July and distributed over the shelf area. As above mentioned, the coastal current with the Pearl River water flows southwestward [62]. Combined with the climatological geostrophic current (Figure 10c,d), one can concludes that the Pearl River water transported by the alongshore current supplies the phytoplankton in the offshore area of A1, which results in a high Chl-a concentration in the winter half year (Figure 3a).

Figure 10e,f show the geostrophic current before and after typhoon Mujigae (2015), which reversed the geostrophic current direction, compared with the climatological geostrophic current in July (Figure 10c). The climatological geostrophic current in October (Figure 10d) flows southwestward at a velocity as high as 0.3 m s^{-1}.

Lü, et al. [28] found no Chl-a bloom in the nearshore and shelf areas in mid-latitude regions during a typhoon event with the characteristics of coastal downwelling and insufficient sunshine. Here, we find that Chl-a increase during three late autumn typhoons (Nesat (2011), Mujigae (2015) and Khanun (2017)) landed in the NWSCS and aim to investigate the mechanisms. However, this study area sits in the low latitudes with the sufficient sunshine. The mixing effect (Section 5.2) is strong in the monsoon season in the offshore area, especially during typhoon periods, implying that the downwelling effect is weak. Thus, the case of no Chl-a bloom is seldom seen.

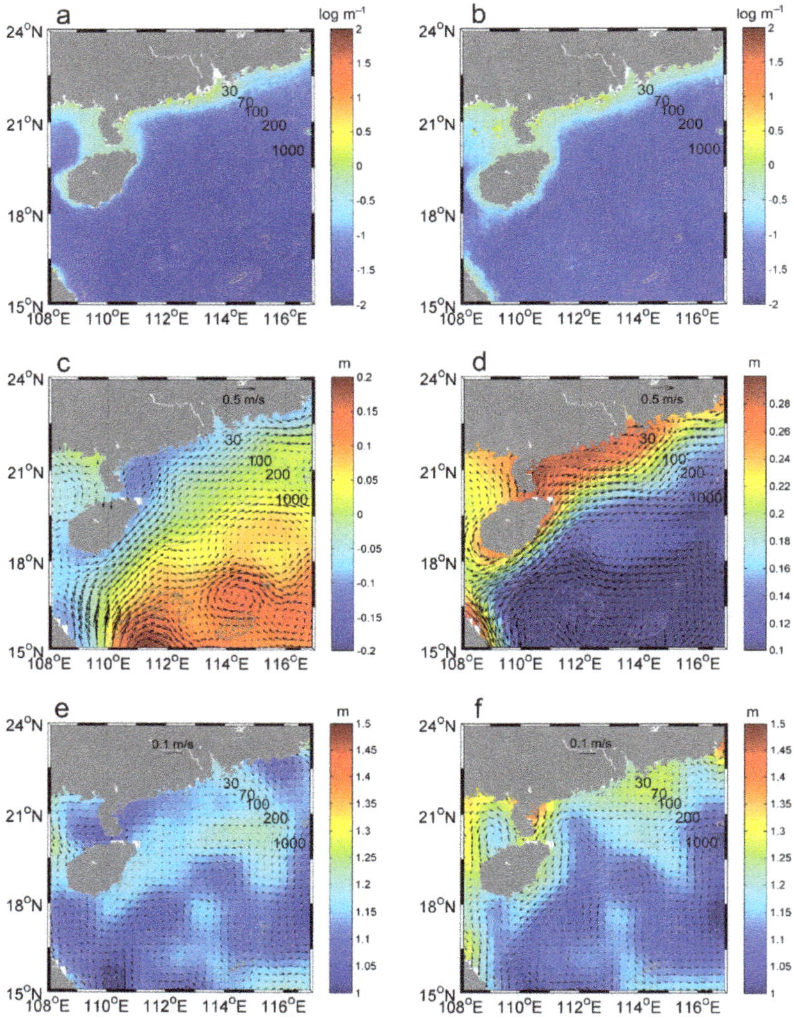

Figure 10. CDOM a_{CDOM} (**a**,**b**), monthly mean SLA (**c**,**d**), and daily SLA (**e**,**f**) in the NWSCS. (**a**) and (**c**) climatological data in July from 2010 to 2019. (**b**) and (**d**) climatological data in October from 2010 to 2019. (**e**) SLA on 24 September 2015. (**f**) SLA on 4 October 2015. Arrows in (**c**–**f**) represent the geostrophic currents. Numerals on the isobaths are in m.

Furthermore, Huang, et al. [63] found that the offshore area of A1 was mainly impacted by the Pearl River water, and nutrition are sufficient to feed phytoplankton in winter. Thus, the balance of nutrition could explain why late autumn typhoons only had a slight effect on the Chl-a concentration in the offshore of A1.

The water on the offshore and shelf area originated from the same source [19], and the coefficient of Chl-a concentration for the offshore and shelf areas of A1 are similar, about 6 mg m^{-3}. The Chl-a increments in the offshore area (1.4 mg m^{-3}) and the shelf area (1.3 mg m^{-3}) are almost the same.

In A2, unusually high TSS concentrations were observed in the offshore area for three cases with a weaker sea surface wind (~100 km away from the typhoon track).

Zheng, et al. [64] found that the upper layer current turned southwestward after typhoon passage in the shelf area of A2. Figure 10e,f show the same result, i.e., the current transported the shelf water with high TSS from A1 into the offshore and shelf areas of A2, which induced the high TSS in A2. The transported TSS induced an increase of Chl-a concentration in the shelf area of A2 (Figure 10d).

5.2. Upwelling Effect and Mixing

Typhoon with strong winds would induce upwelling and mixing, which are important to nutrient transport. Figure 11 shows the Ekman pumping velocity on 2, 3, and 4 October 2015. The upwelling before typhoon passage was weak (about 1×10^{-6} m/s) in A1 and A2. On 4 October, when typhoon Mujigae (2015) reached A1, upwelling became stronger. The upwelling velocity for the whole layer was as high as 2×10^{-5} m s^{-1}, representing an increase of twenty times. Because A2 is far away from the typhoon track, the upwelling effect changed little. Wang, et al. [65] pointed out that the pumping velocity in the upper layer was as high as 1×10^{-4} m s^{-1} during typhoon Washi (2005) using mooring observation data. The Ekman pumping velocity for the other two typhoons was almost the same. The ocean surface cooling reached ~5 °C after the typhoon events [16,17].

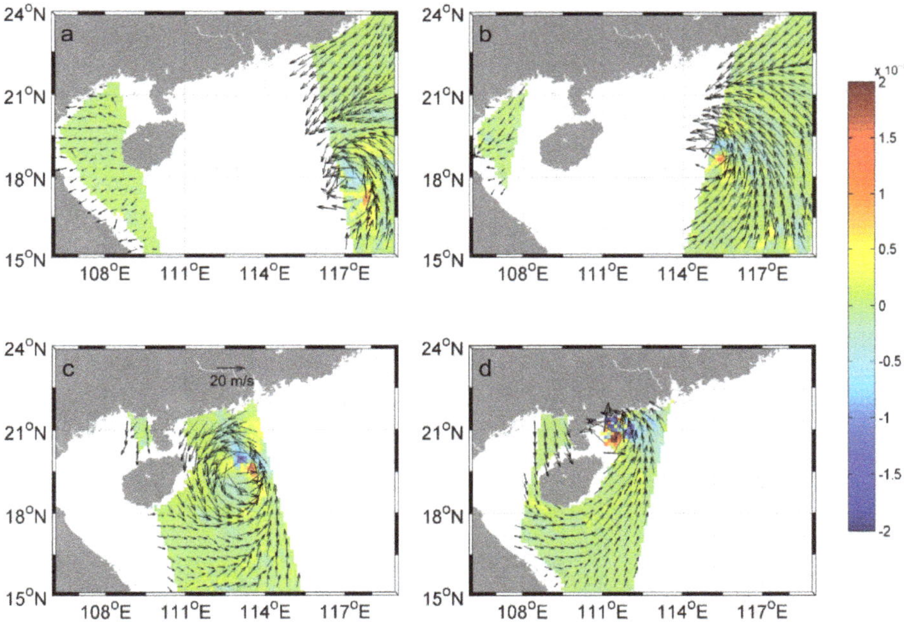

Figure 11. Sea surface wind (arrows) and Ekman pumping velocity (colored) in the NWSCS. The unit of velocity (color bar) is m s^{-1}. (**a–d**) were captured at 13:45 UTC on 2 October, 2:16 UTC on 3 October, 13:24 UTC on 3 October, and 1:55 UTC on 4 October 2015, respectively.

In addition to Ekman pumping, the mixing effect during the typhoon event is also important. The climatological wind speed in October is about 10 m s^{-1}, whereas about 3 m s^{-1} in July. The mixing forced by the prevailing wind is another reason for the CDOM distribution in the shelf area (Figure 10a,b). A high nutrient concentration has been observed in the bottom water (~40 m) of the offshore area (in A1) [66]. The nutrients would be transported to the surface layer by the mixing effect together with the Ekman pumping. Moreover, wind-induced upper water mixing during the winter in the NWSCS may also provide subsurface nutrients as the ultimate marine end-member for offshore surface waters [67].

6. Conclusions

In this study, we investigate the response of TSS concentration and Chl-a blooms to late autumn typhoon events over the NWSCS. Three cases of typhoons Nesat (2011), Mujigae (2015) and Khanun (2017) in October are examined using satellite observations. The major findings and results are summarized as follows.

The high TSS and Chl-a concentrations are mainly concentrated in the offshore area throughout the year. They concurrently increase after September. The mean TSS and Chl-a concentrations in October are 1.4 mg L^{-1} and 3.1 mg m^{-3}, respectively. In the upwelling area (A2), in contrast to high Chl-a, a considerably low TSS concentrations (<0.1 mg L^{-1}) occur from April to August in the offshore area. High TSS concentrations (>1.5 mg L^{-1}) accompanied by high Chl-a concentrations (>1 mg m^{-3}) occur in March and October.

The mechanisms of Chl-a concentration increase, induced by late autumn typhoon events vary throughout the study area. The alongshore currents play different roles in Chl-a blooming on the track and adjacent area, which would reduce the Chl-a blooming in the NWSCS during late autumn typhoon events. Chl-a concentrations on the typhoon tracks increase 1.4 mg m^{-3} (23%), with TSS concurrently increasing by 13.6 mg L^{-1} (280%) in the offshore area. The Chl-a bloom is restrained by the alongshore current from the Pearl River Estuary with the sufficient nutrition. However, in the upwelling area, the Chl-a bloom in the shelf area is mainly associated with TSS transported from the typhoon track. The empirical analysis is applied to the time series TSS and Chl-a concentration data. The results show that the Rayleigh function describe the TSS and Chl-a variation process during typhoon events well. With three late autumn typhoons cases, this study contributes toward further evaluation of typhoon-induced biological responses.

Author Contributions: All authors have made significant contributions to this research. J.L., Z.L. and H.Z. analyzed the data; L.X. and M.L. provided important insights and suggestions on this research; J.L. and Q.Z. prepared the manuscript. All authors have read and agreed to the published version of the manuscript.

Funding: Supported by the National Natural Science Foundation of China (41706025, 41776034, 41476009), the Matched Grant of Guangdong Ocean University (P17263), the Innovation Team Project of Guangdong Ocean University (CXTD2019001, 570119020), College Students' innovation and Entrepreneurship Project (CXXL2019151), the Guangdong Provincial College Innovation Team Project (2019KCXTF021), the First-class Discipline Plan of Guangdong Province (231419012, 231919030), and the Open Fund of the Key Laboratory of Ocean Circulation and Waves, Chinese Academy of Sciences, the Program for Scientific Research Start-up Funds of Guangdong Ocean University.

Institutional Review Board Statement: Not applicable.

Informed Consent Statement: Not applicable.

Data Availability Statement: Publicly available datasets were analyzed in this study. This data can be found here: Typhoon track is downloaded from http://tcdata.typhoon.org.cn (accessed on 17 July 2021). Satellite Ocean Color Data are downloaded from http://oceandata.sci.gsfc.nasa.gov/ (accessed on 17 July 2021). Sea surface wind data and SLA data are downloaded from http://marine.copernicus.eu/ (accessed on 17 July 2021).

Acknowledgments: The authors are grateful to the anonymous reviewers for their valuable suggestions and comments.

Conflicts of Interest: The researchers claim no conflict of interests.

References

1. Hu, J.; Kawmura, H. Detection of cyclonic eddy generated by looping tropical cyclone in the northern South China Sea: A case study. *Acta Oceanol. Sin.* **2004**, *232*, 213–224.
2. Zheng, Q.; Xie, L.; Xiong, X.; Hu, X.; Chen, L. Progress in research of submesoscale processes in the South China Sea. *Acta Oceanol. Sin.* **2020**, *39*, 1–13. [CrossRef]
3. Guan, S.; Wei, Z.; Huthnance, J.; Tian, J.; Wang, J. Observed upper ocean response to typhoon Megi (2010) in the Northern South China Sea. *J. Geophys. Res. Oceans* **2014**, *119*, 3134–3157. [CrossRef]
4. Sun, Z.; Hu, J.; Zheng, Q.; Gan, J. Comparison of typhoon-induced near-inertial oscillations in shear flow in the northern South China Sea. *Acta Oceanol. Sin.* **2015**, *45*, 38–45. [CrossRef]
5. Lu, Z.; Wang, G.; Shang, X. Strength and Spatial Structure of the Perturbation Induced by a Tropical Cyclone to the Underlying Eddies. *J. Geophys. Res. Oceans* **2020**, *125*, e2020JC016097. [CrossRef]
6. Sun, J.; Wang, G.; Xiong, X.; Hui, Z.; Hu, X.; Ling, Z.; Long, Y.; Yang, G.; Guo, Y.; Ju, X.; et al. Impact of warm mesoscale eddy on tropical cyclone intensity. *Acta Oceanol. Sin.* **2020**, *39*, 1–13. [CrossRef]
7. Shang, S.; Li, L.; Sun, F.; Wu, J.; Hu, C.; Chen, D.; Ning, X.; Qiu, Y.; Zhang, C.; Shang, S. Changes of temperature and bio-optical properties in the South China Sea in response to Typhoon Lingling, 2001. *Geophys. Res. Lett.* **2008**, *35*, L10602. [CrossRef]
8. Zhao, H.; Pan, J.; Han, G.; Devlin, A.T.; Zhang, S.; Hou, Y. Effect of a fast-moving tropical storm Washi on phytoplankton in the northwestern South China Sea. *J. Geophys. Res. Oceans* **2017**, *122*, 3404–3416. [CrossRef]
9. Zhang, S.; Xie, L.; Hou, Y.; Zhao, H.; Qi, Y.; Yi, X. Tropical storm-induced turbulent mixing and chlorophyll-a enhancement in the continental shelf southeast of Hainan Island. *J. Mar. Syst.* **2014**, *129*, 405–414. [CrossRef]
10. Liu, Y.; Tang, D.; Morozov, E. Chlorophyll Concentration Response to the Typhoon Wind-Pump Induced Upper Ocean Processes Considering Air–Sea Heat Exchange. *Remote Sens.* **2019**, *11*, 1825. [CrossRef]
11. Yasuki, N.; Suzki, K.; Tsuda, A. Responses of lower trophic-level organisms to typhoon passage on the outer shelf of the East China Sea: An incubation experiment. *Biogeosic. Discuss.* **2013**, *10*, 6605–6635.
12. Li, J.; Zheng, Q.; Li, M.; Li, Q.; Xie, L. Spatiotemporal Distributions of Ocean Color Elements in Response to Tropical Cyclone: A Case Study of Typhoon Mangkhut (2018) Past over the Northern South China Sea. *Remote Sens.* **2021**, *13*, 687. [CrossRef]
13. Wang, Y. Composite of Typhoon-Induced Sea Surface Temperature and Chlorophyll-a Responses in the South China Sea. *J. Geophys. Res. Oceans* **2020**, *125*, e2020JC016243. [CrossRef]
14. Wang, G.; Su, J.; Ding, Y.; Chen, D. Tropical cyclone genesis over the south China sea. *J. Mar. Syst.* **2007**, *68*, 318–326. [CrossRef]
15. Huynh, H.; Alvera-Azcarate, A.; Beckers, J.M. Analysis of surface chlorophyll a associated with sea surface temperature and surface wind in the South China Sea. *Ocean Dyn.* **2020**, *70*, 139–161. [CrossRef]
16. Wang, T.; Zhang, S.; Chen, F.; Ma, Y.; Jiang, C.; Yu, J. Influence of sequential tropical cyclones on phytoplankton blooms in the northwestern South China Sea. *Chin. J. Oceanol. Limnol.* **2020**. [CrossRef]
17. Shi, Y.; Xie, l.; Wang, L.; Zheng, M.; Shen, Y. Impacts of Typhoon Mujigea on Sea Surface Temperature and Chlorophyll-a Concentration in the Coastal Ocean of Western Guangdong. *J. Guangdong Ocean Univ.* **2017**, *37*, 49–58. (In Chinese)
18. Shi, Y.; Xie, L.; Zheng, Q.; Zhang, S.; Li, J. Unusual coastal ocean cooling in the northern South China Sea by a katabatic cold jet associated with Typhoon Mujigea. *Acta Oceanol. Sin.* **2019**, *38*, 62–75. [CrossRef]
19. Ding, Y.; Yao, Z.; Zhou, L.; Bao, M.; Zang, Z. Numerical modeling of the seasonal circulation in the coastal ocean of the Northern South China Sea. *Front. Earth Sci.* **2018**, *14*, 90–109. [CrossRef]
20. Xie, L.L.; Cao, R.X.; Shang, Q.T. Progress of Study on Coastal Circulation near the Shore of Western Guangdong. *J. Guangdong Ocean Univ.* **2012**, *32*, 94–98. (In Chinese)
21. Ding, Y.; Bao, X.; Yao, Z.; Zhang, C.; Wan, K.; Bao, M.; Li, R.; Shi, M. A modeling study of the characteristics and mechanism of the westward coastal current during summer in the northwestern South China Sea. *Ocean Sci. J.* **2017**, *52*, 11–30. [CrossRef]
22. Li, R.; Chen, C.; Xia, H.; Beardsley, R.C.; Shi, M.; Lai, Z.; Lin, H.; Feng, Y.; Liu, C.; Xu, Q.; et al. Observed wintertime tidal and subtidal currents over the continental shelf in the northern South China Sea. *J. Geophys. Res. Oceans* **2014**, *119*, 5289–5310. [CrossRef]
23. Zheng, M.; Li, M.; Xie, L.; Hong, Y.; He, Y.; Zong, X. Observation of hydrographic characteristics of northwestern shelf of the South China Sea in winter 2012. *Oceanol. Et Limnol. Sin.* **2018**, *49*, 734–745. (In Chinese) [CrossRef]
24. Shan, G.; Hui, W.; Gui-Mei, L.; Liang-Min, H. The statistical estimation of the vertical distribution of chlorophyll a concentration in the South China Sea. *Acta Oceanol. Sin.* **2010**, *5*, 13–26.
25. Liao, X.; Dai, M.; Gong, X.; Liu, H.; Huang, H. Subsurface chlorophyll a maximum and its possible causes in the southern South China Sea. *J. Trop. Oceanogr.* **2018**, *37*, 45–56. [CrossRef]
26. Ravichandran, M.; Girishkumar, M.S.; Riser, S. Observed variability of chrolophyll-a using Argo profiling floats in the southeastern Arabian Sea. *Deep Sea Res. Part I Oceanogr. Res. Pap.* **2012**, *65*, 15–25. [CrossRef]
27. Xie, L.L.; Zhang, S.W. Overview of studies on Qiongdong upwelling. *J. Trop. Oceanogr.* **2012**, *31*, 35–41. (In Chinese)
28. Lü, H.; Ma, X.; Wang, Y.; Xue, H.; Chai, F. Impacts of the unique landfall Typhoons Damrey on chlorophyll-a in the Yellow Sea off Jiangsu Province, China. *Reg. Stud. Mar. Sci.* **2020**, *39*, 101394. [CrossRef]
29. Zheng, Q.; And, G.F.; Song, Y.T. Introduction to special section: Dynamics and Circulation of the Yellow, East, and South China Seas. *J. Geophys. Res. Oceans* **2006**, *111*, C11. [CrossRef]

30. Hu, J.; Kawamura, H.; Li, C.; Hong, H.; Jiang, Y. Review on Current and Seawater Volume Transport through the Taiwan Strait. *J. Oceanogr.* **2010**, *66*, 591–610. [CrossRef]
31. Hu, J.; Wang, X.H. Progress on upwelling studies in the China seas. *Rev. Geophys.* **2016**, *54*, 653–673. [CrossRef]
32. Shi, W.; Huang, Z.; Hu, J. Using TPI to Map Spatial and Temporal Variations of Significant Coastal Upwelling in the Northern South China Sea. *Remote Sens.* **2021**, *13*, 1065. [CrossRef]
33. Xie, L.; Pallas-Sanz, E.; Zheng, Q.; Zhang, S.; Zong, X.; Yi, X.; Li, M. Diagnosis of 3-D vertical circulation in the upwelling and frontal zones east of Hainan Island, China. *J. Phys. Oceanogr.* **2017**, *47*, 755–774. [CrossRef]
34. Lu, X.; Yu, H.; Ying, M.; Zhao, B.; Zhang, S.; Lin, L.; Bai, L.; Wan, R. Western North Pacific Tropical Cyclone Database Created by the China Meteorological Administration. *Adv. Atmos. Sci.* **2021**, *38*, 690–699. [CrossRef]
35. Ying, M.; Zhang, W.; Yu, H.; Lu, X.; Feng, J.; Fan, Y.; Zhu, Y.; Chen, D. An Overview of the China Meteorological Administration Tropical Cyclone Database. *J. Atmos. Ocean. Technol.* **2014**, *31*, 287–301. [CrossRef]
36. Kahru, M.; Kudela, R.M.; Lorenzo, E.; Manzano-Saraba, M.; Mitchell, B.G. Trends in the surface chlorophyll of the California Current: Merging data from multiple ocean color satellites. *Deep Sea Res. Part II Top. Stud. Oceanogr.* **2012**, *77–80*, 89–98. [CrossRef]
37. Teodoro, A.C.; Veloso-Gomes, F. Quantification of the Total Suspended Matter concentration around the sea breaking zone from in situ measurements and TERRA/ASTER data. *Mar. Georesour. Geotechnol.* **2007**, *25*, 67–80. [CrossRef]
38. Teodoro, A.C.; Veloso-Gomes, F.; Goncalves, H. Retrieving TSM Concentration from Multispectral Satellite Data by Multiple Regression and Artificial Neural Networks. *IEEE Trans. Geosci. Remote Sens.* **2007**, *45*, 1342–1350. [CrossRef]
39. Miller, R.; McKee, B. Using MODIS Terra 250 m imagery to map concentrations of total suspended matter in coastal waters. *Remote Sens. Environ.* **2004**, *93*, 259–266. [CrossRef]
40. Tassan, S. An Improved In-Water Algorithm for the Determination of Chlorophyll and Suspended Sediment Concentration from Thematic Mapper Data in Coastal Waters. *Int. J. Remote Sens.* **1993**, *14*, 1221–1229. [CrossRef]
41. Zhang, M.; Tang, J.; Dong, Q.; Song, Q.; Ding, J. Retrieval of total suspended matter concentration in the Yellow and East China Seas from MODIS imagery. *Remote Sens. Environ.* **2010**, *114*, 392–403. [CrossRef]
42. Asaoka, S.; Nakada, S.; Umehara, A.; Ishizaka, J.; Nishijima, W. Estimation of spatial distribution of coastal ocean primary production in Hiroshima Bay, Japan, with a geostationary ocean color satellite. *Estuar. Coast. Shelf Sci.* **2020**, *244*, 106897. [CrossRef]
43. Nakada, S.; Kobayashi, S.; Hayashi, M.; Ishizaka, J.; Akiyama, S.; Fuchi, M.; Nakajima, M. High-resolution surface salinity maps in coastal oceans based on geostationary ocean color images: Quantitative analysis of river plume dynamics. *J. Oceanogr.* **2018**, *74*, 287–304. [CrossRef]
44. Enriquez, A.; Friehe, C. Effects of Wind Stress and Wind Stress Curl Variability on Coastal Upwelling. *J. Phys. Oceanogr.* **1995**, *25*, 1651–1671. [CrossRef]
45. Chen, X.; Pan, D.; He, X.; Bai, Y.; Wang, D. Upper ocean responses to category 5 typhoon Megi in the western north Pacific. *Acta Oceanol. Sin.* **2012**, *1*, 51–58. [CrossRef]
46. Hellerman, S.; Rosenstein, M. Normal Monthly Wind Stress over the World Ocean with Error Estimates. *J. Phys. Oceanogr.* **1983**, *13*, 1093–1104. [CrossRef]
47. Garratt, J.R. Review of Drag Coefficients over Oceans and Continents. *Mon. Weather Rev.* **1977**, *105*, 915–929. [CrossRef]
48. Pan, G.; Chai, F.; Tang, D.; Wang, D. Marine phytoplankton biomass responses to typhoon events in the South China Sea based on physical-biogeochemical model. *Ecol. Model.* **2017**, *356*, 38–47. [CrossRef]
49. Zheng, G.; Tang, D. Offshore and nearshore chlorophyll increases induced by typhoon winds and subsequent terrestrial rainwater runoff. *Mar. Ecol. Prog. Ser.* **2007**, *333*, 61–74. [CrossRef]
50. Ye, H.J.; Sui, Y.; Tang, D.L.; Afanasyev, Y.D. A subsurface chlorophyll a bloom induced by typhoon in the South China Sea. *J. Mar. Syst.* **2013**, *128*, 138–145. [CrossRef]
51. Zhao, H.; Tang, D.; Wang, D. Phytoplankton blooms near the Pearl River Estuary induced by Typhoon Nuri. *J. Geophys. Res.* **2009**, *114*, C12027. [CrossRef]
52. Wang, Z.; Li, W.; Zhang, K.; Agrawal, Y.C.; Huang, H. Observations of the distribution and flocculation of suspended particulate matter in the North Yellow Sea cold water mass. *Cont. Shelf Res.* **2020**, *204*, 104187. [CrossRef]
53. Chen, Y.Q.; Tang, D.L. Remote Sensing Analysis of Impact of Typhoon on Environment in the Sea Area South of Hainan Island. *Procedia Environ. Sci.* **2011**, *10*, 1621–1629.
54. Yu, X.; Xu, J.; Long, A.; Li, R.; Shi, Z.; Li, Q.P. Carbon-to-chlorophyll ratio and carbon content of phytoplankton community at the surface in coastal waters adjacent to the Zhujiang River Estuary during summer. *Acta Oceanol. Sin.* **2020**, *39*, 123–131. [CrossRef]
55. Hu, B.; Wang, P.; Bao, T.; Qian, J.; Wang, X. Mechanisms of photochemical release of dissolved organic matter and iron from resuspended sediments. *J. Environ. Sci.* **2021**, *104*, 288–295. [CrossRef] [PubMed]
56. Southwell, M.W.; Kieber, R.J.; Mead, R.N.; Avery, G.B.; Skrabal, S.A. Effects of sunlight on the production of dissolved organic and inorganic nutrients from resuspended sediments. *Biogeochemistry* **2010**, *98*, 115–126. [CrossRef]
57. Shank, G.C.; Evans, A.; Jaffé, R.; Yamashita, Y. Influence of solar radiation on DOM release from resuspended Florida Bay sediments. In Proceedings of the AGU Fall Meeting Abstracts, San Francisco, CA, USA, 14–18 December 2019.
58. Schiebel, H.N.; Wang, X.; Chen, R.F.; Peri, F. Photochemical Release of Dissolved Organic Matter from Resuspended Salt Marsh Sediments. *Estuaries Coasts* **2015**, *38*, 1692–1705. [CrossRef]

59. Bai, Y.; Su, R.; Han, X.; Zhang, C.; Shi, X. Investigation of seasonal variability of CDOM fluorescence in the southern Changjiang River Estuary by EEM-PARAFAC. *Acta Oceanol. Sin.* **2015**, *34*, 1–12. [CrossRef]
60. Huang, C.; Chen, F.; Zhang, S.; Chen, C.; Meng, Y.; Zhu, Q.; Song, Z. Carbon and nitrogen isotopic composition of particulate organic matter in the Pearl River Estuary and the adjacent shelf. *Estuar. Coast. Shelf Sci.* **2020**, *246*, 107003. [CrossRef]
61. Lao, Q.; Chen, F.; Liu, G.; Chen, C.; Jin, G.; Zhu, Q.; Wei, C.; Zhang, C. Isotopic evidence for the shift of nitrate sources and active biological transformation on the western coast of Guangdong Province, South China. *Mar. Pollut. Bull.* **2019**, *142*, 603–612. [CrossRef]
62. Yang, Y.; Yan-Dong, X.U.; Wang, F.Y.; Wei, X. A Numerical Hydrodynamic and Transport Model in the West Coast of Guangdong Province. *Sci. Technol. Eng.* **2015**, *19*, 86–91.
63. Huang, Y.; Chen, F.; Zhao, H.; Zeng, Z.; Chen, J. Concentration distribution and structural features of nutrients in the northwest of the South China Sea in winter 2012. *J. Appl. Oceanogr.* **2015**, *34*, 310–316. (In Chinese)
64. Zheng, M.; Xie, L.; Zheng, Q.; Li, M.; Li, J. Volume and Nutrient Transports Disturbed by the Typhoon Chebi (2013) in the Upwelling Zone East of Hainan Island, China. *J. Mar. Sci. Eng.* **2021**, *9*, 324. [CrossRef]
65. Wang, L.; Xie, L.; Zheng, Q.; Li, J.; Li, M.; Hou, Y. Tropical cyclone enhanced vertical transport in the northwestern South China Sea I: Mooring observation analysis for Washi (2005). *Estuar. Coast. Shelf Sci.* **2020**, *235*, 106599. [CrossRef]
66. Jiang, C.; Cao, R.; Lao, Q.; Chen, F.; Zhang, S.; Bian, P. Typhoon Merbok induced upwelling impact on material transport in the coastal northern South China Sea. *PLoS ONE* **2020**, *15*, e0228220. [CrossRef]
67. Wong, G.; Pan, X.; Li, K.-Y.; Shiah, F.-K.; Ho, T.-Y.; Guo, X. Hydrography and nutrient Dynamics in the Northern South China Sea Shelf-sea (NoSoCS). *Deep Sea Res. Part II Top. Stud. Oceanogr.* **2015**, *117*, 23–40. [CrossRef]

remote sensing

MDPI

Article

Spatiotemporal Variability of Chlorophyll-a and Sea Surface Temperature, and Their Relationship with Bathymetry over the Coasts of UAE

Khalid A. Hussein [1,2,*], Khameis Al Abdouli [3], Dawit T. Ghebreyesus [4], Pakorn Petchprayoon [5], Naeema Al Hosani [1] and Hatim O. Sharif [4]

1 Geography and Urban Sustainability Department, College of Humanities and Social Sciences, United Arab Emirates University, Al Ain P.O. Box 15551, United Arab Emirates; naeemam@uaeu.ac.ae
2 Cooperative Institute for Research in Environmental Sciences (CIRES), University of Colorado, Boulder, CO 80309, USA
3 Business Continuity Management & Integrated Emergency Management, Rabdan Academy, Abu Dhabi P.O. Box 114646, United Arab Emirates; kabdouli@ra.ac.ae
4 Department of Civil and Environmental Engineering, University of Texas at San Antonio, San Antonio, TX 78249, USA; dawit.ghebreyesus@my.utsa.edu (D.T.G.); hatim.sharif@utsa.edu (H.O.S.)
5 Geo-Informatics and Space Technology Development Agency (GISTDA), Bangkok 10210, Thailand; pakorn@gistda.or.th
* Correspondence: khalid.hussein@uaeu.ac.ae

Citation: Hussein, K.A.; Al Abdouli, K.; Ghebreyesus, D.T.; Petchprayoon, P.; Al Hosani, N.; O. Sharif, H. Spatiotemporal Variability of Chlorophyll-a and Sea Surface Temperature, and Their Relationship with Bathymetry over the Coasts of UAE. *Remote Sens.* **2021**, *13*, 2447. https://doi.org/10.3390/rs13132447

Academic Editors: Chung-Ru Ho, Xiaofeng Li and Antony K. Liu

Received: 19 May 2021
Accepted: 18 June 2021
Published: 23 June 2021

Publisher's Note: MDPI stays neutral with regard to jurisdictional claims in published maps and institutional affiliations.

Abstract: The catastrophic implication of harmful algal bloom (HAB) events in the Arabian Gulf is a strong indication that the study of the spatiotemporal distribution of chlorophyll-a and its relationship with other variables is critical. This study analyzes the relationship between chlorophyll-a (Chl-a) and sea surface temperature (SST) and their trends in the Arabian Gulf and the Gulf of Oman along the United Arab Emirates coast. Additionally, the relationship between bathymetry and Chl-a and SST was examined. The MODIS Aqua product with a resolution of 1×1 km^2 was employed for both chlorophyll-a and SST covering a timeframe from 2003 to 2019. The highest concentration of chlorophyll-a was seen in the Strait of Hormuz with an average of 2.8 mg m^{-3}, which is 1.1 mg m^{-3} higher than the average for the entire study area. Three-quarters of the study area showed a significant correlation between the Chl-a and SST. The shallow (deep) areas showed a strong positive (negative) correlation between the Chl-a and SST. The results indicate the presence of trends for both variables across most of the study area. SST significantly increased in more than two-thirds of the study area in the summer with no significant trends detected in the winter.

Keywords: Arabian Gulf; Gulf of Oman; MODIS; algal blooms; chlorophyll-a; SST; bathymetry

1. Introduction

The Arabian Sea is one of the most essential bodies of water not only for the local economy, but also for the global one, because it serves as a route to a significant portion of the world's oil supply. The ecosystems of the Arabian seas (Arabian Gulf (thereafter AG), Gulf of Oman (thereafter GO), and Arabian Sea) are fragile, and susceptible to pollution. Among these pollutants are algal blooms, particularly red tide [1–3]. The bloom's growth and biomass depend on the availability of nutrients in the surface layer. Therefore, the processes by which the nutrients reach the surface are of crucial importance. The main source of nutrients to the surface layer is the deep water, which is rich in nutrients [4]. The transfer of these deep nutrients is affected by wind-induced or thermohaline upwelling, vertical diffusion, deepening of the surface layer, and vertical overturning [4]. In the Arabian Sea, the transfer of nutrients is related to the summer (southwest) and winter (northeast) monsoon seasons. The distinct direction of the summer monsoon from the southwest, which is almost parallel to the Oman coastline in the northern Arabian Sea, produces a strong coastal upwelling system that highly contributes to bringing the nutrient-rich deep water to the surface and supporting phytoplankton blooms [5–7]. The northeast monsoon drives convective mixing in the northern Arabian Sea, resulting in an upward transport of nutrients from the base of the mixed layer and upper thermocline [2,8–10]. These processes make the conditions conducive for phytoplankton growth and development in the AG and GO all year around. The timely identification of the location and extent of the blooms is crucial for assessing and managing the coastal environment as well as forecasting and mitigating their negative impact [11].

Mapping, monitoring, and forecasting algal blooms in an efficient manner is critical for mitigating their impacts. However, monitoring of algal blooms using traditional methods, such as near coastal line and shipboard measurements, is very difficult because of spatial and temporal data gaps. These problems can be addressed using remote sensing data, which offer a supplement to local measurements by providing comprehensive coverage of large areas, which are reliable data and are regularly updated.

Satellite ocean color data, remote sensing techniques, and algorithms are widely used for the detection, measuring, mapping, monitoring, modeling, and managing of phytoplankton blooms because satellite earth observation derived from various sensors provides a synoptic view of the ocean, both spatially and temporally [12]. The main limitation of these sensors is their inability to penetrate clouds, which makes their data limited to only clear-sky conditions [13,14]. To fill in the gaps of remote sensing data, several interpolation techniques are employed. One of these techniques is the data interpolating empirical orthogonal function (DINEOF) method, which is used to reconstruct the monthly mean datasets [15]. Other interpolation techniques are also used which are simpler and computationally less expensive. These techniques can be very useful, especially in regions and/or times where clouds do not cover a significant portion of the study area [16].

Numerous studies have been conducted to investigate and assess the spatial and temporal distribution of phytoplankton and red tides from remotely sensed data. For example, Brewin [17] used MODIS/Aqua data to assess the spatial and temporal distribution of Chl-a in the Red Sea. The operational Chl-a algorithm, National Aeronautics and Space Administration (NASA) OC3, and the Color Index (CI) algorithm developed by Hu [18] were employed in the study. The OC3 algorithm is a polynomial function that relates the remote sensing reflectance at wavelengths 443, 488, and 547 nm to the Chl-a concentration. The CI is defined as the difference between the reflectance in the green region and the blue and red regions of the visible spectrum. Their results revealed that OC3 and CI-derived Chl-a concentrations were comparable to the in situ measurements and to other areas in the global ocean.

Nezlin [19] and Tang [20] investigated the seasonal and inter-annual variations of surface Chl-a concentration and their causes in the Black Sea and southwest of the Luzon Strait in the South China Sea, respectively, from CZCS data collected during the period 1978 to 1986. They concluded that remotely sensed data are useful in detecting Chl-a concentration over large areas. MERIS and MODIS data have been used by Gurlin [21] to estimate Chl-a concentrations in turbid water of the Fremont Lakes State Recreation Area in Nebraska, USA. Gower [22] and Gower [23] studied the global algal blooms from MERIS data using the maximum Chl-a Index (MCI).

Cannizzaro [24] used SeaWifs and MODIS data for the detection of the toxic dinoflagellate, Karenia brevis, in the Gulf of Mexico. Hu [25] used the Floating Algae Index (FAI) to characterize the cyanobacteria (Microcystis aeruginosa) blooms primarily in Taihu Lake, China, using MODIS time series of nine years (2000 to 2008). Anderson [26] demonstrated the combined use of the empirical harmful algal blooms (HABs) models, MODIS/Aqua data, and a regional ocean model for the prediction of the toxic Pseudo-nitzschia bloom events in the Santa Barbara Channel.

SST is one of the main factors that affects the growth of phytoplankton in oceans, especially at an optimum temperature when the correlation is significantly high [27]. However, as Nurdin [28] reported, an excessive increase in SST would hinder the growth of phytoplankton. Another factor that affects the growth of phytoplankton is the amount of nutrients loaded with the freshwater from river discharges. Jutla [29] found a positive correlation between seasonal river discharges, SST, and Chl-a and vice versa in the coastal Bay of Bengal region. Seawater current is also one of the main factors that drive the Chl-a concentration in the water bodies. Kouketsu [30] and Chu [31] suggest that in the Kuroshio Extension the cyclonic eddies are related to high area-averaged Chl-a concentration and anti-cyclonic eddies are often related to low area-averaged Chl-a.

The existence of large spatial and temporal gaps in in situ measurements hamper the complete understanding of Chl-a behavior. Our work utilizes satellite data that provide regular long-term temporal and spatial continuity to comprehend the pattern and change of Chl-a characteristics in both space and time. The main goal of this study is to examine the spatiotemporal variability of Chl-a and other oceanography variables over the AG and GO for the period span between 2003 and 2019. The spatiotemporal analysis elucidates the impact of the SST on the growth of phytoplankton over the region. Additionally, we investigated the variability of both SST and Chl-a over the coastal areas of the UAE using the empirical orthogonal function (EOF). The objectives of this research are to (i) conduct frequency analysis of the mode of the variability in SST and Chl-a and their relationship with regional wind circulations, and (ii) investigate the presence of trends in both variables and their seasonal decomposition.

2. Study Area and Dataset

2.1. Study Area

The study area is shown in Figure 1. The area covers the AG and GO along the UAE coasts (1318 km). The AG is located in the Middle East between latitude 24.0° N and 30.0° N and longitude 48.0° E and 56.5° E. The AG is separated from the northern Indian Ocean by the Strait of Hormuz and the GO [3,6]. The AG is 990 km long with a maximum width of 338 km and an average depth of 36 m for much of the Arabian coast and 60 m depth along the Iranian coast [32,33]. The GO is situated between 22.0° N to 26.0° N and 56.5° E to 61.7° E. The GO is 320 km wide between Ra's Al-Ḩadd in Oman and Gwādar Bay on the Pakistan–Iran border. It is 560 km long and connects with AG through the Strait of Hormuz [34]. Although the AG is located entirely north of the Tropic of Cancer, its climate is tropical in the summer and temperate in the winter (Reynolds, 1993). The climate of the AG has two main seasons: winter (December to March) and summer (June to September), and two transition periods, fall (October to November), and spring (April to May) [35]. In the summer, the air temperature reaches up to 51°C with an average of 41 °C, while in winter the air temperature drops to as low as 15 °C [33]. Due to the surrounding arid climate, evaporation surpasses the combination of precipitation and runoff resulting in hypersaline water mass production [36]. The climate of the GO and the northern Arabian Sea is significantly influenced by the summer and winter monsoons driven by land–sea latent heat differences. The summer monsoon occurs from July to September and the winter monsoon from November to April [5,7]. The SST is a considerably variable in both Gulfs due to the effects of the surrounding landmass and air temperatures [37].

2.2. Dataset

2.2.1. Chl-a Data

To characterize the spatial and temporal distribution of algal blooms along the coast of the UAE, daily remotely sensed Chl-a concentration and SST were obtained for the period between 2003 and 2019. Level 2 product with a spatial resolution of 1 × 1 km^2 of the Chl-a concentration from MODIS onboard Aqua satellite were downloaded from the NASA MODIS standard products at https://oceancolor.gsfc.nasa.gov/cgi/browse.pl. These data are in the netCDF-4 format (.nc), which contains multi-object files [38]. The Band Select of Data Conversion tool from Sentinel Application Platform (SNAP) was used to extract the products of both Chl-a and SST and mosaicked using the Geospatial Data Abstraction Library (GDAL) merge tool by pyQGIS. The Chl-a data contain gaps mainly due to the inability of the sensors to perpetrate through clouds. From the study period (1 January 2003 to 31 December 2019), out of 6208 days, 6149 daily imageries were available in the archive. Out of the available daily images, 217 images were found to be covered by clouds for more than 75% of the study area. The temporal distribution showed that the daily images were missing around 5% of the study area every day before January 2018 (Figure 2B). The spatial distribution suggested that the area that failed to be covered consistently was the northwestern tip (Figure 2A). The areal coverage drops to below 75% during only a few days for a small number of months. Due to a lack of sufficient data, the areas with a dataset that are missing more than 75% (~5% of the study area) of their observations were masked out before the analysis was conducted. The areal average amount of missing data over the entire Arabian Sea was recorded as 16.3%. The month of July had most of the missing data, including on seven occasions wherein the daily images failed to cover more than 25% of the study area. Moradi [39] also suggested that the data of July included the highest missing values in the region followed by August and June.

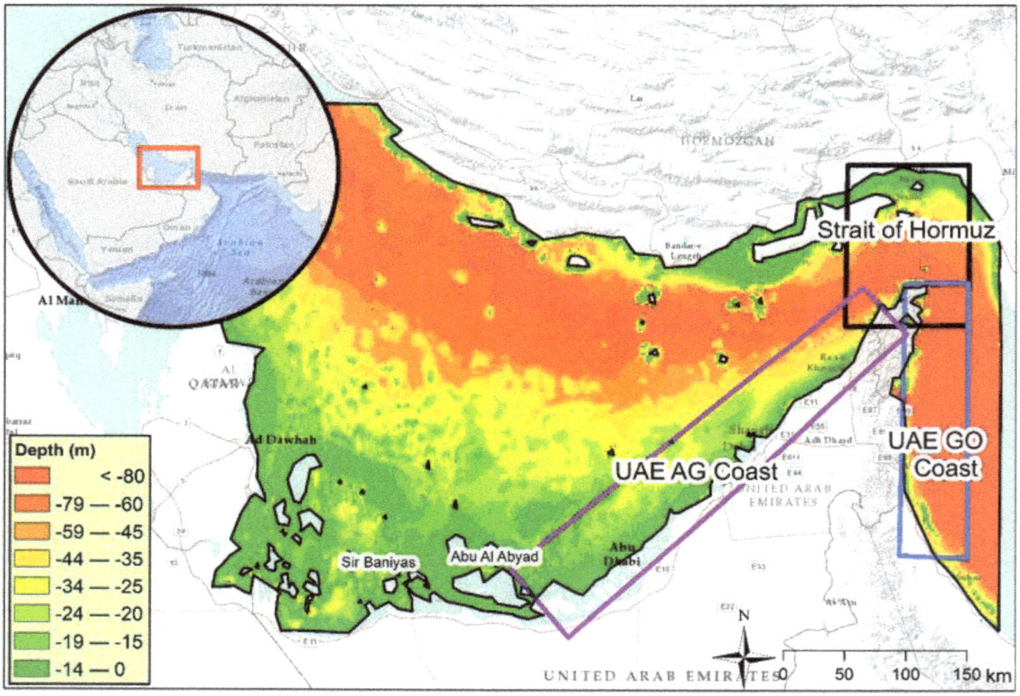

Figure 1. Map of the study area with bathymetric data of the Arabian and Oman Gulfs and the location of the three sections (UAE Arabian Gulf (AG) Coast, UAE Gulf of Oman (GO), and Strait of Hormuz) that were used for the empirical orthogonal function (EOF) analysis.

2.2.2. SST Data

Level 2 product with a spatial resolution of 1×1 km^2 of the SST from MODIS onboard Aqua satellite was downloaded from the NASA MODIS standard products at https://oceancolor.gsfc.nasa.gov/cgi/browse.pl. For the MODIS data, thermal channels 31 (10.780 to 11.280 µm) and 32 (11.770 to 12.270 µm) are particularly suited to estimate the surface temperature [40]. The MODIS sea surface temperature data have been widely validated for open waters and therefore are widely accepted as accurate [37,41–45].

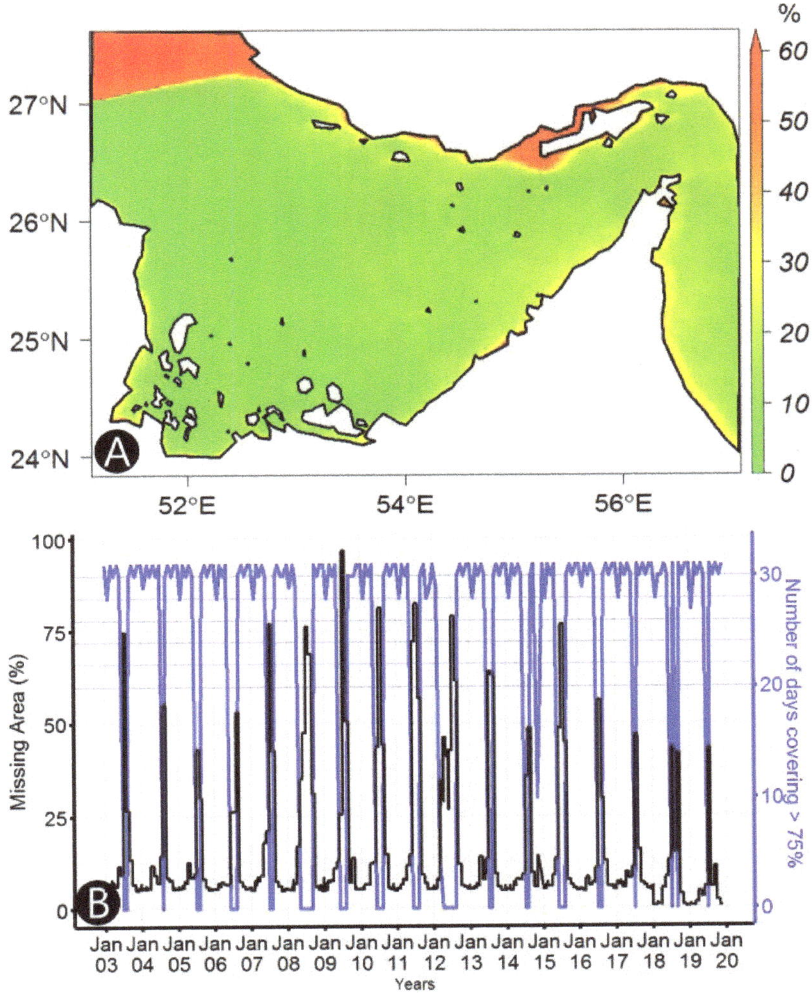

Figure 2. (**A**) Spatial distribution of the missing data of the Chl-a concentration across the study area. (**B**) Temporal distribution of the missing data of the Chl-a concentration (black line represents a daily fraction of missing area and the blue line represents the number of days in a month with data coverage fraction greater than 75%).

The SST data have relatively better coverage than the Chl-a data described above. The study spanned for a period (1 January 2003 to 31 December 2019) of 6208 days, out of which 6144 daily images were available in the archive. The notable missing data from the archive is that only 10 days of data were available for the months of November and December 2014. However, only 19 daily images out of the available 6144 imageries had a missing area of more than 75% of the study area (mainly due to clouds). The temporal distribution of the missing data shows that the cloud coverage is much higher in the winter months (from November to April), as shown in Figure 3B. The areal coverage drops to below 75% during only a few days for a small number of months. The spatial distribution of the missing data indicates that the northwestern tip of the study area is the area with the most missing data (~6%). The areal average of missing data was 2.4%. The amount of missing data decreases as you move from the northwestern to the southeastern corner (Figure 3A). Interestingly, from January 2018 to December 2019, there was no significant missing data (images covered more than 98% of the study area). That is likely due to an enhancement of the product processing algorithm.

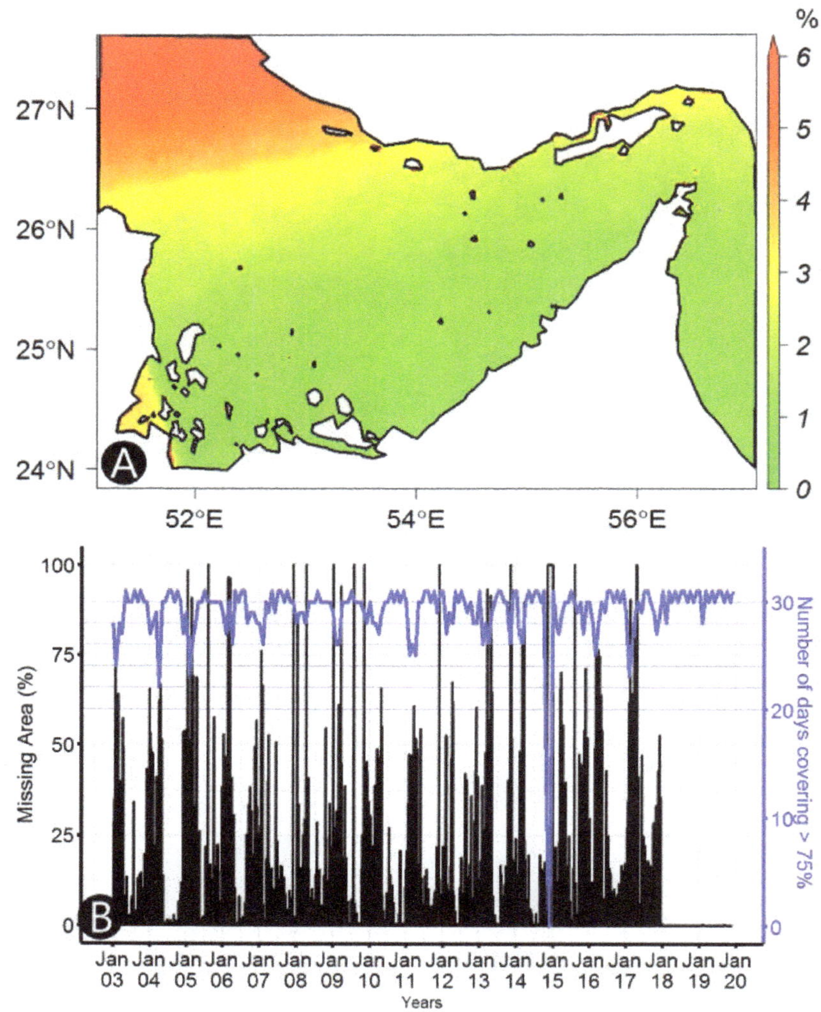

Figure 3. (**A**) Spatial distribution of the missing data of SST across the study area. (**B**) Temporal distribution of the missing data of SST (black line represents a daily fraction of missing area and the blue line represents the number of days in a month with data coverage fraction greater than 75%).

2.2.3. Bathymetry Data

The bathymetric data, the Global Relief Model referred to as ETOPO1, which is an improved model of the ETOPO2v2 Global Relief Model, were used in the study. The data are developed by the National Geophysical Data Center (NGDC) of the National Oceanic and Atmospheric Administration (NOAA). The ETOPO1 has two versions—Ice Surface and Bedrock. The Ice Surface version includes the top of the ice sheets (Antarctica and Greenland), while the Bedrock version depicts the base of the ice sheets [46]. For this study, the Bedrock version was used. The vertical datum is referenced from the mean sea level and the World Geodetic System of 1984 (WGS 84) datum was used as a horizontal datum. The spatial resolution of the data is one arcminute with global coverage. The bathymetry of the study area shows a shallow AG and a much deeper GO (Figure 1).

3. Methodology

3.1. Filling Missing Data

Cloud cover significantly obscures surface information; therefore, it is very important to retrieve the Chl-a and SST under overcast skies. The biggest challenge in retrieving the Chl-a and SST is to eliminate cloud contamination. Figure 4 presents how the missing data values and gaps of Chl-a and SST are filled. To fill the missing values of Chl-a, the daily Chl-a data for the study period (2003 to 2019) are used to develop the monthly composites of each Julian day. Then, the missing data values and gaps of the daily data are filled with the corresponding monthly composite values. MODIS cloud-free data composite image (SST, monthly composite product) was employed to fill in these missing pixels' values of SST. This method was used because the amount of missing data is not as significant relative to the other regions of the world where cloud cover is a major issue. For example, Li [47] found that only 2872 daily snapshots were useful out of 3653 imageries in the Gulf of Maine. However, in this study, only 217 days of Chl-a had missing data covering more than 75% of the study area out of 6149 obtained daily images. The northwestern part of the study area was found to have significant gaps in the data. For this reason, the area which covers ~5% of the total study area was masked from the analysis. The average missing data across the study period for Chl-a was around 10%. Conversely, the average missing data of SST is ~2% of the study area after excluding the northwestern part of the study area.

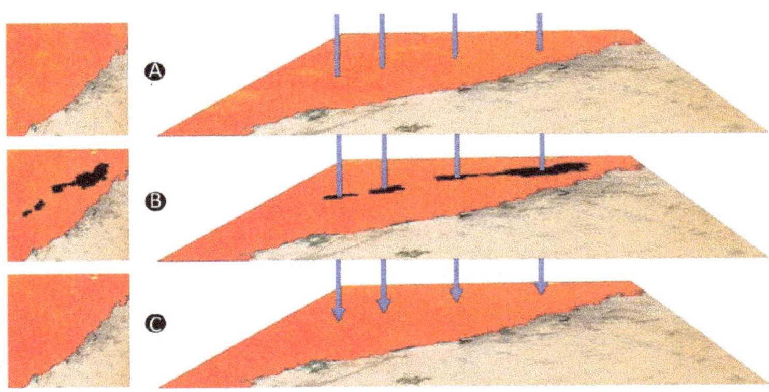

Figure 4. Presents how the data values of Chl-a and SST are filled. (**A**) Monthly composites of the Julian day. (**B**) Original daily data with missing values and gaps (black spots are missing values). (**C**) Filled data.

3.2. Empirical Orthogonal Function (EOF) Analysis

The primary application of EOF is that it helps in understanding the spatial patterns of variability in spatiotemporal data by examining the EOF coefficient maps. Secondly, it can be used to reduce the dimension of the components by using the optimal number of components that explain the majority of the variability in the spatiotemporal dataset [48,49]. The EOFs analysis was conducted over the coastal areas of the UAE, the AG Coast, Strait of Hormuz, and GO Coast (Figure 1) to examine the spatial variability of the SST and Chl-a concentration.

The raw data matrix F is arranged in a matrix format $M \times N$, where M is the time series dimension and N is the space dimension. The covariance matrix R is calculated using Equation (1). Then, the eigenvalues are solved with Equation (2), which provides the information about the amount of variability explained by each component [48]. The highest three components were selected for this study as they explain more than three-quarters of the variability in both Chl-a and SST.

$$R = F^t F \tag{1}$$

$$RC = C\Delta \tag{2}$$

Δ is a diagonal matrix containing eigenvalues λ_i of R where i is the length of the time series ranges from 1 to p (size of M). The column vectors of C are the eigenvectors of R that corresponds to the eigenvalues λ_i which contain information about the spatial distribution of the eigenvalues.

The raw data can be reconstructed from the EOFs and the eigenvectors using the following:

$$F = \sum_{i=1}^{p} \vec{c}_i(EOF_i) \tag{3}$$

The amount of variability explained by one EOFs component a can be estimated as a fraction of the total variability using Equation (4).

$$\sigma_a = \frac{\lambda_a}{\sum_{i=1}^{p} \lambda_i} \tag{4}$$

3.3. Correlation Analysis

The Pearson correlation coefficient (PCC) statistical tool was used to evaluate the impact of SST on Chl-a concentration. If the value approaches $1/-1$, it indicates that the relationship is strongly positive/negative, and if the coefficient is closer to 0, it indicates that the relationship between the variables is weak. Cross-correlation was conducted to assess the possible lag time of the impact of the SST over the formation of the Chl-a concentration. The mathematical formula used is obtained from Pearson [50]:

$$r = \frac{1}{n-1} \sum_{i=1}^{n} \left[\left(\frac{x_i - \overline{x}}{S_x} \right) \left(\frac{y_i - \overline{y}}{S_y} \right) \right] \tag{5}$$

where n is the sample size, x_i and y_i are records of the variables (SST and Chl-a in this case), $\overline{x}$ and $\overline{y}$ are the average values of the variables, and S_x and S_y are the standard deviations of the variables.

3.4. Correlated Seasonal Mann–Kendal Trend Test

The corrected seasonal Mann–Kendal trend test was used to investigate the presence of a significant trend in the data. This test was a modified version of the original Mann–Kendal test to accommodate seasonally correlated data. The adjustment was used by Hirsch [51] and Libiseller [52] to reduce the seasonal autocorrelation in the dataset. The Mann–Kendall scores are first computed for each month separately as follows:

$$S_i = \sum_{k=1}^{n_i-1} \sum_{j=k+1}^{n_i} sgn(x_{ij} - x_{ik}) \tag{6}$$

where $sgn()$ is a sign function obtaining the sign of real number, x_{ij} and x_{ik} are monthly series values for the periods k and j, respectively, and i represent the month. The variance for each month is given by:

$$\text{Var}(S_i) = \frac{n_i(n_i-1)(2n_i+5) - \sum_{p=1}^{g_i} t_{ip}(t_{ip}-1)(2t_{ip}+5)}{18} \tag{7}$$

where g_i is the number of tied groups for the ith month and t_{ip} is the number of observations in the pth group for the ith month. Then, the Mann–Kendall score and variance for the entire series are computed as follows:

$$S' = \sum_{i=1}^{m} S_i \quad (8)$$

$$\mathrm{Var}(S') = \sum_{i=1}^{m} \mathrm{Var}(S_i) \quad (9)$$

where S_i is the Mann–Kendall score of an individual month and m, the number of months in this study, is 12. Similarly, $\mathrm{Var}(S')$ is the variance of individual months. The seasonal adjusted Mann–Kendall test statistics for the series (Z_{SK}) is given by:

$$Z_{\mathrm{MK}} = \begin{cases} \frac{S'-1}{\sqrt{\mathrm{VAR}(S')}} & \text{if } S' > 0 \\ 0 & \text{if } S' = 0 \\ \frac{S'+1}{\sqrt{\mathrm{VAR}(S')}} & \text{if } S' < 0 \end{cases} \quad (10)$$

Finally, for the areas with a significant trend, the magnitude of the trend was computed using a linear model ($y = \alpha + \beta x$). Moreover, trend analysis was conducted over the summer and winter months separately to assess the influence of seasonality.

4. Results and Discussion

4.1. Spatiotemporal Distribution of Chl-a and SST

The spatial distribution of the long-term average of Chl-a concentration was unevenly distributed across the Arabian Sea. The Chl-a concentration was high in the coastal areas and the Strait of Hormuz (Figure 5A). The coastal hotspots of Chl-a concentration are usually created due to the loading of nutrients with the discharge from the Wadis and artificial loading of nutrients from agricultural and aquaculture activities around the shores [3,5]. The areal average concentration of Chl-a in the Strait of Hurmuz was 2.8 mg m^{-3}, whereas the areal average concentration across the entire study area was 1.7 mg m^{-3}. The seasonal distribution shows that February and March are the months with the highest Chl-a concentration, especially in the Strait of Hormuz and GO (Appendix A). The main reasons for such a high concentration of Chl-a in the Strait of Hormuz are seasonal upwelling, mixing of the AG and GO, and the high concentration of pollutants and river discharge from the northern coast [53]. Over the study period of 17 years, 2008 and 2009 showed peak concentration of Chl-a with an average concentration of 2.4 mg m^{-3} and 2.1 mg m^{-3}, respectively (Appendix C). This period includes the red tide events that were reported by Richlen [54]. Moreover, the seasonal mean distribution showed a distinct pattern between winter and summer. The winter had a higher concentration of Chl-a, which was clearly observed in the Strait of Hormuz (Figure 5C). However, in summer, the coastal areas exhibited a relatively high concentration of Chl-a (Figure 5E).

Unlike Chl-a concentration, the spatial distribution of the average SST for the period 2003 to 2019 shows a uniform linear increase in the west–east direction, as shown in Figure 5B. The GO experienced an average SST of about 26 °C, which makes it the hottest region in the study area. The areal average SST over the entire study area was around 25 °C. A difference of ~3 °C was observed between the hottest region (GO) and the coldest region (northwestern AG) in the long-term average of SST. The monthly distribution of SST showed very little spatial variability (Appendix B). The annual average suggests that the hottest years were 2018 and 2019 with an average SST of 26. 3 °C and 26. 4 °C, respectively (Appendix D). The AG and GO experience different winter and summer temperature patterns. The southern AG was warmer than GO in the summer (Figure 5F), whereas the GO was warmer than AG in the winter (Figure 5D).

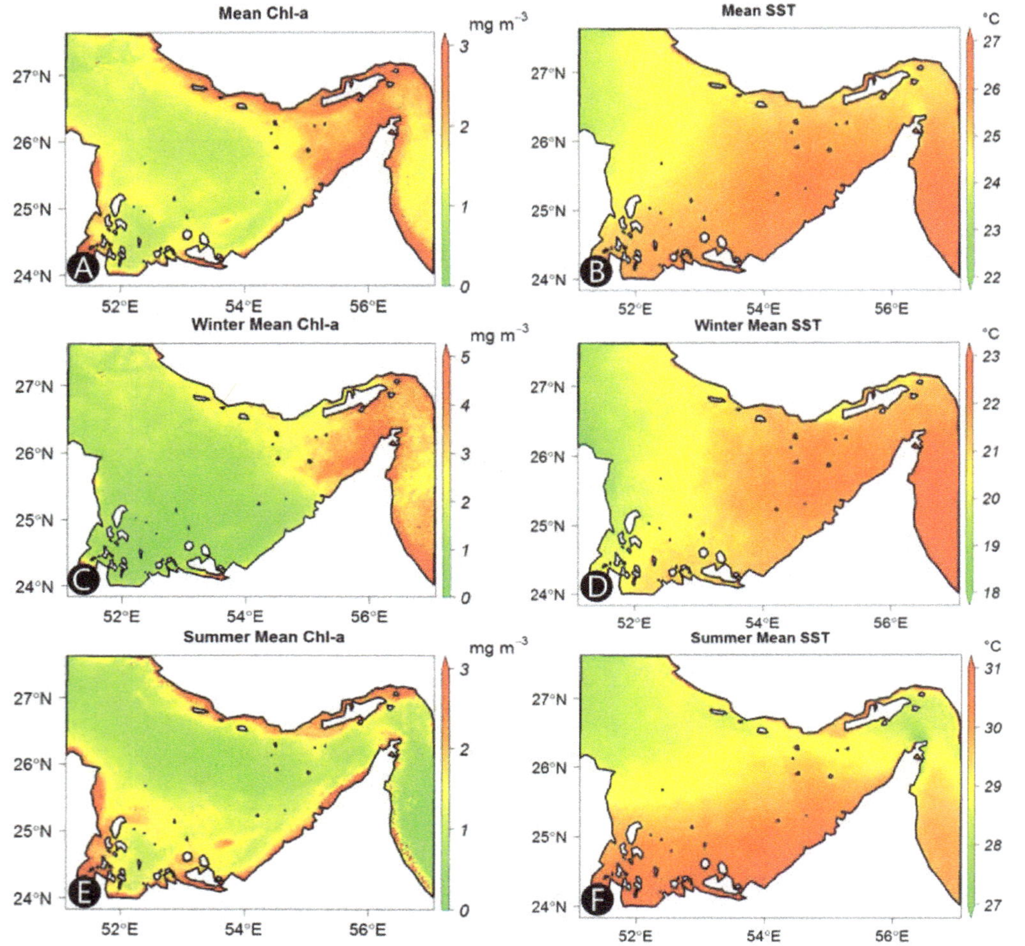

Figure 5. Spatial distribution of the long-term average of (**A**) Chl-a concentration, (**B**) SST, (**C**) long-term winter Chl-a average, (**D**) long-term winter SST average, (**E**) long-term summer Chl-a, and (**F**) long-term summer SST average across the Arabian Sea (Arabian and Oman Gulfs) during the study period (1 January 2003 to 31 December 2019).

The temporal distribution of the Chl-a concentration suggests that different parts of the Arabian Sea express different seasonal variability. The shape of the seasonal cycle appears to be a smooth sinusoidal curve with relatively smaller amplitude in the case of the AG coast of UAE and a more pointed shape with a higher variability for both the Strait of Hormuz and the GO. The UAE's coast across the AG experiences small seasonal variability with the peak concentration seen in November and the lowest concentration observed in May (Figure 6A). The highest variability is seen in the time series of the GO with the peak concentration observed in February and the lowest reported in May (Figure 6C). In the summer of 2012 (April, May, and June), the entire region experienced the lowest concentration of Chl-a (Figure 6). After that point, the Chl-a concentration was above normal in winter and below normal in the summer, especially in the Strait of Hormuz and the GO.

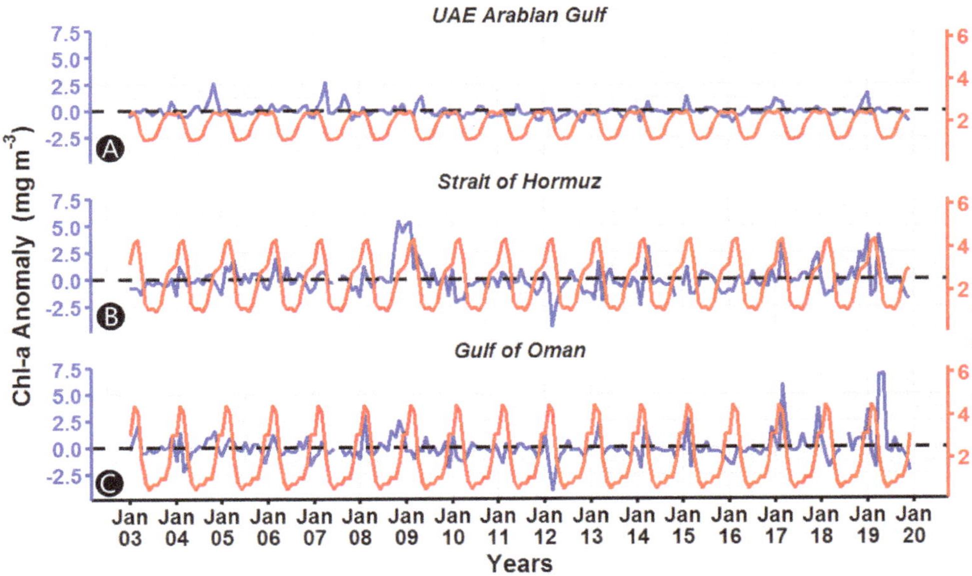

Figure 6. The time series of Chl-a concentration across different sections of the Arabian Sea (**A**) UAE Arabian Gulf coast, (**B**) Strait of Hurmuz, and (**C**) Gulf of Oman coast (the red line is the seasonal cycle, which is the average of each month; the blue line is the anomaly, which is the difference between monthly average and seasonal cycle).

The temporal distribution of SST reveals different behaviors among the three sample regions. The UAE's AG coast showed a smooth sinusoidal seasonal cycle with the highest variability between winter and summer (Figure 7). The SST ranges from 19 °C in January to as high as 31 °C in September. The GO showed a bimodal seasonal cycle with peaks in June (30 °C) and September (30 °C) and February as the coldest month with 21 °C (Figure 7). This bimodal cycle is due to the decrease of the temperature during the southwest monsoon that occurs from June to September [55]. The time series showed that the Arabian Sea, in general, experienced cooler than usual winters between 2005 and 2008. Towards the end of the study period (2014 to 2019), the summers became hotter than the typical summer. The northeast monsoon is the main reason for cool SST across the entire AG and the GO from November to March [56].

4.2. Variability in Chl-a and SST

The best way to display the EOFs components as meaningful indicators is to represent them as homogeneous correlation maps. The homogenous correlation map of the EOFs' first component is the correlation of the raw data with the expansion coefficient of the first component of EOFs [57]. With regard to the mode of variability of the Chl-a concentration, the first three components captured 74% of the variability on average. The first component, which explains 42% of the variability, had a strong relationship in the Strait of Hormuz and GO, as shown in Figure 8A. The first component is related to the northeast monsoon winds that move heat from the surface of the Arabian Sea, which occurs from early November to March. Lower solar radiation and increased salinity create convective mixing that drives upward transport of nutrients [2]. The availability of nutrients with optimal atmospheric conditions results in excessive growth of phytoplankton biomass. The second component, with an average variability of 24%, had the reverse impact of the first component with the western coast of the UAE affected significantly more than the rest of the area (Figure 8B). The spikes of Chl-a concentration over the coast of UAE (AG) at the end of 2004, 2007, and 2018 to 2019 were related to the second component (Figure 8D). The third component, responsible for 8% of the variability, was highly related to the Strait of Hormuz, which captured the peaks in Chl-a concentration observed in 2008 to 2009 (Figure 8D). The 2008 to 2009 algal blooms were catastrophic to the infrastructure of the countries in the AG, especially in the water supply system and tourism industry. The blooms dissipated in August 2009 about nine months after they first appeared on the coast [54].

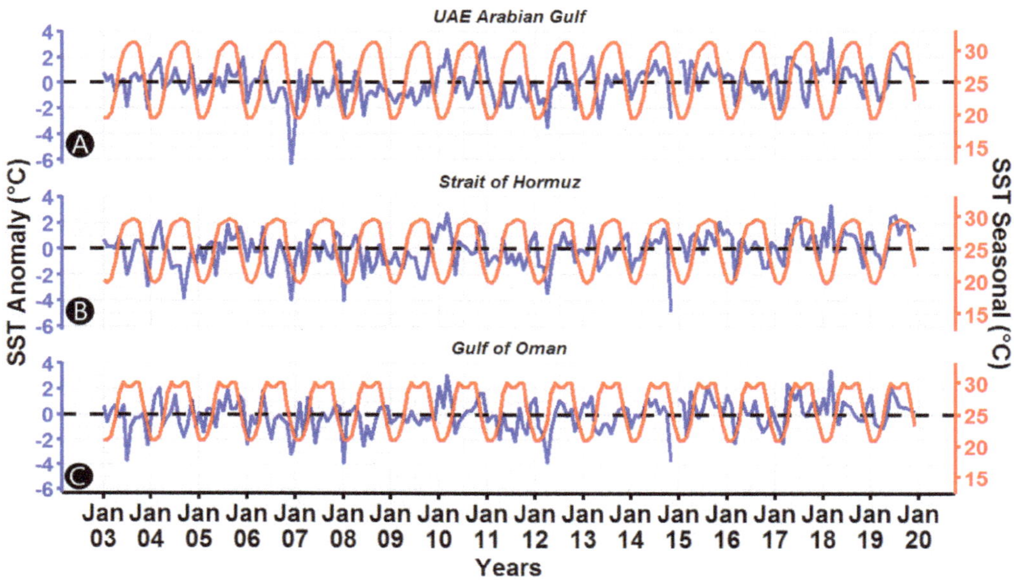

Figure 7. The time series of SST across different sections of the Arabian Sea (**A**) UAE Arabian Gulf coast, (**B**) Strait of Hurmuz, and (**C**) Gulf of Oman coast (the red line is the seasonal cycle, which is the average of each month; the blue line is the anomaly, which is the difference between monthly average and seasonal cycle).

The first three EOFs modes of SST captured more than 96% of the variability in the dataset. The first EOF component, which represents the annual seasonal component of SST, accounts for more than 94% of the variability, as shown in Figure 9A. The entire study area showed a strong homogeneous correlation coefficient of more than 0.9. This means that the EOF first component is highly influenced by the annual periodicity that reaches its peak in summer and its lowest in winter. The spatial variability is very small, indicating that the seasonal variability is uniform across the entire area. This shows that the annual variability of the SST (that represents 95% of the total variability) showed a very small spatial variability across the coasts of the UAE. This is evident in the fact that the entire UAE coast demonstrates an interquartile range of only 0.7 °C in the long-term average SST. However, the second EOF component of SST, accounting for 1.07% of the total variance, showed a significant spatial variability (Figure 9B). The UAE's AG coast is positively correlated, whereas the GO coast is negatively correlated. The second component seems to capture the impact of the southwest monsoon with a spatial variability that is oriented in the east–west direction. The southwest monsoon decreases the temperature of the GO, causing a bimodal cycle. The southwest monsoon does not have a significant impact on the SST of the AG; on the contrary, SST increases during that period. This result is in line with the findings of Nandkeolyar [56]. The third component of SST also revealed a significant spatial variability, whereas the Strait of Hormuz is negatively correlated and the rest of the coasts are positively correlated. The spatial variability is oriented in the north–south direction (Figure 9C).

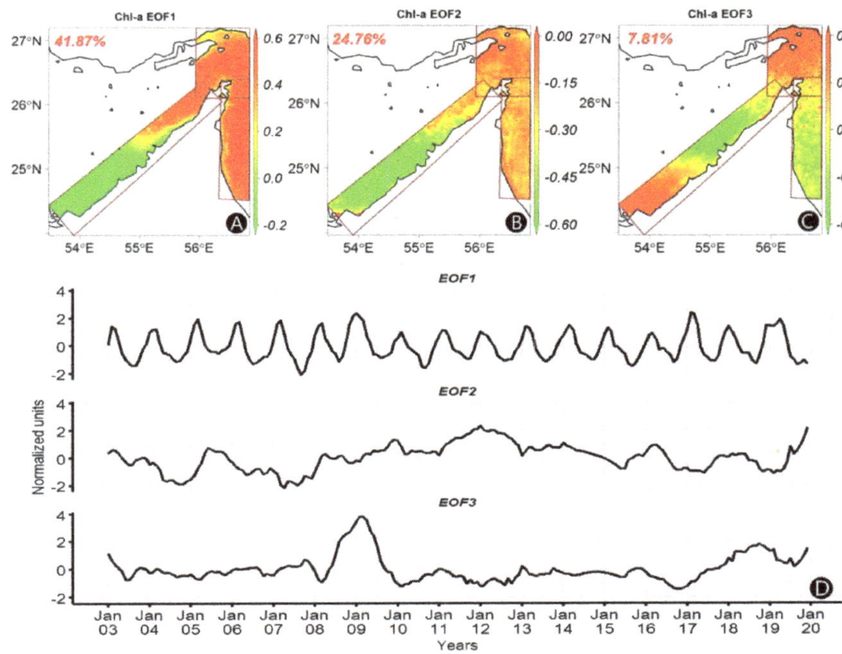

Figure 8. Spatial distribution of the homogenous correlation map of the first three EOFs components of the Chl-a concentration ((**A**–**C**), respectively), and their time-series component (**D**).

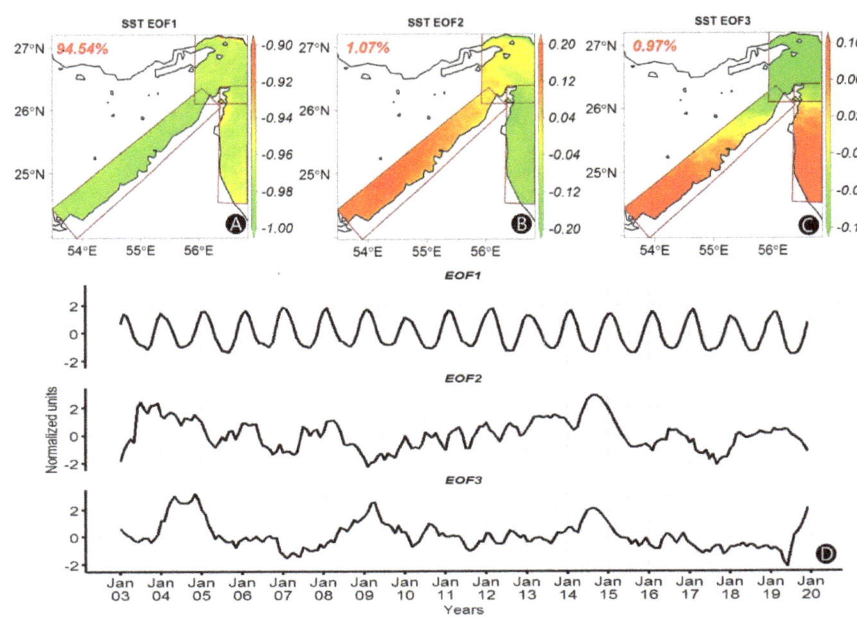

Figure 9. Spatial distribution of the homogenous correlation map of the first three EOFs components of the SST ((**A**–**C**), respectively), and their time-series component (**D**).

4.3. Correlation of Chl-a and SST

The correlation coefficient of Chl-a and SST indicates that around 75% of the study area exhibits significant correlation as shown in Figure 10A. The coastal area of western UAE showed a significant positive correlation which suggests that the SST affects the concentration of Chl-a. However, more than half of the study area indicated that SST negatively influenced the Chl-a concentration in the AG and GO, whereas more than a quarter of the area showed a positive relationship. The southern coast of the AG showed a significant positive correlation coefficient of 0.44 on average and standard deviation (SD) of 0.15. The areas with positive correlation had an average Chl-a of 1.36 mg m^{-3} (SD of 0.51 mg m^{-3}) and an average SST of 25.39 °C (SD of 0.61 °C). On the other hand, the negatively correlated areas had an average correlation coefficient of −0.33 (SD of 0.09). The negatively correlated areas have shown a compact distribution of the correlation coefficient despite covering an area almost twice the size of the positively correlated areas. This shows that the variability in the negatively correlated areas is small relative to the positively correlated places. The average Chl-a of the negatively correlated areas was 1.66 mg m^{-3} (SD of 0.66 mg m^{-3}) and the average SST was 25.18 °C (SD of 0.77 °C). Additionally, cross-correlation analysis revealed that the best correlation between Chl-a and SST was found without any lag, i.e. the largest area with a significant correlation coefficient (Figure 10B).

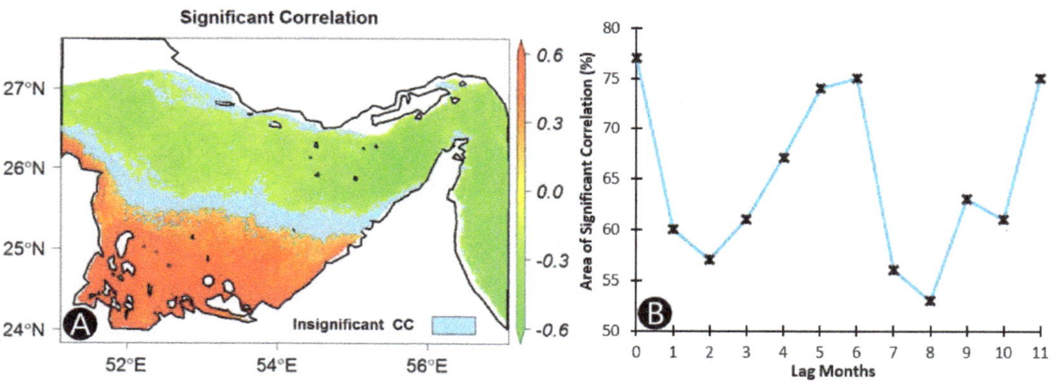

Figure 10. (**A**) The spatial distribution of the correlation coefficients over the Arabian and Oman Gulfs. (**B**) The relationship between the percentages of the area with a significant correlation coefficient and the time lag of the SST (zero meaning without any lag).

The spatial distribution of the correlation indicates that the UAE's AG coast showed a positive correlation between the Chl-a concentration and the SST and especially coasts near the Abu Al Abyad and Sir Baniyas islands. However, the northeastern coast of the UAE (from Dubai northward) showed a significant negative correlation covering around one-third of the study area. Along the coasts of the GO and the Strait of Hormuz, the correlation was almost uniform, with more than 80% of the area showing a negative correlation. Detailed summary statistics of the correlation in the three regions are shown in Table 1.

Moreover, the relationship between the Chl-a concentration and the SST was highly dependent on the bathymetry of the seawaters. This relationship is due to the difference in gaining the solar heat between the shallow (warmer) and deeper portions (colder) of the sea. The deeper sea areas have greater thermal memory; in turn, they require a longer time to heat and never reach the optimum temperature for algal blooms (Chl-a) growth. Therefore, the surface water in the middle (deep) sea generally gains lower temperature than the surface water near the shore [16]. All the areas that showed a significant positive correlation were in the shallow coastal areas. On the contrary, the deeper areas seem to have an inverse relationship between Chl-a and SST (Figure 11). In the areas where the depth of the sea is less than 20 m below sea level, the average correlation coefficient between Chl-a concentration and SST was 0.43, whereas an average correlation of −0.34 was found in waters deeper than 40 m. Moreover, the areas that did not exhibit a significant correlation have an average depth of 33 m and a median of 28 m below the sea level. This means an increase in SST increases the concentration of Chl-a in shallow water with less than 20 m depth, while an increase in SST tends to decrease the concentration of Chl-a in the deeper waters below 40 m sea level. The full relationship between the bathymetry and the correlation of the Chl-a and SST concentration is shown in Figure 11B.

Table 1. Basic statistics of the variables in the three main regions of interest.

	UAE-Arabian Gulf		Strait of Hormuz		UAE-Gulf of Oman	
	Average	Median	Average	Median	Average	Median
SST (°C)	26.0	26.0	25.2	25.1	26.2	26.2
Chl-a (mg m^{-3})	2.00	1.8	2.8	2.6	2.3	2.0
Depth (m)	−22.8	−20.0	−56.7	−60.0	−104.7	−96.0
Correlation Coefficient (CC)	0.09	0.27	−0.36	−0.36	−0.41	−0.42
Area with positive CC (%)	47.60		0.01		0.00	
Area with negative CC (%)	30.77		80.68		81.85	

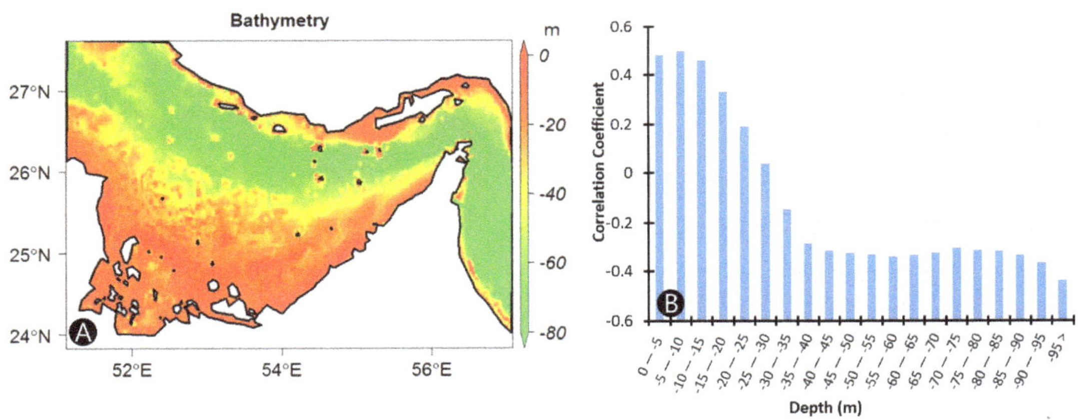

Figure 11. (**A**) Bathymetry of the Arabian Gulf and Gulf of Oman. (**B**) The relationship between the correlation coefficient and the depth of the seawater.

The relationship between bathymetry, Chl-a, and SST follows a U-shaped curve with the shallow and very deep seas having higher Chl-a and SST, respectively, as shown in Figure 12. The shallow waters experienced the highest Chl-a concentration in terms of average and median values. The deeper areas were observed to have the highest SST (Figure 12). Eventually, as the depth increases, both Chl-a and SST start to decline rapidly until 20 m below sea level. Then, the Chl-a concentration remains relatively stable while SST decreases until it reaches 24.7 °C at a depth of 70 m. Then, the SST begins to increase, reaching more than 26 °C at the depth of >95 m while the Chl-a also rises but at a lower rate, reaching 1.8 mg m^{-3} from 1.4 mg m^{-3}. The deepest region is the GO with the highest SST was also one of the areas with a high concentration of Chl-a.

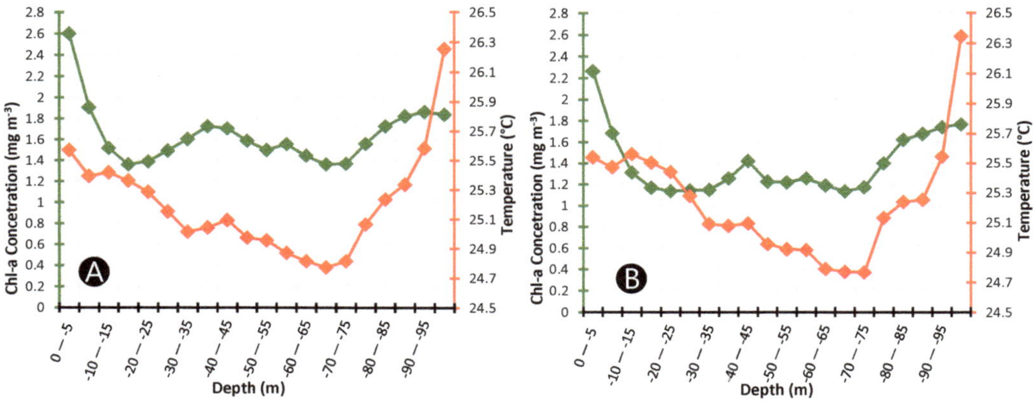

Figure 12. The relationship between the Chl-a and SST with respect to the depth of the seawaters using (**A**) the long-term average and (**B**) the long-term median of Chl-a and SST.

4.4. Trend Analysis

Mann–Kendal trend analysis was carried out to investigate the possibility of a significant trend in both variables (Chl-a and SST) over the span of 17 years. The Chl-a showed a decreasing trend in most areas, except on the coast of Abu Al Abyad Island, which is located in the western part of the UAE. This is mainly due to nutrient leaching from the orchards' soil and aquafarming drainage that contains nutrients useful for algae growth. Overall, 21% of the study area had a significant trend in Chl-a concentration. The majority (95%) of this area experienced a decreasing trend in the concentration of the Chl-a with an average of -0.28 mg m^{-3} per decade rate of decline (Figure 13A). The decreasing trend appears to increase in areas with higher average Chl-a concentration (Figure 13B). This suggests that the concentration of Chl-a is decreasing and at a higher rate in areas with a relatively high concentration during the last two decades. However, the areas with the highest concentration of Chl-a (Strait of Hormuz and the GO) did not experience a significant trend. In the places where the trend was increasing (Abu Abyad Island), the trend rate was increased as the average concentration increased. This is mainly due to the agricultural and aquaculture activities involved in Abu Abyad Island and the results suggest that the activities have increased over time (Figure 13B).

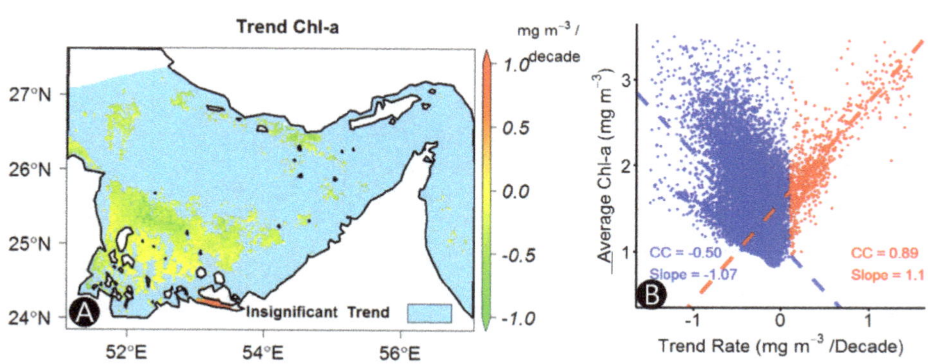

Figure 13. (**A**) Spatial distribution of the estimated trend using linear regression over the Arabian and Oman Gulfs for Chl-a and (**B**) scatter plot of the estimated rate of trend versus the long-term average of Chl-a concentration. Red points represent a positive trend and blue points represent a negative trend (n = 35,358).

The trend test indicated that 52% of the study area, mainly located in the northern part, experienced a significant SST-positive trend (Figure 14). The average rate of increase in these regions is estimated as 0.91 °C per decade. Most of the areas that showed an increasing trend are places with relatively lower mean SST. As the long-term average SST decreases, the rate of trend increases sharply, as shown in Figure 14B. This means that the cooler regions of the AG are experiencing an increase in temperature at an alarming rate.

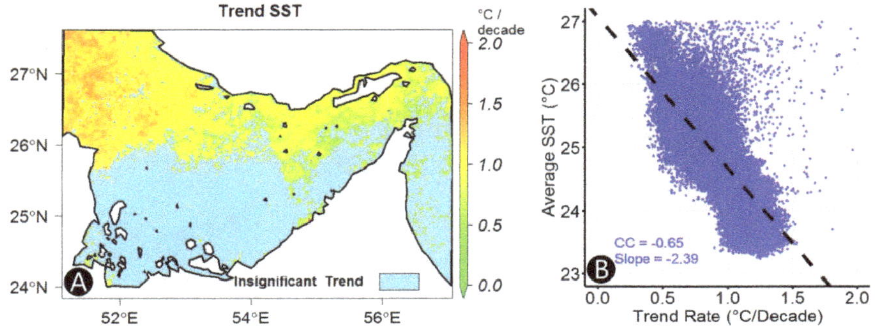

Figure 14. (**A**) Spatial distribution of the estimated trend using a linear regression model over the Arabian and Oman Gulfs for SST and (**B**) scatter plot of the estimated rate of trend versus the long-term average of SST (n = 85,693).

The trend in both the winter and summer seasons has been further analyzed to investigate the seasonality of the trends. The time series is divided into two six-month periods of summer and winter. Summer months (May to October) are categorized as very hot and humid and the winter months (November to April) are characterized by relatively cooler months (Appendix B). The seasonal trend analysis of the Chl-a indicated that 18% and 13% of the study area have a significant trend for the months of summer and winter, respectively. Less than 1% of the area with a significant trend showed a positive trend. The Abu Abyad Island and its surroundings showed an increasing trend in both seasons. This further supports the aforementioned reasoning that the higher concentration of Chl-a near the island is not related to climatological phenomena but to activities on the island. The spatial distribution showed that the trend in the winter is concentrated in the coastal area located between Qatar and UAE, whereas in summer, the trend is experienced further from the seashore (Figure 15A,B). The rate of decline was higher during summer with an average rate of -0.41 mg m^{-3} compared to the average rate of -0.22 mg m^{-3} in winter. Similar to the results of the trend analysis, the areas with higher average Chl-a concentration (Strait of Hormuz and GO as shown in Figure 2B) did not show a significant trend in both seasons over the last two decades.

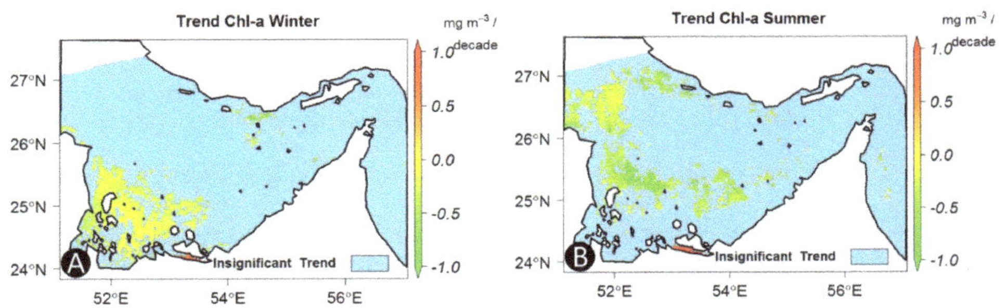

Figure 15. Spatial distribution of the seasonal trend of Chl-a in (**A**) winter (November to April) and (**B**)summer (May to October).

The seasonal trend analysis of the SST showed that the winter months had no significant trend over the entire area (Figure 16A). On the contrary, the summer was the dominant season of the trend. The summer months demonstrated an increasing trend in more than two-thirds of the study area (Figure 16B). The results indicated that the summer months are becoming hotter at a rate of higher than 1.0 °C per decade in half of the area. Around 20% of the study area (almost all of them located in the northeastern tip) exhibits an increasing trend rate higher than 1.5 °C per decade. The regions that are warming at a higher rate are the areas with relatively lower average SST. Previous studies also reached a similar conclusion, which indicates that the summer months are becoming hotter at a much higher rate [56,58]. Piontkovski [58] showed that the trend of SST in June and July was more than double the trend of average annual SST.

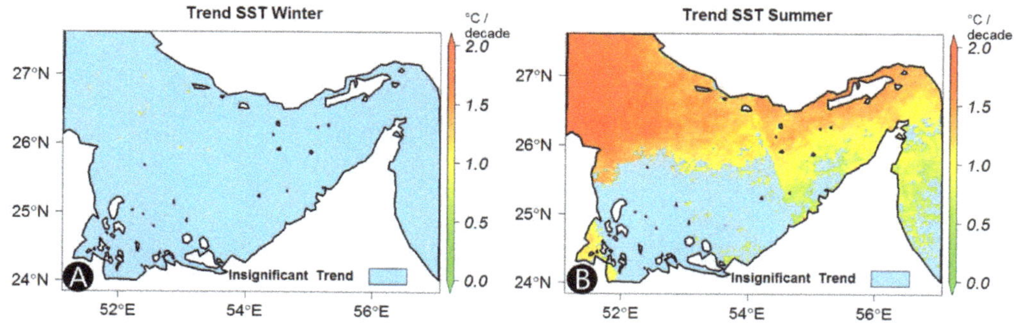

Figure 16. Spatial distribution of the seasonal trend of SST in (**A**) winter (November to April) and (**B**) summer (May to October).

5. Summary and Conclusions

The ecosystems of the Arabian seas (Arabian Gulf, Gulf of Oman, and Arabian Sea) are fragile and susceptible to pollution. Among these pollutants are algal blooms. The most effective approach for estimating the Chl-a concentration and assessing the spatiotemporal distribution of algal blooms is the employment of remotely sensed data and remote sensing techniques. This study analyzed the spatiotemporal variability of the Chl-a concentration and SST in the Arabian Gulf and the Gulf of Oman along the UAE coasts. The correlation between the Chl-a and SST is also investigated as it sheds light on the impact of the SST on the growth of phytoplankton. The variability of both Chl-a and SST is also examined using the empirical orthogonal function (EOF) analysis, which helps in understanding the impact of major wind currents in the area.

The spatial distribution of the Chl-a concentration showed that the highest concentration was observed in the Strait of Hormuz with an average of 2.8 mg m^{-3}, which is 1.1 mg m^{-3} higher than the average for the entire study area. The Gulf of Oman was also the hottest region with an average of 26 °C, which is one degree hotter than the average of the total area. Moreover, SST showed a uniform gradient in the northwest to southeast direction. The summer months (May to October) were the hottest months, with an average of ~31 °C, whereas in the winter months (November to April), the SST reached as low as ~19 °C.

The first EOF component of Chl-a is related to the northeast monsoon winds (November to March), which cools the sea surface. The Chl-a concentration increased in the Strait of Hormuz and the Gulf of Oman due to the availability of nutrients in addition to the optimal atmospheric conditions. The spikes in concentration over the coast of UAE (AG) at the end of 2004, 2007, and 2018 to 2019 were related to the second component.

Three quarters of the study area experienced a significant correlation between the Chl-a and SST. The coastal areas of western UAE and Qatar showed a significant positive correlation, which suggests that the SST affects the concentration of Chl-a. However, more than half of the study area indicated that SST negatively influenced the Chl-a concentration in the Arabian Gulf and the Gulf of Oman, especially areas in the deep sea. Furthermore, the correlation coefficient and the bathymetry of the seas showed a strong relationship. The shallow areas had a strong positive correlation between the SST and Chl-a, whereas the deeper areas were inclined to have a negative correlation.

Lastly, trend analysis was carried out to investigate the presence of significant trends using the correlated seasonal Mann–Kendal trend test. The Chl-a data showed the presence of a trend in just 21% of the study area, of which 95% indicated a decreasing trend. Most of the area with a decreasing trend is located in the southern region, which is closer to the coasts of the UAE and Qatar. The rate at which the trend is decreasing is also related to the average Chl-a concentration. Higher average values of Chl-a concentration are associated with a higher rate of decline, and vice versa. The SST also showed the presence of a significant trend in more than 52% of the study area. However, in this case, an increasing trend is observed. Similarly, the rate of trend showed an inverse relationship with the average SST, the higher the average SST, the smaller the rate of increase, and vice-versa.

The main limitations of the study are the missing data due to cloud cover and the relatively short period of the dataset (2003 to 2019). Even though a mature technique of filling the data was followed, a relatively large amount of missing data can cause uncertainty in the results. The conclusions of this research were similar to previously conducted studies. However, the authors feel that these limitations are worth mentioning.

Author Contributions: K.A.H. and K.A.A. guided this research and contributed significantly to preparing the manuscript for publication; K.A.H., P.P., K.A.A., N.A.H. and H.O.S. developed the research methodology; K.A.H., K.A.A., P.P. and D.T.G. downloaded and processed the remote sensing products; D.T.G. and P.P. developed the scripts used in the analysis. K.A.H., K.A.A., N.A.H., H.O.S. and D.T.G. prepared the first draft; K.A.H., D.T.G., K.A.A. and H.O.S. performed the final overall proofreading of the manuscript. All authors have read and agreed to the published version of the manuscript.

Funding: This research was partially funded through the UAEU Research and sponsored Projects Office (Grant: G00002687).

Data Availability Statement: Publicly available datasets were analyzed in this study. Chlorophyll-a Level 2 data can be found at [https://oceancolor.gsfc.nasa.gov/cgi/browse.pl?sen=amod]. Publicly available datasets were analyzed in this study. SST Level 2 data presented in this study are openly available at [https://podaac.jpl.nasa.gov/dataset/MODIS_A-JPL-L2P-v2019.0] at [doi: 10.5067/GHMDA-2PJ19].

Acknowledgments: The United Arab Emirates University—Research Affairs is gratefully acknowledged for their support. Gratitude is also extended to Marwan Almubarak and Rowan and Linda Hussein for following up with editing.

Conflicts of Interest: The authors declare no conflict of interest.

Appendix A

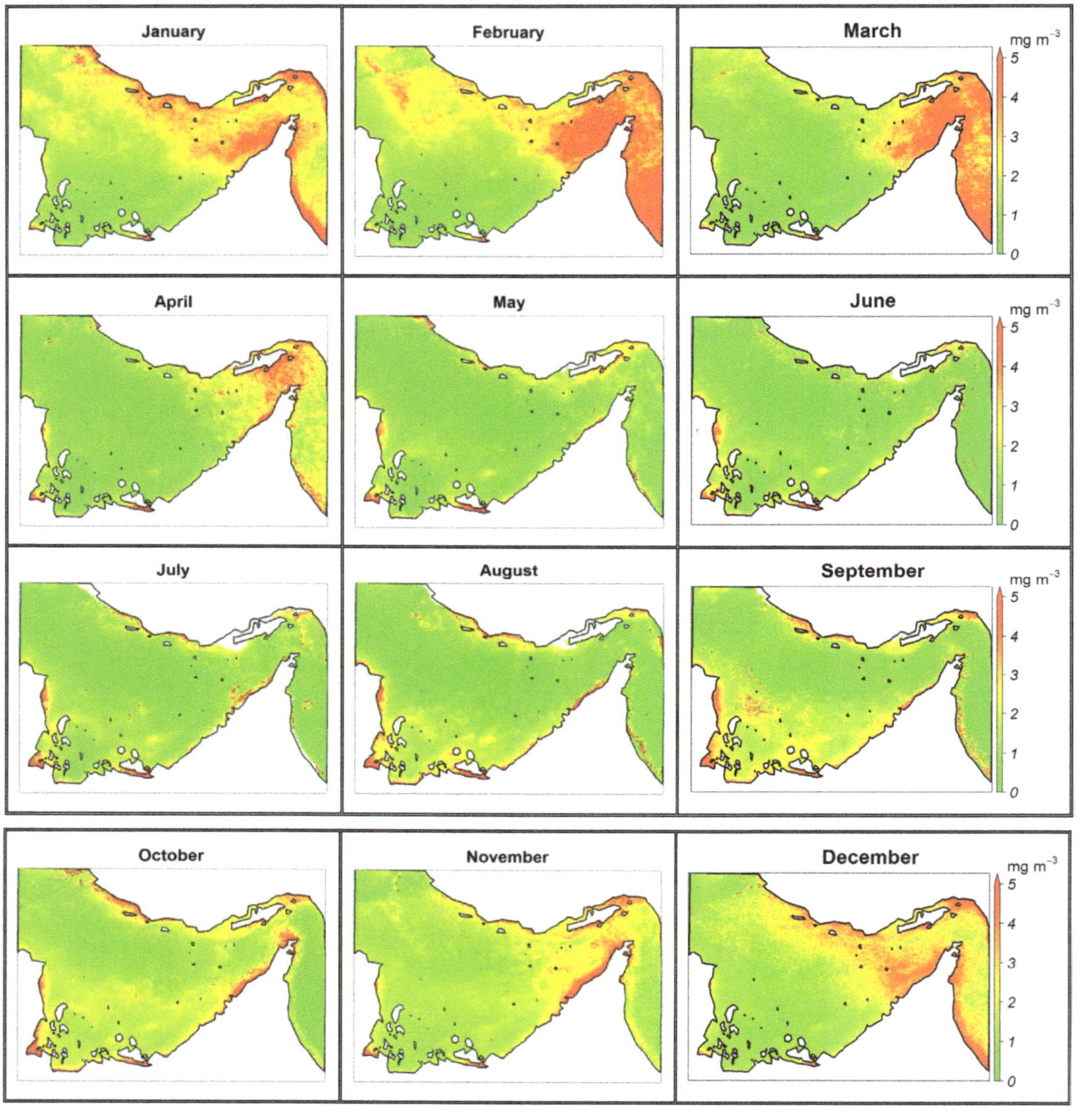

Figure A1. The spatial distribution of the average monthly Chl-a concentration across the Arabian and Oman Gulfs.

Appendix B

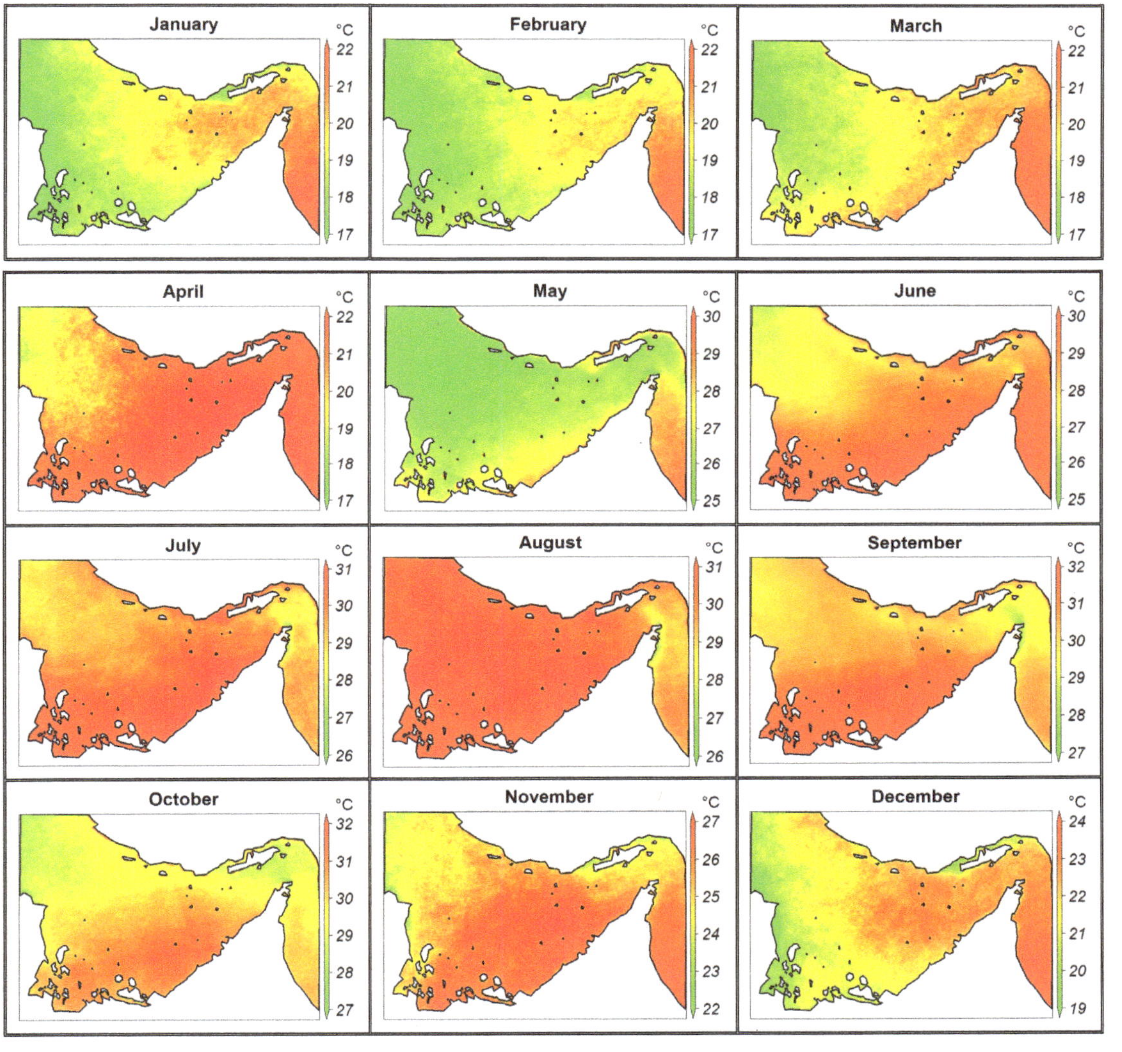

Figure A2. The spatial distribution of the average monthly sea surface temperature (SST) across the Arabian and Oman Gulfs (N.B. the color bar scale varies from figure to figure).

Appendix C

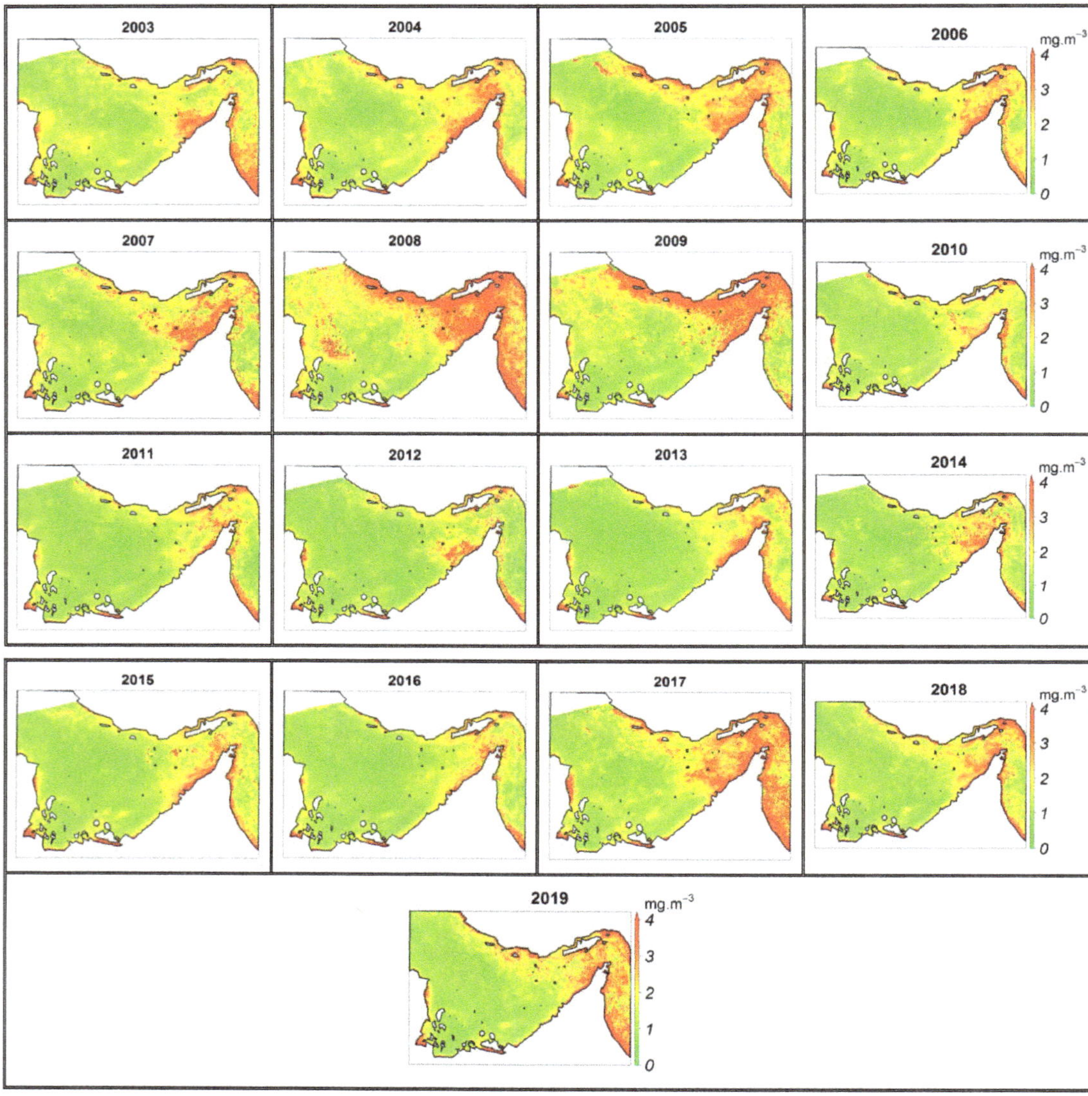

Figure A3. The spatial distribution of the annual average Chl-a concentration across the Arabian and Oman Gulfs.

Appendix D

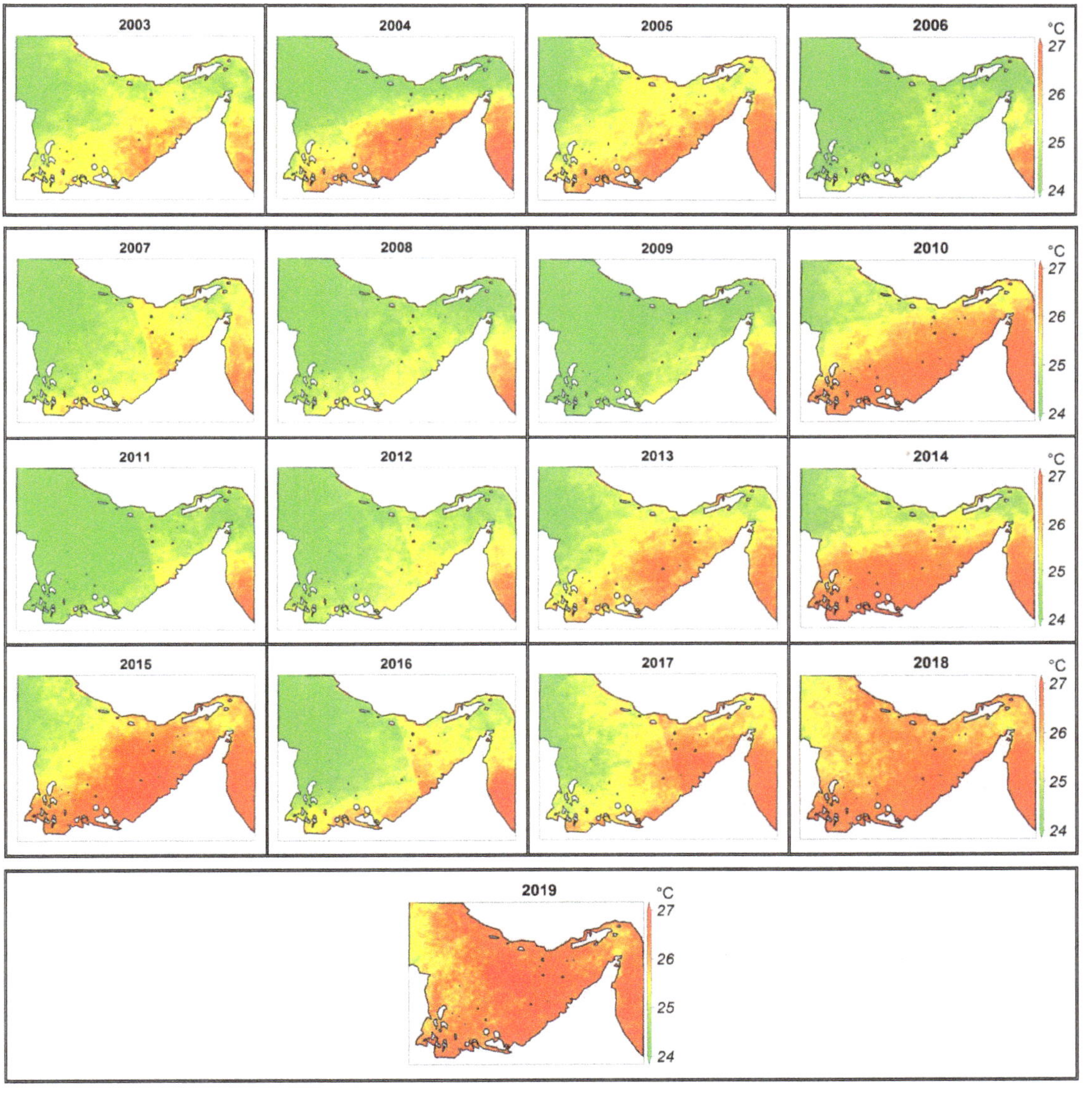

Figure A4. The spatial distribution of the annual average SST across the Arabian and Oman Gulfs.

References

1. Al Shehhi, M.R.; Gherboudj, I.; Ghedira, H. An overview of historical harmful algae blooms outbreaks in the Arabian Seas. *Mar. Pollut. Bull.* **2014**, *86*, 314–324. [CrossRef]
2. do Rosario Gomes, H.; Goes, J.I.; Matondkar, S.P.; Parab, S.G.; Al-Azri, A.R.; Thoppil, P.G. Blooms of Noctiluca miliaris in the Arabian Sea—An in situ and satellite study. *Deep. Sea Res. Part I Oceanogr. Res. Pap.* **2008**, *55*, 751–765. [CrossRef]
3. Zhao, J.; Ghedira, H. Monitoring red tide with satellite imagery and numerical models: A case study in the Arabian Gulf. *Mar. Pollut. Bull.* **2014**, *79*, 305–313. [CrossRef]
4. Sastry, J.; d'Souza, R. Upwelling and upward mixing in the Arabian Sea. *Indian J. Geo Mar. Sci.* **1972**, *1*, 17–27.

5. Wiggert, J.; Jones, B.; Dickey, T.; Brink, K.; Weller, R.; Marra, J.; Codispoti, L. The Northeast Monsoon's impact on mixing, phytoplankton biomass and nutrient cycling in the Arabian Sea. *Deep. Sea Res. Part II Top. Stud. Oceanogr.* **2000**, *47*, 1353–1385. [CrossRef]
6. Barzandeh, A.; Eshghi, N.; Hosseinibalam, F.; Hassanzadeh, S. Wind-driven coastal upwelling along the northern shoreline of the Persian Gulf. *Boll. Geofis. Teor. Appl.* **2018**, *59*, 301–312.
7. Chaichitehrani, N.; Allahdadi, M.N. Overview of wind climatology for the Gulf of Oman and the northern Arabian Sea. *Am. J. Fluid Dyn.* **2018**, *8*, 1–9.
8. Banse, K. Seasonality of phytoplankton chlorophyll in the central and northern Arabian Sea. *Deep Sea Res. Part A Oceanogr. Res. Pap.* **1987**, *34*, 713–723. [CrossRef]
9. Kumar, S.P.; Ramaiah, N.; Gauns, M.; Sarma, V.V.S.S.; Muraleedharan, P.M.; Raghukumar, S.; Dileep Kumar, M.; Madhupratap, M. Physical forcing of biological productivity in the Northern Arabian Sea during the Northeast Monsoon. *Deep Sea Res. Part II Top. Stud. Oceanogr.* **2001**, *48*, 1115–1126. [CrossRef]
10. Banse, K.; English, D.C. Comparing phytoplankton seasonality in the eastern and western subarctic Pacific and the western Bering Sea. *Prog. Oceanogr.* **1999**, *43*, 235–288. [CrossRef]
11. Palanisamy, S.; Ahn, Y.-H.; Ryu, J.-H.; Moon, J.-E. Application of optical remote sensing imagery for detection of red tide algal blooms in Korean waters. In Proceedings of the 2005 IEEE International Geoscience and Remote Sensing Symposium—IGARSS'05, Seoul, Korea, 25–29 July 2005; IEEE: New York, NY, USA, 2005.
12. Blondeau-Patissier, D.; Gower, J.F.; Dekker, A.G.; Phinn, S.R.; Brando, V.E. A review of ocean color remote sensing methods and statistical techniques for the detection, mapping and analysis of phytoplankton blooms in coastal and open oceans. *Prog. Oceanogr.* **2014**, *123*, 123–144. [CrossRef]
13. Prigent, C.; Aires, F.; Rossow, W.B. Land surface skin temperatures from a combined analysis of microwave and infrared satellite observations for an all-weather evaluation of the differences between air and skin temperatures. *J. Geophys. Res. Atmos.* **2003**, *108*. [CrossRef]
14. Miles, T.; He, R. Seasonal surface ocean temporal and spatial variability of the South Atlantic Bight: Revisiting with MODIS SST and Chl-a imagery. *Cont. Shelf Res.* **2010**, *30*, 1951–1962. [CrossRef]
15. Alvera-Azcárate, A.; Barth, A.; Beckers, J.M.; Weisberg, R.H. Multivariate reconstruction of missing data in sea surface temperature, chlorophyll, and wind satellite fields. *J. Geophys. Res. Oceans* **2007**, *112*. [CrossRef]
16. Petchprayoon, P. *Analysis of Climate Change Impacts on Surface Energy Balance of Lake Huron (Estimation of Surface Energy Balance Components: Remote Sensing Approach for Water–Atmosphere Parameterization)*; University of Colorado at Boulder: Boulder, CO, USA, 2015.
17. Brewin, R.J.; Raitsos, D.E.; Pradhan, Y.; Hoteit, I. Comparison of chlorophyll in the Red Sea derived from MODIS-Aqua and in vivo fluorescence. *Remote Sens. Environ.* **2013**, *136*, 218–224. [CrossRef]
18. Hu, C.; Lee, Z.; Franz, B. Chlorophyll aalgorithms for oligotrophic oceans: A novel approach based on three-band reflectance difference. *J. Geophys. Res. Oceans* **2012**, *117*. [CrossRef]
19. Nezlin, N.P.; Kostianoy, A.G.; Grégoire, M. Patterns of seasonal and interannual changes of surface chlorophyll concentration in the Black Sea revealed from the remote sensed data. *Remote Sens. Environ.* **1999**, *69*, 43–55. [CrossRef]
20. Tang, D.-L.; Ni, I.-H.; Kester, D.R.; Müller-Karger, F.E. Remote sensing observations of winter phytoplankton blooms southwest of the Luzon Strait in the South China Sea. *Mar. Ecol. Prog. Ser.* **1999**, *191*, 43–51. [CrossRef]
21. Gurlin, D.; Gitelson, A.A.; Moses, W.J. Remote estimation of chl-a concentration in turbid productive waters—Return to a simple two-band NIR-red model? *Remote Sens. Environ.* **2011**, *115*, 3479–3490. [CrossRef]
22. Gower, J.F.; King, S.A. Distribution of floating Sargassum in the Gulf of Mexico and the Atlantic Ocean mapped using MERIS. *Int. J. Remote Sens.* **2011**, *32*, 1917–1929. [CrossRef]
23. Gower, J.; King, S.; Goncalves, P. Global monitoring of plankton blooms using MERIS MCI. *Int. J. Remote Sens.* **2008**, *29*, 6209–6216. [CrossRef]
24. Cannizzaro, J.P.; Carder, K.L.; Chen, F.R.; Heil, C.A.; Vargo, G.A. A novel technique for detection of the toxic dinoflagellate, Karenia brevis, in the Gulf of Mexico from remotely sensed ocean color data. *Cont. Shelf Res.* **2008**, *28*, 137–158. [CrossRef]
25. Hu, C.; Lee, Z.; Ma, R.; Yu, K.; Li, D.; Shang, S. Moderate resolution imaging spectroradiometer (MODIS) observations of cyanobacteria blooms in Taihu Lake, China. *J. Geophys. Res. Oceans* **2010**, *115*. [CrossRef]
26. Anderson, C.R.; Kudela, R.M.; Burrell, C.T.; Langlois, G.; Goodman, J.; Benitez-Nelson, C.; Sekula-Wood, E.; Chao, Y.; Siegel, D.A. Detecting toxic diatom blooms from ocean color and a regional ocean model. *Geophys. Res. Lett.* **2011**, *38*. [CrossRef]
27. Kavak, M.T.; Karadogan, S. The relationship between sea surface temperature and chlorophyll concentration of phytoplanktons in the Black Sea using remote sensing techniques. *J. Environ. Biol.* **2012**, *33*, 493. [PubMed]
28. Nurdin, S.; Mustapha, M.A.; Lihan, T. The relationship between sea surface temperature and chlorophyll-a concentration in fisheries aggregation area in the archipelagic waters of Spermonde using satellite images. In *AIP Conference Proceedings*; American Institute of Physics: New York, NY, USA, 2013.
29. Jutla, A.S.; Akanda, A.S.; Griffiths, J.K.; Colwell, R.; Islam, S. Warming oceans, phytoplankton, and river discharge: Implications for cholera outbreaks. *Am. J. Trop. Med. Hyg.* **2011**, *85*, 303–308. [CrossRef] [PubMed]
30. Kouketsu, S.; Kaneko, H.; Okunishi, T.; Sasaoka, K.; Itoh, S.; Inoue, R.; Ueno, H. Mesoscale eddy effects on temporal variability of surface chlorophyll a in the Kuroshio Extension. *J. Oceanogr.* **2016**, *72*, 439–451. [CrossRef]

31. Chu, P.C.; Kuo, Y.-H. Biophysical variability in the Kuroshio Extension from Altimeter and SeaWiFS. In Proceedings of the OCEANS 2009, Biloxi, MS, USA, 26–29 October 2009; IEEE: New York, NY, USA, 2009.
32. Emery, K.O. Sediments and water of Persian Gulf. *AAPG Bull.* **1956**, *40*, 2354–2383.
33. Vaughan, G.O.; Al-Mansoori, N.; Burt, J.A. The arabian gulf. In *World Seas: An Environmental Evaluation*; Elsevier: Amsterdam, The Netherlands, 2019; pp. 1–23.
34. The Editors of Encyclopaedia Britannica. Gulf of Oman. In Encyclopedia Britannica. 2020. Available online: https://www.britannica.com/place/Gulf-of-Oman (accessed on 9 March 2021).
35. Al Senafi, F.; Anis, A. Shamals and climate variability in the Northern Arabian/Persian Gulf from 1973 to 2012. *Int. J. Climatol.* **2015**, *35*, 4509–4528. [CrossRef]
36. Nezlin, N.P.; Polikarpov, I.G.; Al-Yamani, F.Y.; Rao, D.S.; Ignatov, A.M. Satellite monitoring of climatic factors regulating phytoplankton variability in the Arabian (Persian) Gulf. *J. Mar. Syst.* **2010**, *82*, 47–60. [CrossRef]
37. Noori, R.; Tian, F.; Berndtsson, R.; Abbasi, M.R.; Naseh, M.V.; Modabberi, A.; Soltani, A.; Kløve, B. Recent and future trends in sea surface temperature across the Persian Gulf and Gulf of Oman. *PLoS ONE* **2019**, *14*, e0212790. [CrossRef]
38. Dwyer, J.; Schmidt, G. The MODIS reprojection tool. In *Earth Science Satellite Remote Sensing*; Springer: Berlin/Heidelberg, Germany, 2006; pp. 162–177.
39. Moradi, M.; Kabiri, K. Spatio-temporal variability of SST and Chlorophyll-a from MODIS data in the Persian Gulf. *Mar. Pollut. Bull.* **2015**, *98*, 14–25. [CrossRef] [PubMed]
40. Wan, Z. *MODIS Land-Surface Temperature Algorithm Theoretical Basis Document (LST ATBD)*; Institute for Computational Earth System Science: Santa Barbara, CA, USA, 1999; Volume 75.
41. Crosman, E.T.; Horel, J.D. MODIS-derived surface temperature of the Great Salt Lake. *Remote Sens. Environ.* **2009**, *113*, 73–81. [CrossRef]
42. Hereher, M.E. Assessment of climate change impacts on sea surface temperatures and sea level rise—The Arabian Gulf. *Climate* **2020**, *8*, 50. [CrossRef]
43. Oesch, D.; Jaquet, J.M.; Klaus, R.; Schenker, P. Multi-scale thermal pattern monitoring of a large lake (Lake Geneva) using a multi-sensor approach. *Int. J. Remote Sens.* **2008**, *29*, 5785–5808. [CrossRef]
44. Oesch, D.C.; Jaquet, J.M.; Hauser, A.; Wunderle, S. Lake surface water temperature retrieval using advanced very high resolution radiometer and Moderate Resolution Imaging Spectroradiometer data: Validation and feasibility study. *J. Geophys. Res. Oceans* **2005**, *110*. [CrossRef]
45. Reinart, A.; Reinhold, M. Mapping surface temperature in large lakes with MODIS data. *Remote Sens. Environ.* **2008**, *112*, 603–611. [CrossRef]
46. Amante, C.; Eakins, B.W. *ETOPO1 Arc-Minute Global Relief Model: Procedures, Data Sources and Analysis*; National Geophysical Data Center: Boulder, CO, USA, 2009; Volume 10, p. V5C8276M.
47. Li, Y.; He, R. Spatial and temporal variability of SST and ocean color in the Gulf of Maine based on cloud-free SST and chlorophyll reconstructions in 2003–2012. *Remote Sens. Environ.* **2014**, *144*, 98–108. [CrossRef]
48. Monahan, A.H.; Fyfe, J.C.; Ambaum, M.H.; Stephenson, D.B.; North, G.R. Empirical orthogonal functions: The medium is the message. *J. Clim.* **2009**, *22*, 6501–6514. [CrossRef]
49. Wikle, C.K.; Zammit-Mangion, A.; Cressie, N. *Spatio-Temporal Statistics with R*; CRC Press: Boca Raton, FL, USA, 2019.
50. Pearson, K., VII. Note on regression and inheritance in the case of two parents. *Proc. R. Soc. Lond.* **1895**, *58*, 240–242.
51. Hirsch, R.M.; Slack, J.R.; Smith, R.A. Techniques of trend analysis for monthly water quality data. *Water Resour. Res.* **1982**, *18*, 107–121. [CrossRef]
52. Libiseller, C.; Grimvall, A. Performance of partial Mann–Kendall tests for trend detection in the presence of covariates. *Environ. Off. J. Int. Environ. Soc.* **2002**, *13*, 71–84. [CrossRef]
53. Nezlin, N.P.; Polikarpov, I.G.; Al-Yamani, F. Satellite-measured chlorophyll distribution in the Arabian Gulf: Spatial, seasonal and inter-annual variability. *Int. J. Oceans Oceanogr.* **2007**, *2*, 139–156.
54. Richlen, M.L.; Morton, S.L.; Jamali, E.A.; Rajan, A.; Anderson, D.M. The catastrophic 2008–2009 red tide in the Arabian gulf region, with observations on the identification and phylogeny of the fish-killing dinoflagellate Cochlodinium polykrikoides. *Harmful Algae* **2010**, *9*, 163–172. [CrossRef]
55. Prasad, T.; Ikeda, M. A numerical study of the seasonal variability of Arabian Sea high-salinity water. *J. Geophys. Res. Oceans* **2002**, *107*, 18-1–18-12. [CrossRef]
56. Nandkeolyar, N.; Raman, M.; Kiran, G.S. Comparative analysis of sea surface temperature pattern in the eastern and western gulfs of Arabian Sea and the Red Sea in recent past using satellite data. *Int. J. Oceanogr.* **2013**, *2013*, 501602. [CrossRef]
57. Björnsson, H.; Venegas, S. A manual for EOF and SVD analyses of climatic data. *CCGCR Rep.* **1997**, *97*, 112–134.
58. Piontkovski, S.; Chiffings, T. Long-term changes of temperature in the Sea of Oman and the western Arabian Sea. *Int. J. Oceans Oceanogr.* **2014**, *8*, 53–72.

remote sensing

MDPI

Article

Long-Term Changes in the Unique and Largest Seagrass Meadows in the Bohai Sea (China) Using Satellite (1974–2019) and Sonar Data: Implication for Conservation and Restoration

Shaochun Xu [1,2,3,4,5,†], Shuai Xu [1,2,3,4,5,†], Yi Zhou [1,2,3,4,5,*], Shidong Yue [1,2,3,4,5], Xiaomei Zhang [1,2,3,4], Ruiting Gu [1,2,3,4,5], Yu Zhang [1,2,3,4,5], Yongliang Qiao [6] and Mingjie Liu [1,2,3,4,5]

1 CAS Key Laboratory of Marine Ecology and Environmental Sciences, Institute of Oceanology, Chinese Academy of Sciences, Qingdao 266071, China; xushaochun@qdio.ac.cn (S.X.); xushuai16@mails.ucas.edu.cn (S.X.); yueshidong17@mails.ucas.ac.cn (S.Y.); zhangxiaomei@qdio.ac.cn (X.Z.); guruiting15@mails.ucas.edu.cn (R.G.); zhangyu185@mails.ucas.ac.cn (Y.Z.); liumingjie20@mails.ucas.ac.cn (M.L.)
2 Laboratory for Marine Ecology and Environmental Science, Qingdao National Laboratory for Marine Science and Technology, Qingdao 266237, China
3 Center for Ocean Mega-Science, Chinese Academy of Sciences, Qingdao 266071, China
4 CAS Engineering Laboratory for Marine Ranching, Institute of Oceanology, Chinese Academy of Sciences, Qingdao 266071, China
5 University of Chinese Academy of Sciences, Beijing 100049, China
6 Qingdao University of Science and Technology, Qingdao 266000, China; qiaoyongliang@qdio.ac.cn
* Correspondence: yizhou@qdio.ac.cn
† These authors contributed equally to this work.

Citation: Xu, S.; Xu, S.; Zhou, Y.; Yue, S.; Zhang, X.; Gu, R.; Zhang, Y.; Qiao, Y.; Liu, M. Long-Term Changes in the Unique and Largest Seagrass Meadows in the Bohai Sea (China) Using Satellite (1974–2019) and Sonar Data: Implication for Conservation and Restoration. *Remote Sens.* **2021**, *13*, 856. https://doi.org/10.3390/rs13050856

Academic Editor: Maria Laura Carranza

Received: 14 January 2021
Accepted: 18 February 2021
Published: 25 February 2021

Abstract: Seagrass meadows play critical roles in supporting a high level of biodiversity but are continuously threatened by human activities, such as sea reclamation. In this study, we reported on a large seagrass (Zostera marina L.) meadow in Caofeidian shoal harbor in the Bohai Sea of northern China. We evaluated the environmental impact of sea reclamation activities using Landsat imagery (1974–2019) by mapping seagrass meadow distribution changes. ISODATA was adopted for the unsupervised classification and mapping of seagrass beds. The error matrix developed using the in situ data obtained from acoustic surveys for Landsat 8OLI image classification was 87.20% accurate. The maps showed rapidly increasing changes in seagrass meadows as the amount of reclaimed land increased. Some seagrass meadows experienced large-scale changes, and sea reclamation has been suggested as the main factor responsible for habitat loss, which results from physical damage, excessive sedimentation, and increased turbidity caused by reclamation. In addition, habitat degradation may have resulted from three storm surges induced by typhoons in 1992–1998. Fortunately, land reclamation, forming an artificial "longshore bar", buffers seagrass meadows from wave actions, providing relatively sheltered conditions, which has allowed a large habitat increase since 2012. These were the largest eelgrass meadows (3,217.32 ha), with a peripheral area of ~100 km^2, in the Bohai Sea of northern China in 2019. However, the existing largest eelgrass beds in China are threatened by trawling, clam harvesting (especially clam sucking), channel dredging, and culture pond construction. Our work will help coastal managers monitor the environmental impacts of reclamation activities on seagrass meadows on a large spatio-temporal scale and will also provide information for seagrass restoration using artificial "longshore bars".

Keywords: seagrass; *Zostera marina* L.; remote sensing; reclamation; spatial and temporal changes

1. Introduction

Seagrasses form productive plant communities, providing habitats, foods, and nurseries for a variety of marine organisms [1–7]. However, with multiple stressors threatening these aquatic plants, seagrass meadows have been declining worldwide since 1990 at a rate of 7% per annum [8]. Many studies have been conducted to effectively understand changes in seagrass habitats' spatial distributions and temporal fluctuations [9–11]. There are 72 species of seagrass in six families and 14 genera worldwide [12]. However, five species within three genera are found in temperate northern China according to the national seagrass resource survey (2015–2020), but only the seagrass *Zostera marina* L. (eelgrass) is found at our study site. Seagrass monitoring and protection programs exist in North America, Europe, and Australia, and many countries have established seagrass monitoring networks (e.g., Seagrass Watch and Seagrass Net). However, limited investigations have been performed in northern China. Although there are preliminary investigations in Shandong Province, northern China, most research has been carried out in the past decade [13,14]. At present, seagrass research mainly focuses on its distribution in China, and the national seagrass resource survey (2015–2020) will be finished in 2021. Seagrass distributions in China have only been conducted using Sonar scanning [15], scuba diving, and direct sampling at limited depths and areas during low tides. Consequently, there is little information on the historical seagrass distribution in China. Even though underwater applications of satellite remote sensing are subject to several limitations, including cloud cover, wavelength-specific water column attenuation, and spatial and spectral resolutions of the sensors [16], satellite remote sensing is an effective tool for monitoring and managing seagrass beds, and it is now used to complement conventional surveying methods because of the large coverage area, rapidity, and repeatability of observations [17]. Compared with in situ sampling surveys, mapping seagrass using remote sensing provides a more spatially comprehensive representation of seagrass distributions [10,18,19]. Remote-sensing techniques have been used successfully to analyze the spatial extents of seagrass areas [18–23].

Seagrass ecosystems are disappearing at an alarming rate worldwide [24] because of natural disturbances and anthropogenic activities [8,25,26]. A large number of seagrass ecosystems have contracted sharply or disappeared in China according to the national seagrass resource survey (2015–2020), and this is mainly attributed to human disturbances caused by fishing, aquaculture, and land reclamation [27]. In general, land reclamation results in the complete, irrecoverable removal of seagrass meadows from natural habitats [28]. Owing to urban construction, as well as port and industrial construction, China's total reclamation area increased from 8241 km^2 to 13,380 km^2 during 1990–2008, with an annual increase of 285 km^2. Therefore, with the rapid development of harbors and ports, assessing the impact of reclamation activities on seagrass meadow distributions has become more and more important. However, data on the impacts of land reclamation on seagrasses in China are limited. Land reclamation was conducted in Caofeidian coastal waters from 2002 to 2012, covering an area of >200 km^2 [29]. According to local fishermen, there was abundant *Z. marina* distributed in the reclamation area before 2003. In addition, coastal ecosystems are influenced by the extreme climatological events, such as typhoons [11,30], and our study area was severely affected by storm surges induced by typhoons [31].

Seagrass monitoring has been conducted using conventional survey methods with GPS that provides limited information on spatial patterns and temporal dynamics in the habitats. These methods do not provide historical distribution information. However, satellite remote sensing provides more frequent data on seagrass distributions over wide ranges of temporal and geographical scales [11,18–23]. Understanding changes in eelgrass distributions during the last 46 years and analyzing its natural and man-made causes are essential for the conservation, management, and restoration of *Z. marina* beds. This study aimed to monitor seagrass distribution changes in the study area using historical satellite images from 1974 to 2019 and to elucidate the causes of these changes. We mapped changes in the eelgrass distribution over the last 46 years and analyzed their natural and man-made causes using remote sensing, sonar, and in situ survey methods. Our work provides new seagrass distribution data from temperate northern China and fundamental information for implementing appropriate seagrass restoration and conservation strategies.

2. Materials and Methods

2.1. Study Sites

The Bohai Sea is a semi-enclosed sea on the northern coast of China. It is often divided into Liaodong Bay, Bohai Bay, Laizhou Bay, and the Central Area. The Bohai Sea is connected to the Yellow Sea by the Bohai Strait [32]. As a semi-enclosed sea, there is a low water exchange rate in the Bohai Sea, and it takes more than 20 years to renew 90% of the water in the Bohai Sea, leading to a weak self-purification capacity and a low environmental carrying capacity [32]. In recent years, with rapid industrial and agricultural development, port construction, and increased urbanization, the impacts of human activities on the marine ecological environment have increased. The Bohai Sea is seriously polluted, and the functions of the marine ecosystem are degraded and damaged [32]. The Caofeidian shoal used to be in the Luanhe River Delta [33,34], but owing to erosion by tides and waves, a barrier island lagoon system eventually formed (Figure 1). The system consists of the Caofeidian islands, including Caofeidian, Yaotuo, Getuo, Longdao, and other sand islands and sandbars, with a zonal distribution [35]. The width of the tidal flat in this system is approximately 15–25 km, and the lagoon experiences irregular semidiurnal mixed tides [36]. The average water depth of lagoon is approximately 0–5 m, with a maximum water depth of 22 m [37]. At low tide, a large area of the tidal flat is exposed, while at high tide, only the barrier islands are visible. However, the barrier islands are severely affected by erosion resulting from storm surges, waves, and currents [38], resulting in morphological changes to the Caofeidian islands [39], with some islands even disappearing. Moreover, since 2003, large-scale land reclamation from the waters of Caofeidian has caused extreme damage to the barrier island lagoon system. Reclamation directly destroys the habitats of intertidal organisms and has large impacts on intertidal communities. The salinity ranged from 31.4 to 34.6 psu at the study site in the Bohai Sea. The sediments at the study site were mainly (93.28 ± 4.52%) composed of sand (0.063–2.0 mm) [40].

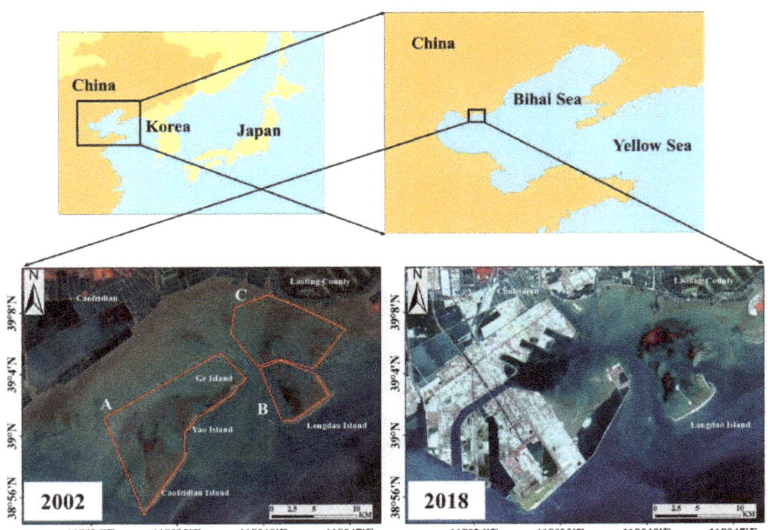

Figure 1. The study area in the northern coast of the China (2002 and 2018). On the basis of changes in the spatial distributions of the seagrass beds, the study area was divided into three areas (**A–C**) for research (2002).

2.2. Sonar Survey of the Seagrass Distribution in 2018

A sonar survey method, using the BioSonics MX echosounder (Biosonics Inc., Seattle, WA, USA) from August 5 to 13, 2018, was used to acquire sonar data, and it was applied to areas B and C. The echosounder, with a 204.8 kHz transducer, was set to operate at pulse rate of 0.4 ms and 5 pings s^{-1}. The ship route is parallel, with a spacing of ~500 m, and the ship speed was set to at 3–4 knots (Figure 2). To acquire information regarding the seagrass status (presence/absence, coverage, canopy height, and water depth), sonar data were analyzed using BioSonics Visual Habitat software (BioSonics Inc.). After mask creation, the interpolation analysis using Topo to Raster was applied using ArcMap 10.2.2 software (ESRI Inc., Redlands, CA, USA). The mask was selected as a type of boundary without a field. The parameters of plant coverage, canopy height, and water depth were selected as fields, respectively, and the type was selected as PointElevation. Vegetation with a canopy height > 0.2 m was considered seagrass.

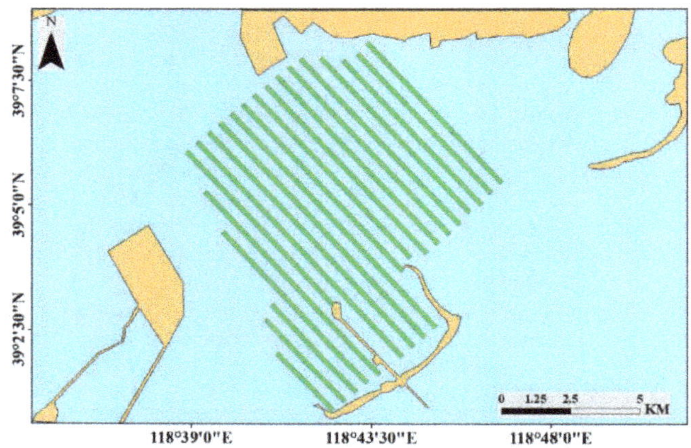

Figure 2. The ship route, with a spacing of ~500 m, for the acoustic survey of the seagrass distribution.

2.3. Seagrass Distribution Changes Inferred from Satellite Remote-Sensing Data

2.3.1. Satellite Image Analysis

In total, 29 satellite images from Landsat Multi-spectral Scanner (MSS)/Thematic Mapper (TM)/Enhanced Thematic Mapper (ETM+)/Operational Land Imager (OLI) were used to examine the long-term changes in the seagrass beds from 1974 to 2019 in the study area (Table S1). The satellite images that were taken during the period (May–September) in which seagrasses with high biomasses were selected for analysis. In addition, satellite images, taken during low tide (<1 m, Table S1), were selected and analyzed to determine the minimum distortion of the seagrass spectrum by the water column [41,42]. The seagrass beds in the study area were easily visible, because the canopy was exposed above the water level during low tide (Figure 3).

The two visible bands (green and red) and the near-infrared band were used to analyze the Landsat images. In addition, wedge-shaped gaps in some Landsat 7ETM+ images were filled [43]. Geometric and radiometric corrections were applied using ENVI 5.2 image processing software (EXELIS Inc., McLean, VA, USA). The FLAASH module was used to make atmospheric corrections [44]. Three masks were created to eliminate the unvegetated areas and to focus on the three specific areas of interest, A, B, and C (Figure 1). The unsupervised classification using ISODATA within ArcMap 10.2.2 software was applied to analyze for seagrass. In addition, because sea reclamations were easily visible in satellite images, to understand the temporal and spatial distributions of the reclamation from 2002 to 2014 in the study area, visual observations were used.

Figure 3. Field photo of a seagrass bed at the study site in the Bohai Sea (Photo by Xiaomei Zhang; tide height of 0.73 m on 8 June 2019, acquired at Shaleitien Tao Station using wxtide47 software).

2.3.2. Accuracy Assessment

Field data are usually used to assess classification accuracy, but there was no synchronous related ground truth data for the past 46 years. Consequently, we used the two visible bands (green and red) and the near-infrared band (described in detail in the satellite image analysis section) to determine the seagrass beds in satellite images lacking in situ field data. A Landsat 8OLI (29 June 2018) image was taken within 7 weeks of the in situ data obtained by the acoustic survey, and the acoustic data were used to assess the accuracy of the classifications of the Landsat 8OLI image using those bands. We used raw points acquired from acoustic data as verification points [45]. Points with canopy heights $\geq$ 0.2 m were defined as in the vegetated class; otherwise, they were defined as in the unvegetated class. The accuracy of the classification was assessed based on an error matrix (overall accuracy) [45], and it was calculated using the ratio of the number of points named to the same class (unvegetated or vegetated) in both the remotely derived classification and the acoustic data, divided by the total number of compared verification points. Depending upon the application, classification accuracy levels of $\leq$60, 70, and $\geq$80% may be unacceptable, marginally acceptable, and acceptable, respectively [20].

2.4. Field Survey of Eelgrass Population

To investigate the biological parameters of eelgrass at the six sites (Figure 4), three sediment cores were randomly collected at each site to investigate the density (shoots m^{-2}) and height (cm) of vegetative shoots, as well as the above- and below-ground biomasses [g m^{-2} dry weight (DW)] of total shoots on 21 August 2019. Samples from the study site were filtered through a 5-mm sieve carefully in situ, and the shoots were kept. The number of shoots was determined in the laboratory, and shoot height and both above- and below-ground biomasses (DW) were measured.

Water temperature (°C), salinity (ppt), dissolved oxygen content (DO, mg L^{-1}), depth (m), pH level, and chlorophyll content (μg L^{-1}) were measured using a multi-parameter water quality sonde (YSI 6600, USA) at the six sites during the seagrass sampling period (Figure 4).

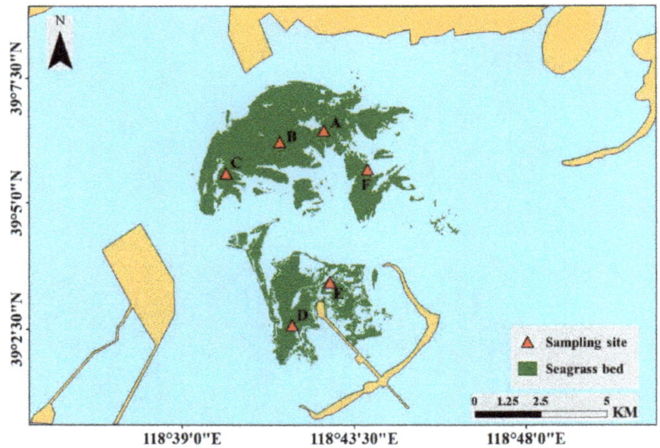

Figure 4. Field survey sites of eelgrass biological parameters in the study area. The red triangles (**A–F**) represent seagrass sampling sites.

2.5. Statistical Analyses

Results are presented as means ± SDs. Canopy height, and plant coverage of Z. marina over depths were modelled using generalized additive models (GAMs), which are an extension of generalized linear models that do not require the assumption of a particular response variable distribution along the environmental gradient [46]. The GAMs were built in R 4.0.2 [47] using the package "mgcv". For remote-sensing data, Pearson's correlation coefficient between seagrass area and sea reclamation area was determined. Pearson's correlation analyses were considered significant at a probability level of $p < 0.05$.

3. Results

3.1. Sonar Survey of Seagrass Distribution in 2018

As illustrated in Figure 5A, we determined the seagrass distribution using canopy height data. The seagrass bed area was 2,916.81 ha in August 2018, and the peripheral area was ~100 km^2. Similarly, we obtained plant percentage cover (Figure 5B) and water depth (Figure 5C) data. As illustrated in Figure 5B, the plant percentage cover map was very similar to the seagrass distribution map (Figure 5A). Most of the seagrasses were distributed at a depth of <3 m. There was a divide between seagrass beds (Figure 5A,B), and the general distribution pattern was caused by deeper water (>6 m) in the divide.

The variations in canopy height and *Z. marina* coverage over depths were analyzed. In response to water depth, canopy height and plant coverage exhibited the same trends, which are shown in Figure 6. The canopy height ($p < 0.001$) and plant coverage ($p < 0.001$) significantly increased with depth in shallow zones (<2.5 m) but decreased in deeper zones (>2.5 m) (Figure 6).

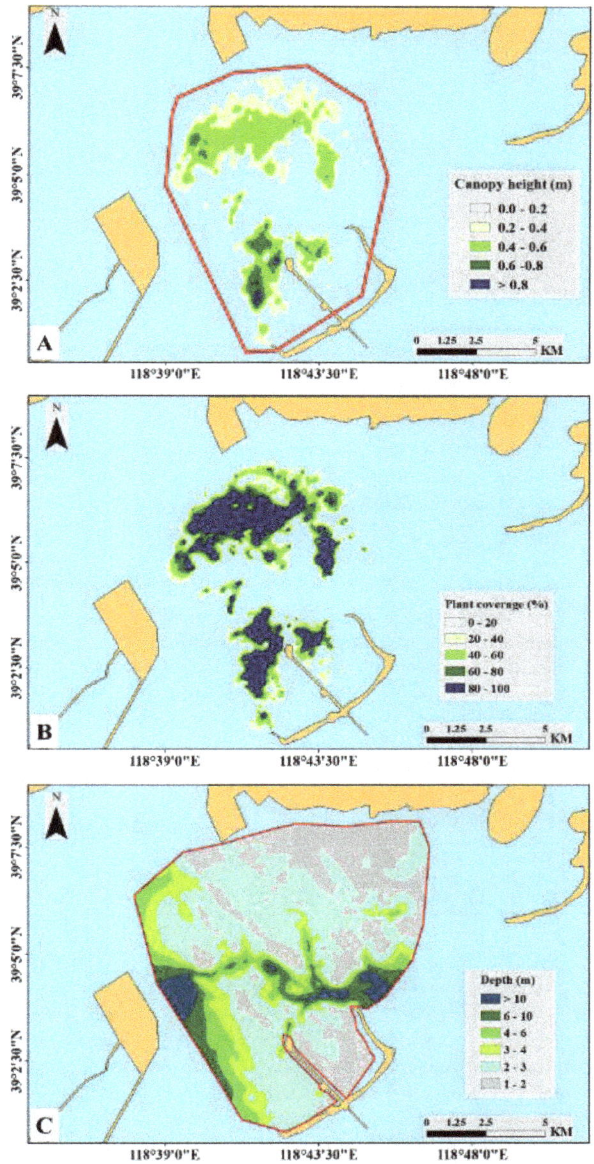

Figure 5. Eelgrass distribution map of plant canopy height (**A**), coverage (**B**), and water depth (**C**) generated from the interpolation analysis of the study area in August 2018. The area marked by the red polygon in A is the peripheral area of the seagrass beds. The red polygon in C marks the area used for the interpolation analysis.

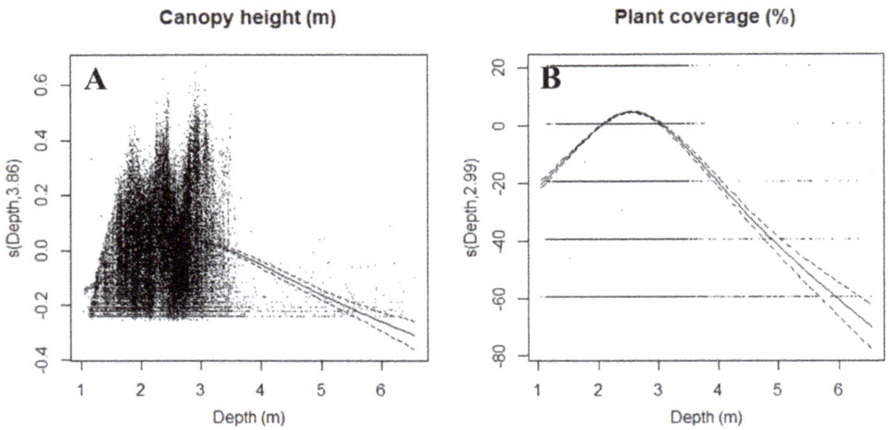

Figure 6. Response curves (thick, solid lines) of *Zostera marina* L. canopy height (**A**) and coverage (**B**) to water depth. Fitted model plots for depth corresponding to the reduced GAM for canopy height (m) and plant coverage (%). The solid lines show the GAM regressions fitted to each relationship, with dotted lines indicating the confidence interval of the mean trend line.

3.2. Seagrass Distribution Changes Inferred from Satellite Remote-Sensing Data

The accuracy of the classification, at 87.20%, was acceptable (Table 1) [20]. Two major seagrass beds were classified on the Landsat 8OLI image (Figure 7), which was consistent with the sonar results (Figure 5A). Thus, it was possible to classify seagrass beds with a high accuracy using Landsat 8OLI image-based classifications.

Table 1. The classification-accuracy assessment performed using verification points from acoustic data.

Verification Points from Acoustic Data	No. Points	No. Points Same Class to Classification	No. Points Different Class to Classification	Accuracy of Classification
Total verification points	128,619	112,155	16,464	87.20%
Vegetated class points	37,153	28,659	8494	77.14%
Unvegetated class points	91,466	83,496	7970	91.29%

The unsupervised classification (ISODATA) method applied to the Landsat 8OLI images produced a reliable accuracy level for this study area, and it was also used to analyze Landsat MSS/TM/ETM+ images from 1974 to 2019. There was a divide between seagrass beds, and the general distribution pattern (A, B, and C) was maintained consistently over the past 46 years (Figure 1). The area of seagrass beds ranged from 937.71 to 4,527.54 ha and averaged 2,487.65 $\pm$ 945.81 ha from 1974 to 2019 (Figures 8 and 9). The beds were largest in 1986 and smallest in 2009. There were, in general, declines in the seagrass bed areas from 1986 to 2009, and a large increase occurred in 2013. In A, there were large declines in the seagrass bed area in 1992–1998 and 2000–2011, and eventually, seagrass disappeared in 2011. Seagrass began to colonize B from 1979, and the seagrass bed area averaged 861.13 $\pm$ 356.29 ha from 1979 to 2019. Seagrass began to colonize C from 1984, and the seagrass bed area averaged 158.86 $\pm$ 149.42 ha from 1984 to 2012, and there was a large increase in 2013.

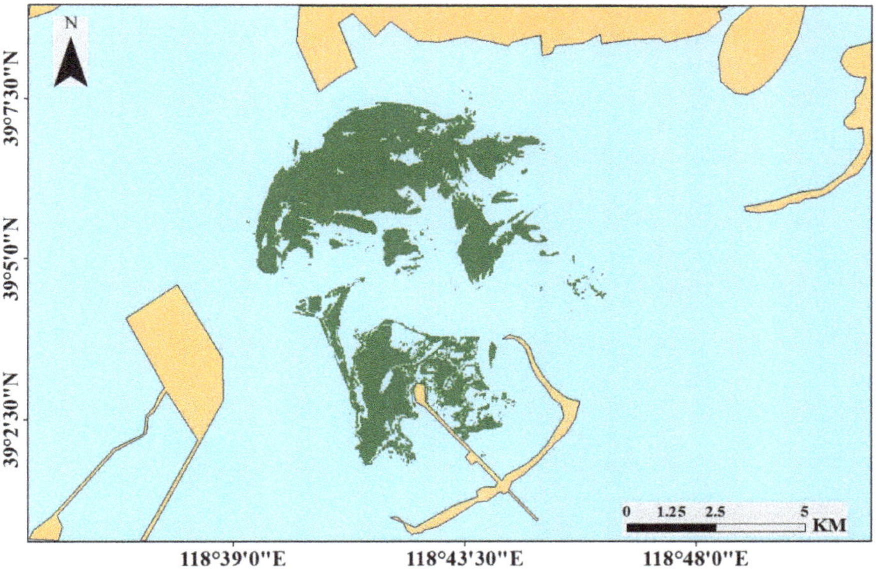

Figure 7. Seagrass distribution acquired from unsupervised classification (ISODATA) using Landsat 8OLI (29 June 2018) images.

The total area of reclamation reached 205.21 km^2 from 2002 to 2014 (Figures 10 and 11). In the years following 2003, when the infrastructure construction began, the reclamation area in the study region substantially increased. The reclamation area increased significantly in 2008–2011, but it decreased in the following years. As the reclamation area increased, the intertidal zone area in the study region decreased annually, and seagrass beds also underwent a considerable reduction (Figures 8 and 9). Pearson's correlation coefficient between the seagrass area of A and sea reclamation area was determined and revealed a significant negative correlation ($p < 0.05$, $R = -0.8588$).

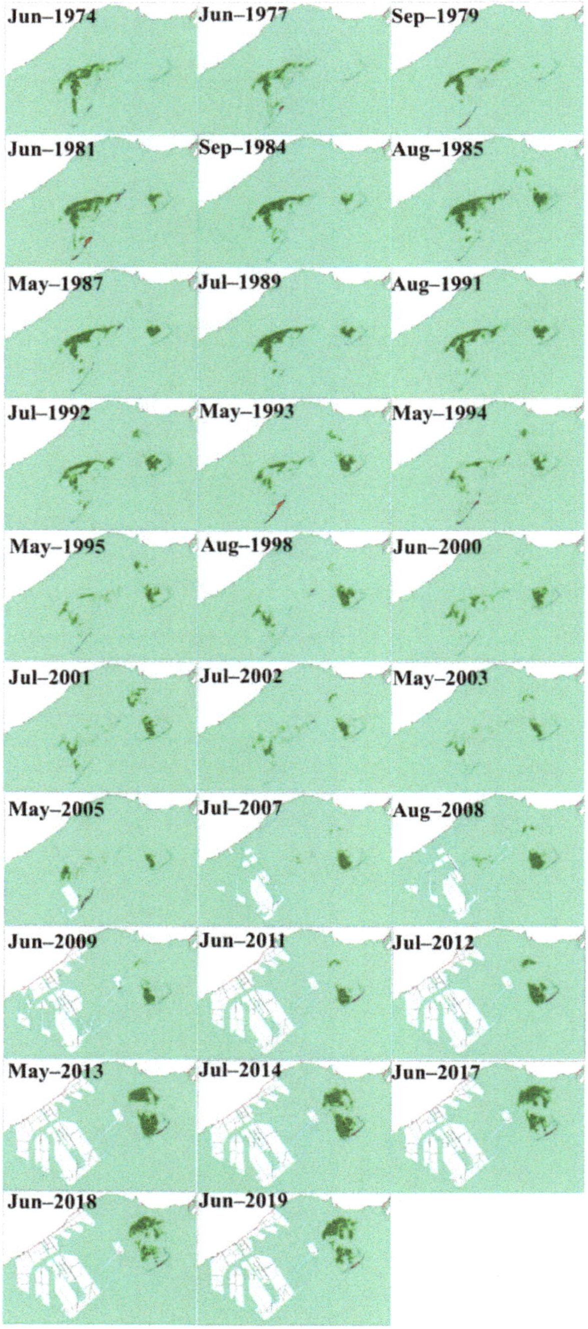

Figure 8. Changes in the spatial distribution of seagrass beds in the study area from 1974 to 2019. In the satellite images, the seagrass beds and unvegetated seawater are represented by green- and cyan-colored pixels, respectively.

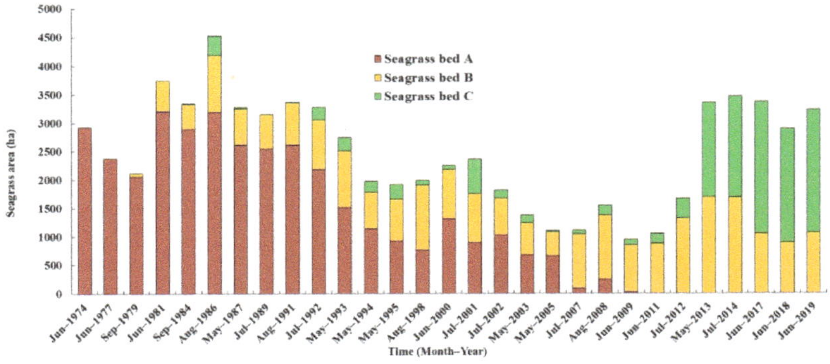

Figure 9. Changes in the areas (ha) of seagrass beds (**A–C**) in the study area from 1974 to 2019 as estimated by a satellite-image-based analysis.

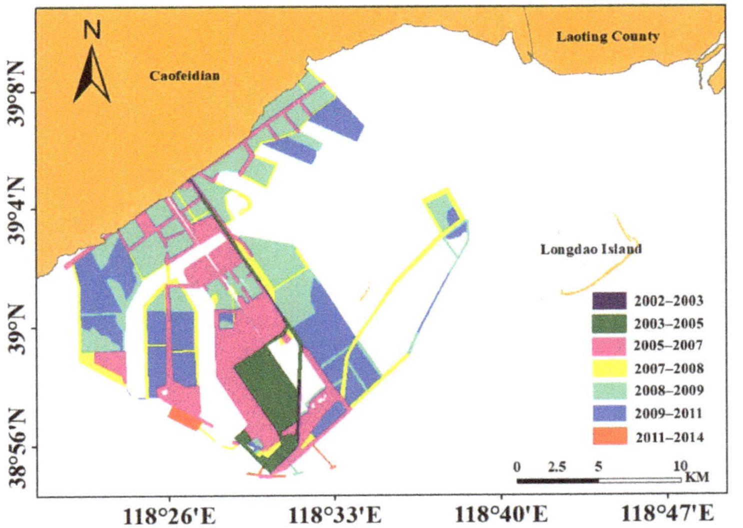

Figure 10. Spatial-temporal distribution of land reclamation in the study area from 2002 to 2014.

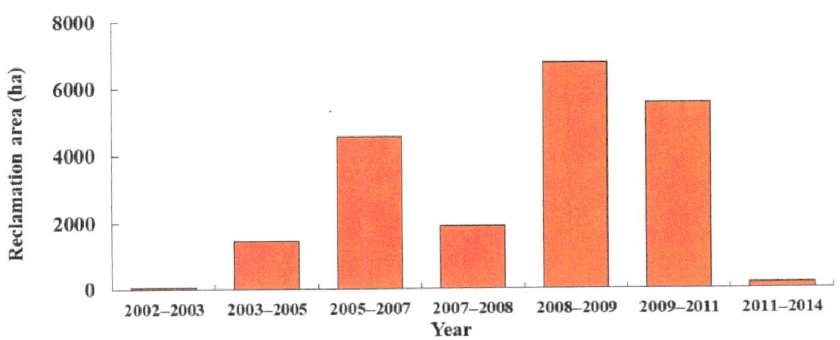

Figure 11. Reclamation area (ha) in the study region from 1974 to 2019 as estimated by satellite-image-based analysis.

3.3. Field Survey of Eelgrass Population

Salinity (31.11–32.08 psu) and pH (8.40-–8.58) varied slightly within sites. The dissolved oxygen levels of sites C (5.43 ± 0.19 mg L^{-1}) and D (5.09 ± 0.22 mg L^{-1}) were higher than those of other sites (2.93 ± 0.61 mg L^{-1}–3.90 ± 0.14 mg L^{-1}). This may occur because the former sites have strong water exchange capacities, being in a channel having a depth of more than 10 m. The chlorophyll content changes within sites were mostly slight (2.00 ± 0.36 μg·L^{-1} m^{-2}–6.49 ± 0.27 μg·L^{-1} m^{-2}).

Total biomass, plant height, and total shoot density varied slightly within the sites on 21 August 2019 (Table 2).

Table 2. Total biomass, belowground biomass, plant height, and shoot density values of eelgrasses at six sites in the study area on 21 August 2019.

Site	Total Biomass (g m^{-2} DW)	Belowground Biomass Ratio to Total Biomass (%)	Plant Height (cm)	Total Shoot Density (shoots m^{-2})
A	189.35 ± 54.05	38.26 ± 8.77	56.71 ± 20.93	152.84 ± 20.36
B	352.85 ± 160.24	37.72 ± 15.38	70.55 ± 15.50	305.68 ± 146.84
C	214.19 ± 37.68	26.82 ± 1.71	78.44 ± 9.37	141.08 ± 35.27
D	263.49 ± 177.27	26.29 ± 8.88	92.11 ± 12.03	211.63 ± 153.74
E	229.74 ± 12.36	31.81 ± 7.39	68.86 ± 10.56	235.14 ± 40.73
F	134.66 ± 18.90	37.37 ± 1.35	42.00 ± 10.68	223.38 ± 53.88

4. Discussion

Remote sensing is a useful tool for long-term monitoring of large seagrassbeds [10,11,48,49] that is able to estimate past changes [9]. In this study, we analyzed 46 years of seagrass distribution changes beginning in 1974 using 29 satellite images and the accuracy of classification was 87.20%, which is acceptable [20].

The seagrass distribution in this study area underwent large changes over the past 46 years. Seagrass distribution changes are usually affected by many factors, and coastal development, degraded water quality, and climate change are the main reasons for the loss of global seagrass beds [8]. In particular, extreme climatic events, such as typhoons and tsunamis, are involved in the large-scale die-off of seagrass beds [30,50,51]. Typhoons greatly influence sediment erosion and burial dynamics, which are the main reasons for seagrass loss [52]. In our study area, there were three storm surges induced by typhoons (September 1992, August 1994, and August 1997) in 1992–1998, accounting for 50% of the storm-surge-related disasters since 1960 [31]. The typhoon-related storm surge that occurred in September 1992 was the largest in the study area since 1949 [31], and it killed 208 people. Consequently, the seagrass bed loss in A from 1992 to 1998 may be related to the three storm surges induced by typhoons, which led to physical disturbances, such as shoot and rhizome uprooting, and damage from sediment erosion or burial. Seagrass beds are able to recover in large disturbed areas [53], and in particular, eelgrass grows quickly through both sexual and asexual reproduction [54]. However, seagrass beds in our study area did not recover and continually decreased. It is possible that the shoots, rhizomes, and seeds in the sediment were completely uprooted and removed by typhoon-related storm surges [11]. In addition, the islands of Caofeidian, providing shelter from oceanic waves and forming the ideal lagoon habitats for *Z. marina*, suffered severe erosion induced by typhoon-related storm surges, waves, and currents [55], and some islands disintegrated, which may have resulted in the severe erosion of seagrass habitats by waves and currents [39]. Lewis [56] reported that seagrasses have substantially retreated in the Tampa Bay (FL, USA) because of the disappearance of longshore bars, which may be particularly vulnerable to storm events. The reduction or disappearance of sand bars in our study area is related to storm surges [57]. According to records, the storm surge wiped out Dawang Island, which was ~10 km from Longdao Island, in 1949 [58].

Reclamation in seagrass habitats might unexpectedly place previously unaffected coastal ecosystems at risk [59]. In the present study, reclamation was the chief reason for seagrass loss in area A, because 205.21 km^2 of the seagrass bed was directly and irreversibly buried by reclaimed land (Figures 10 and 11). However, there has been a large increase in seagrass beds in area C since 2012, when the main land reclamation was completed. Lu et al. [36] reported that the mean velocity of ebbing tides in C decreased by 2 to 10% after reclamation, resulting in a slight deposition, of 0.01–0.10 m, after one year. The reduced wave energies in C may provide a favorable environment for eelgrass expansion through sexual and asexual reproduction. Therefore, the land reclamation, forming an artificial "longshore bar", buffers seagrass beds from wave actions, providing relatively sheltered conditions. Lewis [56] reported that longshore bars maximize the potential for seagrass regrowth. In addition, eelgrasses, as larger species, show greater resilience to reclamation-induced stressors than other smaller species [60]. This allowed the large increase in the seagrass area of C to occur since 2012. During 2003–2011, the seagrass area in C remained at a low level of 109.49 $\pm$ 64.08 ha, likely resulting from excessive sedimentation and increased turbidity caused by reclamation.

Seagrass meadows are declining at an unprecedented rate, and there is a general lack of environmental protection for seagrasses in China. As Australia and the USA did decades ago [61–63], seagrass needs to be included in the Chinese conservation agenda. In recent years, questionnaires indicate that the existing largest eelgrass beds in China are being threatened by trawling, clam harvesting (especially clams sucking), dredging channels, and the construction of culture ponds (Figure 12). Unfortunately, there were no seagrasses at site C (Figure 4) in June 2020 as a result of clam sucking, which removes sediments from a depth of >30 cm, resulting in great damage to seagrass beds. In addition, in recent years, sand excavation for building construction in the study area has also negatively affected seagrass habitats.

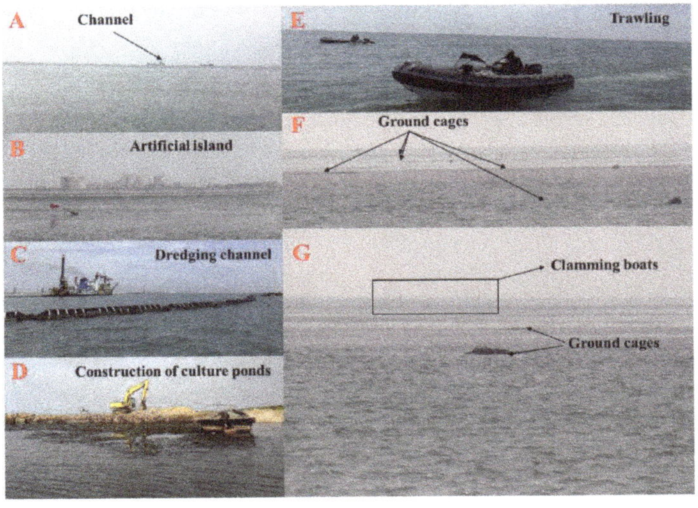

Figure 12. Existing seagrass beds are subjected to threats resulting from artificial factors, especially anthropogenic causes, such as (**A**) boating in channels near seagrass beds, (**B**) construction of artificial islands and industrial activities on the islands, (**C**) dredging of channels, (**D**) construction of culture ponds, (**E**) trawling, (**F**) fish collection using ground cages, and (**G**) harvesting clams using specific boats.

In view of the main threats to the largest eelgrass beds in China, we raise the following suggestions for seagrass protection and management: (1) it should be forbidden to use clam sucking boats for clam harvesting in seagrass beds; (2) it is prohibited to illegally fish within the banned fishing periods, but illegal fishing is especially popular in seagrass beds; (3) it is necessary to increase the fishermen's and policymakers' understanding of the ecological significances of seagrass beds; (4) environmental impact assessments of construction projects should include the effects on seagrass; (5) seagrass reserves need to be established to protect the largest eelgrass beds in China.

5. Conclusions

We described the long-term changes (>40 years) of a seagrass meadow using historical satellite images. We made accuracy assessments of the satellite image classifications using acoustic data and found that the classification accuracy, at 87.20%, was acceptable. The spatial and temporal changes were mainly driven by sea reclamation since 2002, and the large changes during 1992–1998 may have been affected by the storm surges induced by typhoons. However, the existing largest eelgrass beds are being threatened by human activities, and the negative impacts of those factors needs further investigation.

Supplementary Materials: The following are available online at https://www.mdpi.com/2072-4292/13/5/856/s1, Table S1: Remote-sensor data of eelgrass habitats in the study area.

Author Contributions: S.X. (Shaochun Xu): investigation, data curation, writing the original draft, software, visualization; S.X. (Shuai Xu): investigation; Y.Z. (Yi Zhou): funding acquisition, writing, review and editing. S.Y., X.Z., R.G., Y.Z. (Yu Zhang), Y.Q., M.L., Y.Z. (Yunling Zhang): investigation. All authors have read and agreed to the published version of the manuscript.

Funding: This research was funded by the National Key R & D Program of China (2019YFD0901300), the National Science & Technology Basic Work Program (2015FY110600), the Key Research Project of Frontier Sciences of CAS (QYZDB-SSW-DQC041-1), and the Taishan Scholars Program (Distinguished Taishan Scholars).

Institutional Review Board Statement: Not applicable for studies not involving humans.

Informed Consent Statement: Not applicable for studies not involving humans or animals.

Data Availability Statement: The data that support the findings of this study are available from the author upon reasonable request.

Acknowledgments: We would like to thank Zhenhai Zhang, Y.Z. (Yunling Zhang), Chun Li, Min Ye, Zhiwei Zhang, and Chen Cui for their help in the field survey.

Conflicts of Interest: The authors declare that they have no known competing financial interests or personal relationships that could have appeared to influence the work reported in this paper.

References

1. Costanza, R.; d'Arge, R.; de Groot, R.; Farber, S.; Grasso, M.; Hannon, B.; Limburg, K.; Naeem, S.; O'Neill, R.V.; Paruelo, J.; et al. The value of the world's ecosystem services and natural capital. *Nature* **1997**, *387*, 253–260. [CrossRef]
2. Jackson, E.L.; Rowden, A.A.; Attrill, M.J.; Bossey, S.J.; Jones, M.B. The importance of seagrass beds as a habitat for fishery species. *Oceanogr. Mar. Biol.* **2001**, *39*, 269–303.
3. Duffy, J.E. Biodiversity and the functioning of seagrass ecosystems. *Mar. Ecol. Prog. Ser.* **2006**, *311*, 233–250. [CrossRef]
4. Verweij, M.C.; Ivan, N.; Ingmar, H.; Ruseler, S.M.; Mason, P.R.D. Seagrass nurseries contribute to coral reef populations. *Limnol. Oceanogr.* **2008**, *53*, 1540–1547. [CrossRef]
5. Barbier, E.B.; Hacker, S.D.; Kennedy, C.; Koch, E.W.; Stier, A.C.; Silliman, B.R. The value of estuarine and coastal ecosystem services. *Ecol. Monogr.* **2011**, *81*, 169–193. [CrossRef]
6. Tol, S.J.; Coles, R.G.; Congdon, B.C. *Dugong dugon* feeding in tropical Australian seagrass meadows: Implications for conservation planning. *PeerJ* **2016**, *4*, e2194. [CrossRef]
7. Taylor, M.D.; Fry, B.; Becker, A.; Moltschaniwskyj, N. Recruitment and connectivity influence the role of seagrass as a penaeid nursery habitat in a wave dominated estuary. *Sci. Tot. Environ.* **2017**, *584–585*, 622–630. [CrossRef] [PubMed]
8. Waycott, M.; Duarte, C.M.; Carruthers, T.J.; Orth, R.J.; Dennison, W.C.; Olyarnik, S.; Calladine, A.; Fourqurean, J.W.; Heck, K.L.; Hughes, A.R. Accelerating loss of seagrasses across the globe threatens coastal ecosystems. *Proc. Natl. Acad. Sci. USA* **2009**, *106*, 12377–12381. [CrossRef]
9. Mumby, P.J.; Green, E.P.; Edwards, A.J.; Clark, C.D. The cost-effectiveness of remote sensing for tropical coastal resources assessment and management. *J. Environ. Manag.* **1999**, *55*, 157–166. [CrossRef]

10. Lyons, M.B.; Roelfsema, C.M.; Phinn, S.R. Towards understanding temporal and spatial dynamics of seagrass landscapes using time-series remote sensing. *Estuar. Coast. Shelf Sci.* **2013**, *120*, 42–53. [CrossRef]
11. Kim, K.; Choi, J.-K.; Ryu, J.-H.; Jeong, H.J.; Lee, K.; Gil Park, M.; Kim, K.Y. Observation of typhoon-induced seagrass die-off using remote sensing. *Estuar. Coast. Shelf Sci.* **2015**, *154*, 111–121. [CrossRef]
12. Short, F.T.; Short, C.A.; Novak, A. Seagrasses. In *The Wetland Book: II: Distribution, Description and Conservation*; Finlayson, C.M., Milton, G.R., Prentice, R.C., Davidson, N.C., Eds.; Springer Science: Dordrecht, The Netherlands, 2016.
13. Guo, D.; Zhang, P.D.; Zhang, X.M.; Li, W.T.; Zhang, X.J.; Yuan, Y.D. Preliminary investigation and study on seagrass species of inshore areas in Shandong Province. *Trans. Oceanol. Limnol.* **2010**, *2*, 17–21, (In Chinese with English Abstract).
14. Liu, B.J. Foundational Studies on Resource Investigation and Ecological Restoration of Eelgrass in Typical Bays in Shandong. Master's Thesis, The University of Chinese Academy of Sciences, Beijing, China, 2012.
15. Xu, S.C.; Xu, S.; Zhou, Y.; Yue, S.D.; Qiao, Y.L.; Liu, M.J.; Gu, R.T.; Song, X.Y.; Zhang, Y.; Zhang, X.M. Sonar and in situ surveys of eelgrass distribution, reproductive effort, and sexual recruitment contribution in a eutrophic bay with intensive human activities: Implication for seagrass conservation. *Mar. Pollut. Bull.* **2020**, *161*, 111706. [CrossRef] [PubMed]
16. Call, K.A.; Hardy, J.T.; Wallin, D.O. Coral reef habitat discrimination using multivariate spectral analysis and satellite remote sensing. *Int. J. Remote Sens.* **2003**, *24*, 2627–2639. [CrossRef]
17. Hossain, M.S.; Bujang, J.S.; Zakaria, M.H.; Hashim, M. The application of remote sensing to seagrass ecosystems: An overview and future research prospects. International Journal of Remote Sensing. *Int. J. Remote Sens.* **2014**, *36*, 61–113. [CrossRef]
18. Barrell, J.; Grant, J. High-resolution, low-altitude aerial photography in physical geography: A case study characterizing eelgrass (*Zostera marina* L.) and blue mussel (*Mytilus edulis* L.) landscape mosaic structure. *Prog. Phys. Geog.* **2015**, *39*, 440–459. [CrossRef]
19. Barrell, J.; Grant, J.; Hanson, A.; Mahoney, M. Evaluating the complementarity of acoustic and satellite remote sensing for seagrass landscape mapping. *Int. J. Remote Sens.* **2015**, *36*, 4069–4094. [CrossRef]
20. Ferguson, R.L.; Korfmacher, K. Remote sensing and GIS analysis of seagrass meadows in North Carolina, USA. *Aquat. Bot.* **1997**, *58*, 241–258. [CrossRef]
21. Howari, F.M.; Jordan, B.R.; Bouhouche, N.; Wyllie-Echeverria, S. Field and remote-sensing assessment of mangrove forests and seagrass beds in the northwestern part of the United Arab Emirates. *J. Coast. Res.* **2009**, *251*, 48–56. [CrossRef]
22. Barillé, L.; Robin, M.; Harin, N.; Bargain, A.; Launeau, P. Increase in seagrass distribution at Bourgneuf Bay (France) detected by spatial remote sensing. *Aquat. Bot.* **2010**, *92*, 185–194. [CrossRef]
23. Knudby, A.; Nordlund, L. Remote sensing of seagrasses in a patchy multi-species environment. *Int. J. Remote Sens.* **2011**, *32*, 2227–2244. [CrossRef]
24. Short, F.T.; Polidoro, B.; Livingstone, S.R.; Carpenter, K.E.; Bandeira, S.; Bujang, J.S.; Calumpong, H.P.; Carruthers, T.J.B.; Coles, R.G.; Dennison, W.C.; et al. Extinction risk assessment of the world's seagrass species. *Biol. Conserv.* **2011**, *144*, 1961–1971. [CrossRef]
25. Green, E.P.; Short, F.T. *World Atlas of Seagrasses*; University of California Press: Berkeley, CA, USA, 2003.
26. Ferwerda, J.G.; Leeuw, J.; Atzberger, C.; Vekerdy, Z. Satellite-based monitoring of tropical seagrass vegetation: Current techniques and future developments. *Hydrobiologia* **2007**, *591*, 59–71. [CrossRef]
27. Zheng, F.Y.; Qiu, G.L.; Fan, H.Q.; Zhang, W. Diversity, distribution and conservation of Chinese seagrass species. *Biodivers. Sci.* **2013**, *21*, 517–526, (In Chinese with English Abstract).
28. Meinesz, A.; Lefevre, J.R.; Astier, J.M. Impact of coastal development on the infralittoral zone along the southeastern Mediterranean shore of continental France. *Mar. Pollut. Bull.* **1991**, *23*, 343–347. [CrossRef]
29. Wu, Y.; Yang, W.B.; Wang, L.; Zhang, Z.; Li, J.L. Primary study on the spatial-temporal changes and the effect of sea reclamation in Caofeidian based on remote sensing. *Trans. Oceanol. Limnol.* **2013**, *1*, 153–158, (In Chinese with English Abstract).
30. Orth, R.J.; Carruthers, T.J.; Dennison, W.C.; Duarte, C.M.; Fourqurean, J.W.; Heck, K.L.; Hughes, A.R.; Kendrick, G.A.; Kenworthy, W.J.; Olyarnik, S. A global crisis for seagrass ecosystems. *Bioscience* **2006**, *56*, 987–996. [CrossRef]
31. Zheng, L.P.; Guo, L.X.; Zhang, W.Y.; Hou, S.X. Analysis on the causes of storm surge in coastal area of Tangshan. *Ann. Meet. China Meteorol. Soc.* **2006**, 1487–1489. (In Chinese)
32. Li, B.L.; Zhao, Y.H.; Yang, K.; Zhao, B. Current environmental status and the protection of the Bohai Sea. *Ocean Dev. Manag.* **2016**, *33*, 59–62.
33. Gao, S.M.; Li, Y.F.; An, F.T.; Li, F.X. The formation of sand bars on the Luanhe River Delta and the change of the coast line. *Acta Oceanol. Sin.* **1981**, *2*, 102–113, (In Chinese with English Abstract).
34. Wang, Y.; Fu, G.H.; Zhang, Y.Z. River-sea interactive sedimentation and plain morphological evolution. *Quat. Sci.* **2007**, *27*, 674–689, (In Chinese with English Abstract).
35. Zhang, R.S.; Li, K.P. Evolution of tidal inlet-drainage basin system on Luanhe River Delta. *Ocean Eng.* **1996**, *14*, 45–52, (In Chinese with English Abstract).
36. Lu, Y.J.; Zuo, L.Q.; Ji, R.Y.; Zhang, J.S. Effect of development of Caofeidian harbor area in Bohai bay on hydrodynamic sediment environment. *Adv. Water Sci.* **2007**, *18*, 793–800, (In Chinese with English Abstract).
37. Yin, Y.H. Thoughts on large area reclamation of Caofeidian shoal in Tangshan, Hebei Province. *Mar. Geol. Lett.* **2007**, *23*, 1–10. (In Chinese)
38. Gu, J.Q.; Zhang, B.H. Evolution, development and utilization of sedimentary Sand Island in Eastern Hebei Province. *Geogr. Territ. Res.* **1997**, *13*, 36–40. (In Chinese)

39. Tian, H.L.; Liu, X.H.; Shi, Y.J.; Li, Q.C.; Wang, Y.X. Study on the morphologic change of Caofeidian islands. *Mar. Sci. Bull.* **2015**, *34*, 695–702, (In Chinese with English Abstract).
40. Xu, S.C.; Wang, P.M.; Zhou, Y.; Wang, F.; Zhang, X.M.; Gu, R.T.; Xu, S.; Yue, S.D.; Zhang, Y.; Suonan, Z.X. Sheaths of *Zostera marina* L. as ecological indicators of shoot length and the elemental stoichiometry of aboveground tissues. *Mar. Pollut. Bull.* **2020**, *160*, 111545. [CrossRef]
41. Zimmerman, R.C.; Dekker, A.G. Aquatic optics: Basic concepts for understanding how light affects seagrasses and makes them measurable from space. In *Seagrasses: Biology, Ecology, and Conservation*; Larkum, A.W.D., Orth, R.J., Duarte, C.M., Eds.; Springer: Dordrecht, The Netherlands, 2006.
42. Wicaksono, P.; Hafizt, M. Mapping seagrass from space: Addressing the complexity of seagrass LAI mapping. *Eur. J. Remote Sens.* **2013**, *46*, 18–39. [CrossRef]
43. Scaramuzza, P.; Micijevic, E.; Chander, G. SLC Gap-Filled Products Phase One Methodology. 2004. Available online: http://landsat.usgs.gov/documents/SLC_Gap_Fill_Methodology.pdf (accessed on 1 January 2020).
44. Research Systems Inc. *ENVI FLAASH User's Guide*; ENVI FLAASH Version 4.1; Research Systems Inc.: Norwalk, CT, USA, 2004; pp. 1–80.
45. Foody, G.M. Status of land cover classification accuracy assessment. *Remote Sens. Environ.* **2002**, *80*, 185–201. [CrossRef]
46. Lepš, J.; Šmilauer, P. *Multivariate Analysis of Ecological Data Using CANOCO*; Cambridge University Press: Cambridge, UK, 2003.
47. R Core Team. *R: A Language and Environment for Statistical Computing*; R Foundation for Statistical Computing: Vienna, Austria, 2016.
48. Cuttriss, A.K.; Prince, J.B.; Castley, J.G. Seagrass communities in southern Moreton Bay, Australia: Coverage and fragmentation trends between 1987 and 2005. *Aquat. Bot.* **2013**, *108*, 41–47. [CrossRef]
49. Mohammad, S.H.; Mazlan, H.; Japar, S.B.; Muta, H.Z.; Aidy, M.M. Assessment of the impact of coastal reclamation activities on seagrass meadows in Sungai Pulai estuary, Malaysia, using Landsat data (1994–2017). *Int. J. Remote Sens.* **2018**, *40*, 3571–3605.
50. Seddon, S.; Connolly, R.M.; Edyvane, K.S. Large-scale seagrass dieback in northern Spencer Gulf, South Australia. *Aquat. Bot.* **2000**, *66*, 297–310. [CrossRef]
51. Skinner, M.A.; Courtenay, S.C.; McKindsey, C.W. Reductions in distribution, photosynthesis, and productivity of eelgrass *Zostera marina* associated with oyster *Crassostrea virginica* aquaculture. *Mar. Ecol. Prog. Ser.* **2013**, *486*, 105–119. [CrossRef]
52. Cabaço, S.; Santos, R.; Duarte, C.M. The impact of sediment burial and erosion on seagrasses: A review. Estuar. *Coast. Shelf Sci.* **2008**, *79*, 354–366. [CrossRef]
53. Greve, T.M.; Krause-Jensen, D.; Rasmussen, M.B.; Christensen, P.B. Means of rapid eelgrass (*Zostera marina* L.) recolonisation in former dieback areas. *Aquat. Bot.* **2005**, *82*, 143–156. [CrossRef]
54. Rasheed, M.A.; McKenna, S.A.; Carter, A.B.; Coles, C.R. Contrasting recovery of shallow and deep water seagrass communities following climate associated losses in tropical north Queensland, Australia. *Mar. Pollut. Bull.* **2014**, *83*, 491–499. [CrossRef] [PubMed]
55. Gao, W.M.; Yang, J.X. Analysis on evolution of offshore bars in the north area of the Bohai Sea Bay using remote sensing. *Geogr. Geo Inf. Sci.* **2006**, *22*, 41–44, (In Chinese with English Abstract).
56. Lewis, R.R. The potential importance of the longshore bar system to the persistence and restoration of Tampa Bay seagrass meadows. In Proceedings of the Conference on Seagrass Management: It's Not Just Nutrients, St. Petersburg, FL, USA, 22–24 August 2000.
57. Hebei Coastal Zone Resources Editorial Committee. *Coastal Resources of Hebei Province;* Hebei Province Science and Technology Press: Shijiazhuang, China, 1988; pp. 94–95. (In Chinese)
58. Gong, R.; Mu, Z.Y. Harm and control of storm surge in Hebei Coastal Area. *Geogr. Territ. Res.* **1994**, *10*, 21–24. (In Chinese)
59. Ghazali, N.; Hisham, M. Coastal Erosion and Reclamation in Malaysia. *Aquat. Ecosyst. Health Manag.* **2006**, *9*, 237–247. [CrossRef]
60. Erftemeijer, P.L.A.; Lewis, R.R.R. Environmental impacts of dredging on seagrasses: A review. *Mar. Pollut. Bull.* **2006**, *52*, 1553–1572. [CrossRef]
61. Schemske, D.W.; Husband, B.C.; Ruckelshaus, M.H.; Goodwillie, C.; Parker, I.M.; Bishop, J.G. Evaluating approaches to the conservation of rare and endangered plants. *Ecology* **1994**, *75*, 584–606. [CrossRef]
62. Fonseca, M.S.; Kenworthy, W.J.; Thayer, G.W. Guidelines for the conservation and restoration of seagrasses in the United States and adjacent Waters. In *Decision Analysis Series No. 12. U.S. Department of Commerce—NOAA, Maryland*; NsCO Program Ed.; NOAA/National Centers for Coastal Ocean Science: Silver Spring, MD, USA, 1998.
63. Lee-Long, W.J.; Coles, R.G.; McKenzie, L.J. Issues for seagrass conservation management in Queensland. *Pac. Conserv. Biol.* **2000**, *5*, 321–328. [CrossRef]

remote sensing

Communication

Extensive Marine Heatwaves at the Sea Surface in the Northwestern Pacific Ocean in Summer 2021

Hiroshi Kuroda [1,*] and Takashi Setou [2]

1 Fisheries Resources Institute (Kushiro), Japan Fisheries Research and Education Agency, Kushiro 085-0802, Hokkaido, Japan
2 Fisheries Resources Institute (Yokohama), Japan Fisheries Research and Education Agency, Yokohama 236-8648, Kanagawa, Japan; setou@affrc.go.jp
* Correspondence: kurocan@affrc.go.jp; Tel.: +81-154-92-1723

Abstract: In July–August 2021, intense marine heatwaves (MHWs) occurred at the sea surface over extensive areas of the northwestern Pacific Ocean, including the entire Sea of Japan and part of the Sea of Okhotsk. In extent and intensity, these MHWs were the largest since 1982, when satellite measurements of global sea surface temperatures started. The MHWs in summer 2021 were observed at the sea surface and occurred concomitantly with a stable shallow oceanic surface boundary layer. The distribution of the MHWs was strongly related to heat fluxes at the sea surface, indicating that the MHWs were generated mainly by atmospheric forcing. The MHWs started to develop after around 10 July, concurrent with an extreme northward shift of the atmospheric westerly jet. The MHWs developed rapidly under an atmospheric high-pressure system near the sea surface, associated with a northwestward expansion of the North Pacific Subtropical High. The MHWs exhibited peaks around 30 July to 1 August. Subsequently, following the southward displacement of the westerly jet, the MHWs weakened and then shrank abruptly, synchronously with rapid deepening of the oceanic surface boundary layer. By 18 August, the MHWs had disappeared.

Keywords: marine heatwaves; sea surface temperatures; summer 2021; northwestern Pacific Ocean; westerly jet; North Pacific Subtropical High

Citation: Kuroda, H.; Setou, T. Extensive Marine Heatwaves at the Sea Surface in the Northwestern Pacific Ocean in Summer 2021. *Remote Sens.* **2021**, *13*, 3989. https://doi.org/10.3390/rs13193989

Academic Editors: Chung-Ru Ho, Antony K. Liu and Xiaofeng Li

Received: 6 September 2021
Accepted: 3 October 2021
Published: 5 October 2021

1. Introduction

As generally defined, an extreme event occurs when the values of a weather or climate variable either exceed a threshold near the upper end of an observed range, or are lower than a threshold near the lower end [1]. Many extreme weather and climate events result from natural climate variability, but some extreme events have occurred as a result of anthropogenic climate change and ongoing global warming (e.g., [2–4]). Marine heatwaves (MHWs) are an example of an extreme climate in an oceanic system [5–7]. In the last century, from 1925 to 2016, the global average frequency and duration of MHWs increased by 34% and 17% respectively, resulting in a 54% increase in annual marine heatwave days [8]. According to Oliver [9], an increase in mean sea surface temperature (SST), rather than its variability, has been the dominant driver of the increasing frequency of MHW days over approximately two-thirds of the world's ocean, and it has been the dominant driver of changes in MHW intensity over approximately one-third of the ocean. Moreover, it has been projected that these historical trends in MHW properties will continue over the coming decades under global warming caused by anthropogenic climate change [10]. Furthermore, MHWs have devastatingly affected marine ecosystems in the past, and they are expected to do so in the future, through abrupt changes in biological habitat, mortality, reproduction, community structure, and so on [10–15].

Most published studies of MHWs in the North Pacific Ocean have investigated conditions in northeastern waters. A MHW known as "the blob," which occurred persistently in 2014–2016 in the northeastern Pacific, is especially famous (e.g., [16]). In contrast, a search for studies conducted in the northwestern Pacific, particularly east of Japan, with the keyword "MHW," yielded only three scientific studies [17–19]: one conducted in subtropical waters [17] and the others in subarctic waters [18,19]. Miyama et al. [18] reported the occurrence of MHWs in summer during 2010–2016 near Hokkaido, northern Japan, which they attributed to warm mesoscale eddies that had become detached from the subtropical Kuroshio Extension and had been advected into the subarctic Oyashio region every summer of that period. These MHWs, which were localized off Hokkaido, also affected the social economy through their effects on fisheries, at least, for instance, fishing grounds of Pacific saury shifted farther offshore [20], and the catch of yellowtail in coastal waters by fixed fishing nets increased [18].

Extensive and intense MHWs occurred in July–August 2021 over the northwestern Pacific, including the entire Sea of Japan and part of the Sea of Okhotsk. In extent and intensity, these MHWs were the largest since 1982, when satellite measurements of global sea surface temperatures (SSTs) started. As the first scientific report, this study aims to describe features of these MHWs together with the accompanying atmospheric conditions and to promote future studies.

2. Materials and Methods

We analyzed gridded "Merged satellite and in situ data Global Daily Sea Surface Temperature" (MGDSST) datasets, comprising data from 1 January 1982 to 31 August 2021. These datasets, which were compiled by the Japan Meteorological Agency, have a spatial resolution of 0.25° (latitude) × 0.25° (longitude) and a daily temporal resolution [21].

We also used the "Roemmich-Gilson Argo Climatology," a global monthly dataset of seawater temperature in the subsurface created by the Scripps Institute of Oceanography [22] with a horizontal resolution of 1° (latitude) × 1° (longitude). We included data from January 2004 to July 2021 in our analysis.

We also used global atmospheric products at multiple vertical levels from the Japanese 55-year Reanalysis dataset (JRA-55), the second global atmospheric reanalysis project of the Japan Meteorological Agency [23]. These data have a horizontal resolution of 1.25° (latitude) × 1.25° (longitude). In our analysis, we used 3-hourly or 6-hourly outputs for the period from 1 January 1990 to 31 August 2021.

To infer the oceanic surface boundary layer conditions during the period of MHWs in 2021, numerical experiments were conducted with realistic ocean circulation models that are part of the operational ocean forecast system of the Japan Fisheries Research and Education Agency, "FRA-ROMS" [24]. The ocean circulation models, which are based on the Regional Ocean Modeling System (ROMS) [25–27], consist of a 1/2° resolution parent model for the North Pacific and a 1/10° resolution child model for the northwestern Pacific, connected by one-way nesting to simulate dominant basin-scale and mesoscale variations around Japan [28]. Daily mean reanalysis data for 1 July 2021, assimilated by using a three-dimensional variational scheme, were utilized for the initial conditions of 2-month hindcast and sensitivity experiments conducted without further data assimilation. To sequentially estimate momentum and heat fluxes at the sea surface during each model run, 31 sets of atmospheric elements near the sea surface (i.e., wind vectors at 10 m, air temperature and specific humidity at 2 m above the sea surface, air pressure, precipitation, and net shortwave and downward longwave radiation at the sea surface) were derived from the 3-hourly outputs of the JRA-55 system for July–August of 1990–2019 and 2021.

Climatological daily means were estimated for each element (e.g., SST, air temperature, sea level pressure), except subsurface ocean temperature, by averaging their data for 1990–2019. This 30-year averaging period, which was selected by referring to Hobday et al. [5], does not include the 1980s, during which a typical cold SST regime prevailed around Japan. Therefore, because the climatological averaging period excluded data from the 1980s, MHWs in recent years were less frequently detected than would have been the case had the selected climatological period included the 1980s. Further, because of limited data availability, climatological monthly means of the Argo-based subsurface ocean temperature were estimated by averaging data over just 15 years (2004–2018) instead of 30 years. Throughout this study, an anomaly was defined as a difference from the climatological mean.

The MHWs were detected from daily SST data and categorized according to intensity following Hobday et al. [6]. Intensity categories were defined based on the SST difference, *Tdiff*, between the 30-year climatological average (*Tavg*) and the 90th percentile value for each grid point. If the SST on a given day was $\geq(Tavg + N \times Tdiff)$ and $<(Tavg + (N + 1) \times Tdiff)$, it was categorized into intensity category *N*. In addition, the MHW index was defined as $(SST - Tavg)/Tdiff$. Note that in this paper, the duration of MHWs is referred to as their age.

3. Results

3.1. Surface Ocean Conditions

3.1.1. Sea Surface Temperatures

Figure 1a,b show intensity and age respectively, of MHWs on 30 July 2021, when the MHWs were at their maximum strength and extent, as explained later. The MHWs in the northwestern Pacific occurred within a zonal band, between 38°N and 50°N and extending from 120°E to 170°W, including the marginal seas, that is, the northern part of the Yellow Sea, the whole of the Sea of Japan, and the southern half of the Sea of Okhotsk. The MHW intensity on 30 July was classified into categories 1–4. Some areas of the MHWs corresponded to large SST anomalies > 6 °C (Figure 1d). The MHWs exhibited maximum intensity (i.e., category 4) on 30 July in the northern part of the Sea of Japan and tended to weaken gradually toward the east. The age of the MHWs in the northwestern Pacific on 30 July was within 1–30 days: it was longest (~30 days) in the northern part of the Sea of Japan and tended to be shorter further east.

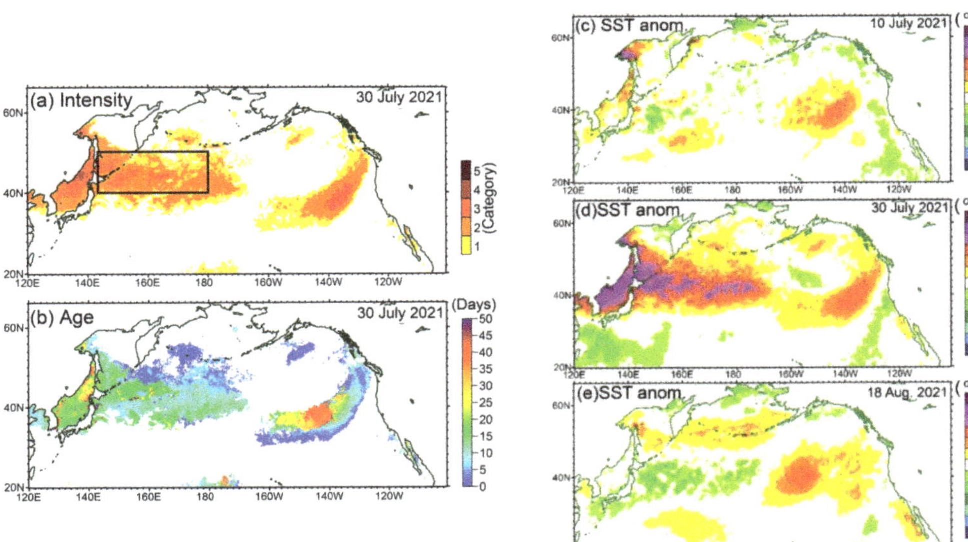

Figure 1. (**a**) Intensity and (**b**) age of marine heatwaves (MHWs) on 30 July 2021. In (**a**), the rectangle indicates the area within 143°E–180° and 40°–50°N, the focal area of our analysis. SST anomalies are shown on (**c**) 10 July, (**d**) 30 July, and (**e**) 18 August 2021.

It is worth noting that intense MHWs appeared to start on the northwestern shelf of the Sea of Okhotsk around 10 July 2021 and subsequently expand from there (Figure 1c,d). However, the SST data analyzed in this study did not strictly discriminate between SSTs in the presence or absence of sea ice; thus, if sea ice remained on the northwestern shelf at the beginning of summer, the estimated MHWs would include large uncertainties. Therefore, whether the above feature is a real phenomenon is questionable.

We focused on MHWs east of Japan in the northwestern Pacific and excluded from our analysis the Sea of Japan, where MHWs that were more intensive than those in 2021 occurred locally after 1982 (e.g., January 2020). Thus, we estimated daily time series of SST anomalies and the MHW index averaged over the rectangular area within 143°E–180° and 40°–50°N (Figure 2). Both time series show historical maxima in July–August 2021 (Figure 2a). The spatial mean of the daily MHW index exceeded 1.0 on a total of 207 days during 1982–2021, and 23 of these days were in July–August 2021. The spatial mean also exceeded 1.5 on 15 days during 1982–2021, of which 14 were in July–August 2021. These results indicate that the MHWs in summer 2021 were the most intense and extensive in the northwestern Pacific during the last four decades.

The transition to MHWs in summer 2021 can be interpreted as follows (Figure 2b). The MHWs started to develop on around 10 July 2021 (Figure 1c), when the spatial mean SST anomaly was close to zero (Figure 2b). The MHWs reached maximum intensity and extent around 30 July to 1 August (Figure 1d), and then immediately started to weaken. By 18 August (Figure 1e), the spatial mean SST anomaly was again near zero (Figure 2b) and the MHWs had almost entirely disappeared.

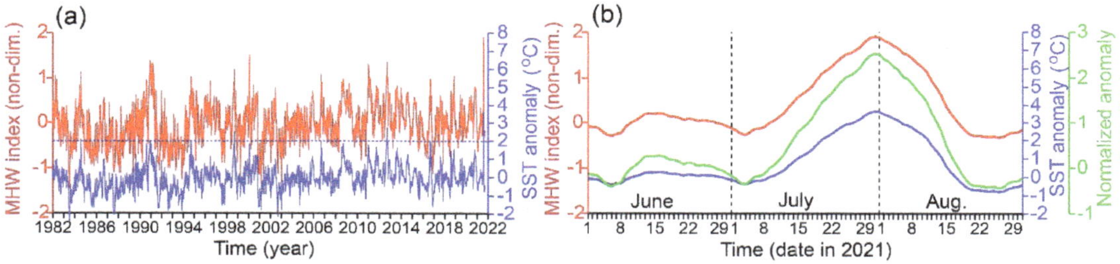

Figure 2. Time series of the daily SST anomaly (blue lines, corresponding to the right blue axes) and MHW index (red lines, corresponding to the left red axes) during (**a**) 1982–2021 and (**b**) June–August 2021. The SST anomaly and MHW index were estimated at each grid point and then spatially averaged over 143°E–180° and 40°–50°N (rectangle in Figure 1a). As indices for evaluating the variability of the daily SST anomaly, the +3 standard deviation line (dashed blue line) is shown in (**a**), and the daily SST anomaly normalized by the standard deviation at each grid point and spatially averaged over the focal area (green line, corresponding to the right green axis) is shown in (**b**).

3.1.2. Heat Fluxes at the Sea Surface

Net heat flux anomalies at the sea surface during 10–30 July 2021 were mostly positive within the focal area, that is, the central area of MHWs (Figure 3a); thus, the sea surface was heated by the atmosphere, relative to the climatology. The distribution of net heat flux anomalies (Figure 3a) also correlated with that of SST anomaly differences between 10 and 30 July (Figure 3b) ($r = 0.69$), suggesting that the MHWs were attributable primarily to increased net heat fluxes at the sea surface over the northwestern Pacific.

The net heat fluxes were decomposed into four components: net shortwave radiation, net longwave radiation, latent heat, and sensible heat (Figure 4). Within our focal area, net shortwave radiation anomalies exhibited three positive maxima, centered around (145°E, 40°N), (170°E, 40°N), and (160°E, 45°N) (Figure 4b). These three positive maxima corresponded spatially to three negative minima of the total cloud cover and low cloud cover anomalies (see Section 3.3.1). Further, the spatial patterns of positive and negative net shortwave radiation anomalies in the focal area (Figure 4b) were roughly opposite to those of the other three components (Figure 4d,f,h), indicating that negative net shortwave radiation anomalies in the focal area were compensated by positive anomalies of the other three components. As a result, net heat flux anomalies at the sea surface were positive over the area of MHWs (Figure 3a).

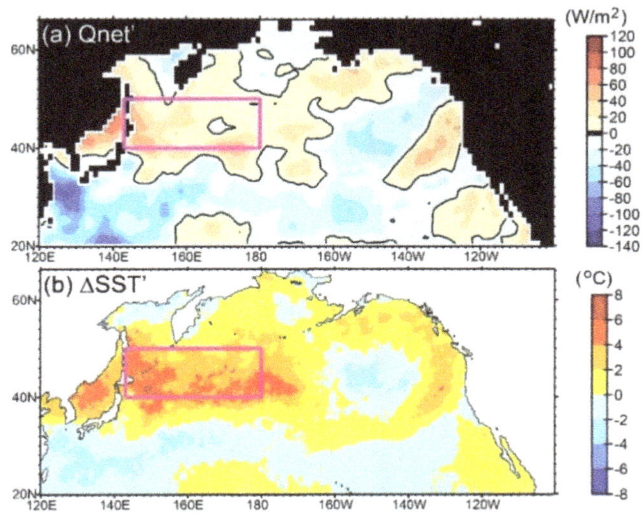

Figure 3. (**a**) Net heat flux anomalies (Qnet′) at the sea surface during 10–30 July 2021. (**b**) SST anomaly differences between 10 and 30 July 2021 (ΔSST′). The pink rectangle indicates the focal area (143°E–180°, 40°–50°N).

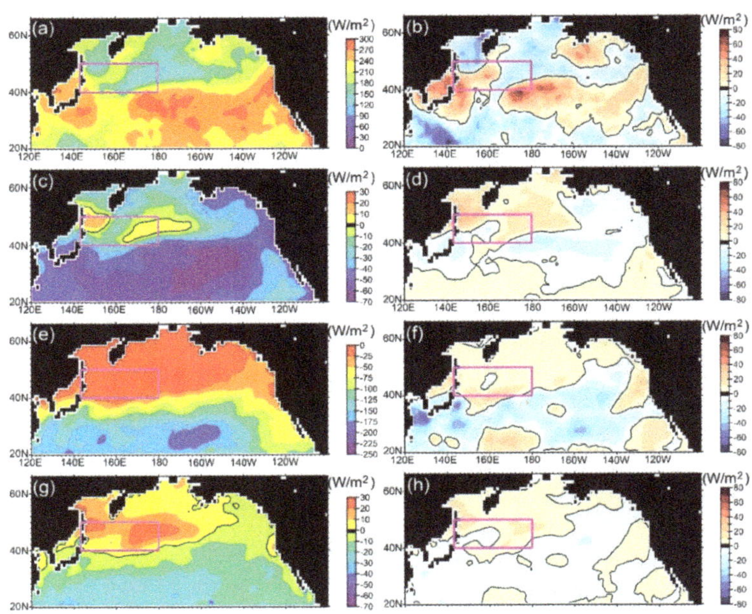

Figure 4. (**a**) Net shortwave radiation, (**c**) net longwave radiation, (**e**) latent heat, and (**g**) sensible heat flux averaged over 10–30 July 2021. Anomalies of (**b**) net shortwave radiation, (**d**) net longwave radiation, (**f**) latent heat, and (**h**) sensible heat flux.

Note that positive anomalies of net longwave radiation and sensible heat around the focal area (Figure 4d,h) corresponded to positive values of the respective components (Figure 4c,g). The positive values thus indicate that the sea surface was actually heated. Although MHWs with high SSTs were expected to cause large upward longwave radiation (i.e., negative heat fluxes), the net longwave radiation was nevertheless positive in some areas within the MHW area, for example, around the northwestern corner of the focal area (Figure 4c). This result suggests that strong downward longwave radiation from clouds exceeded the large upward longwave radiation from the sea surface. In fact, as described in Section 3.3.1, dense clouds tended to cover areas with positive net longwave radiation (Figure 4c). In any case, it should be emphasized that there were inter-regional differences in the origin of positive net heat flux anomalies over the area of MHWs; that is, the SST heating process differed inter-regionally within the area of MHWs.

3.2. Subsurface Ocean Conditions

3.2.1. Subsurface Sea Temperatures

In July 2021, temperatures in the subsurface at 20 m depth, just below the oceanic surface boundary layer, exhibited positive anomalies, mostly over the whole of the focal area (Figure 5c), although relatively large positive anomalies (>1 °C) were limited to its southern half. These relatively large positive anomalies were also distributed continuously after March 2021 (e.g., Figure 5a–c). In addition, subsurface temperatures at 20 m depth spatially averaged over the focal area reached a local maximum not in July 2021 but in May 2021 (Figure 5d), after which they decreased slightly until July 2021. Moreover, the positive anomalies in July 2021 were not the largest during 2004–2021. These features of the subsurface temperature were thus clearly different from the corresponding SST features. Hence, the main factors controlling the positive temperature anomalies in July 2021 seemed to differ between 20 m depth and the sea surface, and the MHWs that developed dramatically in summer 2021 were mostly limited to the sea surface (<20 m depth). With respect to this point, however, the positive subsurface temperature anomalies apparent after March 2021 might have contributed to the formation or maintenance of stratified conditions in the oceanic surface boundary layer in July 2021, as described in the next section.

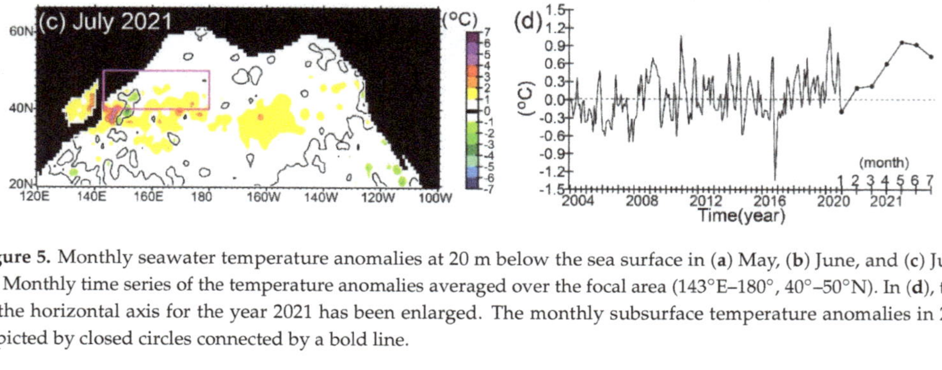

Figure 5. Monthly seawater temperature anomalies at 20 m below the sea surface in (**a**) May, (**b**) June, and (**c**) July 2021. (**d**) Monthly time series of the temperature anomalies averaged over the focal area (143°E–180°, 40°–50°N). In (**d**), the scale of the horizontal axis for the year 2021 has been enlarged. The monthly subsurface temperature anomalies in 2021 are depicted by closed circles connected by a bold line.

3.2.2. Oceanic Surface Boundary Layer

In this study, the depth of the oceanic surface boundary layer, defined as the shallowest depth where the bulk Richardson number is equal to its critical value, was derived from model simulations instead of the more conventional mixed layer depth (i.e., the depth at which the water density in the upper layer exceeds a threshold value, typically 0.125 kg m^{-3}, relative to the 10 m density), primarily because the mixed layer was very thin during summer 2021 and could not be accurately evaluated by the conventional mixed layer depth.

Hindcast and sensitivity experiments conducted with realistic ocean circulation models revealed that thickness anomalies of the oceanic surface boundary layer during 10–30 July 2021 were mainly negative within the focal area, except for an area of positive anomalies that extended southwest–northeast, from (165°E, 40°N) to (180°, 46°N) (Figure 6a). The thickness of the oceanic surface boundary layer remained small in the focal area during the period from 13 July to 5 August 2016 (Figure 6b), during most of which MHWs were rapidly developing (Figure 2b). From 7 August, however, the thickness increased abruptly, at the same time that the MHWs were weakening rapidly (Figure 2b). These results suggest that positive net heat fluxes (Figure 3a) and relatively weak wind stress (as explained in Section 3.3.1) near the sea surface stably maintained strong stratification, suppressed vertical convection, and accelerated the SST increase, that is, the rapid development of MHWs.

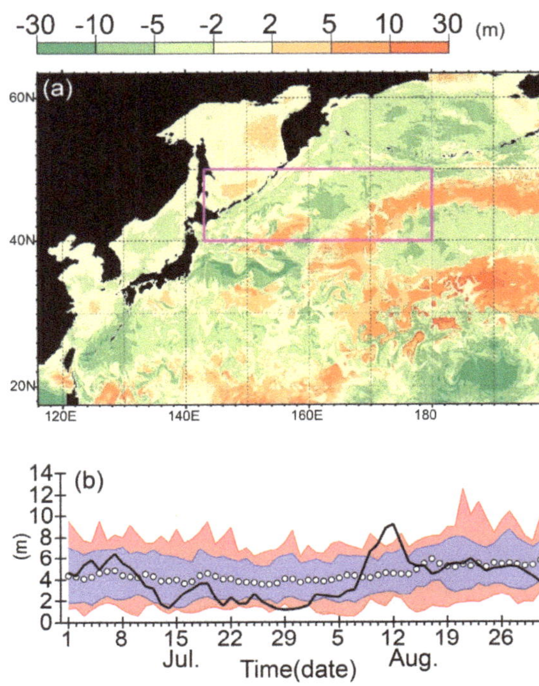

Figure 6. (**a**) Thickness anomalies of the oceanic surface boundary layer during 10–30 July 2021, estimated by a mixed layer model based on K-profile parameterization in the 1/10° ocean circulation model. Climatological means were estimated by averaging the thickness over 30 sensitivity simulations based on external forcings in 1990–2019. (**b**) Time series of oceanic surface boundary layer thickness spatially averaged over the focal area (143°E–180° and 40°–50°N; the purple rectangle in (**a**)). The bold black line shows the daily time series in 2021. Open circles denote daily means of the 30 simulations. Blue shading shows the 10th to 90th percentile range. Red shading shows the range, beyond these percentiles, between the minimum and maximum.

3.3. Atmospheric conditions

3.3.1. Conditions Near the Sea Surface

Sea level pressure anomalies during 10–30 July 2021 were widely positive in the northwestern Pacific, except in the northern Sea of Okhotsk and the western Bering Sea (Figure 7b). These positive anomalies were attributable to a northwestward expansion of the North Pacific Subtropical High (Figure 7a). The distribution of sea level pressure anomalies was also similar to MHW distribution (Figure 1a), although the distribution of air temperature anomalies near the sea surface (Figure 7c) more closely resembled the MHW distribution (Figure 1a). Local maxima of the air temperature anomalies exceeded 3 °C.

As mentioned in Section 3.1.2, the distributions of positive and negative anomalies of total and low cloud cover (Figure 7e,f) were almost opposite to those of positive and negative net shortwave radiation anomalies (Figure 4b). In addition, in the focal area, the positive anomalies of total and low cloud cover (Figure 7e,f) corresponded to both positive values and positive anomalies of net longwave radiation (Figure 4c,d), where strong downward longwave radiation from clouds to the sea surface exceeded the large upward longwave radiation attributed to MHWs. Moreover, as mentioned in Section 3.2.2, negative or small positive (<0.02 Pa) wind stress magnitude anomalies were distributed in the western half of the focal area (Figure 7d). In contrast, in the eastern half of the focal area, relatively large positive anomalies (0.02–0.06 Pa) were distributed in an area extending from southwest to northeast, corresponding to the area of positive oceanic surface boundary layer anomalies (Figure 6a).

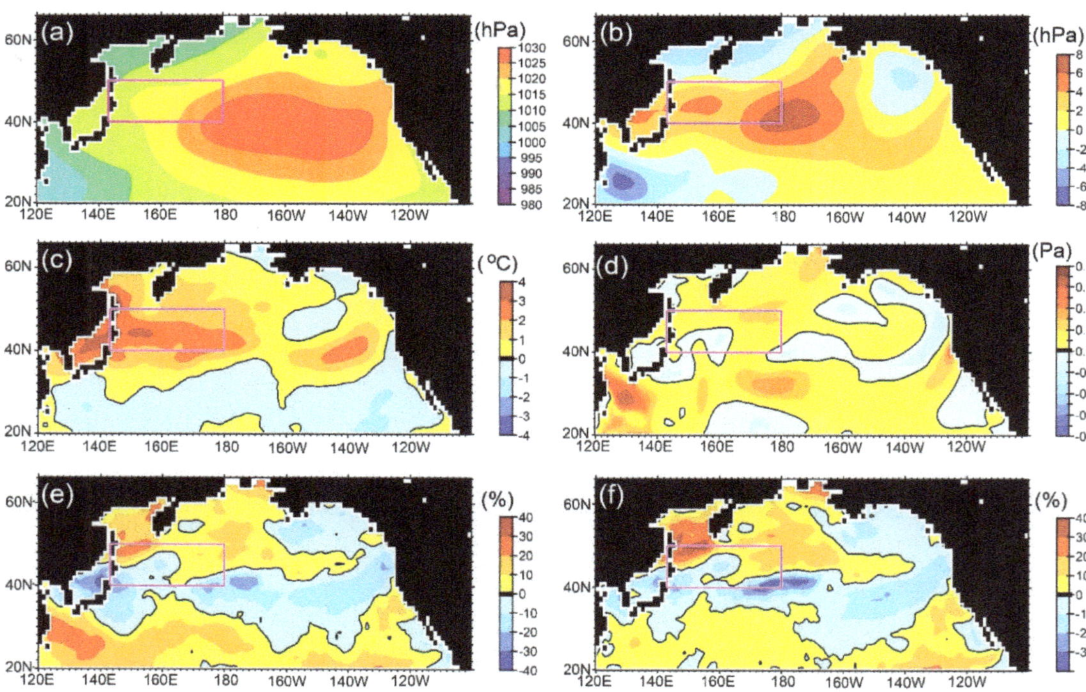

Figure 7. (**a**) Sea level pressure during 10–30 July 2021. Anomalies during 10–30 July 2021: (**b**) sea level pressure, (**c**) air temperature at 2 m above the sea surface, (**d**) magnitude of wind stress at 10 m above the sea surface, (**e**) total cloud cover, and (**f**) low cloud cover.

3.3.2. Variation of the Westerly Jet at 200 hPa

Here, we focus on the westerly jet variation at 200 hPa as a potential driver or trigger of the atmospheric variations and MHWs in summer 2021. Maps of eastward wind velocity at 200 hPa averaged over 10–30 July (Figure 8a) revealed that the strong eastward wind (>20 m s^{-1}) associated with the westerly jet [29] migrated northward, largely to the north of Japan, and skirted the focal area. The position of the velocity maximum reached 55°N at around 140°E. The velocity maximum between 120°E and 180° during 10–30 July 2021 was clearly located north of the climatological mean position (Figure 8b).

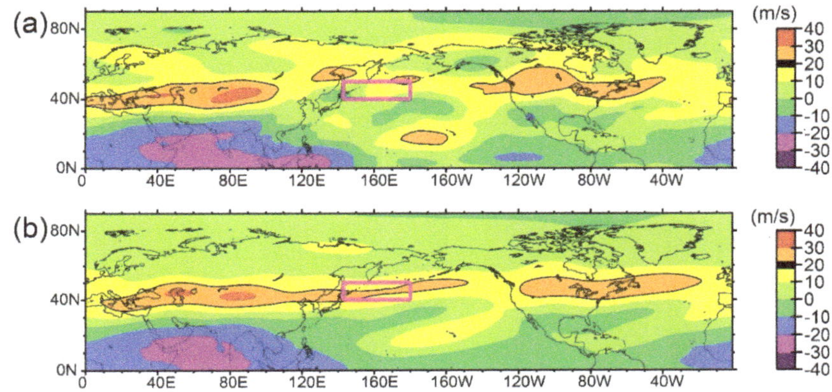

Figure 8. Eastward wind velocity at 200 hPa averaged over 10–30 July (**a**) in 2021 and (**b**) 1990–2019. The purple rectangle indicates the focal area (143°E–180°, 40°–50°N). Strong velocities of >20 m s^{-1} are associated with the westerly jet.

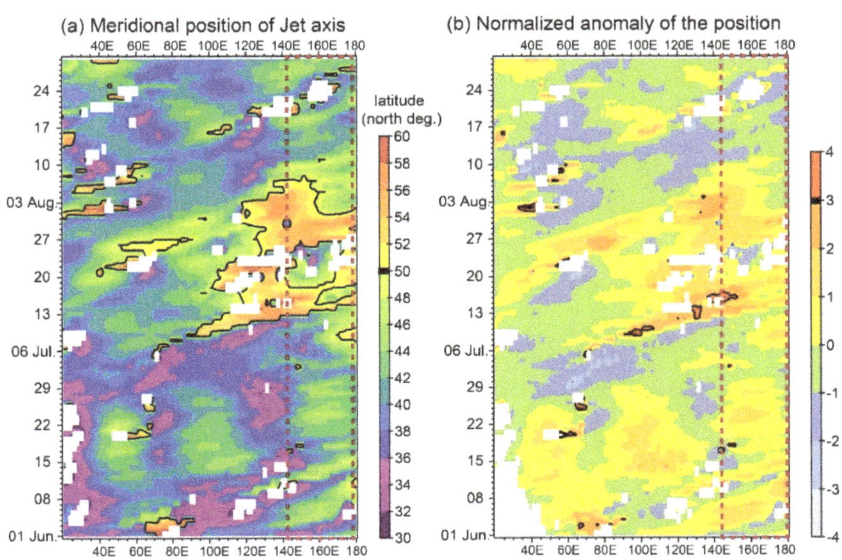

Figure 9. (**a**) Daily meridional position of the westerly jet axis during June–August 2021. The jet axis was defined as the velocity maximum at 200 hPa between 30°N and 60°N. When the velocity maximum was at either 30°N or 60°N, the axis position was not specified (i.e., white areas). (**b**) Normalized anomalies of the axis position. The daily mean axis position (Y) was normalized by the 30-year daily mean (M) and the standard deviation (S): i.e., $(Y - M)/S$. The values of 50°N in panel (**a**) and 3 in panel (**b**) are emphasized by bold contours. The red dashed rectangle denotes the area within 143°E–180°.

The northward shift of the westerly jet began suddenly at the beginning of July: large northward shifts of the westerly jet to the north of 50°N started at 70°E on 5 July, propagated eastward, and reached 120°–140°E around Japan on 11–15 July (Figure 9a). After 15 July, the meridional position of the jet in 143°E–180° stayed north of 50°N until the beginning of August, although it was intermittently disturbed (e.g., by typhoon events in late July): 50°N corresponded to the northern boundary of the MHWs in the northwestern Pacific (Figure 1a). After around 1–8 August, the axial position of the jet was displaced to the south. Simultaneously, the MHWs in the northwestern Pacific started to weaken and shrink.

The large northward shift of the westerly jet identified in early to mid-July exceeded the climatological mean plus 3 standard deviations (Figure 9b); therefore, it was interpreted as a rare event. This result suggests that the abnormally large northward shift of the westerly jet triggered warm atmospheric conditions near the sea surface and thus caused the strongest, most extensive MHWs yet recorded over the northwestern Pacific.

4. Discussion

In this article, we reported on the most extensive and intense MHWs in the historical record of the last 40 years, which occurred in the northwestern Pacific during July–August 2021. However, we mainly limited our explanation to certain features: SST, the oceanic surface boundary layer, heat flux at the sea surface, and atmospheric conditions. We only briefly mentioned the redistribution of heat from the sea surface to the ocean subsurface by ocean advection and diffusion. In addition, although differences in SST anomalies were positively correlated with net heat flux anomalies ($r = 0.69$, Figure 3), anomalies of SST and net heat flux were not strongly correlated with each other. In considering these problems, the uncertainties of the SST and heat flux data should first of all be meticulously validated by comparing different datasets with each other and with in situ data (e.g., surface buoy and vessel data). Moreover, it should be noted that the core area of the MHWs was at 43°–46°N (not shown), whereas the net heat flux anomalies did not show corresponding maxima at 43°–46°N. Additionally, as described in Section 3.1.2, SST heating processes differed inter-regionally within the area of MHWs. Hence, we anticipate that a feedback process from the ocean to the atmosphere contributed to localized enhancement of MHWs. Air–sea processes that might be involved in the enhancement of MHWs at mid-latitudes include SST–cloud feedback (e.g., [30–32]), surface heat flux feedback dominated by turbulent flux (e.g., [33]), or an air–sea process related to the imbalance between incoming and outgoing fluxes through the sea surface (e.g., [34]). A useful first step to clarify such processes would be to conduct numerical experiments with a coupled air–sea model.

In this article, we also focused mainly on the development of MHWs in the northwestern Pacific, but the MHWs also decayed rapidly after the westerly jet was displaced southward at the beginning of August 2021. Therefore, we provide here a brief description of atmospheric conditions near the sea surface during 1–18 August 2021. In this period, the North Pacific Subtropical High retreated southward, and the Okhotsk High strengthened (Figure 10a). The sea level pressure valley between the two high-pressure systems was located in the southeastern half of the focal area (143°E–180°, 40°–50°N) (Figure 10a,b). Air temperature greatly decreased over the northwestern Pacific after 1 August (Figure 10c), and the negative net heat flux anomalies at the sea surface in the focal area (Figure 10d) contributed to sea surface cooling. Simultaneously, as shown by hindcast experiments, the oceanic surface boundary layer deepened (Figure 6b), and this deepening implies that entrainment of subsurface water into the surface boundary layer accelerated cooling at the sea surface. As a consequence, the extensive and intense MHWs had disappeared by 18 August.

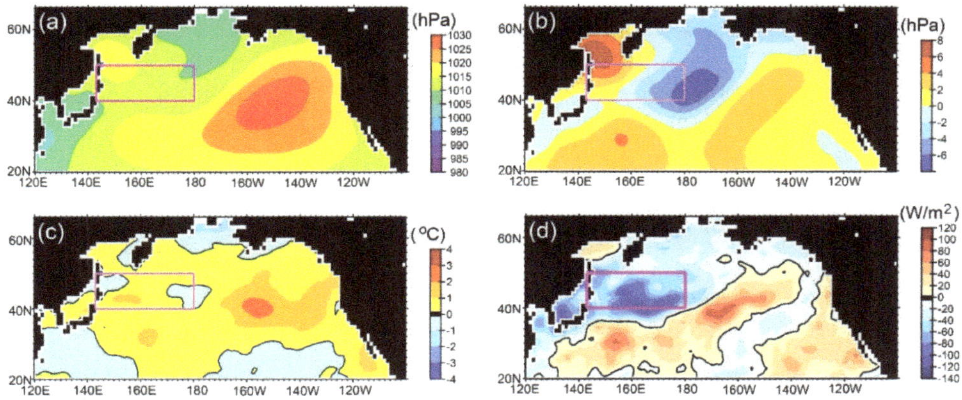

Figure 10. (**a**) Sea level pressure during 1–18 August 2021. Anomalies during the same period: (**b**) sea level pressure, (**c**) air temperature at 2 m above the sea surface, and (**d**) net heat flux at the sea surface.

5. Conclusions

In July–August 2021, the largest and most intense MHWs of the satellite era occurred at the sea surface over extensive areas of the northwestern Pacific Ocean. These were observed at the sea surface and were accompanied by a stable shallow oceanic surface boundary layer. Their spatial relationship to heat fluxes at the sea surface indicates that the MHWs were generated mainly by atmospheric forcing associated with northwestward expansion of the North Pacific Subtropical High and northward displacement of the westerly jet. Additional air–sea studies are needed to understand inter-regional differences in SST heating processes within the area of MHWs.

Moreover, the MHWs in summer 2021 were limited to the vicinity of the sea surface (<20 m depth), where primary production is especially high (e.g., [35–37]). In fact, some studies have reported impacts of MHWs on lower trophic levels of marine ecosystems (e.g., [38,39]). Commercially important fisheries' resources include Japanese sardine *Sardinops melanostictus* and Pacific saury *Cololabis saira*, which utilize the northwestern Pacific waters around the focal area as a nursery ground in summer (e.g., [40,41]). Hence, for sustainable management of ecosystems in the northwestern Pacific Ocean, physical and biological studies are essential, not only to evaluate influences of the 2021 MHWs but also to project the influences of future MHWs on marine ecosystems.

Author Contributions: Conceptualization, H.K.; methodology, H.K. and T.S.; software, H.K. and T.S.; validation, H.K and T.S.; investigation, H.K.; writing—original draft preparation, H.K.; writing—review and editing, T.S.; visualization, H.K.; funding acquisition, H.K. All authors have read and agreed to the published version of the manuscript.

Funding: This research was supported by KAKENHI, Grant Number 19K06216, from the Ministry of Education, Culture, Sports, Science and Technology, and by the Environment Research and Technology Development Fund (4-2102) of the Environmental Restoration and Conservation Agency of Japan. This work was also partially supported by the Fisheries Agency (promotion project to precisely estimate fish stock size around Japan (no grant number)), but the contents of this study do not necessarily reflect the views of the Fisheries Agency.

Data Availability Statement: All the data analyzed in this study are publicly available. SST data were collected from a data server of NEAR-GOOS (https://ds.data.jma.go.jp) (accessed on 2 September 2021). JRA-55 data were downloaded from the University Corporation for Atmospheric Research data server (https://rda.ucar.edu) (accessed on 4 September 2021). Gridded Argo-based seawater temperatures in the subsurface were obtained from Scripps Institute of Oceanography (http://sio-argo.ucsd.edu/RG_Climatology.html) (accessed on 17 August 2021). Daily reanalysis data from

FRA-ROMS are available at http://fm.dc.affrc.go.jp/fra-roms after user registration (accessed on 15 August 2021). The source code of the ocean circulation models based on Rutgers University's ROMS is available at https://www.myroms.org (accessed on 15 August 2021).

Acknowledgments: Numerical experiments based on FRA-ROMS were conducted using the Scientific Computing System in the Agriculture, Forestry and Fisheries Research Information Technology Center. We extend a special thanks to associate editors, special issue editors, and three reviewers for many constructive and fruitful comments.

Conflicts of Interest: The authors declare no conflict of interest.

References

1. Seneviratne, S.I.; Nicholls, N.; Easterling, D.; Goodess, C.M.; Kanae, S.; Kossin, J.; Luo, Y.; Marengo, J.; McInnes, K.; Rahimi, M.; et al. Changes in Climate Extremes and their Impacts on the Natural Physical Environment. In *Managing the Risks of Extreme Events and Disasters to Advance Climate Change Adaptation*, 1st ed.; Field, C.B., Barros, V., Stocker, T.F., Qin, D., Dokken, D.J., Ebi, K.L., Mastrandrea, M.D., Mach, K.J., Plattner, G.-K., Allen, S.K., et al., Eds.; A Special Report of Working Groups I and II of the Intergovernmental Panel on Climate Change (IPCC); Cambridge University Press: Cambridge, UK; New York, NY, USA, 2012; pp. 109–230. Available online: https://www.ipcc.ch/site/assets/uploads/2018/03/SREX-Chap3_FINAL-1.pdf (accessed on 5 September 2021).
2. Fischer, E.M.; Knutti, R. Anthropogenic contribution to global occurrence of heavy-precipitation and high-temperature extremes. *Nat. Clim. Chang.* **2015**, *5*, 560–564. [CrossRef]
3. Easterling, D.R.; Kunkel, K.E.; Wehner, M.F.; Sun, L. Detection and attribution of climate extremes in the observed record. *Weather. Clim. Extrem.* **2016**, *11*, 17–27. [CrossRef]
4. Mitchell, J.F.B.; Lowe, J.; Wood, R.A.; Vellinga, M. Extreme events due to human-induced climate change. *Phil. Trans. R. Soc. A* **2006**, *364*, 2117–2133. [CrossRef]
5. Hobday, A.J.; Alexander, L.V.; Perkins, S.E.; Smale, D.A.; Straub, S.C.; Oliver, E.C.J.; Benthuysen, J.A.; Burrows, M.T.; Donat, M.G.; Feng, M.; et al. A hierarchical approach to defining marine heatwaves. *Prog. Oceanogr.* **2016**, *141*, 227–238. [CrossRef]
6. Hobday, A.J.; Oliver, E.C.J.; Sen Gupta, A.; Benthuysen, J.A.; Burrows, M.T.; Donat, M.G.; Holbrook, N.J.; Moore, P.J.; Thomsen, M.S.; Wernberg, T.; et al. Categorizing and naming marine heatwaves. *Oceanography* **2018**, *31*, 162–173. [CrossRef]
7. Holbrook, N.J.; Scannell, H.A.; Sen Gupta, A.; Benthuysen, J.A.; Feng, M.; Oliver, E.C.J.; Alexander, L.V.; Burrows, M.T.; Donat, M.G.; Hobday, A.J.; et al. A global assessment of marine heatwaves and their drivers. *Nat. Commun.* **2019**, *10*, 2624. [CrossRef]
8. Oliver, E.C.J.; Donat, M.G.; Burrows, M.T.; Moore, P.J.; Smale, D.A.; Alexander, L.V.; Benthuysen, J.A.; Feng, M.; Sen Gupta, A.; Hobday, A.J.; et al. Longer and more frequent marine heatwaves over the past century. *Nat. Commun.* **2018**, *9*, 1324. [CrossRef]
9. Oliver, E.C.J. Mean warming not variability drives marine heatwave trends. *Clim. Dyn.* **2019**, *53*, 1653–1659. [CrossRef]
10. Oliver, E.C.J.; Burrows, M.T.; Donat, M.G.; Sen Gupta, A.; Alexander, L.V.; Perkins-Kirkpatrick, S.E.; Benthuysen, J.A.; Hobday, A.J.; Holbrook, N.J.; Moore, P.J.; et al. Projected marine heatwaves in the 21st century and the potential for ecological impact. *Front. Mar. Sci.* **2019**, *6*, 734. [CrossRef]
11. Straub, S.C.; Wernberg, T.; Thomsen, M.S.; Moore, P.J.; Burrows, M.T.; Harvey, B.P.; Smale, D.A. Resistance, extinction, and everything in between—The diverse responses of seaweeds to marine heatwaves. *Front. Mar. Sci.* **2019**, *6*, 763. [CrossRef]
12. Jacox, M.G.; Alexander, M.A.; Bograd, S.J.; Scott, J.D. Thermal displacement by marine heatwaves. *Nature* **2020**, *584*, 82–86. [CrossRef] [PubMed]
13. Piatt, J.F.; Parrish, J.K.; Renner, H.M.; Schoen, S.K.; Jones, T.T.; Arimitsu, M.L.; Kuletz, K.J.; Bodenstein, B.; García-Reyes, M.; Duerr, R.S.; et al. Extreme mortality and reproductive failure of common murres resulting from the northeast Pacific marine heatwave of 2014-2016. *PLoS ONE* **2020**, *15*, e0226087. [CrossRef]
14. Cheung, W.W.L.; Frölicher, T.L. Marine heatwaves exacerbate climate change impacts for fisheries in the northeast Pacific. *Sci. Rep.* **2020**, *10*, 6678. [CrossRef]
15. Gómez-Gras, D.; Linares, C.; Dornelas, M.; Madin, J.S.; Brambilla, V.; Ledoux, J.-B.; López-Sendino, P.; Bensoussan, N.; Garrabou, J. Climate change transforms the functional identity of Mediterranean coralligenous assemblages. *Ecol. Lett.* **2021**, *24*, 1038–1051. [CrossRef]
16. Di Lorenzo, E.; Mantua, N. Multi-year persistence of the 2014/15 North Pacific marine heatwave. *Nat. Clim. Chang.* **2016**, *6*, 1042–1047. [CrossRef]
17. Hayashi, M.; Shiogama, H.; Emori, S.; Ogura, T.; Hirota, N. The northwestern Pacific warming record in August 2020 occurred under anthropogenic forcing. *Geophys. Res. Lett.* **2021**, e2020GL090956. [CrossRef]
18. Miyama, T.; Minobe, S.; Goto, H. Marine heatwave of sea surface temperature of the Oyashio region in summer in 2010–2016. *Front. Mar. Sci.* **2021**, *7*, 576240. [CrossRef]
19. Kuroda, H.; Taniuchi, Y.; Kasai, H.; Nakanowatari, T.; Setou, T. Co-occurrence of marine extremes induced by tropical storms and an ocean eddy in summer 2016: Anomalous hydrographic conditions in the Pacific shelf waters off southeast Hokkaido, Japan. *Atmosphere* **2021**, *12*, 888. [CrossRef]

20. Kuroda, H.; Yokouchi, K. Interdecadal decrease in potential fishing areas for Pacific saury off the southeastern coast of Hokkaido, Japan. *Fish. Oceanogr.* **2017**, *26*, 439–454. [CrossRef]
21. Kurihara, Y.; Sakurai, T.; Kuragano, T. Global daily sea surface temperature analysis using data from satellite microwave radiometer, satellite infrared radiometer and in-situ observations. *Weather. Serv. Bull.* **2006**, *73*, S1–S18. (In Japanese)
22. Roemmich, D.; Gilson, J. The 2004-2008 mean and annual cycle of temperature, salinity, and steric height in the global ocean from the Argo Program. *Prog. Oceanogr.* **2009**, *82*, 81–100. [CrossRef]
23. Kobayashi, S.; Ota, Y.; Harada, Y.; Ebita, A.; Moriya, M.; Onoda, H.; Onogi, K.; Kamahori, H.; Kobayashi, C.; Endo, H.; et al. The JRA-55 reanalysis: General specifications and basic characteristics. *J. Meteor. Soc. Jpn.* **2016**, *93*, 5–48. [CrossRef]
24. Kuroda, H.; Setou, T.; Kakehi, S.; Ito, S.; Taneda, T.; Azumaya, T.; Inagake, D.; Hiroe, Y.; Morinaga, K.; Okazaki, M.; et al. Recent advances in Japanese fisheries science in the Kuroshio-Oyashio region through development of the FRA-ROMS ocean forecast system: Overview of the reproducibility of reanalysis products. *Open J. Mar. Sci.* **2017**, *7*, 62–90. [CrossRef]
25. Shchepetkin, A.F.; McWilliams, J.C. A method for computing horizontal pressure-gradient force in an oceanic model with a nonaligned vertical coordinate. *J. Geophys. Res.* **2003**, *108*, 3090. [CrossRef]
26. Shchepetkin, A.F.; McWilliams, J.C. The regional oceanic modeling system (ROMS): A split-explicit, free-surface, topography-following-coordinate oceanic model. *Ocean Modell.* **2005**, *9*, 347–404. [CrossRef]
27. Haidvogel, D.B.; Arango, H.; Budgell, W.P.; Cornuelle, B.D.; Curchitser, E.; Di Lorenzo, E.; Fennel, K.; Geyer, W.R.; Hermann, A.J.; Lanerolle, L.; et al. Ocean forecasting in terrain-following coordinates: Formulation and skill assessment of the Regional Ocean Modeling System. *J. Comput. Phys.* **2008**, *227*, 3595–3624. [CrossRef]
28. Kuroda, H.; Setou, T.; Aoki, K.; Takahashi, D.; Shimizu, M.; Watanabe, T. A numerical study of the Kuroshio-induced circulation in Tosa Bay, off the southern coast of Japan. *Cont. Shelf Res.* **2013**, *53*, 50–62. [CrossRef]
29. Schiemann, R.; Lüthi, D.; Schär, C. Seasonality and interannual variability of the westerly jet in the Tibetan Plateau region. *J. Clim.* **2009**, *22*, 2940–2957. [CrossRef]
30. Clement, A.M.; Burgman, R.; Norris, J.R. Observational and model evidence for positive low-level cloud feedback. *Science* **2009**, *325*, 460–464. [CrossRef]
31. Wu, R.; Kinter, J.L. Atmosphere-ocean relationship in the midlatitude North Pacific: Seasonal dependence and east-west contrast. *J. Geophys. Res.* **2010**, *115*, D06101. [CrossRef]
32. Schmeisser, L.; Bond, N.A.; Siedlecki, S.A.; Ackerman, T.P. The role of clouds and surface heat fluxes in the maintenance of the 2013–2016 Northeast Pacific marine heatwave. *J. Geophys. Res. Atmos.* **2019**, *124*, 10772–10783. [CrossRef]
33. Frankignoul, C.; Kestenare, E. The surface heat flux feedback. Part I: Estimates from observations in the Atlantic and the North Pacific. *Clim. Dyn.* **2002**, *19*, 633–647. [CrossRef]
34. Pinault, J.-L. Modulated response of subtropical gyres: Positive feedback loop, subharmonic modes, resonant solar and orbital forcing. *J. Mar. Sci. Eng.* **2018**, *6*, 107. [CrossRef]
35. Shiomoto, A.; Sasaki, K.; Shimoda, T.; Matsumura, S. Primary Productivity in the Offshore Oyashio in the Spring and Summer 1990. *J. Oceanogr.* **1994**, *50*, 209–222. [CrossRef]
36. Shiomoto, A.; Ishida, Y.; Tamaki, T.; Yamanaka, Y. Primary production and chlorophyll *a* in the northwestern Pacific Ocean in summer. *J. Geophys. Res.* **1998**, *103*, 24651–24661. [CrossRef]
37. Kameda, T.; Matsumura, S. Chlorophyll biomass off Sanriku, northwestern Pacific estimated by Ocean Color and Temperature Scanner (OCTS) and a vertical distribution model. *J. Oceanogr.* **1998**, *54*, 509–516. [CrossRef]
38. Hayashida, H.; Matear, R.J.; Strutton, P.G. Background nutrient concentration determines phytoplankton bloom response to marine heatwaves. *Glob Change Biol.* **2020**, *26*, 4800–4811. [CrossRef] [PubMed]
39. Montie, S.; Thomsen, M.; Rack, W.; Broady, P. Extreme summer marine heatwaves increase chlorophyll *a* in the Southern Ocean. *Antarct. Sci.* **2020**, *32*, 508–509. [CrossRef]
40. Okunishi, T.; Ito, S.; Ambe, D.; Takasuka, A.; Kameda, T.; Tadokoro, K.; Setou, T.; Komatsu, K.; Kawabata, A.; Kubota, H.; et al. A modeling approach to evaluate growth and movement for recruitment success of Japanese sardine (*Sardinops melanostictus*) in the western Pacific. *Fish. Oceanogr.* **2012**, *21*, 44–57. [CrossRef]
41. Miyamoto, H.; Vijai, D.; Kidokoro, H.; Tadokoro, K.; Watanabe, T.; Fuji, T.; Suyama, S. Geographic variation in feeding of Pacific saury *Cololabis saira* in June and July in the North Pacific Ocean. *Fish. Oceanogr.* **2020**, *29*, 558–571. [CrossRef]

remote sensing

MDPI

Article

Long-Term Changes and Factors That Influence Changes in Thermal Discharge from Nuclear Power Plants in Daya Bay, China

Zhihua Zhang [1,2], Difeng Wang [2,3,*], Yinhe Cheng [1] and Fang Gong [2,3]

1 School of Marine Technology and Geomatics, Jiangsu Ocean University, Lianyungang 222005, China; 2019220288@jou.edu.cn (Z.Z.); chengyh@jou.edu.cn (Y.C.)
2 State Key Laboratory of Satellite Ocean Environment Dynamics, Second Institute of Oceanography, Ministry of Natural Resources of the People's Republic of China, Hangzhou 310012, China; gongfang@sio.org.cn
3 Southern Marine Science and Engineering Guangdong Laboratory (Guangzhou), Guangzhou 511400, China
* Correspondence: dfwang@sio.org.cn

Abstract: Thermal discharge (i.e., warm water) from nuclear power plants (NPPs) in Daya Bay, China, was analyzed in this study. To determine temporal and spatial patterns as well as factors affecting thermal discharge, data were acquired by the Landsat series of remote-sensing satellites for the period 1993–2020. First, sea surface temperature (SST) data for waters off NPPs were retrieved from Landsat imagery using a radiative transfer equation in conjunction with a split-window algorithm. Then, retrieved SST data were used to analyze seasonal and interannual changes in areas affected by NPP thermal discharge, as well as the effects of NPP installed capacity, tides, and wind field on the diffusion of thermal discharge. Analysis of interannual changes revealed an increase in SST with an increase in NPP installed capacity, with the area affected by increased drainage outlet temperature increasing to different degrees. Sea surface temperature and NPP installed capacity were significantly linearly related. Both flood tides (peak spring and neap) and ebb tides (peak spring and neap) affected areas of warming zones, with ebb tides having greater effects. The total area of all warming zones in summer was approximately twice that in spring, regardless of whether winds were favorable (i.e., westerly) or adverse (i.e., easterly). The effects of tides on areas of warming zones exceeded those of winds.

Keywords: Daya Bay Nuclear Power Plants; thermal discharge; long-term changes; Landsat; radiative transfer equation; split-window algorithm; power plant installed capacity; flood tide; ebb tide; wind field

Citation: Zhang, Z.; Wang, D.; Cheng, Y.; Gong, F. Long-Term Changes and Factors That Influence Changes in Thermal Discharge from Nuclear Power Plants in Daya Bay, China. *Remote Sens.* **2022**, *14*, 763. https://doi.org/10.3390/rs14030763

Academic Editors: Chung-Ru Ho, Xiaofeng Li and Antony K. Liu

Received: 3 January 2022
Accepted: 1 February 2022
Published: 7 February 2022

1. Introduction

In recent years, the increasing energy demand of coastal cities has increased both the number and scale of coastal thermal power plants and NPPs. In NPPs, only 30% to 35% of nuclear energy is converted to electrical energy [1], and most of the remaining energy is discharged as thermal energy in cooling water. The discharge of warm water rapidly increases the temperature of surrounding waters, which in turn directly or indirectly affects the growth and reproduction of aquatic organisms [2].

To monitor thermal discharge, many studies have examined algorithms to retrieve sea surface temperatures (SSTs) from thermal infrared remote sensing. In 1975, McMillin [3] first introduced a split-window algorithm (SWA) based on the radiative transfer equation (RTE), which was relatively effective at retrieving SST [4]. Liu and Zhou [5] introduced a multichannel SWA to retrieve SST data for the Yellow and East China seas, and the retrieved data adequately reflected SST distribution patterns. Rozenstein et al. [6] and Chen et al. [7] revised the SWA to use data acquired by the Advanced Very-High-Resolution Radiometer instruments onboard the United States National Oceanic and Atmospheric Administration

(NOAA) family of satellites in order to retrieve SSTs from Landsat satellite data. Ai et al. [8] presented a new SWA-based SST retrieval model and validated its reliability by comparing SST data for the Bohai Sea with those extracted from a MODIS SST product.

Furthermore, clarifying the influencing factors of the warm drainage of nuclear power plants is of great significance for predicting its changing trend and evaluating and finding ways to speed up the hydrodynamic exchange of warm drainage to minimize its ecological impact. The velocity and direction of tide, wind, water depth, and installed capacity are considered to be the main factors affecting thermal discharge [9–12]. Hence, it is crucial to investigate factors and mechanisms that influence changes in thermal discharge from NPPs.

Many studies have recently examined the effects of NPPs on Daya Bay in China. For example, nutrients, phytoplankton, and zooplankton in Daya Bay show marked responses to thermal discharge from NPPs [13]. However, changes in Daya Bay based on time-series data covering more than two decades have not been investigated. High-accuracy remote-sensing time-series data can be used to monitor the zones affected by thermal discharge, as well as to identify patterns of change in and factors influencing thermal discharge. Hence, in this study, data reflecting changes in thermal discharge in waters surrounding NPPs in Daya Bay from 1993 to 2020 were retrieved from remote sensing imagery, acquired by Landsat satellites. Those data were then used to analyze the seasonal and interannual temporal and spatial distribution patterns of thermal discharge. In addition, the effects of tides and the wind field on diffusion of thermal discharge were examined. The results will provide an important reference for protection of the ecology and environments in waters off coastal NPPs.

2. Data

2.1. Study Area

In the northern part of the South China Sea, Daya Bay (23°31′12″ N to 24°50′00″ N, 113°29′42″ E to 114°49′42″ E) encompasses ~600 km^2 [14,15] and includes more than 50 islands (Figure 1). The bay is very shallow, with an average depth of 11 m and a maximum depth of 21 m, and its waters are cold with high salinity in winter and hot with low salinity in summer [16]. A monsoon and oceanic climate prevails, with four distinct seasons and an annual average temperature of ~22 °C. Wind over Daya Bay is dominated by northeasterlies in winter and southwesterlies in summer, with relatively high speeds (monthly average: 5.0–5.4 m/s) in spring and early summer and relatively low speeds (monthly average: 4.6–4.8 m/s) in the remainder of summer and winter [17]. The Daya Bay NPP (DBNPP), with an installed capacity of 1968 MW, and the Lingao NPP (LNPP), with an installed capacity of 4152 MW, began operating in 1994 and 2002, respectively.

2.2. Landsat Data

Landsat Level-1 imagery (spatial resolution: 30 m) was acquired for the period 1993 to 2020. Specifically, 125 good-quality Landsat-5, -7, and -8 images, each with cloud coverage less than 20%, were downloaded from the website of the United States Geological Survey (https://earthexplorer.usgs.gov/, accessed on 2 January 2022). Each image was acquired at approximately 0245 Greenwich Mean Time (GMT) when the satellite flew over Daya Bay. The temperature data contained in those images spanned almost the entire range of temperatures over the period, from low temperatures in winter to high temperatures in summer, and therefore were representative of seasons in the area. Figure 2 shows the amount of Landsat images in each season.

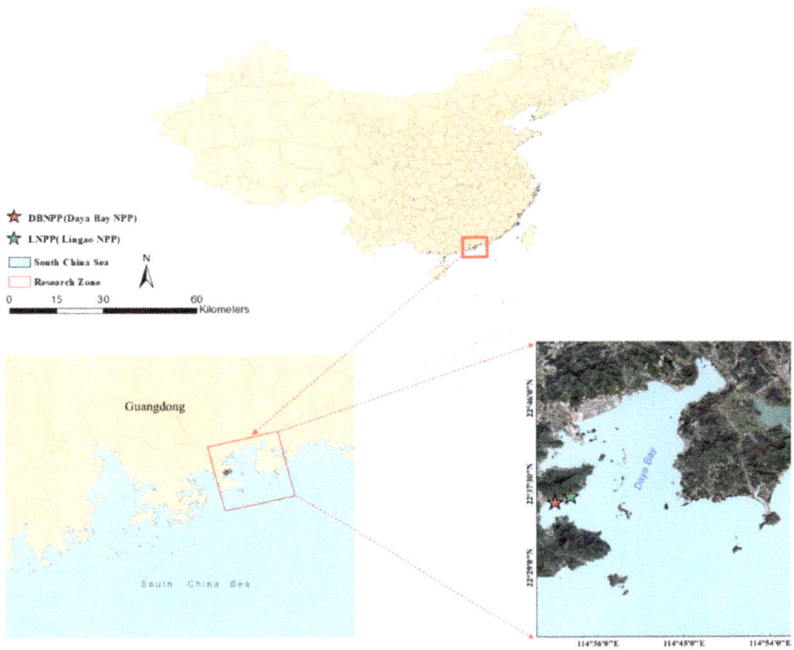

Figure 1. Geographic location of Daya Bay, China.

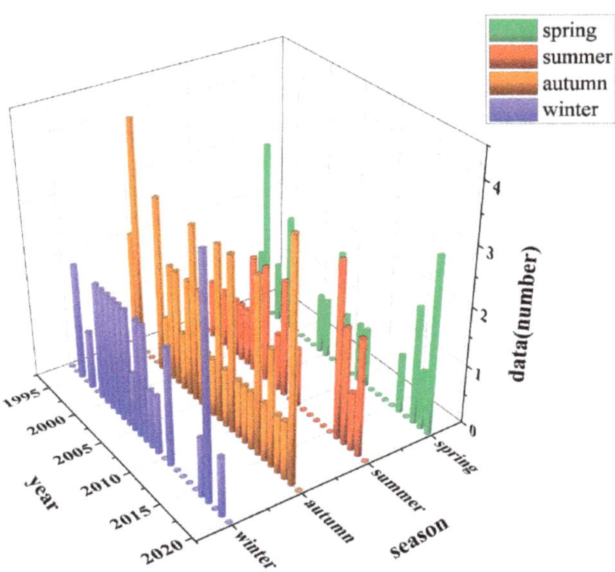

Figure 2. Amount of Landsat images of Daya Bay in spring (green), summer (red), autumn (orange), and winter (blue) from 1993 to 2020.

2.3. MODIS Data

MODIS is an important sensor onboard the Terra and Aqua satellites. Each day, the Terra satellite flies over the study area at approximately 1030 and 2230 GMT, whereas the Aqua satellite flies over at approximately 0130 and 1330 GMT. A MODIS Level-2 SST product that provides data with 1-km resolution and long-term quality control was obtained from the official website of the United States National Aeronautics and Space Administration (NASA, http://oceancolor.gsfc.nasa.gov/, accessed on 2 January 2022).

2.4. Tidal Data and Processing

Daily tidal level data were collected at a Daya Bay meteorological station (22°35′ N, 114°31′ E) from 1993 to 2020. A tide table was produced based on the tidal height at each time point to determine the tidal state of waters when the satellites flew over the study area. In addition, average areas of warming zones with different levels of increase in SST in each tidal state were compared in order to analyze the effects of tides on the diffusion of thermal discharge.

2.5. Wind Field Data and Processing

To analyze the effects of the wind field on diffusion of thermal discharge, average direction and speed data were determined for the wind field over a zone approximately 15 km offshore (22°45′N to 23°12′N, 114°36′E to 115°00′E). Those data were extracted from a Cross-Calibrated Multi-Platform Level-3 product (www.remss.com/measurements/ccmp, accessed on 2 January 2022) that provided ocean surface wind data (i.e., data for wind fields 10 m above the ocean surface) with a spatial resolution of 0.25° × 0.25°. Westerlies (including southwesterlies and northwesterlies), which promote diffusion of thermal discharge, were defined as favorable winds, whereas easterlies (including northeasterlies and southeasterlies), which limit diffusion of thermal discharge, were defined as adverse winds.

2.6. Installed Capacity Data for Nuclear Power Plants in the Study Area

Installed capacity data for the NPPs in Daya Bay (Table 1) were extracted from their introductions given by the South China Nuclear and Radiation Oversight Station, Ministry of Ecology and Environment of China (https://scro.mee.gov.cn, accessed on 2 January 2022).

Table 1. Installed capacity data for the Daya Bay (DB) and Lingao (L) nuclear power plants (NPPs) in Daya Bay, China.

Scheme	Cooling Water Flow Rate (m^3 s^{-1})	Total Installed Capacity (MW)	Installed Capacities in Different Periods (MW)
DBNPP	319	6120	1968
LNPP	329		Phase 1: 1980 Phase 2: 2172

3. Methods

3.1. Data Preprocessing

Landsat imagery was subjected to preprocessing procedures such as radiometric calibration and conversion of gray values to radiance values with a physical meaning [18], i.e., top-of-atmosphere radiance, $L_{(\lambda)}$, values, which are input data for temperature inversion. $L_{(\lambda)}$ was calculated using the following equation:

$$L_{(\lambda)} = DN \times G_{ain} + h \quad (1)$$

where DN is the original value recorded by the sensor, G_{ain} is the absolute calibration coefficient, and h is the offset. G_{ain} and h were obtained directly from the parameter file of Landsat data [19].

3.2. Sea Surface Temperature Retrieval

Sea surface temperatures were retrieved from Landsat-5 and -7 data using the RTE [20] and from Landsat-8 data using a SWA [21,22].

Because it is difficult to obtain accurate and real-time atmospheric information, and to avoid dependence on atmospheric parameters, a split-window algorithm has been developed based on the difference in atmospheric absorption between two adjacent channels (center wavelengths of approximately 11 μm and 12 μm) in the atmospheric window [23]. The influence of the atmosphere is eliminated through combinations of the measured values of the two channels. Therefore, atmospheric correction preprocessing of Landsat-8 data is no longer necessary [24,25].

For Landsat 5/7, the RTE algorithm was used. The following equations were used:

$$L(\lambda) = \left[(\varepsilon) \times B(T_s) + (1-\varepsilon)L_{\downarrow}\right]\tau + L_{\uparrow} \tag{2}$$

$$B(T_s) = \frac{\left[L(\lambda) - L_{\uparrow} - \tau \times (1-\varepsilon)L_{\downarrow}\right]}{\tau - \varepsilon} \tag{3}$$

$$T_s = K_2 / ln(K_1 / L(\lambda) + 1) \tag{4}$$

where $L(\lambda)$ is the radiance of the top layer of the atmosphere received by the sensor; $B(T_s)$ is the blackbody thermal radiance; ε is the surface emissivity (0.992); and T_s is the sea surface temperature. $L_{\downarrow}$ and $L_{\uparrow}$ are the atmospheric downward and upward radiation, respectively, and τ is the atmospheric transmittance. The three parameters were obtained through the NASA official website (https://atmcorr.gsfc.nasa.gov, accessed on 2 January 2022). Values of K_1 and K_2 were obtained from the header file of Landsat. For Landsat-5, K_1 = 607.76 $W \cdot m^{-2} \cdot sr^{-1} \cdot \mu m^{-1}$, and K_2 = 1260.56 K; and for Landsat-7, K_1 = 666.09 $W \cdot m^{-2} \cdot sr^{-1} \cdot \mu m^{-1}$, and K_2 = 1282.71 K [26,27].

3.3. Sea Surface Temperature Accuracy Evaluation

The accuracy of SST data retrieved from Landsat imagery was evaluated in a comparison with a MODIS Aqua Level-2 SST product (resolution: 1 km) for the period 2003 to 2020 (see Table A1 of Appendix A for matching files). Landsat and Aqua satellites fly over the study area at approximately 0245 and 0530 GMT, respectively. According to the analysis by Li et al. [28] of daily global SST variations, SST in the South China Sea varies by only 0.2 °C to 0.4 °C each day. Therefore, SST in the study area was considered to be approximately constant for three hours. The following procedure was used in matching datasets. First, data retrieved from Landsat imagery were resampled to the same resolution as that of MODIS images. Then, Landsat and MODIS images of the waters off the NPPs in Daya Bay at the same latitudinal and longitudinal coordinates, as well as corresponding SSTs, were extracted for same-point validation. The accuracy analysis was based on 51 Landsat images and the same number of matching MODIS images (Figure 3). Data retrieved from the Landsat images were significantly linearly related to the matching data extracted from the MODIS SST product. The relation was described by the equation y = 1.058x − 2.269, with a correlation coefficient (R) of 0.94. Therefore, data retrieved from multisource Landsat imagery in this study were considered suitable to calculate increases in SSTs in waters off the NPPs in Daya Bay over a long period.

3.4. Criterion for Determining Increases in Sea Surface Temperatures

The background temperature selected in an investigation of the thermal discharge from an NPP must be close to the natural temperature in the area in the absence of the NPP [29]. Because Daya Bay is semi-closed, an adjacent-zone substitution method was adopted to determine a reference temperature, T_r. Specifically, the average temperature in a square zone with an area of 8 km × 8 km, located approximately 10 km from the LNPP, was selected as T_r to determine SST increases (Figure 4).

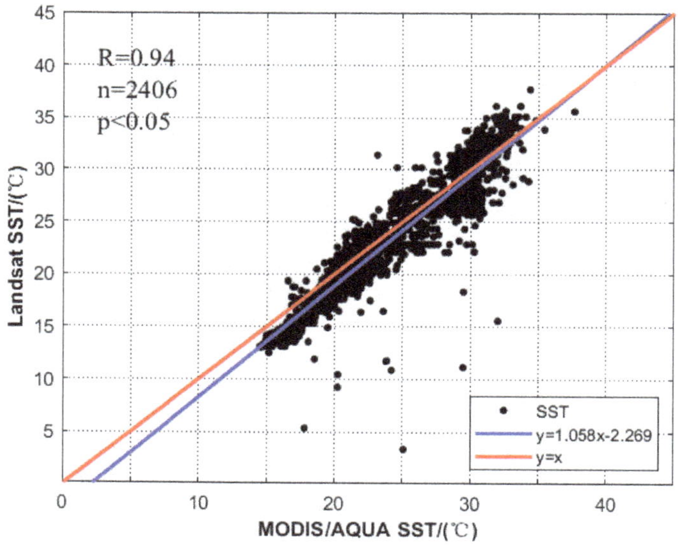

Figure 3. Relation between Landsat sea surface temperature (SST, °C) and MODIS SST product. Blue line is the regression line (y = 1.058x − 2.269); red line is the 1:1 line (y = x).

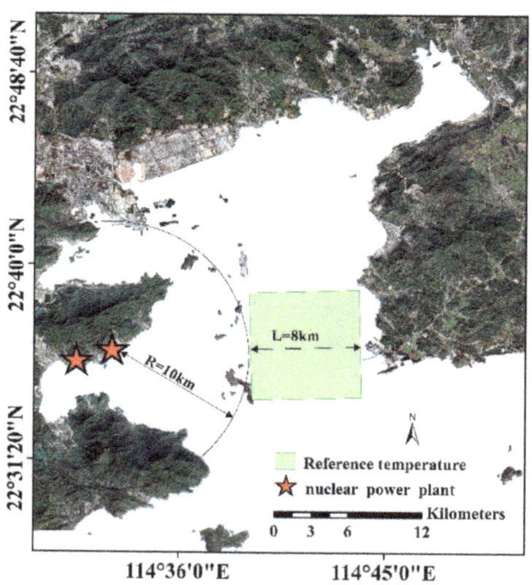

Figure 4. Reference zone for background temperature (T_r) to determine increases in sea surface temperature.

3.5. Methods for Analyzing Sea Surface Temperature Increases

Contours of increases in SST were plotted for each season (spring: March–May; summer: June–August; fall: September–November; winter: December–February of the following year) [30]. The average total area of all warming zones with SST increases ranging from 2 °C to greater than 7 °C (A_{total}) and average total area of warming zones with SST increases at different levels were calculated for each season.

Similarly, annual contours of increases in SST were plotted for the period 1993 to 2020 (excluding 2012, 2013, 2014, 2015, and 2020, because data were available for only one season in each of those years and therefore were not representative) in order to analyze patterns of interannual changes in thermal discharge. To examine distributions of temperature increases more clearly and to facilitate subsequent analysis and research, different temperature levels were set. Table 2 summarizes the range for each level of SST increase. For convenience in description, average total areas of warming zones with SST increases of 2–3 °C, 3–4 °C, 4–5 °C, 5–6 °C, 6–7 °C, and >7 °C are denoted by $A_{+2°C}$, $A_{+3°C}$, $A_{+4°C}$, $A_{+5°C}$, $A_{+6°C}$, and $A_{+7°C}$, respectively.

Table 2. Temperature ranges (°C) for different levels of increases in sea surface temperature (SST).

Range of SST Increases (>T_r)	Level (>T_r)
<2 °C	<2 °C
[+2 °C, +3 °C]	+2 °C
[+3 °C, +4 °C]	+3 °C
[+4 °C, +5 °C]	+4 °C
[+5 °C, +6 °C]	+5 °C
[+6 °C, +7 °C]	+6 °C
>7 °C	+7 °C

4. Results

4.1. Seasonal Changes in Distribution Pattern of Thermal Discharge

Figure 5 shows seasonal contours of SST increases, plotted on the basis of seasonal averages of data retrieved from Landsat Level-1 imagery for the period 1993–2020. Table 3 shows areas of warming zones in each season. In each season, the total areas of warming zones with low SST increases (at +2 °C and +3 °C levels) were relatively large and displayed an outward fan-shaped diffusion pattern, whereas those with high SST increases (at +6 °C and +7 °C levels) were relatively small and concentrated primarily near discharge outlets of the NPPs along the shore of Daya Bay. Summer had the largest A_{total} (total area with temperature increase from 2 °C to 7 °C) at 31.58 km^2, with $A_{+2°C}$ accounting for the largest proportion (49.6%; 15.67 km^2). The smallest A_{total} was in winter (7.89 km^2). The average total area of warming zones with SST increases at each of the +2 °C to +7 °C levels in summer was approximately four times that in winter and approximately twice that in spring and fall. The seasonal pattern was similar for A_{total}.

Table 3. Total area (A_{total}) of warming zones and areas of warming zones at different temperature levels ($A_{+2°C}$ to $A_{+7°C}$) in each season for the period 1993–2020.

Season\Area (km^2)	$A_{+2°C}$	$A_{+3°C}$	$A_{+4°C}$	$A_{+5°C}$	$A_{+6°C}$	$A_{+7°C}$	A_{total}
Spring	8.83	3.6	1.8	1.06	0.44	0.3	16.03
Summer	15.67	9.49	3.29	1.37	0.79	1.0	31.58
Fall	9.4	3.51	1.69	0.64	0.55	0.56	16.35
Winter	4.23	1.84	0.94	0.4	0.29	0.19	7.89

4.2. Interannual Changes in Thermal Discharge

Because Landsat sensors are affected by factors such as clouds, Landsat data suitable for retrieval of SSTs are limited. Therefore, images that contained as much data as possible were selected to analyze interannual changes in thermal discharge. Specifically, images of at least two seasons in each year (except 1993) were used in the analysis.

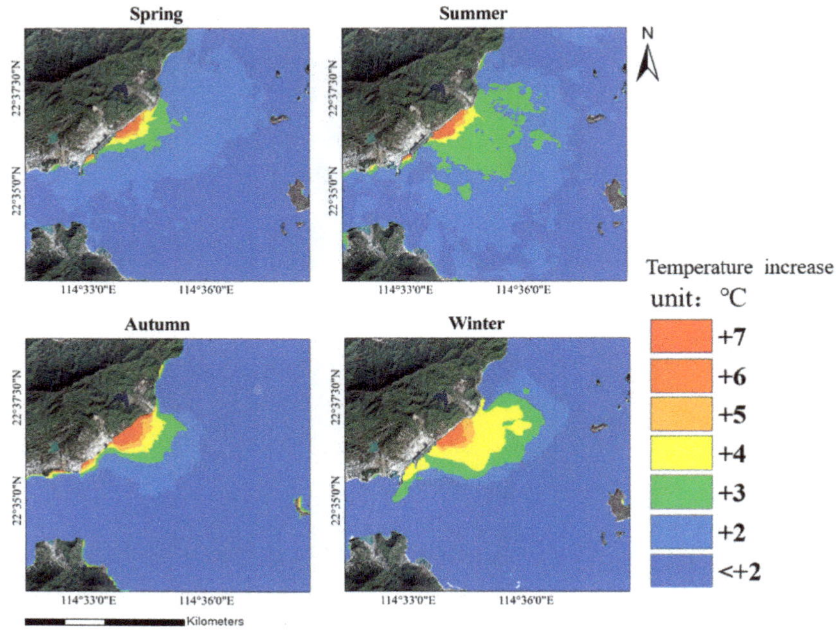

Figure 5. Seasonal changes (spring, summer, autumn, winter) in sea surface temperature (°C) contours for 1the period 993–2020.

Figures 6 and 7 show the interannual contours of SST increases and the interannual changes in areas of warming zones, respectively. In 1993, there were no thermal discharge-affected zones. Because the No. 1 unit of the DBNPP began operation on May 6, 1994, warm water was discharged within an extremely small zone in that year. A notable diffusion of thermal discharge began in 2000, with SST increases occurring primarily at +2 °C and +3 °C levels and at locations approximately 1.5 to 1.7 km east of the discharge outlet. Later, in 2004, with intensification of operations and an increase in installed capacity (first phase of LNPP became operational in January 2003), the zones affected by thermal discharge expanded, with SST increases occurring primarily at +2 °C to +5 °C levels. In addition, $A_{+2°C}$, $A_{+3°C}$, $A_{+4°C}$, $A_{+5°C}$, and A_{total} in 2004 were approximately three times those in 2000. The warming zones in 2004 were distributed 3.5 to 3.8 km east of the discharge outlets, and SST increases at +6 °C and +7 °C levels became increasingly prominent (at locations ~0.5 km east of the discharge outlets). The second phase of LNPP was completed and began commercial service on 15 July 2010, resulting in a notable expansion of zones affected by thermal discharge. Increases in SST were at +2 °C to +4 °C levels in most of the nearby waters and were at +5 °C to +7 °C levels near the discharge outlets. In 2019, the effects of thermal discharge peaked, with A_{total} also reaching its maximum (24.28 km^2). Increases in SST occurred primarily at +2 °C to +5 °C levels, but there were also notable increases in $A_{+6°C}$ and $A_{+7°C}$.

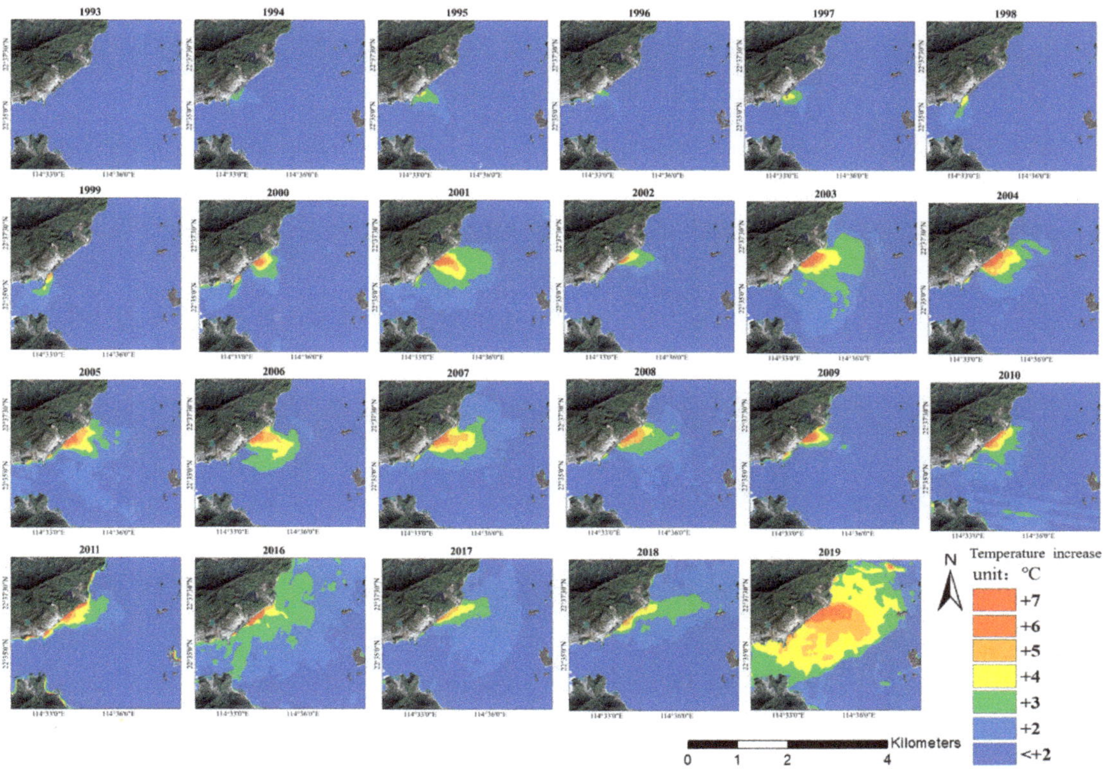

Figure 6. Interannual changes in sea surface temperature (°C) contours of areas of warming zones from the period 1993 to 2019.

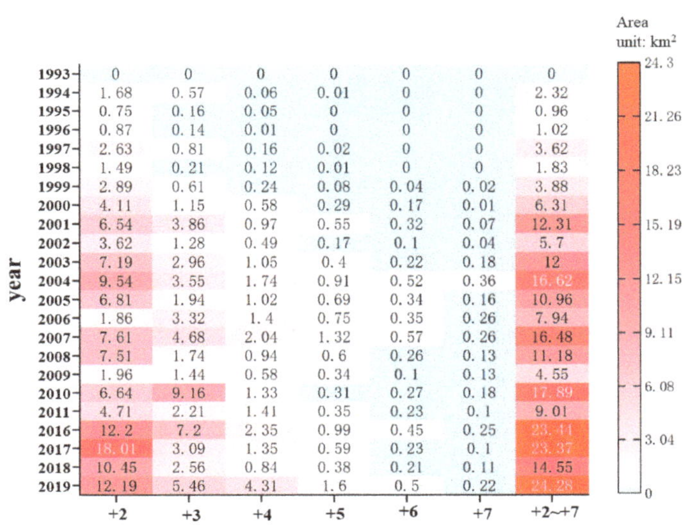

Figure 7. Interannual changes in areas of warming zones for the period 1993 to 2019.

5. Discussion

5.1. Effects of Installed Capacity of Nuclear Power Plants on Thermal Discharge

To analyze the effects of changes in installed capacity of the NPPs on thermal discharge, three test zones and three control zones, each with an area within 30 km^2, were selected within different distance intervals (i.e., 0–2 km, 2–5 km, and 5–10 km) from the shoreline within the study area. The test zones were A, B, and C, respectively, and the corresponding control zones were D, E, and F (Figure 8).

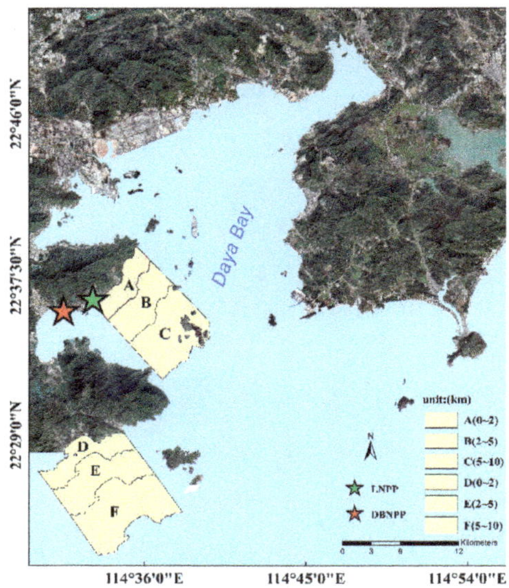

Figure 8. Schematic of test zones (A, B, C) and control zones (D, E, F), which are used to analyze the effects of changes in installed capacity of the NPPs on thermal discharge.

With SSTs in test zone C and control zone F as references, interannual changes in SSTs were calculated based on MODIS Level-2 SST data for the period 2003 to 2020. Let Δ_{AC} and Δ_{BC} be the annual average SST in zone A minus that in zone C and the annual average SST in zone B minus that in zone C, respectively, with Δ_{DF} and Δ_{EF} similarly defined. The following equations define Δ_{AC}, Δ_{BC}, Δ_{DF}, and Δ_{EF}:

$$\Delta_{AC} = SST_A - SST_C \quad (5)$$

$$\Delta_{BC} = SST_B - SST_C \quad (6)$$

$$\Delta_{DF} = SST_D - SST_F \quad (7)$$

$$\Delta_{EF} = SST_E - SST_F \quad (8)$$

where SST_A, SST_B, SST_C, SST_D, SST_E, and SST_F are the annual average SSTs in zones A, B, C, D, E, and F, respectively. Annual rates of increase in SST in zones A and B (R_A and R_B, respectively) were defined as follow [31]:

$$R_A = \Delta_{AC} - \Delta_{DF} \quad (9)$$

$$R_B = \Delta_{BC} - \Delta_{EF} \quad (10)$$

Figure 9 shows the relations between the installed capacity of the NPPs and increases in SST, determined based on changes in R_A and R_B.

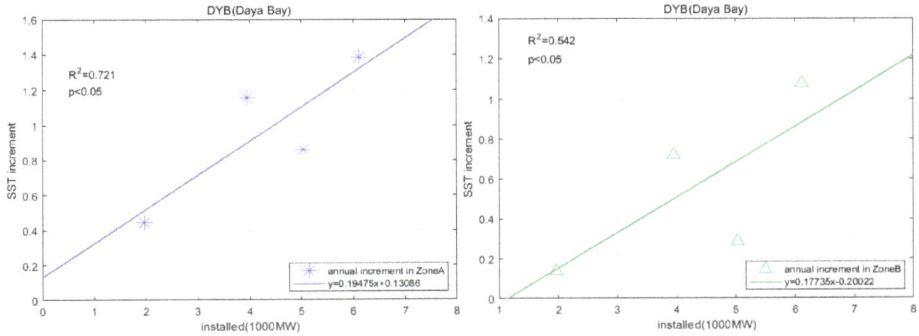

Figure 9. Relation between installed capacity of nuclear power plants and increase in sea surface temperature (SST).

There was a significant linear relation between the increase in SST and the installed capacity of NPPs in each of zones A and B, with the relation stronger in zone A than in zone B (Figure 9). Thus, an increase in installed capacity of the NPPs led to a greater increase in SST in zone A (i.e., within 2 km from the shoreline) than in zone B (i.e., within 2–5 km from the shoreline). In summary, an increase in installed capacity resulted in more heat released into the sea, which in turn resulted in a larger increase in SST over a larger area. This conclusion is consistent with that of Lin et al. [31].

5.2. Effects of Tides on Thermal Discharge

Tides in Daya Bay are irregular and semidiurnal, with an average height of 1.01 m and a maximum height of 2.5 m [32]. Tidal flow rate in Daya Bay decreases gradually from the mouth to the north and is relatively low at the head. Weak tidal currents dominate the waters of Daya Bay, except for those in the east, where there are relatively strong tidal currents. Specifically, tidal flow rates are approximately 30 and 20 cm/s in the eastern and western waters of Daya Bay, respectively, and range from 5 to 10 cm/s in waters near the NPPs [16].

Figure 10 shows areas of warming zones corresponding to different tidal states (peak spring flood tides (PSFTs), peak spring ebb tides (PSETs), peak neap flood tides (PNFTs), and peak neap ebb tides (PNETs)) for the period 1993–2020. Total area of warming zones with SST increases at each of the +2 °C to +7 °C levels, and A_{total} was greater during ebb tides (ETs) than during flood tides (FTs). Values of $A_{+2°C}$, $A_{+4°C}$, $A_{+5°C}$, and A_{total} during PSETs compared with those during PSFTs were approximately two times higher. Those areas during PNETs were also greater than those during PNFTs. The largest and smallest A_{total} values (29.23 and 14.5 km^2, respectively) occurred during PNETs and PSFTs, respectively, and $A_{+6°C}$ and $A_{+7°C}$ during PSETs were approximately six times those during PSFTs.

Table 4 shows the variation in areas of warming zones with tidal state and season (spring, summer, fall, and winter) for the period 1993–2020. The A_{total} was greater during peak ETs (PETs), both PSETs and PNETs, than during peak FTs (PFTs) in each season, except in fall, when A_{total} was smaller during PNETs than during PNFTs. The A_{total} during PETs was approximately 1.5 times that during PFTs in spring and summer, whereas A_{total} during PETs was approximately twice that during PFTs in winter. The largest and smallest A_{total} values (56.19 and 6.82 km^2, respectively) appeared during PNETs in summer and PNFTs in winter, respectively. In summary, the effects of ETs on areas of warming zones exceeded those of FTs [33]. In addition, as shown in Figure 11, tides affected the direction of thermal discharge.

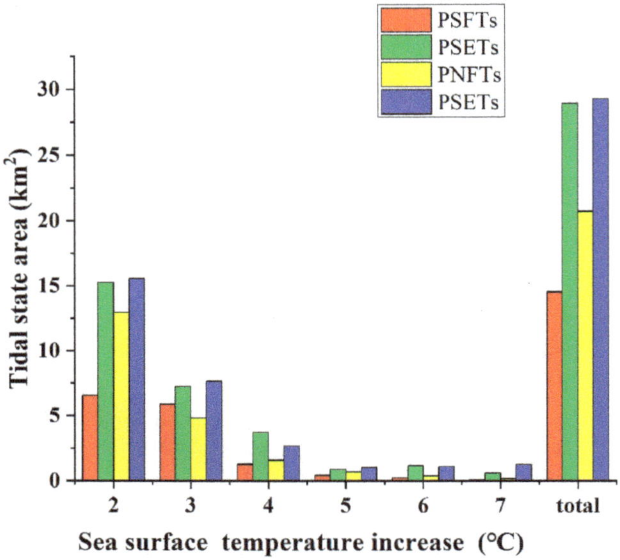

Figure 10. Total area of warming zones and areas of warming zones at different levels of sea surface temperature increase in different tidal states.

Table 4. Total area (A_{total}) of warming zones and areas of warming zones at different levels of sea surface temperature increase ($A_{+2°C}$ to $A_{+7°C}$) in different tidal states and seasons.

Season\Area (km^2)	Tidal State		$A_{+2°C}$	$A_{+3°C}$	$A_{+4°C}$	$A_{+5°C}$	$A_{+6°C}$	$A_{+7°C}$	A_{total}
Spring	STs	FTs	12.22	3.15	2.28	0.62	0.71	0.27	19.25
		ETs	18.14	9.36	3.21	1.93	0.65	0.25	33.54
	NTs	FTs	13.80	3.60	1.53	0.55	0.33	0.19	20.0
		ETs	11.81	7.01	5.35	3.32	1.5	1.07	30.06
Summer	STs	FTs	7.81	13.01	1.29	0.60	0.26	0.10	23.07
		ETs	7.88	11.36	12.97	0.51	0.42	0.17	33.31
	NTs	FTs	16.74	8.95	3.63	1.02	0.46	0.29	31.09
		ETs	36.0	17.85	1.91	0.57	0.45	0.26	56.19
Fall	STs	FTs	3.67	2.60	1.21	0.39	0.38	0.18	8.43
		ETs	20.42	6.77	1.75	0.69	0.37	0.17	30.17
	NTs	FTs	15.58	3.79	0.69	0.65	0.28	0.16	21.15
		ETs	7.68	1.93	0.96	0.62	0.23	0.18	11.6
Winter	STs	FTs	6.01	1.89	1.09	0.25	0.04	0.02	9.3
		ETs	13.39	1.6	1.06	0.37	0.05	0.01	16.48
	NTs	FTs	2.73	2.21	0.80	0.51	0.48	0.09	6.82
		ETs	7.5	2.71	1.67	0.33	0.12	0.16	12.49

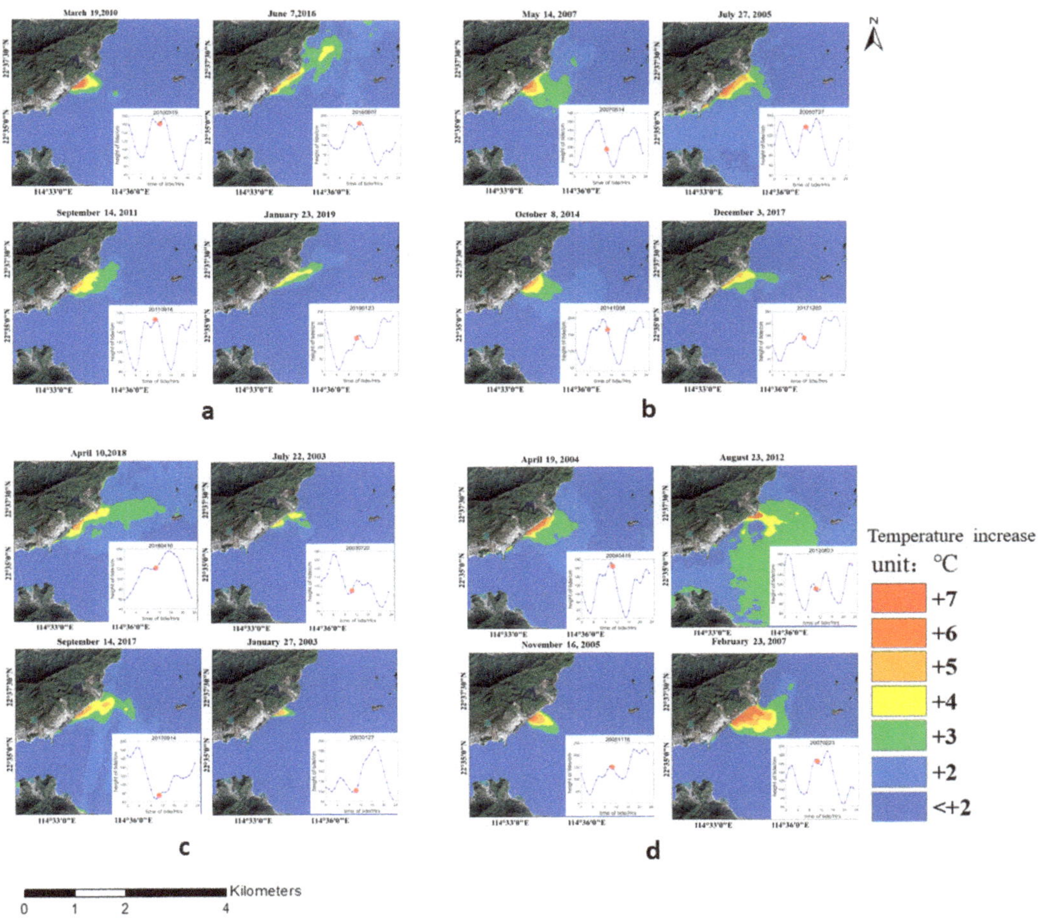

Figure 11. Areas of warming zones in different tidal states (images are for typical tidal states in different seasons): (**a**) peak spring flood tides; (**b**) peak spring ebb tides; (**c**) peak neap flood tides; (**d**) peak neap ebb tides.

5.3. Effects of Monsoons on Thermal Discharge

A marked monsoon climate prevails in Daya Bay. Figure 12 shows wind speeds and directions extracted from wind field data. In winter, the only winds over Daya Bay were easterlies (average speed: 7.7 m/s). In summer, most (approximately 60%) winds were from the west, with an average speed of only 3.4 m/s, whereas the rest (approximately 40%) originated from the east, with an average speed of 4.1 m/s. Figure 13 shows areas of warming zones corresponding to different wind speeds. In summer, A_{total} was approximately twice that in spring, regardless of whether winds were favorable (westerly) or adverse (easterly). In addition, in summer, A_{total} was greater with adverse winds than with favorable winds, which might be because the effects of tides on thermal discharge exceeded those of winds [34]. Under adverse winds, the areas of warming zones with low SST increases (i.e., $A_{+2^{\circ}C}$ and $A_{+3^{\circ}C}$), those of warming zones with high SST increases (i.e., $A_{+6^{\circ}C}$ and $A_{+7^{\circ}C}$), and A_{total} in summer were approximately five times those in winter. Several factors could explain those results. In summer, the waters of Daya Bay have inherently relatively higher temperatures, resulting in relatively poor conditions for seawater exchange. Exchange between seawater within Daya Bay and open seawater occurs primarily through the mouth of the bay. Moreover, the tidal range is small in summer, and changes in tidal currents are controlled by tides [6]. In spring, A_{total} under favorable winds differed from that under adverse winds by 1.35 km^2, and in fall, A_{total} under favorable winds was 70% greater than that under adverse winds.

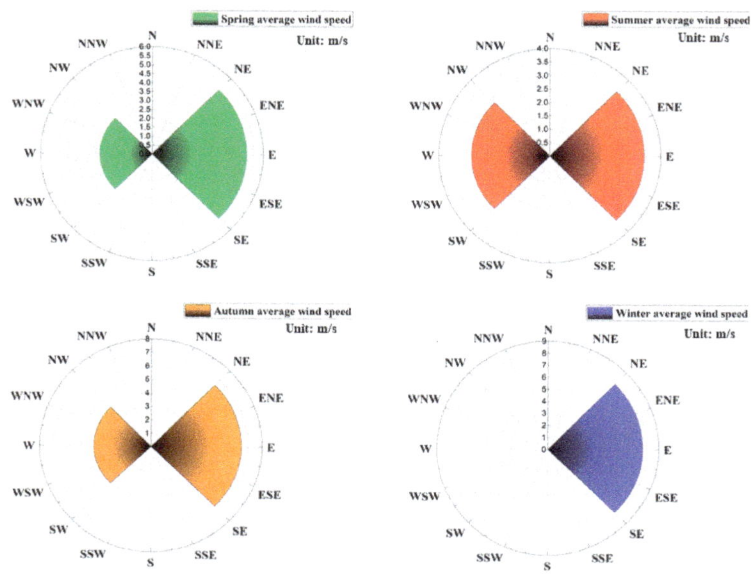

Figure 12. Average wind speed (m/s) in different seasons over Daya Bay, China.

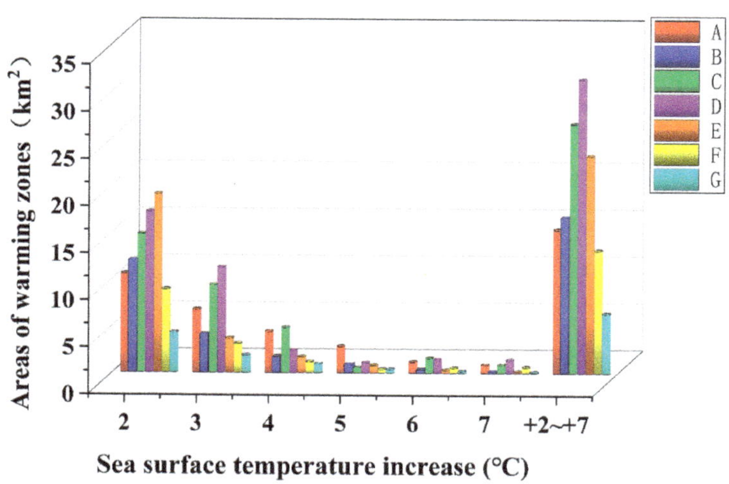

Figure 13. Areas of warming zones corresponding to different wind direction in different seasons (A, B, C, D, E, F, and G are spring westerlies, spring easterlies, summer westerlies, summer easterlies, autumn westerlies, autumn easterlies, and winter easterlies, respectively).

Table 4, Figures 10 and 13 were combined to produce Table 5. In spring, total area of warming zones with SST increases at each of the +2 °C to +7 °C levels and A_{total} during PETs (PSETs and PNETs) were twice the respective seasonal averages, whereas $A_{+3°C}$ and A_{total} under favorable winds were only ~30% greater than the respective seasonal averages. In summer, during PETs (PSETs and PNETs), $A_{+3°C}$ and A_{total} values were ~50% greater than those of respective seasonal averages. However, under favorable winds in summer, $A_{+4°C}$ and $A_{+6°C}$ were, to some extent, greater than the respective seasonal averages, whereas areas at other temperatures were below respective seasonal averages. In fall, during PETs, $A_{+2°C}$, $A_{+3°C}$, and A_{total} values were approximately twice those of respective seasonal averages. However, under favorable winds in fall, only $A_{+2°C}$ was approximately twice the seasonal average, whereas areas at other temperatures differed only slightly from respective seasonal averages. In winter, when all winds over Daya Bay are adverse, $A_{+2°C}$ and A_{total} values during ETs were approximately twice those of respective seasonal averages.

Table 5. Changes in total area (A_{total}) of warming zones and areas of warming zones at different levels of sea surface temperature increase ($A_{+2°C}$ to $A_{+7°C}$) under different tide and wind conditions.

Season	Conditions	$A_{+2°C}$ (km^2)	$A_{+3°C}$ (km^2)	$A_{+4°C}$ (km^2)	$A_{+5°C}$ (km^2)	$A_{+6°C}$ (km^2)	$A_{+7°C}$ (km^2)	A_{total} (km^2)
Spring	Average	8.83	3.6	1.8	1.06	0.44	0.3	16.03
	ETs	14.98	8.18	4.28	2.62	1.01	0.66	31.73
	Favorable winds	9.51	4.72	2.38	1.81	1.2	0.87	20.28
Summer	Average	15.67	9.49	3.29	1.37	0.79	1.0	31.58
	ETs	21.95	14.61	5.4	1.54	1.04	0.6	45.14
	Favorable winds	14.74	9.31	4.81	0.57	1.58	0.86	26.49
Fall	Average	9.4	3.51	1.69	0.64	0.55	0.56	16.35
	ETs	20.42	6.77	1.75	0.69	0.37	0.17	30.17
	Favorable winds	18.97	3.64	1.7	0.82	0.32	0.17	23.1
Winter	Average	4.23	1.84	0.94	0.4	0.29	0.19	7.89
	ETs	10.45	2.16	1.37	0.35	0.09	0.09	14.51

In summary, the effects of tides on seasonal areas of warming zones exceeded those of favorable winds, suggesting that the effects of tides on the diffusion of thermal discharge surpass those of winds. In addition, the shapes and distributions of warming zones depended primarily on tides instead of winds.

6. Conclusions

Seasonal and interannual changes and the factors influencing changes in thermal discharge from the NPPs in Daya Bay were examined for the period 1993–2020. The conclusions are summarized below.

(1) As indicated by an R^2 value of 0.89, the SST inversion algorithm of Landsat-5 and -7 imagery could adequately retrieve temperature-increase data for Daya Bay.

(2) Temporal and spatial analyses of the retrieved time series data from the period 1993 to 2020 revealed that the range of warming zones has expanded to a certain extent. In terms of interannual changes, SSTs increased as the installed capacity of the NPPs increased. There was a relatively significant linear relation between SST and the installed capacity of the NPPs. An increase in installed capacity resulted in more heat released into the sea, which in turn resulted in a considerable increase in SST over a relatively large area. Warming zones with SST increases at +5 °C to +7 °C levels have remained near discharge outlets of the NPPs since 2000. In 2019, the effects of thermal discharge from the NPPs peaked, and A_{total} also reached its maximum (24.28 km^2). Increases in SST occurred primarily at the +2 °C to +5 °C levels, but there was also a notable increase in $A_{+6°C}$ and $A_{+7°C}$.

(3) The effects of tides on areas of warming zones exceeded those of winds. Specifically, both FTs (PSFTs and PNFTs) and ETs (PSETs and PNETs) affected areas of warming zones.

The highest and lowest values of A_{total} (29.23 and 14.5 km^2, respectively) occurred during PNETs and PSFTs, respectively. A_{total} was greater during PETs (both PSETs and PNETs) than during PFTs in each season, except in fall, when A_{total} was smaller during PNETs than during PNFTs. Therefore, the effects of ETs on areas of warming zones exceeded those of FTs. Favorable winds promoted the diffusion of thermal discharge, whereas adverse winds inhibited it and could even alter the original direction of diffusion. Under either favorable (westerly) or adverse (easterly) winds, A_{total} in summer was approximately twice that in spring. In addition, warming zones in summer were larger under adverse winds than under favorable winds.

Therefore, if we want to minimize the warming zone area, we must focus on the dynamic environment first. For example, reducing land reclamation in the Daya Bay and reducing seabed sedimentation may be more effective approaches.

Author Contributions: Conceptualization, D.W.; methodology, Z.Z.; software, validation, Z.Z.; formal analysis, Z.Z. and F.G.; resources, F.G. and D.W.; writing—original draft preparation, Z.Z.; writing—review and editing, D.W. and Y.C.; supervision, project administration, and funding acquisition, D.W. All authors have read and agreed to the published version of the manuscript.

Funding: This research was funded by the National Key R&D Program of China, grant nos. 2017YFC 1405300 and 2018YFB0505005; the Key Research and Development Plan of Zhejiang Province, contract no. 2017C03037; the National Natural Science Foundation of China, contract nos. 41476157 and 41776029; and the Marine Science and Technology Cooperation Project between the Maritime Silk Route and island countries based on marine sustainability.

Data Availability Statement: Not applicable.

Acknowledgments: The authors thank the USGS (United States Geological Survey) and NASA (National Aeronautics and Space Administration) for excellent Landsat and MODIS data, respectively. We also thank the satellite ground station and the satellite data processing and sharing center of SOED/SIO for help with data processing. Our deepest gratitude goes to the editors and reviewers for their careful work and thoughtful suggestions.

Conflicts of Interest: The authors declare no conflict of interest.

Appendix A

Table A1. Matching MODIS Sea surface temperature (SST) and Landsat files.

MODIS Product File	Landsat File	MODIS Product File	Landsat File
AQUA_MODIS.20030722T051506.L2.SST.nc	LT05_L1TP_121044_20030722_20161205_01_T1	AQUA_MODIS.20030823T051506.L2.SST.nc	LT05_L1TP_121044_20030823_20161204_01_T1
AQUA_aODIS.20031028T050505.L2.SST.nc	LT05_L1TP_121044_20031026_20161204_01_T1	AQUA_MODIS.20031213T052006.L2.SST.nc	LT05_L1TP_121044_20031213_20161204_01_T1
AQUA_MODIS.20040213T053006.L2.SST.nc	LT05_L1TP_121044_20040215_20161202_01_T1	AQUA_MODIS.20040422T055006.L2.SST.nc	LT05_L1TP_121044_20040419_20161201_01_T1
AQUA_MODIS.20040926T052006.L2.SST.nc	LT05_L1TP_121044_20040926_20161129_01_T1	AQUA_MODIS.20030823T051506.L2.SST.nc	LT05_L1TP_121044_20041012_20161129_01_T1
AQUA_MODIS.20041013T060005.L2.SST.nc	LT05_L1TP_121044_20041012_20161129_01_T1	AQUA_MODIS.20041129T052006.L2.SST.nc	LT05_L1TP_121044_20041129_20161128_01_T1
AQUA_MODIS.20041215T052006.L2.SST.nc	LT05_L1TP_121044_20041215_20161127_01_T1	AQUA_MODIS.20050116T052006.L2.SST.nc	LT05_L1TP_121044_20050116_20161128_01_T1
AQUA_MODIS.20050306T060006.L2.SST.nc	LT05_L1TP_121044_20050305_20161128_01_T1	AQUA_MODIS.20051013T053006.L2.SST.nc	LT05_L1TP_121044_20051015_20161124_01_T1
AQUA_MODIS.20061019T060508.L2.SST.nc	LT05_L1TP_121044_20061018_20161119_01_T1	AQUA_MODIS.20061205T052008.L2.SST.nc	LT05_L1TP_121044_20061205_20161117_01_T1
AQUA_MODIS.20061221T052008.L2.SST.nc	LT05_L1TP_121044_20061221_20161118_01_T1	AQUA_MODIS.20070207T052008.L2.SST.nc	LT05_L1TP_121044_20070207_20161117_01_T1
AQUA_MODIS.20070223T052007.L2.SST.nc	LT05_L1TP_121044_20070223_20161116_01_T1	AQUA_MODIS.20070514T052007.L2.SST.nc	LT05_L1TP_121044_20070514_20161115_01_T1
AQUA_MODIS.20071005T052007.L2.SST.nc	LT05_L1TP_121044_20071005_20161110_01_T1	AQUA_MODIS.20080516T052006.L2.SST.nc	LT05_L1TP_121044_20080516_20161031_01_T1
AQUA_MODIS.20081023T052506.L2.SST.nc	LT05_L1TP_121044_20081023_20161029_01_T1	AQUA_MODIS.20081210T052506.L2.SST.nc	LT05_L1TP_121044_20081210_20161028_01_T1
AQUA_MODIS.20090111T052508.L2.SST.nc	LT05_L1TP_121044_20090111_20161028_01_T1	AQUA_MODIS.20091010T052508.L2.SST.nc	LT05_L1TP_121044_20091010_20161019_01_T1
AQUA_MODIS.20101029T052007.L2.SST.nc	LT05_L1TP_121044_20101029_20161012_01_T1	AQUA_MODIS.20101130T052008.L2.SST.nc	LT05_L1TP_121044_20101130_20161012_01_T1
AQUA_MODIS.20110101T052008.L2.SST.nc	LT05_L1TP_121044_20110101_20161011_01_T1	AQUA_MODIS.20110202T052007.L2.SST.nc	LT05_L1TP_121044_20110202_20161010_01_T1
AQUA_MODIS.20110407T052008.L2.SST.nc	LT05_L1TP_121044_20110407_20161208_01_T1	AQUA_MODIS.20110914T052007.L2.SST.nc	LT05_L1TP_121044_20110914_20161006_01_T1
AQUA_MODIS.20121009T052509.L2.SST.nc	LE07_L1TP_121044_20121010_20161128_01_T1	AQUA_MODIS.20131005T052009.L2.SST.nc	LC08_L1TP_121044_20131005_20170429_01_T1
AQUA_MODIS.20141008T052008.L2.SST.nc	LC08_L1TP_121044_20141008_20180205_01_T1	AQUA_MODIS.20141125T052000.L2.SST.nc	LC08_L1TP_121044_20141125_20170417_01_T1
AQUA_MODIS.20150925T052009.L2.SST.nc	LC08_L1TP_121044_20150925_20170403_01_T1	AQUA_MODIS.20160623T052009.L2.SST.nc	LC08_L1TP_121044_20160623_20170323_01_T1
AQUA_MODIS.20161216T052010.L2.SST.nc	LC08_L1TP_121044_20161216_20180205_01_T1	AQUA_MODIS.20170102T060511.L2.SST.nc	LC08_L1TP_121044_20170101_20170312_01_T1
AQUA_MODIS.20171101T052010.L2.SST.nc	LC08_L1TP_121044_20171101_20171109_01_T1	AQUA_MODIS.20171117T052010.L2.SST.nc	LC08_L1TP_121044_20171117_20171122_01_T1
AQUA_MODIS.20171203T052001.L2.SST.nc	LC08_L1TP_121044_20171203_20171207_01_T1	AQUA_MODIS.20171219T052001.L2.SST.nc	LC08_L1TP_121044_20171219_20171224_01_T1
AQUA_MODIS.20180309T052001.L2.SST.nc	LC08_L1TP_121044_20180309_20180320_01_T1	AQUA_MODIS.20181003T052001.L2.SST.nc	LC08_L1TP_121044_20181003_20181010_01_T1
AQUA_MODIS.20190123T052001.L2.SST.nc	LC08_L1TP_121044_20190123_20190205_01_T1	AQUA_MODIS.20190312T052001.L2.SST.nc	LC08_L1TP_121044_20190312_20190325_01_T1
AQUA_MODIS.20190920T052000.L2.SST.nc	LC08_L1TP_121044_20190920_20190926_01_T1	AQUA_MODIS.20191022T052001.L2.SST.nc	LC08_L1TP_121044_20191022_20191030_01_T1
AQUA_MODIS.20191123T052001.L2.SST.nc	LC08_L1TP_121044_20191123_20191203_01_T1		

References

1. Liang, S.S.; Zhang, B.; Li, J.S.; Zhang, H.; Shen, Q. Distribution of therm-water pollution of nuclear powerplant using the thermal infrared Band of HJ-IRS data-taking Daya Bay as an example. *Remote Sens. Inf.* **2012**, *2*, 43–48.
2. Verones, F.; Hanafiah, M.M.; Pfister, S.; Huijbregts, M.A.J.; Pelletier, G.J.; Koehler, A. Characterization factors for thermal pollution in freshwater aquatic environments. *Environ. Sci. Technol.* **2010**, *44*, 9364–9369. [CrossRef] [PubMed]
3. McMillin, L.M. Estimation of sea surface temperatures from two infrared window measurements with different absorption. *J. Geophys. Res.* **1975**, *80*, 80–82. [CrossRef]
4. Loncan, L.; de Almeida, L.B.; Bioucas-Dias, J.M.; Briottet, X.; Chanussot, J.; Dobigeon, N.; Fabre, S.; Liao, W.; Licciardi, G.A.; Simoes, M.; et al. Hyperspectral pansharpening: A review. *IEEE Geosci. Remote Sens. Mag.* **2015**, *3*, 27–46. [CrossRef]
5. Liu, L.; Zhou, J. Using MODIS imagery to map sea surface temperature. *Geospat. Inf.* **2006**, *4*, 7–9.
6. Rozenstein, O.; Qin, Z.; Derimian, Y.; Karnieli, A. Derivation of land surface temperature for Landsat-8 TIRS using a split window algorithm. *Sensors* **2014**, *14*, 5768–5780. [CrossRef]
7. Chen, H.; Zhu, L.; Li, J.; Fan, X. A comparison of two mono-window algorithms for retrieving sea surface temperature from Landsat8 data in coastal water of Hongyan River nuclear power station. *Remote Sens. Land Resour.* **2018**, *30*, 45–53.
8. Ai, B.; Wen, Z.; Jiang, Y.; Gao, S.; Lv, G. Sea surface temperature inversion model for infrared remote sensing images based on deep neural network. *Infrared Phys. Technol.* **2019**, *99*, 231–239. [CrossRef]
9. Zhang, H.R.; Zhao, Y.; Yang, H.; Chen, B.R.; Ding, J.; Dai, G.X.; Chuan, P.J. Study on the temperature rise characteristics and influence effects of thermal discharge from coastal power plant in Xiangshan Bay. *J. Shanghai Ocean Univ.* **2013**, *22*, 274–281.
10. Jia, H.L.; Zheng, S.; Xie, J.; Ying, X.M.; Zhang, C.P. Influence of geographic setting on thermal discharge from coastal power plants. *Mar. Pollut. Bull.* **2016**, *111*, 106–114. [CrossRef]
11. Liu, R.; Wang, Y.G.; Huang, H.M.; Hua, X. Research on effect of water depthand flow intensity in coastal power plant outfall on warming area. *J. Waterway Harbor* **2017**, *38*, 26–30.
12. Lentz, S.J.; Largier, J. The influence of wind forcing on the Chesapeake Bay buoyant coastal current. *J. Phys. Oceanogr.* **2006**, *36*, 1305–1316. [CrossRef]
13. Jiang, R.; Wang, Y. Modeling the ecosystem response of the semi-closed Daya Bay to the thermal discharge from two nearby nuclear power plants. *Ecotoxicology* **2020**, *29*, 736–750. [CrossRef]
14. Zhang, X.; Zhang, J.; Shen, Y.; Zhou, C.; Huang, X. Dynamics of alkaline phosphatase activity in relation to phytoplankton and bacteria in a coastal embayment Daya Bay, South China. *Mar. Pollut. Bull.* **2018**, *131*, 736–744. [CrossRef] [PubMed]
15. Ye, Y.; Chen, K.; Zhou, Q.; Xiang, P.; Huo, Y.; Lin, M. Impacts of thermal discharge on phytoplankton in Daya Bay. *J. Coast. Res.* **2019**, *83*, 135–147. [CrossRef]
16. Wu, C.; Wang, Q.; Yang, Z.; Wang, W. Monitoring heated water pollution of the DaYaWan nuclear power plant using TM images. *Int. J. Remote Sens.* **2007**, *28*, 885–890. [CrossRef]
17. Liu, M.; Yin, X.; Xu, Q.; Chen, Y.; Wang, B. Monitoring of fine-scale warm drain-off water from nuclear power stations in the Daya Bay based on Landsat 8 data. *Remote Sens.* **2020**, *12*, 627. [CrossRef]
18. Fu, J.; Chen, C.; Guo, B.; Chu, Y.; Zheng, H. A split-window method to retrieving sea surface temperature from Landsat 8 thermal infrared remote sensing data in offshore waters. *Estuar. Coast. Shelf Sci.* **2020**, *236*, 106626. [CrossRef]
19. Schott, J.R.; Volchok, W.J. Thematic Mapper thermal infrared calibration. *Photogramm. Eng. Remote Sens.* **1985**, *51*, 1351–1357.
20. Sobrino, J.A.; Jiménez-Muñoz, J.C.; Paolini, L. Land surface temperature retrieval from Landsat TM 5. *Remote Sens. Environ.* **2004**, *90*, 434–440. [CrossRef]
21. Sekertekin, A.; Bonafoni, S. Land surface temperature retrieval from Landsat 5, 7, and 8 over rural areas: Assessment of different retrieval algorithms and emissivity models and toolbox implementation. *Remote Sens.* **2020**, *12*, 294. [CrossRef]
22. Yu, X.; Guo, X.; Wu, Z. Land surface temperature retrieval from Landsat 8 TIRS-comparison between radiative transfer equation-based method, split window algorithm and single channel method. *Remote Sens.* **2014**, *6*, 9829–9852. [CrossRef]
23. Sobrino, J.; Li, Z.; Stoll, M.; Becker, F. Multi-channel and multi-angle algorithms for estimating sea and land surface temperature with ATSR data. *Int. J. Remote Sens.* **1996**, *17*, 2089–2114. [CrossRef]
24. Arabi Aliabad, F.; Zare, M.; Ghafarian Malamiri, H. A comparative assessment of the accuracies of split-window algorithms for retrieving of land surface temperature using Landsat 8 data. *Model. Earth Syst. Environ.* **2021**, *7*, 2267–2281. [CrossRef]
25. Kerr, Y.H.; Lagouarde, J.P.; Imbernon, J. Accurate land surface temperature retrieval from AVHRR data with use of an improved split window algorithm. *Remote Sens. Environ.* **1992**, *41*, 197–209. [CrossRef]
26. Qin, Z.; Karnieli, A.; Berliner, P. A mono-window algorithm for retrieving land surface temperature from Landsat TM data and its application to the Israel-Egypt border region. *Int. J. Rem. Sens.* **2001**, *22*, 3719–3746. [CrossRef]
27. Jimenez-Munoz, J.C.; Cristobal, J.; Sobrino, J.A.; Soria, G.; Ninyerola, M.; Pons, X.; Pons, X. Revision of the Single-Channel Algorithm for Land Surface Temperature Retrieval from Landsat Thermal-Infrared Data. *IEEE Trans. Geosci. Remote Sens.* **2009**, *47*, 339–349. [CrossRef]
28. Li, X.; Ling, T.; Zhang, Y.; Zhou, Q. A 31-year global diurnal sea surface temperature dataset created by an ocean mixed-Layer model. *Adv. Atmos. Sci.* **2018**, *35*, 1443–1454. [CrossRef]
29. Zhou, Y.; Gong, C.; Kuang, D.; Hu, Y. Research on the distribution of temperature and drainage of nuclear power plants based on the thermal infrared band data of environmental disaster mitigation satellites. *J. Infrared Millim. Waves* **2012**, *31*, 544–549. [CrossRef]

30. Wang, R.; Yang, H.; Zhu, L.; Wu, C.; Chen, Y. Application of temperature rise envelop in thermal discharge from nuclear power plant. *Environ. Monit. Manag. Technol.* **2020**, *32*, 49–52. [CrossRef]
31. Lin, J.; Zou, X.; Huang, F.; Yao, Y. Quantitative estimation of sea surface temperature increases resulting from the thermal discharge of coastal power plants in China. *Mar. Pollut. Bull.* **2021**, *164*, 112020. [CrossRef] [PubMed]
32. Chen, C.S.; Liu, H.D.; Beardsley, R.C. An unstructured grid, finite-volume, three-dimensional, primitive equations ocean model: Application to coastal ocean and estuaries. *J. Atmos. Ocean. Technol.* **2003**, *20*, 159–186. [CrossRef]
33. Wang, G.L.; Xiong, X.J. Distribution and variation of warm water discharge in the coastal area of Tianwan. *Adv. Mar. Sci.* **2013**, *31*, 69–74.
34. Xie, F.; Liu, H.; Huang, H.; Song, X. Effects of thermal discharge and nutrients input on size structure of phytoplankton in Daya Ba. *J. Trop. Oceanogr.* **2008**, *37*, 55–64. [CrossRef]

remote sensing

MDPI

Article

Enhanced Turbulent Mixing in the Upper Ocean Induced by Super Typhoon Goni (2015)

Mengtian Qiao [1,2], Anzhou Cao [1,2,*], Jinbao Song [1], Yunhe Pan [1] and Hailun He [2]

1 Ocean College, Zhejiang University, Zhoushan 316021, China; 12134027@zju.edu.cn (M.Q.); songjb@zju.edu.cn (J.S.); 21934020@zju.edu.cn (Y.P.)
2 State Key Laboratory of Satellite Ocean Environment Dynamics, Second Institute of Oceanography, Ministry of Natural Resources, Hangzhou 310012, China; hehailun@sio.org.cn
* Correspondence: caoanzhou@zju.edu.cn

Abstract: Based on the satellite-observed sea surface temperature (SST) data, high-resolution Argo observations and hybrid coordinate model (HYCOM) reanalysis results, this study examined the upper ocean response to Super Typhoon Goni in 2015 in the western north Pacific and revealed the significant diapycnal diffusivity enhancement in the upper ocean induced by Goni. Results indicate that the maximum SST cooling caused by Goni was 7.7 °C, which is greater than the SST cooling caused by most typhoons. The severe SST cooling was related to the enhancement of turbulent mixing induced by Goni. To the right of the typhoon track, the diapycnal diffusivity enhancement in the upper ocean caused by Goni could reach three orders of magnitude, from O (10^{-5} m^2/s) to O (10^{-2} m^2/s) and could last at least 9 days after the passage of Goni. In contrast, the diapycnal diffusivity to the left of the typhoon track did not show significant variations. The enhancement of turbulent mixing was found to be consistent with Goni-induced near-inertial kinetic energy calculated from the HYCOM reanalysis results, which suggests that the enhanced turbulent mixing was caused by Goni-induced near-inertial waves.

Keywords: turbulent mixing; upper ocean response; Super Typhoon Goni; satellite observations; sea surface temperature; HYCOM reanalysis results

Citation: Qiao, M.; Cao, A.; Song, J.; Pan, Y.; He, H. Enhanced Turbulent Mixing in the Upper Ocean Induced by Super Typhoon Goni (2015). *Remote Sens.* **2022**, *14*, 2300. https://doi.org/10.3390/rs14102300

Academic Editors: Chung-Ru Ho, Xiaofeng Lin and Antony K. Liu

Received: 28 March 2022
Accepted: 8 May 2022
Published: 10 May 2022

1. Introduction

Tropical cyclones (TCs), commonly known as typhoons in the western Pacific, are disastrous weather systems generated and developed in the tropical ocean. According to the tropical cyclone classification issued by the China Meteorological Administrationtropical cyclone database (https://tcdata.typhoon.org.cn/, accessed on 20 April 2021), TCs can be classified into different categories based on the maximum wind speed near the center, i.e., tropical depression (10.8–17.1 m/s), tropical storm (17.2–24.4 m/s), several tropical storm (24.5–32.6 m/s), typhoon (32.7–41.4 m/s), severe typhoon (41.5–50.9 m/s) and super typhoon (≥51.0 m/s). In recent years, a lot of studies have focused on the interaction between the upper ocean and TCs. When a TC passes over the ocean, the bottom atmosphere draws energy and moisture from the warm upper ocean to maintain or increase its intensity [1,2]. At the same time, the intense wind stress of a TC can penetrate 100–200 m depths of the upper ocean and generate strong near-inertial internal currents. The strong near-inertial currents [O (1 m/s)] cause enhanced turbulent mixing through shear instability, which brings the cold water below the mixed layer to the sea surface and decreases the sea surface temperature (SST), ranging from 1 to 6 °C; in turn, the SST cooling inhibits the heat exchange between the air-sea interface and hence, limits the intensity of the development of a TC [3–14]. The oceanic response to a TC is affected by the intensity, size and translation speed of the TC [5,15–20]. On the global scale, TCs are responsible for 1.87 PW (11.05 W/m^2) of heat transfer annually from the ocean to the atmosphere [21].

Thus, more knowledge and a better understanding of the dynamic and thermal response of the upper ocean to a TC is urgently required to improve TC forecasting [2,22–24].

There are three major processes that control the SST cooling, i.e., oceanic diapycnal mixing, advection (mostly upwelling) and air-sea heat exchange [4]. It is demonstrated that nearly 75–90% of the TC-induced SST cooling is caused by diapycnal mixing, while the upwelling and air-sea heat exchange contribute less in the open ocean [4,8,25–27]. It is reported that approximately 15% of the peak ocean heat transport may be associated with the vertical mixing induced by TCs [28]. In addition, SST cooling is also associated with the ocean background conditions [29], such as mesoscale eddies and barrier layers [30]. The TC-induced SST cooling has significant asymmetry, which is greater to the right of the TC track than to the left in the northern hemisphere, owing to the asymmetry of TC-induced near-inertial currents and vertical mixing [4,31]. With the development of observation technology and research approaches, satellite remote sensing, in situ observations and numerical simulations have been widely used in the studies of oceanic response to TCs. For example, Zhang et al. [14] studied the upper ocean response to typhoon Kalmaegi in 2014, based on an array of buoys and moorings and a numerical model. Guan et al. [32] used satellite remote sensing observations to investigate SST cooling, which was induced by four typhoons in the Yellow Sea and the Bohai Sea in 2019, and explored the possible mechanisms. Combining the hybrid coordinate model (HYCOM) reanalysis results and moored observations, Cao et al. [33] and Yang et al. [34] explored the dynamical response of the upper South China Sea to Megi in 2014 and Noul in 2020, respectively.

Although the turbulent mixing is known to play a leading role in SST cooling caused by TCs, it is difficult to quantify the diapycnal diffusivity by conventional ship-based observations because of the extremely dangerous conditions under TCs, which limits our understanding to some extent. Fortunately, the temperature and salinity profiles measured by Argo floats provide us an opportunity to estimate the diapycnal diffusivity and understand the turbulent mixing induced by TCs [35]. In this study, based on 42 high-resolution temperature and salinity profiles measured by 6 Argo floats andsimultaneoussatellite-observed SST data and HYCOM reanalysis results, we investigated the dynamical and thermal response of the upper ocean to Super Typhoon Goni in 2015. The enhancement of turbulent mixing induced by Goni was quantified by estimating the diapycnal diffusivity based on the fine-scale parameterization method. The remainder of the paper is organized as follows. The data and analysis methodology are introduced in Section 2. In Section 3, the dynamical and thermal response of the upper ocean to Goni is shown. Finally, a discussion and conclusions complete the paper in Sections 4 and 5, respectively.

2. Data and Methodology

2.1. Super Typhoon Goni

Goni was a super typhoon that occurred in 2015 in the western Pacific. According to the best track data from the China Meteorological Administration tropical cyclone database (https://tcdata.typhoon.org.cn/, accessed on 20 April 2021, [36,37]). Goni first developed as a tropical storm east of the Mariana Islands (13°00′ N, 148°20′ E) at 1700 UTC on 15 August 2015. Thereafter, it quickly intensifiedinto a super typhoon with amaximum sustained wind speed of 52 m/s on 17 August. It weakened to a severe typhoon on 18 August but upgraded to a super typhoon on 19 August again. It soon weakened and continuedas a severe typhoon for three days, at about 150 km east of the Luzon Strait. Then, it moved to the northeast after 23 August through the Okinawa Trough. Figure 1 shows the track and intensity of Goni.

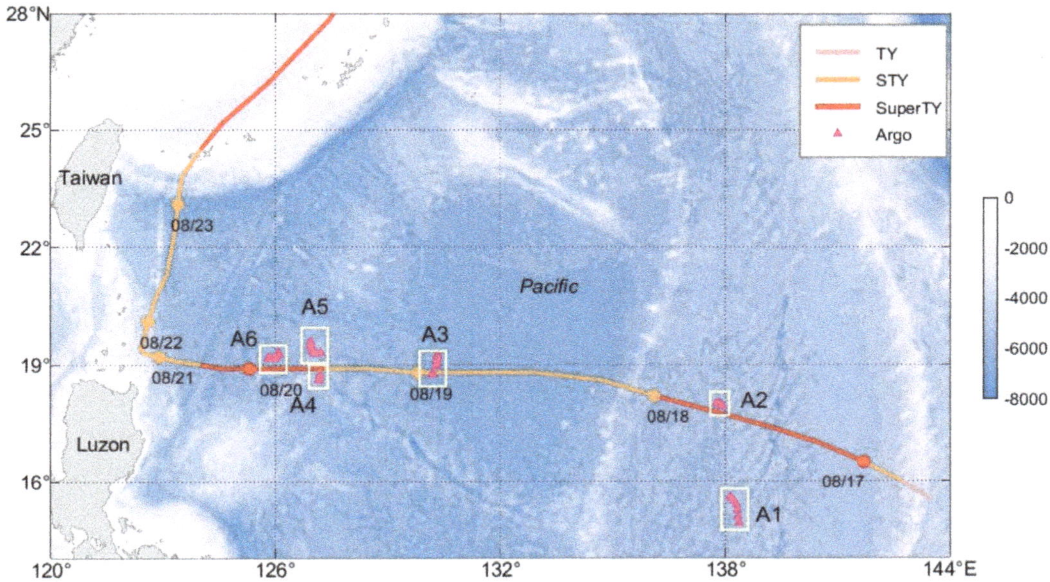

Figure 1. Bathymetry (shading, unit: m) of the study area, track and intensity of Goni in August 2015 (colored lines) and positions of Argo profiles (rose triangle, A1–A6). The time at 0000 UTC from 17 to 23 August is labeled. In the legend, TY, STY and SuperTY are abbreviations fortyphoon, severe typhoon and super typhoon, respectively.

2.2. Data

The temperature and salinity profile data observed by the Argo floats were obtained from the China Argo Real-time Data Centre (https://www.argo.org.cn/, accessed on 24 April 2021). To explore the turbulent mixing of the upper ocean caused by Goni, we searched Argo floats from 7 to 28 August 2015, which meet the following requirements: (1) The Argo float had observation profiles both before and after the passage of Goni; (2) The position of the Argo float was within 300 km from Goni's track; (3) Temperature and salinity profile data hada high vertical resolution (about 2 m). As shown in Figure 1, a total of 42 profiles measured by 6 Argo floats (A1–A6) were selected. Table 1 lists the detailed information of these Argo floats and profiles.

Table 1. Information onArgo floats and profiles.

Argo Float	WMO Number	Number of Profiles	Distance to Goni's Center (km)	Time of Typhoon Passed (mm/dd)	Observation Time of Profiles (mm/dd)
A1	2901199	7	227–295	08/17	08/09, 08/12, 08/15, 08/18, 08/21, 08/24, 08/26
A2	2901543	6	21–27	08/17	08/09, 08/12, 08/15, 08/20, 08/23, 08/26
A3	2901579	8	4–53	08/19	08/08, 08/10, 08/14, 08/16, 08/18, 08/20, 08/25, 08/27
A4	5904317	4	18–35	08/19	08/09, 08/14, 08/19, 08/24
A5	2901494	7	42–78	08/19	08/12, 08/14, 08/17, 08/20, 08/23, 08/26, 08/29
A6	2901578	10	30–47	08/19	08/09, 08/11, 08/13, 08/15, 08/17, 08/20, 08/22, 08/24, 08/26, 08/28

We downloaded the wind speed and direction data from the high-resolution Climate Forecast System, version 2 (CFSv2, https://rda.ucar.edu/, accessed on 9 May 2021) from the National Centers for Environmental Prediction (NCEP), which hasa spatial resolution of 38 km and a temporal interval of 6 h. As shown in Figure 2a–c, A2–A6 are nearly inside the typhoon center, but A1 is at the edge of the typhoon center. The translation speed of Goni is shown in Figure 2d.

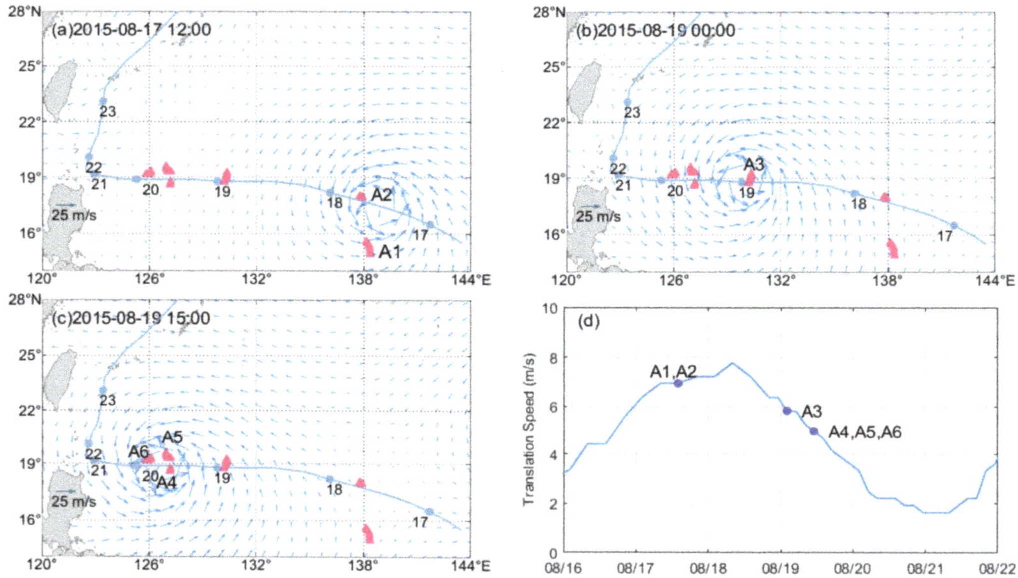

Figure 2. The wind speed and direction at (**a**) 1200 UTC on 17 August, (**b**) 0000 UTC on 19 August and (**c**) 1500 UTC on 19 August, respectively. (**d**) The translation speed of Goni from 16 to 22 August.The six Argo floats (A1–A6) are labeled.

We also used the microwave and infrared merged optimally interpolated SST data, which are provided by the remote sensing systems (https://www.remss.com/, accessed on 9 May 2021). The SST data have a spatial resolution of 1/4° and a temporal interval of one day. Moreover, the horizontal velocities derived from the HYCOM reanalysis results (GLBb0.08-53.X, https://www.hycom.org/, accessed on 21 May 2021) from 14 to 28 August with a spatial resolution of 1/12.5° and a temporal interval of 3 h were also used in this study. Appendix A showsthe comparison between the satellite-observed and HYCOM-simulated SST cooling, which validates thereasonability of the HYCOM reanalysis results.

2.3. Methodology

2.3.1. Gregg–Henyey–PolzinParameterization

Based on the internal wave–wave interaction theory [38], the Gregg–Henyey–Polzin (GHP) parameterization was used to estimate the diffusivity K:

$$K = K_0 \frac{\langle \xi_z^2 \rangle^2}{_{GM}\langle \xi_z^2 \rangle^2} h_2(R_\omega) j(f/N) \quad (1)$$

where $K_0 = 5 \times 10^{-6}$ m^{2}s^{-1}, $<\xi_z{}^2>$ and $_{GM}<\xi_z{}^2>$ are the strain variance derived from the observations and the Garrett–Munk model spectrum [39], respectively, and $h_2(R_\omega)$ and $j(f/N)$ are the correction items of internal wave structure and latitude:

$$h_2(R_\omega) = \frac{1}{6\sqrt{2}} \frac{R_\omega(R_\omega + 1)}{\sqrt{R_\omega - 1}} \tag{2}$$

$$j(f/N) = \frac{f\text{arccosh}(N/f)}{f_{30}\text{arccosh}(N_0/f_{30})} \tag{3}$$

where R_ω represents the shear/strain variance ratio, which was suggested to be a constant of 7 in the western north Pacific [15], f and N are the Coriolis and buoyancy frequencies, $f_{30} = f$ (30°), and $N_0 = 5.2 \times 10^{-3}$ rads^{-1}.

All the temperature and salinity profiles measured by the Argo floats were broken into 300-m segments to evaluate the strain spectra and then the segment-averaged diffusivity. According to [38,40,41], the GHP parameterization is not applicable in the upper ocean, because the strain spectrum may be contaminated due to great depth variability in the background stratification. Therefore, the temperature and salinity data in the upper 100 m were not used. The strain was calculated by:

$$\xi_z = \frac{N^2 - \overline{N^2}}{\overline{N^2}} \tag{4}$$

where $N^2(z) = -\frac{g}{\rho_0}\frac{d\sigma}{dz}$, obtained by the vertical difference of the potential density, and σ is the potential density. $\overline{N^2}$ is the mean value of stratification squared obtained by quadratic fitting of the potential density for each segment. Based on the multi-taper technique, Fourier transform gives the spectral representation $\varphi(k)$ for each segment [40,42,43]. Strain variance is determined by integrating $\varphi(k)$ from the lowest resolved wavenumber $k_{min} = 2\pi/150$ rads^{-1} to the maximum wavenumber k_{max} which satisfies

$$\left\langle \xi_z^2 \right\rangle = \int_{k_{min}}^{k_{max}} \varphi(k)dk = 0.1 \tag{5}$$

The strain corresponding to Garrett–Munk model spectrum is calculated as

$$_{GM}\left\langle \xi_z{}^2 \right\rangle = \frac{\pi E_0 b j_*}{2} \int_{k_{min}}^{k_{max}} \frac{k^2}{(k + k_*)^2} dk \tag{6}$$

where $E_0 = 6.3 \times 10^{-5}$, b = 1300 m is the scale depth of the thermocline, $j_* = 3$ is the reference mode number, $k_* = (\pi j_* N)/(bN_0)$ is the reference wave number [44].

2.3.2. Near-Inertial Kinetic Energy (NIKE)

In this study, the power spectral analysis was performed on the HYCOM horizontal velocities at 19.5° N, 132° E in August 2015. Based on the result of the power spectral analysis, the fourth-order Butterworth filter was adopted to extract the near-inertial velocities [33] with a cutoff frequency of [0.53, 0.87] cpd, corresponding to 0.80–1.30 times the local Coriolis frequency. Thereafter, the NIKE was calculated as

$$\text{NIKE} = \frac{1}{2}\rho_0\left(u_f{}^2 + v_f{}^2\right) \tag{7}$$

where ρ_0 = 1024 kgm^{-3} is the seawater density, u_f and v_f are the zonal and meridional near-inertial velocities, respectively.

3. Ocean Response to Super TyphonGoni

3.1. *Satellite-Observed SST Cooling*

Based on the satellite-observed data, the SST evolution during the passage of Goni from 16 to 28 August is shown in Figure 3a–k. On 16 August, when Goni did not enter the domain (Figure 3a), the SST was generally higher than 29 °C and the highest SST exceeding 31 °C appeared to the east of Luzon Island. On 17 August (Figure 3b), Goni entered the Philippine Sea, and the SST to the right of the typhoon track was cooled slightly. From 18 to 21 August (Figure 3c–f), as Goni moved westward, the range of SST cooling also moved westward and expanded. The SST cooling was enhanced on 22 and 23 August (Figure 3g,h) when Goni was about to leave this region. The lowest SST (smaller than 25 °C) appeared at 150 km east of the Luzon Strait, which was the location Goni turned northward. After 24 August when Goni left the domain, the SST was gradually heating (Figure 3i–k). On 28 August, the SST almost shared the same pattern as that of before the passage of Goni, except for the region to the east of the Luzon Strait and Luzon Island. Figure 3l displays the maximum SST cooling caused by Goni, which was calculated as the difference between the minimum SST from 17 to 24 August and the SST on 16 August. As shown, Goni caused significant SST cooling in the domain. The maximum SST cooling was 7.7 °C, appearing at about 150 km east ofthe Luzon Strait.

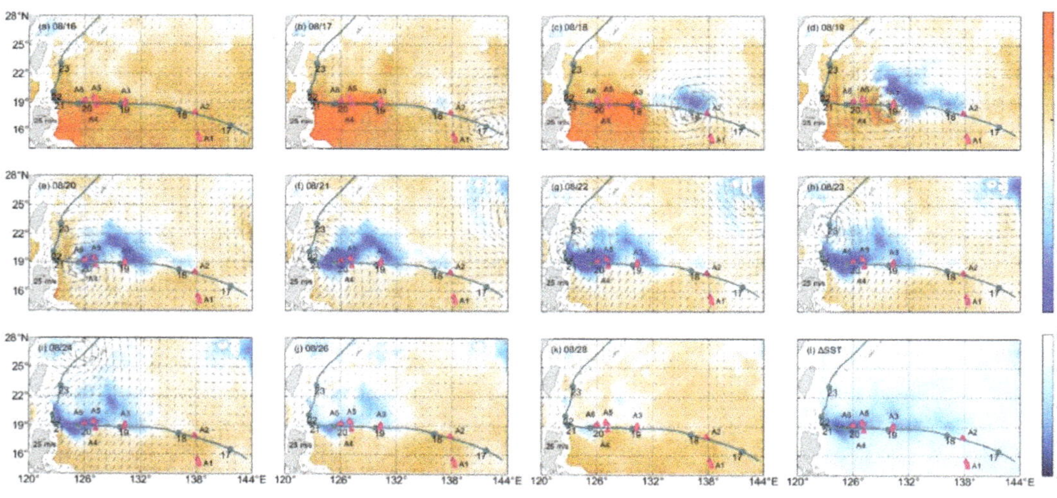

Figure 3. (**a**–**k**) SST evolution during the passage of Goni from 16 to 28 August. The gray quivers denote the CFSv2 winds at 0000 UTC. (**l**) The maximum SST cooling caused by Goni.

3.2. *Goni-Induced Mixing*

The diapycnal diffusivity was calculated by GHP parameterization to estimate Goni-induced mixing. As shown in Figure 2a, at 1200 UTC on 17 August, Goni passed over Argo floats A1 and A2. At the same time, it developed into a super typhoon with amaximum wind speed exceeding 51 m/s. Argo float A1 was located to the left of the typhoon track, about 200–300 km away from the typhoon center. Argo float A2 was about 20 km away to the right of the typhoon track. Figure 4a shows the temperature profiles observed by Argo float A1 and Figure 4b displays the corresponding diapycnal diffusivity from 100 m to 1000 m. The values in the brackets represent the observation time relative to the time Goni passed. Hence, negative and positive values correspond to the time before and after the passage of Goni, respectively. As shown, the temperature was nearly unchanged at Argo float A1 during the passage of Goni. This is consistent with SST cooling at Argo float A1, which was very small and close to zero (Figure 3l). At the same time, the diapycnal diffusivity was generally on the level of 10^{-5} m^2/s, the same order of background value of the abyssal ocean [38,45], and did not show significant variations during the passage of Goni (Figure 4b). All these results suggest that Goni did not enhance the turbulent mixing at Argo float A1.

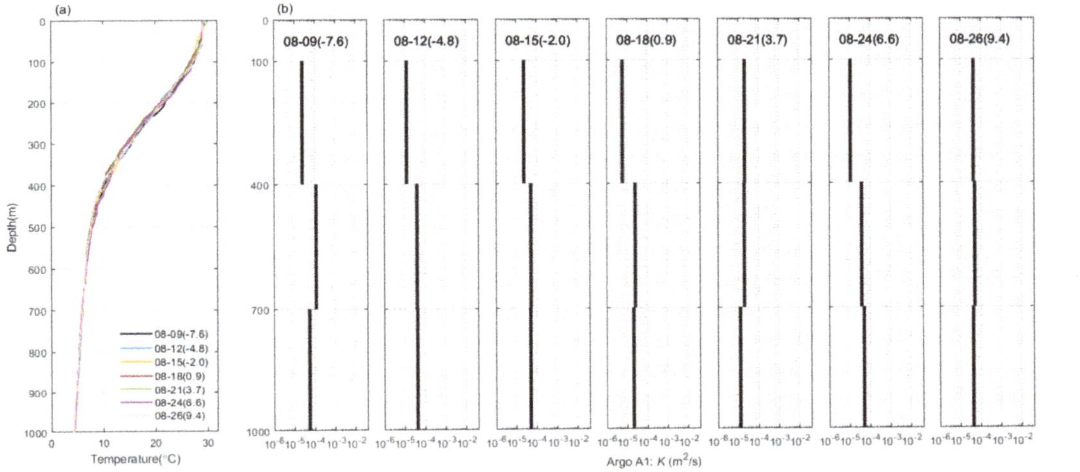

Figure 4. (**a**) Temperature profiles observed by Argo float A1 and (**b**) estimated diapycnal diffusivity based on the GHP parameterization. In each subfigure, the values in the brackets represent the observation time relative to the time Goni passed (unit: day).

Figure 5 displays the temperature profiles and diapycnal diffusivity at Argo float A2. Compared with Argo float A1, we can find a slight cooling of water temperature and a slight enhancement of diapycnal diffusivity at Argo float A2, especially in the upper 400 m. On 12 August, about 5.6 days before the passage of Goni, the diapycnal diffusivity was O (10^{-4} m^2/s). It increased to O (10^{-3} m^2/s) on 15 August and lasted to 20 August. After 23 August, the diapycnal diffusivity dropped to below 10^{-4} m^2/s.

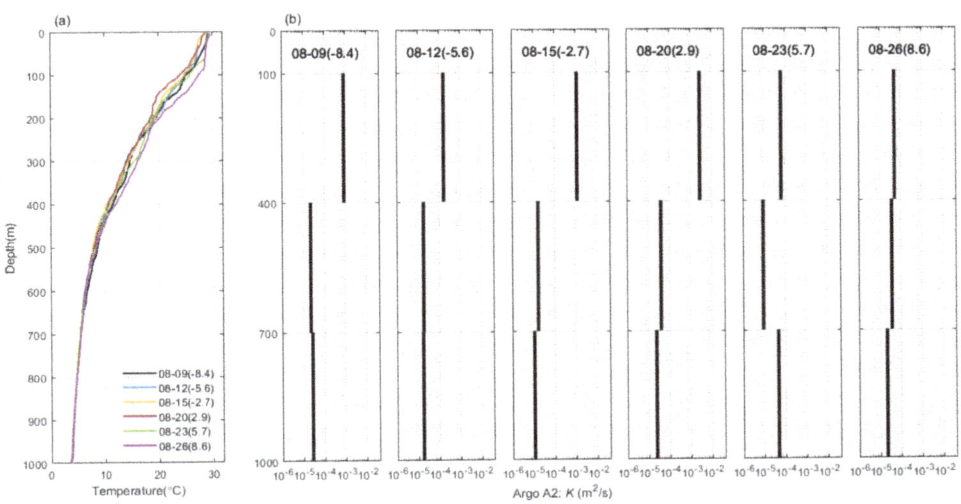

Figure 5. Same as Figure 4 but for Argo float A2.

At 0000 UTC on 19 August, Goni passed over Argo float A3 when it was a severe typhoon with the maximum wind speed exceeding 41 m/s. Argo float A3 was located about 4–50 km away to the right of the typhoon track (Figure 2b). Figure 6a shows the temperature profiles at Argo float A3, from which a continuous temperature cooling lastingfor more than 8 days in the upper 600 m can be detected. The maximum temperature cooling appeared around 150 m in depth, which exceeded 5 °C. Moreover, it is found that Goni caused significantly enhanced turbulent mixing at Argo float A3 (Figure 6b). Before the passage of Goni, the diapycnal diffusivity was generally below 10^{-4} m^2/s. Only one day after the passage of Goni, the diapycnal diffusivity at 100–400 m was increased to 5×10^{-4} m^2/s, which was nearly amplified by one order of magnitude. With time going on, the diapycnal diffusivity at 100–400 m was continuously increased. On 27 August, which was about 9 days after the passage of Goni, the diapycnal diffusivity was as high as 5×10^{-2} m^2/s. In contrast, the diapycnal diffusivity below a 400 m depth did not show significant enhancement and was generally atthe levels of 10^{-6} and 10^{-5} m^2/s.

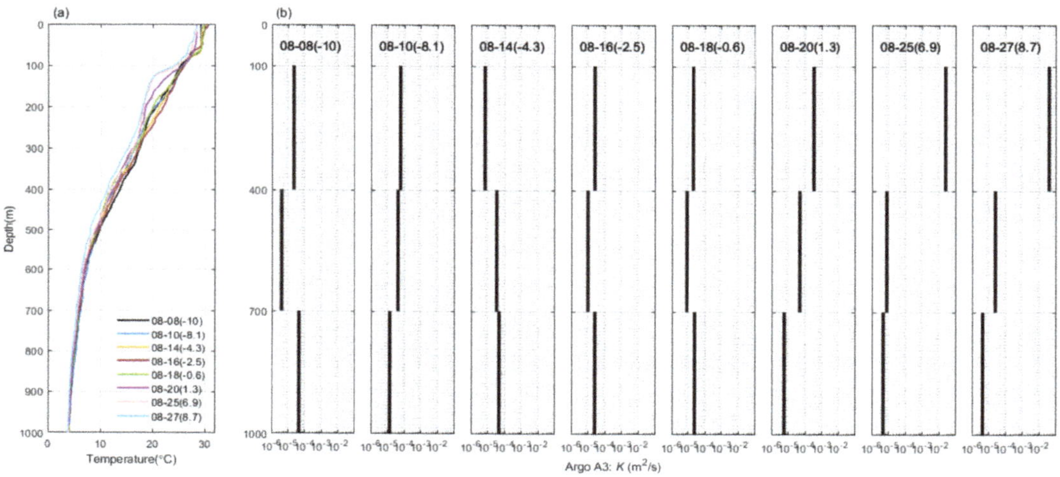

Figure 6. Same as Figure 4 but for Argo float A3.

As is shown in Figure 2c, at 1500 UTC on 19 August, Goni passed over Argo floats A4, A5 and A6 when it intensified again to a super typhoon with the maximum wind speed exceeding 51 m/s. Argo float A4 was about 30 km away to the left of the typhoon track, whereas Argo floats A5 and A6 were located 55 km and 35 km to the right of the typhoon track, respectively.

Figure 7a shows the temperature profiles at Argo float A4. It is easy to find that Goni caused temperature cooling in the upper 40 m but warming at 40–100 m five hours after its passage. However, the turbulent mixing was not significantly enhanced at the same time (Figure 7b). On 24 August (five days after the passage of Goni), the temperature nearly became the same as that before the passage of Goni.

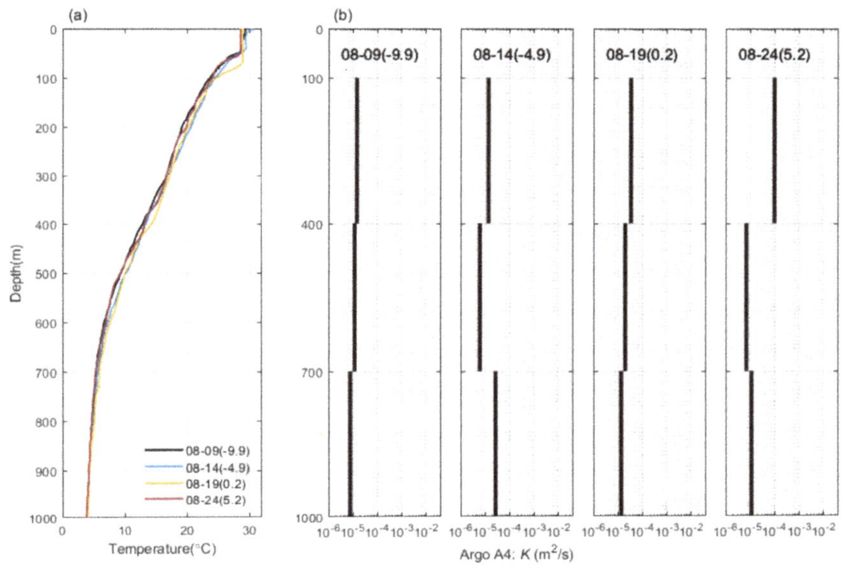

Figure 7. Same as Figure 4 but for Argo float A4.

Figure 8 shows the temperature profiles and diapycnal diffusivity at Argo float A5. It is clearly shown that after the passage of Goni, the temperature near the surface had a significant cooling of approximately 5 °C, which is consistent with the satellite observations shown in Figure 3l. At the same time, the turbulent mixing was remarkably enhanced: Before the typhoon, the diapycnal diffusivity was below 10^{-4} m^2/s; however, the diapycnal diffusivity at 100–400 m was rapidly increased to 3.5×10^{-4} m^2/s about one day after the typhoon (20 August); with time going on, the diapycnal diffusivity was gradually increased; about 4 days after the passage of Goni (23 August), the diapycnal diffusivity at 100–400 m was higher than 10^{-2} m^2/s and this phenomenon lasted to 29 August, approximately 10 days after the passage of Goni. Moreover, it is also found that the enhanced turbulent mixing was concentrated in the upper 400 m, whereas the diapycnal diffusivity below 400 m depth did not have a significant change.

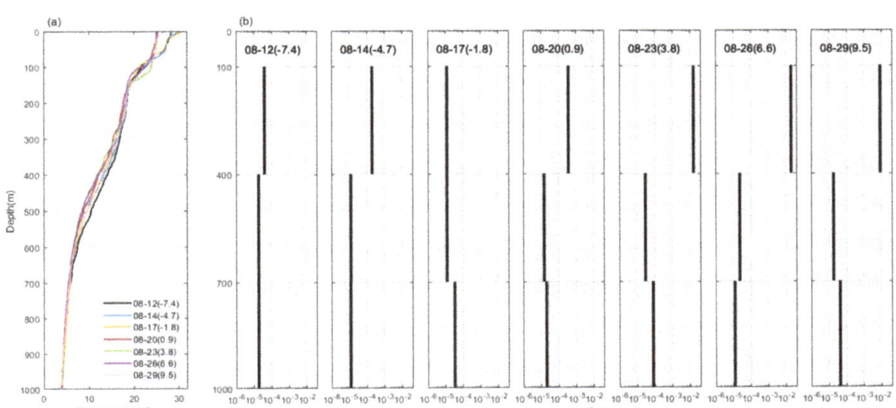

Figure 8. Same as Figure 4 but for Argo float A5.

As for Argo float A6, the maximum temperature cooling was greater than 7 °C (Figure 9a), which was even larger than the satellite-observed SST cooling (Figure 3l). The temperature cooling in the upper 80 m almost lasted approximately 9 days after the passage of Goni. However, the turbulent mixing at 100–400 m at Argo float A6 showed an interesting variation (Figure 9b). Before 15 August, the diapycnal diffusivity was generally below 10^{-4} m^2/s. On 17 August (approximately 3 days before the passage of Goni), the diapycnal diffusivity was slightly increased to be a little larger than 10^{-4} m^2/s. In the three days after the passage of Goni, the diapycnal diffusivity was rapidly increased to 1.5×10^{-2} m^2/s. Then, it rapidly decreased to 10^{-5} m^2/s on 24 August. After 26 August, the diapycnal diffusivity was increased again. On 28 August (approximately 9 days after the passage of Goni), the diapycnal diffusivity at 100–400 m became 10^{-2} m^2/s.

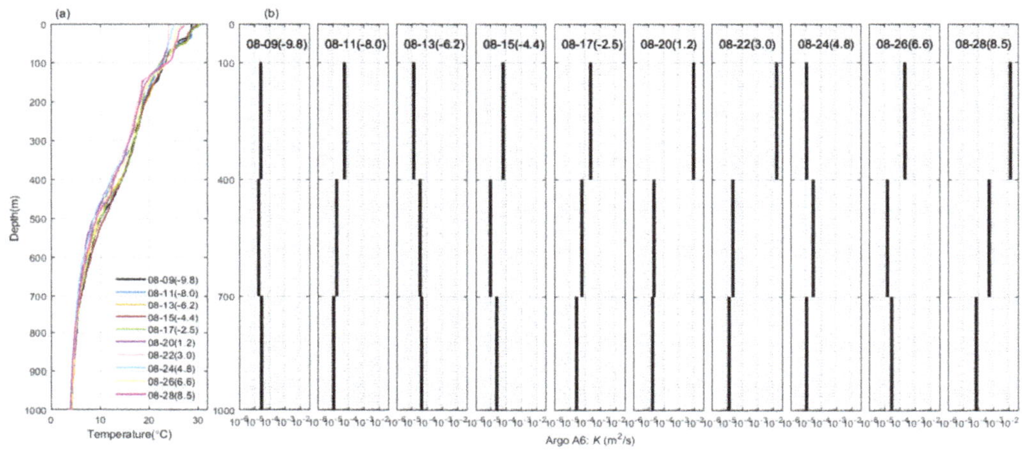

Figure 9. Same as Figure 4 but for Argo float A6.

To further understand Goni's influence on the turbulent mixing in the upper ocean, Figure 10 shows the diapycnal diffusivity at 100–400 m as a function of the observation time relative to the time when Goni passed. It is found that the diapycnal diffusivity at Argo float A1 nearly showed no enhancement after the passage of Goni. A similar result is found at Argo float A4, although Argo float A4 was much closer to Goni's track than A1. All the estimated diapycnal diffusivity at Argo floats A1 and A4 werebelow 10^{-4} m^2/s. We speculate that the cause is related to the locations of Argo floats A1 and A4, both of which were to the left of the typhoon track (Figure 1). In contrast, the diapycnal diffusivity showed enhancement in various degrees at the four Argo floats (A2, A3, A5 and A6) to the right of the typhoon track (Figure 1). At Argo float A2, the diapycnal diffusivity caused by Goni was increased from 10^{-4} m^2/s to 5×10^{-3} m^2/s; whereas at Argo floats A3, A5 and A6, the diapycnal diffusivity caused by Goni was enhanced by at least three orders of magnitude, from 10^{-5} m^2/s to more than 10^{-2} m^2/s. Moreover, at Argo floats A3 and A5, the diapycnal diffusivity was nearly increased continuously until approximately 9 days after the passage of Goni; while the diapycnal diffusivity at Argo floats A2 and A6 showed a rapid decrease 4 days after the passage of Goni. The diapycnal diffusivity at Argo float A6 was increased again 6 days after the passage of Goni, while it almost kept invariant at Argo float A2.

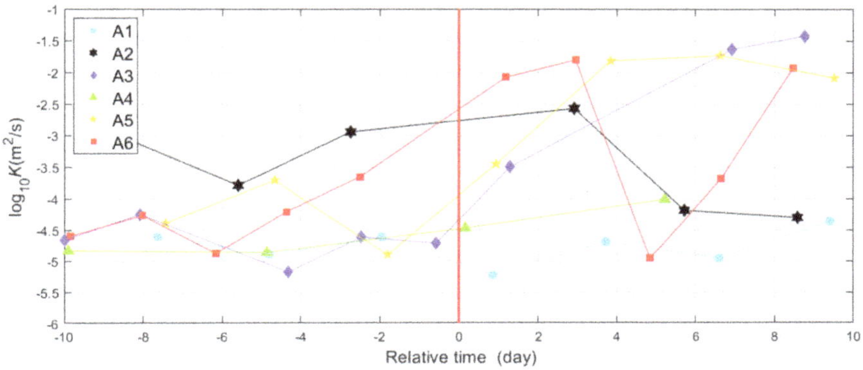

Figure 10. Diapycnal diffusivityversus the observation time relative to the time Goni passed.

3.3. Goni-Induced NIKE

To further study Goni's effect on the turbulent mixing in the upper ocean, Figure 11 illustrates the depth-integrated NIKE from the sea surface to the sea bottom of the HYCOM reanalysis results at 1200 UTC from 16 to 28 August. From 16 to 17 August, before Goni passed, the NIKE was at a low level. On 18 August when Goni's center reached 136.1° E, 18.2° N, only slight NIKE appeared to the northeast of Goni's center and the maximum NIKE was approximately 33 kJ/m^2. From 19 to 21 August, the NIKE was increased, and at the same time, the range of strong NIKE was gradually expanded (strong NIKE, exceeding 30 kJ/m^2, could reach about 330 km to the right of the typhoon track). On 21 August, the NIKE reached the maximum, which was greater than 80 kJ/m^2. Thereafter, both the NIKE and the area with strong NIKE were decreased. We also note, however, that the NIKE at 127–131° E and 18–22° N was still significant ($\geq$30 kJ/m^2) on 28 August, whereas the NIKE outside the region was quickly damped to below 20 kJ/m^2. Moreover, strong NIKE was found to be concentrated to the right of the track typhoon. To the left of the typhoon track, the depth-integrated NIKE was generally below 20 kJ/m^2 during the passage of Goni. Combining these results with the estimated diapycnal diffusivity at the 6 Argo floats, we can conclude that the strong (weak) turbulent mixing at Argo floats A2, A3, A5 and A6 (A1 and A4) was related to the strong (weak) NIKE to the right (left) of the typhoon track.

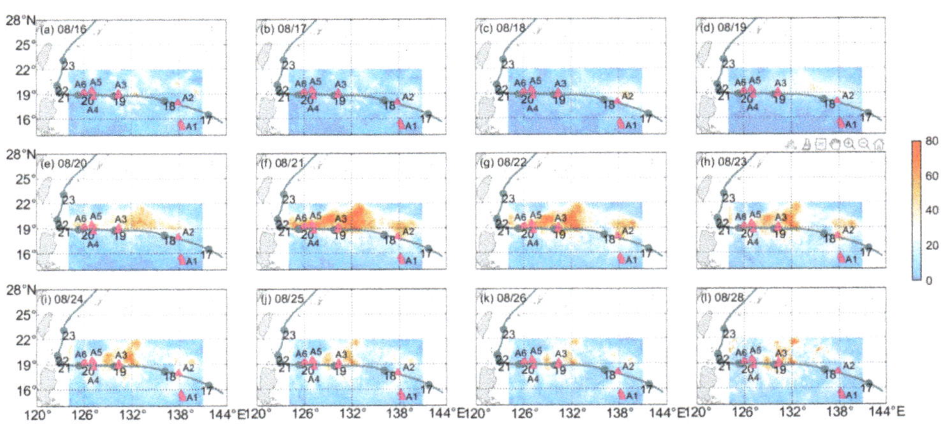

Figure 11. Depth-integrated NIKE (shading, unit: kJ/m^2) during the passage of Goni (**a**–**l**) from 16 to 28 August.

Since Argo floats A3, A5 and A6 were close to the typhoon track which was nearly along 19° N, Figure 12 illustrates the NIKE along 19.12° N from 16 to 28 August. Before 17 August, the NIKE at 0–800 m was generally below 50 J/m^3. On 18 August, strong NIKE exceeding 100 J/m^3 appeared in the upper 100 m between 132° E and 133° E. As Goni moved westward, the region with strong NIKE gradually expanded westward. From 19 to 23 August, strong NIKE exceeding 300 J/m^3 was concentrated in the upper 100 m. With the increase indepth, the NIKE decreased significantly, which is consistent with [33]. At 100–400 m depth, the NIKE was generally on the level of 50 J/m^3, whereas below 400 m depth, the NIKE could be one order of magnitude smaller than that at 100–400 m depth. This result can account for the enhanced turbulent mixing at 100–400 m and nearly invariant diapycnal diffusivity below 400 m depth at the Argo floats (Figures 5, 6, 8 and 9). Furthermore, although Goni had left the domain on 24 August, it can be detected from Figure 12 that the strong NIKE at 100–400 m depth could last to 28 August, which can account for the enhanced turbulent mixing at Argo floats A3, A5 and A6 on 27–29 August (Figures 6 and 8–10).

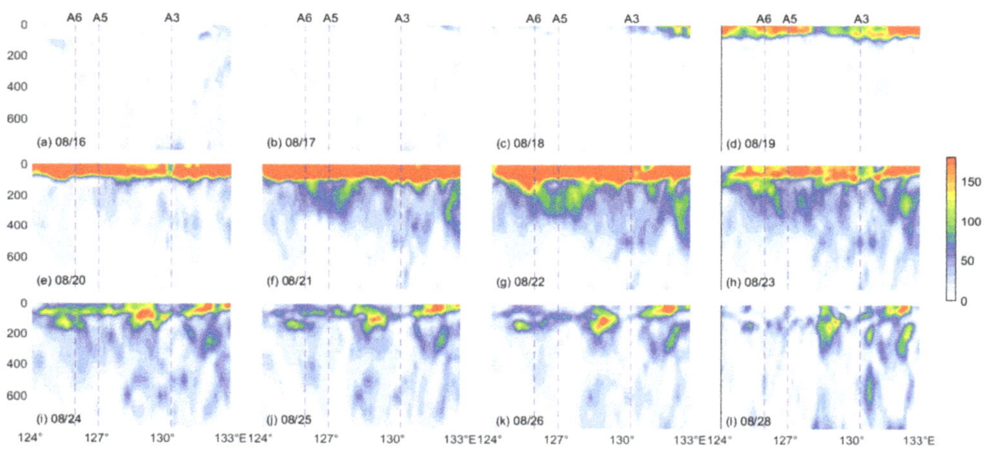

Figure 12. Profiles of NIKE (shading, unit: J/m^3) along 19.12° N (**a**–**l**) from 16 to 28 August.

4. Discussion

Satellite observations indicate that Super Typhoon Goni caused significant SST cooling in the western Pacific, which was mainly concentrated to the right of the typhoon track. This is consistent with the rightward biased feature of the ocean's response to a typhoon in the northern hemisphere [46]. The SST cooling induced by Goni could reach 660 km away from the typhoon track and the maximum SST cooling was 7.7 °C, which exceeded the SST cooling (1–6 °C) caused by most typhoons [4,8] and that (4.2 °C) caused by another Super Typhoon, Megi, in 2010 [27]. Moreover, it is found that the maximum SST cooling generally occurred about one day after the passage of Goni, and Goni-induced SST cooling could last for more than one week. This result is consistent with [47] that stronger SST cooling corresponds to a longer recovery time.

Previous studies have demonstrated that SST cooling is mainly related to the enhanced turbulent mixing caused by the typhoon [4,8,22,25,27]. In this study, six Argo floats with a high vertical resolution fortunately captured the temperature cooling and turbulent mixing enhancement induced by Goni in the upper ocean. At the four Argo floats (A2, A3, A5 and A6) to the right of the typhoon track, the temperature in the mixed layer was cooled to different degrees, and the diapycnal diffusivity was significantly enhanced. The consistency between the temperature cooling and diapycnal diffusivity enhancement indicates their correlation. It is interesting to find that at Argo floats A3, A5 and A6, the diapycnal diffusivity enhancement could reach three orders of magnitude, from O (10^{-5} m^2/s) to O (10^{-2} m^2/s), which, to the best of our knowledge, has not been reported. This diapycnal diffusivity enhancement, caused by Goni, is much greater than that caused by Super Typhoon Tembin in 2012 [35]. It is also found that the diapycnal diffusivity enhancement at Argo floats A3, A5 and A6 was mainly concentrated in the upper ocean, to be specific, at 100–400 m depth. Meanwhile, the diapycnal diffusivity below 400 m depth was generally atthe level of 10^{-5} m^2/s and did not exhibit significant variations before and after the passage of Goni. Moreover, the enhanced turbulent mixing in the upper ocean at Argo floats A3, A5 and A6 could last 9 days after the passage of Goni (Figure 10). In contrast, at the two Argo floats (A1 and A4) to the left of the typhoon track, either the observed temperature profiles or the estimated diapycnal diffusivity did not show significant variations (Figures 4 and 7) during the passage of Goni. Especially, the diapycnal diffusivity before and after the passage of Goni at Argo floats A1 and A4 was always on the level of 10^{-5} m^2/s, the same order of background value in the abyssal ocean [38,45].

The HYCOM reanalysis results further reveal that the enhanced turbulent mixing in the upper ocean was related to Goni-induced NIKE. Results show that the depth-integrated NIKE to the right of the typhoon could reach 80 kJ/m^2, whereas it was very small to the left (generally below 20 kJ/m^2). This result can explain the strong turbulent mixing at Argo floats A2, A3, A5 and A6 to the right of the typhoon track, and the weak turbulent mixing at Argo floatsA1 and A4 to the left. Moreover, the profiles of NIKE along 19.12° N indicate that the strongest NIKE exceeding 300 J/m^3 was concentrated in the upper 100 m. With the increase indepth, the NIKE was rapidly decreased: At 100–400 m depth, the NIKE was generally atthe level of 50 J/m^3, whereas below 400 m depth, the NIKE was nearly one order of magnitude smaller than that at 100–400 m depth. This result is generally consistent with [33] and could account for the enhanced turbulent mixing at 100–400 m and the nearly invariant diapycnal diffusivity below 400 m depth. Moreover, strong NIKE was found to last to 28 August, which is consistent with the duration of enhanced turbulent mixing in the upper ocean at Argo floats A3, A5 and A6. In a word, Goni-induced near-inertial waves caused a significant enhancement of turbulent mixing in the upper ocean, which finally led to severe SST cooling. Because Goni-induced near-inertial waves existed for more than one week after the passage of Goni (Figures 11 and 12), both an enhanced turbulent mixing and SST cooling lasted for more than one week (Figures 3 and 10).

5. Conclusions

Based on the satellite remote sensing, Argo measurements and HYCOM reanalysis results, this study investigates the oceanic dynamical and thermal response to Super Typhoon Goni in 2015 and highlights the enhanced turbulent mixing in the upper ocean caused by Goni. Results indicate that the super typhoon caused significant near-inertial waves in the upper ocean, which further enhanced the turbulent mixing. To the right of the typhoon track, the diapycnal diffusivity enhancement in the upper ocean, caused by Goni, could reach three orders of magnitude, from O (10^{-5} m^2/s) to O (10^{-2} m^2/s) and last at least 9 days after the passage of Goni. In contrast, the diapycnal diffusivity to the left of the typhoon track did not show significant variations. As a result, the maximum SST cooling caused by Goni was 7.7 °C, which is greater than the SST cooling caused by most typhoons, and the SST cooling exhibitedan apparent rightward biased feature. Because Goni-induced near-inertial waves existed for more than one week after the passage of Goni, both enhanced turbulent mixing and SST cooling lasted for more than one week.

This study reveals the significant SST cooling and diapycnal diffusivity enhancement in the upper ocean induced by Super Typhoon Goni; however, there still exists a problem thatis not solved, i.e., what causes the rapid decrease and reinforcement of diapycnal diffusivity at Argo float A6 from 22 to 26 August? This process is worthy to be investigated in the future.

Author Contributions: Conceptualization, A.C.; methodology, M.Q. and Y.P.; formal analysis, M.Q. and Y.P.; writing—original draft preparation, M.Q.; writing—review and editing, M.Q., A.C., J.S., Y.P. and H.H.; supervision, A.C. and J.S.; funding acquisition, A.C. and J.S. All authors have readand agreed to the published version of the manuscript.

Funding: This research was funded by the National Natural Science Foundation of China (Grant numbers: 42176002 and 41830533) and the open fund of the State Key Laboratory of Satellite Ocean Environment Dynamics, Second Institute of Oceanography, Ministry of Natural Resources (Grant number: QNHX2218).

Data Availability Statement: The best track data of Super Typhoon Goni were obtained from the China Meteorological Administration tropical cyclone database (https://tcdata.typhoon.org.cn/, accessed on 20 April 2021). The wind data were downloaded from the high-resolution Climate Forecast System, version 2 (https://rda.ucar.edu/, accessed on 9 May 2021). The Argo float data were obtained from the China Argo Real-time Data Centre (https://www.argo.org.cn/, accessed on 24 April 2021). The SST data were downloaded from the Remote Sensing Systems (http://www.remss.com/, accessed on 9 May 2021). The HYCOM reanalysis results (GLBb0.08-53.X) were obtained from https://www.hycom.org/ (accessed on 21 May 2021).

Conflicts of Interest: The authors declare no conflict of interest.

Appendix A

Figure A1 shows the Goni-induced SST cooling maps from HYCOM reanalysis data and satellite observations. On the whole, the HYCOM-simulated SST cooling is generally consistent with that from the satellite observations: the maximum SST cooling caused by Goniwas approximately −8 °C, which appeared to the right of Goni's track. In addition, the scatters of HYCOM-simulated and satellite-observed SST cooling are almost distributed along the line $y = x$ (Figure A2), which indicates the consistency between them again. The slight difference between them may be attributed to the different temporal intervals of HYCOM reanalysis results (3 h) and satellite observations (1 day). Based on the aforementioned results, we can conclude that the HYCOM reanalysis data reasonably and reliably reproduces the oceanic thermal response to Super Typhoon Goni.

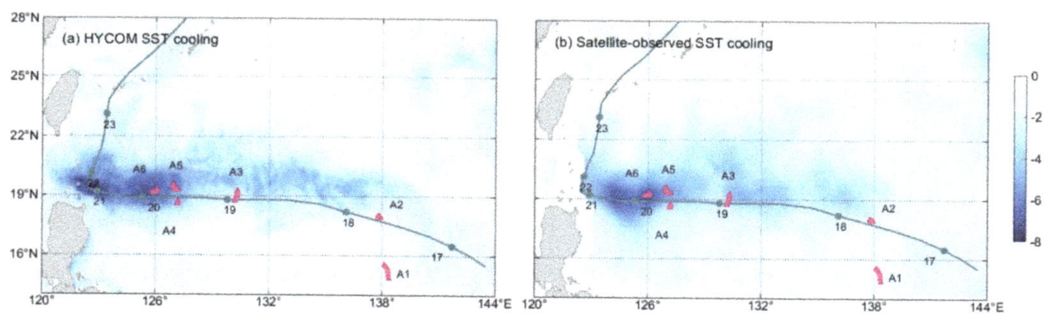

Figure A1. SST cooling caused by Goni corresponds to the (**a**) HYCOM reanalysis results and (**b**) satellite observations, respectively.

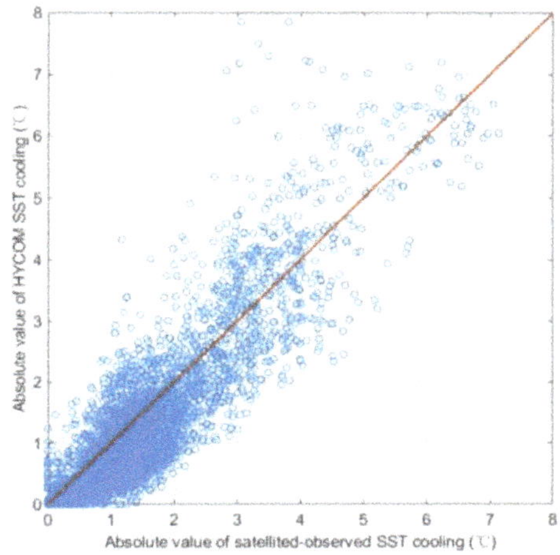

Figure A2. Comparison between the HYCOM-simulated and satellite-observed SST cooling caused by Goni.

References

1. Emanuel, K.A. Thermodynamic control of hurricane intensity. *Nature* **1999**, *401*, 665–669. [CrossRef]
2. Emanuel, K.A. Tropical cyclones. *Annu. Rev. Earth Planet. Sci.* **2003**, *31*, 75–104. [CrossRef]
3. Sanford, T.B.; Price, J.F.; Girton, J.B.; Webb, D.C. Highly resolved observations and simulations of the ocean response to a hurricane. *Geophys. Res. Lett.* **2007**, *34*. [CrossRef]
4. Price, J.F. Upper Ocean Response to a Hurricane. *J. Phys. Oceanogr.* **1981**, *11*, 153–175. [CrossRef]
5. Chang, S.; Anthes, R. Numerical simulations of the ocean's nonlinear baroclinic response to translating hurricanes. *J. Phys. Oceanogr.* **1978**, *8*, 468–480. [CrossRef]
6. Lin, I.I.; Black, P.; Price, J.F.; Yang, C.Y.; Chen, S.S.; Lien, C.C.; Harr, P.; Chi, N.H.; Wu, C.C.; D'Asaro, E.A. An ocean coupling potential intensity index for tropical cyclones. *Geophys. Res. Lett.* **2013**, *40*, 1878–1882. [CrossRef]
7. Black, W.J.; Dickey, T.D. Observations and analyses of upper ocean responses to tropical storms and hurricanes in the vicinity of Bermuda. *J. Geophys. Res.* **2008**, *113*, C08009. [CrossRef]
8. D'Asaro, E.A.; Sanford, T.B.; Niiler, P.P.; Terrill, E.J. Cold wake of hurricane Frances. *Geophys. Res. Lett.* **2007**, *34*, L15609. [CrossRef]
9. Hsu, H.H.; Hung, C.H.; Lo, A.K.; Wu, C.C.; Hung, C.W. Influence of tropical cyclones on the estimation of climate variability in the tropical western North Pacific. *J. Clim.* **2008**, *21*, 2960–2975. [CrossRef]
10. Lin, I.I.; Wu, C.; Chiang, J.C.H.; Sui, C.; Lin, I.; Liu, W.T. Satellite observations of modulation of surfacewinds by typhoon-induced upper ocean cooling. *Geophys. Res. Lett.* **2003**, *30*, 1131. [CrossRef]
11. Meyers, P.C.; Shay, L.K.; Brewster, J.K.; Jaimes, B. Observed ocean thermal response to Hurricanes Gustav and Ike. *J. Geophys. Res. Ocean.* **2016**, *121*, 162–179. [CrossRef]
12. Wu, R.; Li, C. Upper ocean response to the passage of two sequential typhoons. *Deep. Sea Res. Part I Oceanogr. Res. Pap.* **2018**, *132*, 68–79. [CrossRef]
13. Yue, X.X.; Zhang, B.; Liu, G.Q.; Li, X.F.; Zhang, H.; He, Y.J. Upper Ocean Response to Typhoon Kalmaegi and Sarikain the South China Sea from Multiple-Satellite Observations and Numerical Simulations. *Remote Sens.* **2018**, *10*, 348. [CrossRef]
14. Zhang, H.; Chen, D.; Zhou, L.; Liu, X.; Ding, T.; Zhou, B. Upper ocean response to Typhoon Kalmaegi (2014). *J. Geophys. Res. Ocean.* **2016**, *121*, 6520–6535. [CrossRef]
15. Emanuel, K.; Desautels, C.; Holloway, C.; Korty, R. Environmental Control of Tropical Cyclone Intensity. *J. Atmos. Sci.* **2004**, *61*, 843–858. [CrossRef]
16. Zhu, T.; Zhang, D.L. The impact of the storm-induced SST cooling on hurricane intensity. *Adv. Atmos. Sci.* **2006**, *23*, 14–22. [CrossRef]
17. Samson, G.; Giordani, H.; Caniaux, G.; Roux, F. Numerical investigation of an oceanic resonant regime induced by hurricane winds. *Ocean Dyn.* **2009**, *59*, 565–586. [CrossRef]
18. Wang, G.; Wu, L.; Johnson, N.C.; Ling, Z. Observed three-dimensional structure of ocean cooling induced by Pacific tropical cyclones. *Geophys. Res. Lett.* **2016**, *43*, 7632–7638. [CrossRef]
19. Lin, S.; Zhang, W.Z.; Shang, S.P.; Hong, H.S. Ocean response to typhoons in the western north pacific: Composite results from argo data. *Deep. Sea Res. Part I Oceanogr. Res. Pap.* **2017**, *123*, 62–74. [CrossRef]

20. Zhang, H.; He, H.; Zhang, W.Z.; Tian, D. Upper ocean response to tropical cyclones: A review. *Geosci. Lett.* **2021**, *8*, 1. [CrossRef]
21. Cheng, L.; Zhu, J.; Sriver, R.L. Global representation of tropical cyclone-induced short-term ocean thermal changes using Argo data. *Ocean Sci.* **2015**, *11*, 2831–2878. [CrossRef]
22. Huang, P.; Lin, I.I.; Chou, C.; Huang, R.H. Change in ocean subsurface environment to suppress tropical cyclone intensification under global warming. *Nat. Commun.* **2015**, *6*, 7188. [CrossRef] [PubMed]
23. Jaimes, B.; Shay, L.K.; Uhlhorn, E.W. Enthalpy and momentum fluxes during Hurricane Earl relative to underlying ocean features. *Mon. Weather Rev.* **2015**, *143*, 111–131. [CrossRef]
24. Schade, L.R.; Emanuel, K.A. The ocean's effect on the intensity of tropical cyclones: Results from a simple coupled atmosphere-ocean model. *J. Atmos. Sci.* **1997**, *56*, 642–651. [CrossRef]
25. Jacob, S.D.; Shay, L.K.; Mariano, A.J.; Black, P.G. The 3D oceanic mixed layer response to hurricane Gilbert. *J. Phys. Oceanogr.* **2000**, *30*, 1407–1429. [CrossRef]
26. Huang, P.; Sanford, T.B.; Imberger, J. Heat and Turbulent Kinetic Energy Budgets for Surface Layer Cooling Induced by the Passage of Hurricane Frances (2004). *J. Geophys. Res. Ocean.* **2009**, *114*. [CrossRef]
27. Guan, S.D.; Zhao, W.; Huthnance, J.; Tian, J.W.; Wang, J.H. Observed upper ocean response to typhoon Megi (2010) in the Northern South China Sea. *J. Geophys. Res. Ocean.* **2014**, *119*, 3134–3157. [CrossRef]
28. Sriver, R.L.; Huber, M. Observational evidence for an ocean heat pump induced by tropical cyclones. *Nature* **2007**, *447*, 577–580. [CrossRef]
29. Jaimes, B.; Shay, L.K. Mixed layer cooling in mesoscale oceanic eddies during hurricanes Katrina and Rita. *Mon. Weather Rev.* **2009**, *137*, 4188–4207. [CrossRef]
30. Balaguru, K.; Chang, P.; Saravanan, R.; Leung, L.R.; Xu, Z.; Li, M.; Hsieh, J.S. Ocean barrier layers' effect on tropical cyclone intensification. *Proc. Natl. Acad. Sci. USA* **2012**, *109*, 14343–14347. [CrossRef]
31. Sun, J.; Oey, L.Y.; Chang, R.; Xu, F.; Huang, S.M. Ocean response to typhoonNuri (2008) in western Pacific and South China Sea. *Ocean Dyn.* **2015**, *65*, 735–749. [CrossRef]
32. Guan, S.D.; Zhao, W.; Sun, L.; Zhou, C.; Liu, Z.; Hong, X.; Zhang, Y.H.; Tian, J.W.; Hou, Y.J. Tropical cyclone-induced sea surface cooling over the yellow sea and bohai sea in the 2019 pacific typhoon season. *J. Mar. Syst.* **2021**, *217*, 103509. [CrossRef]
33. Cao, A.Z.; Guo, Z.; Pan, Y.H.; Song, J.B.; He, H.L.; Li, P.L. Near-Inertial Waves Induced by Typhoon Megi(2010) in the South China Sea. *J. Mar. Sci. Eng.* **2021**, *9*, 440. [CrossRef]
34. Yang, B.; Hu, P.; Hou, Y.J. Observed Near-Inertial Waves in the Northern South China Sea. *Remote Sens.* **2021**, *13*, 3223. [CrossRef]
35. Guan, S.D.; Liu, Z.; Song, J.B.; Hou, Y.J.; Feng, L.Q. Upper ocean response to Super Typhoon Tembin (2012) explored using multiplatform satellites and Argo float observations. *Int. J. Remote Sens.* **2017**, *38*, 5150–5167. [CrossRef]
36. Ying, M.; Zhang, W.; Yu, H.; Lu, X.Q.; Feng, J.X.; Fan, Y.X.; Zhu, Y.T.; Chen, D.Q. An Overview of the China Meteorological Administration Tropical Cyclone Database. *J. Atmos. Ocean. Technol.* **2014**, *31*, 287–301. [CrossRef]
37. Lu, X.Q.; Yu, H.; Ying, M.; Zhao, B.K.; Zhang, S.; Lin, L.M.; Bai, L.N.; Wan, R.J. Western North Pacific tropical cyclone database created by the China Meteorological Administration. *Adv. Atmos. Sci.* **2021**, *38*, 690–699. [CrossRef]
38. Kunze, E.; Firing, E.; Hummon, J.M.; Chereskin, T.K.; Thurnherr, A.M. Global abyssal mixing inferred from lowered ADCP shear and CTD strain profiles. *J. Phys. Oceanogr.* **2006**, *36*, 1553–1576. [CrossRef]
39. Garrett, C.; Munk, W. Space-time scales of internal waves: A progress report. *J. Geophys. Res.* **1975**, *80*, 291–297. [CrossRef]
40. Jing, Z.; Wu, L.X.; Li, L.; Liu, C.Y.; Liang, X.; Chen, Z.H.; Hu, D.X.; Liu, Q.Y. Turbulent diapycnal mixing in the subtropical northwestern Pacific: Spatial-seasonal variations and role of eddies. *J. Geophys. Res.* **2011**, *116*, C10028. [CrossRef]
41. Whalen, C.B.; MacKinnon, J.A.; Talley, L.D. Large-scale impacts of the mesoscale environment on mixing from wind-driven internal waves. *Nat. Geosci.* **2018**, *11*, 842–847. [CrossRef]
42. Jing, Z.; Wu, L.X. Seasonal variation of turbulent diapycnal mixing in the northwestern Pacific stirred by wind stress. *Geophys. Res. Lett.* **2010**, *37*, L23604. [CrossRef]
43. Wu, L.; Jing, Z.; Riser, S.; Visbeck, M. Seasonal and spatial variations of Southern Ocean diapycnal mixing from Argo profiling floats. *Nat. Geosci.* **2011**, *4*, 363–366. [CrossRef]
44. Gregg, M.C.; Kunze, E. Shear and strain in Santa Monica Basin. *J. Geophys. Res.* **1991**, *96*, 16709–16719. [CrossRef]
45. Munk, W.; Wunsch, C. Abyssal Recipes II: Energetics of Tidal and Wind Mixing. *Deep. Sea Res. Part IOceanogr. Res. Pap.* **1998**, *45*, 1977–2010. [CrossRef]
46. Price, J.F.; Sanford, T.B.; Forristall, G.Z. Forced stage response to a moving hurricane. *J. Phys. Oceanogr.* **1994**, *24*, 233–260. [CrossRef]
47. Ling, Z.; Chen, Z.; Wang, G.; He, H.; Chen, C. Recovery ofTropical Cyclone Induced SSTCooling Observed by Satellite in theNorthwestern Pacific Ocean. *Remote Sens.* **2021**, *13*, 3781. [CrossRef]

remote sensing

Article

Bayesian Sea Ice Detection Algorithm for CFOSAT

Zhen Li *, Anton Verhoef and Ad Stoffelen

Royal Netherlands Meteorological Institute, 3731 GA De Bilt, The Netherlands; anton.verhoef@knmi.nl (A.V.); ad.stoffelen@knmi.nl (A.S.)
* Correspondence: zhen.li@knmi.nl

Abstract: This paper describes the adaptation of the Bayesian sea ice detection algorithm for the rotating fan-beam scatterometer CSCAT onboard the China–France Oceanography Satellite (CFOSAT). The algorithm was originally developed and applied for fixed fan-beam and rotating pencil-beam scatterometers. It is based on the probability of the wind and ice backscatter distances from the measurements to their corresponding geophysical model functions (GMFs). The new rotating Ku-band fan-beam design introduces very diverse geometry distributions across the swath, which leads to three main adaptations of the algorithm: (1) a new probability distribution function fit for the backscatter distances over open sea; (2) a linear ice GMF as a function of incidence angle; (3) the separation of outer swath wind vector cells ((WVCs) number 1, 2, 41, 42) from the other WVCs to form two sets of probability distribution function fits for these two WVC groups. The results are validated against sea ice extents from the active microwave ASCAT and the passive microwave SSMI. The validation shows good agreement with both instruments, despite the discrepancies with SSMI during the melting season, and this discrepancy is caused by the lower sensitivity of the passive microwave to detect the ice at a low concentration with a mixed water/ice state, while the scatterometer is more tolerant regarding this situation. We observed that the sea-ice GMF regression between HH and VV sea-ice backscatter at low and high incidence angles decorrelates at around -12 dB (28°) and -20 dB (50°) and an experiment with truncated backscatter values at these incidence angles is executed, which significantly improves the year-long average sea ice extents. In conclusion, the adapted algorithm for CSCAT works effectively and yields consistent sea ice extents compared with active and passive microwave instruments. As such, it can, in principle, contribute to the long-term global scatterometer sea ice record, and as the algorithm was adapted for a rotating fan-beam scatterometer, it also can serve as a guideline for the recently launched, dual-frequency, rotating fan-beam scatterometer WindRAD.

Keywords: sea ice; Bayesian algorithm; CFOSAT; scatterometer

Citation: Li, Z.; Verhoef, A.; Stoffelen, A. Bayesian Sea Ice Detection Algorithm for CFOSAT. *Remote Sens.* **2022**, *14*, 3569. https://doi.org/10.3390/rs14153569

Academic Editors: Xiaofeng Li, Chung-Ru Ho and Antony K. Liu

Received: 12 May 2022
Accepted: 21 July 2022
Published: 25 July 2022

1. Introduction

Sea ice plays an important role in global climate change, and the polar regions are a central focus of climate studies due to the significant changes that have been observed by satellites over time [1–4]. Satellite observations have been supporting the growing interest in polar regions in recent decades and the daily long-term historical sea ice extent records have been assessed and mediated by passive microwave sensors [5]. Active microwave scatterometer methods have also been developed to map sea ice extents, etc., and proven to be powerful [6–9]. The Bayesian sea ice detection algorithm developed for scatterometer was first applied on C-band fixed fan-beam European Remote Sensing satellite (ERS) data [10], then on Ku-band rotating pencil-beam Quick Scatterometer (USA QuikSCAT) data [9] and finally on the European C-band fixed fan-beam Advanced Scatterometer (ASCAT) data [11]. An independent record of sea ice extents has been produced from intercalibrated scatterometer data from 1992 to 2016 with ERS, QuikSCAT and ASCAT [12,13]. Bayesian sea ice detection, applied using the above scatterometers, demonstrated excellent agreement with passive microwave records during freezing seasons, and more sensitivity during melting seasons, as compared to passive microwave [9]. This is caused by a low concentration and mixed water/ice conditions [14–16]. In this context, Bayesian sea ice detection for scatterometers can serve as a nice complement to passive microwave products, especially for mixed and saturated water/ice surface states.

In late 2018 the China-France Oceanography SATellite (CFOSAT) was launched with a new type Ku-band scatterometer, CFOSAT Scatterometer (CSCAT), with a unique design of a rotating fan-beam antenna [17–19]. The aim of this paper is to adapt the Bayesian sea ice detection algorithm to the rotating fan-beam CSCAT, investigate sea ice mapping capabilities, generate sea ice extents and validate these using both active and passive microwave radiometers. The adapted Bayesian sea ice detection algorithm for CSCAT provides the possibility of continuous scatterometer sea ice records and more diverse ice backscatter information and can also serve as a guideline for another recently launched dual-frequency rotating fan-beam Wind Radar scatterometer (WindRAD) on board FY-3E (Feng Yun-3E) [20].

Section 2 describes the details of the Bayesian algorithm adaptation. Section 3 describes the sea ice detection results of CSCAT, including a discussion and cross-validation of the sea ice extent and sea ice edges among CSCAT, ASCAT, and the Special Sensor Microwave/Imager (SSMI). Conclusions are given in Section 4.

2. Algorithm Description and Adaptation

Ocean-surface wind speed and wind-direction retrieval are the prime purpose of scatterometers. However, they have also been used to detect and characterize sea ice [7,21]. The sea ice detection method we propose here is an adapted version of the existing algorithm developed for pencil-beam scatterometers such as QuikSCAT [9]. CSCAT and QuikSCAT both have rotating beams at the Ku-band radar frequency. The differences are as follows: CSCAT has a rotating fan-beam, emitting alternating HH/VV polarized pulses, whereas QuikSCAT has two rotating pencil or spot beams, one with HH polarization and the other with VV polarization. CSCAT flies in a sun-synchronous near-circular orbit at an altitude of 519 km. It can provide global wind field coverage within 3 days. The rotating fan-beam design results in multiple overlapping views with diverse incidence and azimuth angles in each individual Wind Vector Cell (WVC). At the same time, this design also yields unbalanced geometry diversity across the swath: outer-swath WVCs contain little diversity, where only side-looking azimuth angles and high incidence angles are available; sweet-swath WVCs contain the most diverse geometries and nadir-swath WVCs contain mainly fore/aft-looking azimuth angles, together with a large range of incidence angles [19]. A rotating pencil-beam scatterometer, such as QuikSCAT, on the other hand, has two fixed incidence angles of 46° for the HH polarization beam and 54° for the VV polarization beam and yields four views for each WVC in most parts of the swath. In the outer swath, only VV polarization measurements are available.

Section 2.1 provides a summary of the Bayesian algorithm for the pencil-beam scatterometer QuikSCAT, which is followed by a description of the adaptation made for CSCAT in Section 2.2. A detailed description of the QuikSCAT algorithm and its validation can be found in [9].

2.1. The Bayesian Sea Ice Detection Algorithm for QuikSCAT

The distribution of the backscatter values, from open water on the one hand and from the sea ice surface on the other hand, occupies distinct sectors in the backscatter measurement space with fore and aft HH and VV backscatter values serving as an axis in a 4-dimensional (4D) measurement space. The surface scattering caused by the wind over open water shows azimuthal anisotropy and a conical surface [10] in 4D, whereas the scattering from the sea ice slab is azimuth-invariant with stronger returns, particularly for HH, and the ice backscatter geophysical model function (GMF) is a linear model of HH and VV backscatter values as a function of sea ice age, thickness or roughness [9,11,12]. The Bayesian ice probability $p(ice|\sigma^{\circ})$ algorithm combines the prior knowledge of the sea ice probability for a specific location, $p_0(ice)$, with the newly available satellite information, based on conditional probabilities and modelled as a function of the distance to the ocean-wind GMF or sea-ice GMF, respectively:

$$p(ice|\sigma^\circ) = \frac{p(\sigma^\circ|ice)p_0(ice)}{p(\sigma^\circ|ice)p_0(ice) + p(\sigma^\circ|wind)p_0(wind)} \tag{1}$$

where $p(\sigma^\circ|wind)$ is the conditional probability of σ°s given wind (in the case where we would measure wind over open sea), i.e., following the wind σ° distribution around the ocean GMF; $p(\sigma^\circ|ice)$ is the conditional probability of σ° given ice (in the case where we would measure over ice), i.e., following the typical ice σ° distribution around the sea-ice GMF in measurement space. Note that $p_0(wind) = 1 - p_0(ice)$. The normalized measures of distance between the observed backscatter values and GMFs are derived by maximum likelihood estimates (MLEs):

$$p(\sigma^\circ|wind) = p(MLE_{wind}) \tag{2}$$

$$p(\sigma^\circ|ice) = p(MLE_{ice}) \tag{3}$$

$p_0(ice)$ and $p_0(wind)$ are a-priori probabilities; they are initialized as $p_0(ice) = 0.50$ and $p_0(wind) = 1 - p_0(ice)$ and updated after every orbital pass with the previous posterior $p(ice|\sigma^\circ)$. The posterior sea ice probability is spatially smoothed once a day and $p_0(ice)$ is relaxed for the next day's processing a priori:

$$p_0(\text{ice}) = \begin{cases} 0.50 & if\ p(ice|\sigma^\circ) > 0.30 \\ 0.15 & if\ p(ice|\sigma^\circ) < 0.30 \end{cases} \tag{4}$$

The relaxation setting aims to avoid saturation in the Bayesian filter. These settings maximize the quality of the prior information regarding sea ice detection and suppress the rain contamination effect. A sea ice coverage map is produced daily on a 12.5 km polar stereographic grid with a 55% threshold to the posterior probability $p(ice|\sigma^\circ)$, i.e., each pixel with a posterior probability above 55% is considered to be covered with ice. The 55% threshold is chosen to have the best match with the 15% sea ice concentration edge derived from passive microwave radiometers during the sea ice growing season. The sea-ice backscatter normalized values are plotted in the sea ice maps, indicating the sea ice roughness, age, or thickness.

2.2. The Adapted Bayesian Ice Detection for CSCAT

As described at the beginning of Section 2, CSCAT differs from QuikSCAT by its rotating fan-beam. This important feature leads to the diverse geometries distributed across the swath, and this diversity also causes the probability distribution of MLE_{wind} and MLE_{ice} to differ from QuikSCAT. New probability distribution fits for $p(MLE_{wind})$ and $p(MLE_{ice})$ are needed to derive $p(\sigma^\circ|wind)$ (Equation (2)) and $p(\sigma^\circ|ice)$ (Equation (3)).

Liu et al. [22] describe a sea ice detection method for CSCAT as well, but in comparison to this work it is simplified by only using HH and VV polarized beams from two azimuth angles and an incidence angle of 40 degrees, which is a rather direct adaptation from the method as applied for pencil-beam scatterometers. Our implementation includes all the measurements and classifies them into different groups to find corresponding and suitable probability distribution functions, and it also diagnoses the incidence angle dependency influences, which is a tailor-made version for CSCAT. Our method thereby follows earlier implementations of the Bayesian sea ice detection method for ERS and ASCAT, which [22] did not consider.

2.2.1. Probability Distribution of $p(\sigma^\circ|\text{wind})$

The wind inversion computes the minimum squared distances, called the maximum likelihood estimator (MLE) [23], and this is implemented in the CFOSAT Wind Data Processor (CWDP):

$$MLE_{wind} = \sum_{i=1}^{N} \frac{(\sigma^\circ{}_{obs,i} - \sigma^\circ{}_{wind,i})^2}{var[\sigma^\circ{}_{wind,i}]} \tag{5}$$

where $\sigma^\circ{}_{obs,i}$ is the measured σ°, and $\sigma^\circ{}_{wind,i}$ is computed from the Ku-band GMF for a given wind speed, wind direction, azimuth and incidence angle, i is the view number, N is the number of views in a WVC, and $var[\sigma^\circ{}_{wind,i}]$ is the expected Gaussian observation noise for view i. MLE_{wind} is normalized with $var[\sigma^\circ{}_{wind,i}]$ to make sure that the variance in backscatter values around the GMF equals unity. In this way, MLE_{wind} is expressed as a sum of the squares of N standard normal random variables. For QuikSCAT, the number of views is four for all the WVCs; hence, $MLE_{wind,QuikSCAT}$ is the squared distance of the four-dimensional wind backscatter vector from a two-dimensional ocean GMF, varying with wind speed and direction. Therefore, the probability of $MLE_{wind,QuikSCAT}$ can be expressed as a chi-square distribution with two independent degrees of freedom [9]. However, for CSCAT, the number of views is variable across the swath [19]. Hence, the probability distribution of the MLE_{wind} is multi-dimensional, with two degrees of freedom, and cannot be simply classified into a chi-square distribution, although it can be empirically derived. The probability of MLE_{wind} as a function of WVC has been tested and the distribution per WVC looks very similar, so it is not necessary to perform $p(MLE_{wind})$ per WVC. Figure 1 shows the observed probability distribution of MLE_{wind} in blue. The best analytical fit is an inverse gamma distribution, as in Equation (6):

$$p(MLE_{wind}) = \frac{x^{-\alpha-1}}{\Gamma(\alpha)} e^{-\frac{1}{x}} \frac{1}{scale} \tag{6}$$

where $x = \frac{MLE_{wind} - loc}{scale}$, $loc = -0.22, \alpha = 0.44, scale = 4.81$, $\Gamma(\alpha)$ is the gamma function. The observed distribution of MLE_{wind} to the ocean GMF closely agrees with this inverse gamma distribution, as shown in Figure 1. Thus, $p(\sigma^\circ|wind)$ can be expressed with the function proposed in Equation (6).

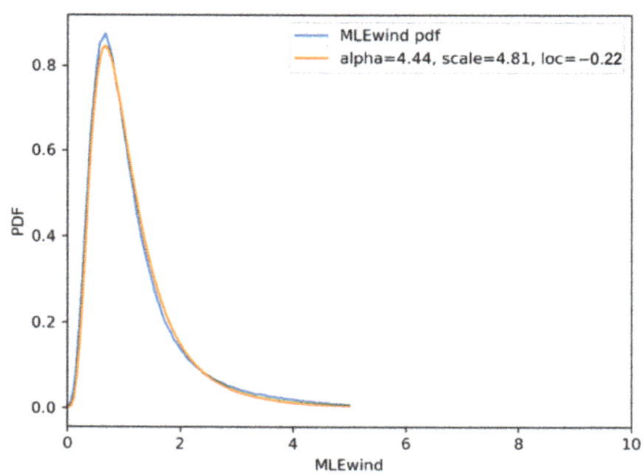

Figure 1. Probability distribution of MLE_{wind} (normalized to unit area): blue is the observed $p(MLE_{wind})$, orange is the fitted probability function derived from an inverse gamma distribution as in Equation (6).

2.2.2. Probability Distribution of $p(\sigma^{\circ}|\text{ice})$

Statistical knowledge is needed to derive the Ku-band sea-ice GMF for CSCAT as a function of incidence angle; hence, the actual distribution of the sea-ice backscatter values in the measurement space is derived by selecting the data located over the Arctic ice shelf with a latitude larger than 70°, a Sea Surface Temperature (SST, from ECMWF) lower than −1 °C and ice flags from the level-1b dataset if applicable. As described in Section 2.1, the sea-ice GMF is a linear function, which we can write as:

$$\sigma^{\circ}{}_{VV,ice} = \sigma^{\circ}{}_{HH,ice} \times slope + offset \quad (7)$$

The slope and offset are estimated using a 2D symmetric regression with VV against HH. Note that this symmetric regression only provides a reasonable result when VV and HH have similar noise characteristics, which might lead to inaccurate regression at incidence angles with different noise characteristics (particularly at low and high incidence angle). The σ°s over sea ice (as described at the beginning of 2.2.2) are binned into incidence angle bins, with a 1° interval for VV and HH, respectively, and then 1-degree Polyfit is applied on the VV and HH σ°s in the same incidence angle bin to derive the slope and offset in Equation(7) for every incidence angle bin (this method is referred to as all_inc from now on). Figure 2 shows the seasonal dependence of the sea-ice GMF slopes over the year 2019 in the Arctic. There is an obvious dispersion among different incidence angles (note that there was an instrument restart in July and a change to a redundant channel by the end of the year). The seasonal dependence is caused by the presence of mixed ice-water areas during the melting season in the summer, which negatively biases the slopes for all incidence angles. The slopes generally increase with increasing incidence angle, except around the incidence angle bin of 50°, most probably due to the much higher noise level in the backscatter values and the unsymmetric noise characteristics between VV and HH at a high incidence angle, as we described earlier. The dispersion of the slopes for different incidence angles becomes much narrower after the instrument restarted in July. One possible reason for this is that it is the melting season, meaning that the regression results are not the same as in the January to March period due to the water/ice mixture. Another reason is that σ°s, on average, started to drift after the instrument restarted. At the beginning of November, the slope dispersion among the different incidence angles returns and the values become generally larger than at the beginning of the year because the level 1b data processing version was updated. Fortunately, the data were stable when the sea ice was formed and stable in the Arctic from January to March, we can use this period to derive the sea-ice GMF.

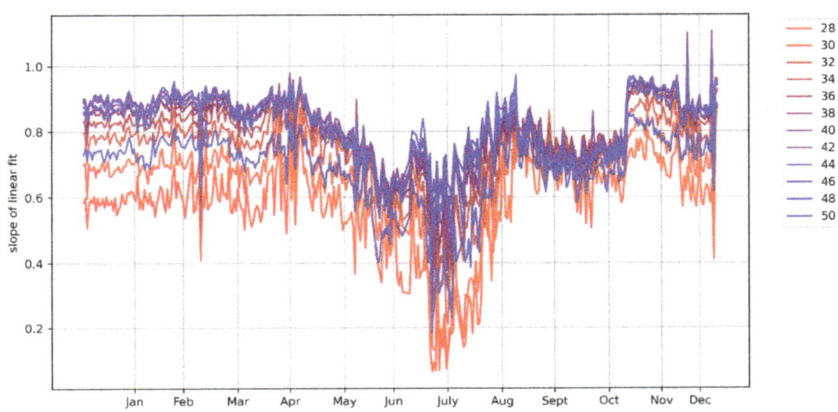

Figure 2. Daily slopes in the linear ice model in the Arctic per incidence angle in the year 2019 (a 2° incidence interval is used for a clearer look).

The mean slopes per incidence angle from the Arctic winter (from January to March) are taken as representative of pure-ice backscatter for the sea-ice GMF (Equation (7)). The scatter plots of the sea-ice backscatter values, together with the corresponding sea-ice GMF per selected incidence angle, are shown in Figure 3. The distribution of the distances between sea-ice backscatter values and the corresponding GMF is Gaussian, which is shown per selected incidence angle in Figure 4. The standard deviations are the largest at the lowest incidence angle of 28° and the highest incidence angle of 50°, where the standard deviation is around 2. The values at the other incidence angles are mainly between 1.0 and 1.2. This can also be observed in Figure 3: the regressions at 28° and 50° do not follow the scatter plot's highest density at high σ° values due to the decorrelation of the VV and HH backscatter at around −12 dB (28° and −20 dB (50°). These limits are most pronounced in VV and a broader decorrelated cloud of points is visible in HH, which suggests that the noise is more disturbing for HH. As we described earlier, the regression might provide inaccurate results if the VV and HH beams are uncorrelated and contain asymmetric noise characteristics, as shown here at 28° and 50°. Hence, we exclude dB values below −10 dB for 28° and 50° (referred to as truncated_inc). Figure 5 gives the averaged slope as a function of the incidence angle for truncated_inc with an increased slope at 28° (from 0.56 to 0.62) and 50° (from 0.71 to 0.77), and Figure 6 illustrates the sea-ice GMF with truncated_inc as a function of incidence angle. For pencil-beam instruments (e.g., SeaWinds), as the VV beam has a fixed incidence angle of 46° and the HH beam has a fixed incidence angle of 54°, its sea-ice GMF has a slope of around 1. The slope for CSCAT increases with increasing incidence angle, reaches saturation at a higher incidence angle 46° with value 0.91, and then goes down. In comparison with SeaWinds, it seems that the slope saturates at higher incidence angles and then tends to go downward.

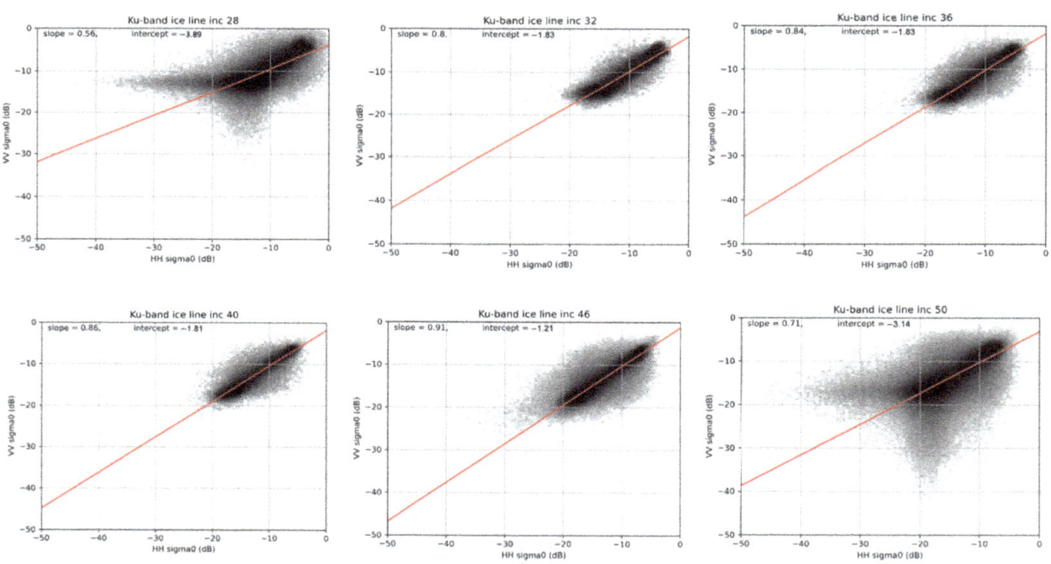

Figure 3. The distribution of the sea-ice backscatter measurements and the corresponding sea-ice GMF (red line). Selected incidence angles from low to high are shown.

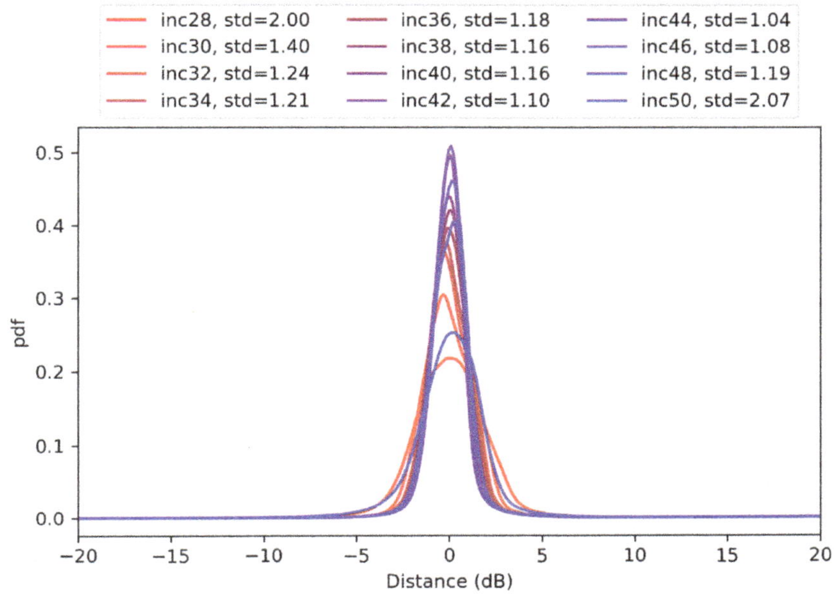

Figure 4. Gaussian distribution of the distance between sea-ice backscatter and the corresponding sea-ice GMF per incidence angle with standard deviation (red to blue indicates incidence angle from low, 28°, to high, 50°).

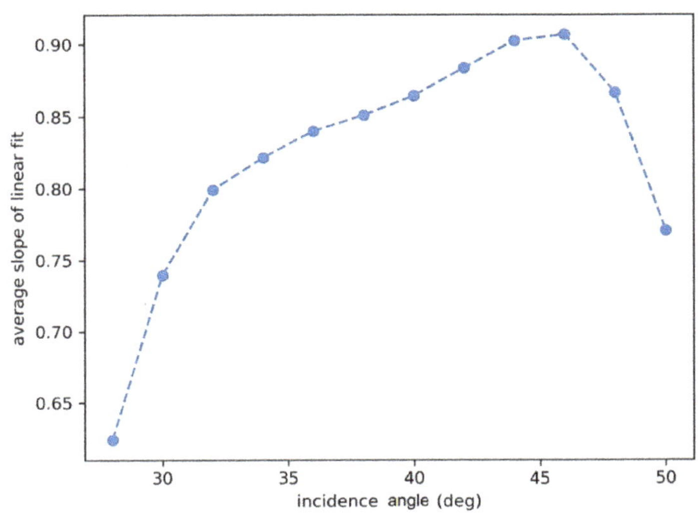

Figure 5. The average slopes from Jan to March in the Arctic as a function of incidence angle (truncated_inc).

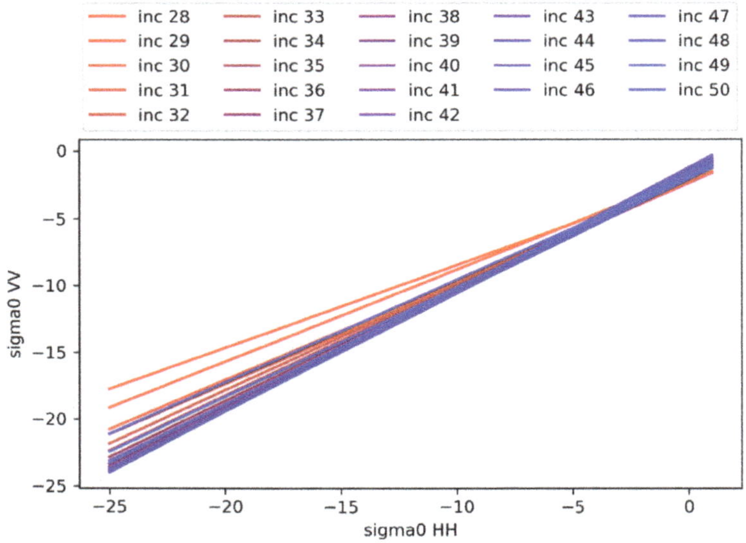

Figure 6. Illustration of the sea-ice GMFs as a function of incidence angle (truncated_inc).

The sea-ice GMFs (slope and offset) are constructed as a function of incidence angle (bin size is 1°) from the HH and VV backscatter measurements over sea ice (Figure 6). Therefore, the minimum squared distance from sea-ice backscatter to sea-ice GMF in each WVC can be expressed as follows:

$$MLE_{ice} = \sum_{i=1}^{N} \frac{(\sigma^{\circ}{}_{obs,i} - \sigma^{\circ}{}_{ice,i})^2}{var[\sigma^{\circ}{}_{ice,i}]} \tag{8}$$

where $\sigma^{\circ}{}_{obs,i}$ is the measured σ° HH and VV pair, i is the pair number, N is the number of pairs in one WVC and $\sigma^{\circ}{}_{obs,i} - \sigma^{\circ}{}_{ice,i}$ is the distance from the HH and VV pair to the sea-ice GMF. $var[\sigma^{\circ}{}_{ice,i}]$ is the squared standard deviation of the corresponding sea-ice backscatter distances' distribution against the sea-ice GMF, as described above.

For QuikSCAT, the probability distribution of MLE_{ice} is modelled as a chi-square distribution with three independent degrees of freedom, since four backscatter measurements are present in all WVCs and the sea-ice GMF is one-dimensional. For CSCAT, the probability distribution of MLE_{ice} cannot be modelled in the same way due to the varying number of views and HH/VV pairs in different WVCs across the swath. Probability distributions of MLE_{ice} in some selected WVCs are shown in Figure 7. The selected WVCs are representative across the swath for an outer-swath WVC, a sweet-swath WVC and a nadir-swath WVC. The distribution of MLE_{ice} in outer-swath WVCs is quite different from the distributions in sweet and nadir WVCs; thus, the WVCs across the swath are grouped into two: one is the 'outer group' of numbers 1, 2, 41 and 42; the other one is the 'rest group' of numbers 3 to 40. The outer group contains only high-incidence-angle observations (Figure 8a) due to the incidence angle distribution across the swath (Figure 8b). The incidence angles become more mixed for WVCs closer to the sweet and nadir swath; this causes the quite different $p(MLE_{ice})$ distributions in the outer group as compared to the rest group. Thus, the $p(MLE_{ice})$ is empirically and separately fitted for the outer group and the rest group (Figure 9). The best probability distribution fit for both groups is a chi-square distribution, but with different parameter values:

$$p(MLE_{ice}) = \frac{(MLE_{ice} - loc)^{\frac{k}{2}-1}}{2^{\frac{k}{2}}\Gamma(k/2)} e^{-\frac{MLE_{ice}-loc}{2}} \quad (9)$$

for the outer group: $k = 3.35$, $loc = -0.1$ and for the rest group: $k = 1.5$, $loc = -0.2$. The observed distribution of MLE_{ice} closely agrees with the chi-square distribution, as shown in Figure 9. Thus, $p(\sigma^{\circ}|ice)$ can be expressed as in Equation (9). Note: the fittings shown here are for all_inc., while $p(MLE_{ice})$ for truncated_inc: outer group is equal to Equation (9) + 0.01; while the rest group is the same setting as Equation (9).

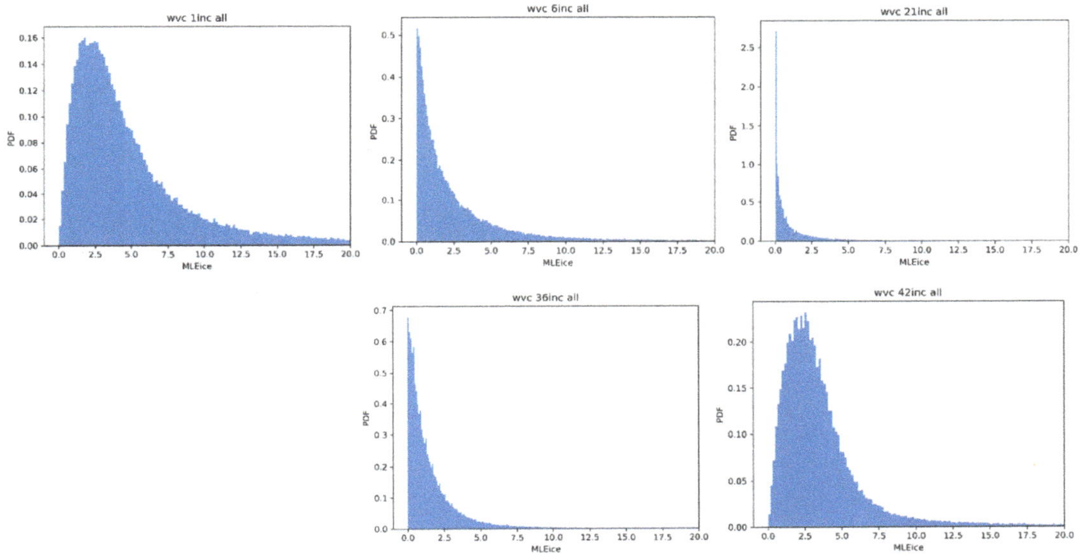

Figure 7. The probability distribution of MLE_{ice} in the selected WVCs across the swath: outer WVCs number 1 and 42; sweet WVCs number 6 and 36; nadir WVC number 21.

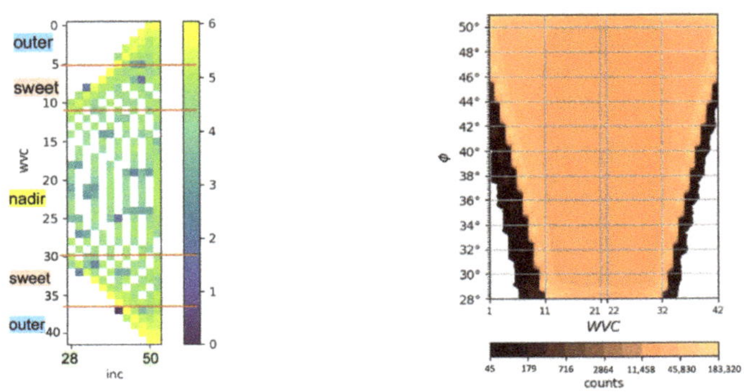

(**a**) (**b**)

Figure 8. (**a**) the number of incidence angle HH/VV pairs across the swath (the number is logarithmic to make the plot clearer); (**b**) the distribution of the incidence angle across the swath.

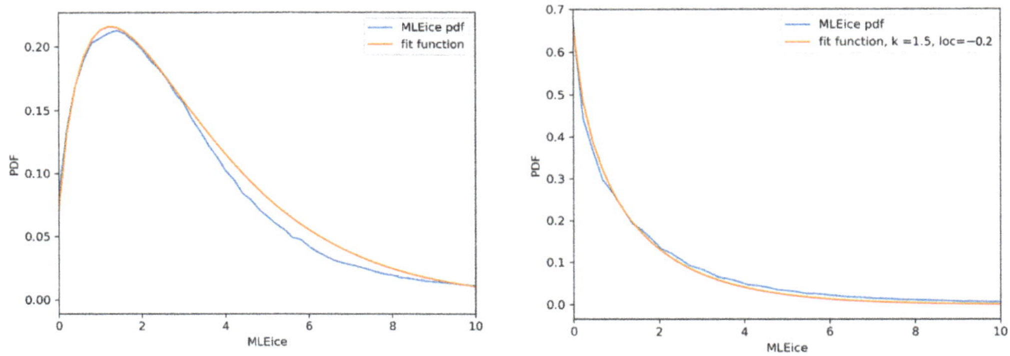

(a) (b)

Figure 9. Probability distribution of MLE_{ice} (normalized to unit area): blue is the observed $p(MLE_{ice})$, orange is the fitted probability function derived from chi-square distribution Equation (9). (**a**) $p(MLE_{ice})$ of outer group (WVC numbers 1, 2, 41 and 42); (**b**) $p(MLE_{ice})$ of rest group (WVC numbers from 3 to 40).

The Bayesian sea ice detection algorithm in Equation (1) can now be solved with the newly constructed $p(\sigma^{\circ}|wind)$, $p(\sigma^{\circ}|ice)$ and the prior $p_0(ice)$ and $p_0(wind)$.

3. Results and Discussion

Two data sources from the year of 2019 were applied to validate the adapted Bayesian sea ice detection algorithm for CSCAT. One data source is the sea ice extent produced from ASCAT with the KNMI (Royal Netherlands Meteorological Institute) Bayesian sea ice detection algorithm [11]. The other data source is the sea ice extent derived from the ice concentration data generated by the NASA team's NT algorithm from the passive microwave radiometer SSMI using a 15% isoline [24]. The all_inc method was first used for the validation, and then the truncated_inc method was applied for comparison.

Figure 10 shows the sea ice maps filled with normalized sea-ice backscatter values, which proxies as ice age or thickness, as well as the geographical behavior of sea ice extent on 10 January 2019 for CSCAT, SSMI (15% concentration threshold), and ASCAT in the Arctic (winter) and Antarctic (summer). All maps are masked with the same land and polar gap mask. As expected, the contrast between multi-year ice (bright) and first-year ice (dimmer) is stronger for CSCAT than ASCAT, while ASCAT gives a stronger response at the edge of the ice area. These are caused by their different frequencies. CSCAT operates at Ku-band, with a shorter wavelength than the ASCAT C-band. The shorter wavelength has a lower response on the rafted sea ice on the edge and is more responsive to rough multi-year ice. The three datasets in general agree well with each other in the Arctic (winter), while there are more disagreements in the Antarctic (summer) between the scatterometer instruments (CSCAT and ASCAT) on the one hand and passive microwave SSMI on the other hand. The ice edges of the three instruments (Figure 11) clearly show the discrepancy between active scatterometer and passive microwave in summer (Antarctic) because the sea ice state is mixed with open water and the passive microwave instrument has difficulties detecting this.

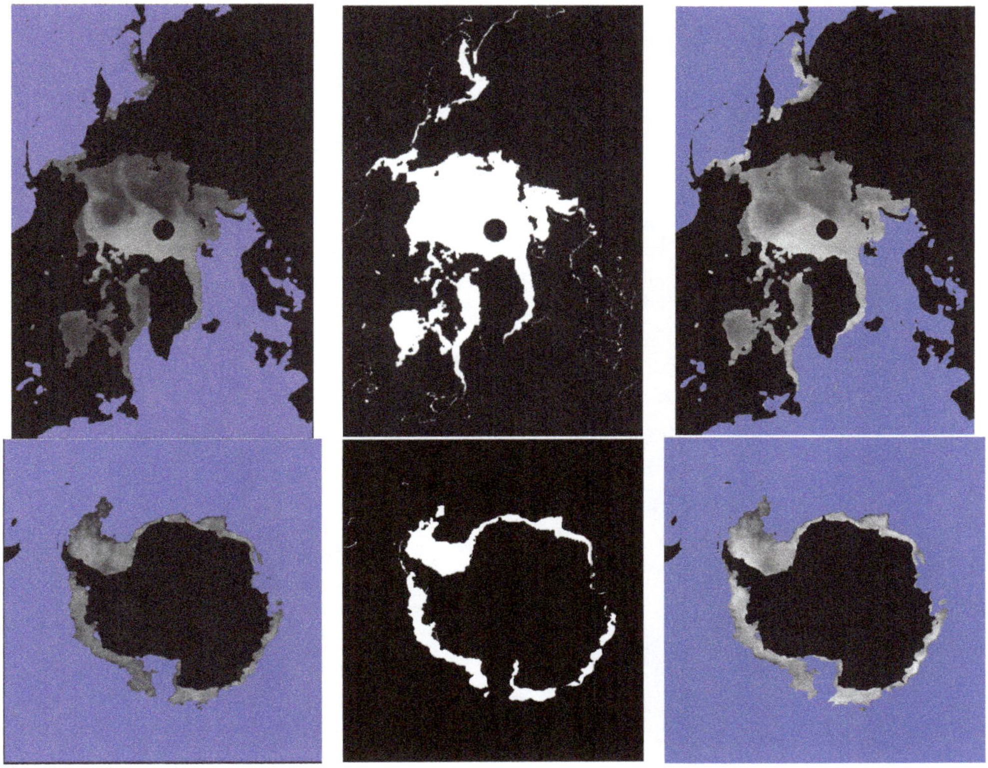

(**a**) (**b**) (**c**)

Figure 10. Sea ice extents in the Arctic (upper panel) and Antarctic (lower panel) on 10 January 2019 with the same land mask and polar gap mask: (**a**) CSCAT; (**b**) SSMI; (**c**) ASCAT; the gray scale for (**a**) and (**c**) indicates sea ice age/brightness. The grey scale for (**a**) and (**c**) represents sea ice normalized VV backscatter and map griding is 12.5 km.

Daily Arctic and Antarctic ice extents were produced over the year 2019 (Figure 12). We observe a large increase in sea ice extent in the Arctic from mid-July to Aug for CSCAT (indicated in Figure 12 with a rectangular box), which was caused by an instrument restart in mid-July. The restart was necessary to correct a mis-registration of the time in the level0 data and a stabilization period was needed afterwards as we learned from communication with National Satellite Ocean Application Service (NSOAS), China. This restart did not have the same impact in the Antarctic because it was wintertime. Thus, the sea ice formed solidly and the wind speed was much higher than in the Arctic. A lower wind speed corresponds to lower backscatter values, which causes a noisier MLE_{wind}, and can reduce the ice screening skill in the Arctic summer; thus, together with the instrument restart incident, the impact was obverse in the Arctic. Apart from this interruption, CSCAT and ASCAT in general agree with each other well throughout the year. Both in the Arctic and the Antarctic, the two scatterometers show better agreement in summer than in winter. The agreement between CSCAT and ASCAT in winter (starting from spring) is less as compared to QuikSCAT and ASCAT [11]. A possible reason for this is that CSCAT data are much noisier than QuikSCAT due to the rotating fan-beam design and the CSCAT asymmetric noise characteristics, not fitting the sea-ice GMF well at both low and high incidence angles, which might lead to misdetection during the high ice extents season. However, the average difference (excluding the instrument restart period) is relatively small,

0.18 million km^2 in the Arctic and 0.13 million km^2 in the Antarctic, and the truncated_inc method further improves the agreement between CSCAT and ASCAT (Figure 13). The result shows the capability of the newly developed Bayesian sea ice detection algorithm for rotating fan-beam scatterometers and the possibility of adding new, consistent data to the existing scatterometer sea ice record. The difference between CSCAT and SSMI was smallest during the autumn and winter months for both the Arctic and Antarctic, whereas the larger discrepancies begin from spring's rapid sea ice melting and summer's rapid sea ice advance period. This is a typical discrepancy between the active and passive radiometers, associated with the low sensitivity of the passive microwave radiometer to melting/mixed sea ice conditions [14]. This discrepancy also indicates that the active microwave scatterometers are more sensitive and proficient at detecting or distinguishing the sea ice in mixed sea ice and water states during the spring and summer seasons.

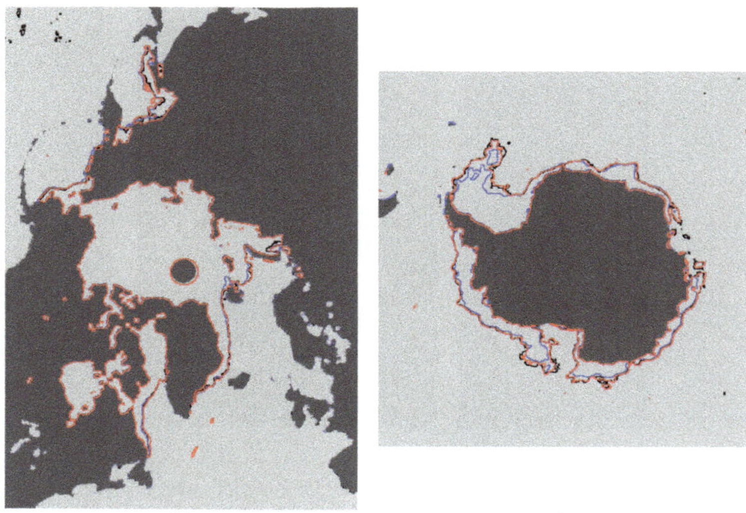

(a) (b)

Figure 11. Ice-edge map on the 10 January 2019 for the (**a**) Arctic and (**b**) Antarctic, CSCAT black line, SSMI blue line, ASCAT red line.

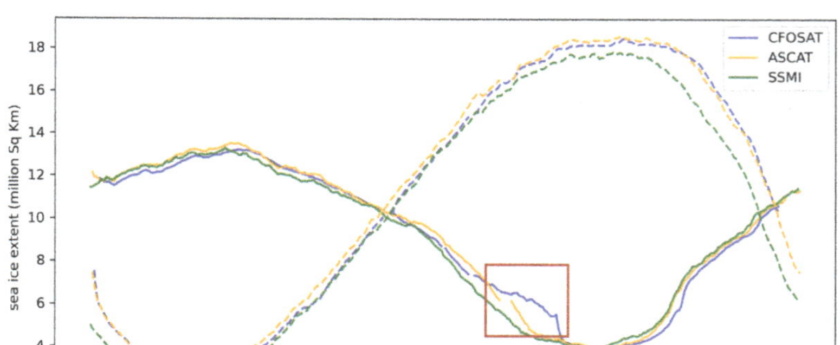

Figure 12. Daily Arctic (solid lines) and Antarctic (dashed lines) sea ice extents in 2019 from CFOSAT (blue), ASCAT (orange), and SSMI (green); the red rectangular area marks the period when CSCAT instrument restarted.

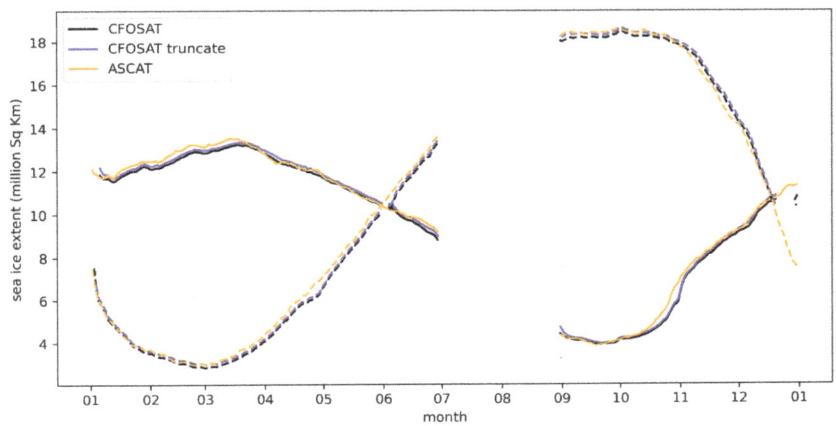

Figure 13. Daily Arctic (solid lines) and Antarctic (dashed lines) sea ice extents in 2019 from CFOSAT all_inc (black), CFOSAT truncated_inc (blue), ASCAT (orange), CSCAT instrument restart period is excluded here.

As we describe in Section 2, the regressions at 28° and 50° do not follow the highest density of scattered data points at high σ° values due to the decorrelation of the VV and HH at around −12 dB (28°) and −20 dB (50°). Hence, the truncated_inc test is proposed and applied. The sea ice extents are similar to those from the all_inc method and agree well with ASCAT and SSMI. If we take ASCAT as a reference, the difference in the year-long average sea ice extent (excluding the instrument restart period) between truncated_inc and ASCAT is 0.05 million km^2 in the Arctic and 0.01 million km^2 in the Antarctic, which reduces the differences by 72% and 92%, respectively, as compared to all_inc (Table 1).

Table 1. Sea ice extent year-long average difference between ASCAT and all_inc, truncated_inc and exclude_inc, respectively, in the Arctic and Antarctic (during the sea-ice GMF construction, all_inc: all the sea-ice backscatter values are included; truncated_inc: sea-ice backscatter values below −10 dB are not considered at incidence 28° and 50°; exclude_inc: sea-ice backscatter values at incidence 50° are excluded).

Average Sea Ice Extent Difference Compared to ASCAT Sea Ice Extent	All_Inc	Truncated_Inc	Exclude_Inc
Arctic (million km^2)	0.18	0.05	0.13
Antarctic (million km^2)	0.13	0.01	0.08

The sea ice retrieval at outer-swath WVCs is only influenced by high incidence angles of around 50°, and corresponds to a larger standard deviation in the backscatter values against the linear ice model. This may cause more uncertainty in the retrieval. We simply exclude incidence angles larger than 49° from the sea ice retrieval and refer to this as exclude_inc. This leads to a narrower usable swath width by about two WVCs (i.e., 50 km). However, the swath width is still wide enough to obtain daily coverage in the polar regions. The sea ice extent results show that the year-long average sea ice extent difference (excluding the instrument restart period) between exclude_inc and ASCAT is 0.13 million km^2 in the Arctic and 0.08 million km^2 in the Antarctic, which reduces the differences by 28% and 38%, respectively, as compared to all_inc (Table 1).

This indicates that the sea ice algorithm skill improves when using the truncated_inc and exclude_inc methods. The truncated_inc method shows results that are more consistent with ASCAT as compared to exclude_inc, without reducing the usable swath width.

Figure 13 shows the daily ice extents of all_inc, truncated_inc and ASCAT in 2019, excluding the instrument restart period (exclude_inc is not shown because the difference between exclude_inc and all_inc is smaller than the difference between truncated_inc and all_inc, which makes it difficult to see), and truncated_inc is usually closer and more consistent to ASCAT than all_inc. Thus, we suggest applying the truncated_inc method as the final Bayesian sea ice detection algorithm.

4. Conclusions

In this paper, the details of the Bayesian sea ice detection algorithm's adaptation to CSCAT onboard CFOSAT are exploited. The main adaptations include: (1) a new fit for the probability distribution of $p(\sigma^{\circ}|wind)$; (2) the introduction of an incidence angle dependency to the linear ice model; (3) the identification of the distinct probability distribution $p(\sigma^{\circ}|ice)$ difference between the outer swath WVCs (number 1, 2, 41, 42) and the other WVCs, and the construction of different probability fits for these two groups. The performance of the new algorithm is validated against active (ASCAT) and passive (SSMI) microwave data at a global and seasonal level. CSCAT provides consistent sea ice extents, which agree well with ASCAT and SSMI, despite the instrument restart and stabilization period after late July, 2019. There was a larger discrepancy between CSCAT and SSMI during the sea ice's fast advancing and retreat episodes, caused by the well-known passive microwave issue of identifying mixed sea-ice and open-water conditions. The scatterometer appears again more sensitive to the detection of low ice concentration and water/ice mixed situations as compared to passive microwave instruments. We observed a larger standard deviation in the ice backscatter compared to the linear ice model for both the lowest and highest incidence angles. Therefore, two extra experiments were conducted: truncating the sea-ice backscatter values below -10 dB at incidences of 28° and 50° (truncated_inc) and excluding high-incidence angles (>49°, exclude_inc). Both tests reduce the year-long average difference in sea ice extents between CSCAT and ASCAT, and the truncated_inc method shows the largest improvement, with 72% and 92% reductions in the difference in the Arctic and the Antarctic, respectively. The truncated_inc test did not reduce the usable swath width, whereas the exclude_inc filter reduces the usable swath width by about 50 km. We recommend using the exclude_inc mode, as swath width is not critical for CSCAT. Overall, the adapted Bayesian sea ice detection for the Ku-band rotating fan-beam CSCAT instrument shows consistency with other active and passive microwave instruments. The active instrument shows more inclusive ice detection at water/ice mixed seasons compared to the passive microwave instruments, as expected. The Bayesian sea ice detection algorithm has now been successfully implemented for all scatterometer types; its performance appears consistent and of high quality. As CSCAT is the first rotating fan-beam scatterometer in orbit, the adapted algorithm can also serve as a guideline for the recently launched dual-frequency rotating fan-beam Wind Radar scatterometer (WindRAD). For future elaboration, CSCAT has the potential to further identify sea ice types: first-year ice and multi-year ice and to explore/inter-compare with the sea ice product from the SWIM instrument, which is also onboard CFOSAT. In addition, it is possible to provide a long-term global scatterometer sea ice record in the case of a stable CSCAT instrument.

Author Contributions: Conceptualization, Z.L. and A.V.; methodology, Z.L. and A.V.; software, Z.L. and A.V.; validation, Z.L.; formal analysis, Z.L.; investigation, Z.L.; resources, A.S.; data curation, Z.L.; writing—original draft preparation, Z.L.; writing—review and editing, A.V. and A.S.; visualization, Z.L.; supervision, A.S. and A.V.; project administration, A.S., Z.L. and A.V.; funding acquisition, A.S., A.V. and Z.L. All authors have read and agreed to the published version of the manuscript.

Funding: This research was supported by EUMETSAT OSI SAF CDOP4 (Darmstadt, Germany) and CNES CF-GSFR-CR-2410-CNES (Toulouse, France).

Data Availability Statement: The data used in the paper are available with the permeant URLs: (the URLs will be obtained upon the paper acceptance from KNMI data platform), which can be accessed via API through KNMI open data platform. There is an introduction on the webpage showing how to connect and download the data through API; it is also possible to browse the dataset and retrieve specific files. The CFOSAT Wind Data Processor (CWDP) used in the study is available at EUMETSAT NWP SAF website: https://nwp-saf.eumetsat.int/site/software/scatterometer/cwdp/(accessed on 23 April 2022), registration is needed to access the software.

Acknowledgments: We acknowledge the support of the CFOSAT team, CNES and NSOAS in providing the data, status and mission information. The support from the EUMETSAT Ocean and Sea Ice Satellite Application Facility is further much appreciated.

Conflicts of Interest: The authors declare no conflict of interest.

References

1. Comiso, J.C.; Parkinson, C.L.; Gersten, R.; Stock, L. Accelerated decline in the Arctic sea ice cover. *Geophys. Res. Lett.* **2008**, *35*, 1–6. [CrossRef]
2. Stroeve, J.C.; Serreze, M.C.; Holland, M.M.; Kay, J.E.; Malanik, J.; Barrett, A.P. The Arctic's rapidly shrinking sea ice cover: A research synthesis. *Clim. Change* **2012**, *110*, 1005–1027. [CrossRef]
3. Stroeve, J.; Notz, D. Changing state of Arctic sea ice across all seasons. *Environ. Res. Lett.* **2018**, *13*, 103001. [CrossRef]
4. Parkinson, C.L. A 40-y record reveals gradual Antarctic sea ice increases followed by decreases at rates far exceeding the rates seen in the Arctic. *Proc. Natl. Acad. Sci. USA* **2019**, *116*, 14414–14423. [CrossRef] [PubMed]
5. Cavalieri, D.J.; Parkinson, C.L.; Gloersen, P.; Comiso, J.C.; Zwally, H.J. Deriving long-term time series of sea ice cover from satellite passive-microwave multisensor data sets. *J. Geophys. Res. Ocean.* **1999**, *104*, 15803–15814. [CrossRef]
6. Haarpaintner, J.; Tonboe, R.T.; Long, D.G.; Van Woert, M.L. Automatic detection and validity of the sea-ice edge: An application of enhanced-resolution QuikScat/SeaWinds data. *IEEE Trans. Geosci. Remote Sens.* **2004**, *42*, 1433–1443. [CrossRef]
7. Anderson, H.S.; Long, D.G. Sea ice mapping method for SeaWinds. *IEEE Trans. Geosci. Remote Sens.* **2005**, *43*, 647–657. [CrossRef]
8. Haarpaintner, J.; Spreen, G. Use of enhanced-resolution QuikSCAT/SeaWinds data for operational ice services and climate research: Sea ice edge, type, concentration, and drift. *IEEE Trans. Geosci. Remote Sens.* **2007**, *45*, 3131–3137. [CrossRef]
9. Belmonte Rivas, M.; Stoffelen, A. New bayesian algorithm for sea ice detection with QuikSCAT. *IEEE Trans. Geosci. Remote Sens.* **2011**, *49*, 1894–1901. [CrossRef]
10. Stoffelen, A.; Verspeek, J.; Haan, S. De C-Band Sea Ice Model. In *2004 Envisat & ERS Symposium*; Lacoste, H., Ouwehand, L., Eds.; European Space Agency: Paris, France, 2005; Volume ESA SP-572, p. 143.1.
11. Belmonte Rivas, M.; Verspeek, J.; Verhoef, A.; Stoffelen, A. Bayesian sea ice detection with the advanced scatterometer ASCAT. *IEEE Trans. Geosci. Remote Sens.* **2012**, *50*, 2649–2657. [CrossRef]
12. Otosaka, I.; Rivas, M.B.; Stoffelen, A. Bayesian sea ice detection with the ERS scatterometer and sea ice backscatter model at C-band. *IEEE Trans. Geosci. Remote Sens.* **2018**, *56*, 2248–2254. [CrossRef]
13. Belmonte Rivas, M.; Otosaka, I.; Stoffelen, A.; Verhoef, A. A scatterometer record of sea ice extents and backscatter: 1992–2016. *Cryosphere* **2018**, *12*, 2941–2953. [CrossRef]
14. Markus, T.; Cavalieri, D.J. An Enhancement of the NASA Team Sea Ice Algorithm. *IEEE Trans. Geosci. Remote Sens.* **2000**, *38*, 1387–1398. [CrossRef]
15. Cavalieri, D.J.; Markus, T.; Hall, D.K.; Ivanoff, A.; Glick, E. Assessment of AMSR-E antarctic winter sea-ice concentrations using aqua MODIS. *IEEE Trans. Geosci. Remote Sens.* **2010**, *48*, 3331–3339. [CrossRef]
16. Meier, W.N.; Stroeve, J. Comparison of sea-ice extent and ice-edge location estimates from passive microwave and enhanced-resolution scatterometer data. *Ann. Glaciol.* **2008**, *48*, 65–70. [CrossRef]
17. Li, Z.; Stoffelen, A.; Verhoef, A. A Generalized Simulation Capability for Rotating Beam Scatterometers. *Atmos. Meas. Tech.* **2019**, *12*, 3573–3594. [CrossRef]
18. Liu, J.; Lin, W.; Dong, X.; Lang, S.; Yun, R.; Zhu, D.; Zhang, K.; Sun, C.; Mu, B.; Ma, J.; et al. First Results from the Rotating Fan Beam Scatterometer Onboard CFOSAT. *IEEE Trans. Geosci. Remote Sens.* **2020**, *58*, 8793–8806. [CrossRef]
19. Li, Z.; Stoffelen, A.; Verhoef, A.; Verspeek, J. Numerical Weather Prediction Ocean Calibration for the Chinese-French Oceanography Satellite Wind Scatterometer and Wind Retrieval Evaluation. *Earth Sp. Sci.* **2021**, *8*, e2020EA001606. [CrossRef]
20. Zhang, P.; Hu, X.; Lu, Q.; Zhu, A.; Lin, M.; Sun, L.; Chen, L.; Xu, N. FY-3E: The First Operational Meteorological Satellite Mission in an Early Morning Orbit. *Adv. Atmos. Sci.* **2022**, *39*, 1–8. [CrossRef]
21. Remund, Q.P.; Long, D.G. Sea ice extent mapping using Ku band scatterometer data. *J. Geophys. Res. Ocean.* **1999**, *104*, 11515–11527. [CrossRef]
22. Liu, L.; Dong, X.; Lin, W.; Lang, S.; Wang, L. Polar Sea Ice Detection with the CFOSAT Scatterometer. In Proceedings of the 2021 IEEE International Geoscience and Remote Sensing Symposium IGARSS, Brussels, Belgium, 11–16 July 2021; pp. 5645–5648. [CrossRef]
23. Stoffelen, A.; Portabella, M. On Bayesian Scatterometer Wind Inversion. *IEEE Trans. Geosci. Remote Sens.* **2006**, *44*, 1523–1533. [CrossRef]
24. Cavalieri, D.J.; Parkinson, C.L.; Gloersen, P.; Zwally, H.J. *Sea Ice Concentrations from Nimbus-7 SMMR and DMSP SSM/I-SSMIS Passive Microwave Data, Version 1*; [NSIDC-0051]; National Snow & Ice Data Center: Boulder, CO, USA, 1996.

remote sensing

MDPI

Article

Unusual Fish Assemblages Associated with Environmental Changes in the East China Sea in February and March 2017

Wenxiang Ding, Caiyun Zhang *, Jianyu Hu and Shaoping Shang

State Key Laboratory of Marine Environmental Science, Key Laboratory of Underwater Acoustic Communication and Marine Information Technology Ministry of Education, College of Ocean and Earth Sciences, Xiamen University, Xiamen 361102, China; wenxiangding2015@163.com (W.D.); hujy@xmu.edu.cn (J.H.); spshang@xmu.edu.cn (S.S.)

* Correspondence: cyzhang@xmu.edu.cn

Abstract: Two large fish assemblages were recorded in the overwintering fishing grounds of the East China Sea in February and March 2017. In this study, available time series of satellite-derived sea surface temperature, wind, chlorophyll *a*, and reanalysis data were used to explore the relationships between the observed large fish aggregations and environmental factors. The bottom waters of the fishing grounds were abnormally warm in winter 2017, and then experienced significant cooling due to the eastward movement of the Yellow Sea Cold Current, which was driven by the increased northwesterly wind from January to mid-March 2017. Fishing areas in the affected region, including No. 1891, which was abnormally warm, and No. 1592, which had a strong thermal front and high chlorophyll *a* concentration, might have provided suitable environments for the warm-temperature fish, resulting in the observed large fish assemblages. The abnormal temperature changes between winter and early spring 2017 may have been associated with changes in local ocean circulation.

Keywords: fish assemblage; temperature; environmental change; Yellow Sea coastal current; East China Sea

Citation: Ding, W.; Zhang, C.; Hu, J.; Shang, S. Unusual Fish Assemblages Associated with Environmental Changes in the East China Sea in February and March 2017. *Remote Sens.* **2021**, *13*, 1768. https://doi.org/10.3390/rs13091768

Academic Editors: Antony Liu and Jorge Vazquez

Received: 2 March 2021
Accepted: 29 April 2021
Published: 1 May 2021

1. Introduction

Changes in fishery resources are not only associated with increased anthropogenic fishing pressure, but are also closely related to fluctuations in marine environmental factors, such as winds, sea surface temperatures (SST), and ocean currents [1–3]. Of these, temperature has the most significant effect on fish activity [4]. Changes in temperature directly or indirectly affect fish migration behaviors, growth, and habitat distributions [5–7]. Eveson et al. [8] demonstrated that water temperature is the environmental variable most commonly used to forecast tuna fishing grounds. The strong relationships between temperature and fish abundance are well documented [9]. In addition, several studies investigated the effects of abrupt changes in temperature on fishing grounds and fishery resources, e.g., [4,10]. For example, some fish species shifted northward during the 2012 ocean heat wave in the Northwest Atlantic [4], while a cold event in 2008 contributed to a massive fish die-off near the Penghu Island, Taiwan [10].

The East China Sea (ECS) is a marginal sea in the western Pacific Ocean that has a vast shallow continental shelf (Figure 1). The Yangtze River carries a large volume of terrestrial material into the ECS, and the southeastern ECS is strongly affected by the warm Kuroshio current [11,12]. The ECS circulation is influenced by the East Asian monsoon: Northerly winds prevail in winter, while southerly winds prevail in summer [13]. The primary currents affecting the ECS include the Kuroshio, Taiwan Warm Current (TWWC), Yellow Sea Warm Current (YSWC), Yellow Sea Coastal Current (YSCC), and Zhe-Min Coastal Current [1,14]. Interactions among bottom topography, large freshwater discharge, monsoon winds, and the Kuroshio intrusion led to the development of particular circulation regimes and distinct water masses [1]. Ocean parameters, such as temperature and salinity, may change abruptly between water masses [15]. Fronts, which are defined as boundaries between water masses, are generally characterized by strong mixing and increased bioproductivity [16–18]. Thus, frontal zones often coincide with fishery grounds in the ECS [19].

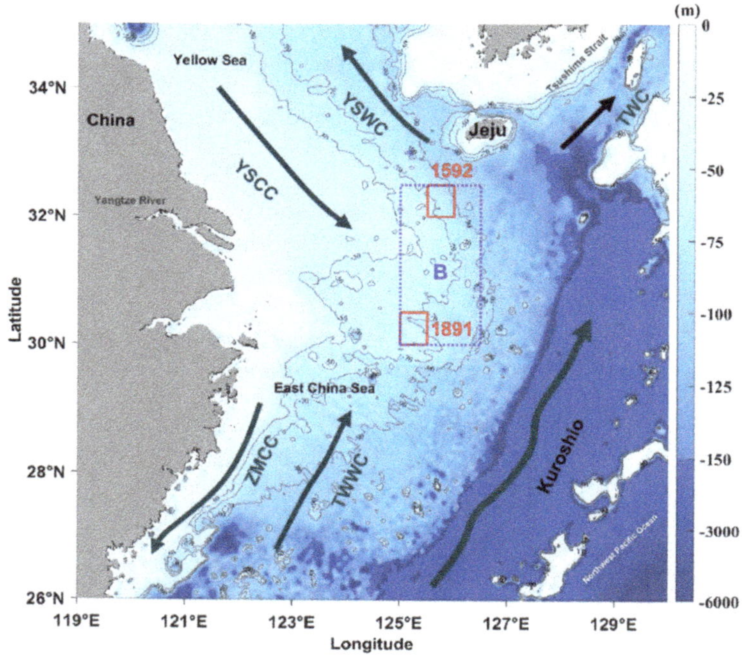

Figure 1. Map showing the study area and water circulation in winter. The two red boxes outline the two fishing areas (Nos. 1592 and 1891). Area B, which includes Nos. 1592 and 1891, is boxed by blue dotted lines. The light gray lines represent isobaths. The bathymetric data were obtained from the General Bathymetric Chart of the Oceans (GEBCO_2020) grid data from the British Oceanographic Data Centre. TWWC: Taiwan Warm Current; ZMCC: Zhe–Min Coastal Current; YSCC: Yellow Sea Coastal Current; YSWC: Yellow Sea Warm Current; TWC: Tsushima Warm Current.

Being one of the most important fishing grounds in the west Pacific, the ECS is biologically diverse and rich in fishery resources, with over 700 recorded species of fish, more than 100 species of crustaceans, and 69 species of cephalopods [20–22]. All fish species are considered inshore, offshore, or migratory [23]. Migratory species include the small yellow croaker (*Larimichthys polyactis*), hairtail, red seabream, and Pacific herring [23]. These species spawn and mature in shallow nearshore waters, then migrate for food and overwintering far offshore [23]. Previous studies show that the abundances and spatial distributions of fishery resources in the ECS are closely related to variations in marine environmental variables [24–26]. For example, due to the influence of various currents and water masses, primary productivity in the ECS is high during the summer [27]. Thus, the ECS is an important spawning and nursery ground for commercially valuable fish such as the small yellow croaker, hairtail, pomfret, and white Chinese croaker [26].

ECS fisheries began to be heavily exploited in the 1980s [28,29], and the biodiversity and abundance of commercial fish species decreased significantly since that time [30]. In the 1990s, the abundance of large-sized, commercially high-value fish species, such as hairtail, small yellow croaker, and silver pomfret, declined to less than 50% of levels in the 1980s, and low-value species began to dominate the catch [29,31]. In addition, although fishing power increased by a factor of 7.6 between the 1960s and the 1990s, the catch per unit effort in the ECS declined by a factor of 3 over the same period [23,32].

In late February 2017, a large number of small yellow croaker (*L. polyactis*) appeared at the surface of fishing area No. 1891 in the ECS (Figure 2a). Fishermen caught more than 9000 kg of *L. polyactis* in about two hours (available online: http://slide.news.sina.com.cn/s/slide_1_2841_108157.html#p=1, accessed on 2 March 2021). Two weeks later, in mid-March 2017, fishermen caught more than 90,000 kg of *Collichthys sp.* and *L. polyactis* near fishing area No. 1592 in about ten days (Figure 2b) (available online: https://www.sohu.com/a/128766304_115864, accessed on 2 March 2021). Large fish catches over short periods such as these are very rare in the ECS, especially as fishery resources are currently reduced due to overfishing. Although some previous studies explored long-term variations in fishery resources and the relationship between these variations and climate change [33–36], investigations of the environmental factors associated with the sudden appearance of abundant fish assemblages in the ECS are limited.

(a) (b)

Figure 2. Fishermen catching large quantities of fish in the ECS. *Larimichthys polyactis* caught in fishing area No. 1891 in late-February 2017 (**a**, available online: http://slide.news.sina.com.cn/s/slide_1_2841_108157.html#p=1, accessed on 2 March 2021); *Collichthys sp.* and *L. polyactis* caught in fishing area No. 1592 in mid-March 2017 (**b**, available online: https://www.sohu.com/a/128766304_115864, accessed on 2 March 2021).

Therefore, to investigate the drivers of the large fish assemblages recorded between February and March 2017, we utilized time series data for satellite-derived sea surface temperature (SST), wind, and Chl *a* concentration, and we used reanalyzed data including data covering temperature and geostrophic currents to investigate variations in marine environmental factors across the fishing grounds, particularly Nos. 1891 and 1592. This study aimed to explore possible mechanisms associated with the large fish accumulations observed in February and March 2017. Usually, large fish assemblages are rare and difficult to predict. However, an improved understanding of the complex relationships among environmental factors in marine ecosystems will help to clarify the responses of fishery resources to marine environmental change, and will inform the development of fishery policies that anticipate future changes in resource availability due to abrupt shifts in marine conditions.

2. Materials and Methods

2.1. Satellite Data and Processing

Daily and monthly global Operational Sea Surface Temperature and Sea Ice Analysis (OSTIA) data from 2011 to 2019 were obtained from the European Copernicus program (http://marine.copernicus.eu, accessed on 2 March 2021), and Advanced Scatterometer (ASCAT)-derived wind-speed and wind-direction data for the same period were obtained from Remote Sensing Systems (http://www.remss.com, accessed on 2 March 2021). The spatial resolution of the OSTIA global sea surface temperature (SST) data is 5 km, and that of the ASCAT-derived wind data is 25 km. Monthly Chl *a* data from the Moderate Resolution Imaging Spectroradiometer (MODIS)/Aqua were obtained from the NASA Goddard Space Flight Center (https://oceancolor.gsfc.nasa.gov, accessed on 2 March 2021) and have a spatial resolution of 4 km. Daily Chl *a* concentration data, derived from geostationary satellite Himawari-8, were obtained from the P-Tree System (Japan Aerospace Exploration Agency, https://www.eorc.jaxa.jp/ptree/index.html, accessed on 2 March 2021) and have a spatial resolution of 5 km.

Temporal changes in the main marine environmental factors in fishing areas No. 1592 (125.5°E–126.0°E, 32°N–32.5°N) and No. 1891 (125.0°E–125.5°E, 30°N–30.5°N; Figure 1) were analyzed, as the abnormal fish assemblages of 2017 were observed in these areas. We also established a larger study site surrounding fishing areas Nos. 1592 and 1891 (area B; 125°E–126.5°E, 30°N–32.5°N) in order to investigate the main environmental parameters of the overwintering fishing grounds.

Gradient methods are commonly used to recognize and define oceanic fronts. The amplitude *grad (i, j)* of the SST gradient was computed at each grid point (*i, j*) in the ECS using the central difference method of Wall et al. [37] as follows:

$$grad(i,j) = \sqrt{(\frac{SST_{i+1,j} - SST_{i-1,j}}{2\Delta x})^2 + (\frac{SST_{i,j+1} - SST_{i,j-1}}{2\Delta y})^2} \quad (1)$$

where Δx and Δy are the distances in kilometers between pairs of neighboring grid points in the east-west direction and the north-south direction, respectively.

Here, we defined thermal fronts as those where the SST gradient was ≥0.025 °C/km. In this way, the weak front at the southern edge of the YSCC was manifested. The frontal intensity was then calculated as the average SST gradient of the pixels with gradients greater than the threshold. We also calculated frontal intensity using other frontal thresholds (e.g., gradient thresholds of ≥0.01, ≥0.015, and ≥0.02 °C/km) and compared our results with those generated using a gradient threshold of ≥0.025 °C/km (Figure 3). We found no significant differences among daily frontal intensities calculated based on different thresholds for fishing areas No. 1891 and 1592 between January and May 2017 (Figure 3). The correlation coefficient (R^2) was >0.74 for fishing area No. 1891 and >0.96 for fishing area No. 1592.

Nighttime light remote sensing data were previously shown to be an effective proxy with which to monitor fishing activity and evaluate fishery resources [38,39]. Waluda et al. [40] used satellite-derived nighttime lights to show that fish catch was positively correlated with estimated fishing extent. Indeed, light intensity at night affects fish activity levels and aggregation [41,42]. Because monthly fish catch data at the study sites are not available, we used nighttime light intensity to approximate the temporal changes in fishery activity levels. Higher light intensities might imply increases in fishing effort or fish abundance. The monthly average radiance composite images derived from the Visible Infrared Imaging Radiometer Suite (VIIRS) Day/Night Band (DNB) for February and March from 2013 to 2019 were obtained from the Earth Observations Group (https://payneinstitute.mines.edu/eog/nighttime-lights/, accessed on 2 March 2021), with a spatial resolution of 15 arc-seconds (~500 m). DNB radiance data affected by stray light, lightning, lunar illumination, and cloud cover are excluded from the DNB datasets before averaging [43]. We used the average monthly radiance data greater than zero in area B to analyze temporal changes in nighttime light intensity.

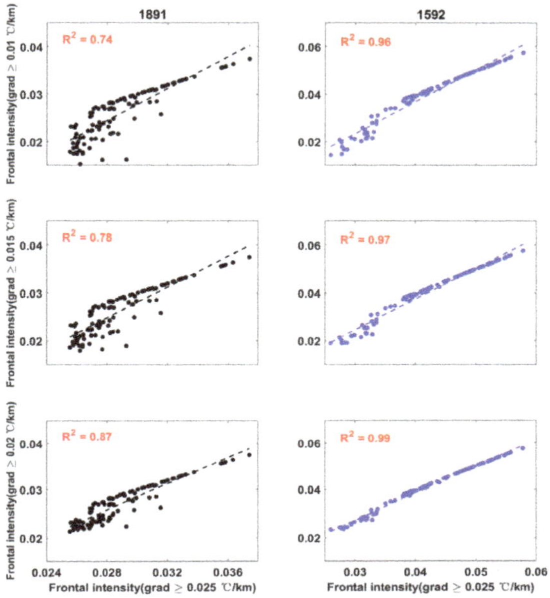

Figure 3. Daily frontal intensities from January to May 2017 calculated using different frontal thresholds for fishing areas No. 1891 (left panels) and No. 1592 (right panels).

2.2. *Temperature and Geostrophic Currents from ARMOR3D*

The Multi Observation Global Ocean ARMOR3D L4 multi-year reprocessed (REP) weekly data from 2011 to 2019 were obtained from the E.U. Copernicus Marine Service (http://marine.copernicus.eu, accessed on 2 March 2021). The product used in this study was MULTIOBS_GLO_PHY_TSUV_3D_MYNRT_015_012. The ARMOR3D product consists of global 3D temperature, salinity, geopotential height, geostrophic currents, and 2D mixed layer depth defined on a 1/4° regular grid, from the surface down to a depth of 5500 m [44,45]. The temperature and salinity observation-based product uses statistical methods to combine remote sensing observations (sea level anomaly and SST) with in situ vertical profiles of temperature and salinity (derived mainly from Argo floats but also from other sources including Conductivity-Temperature-Depth (CTDs) and eXpandable-Bathy-Thermograph (XBTs)) [45]. The global 3D geostrophic circulation was estimated by merging altimetric data and a synthetic 3D thermohaline field using the thermal wind equation with the reference level set at the surface [44].

3. Results

3.1. *Temperature*

The spatial distribution of average SST in the ECS from February to March clearly showed a cold-water tongue, represented by a 12.5 °C isotherm, extending southeastward from the Yellow Sea (Figure 4). Fishing areas Nos. 1891 and 1592 were located at the edge of this cold-water tongue. Unlike the 12.5 °C isotherm observed each year from 2011 to 2019 (but excluding 2011 and 2017), the 12.5 °C isotherm of 2017 shifted northwards or eastwards (Figure 4).

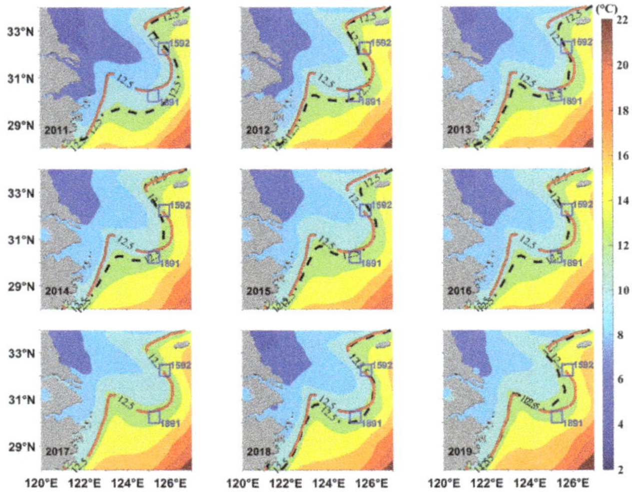

Figure 4. The spatial distributions of average OSTIA SST in February and March in the East China Scheme from 2011 to 2019. The overlaid solid red line represents the 12.5 °C isotherm based on the average SST from February to March in 2017, while the dashed black line represents the 12.5 °C isotherm based on average SST from February to March in the year indicated.

Daily variations in SST between January and April from 2011 to 2019 indicated that relatively high SST took place early in January in both fishing areas Nos. 1891 and 1592. Then, the SSTs decreased until mid- or late-March (Figure 5). The SST of No.1891 was warmer in 2017 than in other years on the same days before mid-March. For No. 1592, warm SST appeared in early January, and it was warmer than the SSTs of other years excepting 2019. After that, the SST decreased and was <12 °C in mid-March 2017 (Figure 5).

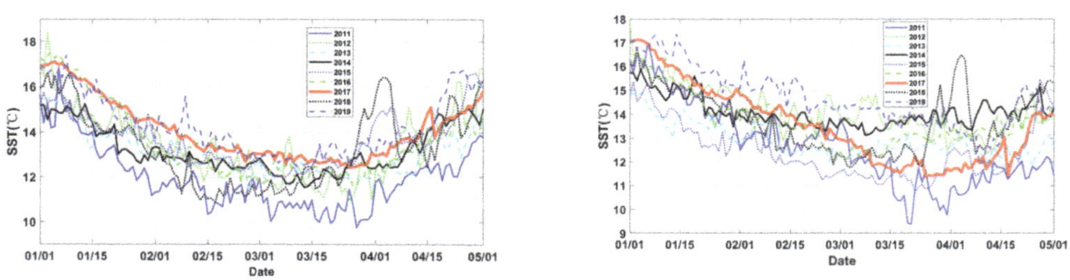

(**a.1891**) (**b.1592**)

Figure 5. Daily SST between January and April 2011–2019 for (**a**) fishing area No. 1891 and (**b**) fishing area No. 1592. The thickened red lines correspond to 2017.

The vertical distributions of temperature derived from ARMOR3D for area B between 2011 and 2019 are shown in Figure 6. The upper 50 m of the water column in area B was relatively cold and vertically mixed; the bottom layer was warmer, possibly due to the influence of the TWCC [46]. Compared with other years, the bottom layer (60–80 m) in early January 2017 was relatively warm (>18 °C), warmer than any other years except for 2016 and 2019 (Figures 6 and 7). Then, water temperature decreased, and relatively cold water (<14 °C) appeared in March 2017 (Figure 6). We calculated the differences in water temperatures between the week with the coldest average water temperature (the 11th week, March 12–18) and the week with the warmest average temperature (the 1st week, 1–7 January) for the upper layer (0–20 m) and the bottom layer (60–80 m) (Figure 7). We found that the magnitude of cooling in area B from early January to mid-March 2017 was generally greater than that in any of the other nine years. This was especially dramatic in the bottom layer, the temperature of which decreased by >4 °C between early January to mid-March 2017, a greater decrease than that in any other year (Figure 7).

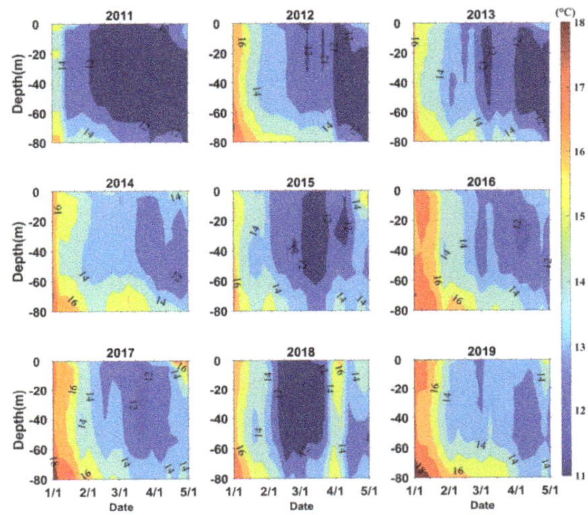

Figure 6. Vertical distributions of temperature between January and April 2011–2019 for area B.

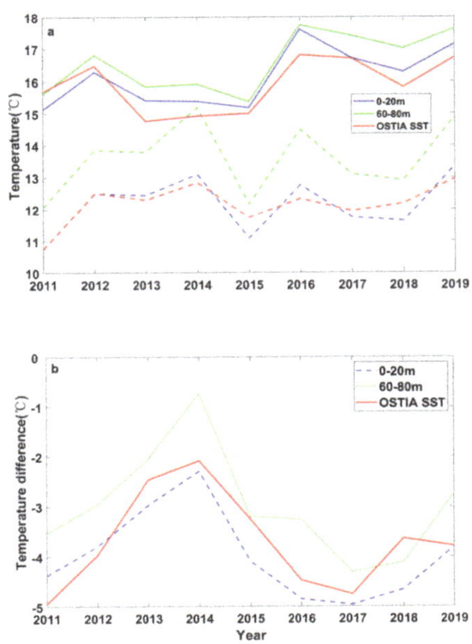

Figure 7. Temperatures in the 1st week (solid lines) and 11th week (dashed lines) in the upper layer (0–20 m) and the bottom layer (60–80 m) in area B. Both temperatures are graphed in (**a**); differences between the 1st and 11th week are shown in (**b**). The OSITA SSTs are superimposed for comparison.

3.2. Thermal Fronts

The spatial distribution of the SST gradient in the ECS is shown in Figure 8. A high SST gradient was particularly noticeable at the northern edge of the cold-water tongue, where the cold YSCC meets the warm YSWC. Fishing area No. 1592 is located inside the northern strong thermal front, and fishing area No. 1891 is located at the edge of the southern weak thermal front (Figure 8a). The frontal intensity in the fishing area No. 1592 was stronger than the multi-year average, while the intensity of the thermal front in most of fishing area No. 1891 was weaker than in normal years (Figure 8b).

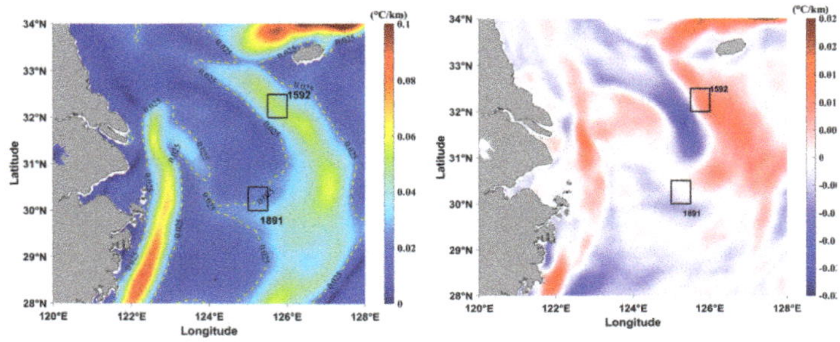

(a) (b)

Figure 8. (**a**) The spatial distribution of the average SST gradient and (**b**) the gradient anomaly during February and March 2017. The gradient anomaly was obtained by subtracting the average gradients for February and March 2011−2019 from the gradients for February and March 2017.

Daily changes in frontal intensity between 1 January 2017 and 1 May 2017 indicate that the frontal intensity in fishing area No. 1891 was weaker than that in fishing area No. 1592 (Figure 9). The strongest frontal intensity in fishing area No. 1891 appeared in late January. Frontal intensity remained relatively strong through mid-February and became relatively weak in March. The strongest frontal intensity in fishing area No. 1592 appeared in mid-February, about one month later than that in fishing area No. 1891. Thus, in neither area did the appearance of the unusually large fish assemblage exactly coincide with the peak of frontal intensity. However, the large fish assemblage did appear in fishing area No. 1592 at a point when frontal intensity was relatively strong (Figure 9). Compared with multi-year average, the frontal intensity in fishing area No. 1891 was weaker on most days in February and March 2017. However, the frontal intensity in fishing area No. 1592 was higher than average between January and mid-March 2017 (Figure 9).

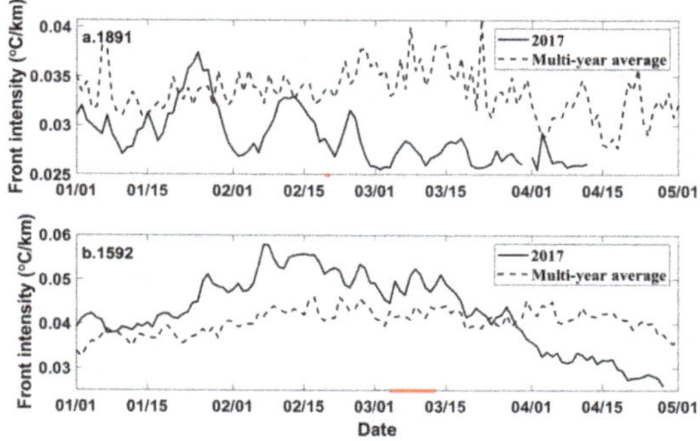

Figure 9. (**a**) Daily fluctuations in frontal intensity for fishing area No. 1891 and (**b**) fishing area No. 1592 from January to April 2017. The solid and dashed lines indicate the frontal intensity in 2017 and the average frontal intensity in 2011−2019, respectively. The bold red lines on each x-axis indicate the time periods at which the large fish assemblages were observed in each area (as shown Figure 2).

3.3. *Chl a*

Analysis of the spatial distribution of MODIS Chl *a* concentration in February and March 2017 showed that nearshore Chl *a* was greater than offshore Chl *a* (Figure 10a). Chl *a* concentrations were greater than 1.0 mg/m^3 in both fishing areas (Nos. 1891 and 1592). In 2017, Chl *a* levels greater than the multi-year average mostly occurred in or near thermal fronts (Figure 10b), and these locations also had stronger than average frontal intensities (Figure 8b). Indeed, the Chl *a* concentration in fishing area No. 1592 was 0.21 mg/m^3 (11.9%) higher than the multi-year average.

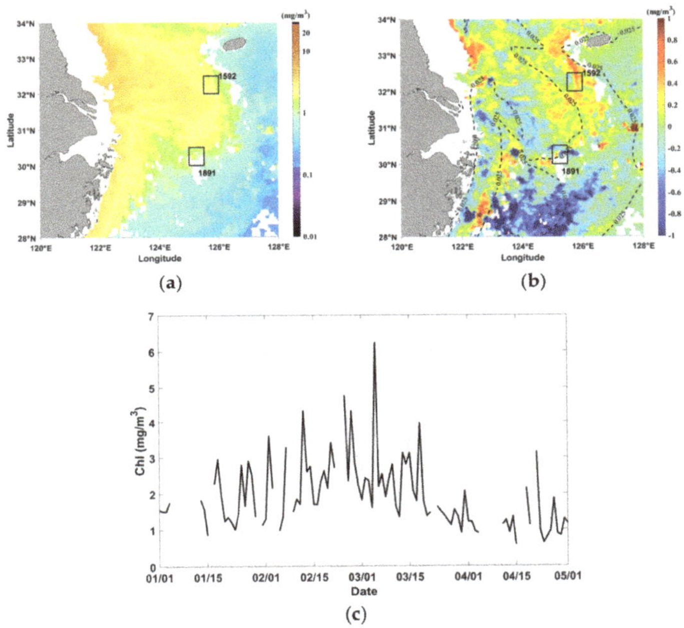

Figure 10. (**a**) The spatial distribution of average MODIS-derived Chl *a* concentrations and (**b**) Chl *a* anomalies in area B, overlaid with an SST gradient of 0.025 °C/km in February and March 2017. (**c**) Daily changes in Himawari-8-derived Chl *a* concentrations in area B. Chl *a* anomaly data were calculated by subtracting the average Chl *a* concentration in February and March 2011−2019 from the Chl *a* concentration in February and March 2017.

Because several datapoints in the daily Himawari-8-derived Chl *a* dataset for fishing areas Nos. 1891 and 1592 are missing due to bad weather or cloud cover, we used the daily changes in Chl *a* concentration in area B as a whole (including both fishing areas; Figure 10c). Chl *a* concentrations were relatively high from early February to mid-March, consistent with the appearance of relatively intense fronts (Figure 9).

3.4. *Wind*

Northerly winds prevailed in the ECS during February and March 2017 (Figure 11a). The polar wind chart indicates that the wind field in 2017 was mainly northerly and northwesterly, in contrast to the multi-year average wind field, which was dominated by northerly and northeasterly winds (Figure 11b). The frequencies of northerly and northwesterly winds in February and March 2017 were greater than the multi-year average, while the frequencies of northeasterly winds were significantly lower. Furthermore, the speeds of northerly and northwesterly winds in February and March 2017 were significantly stronger than the multi-year average, while the northeasterly winds were weaker (Figure 11b).

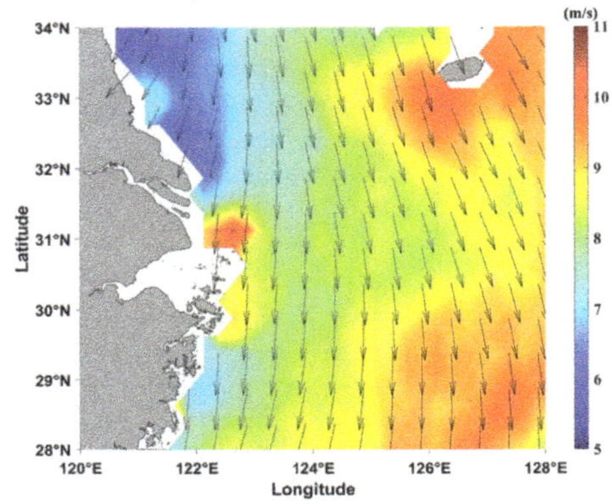

(a)

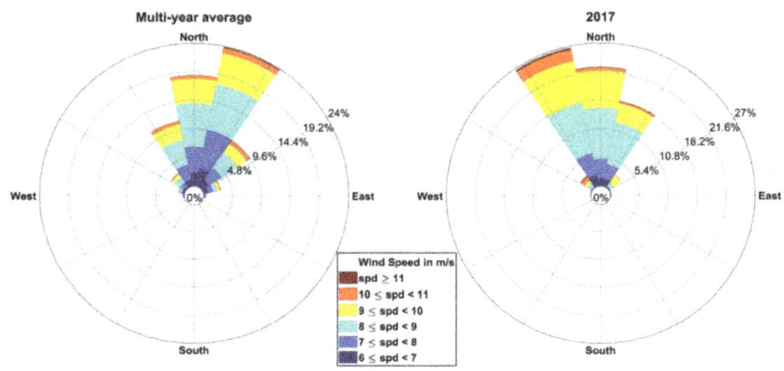

(b)

Figure 11. (**a**) The spatial distribution of the average wind field in February and March 2017. The color represents wind speed, and the black arrows indicate wind direction. (**b**) The polar wind chart showing the multi-year average wind for 2011−2019 (left panel) and the average wind in February and March 2017 (right panel).

4. Discussion

During February and March 2017, large fish assemblages were observed in fishing areas Nos. 1891 and 1592 (Figure 2). We used monthly average nighttime light radiance data to demonstrate that fish abundance was also high in other locations in area B (Figure 12a). Nighttime light radiance in area B was high in 2017 and 2018 (Figure 12a). The average radiance in February and March 2017 was 38% higher than that of 2013–2016. High radiance levels correspond to higher light intensities, indicating that greater fishing fleets and fish abundance were at sea [40]. Indeed, it was previously reported that many fisheries worldwide use artificial light to attract pelagic fish and increase fish catches, e.g., [41,47]. Our results suggested that there might have been a considerable number of fishing fleets operating near the fishing grounds of area B due to the relative abundance of fish during February and March 2017 (Figure 12a).

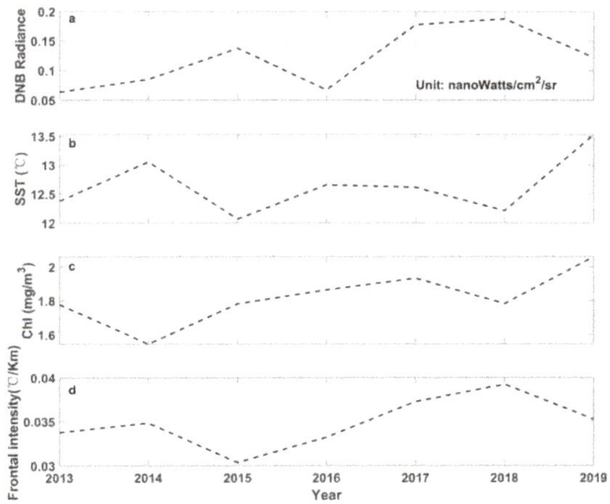

Figure 12. (**a**) Average DNB radiance, (**b**) SST, (**c**) Chl *a*, and (**d**) frontal intensity in February and March 2013–2019 in area B (the area outlined with a dashed blue line in Figure 1).

L. polyactis and *Collichthys sp.*, which are economically important fishery species in China, are warm-temperate demersal fish [48,49]. Generally, the spawning grounds of these fish are located on the inner shelf of the ECS or its estuaries [50]. Due to the southeastward intrusion of the cold YSCC in the winter, the nearshore environment becomes unsuitable for the growth of these species; both species thus migrate to warm offshore areas and overwinter at depths of 40–80 m [51–53]. Previous studies showed that the main winter fishing grounds in the ECS are located at the southern shelf area, where the YSCC and TWWC meet, or at the eastern slope area, where the YSCC and YSWC meet [50]. The two fishing areas examined in this study, Nos. 1891 and 1592, are located in these mixed waters (Figure 1). These fishing grounds are influenced by various different water masses and consequently have strong temperature and salinity gradients [15]. These fishing grounds also include areas with high abundances of small- and medium-sized copepods, which provide a rich food source for *L. polyactis* and *Collichthys sp.* [54]. Therefore, *L. polyactis* and *Collichthys sp.* tend to migrate and gather in these fishing areas for the winter.

Environmental changes strongly affect the paths, compositions, and durations of fish migrations for feeding or overwintering [55–57]. In January 2017, the bottom water became relatively warm (>18 °C). The water temperature in this layer then decreased due to the eastward movement of the cold YSCC. The decrease in the temperature of the bottom layer was especially dramatic in 2017 (Figure 7). Liu et al. [58] indicated that *L. polyactis* tolerate a wide range of temperatures (6–26 °C). However, as both *L. polyactis* and *Collichthys sp.* are warm-temperate demersal fish, they might tend to migrate to warmer water [50]. Indeed, *L. polyactis* and *Collichthys sp.* may not have been able to the tolerate the continuous cooling of the bottom layer in February and March 2017, and these fish may have attempted to migrate to the surrounding warm water.

Fishing areas Nos. 1891 and 1592 are located at the edge of the cold-water tongue, and are warmer than the surrounding water that is influenced by the YSCC (Figure 4). Fishing area No. 1891 was warmer than average during January and March 2017 (Figure 5a), although frontal intensity was weak, and Chl *a* concentrations were not high (Figures 8 and 10). Several studies suggested that fish abundance increases with temperature, e.g., [9,59,60]. Therefore, the warmth of fishing area No. 1891 in February 2017 might have provided favorable conditions for warm-temperature fish, causing the observed large aggregation of fish in this fishing area (Figure 2). However, the factors leading to the large aggregation of fish on the sea surface in fishing area No. 1891 remain unclear due to limited data. Additional observations, combined with fish behavior analyses, are necessary to understand better the factors that generated this aggregation at the surface.

In contrast to fishing area No. 1891, fishing area No. 1592 had stronger fronts (Figures 8 and 9) and higher Chl a concentrations (Figure 10) in 2017 as compared to other years. Usually, a strong front is characterized by abundant nutrients and high primary productivity [61–64], which provides a suitable habitat for fish [48,65]. Figure 12 shows the interannual variation in nighttime light intensity, average SST, front intensity, and Chl a concentrations for area B over the period 2013–2019. The two years with the greater light intensity (2017 and 2018) also seem to have high frontal intensity (Figure 12a,d). These results further suggested that fishing grounds with high frontal intensity attracted more fish schools. Consequently, fish abundance was high in these frontal areas, and more fishing boats were present.

Fishery resources in the ECS have been gradually increasing as a result of recent moratoriums on summertime fishing [49]. As fishing grounds recover, large fish assemblages may reappear in the ECS. It is possible that the particular environmental conditions identified in winter and spring 2017, which provided a suitable environment for migratory fish and which may have played an important role in the large fish aggregations of that year, were also present in other years. These conditions included the relatively warm temperature of the bottom layer in January (Figure 6), followed by a significant cooling of the bottom layer between January and March (Figures 6 and 7), while the surrounding waters remained warm; these conditions also included the observed strong frontal intensity and high Chl *a* concentrations. Long-term time series of multi-disciplinary data, including historical catch data, should be collected to investigate the complex relationships between environmental change and variations in fishery resources further.

Numerous studies previously showed that increases in ECS temperatures are associated with global warming [66,67], and average winter SST in the ECS has increased an average of ~0.3 °C/decade, which far exceeds the globally averaged rate of ocean surface warming [68]. Indeed, the ECS experienced record-high temperatures in the winter months of 2017 [69], which was consistent with the high temperatures recorded in January 2017 (Figures 5 and 6).

The significant cooling from January to mid-March 2017 might be linked to the eastward movement of cold YSCC (Figures 4 and 13a). Circulation in the ECS and the Yellow Sea is controlled by the East Asian monsoon [70,71]. As shown in Figure 13a, the Kuroshio, TWC, and TWWC have greater northward velocities than other surrounding currents. The frequent and strong northerly and westerly winds (Figure 11) in February and March 2017 enhanced the YSCC (Figure 13b), which might have driven more cold water from the Yellow Sea into the northern ECS, causing the cold-water tongue of the YSCC to move further eastward in February and March (Figures 4 and 13b). At the same time, the warm TWWC in the south moved abnormally northward (Figure 13b), raising the SST of fishing area No. 1891 above the multi-year average (Figures 4 and 5). The strengthened YSCC reduced SSTs, increasing the temperature difference between the YSCC and the YSWC. This increased frontal intensity between the cold and warm water masses (Figure 8) led to an increase in Chl *a* concentrations (Figure 10). As demonstrated by Weisberg et al. [72], ocean circulation unites nutrients with light, fueling primary productivity and higher trophic level interactions. Thus, ocean currents may play an important role in the regulation of phytoplankton growth and fishery resources.

Over recent decades, ocean temperatures increased steadily due to climate change, and these increases strongly impacted the marine ecosystem [73–75]. The ECS was previously identified as a critical area of significant ocean warming, e.g., [76–78]. Abrupt marine environmental changes are expected to become more common as climate change progresses [35]. Warming trends and/or abnormal temperature changes may affect the abundance, mortality, growth, and distribution of marine fish [33,34,79]. Here, we showed that the large fish assemblage in the ECS winter fishing ground in February and March 2017 may have been associated with abnormal temperature changes associated with the eastward movement of the YSCC. This event provided a rare case study that exemplified how complex relationships among climate-driven physical changes and biological processes may affect fishery resources. Future fishery management programs must develop strategies to adapt to similar environmental events that will become more frequent due to climate change.

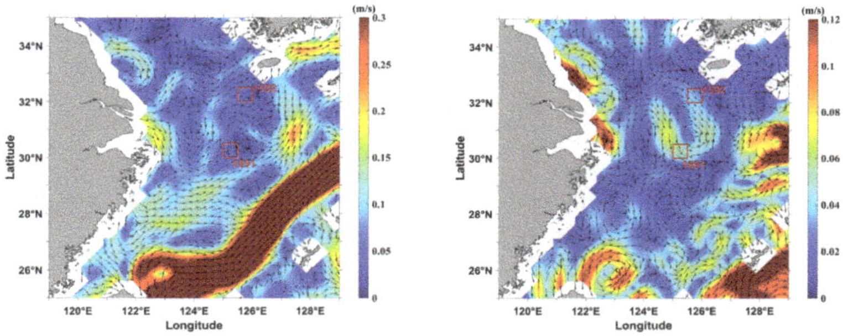

(**a**) (**b**)

Figure 13. (**a**) The distribution of average sea surface currents (SSCs) and (**b**) SSC anomalies in February and March 2017. Colors correspond to current speed, while arrows indicate current direction. SSC anomaly data were calculated by subtracting the average SSC during February and March 2011–2019 from the SSC in February and March 2017.

5. Conclusions

Large fish assemblages were reported in the winter fishing grounds of the ECS during February and March 2017. The bottom waters of these fishing grounds were abnormally warm in winter 2017. This layer then cooled substantially from January to mid-March 2017 due to the eastward movement of the YSCC associated with the increased northwesterly winds. Warm-temperature demersal fish, such as the small yellow croaker (*Larimichthys polyactis*) and *Collichthys sp.*, might tend to migrate to fishing areas with warm temperature, strong thermal front, and high Chl *a* concentrations; these conditions were found in fishing areas Nos. 1891 and 1592 during February and March 2017. The observed abnormal temperature changes in winter and early spring 2017 might reflect the effects of changes in local ocean circulation.

This preliminary study proposes some possible mechanisms underlying the formation of two large fish assemblages based on satellite remote sensing and reanalysis data. Long-term time series data, obtained using a variety of coastal ocean observing systems [80], in conjunction with numerical models, are required to clarify and explain these physical mechanisms further. However, this rare case study demonstrates how fishery resources respond to climate-driven physical and biological changes. Thus, fishery management programs should carefully consider ecological responses to similar abnormal changes in marine environment.

Author Contributions: Conceptualization, C.Z. and W.D.; methodology, C.Z.; validation, W.D. and C.Z.; formal analysis, C.Z. and W.D.; investigation, W.D.; resources, C.Z. and W.D.; data curation, W.D.; writing—original draft preparation, W.D.; writing—review and editing, C.Z., W.D., J.H. and S.S.; visualization, W.D.; supervision, C.Z.; project administration, C.Z.; funding acquisition, C.Z. All authors have read and agreed to the published version of the manuscript.

Funding: This research was funded by the Key Program of NSF-China (NSFC) (Grant No. U1805241) and the National Key Research and Development Plan of China (Grant Nos. 2016YFE0202100, 2018YFC1406302).

Data Availability Statement: Publicly available data sets were analyzed in this study. OSTIA and ARMOR3D data were downloaded from Europe Copernicus program (http://marine.copernicus.eu, accessed on 2 March 2021). ASCAT wind and MODIS ocean color data were produced by Remote

Sensing Systems (http://www.remss.com, accessed on 2 March 2021) and NASA Goddard Space Flight Center (https://oceancolor.gsfc.nasa.gov, accessed on 2 March 2021), respectively. VIIRS DNB radiance data were downloaded from the Earth Observations Group (https://payneinstitute.mines.edu/eog/nighttime-lights/, accessed on 2 March 2021). Himawari-8 Chl a concentration were downloaded from the P-Tree System (https://www.eorc.jaxa.jp/ptree/index.html, accessed on 2 March 2021).

Acknowledgments: Anonymous reviewers contributed substantially to the improvement of this paper.

Conflicts of Interest: The authors declare no conflict of interest.

References

1. Lie, H.-J.; Cho, C.-H. Recent advances in understanding the circulation and hydrography of the East China Sea. *Fish. Oceanogr.* **2002**, *11*, 318–328. [CrossRef]
2. Drinkwater, K.F.; Tremblay, M.J.; Comeau, M. The influence of wind and temperature on the catch rate of the American lobster (*Homarus americanus*) during spring fisheries off eastern Canada. *Fish. Oceanogr.* **2006**, *15*, 150–165. [CrossRef]
3. Winters, D.B. Offshore Wind Energy Impacts on Fisheries: Investigating Uncharted Waters in Research and Monitoring. *Fisheries* **2018**, *43*, 343–344. [CrossRef]
4. Mills, K.E.; Pershing, A.J.; Brown, C.J.; Chen, Y.; Chiang, F.-S.; Holland, D.S.; Lehuta, S.; Nye, J.A.; Sun, J.C.; Thomas, A.C.; et al. Fisheries Management in a Changing Climate: Lessons From the 2012 Ocean Heat Wave in the Northwest Atlantic. *Oceanography* **2013**, *26*, 191–195. [CrossRef]
5. Sambou, O.S.; Kang, B.; Xu, H.X.; Zhou, Y.D.; Panhwar, S.K. Fish assemblage in the Hairtail Protected Area, East China Sea in relation to environmental variables. *Cah. Biol. Mar.* **2020**, *61*, 279–289. [CrossRef]
6. Huang, M.R.; Ding, L.Y.; Wang, J.; Ding, C.Z.; Tao, J. The impacts of climate change on fish growth: A summary of conducted studies and current knowledge. *Ecol. Indic.* **2021**, *121*, 106976. [CrossRef]
7. Xu, Z.L.; Chen, J.J. Analysis on migratory routine of *Larimichthy polyactis*. *J. Fish. Sci. China* **2009**, *16*, 931–940.
8. Eveson, J.P.; Hobday, A.J.; Hartog, J.R.; Spillman, C.M.; Rough, K.M. Seasonal forecasting of tuna habitat in the Great Australian Bight. *Fish. Res.* **2015**, *170*, 39–49. [CrossRef]
9. Methven, D.A.; Piatt, J.F. Seasonal abundance and vertical distribution of capelin (*Mallotus villosus*) in relation to water temperature at a coastal site off eastern Newfoundland. *ICES J. Mar. Sci.* **1991**, *48*, 187–193. [CrossRef]
10. Chang, Y.; Lee, M.-A.; Lee, K.-T.; Shao, K.-T. Adaptation of fisheries and mariculture management to extreme oceanic environmental changes and climate variability in Taiwan. *Mar. Policy* **2013**, *38*, 476–482. [CrossRef]
11. Hu, X.M.; Xiong, X.J.; Qiao, F.L.; Guo, B.H.; Lin, X.P. Surface current field and seasonal variability in the Kuroshio and adjacent regions derived from satellite-tracked drifter data. *Acta Oceanol. Sin.* **2008**, *27*, 11–29.
12. Zuo, J.; Song, J.; Yuan, H.; Li, X.; Li, N.; Duan, L. Impact of Kuroshio on the dissolved oxygen in the East China Sea region. *J. Oceanol. Limnol.* **2018**, *37*, 513–524. [CrossRef]
13. Yang, X.M.; Yao, T.D. The progress on the Asian monsoon study. *Chin. J. Nat.* **1999**, *6*, 3–5.
14. Ichikawa, H.; Beardsley, R.C. The Current System in the Yellow and East China Seas. *J. Oceanogr.* **2002**, *58*, 77–92. [CrossRef]
15. Park, S.; Chu, P.C. Thermal and haline fronts in the Yellow/East China Seas: Surface and subsurface seasonality comparison. *J. Oceanogr.* **2006**, *62*, 617–638. [CrossRef]
16. Chen, C.T.A. Chemical and physical fronts in the Bohai, Yellow and East China seas. *J. Mar. Syst.* **2009**, *78*, 394–410. [CrossRef]
17. He, S.; Huang, D.; Zeng, D. Double SST fronts observed from MODIS data in the East China Sea off the Zhejiang–Fujian coast, China. *J. Mar. Syst.* **2016**, *154*, 93–102. [CrossRef]
18. Chang, Y.; Lee, M.A.; Shimada, T.; Sakaida, F.; Kawamura, H.; Chan, J.W.; Lu, H.J. Wintertime high-resolution features of sea surface temperature and chlorophyll-a fields associated with oceanic fronts in the southern East China Sea. *Int. J. Remote Sens.* **2008**, *29*, 6249–6261. [CrossRef]
19. Huang, D.; Zhang, T.; Zhou, F. Sea-surface temperature fronts in the Yellow and East China Seas from TRMM microwave imager data. *Deep. Sea Res. Part II Top. Stud. Oceanogr.* **2010**, *57*, 1017–1024. [CrossRef]
20. Zhao, C.Y. *Marine Fishery Resource of China*; Zhejiang Science and Technology Press: Hangzhou, China, 1990. (In Chinese)
21. Zheng, J.Y.; Chen, X.Z.; Cheng, J.H. *Fisheries Resource and Environment of Continental Shelf in East China Sea*; Shanghai Science and Technology Press: Shanghai, China, 2003. (In Chinese)
22. Cheng, J.; Cheung, W.W.; Pitcher, T.J. Mass-balance ecosystem model of the East China Sea. *Prog. Nat. Sci.* **2009**, *19*, 1271–1280. [CrossRef]
23. Kang, J.-S. Analysis on the development trends of capture fisheries in North-East Asia and the policy and management implications for regional co-operation. *Ocean Coast. Manag.* **2006**, *49*, 42–67. [CrossRef]
24. Zheng, B.; Chen, X.J.; Li, G. Relationship between the resource and fishing ground of mackerel and environmental factors based on GAM and GLM models in the East China Sea and Yellow Sea. *J. Fish. China* **2008**, *32*, 379–386.
25. Li, G.; Chen, X.; Lei, L.; Guan, W. Distribution of hotspots of chub mackerel based on remote-sensing data in coastal waters of China. *Int. J. Remote. Sens.* **2014**, *35*, 4399–4421. [CrossRef]

26. Lin, H.-Y.; Chiu, M.-Y.; Shih, Y.-M.; Chen, I.-S.; Lee, M.-A.; Shao, K.-T. Species composition and assemblages of ichthyoplankton during summer in the East China Sea. *Cont. Shelf Res.* **2016**, *126*, 64–78. [CrossRef]
27. Liu, K.-K.; Chao, S.-Y.; Lee, H.-J.; Gong, G.-C.; Teng, Y.-C. Seasonal variation of primary productivity in the East China Sea: A numerical study based on coupled physical-biogeochemical model. *Deep. Sea Res. Part II Top. Stud. Oceanogr.* **2010**, *57*, 1762–1782. [CrossRef]
28. Zhang, C.-I.; Seo, Y.-I.; Kang, H.-J.; Lim, J.-H. Exploitable carrying capacity and potential biomass yield of sectors in the East China Sea, Yellow Sea, and East Sea/Sea of Japan large marine ecosystems. *Deep. Sea Res. Part II Top. Stud. Oceanogr.* **2019**, *163*, 16–28. [CrossRef]
29. Teh, L.S.L.; Cashion, T.; Cheung, W.W.L.; Sumaila, U.R. Taking stock: A Large Marine Ecosystem perspective of socio-economic and ecological trends in East China Sea fisheries. *Rev. Fish Biol. Fish.* **2020**, *30*, 269–292. [CrossRef]
30. Jiang, Y.Z.; Cheng, J.H.; Li, S.F. Temporal changes in the fish community resulting from a summer fishing moratorium in the northern East China Sea. *Mar. Ecol. Prog. Ser.* **2009**, *387*, 265–273. [CrossRef]
31. Chen, W.Z.; Zheng, Y.Z.; Chen, Y.Q.; Mathews, C.P. An assessment of fishery yields from the East China Sea Ecosystem. *Mar. Fish. Rev.* **1997**, *59*, 1–7.
32. FAO. *The State of World Fisheries and Aquaculture*; FAO: Rome, Italy, 1997; p. 125.
33. Edwards, M.; Richardson, A.J. Impact of climate change on marine pelagic phenology and trophic mismatch. *Nat. Cell Biol.* **2004**, *430*, 881–884. [CrossRef]
34. Perry, A.L.; Low, P.J.; Ellis, J.R.; Reynolds, J.D. Climate Change and Distribution Shifts in Marine Fishes. *Science* **2005**, *308*, 1912–1915. [CrossRef]
35. Tebaldi, C.; Hayhoe, K.; Arblaster, J.M.; Meehl, G.A. Going to the Extremes. *Clim. Chang.* **2006**, *79*, 185–211. [CrossRef]
36. Sukgeun, J.; Hyung, K.C. Fishing vs. Climate Change: An example of filefish (*Thamnaconus modestus*) in the northern east China sea. *J. Mar. Sci. Technol.* **2013**, *21*, 15–22. [CrossRef]
37. Wall, C.C.; Muller-Karger, F.E.; Roffer, M.A.; Hu, C.; Yao, W.; Luther, M.E. Satellite remote sensing of surface oceanic fronts in coastal waters off west–central Florida. *Remote. Sens. Environ.* **2008**, *112*, 2963–2976. [CrossRef]
38. Straka, W.; Seaman, C.J.; Baugh, K.; Cole, K.; Stevens, E.; Miller, S.D. Utilization of the Suomi National Polar-Orbiting Partnership (NPP) Visible Infrared Imaging Radiometer Suite (VIIRS) Day/Night Band for Arctic Ship Tracking and Fisheries Management. *Remote. Sens.* **2015**, *7*, 971–989. [CrossRef]
39. Cozzolino, E.; Lasta, C.A. Use of VIIRS DNB satellite images to detect jigger ships involved in the Illex argentinus fishery. *Remote. Sens. Appl. Soc. Environ.* **2016**, *4*, 167–178. [CrossRef]
40. Waluda, C.M.; Griffiths, H.J.; Rodhouse, P.G. Remotely sensed spatial dynamics of the Illex argentinus fishery, Southwest At-lantic. *Fish. Res.* **2008**, *91*, 196–202. [CrossRef]
41. Cooke, S.; Lennox, R.; Bower, S.; Horodysky, A.; Treml, M.; Stoddard, E.; Donaldson, L.; Danylchuk, A. Fishing in the dark: The science and management of recreational fisheries at night. *Bull. Mar. Sci.* **2017**, *93*, 519–538. [CrossRef]
42. Hammerschlag, N.; Meyer, C.; Grace, M.; Kessel, S.; Sutton, T.; Harvey, E.; Paris-Limouzy, C.; Kerstetter, D.; Cooke, S. Shining a light on fish at night: An overview of fish and fisheries in the dark of night, and in deep and polar seas. *Bull. Mar. Sci.* **2017**, *93*, 253–284. [CrossRef]
43. Elvidge, C.D.; Baugh, K.; Zhizhin, M.; Hsu, F.C.; Ghosh, T. VIIRS night-time lights. *Int. J. Remote. Sens.* **2017**, *38*, 5860–5879. [CrossRef]
44. Mulet, S.; Rio, M.-H.; Mignot, A.; Guinehut, S.; Morrow, R. A new estimate of the global 3D geostrophic ocean circulation based on satellite data and in-situ measurements. *Deep. Sea Res. Part II Top. Stud. Oceanogr.* **2012**, *77–80*, 70–81. [CrossRef]
45. Guinehut, S.; Dhomps, A.-L.; Larnicol, G.; Le Traon, P.-Y. High resolution 3-D temperature and salinity fields derived from in situ and satellite observations. *Ocean Sci.* **2012**, *8*, 845–857. [CrossRef]
46. Kim, S.H.; Choi, B.K.; Kim, E. Study on the Behavior of the Water Temperature Inversion Layer in the Northern East China Sea. *J. Mar. Sci. Eng.* **2020**, *8*, 157. [CrossRef]
47. Solomon, O.O.; Ahmed, O.O. Fishing with light: Ecological consequences for coastal habitats. *Int. J. Fish. Aquat. Stud.* **2016**, *4*, 474–483.
48. Du, J.-L.; Yang, S.-L.; Feng, H. Recent human impacts on the morphological evolution of the Yangtze River delta foreland: A review and new perspectives. *Estuarine, Coast. Shelf Sci.* **2016**, *181*, 160–169. [CrossRef]
49. Lin, L.; Liu, Z.; Jiang, Y.; Huang, W.; Gao, T. Current status of small yellow croaker resources in the southern Yellow Sea and the East China Sea. *Chin. J. Oceanol. Limnol.* **2011**, *29*, 547–555. [CrossRef]
50. Chen, J.-J.; Xu, Z.-L.; Chen, X.-Z. The spatial distribution pattern of fishing ground for small yellow croaker in China Seas. *J. Fish. China* **2010**, *34*, 236–244. [CrossRef]
51. Liu, X. *The Research of Small Yellow Croaker (Larimichthys polyactis) Geographic Race and Gonad*; Science Press: Beijing, China, 1962; pp. 35–70.
52. Li, X.D. A preliminary study on the division of water-system and fishing grounds during winter in the south of Huanghai Sea and the East China Sea. *Mar. Forecast. Serv.* **1985**, *2*, 67–72.
53. Liu, Z.L.; Yuan, X.W.; Yang, L.L.; Yan, L.P.; Tian, Y.J.; Chen, J.H. Effect of climate change on the fisheries community pattern in the overwintering ground of open waters of northern East China Sea. *Chin. J. Appl. Ecol.* **2015**, *26*, 901–911. [CrossRef]

54. Xu, Z.L.; Wang, R.; Chen, Y.Q. Study on ecology of meso-small pelagic copepods in the Southern Yellow Sea and the East Chi-na Sea I. Quantitative distribution. *J. Fish. China* **2003**, *27*, 1–8. [CrossRef]
55. Yatsu, A.; Watanabe, T.; Ishida, M.; Sugisaki, H.; Jacobson, L.D. Environmental effects on recruitment and productivity of Japanese sardine Sardinops melanostictus and chub mackerel Scomber japonicus with recommendations for management. *Fish. Oceanogr.* **2005**, *14*, 263–278. [CrossRef]
56. Yukami, R.; Ohshimo, S.; Yoda, M.; Hiyama, Y. Estimation of the spawning grounds of chub mackerel Scomber japonicus and spotted mackerel Scomber australasicus in the East China Sea based on catch statistics and biometric data. *Fish. Sci.* **2009**, *75*, 167–174. [CrossRef]
57. Guo, A.; Yu, W.; Chen, X.J.; Qian, W.G.; Li, R.C. Relationship between spatio-temporal distribution of chub mackerel Scomber japonicus and net primary production in the coastal waters of China. *Acta Oceanol. Sin.* **2018**, *40*, 42–52. [CrossRef]
58. Liu, F.; Chu, T.; Wang, M.; Zhan, W.; Xie, Q.; Lou, B. Transcriptome analyses provide the first insight into the molecular basis of cold tolerance in Larimichthys polyactis. *J. Comp. Physiol. B* **2019**, *190*, 27–34. [CrossRef]
59. O'Gorman, E.J.; Ólafsson, Ó.P.; Demars, B.O.L.; Friberg, N.; Guðbergsson, G.; Hannesdóttir, E.R.; Jackson, M.C.; Johansson, L.S.; McLaughlin, Ó.B.; Ólafsson, J.S.; et al. Temperature effects on fish production across a natural thermal gradient. *Glob. Chang. Biol.* **2016**, *22*, 3206–3220. [CrossRef]
60. Friedland, K.D.; Langan, J.A.; Large, S.I.; Selden, R.L.; Link, J.S.; Watson, R.A.; Collie, J.S. Changes in higher trophic level productivity, diversity and niche space in a rapidly warming continental shelf ecosystem. *Sci. Total. Environ.* **2020**, *704*, 135270. [CrossRef]
61. Moser, H.G.; Smith, P.E. Larval fish assemblages of the California Current region and their horizontal and vertical distribu-tions across a front. *Bull. Mar. Sci.* **1993**, *53*, 645–691.
62. Sournia, A. Pelagic biogeography and fronts. *Prog. Oceanogr.* **1994**, *34*, 109–120. [CrossRef]
63. Okazaki, Y.; Nakata, H. Effect of the mesoscale hydrographic features on larval fish distribution across the shelf break of East China Sea. *Cont. Shelf Res.* **2007**, *27*, 1616–1628. [CrossRef]
64. Bost, C.; Cotté, C.; Bailleul, F.; Cherel, Y.; Charrassin, J.; Guinet, C.; Ainley, D.; Weimerskirch, H. The importance of oceanographic fronts to marine birds and mammals of the southern oceans. *J. Mar. Syst.* **2009**, *78*, 363–376. [CrossRef]
65. Kindong, R.; Wu, J.; Gao, C.; Dai, L.; Tian, S.; Dai, X.; Chen, J. Seasonal changes in fish diversity, density, biomass, and assemblage alongside environmental variables in the Yangtze River Estuary. *Environ. Sci. Pollut. Res.* **2020**, *27*, 25461–25474. [CrossRef]
66. Yeh, S.W.; Kim, C.H. Recent warming in the yellow/East China Sea during winter and the associated atmospheric circulation. *Cont. Shelf Res.* **2010**, *30*, 1428–1434. [CrossRef]
67. Cai, R.; Tan, H.; Kontoyiannis, H. Robust surface warming in offshore China seas and its relationship to the east Asian mon-soon wind field and ocean forcing on interdecadal time scales. *J. Clim.* **2017**, *30*, 8987–9005. [CrossRef]
68. Cai, R.; Tan, H.; Qi, Q. Impacts of and adaptation to inter-decadal marine climate change in coastal China seas. *Int. J. Clim.* **2015**, *36*, 3770–3780. [CrossRef]
69. Qi, L.; Hu, C.; Wang, M.; Shang, S.; Wilson, C. Floating Algae Blooms in the East China Sea. *Geophys. Res. Lett.* **2017**, *44*, 11–501. [CrossRef]
70. Su, J.L. A review of circulation dynamics of the coastal oceans near China. *Acta Oceanol. Sin.* **2001**, *4*, 1–16.
71. Guan, B.; Fang, G. Winter counter-wind currents off the southeastern China coast: A review. *J. Oceanogr.* **2006**, *62*, 1–24. [CrossRef]
72. Weisberg, R.H.; Zheng, L.; Liu, Y. Basic tenets for coastal ocean ecosystems monitoring. In *Coastal Ocean Observing Systems*; Academic Press: Cambridge, MA, USA, 2015; pp. 40–57. [CrossRef]
73. IPCC. *Climate Change 2007: The Physical Science Basis, Contribution of Working Group I to the Fourth Assessment Report of the Intergovernmental Panel on Climate Change*; Solomon, S., Qin, D., Manning, M., Chen, Z., Marquis, M., Averyt, K.B., Tignor, M., Miller, H.L., Eds.; Cambridge University Press: New York, NY, USA, 2007.
74. Strüssmann, C.A.; Conover, D.O.; Somoza, G.M.; Miranda, L.A. Implications of climate change for the reproductive capacity and survival of New World silversides (family Atherinopsidae). *J. Fish Biol.* **2010**, *77*, 1818–1834. [CrossRef] [PubMed]
75. Brander, K.M. Cod Gadus morhua and climate change: Processes, productivity and prediction. *J. Fish Biol.* **2010**, *77*, 1899–1911. [CrossRef] [PubMed]
76. Belkin, I.M. Rapid warming of Large Marine Ecosystems. *Prog. Oceanogr.* **2009**, *81*, 207–213. [CrossRef]
77. Liu, Q.; Zhang, Q. Analysis on long-term change of sea surface temperature in the China Seas. *J. Ocean Univ. China* **2013**, *12*, 295–300. [CrossRef]
78. Bao, B.; Ren, G. Climatological characteristics and long-term change of SST over the marginal seas of China. *Cont. Shelf Res.* **2014**, *77*, 96–106. [CrossRef]
79. Nye, J.; Link, J.; Hare, J.; Overholtz, W. Changing spatial distribution of fish stocks in relation to climate and population size on the Northeast United States continental shelf. *Mar. Ecol. Prog. Ser.* **2009**, *393*, 111–129. [CrossRef]
80. Liu, Y.G.; Kerkering, H.; Weisberg, R. *Coastal Ocean Observing Systems*; Academic Press: London, UK, 2015.

remote sensing

MDPI

Article

On Characteristics and Mixing Effects of Internal Solitary Waves in the Northern Yellow Sea as Revealed by Satellite and In Situ Observations

Heping Liu [1], Wei Yang [1,*], Hao Wei [1], Chengfei Jiang [1], Changgen Liu [2] and Liang Zhao [3]

[1] School of Marine Science and Technology, Tianjin University, Tianjin 300072, China; peace_liu@tju.edu.cn (H.L.); hao.wei@tju.edu.cn (H.W.); cfjiang@tju.edu.cn (C.J.)
[2] School of Mechanical Engineering, Tianjin University, Tianjin 300350, China; lchg@tju.edu.cn
[3] College of Marine and Environmental Sciences, Tianjin University of Science and Technology, Tianjin 300457, China; zhaoliang@tust.edu.cn
* Correspondence: wei_yang@tju.edu.cn; Tel.: +86-17695508926

Abstract: This study examines the characteristics, statistics, and mixing effects of internal solitary waves (ISWs) observed in the northern Yellow Sea (YS) during the summers of 2018 and 2019. The mooring stations are located between offshore islands with rough topographic features. Throughout the observation period, the ISWs with vertical displacements of up to 10 m induced prevailing high-frequency (3–10 min period) temperature variations. Synthetic aperture radar (SAR) images showed that the observed ISWs propagate in zonal directions generated around the islands where internal-tide-generating body force is strong. The estimated ISW propagation speed ranges from 0.16 to 0.25 m s^{-1}, which agrees with the Korteweg-de Vries (KdV) model. The ISW intensity exhibits a clear spring-neap cycle corresponding to the local tidal forcing. The constant occurrence of ISWs at low tide suggests an important generation site where the ISWs are tidally generated. The ray-tracing result indicates that this generation site appears to be located at a strait between Dahao and Xiaohao islands. A generalized KdV model successively reproduces the propagation process from the generation site to the mooring station. Following the passage of ISWs, microstructure profiling observations reveal a high turbulent kinetic energy dissipation rate (10^{-6} W kg^{-1}). The prevalence of ISWs in the study area is believed to play a crucial role in regulating vertical heat and nutrient transport, thereby modulating the biogeochemical cycle.

Keywords: SAR; internal solitary waves; turbulence; Yellow Sea

Citation: Liu, H.; Yang, W.; Wei, H.; Jiang, C.; Liu, C.; Zhao, L. On Characteristics and Mixing Effects of Internal Solitary Waves in the Northern Yellow Sea as Revealed by Satellite and In Situ Observations. *Remote Sens.* **2022**, *14*, 3660. https://doi.org/10.3390/rs14153660

Academic Editors: Xiaofeng Li, Chung-Ru Ho and Antony K. Liu

Received: 28 June 2022
Accepted: 26 July 2022
Published: 30 July 2022

Publisher's Note: MDPI stays neutral with regard to jurisdictional claims in published maps and institutional affiliations.

1. Introduction

Internal solitary waves (ISWs) are ubiquitous in world coastal oceans; they are long-lived features and can travel tens/hundreds of kilometers before dissipating [1,2]. These waves can generate strong reverse flow across the pycnocline, which induces the intense turbulent vertical exchange of heat and nutrients, regulating the biogeochemical cycle [3–5]. Therefore, comprehending the characteristics, statistics, and mixing effects of ISWs is crucial to understanding how marine ecosystems behave.

ISWs can be produced by a variety of mechanisms, including the lee wave mechanism [6], a transformation of the internal tide [7,8], or resonance in the transcritical flow [9], depending on the local stratification, topographic features, and tidal amplitudes. The lee wave mechanism has traditionally been used to explain the generation of ISWs at an underwater sill or bank [6]. The isopycnal disturbance on the topography exhibits a large lee wave when the flow reaches a supercritical flow condition (the Froude number $F_r = U/c_0 > 1$, where U is the tidal current and c_0 is the first-mode linear internal wave phase speed). The lee wave moves upstream as the tidal flow slows, evolving into several ISWs. Therefore, locally generated ISWs are often associated with tides. Kozlov et al. found a similar distance between adjacent ISW packages in the White Sea that matched the M_2 wavelength value based on synthetic aperture radar (SAR) images [10]. According to statistical findings from years of observations in the Strait of Georgia, the ISW packets tended to appear at the specific tidal phase with similar time intervals [11], which emphasized the importance of tidal forcing in generating ISWs in the ocean.

ISWs can cause significant vertical displacements across the pycnocline and exacerbate turbulence [12,13]. In the investigation of Oregon's continental shelf, Moum et al. revealed the evolving nature of interfaces through microstructure observations. As ISWs propagate shoreward, the interfaces may become unstable and break, creating turbulent flow [14]. Lee et al. investigated the generation and longevity of nonlinear internal waves in the northern East China Sea [4], suggesting that shear instability was the generation mechanism for the observed turbulence. The turbulent mixing can modify the transport of particulates and nutrients, influencing the distribution and recruitment of various littoral larval species [15]. These dynamical and ecological consequences provide additional motivation to investigate the causes of internal turbulent mixing and their effects on ecology and aquaculture.

The Yellow Sea (YS) is a temperate shelf sea in the northwest Pacific Ocean with a mean depth of 44 m. It connects to the East China Sea in the south and the Bohai Sea in the north. Limited observations have revealed the existence of ISWs, wind–induced near–inertial waves, and semidiurnal internal tides there [12,16–19]. Hsu et al. examined the distribution of ISWs in the YS based on SAR images onboard the satellites ERS-1 and ERS-2 [16]. They found that the ISWs were widely distributed along the coast of the YS, with the islands west of the Korea Peninsula being the local hot spots for ISWs due to their topographic features. After their generation on the west coast of the Korean Peninsula, ISWs propagated into the YS. Their phase speed increased nonlinearly with the water depth, averaging a value of approximately 0.1–0.7 m/s [20]. A recent numerical study showed that the simulated internal tides in the YS have an excellent spatial consistency with the satellite-detected ISWs because both the internal tides and ISWs are of tidal origin [21]. Moreover, Liu et al. observed that ISWs with a vertical displacement of 4–5 m had induced intensified turbulence in the pycnocline by creating small–scale shear instabilities based on microstructure observations in the local shelf break of the southern YS [12].

Only limited studies have looked into the ISWs in the northern YS. Lin et al. used numerical simulations to explore that energetic internal tides were generated near the coast around the northern YS [22], with a maximum baroclinic energy flux of 45 W/m. Although the baroclinic energy flux is much lower in the northern YS due to the shallow bathymetry, there are still significant large values at the topographic changes, which indicate the local generation of ISWs and internal tides. Alpers et al. have confirmed the presence of ISW packets in the northern YS through satellite imagery [17]. These images show the characteristics of ISWs in the northern YS, which appear frequently in the summer and seldom have wavelengths longer than 1000 m due to the shallow water. However, it is still challenging to determine the variability, property, and mixing effects of the ISWs because satellite observations alone cannot shed light on the features of ISWs in the interior of oceans. In situ observations are important methods to reveal the characteristics and mixing effects of ISWs [23,24]. Although the combination of satellite observations and in situ observations has been widely used [25,26], the characteristics of ISWs and their effects on turbulent mixing are still unclear in the northern YS due to the lack of in situ observations.

In this study, we conducted long-term high-frequency temperature observation for two mooring stations in the northern YS. SAR images and mooring observations were combined to analyze the statistical characteristics of the ISWs. The effect of ISWs on turbulent mixing was also discussed. This paper is organized as follows. Section 2 describes the observation data and methods. In Section 3, the results of the in situ hydrographic observations and satellite SAR investigations are presented. The correlation between ISWs and tide, potential generation sites, and the influence of ISWs on turbulence are discussed in Section 4. Conclusions are summarized in Section 5.

2. Materials and Methods

2.1. Mooring Observation

The data for this study were collected at two mooring stations in the northern YS from 29 June to 18 August 2018 and 5 June to 18 August 2019 (Figure 1b). The mooring stations (Stns.) A4 (39.03°N 122.98°E) and B4 (39.04°N 122.96°E) are located between two offshore islands (Zhangzi and Haiyang islands) in water with a mean depth of about 40 m. The T-chain is used to measure temperature at the mooring stations, equipped with four temperature sensors (T, ONT18S) with a frequency of 1 Hz at Stn. A4. At Stn. B4, the T-chain is equipped with two Conductivity-Temperature-Depth instruments (CTD, RBR420), two Temperature-Depth sensors (TD, RBR duet), and two temperature sensors (T, ONT18S). Temperature measurements are valid from 1.5 m to 25 m above the bottom. Table 1 summarizes information on the corrected sensor depths and temporal resolution. Figure 1c depicts the instrument configuration diagram.

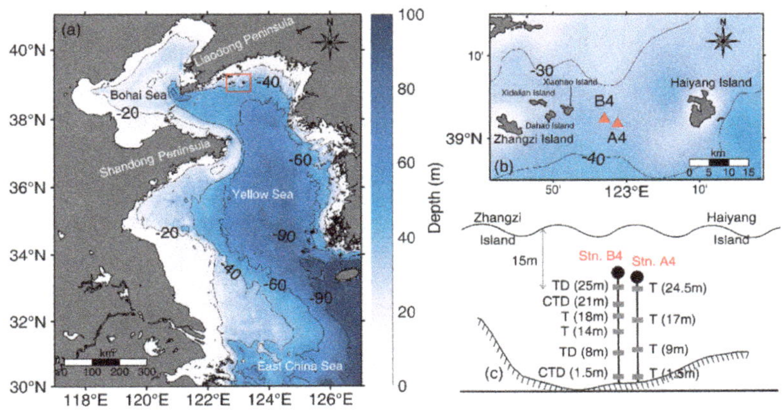

Figure 1. (**a**) Bathymetry of the Yellow Sea. (**b**) Inset of the outlined square in (**a**), indicating the location of the mooring station (red triangle). (**c**) Instrument configuration schematic of the T-chain at Stns. A4 and B4.

Table 1. Stations, sensors, frequency, and sensor height of the mooring observations.

Stations	Depth (m)	Sensors	Sampling Period (s)	Sensor Height (m)
A4	39	T	1	1.5
		T	1	9
		T	1	17
		T	1	24.5
B4	38.8	CTD	300	1.5
		TD	10	8
		T	10	14
		T	10	18
		CTD	10	21
		TD	10	25

We calculated the vertical velocity (w) from vertically densely spaced temperature sensor data after converting the series $T(z, t)$ to isotherm displacements $\eta(z, t)$ [27]. This study selects a set of isotherms with constant mean spacing (0.5 °C). For each isotherm, w is calculated by dividing the displacement of the isotherms in adjacent time by the time interval

$$w = \frac{\eta}{t} \tag{1}$$

where η is the isotherm displacements in adjacent time and t is the time interval. The velocity of each layer is acquired by linear interpolation.

2.2. VMP Observations

We deployed the Vertical Microstructure Profiler-(VMP-) at Stn. B4 on 14–15 July 2019, to investigate the role of ISWs in inducing turbulent mixing. The VMP was activated once every hour for 22 h. The VMP is outfitted with two high-resolution frequency shear probes, one high-frequency temperature probe (FP07), and one high-frequency conductivity probe (SBE7), all with sampling rates of 512 Hz. The turbulent kinetic energy dissipation rate was calculated using microstructure shear. Using the assumption of isotropic turbulence, the turbulent kinetic energy dissipation rate ε was calculated as follows:

$$\varepsilon = \frac{15}{2}v\overline{\left(\frac{\partial u}{\partial z}\right)^2} = \frac{15}{2}v\int_{k_1}^{k_2}\phi(k)dz \tag{2}$$

where v is the kinematic molecular viscosity; $\overline{(\partial u/\partial z)^2}$ is the variance of the vertical shear; $\varphi(k)$ is the power spectrum of velocity shear; and k_1 and k_2 are the lower and upper limits of the wavenumber for integration, respectively. The calculation of ε is based on the ODAS Matlab Library Manual program by Rockland Scientific International (RSI) Inc. Previous studies have extensively described the corresponding data processing theories and methods (Gregg, 1999). For details of the calculation methods, please refer to Yang et al. [28] and Xu et al. [18]. The shear spectrum is calculated over consecutive segments of 2 m with a 50% overlap. As a result, we could obtain vertical profiles of ε with a vertical resolution of 1 m.

2.3. SAR Images

The ISW—induced currents modify sea surface roughness, which can be imaged using satellite images via Bragg backscattering [29–31]. Synthetic aperture radar (SAR) has long been a principal sensor in the observation of ISWs, because of its all-day, all-weather imaging capability [32]. SAR maps the sea surface roughness through Bragg scattering from the capillary waves and short gravity waves. The relation is

$$\Lambda = \frac{\lambda}{2\sin\theta} \tag{3}$$

where Λ and λ are the wavelength of capillary waves and radar, respectively, and θ is the local incidence angle. In satellite SAR images, the ISWs typically appear as a pair of bright and dark stripes, corresponding to the rougher (convergence) and smoother (divergence) surface zones, respectively [30,32]. In this study, we used the C-band SAR of Sentinel-1 and GF-3 satellites launched by the European Space Agency on 3 April 2014 and China Aerospace Science and Technology Corporation on 10 August 2016, respectively. Both satellites can continuously image ISWs at 10 m resolution using HH-polarization. There are 59 images taken in the study area during the observation periods in 2018 and 2019, 11 of which show clear signatures of ISW packets.

2.4. The KdV Equation

The evolution of ISWs is generally influenced by topography and horizontally variable stratification [33,34]. For shallow water approximation, the KdV equation and its generalizations are widely used for the investigation of the propagation and transformation of ISWs in the shelf and coastal oceans [10,33]. To overcome the lack of direct observations of continuous stratification in the northern YS, we used the generalized Korteweg-de Vries (gKdV) equation to simulate the propagation of ISWs in the varied topography [10,35] as follows:

$$A_t + C_0(x)A_x + \alpha(x)AA_x + \alpha_1(x)A^2A_x + \beta(x)A_{xxx} + \left(\frac{C_0(x)}{2\gamma(x)}\right)\gamma_x A - \zeta A_{xx} = 0 \quad (4)$$

where $A(x, t)$ is the interface amplitude; $C_0(x)$ is the linear wave speed; and parameters $\alpha(x)$, $\alpha_1(x)$, $\beta(x)$, $\gamma(x)$, and ζ are the coefficients for the nonlinear, higher–order nonlinear (cubic), dispersion, transformation, and dissipation effects, respectively.

The numerical model based on the gKdV equation was discretized by the Predictor–Evaluation-Corrector-Evaluation (PECE) method and iterated by the implicit corrector scheme [36], which ensures convergence and computational efficiency of the iterative process. There is a steady summer pycnocline between the surface and intermediate water masses along the ISW propagation trajectory at depths of 5–10 m. Segur and Hammack [37] and Koop and Butler [38] found that the KdV model can be simplified to a two-layer model and predicted the solitary waves with remarkable accuracy. Thus, a two-layer system of upper quasi-homogeneous and lower layers with thicknesses h_1 and h_2, respectively, can be used [10,33]:

$$C_0 = \sqrt{\Delta\rho g h_1 h_2/\rho(h_1 + h_2)} \quad (5)$$

$$\alpha = [3(h_1 - h_2)/2(h_1 h_2)]C_0 \quad (6)$$

$$\alpha_1 = \left[-3\left(h_1^2 + h_2^2 + 6h_1h_2\right)/8(h_1h_2)^2\right]C_0 \quad (7)$$

$$\beta = (h_1 h_2/6)C_0 \quad (8)$$

$$\gamma = \left\{C_0(0)C_0^{-1}\left[h_1^{-1}(0) + h_2^{-1}(0)\right]/\left[h_1^{-1} + h_2^{-1}\right]\right\}^3 \quad (9)$$

where g is the acceleration of gravity; ρ_1, h_1 and ρ_2, h_2 are the density and thickness of upper and lower layers, respectively.

3. Results

This section is divided into subheadings. It should provide a concise and precise description of the experimental results, their interpretation, as well as the experimental conclusions that can be drawn.

3.1. Temperature Variations

Figures 2 and 3 show the temperature variation observed by the T-chain at Stns. A4 and B4, respectively. These two stations shared many characteristics due to their close distance and similar time of the year for observation. The observed temperature shows a warming trend at both stations. The temperature measured by the lowest sensors increased from 10 to 18 °C and from 8 to 20 °C at Stns. A4 and B4, respectively. The observed part of the water column is usually stratified, with the largest vertical temperature difference reaching 11.2 °C and 12.8 °C at Stns. A4 and B4, respectively. There was a significant increase in temperature throughout the water column on 13 August at Stn. B4 (Figure 3a,b). The bottom temperature changed from 15 to 19 °C, after which the temperature fluctuated and dropped gradually. The surface elevation demonstrates that semidiurnal tides dominate at both stations with a distinct spring–neap cycle (Figures 2a and 3a).

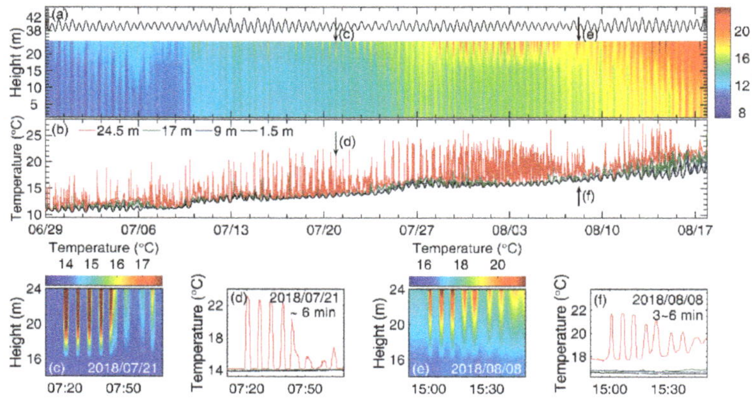

Figure 2. (**a**) Time–depth variations of observed temperature at Stn. A4. The black curve represents sea surface elevation. (**b**) Temporal variation of temperature as monitored by four temperature sensors. (**c**–**f**) Close views of the high-frequency variation of the temperature.

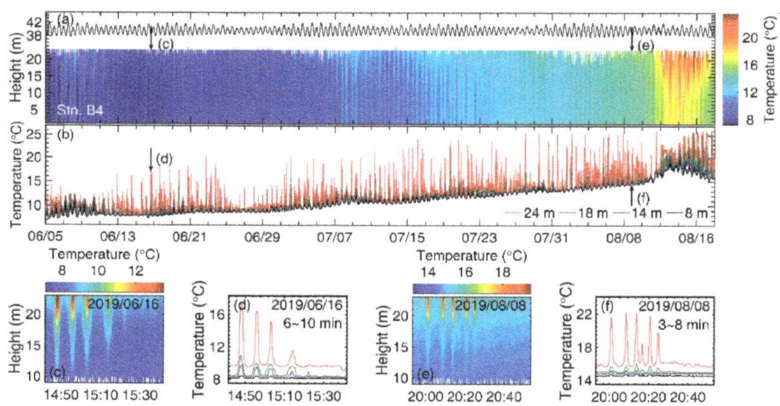

Figure 3. The same as in Figure 2 but for Stn. B4.

One notable phenomenon observed at both stations is prevailing temperature spikes that last throughout the observation period. Close-view examples of the temperature spikes show that these spikes have frequencies ranging from 3–10 min (Figures 2c,e and 3c,e). When the bottom layer is well-mixed, temperature spikes may appear only in the upper sensors (Figure 2d,f). The uppermost temperature sensor increased from 14 to 23 °C in 100 s with a period of 5 min (Figure 2c). Figures 2c,e and 3c,e show that these temperature spikes are caused by the prevailing high-frequency ISWs of depression in this area. The vertical displacement of the ISWs is about 8–10 m (Figure 2c). Additionally, the leading waves have the largest vertical displacement of 13 m, corresponding to the highest temperature variation of 7 °C, as observed in Figure 3d,f. The observed prevailing ISW packets in the study area are surprising. This is because, although the previous SAR imagery studies have revealed the presence of ISW packets around offshore islands in the northern YS, this region has never been considered a hot spot for ISWs [17]. Given the prevalence of the observed ISWs, they may have significant biogeochemical implications. Next, the properties of the ISWs are examined.

3.2. Spectral Estimates

The frequency spectrum of temperature is calculated to examine the energy content as a function of frequency. The temperature in the uppermost layer is selected to calculate the frequency spectrum. Figure 4c,d depicts the frequency spectrum calculated from mooring temperature observations at Stns. A4 and B4, respectively. The frequency spectra are dominated by near–inertial/diurnal (~1 cycle per day (cpd)), semidiurnal (~2 cpd), and higher tidal harmonics at both stations. The strongest peak corresponds to the frequency of the M_2 tide. The spectra generally decrease with increasing frequency, except where there seems to exist a spectral bump at ~144 to 480 cpd (a period of ~3 to 10 min). Figures 2 and 3 depict that the spectral bump is associated with the high–frequency temperature variations.

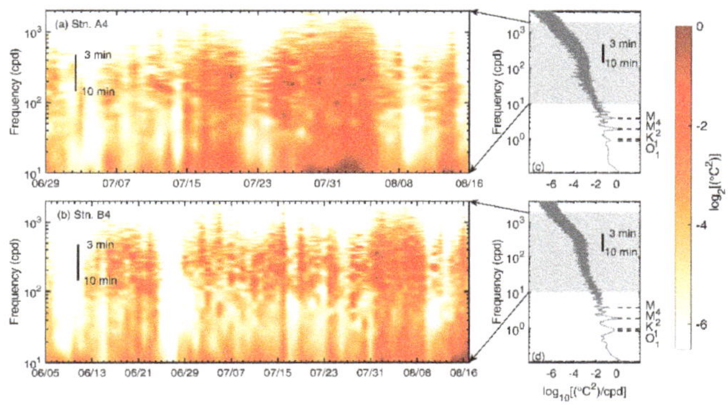

Figure 4. Temperature–frequency spectra of daily temperature variations at (**a**) Stn. A4 and (**b**) Stn. B4. The frequency spectra during the two–month–long observations at (**c**) Stn. A4 and (**d**) Stn. B4. For reference, the frequencies of the essential components are indicated by horizontal dashed lines.

The high-frequency region of the spectra was then examined to obtain a better look at the spectral bump. In order to examine the temporal variation of the high-frequency motions, we calculated the frequency spectrum for each day by dividing the temperature variation into daily segments. Figure 4a,b shows that high spectral energy appears between 3 and 10 min periods, which corresponds to the spectral bump. The spectral bump almost lasts throughout the observation period. This is consistent with the observed frequent temperature spikes, indicating that ISWs are common in the study area.

3.3. Close View of the ISWs

The SAR images and mooring observations are combined to examine the properties of ISWs. For example, we selected two typical SAR images (10 m resolution, HH-polarization) with clear ISWs acquired on 11 July and 8 August 2019. The in situ mooring observations are analyzed alongside the SAR images to provide additional information about propagation details. Figure 5d depicts the capture of two ISW packets (ISW A and ISW B) at 3.4 and 7.9 km from Stn. B4. ISW A arrives at Stn. B4 approximately 4 h after the imaging time, causing significant temperature variations. Within 3 min, the temperature measured by the uppermost sensor rises from 13 to 22 °C. Close inspection reveals that ISW A has a vertical displacement of 10 m and a period of about 10 min (Figure 5c). The ISW propagation speed is estimated to be 0.25 m s^{-1} based on relative packet positions and arrival times. ISW B originates on Haiyang Island and has a propagation speed of 0.16 m s^{-1}.

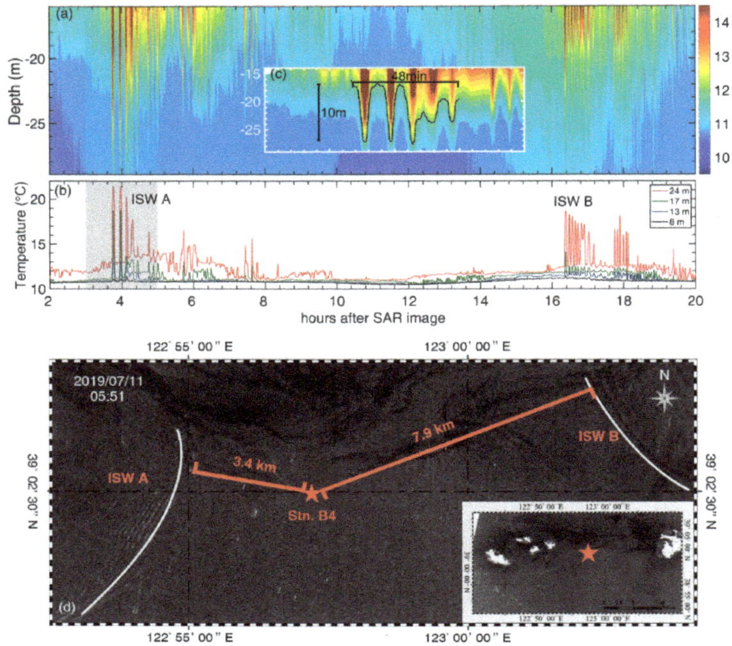

Figure 5. (**a**) Time–depth variations of temperature from 11 July 2019 at 7:51 to 12 July 2019 at 3:51 (UTC + 8). (**b**) The corresponding temporal variations of temperature. (**c**) Inset of the shaded area in (**a**) showing the typical ISWs marked with a black line. (**d**) SAR image was acquired over the northern YS on 11 July 2019 (05:51 UTC + 8). Red stars denote mooring stations. Notably, the x-axis has been changed to hours after this SAR image time.

The other SAR image was acquired on 8 August 2019, at 17:46, capturing at least four ISW packets (Figure 6d). Three of the four packets propagate from west to east in a similar direction. ISW C first passes through the mooring station, causing high–frequency temperature variations for 28 min. This ISW has an amplitude of approximately 9 m and a period of 7 min (Figure 6c). The propagation speed of ISW C is estimated to be 0.21 m s^{-1}. The propagation speeds of ISWs A and B are estimated to be 0.21 and 0.17 m s^{-1}, respectively, based on the distance revealed by SAR images and arrival time. Wavelengths appear to be variable during the propagation of the ISW. ISW A has the largest wavelength at about 500 m, and ISWs B and C have relatively small wavelengths of less than 200 m. In the same ISW packet, the wavelengths of ISWs are also different. As shown in ISW B and ISW C in Figure 6, the wavelengths are significantly larger in the southern part of the waves.

The linear phase speed at any frequency depends on the vertical density profile of the water column, based on the KdV equation for internal waves [39]. The two-layer model, consisting of the upper layer with thickness h_1 and density ρ_1, and the lower layer with thickness $h_2 = H - h_1$ and density $\rho_2 > \rho_1$, is widely considered for the simulation of internal waves [40,41]. In this case, assuming the ISWs are of the KdV type, the speeds of propagation of ISWs are generally calculated from Equation (10):

$$c = \sqrt{\frac{g(\rho_2 - \rho_1)h_1 h_2}{\rho_2 h_1 + \rho_1 h_2}} + \frac{\alpha \eta_0}{3} \tag{10}$$

where g is the acceleration due to gravity; η_0 is the vertical displacement of ISWs; ρ_1 and ρ_2 are the upper layer and lower layer density; h_1 and h_2 are the upper layer and lower layer thickness, respectively. The parameter α is calculated using Equation (5). At the mooring station, the phase speed calculated according to Equation (10) is about 0.25 m s^{-1}, consistent with estimates from satellite observations.

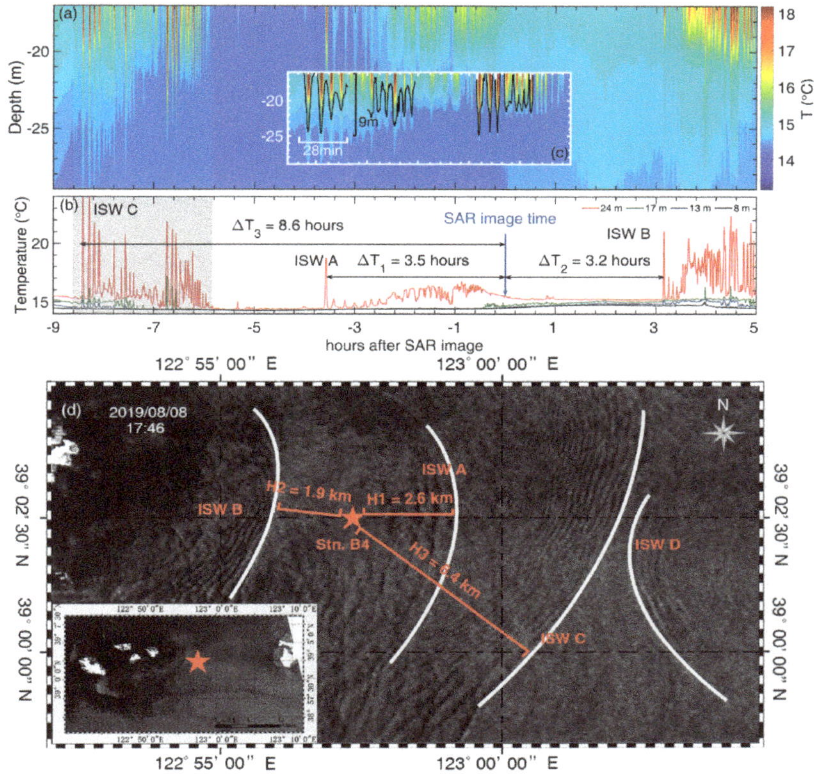

Figure 6. Same as in Figure 5 but for another typical SAR image. The SAR image (**a–d**) was acquired on 8 August 2019 (17:46 UTC + 8).

4. Discussion

4.1. Statistics of ISWs

Mooring observations indicate that ISWs are prevailing phenomena in our study area that exist throughout the observation period in summer. We performed a statistical analysis on the time of appearance of ISWs to the tidal phases to investigate their characteristics and generation mechanisms. Wave packets were defined as ISWs with vertical displacements greater than 5 m. At Stns. A4 and B4, a total of 125 and 277 ISW packets were observed, respectively. The surface elevation shows a clear spring–neap cycle at both stations (Figures 2a and 3a). The occurrence of ISWs is binned by their time relative to the nearest maximum of the fortnightly cycle of tidal magnitudes and by their time relative to the low tide. At Stn. A4, more ISWs seem to be observed during spring tide than during neap tide (Figure 7c). However, ISWs do not appear more frequently at Stn. B4 during spring tide (Figure 7f), showing no obvious correlation with tidal amplitude. The one-hour and one-day root-mean-square (RMS) vertical velocity (σ_w) are then calculated (Figure 7b,e). The (σ_w) shows a clear fortnightly cycle which agrees well with the variation of RMS tidal amplitude (σ_H).

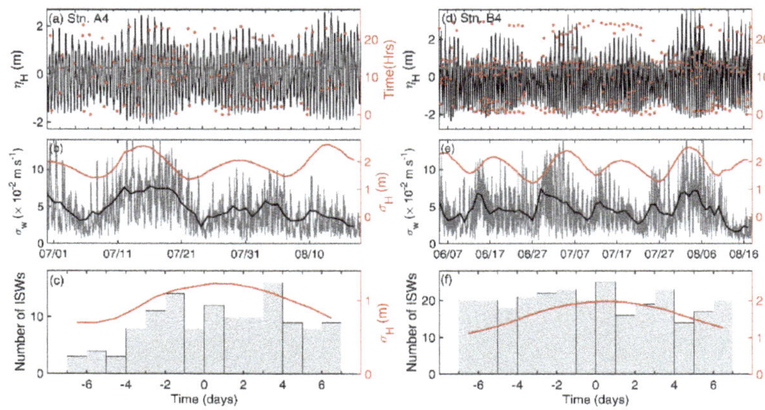

Figure 7. Statistical characteristics of ISWs. (**a**,**d**) The temporal variation of surface elevation (black line) and occurrence of ISWs (red points) relative to the time of low tide. (**b**,**e**) RMS vertical velocity in one-hour (grey curves) and one–day intervals (black curves). The red curve represents the 1.8 × RMS variations of the sea surface. (**c**,**f**) The number of observed ISWs during a fortnightly cycle, binned by days before/after the time of maximum tidal range.

Variations in both the occurrence frequency and intensity of the ISWs can affect the (σ_w). Previous research found that the likelihood of ISW packets peaked around the strongest tides in the spring–neap tide [11,17,42]. However, the likelihood of ISWs shows no apparent correlation with the spring–neap tide in the present study (Figure 7c,f). Thus, we suggest that the fortnightly variations of (σ_w) are responsible for modulating the ISW intensity by the spring–neap tide.

The occurrence of the ISW packet was examined by comparing it with the flood–ebb cycle of the local tides. At both stations, we first calculated the average value of the surface elevation for 25 h, showing a clear dominant semidiurnal cycle (Figure 8). The occurrence of ISWs was classified according to the local tidal phases. Figure 8 shows that most ISWs are observed at both stations during low tides. This relationship appears to be stronger at Stn. B4. In other regions, the occurrence of ISWs at a particular tidal phase has been interpreted as evidence that their generation mechanisms are connected to tidal phases [11]. This strengthens the case that tides cause the observed ISWs.

Figure 9 depicts a typical example of the phase–locked ISW packets at Stn. B4. ISW packets always occurred approximately one hour later than the previous day (marked by red arrows). Although the ISW packets appear at irregular times of the day, they occur at every low–tide phase. Furthermore, the intensity of ISWs occurring during spring tides (19–22 July) tends to be greater than during neap tides (12–18 July), consistent with the statistical results. The occurrence of ISWs at the fixed low–tide phase suggests that tide–topography interactions generate ISWs.

4.2. Generation Sites of ISWs

Next, we investigated the generation and propagation of ISWs. Figure 10a depicts an ISW distribution map of the northern YS based on the SAR images collected during the observation period. ISWs propagate in various directions, implying multiple generation sites. Specifically, the observed ISWs at the mooring station are generated at the rough topography around the island rather than being locally generated. Many ISWs propagating eastwards seem to originate from Zhangzi Island. A small number of westward-propagating ISWs also appear, which seem to have been generated around Haiyang Island.

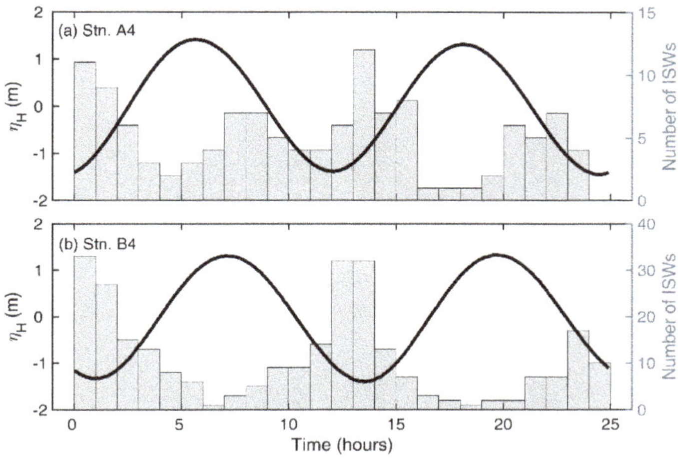

Figure 8. The occurrence of ISWs relative to the tidal phase at (**a**) Stn. A4 and (**b**) Stn. B4. Black curves represent the time series of the tidal elevation simulated using the Ohio State University Tidal Prediction Software (OTPS). The number of ISWs is indicated by vertical bars.

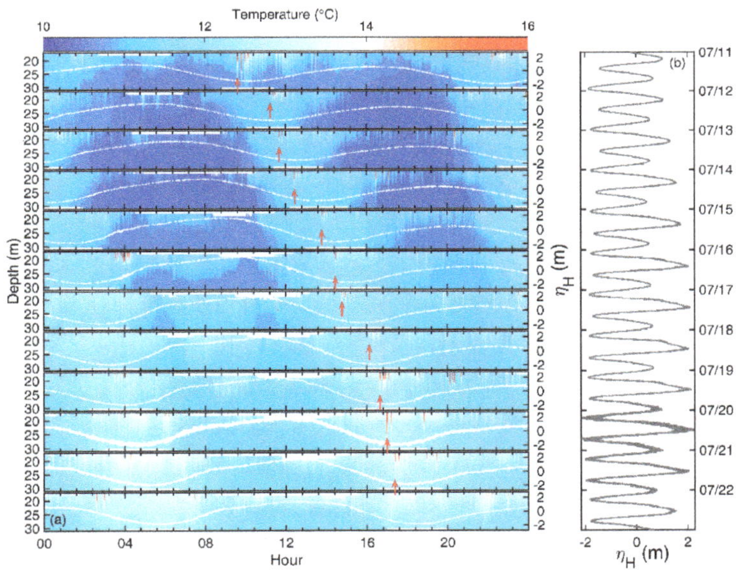

Figure 9. (**a**) Stack plots of the temperature at Stn. B4 from 11–22 July 2019 (UTC + 8). Red arrows indicate the arrival times of the ISW packets, and the white line indicates the surface elevation. (**b**) The corresponding time series of the barotropic tide at Stn. B4.

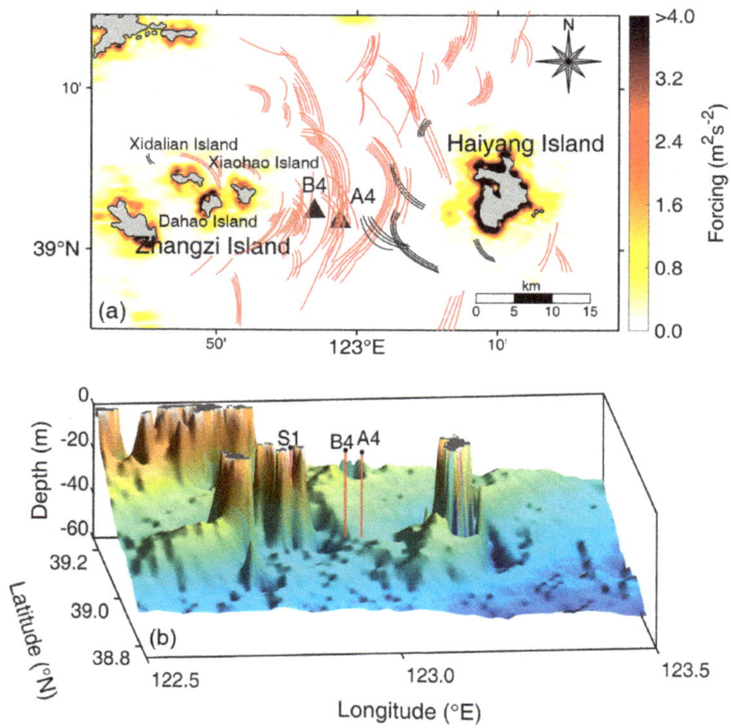

Figure 10. (**a**) The spatial distribution of the depth-integrated internal-tide-generating body force. The superposed curves represent the ISWs detected by SAR images during the observation periods of 2018 and 2019. ISWs propagating eastward and westward are shown in red and black, respectively. (**b**) The three-dimensional topography shows the surrounding study area. The ISW generation site, as traced by the ray-tracing method, is represented by S1. The locations of mooring stations are also indicated.

In order to investigate the generation sites of the ISWs, we examined the distribution of internal-tide-generating body force in the northern YS (Figure 10). The depth-integrated internal–tide–generating body force is calculated as follows (Baines, 1982):

$$F = \frac{1}{\omega}\frac{1}{H^2}\int_H^0 N^2(z)z\mathrm{d}z \cdot \sqrt{(Q_x\frac{\mathrm{d}H}{\mathrm{d}x})^2 + (Q_y\frac{\mathrm{d}H}{\mathrm{d}y})^2} \tag{11}$$

where ω is the M_2 tidal frequency; H is the water depth; $N^2(z)$ is the squared buoyancy frequency; Q_x and Q_y are the zonal and meridional components of the tidal transport calculated by $(Q_x, Q_y) = (uH, vH)$; u and v are the zonal and meridional components of the barotropic tidal current; and $\frac{\mathrm{d}H}{\mathrm{d}x}$ and $\frac{\mathrm{d}H}{\mathrm{d}y}$ are the zonal and meridional bottom slopes. The bottom topography is based on the 0.25′ grid general bathymetric chart of the oceans (GEBCO_2019) provided by the Intergovernmental Oceanographic Commission (IOC) and the International Hydrographic Organization (IHO). The OSU Tidal Prediction Software (OTPS) forecasts the zonal and meridional components of tidal velocity. Potential hot spots of internal wave generation sites around the world ocean have been examined using the spatial distribution of the F [43–45]. $F > 0.25$ m^2 s^{-2} is usually regarded as the critical value for the generation of ISWs [44,46]. Figure 10a depicts that large F tends to be located around islands with striking topographic features.

That energetic internal tides were generated close to the offshore islands around the northern YS is consistent with recent numerical simulations [22]. Both SAR images and calculated F indicate that the ISWs are generated at these offshore islands and then propagate to the mooring stations.

Statistical analysis showed that the ISWs are related to tidal phases, with most ISWs occurring during low tide (Figures 8 and 9). As a result, even though the SAR images show multiple ISW generation sites around the islands, there should be a major ISW generation site with tidally-generated ISWs reaching the mooring station periodically. Figure 9 depicts an example of ISWs fixed to local low tides, implying that they are generated tidally at a fixed location. The data from SAR images are then combined with in situ observations and the KdV model to provide additional insights into the generation and propagation details.

The ray-tracing method was used to find the major generation sites of ISWs in the northern YS [47–49]. Assuming that the ISWs propagate with a circular wave crest, the generation site of ISWs is defined as the intersection of two perpendicular lines across the wave crest. Figure 11a depicts an ISW packet captured on 11 July 2019, corresponding to the first ISW packet that appeared at low tide in Figure 9. The observed ISW packet from satellite images traces the generation site S1 to the straits between Dahao and Xiaohao islands, approximately 10.1 km from Stn. B4 (Figures 10b and 11a). The generalized KdV model was then used to simulate the propagation of ISWs. The density is referred from the WOA18 database, showing that the upper and lower layer densities in the study area are approximately 1023 and 1024 kg m^{-3}, respectively. We then simulated the propagation of ISWs with the generalized KdV model. The real topography from GEBCO was used for simulation. The density was referred to in the WOA18 database, showing that the upper and lower layer densities are about 1023 and 1024 kg m^{-3} in the study area. The depth of pycnocline is 5 m. After that, Equation (4) is numerically solved on a regular grid with Δx = 10 m, Δt = 10 s, the viscosity coefficient ζ = 0.0025 m^2 s^{-2} [10], and a total integration time of 20 h. At x = 0 km, a train of two consecutive waves with vertical displacements of −5 and −4 m appears.

Figure 11b shows that the ISW packet undergoes several changes as they propagate seaward from S1. Their vertical displacements are significantly reduced, and they become broader and flatter. The ISW packet reaches the position corresponding to the SAR image after ~8.2 h of propagation from S1 (Figure 11a). The propagation from S1 to B4 takes about 12 h in total, with a mean propagating speed of ~0.25 m s^{-1}, consistent with the velocity estimated from the SAR image and mooring observations (Figure 5d). Figure 11c shows the barotropic tide at S1 when ISWs are generated. At low tide, when the direction of tidal velocities is toward the northeast, ISWs are generated in the straits between the islands. Although the site between the islands was estimated to be a critical ISW generation site, ISWs coming from different directions and at different times indicate the existence of other source regions. Furthermore, although we have shown that the ISWs are tidally generated at rough topography, the exact generation mechanism of the ISWs also awaits further investigation in the future based on numerical simulations. Nevertheless, the prevalence and properties of ISWs originating from offshore islands in the northern YS are documented for the first time based on comprehensive in situ observations and SAR images.

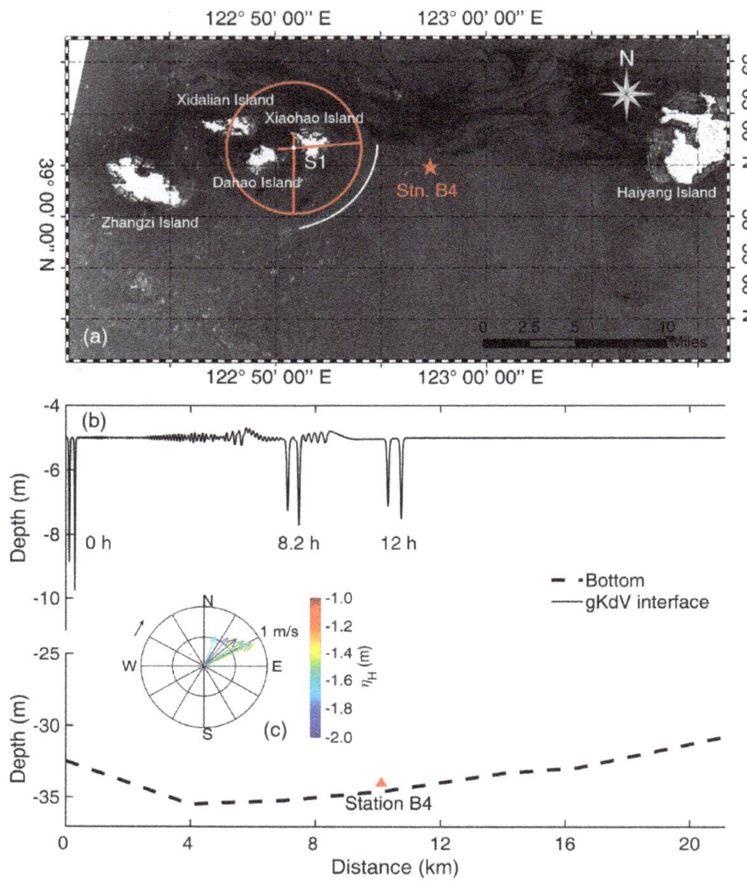

Figure 11. (**a**) SAR image was taken on 11 July 2019 (05:51 UTC + 8). The red circle indicates the estimated generation site (S1) traced by the ray-tracing method. The red star denotes location of Stn. B4. (**b**) The propagation process of ISWs simulated by the gKdV model. The dashed line indicates the topography. (**c**) The corresponding tidal current at S1 at the moment of ISW generation. Colors represent the surface elevation, and the arrow indicates the direction of tidal variations.

4.3. Influences of ISWs on Internal Turbulence

Turbulent mixing in shelf seas is a critical process that controls diapycnal nutrient transport, and it plays a vital role in biological production and carbon cycles [50,51]. The one-day VMP observations enable us to investigate the impact of ISWs on internal turbulence. Figure 12b shows the time–depth evolution of the observed ε and temperature. Previous studies showed that the energy input, conversion, and radiation rates increase monotonically with ε [52]. The largest ε (~10^{-6} W kg^{-1}) occurred in the well–mixed bottom boundary layer (BBL). One of the most noticeable features here is the presence of two peaks near the pycnocline. At the 4th hour after 13:00 on 14 July 2019, the ε reached the first peak with a maximum value of 10^{-7} W kg^{-1}. About eleven hours later, the ε reached the other peak with a larger value of 10^{-6} W kg^{-1}.

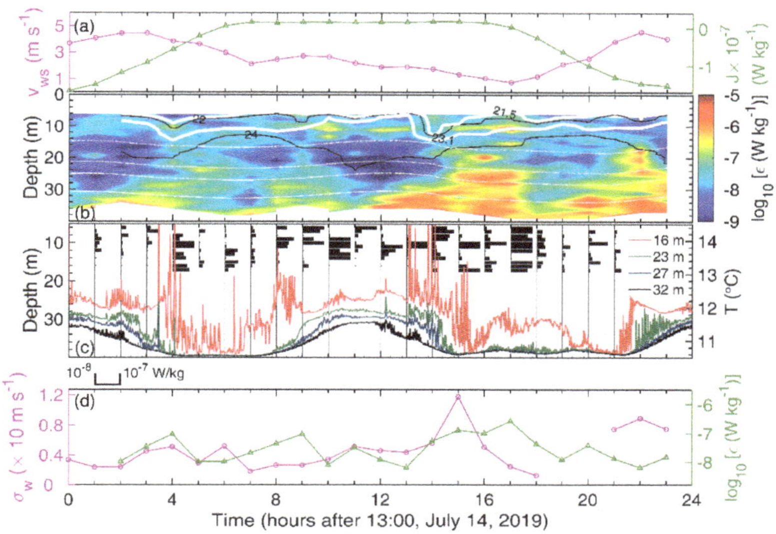

Figure 12. (**a**) Temporal variation of the wind speed (purple line) and buoyancy flux (green line). (**b**) Time-depth variation of ε with the isotherms (solid lines) superposed. White lines represent pycnocline boundaries. (**c**) The replotted ε around the pycnocline (black bars) and temperature. (**d**) Temporal variation of the averaged ε within the pycnocline (green line) and 1 h RMS vertical velocity (purple line).

It is widely believed that turbulence in the surface boundary layer (SBL) in shelf seas can be strongly influenced by wind forcing. In order to investigate the impact of wind on the two ε peaks, we investigated the variation of wind speed at 10 m above the sea surface from the fifth generation European Centre for Medium-Range Weather Forecasts (ECMWF)'s atmospheric reanalysis of the global climate (ERA5). Figure 12a presents that the wind speed is less than 5 m s^{-1} during the observation period, suggesting that the wind may not be the cause of turbulence. Surface buoyancy fluxes do not dramatically change when ε is intensified, which shows that changes in surface buoyancy fluxes do not cause internal turbulence. Kay and Jay found that substantial internal turbulence occurred even during periods of weak near–bed shear [53], suggesting that internal processes must contribute significantly to the internal turbulence. Figure 12c depicts the variations in internal temperature and turbulence. The temperature variations show a semidiurnal period, consistent with the M_2 tide. Around four hours (17:00 14 July 2019) and thirteen hours (02:00 15 July 2019) after the beginning of the observations, the temperature shows continuous high-frequency variations. Furthermore, we selected the depths of the σ_θ = 22.5 and 20.9 kg m^{-3} isopycnals to represent the lower and upper boundaries of the pycnocline, respectively (as shown by the white contours in Figure 12b). The two ε peaks in the pycnocline correspond to the high-frequency temperature variations (Figure 12c). The average ε in the pycnocline also shows a similar trend as the one-day RMS vertical velocity (Figure 12d). The largest root-mean-square vertical velocity appears at 15 h, corresponding to the long-lasting strong turbulence from 15 to 17 h. Although their largest peaks do not exactly coincide with each other, we suggest that the sustaining strong turbulence is related to the high-frequency temperature variation event which has induced a long-lasting influence. We therefore suggest that ISWs cause the observed internal turbulence.

The present observations demonstrate that ISWs are quite prevalent in the study area, originating at rough topographic features near the offshore island. The one-day VMP observation shows how passing ISWs can significantly alter internal turbulence in the northern YS. Moreover, the frequent turbulent mixing induced by ISWs can increase vertical nutrient supply and influence the local biogeochemical cycle [54–56]. Furthermore, the areas around the Changshan Islands are crucial aquaculture regions in China [57,58]. The presence of ISWs, which cause increased vertical flux and high–frequency temperature variations during warm seasons, may have significant ecological consequences that must be investigated.

5. Conclusions

The properties, generation mechanisms, statistics, and mixing effects of ISWs are investigated using mooring observations and SAR images from the northern YS. According to the spectral analysis, previous ISWs induced numerous high-frequency temperature spikes with periods ranging from 3 to 10 min. These ISWs are mainly depression type, with vertical displacements of up to 10 m. Dramatic vertical displacements can lead to nearly 9 °C temperature variations within the pycnocline. We estimated the propagation velocity of ISWs to be 0.16 to 0.25 m s^{-1} by combining mooring observations and SAR images, which is consistent with the theoretical value. According to the statistical analysis, ISWs at the mooring stations do not indicate a clear preference for spring tides, but the RMS vertical velocity of ISWs (ISW intensity) exhibits a spring-neap tidal cycle and is linearly proportional to the barotropic tidal height. Furthermore, while ISWs can be found at the mooring stations at any tidal phase, they are far more prevalent during low tides.

These findings indicate that the ISWs are tidal in origin, with a major generation site. SAR images obtained during the observation period show that most ISWs in this region propagate in the zonal direction, which seems to originate from the Zhangzi and Haiyang islands. This is supported by the horizontal distribution of the depth-integrated internal-tide-generating body force, which shows consistent features of high values around the offshore islands. The ray–tracing method was used to locate the major generation site, which appeared to be located in the straits surrounding Zhangzi Island. The KdV model was used to simulate the propagation of the ISWs from the generation site to the mooring station. The ISWs take ~12 h to propagate from the generation site to the mooring station, which coincides with the observational estimates.

The ε showed two large values (~10^{-6} W kg^{-1}) when there were temperatures with high–frequency variations caused by ISW packets. The similar variation trends between ε and vertical velocity confirmed that ISWs cause internal turbulence. Turbulent mixing induced by frequent ISWs can increase the vertical nutrient supply and influence the biogeochemical cycle. Considering that the regions around these offshore islands are now important bottom–aquaculture farms in China, the influence of the prevailing ISWs on aquaculture remains to be examined.

Author Contributions: Conceptualization, H.L. and W.Y.; methodology, H.L.; software, C.J.; validation, H.L., W.Y., H.W., C.L., L.Z. and C.J.; formal analysis, H.L.; investigation, H.L.; resources, W.Y.; data curation, H.L. and W.Y.; writing—original draft preparation, H.L.; writing—review and editing, W.Y., H.W., L.Z. and C.L.; visualization, H.L., W.Y., H.W., L.Z. and C.J.; supervision, W.Y.; project administration, W.Y.; funding acquisition, W.Y. All authors have read and agreed to the published version of the manuscript.

Funding: This study was supported by the National Natural Science Foundation of China (41906006), the Tianjin Natural Science Foundation (21JCYBJC00500), and the National Key Research and Development Program of China (2018YFD0900704).

Data Availability Statement: The SAR images from Sentinel–1 and GF–3 satellites were obtained online (https://scihub.copernicus.eu/dhus/#/home (accessed on 1 December 2020) and https://osdds.nsoas.org.cn/#/ (accessed on 1 December 2020)). The ERA5 data were obtained online (https://cds.climate.copernicus.eu/cdsapp#!/home, accessed on 1 May 2021). Mooring and VMP data analyzed in this study are available at https://zenodo.org/record/6609695 (accessed on 30 May 2022).

Acknowledgments: The acquisition of mooring data was supported by Laboratory of Ocean Dynamics and Ecology, LODE. We are grateful to the entire crew for their expertise and hard work. The authors thank the editor and three anonymous reviewers for their constructive suggestions to improve the quality of the manuscript.

Conflicts of Interest: The authors declare no conflict of interest.

References

1. Helfrich, K.R.; Melville, W.K. Long nonlinear internal waves. *Annu. Rev. Fluid Mech.* **2006**, *38*, 395–425. [CrossRef]
2. Jackson, C. Internal wave detection using the Moderate Resolution Imaging Spectroradiometer (MODIS). *J. Geophys. Res. Atmos.* **2007**, *112*, C11012. [CrossRef]
3. Duda, T.F.; Lynch, J.F.; Irish, J.D.; Beardsley, R.C.; Ramp, S.R.; Chiu, C.S.; Tang, T.Y.; Yang, Y.J. Internal tide and nonlinear internal wave behavior at the continental slope in the northern south China Sea. *IEEE J. Ocean. Eng.* **2005**, *29*, 1105–1130. [CrossRef]
4. Lee, J.H.; Lozovatsky, I.; Jang, S.T.; Chan, J.J.; Fernando, H. Episodes of nonlinear internal wave in the northern East China Sea. *Geophys. Res. Lett.* **2006**, *33*. [CrossRef]
5. Dong, J.; Zhao, W.; Chen, H.; Meng, Z.; Shi, X.; Tian, J. Asymmetry of internal waves and its effects on the ecological environment observed in the northern South China Sea. *Deep-Sea Res. Part I* **2015**, *98*, 94–101. [CrossRef]
6. Maxworthy, T. A note on the internal solitary waves produced by tidal flow over a three-dimensional ridge. *J. Geophys. Res. Oceans* **1979**, *84*, 338–346. [CrossRef]
7. Holloway, P.E. Internal hydraulic jumps and solitons at a shelf break region on the Australian North West Shelf. *J. Geophys. Res. Atmos.* **1987**, *92*, 5405–5416. [CrossRef]
8. Sandstrom, H.; Elliot, J.A.; Cchrane, N.A. Observing groups of solitary internal waves and turbulence with BATFISH and echo-sounder. *J. Phys. Oceanogr.* **1989**, *19*, 987–997. [CrossRef]
9. Melville, W.K.; Helfrich, K.R. Transcritical two-layer flow over topography. *J. Fluid Mech.* **1987**, *178*, 31–52. [CrossRef]
10. Kozlov, I.; Romanenkov, D.; Zimin, A.; Chapron, B. SAR observing large-scale nonlinear internal waves in the White Sea. *Remote Sens. Environ.* **2014**, *147*, 99–107. [CrossRef]
11. Li, L.; Pawlowicz, R.; Wang, C. Seasonal variability and generation mechanisms of nonlinear internal waves in the strait of Georgia. *J. Geophys. Res. Oceans* **2018**, *123*, 5706–5726. [CrossRef]
12. Liu, Z.; Wei, H.; Lozovatsky, I.D.; Fernando, H.J.S. Late summer stratification, internal waves, and turbulence in the Yellow Sea. *J. Mar. Syst.* **2009**, *77*, 459–472. [CrossRef]
13. Bourgault, D.; Janes, D.C.; Galbraith, P.S. Observations of a large-amplitude internal wave train and its reflection off a steep slope. *J. Phys. Oceanogr.* **2011**, *41*, 586–600. [CrossRef]
14. Moum, J.N.; Farmer, D.M.; Smyth, W.D.; Armi, L.; Vagle, S. Structure and generation of turbulence at interfaces strained by internal solitary waves propagating shoreward over the continental shelf. *J. Phys. Oceanogr.* **2003**, *33*, 2093–2112. [CrossRef]
15. Scotti, A.; Pineda, J. Obervation of very large and steep waves of elevation near the Massachusetts coast. *Geophys. Res. Lett.* **2004**, *31*. [CrossRef]
16. Hsu, M.; Liu, A.K.; Liu, C. A study of internal waves in the China Seas and Yellow Sea using SAR. *Cont. Shelf Res.* **2000**, *20*, 389–410. [CrossRef]
17. Alpers, W.; He, M.X.; Zeng, K.; Guo, L.F.; Li, X.M. The distribution of internal waves in the East China Sea and the Yellow Sea studied by multi-sensor satellite images. In Proceedings of the IEEE International Geoscience & Remote Sensing Symposium, Seoul, Korea, 29 July 2005. [CrossRef]
18. Xu, P.; Yang, W.; Zhu, B.; Wei, H.; Zhao, L.; Nie, H. Turbulent mixing and vertical nitrate flux induced by the semidiurnal internal tides in the southern Yellow Sea. *Cont. Shelf Res.* **2020**, *208*, 104240. [CrossRef]
19. Yang, W.; Wei, H.; Liu, Z.; Li, G. Intermittent intense thermocline shear associated with wind-forced near-inertial internal waves in a summer stratified temperate shelf sea. *J. Geophys. Res. Oceans* **2021**, *126*, e2021JC017576. [CrossRef]
20. Kim, H.; Son, Y.B.; Jeong, J.; Jo, Y. Comparison of internal waves in various ocean fields around the Korean Peninsula. *J. Coast. Res.* **2018**, *85*, 466–470. [CrossRef]
21. Liu, K.; Sun, J.; Guo, C.; Yang, Y.; Yu, W.; Wei, Z. Seasonal and spatial variations of the M2 internal tide in the Yellow Sea. *J. Geophys. Res. Oceans* **2019**, *124*, 1115–1138. [CrossRef]
22. Lin, F.; Asplin, L.; Wei, H. Summertime M2 internal tides in the Northern Yellow Sea. *Front. Mar. Sci.* **2021**, *8*, 798504. [CrossRef]
23. Lien, C.R. Energy of nonlinear internal waves in the South China Sea. *Geophys. Res. Lett.* **2005**, *32*, 215–236. [CrossRef]
24. McSweeney, J.M.; Lerczak, J.A.; Barth, J.A.; Becherer, J.; Colosi, J.A.; MacKinnon, J.A.; MacMahan, J.H.; Moum, J.N.; Pierce, S.D.; Waterhouse, A.F. Observations of shoaling nonlinear internal bores across the Central California inner shelf. *J. Phys. Oceanogr.* **2020**, *50*, 111–132. [CrossRef]
25. Lai, Z.; Jin, G.; Huang, Y.; Chen, H.; Shang, X.; Xiong, X. The generation of nonlinear internal waves in the South China Sea: A three-dimensional, nonhydrostatic numerical study. *J. Geophys. Res. Oceans* **2019**, *124*, 8949–8968. [CrossRef]
26. Bai, X.; Liu, Z.; Zheng, Q.; Hu, J.; Lamb, K.G.; Cai, S. Fission of shoaling internal waves on the Northeastern Shelf of the South China Sea. *J. Geophys. Res. Oceans* **2019**, *124*, 4529–4545. [CrossRef]
27. Pinkel, R. Observations of the near-surface internal wavefield. *J. Phys. Oceanogr.* **1981**, *11*, 1248–1257. [CrossRef]

28. Yang, W.; Wei, H.; Zhao, L.; Zhang, J. Turbulence and vertical nitrate flux adjacent to the Changjiang Estuary during fall. *J. Mar. Syst.* **2020**, *212*, 103427. [CrossRef]
29. Valenzuela, G.R. Theories for the interaction of electromagnetic and oceanic waves—A review. *Bound.-Lay. Meteorol.* **1978**, *13*, 61–85. [CrossRef]
30. Alpers, W. Theory of radar imaging of internal waves. *Nature* **1985**, *314*, 245–247. [CrossRef]
31. Chang, M.H.; Lien, R.C.; Yang, Y.J.; Tang, T.Y.; Wang, J. A composite view of surface signatures and interior properties of nonlinear internal waves: Observations and applications. *J. Atmos. Ocean. Technol.* **2008**, *25*, 523–531. [CrossRef]
32. Fu, L.L.; Holt, B. Internal waves in the Gulf of California—Observations from a spaceborne radar. *J. Geophys. Res. Oceans* **1984**, *89*, 2053. [CrossRef]
33. Liu, A.K.; Chang, Y.S.; Hsu, M.K.; Liang, N.K. Evolution of nonlinear internal waves in the East and South China Seas. *J. Geophys. Res. Oceans* **1998**, *103*, 7995–8008. [CrossRef]
34. Zhao, Z. Satellite observation of internal solitary waves converting polarity. *Geophys. Res. Lett.* **2003**, *30*. [CrossRef]
35. Grimshaw, R.; Pelinovsky, E.; Talipova, T. Modelling internal solitary waves in the Coastal Ocean. *Surv. Geophys.* **2007**, *28*, 273–298. [CrossRef]
36. Djidjeli, K.; Price, W.G.; Twizell, E.H.; Wang, Y. Numerical methods for the solution of the third- and fifth-order dispersive Korteweg-de Vries equations. *J. Comput. Appl. Math.* **1995**, *58*, 307–336. [CrossRef]
37. Segur, H.; Hammack, J.L. Soliton models of long internal waves. *J. Fluid Mech.* **1982**, *118*, 285–304. [CrossRef]
38. Koop, C.G.; Butler, G. An investigation of internal solitary waves in a two-fluid system. *J. Fluid Mech.* **1981**, *112*, 225–251. [CrossRef]
39. Apel, J.R.; Ostrovsky, L.A.; Stepanyants, Y.A.; Lynch, J.F. Internal solitons in the ocean and their effect on underwater sound. *J. Acoust. Soc. Am.* **2007**, *121*, 695–722. [CrossRef]
40. Liu, B.; Yang, H.; Zhao, Z.; Li, X. Internal solitary wave propagation observed by tandem satellites. *Geophys. Res. Lett.* **2014**, *41*, 2077–2085. [CrossRef]
41. Shen, H.; Perrie, W.; Johnson, C.L. Predicting internal solitary waves in the gulf of maine. *J. Geophys. Res. Oceans* **2020**, *125*, e2019JC015941. [CrossRef]
42. Colosi, J.A.; Kumar, N.; Suanda, S.H.; Freismuth, T.M.; MacMahan, J.H. Statistics of internal tide bores and internal solitary waves observed on the inner continental shelf off Point Sal, California. *J. Phys. Oceanogr.* **2018**, *48*, 123–143. [CrossRef]
43. Sherwin, T.J.; Vlasenko, V.I.; Stashchuk, N.; Jeans, D.R.G.; Jones, B. Along-slope generation as an explanation for some unusually large internal tides. *Deep-Sea Res. Part I: Oceanogr. Res. Pap.* **2002**, *49*, 1787–1799. [CrossRef]
44. Li, X.; Zhao, Z.; Han, Z.; Xu, L. Internal solitary waves in the East China Sea. *Acta Oceanol. Sin.* **2008**, *27*, 51–59. [CrossRef]
45. Bai, X.; Liu, Z.; Li, X.; Hu, J. Generation sites of internal solitary waves in the southern Taiwan Strait revealed by MODIS true-colour image observations. *Int. J. Remote Sens.* **2014**, *35*, 4086–4098. [CrossRef]
46. Lozovatsky, I.; Liu, Z.; Fernando, H.; Armengol, J.; Roget, E. Shallow water tidal currents in close proximity to the seafloor and boundary-induced turbulence. *Ocean Dyn.* **2012**, *62*, 177–191. [CrossRef]
47. Da Silva, J.C.B.; Buijsman, M.C.; Magalhaes, J.M. Internal waves on the upstream side of a large sill of the Mascarene Ridge: A comprehensive view of their generation mechanisms and evolution. *Deep Sea Res. Part I Oceanogr. Res. Pap.* **2015**, *99*, 87–104. [CrossRef]
48. Sun, L.; Zhang, J.; Meng, J. A study of the spatial-temporal distribution and propagation characteristics of internal waves in the Andaman Sea using MODIS. *Acta Oceanol. Sin.* **2019**, *38*, 121–128. [CrossRef]
49. Zhang, X.; Li, X.; Zhang, T. Characteristics and generations of internal wave in the Sulu Sea inferred from optical satellite images. *J. Oceanol. Limnol.* **2020**, *38*, 1435–1444. [CrossRef]
50. Rippeth, T.P.; Palmer, M.R.; Simpson, J.H.; Fisher, N.R.; Sharples, J. Thermocline mixing in summer stratified continental shelf sea. *Geophys. Res. Lett.* **2005**, *32*, 159–171. [CrossRef]
51. Sharples, J.; Moore, M.C.; Hickman, A.E.; Holligan, P.M.; Simpson, J.H. Internal tidal mixing as a control on continental margin ecosystems. *Geophys. Res. Lett.* **2009**, *36*, L23603. [CrossRef]
52. Chen, Z.; Xie, J.; Xu, J.; Zhan, J.; Cai, S. Energetics of nonlinear internal waves generated by tidal flow over topography. *Ocean Model.* **2013**, *68*, 1–8. [CrossRef]
53. Kay, D.J. Interfacial mixing in a highly stratified estuary 1. Characteristics of mixing. *J. Geophys. Res. Oceans* **2003**, *108*. [CrossRef]
54. Williams, C.; Sharples, J.; Mahaffey, C.; Rippeth, T. Wind-driven nutrient pulses to the subsurface chlorophyll maximum in seasonally stratified shelf seas. *Geophys. Res. Lett.* **2013**, *40*, 5467–5472. [CrossRef]
55. Shroyer, E.L.; Benoit-Bird, K.J.; Nash, J.D.; Moum, J.N. Stratification and mixing regimes in biological thin layers over the Mid-Atlantic Bight. *Limnol. Oceanogr.* **2014**, *59*, 1349–1363. [CrossRef]
56. Sharples, J.; Zeldis, J.R. Variability of internal tide energy, mixing and nitrate fluxes in response to changes in stratification on the northeast New Zealand continental shelf. *N. Z. J. Mar. Freshw.* **2019**, *55*, 51–58. [CrossRef]
57. Zhao, Y.; Zhang, J.; Lin, F.; Ren, J.S.; Sun, K.; Liu, Y.; Wu, W.; Wang, W. An ecosystem model for estimating shellfish production carrying capacity in bottom culture systems. *Ecol. Model.* **2019**, *393*, 1–11. [CrossRef]
58. Nan, X.; Wei, H.; Zhang, H.; Nie, H. Factors influencing the interannual variation in biomass of bottom-cultured yesso scallop (*Patinopecten yessoensis*) in the Changhai Sea Area, China. *Front. Mar. Sci.* **2022**, *8*, 798359. [CrossRef]

remote sensing

MDPI

Article

Satellite Investigation of Semidiurnal Internal Tides in the Sulu-Sulawesi Seas

Xiaoyu Zhao [1,2], Zhenhua Xu [1,3,4,5,*], Ming Feng [6], Qun Li [7], Peiwen Zhang [1,2,4], Jia You [1,2,4], Song Gao [8] and Baoshu Yin [1,2,3,4,5]

1 CAS Key Laboratory of Ocean Circulation and Waves, Institute of Oceanology, Chinese Academy of Sciences, Qingdao 266071, China; zhaoxiaoyu@qdio.ac.cn (X.Z.); zpw@qdio.ac.cn (P.Z.); youjia@qdio.ac.cn (J.Y.); bsyin@qdio.ac.cn (B.Y.)
2 University of Chinese Academy of Sciences, Beijing 100049, China
3 Pilot National Laboratory for Marine Science and Technology, Qingdao 266237, China
4 Center for Ocean Mega-Science, Chinese Academy of Sciences, Qingdao 266071, China
5 CAS Engineering Laboratory for Marine Ranching, Institute of Oceanology, Chinese Academy of Sciences, Qingdao 266071, China
6 CSIRO Oceans and Atmosphere, Indian Ocean Marine Research Centre, Crawley, WA 6009, Australia; ming.feng@csiro.au
7 Polar Research Institute of China, Shanghai 200136, China; liqun@pric.org.cn
8 North China Sea Marine Forecasting Center of State Oceanic Administration, Qingdao 266061, China; gaosong@ncs.mnr.gov.cn
* Correspondence: xuzhenhua@qdio.ac.cn; Tel.: +86-532-8289-8991

Abstract: The mode-1 semidiurnal internal tides that emanate from multiple sources in the Sulu-Sulawesi Seas are investigated using multi-satellite altimeter data from 1993–2020. A practical plane-wave analysis method is used to separately extract multiple coherent internal tides, with the nontidal noise in the internal tide field further removed by a two-dimensional (2-D) spatial band-pass filter. The complex radiation pathways and interference patterns of the internal tides are revealed, showing a spatial contrast between the Sulu Sea and the Sulawesi Sea. The mode-1 semidiurnal internal tides in the Sulawesi Sea are effectively generated from both the Sulu and Sangihe Island chains, forming a spatially inhomogeneous interference pattern in the deep basin. A cylindrical internal tidal wave pattern from the Sibutu passage is confirmed for the first time, which modulates the interference pattern. The interference field can be reproduced by a line source model. A weak reflected internal tidal beam off the Sulawesi slope is revealed. In contrast, the Sulu Island chain is the sole energetic internal tide source in the Sulu Sea, thus featuring a relatively consistent wave and energy flux field in the basin. These energetic semidiurnal internal tidal beams contribute to the frequent occurrence of internal solitary waves (ISWs) in the study area. On the basis of the 28-year consistent satellite measurements, the northward semidiurnal tidal energy flux from the Sulu Island chain is 0.46 GW, about 25% of the southward energy flux. For M_2, the altimetric estimated energy fluxes from the Sulu Island chain are about 80% of those from numerical simulations. The total semidiurnal tidal energy flux from the Sulu and Sangihe Island chains into the Sulawesi Sea is about 2.7 GW.

Keywords: semidiurnal internal tides; the Sulu-Sulawesi Seas; sea surface height; plane wave fit method; energy flux

Citation: Zhao, X.; Xu, Z.; Feng, M.; Li, Q.; Zhang, P.; You, J.; Gao, S.; Yin, B. Satellite Investigation of Semidiurnal Internal Tides in the Sulu-Sulawesi Seas. *Remote Sens.* **2021**, *13*, 2530. https://doi.org/10.3390/rs13132530

Academic Editors: Xiao-Feng Li, Chung-Ru Ho and Antony K. Liu

Received: 15 May 2021
Accepted: 23 June 2021
Published: 28 June 2021

1. Introduction

Internal tides are widespread in the stratified ocean and act as a significant part in the energy cascade of multiscale oceanic processes. They are generated by the flux of barotropic tidal currents over complex bathymetries, such as the seamounts, trenches, ridges, and continental slopes. The globally integrated barotropic into baroclinic conversion rate in the deep ocean is estimated at ~1 TW [1]. The long-range radiation of internal tides in the open ocean is relevant to low modes that emanate from the generation sites [2]. The breaking of internal tides can cause energetic tidal mixing, which is essential to drive the large-scale meridional overturning circulation (MOC) and, subsequently, to affect the climate variability [3].

The propagation of internal tides redistributes the baroclinic tidal energy across the ocean basin, which is closely related to the inhomogeneous distribution of deep-sea mixing [4]. In particular, the semidiurnal internal tides dominate the tidal energy transfer process in the deep ocean and balance the global internal tide energy budget [5]. A correct characterization of the propagation paths and directions of the internal tidal energy flux will significantly contribute to parameterizing realistic diapycnal mixing in the ocean models [6]. However, the horizontal inhomogeneity in the energy flux density largely modulated by the interference of internal tides remains unknown [7]. Both the numerical models and satellite altimetry have revealed the widespread presence of multiwave interferences in the open ocean [8]. Therefore, it is crucial to regionally characterize multisource internal tides among the world ocean's hotspots [9–12], based on a combination of simulations and multi-platform observations.

The Indonesian Archipelago acts as the only tropical connection of the Pacific and Indian Oceans and features the most complicated topography among the world's regional seas. It is identified as an energetic internal tide generation field in the global ocean based on in situ measurements, numerical model simulations, and satellite altimeters [4,8,13]. The internal tide source regions include the Manipa, Lifamatola, Lombok, and Ombai Straits, and the Sulu and Sangihe Island chains et al. Previous studies showed that internal tides in different sea areas interact with each other, and internal tide fractions of different cycles interact nonlinearly, forming extremely complex and variable internal tide fields [14,15]. Koch-Larrouy et al. [16] showed numerically that tidal mixing has a considerable effect on the tropical climate system. The tidal mixing influences the SST patterns in the Indonesian Seas and changes the character of the Indonesian Throughflow (ITF) [17,18]. To date, much work on internal tides in the Indonesian Archipelago has been made based on numerical simulations. However, the verification of models remains limited due to insufficient field observations.

In the last two decades, satellite altimetry has been widely used as an important method to observe internal tides [19–21]. A recent M_2 internal tide prediction model on a global scale without blind directions was constructed using multiple satellite altimeter measurements [22]. In addition, Zaron [21] filtered the main mesoscale noise and proposed global internal tide fields for the diurnal constituent (O_1 and K_1) and the semidiurnal constituent (M_2 and S_2). The radiation and propagation of internal tides have been investigated in numerous marginal oceans by satellite observations [23,24]. In the regions where in situ measurements are lacking, satellite altimeter measurements can be utilized to reveal a broad distribution of internal tide signals. Note that most existing satellite altimeter measurements focused on mid-latitude open seas where the M_2 internal tides can usually travel thousands of kilometers across ocean basins.

The Sulu-Sulawesi Seas are characterized as energetic generation sources of the semidiurnal internal tides in the semi-closed sea basin near the equator [25]. As Figure 1a shows, the two basins are deeper than 4000 m and exhibit steep topography at the boundary, including the Sulu and Sangihe Island chains. The topographic conditions are critical for the M_2 internal tides (Figure 1b), as are the S_2 internal tides. According to the TOPEX/POSEIDON global tidal model (TPXO) [26], the semidiurnal M_2 and S_2 barotropic tidal currents are energetic in the two island chains. Thus, the Sulu Island chain radiates intense internal tides northward into the Sulu Sea and southward into the Sulawesi Sea, enabling a north-south asymmetry pattern between the two basins. Nagai and Hibiya [27] launched numerical experiments to estimate that the M_2 barotropic energy to baroclinic energy conversion rate in two island chains is 18.4 GW, and 50–70% of their energy is dissipated near the generation sites. The work suggests that ocean models need to consider the energy dissipation far from the generation sources caused by internal tide propagation [27]. A coherent portion of the semidiurnal internal tides near two island chains is dominant, making the altimetry measurements more reliable and suitable [28]. The semidiurnal internal tides lose coherence in propagation due to the background flow.

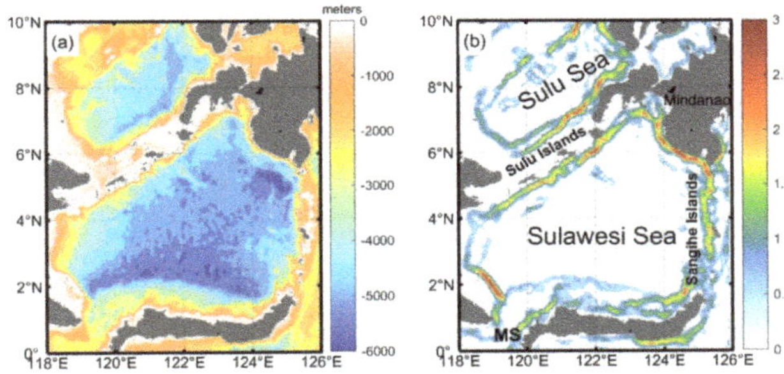

Figure 1. (**a**) Bathymetry of the Sulu-Sulawesi Seas. (**b**) Slope criticality γ/c. ($c = [(\omega^2 - f^2)/(N^2 - \omega^2)]^{1/2}$), which indicates the internal tide propagation angle to the horizontal, and γ is the bottom gradient.) The study region area is 0° to 10°N, 118° to 126°E near the equator. MS: Makassar Strait. The Sulawesi Sea is semi-closed, connected with the Makassar Strait to the south and the Western Pacific to the east. The Sulu Island chain forms the southern edge of the Sulu Sea and the northern edge of the Sulawesi Sea.

In this paper, to clarify semidiurnal internal tides in the Sulu-Sulawesi Seas, we are motivated to extract the internal tides using a practical plane-wave analysis method [22], which resolves multiwave interference well. From a global view, satellite altimetry shows that the Sulu Island chain radiates the transbasin coherent M_2 internal tidal beams [8,29]. However, the multidirectional internal tidal beams are usually represented by the strongest beam in the small-scale basin and become undetected. The decomposition of the multi-directional wave fields allows the covered internal tides to be exposed. Accumulating multiyear satellite altimeter data, multi-satellite altimetry with denser ground tracks can resolve the internal tide with a short wavelength. In Section 2, the satellite altimeter data and plane-wave analysis method will be described. We will propose the semidiurnal internal tide field of the Sulu-Sulawesi Seas in Section 3, with a focus on the interference of multisource internal tides. The interference patterns are explained by a line source model. Section 4 discusses the reflection and energy of internal tides. Section 5 summarizes the results and conclusions.

2. Data and Methods

2.1. Satellite Altimeter Data

In this paper, we used the sea surface height (SSH) observations combined by multiple altimeter satellites, including TOPEX/Poseidon (TP), Jason 1, 2, and 3 (J1, J2, and J3), Geosat Follow-On (GFO), Envisat (EN), and European Remote Sensing Satellite 1 and 2 (ERS-1 and ERS-2). The satellites usually take over and continue the previous satellite missions. Thus, a series of satellites have the same parameters, including the ground track and the repeated cycle (Figure 2a). According to satellite ground tracks, the SSH data are divided into four sets, namely TPJ (TP, J1, J2, and J3), TPT (TP tandem mission), ERS (ERS-1 and ERS-2), and GFO (Figure 2c). Standard corrections for geophysical effects, surface wave bias, and atmospheric effects are applied to process the SSH measurements. Global Ocean Tide 4.7 (GOT4.7) was used to correct the barotropic tide and loading tide. The noise measurement error on the same scale in different satellites is less than the amplitude of internal tides in our survey region. Our study area extends from 118°E, 0° to 126°E, 10°N to contain the Sulu-Sulawesi Seas. To avoid the issues of tide aliasing in marginal seas caused by satellite altimetry measurements, measurements of water depths less than 400 m were eliminated. The reference depth is determined based on the sensitivity analysis of different depths. The setting of a threshold is necessary because the quality of satellite altimeter data in coastal areas is not good enough. The Copernicus Marine and Environment Monitoring Service (CMEMS) is responsible for processing and distributing SSH measurements. The satellite SSH products were downloaded on 16 July 2020. A global mode-1 M_2 internal tide field was constructed using a similar dataset [8].

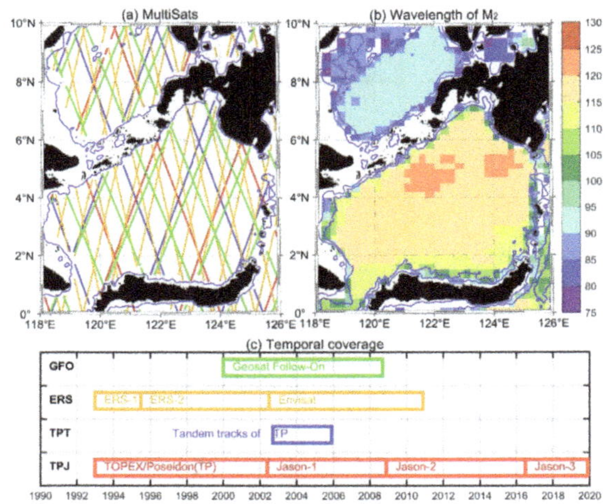

Figure 2. Satellite altimeter data and wavelength of M_2 internal tide. (**a**) Tracks of multiple satellites (MultiSats). The colors correspond to the boxes in (**c**). (**b**) The wavelength of the mode-1 M_2 internal tide is calculated from annual-mean ocean stratification in the World Ocean Atlas 2018 (WOA2018). (**c**) The duration of satellite altimeter observation is from 1993 to 2020.

These datasets have different temporal lengths and spatial coverage. The repeat periods of ERS, GFO, and TPJ are 35, 17, and 10 days. Based on the Rayleigh criterion, the M_2 and S_2 internal tides can be reliably separated by the long enough datasets. The ground tracks of TPJ (TPT), GFO, and ERS are 254, 488, and 1002. The datasets' along-track resolution is about 6–7 km, sufficient to resolve internal tides. Noise signals and noise-affected short wavelengths are removed from measurements in the along-track noise filtering with a cutoff wavelength of 65 km. Figure 2b shows the theoretical wavelengths of mode-1 M_2 internal tides in the Sulu-Sulawesi Sea, which range from 75 km to 130 km. Because of the distinct stratification caused by the background flow, the wavelengths between the two basins are different. This cutoff wavelength is less than the theoretical minimum wavelength of mode-1 M_2 internal tides and denotes the minimum wavelength associated with the dynamical scales to be statistically resolved by the altimetry.

2.2. *Two-Dimensional Plane-Wave Fit*

Internal tides were extracted by a 2-D plane-wave fit method, using satellite altimeter data with a large spatial coverage. Plane-wave fitting is an extension of traditional point-wise harmonic analysis. This method was first proposed by Ray and subsequently promoted by Zhao [30]. Through plane-wave fitting, the propagation direction and amplitude of the internal tide can be determined by solving multiple stable internal tidal waves in any direction on the plane.

$$\eta(a, \phi, \theta) = a\cos(k_0 x \cos\theta + k_0 y sin\theta - \omega_0 t - \phi), \quad (1)$$

where k_0 and ω_0 are the tidal frequency and wavenumber, respectively, and x and y are the east and north in the Cartesian coordinates. The expected output, including tidal wave amplitude a, phase ϕ, and propagation direction θ, is obtained by fitting plane waves to satellite SSH data. The buoyancy frequency N, wavenumber k_0, and phase velocity C_n are calculated based on WOA2018 [31]. The WOA2018 is a collection of objectively analyzed ocean water quality parameters which has been widely used in ocean models and the corroboration of satellite data.

An example is shown in Figure 3 to illustrate the process of extracting mode-1 M_2 internal tides. Internal tides are extracted using over 1×10^5 SSH data in a selected window centered at 3.6°N, 121.8°E (Figure 3a). First, using the least-square fitting, the amplitude and phase of a plane-wave are determined in each compass direction with a 1° angular resolution. When we plotted the amplitudes in polar coordinates as a function of direction, the lobes appeared and indicated the internal tidal beams. The amplitude and direction of the first M_2 internal tide are determined by the largest lobe (Figure 3d). Then, the phase and direction corresponding to the maximum amplitude is determined. The amplitude maximum corresponds to the residual minimum in the same direction (Figure 3g). When the first internal tide wave is identified, its signal can be reproduced and then removed from the original measurements. The extraction process above is repeated (Figure 3e,f). The residual variance also varies with the direction (Figure 3g–i). The new maximum amplitude tidal wave can then be considered as another tidal wave direction. In this study, five internal tidal waves are extracted in a fitting window of 180 km by 180 km at each grid point, based on a sensitivity analysis of different sizes (Figure 3b,c). The size of the window is about one and a half wavelength of mode-1 M_2 internal tides in the Sulu-Sulawesi Seas. The last two internal tide waves are too small, so only the first three waves are presented in Figure 3.

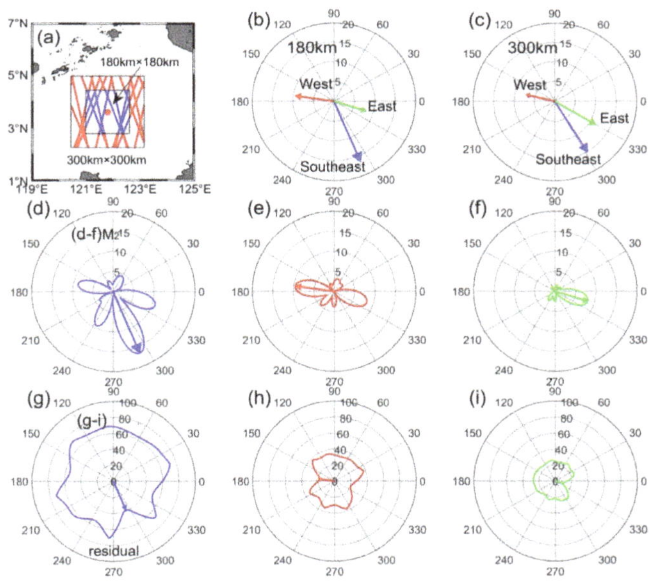

Figure 3. An example illustrates the plane-wave fit method. (**a**) A 180-km wide fitting window and a 300-km wide fitting window are centered at 3.6°N, 121.8°E, displaying multi-satellite SSH measurements in this region. (**b**) Three determined internal tidal waves in a fitting window of 180 km by 180 km. (**c**) As in (**b**), but for a fitting window of 300 km by 300 km. (**d**) Amplitudes (mm) show as a function of direction obtained in the plane-wave fit. The amplitude and direction of the first M_2 internal tide are determined by the largest lobe. (**e**), (**f**) As in (**d**), but for westward and eastward waves. After removing the first internal tidal wave from the original measurements, the second M_2 internal tidal wave is determined by repeating the procedure. (**g**) Residual variance (mm^2) vs. direction in the plane-wave fit. (**h**), (**i**) As in (**g**), but for westward and eastward waves.

The plane-wave fit method is applied at a regular grid resolution of one-tenth degree (0.1° longitude by 0.1° latitude) in the Sulu-Sulawesi Seas. The five extracted waves get summed in each grid point. Besides, the internal tide solution exists in water deeper than 400 m. Figure 4a presents the map of the mode-1 M_2 internal tide field. From the SSH pattern seen in satellite altimetry, the generation sites for the internal tides appear to be the Sulu and Sangihe Island chains, favoring a spatially inhomogeneous SSH field in the Sulu-Sulawesi Seas. An interference pattern means that mooring measurements at different locations vary greatly. Neighboring fitting windows are mostly overlapped, due to the large window and small grid, producing a smooth amplitude, phase, and direction. However, because mesoscale eddies contaminate the internal tide solution in the Sulu-Sulawesi Seas [32,33], some SSH perturbations exist in the internal tide field. A 2-D band-pass filter is required to clean the internal tide field.

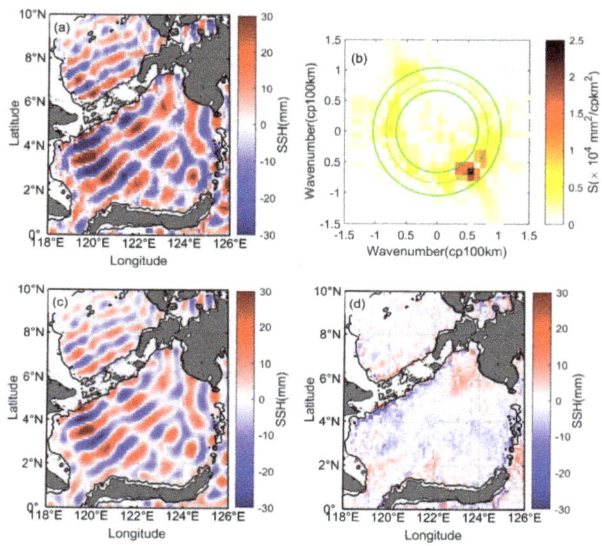

Figure 4. Mapping the M_2 internal tide from satellite altimetry. (**a**) The five-wave superimposed M_2 internal tide SSH field was constructed by plane-wave analysis from along-track altimeter measurements. (**b**) 2-D wavenumber spectrum of (**a**). The theoretical wavenumber is indicated by the dotted-line circle, and the filter's cutoff wavenumbers are indicated by the solid-line circles. They are [4/5 8/5] times the theoretical wavenumber. (**c**) The M_2 internal tide SSH field, obtained from the 2-D wavenumber band-pass filtering of (**a**). (**d**) Nontidal noise, obtained from the difference between (**a**,**c**).

2.3. Two-Dimensional Bandpass Filter

The nontidal noise in the M_2 internal tide field is removed by a 2-D band-pass filter (Figure 4). It processes the internal tide field by preserving the internal tides that meet the theoretical wavelengths. Figure 4b shows the two-dimensional wavenumber spectrum conducted by processing the noisy internal tide field on a regular spatial grid. It shows a nearly circular energy pattern indicated by a dotted circle corresponding to a wavenumber of 8×10^{-3} cpkm for the mode-1 M_2 internal tide. Its reciprocal corresponds to the theoretical wavelength (Figure 2b). Ray and Zaron [29] constructed a global map of modal wavelengths by applying a two-dimensional wavenumber spectrum. The cutoff wavenumbers are indicated by solid circles, which are chosen as [4/5 8/5] times the theoretical wavenumber. Figure 4c shows 2-D band-pass filtered internal tide field with a smooth pattern. Figure 4d gives a residual field representing the nontidal noise acquired from the difference between Figure 4a,c. Because the wavenumber varies more dramatically in small basins, the wavenumber band is broader than the similar band shown by Zhao [22], who fully demonstrated that the 2-D filter does not lead to blind directions. The results will be discussed in Section 3.1.

2.4. Phase Speed and Energy Flux

The horizontal scale of internal tide is much larger than the mean ocean depth. The variable separation method is widely applied to simplify with hydrostatic approximation [34]. A sum of normal vertical modes in a stratified ocean expresses internal tides. We study the first mode of internal tides in this paper. The buoyancy frequency profile $N(z)$ decides the modal structure in the vertical. $\Phi(z)$ presents the vertical displacement structure. The following eigenvalue equation determines the modal structure [35]

$$\frac{d^2\Phi(z)}{dz^2} + \frac{N^2(z)}{c^2}\Phi(z) = 0, \tag{2}$$

which satisfies the boundary condition $\Phi(-H) = \Phi(0) = 0$, where H is the ocean depth and c is the eigenvalue speed. The buoyancy frequency profiles are calculated from WOA2018.

In a non-rotating fluid, the eigenvalue speed c is the phase speed. Considering the effect of the Earth's rotation (Ω), the dispersion relation of an internal wave is

$$\omega^2 = k^2c^2 + f^2, \tag{3}$$

where k is the wavenumber, and the inertial frequency is expressed as $f \equiv 2\Omega$ sin(latitude). Equation (3) suggests that the internal tide's wavelength usually increases with latitude, with constant depth and stratification. Before we apply the plane-wave fit method, we will calculate the internal tide's wavenumber as an input parameter. The phase velocity c_p is calculated from c,

$$c_p = \frac{\omega}{(\omega^2 - f^2)^{1/2}}c, \tag{4}$$

where ω stands for the tidal frequencies. Equation (4) also suggests that the internal tide's phase speed increases with latitude, which causes the refraction of internal tide. Due to a tidal frequency $\omega > f$, phase velocity c_p is always greater than c.

Using the vertical modal structures $\Phi(z)$, The SSH amplitude can deduce the interior displacement of internal tides. Then, the energy flux of the mode-1 internal tide can be vertically integrated by

$$\begin{aligned} F &= \tfrac{1}{2}\int_{-H}^{0} u(z)p(z)dz \\ &= \tfrac{1}{2}a^2F_n(f,\omega,N,H), \end{aligned} \tag{5}$$

where $F_n(H, N, \omega, f)$ characterizes the depth-integrated energy flux per unit amplitude of 1 cm at the surface. It is a function of the inertial frequency f, the tidal frequency ω, ocean stratification $N(z)$, and water depth H, and if F_n is determined, the energy flux will be proportional to the amplitude squared. We refer readers to the papers [8,36], which described the calculation procedure in detail.

3. Results

3.1. Decomposed M_2 Internal Tides

This study reveals the constructed M_2 internal tide field and its three decomposed components in the Sulu-Sulawesi Seas (Figure 5). The black arrows indicate depth-integrated fluxes, which point away from where they are generated. The M_2 internal tide field can be separated into three components by the plane-wave fit method due to their different propagation directions (Figure 5b–d). They are southeastward (270–320°), non-dominant directional (−40–50° and 250–270°), and the westward component (90–250°), respectively. We select the internal tides with the maximum amplitude in the chosen directions in each grid point. Unlike Luzon Strait, where multidirectional internal tides propagate away and are observed by satellite altimetry, the Sulu Island chain radiates multidirectional internal tides, which are represented by the biggest internal tide in the small sea basin. Multidirectional decomposition allows the originally covered internal tides to be exposed. The decomposed components have stable phases and wavefronts, compared to the original M_2 internal tide field. Multiple waves superpose to form an intricate interference pattern. For example, three tidal beams are marked by green lines and dots (Figure 5a). We refer to the M_2 internal tide field simulated by two global numerical models, including GOLD [13] and STORMTIDE [37]. Both models show the existence of the three tidal beams. We will explain the formation of three internal tidal beams in Section 3.3.

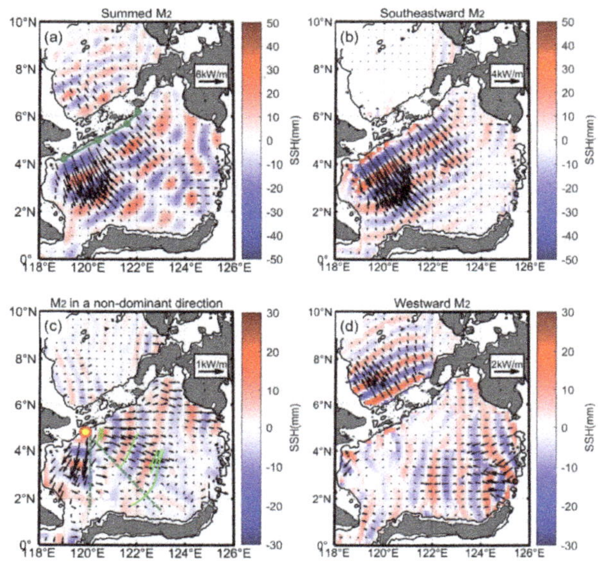

Figure 5. The M_2 internal tide field is decomposed in the Sulu-Sulawesi Seas. Black lines indicate isobathic contours of 400 m. (**a**) The five-wave-summed M_2 internal tide field. Three internal tidal beams are marked by green lines and dots. The internal tide field can be divided into three components. (**b**) The southeastward component (270–320°). (**c**) The M_2 internal tides in a non-dominant direction consist of the eastward component (−40–50°) and the southward component (250–270°). Two dash green lines indicate a 270° direction and a 320° direction. The yellow dot marks out the Sibutu Passage. The green curves indicate ISWs. (**d**) The westward component (90–250°). The depth-integrated energy fluxes are shown as black arrows.

The internal tidal wave with the maximum amplitude is separated in the direction (270–320°) on every grid point. Figure 5b displays the southeastward internal tide from the Sulu Island chain. When the internal tide propagates toward the Sulawesi Sea, its isophase lines are almost parallel to the shoreline of the Sulu Island chain. We believe that the feature is formed by the interference of internal tides with similar strength from multiple point sources in the chain. The transbasin internal tide propagates across almost the entire basin with isophase contours, arriving at the Sulawesi continental slope. The fate of the internal tide reaching the continental slope is an open question. We will investigate it in the discussion part.

A group of the M_2 internal tides emanate from the Sulu Island chain into the Sulawesi Sea in a non-dominant direction (Figure 5c). They consist of the eastward component (-40–50°) and the southward component (250–270°). The yellow dot indicating the Sibutu passage as a hypothetic point source radiates a cylindrical internal tidal wave with a missing piece. The missing piece is outlined by two dash green lines. The amplitude of the southeastward internal tide is about 40 mm, and the amplitude of a cylindrical internal tidal wave is about 15 mm. The eastward ISWs observed by Liu and D'Sa are spatially coincident with the internal tidal wavefronts (Figure 5c). For the first time, the cylindrical M_2 internal tide in the Sulawesi Sea is reported. The recognition of this covered beam benefits from the plane-wave fit method. Here we explain the missing piece of the cylindrical internal tide. The M_2 internal tides are separated from the point source and the other source over the Sulu Island chain due to different propagating directions. In contrast, we cannot separate them where their propagation directions become close. The missing piece corresponds to the dominant direction shown in Figure 5b, enhancing the strength of the southeastward internal tide. In addition, the eastward M_2 internal tides start from the continental slope in the western Sulu Sea along 119°E ranging from 6 to 9°N (Figure 5c). The weak eastward internal tide propagates along the Sulu Island chain and significantly modulates the dominant northward internal tide from the Sulu Island chain.

Figure 5d shows the westward internal tides from Sulu and Sangihe Island chains. The northwestward internal tides emanate from the Sulu Island chain into the Sulu Sea, propagating in almost the same direction. The internal tide causes a relatively consistent wave and flux pattern in the basin, which agrees well with HYCOM results around the Sulu Sea [38,39]. The Sangihe Island chain is another strong generation source that radiates westward internal tides [40]. The northern and southern sections of the Sangihe Island chain radiate two westward internal tidal beams into the Sulawesi Sea. Two internal tidal beams propagate at an angle and interfere with each other.

3.2. The Dominant Tidal Beam and Internal Solitary Waves

The relation of internal tides and ISWs in the Sulu-Sulawesi Seas is explored by investigating their spatial distribution. The internal tides evolve nonlinearly to generate ISWs from the Sulu Island chain [41–43]. Field measurement [44] and satellite observations have confirmed this relation [45]. Figure 6a shows the internal tides superposed with ISWs observed by Liu and D'Sa in 2019 [45]. The relation is revealed by their spatial distribution. The ISWs that are indicated by black curves are spatially coincident with the internal tidal wavefronts (Figure 6a).

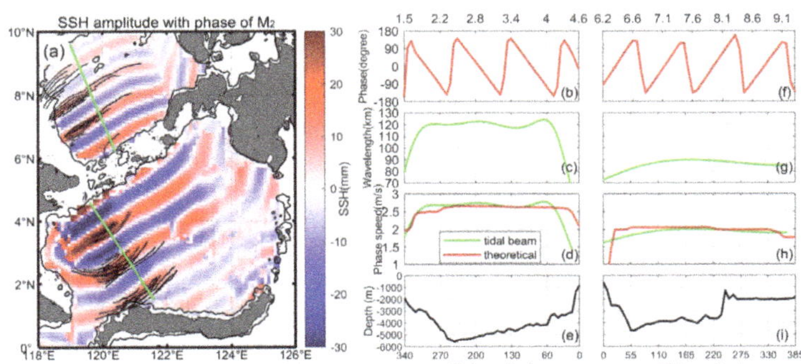

Figure 6. (**a**) M_2 internal tides and ISWs in Sulu-Sulawesi Seas. Black curves indicate ISWs. The green lines indicate the paths of two M_2 beams. (**b–i**) The M_2 internal tides propagate along a definite direction: (**b–e**) in the Sulawesi Sea and (**f–i**) in the Sulu Sea. Internal tides phase (Figure 6b,f). Wavelength (Figure 6c,g). Phase speed (Figure 6d,h). The theoretical and observational values are shown as red and green curves, respectively. Submarine topography (Figure 6e,i).

Liu and D'Sa [45] reported that ISWs are observed to propagate into the Sulu Sea and the Sulawesi Sea, showing the behavior of a spring-neap tidal cycle. The ISWs in the Sulawesi Sea are less frequently observed compared to those in the Sulu Sea. Zhang et al. [43] further confirmed that ISWs are mainly observed in the shallower western zones rather than the deeper eastern areas in the Sulu Sea. Tessler et al. [42] estimated the observed energy of the waves, maintaining baroclinic tidal mixing rates. A large number of ISWs accompany the internal tides, which indicates that their common generation source is the Sulu Island chain.

The green lines suggest the propagation paths of two M_2 beams in the Sulu- Sulawesi Seas. The along-beam phase speed of M_2 internal tide is examined by comparing the theoretical and altimetric results. Many factors affect the phase speed, including the depth and latitude. We calculated these factors along the two green lines, and the calculation procedure is illustrated in Figure 6b–i. First, the along-beam phase usually increases with the propagation distance in the Sulu-Sulawesi Sea, respectively (Figure 6b,f). Then, the polynomial fitting is employed to smooth the along-beam phase. The wavelength is calculated from the relation of phase gradient to distance (Figure 6c,g). The graphs show that the wavelength is mainly affected by the basin's depth rather than the Coriolis parameter near the equator. Finally, the phase speed is deduced from the relation of frequency and wavelength (Figure 6d,h). For comparison, light green curves indicate the theoretical phase speeds extracted from WOA2018 [31]. Due to the rapid decrease in the depth of submarine topography, the low values of the theoretical phase speed appear at starting and ending points (Figure 6e,i); meanwhile, the satellite-derived phase speeds coincide with the theoretical phase speeds.

3.3. *Sulu Sea vs. Sulawesi Sea*

Semidiurnal internal tides display an apparent contrast between the Sulu Sea and the Sulawesi Sea. The Sulu Island chain radiates the semidiurnal internal tides northward into the Sulu Sea and southward into the Sulawesi Sea. The amplitude of the northward M_2 internal tide is about 20 mm, about half of the southward one. The northward internal tide from the Sulu Island chain propagates in almost the same direction. However, multidirectional internal tides from the Sulu Island chain are superposed on the westward internal tide emanating from the Sangihe Island chain, forming a complex interference pattern in the Sulawesi Sea. The wavelength of the internal tide in the Sulawesi Sea is longer than that in the Sulu Sea. In summary, internal tides in the Sulu Sea and the Sulawesi Sea have different strengths, directions, and wavelengths. The reason for this contrast is that the number and distribution of their generation sources are distinct. Thus, the internal tides in the Sulawesi Sea have significant spatial inhomogeneity due to multisource interference. Next, we will further investigate the distinctions of internal tides between the two basins in detail from their multiple generation sources and multiwave interference process.

3.3.1. Comparison of Multiple Generation Sources

The multidirectional internal tides emanate away from two island chains, suggesting the presence of multiple sources in the boundaries. The phenomenon captured by the satellite altimeter is clarified in this section. Previous studies have investigated the barotropic tide in the Sulu-Sulawesi Seas by mooring measurements, satellite altimetry, and numerical models [46]. The model product is generated by a TPXO developed by Egbert and Erofeeva [26]. Figure 7a shows M_2 barotropic tidal ellipses and volume transport in the Sulu-Sulawesi Seas. The M_2 tide enters the Sulawesi Sea via the Sulu Island chain (Figure 7b). Strong tidal currents mainly occur in the eastern section of the Sibutu Passage in the Sulu Island chain, corresponding to the hypothetic point source. Affected by the rugged terrain, the tidal currents follow multiple directions, favoring a multidirectional internal tide generation. In contrast, the tidal currents mainly flow across the Sangihe Island chain (Figure 7c). The southern section of the Sangihe Island chain is the main channel of volume transport.

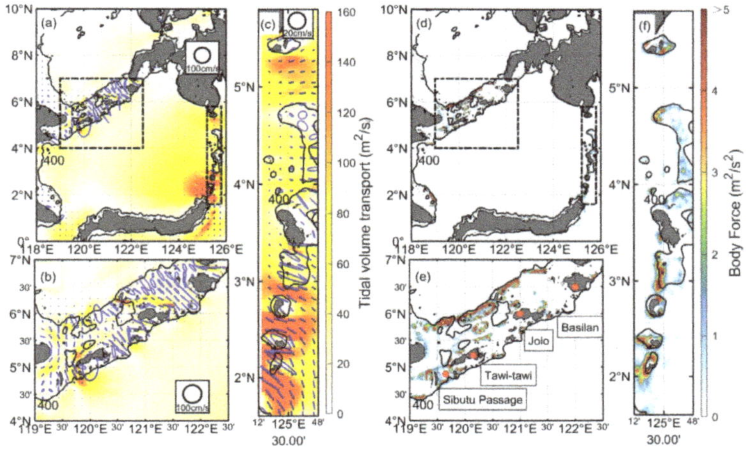

Figure 7. M_2 barotropic tide from TPXO and body force. Isobathic contours of 400 m are shown as black lines. (**a**) M_2 barotropic tidal ellipses and volume transport. Tidal ellipses are shown in blue. Two dotted boxes indicate two island chains. (**b**,**c**) are partial enlargements of (**a**). (**b**) Tidal ellipses and volume transport in the Sulu Islands chain. (**c**) As in (**b**), but in the Sangihe Islands chain. (**d**) Barotropic tidal body force. (**e**,**f**) are also partial enlargements of (**d**). (**e**) Barotropic tidal body force in the Sulu Islands chain. The four red dots represent Basilan Island, Jolo Island, Tawi-tawi Island, and Sibutu Passage. (**f**) As in (**e**), but in the Sangihe Islands chain. The dark red in the colorbar indicates values greater than five.

The barotropic tidal body force is used to reveal generation sites. The calculation of the barotropic tidal body force is expressed as

$$F = \frac{Q\nabla H}{\omega H^2}\int_{-H}^{0} N^2 z dz, \quad (6)$$

where H presents the basin depth, ∇H is the bottom gradient, Q is the barotropic tidal volume transport from TPXO, ω is the M_2 tidal frequency, and N^2 is the buoyancy frequency calculated from stratification from WOA2018. Figure 7d shows the barotropic tidal body force in the Sulu-Sulawesi Seas. In particular, Sibutu Passage, near Tawi-tawi Island, and Jolo Island, along the Sulu Island chain, are strong sources (Figure 7e). The body force shows that the generation sites are on the northern and southern sections of the Sangihe Island chain (Figure 7f). It reveals that strong conversion sites are scattered around the two island chains, which is consistent with previous work [43].

3.3.2. Comparison of Multiwave Interference Process

The decomposed internal tidal waves are revealed by the satellite altimeter observation. Their interference mechanism should be taken into account to interpret the complex structures of the SSH pattern. To illustrate the SSH pattern observed by satellites, an ideal line source model is employed to simulate the interference between multidirectional internal tidal waves. The line source model was proposed by St. Laurent et al. [47] and subsequently promoted by Rainville et al. [7]. The model uses a zero-width ridge as the source and a sinusoidal barotropic tidal current perpendicular to the ridge. Rainville et al. [7] noted that the simple model could describe the surface elevations caused by baroclinic tides:

$$\zeta(t,r,\theta) = \zeta_0\left(\frac{r_0}{r}\right)^{\beta} exp(ik_r r - i\omega t + \phi_0) \quad for\ |\theta - \theta_0| < a_0/2r_0\,, \quad (7)$$

where ζ_0, r_0, β, a_0 are the wave amplitude, the source radius, the attenuation coefficient, and the "arc length". ω, k_r, and ϕ_0 are respectively the M_2 frequency, wavenumber, and phase. The internal tidal wave is limited to the distance within $r > r_0$ and the direction between $\theta_0 \pm (a_0/2r_0)$. The altimetric results determine the parameters applied in the line source model (Table 1). The line source model clarifies the formation of interference by characterizing the plane waves. The phase Φ_0 of the internal tides is associated with the phase of the barotropic tides at the generation sites [47]. The wavelength for mode-1 M_2 internal tides is obtained by averaging the theoretical wavelength from WOA2018.

Table 1. Parameters are used to reproduce the interference in the Line Source Model.

Source [1]	Elevation (mm)	Radius (km): R_0	Direction (°): θ
PS	20	100	305 ± 85
LS_1	30	2000 [2]	302 ± 8
LS_2	15	150	200 ± 40
LS_3	15	150	170 ± 30
LS_4	20	800	112 ± 12
LS_5	5	500	20 ± 20

[1] PS is a point source in the Sulu Island chain. LS_1 is a line source in the Sulu Island chain that radiates the southeastward internal tide. LS_2 and LS_3 are line sources located at the northern and southern sections of the Sangihe Island chain. LS_4 is a line source located at the continental slope in the western Sulu Sea. LS_5 is a line source in the Sulu Island chain that radiates the southeastward internal tide. [2] A flat line source is produced by a small angle and a large radius.

Internal tide generation sites along the two island chains are not uniform. The line source model described a wave propagation that can have various levels of source numbers. For example, a few sources at the major generation sites can be superposed to create a line source, and larger amounts of sources along island chains also create a line source with a similar spatial pattern. Here we retain only the six dominant sources in the line source model (Figure 8). Figure 8a shows that a line source is used to characterize the smooth internal tide. Figure 8b shows the Sibutu Passage as a point source. The superposition of a line source and a point source presents spatial variability and three strong beams (Figure 8c). The northern and southern sections of Sangihe Island chain as two sources are shown in Figure 8d,e. Figure 8f shows the interference of M_2 internal tides from the two sources. The interference between M_2 internal tides from the continental slope in the eastern Sulu Sea and Sulu Island chain is not shown here. Based on the superpose principle, due to the destructive effects of interference, the SSH becomes weak between the beams. Internal tidal waves form nodes and antinodes not only in the SSH field but also in the energy flux.

The interference pattern caused by several sources in the LSM explains the complex structures of baroclinic SSH observed by satellite altimetry. Interference patterns in the Sulu-Sulawesi Sea are present by the line source model and satellite altimetry (Figure 9). Different from the decomposed internal tidal fields in Figure 5, the internal tides observed in each grid point get summed in the chosen direction (Figure 9a–c). To demonstrate the spatial pattern of the M_2 internal tide field modulated by multiwave interference, the Sulawesi Sea is divided into two fields. The first field in the western basin is predominantly influenced by internal tides from the Sulu Island chain. The second field in the eastern basin is influenced by internal tides from the Sangihe Island chain. Two island chains as the boundary of the Sulawesi Sea radiate internal tides, contributing to the interference pattern. Finally, internal tides from both island chains superposed together to construct the spatial pattern in the Sulu-Sulawesi Sea.

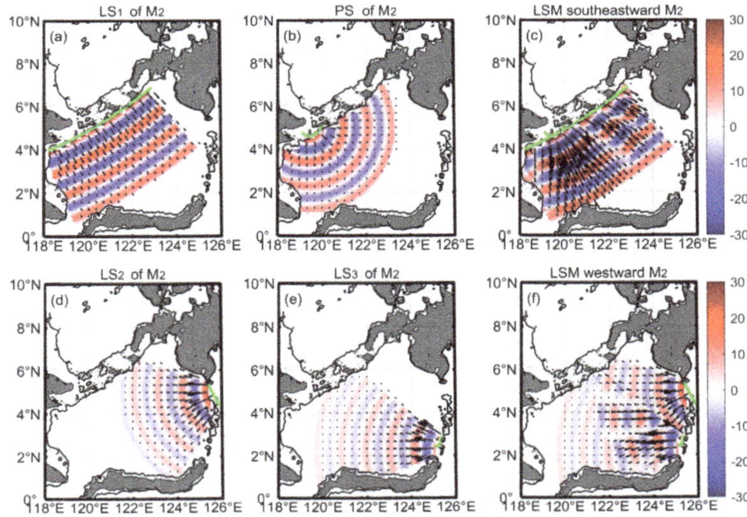

Figure 8. The line sources in the ideal line source model. (**a**) A line source along the Sulu Island chain. (**b**) A point source is located on the Sibutu Passage. (**c**) Interference pattern in the Sulawesi Sea superimposed to (**a**,**b**). The line source is located on (**d**) the northern section and (**e**) the southern section of the Sangihe Island chain. (**f**) Interference pattern in the Sulawesi Sea superimposed to (**d**,**e**). Parameters of LSM are listed in Table 1.

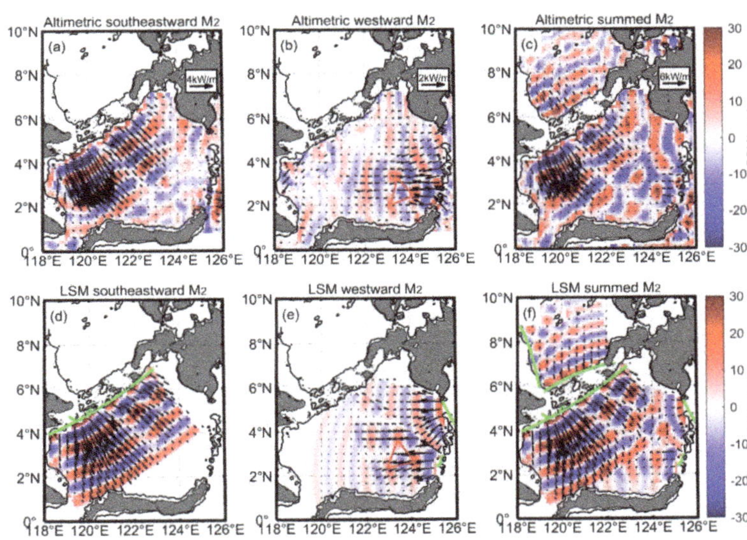

Figure 9. Interference patterns in the Sulawesi Sea by satellite altimetry and the line source model. (**a**) Altimetric M_2 internal tide from the Sulu Island chain. (**b**,**c**) as in (**a**), for the M_2 internal tide from the Sangihe Island chain and both island chains. (**d**) M_2 internal tide from the Sulu Island chain in the ideal line source model. (**e**,**f**) as in (**d**), for the M_2 internal tide from the Sangihe Island chain and both island chains. The red triangle represents the enhancement area. The green curves indicate generation sources.

In the western basin of the Sulawesi Sea, a cylindrical internal tidal wave from the Sibutu Passage significantly modulates the dominant internal tides from the Sulu Island chain. The interference shapes three internal tidal beams with elevated energy flux magnitude (Figure 9a). The three internal tidal beams with various strengths are all presented by satellite altimetric observation and the ideal line source model (Figure 9a,d). This interference pattern has been suggested by global satellite observations [8]. This feature is similar to that at the Hawaiian Ridge, where internal tides interfere to form distinct beams [7]. We can conclude that the weak internal tide modulates the SSH field, contributing to the inhomogeneous distribution of M_2 internal tides. In the eastern basin of the Sulawesi Sea, the internal tides are generated at the northern and southern sections of the Sangihe Island chain. One can see that an enhanced central internal tidal beam is formed at the intersection of their propagation paths (Figure 9b,e). In the whole sea basin, the complex SSH pattern from the idea line source model is consistent with the satellite altimetric observation. The internal tide in the Sulu Sea propagates about four wavelengths (Figure 9f), while the internal tide in the Sulawesi Sea propagates about three wavelengths for approximate distance. Eventually, the sources are summed to be compared to the SSH for satellite altimetric observations (Figure 9c). A spatially inhomogeneous SSH field is shaped by internal tides from the Sulu and Sangihe Island chains. Semidiurnal internal tides display complex and distinct geographical variations between the Sulu and Sulawesi Seas. Many simplifications are made in the line source model, including tidal beams perpendicular to the source ridge, coherent and two-dimensional [47]. Furthermore, the line source model neglects the topography-scattering effects to characterize the interference of plane waves ideally. An ideal line source model applied in the small basin should take into account neighboring sources and the short propagation distance.

3.4. S_2 Internal Tides

The S_2 internal tide field in the Sulu-Sulawesi Seas is constructed by fitting plane waves (Figure 10a). The specific frequency and wavelength of the S_2 internal tide are input parameters in the extraction process. Like point harmonic analysis, plane-wave analysis can extract multi-frequency internal tide signals in one step. The wavenumber band of the 2-D band-pass filter processing the S_2 internal tide field is similar to that of the M_2 internal tide. Because of tidal aliasing, the S_2 tidal signals could not be extracted from ERS satellite altimeter data which are excluded in the extraction process. Although the amount of data is reduced, the same fitting window as that of the M_2 internal tide is still used after different window size attempts to investigate the spatial propagation characteristics of the S_2 internal tide.

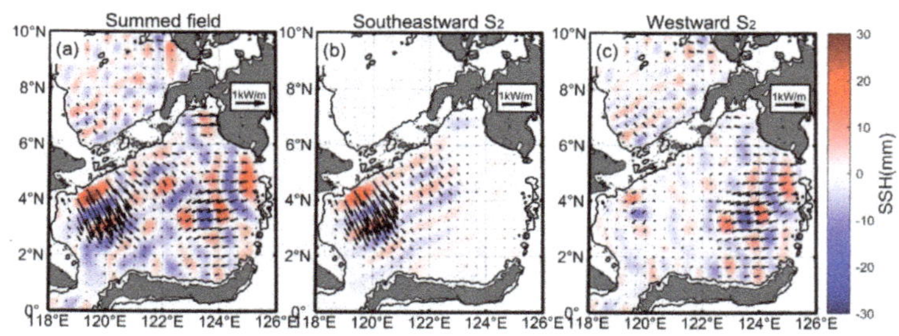

Figure 10. The mode-1 S_2 internal tide field is decomposed. (**a**) The five-wave-summed S_2 internal tide field. The internal tide field can be divided into two components. (**b**) The southward component (250–330°). (**c**) The westward component (90–270°). The S_2 internal tidal energy fluxes are shown as black arrows in (**a**–**c**). Isobathic contours of 400 m are indicated as black lines, respectively.

Figure 10a shows the superposition map of mode-1 S_2 internal tides. Consistent with the M_2 internal tide field, the Sulu and Sangihe Island chains are also essential generation sources for the S_2 internal tides. Several internal tidal beams are shaped by multiwave interference. Then, we separated the southward and westward components, respectively. Figure 10b shows the S_2 internal tidal beams propagated from the Sulu Island chain into the Sulawesi Sea. They propagate about three wavelengths and become undetected by satellite altimeters. The western internal beam is stronger than the eastern one, which indicates that the Sibutu Passage is also an energetic generation site. The SSH ratio of S_2 to M_2 is 0.46, consistent with their ratio in the global ocean [48]. However, differently from M_2 internal tides, we find the lack of a stable and continuous eastward S_2 internal tide signal. We deduce that the eastward S_2 internal tide is very weak and becomes undetectable.

Westward S_2 internal tides emanate from the Sulu and Sangihe Island chains (Figure 10c). On the one hand, two internal tidal beams propagate from the Sangihe Island chain. They interfere with each other and enhance the S_2 internal tidal amplitude and energy flux. The westward S_2 beams travel a shorter distance than the M_2 beams because nontidal noise disturbs the weak S_2 internal tide more easily. On the other hand, the northward internal tidal beam emanates from the Sulu Island chain into the Sulu Sea. The internal tide passes through the basin and reaches the other side due to the stable ocean environment.

4. Discussion

4.1. Reflection on the Sulawesi Continental Slope

In this section, we investigate the fate of the transbasin M_2 internal tide in the Sulawesi Sea (Figure 11a). Theoretically, a small fraction of the M_2 internal tide is expected to reflect into the Sulawesi sea, since the Sulawesi slope is supercritical to M_2 (Figure 1b). The incident and reflected M_2 internal tides have been investigated in marginal seas by satellite altimeters, such as the Tasman Sea [49]. The incident and reflected internal tides can be separated according to their propagation directions. This separation is attributed to the multiwave internal tidal field decomposed by the plane-wave fitting method (Figure 3). Here, we choose waves in the direction of 280°–330° as the incident internal tides and 0°–70° as the reflected internal tides (Figure 11).

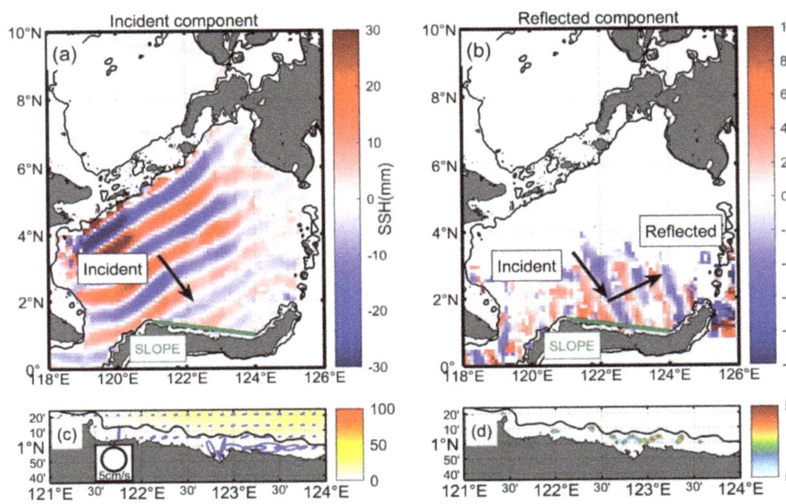

Figure 11. The M_2 internal tide reflects in the Sulawesi Sea. (**a**) Incident part (ranging 280–330°). (**b**) Reflected part (ranging 0–70°). The Sulawesi slope is denoted as the green line. The incident and reflected waves are shown as two black arrows, respectively. The ratio of the incident and reflected SSH amplitudes near the Sulawesi slope presents reflectivity. (**c**) M_2 barotropic tide ellipse and volume transport in the Sulawesi slope. (**d**) Barotropic tidal body force in the Sulawesi slope.

Figure 11 shows that the dominant M_2 internal tidal beam hits the Sulawesi slope and reflects into the Sulawesi Sea. Snell's law is used to examine the incident and reflected waves. The black arrow with a 300° incidence angle denotes the incident internal tidal wave. A black arrow toward 28° denotes the reflected wave. The green line denotes the continental slope as the wall. Satellite altimetric observation agrees well with the law of reflection, indicating that the northeastward tidal beam is reflected from the Sulawesi slope. When the dominant internal tidal beam reaches the slope, the amplitude of the incident part is about 6 mm. The amplitude of the reflected wave is about 4 mm. The energy flux is approximately proportional to the SSH squared in Equation (5). Thus, the reflectivity from the incident and reflected energy flux is about 45%. Our estimation may be affected by many factors.

In addition, we need to verify that the northeastward internal tide is not locally generated. We check the tidal ellipses, volume transport, and barotropic tidal body force on the Sulawesi slope. The direction of the tidal current is not consistent with the propagation direction of the reflected internal tide. Therefore, it can be concluded that the northeastward internal tide is formed by reflection rather than local generation. A key point of this paper is to note the existence of the reflection phenomenon but not to estimate the reflectivity accurately. A quantitative estimation of reflectivity requires further field observations and numerical models with a suitable method in the Sulawesi Sea [50].

4.2. Internal Tide Energetics

In this section, we estimate the semidiurnal internal tidal energy fluxes from two island chains. The total energy fluxes are integrated along a section across the whole sea basin parallel to the two island chains. For the southeastward M_2 internal tide from the Sulu Island chain, we acquire a total energy flux of 2.07 GW. Similarly, we estimate the northwestward energy flux from the Sulu Island chain and westward energy flux from the Sangihe Island chain. The total energy fluxes for the northwestward and westward internal tides are 0.41 and 0.28 GW, respectively. Thus, the total M_2 tidal energy flux from the Sulu Island chain is 2.48 GW (Figure 12a). It is about 80% of 2.8 GW from the numerical simulation by Nagai and Hibiya [27]. The westward internal tide is smaller than the model result because only the energy flux in the Sulawesi Sea is considered [51]. Likewise, the energy flux for each direction of the S_2 internal tides is estimated. The energy fluxes for the northwestward, westward, and southeastward internal tides are 0.04, 0.09, and 0.25 GW, respectively. The total semidiurnal tidal energy fluxes from the Sulu and Sangihe Island chains into the Sulawesi Sea are about 2.7 GW, measured by satellite altimetry.

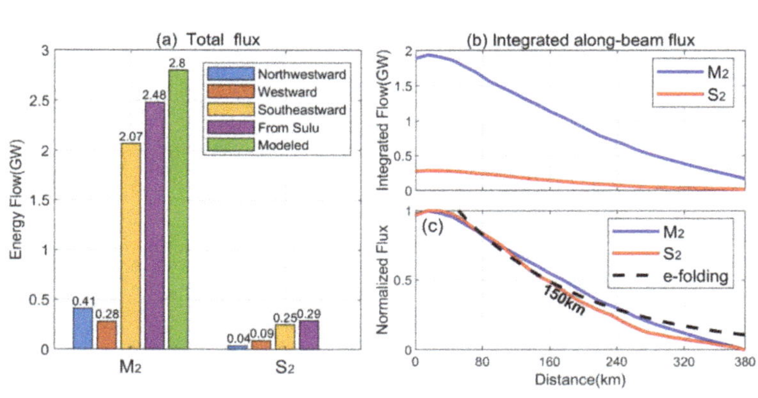

Figure 12. The total energy fluxes and the decay rate of the southeastward internal tide. (**a**) The total internal tidal energy flux from two island chains. (**b**) The decay of the southeastward internal tide. (**c**) Internal tides energy flux after normalization. The E-folding scale of 150 km is indicated as the dashed curve.

Furthermore, we estimate the decay rate of the dominant internal tide (Figure 5b). We sample the internal tide from the Sulu Island chain to the Sulawesi slope using 30 equidistant sections. We integrate them along each section, which is perpendicular to the energy fluxes. The blue line and red line in Figure 12b respectively denote the M_2, and S_2 energy flux obtained using plane-wave fitting. Figure 12c indicates the M_2 and S_2 internal tides energy flux after normalization. Even though their amplitudes are different, both of them decay at close rates. The E-folding decay scale of 150 km is indicated as the dashed curve. The short *e*-folding scale may be the result of either undetectable incoherent internal tides or real dissipation.

5. Conclusions

In this paper, we apply the plane-wave analysis method to construct the regional semidiurnal internal tidal field in the Sulu-Sulawesi Seas from multi-satellite altimeter data. We separately decompose the internal tidal field due to different propagation directions. The components obtained from the decomposition contribute to a better understanding of the formation and propagation of internal tidal beams. The multidirectional internal tides and their interference are revealed in the Sulu-Sulawesi Seas. We presented the geographic distribution of the semidiurnal internal tidal beams from the Sulu and Sangihe Island chains. The internal tides in the Sulawesi Sea dissipate in the semi-closed basin, causing strong tidal mixing.

Our main scientific findings are as follows. (1) The constructed M_2 internal tidal field is decomposed into three components, which have different propagation directions. The interference pattern is well reproduced by the line source model. (2) The Sibutu passage in the Sulu Island chain acts as a point source that radiates a cylindrical internal tidal wave. (3) The ISWs spatially correspond to the internal tidal wavefronts. The phase speed derived from the satellite is in good agreement with the theoretical phase speed. The above two results confirm that the analysis method can be applied to local small sea basins. (4) The northward M_2 tidal energy flux from the Sulu Island chain is 0.41 GW, about 20% of the southward energy flux. The total energy fluxes from satellite altimeters are about 80% of those from numerical simulations in the Sulu Island chain. (5) The M_2 and S_2 internal tides have a similar decay rate, although the SSH ratio of S_2 to M_2 is 0.45. (6) The total semidiurnal tidal energy fluxes from the Sulu and Sangihe Island chains into the Sulawesi Sea are about 2.7 GW. It is important to note that these values are all at the lower limits because they are multi-year coherent results. The plane-wave analysis only extracts the temporally coherent internal tide. The internal tides become incoherent due to the variation of the ocean environment, such as mesoscale eddies and circulation [52].

The present study focuses on the coherent internal tides over 20 years. However, recent studies revealed that the internal tides show the spatial distribution and strong seasonal variation in the Indonesian Archipelago and adjacent seas [53–55]. In addition, the Sulawesi Sea is the main channel of ITF, which affects the generation and propagation of internal tides. The propagation direction of internal tides from the Sulu Island chain is almost perpendicular to the direction of ITF in the Sulawesi Sea. In contrast, the internal tides propagate from the Sangihe Island chain in the same direction as the ITF, weakening the westward internal tides observed by satellite altimetry. The Sulu-Sulawesi Seas are also identified to be the source of the energetic generation of diurnal internal tides. The investigation of these tides using multi-satellite altimeter is underway.

Author Contributions: Conceptualization, X.Z., Z.X. and Q.L.; methodology, X.Z. and Z.X.; software, X.Z.; writing—original draft preparation, X.Z.; writing—review and editing, Z.X., M.F., Q.L., P.Z., J.Y., S.G., B.Y. and X.Z. All authors have read and agreed to the published version of the manuscript.

Funding: This study was supported by the Strategic Pioneering Research Program of CAS, the National Natural Science Foundation of China, the National Key Research and Development Program of China (XDB42000000, XDA22050202, 92058202, 91858103, 2017YFA0604102, 2016YFC1401404), the CAS Key Research Program of Frontier Sciences, the Key Deployment Project of Centre for Ocean Mega-Research of Science, and the project jointly funded by the CAS and CSIRO (QYZDB-SSW-DQC024, COMS2020Q07, 133244KYSB20190031).

Institutional Review Board Statement: Not applicable.

Informed Consent Statement: Not applicable.

Data Availability Statement: The datasets presented in this study are publicly available. The altimetric sea surface height measurements are produced by Archiving, Validation, and Interpolation of Satellite Oceanographic Data (AVISO) and distributed by the Copernicus Marine and Environment Monitoring Service (CMEMS) via https://marine.copernicus.eu/, (accessed on 25 June 2021). The World Ocean Atlas 2018 is produced and made available by NOAA National Centers for Environmental Information via https://accession.nodc.noaa.gov/NCEI-WOA18, (accessed on 25 June 2021). The TPXO tide model is from the Oregon State University via https://www.tpxo.net/global/tpxo9-atlas, (accessed on 25 June 2021). The internal tide products presented in this paper are available upon request.

Acknowledgments: We would like to thank the editors and three anonymous reviewers whose constructive comments have significantly improved the manuscript.

Conflicts of Interest: The authors declare no conflict of interest. The funders had no role in the design of the study; in the collection, analyses, or interpretation of data; in the writing of the manuscript, or in the decision to publish the results.

References

1. Munk, W.; Wunsch, C. Abyssal recipes II: Energetics of tidal and wind mixing. *Deep Sea Res. Part I Oceanogr. Res. Pap.* **1998**, *45*, 1977–2010. [CrossRef]
2. Alford, M.H. Redistribution of energy available for ocean mixing by long-range propagation of internal waves. *Nature* **2003**, *423*, 156–162. [CrossRef] [PubMed]
3. Neale, R.; Slingo, J. The Maritime Continent and Its Role in the Global Climate: A GCM Study. *J. Clim.* **2003**, *16*, 834–848. [CrossRef]
4. Alford, M.H.; Zhao, Z. Global Patterns of Low-Mode Internal-Wave Propagation. Part I: Energy and Energy Flux. *J. Phys. Oceanogr.* **2007**, *37*, 1829–1848. [CrossRef]
5. Egbert, G.D.; Ray, R.D. Semi-diurnal and diurnal tidal dissipation from TOPEX/Poseidon altimetry. *Geophys. Res. Lett.* **2003**, *30*, 1907. [CrossRef]
6. MacKinnon, J.A.; Alford, M.H.; Ansong, J.K.; Arbic, B.K.; Barna, A.; Briegleb, B.P.; Bryan, F.O.; Buijsman, M.C.; Chassignet, E.P.; Danabasoglu, G.; et al. Climate Process Team on Internal Wave-Driven Ocean Mixing. *Bull. Am. Meteorol. Soc.* **2017**, *98*, 2429–2454. [CrossRef]
7. Rainville, L.; Johnston, T.M.S.; Carter, G.S.; Merrifield, M.A.; Pinkel, R.; Worcester, P.F.; Dushaw, B.D. Interference Pattern and Propagation of the M2 Internal Tide South of the Hawaiian Ridge. *J. Phys. Oceanogr.* **2010**, *40*, 311–325. [CrossRef]
8. Zhao, Z.; Alford, M.H.; Girton, J.B.; Rainville, L.; Simmons, H.L. Global Observations of Open-Ocean Mode-1 M2 Internal Tides. *J. Phys. Oceanogr.* **2016**, *46*, 1657–1684. [CrossRef]
9. Chang, H.; Xu, Z.; Yin, B.; Hou, Y.; Liu, Y.; Li, D.; Wang, Y.; Cao, S.; Liu, A.K. Generation and Propagation of M2 Internal Tides Modulated by the Kuroshio Northeast of Taiwan. *J. Geophys. Res. Ocean.* **2019**, *124*, 2728–2749. [CrossRef]
10. Wang, Y.; Xu, Z.; Yin, B.; Hou, Y.; Chang, H. Long-Range Radiation and Interference Pattern of Multisource M2 Internal Tides in the Philippine Sea. *J. Geophys. Res. Ocean.* **2018**, *123*, 5091–5112. [CrossRef]
11. Xu, Z.; Liu, K.; Yin, B.; Zhao, Z.; Wang, Y.; Li, Q. Long-range propagation and associated variability of internal tides in the South China Sea. *J. Geophys. Res. Ocean.* **2016**, *121*, 8268–8286. [CrossRef]
12. Zhao, C.; Xu, Z.; Robertson, R.; Li, Q.; Wang, Y.; Yin, B. The Three-Dimensional Internal Tide Radiation and Dissipation in the Mariana Arc-Trench System. *J. Geophys. Res. Ocean.* **2021**, *126*. [CrossRef]
13. Simmons, H.L.; Hallberg, R.W.; Arbic, B.K. Internal wave generation in a global baroclinic tide model. *Deep Sea Res. Part II Top. Stud. Oceanogr.* **2004**, *51*, 3043–3068. [CrossRef]
14. Robertson, R. Interactions between tides and other frequencies in the Indonesian seas. *Ocean Dyn.* **2011**, *61*, 69–88. [CrossRef]
15. Robertson, R.; Ffield, A. Baroclinic tides in the Indonesian seas: Tidal fields and comparisons to observations. *J. Geophys. Res.* **2008**, *113*, C07031. [CrossRef]
16. Koch-Larrouy, A.; Lengaigne, M.; Terray, P.; Madec, G.; Masson, S. Tidal mixing in the Indonesian Seas and its effect on the tropical climate system. *Clim. Dyn.* **2010**, *34*, 891–904. [CrossRef]

17. Ffield, A.; Gordon, A.L. Tidal Mixing Signatures in the Indonesian Seas. *J. Phys. Oceanogr.* **1996**, *26*, 1924–1937. [CrossRef]
18. Hatayama, T.; Awaji, T.; Akitomo, K. Tidal currents in the Indonesian Seas and their effect on transport and mixing. *J. Geophys. Res. Ocean.* **1996**, *101*, 12353–12373. [CrossRef]
19. Ray, R.D.; Mitchum, G.T. Surface manifestation of internal tides in the deep ocean: Observations from altimetry and island gauges. *Prog. Oceanogr.* **1997**, *40*, 135–162. [CrossRef]
20. Tian, J.; Zhou, L.; Zhang, X.; Liang, X.; Zheng, Q.; Zhao, W. Estimates of M2 internal tide energy fluxes along the margin of Northwestern Pacific using TOPEX/POSEIDON altimeter data. *Geophys. Res. Lett.* **2003**, *30*, 17. [CrossRef]
21. Zaron, E.D. Baroclinic Tidal Sea Level from Exact-Repeat Mission Altimetry. *J. Phys. Oceanogr.* **2019**, *49*, 193–210. [CrossRef]
22. Zhao, Z. Mapping Internal Tides from Satellite Altimetry without Blind Directions. *J. Geophys. Res. Ocean.* **2019**, *124*, 8605–8625. [CrossRef]
23. Jithin, A.K.; Subeesh, M.P.; Francis, P.A.; Ramakrishna, S. Intensification of tidally generated internal waves in the north-central Bay of Bengal. *Sci. Rep.* **2020**, *10*, 6059. [CrossRef] [PubMed]
24. Zhao, Z. Southward Internal Tides in the Northeastern South China Sea. *J. Geophys. Res. Ocean.* **2020**, *125*. [CrossRef]
25. Koch-Larrouy, A.; Madec, G.; Bouruet-Aubertot, P.; Gerkema, T.; Bessières, L.; Molcard, R. On the transformation of Pacific Water into Indonesian Throughflow Water by internal tidal mixing. *Geophys. Res. Lett.* **2007**, *34*. [CrossRef]
26. Egbert, G.D.; Erofeeva, S.Y. Efficient Inverse Modeling of Barotropic Ocean Tides. *J. Atmos. Ocean. Technol.* **2002**, *19*, 183–204. [CrossRef]
27. Nagai, T.; Hibiya, T. Internal tides and associated vertical mixing in the Indonesian Archipelago. *J. Geophys. Res. Ocean.* **2015**, *120*, 3373–3390. [CrossRef]
28. Buijsman, M.C.; Arbic, B.K.; Richman, J.G.; Shriver, J.F.; Wallcraft, A.J.; Zamudio, L. Semidiurnal internal tide incoherence in the equatorial Pacific. *J. Geophys. Res. Ocean.* **2017**, *122*, 5286–5305. [CrossRef]
29. Ray, R.D.; Zaron, E.D. M2 Internal Tides and Their Observed Wavenumber Spectra from Satellite Altimetry. *J. Phys. Oceanogr.* **2016**, *46*, 3–22. [CrossRef]
30. Ray, R.D.; Cartwright, D.E. Estimates of internal tide energy fluxes from Topex/Poseidon Altimetry: Central North Pacific. *Geophys. Res. Lett.* **2001**, *28*, 1259–1262. [CrossRef]
31. Locarnini, R.A.; Mishonov, A.V.; Baranova, O.K.; Boyer, T.P.; Locarnini, R.A. *World Ocean Atlas 2018, Volume 1: Temperature*; NOAA Atlas NESDIS: Silver Spring, MD, USA, 2019.
32. Hao, Z.; Xu, Z.; Feng, M.; Li, Q.; Yin, B. Spatiotemporal Variability of Mesoscale Eddies in the Indonesian Seas. *Remote Sens.* **2021**, *13*, 1017. [CrossRef]
33. He, Y.; Feng, M.; Xie, J.; Liu, J.; Chen, Z.; Xu, J.; Fang, W.; Cai, S. Spatiotemporal Variations of Mesoscale Eddies in the Sulu Sea. *J. Geophys. Res. Ocean.* **2017**, *122*, 7867–7879. [CrossRef]
34. Wunsch, C. Internal tides in the ocean. *Rev. Geophys.* **1975**, *13*, 167–182. [CrossRef]
35. Gill, A. *Atmosphere-Ocean Dynamics*; Academic Press: Cambridge, MA, USA, 1982.
36. Zhao, Z.; Alford, M.H. New Altimetric Estimates of Mode-1 M2 Internal Tides in the Central North Pacific Ocean. *J. Phys. Oceanogr.* **2009**, *39*, 1669–1684. [CrossRef]
37. Muller, M. On the space- and time-dependence of barotropic-to-baroclinic tidal energy conversion. *Ocean Model.* **2013**, *72*, 242–252. [CrossRef]
38. Hurlburt, H.; Metzger, J.; Sprintall, J.; Riedlinger, S.; Arnone, R.; Shinoda, T.; Xu, X. Circulation in the Philippine Archipelago Simulated by 1/12° and 1/25° Global HYCOM and EAS NCOM. *Oceanography* **2011**, *24*, 28–47. [CrossRef]
39. Metzger, E.J.; Hurlburt, H.E.; Xu, X.; Shriver, J.F.; Gordon, A.L.; Sprintall, J.; Susanto, R.D.; van Aken, H.M. Simulated and observed circulation in the Indonesian Seas: 1/12° global HYCOM and the INSTANT observations. *Dyn. Atmos. Ocean.* **2010**, *50*, 275–300. [CrossRef]
40. Nugroho, D.; Koch-Larrouy, A.; Gaspar, P.; Lyard, F.; Reffray, G.; Tranchant, B. Modelling explicit tides in the Indonesian seas: An important process for surface sea water properties. *Mar. Pollut. Bull.* **2018**, *131*, 7–18. [CrossRef]
41. Chen, G.-Y.; Su, F.-C.; Wang, C.-M.; Liu, C.-T.; Tseng, R.-S. Derivation of internal solitary wave amplitude in the South China Sea deep basin from satellite images. *J. Oceanogr.* **2011**, *67*, 689–697. [CrossRef]
42. Tessler, Z.D.; Gordon, A.L.; Jackson, C.R. Early Stage Soliton Observations in the Sulu Sea. *J. Phys. Oceanogr.* **2012**, *42*, 1327–1336. [CrossRef]
43. Zhang, X.; Li, X.; Zhang, T. Characteristics and generations of internal wave in the Sulu Sea inferred from optical satellite images. *J. Oceanol. Limnol.* **2020**, *38*, 1435–1444. [CrossRef]
44. Apel, J.R.; Holbrook, J.R.; Liu, A.K.; Tsai, J.J. The Sulu Sea Internal Soliton Experiment. *J. Phys. Oceanogr.* **1985**, *15*, 1625–1651. [CrossRef]
45. Liu, B.; D'Sa, E.J. Oceanic Internal Waves in the Sulu–Celebes Sea Under Sunglint and Moonglint. *IEEE Trans. Geosci. Remote Sens.* **2019**, *57*, 6119–6129. [CrossRef]
46. Ray, R.D.; Egbert, G.D.; Erofeeva, S. A brief overview of tides in the Indonesian Seas. *Oceanography* **2005**, *18*, 74–79. [CrossRef]
47. St. Laurent, L.; Stringer, S.; Garrett, C.; Perrault-Joncas, D. The generation of internal tides at abrupt topography. *Deep Sea Res. Part I Oceanogr. Res. Pap.* **2003**, *50*, 987–1003. [CrossRef]
48. Zhao, Z. The Global Mode-1 S2 Internal Tide. *J. Geophys. Res. Ocean.* **2017**, *122*, 8794–8812. [CrossRef]

49. Zhao, Z.; Alford, M.H.; Simmons, H.L.; Brazhnikov, D.; Pinkel, R. Satellite Investigation of the M2 Internal Tide in the Tasman Sea. *J. Phys. Oceanogr.* **2018**, *48*, 687–703. [CrossRef]
50. Wang, S.; Cao, A.; Chen, X.; Li, Q.; Song, J.; Meng, J. Estimation of the Reflection of Internal Tides on a Slope. *J. Ocean Univ. China* **2020**, *19*, 489–496. [CrossRef]
51. Hermansyah, H.; Atmadipoera, A.S.; Prartono, T.; Jaya, I.; Syamsudin, F. Energetics of internal tides over the Sangihe-Talaud ridge—Sulawesi sea. *J. Phys. Conf. Ser.* **2019**, *1341*, 082001. [CrossRef]
52. Xu, Z.; Wang, Y.; Liu, Z.; McWilliams, J.C.; Gan, J. Insight into the Dynamics of the Radiating Internal Tide Associated with the Kuroshio Current. *J. Geophys. Res. Ocean.* **2021**, *126*. [CrossRef]
53. Song, P.; Chen, X. Investigation of the Internal Tides in the Northwest Pacific Ocean Considering the Background Circulation and Stratification. *J. Phys. Oceanogr.* **2020**, *50*, 3165–3188. [CrossRef]
54. Xu, Z.; Yin, B.; Hou, Y.; Liu, A.K. Seasonal variability and north–south asymmetry of internal tides in the deep basin west of the Luzon Strait. *J. Mar. Syst.* **2014**, *134*, 101–112. [CrossRef]
55. Xu, Z.; Yin, B.; Hou, Y.; Xu, Y. Variability of internal tides and near-inertial waves on the continental slope of the northwestern South China Sea. *J. Geophys. Res. Ocean.* **2013**, *118*, 197–211. [CrossRef]

remote sensing

MDPI

Article

Distinct Variability between Semidiurnal and Diurnal Internal Tides at the East China Sea Shelf

Weidong Wang [1,2,3,4], Robin Robertson [5], Yang Wang [1,2,3,4], Chen Zhao [1,2,3,4], Zhanjiu Hao [1,2,3,4], Baoshu Yin [1,2,3,4,6] and Zhenhua Xu [1,2,3,4,*]

1 CAS Key Laboratory of Ocean Circulation and Waves, Institute of Oceanology, Chinese Academy of Sciences, Qingdao 266071, China; wangweidong@qdio.ac.cn (W.W.); wangyang@qdio.ac.cn (Y.W.); zhaochen@qdio.ac.cn (C.Z.); haozhanjiu@qdio.ac.cn (Z.H.); bsyin@qdio.ac.cn (B.Y.)
2 Pilot National Laboratory for Marine Science and Technology, Qingdao 266237, China
3 Center for Ocean Mega-Science, Chinese Academy of Sciences, Qingdao 266071, China
4 University of Chinese Academy of Sciences, Beijing 100049, China
5 China-Asean College of Marine Science, Xiamen University Malaysia, Sepang 43900, Malaysia; robin.robertson@xmu.edu.my
6 CAS Engineering Laboratory for Marine Ranching, Institute of Oceanology, Chinese Academy of Sciences, Qingdao 266071, China
* Correspondence: xuzhenhua@qdio.ac.cn; Tel.: +86-532-8288-8991

Abstract: Breaking internal tides and induced mixing are critical to shelf dynamics, including heat and mass exchanges. Spatiotemporal variability of internal tides and modulation factors for the southern East China Sea shelf were examined based on a combination of a three-month mooring velocity and satellite altimeter data. Semidiurnal and diurnal internal tides exhibited distinct temporal trends, with the semidiurnal internal tides enhanced by an order of magnitude during the latter half of the record, while the diurnal internal tides followed quasi spring-neap cycles with a generally stable intensity except for two specific periods of strengthening. These internal tides probably originated remotely over the shelf-slope area northeast of Taiwan. Time-varying stratification was the most important factor for the internal tidal magnitude. In addition, varying background currents influenced the diurnal critical latitude band, which explains the slightly enhanced diurnal internal tides during the two periods. Although both semidiurnal and diurnal internal tides were mode-1 dominated, the semidiurnal internal tides were surface intensified while the diurnal tides were bottom intensified. The proportion of higher mode internal tides increased during robust eddy activities. Stronger background vertical shear corresponded to high-frequency events and energy transfers from tidal frequencies to high frequencies associated with turbulent mixing.

Keywords: internal tides; spatiotemporal variation; East China Sea; modal structure; energy cascade

Citation: Wang, W.; Robertson, R.; Wang, Y.; Zhao, C.; Hao, Z.; Yin, B.; Xu, Z. Distinct Variability between Semidiurnal and Diurnal Internal Tides at the East China Sea Shelf. *Remote Sens.* **2022**, *14*, 2570. https://doi.org/10.3390/rs14112570

Academic Editors: Xiaofeng Li, Chung-Ru Ho and Antony K. Liu

Received: 27 March 2022
Accepted: 19 May 2022
Published: 27 May 2022

1. Introduction

Internal tides are generated when barotropic tides in a stratified water column flow over abrupt topography, such as a continental shelf edge, subsurface ridge, sill, or seamount [1]. Previous analysis indicates that open-ocean low-mode internal tides lose up to 60% of their energy as they impinge onto the continental shelf [2]. Along the continental margins, internal tides can induce turbulent mixing, playing a pivotal role in mass and heat transports, biological production, and possibly even shaping the continental slope [3]. Deciphering the formation, structure, and variability of internal tides at the continental shelf is of significance in understanding the coastal regions and their energy exchange with the open ocean.

Global maps of mode-1 internal tides have been estimated by numerical models [4,5] and satellite observations [6–8]. Nevertheless, some pieces are still missing over the continental shelves and in coastal regions, and the accuracy decreases near land. Accurately characterizing the internal tide is challenging for numerical models due to the highly variable stratification and complex topography. Noise contamination near land prevents internal tides from being identified in altimeter results in continental shelf areas [6,8,9]. All of these make in situ observations indispensable to accurately characterize internal tides in shallow coastal waters.

The East China Sea (ECS) is considered to be the second-largest M_2 internal tidal generating site among global shelf regions, owing to the strong tidal currents that are perpendicular to the steep continental slope [1]. In the ECS, strong internal tides are effectively generated by multiple sources, including the Ryukyu Island chain, Tokara Straits, and the continental shelf break. Furthermore, these waves can propagate both onshore and offshore [10]. Previous studies primarily focused on offshore propagating tidal beams, particularly their propagation path, their complicated multiple-source interference patterns in the deep waters of the western Pacific Ocean [10–12], and the energetics and variability modulated by the Kuroshio at the mouth of a canyon northeast of Taiwan [13–15]. Synthetic Aperture Radar (SAR) images illuminated the manifestation of internal solitary waves northeast of Taiwan [16–18], an indicator of shoreward internal tidal activities. Consequently, inshore propagating internal tides from these strong sources and their influence requires further investigation over the inner continental shelf.

Temporal variation is a common feature of internal tides, with multiple factors leading to a range of time scales from days to several years. Time-varying stratification, both seasonal and by the spring-neap cycle, can significantly affect the energetics and turbulent dissipation of internal tides [19]. Seasonal variability of semidiurnal and diurnal internal tides in the northern South China Sea (SCS) are subject to changes in the corresponding barotropic tide in the Luzon Strait [20–22]. Over a longer interannual period, ENSO events affect the intensity and modal structure of diurnal and semidiurnal internal tides through stratification changes [23,24]. Background currents and eddies strongly modify the generation, propagation, and evolution of internal tides [25,26], by regulating the energy source [13], altering the phase speed of refraction [27,28], scattering to higher modes [29,30], adding relative vorticity to the system [31,32], and dephasing the internal tides, so they become nonstationary [33–35]. The ECS shelf features multi-scale subtidal frequency processes: the monthly changing Kuroshio front [36] and its intrusion branch [37], the seasonal varying Taiwan Warm Current [38], and vigorous eddies [39,40]. The temporal characteristics of internal tides under these complex dynamical processes over the ECS shelf remain unknown.

The ECS generates both semidiurnal and diurnal internal tides [10,41]. Although they share a similar generation mechanism, they differ in frequency, wavelength, and propagation directions, both horizontal and vertical. Numerical simulation results suggest that diurnal and semidiurnal tidal beams originating from the Luzon Strait are quite different in propagation paths and interference patterns [11,28]. The generation and propagation of internal tides are strongly latitude dependent [42]. Internal tides tend to resonate in the critical latitude zones and cannot freely propagate poleward of their critical latitude [31]. For diurnal internal tides, the critical latitudes for the O_1 and K_1 tides are 27.6° and 30°, respectively. Recently, the role of low-frequency flows has been recognized in modulating and broadening the impact range of critical latitudes [32]. Particularly, the southern ECS is located in the latitude range affected by the diurnal critical latitudes. Until now, there is no literature describing the diurnal internal tides in the ECS shelf region. The ECS, with its wide continental shelf with a steep continental shelf break and a time-varying background current, is a complicated region for internal tidal dynamics. The behavior of diurnal and semidiurnal tides, their different characteristics, and the exact impacts of the critical latitudes on diurnal internal tidal variability remains unknown.

Therefore, mooring observations combined with altimeter data for the ECS shelf region were employed to explore the vertical structure and temporal variability of internal tides. The contributions of the relative vorticity and stratification induced by background flow in modifying the temporal variations of semidiurnal and diurnal internal tides were assessed.

The remainder of this paper is organized as follows: Section 2 describes the characteristics of the in situ velocities and satellite data, the background barotropic tides, and the methods used in this paper. Section 3 provides temporal variability of semidiurnal and diurnal internal tides. The source region and travel times of internal tides, the roles of stratification, and the relative vorticity associated with background conditions are discussed in Section 4. Section 5 summarizes the results and conclusions.

2. Materials and Methods

2.1. Mooring Observations

From 29 May 2014 to 2 September 2014, a mooring was deployed on the western ECS shelf. It was located at 26°35′N, 121°09′E, at a water depth of 72 m (Figure 1a). A nearly three-month-long time series was obtained from an upward-looking 300 kHz Acoustic Doppler Current Profiler (ADCP) mounted on the sea bottom. The vertical sampling interval was set to 2 m, with a time interval of 30 min. To eliminate near-surface contamination from noise reflection, the upper two layers of data were removed. Additionally, occasional isolated spikes were eliminated and replaced through interpolation. Thus, the depth ranges for the available current data ranged from 8 m to 68 m, with a precision of ±0.5 cm/s. Since nearly the entire water column was sampled, the barotropic current was defined as the depth-averaged flow. The residual baroclinic anomalies were determined by subtracting the barotropic velocities from the total velocities. Using a first-order Butterworth filter, these baroclinic anomalies were divided into 7-day low-pass sections, diurnal internal tides (frequency bound: (0.9, 1.1) K_1), semidiurnal internal tides (frequency bound: (0.9, 1.1) M_2). Tidal harmonic analysis was conducted on the velocity records to obtain the phase-locked internal tides using the UTide toolbox [43]. We also employed an empirical orthogonal function (EOF) method to characterize the detailed modal structure of the baroclinic signals. Although this method is based on data statistics, it can demonstrate a reasonable modal structure of baroclinic tides, and it is widely used in internal wave analysis [44–46]. Considering the variable stratifications and currents, we used 14-day moving overlapped EOFs to analyze the modal content.

2.2. Satellite Altimetric Data

To estimate the influence of background currents and sea level height variations on internal tides, gridded sea level anomaly (SLA) and geostrophic velocities were obtained from Archiving, Validation, and Interpretation of Satellite Oceanographic Data (AVISO). The delayed-time data are all satellite mission merged, produced by Ssalto/Duacs, and distributed by the Copernicus Marine Environment Monitoring Service (CMEMS). They range from 119°E to 124°E, 24°N to 29°N with a 1/4° by 1/4° spatial resolution and daily temporal resolution simultaneous with the mooring observation period.

2.3. Body Force Calculation

The generation sites of the internal tides can be further examined by calculating the barotropic tidal force. This has been widely used in prior studies to identify possible internal tidal generation hot spots [10,47,48]. The formula used here follows [48] according to [1]. The depth-integrated body force F is calculated as:

$$F = -\frac{\vec{Q}\nabla H}{\omega H^2}\int_{-H}^{0} zN^2(z)dz \tag{1}$$

where ω is the tidal angular frequency (rad s^{-1}), z is the vertical coordinate (z = 0 at sea surface, upward positive), $N(z)$ is the local buoyancy frequency calculated by the World Ocean Atlas (WOA18) salinity and temperature, Q is the barotropic tidal volume transport extracted from the Oregon State University (OSU) TOPEX/Poseidon global tidal model (TPXO 7.2) [49]. H presents the local water depth, and ∇H is the bottom slope. The bathymetry used here is from the Smith and Sandwell database at a spatial sampling interval of 1 arc-minute [42]. This widely-used database is derived from satellite and ship depth soundings [50].

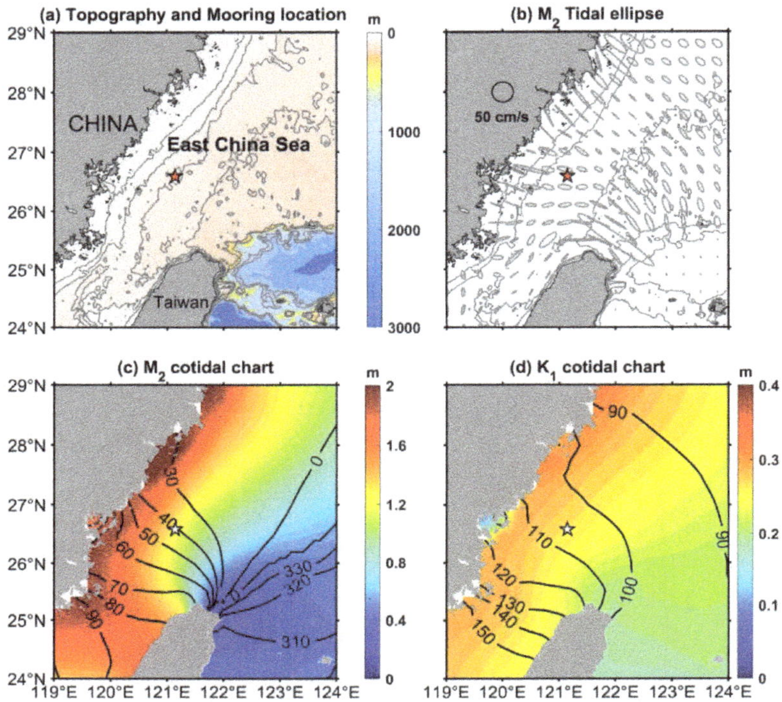

Figure 1. (**a**) Bathymetry for the East China Sea Shelf (color), with the mooring location (red star). The solid gray lines indicate the isobathic contours of 30, 50, 70, 100, 200, 800, and 1000 m (Smith and Sandwell 1/60° by 1/60°). (**b**) M_2 tidal ellipse obtained from TPXO 7.2. Tidal ellipses (with major axes scaled to unit) are shown in gray at a grid spacing of 0.3° by 0.3°. The contours are the isobaths of 30, 50, 100, 200, and 1000 m. (**c**) M_2 cotidal chart based on TPXO 7.2. The contours indicate the co-phase lines with interval of 10°, except at 330°, 0°, and 30° with interval of 30°. (**d**) K_1 cotidal chart based on TPXO 7.2.

2.4. Critical Latitude and Effective Latitude

According to linear wave theory, the frequency of internal tides must be between the local Coriolis frequency and the buoyancy frequency ($f \le \omega \le N$). The critical latitude is generally defined as the latitude where the tidal frequency equals the local inertial frequency. According to linear internal wave theory, internal tides are trapped and cannot freely propagate poleward of their critical latitude. This effect occurs in polar regions for semidiurnal internal tides (M_2: 74.45° and S_2: 85.7°) and in temperate regions for diurnal tides (K_1: 30° and O_1: 27.6°) in both the northern and southern hemispheres. However, the added relative vorticity from background currents can shift the effective region of a critical latitude, potentially up to several degrees [32,51–53]. Consequently, it is insufficiently rigorous to only take Coriolis frequency, f, into account. This should be the combination of the planetary vorticity (f) and relative vorticity ($\zeta = \frac{\partial u}{\partial y} - \frac{\partial v}{\partial x}$) induced by the background flow. This complicates the concept dramatically. The Coriolis frequency, f, is essentially modified as an effective Coriolis frequency ($f_{eff} = f + \zeta/2$) when describing motions on a reference frame rotating with the Earth [54]. Applying a traditional approximation [55], Coriolis frequency can be defined as $f = 2\Omega sin\theta$, where Ω denotes the Earth's angular velocity (7.29×10^{-5} rad s^{-1}) and θ is latitude. Similarly, a one-to-one correspondence between the effective Coriolis (f_{eff}) and the effective latitude becomes $f_{eff} = 2\Omega sin\theta_{eff}$, which is defined θ_{eff} as effective latitude. We introduce effective latitude for straightforward comparison with the K_1 and O_1 critical latitudes here.

2.5. Barotropic Tides

The barotropic tides in the ECS are subject to the northwestern Pacific tides. Barotropic tides in the northwestern Pacific propagate northwestward, passing through the Ryukyu Island chain and Okinawa Trough, and then reaching the ECS continental shelf. Affected by the Coriolis force, topography, and the coastline, the tides bifurcate, rotate, and shape several cyclonic amphidromic systems in the SCS and ECS. One of these amphidromic systems occupies our survey region. Both semidiurnal and diurnal tides enter the ECS and rotate counterclockwise around the northeast corner of Taiwan, finally entering the Taiwan Strait (Figure 1c,d). The four major tidal constituents dominate, with two main semidiurnal constituents M_2 and S_2, and two diurnal constituents O_1 and K_1, respectively (Table 1). As the tides shoal and narrow over the wide ECS inner continental shelf, stronger tidal flows and higher tidal elevations develop. The tidal ellipses show spatial variation, with stronger tidal currents northeast of Taiwan (Figure 1b). Generally, the major axis of M_2 and K_1 are aligned in the cross-isobaths direction. When the tidal flow encounters the featured topography, such as the northern continental shelf or Mien-Hua Canyon, the tidal ellipses alter to rectilinear. This reversing tidal flow here with the large topography gradient in shelf-slope favors the generation of internal tides [14]. According to the cotidal chart (Figure 1c,d), the barotropic tidal phase lines near our measurement locations are perpendicular to the slope, indicating that the tidal phase and tidal cycles are nearly the same between the mooring and potential generation sites. The detailed generation sites are estimated in Section 3.3.

Table 1. Ellipse properties of the major semidiurnal and diurnal barotropic tidal constituents at the mooring location.

Constituent	Major, cm/s	Minor, cm/s	Inclination, Deg	Phase [1], Deg
M_2	55.0	18.9	176.5	336
S_2	16.8	7.5	169.5	6.3
K_1	5.05	2.28	170	34
O_1	4.8	0.72	11	17

[1] The 'Phase' indicated in table is Greenwich Phase.

3. Results

3.1. Spectral Characteristics and Tidal Ellipses

Power spectra were used to explore the frequency distribution of the baroclinic energy. There were several significant peaks at tidal frequencies (K_1 at 23.93 h, M_2 at 12.42 h, and S_2 at 12 h), while the power density was less prominent for O_1 (25.82 h) or the inertial frequency (26.74 h). The baroclinic energy was concentrated at the semidiurnal and diurnal tidal bands (Figure 2a), particularly the semidiurnal bands with spectral peaks nearly five times larger than the diurnal peaks. For the semidiurnal constituents, M_2 was much larger than S_2 according to the spectral peaks and internal tidal ellipses; however, they had similar vertical structures (Figure 2b). These results were consistent with the previous numerical simulations and altimeter estimates and underscore the dominance of M_2 and K_1 internal tides in the western Pacific margin [4,8]. Consequently, we choose M_2 and K_1 as the two major constituents to represent the semidiurnal and diurnal internal tidal generation, respectively.

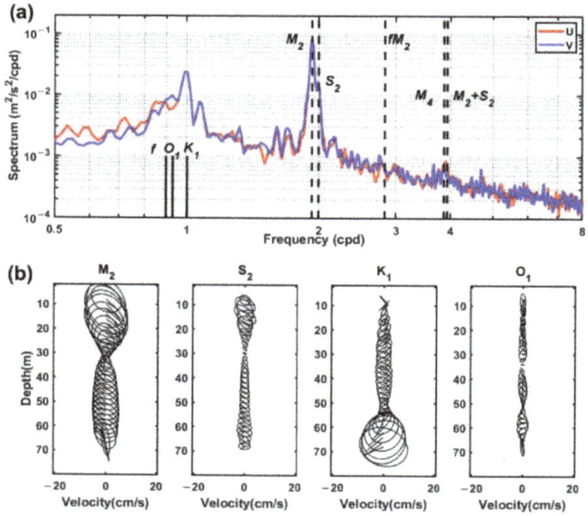

Figure 2. (**a**) Full-depth averaged power spectra of the baroclinic zonal (red line) and meridional (blue line) velocities during the deployment period. The vertical dashed lines indicate the local Coriolis frequency f, four major tidal constituents (O_1, K_1, M_2, S_2), and some nonlinear couplings (fM_2, M_4, $M_2 + S_2$). (**b**) The internal tidal current ellipses for the four major tidal constituents (M_2, S_2, K_1, and O_1) at different depths.

The semidiurnal and diurnal internal tides had different vertical structures (Figure 2b). To investigate the variance and amplitude of the four major internal tidal constituents, harmonic analysis was performed on each layer's baroclinic anomalies over the entire period (Figure 2b). It should be noted that harmonic analysis only manifests the coherent part (phase-locked to astronomical tide) of the internal tide. There was a remarkable difference, as M_2 intensified near the surface, whereas K_1 intensified directly above the bottom. Compared to the M_2 baroclinic currents, the K_1 coherent internal tide was of comparable magnitude (approximately $\pm$ 10 cm/s) at their intensified depths. This was quite different from the results that the observed diurnal internal tides were very weak over the continental slope [14,56]. With increasing depth, the M_2 amplitude first weakened to a minimum, and the inclination of the ellipse oscillated nearly 180° out of phase near two layers at 35 m. The enhanced K_1 changed little in amplitude at the surface and mid-water column, but was dramatically enhanced below the oscillating layer at 55 m. The inclinations of tidal ellipses significantly changed and exhibited the clockwise rotation common in the Northern Hemisphere [20,21]. The number of oscillating layers indicated the modal structure of the internal tide [20,45]. The singular oscillating layer for M_2, K_1, and S_2 suggested the dominance of mode-1, while O_1 showed multimodal structure as indicated by several reversing layers.

3.2. Distinct Variability of Semidiurnal and Diurnal Internal Tides

The temporal variations of semidiurnal baroclinic velocities can be seen in Figure 3. Overall, semidiurnal internal tides were quite weak before about 17 July. They appeared near the bottom for the first 14 days and then became calm after more than one month. The semidiurnal baroclinic velocities were dramatically enhanced to 30–40 cm/s from 17 July (Figure 3a–c). This velocity magnitude enhancement occurred throughout the entire water column, and in both the zonal and meridional directions. This indicates a northwest-southeast current, consistent with the major axes of the M_2 and S_2 tidal ellipses (Figure 2b). The tidal ellipse characteristics at the mooring disagreed with the local tidal ellipse (east-west direction) but agreed with the tidal ellipses on the corrugated continental slope and upper continental shelf (Figures 1b and 2b). During the enhancement period, semidiurnal internal tides followed a spring-neap tide cycle, roughly phase-locked to the local M_2-S_2 spring-neap cycle. Three spring peaks were observed in the record (Figure 3c). The phase difference to the local barotropic (M_2 + S_2) tidal cycles was less than 2 days. These results indicated the semidiurnal internal tides may not be locally generated near the mooring, but possibly nearby on the shelf.

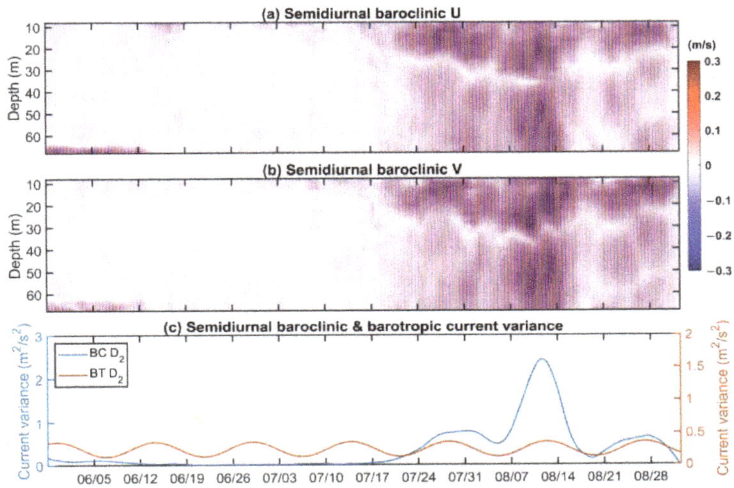

Figure 3. Time series of semidiurnal band baroclinic velocities (color, unit: m/s) derived from mooring observations from 29 May 2014 to 2 September 2014: (**a**) zonal component, (**b**) meridional component. (**c**) The vertically integrated current variance of baroclinic (blue) and local barotropic (M_2 + S_2) (red) tides predicted by TPXO 7.2.

Unlike the dramatic change of the semidiurnal internal tides before and after 17 July, the diurnal baroclinic velocity changed less and exhibited quasi-spring-neap cycles during the full record (Figure 4). Both the timing and magnitude varied for each spring-neap cycle (Figure 4c). A majority of the internal tidal peaks lagged the barotropic tide by 5–7 days at spring tides. The timing of the second spring tide was delayed by nearly 10 days and persisted longer than the others. Compared with the strong semidiurnal internal tide, the diurnal tide was weaker, both in depth-integrated magnitude and maximum baroclinic speed.

The semidiurnal current field was divided into two vertical sections: subsurface and lower middle. A zero-crossing point fluctuated over time between 20 and 36 m (Figure 3a,b). Near-bottom intensification featured prominently in the diurnal field, with the zero-crossing point ranging between 52 and 60 m. In previous observational studies, the reversal depth was usually at the thermocline, and the velocity structure was associated with time-varying stratification [19]. As the energy source, the barotropic tides were stable during the mooring period (Figures 3c and 4c).

Stratification is one of the key factors that influence internal tide generation and radiation [57]. To further demonstrate the temporal and vertical internal tidal structure, the buoyancy frequencies were calculated from monthly mean T (temperature) and S (salinity) extracted from WOA18 data (Figure 5). In Figure 5b, the stratification in the slope region was quite weak in June but enhanced and formed pycnocline layers at 70 m in July and August. In contrast, the local stratification at the mooring featured different trends (Figure 5a). At depths of 10 m and 40 m, the double N^2 peaks indicated the positions of two pycnocline layers. The 10 m and 40 m stratification weakened during the mooring period, while the benthic stratification increased. Different from the extremely weak abyssal stratification in the deep ocean, the benthic stratification over the shelf was comparable to that in the upper pycnocline. From Equation (1), the larger z near the seafloor resulted in a larger influence of the benthic N^2 in generation. This suggests the varying stratification was potentially associated with the strength of internal tides, especially where the benthic stratification changed in the shelf region.

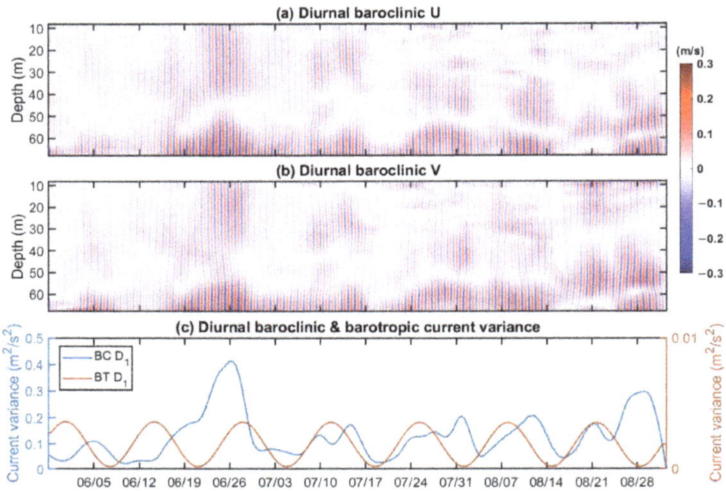

Figure 4. Same as in Figure 3 but for the diurnal baroclinic current and $K_1 + O_1$ barotropic tides.

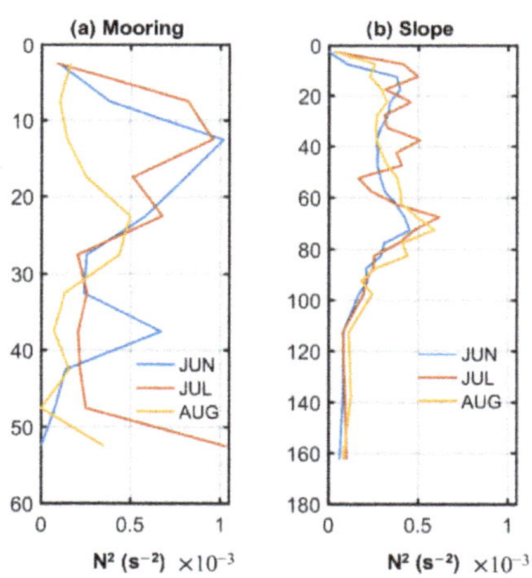

Figure 5. Comparison of monthly buoyancy frequency profiles from mooring (**a**) and slope (**b**).

3.3. Internal Tide Generation Sites

Potential internal tidal generation sites can be determined from the spatial pattern of the barotropic tidal body force. All of the high body forces were distributed near the canyon and ridges northeast of Taiwan, where the topography changes abruptly (Figure 6). Previous studies have found that the nearby continental slope and submarine canyons were major generation sites, and a portion of M_2 internal tidal energy may spread to the continental shelf over a distance of one or two wavelengths [10,13]. Many discrete medium and weak M_2 internal tidal energy sources were also found in the shelf region (Figure 6a). Other numerical results suggested the shelf as an M_2 internal tidal energy source [41]. Considering body forces near the observation point are very small, they are unlikely to generate noticeable internal tides; therefore, the internal tide at the mooring location most likely radiated from northeast of Taiwan (Figure 6 black box). In our area of concern, the semidiurnal tidal body force was obviously larger than that of diurnal tides, and the integrated body force of semidiurnal tides was more than four times that of diurnal tides (Figure 6 and Table 2). The proportion of the shelf region was around 20% of the total for both M_2 and K_1 internal tides. The K_1 slope generation is nearly equivalent to M_2 generation over the shelf. Meanwhile, diurnal baroclinic velocity was also comparable to its semidiurnal counterpart in terms of the order of magnitude. The association between the shelf-slope generation and distinct internal tides features will be further explored in the following sections.

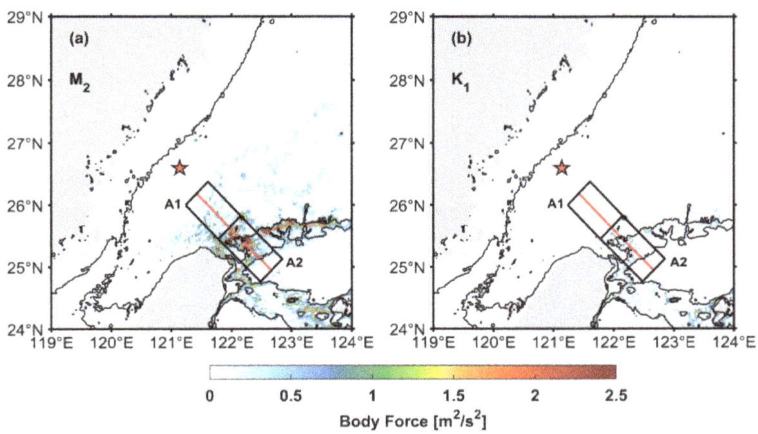

Figure 6. Body forces for (**a**) semidiurnal (M_2) and (**b**) diurnal (K_1) internal tides. The black solid lines indicate the 50, 200, and 1000 m isobaths. The red star indicates the mooring location. The red solid line indicates the possible propagation path of a remotely generated internal tide. The black box indicates the area of integration in Table 2. A1 (the upper box) represents the shelf generation region, A2 (the lower box) represents the slope generation region.

Table 2. Integration of M_2 and K_1 body forces in the generation domain (black box shown in Figure 5) for each month.

	June	July		August	
	Value (m^2/s^2)	Value (m^2/s^2)	Increasing Rate [1] (%)	Value (m^2/s^2)	Increasing Rate [2] (%)
M_2 all	4462	5281	18.35	5486	22.95
M_2 in A1	893	1117	25.22	1230	51.12
M_2 in A2	3569	4164	16.67	4256	19.25
K_1 all	1104	1297	17.50	1348	22.09
K_1 in A1	213	266	24.88	293	37.56
K_1 in A2	891	1031	15.71	1055	18.41

[1,2] Increasing rates indicate July-on-June ratio, August-on-June ratio, respectively.

The observations indicated considerably more semidiurnal internal tide energy at the moorings later in the record after 17 July. To investigate this change, we applied monthly means of WOA18 stratifications to the body force calculations. Background stratification can affect internal tidal generation, leading to temporal variability in the baroclinic tidal energy, which will also be reflected in the body force (Table 2). Stratification both at the mooring (Figure 5) and over the slope northeaster of Taiwan changed significantly. Generally, the stratification was stronger during both July and August compared to June, so the body forces in July and August also increased, with the largest increase exceeding 20% in August. However, the changing stratification trends for the shelf and slope were not exactly similar. We further investigated the body force in the shelf (A1 in Figure 6) and slope regions (A2 in Figure 6). The M_2 body force in A1 (shelf region) had greater variability than that in A2 (slope region). The M_2 internal tide generation in the shelf region increased ~25% in July over June and ~51% in August over June. The increasing rates in the slope region were all below 20%. Therefore, the varying stratification contributed significantly to the variability of semidiurnal internal tide generation, especially the shelf generation. The occurrence of strong semidiurnal internal tides possibly resulted from the preferable stratification conditions for generation over the shelf region.

3.4. Modal Contribution Modulated by Background Current

Through moving overlapped EOF analysis, the depth-integrated velocity variance for the lowest three modes of the semidiurnal and diurnal internal tides can be seen in the stacked histograms (Figure 7e,g). Modes higher than three accounted for less than 3% energy in total out of 8 modes (not shown). Semidiurnal internal tides at the early stages were not analyzed due to their low energies. For both the semidiurnal and diurnal internal tides, the first mode dominated and provided over 85% of the energy most of the time. The second and third modes made minor contributions. These results are consistent with the former analysis of tidal current ellipses and baroclinic velocity fields. However, there was still a strong temporal dependence in the energy proportions (Figure 7f,h). Vertical modes of internal tides can be modified by background current shear [30]. Our results show that low-frequency shear was not strong in the early stage, and the proportion of first mode diurnal internal tides was relatively stable, accounting for ~90%. After 17 July, the intensified shear was accompanied by intermittent increases of mode-2 proportions for both semidiurnal and diurnal internal tidal constituents. Especially around 14 and 28 August, low-frequency shear was significantly enhanced in the shallow layer, and the proportion of the mode-2 and 3 energy exceeded ~30% for semidiurnal and diurnal internal tides. It is likely that the energy of the mode-2 and 3 internal tides was transferred through the mode-1 internal tide because the total energy remained relatively constant and still followed the quasi spring-neap variability. Furthermore, the high mode increase occurred during some spring tides and during times of weaker internal tides.

Based on observations and numerical simulations, previous studies found that eddy-wave interactions could result in energy transfers from the mode-1 internal tides to higher modes [27,29]. Eddies frequently occur in this region according to AVISO data, so eddy-wave interactions are possible. During the three eddy periods (Figure 7a–c), the mooring was located on the edge of a strong lateral shear area, and the mode-2 and mode-3 proportions increased.

3.5. Diurnal Tidal Critical Latitude Effect

Where and when diurnal critical latitude effects occur may shift through the addition of positive or negative relative vorticity associated with varying circulation conditions. Relative vorticity exists in eddies, at the edges and meanders of western boundary currents, and in the flow along the continental slope. Observation and simulation results indicated a shift of the diurnal critical latitude(s), potentially of up to several degrees, when it encounters a mesoscale current [32]. Mesoscale motions are energetic enough to evoke strong vorticity for constructing a complicated field of effective latitude (Figure 8a–c). The Kuroshio front and its intrusion bifurcation beam, the Taiwan Warm Current (TWC), together shifted the effective latitude in the area northeast of Taiwan Island [58]. The circulation pattern in summer created a prominent positive shift of effective latitude. The vorticity was sufficient to shift the effective latitude of the shelf slope area from geographical 25°N–26°N to beyond 27.6° and even 30°. This made an area geographically

2–5° equatorward of the diurnal critical latitude, also under the influence of the diurnal critical latitude(s). This area covered a large part of the diurnal internal tidal generation sites (Figure 6). In this context, O_1 and K_1 critical latitude effects influenced the generation and propagation of diurnal internal tides by various degrees.

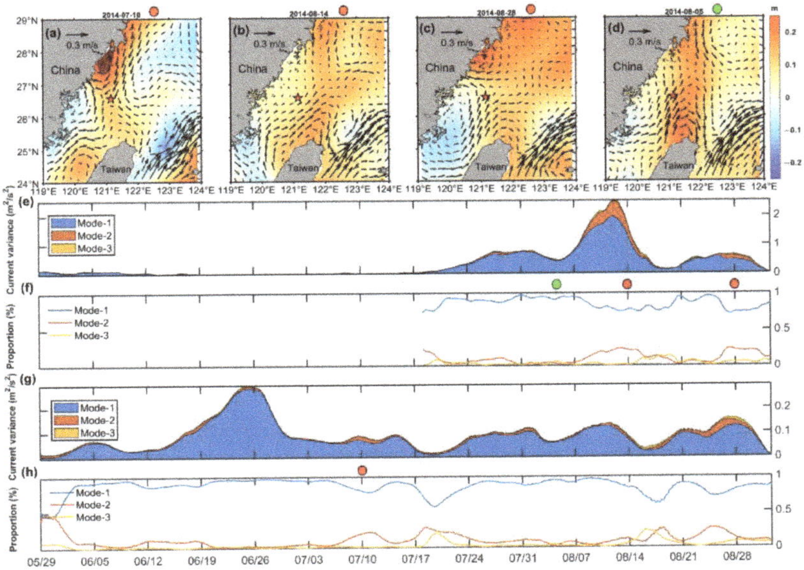

Figure 7. SLA and geostrophic currents during (**a**–**c**) three eddy periods and (**d**) a calm time. Time series of depth-integrated (**e**) semidiurnal and (**g**) diurnal baroclinic current variance in first three modes (stacked colors) by 14-day moving overlapped EOFs, and their respective proportions of the first eight EOF modes (**f**,**h**).

There was a good correspondence between the temporal variations of the diurnal internal tides and the effective latitude near the generation sites. To further evaluate the role of the diurnal critical latitude in modulating internal tidal propagation, the daily shifts of the effective latitude on the propagation path (Figure 8d) were connected with the time series of diurnal internal tides (Figure 8e). Critical latitude effects can cause remarkable changes in the intensity of remote diurnal internal tides. When the cyclonic eddy was weak during two specific periods (12–16 June and 21–28 August), the effective latitude deviated less from the local latitude. Unlimited by the critical latitude restrictions, all along-slope-generated internal tides freely propagated, resulting in a larger observed magnitude and a wider spring-tide peak (Figure 8e).

At other times, the effective latitude value frequently exceeded 27.6° and 30° (red areas in Figure 8d). The red areas correspond to the lower part of the red line in Figures 6 and 8a–c, which were both major generation sites and beginning points of propagation for diurnal internal tides. Consequently, the critical latitude effects actually work in this area during these times. Considering that the effective area of critical latitudes and the generation sites in the shelf-slope region heavily overlapped, it was unlikely to cover all source regions (Figures 6 and 8). The effective area of O_1 critical latitude was larger than K_1 (Figure 8a–d). Therefore, the portion of K_1 generated beyond the effective area of K_1 critical latitude could freely propagate, while most of the O_1 internal tides were expected to be trapped. Observed diurnal internal tides were not so strong at that time. Spring-neap cycles were also modified by the effects on these two tidal constituents, O_1 and K_1, as the superposition of O_1 and K_1 dominated the spring-neap cycle.

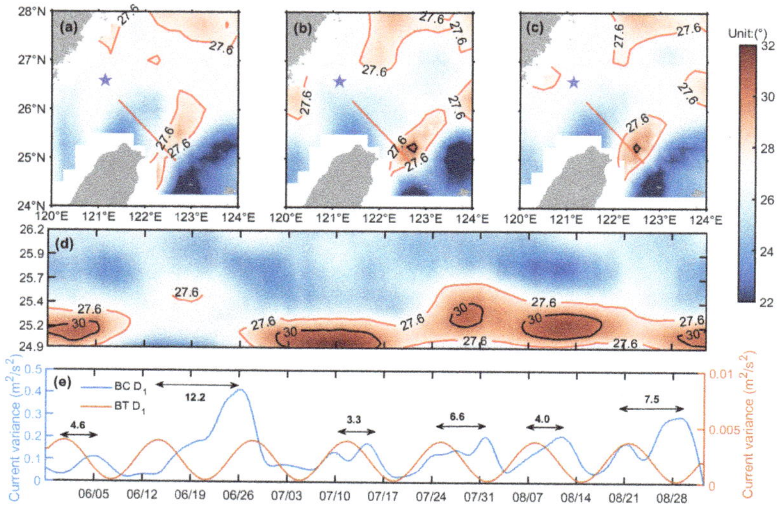

Figure 8. Monthly-mean effective latitude (θ_{eff}) maps for June (**a**), July (**b**), and August (**c**) were calculated based on the effective Coriolis frequency ($f_{eff} = f + \zeta/2$) and the formula ($f_{eff} = 2\Omega sin\theta_{eff}$). (**d**) Hovmöller Diagrams of effective latitude along the propagation path of diurnal internal tide (red line shown in Figures 6 and 8a–c. The contours indicate the effective latitude. *X*-axis represents time, and *Y*-axis the geographical latitude. The red and black lines represent the critical latitudes for the O_1 (27.6°) and K_1 (30°) internal tides, respectively. (**e**) The time series of diurnal internal tide. The vertically integrated current variance of diurnal baroclinic (blue) and barotropic (K_1 + O_1) (red) tides are the same as in Figure 4c. Time lag (unit: days) of each baroclinic peak to the barotropic peak are labeled and remarked with black double arrows.

3.6. High-Frequency Internal Wave and Energy Cascades

Wavelet analysis was applied to the time series of the baroclinic velocity anomalies at 60 m depth, and the baroclinic energy distribution in frequency and time is presented in Figure 9a. From the magnitudes in the scalogram, baroclinic energy was concentrated in the diurnal (1 cpd) and semidiurnal (2 cpd) bands. The wavelet results also showed vigorous energy and variability at these tidal bands, consistent with the baroclinic velocity record and the mode histograms. The periods of energetic semidiurnal and diurnal internal tides revealed in Figure 9a were consistent with former results.

Apart from internal tides, intermittent high-frequency internal waves occurred and responded significantly to the low-frequency vertical velocity shear. High-frequency internal wave (>4 cpd) pulses occurred on 10, 17, 24, and 31 July and were enhanced from 7 to 21 August (Figure 9a). The occurrences of these high-frequency signals coincided with the strong vertical shear periods (Figure 9). These periods were observed to be deep-reaching and could potentially catalyze nonlinear wave–wave interactions at 60 m. Moreover, when both the semidiurnal and diurnal internal tides were energetic during these periods, the high-frequency waves were also more powerful and continuous in the frequency domain from 2 cpd to above 10 cpd. This indicates that when low-frequency shear flows encountered the strong internal tides, they drew internal tidal energy to cascade from low to higher frequencies.

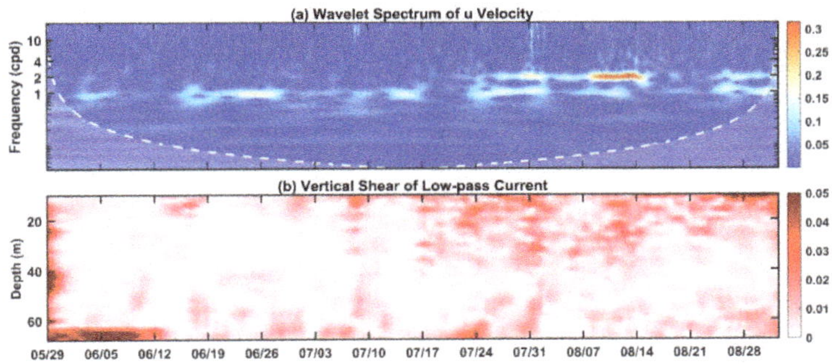

Figure 9. (**a**) Wavelet power spectra of baroclinic zonal component during observation period at 60 m depth. The white dashed line indicates the cone of influence. The white dashed lines and the shaded regions below are suspect and potentially influenced by edge effects. The frequency unit is cycles per day (cpd), and the spectrum uses L1 normalization to show a more accurate representation of the signal. (**b**) The vertical profile of 7-day low-pass vertical shear variance (unit: $1/s^2$).

4. Discussion

Both semidiurnal and diurnal internal tides have been observed over the ECS continental shelf. Although semidiurnal internal tides dominated, the diurnal signal was also significant, particularly in benthic layers. The phases of the diurnal baroclinic velocity anomalies were obviously associated with the tidal cycles, apparently indicating generation through a barotropic flow-topography interaction. The results were quite different from previous slope observations. They suggested either the diurnal internal tides were an order of magnitude weaker with a secondary influence [56] or associated diurnal-band internal waves with a parametric subharmonic instability (PSI) mechanism that transferred energy from semidiurnal internal tides to half their frequency, also diurnal band, but mainly regarded as near-inertial wave [15].

Similar mode-1 structures of two semidiurnal internal tidal constituents, M_2 and S_2, reinforced the dominance of mode-1 and the magnitude of semidiurnal internal tides. Multimodal O_1 tides were found in our results and other ECS slope studies [56]. In the SCS, another marginal sea of the western Pacific, O_1, usually exhibited a mode-1 structure [21]. Compared with SCS, O_1 is closer to the local Coriolis frequency in the ECS, which is regarded as the lower limit for the O_1 internal tide. However, the background flow induced sufficient positive relative vorticity to the Coriolis frequency (Figure 8d) for O_1 to be sub-inertial or inertial. Internal tides will not propagate freely in proximity to their critical latitude, and their vertical wave number tends to be infinite, resulting in a multimodal O_1 structure [32].

Although the generation sites were indicated by the body force distribution, specific source regions require further investigation. It is evident that spot bands of semidiurnal internal tide generation existed over both the shelf and slope, while the diurnal internal tide was only generated in a narrow slope area (Figure 6). The time lags between semidiurnal internal tides and semidiurnal barotropic velocities (~1–2 days) were smaller than those between the diurnal internal tides and the diurnal barotropic velocities (~3–12 days) (Figures 3c and 8e). To estimate the energy source(s) and evaluate the propagation of internal tides, the eigenspeed (C_n) of normal modes (n) can be determined by solving the Taylor–Goldstein equation with zero background flow [59]:

$$\frac{d^2\Phi(z)}{dz^2} + \frac{N^2(z)}{c_n^2} = 0, \tag{2}$$

where $\Phi(z)$ is the vertical displacement in baroclinic modes, which is subjected to buoyancy frequency profile $N^2(z)$. The Eigenvalue equation can be numerically solved [60]. We focus on the mode-1 internal tides. Due to the Earth's rotation, the phase speed C_p can be calculated from eigenspeed C as follows [33]:

$$C_p = \frac{\omega}{\sqrt{\omega^2 - f^2}} C, \tag{3}$$

where ω is the tidal frequency, f is the Coriolis frequency. Equation (3) When we replace the f in Equation (3) with f_{eff}, we can get phase velocity with the background currents. The effect of background current was taken into consideration by modified effective Coriolis frequency:

$$C_{peff} = \frac{\omega}{\sqrt{\omega^2 - f_{eff}^2}} C, \tag{4}$$

The spring-neap cycle and energy of the semidiurnal internal tide propagate at the group velocity [61]. The group velocity C_g and phase velocity C_p can be calculated as follows:

$$C_p = \frac{\omega}{k} \tag{5}$$

$$C_g = \frac{d\omega}{dk} = \frac{\omega_1 - \omega_2}{k_1 - k_2} \tag{6}$$

On the basis of the phase velocity, the wave number k of two participating tidal constituents can be obtained. The group velocity can be calculated as the frequency and wave number differences between the tidal constituents (M_2–S_2 in semidiurnal and K_1–O_1 in diurnal). Similarly, the effect of background current can be taken into account as:

$$C_{peff} = \frac{\omega}{k_{eff}} \tag{7}$$

$$C_{geff} = \frac{d\omega}{dk_{eff}} = \frac{\omega_1 - \omega_2}{k_{eff1} - k_{eff2}} \tag{8}$$

The internal tidal travel times are calculated by cumulating the time at group velocity along the propagation path. Travel time for K_1 + O_1 from the source area is close to twice that of M_2 + S_2, that is, for the same distance, it takes longer for the diurnal spring-neap phase to travel compared to the semidiurnal spring-neap (Figure 10a). From near shelf to far slope, the internal tides propagated faster and faster. One reason is the latitude variation. In general, internal tides at higher latitudes propagate slower. However, such a sharp change is not indicated in the narrow latitude range in the former study [61]. Stratification may reinforce this change. The strongest stratification in August slows down the propagation of internal tides.

The role of stratification and background currents in modifying the travel time is also shown in Figure 10. Lines in different colors represented the effect of stratification for different months. Dashed lines represent the monthly mean horizontal flow. Compared with semidiurnal internal tides, diurnal internal tides are more susceptible to the influence of stratification and currents. The influence of these two factors was comparable to diurnal internal tides (~1 day). The stratification changes slightly influenced the semidiurnal travel times (several hours). While the dashed lines nearly overlapped with the solid lines, this indicated the relative vorticity of mean flow made a negligible change on semidiurnal propagation. Previous study found the different Kuroshio paths in Luzon Strait could alter the radiation pattern and magnitude of semidiurnal internal tides [62]. In present study, background current changed not so much in these several months, and less regulated the semidiurnal internal tides.

The travel times along the propagation path are presented in Figure 10b. The observed time lags to barotropic tidal cycles combined with travel times indicated the internal tides' energy source in the generation map. It can be expected that semidiurnal internal tides originate from the nearby inside shelf and diurnal internal tides far from the shelf-slope and even the slope region when the critical latitude effects are weak.

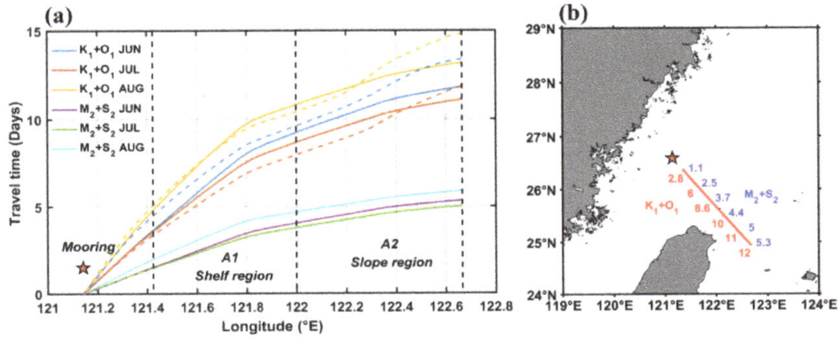

Figure 10. (**a**) Cumulative travel time of diurnal ($K_1 + O_1$) internal tides and semidiurnal ($M_2 + S_2$) internal tides along the propagation path of internal tide. The path is the same as the one shown in (Figures 6, 8, and 10b), except ahead of the starting point to mooring location. The travel times are estimated by group velocity of $K_1 + O_1$ and $M_2 + S_2$ waves using WOA 18 monthly-mean stratification. The dash lines indicate the travel times considering the monthly mean background current of the month with the same color as solid lines. The mooring location, A1 and A2 demarcation lines (same as Figure 6) are marked for reference. (**b**) The schematic of the travel times along the propagation path. Semidiurnal and diurnal internal tide travel times in August (Unit: days) are labeled in blue and red as reference, respectively.

Moreover, semidiurnal and diurnal internal tidal wavelengths were estimated to be 10–20 km (Figure 11) and 70–80 km (Figure 12), respectively. Bathymetry criticality along the propagation path is subcritical for both M_2 (Figures 11 and 13a) and K_1 (Figures 12 and 13b) internal tides. The topographic slope is not a limit for M_2 and K_1 generated on the slope shoaling on the shelf; however, due to the large bottom friction on the shelf, it is still highly unlikely that the semidiurnal internal tides from the slope area reach the mooring site after ~10 bottom-surface reflections, without any signs of damping [2,63]. Therefore, the observed diurnal internal tides should be remotely generated on the slope, while the semidiurnal internal tides were generated on the shelf near the observation site.

The background vorticity has a greater impact on the propagation speed of the K_1 internal tide. The tidal frequency of the K_1 tide is closer to f, and by Equation (3), the diurnal phase velocity and group velocity are more sensitive to the change from f to f_{eff}. In addition, the propagation directions of diurnal internal tides were more horizontal than those of semidiurnal internal tides under the same stratification, so that diurnal internal tides were more sensitive to the horizontal circulation. As a result, the spring tide peak was flatter, the duration longer, and the time lags of each tidal cycle were more changed.

Internal tidal energy with respect to mode contents and frequency shift was modulated by the background currents. The decrease of mode-1 internal tidal proportion corresponded to the enhancement of horizontal shear associated with background circulation, indicating the energy scattering from low modes to higher modes. The existence of background low-frequency shear also coincided with strengthened high-frequency energy. Furthermore, the coexistence of semidiurnal and diurnal frequencies more easily facilitated the energy cascade to high-frequency bands (Figure 9). The occurrence of semidiurnal internal tides increased the energy source for the energy cascade and became a jumping board for more wave–wave nonlinear interactions in frequency space. The resultant high-mode and high-frequency internal waves promoted the transfer of tidal energy to turbulence scales, providing more energy for mixing.

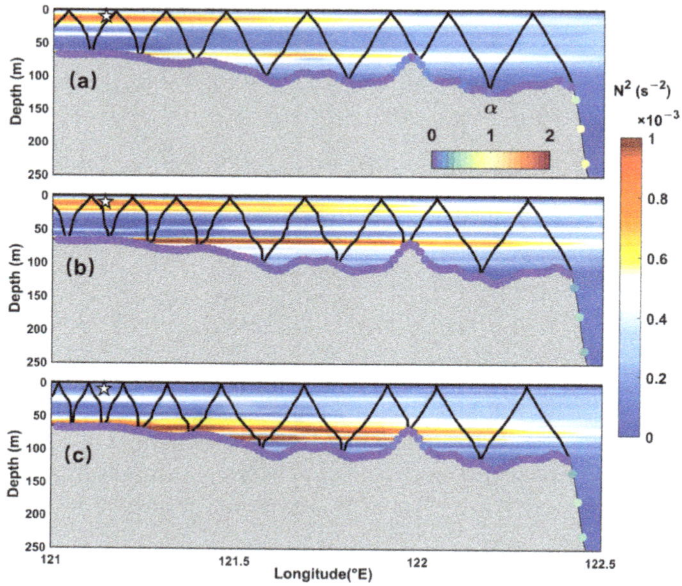

Figure 11. Calculation of the M_2 tidal beam reflections on the continental shelf from the slope area to mooring (white pentagram). The color indicates monthly climatology stratification derived from the WOA18 data. The black solid lines indicate the ray tracks of tidal beams, based on the formula ($\tan\theta = \sqrt{(\omega^2 - f^2)/(N^2 - \omega^2)}$). The color dots indicate the bathymetry criticality $\alpha = \frac{\nabla H}{\tan\theta}$. The ray tracks are presented in (**a**) June, (**b**) July and (**c**) August.

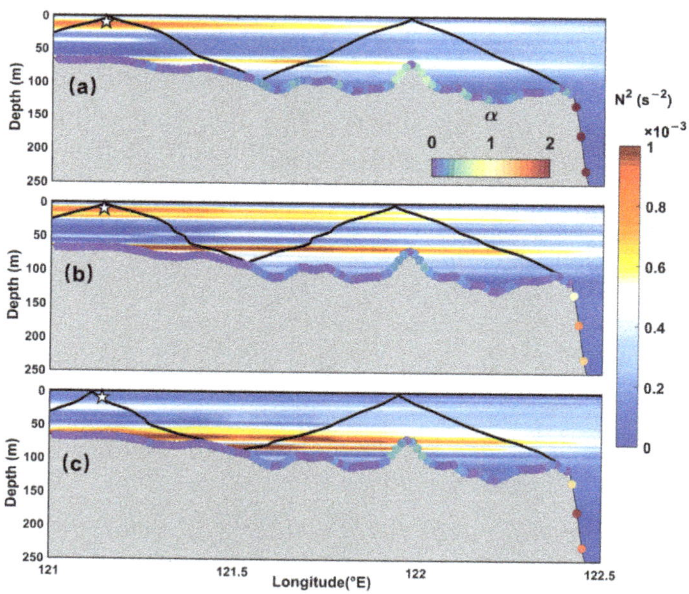

Figure 12. Same as in Figure 11, but for the K_1 tidal beam.

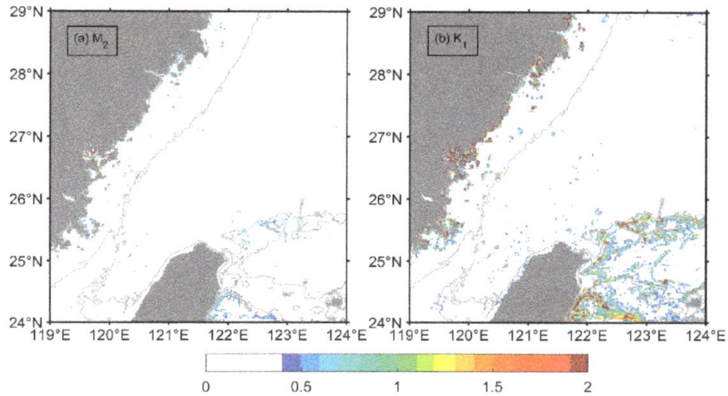

Figure 13. Bathymetry criticality (α) for (**a**) M_2 and (**b**) K_1 tidal constituents.

In addition, the relative vorticity of the ECS background currents was another contributor to the variation of internal tides. We found the mesoscale subtidal circulation near the critical latitude significantly regulated the diurnal internal tides. Previous research pointed out that critical latitude effects may modify the internal wave frequency continuum through nonlinear interactions [31,32]. The effective critical latitude described here is not only a geographic concept, but a geodynamic factor, quantitatively expressing the influence of background flow on shifting the effects of critical latitude. By comparing the effective latitude with critical latitude, we could intuitively estimate critical latitude effects on the internal tide energetics in a realistic ocean.

5. Conclusions

In this study, the temporal variations of internal tides on the ECS continental shelf and their controlling factors were investigated based on mooring observations. We highlighted the different temporal variability of semidiurnal internal tides and diurnal internal tides. Semidiurnal internal tides showed significant enhancement over the last three spring-neap cycles between early summer and late summer. The diurnal internal tidal amplitudes were relatively stable during the mooring period. M_2 and K_1 internal tides were the largest constituents, and both were mode-1dominated. The M_2 internal tidal amplitude intensified near the surface while that of K_1 intensified above the bottom.

The generation sources of semidiurnal (M_2) and diurnal (K_1) internal tides were revealed by distribution maps of body force. The shelf-slope region northeast of Taiwan was the major source region, as preceding research concluded. A majority of M_2 and K_1 internal tides were generated over the slope. M_2 internal tides were also locally generated inside the inner ECS shelf, which has been rarely mentioned before. Nevertheless, the local source sites are still away from the observation site. The distribution of generation sources, coupled with the time lag between baroclinic and barotropic tides, indicated that the observed internal tides propagated from a remote region. The stratification was stronger in July and August than in June, so the calculated body force changed similarly. In addition, changes in stratifications at the shelf explained a significant fraction of the variability in the semidiurnal internal tides.

The relative vorticity associated with background currents regulated the critical latitude effects by changing the size of the effective area and the time range in which they took effect. As a result, the propagation of the related diurnal internal tides varied with time and space. The increasing diurnal internal tidal magnitudes and the accordance of observed time lag and theoretical travel time on the propagation path consistently confirmed the process at the remote slope identified in observations. In addition, background currents played a vital role in tidal energy transfer, not only between the different vertical modes, but also between internal tides with high-frequency internal waves. The lateral shear resulted in high mode, and the vertical shear facilitated high-frequency internal waves and an energy cascade.

The upcoming Surface Water Ocean Topography (SWOT) mission will measure the sea surface with an unprecedented high spatial resolution and provide us with a wealth of data. However, we need to combine multiple methods with the SWOT mission in order to accurately investigate the energy cascade processes between tidal signals, mesoscale, and sub-mesoscale variability. Therefore, more specific in situ observations and numerical simulations are critical for further investigation.

Author Contributions: Conceptualization, W.W., Y.W. and Z.X.; methodology, W.W., Y.W. and R.R.; validation, Y.W., Z.H. and R.R.; formal analysis, W.W.; resources, B.Y.; writing—original draft preparation, W.W.; writing—review and editing, R.R., Y.W., Z.H. and Z.X.; visualization, W.W., C.Z.; supervision, B.Y. and Z.X.; project administration, B.Y. and Z.X.; funding acquisition, B.Y. and Z.X. All authors have read and agreed to the published version of the manuscript.

Funding: This study was jointly supported by the Strategic Pioneering Research Program of CAS, the National Natural Science Foundation of China (NSFC), the NSFC-Shandong Joint Fund, the National Key Research and Development Program of China (XDA22050202, 92058202, XDB42000000, 42076022, U1806227, 2017YFA0604102), the CAS Key Research Program of Frontier Sciences, the Key Deployment Project of Center for Ocean Mega-Research of Science, and the project jointly funded by the CAS and CSIRO (QYZDB-SSW-DQC024, COMS2020Q07, 133244KYSB20190031).

Data Availability Statement: The datasets for this study are publicly available. The altimetry sea level anomalies and geostrophic velocities are provided by Archiving, Validation, and Interpolation of Satellite Oceanographic (AVISO) data and distributed by the Copernicus Marine Environment Monitoring Service (CMEMS) via https://marine.copernicus.eu/ (accessed on 12 March 2022). The bathymetry data are available from https://topex.ucsd.edu/cgi-bin/get_data.cgi (accessed on 12 March 2022). The TPXO tidal atlas can be downloaded from https://www.tpxo.net/tpxo-products-and-registration (accessed on 12 March 2022). The World Ocean Atlas (WOA18) is available from NOAA National Centers for Environmental Information via https://accession.nodc.noaa.gov/NCEI-WOA18 (accessed on 12 March 2022). The mooring data presented in this study are available from the corresponding author upon reasonable request.

Acknowledgments: The acquisition of mooring data was supported by Western Pacific Ocean System: Structure, Dynamics, and Consequences, WPOS. We are grateful to the entire crew of R/V Kexue I for their expertise and hard work. We are grateful to Jifeng Qi and Fei Yu for data curation. Discussions with my fellows Xiaoyu Zhao and Jia You were very helpful.

Conflicts of Interest: The authors declare no conflict of interest. The funders had no role in the design of the study; in the collection, analyses, or interpretation of data; in the writing of the manuscript, or in the decision to publish the results.

References

1. Baines, P.G. On Internal Tide Generation Models. *Deep Sea Res. Part A Oceanogr. Res. Pap.* **1982**, *29*, 307–338. [CrossRef]
2. Kelly, S.M.; Jones, N.L.; Nash, J.D.; Waterhouse, A.F. The Geography of Semidiurnal Mode-1 Internal-Tide Energy Loss. *Geophys. Res. Lett.* **2013**, *40*, 4689–4693. [CrossRef]
3. Cacchione, D.A.; Pratson, L.F.; Ogston, A.S. The Shaping of Continental Slopes by Internal Tides. *Science* **2002**, *296*, 724–727. [CrossRef]
4. Müller, M. On the Space- and Time-Dependence of Barotropic-to-Baroclinic Tidal Energy Conversion. *Ocean Model.* **2013**, *72*, 242–252. [CrossRef]
5. Simmons, H.L.; Hallberg, R.W.; Arbic, B.K. Internal Wave Generation in a Global Baroclinic Tide Model. *Deep. Res. Part II Top. Stud. Oceanogr.* **2004**, *51*, 3043–3068. [CrossRef]
6. Zhao, Z.; Alford, M.H.; Girton, J.B.; Rainville, L.; Simmons, H.L. Global Observations of Open-Ocean Mode-1 M_2 Internal Tides. *J. Phys. Oceanogr.* **2016**, *46*, 1657–1684. [CrossRef]
7. Carrere, L.; Arbic, B.K.; Dushaw, B.; Egbert, G.; Erofeeva, S.; Lyard, F.; Ray, R.D.; Ubelmann, C.; Zaron, E.; Zhao, Z.; et al. Accuracy Assessment of Global Internal-Tide Models Using Satellite Altimetry. *Ocean Sci.* **2021**, *17*, 147–180. [CrossRef]
8. Zaron, E.D. Baroclinic Tidal Sea Level from Exact-Repeat Mission Altimetry. *J. Phys. Oceanogr.* **2019**, *49*, 193–210. [CrossRef]
9. Zhao, Z. Seasonal Mode-1 M_2 Internal Tides from Satellite Altimetry. *J. Phys. Oceanogr.* **2021**, *51*, 3015–3035. [CrossRef]

10. Niwa, Y.; Hibiya, T. Three-Dimensional Numerical Simulation of M_2 Internal Tides in the East China Sea. *J. Geophys. Res. Ocean.* **2004**, *109*, 1–14. [CrossRef]
11. Wang, Y.; Xu, Z.; Yin, B.; Hou, Y.; Chang, H. Long-Range Radiation and Interference Pattern of Multisource M_2 Internal Tides in the Philippine Sea. *J. Geophys. Res. Ocean.* **2018**, *123*, 5091–5112. [CrossRef]
12. Tian, J.; Zhou, L.; Zhang, X.; Liang, X.; Zheng, Q.; Zhao, W. Estimates of M_2 Internal Tide Energy Fluxes along the Margin of Northwestern Pacific Using TOPEX/POSEIDON Altimeter Data. *Geophys. Res. Lett.* **2003**, *30*, 4–7. [CrossRef]
13. Chang, H.; Xu, Z.; Yin, B.; Hou, Y.; Liu, Y.; Li, D.; Wang, Y.; Cao, S.; Liu, A.K. Generation and Propagation of M_2 Internal Tides Modulated by the Kuroshio Northeast of Taiwan. *J. Geophys. Res. Ocean.* **2019**, *124*, 2728–2749. [CrossRef]
14. Lien, R.C.; Sanford, T.B.; Jan, S.; Chang, M.H.; Ma, B.B. Internal Tides on the East China Sea Continental Slope. *J. Mar. Res.* **2013**, *71*, 151–186. [CrossRef]
15. Yang, W.; Wei, H.; Zhao, L. Parametric Subharmonic Instability of the Semidiurnal Internal Tides at the East China Sea Shelf Slope. *J. Phys. Oceanogr.* **2020**, *50*, 907–920. [CrossRef]
16. Hsu, M.K.; Liu, A.K.; Liu, C. A Study of Internal Waves in the China Seas and Yellow Sea Using SAR. *Cont. Shelf Res.* **2000**, *20*, 389–410. [CrossRef]
17. Zhao, Z.; Liu, B.; Li, X. Internal Solitary Waves in the China Seas Observed Using Satellite Remote-Sensing Techniques: A Review and Perspectives. *Int. J. Remote Sens.* **2014**, *35*, 3926–3946. [CrossRef]
18. Li, X.; Zhao, Z.; Han, Z.; Xu, L. Internal Solitary Waves in the East China Sea. *Acta Oceanol. Sin.* **2008**, *27*, 51–59.
19. Zhao, Z.; Alford, M.H.; Lien, R.C.; Gregg, M.C.; Carter, G.S. Internal Tides and Mixing in a Submarine Canyon with Time-Varying Stratification. *J. Phys. Oceanogr.* **2012**, *42*, 2121–2142. [CrossRef]
20. Xu, Z.; Yin, B.; Hou, Y.; Liu, A.K. Seasonal Variability and North-South Asymmetry of Internal Tides in the Deep Basin West of the Luzon Strait. *J. Mar. Syst.* **2014**, *134*, 101–112. [CrossRef]
21. Cao, A.; Guo, Z.; Lv, X.; Song, J.; Zhang, J. Coherent and Incoherent Features, Seasonal Behaviors and Spatial Variations of Internal Tides in the Northern South China Sea. *J. Mar. Syst.* **2017**, *172*, 75–83. [CrossRef]
22. Liu, J.; He, Y.; Wang, D.; Liu, T.; Cai, S. Observed Enhanced Internal Tides in Winter near the Luzon Strait. *J. Geophys. Res. Ocean.* **2015**, *120*, 6637–6652. [CrossRef]
23. Zhai, R.W.; Chen, G.Y.; Liang, C.R.; Shang, X.D.; Xie, J.S. The Influence of ENSO on the Structure of Internal Tides in the Xisha Area. *J. Geophys. Res. Ocean.* **2020**, *125*, e2019JC015405. [CrossRef]
24. Devlin, A.T.; Jay, D.A.; Talke, S.A.; Zaron, E. Can Tidal Perturbations Associated with Sea Level Variations in the Western Pacific Ocean Be Used to Understand Future Effects of Tidal Evolution? *Ocean Dyn.* **2014**, *64*, 1093–1120. [CrossRef]
25. Li, Q.; Wang, B.; Chen, X.; Chen, X.; Park, J.H. Variability of Nonlinear Internal Waves in the South China Sea Affected by the Kuroshio and Mesoscale Eddies. *J. Geophys. Res. Ocean.* **2016**, *121*, 2098–2118. [CrossRef]
26. Song, P.; Chen, X. Investigation of the Internal Tides in the Northwest Pacific Ocean Considering the Background Circulation and Stratification. *J. Phys. Oceanogr.* **2020**, *50*, 3165–3188. [CrossRef]
27. Huang, X.; Wang, Z.; Zhang, Z.; Yang, Y.; Zhou, C.; Yang, Q.; Zhao, W.; Tian, J. Role of Mesoscale Eddies in Modulating the Semidiurnal Internal Tide: Observation Results in the Northern South China Sea. *J. Phys. Oceanogr.* **2018**, *48*, 1749–1770. [CrossRef]
28. Wang, Y.; Xu, Z.; Hibiya, T.; Yin, B.; Wang, F. Radiation Path of Diurnal Internal Tides in the Northwestern Pacific Controlled by Refraction and Interference. *J. Geophys. Res. Ocean.* **2021**, *126*, e2020JC016972. [CrossRef]
29. Dunphy, M.; Lamb, K.G. Focusing and Vertical Mode Scattering of the First Mode Internal Tide by Mesoscale Eddy Interaction. *J. Geophys. Res. Ocean.* **2014**, *119*, 523–536. [CrossRef]
30. Klymak, J.M.; Alford, M.H.; Pinkel, R.; Lien, R.C.; Yang, Y.J.; Tang, T.Y. The Breaking and Scattering of the Internal Tide on a Continental Slope. *J. Phys. Oceanogr.* **2011**, *41*, 926–945. [CrossRef]
31. Robertson, R.; Dong, J.; Hartlipp, P. Diurnal Critical Latitude and the Latitude Dependence of Internal Tides, Internal Waves, and Mixing Based on Barcoo Seamount. *J. Geophys. Res. Ocean.* **2017**, *122*, 7838–7866. [CrossRef]
32. Dong, J.; Robertson, R.; Dong, C.; Hartlipp, P.S.; Zhou, T.; Shao, Z.; Lin, W.; Zhou, M.; Chen, J. Impacts of Mesoscale Currents on the Diurnal Critical Latitude Dependence of Internal Tides: A Numerical Experiment Based on Barcoo Seamount. *J. Geophys. Res. Ocean.* **2019**, *124*, 2452–2471. [CrossRef]
33. Rainville, L.; Pinkel, R. Propagation of Low-Mode Internal Waves through the Ocean. *J. Phys. Oceanogr.* **2006**, *36*, 1220–1236. [CrossRef]
34. Nash, J.D.; Kelly, S.M.; Shroyer, E.L.; Moum, J.N.; Duda, T.F. The Unpredictable Nature of Internal Tides on Continental Shelves. *J. Phys. Oceanogr.* **2012**, *42*, 1981–2000. [CrossRef]
35. Savage, A.C.; Waterhouse, A.F.; Kelly, S.M. Internal Tide Nonstationarity and Wave-Mesoscale Interactions in the Tasman Sea. *J. Phys. Oceanogr.* **2020**, *50*, 2931–2951. [CrossRef]
36. Liu, Z.; Hou, Y. Kuroshio Front in the East China Sea from Satellite SST and Remote Sensing Data. *IEEE Geosci. Remote Sens. Lett.* **2012**, *9*, 517–520. [CrossRef]
37. Yang, D.; Huang, R.X.; Yin, B.S.; Feng, X.R.; Chen, H.Y.; Qi, J.F.; Xu, L.J.; Shi, Y.L.; Cui, X.; Gao, G.D.; et al. Topographic Beta Spiral and Onshore Intrusion of the Kuroshio Current. *Geophys. Res. Lett.* **2018**, *45*, 287–296. [CrossRef]
38. Lian, E.; Yang, S.; Wu, H.; Yang, C.; Li, C.; Liu, J.T. Kuroshio Subsurface Water Feeds the Wintertime Taiwan Warm Current on the Inner East China Sea Shelf. *J. Geophys. Res. Ocean.* **2016**, *121*, 4790–4803. [CrossRef]

39. Shi, Y.; Yang, D.; He, Y. Numerical Study on Interaction between Eddies and the Kuroshio Current East of Taiwan, China. *J. Oceanol. Limnol.* **2021**, *39*, 388–402. [CrossRef]
40. Yin, Y.; Lin, X.; He, R.; Hou, Y. Impact of Mesoscale Eddies on Kuroshio Intrusion Variability Northeast of Taiwan. *J. Geophys. Res. Ocean.* **2017**, *122*, 3021–3040. [CrossRef]
41. Niwa, Y.; Hibiya, T. Numerical Study of the Spatial Distribution of the M_2 Internal Tide in the Pacific Ocean. *J. Geophys. Res. Ocean.* **2001**, *106*, 22441–22449. [CrossRef]
42. Van Haren, H. Shear at the Critical Diurnal Latitude. *Geophys. Res. Lett.* **2007**, *34*, 2–5. [CrossRef]
43. Codiga, D.L. *Unified Tidal Analysis and Prediction Using the UTide Matlab Functions*; Technical Report 2011-01; Graduate School of Oceanography, University of Rhode Island: Narragansett, RI, USA, 2011; p. 59. [CrossRef]
44. Xu, Z.; Yin, B.; Hou, Y. Multimodal Structure of the Internal Tides on the Continental Shelf of the Northwestern South China Sea. *Estuar. Coast. Shelf Sci.* **2011**, *95*, 178–185. [CrossRef]
45. Xu, Z.; Yin, B.; Hou, Y.; Xu, Y. Variability of Internal Tides and Near-Inertial Waves on the Continental Slope of the Northwestern South China Sea. *J. Geophys. Res. Ocean.* **2013**, *118*, 197–211. [CrossRef]
46. Xie, J.; Pan, J.; Jay, D.A. Multimodal Internal Waves Generated over a Subcritical Ridge: Impact of the Upper-Ocean Stratification. *J. Phys. Oceanogr.* **2015**, *45*, 904–926. [CrossRef]
47. Zhao, Z. Southward Internal Tides in the Northeastern South China Sea. *J. Geophys. Res. Ocean.* **2020**, *125*, e2020JC016554. [CrossRef]
48. Bai, X.; Liu, Z.; Li, X.; Hu, J. Generation Sites of Internal Solitary Waves in the Southern Taiwan Strait Revealed by MODIS True-Colour Image Observations. *Int. J. Remote Sens.* **2014**, *35*, 4086–4098. [CrossRef]
49. Egbert, G.D.; Erofeeva, S.Y. Efficient Inverse Modeling of Barotropic Ocean Tides. *J. Atmos. Ocean. Technol.* **2002**, *19*, 183–204. [CrossRef]
50. Smith, W.H.F.; Sandwell, D.T. Global Sea Floor Topography from Satellite Altimetry and Ship Depth Soundings. *Science* **1997**, *277*, 1956–1962. [CrossRef]
51. Kunze, E.; Toole, J.M. Tidally Driven Vorticity, Diurnal Shear, and Turbulence atop Fieberling Seamount. *J. Phys. Oceanogr.* **1997**, *27*, 2663–2693. [CrossRef]
52. Robertson, R. Modeling Internal Tides over Fieberling Guyot: Resolution, Parameterization, Performance. *Ocean Dyn.* **2006**, *56*, 430–444. [CrossRef]
53. Robertson, R. Tidally Induced Increases in Melting of Amundsen Sea Ice Shelves. *J. Geophys. Res. Ocean.* **2013**, *118*, 3138–3145. [CrossRef]
54. Kunze, E. Near-Inertial Wave Propagation in Geostrophic Shear. *J. Phys. Oceanogr.* **1985**, *15*, 544–565. [CrossRef]
55. Gerkema, T.; Zimmerman, J.T.F.; Maas, L.R.M.; Van Haren, H. Geophysical and Astrophysical Fluid Dynamics beyond the Traditional Approximation. *Rev. Geophys.* **2008**, *46*, 1–33. [CrossRef]
56. Yin, Y.; Liu, Z.; Zhang, Y.; Chu, Q.; Liu, X.; Hou, Y.; Zhao, X. Internal Tides and Their Intraseasonal Variability on the Continental Slope Northeast of Taiwan Island Derived from Mooring Observations and Satellite Data. *Remote Sens.* **2021**, *14*, 59. [CrossRef]
57. Kelly, S.M.; Nash, J.D. Internal-Tide Generation and Destruction by Shoaling Internal Tides. *Geophys. Res. Lett.* **2010**, *37*, L23611. [CrossRef]
58. Jan, S.; Chen, C.C.; Tsai, Y.L.; Yang, Y.J.; Wang, J.; Chern, C.S.; Gawarkiewicz, G.; Lien, R.C.; Centurioni, L.; Kuo, J.Y. Mean Structure and Variability of the Cold Dome Northeast of Taiwan. *Oceanography* **2011**, *24*, 100–109. [CrossRef]
59. Gill, A.E.; Adrian, E. *Atmosphere-Ocean Dynamics*; Academic Press: Cambridge, MA, USA, 1982; Volume 30.
60. Wunsch, C. Baroclinic Motions and Energetics as Measured by Altimeters. *J. Atmos. Ocean. Technol.* **2013**, *30*, 140–150. [CrossRef]
61. Zhao, Z. Propagation of the Semidiurnal Internal Tide: Phase Velocity Versus Group Velocity. *Geophys. Res. Lett.* **2017**, *44*, 11942–11950. [CrossRef]
62. Xu, Z.; Wang, Y.; Liu, Z.; McWilliams, J.C.; Gan, J. Insight Into the Dynamics of the Radiating Internal Tide Associated With the Kuroshio Current. *J. Geophys. Res. Ocean.* **2021**, *126*, 1–14. [CrossRef]
63. Ansong, J.K.; Arbic, B.K.; Buijsman, M.C.; Richman, J.G.; Shriver, J.F.; Wallcraft, A.J. Indirect Evidence for Substantial Damping of Low-Mode Internal Tides in the Open Ocean. *J. Geophys. Res. Ocean.* **2015**, *120*, 6057–6071. [CrossRef]

remote sensing

Article

Reconstructing the Three-Dimensional Structure of Loop Current Rings from Satellite Altimetry and In Situ Data Using the Gravest Empirical Modes Method

Thomas Meunier [1,*], Paula Pérez-Brunius [2] and Amy Bower [1]

1 Woods Hole Oceanographic Institution, Woods Hole, MA 02543, USA
2 Centro de Investigacion Cientifica y de Educacion Superior de Ensenada, Ensenada 22860, Mexico
* Correspondence: tmeunier@whoi.edu

Abstract: The three-dimensional structure of Gulf of Mexico's warm-core rings, detaching from the Loop Current, is investigated using satellite altimetry and a large set of ARGO float profiles. Reconstruction of the Loop Current rings (LCRs) vertical structure from sea surface height observations is made possible by the use of the gravest empirical modes method (GEM). The GEMs are transfer functions that associate a value of temperature and salinity for each variable pair {dynamic height; pressure}, and are computed by estimating an empirical relationship between dynamic height and the vertical thermohaline structure of the ocean. Between 1993 and 2021, 40 LCRs were detected in the altimetry and their three-dimensional thermohaline structure was reconstructed, as well as a number of dynamically relevant variables (geostrophic and cyclogeostrophic velocity, relative vorticity, potential vorticity, available potential energy and kinetic energy density, etc.). The structure of a typical LCR was computed by fitting an analytical stream function to the LCRs dynamic height signature and reconstructing its vertical structure with the GEM. The total heat and salt contents and energy of each LCR were computed and their cumulative effect on the Gulf of Mexico's heat, salt and energy balance is discussed. We show that LCRs have a dramatic impact on these balances and estimate that residual surface heat fluxes of -13 W m^{-2} are necessary to compensate their heat input, while the fresh water outflow of the Mississippi river approximately compensates for their salt excess input. An average energy dissipation of O $[10^{-10}–10^{-9}]$ W kg^{-1} would be necessary to balance their energy input.

Keywords: three-dimensional eddy reconstruction; loop current rings; gulf of Mexico; gravest empirical modes

Citation: Meunier, T.; Pérez-Brunius, P.; Bower, A. Reconstructing the Three-Dimensional Structure of Loop Current Rings from Satellite Altimetry and In Situ Data Using the Gravest Empirical Modes Method. *Remote Sens.* **2022**, *14*, 4174. https://doi.org/10.3390/rs14174174

Academic Editors: Xiaofeng Li, Chung-Ru Ho and Antony K. Liu

Received: 1 July 2022
Accepted: 16 August 2022
Published: 25 August 2022

1. Introduction

Although the circulation of the world ocean is dominated by geostrophic turbulence, which is transient by nature, long lived coherent mesoscale eddies can be found in virtually every oceanic basin (e.g., Agulhas rings in the South Atlantic [1,2], Gulf Stream rings in the North Atlantic [3,4], Kuroshio rings in the North Pacific [5,6], Loop Current rings in the Gulf of Mexico [7,8]). Because of their longevity and coherence, these eddies are able to trap and transport tracers (heat, salt, oxygen, plankton, nutrients) far away across basins [2,9,10]. The advent of satellite altimetry in the early 1990s yielded a dramatic increase in the knowledge and understanding of the surface properties of mesoscale eddies [11]. However, energy and tracer transport is by essence a three-dimensional process, as momentum and tracer distribution within mesoscale structures is clearly baroclinic. Understanding and quantifying the role of mesoscale coherent eddies in tracer transports requires a detailed assessment of their vertical structure. Although ship and glider surveys can offer occasional detailed pictures of a limited number of eddies, the observations they provide are too limited in time and space for systematic statistical analysis on a regional or global scale. To address this setback in the availability of a solid statistical description of the three-dimensional properties of mesoscale eddies, a statistical method using jointly in situ observations from ARGO profiling floats and surface observations of sea surface height (SSH) from satellite altimetry were developed [12]. The method consists of automatically detecting mesoscale eddies as closed sea level anomaly (SLA) contours, and searching for ARGO profiles within, and in the vicinity of, the eddy's boundary. The position of the profile is then referenced to the eddy's rotation axis and normalized by the eddy's radius or simply localized by its zonal and meridional distance. Given a sufficient number of profiles, the method allows for the computation of one mean 3-dimensional profile, supposed to be representative of a typical eddy in a given region. The method has since been extensively used in many regions of the ocean ([13–15] in the tropical and sub-tropical Pacific, [16]) in the South China sea, [17] in the Arabian sea, [18] in the South Atlantic, [19] in the Lofoten basin, among many). Although the method, known as composite or co-location method, greatly helped to quantify regional statistical properties of mesoscale eddies, they are of limited use, because they only allow for the computation of one average eddy, and not for the reconstruction of the vertical structure of each eddy spotted in the altimetry. Recently, Meunier et al. [20] proposed an alternative method for the estimation of the heat anomaly carried by Loop Current rings (LCR) in the Gulf of Mexico (GoM), based on satellite altimetry and in situ data. Taking advantage of a convenient linear relationship between the local heat content anomaly and SSH, they were able to estimate the total heat content anomaly of each individual eddy. Their method was limited to vertically integrated quantities, and did not provide a full three-dimensional picture of the eddies structure. It could, thus, not provide any information on the energetics of LCRs.

Over two decades ago, Watts et al. [21] and Sun and Watts [22] proposed a method to estimate the full water column's thermohaline structure from dynamic height observations only. The procedure, known as the Gravest Empirical Modes (GEM), consists in establishing an empirical relationship between dynamic height and temperature and salinity, at a given pressure level, from in situ observations. In the Antarctic circumpolar region, the GEM representation was shown to account for over 97% of the thermohaline variance. Taking advantage of the close relationship between dynamic height and sea surface height, Swart et al. [23] use the GEM methods to reconstruct vertical hydrographic transects in the Antarctic Circumpolar Current from satellite altimetry. More recently, Müller et al. [24] used the GEM method along with satellite altimetry to estimate the heat and fresh water transport by mesoscale eddies in the subpolar north Atlantic. This method is of particular interest because it allows the computation of the thermohaline structure of each individual eddy.

In this study, we follow the procedure of Müller et al. [24], to infer the three-dimensional thermohaline structure of mesoscale eddies, as well as their heat and salt anomalies, and extend it to the computation of other relevant variables, such as geostrophic and cyclogeostrophic velocity, relative vorticity, potential vorticity, as well as kinetic and available potential energy density.

The data used are described in Section 2 and the methods in Section 3. Validation using independent glider observation across an LCR is presented in Section 4. The method is then applied to the 29 years-long AVISO altimetry record in the GoM, where we identified and reconstructed 40 Loop Current rings. The vertical structure of a typical LCR is presented in Section 5 and the statistical properties of LCRs characteristics, with an emphasis on their heat, salt, and energy contents, are presented in Section 6.

2. Data

2.1. In Situ Data

In situ temperature and salinity data were obtained from 6792 ARGO profiles in the Gulf of Mexico. The profiles locations are shown on the map of Figure 1a, showing a wide coverage of the GoM and a homogeneous distribution of the collected data. The color coding corresponds to the local steric height referenced to 2000 dbar computed from the profile data. The presence of the Loop Current (LC) is evident as a tongue of high steric height extending northward from the Yucatan channel and bounded by the West Florida and Louisiana shelves to the East and North, respectively. Sparse high steric height values can also be found in the central and western basin and are the signature of detached LCRs drifting westward. Figure 1b shows a TS diagram of all ARGO profiles. Color coding is the same as in Figure 1a. High steric height water columns are characterized by a warm and salty anomaly between the 1024 and 1026 kg m^{-3} isopycnals at intermediate depths. This water mass corresponds to the Atlantic subtropical underwater (SUW), of Caribbean origin, found in the LC and LCRs [7,25,26].

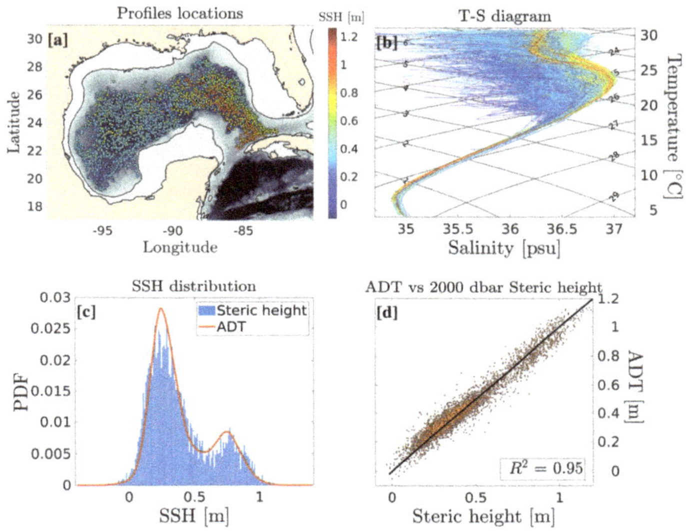

Figure 1. (**a**): Location of all available ARGO profiles in the Gulf of Mexico. The local 2000 dbar-referenced steric height is color coded. (**b**): T-S diagram of all ARGO profiles. The color code is the same as in panel (**a**). The black contours are isopleths of spice and potential density. (**c**): Distributions (normalized PDF) of the 2000 dbar-referenced steric height computed from in situ data (blue bars) and the gridded absolute dynamic topography (ADT) data (orange line). (**d**): 2000 dbar-referenced steric height against ADT.

2.2. Altimetry Data

AVISO gridded absolute dynamic topography (ADT) is used for eddy detection, and as a proxy for dynamic height. The gridded fields have a spatial resolution of $1/4°$ and are available on a daily basis. In situ measured steric height and ADT distributions are compared in Figure 1c. The principle mode of the distribution at about 30 cm corresponds to the Gulf Common Water (GCW), while the secondary mode at about 70 cm corresponds to the SUW. The similarity between the distributions guarantees an unbiased geographic sampling of the ARGO profiling floats. ADT is plotted against the 2000 dbar-referenced steric height in Figure 1d. The clear linear relationship, with a coefficient of determination (R^2) of 0.95, shows that the 2000 dbar-referenced steric height is a solid proxy for dynamic height (i.e., the geopotential at 2000 dbar can be considered as flat).

2.3. Glider Data

For validation purposes, we used recent glider observations across an LCR. The glider repeatedly crossed LCR Poseidon [27] through its center between August and November 2016. It was equipped with a sea bird CT-sail CTD probe, and had mean vertical and horizontal resolution of 2 m and 2 km, respectively. More details on the glider observations can be found in Meunier et al. [27–30].

3. Methods

3.1. The Gravest Empirical Modes (GEM) Method

The gravest empirical mode projection (GEM) [21–24,31] consists in establishing an empirical relationship between the vertical thermohaline structure of the ocean and the dynamic height [21,22]. The procedure used here is straightforward and can be summarized as follows:

- **A.** The steric height relative to 2000-dbar is computed for each in situ profile of temperature and salinity;
- **B**. All profiles are sorted according to their steric height. The sorted ARGO temperature and salinity profiles are shown in Figure 2, where a pattern already emerges, showing the clear relationship between steric height, pressure, and both temperature and salinity;
- **C**. A regular pressure grid is defined ((0–2000 dbar) with a vertical grid-step of 2 dbar) and for each reference pressure value, a spline interpolant is fitted to the functions $T(\eta_{2000})|_p$ and $S(\eta_{2000})|_p$, where T and S are temperature and salinity, η_{2000} is the 2000 dbar-referenced steric height, and p is the pressure at which the variables are evaluated;
- **D**. The relationship between 2000-dbar referenced steric height and dynamic height is assessed by comparing local ADT and η_{2000} to ensure that the empirical relationship obtained from in situ steric height holds when using ADT (Figure 1d).

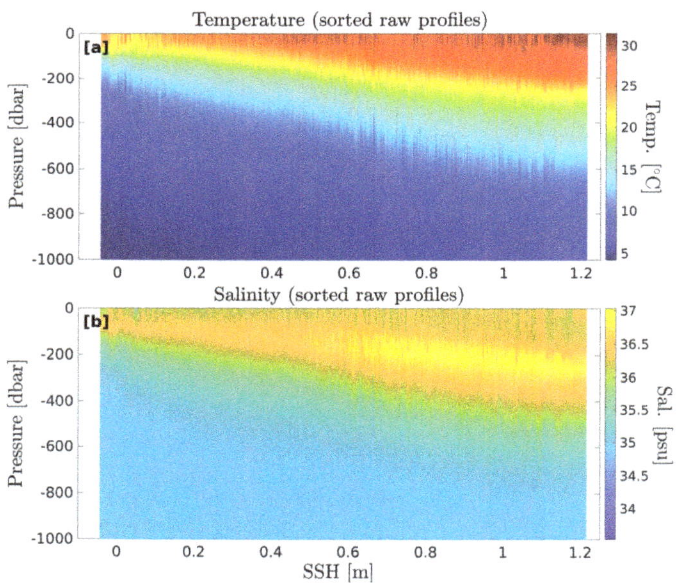

Figure 2. Steric height-sorted raw temperature (**a**) and salinity (**b**) profiles for the whole ARGO dataset. The x-axis represents 2000 dbar-referenced steric heigh (η_{2000}) and the y-axis is pressure. The 2000 dbar-referenced steric height closely matches absolute dynamic topography (ADT), as shown in Figure 1c,d.

For each couple $\{p, \text{ADT}\}$, the GEM fields provide a single value of temperature and salinity, so that the three-dimensional structure of the ocean can be inferred from any two-dimensional map of ADT. In that regard, the GEM can be thought of as a transfer function [31]. Yearly averaged GEM fields are shown in Figure 3 for temperature, salinity, and potential density (referenced to the surface). The salinity maximum between 100 and 300 dbar associated with the SUW is particularly evident in the high end of the dynamic height range. It is accompanied by a weakly stratified and warm temperature anomaly. One striking feature, also evident in the sorted raw profiles sections of Figure 2, is the smooth transition between SUW and GCW across the dynamic height range. Maps of the root mean square (RMS) errors are also provided in Figure 3d–f. For a given variable X, they are computed as $E_{RMS} = \sqrt{\overline{(X_m - X_e)^2}}$, where the overbar represents the ensemble average, X_m is the measured variable (the actual value from the ARGO profile) and X_e is the estimated variable (the value computed using the GEM). The ensemble averages are performed over bins of 2 cm in SSH and 2 m in pressure. Error is maximum near the surface in low dynamic height water columns (GCW), and below and at the edges of, the salinity maximum. This increased uncertainty at the transition between GCW and SUW might be related to the presence of density compensated interleavings, recently reported around LCRs [29], or could as well be the consequence of a lower number of profiles (Figure 1c). The average RMS error is of 0.2 °C, 0.028 psu, and 0.035 kg m^{-3} for temperature, salinity, and potential density, respectively. They are slightly larger than the values reported by Swart et al. [23], using the same methods south of Africa (0.15 °C, 0.02 psu, and 0.02 kg m^{-3}).

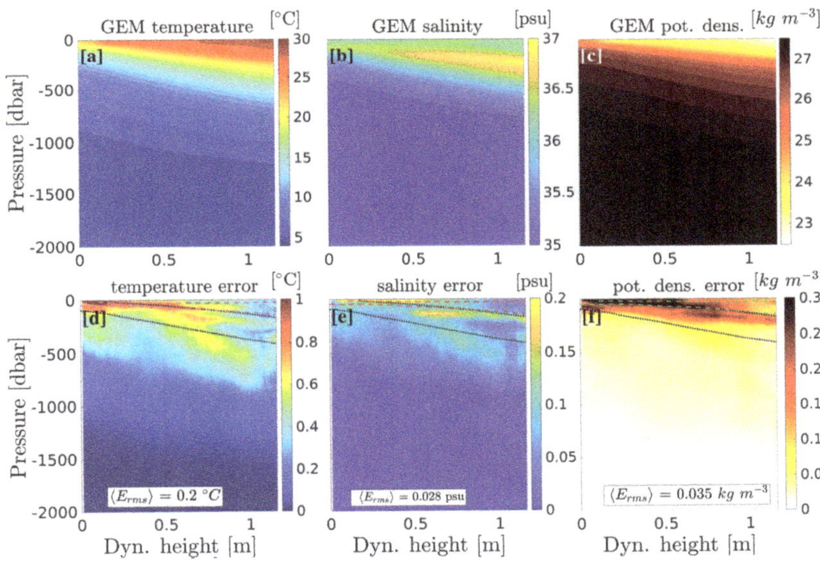

Figure 3. Yearly averaged gravest empirical mode (GEM) fields for temperature (**a**), salinity (**b**), and potential density (**c**). The x-axis represents dynamic height, and the y-axis represents pressure. Dynamic height was shown to be equal to the 2000 dbar-referenced steric height η_{2000} and absolute dynamic topography (ADT) in Figure 1d. The root mean squared (RMS) errors are shown in panels (**d–f**) for temperature, salinity, and potential density, respectively. The black dashed lines represent the limits of the SUW potential density range (1024–1026.5 kg m^{-3}), while the red or green dashed lines represent the depth range of the mixed layer.

To account for the seasonality of surface conditions, which affects the accuracy of the three-dimensional reconstruction in the top 200 dbar, the GEM fields were constructed on a monthly basis. Figure 4 shows a Hovmuller diagram of the temperature, salinity, and potential density GEM for a dynamic height of 80 cm. Seasonal variations of temperature, potential density, and mixed layer depth are evident near the surface. Salinity is less affected by the seasonal cycle. Note that the error was also computed individually for each monthly GEM, and we found no large seasonal variation of the error (range (0.19–0.22) °C for temperature and (0.017–0.019) psu for salinity).

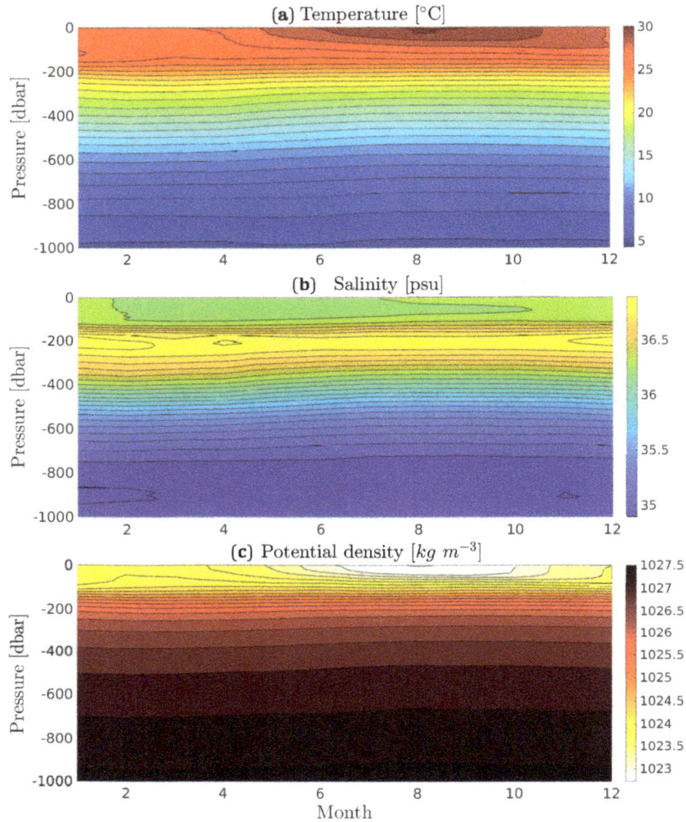

Figure 4. Example of seasonal variation of GEM fields for a given value of sea surface height (SSH; 0.8 m here). The x-axis is the time of year while the y axis is pressure. Temperature is shown in panel (**a**), salinity in panel (**b**), and potential density in panel (**c**).

3.2. Eddy Detection and Edge Definition

The LCR detection method is based on daily ADT maps. First, the edge of the LC is defined as the ADT contour passing through both the Yucatan Channel and the Florida strait, along which the mean velocity is maximum. An LCR is detected when this contour forms a closed loop. However, recently detached LCRs may reattach to the LC, and repeated detachment-reattachment sequences are not rare [32]. Here, we only take into account LCRs that permanently detached from the LC. After detachment, the LCRs edge is defined as the closed ADT contour along which the mean azimuthal velocity is maximum. The maximum velocity contour coincides with a sign-change of the Okubo–Weiss parameter [20,33], so that our edge definition criterion ensures that only the vorticity-dominated portion of the eddy is retained, while the strain-dominated periphery is discarded. For comparison purposes, and because the last closed SSH contour is a commonly used definition for eddies boundaries, the latter was also computed and briefly discussed. Between January 1993 and January 2022, 40 LCRs were detected. A map of their edge contours is shown in Figure 5a. The maximum ADT value at their center is color coded. The location of newly detached LCRs exhibit a large zonal and meridional variability. The same edge contours are plotted in an eddy-centric frame in Figure 5b, showing that LCRs have a wide range of sizes. Figure 5a,b also show a clear tendency for smaller LCRs and for LCRs shed at the tip of the LC (the north-western most eddies) to have a weaker ADT signal. These tendencies are explored further in Figure 5c,d, showing a nearly linear relationship between LCRs maximum ADT and their radius, and between their volume anomaly (surface integral of the ADT anomaly) and their distance from the Yucatan channel.

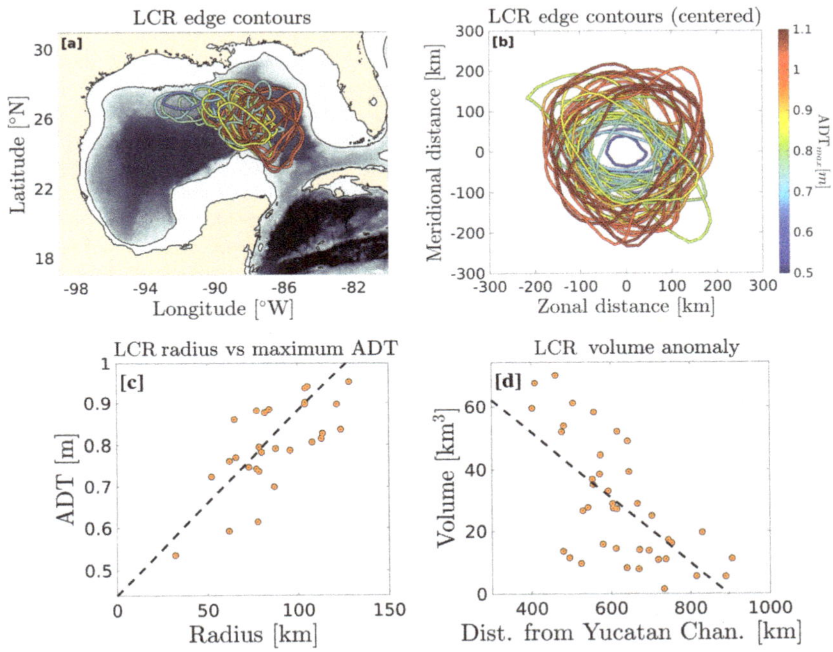

Figure 5. (**a**): Map of the edge contours of the 40 detected Loop Current rings (LCRs) one day after detachment. The maximum absolute dynamic topography (ADT) value within each eddy is color coded. (**b**): Same as (**a**) in an eddy-centric frame: the x and y axis are the distance (in km) from the eddy's rotation axis. The color code is the same as in (**a**). (**c**): Maximum ADT value (at the eddy's center) against eddy's radius. (**d**): Volume anomaly (surface-integral of the ADT anomaly) against distance from the Yucatan channel.

4. Validation Using Independent Observations

To validate the methods, we directly compared glider observations with GEM inferred vertical sections of temperature, salinity, and geostrophic velocity across an LCR. The validation procedure consists of computing a vertical profile from ADT and the GEM at each glider's dive location. For comparison purposes, geostrophic velocity is then computed after applying a Gaussian low-pass filter with vertical and horizontal decorrelation radii of 15 m and 30 km, respectively, and assuming no motion at 1000 dbar [27,34]. Figure 6 shows the two glider sections and the GEM-reconstructed sections. Note that the first section (panels a, c, and e) was performed as the glider was navigating towards the drifting eddy, while in the second section (panels b, d, and f), the glider and the eddy were moving in the same direction. This results in an under (over) estimation of distances and an over (under) estimation of velocity in the first (second) glider section. This bias was discussed in detail by Meunier et al. [27] and a correction method was proposed by Meunier et al. [30]. However, for validation of the GEM fields, we chose to use the uncorrected glider along track coordinates to keep the analysis as straightforward as possible. For the ease of visualisation, the GEM sections are flipped laterally to appear as a mirror image of the glider sections. In both sections, the LCR is obvious as a downward tilting of the isotherms, both in the glider observations and in the GEM-reconstructed fields. The displacement of the isotherms in the GEM sections is in good agreement with the glider observations up to the 25 °C isotherm. However, the homogeneity of temperature in the upper part of the eddy's core (thermostat) is not faithfully reproduced by the GEM method, which exhibits a slightly more stratified structure. Note that LCR Poseidon was an uncommon LCR with an exceptionally thick thermostat [27,30], so we hypothesize that this difference is, in part, related to the exceptional nature of Poseidon. The salinity sections, on the other hand, are not subject to this bias, and the double core structure, consisting of a salinity maximum between 200 and 350 m, and a homogeneous salinity minimum above, is well reproduced by the GEM reconstruction. Note that, in both sections, the GEM-reconstructed eddy is slightly smoother, as expected from the methods, which essentially captures the geostrophic, or *slow* structure of the flow [22]. It should be pointed out that the smoothing of thermohaline gradient has little effect on the geostrophic velocity difference between glider and GEM-derived fields. Indeed, the low-pass filtering required to compute geostrophic velocity from glider observation, removes high wavenumber variability and tends to smooth out gradients, whatever the glider's original resolution. In other words, the small scale variability that the GEM method is unable to capture has to be removed from the glider data anyway. The vertical sections of geostrophic velocity are shown in panels (e) and (f). In both sections, the agreement between the GEM-reconstructed and the glider sections is evident, both in the spatial patterns and in the magnitude of the velocity.

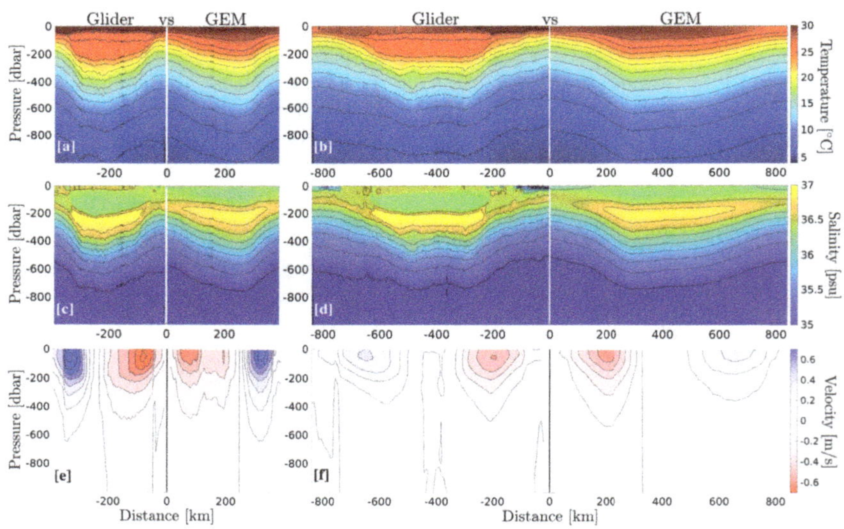

Figure 6. (**a**,**b**): Temperature sections across a Loop Current ring measured by the glider and reconstructed using the GEM. in each panel, the GEM-reconstructed sections are flipped laterally in order to appear as a mirror image of the glider section. (**c**,**d**): Same as (**a**,**b**) for salinity. (**e**,**f**): Same as (**a**,**b**) for geostrophic velocity.

Although the detailed vertical structure of LCRs is of interest, knowledge of the vertical structure of ocean eddies is particularly crucial for the computation of their heat content and transport, which rely on depth-integrated temperature anomaly. Depth-averaged temperature and salinity anomalies, as well as the depth-averaged geostrophic velocity, are shown in Figure 7. In both cross-sections, the glider observation and the GEM-reconstruction are in striking agreement, with a coefficient of determination (R^2) ranging between 0.91 and 0.94, meaning that the GEM method captures over 90 % of the depth averaged velocity, temperature anomaly, and salinity anomaly variance. In particular, one should note that the lateral gradients of depth-averaged variables do not suffer from the over-smoothing that was discernible in the detailed vertical sections. The GEM thus appears to be particularly well-suited to compute integrated variables, such as heat and salt content, or kinetic and available potential energy.

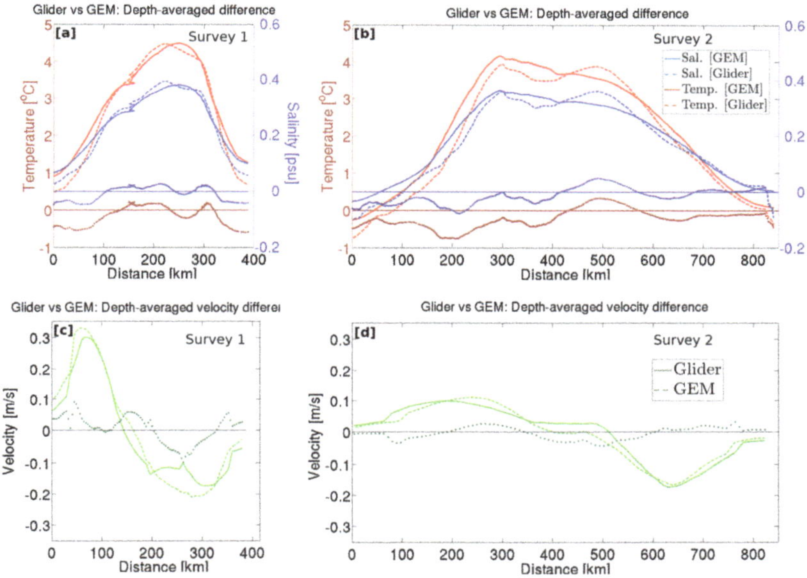

Figure 7. (**a**,**b**): Horizontalprofiles of depth-averaged temperature (red) and salinity (blue) anomalies. The glider observations are plotted as dotted lines, while the GEM-reconstructed profile is plotted as plain lines. The difference between the glider and GEM sections are plotted as dotted lines. (**c**,**d**): Same as (**a**,**b**) for the geostrophic velocity.

5. The 3D Structure of an Average LCR

The 3D reconstruction method was applied to 40 LCRs that detached between 1993 and 2021. Cross-sections of salinity, temperature, and cyclogeostrophic velocity are shown in Figure 8 for 6 selected examples. Maps of ADT 5 days after detachment along with the virtual transect trajectories are shown in the left-hand-side panels. These examples were chosen to represent small, average, and large LCRs in spring/summer conditions and in fall/winter conditions. The reconstructed fields capture well the typical LCR structure, which is characterized by a double core salinity structure, consisting of a fresh anomaly in the top 100 to 150 m, lying over a salty anomaly between 150–200 m and 300 m. The thick homogeneous warm anomaly is also evident between the surface and 200 m, and the isotherms are doming downward throughout the water column. In spring/summer conditions, a shallow thermocline lies over the main LCR structure, while in fall/winter conditions, the mixed layer extends down to the base of the thermostat, and is deeper in the LCR than at its periphery.

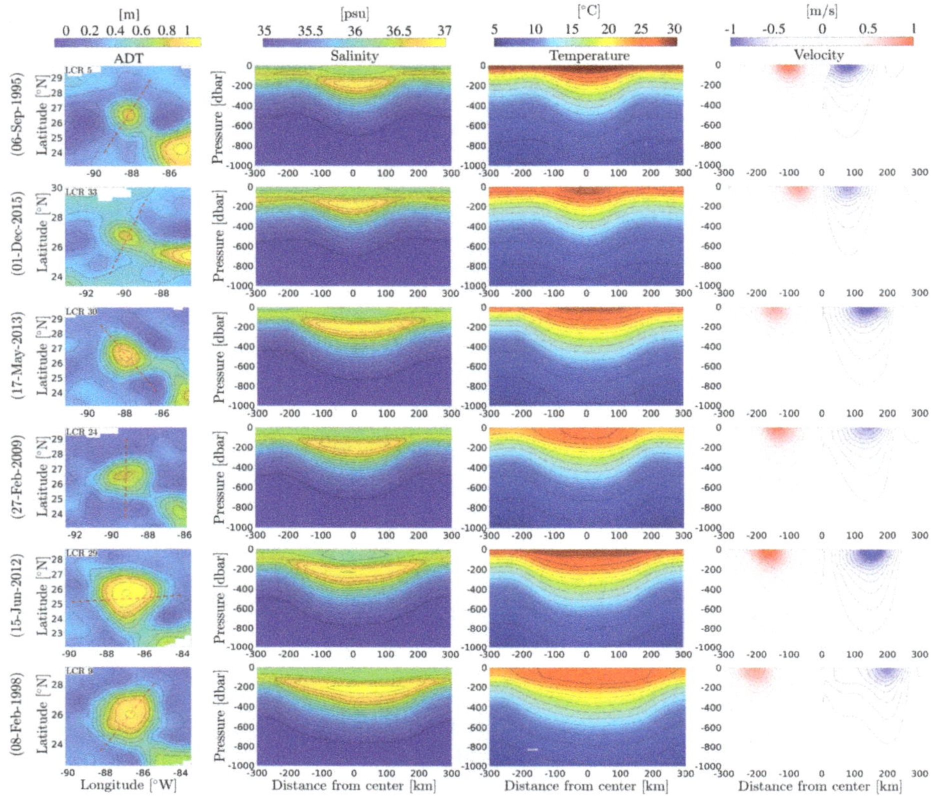

Figure 8. Selected examples of LCRs three-dimensional structure reconstructed from altimetry and the GEM method. (**First row**): maps of absolute dynamic topography (ADT). The dotted red line represents the trajectory of the virtual vertical sections. (**Second row**): Vertical sections of salinity. (**Third row**): Vertical sections of temperature. (**Fourth row**): Vertical sections of cyclogeostrophic velocity.

Geostrophic velocity was computed using the thermal wind relations, using H = 2000 dbar as the level of no motion:

$$\mathbf{u}_g(x,y,z,t) = \frac{g}{\rho_0 f} \int_{-H}^{0} \mathbf{k} \times \boldsymbol{\nabla}\rho(x,y,z,t)dz, \tag{1}$$

where g is the gravity acceleration, ρ_0 is a reference density, f is the Coriolis frequency, $\boldsymbol{\nabla}$ is the horizontal gradient operator, $\mathbf{k}$ is the vertical unit vector, and ρ is in situ density.

Note that the reference can equally be taken as the surface geostrophic velocity inferred from satellite altimetry, yielding exactly similar results since the GEMs were computed assuming that the geopotential is flat at 2000 dbar.

The velocity fields have maxima ranging between 0.6 and 1 m s^{-1} and exhibit intense vertical shear in subsurface, with the velocity dropping by ≈70% in the top 300 m.

Although the statistical properties of LCRs presented in this work (Section 6) are computed using individual GEM reconstructions, it is of interest to determine the average structure of an LCR. To do so, we defined a typical surface signature of LCRs (radial ADT profile) and then used the GEM's transfer functions to reconstruct the vertical thermohaline structure. Each radial ADT profile was fitted to Zhang et al.'s [35] universal stream function, defined as:

$$\psi(r) = \psi_0 + \psi_m(1 - \tilde{r}^2)e^{-\tilde{r}^2}, \quad (2)$$

$$\tilde{r} = \frac{r}{L}, \quad (3)$$

where $\psi(r)$ is the surface stream function, $\tilde{r}$ is the non-dimensional radial coordinate and L is the radial length scale. ψ_0 is the background stream function value outside the eddy, and ψ_m is the amplitude parameter which is equal to the maximum value at the centre of the eddy. In the geostrophic framework considered here, ψ is simply proportional to ADT ($\psi = \frac{g}{f}$ADT). For each radial profile of ADT, the parameters η_0, η_m, and L are determined using least-square fitting. Figure 9a shows the mean profiles of each of the 40 LCRs (gray lines), along with the universal stream function computed using the average parameters of each least-square fit (black line). Figure 9b shows the distribution of the coefficient of determination R^2 between the observed and the fitted profiles. R^2 is a measure of the variance fraction that is reproduced by the analytical stream function. The universal stream function appears to faithfully represent LCRs surface signature, with 37 eddies out of 40 having a coefficient of determination superior to 0.95.

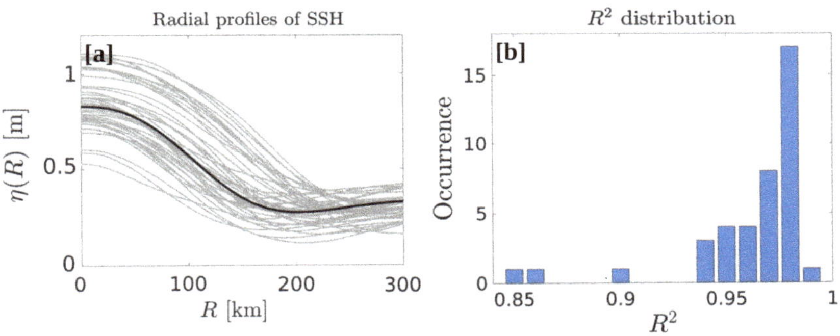

Figure 9. (**a**): Radial profiles of sea surface height (SSH) for the 40 detached Loop Current rings (LCRs) (gray lines) and mean universal SSH profile (Equation (3)) computed using the mean parameters of the 40 least-square fits (black line). (**b**): Coefficient of determination (R^2) of the observed SSH profiles and the fitted universal profiles.

Figure 10 shows values of the ADT anomaly η_m against the radial length scale L. In agreement with Figure 5c, the amplitude of the sea surface height deviation in the LCR's core appears to be approximately proportional to the radial length scale. The parameters chosen for the reference ADT profile are the average of each fitted values, which naturally fall on the linear trend line of Figure 10.

The vertical temperature and salinity fields were then built using the yearly-averaged GEM, and are shown in Figure 11. As in the selected individual examples of Figure 8, the downward doming of the isotherms towards the eddy's center is evident, along with a slight decrease in stratification in the eddy's upper core. The salinity section exhibits the SUW salinity maximum signature near 200 m, and fresher water above. The geostrophic velocity vertical structure (Figure 12a) exhibits well defined velocity maxima of about 0.63 m s^{-1}, with vertical shear reaching 1.5×10^{-3} s^{-1} in the top 200 m.

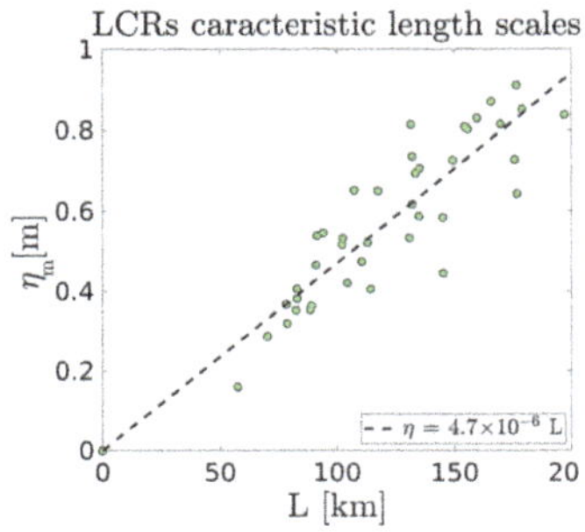

Figure 10. Parameters of the least-square fit to the so-called universal stream function (Equation (3)) for each detected Loop Current ring. The x-axis is the radial length scale and the y-axis is the maximum ADT anomaly η_m.

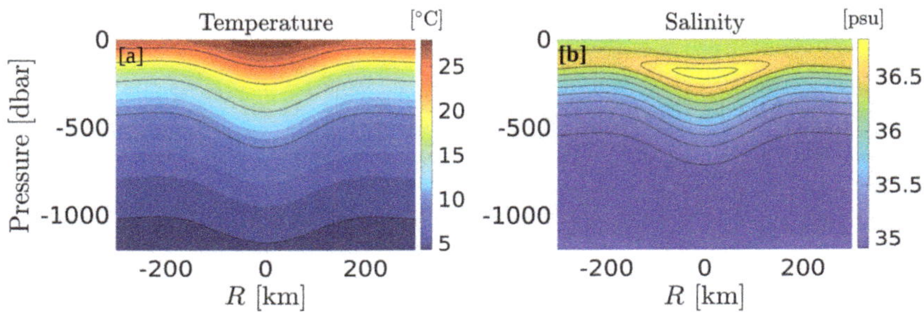

Figure 11. Vertical profiles of temperature (**a**) and salinity (**b**) for the reconstructed average Loop Current ring (LCR). The average LCR is computed using the universal sea surface height (SSH) profile and the Gravest Empirical Mode (GEM) fields. The parameters used in the universal profile are the mean from the 40 least-square fits.

Cyclogeostrophic velocity was also computed for the reference LCR, following Holton [36]. It is the solution of the gradient–wind balance and reads:

$$u_{cg}(r,z) = -\frac{fr}{2} \pm \left(\frac{f^2r^2}{4} + fru_g(r,z)\right). \quad (4)$$

Note that for the computation of the mean LCR's characteristics, the Coriolis frequency is chosen to be constant (Beta plane) and equal to its value at the average eddy-separation latitude. The vertical section of cyclogeostrophic velocity is shown in Figure 12b. Maximum velocity is increased with values reaching 0.76 m s^{-1}. As expected, the impact of including the centrifugal force in the balance has more impacts in the vicinity of the rotation axis, and the mean velocity increase is of $\approx$20 % between the velocity maxima and the rotation axis. Relative vorticity was computed both from the geostrophic and cyclogeostrophic velocities. In cylindrical coordinates, it is defined as:

$$\zeta(r,z) = \frac{1}{r}\frac{\partial(ru_\phi)}{\partial r}, \quad (5)$$

where u_ϕ is the azimuthal velocity. The LCR's relative vorticity signature consists in a bowl of negative relative vorticity and is discernible down to 1000 dbar. It is enclosed within a crown of positive relative vorticity at the eddy's periphery, with a more modest depth extent ($\approx$300 m). As for the azimuthal velocity, cyclogeostrophic vorticity is more intense than geostrophic vorticity, with maximum normalized values reaching $0.51 \times f_0$, and $0.35 \times f_0$, respectively.

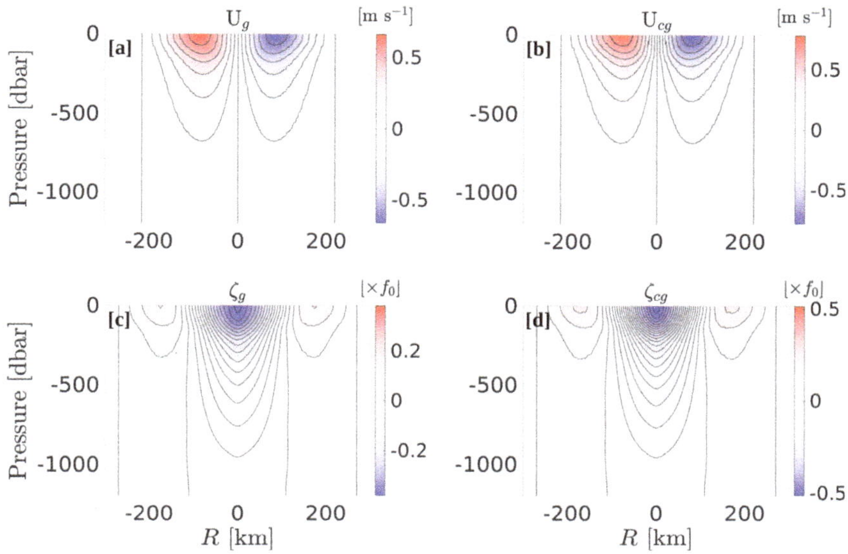

Figure 12. Same as Figure 11 for geostrophic velocity (**a**), cyclogeostrophic velocity (**b**), geostrophic relative vorticity (**c**), and cyclogeostrophic relative vorticity (**d**). The contour interval is 0.1 m s^{-1}. Note that the color bars have a different range in each panel.

A section of Ertel's potential vorticity (PV) is shown in Figure 13a. In cylindrical coordinates, PV is defined as:

$$q(r,z) = \frac{1}{g}(f+\zeta)N^2 - \frac{\partial u_\phi}{\partial z}\frac{\partial \sigma}{\partial r}, \tag{6}$$

where N^2 is the buoyancy frequency, defines as $N^2 = -\frac{g}{\sigma}\frac{\partial \sigma}{\partial z}$. The LCR is obvious as a bowl of extremely low PV in the top 200 m, deflecting the pycnocline downward. Examination of the vertical structure of the buoyancy frequency (Figure 13b) reveals very similar patterns, suggesting that PV is mostly influenced by the LCR's stratification.

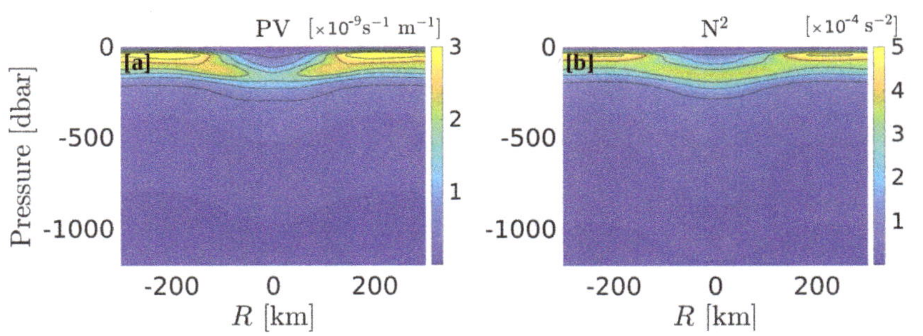

Figure 13. (**a**): Same as Figure 11 for potential vorticity. (**b**): same as (**a**) for the squared buoyancy frequency.

To conclude this description of the vertical structure of an average LCR, the distribution of mechanical energy density is shown in Figure 14. Kinetic energy density

(KED) and available potential energy density (APED) are defined following Holliday and McIntyre [37]:

$$E_k(r,z) = \frac{1}{2}\rho u_\phi^2(r,z), \tag{7}$$

$$E_p(r,z) = -g\int_0^\delta \tilde{\delta}(r,z)\frac{d\overline{\rho}}{dz}(z-\tilde{\delta}(r,z))d\tilde{\delta}, \tag{8}$$

$\overline{\rho}(z)$ is the reference density profile, which is defined as the minimum potential energy profile in the GoM, and obtained by adiabatically sorting all available density measurements, and δ is the isopycnal displacement. The LCR's vertical structure consists of a subsurface bowl of intense APED intensified between 150 and 200 m, where isopycnal displacement and density anomaly are maximum. There is no surface signature, while APED anomaly is evident down to 1200 m. KED exhibits significantly smaller values than APED, and is maximum at the periphery of the eddy, while APED is maximum in the core. When integrated over the whole eddy's volume, we find a ratio of KE/APE $\approx 1/3$, so that energy partition is strongly skewed.

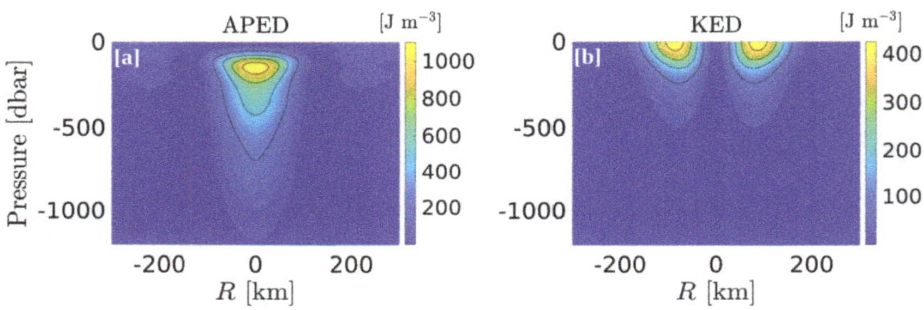

Figure 14. Same as Figure 11 for available potential energy density (**a**), and kinetic energy density (**b**).

6. Heat, Salt, and Energy Statistical Properties

One particularly important application of this three-dimensional individual eddy reconstruction method, is to achieve a statistical representation of LCRs properties, and of their impacts on the GoM's heat, salt, and energy budget. The heat and salt content anomalies associated with each LCR with a boundary $\mathscr{C}$ enclosing a surface $\mathscr{S}$ are defined as:

$$\mathbb{H} = \iint_{\mathscr{S}} \Big(\int_{-H}^{0} \rho_0 C_p \delta T dz \Big) dS \tag{9}$$

$$\mathbb{S} = \iint_{\mathscr{S}} \Big(\int_{-H}^{0} \delta S dz \Big) dS, \tag{10}$$

where dS is a surface element, C_p is the specific heat of sea water, ρ_0 is the mean density, δT is the temperature anomaly, defined as the difference between the temperature $T(x,y,z,t)$ and the GoM's mean profile $\overline{T}(z)$, and δS is salinity anomaly (in kg m^{-3}), defined using the same procedure as for temperature anomaly. Heat and salt contents of the 40 detected eddies are shown on the bar plots of Figure 15a,b for two different eddy boundary criteria (maximum velocity contour and last closed ADT contour) and are compared against each other on Figure 15c. The average heat content of an LCR is of 0.42 and 0.46 ZJ for the maximum velocity contour and the last closed contour boundary criteria, respectively, while the average salt content is of 9.43 and 10.24 billion tons. Heat and salt are extremely variable from one eddy to the other, with a range spanning nearly 2 orders of magnitude ((0.017–1.14) ZJ for heat and (0.38–25.5) billion tons for salt). There is a solid proportionality relationship (R^2 = 0.98) between heat and salt contents (Figure 15c), in agreement with Meunier et al. [20]. The cumulative heat and salt input into the GoM were also computed between 1993 and 2022, and are shown as the orange lines in Figure 15a,b. Despite the large variability of LCRs heat and salt contents, and the lack of periodicity in eddy detachment events, the cumulative heat input grows nearly linearly with time, with a growth rate of 0.60 ZJ per year for heat and 13.5 billion tons per year for salt (coefficient of determination $R^2 = 0.99$ for both heat and salt linear fits). It is also interesting to note that the individual heat contents of LCRs do not grow linearly with their surface area, but rather quadratically (Figure 15d), which might be attributed to the fact that larger eddies also have larger maximum SSH anomalies, hence, not only larger areas, but also larger heat content anomalies per unit area.

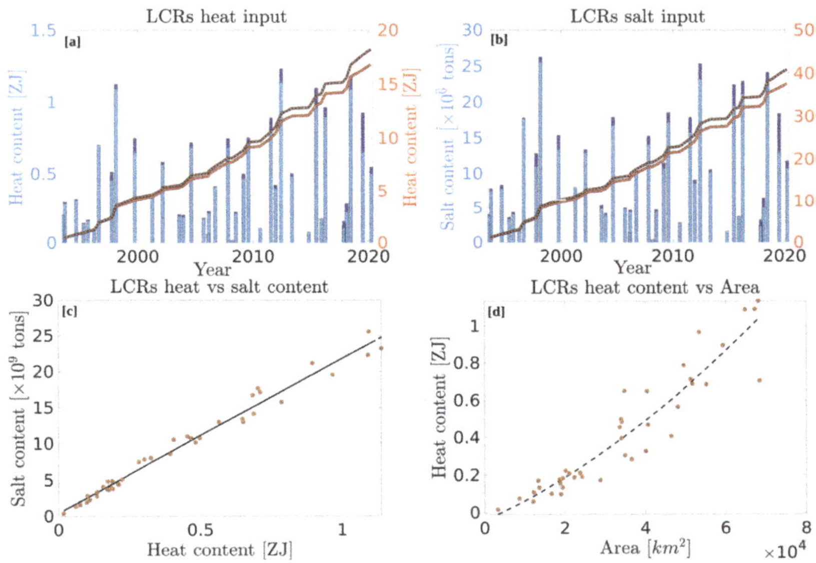

Figure 15. (**a**): Total heat content of each detached Loop Current ring (bar plot). The light blue bars represent values computed using the maximum velocity contour as the LCR's edge, while dark blue bars were obtained using the last closed absolute dynamic topography (ADT) contour. The orange curves, referenced on the right-hand side y-axis, represent the cumulative heat input of the LCRs over time. The plain and dotted lines represent the maximum velocity and the last closed contour criteria, respectively. (**b**): same as panel (**a**) for the salt input. (**c**): Total heat content of each detached LCR against total salt content. (**d**): Total heat content of each LCR against its total area.

The kinetic and potential energy carried by the LCRs were also estimated. Kinetic and available potential energy ($\mathbb{E}_k$ and $\mathbb{E}_p$, respectively) are defined as the volume integral of KED and APED (defined in Equations (7) and (8), respectively):

$$\mathbb{E}_k = \iint_{\mathscr{S}} \Big(\int_{-H}^{0} E_k dz \Big) dS, \tag{11}$$

$$\mathbb{E}_p = \iint_{\mathscr{S}} \Big(\int_{-H}^{0} E_p dz \Big) dS \tag{12}$$

Similar to the heat and salt contents discussed above, the total energy carried by each individual LCR is shown in the bar plot of Figure 16a. The cumulative energy is plotted for KE, APE, and total mechanical energy (TE = KE + APE), while the individual energy contents (bar plot) is only shown for TE, for the sake of clarity. Total mechanical energy has an average of 10.0 (10.9) PJ per eddy when defining the eddies boundaries as the maximum velocity contour (last closed contour). They also exhibit a wide range of values with nearly two orders of magnitudes between the less energetic and the more energetic eddies ((0.15–36.6) PJ). On average, APE is 3.8 times larger than KE. This bias in the energy partition is particularly evident in Figure 16b, which shows KE against APE for each detected LCR. The black line represent equipartition (Burger number unity). Although APE dominates over KE in all LCRs, the ratio between KE and APE (Burger number) decreases as the LCRs total energy increases: large eddies have very small Burger numbers, while smaller eddies can get closer to energy equipartition. The growth of cumulative energy is also nearly linear, with values of 2.95 and 11.2 PJ per year for KE and APE, respectively (R^2 = 0.99 for KE and 0.97 for APE).

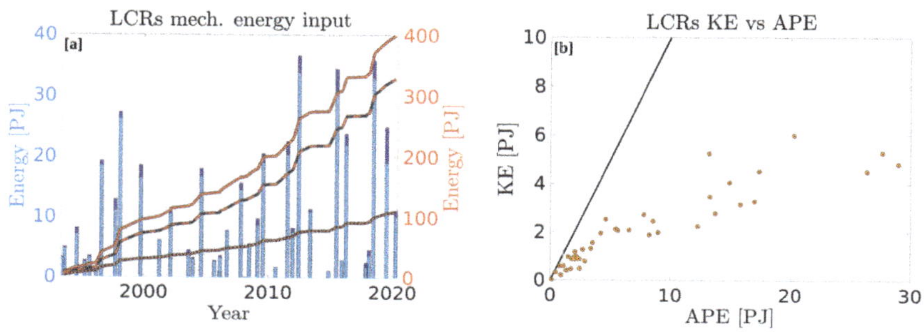

Figure 16. (**a**): Bar graph of the total mechanical energy (sum of the kinetic and available potential energy) of each detached Loop Current rings (LCRs). The light blue bars represent values computed using the maximum velocity contour as the LCR's edge, while dark blue bars were obtained using the last closed absolute dynamic topography (ADT) contour. The orange curves, referenced on the right-hand side y-axis, represent the cumulative energy input of the LCRs over time. The plain, dotted, and dashed lines represent total mechanical energy (TE), available potential energy (APE), and kinetic energy (KE), respectively. (**b**): KE against APE for each of the 40 detached LCRs. The black line represents equal partition of energy (Burger number equals to one).

It is of interest to put these large numbers back into the context of temperature, salinity and energy balance in the Gulf of Mexico. Attempting a full closed budget of the GoM is beyond the scope of this paper, but we can compute a number of meaningful quantities that highlight the importance of LCRs in the GoM's dynamics.

For instance, it is of interest to estimate the residual net surface heat flux that would be necessary to balance the 0.60 ZJ per year heat input of LCRs into the GoM. Under the assumption that the heat carried by each LCR will eventually totally mix with the GoM water, the necessary residual heat fluxes can be simply estimated by dividing this heat growth rate by the surface of the GoM ($\mathcal{A}_{\text{GOM}}$ = 1.58 millions km^2), or equivalently, since the heat growth rate can be considered as linear, by dividing the total heat input of the 40 detected eddies by the time interval (τ = 29 years) multiplied by the surface of the GoM [20]:

$$\mathcal{Q}_{\text{b}} = \frac{\sum_{i=1}^{40} \mathbb{H}_i}{\tau \mathcal{A}_{\text{GOM}}}. \tag{13}$$

Here, we find that a yearly net residual heat flux of -13 W m^{-2} is necessary to compensate LCRs heat input into the GoM. This value is very close from that of Meunier et al. [20] (14 W m^{-2}), using a simple linear relationship between SSH and local heat content. Because the literature reveals a wide range of residual net surface heat flux estimates (between

−24 and +46 W m^{-2} [38–40]), knowing the value necessary to balance LCRs heat could be helpful to calibrate heat flux products.

A similar argument can be used to estimate the necessary fresh water input in the GoM to balance the 13.5 billion tons of salt excess per year carried by LCRs. Following Meunier et al. [20], the necessary flux of fresh water input is:

$$\mathcal{F}_{\mathrm{b}} = \frac{\sum_{i=1}^{40} \mathbb{S}_i}{\tau \overline{S}_{\mathrm{GOM}}}, \tag{14}$$

where $\overline{S}_{\mathrm{GOM}}$ is the average salinity of the GoM. Here, we find that a fresh water flux of 12,000 $m^3\ s^{-1}$ would be necessary for the GoM's mean salinity to remain constant despite the LCRs salt input. This value is also very close from Meunier et al.'s [20] estimate (12,700 $m^3\ s^{-1}$) using a simple linear relationship between SSH and local salt content. It should be pointed out that these values are closely matching Morey et al.'s [41] recent estimates of the Mississippi river outflow (13,000 $m^3 s^{-1}$), suggesting that the opposing effects of LCRs and the Mississippi river on the GoM's salinity approximately cancel each other.

We can similarly estimate the energy dissipation rate that would be necessary to dissipate LCRs energy:

$$\epsilon_{\mathrm{b}} = \frac{\sum_{i=1}^{40} \mathbb{E}_i}{\tau \rho_0 \mathcal{V}}, \tag{15}$$

where $\mathcal{V}$ is the volume of water in which we expect energy to be dissipated. We explore three different hypotheses: (a) energy is homogeneously dissipated within the whole GoM's volume; (b) energy is dissipated within the top 1000 m; and (c) energy is dissipated within the top 500 m. The necessary dissipation rate is respectively of 1.8, 4.5, and 9.0 × 10^{-10} W kg^{-1}. These values are lower than direct microstructure measurements in the vicinity of LCRs by Molodstov et al.'s [42] ((10^{-9}–10^{-8}) W kg^{-1}), but are consistent with their GoM's background values (10^{-10}–10^{-9}) W kg^{-1}, as well as Whalen et al.'s [43] estimates between 250 and 500 m, using fine-scale strain parameterization (≈5 ×10^{-10} W kg^{-1}).

To emphasize the need for the compensation of heat, salt and energy excesses input to the GoM through LCR detachment, it is of interest to investigate what would happen in the absence of surface heat fluxes, fresh water influx, and energy dissipation. To do so, we computed the temperature, salinity, SSH, and energy density rise that would occur if LCRs detachment was not balanced by any process. As for our energy dissipation estimates, we propose three scenarios: the heat, salt, and energy excess are homogeneously redistributed into: (a) the whole GoM volume; (b) the top 2000 m; and (c) the top 1000 m. For each individual LCR, the mean temperature, salinity, and energy density rises read:

$$\Delta T_i = \frac{\mathbb{H}_i}{\rho_0 C_p \mathcal{V}} \tag{16}$$

$$\Delta S_i = \frac{\mathbb{S}_i}{\mathcal{V}} \tag{17}$$

$$\Delta E_i = \frac{\mathbb{E}_i}{\mathcal{V}} \tag{18}$$

where $\mathcal{V}$ represents weather the full water volume of the GoM, or that of the top 2000 or 1000 m. The equivalent sea level rise is computed as the difference between the steric height associated with the mean GoM temperature and salinity ($\overline{T}$ and $\overline{S}$), and the steric height associated with the hypothetical increased temperature and salinity ($\overline{T} + \Delta T$ and $\overline{S} + \Delta S$). The hypothetical (unbalanced) impacts of each individual LCR, as well as their cumulative impacts over time on temperature, salinity, SSH, and energy density are shown in Figure 17. For the sake of clarity, only hypothesis (a) (redistribution of tracers over the entire GoM volume) is shown in the bar graphs, while the three scenarios are plotted for the cumulative effects. Because of the heterogeneity of LCRs heat, salt, and energy content, the hypothetical unbalanced response of the GoM to individual eddies is highly variable. On average, mixing of a mean LCR into the whole GoM, the top 2000 m, or the top 1000 m would result in a rise of 0.04, 0.06, and 0.10 °C of the GoM's mean temperature, respectively (Figure 17a). For the largest LCRs, these values reach up to 0.12, 0.16, and 0.28 °C. Looking at the cumulative effects of LCRs, between 1993 and 2021, if unbalanced by surface heat fluxes, the GoM's mean temperature would have risen by 1.71, 2.42, and 4.23 °C in the whole GoM, top 2000 m, and top 1000 m mixing scenarios, respectively.

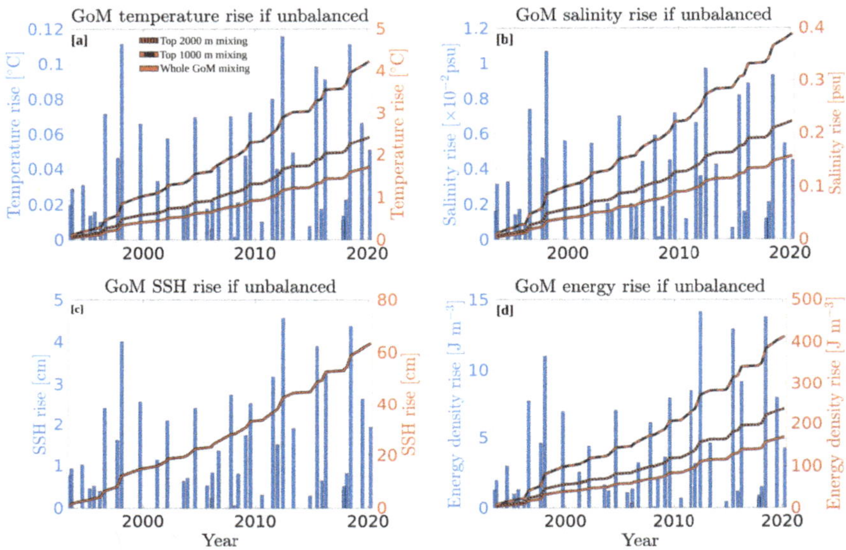

Figure 17. Hypothetical evolution of temperature (**a**), salinity (**b**), sea surface height (**c**), and energy density (**d**) if the heat, salt, and energy inputs of Loop Current rings into the Gulf of Mexico (GoM) were not balanced at all. The bar plots represent the effect of individual eddies while the orange lines represent their cumulative effects over time. The plain line represents the scenario where the excess heat, salt and energy are mixed homogeneously within the entire GoM's volume. The dashed and dotted lines represent scenarios where the excess of heat, salt and energy are mixed within the top 2000 and 1000 m, respectively.

Using similar arguments, in the absence of fresh water input, the average salinity of the GoM would rise by 0.0039 psu after mixing an average LCR within the whole GoM volume (0.0056 and 0.010 psu if the LCRs mixes with the GoM's top 2000 and 1000 m water mass), while the largest individual LCRs could yield a mean salinity increase of 0.011, 0.015, and 0.026 for in the three scenarios (Figure 17b). If unbalanced, the salt excess of LCRs would have induced a salinity rise of 0.16, 0.22, or 0.39 psu depending on the mixing depth, between 1993 and 2021.

Although the halosteric effect associated with the salinity increase would partially compensate the thermosteric effects due to temperature increase, in the absence of balancing processes, LCRs would have caused a sea level rise of 63 cm over the 29 year study period (Figure 17c).

If energy was not dissipated, energy density in the GoM would have increased by 167, 236, or 413 J m^{-3}, depending on the scenario. As an illustrative reference, these levels of energy density would be equivalent to the kinetic energy density of currents of 0.57, 0.68, and 0.89 m s^{-1} over the full water column, top 2000, and top 1000 m of the entire GoM. Obviously, such temperature, salinity, SSH, and energy density increase is not observed, and these hypothetical scenarios are presented to highlight both the crucial importance of LCRs in the GoM's dynamics and budgets, as well as the evident need to accurately measure the surface heat fluxes and fresh water inputs when modelling a semi-enclosed basin with such important advective fluxes.

7. Conclusions

In this work, we applied the GEM method [21,22] to satellite altimetry data, similarly to Swart et al. [23], Stendardo et al. [31], and Müller et al. [24], to reconstruct the three-dimensional structure of individual LCRs in the GoM.

Although the joint use of the GEM and satellite altimetry to infer heat and salt contents of mesoscale eddies was first proposed by Müller et al. [24], here, we extended the method to the computation of the full three-dimensional velocity, vorticity and energy density structure of mesoscale eddies.

The method was validated using independent glider observations, showing that the GEM-reconstruction was able to represent accurately the vertical structure of temperature, salinity, and geostrophic velocity of LCRs, especially when comparing depth-integrated variables.

The application of this three-dimensional reconstruction procedure allowed the success of two primary goals: 1. determine the typical structure of LCRs by computing their average thermohaline and dynamical structure; 2. estimate statistical properties of LCRs heat, salt, and energy contents, as well as their cumulative effect.

Consistent with previous ship and glider observations of individual LCRs [7,27], the typical LCR is characterized by a warm temperature anomaly with weaker stratification, and a double core salinity structure, with a fresher anomaly near the surface and a high salinity anomaly between 150 and 300 m. Although typical LCRs are large eddies, we found that the gradient–wind balanced velocity was significantly larger than the geostrophic velocity ($\approx$+20% between the rotation axis and the maximum velocity radius), similar to Meunier et al.'s [30] recent observation in LCR Poseidon. This results in an increased relative vorticity, reaching half of the Coriolis frequency, showing that average LCRs (medium size) are significantly non-linear eddies with a Rossby radius of 0.5. The average LCR's PV structure consists of a bowl of low PV deflecting the main pycnocline's high PV strip downwards, and is mostly controlled by density stratification. It should be pointed out that the average LCR computed here has a weaker PV anomaly, with a lesser vertical extension, than the recent observations of Meunier et al. [30] of LCR Poseidon. We stress this is related to the exceptionally thick thermostadt observed in Poseidon, while the work is focused on describing an average LCR. However, the GEM-reconstructed energy density structure of the mean LCR exhibited a similar pattern than Meunier et al.'s [30] direct observations, with a clear dominance of APE over KE. However, it should be pointed out that the smoothing of the thermostat by the GEM reconstruction, as compared to the glider observations (Section 4) might slightly bias our estimates of APE and KE. For the two available glider sections, we found that the GEM-reconstructed eddy's APE was about 10% smaller than the glider-measured eddy. Similarly, KE was reduced by about 9% and potential enstrophy (volume integral of the squared PV) by 12.5 and 22% depending on the glider section.

By detecting and studying a large number of LCRs (40), we were able to assess statistical properties of their heat, salt, and energy contents, as well as their cumulative effects on the GoM. One particularly striking characteristic of LCRs is their heterogeneity: the ratio of standard deviation over the mean value for LCRs heat and salt contents is of 0.76 and 0.73, respectively. They, thus, have a very variable impact on heat and salt input into the GoM: the cumulative effect of the 20% largest eddies contribute to half of the total heat and salt input between 1993 and 2021.

As an illustration of the importance of LCRs to the heat, salt, and energy inputs of LCRs in the GoM, we computed the temperature, salinity, SSH, and energy density rises that would occur in the absence of balancing processes and showed that, in the hypothesis that LCRs would eventually mix homogeneously within the entire GoM's volume, the mean temperature and salinity would have increased by nearly 2 °C and 0.15 psu, respectively, between 1993 and 2021, causing a ≈60 cm sea level rise. Over the same period, the energy density would have increased to a level equivalent to mean barotropic currents of nearly 0.6 m s^{-1} over the whole GoM. Another way to appreciate these numbers is to estimate the time it would take for the GoM's mean temperature and salinity (7.7 °C and 35.2 psu for the top 2000 m for the 2010–2020 period) to reach Caribbean values (9.9 °C and 35.36 psu for the same depth range and period). Here, we find that if LCRs heat and salt inputs were not balanced, it would only take 25 years for the GoM to have pure SUW properties. Another particularly striking result is that one single unbalanced large LCR would be able to increase the GoM's mean temperature by over 0.1 °C, yielding a sea level rise of nearly 5 cm.

Obviously, the heat, salt, and energy budgets in the Gulf of Mexico do not fall down to an inevitable accumulation of LCRs input, and the numbers presented in the last paragraph are only intended to emphasize the large impact of LCRs, as well as the need for compensating processes. Because the GoM is a semi-enclosed basin, whose entrance (the Yucatan channel) and exit (the Florida strait) are directly connected by the Loop Current, a straightforward model for volume-integrated budgets is that the heat input of LCRs can be balanced by an outward advective heat flux and surface heat fluxes, their salt input by an outward advective salt flux and fresh water input, and their energy by an outward advective energy flux, energy dissipation, and wind stress work. Because the Florida strait is shallower than the Yucatan channel, and the warm and salty anomaly associated with the SUW reach deeper depths than the strait's depth [44,45], the only possible advective heat flux to partially compensate for the LCRs input would take place in the deeper Yucatan channel. However, Bunge et al. [46] and Candela et al. [47] showed that, despite bursts of outflow through the deep Yucatan channel, mostly related to mass conservation as the Loop Current grows in the GoM, the long term average deep transport is near-zero. Rivas et al. [48] estimated that the advective heat and salt flux through the Yucatan channel were of −30 GW and 1.1 tons per second, respectively. Here, we find that the 29 year trends in LCR heat and salt input are of 19,100 GW and 427 tons of salt per second, so that the advective heat and salt fluxes through the deep Yucatan channel are several orders of magnitudes too small to balance LCRs. LCRs heat, salt, and energy thus must be entirely compensated by surface heat fluxes, fresh water input (river outflow plus precipitation minus evaporation), and energy dissipation. From this very simple remaining balance, we found that net residual heat fluxes of −13 W m^{-2} were necessary to keep the GoM's mean temperature constant, in good agreement with Meunier et al.'s [20] estimate. This number could be useful to calibrate and validate heat flux products in the GoM, as well as regional model configurations.

We also estimated the necessary fresh water input to be of 12,000 m^3 s^{-1}: 5 % less than Meunier et al.'s [20] estimates, and still closely matching the Mississippi river outflow. Note that a fully closed salinity budget of the GoM should also include evaporation and precipitation, which is expected to be a fresh water loss in the GoM, hence requiring larger river outflows for balance to be reached. However, as mentioned above, the scope of this paper is not to make a full budget analysis, but rather to quantify the impact of LCRs and highlight possible balancing processes. Similarly, an attempt to close the GoM's energy budget is beyond the scope of this paper, and would require a careful computation of the wind work and of the buoyancy fluxes through Ekman pumping, using the relative wind (wind minus current velocity) [49–51]. In fact, energy loss of LCRs through relative wind work and energy transfer from APE to KE through Ekman buoyancy fluxes are to be expected for a mesoscale eddy subject to wind forcing [51], and are currently under investigation. However, our simple scaling of the order of magnitude of an equivalent energy dissipation rate ϵ is in good agreement with values observed in the GoM [42,43].

The results reported here highlight the possibility and the utility to reconstruct the three-dimensional structure of individual mesoscale eddies (as opposed to the computation of one single mean composite eddy [12]) from satellite altimetry and the GEM method.

The application of the method used and described in the present paper to other regions of the ocean could help to elucidate the role of coherent mesoscale eddies in basin-scale heat, salt, and energy exchange. However, one should note that the method is not expected to be accurate everywhere in the ocean, since more complex hydrographic conditions may exist and make the computation of a reliable GEM field more difficult. These results also emphasize the crucial role of LCRs in the GoM, and suggests that more research is necessary to elucidate the processes controlling LCRs (and mesoscale coherent eddies in general) mixing and decay processes.

Author Contributions: Conceptualization, T.M.; methodology, T.M. and P.P.-B.; software, T.M.; validation, T.M.; formal analysis, T.M.; investigation, T.M.; resources, A.B.; data curation, T.M.; writing—original draft preparation, T.M.; writing—review and editing, A.B. and P.P.-B.; visualization, T.M.; supervision, T.M., A.B., and P.P.-B.; project administration, A.B.; funding acquisition, A.B. All authors have read and agreed to the published version of the manuscript.

Funding: This work is part of the LC-floats project, funded by the US National Academy of Sciences through the *Understanding Gulf Ocean Systems* grant 2000010488.

Data Availability Statement: All data used in this study are publicly available. The ARGO data can be retrieved from https://argo.ucsd.edu/ (accessed on 1 July 2022). The absolute dynamic topography data are produced by AVISO, and distributed by Copernicus Marine Service (https://resources.marine.copernicus.eu/ (accessed on 1 July 2022)). Glider data were collected by the GMOG group at CICESE (https://gliders.cicese.mx/ (accessed on 1 July 2022)).

Acknowledgments: The authors are grateful to Enric Pallás-Sanz and Miguel Tenreiro from the GMOG group at CICESE for providing the glider data used for validation of the GEM method.

Conflicts of Interest: To the best of their knowledge, the authors have no conflicts of interest.

References

1. Olson, D.B.; Evans, R.H. Rings of the Agulhas current. *Deep Sea Res. A* **1986**, *33*, 27–42. [CrossRef]
2. Wang, Y.; Beron-Vera, F.J.; Olascoaga, M.J. The life cycle of a coherent Lagrangian Agulhas ring. *J. Geophys. Res. Ocean.* **2016**, *121*, 3944–3954. [CrossRef]
3. Richardson, P. Gulf stream rings. In *Eddies in Marine Science*; Springer: Berlin/Heidelberg, Germany, 1983; pp. 19–45.
4. Liu, Y.; Wilson, C.; Green, M.A.; Hughes, C.W. Gulf Stream Transport and Mixing Processes via Coherent Structure Dynamics. *J. Geophys. Res. Ocean.* **2018**, *123*, 3014–3037. [CrossRef]
5. Li, L.; Nowlin, W.D.; Jilan, S. Anticyclonic rings from the Kuroshio in the South China Sea. *Deep Sea Res. Part I Oceanogr. Res.* **1998**, *45*, 1469–1482. [CrossRef]
6. Sasaki, Y.N.; Minobe, S. Climatological Mean Features and Interannual to Decadal Variability of Ring Formations in the Kuroshio Extension Region. In Proceedings of the AGU Fall Meeting Abstracts, 2014; Volume 2014, p. OS41F-07.
7. Elliott, B.A. Anticyclonic Rings in the Gulf of Mexico. *J. Phys. Oceanogr.* **1982**, *12*, 1292–1309. [CrossRef]
8. Donohue, K.A.; Watts, D.R.; Hamilton, P.; Leben, R.; Kennelly, M. Loop Current Eddy formation and baroclinic instability. *Dyn. Atmos. Ocean.* **2016**, *76*, 195–216. [CrossRef]
9. Beron-Vera, F.J.; Olascoaga, M.J.; Wang, Y.; Triñanes, J.; Pérez-Brunius, P. Enduring Lagrangian coherence of a Loop Current ring assessed using independent observations. *Sci. Rep.* **2018**, *8*, 11275. [CrossRef]
10. Andrade-Canto, F.; Karrasch, D.; Beron-Vera, F.J. Genesis, evolution, and apocalypse of Loop Current rings. *Phys. Fluids* **2020**, *32*, 116603. [CrossRef]
11. Chelton, D.B.; Schlax, M.G.; Samelson, R.M. Global observations of nonlinear mesoscale eddies. *Prog. Oceanogr.* **2011**, *91*, 167–216. [CrossRef]
12. Chaigneau, A.; Le Texier, M.; Eldin, G.; Grados, C.; Pizarro, O. Vertical structure of mesoscale eddies in the eastern South Pacific Ocean: A composite analysis from altimetry and Argo profiling floats. *J. Geophys. Res. Ocean.* **2011**, *116*, C11025. [CrossRef]
13. Yang, G.; Wang, F.; Li, Y.; Lin, P. Mesoscale eddies in the northwestern subtropical Pacific Ocean: Statistical characteristics and three-dimensional structures. *J. Geophys. Res. Ocean.* **2013**, *118*, 1906–1925. [CrossRef]
14. Yang, G.; Yu, W.; Yuan, Y.; Zhao, X.; Wang, F.; Chen, G.; Liu, L.; Duan, Y. Characteristics, vertical structures, and heat/salt transports of mesoscale eddies in the southeastern tropical Indian Ocean. *J. Geophys. Res. Ocean.* **2015**, *120*, 6733–6750. [CrossRef]
15. Keppler, L.; Cravatte, S.; Chaigneau, A.; Pegliasco, C.; Gourdeau, L.; Singh, A. Observed Characteristics and Vertical Structure of Mesoscale Eddies in the Southwest Tropical Pacific. *J. Geophys. Res. Ocean.* **2018**, *123*, 2731–2756. [CrossRef]
16. He, Q.; Zhan, H.; Cai, S.; He, Y.; Huang, G.; Zhan, W. A New Assessment of Mesoscale Eddies in the South China Sea: Surface Features, Three-Dimensional Structures, and Thermohaline Transports. *J. Geophys. Res. Ocean.* **2018**, *123*, 4906–4929. [CrossRef]

17. de Marez, C.; L'Hégaret, P.; Morvan, M.; Carton, X. On the 3D structure of eddies in the Arabian Sea. *Deep Sea Res. Part I Oceanogr. Res.* **2019**, *150*, 103057. [CrossRef]
18. Laxenaire, R.; Speich, S.; Stegner, A. Evolution of the Thermohaline Structure of One Agulhas Ring Reconstructed from Satellite Altimetry and Argo Floats. *J. Geophys. Res. Ocean.* **2019**, *124*, 8969–9003. [CrossRef]
19. Sandalyuk, N.V.; Bosse, A.; Belonenko, T.V. The 3-D Structure of Mesoscale Eddies in the Lofoten Basin of the Norwegian Sea: A Composite Analysis From Altimetry and In Situ Data. *J. Geophys. Res. Ocean.* **2020**, *125*, e16331. [CrossRef]
20. Meunier, T.; Sheinbaum, J.; Pallàs-Sanz, E.; Tenreiro, M.; Ochoa, J.; Ruiz-Angulo, A.; Carton, X.; de Marez, C. Heat Content Anomaly and Decay of Warm-Core Rings: The Case of the Gulf of Mexico. *Geophys. Res. Lett.* **2020**, *47*, e85600. [CrossRef]
21. Watts, D.R.; Sun, C.; Rintoul, S. A Two-Dimensional Gravest Empirical Mode Determined from Hydrographic Observations in the Subantarctic Front. *J. Phys. Oceanogr.* **2001**, *31*, 2186–2209. [CrossRef]
22. Sun, C.; Watts, D.R. A circumpolar gravest empirical mode for the Southern Ocean hydrography. *J. Geophys. Res. Ocean.* **2001**, *106*, 2833–2855. [CrossRef]
23. Swart, S.; Speich, S.; Ansorge, I.J.; Lutjeharms, J.R.E. An altimetry-based gravest empirical mode south of Africa: 1. Development and validation. *J. Geophys. Res. Ocean.* **2010**, *115*, C03002. [CrossRef]
24. Müller, V.; Kieke, D.; Myers, P.G.; Pennelly, C.; Steinfeldt, R.; Stendardo, I. Heat and Freshwater Transport by Mesoscale Eddies in the Southern Subpolar North Atlantic. *J. Geophys. Res. Ocean.* **2019**, *124*, 5565–5585. [CrossRef]
25. Hamilton, P.; Leben, R.; Bower, A.; Furey, H.; Pérez-Brunius, P. Hydrography of the Gulf of Mexico Using Autonomous Floats. *J. Phys. Oceanogr.* **2018**, *48*, 773–794. [CrossRef]
26. Portela, E.; Tenreiro, M.; Pallàs-Sanz, E.; Meunier, T.; Ruiz-Angulo, A.; Sosa-Gutiérrez, R.; Cusí, S. Hydrography of the Central and Western Gulf of Mexico. *J. Geophys. Res. Ocean.* **2018**, *123*, 5134–5149. [CrossRef]
27. Meunier, T.; Pallás-Sanz, E.; Tenreiro, M.; Portela, E.; Ochoa, J.; Ruiz-Angulo, A.; Cusí, S. The vertical structure of a Loop Current Eddy. *J. Geophys. Res. Ocean.* **2018**, *123*, 6070–6090. [CrossRef]
28. Meunier, T.; Tenreiro, M.; Pallàs-Sanz, E.; Ochoa, J.; Ruiz-Angulo, A.; Portela, E.; Cusí, S.; Damien, P.; Carton, X. Intrathermocline Eddies Embedded Within an Anticyclonic Vortex Ring. *Geophys. Res. Lett.* **2018**, *45*, 7624–7633. [CrossRef]
29. Meunier, T.; Pallàs Sanz, E.; Tenreiro, M.; Ochoa, J.; Ruiz Angulo, A.; Buckingham, C. Observations of Layering under a Warm-Core Ring in the Gulf of Mexico. *J. Phys. Oceanogr.* **2019**, *49*, 3145–3162. [CrossRef]
30. Meunier, T.; Pallás Sanz, E.; de Marez, C.; Pérez, J.; Tenreiro, M.; Ruiz Angulo, A.; Bower, A. The Dynamical Structure of a Warm Core Ring as Inferred from Glider Observations and Along-Track Altimetry. *Remote Sens.* **2021**, *13*, 2456. [CrossRef]
31. Stendardo, I.; Rhein, M.; Hollmann, R. A high resolution salinity time series 1993–2012 in the North Atlantic from Argo and Altimeter data. *J. Geophys. Res. Ocean.* **2016**, *121*, 2523–2551. [CrossRef]
32. Leben, R.R. Altimeter-derived loop current metrics. *Wash. Am. Geophys. Union Geophys. Monogr. Ser.* **2005**, *161*, 181–201. [CrossRef]
33. Isern-Fontanet, J.; Font, J.; García-Ladona, E.; Emelianov, M.; Millot, C.; Taupier-Letage, I. Spatial structure of anticyclonic eddies in the Algerian basin (Mediterranean Sea) analyzed using the Okubo Weiss parameter. *Deep Sea Res. Part II Top. Stud. Oceanogr.* **2004**, *51*, 3009–3028. [CrossRef]
34. Rudnick, D.L.; Johnston, T.M.S.; Sherman, J.T. High-frequency internal waves near the Luzon Strait observed by underwater gliders. *J. Geophys. Res. Ocean.* **2013**, *118*, 774–784. [CrossRef]
35. Zhang, Z.; Zhang, Y.; Wang, W.; Huang, R.X. Universal structure of mesoscale eddies in the ocean. *Geophys. Res. Lett.* **2013**, *40*, 3677–3681. [CrossRef]
36. Holton, J.R. *An Introduction to Dynamic Meteorology*; Academic Press: Boston, MA, USA, 1992.
37. Holliday, D.; McIntyre, M.E. On potential energy density in an incompressible, stratified fluid. *J. Fluid Mech.* **1981**, *107*, 221–225. [CrossRef]
38. Etter, P.C. Heat and Freshwater Budgets of the Gulf of Mexico. *J. Phys. Oceanogr.* **1983**, *13*, 2058–2069. [CrossRef]
39. Adem, J.; Villanueva, E.; Mendoza, V. A new method for estimating the seasonal cycle of the heat balance at the ocean surface, with application to the Gulf of Mexico. *Geofísica Int.* **1993**, *32*, 21–34. [CrossRef]
40. Zavala-Hidalgo, J.; PARÉS-SIERRA, A.; Ochoa, J. Seasonal variability of the temperature and heat fluxes in the Gulf of Mexico. *Atmósfera* **2002**, *15*, 81–104.
41. Morey, S.L.; Martin, P.J.; O'Brien, J.J.; Wallcraft, A.A.; Zavala-Hidalgo, J. Export pathways for river discharged fresh water in the northern Gulf of Mexico. *J. Geophys. Res. Ocean.* **2003**, *108*, 1. [CrossRef]
42. Molodtsov, S.; Anis, A.; Amon, R.M.W.; Perez-Brunius, P. Turbulent Mixing in a Loop Current Eddy From Glider-Based Microstructure Observations. *Geophys. Res. Lett.* **2020**, *47*, e88033. [CrossRef]
43. Whalen, C.B.; MacKinnon, J.A.; Talley, L.D. Large-scale impacts of the mesoscale environment on mixing from wind-driven internal waves. *Nat. Geosci.* **2018**, *11*, 842–847. [CrossRef]
44. Candela, J.; Sheinbaum, J.; Ochoa, J.; Badan, A.; Leben, R. The potential vorticity flux through the Yucatan Channel and the Loop Current in the Gulf of Mexico. *Geophys. Res. Lett.* **2002**, *29*, 16-1–16-4. [CrossRef]
45. Sheinbaum, J.; Athié, G.; Candela, J.; Ochoa, J.; Romero-Arteaga, A. Structure and variability of the Yucatan and loop currents along the slope and shelf break of the Yucatan channel and Campeche bank. *Dyn. Atmos. Ocean.* **2016**, *76*, 217–239. [CrossRef]
46. Bunge, L.; Ochoa, J.; Badan, A.; Candela, J.; Sheinbaum, J. Deep flows in the Yucatan Channel and their relation to changes in the Loop Current extension. *J. Geophys. Res. Ocean.* **2002**, *107*, 3233. [CrossRef]

47. Candela, J.; Tanahara, S.; Crepon, M.; Barnier, B.; Sheinbaum, J. Yucatan Channel flow: Observations versus CLIPPER ATL6 and MERCATOR PAM models. *J. Geophys. Res. Ocean.* **2003**, *108*, 3385. [CrossRef]
48. Rivas, D.; Badan, A.; Ochoa, J. The Ventilation of the Deep Gulf of Mexico. *J. Phys. Oceanogr.* **2005**, *35*, 1763. [CrossRef]
49. Gaube, P.; Chelton, D.B.; Samelson, R.M.; Schlax, M.G.; O'Neill, L.W. Satellite Observations of Mesoscale Eddy-Induced Ekman Pumping. *J. Phys. Oceanogr.* **2015**, *45*, 104–132. [CrossRef]
50. Renault, L.; Molemaker, M.J.; Gula, J.; Masson, S.; McWilliams, J.C. Control and Stabilization of the Gulf Stream by Oceanic Current Interaction with the Atmosphere. *J. Phys. Oceanogr.* **2016**, *46*, 3439–3453. [CrossRef]
51. Wilder, T.; Zhai, X.; Munday, D.; Joshi, M. The response of a baroclinic anticyclonic eddy to relative wind stress forcing. *J. Phys. Oceanogr.* **2022**. [CrossRef]

remote sensing

Article

Spatiotemporal Variability of Mesoscale Eddies in the Indonesian Seas

Zhanjiu Hao [1,2], Zhenhua Xu [1,3,4,5,*], Ming Feng [6,7], Qun Li [8] and Baoshu Yin [1,2,3,4,5]

1 CAS Key Laboratory of Ocean Circulation and Waves, Institute of Oceanology, Chinese Academy of Sciences, Qingdao 266071, China; haozhanjiu@qdio.ac.cn (Z.H.); bsyin@qdio.ac.cn (B.Y.)
2 University of Chinese Academy of Sciences, Beijing 100049, China
3 Pilot National Laboratory for Marine Science and Technology, Qingdao 266237, China
4 Center for Ocean Mega-Science, Chinese Academy of Sciences, Qingdao 266071, China
5 CAS Engineering Laboratory for Marine Ranching, Institute of Oceanology, Chinese Academy of Sciences, Qingdao 266071, China
6 CSIRO Oceans and Atmosphere, Indian Ocean Marine Research Centre, Crawley, WA 6009, Australia; ming.feng@csiro.au
7 Centre for Southern Hemisphere Oceans Research, Hobart, TAS 7000, Australia
8 Polar Research Institute of China, Shanghai 200136, China; liqun@pric.org.cn
* Correspondence: xuzhenhua@qdio.ac.cn

Citation: Hao, Z.; Xu, Z.; Feng, M.; Li, Q.; Yin, B. Spatiotemporal Variability of Mesoscale Eddies in the Indonesian Seas. *Remote Sens.* **2021**, *13*, 1017. https://doi.org/10.3390/rs13051017

Academic Editors: Chung-Ru Ho, Antony K. Liu and Xiaofeng Li

Received: 25 January 2021
Accepted: 4 March 2021
Published: 8 March 2021

Abstract: Mesoscale eddies are ubiquitous in the world ocean and well researched both globally and regionally, while their properties and distributions across the whole Indonesian Seas are not yet fully understood. This study investigates for the first time the spatiotemporal variations and generation mechanisms of mesoscale eddies across the whole Indonesian Seas. Eddies are detected from altimetry sea level anomalies by an automatic identification algorithm. The Sulu Sea, Sulawesi Sea, Maluku Sea and Banda Sea are the main eddy generation regions. More than 80% of eddies are short-lived with a lifetime below 30 days. The properties of eddies exhibit high spatial inhomogeneity, with the typical amplitudes and radiuses of 2–6 cm and 50–160 km, respectively. The most energetic eddies are observed in the Sulawesi Sea and Seram Sea. Eddies feature different seasonal cycles between anticyclonic and cyclonic eddies in each basin, especially given that the average latitude of the eddy centroid has inverse seasonal variations. About 48% of eddies in the Sulawesi Sea are highly nonlinear, which is the case for less than 30% in the Sulu Sea and Banda Sea. Instability analysis is performed using high-resolution model outputs from Bluelink Reanalysis to assess mechanisms of eddy generation. Barotropic instability of the mean flow dominates eddy generation in the Sulu Sea and Sulawesi Sea, while baroclinic instability is slightly more in the Maluku Sea and Banda Sea.

Keywords: mesoscale eddies; the Indonesian Seas; sea level anomaly; nonlinearity; barotropic instability; baroclinic instability

1. Introduction

The progress of numerical models and remote sensing techniques, especially satellite altimeter, in the past few decades has greatly advanced our understanding of mesoscale eddies in the ocean, which contain most of the oceanic kinematic energy [1]. Mesoscale eddies have characteristic spatial and temporal scales ranging from tens to hundreds of kilometers and from several days to years, respectively [2]. Both cyclonic (CE) and anticyclonic (AE) eddies are ubiquitous in the ocean [1], serving as a key bridge of energy cascade between large-scale and sub-mesoscale oceanic dynamics [3–5]. Because of their nonlinearity, mesoscale eddies play a vital role in the transport of momentum, mass, heat and biogeochemical properties and further impact tracer budgets and primary production [2,6].

Most existing analyses focused on several hot spots, such as strong meandering currents, the eastern boundary upwelling system and marginal seas [7–13]. However, eddies at low latitudes, where meridional heat transport was found to be significant, are not well studied [14]. To our knowledge, the sparse temporal resolution of altimeter data over the past few decades restricted researchers to only focusing on the long-lived eddies mainly populating middle to high latitudes [15]. In addition, numerical models may not reproduce realistic positions and properties of eddies because of their highly variable nature [16]. Fortunately, the daily-resolution merged sea surface height (SSH) products have been available in recent years, which allow us to study short-lived eddies with a lifetime of several days [15,17]. Hence, more and more investigators have been addressing the spatiotemporal variations of eddies in low latitudes, such as the Tropical Atlantic Ocean and Peruvian oceanic region [18,19]. Based on daily-resolution SSH products, Chen and Han [15] find that an apparent short-lived eddy zonal belt populates the tropical ocean, including the Indo-Pacific warm pool.

The Indonesian Seas, with numerous narrow channels connecting seas of different sizes and depths, provide the sole tropical pathway from the Pacific to the Indian Ocean known as the Indonesian Throughflow (ITF) [20,21]. Extensive studies based on the moorings and numerical models revealed the upper-ocean circulation in this area as schematically shown in Figure 1. In the west route of the ITF, the Mindanao Current (MC) divides into three branches south of the Mindanao Island: one flows eastward into the North Equatorial Countercurrent (NECC); another one flows southwestward firstly and then turns back into the Pacific; and the thirds intrudes into the Sulawesi Sea [22,23]. For the intrusion of MC, a portion turns eastward and flows back into the Pacific along the pathway north of the Sulawesi Island; and the rest flows southward through Makassar Strait, known as the Makassar Strait Throughflow (MST), then flows into the Flores Sea and Banda Sea [24]. Meanwhile, the east route of the ITF is from the Maluku Sea and Halmahera Sea to the Seram Sea then into the Banda Sea [23,25]. Additionally, small amounts of water from the South China Sea (SCS) are carried into the Sulu Sea and Java Sea via the Mindoro Strait and Karimata Strait (KS), respectively [26,27]. The ITF waters eventually exit the Indonesian Seas via the narrow straits of Lesser Sunda Islands and the Timor Passage [20,21]. As one of the significant features in the ocean circulation, mesoscale eddies can exchange energy with background currents through eddy-mean flow interaction [3–5]. Although we presently have a more advanced understanding of the ITF and its intraseasonal-to-decadal variability [20,21], eddies in the Indonesian Seas, significantly contributing to the intraseasonal variation of ITF transport, are poorly understood [28,29].

Mesoscale eddies in the Indonesian Seas exhibit multiscale temporal variations associated with the ITF. For the intraseasonal scale, a 50-day oscillation of horizontal velocity was observed from moorings at the entrance of Sulawesi Sea [30]. Based on a 1.5-layer reduced-gravity model, Qiu et al. [28] pointed out that this intense 50-day oscillation in the Sulawesi basin is a result of baroclinic Rossby wave resonance. The 50-day oscillation signals also exhibit intense interannual variability modulated by active eddy shedding, and enhanced 50-day oscillation can freshen the upper-ocean water mass in the Sulawesi Sea and Makassar Strait [31]. Through an ocean general circulation model (OGCM), three eddies in the Flores Sea were simulated in austral summer when ITF transport is low, and these eddies vary synchronously at an interannual scale, thus named as "Lombok Eddy Train" [32]. In terms of the seasonal cycle, the variation of eddy kinematic energy (EKE) in the Sulawesi Sea has different periods at different depths, and the annual and semi-annual peaks are in the 0–100 m layer and the 100–300 m layer, respectively [33]. In the Sulu Sea, AEs and CEs display an opposite seasonal variability, with more and larger AEs in boreal winter and CEs in boreal summer [34]. However, a comprehensive description of properties, seasonal variability and generation mechanism for eddies in the Indonesian Seas is still lacking.

This study provides a comprehensive statistical description of the spatiotemporal variability of eddies in the Indonesian Seas using long term altimeter data. The mechanisms of eddy generation are investigated through instability analysis based on high-resolution model outputs. We believe that an overall presentation of mesoscale eddies in the Indonesian Seas will not only improve local ocean forecasts but also facilitate a better understanding of their roles in climate and ecosystems. The remainder of this paper is organized as follows: Section 2 describes the details of data and methods used in this study. Section 3 provides a statistical description of eddy properties. The mechanisms of eddy generation are discussed in Section 4. Section 5 provides a summary of the results.

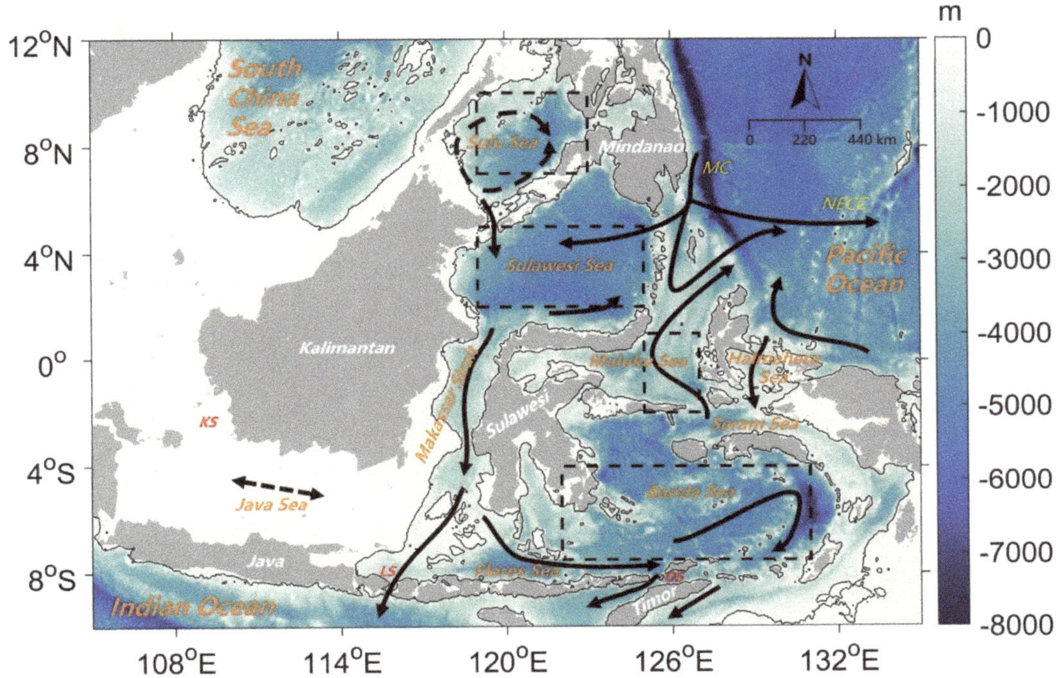

Figure 1. Schematic of the upper-ocean circulation in the Indonesian Seas. Color shading is the bathymetry from ETOPO1. The solid line is the 200 m isobath. KS, LS and OS represent the Karimata Strait, Lombok Strait and Ombai Strait, respectively. MC and NECC represent the Mindanao Current and North Equatorial Countercurrent, respectively. Four dashed boxes are the Sulu Box, Sulawesi Box, Maluku Box and Banda Box from north to south, respectively.

2. Materials and Methods

2.1. Altimeter Data

The delayed-time altimeter data used in this study is a Ssalto/Ducas gridded product of SSH, sea level anomaly (SLA) and geostrophic current from January 1992 to December 2019 provided by Archiving, Validation, and Interpolation of Satellite Oceanographic (AVISO) data and distributed by the Copernicus Marine Environment Monitoring Service (CMEMS). This product has been widely used to detect mesoscale eddies globally [17,35] and regionally [12,19,34,36]. The SLA product was constructed by merging multi-mission satellites since 1992 optimally interpolated onto 0.25° × 0.25° grid with a daily resolution. The geostrophic velocity was computed by the Lagerloef methodology [37] introducing the β-plane approximation in the equatorial band (5°S~5°N), and by the 9-point stencil width ("stencil width" means the number of grid points utilized to estimate the finite difference approximation to the derivative on a grid) methodology outside the equatorial band [38]. Following Chelton et al. [2], the SLA fields are therefore spatially high-pass filtered with half-power filter cutoffs of 20° longitude × 10° latitude to remove steric heating and cooling effects, as well as other large-scale variability of SLA. The readers should be reminded that altimeter data become less accurate over the shelf area due to some aliases from the tides, coastal wave signals and sea-land transition [39]. Thus, the altimeter data in areas shallower than 200 m are masked in this study.

2.2. Bluelink ReANalysis

To study the mechanisms of eddy generation, we use the latest model outputs from Bluelink ReANalysis (BRAN, version of 3p5) to perform an instability analysis. The BRAN model is a multi-year integration of Ocean Forecasting Australian Model (OFAM) assimilating observations of SLA and SST from satellite, sea level data from tide gauges and in situ temperature and salinity profiles by using Bluelink Ocean Data Assimilation System (BODAS) [40]. This model domain covers the Indonesian Seas and surrounding oceans with a horizontal resolution of 0.1° and a vertical resolution of 10 m in the upper 300 m [40]. Compared with existing observations (e.g., moorings, drifters and fields survey) which are not assimilated into the model, BRAN performs well in the Indonesian Seas and surrounding ocean and can especially capture the realistic details of seasonal circulation and its interactions with various topographic features [41–43]. The entire dataset covers the period from January 1994 to August 2016. The daily outputs of sea level, zonal velocity, meridional velocity, potential temperature and salinity in the upper 300 m from January 1994 to December 2015 are used for the instability analysis.

To compare the time-varying signals of SSH between BRAN and AVISO, an empirical orthogonal function (EOF) analysis is performed to both daily BRAN and AVISO SSH data with a daily resolution from 1994 to 2015. The first two modes of BRAN and AVISO reveal good resemblance in both spatial patterns and principal components (PC), and their cumulative explained variances are 64.42% and 63.89%, respectively, which can describe the main characteristics of SSH variations in the Indonesian Seas and surrounding oceans. The spatial patterns and PCs of EOF mode 1 from BRAN and AVISO are displayed in Figure 2. Mode 1, respectively, explains 45.55% and 42.89% of the total variance in the BRAN and AVISO. BRAN and AVISO share a similar spatial pattern for EOF mode 1 (Figure 2a,b). The time series of two PCs also match favorably with a high correlation coefficient of 0.98. The EOF mode 1 physically characterizes the large-scale variations of SSH in response to wind forcing dominantly modulated by the El Niño-Southern Oscillation (ENSO) [31].

The spatial patterns and PCs for EOF mode 2 of SSH from BRAN and AVISO are shown in Figure 3. BRAN captures the EOF mode 2 of observed SSH field faithfully, not only in spatial patterns (Figure 3a,b) but also in PCs (Figure 3c). The variances explained by the EOF mode 2 are 18.87% and 21% for BRAN and AVISO, respectively, and the correlation coefficient between two PCs is 0.98. From the PCs (Figure 3c), it is clear that the EOF mode 2 represents the seasonal SSH variations in the Indonesian Seas and surrounding oceans. For example, Figure 3a,b indicate the surface circulations in the Sulu Sea and southern SCS exhibit a clear seasonal cycle, with an anticyclonic circulation during summer and a cyclonic circulation during winter, which has been proved based on both numerical models and observations [44,45]. The fact that BRAN simulates successfully the first two EOF modes of observed SSH signals is important since the large-scale circulation patterns tend to modulate mesoscale eddy activities.

2.3. Eddy Detection and Tracking Algorithm

Numerous automatic eddy detection algorithms have been developed based on the physical or geometric criteria, and they can be divided into three categories: (1) the physical parameter method, such as the Okubo-Weiss parameter method [46]; (2) the flow geometry method, including the winding-angle method [18,47] and the vector geometry method [48]; (3) the SSH-based method [2,17,35]. However, not all algorithms are suitable for identifying mesoscale eddies in the Indonesian Seas according to three reasons. Firstly, the SSH-based method performs better than the Okubo-Weiss parameter method because of its ability to avoid noise and excess eddy detections [2]. Secondly, the flow geometry algorithms require higher resolution data to get an accurate flow field to identify eddies, and the existing observational data in the Indonesian Seas cannot satisfy this demand [35]. Therefore, we adopted the SSH-based method developed by Faghmous et al. [17] (hereafter JHF15) which has been applied in the Kuroshio Extension Region [10], the Bay of Bengal [36] and the Southeastern Indian Ocean [12]. JHF15 identifies an eddy as a closed SLA contour with a single extreme. It is considered as a parameter-free method in which no empirical parameters are applied and the identified eddies' edges depend only on the single extreme approximation. In this study, both altimeter data and model outputs are used for eddy detection. For each dataset, all identified eddies larger than 9 corresponding grid cells are kept, to avoid some spurious features, and more details of JHF15 refer to Faghmous et al. [17]. The reader should note that the results in Section 3 are only from the altimeter data, while eddy detections from the model outputs are used to assess the capability of BRAN in terms of capturing mesoscale eddies.

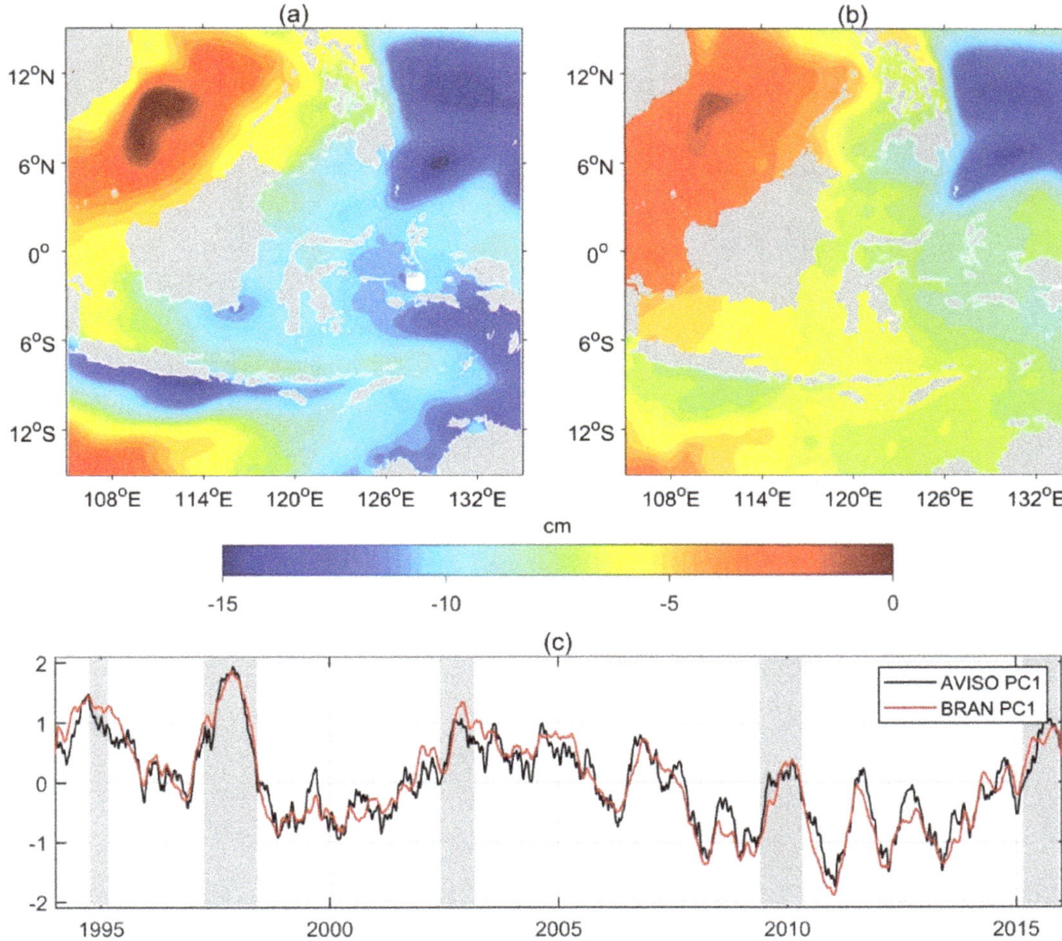

Figure 2. Spatial pattern of empirical orthogonal function mode 1 of sea surface height from (**a**) AVISO and (**b**) BRAN. (**c**) Principal component of empirical orthogonal function mode 1 from AVISO (black line) and BRAN (red line). The gray shadings in (**c**) represent El Niño events.

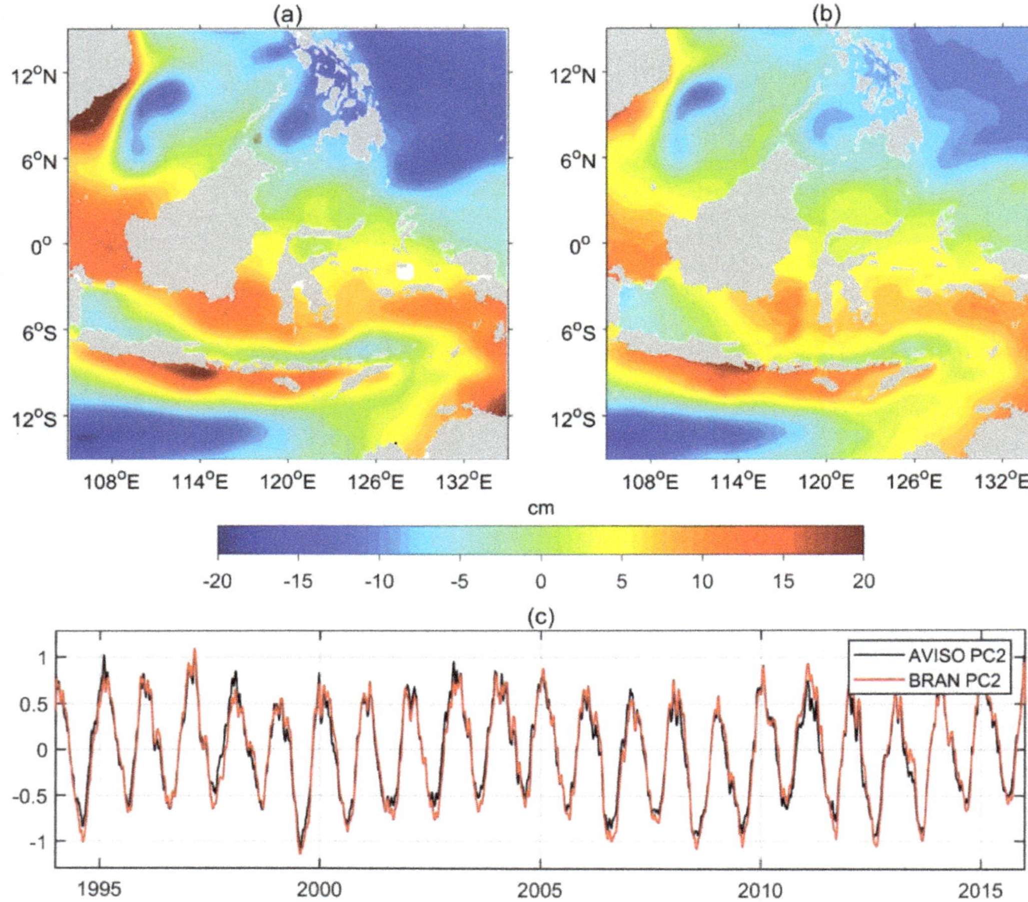

Figure 3. Spatial pattern of empirical orthogonal function mode 2 of sea surface height from (**a**) AVISO and (**b**) BRAN. (**c**) Principal component of empirical orthogonal function mode 2 from AVISO (black line) and BRAN (red line).

After eddies are detected in each daily SLA snapshot, they are tracked by an algorithm developed by Penven et al. [49]. The nondimensional distance of an eddy pair with the same polarity from two consecutive maps is defined as:

$$D_{e_1,e_2} = \sqrt{\left(\frac{\Delta D}{D_0}\right)^2 + \left(\frac{\Delta A}{A_0}\right)^2 + \left(\frac{\Delta a}{a_0}\right)^2}, \quad (1)$$

where ΔD, ΔA and Δa are, respectively, the differences in eddy centroid location, surface area and amplitude between e_1 and e_2 in two consecutive maps; and the characteristic length scale D_0, the characteristic surface area A_0, and the characteristic amplitude a_0 are 25 km, $\pi 60^2$ km^2 and 2 cm, respectively. The smaller D_{e_1,e_2}, the higher similarity of the eddy pair. Due to the sampling errors and measurement noise of satellite altimeter, some eddies may disappear at some time and reappear several time steps later. If an eddy moves into the gap among satellites' orbits, it will also vanish in the SLA map. To solve this problem, we repeat this tracking algorithm for a longer time step, from 2 to 7 days, for those unpaired eddies. We also set the maximum search distance ΔD to 150 km per week to avoid eddies jumping to another track. Following He et al. [34] and Zhang et al. [12], the number of eddy tracks and the number of eddies are different. The former is counted when an eddy track is once identified, while the latter is the total number of eddies along the eddy tracks.

2.4. Definition of Eddy Properties

The amplitude of eddy a is defined as:

$$a = \left| SLA_{center} - \left\langle SLA_{edge} \right\rangle \right|, \tag{2}$$

where $\left\langle SLA_{edge} \right\rangle$ is the mean value of SLA at the eddy edge.

The surface area of eddy A is the area delimited by the outermost contour of SLA, the eddy edge, whereas its apparent radius R corresponds to the radius of an equivalent circular eddy with the same area. Thus, R is calculated as:

$$R = \sqrt{A/\pi}. \tag{3}$$

The mean EKE of eddy is calculated as:

$$\overline{EKE} = \frac{\iint_s \frac{1}{2}\left(u_g'^2 + v_g'^2\right) dxdy}{A}, \tag{4}$$

where u_g' and v_g' are the zonal and meridional components of geostrophic velocity anomaly, respectively, calculated from SLA.

The mean vorticity of eddy is calculated as:

$$\overline{\xi} = \frac{\iint_s \frac{1}{2}\left(\frac{\partial v_g'}{\partial x} - \frac{\partial u_g'}{\partial y}\right) dx\, dy}{A}. \tag{5}$$

2.5. Instability Analysis

The EKE is generally converted from mean-flow kinematic energy (MKE) through barotropic instability and Kelvin-Helmholtz instability or from eddy potential energy (EPE) via baroclinic instability [3–5,50]. Therefore, the energy conversion rates for the above three instabilities are calculated as follows [3,12,50]:

Barotropic conversion rate (BTR) from MKE to EKE via barotropic instability, depending on horizontal shears of mean flow, is defined as:

$$BTR = -\left(\overline{u'u'}\frac{\partial \overline{u}}{\partial x} + \overline{u'v'}\frac{\partial \overline{v}}{\partial x} + \overline{u'v'}\frac{\partial \overline{u}}{\partial y} + \overline{v'v'}\frac{\partial \overline{v}}{\partial y}\right). \tag{6}$$

Kelvin-Helmholtz conversion rate (KHR) from MKE to EKE via Kelvin-Helmholtz instability depending on vertical shears of mean flow and Reynolds stresses, is defined as:

$$KHR = -\left(\overline{u'w'}\frac{\partial \overline{u}}{\partial z} + \overline{v'w'}\frac{\partial \overline{v}}{\partial z}\right). \tag{7}$$

Baroclinic conversion rate (BCR) from EPE to EKE via baroclinic instability is defined as:

$$BCR = -\frac{g\overline{\rho'w'}}{\rho_0}, \tag{8}$$

where ρ_0 is the background density set to 1030 kg/m^3.

The overbars and primes represent time mean (1 month) and anomalies from time mean, respectively.

3. Results

In the Indonesian Seas, a total of 46,676 AEs (Figure 4a) and 47,004 CEs (Figure 4b) are identified from January 1993 to December 2018. Most identified eddies are concentrated on four larger and deeper basins (i.e., the Sulu Sea, Sulawesi Sea, Maluku Sea and Banda Sea). Due to the shallower depth (<200 m) and smaller basin scale, fewer eddies are detected in the rest of the Indonesian Seas. To obtain robust statistical description for eddy properties, eddies with their amplitudes larger than 2 cm are analyzed in our study, with the consideration that the accuracy of SLA gridded data from AVISO is about 2 cm [15,51].

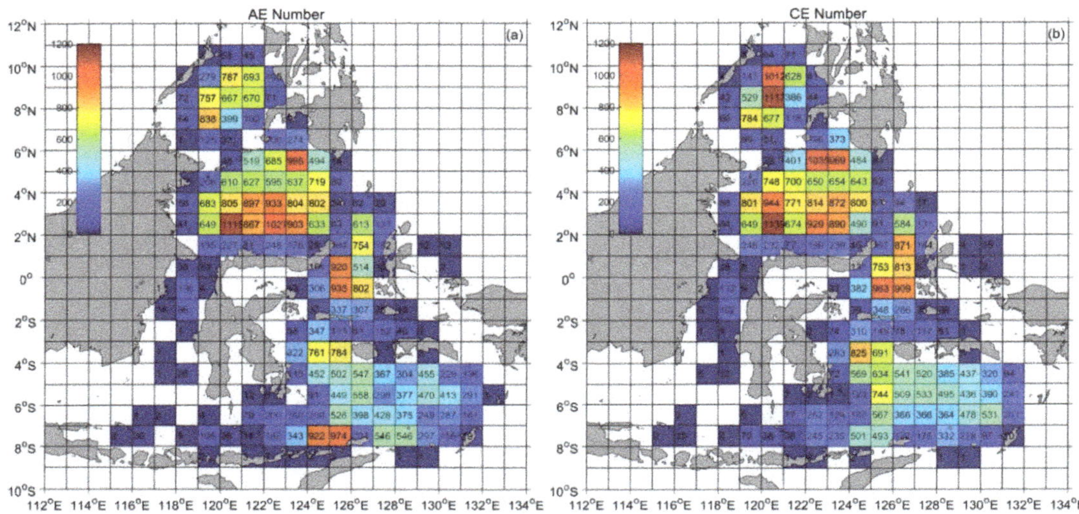

Figure 4. Spatial distribution of the numbers of identified (**a**) anticyclonic (AE) and (**b**) cyclonic (CE) eddies in the Indonesian Seas over the 1993–2018 period.

3.1. Eddy Geneis and Decay

The spatial distributions of the numbers of eddy genesis and decay events for AE and CE are shown in Figure 5. The locations of eddy for generation and decay are the first and the last point in each eddy track, respectively. Because the minimum resolvable temporal scale of gridded SLA product from AVISO is about 10 days [52], we excluded those eddies with a lifespan shorter than 10 days. Consequently, a total of 469 AE tracks and 500 CE tracks are identified in the Indonesian Seas. AE tracks in the Sulu Sea, Sulawesi Sea, Maluku Sea and Banda Sea, respectively, account for 15%, 53%, 7% and 25% of the total AE tracks, while the proportion of CE tracks are 15%, 52%, 7% and 27%, respectively. Hence, no apparent regional preference for AE and CE is found in the Indonesian Seas as is also observed in the Tropical Atlantic Ocean [19]. Due to the removal of SLA data in the area shallower than 200 m, the eddy generation and decay are less frequent near the coast. It is interesting that both AE and CE preferentially generate in the areas north of Ombai Strait (Figure 5a,b), which may result from the interactions of intense mean flow with topographic features [53,54]. However, there are some differences of the geographical pattern for the formation and decay between AE and CE. We find that the generation of AE is concentrated on the southeastern portion of the Sulu Sea (Figure 5a) while CE tend to form in northeastern portion of the Sulu Sea (Figure 5b). Additionally, the larger value of CE formation southwest of the Mindanao Island may partially result from the interaction between eastward currents and the coastline [53,54]. In the Sulawesi Sea, AE usually decays in the central and western part (Figure 5c) while the death of CE concentrates on the western and northeastern part (Figure 5d).

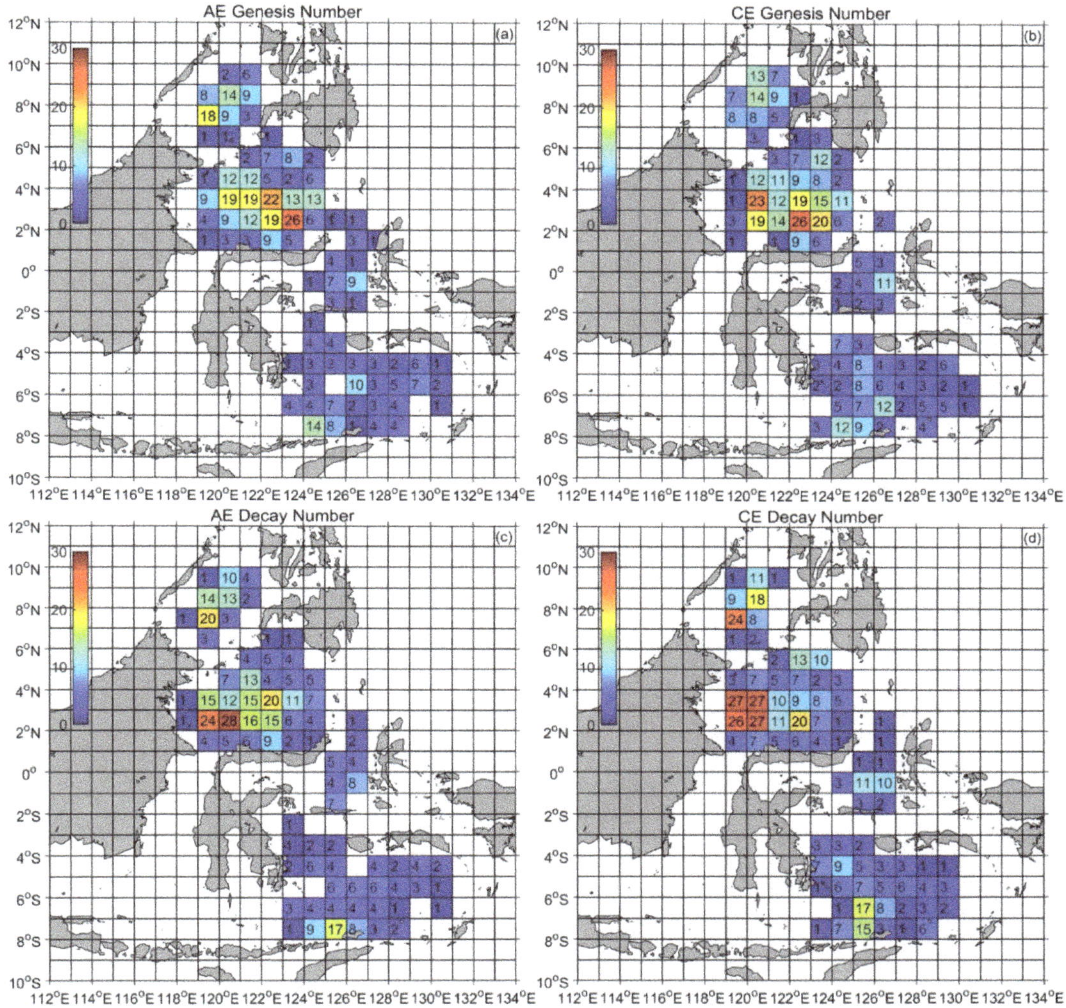

Figure 5. Spatial distribution of eddy genesis and decay events over the 1993–2018 period: (**a**) AE genesis events, (**b**) CE genesis events, (**c**) AE decay events and (**d**) CE decay events. The unit is the number of events.

3.2. Eddy Propagation

The propagation velocities are calculated by a forward difference scheme for eddy centroid displacement every day. The velocities of each eddy track were further filtered using a central moving average of 7 days to reduce the random noise and then averaged in each 0.25° × 0.25° grid, following He et al. [34]. Figure 6 displays the mean propagation velocity fields and the standard deviation (STD) of azimuth of propagation direction relatively to the west for AE and CE. In the Sulu Sea, the AE (Figure 6a) and CE (Figure 6b) translation speed are approximately 2.3 ± 1.0 cm/s and 0.9 ± 0.5 cm/s, respectively. AEs move southeastward at the northeastern half basin because of the advection of mean flow from the Mindoro Strait during boreal spring and summer when AEs are more frequent (Figure 6a). In the Sulawesi Sea, the AE translation speed is approximately 2.5 ± 2.3 cm/s, with a westward propagation in the central basin (Figure 6a). The CE migrations across the entire basin are significantly affected by the intense intrusion of MC (Figure 6b), with a speed of 6.3 ± 2.7 cm/s. In the Maluku Sea, all eddies propagate southwestward, which is parallel to the long axis of basin, at a velocity of 1.9 ± 0.7 cm/s for AE and 1.7 ± 0.6 cm/s for CE (Figure 6a,b). In the Banda Sea, the CEs are advected by the eastward mean flow during austral summer with more CEs (Figure 6b). More frequent AEs move southeastward with mean flow in the austral spring and autumn (Figure 6a). The corresponding propagation speeds of AE and CE in the Banda Sea are 5.7 ± 2.5 cm/s and 3.9 ± 2.5 cm/s (More details refer to Section 3.5 and Appendix A).

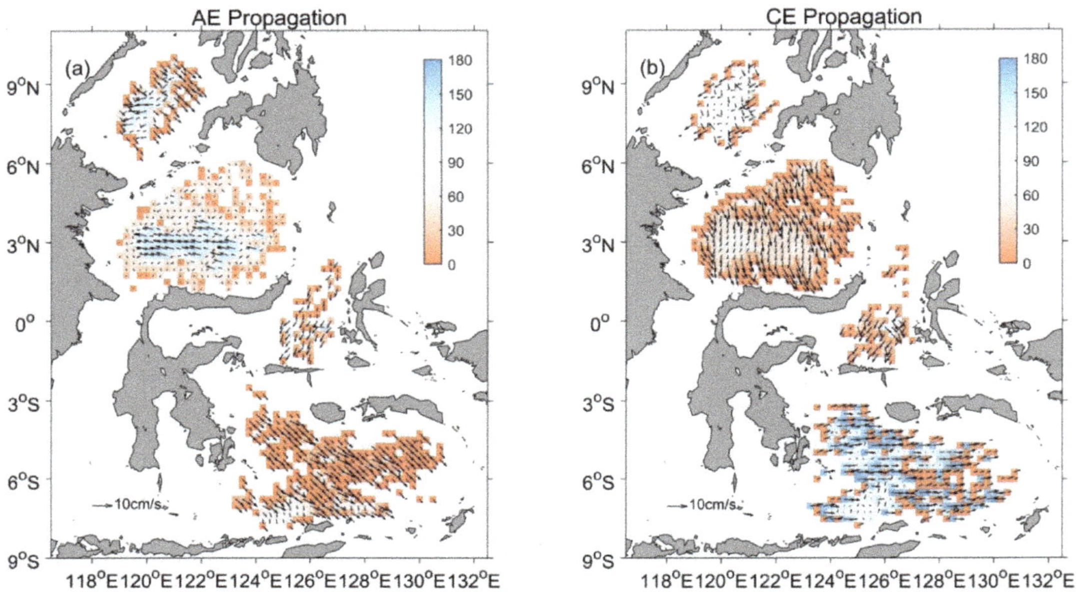

Figure 6. Spatial distribution of propagation velocity vectors (black arrows, unit: cm/s) for (**a**) AE and (**b**) CE over the 1993–2018 period. The color shading represents the standard deviation (STD) of propagation azimuths relative to west at each 0.25° × 0.25° grid (units: degree).

3.3. Eddy Lifespan

In this study, the lifespan of an eddy is defined as the number of days between the genesis and decay of the eddy. The upper-tail cumulative histograms of the eddy lifespan for the Sulu Sea, Sulawesi Sea, Maluku Sea and Banda Sea are presented in Figure 7. The short-lived (10~30 days, referenced in [15]) AEs (CEs) account for 73% (84%), 85% (86%), 100% (94%) and 86% (80%) in the Sulu Sea, Sulawesi Sea, Maluku Sea and Banda Sea, respectively. This suggests that most mesoscale eddies in the Indonesian Seas are short-lived, as also observed in the tropical oceans including the equatorial current systems and tropical oceanic warm pools [15]. The lifespans of AE and CE have a generally similar distribution in all these seas with a rapid decrease within 40 days and a slow decrease beyond 40 days except for the Maluku Sea where the eddy lifespan is mostly shorter than 30 days. However, there are some differences between AE and CE in each sea. In the Sulu Sea (Figure 7a), the portion of AE with lifespan larger than 30 days, 27%, is higher than CE with a percentage of 16%. The distributions of lifespan for eddies are more skewed to high values for AE than CE in the Sulawesi Sea (Figure 7b). For the relatively longevous eddies, CEs are more abundant than AEs in the Banda Sea, with a percentage of 20% (Figure 7d).

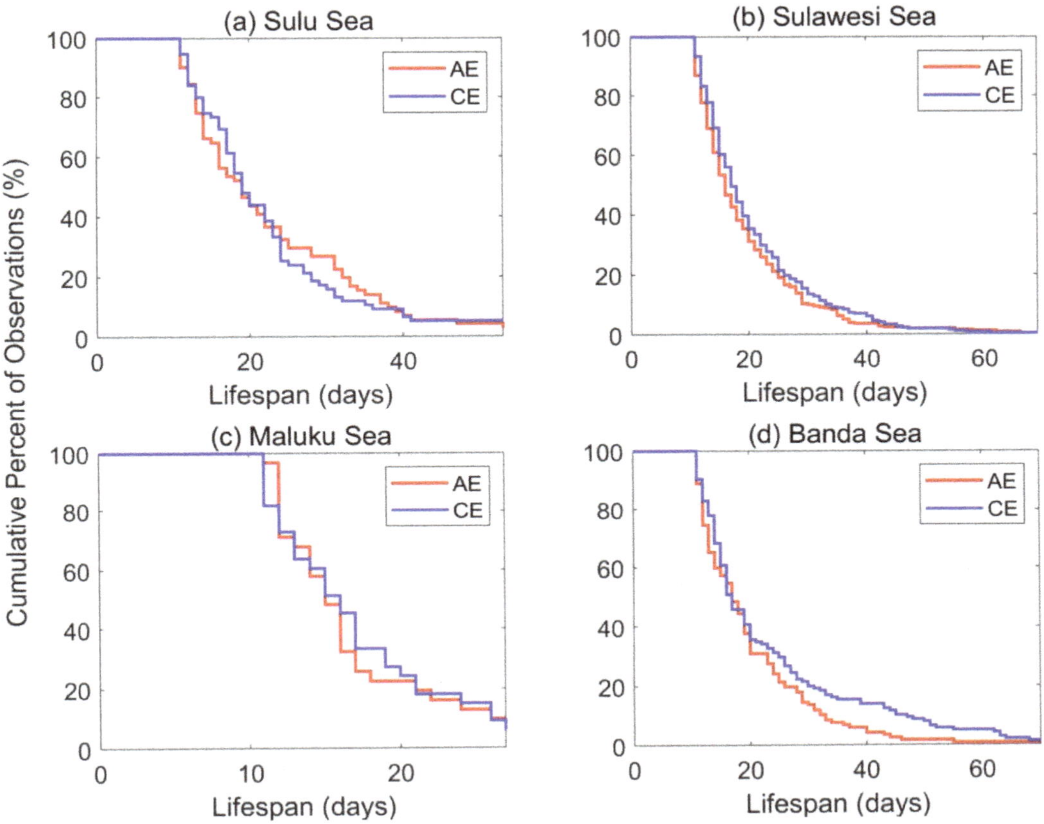

Figure 7. Upper-tail cumulative histograms of the eddy lifespan in the (**a**) Sulu Sea, (**b**) Sulawesi Sea, (**c**) Maluku Sea and (**d**) Banda Sea. The red and blue lines in each panel correspond to AE and CE, respectively.

3.4. Distribution of Eddy Properties

The spatial distributions of amplitude and radius for AEs and CEs averaged in $1^{\circ} \times 1^{\circ}$ grid are shown in Figure 8. Eddies have typical amplitudes of 2-6 cm and radiuses of 50–160 km in the Indonesian Seas. The patterns of amplitude (Figure 8a,b) are generally similar to those of radius (Figure 8c,d), indicating that the amplitude is positively correlated with the radius. Both amplitude (Figure 8a,b) and radius (Figure 8c,d) are larger in the interior basin and smaller near the coast for the Sulu Sea, Sulawesi Sea and Banda Sea. This may be due to the strong energy dissipation near the coast [34], and eddy growth in the central basin resulted from the eddy interactions [55]. Meanwhile the amplitude and radius for AE and CE in the Maluku Sea and Seram Sea are smaller due to small basin scale, as also observed in the Red Sea by Zhan et al. [47]. The large values of eddy radius are observed in the Banda Sea, where the maximum value of 157 km is observed in the central basin and the majority is more than 100 km. Additionally, the amplitude and radius of CE (Figure 8b,d) are stronger than those of AE (Figure 8a,c) in the Sulawesi Sea, where MC intrudes into Sulawesi Sea as a cyclonic loop structure [22].

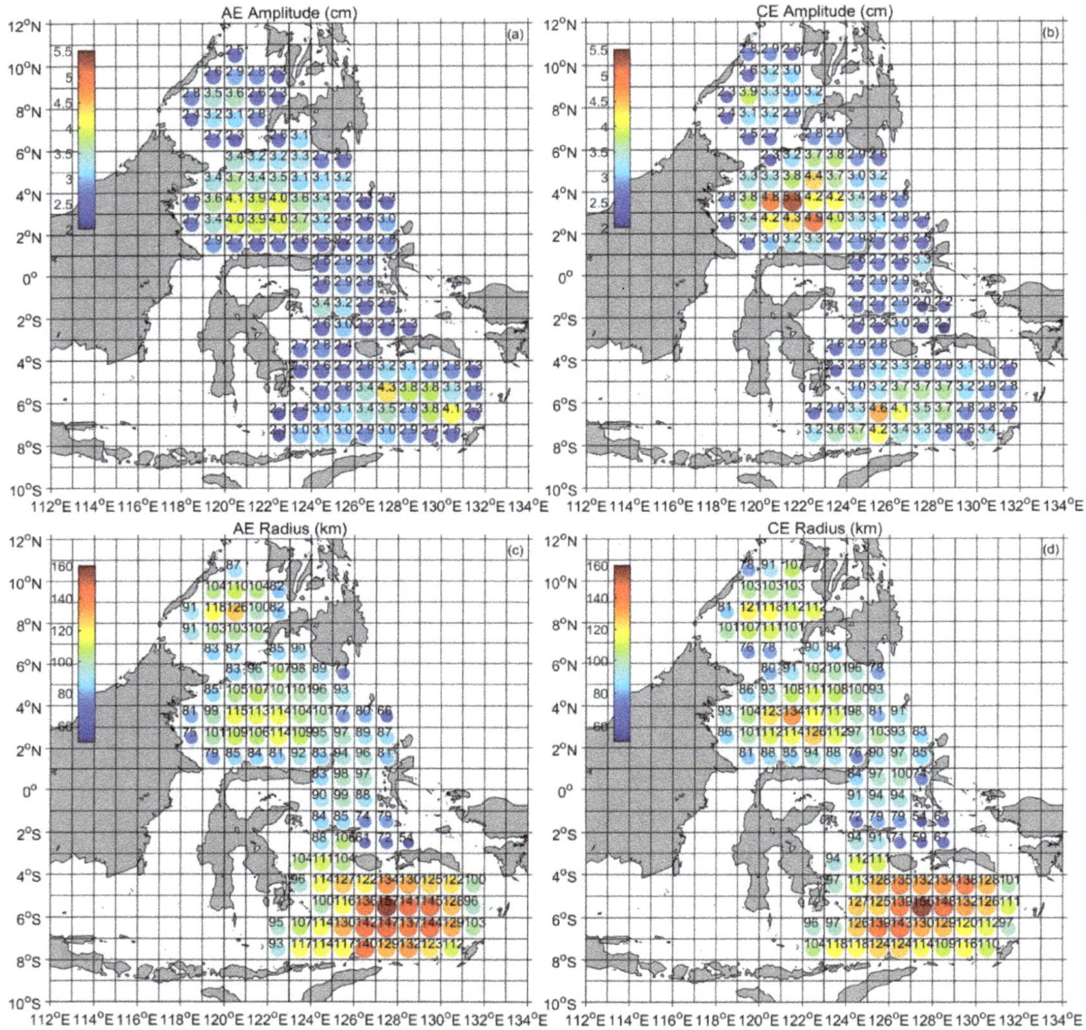

Figure 8. Spatial distribution (color spotted) of mean eddy properties over the 1993–2018 period: (**a**) AE amplitude, (**b**) CE amplitude, (**c**) AE radius and (**d**) CE radius. A value in each grid is averaged from eddies centered within this grid. The units of amplitude and radius are cm and km, respectively.

Figure 9 displays the spatial distributions of mean EKE and mean relative vorticity for AE and CE averaged in $1^\circ \times 1^\circ$ grid. There is no significant difference between AE and CE in the spatial pattern of mean EKE in the Indonesian Seas. The spatial pattern of mean EKE is similar to that of amplitude, which indicates that the tangential speed of eddy is largely determined by amplitude [12,34]. Both the magnitude and variation of mean EKE in the Sulu Sea agree well with the estimate of He et al. [34]. The most energetic eddies are observed in the Sulawesi Sea and Seram Sea with high MKE [42,56]. MKE is converted to EKE through the barotropic pathway [3–5], which will be discussed in Section 4. Although high eddy amplitude is also observed in the central Banda Sea (Figure 8a,b), the value of mean EKE (Figure 9a,b) is smaller than those in the Sulawesi Sea due to larger eddy area (Figure 8c,d). The spatial distribution of mean relative vorticity for AE and CE are generally similar (Figure 9c,d). Eddies with high mean relative vorticity were found in the Sulawesi Sea and Seram Sea, which is similar to the distribution of mean EKE. However, compared with other eddy properties, high value of mean vorticity is located near the coast instead of in the interior basin. On the one hand, the sampling number of eddies is very small in the coastal area. On the other hand, strong lateral friction leads to strong horizontal shear of velocities in the coastal region [55].

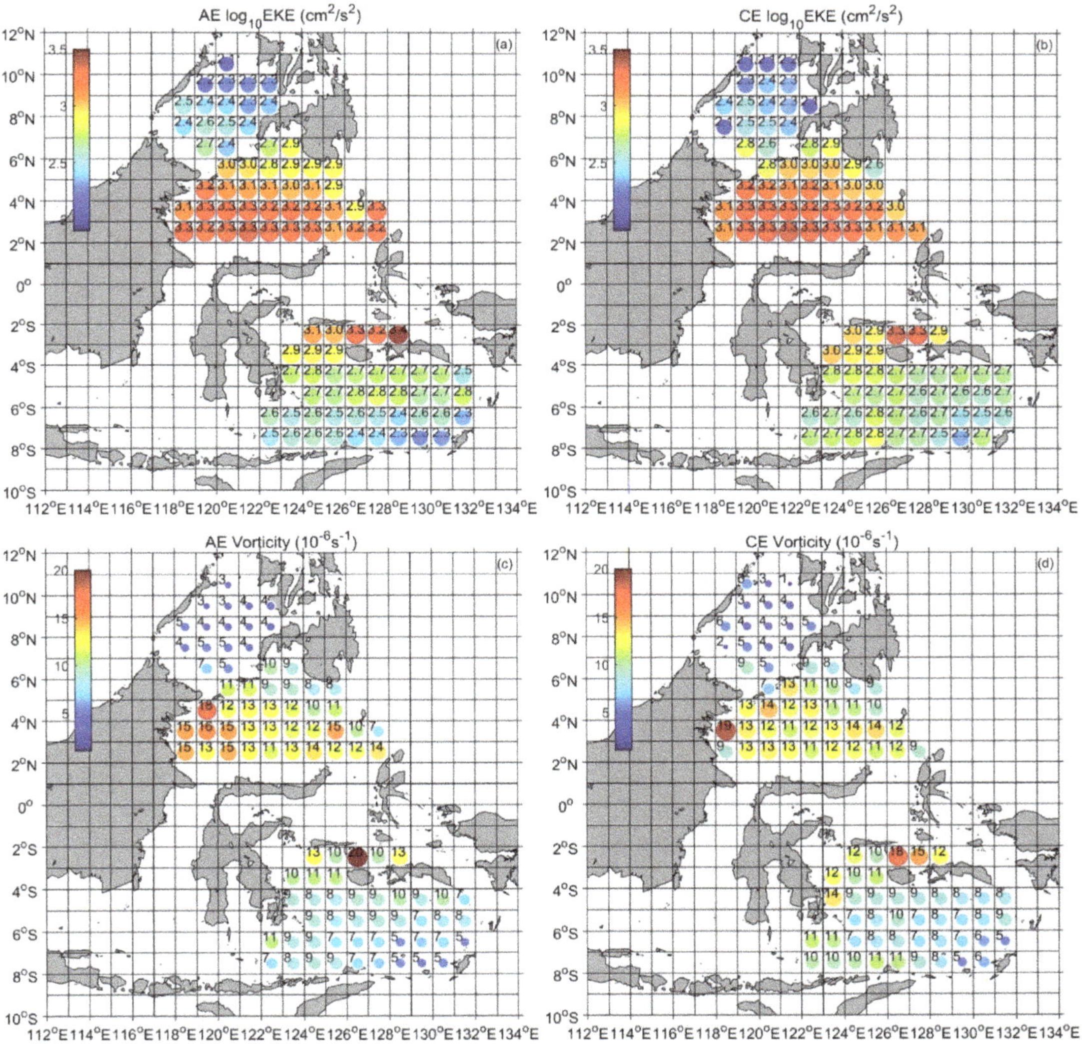

Figure 9. Spatial distribution (color spotted) of mean eddy properties over the 1993–2018 period: (**a**) the common logarithm of mean AE EKE, (**b**) the common logarithm of mean CE EKE, (**c**) mean AE vorticity and (**d**) mean CE vorticity. A value in each grid is averaged from eddies centered within this grid. The units of EKE and vorticity are cm^2/s^2 and 10^{-6}s^{-1}, respectively.

3.5. Seasonal Variability

The mean seasonal cycle of eddy properties in the Sulu Sea is shown in Figure 10. For AE, the mean number (Figure 10a), amplitude (Figure 10b) and radius (Figure 10c) display similar patterns, which are large in boreal summer and small in boreal winter, with peak values of 0.4, 3.7 cm and 123.5 km in August, respectively. The mean number, amplitude and radius of CE display similar patterns but in an opposite phase with AE, with large values in boreal winter and small values in boreal summer (Figure 10a–c). The latitude of eddy centroid presents opposite seasonal variability between AE and CE (Figure 10d). This indicates more AEs than CEs in the northern Sulu Sea from November to February, while more CEs populate in the southern Sulu Sea from August to October. The mean EKE of CE have a seasonal cycle with a bimodal structure (Figure 10e), which reaches the annual maximum of 352.5 cm^2/s^2 in June, and then drops rapidly to the annual minimum of 239.6 cm^2/s^2 in October. From then it increases dramatically to the local maximum of 317.4 cm^2/s^2 in December and decreases to the local minimum of 248.6 cm^2/s^2 in January. While the mean EKE of AE shows a single peak in May with a value of 334.5 cm^2/s^2, and a single trough with a value of 204.2 cm^2/s^2 in March. The mean relative vorticity of CE (Figure 10f) also has a bimodal structure, in which two peaks exist in September and December, with corresponding values of 5.1×10^{-6} s^{-1} and 4.4×10^{-6} s^{-1}, and two troughs appear in February and October with corresponding values of 3.5×10^{-6} s^{-1} and 3.8×10^{-6} s^{-1}. No significant seasonal variation is observed for the mean relative vorticity of AE (Figure 10f). Although the magnitudes of monthly average eddy properties are different from He et al. [34] due to different limitations of amplitude in eddy identification algorithms, the seasonal cycle of eddy characteristics displays almost identical tendency not only for AE but also for CE.

Figure 11 displays the mean seasonal cycle of eddy properties in the Sulawesi Sea. The seasonal variations of mean number (Figure 11a), amplitude (Figure 11b) and radius (Figure 11c) of AE and CE follow the bimodal curves. For AE, the mean number, amplitude and radius reach annual minimum in September, with corresponding values of 0.6, 3.2 cm and 97.6 km, and then achieve annual maximum in boreal winter, with values of 1.0, 4.1 cm and 111.8 km, respectively. From then they attain local minimum in May with corresponding values of 0.7, 3.5 cm and 99.9 km, and reach local maximum in July, with values of 0.8, 3.6 cm and 105.2 km, respectively. While the mean number, amplitude and radius of CE vary in an opposite phase compared with those of AE (Figure 11a–c). It is noted that CEs have larger amplitude than AEs through the whole year in the Sulawesi Sea. This further confirms the fact found above from the spatial perspective, the CE amplitude is larger than AE amplitude. The latitude of eddy centroid between AE and CE presents the opposite tendency in the Sulawesi Sea, with more AEs in the southern Sulawesi Sea from March to September while more CEs in the northern Sulawesi Sea from October to February (Figure 11d). Both the seasonal cycle of mean EKE and mean relative vorticity have bimodal structure for CE (Figure 11e,f). Two peaks appear in January with values of 1816 cm^2/s^2 and 12.9×10^{-6} s^{-1} and May with values of 1688 cm^2/s^2 and 12.3×10^{-6} s^{-1}, respectively, while two troughs exist in April with values of 1480 cm^2/s^2 and 11.3×10^{-6} s^{-1} and September with values of 1493 cm^2/s^2 and 11.2×10^{-6} s^{-1}, respectively. The mean EKE of AE have a similar pattern to that of CE, with two peaks in December and May and two troughs in April and September (Figure 11e). No apparent seasonal variations are observed for the mean relative vorticity of AE in the Sulawesi Sea, with a maximum value of 13.2×10^{-6} s^{-1} and a minimum value of 11.5×10^{-6} s^{-1} (Figure 11f).

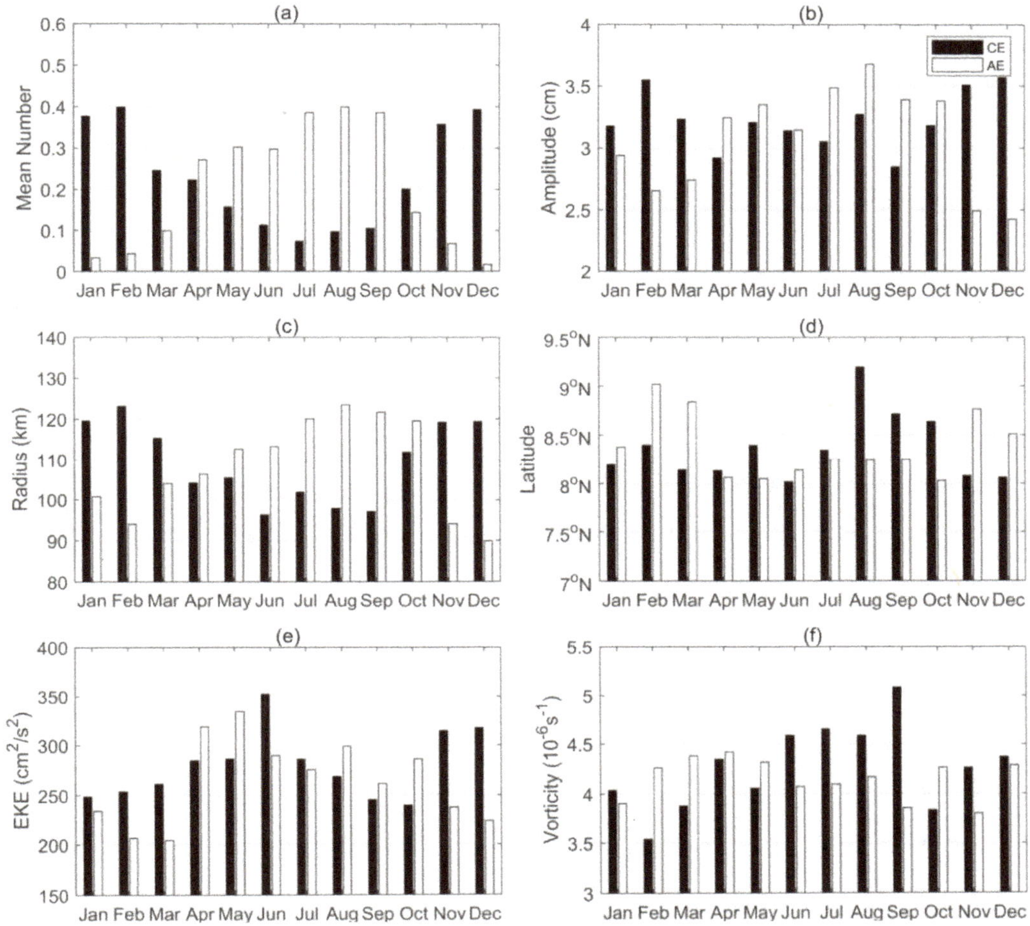

Figure 10. Seasonal cycles of eddy properties (black bars are for CE and white bars are for AE) in the Sulu Sea: (**a**) number of eddies, (**b**) amplitude, (**c**) radius, (**d**) central latitude, (**e**) mean EKE and (**f**) mean vorticity. The units of amplitude, radius, mean EKE and mean vorticity correspond to cm, km, cm^2/s^2 and $10^{-6}s^{-1}$.

The mean seasonal cycle of eddy properties in the Maluku Sea is shown in Figure 12. The monthly mean number of AE and CE are less than 0.2 in the Maluku Sea (Figure 12a). The amplitude of CE has a bimodal curve with annual maximum of 3.1 cm in February, annual minimum of 2.6 cm in July, local maximum of 2.8 cm in October and local minimum of 2.7 cm in December (Figure 12b). While there is no clear seasonal cycle of AE amplitude with the maximum of 3 cm in November and the minimum of 2.7 cm in July (Figure 12b). The seasonal variation of CE radius is similar to that of CE amplitude. Two peaks appear in March and October, with corresponding values of 98.7 km and 103.7 km, and two troughs are in February and May, with corresponding values of 88.51 km and 91.4 km (Figure 12c). Like AE amplitude, the AE radius has no apparent seasonal cycle, with the maximum of 99.8 km in October and the minimum of 87.9 km in January (Figure 12c). AEs in the Maluku Sea prefer to locate south of the equator from July to November and appear north of equator from December to June (Figure 12d). On the contrary, CEs in the Maluku Sea tend to populate south of the equator in the first half of the year and north of the equator in the second half of the year (Figure 12d).

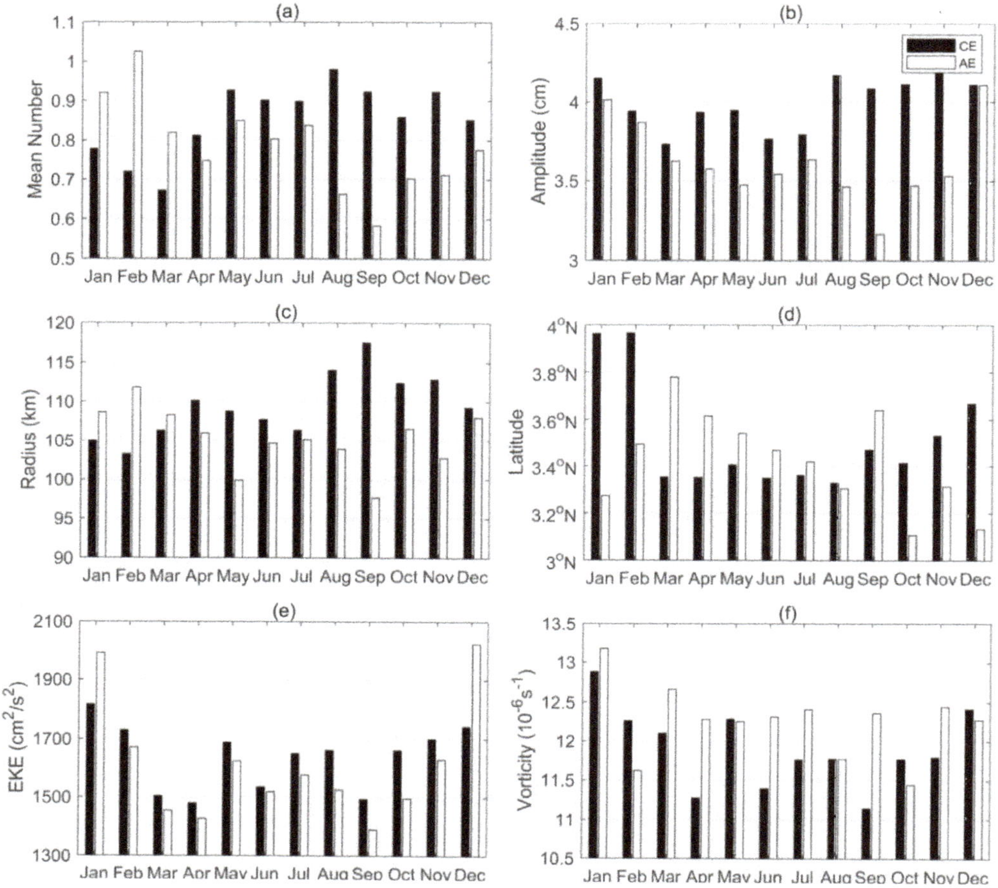

Figure 11. Seasonal cycles of eddy properties (black bars are for CE and white bars are for AE) in the Sulawesi Sea: (**a**) number of eddies, (**b**) amplitude, (**c**) radius, (**d**) central latitude, (**e**) mean EKE and (**f**) mean vorticity. The units of amplitude, radius, mean EKE and mean vorticity correspond to cm, km, cm^2/s^2 and $10^{-6}s^{-1}$.

Figure 13 shows the mean seasonal cycle of eddy properties in the Banda Sea. The mean number (Figure 13a), amplitude (Figure 13b) and radius (Figure 13c) of AE in the Banda Sea exhibit bimodal structure in seasonal variability. They reach the annual maximum in austral autumn with corresponding values of 0.6, 3.6 cm and 139.9 km, then drop to annual minimum in austral winter with the values of 0.3, 3.0 cm and 124.6 km, respectively. After achieving the minima, they increase to a local maximum with corresponding values of 0.4, 3.2 cm and 137.2 km in austral spring and then decrease to a local minimum with the values of 0.2, 3.1 cm and 122.1 km in austral summer, respectively. As shown in Figure 13a,b, the mean number and amplitude of CE hit peak in austral summer with corresponding values of 0.7 and 4.8cm. From then, they decrease gradually to the trough in austral winter with values of 0.3 and 2.9 cm, respectively. Meanwhile, the CE radius reaches the maximum of 143.1 km in January and the minimum of 124.0 km in March (Figure 13c). The latitude of AE and CE shows opposite patterns in the Banda Sea (Figure 13d). AEs tend to seat in the southern Banda Sea during austral winter and in the northern Banda Sea during austral summer. Meanwhile, CEs prefer to populate in the southern Banda Sea for austral summer and in the northern Banda Sea for austral winter. More energetic AEs exist during austral summer with a peak value of 736 cm^2/s^2, and the mean EKE of AE reaches the minimum of 335.2 cm^2/s^2 in austral winter (Figure 13e). The seasonal cycle of mean EKE for CE is a bimodal curve with two peaks in February and June with corresponding values of 672.5 cm^2/s^2 and 585.6 cm^2/s^2, and two troughs in April and October with the values of 509.2 cm^2/s^2 and 488.5 cm^2/s^2, respectively (Figure 13e). No significant seasonal variability of mean relative vorticity for CE is observed, with the maximum of 9.5×10^{-6} s^{-1} in March and the minimum of 8.0×10^{-6} s^{-1} in October (Figure 13f). While the seasonal cycle of mean relative vorticity for AE displays a bimodal structure in the Banda Sea: two peaks exist in March and July with corresponding values of 9.4×10^{-6} s^{-1} and 7.9×10^{-6} s^{-1} and two troughs appear in June and September with values of 7.1×10^{-6} s^{-1} and 6.9×10^{-6} s^{-1}, respectively (Figure 13f).

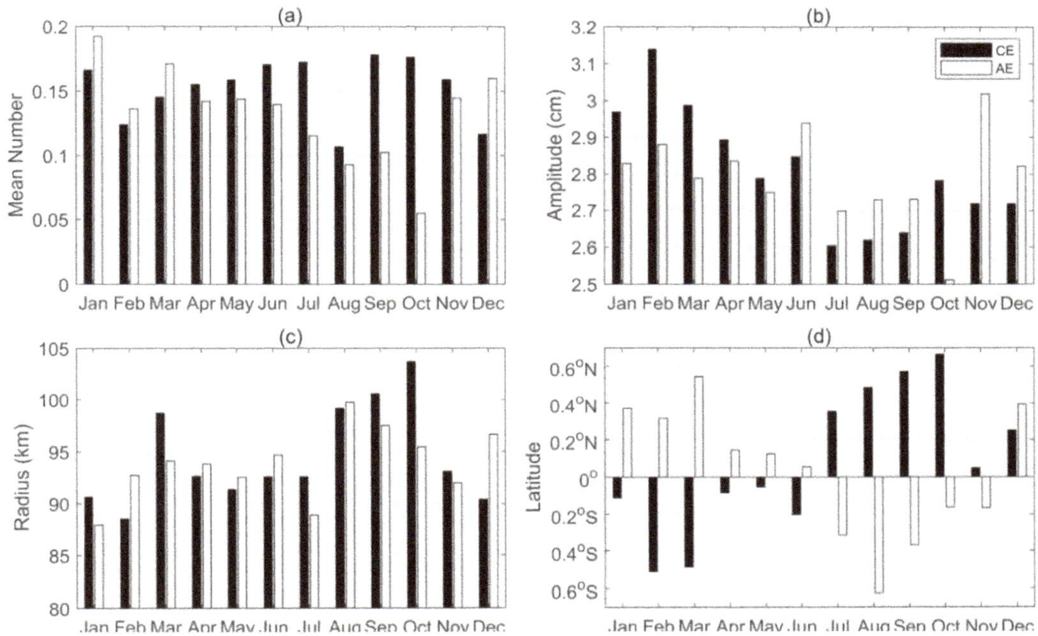

Figure 12. Seasonal cycles of eddy properties (black bars are for CE and white bars are for AE) in the Maluku Sea: (**a**) number of eddies, (**b**) amplitude, (**c**) radius and (**d**) central latitude. The units of amplitude and radius correspond to cm and km.

3.6. Eddy Nonlinearity

The advective nonlinearity parameter is calculated to assess the degree of eddy nonlinearity in the Indonesian Seas. The advective nonlinearity parameter is a nondimensional ratio of U/c, where U is the maximum rotation speed and c is the propagation speed of each eddy [2]. The definition of c is shown in Section 3.2. Figure 14 displays the upper-tail histogram of U/c for the Sulu Sea, Sulawesi Sea and Banda Sea. Most eddies in these three seas are nonlinear by this measure, among which U/c exceeding 1 account for 89% (88%), 100% (100%) and 94% (96%) for AE (CE) in the Sulu Sea, Sulawesi Sea and Banda Sea, respectively. Most highly nonlinear eddies are observed in the Sulawesi Sea, with 48% of the U/c values exceeding 5 and 17% exceeding 10, as observed in major unstable and meandering currents of global ocean [2]. Fewer eddies with high nonlinearity are detected in the Sulu Sea and Banda Sea (Figure 14a,c), where the U/c values above 10 is less than 10%. The distributions of U/c for eddies are more skewed to higher values for AE than for CE in the Sulu Sea, but the opposite is found in the Banda Sea.

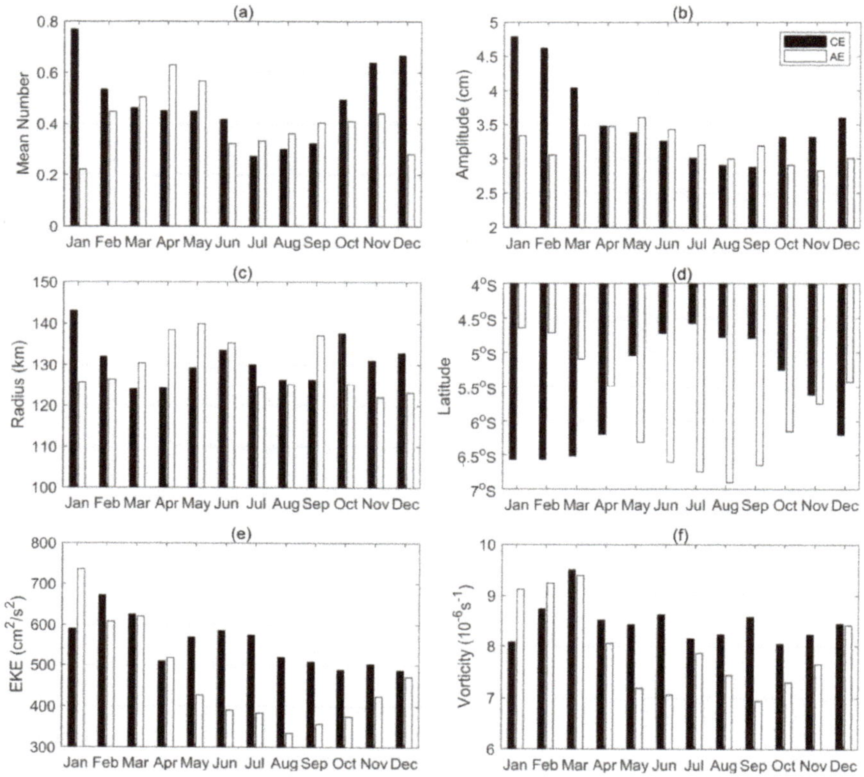

Figure 13. Seasonal cycles of eddy properties (black bars are for CE and white bars are for AE) in the Banda Sea: (**a**) number of eddies, (**b**) amplitude, (**c**) radius, (**d**) central latitude, (**e**) mean EKE and (**f**) mean vorticity. The units of amplitude, radius, mean EKE and mean vorticity correspond to cm, km, cm^2/s^2 and $10^{-6}s^{-1}$.

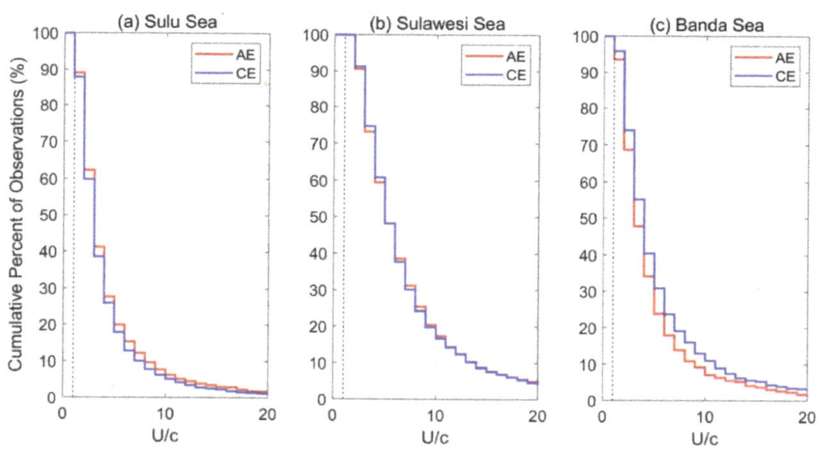

Figure 14. Upper-tail cumulative histograms of U/c in the (**a**) Sulu Sea, (**b**) Sulawesi Sea and (**c**) Banda Sea. The red and blue lines in each panel correspond to AE and CE, respectively. Additionally, the vertical dotted lines in each panel indicate that the value of U/c is equal to 1.

4. Discussion

To investigate the generation mechanism of eddies in the Indonesian Seas, we calculated the BTR, KHR and BCR, respectively, using BRAN outputs from 1994 to 2015. Due to the complex current patterns and the different spatial distribution of eddy genesis in each basin, we select four subregions in the Sulu Sea, Sulawesi Sea, Maluku Sea and Banda Sea (Figure 1, dashed boxes). These four subregions are defined in the Appendix A. We also find that BRAN can reasonably reproduce mesoscale eddies in the Indonesian Seas, and more details of evaluation refer to Appendix B. The seasonal cycles of volume integral BTR, KHR and BCR in each box are displayed in Figure 15. In these four regions, the volumetric integral BCR is positive throughout the year, which means EPE converts to EKE through baroclinic instability. Meanwhile, the volume integration of BTR for each box is positive in most seasons, except for some seasons in the Sulu Box and Maluku Box (Figure 15a,c). The weak negative BTR may result from inverse energy cascade from EKE to MKE [4,5], and the detailed discussions for this beyond the scope of present paper. However, the BTR almost dominates the energy conversion in the Sulawesi box throughout the year (Figure 15b), as in the Gulf Stream [4] and Kuroshio [5]. This result is similar to the conclusions of Yang et al. [33] that the EKE is governed by barotropic instability of mean flow. Compared with BTR and BCR, KHR are much smaller and make little net contribution to EKE in all four boxes. Therefore, barotropic and baroclinic instability dominate eddy generation in the Indonesian Seas. There are some seasonal variations of BTR and BCR, likely associated with the seasonal variations if the mean flow.

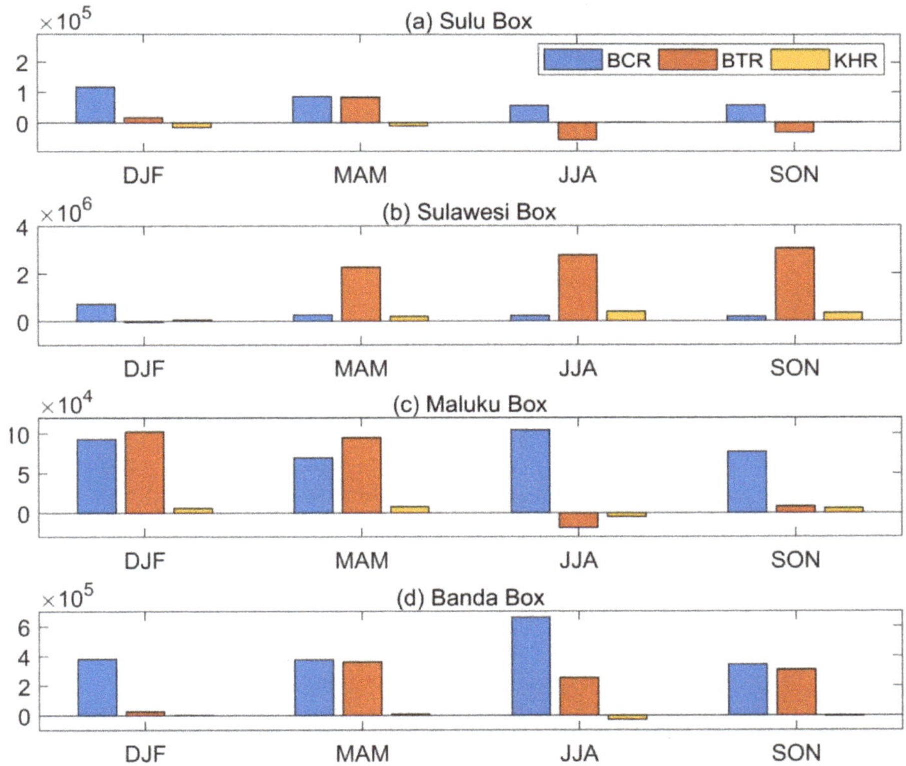

Figure 15. Seasonal variation of volume integral barotropic conversion rate (BTR), baroclinic conversion rate (BCR) and Kelvin-Helmholtz conversion rate (KHR) in the (**a**) Sulu Box, (**b**) Sulawesi Box, (**c**) Maluku Box and (**d**) Banda Box, respectively. The unit is m^5/s^3.

To further determine which instability dominates for individual eddy generation over the 1994–2015 period, the monthly BTR and BCR integrated vertically in the upper 300 m are firstly calculated. Secondly, the corresponding values of BTR and BCR for each eddy genesis event were extracted according to its generation position and month. Thirdly, a comparison between BTR and BCR is performed to assess an eddy genesis event dominated by either barotropic instability (BTR > BCR > 0) or baroclinic instability (BCR > BTR > 0). Figure 16 presents the spatial distribution of the number of eddy genesis events dominated by barotropic and baroclinic instability. In the Sulu Sea, eddy genesis events dominated by barotropic instability account for 48% while those dominated by baroclinic instability are 35% of total eddy generation. This is similar to He et al. [34], who concluded that the instability of mean flow is one of the major eddy generation mechanisms. Barotropic instability dominates eddy generation in the Sulawesi Sea with a percentage of 55%, as in Yang et al [33], while the corresponding value of baroclinic dominance is 26% in the Sulawesi Sea. It is clear that barotropic instability is more frequent in the central Sulawesi Sea where the most intrusion of MC is located (Figure 16a), and eddy generation is dominated by baroclinic instability in the northern Sulawesi Sea (Figure 16b). In the Maluku Sea, 47% of eddy genesis events are dominated by baroclinic instability, while 43% are dominated by barotropic instability. Baroclinic instability in the Banda Sea dominates by 49% of eddy generation (Figure 16b). Additionally, the percentage of eddy generation dominated by barotropic instability is 40% in the Banda Sea (Figure 16a). For the Indonesian Seas as a whole, barotropic instability dominates the eddy generation with a percentage of 49%, which is higher than that dominated by baroclinic instability with a percentage of 35%. This is similar to the previous conclusions that barotropic instability is the dominant mechanisms in EKE generation along intense boundary currents, such as the Gulf Stream and Kuroshio, while baroclinic instability dominates in the broader open ocean [3–5]. A total of 16% of eddies in the Indonesian Seas are not dominated by either barotropic or baroclinic instability, so that costal Kelvin waves and complex topography may also affect eddy generation [57].

In our analysis, eddies with amplitude less than 2 cm are excluded according to the fact that the accuracy of gridded SLA product provided by AVISO is about 2 cm [15,51]. Generally, the variability of eddy amplitude during its lifetime undergoes growing stage, mature stage and decaying stage [15,19]. Inevitably, the eddy amplitudes less than 2 cm exist in the beginning and end of eddy lifespan, which are below the resolving ability of altimeter data [2,51]. There are two possible effects: one is underestimate of eddy lifespan; the other is that locations of eddy generation and decay tracked by eddy tracking algorithm may have some offset. However, our results in the Sulu Sea are similar to that of He et al. [34], especially with same spatial and temporal variability for eddy properties. The eddy amplitudes dramatically increase and decrease in growing stage and decaying stage, respectively, which account for less than 20% of the eddy lifespan [15,19]. Besides, the propagation speed of eddies in the Indonesian Seas is really small, with a mean value below 6 cm/s and identified eddies whose lifespan are less 30 days account for more than 80%. Thus, there is likely no significant difference between the positions detected by eddy tracking algorithm and the actual positions for eddy generation and decay in the region.

Additionally, the reader should be reminded that a free running model cannot reproduce real particular eddy events, while data assimilation inevitably interferes the dynamical or energy balance [40]. The new version of BRAN have used globally balanced forcing and low update cycle of assimilation (4 days) which improved the dynamic imbalance in the previous version [40]. Zhang et al. [12] have proved that BRAN performs well for the instability analysis and the robustness of their conclusions is further proved by a free running model, the eddy-resolving OGCM for the Earth Simulator (OFES). Besides, the magnitude of BTR averaged vertically in the Sulawesi Sea using BRAN outputs in this study is equal to the magnitude, 10^{-3} cm^2/s^3, estimated based on OFES outputs [33]. Although the detailed assessment of the difference of EKE budget analysis between BRAN and a free running model beyond the scope of this paper, future studies about this assessment are on the way.

This paper reports the statistical characteristics and spatiotemporal variations for the eddies across the whole Indonesian Seas for the first time. We believe that our results can be used to assess the performance of numerical models in terms of reproducing mesoscale eddies and to improve local ocean forecasts. Moreover, the significant meridional migrations and the high nonlinearity of eddies revealed in this study indicate considerable eddy transports of heat and salt [29,31], which will be the focus of our ongoing research. In addition, because eddies play a key role in the circulation of semi-closed and enclosed basins [34,47], our conclusions can also be used for the further investigations of eddy-ITF interaction.

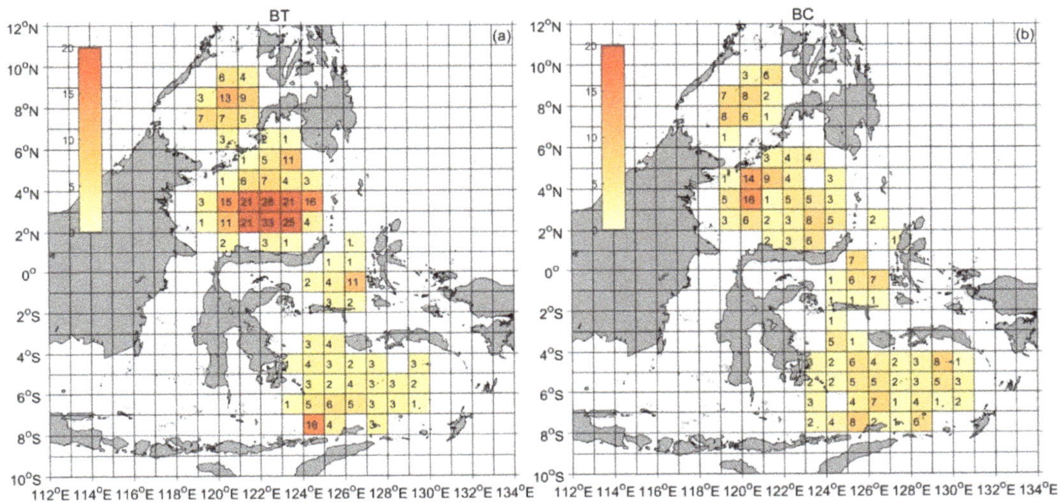

Figure 16. Spatial distribution of eddy genesis events dominated by (**a**) barotropic instability and (**b**) baroclinic instability. The unit is the number of eddy genesis events.

5. Conclusions

In this study, we provided a detailed description for both spatial and temporal variability of eddies in the Indonesian Seas using altimeter data and discussed their generation mechanisms using high-resolution model outputs. A total of 469 AE tracks and 500 CE tracks were identified from January 1993 to December 2018. Most of eddies in the Indonesian Seas are short-lived (below 30 days) with a percentage of 73% (84%), 85% (86%), 100% (94%) and 86% (80%) for AE (CE) in the Sulu Sea, Sulawesi Sea, Maluku Sea and Banda Sea, respectively. There is no geographical preference for AE and CE in above four basins. For the mechanisms of eddy generation, barotropic instability dominates over baroclinic instability in the Sulu Sea and Sulawesi Sea, while baroclinic instability is slightly more important in the Maluku Sea and Banda Sea. Most energetic eddies are observed in the Sulawesi Sea and Seram Sea. The spatial distribution of eddy properties shows high inhomogeneity with typical amplitudes and radiuses of 2–6 cm and 50–160 km, respectively. Because of the intrusion of MC in the Sulawesi Sea, CE has larger amplitude and radius than AE. Limited by basin scale, the eddies in the Maluku Sea show smaller amplitude and radius.

The climatological properties of eddies in the Indonesian Seas exhibit different seasonal variations between AE and CE not only in each basin but also among four major basins. In the Sulu Sea, AEs prevail with relatively large amplitude and radius in the boreal summer, while CEs display an opposite seasonal variability. The mean number, amplitude and radius of eddies in the Sulawesi Sea present bimodal structure in seasonal cycles, with opposite phases between AE and CE. In the Banda Sea, the seasonal cycle of AE properties, including mean number, amplitude and radius, shows a bimodal structure, while CEs are more abundant with larger amplitude during austral summer than the other three seasons. There is no consistent tendency in the seasonal cycle of the mean number, amplitude and radius for both AE and CE in the Maluku Sea. It is interesting that every basin in the Indonesian Seas displays a reversal meridional distribution of eddy polarity with season. Nevertheless, our results only focused on a comprehensive description of eddy properties at the surface due to the limitation of altimeter data, future studies should explore the three-dimensional structure of eddies in the Indonesian Seas and their contributions to the ITF dynamics and thermodynamics.

Supplementary Materials: The following are available online at https://www.mdpi.com/2072-4292/13/5/1017/s1, Figure S1: Spatial distribution of amplitude and radius of AE and CE over the 1994–2015 period from BRAN. Figure S2: Spatial distribution of mean EKE and mean relative vorticity of AE and CE over the 1994–2015 period from BRAN. Figure S3: Seasonal cycles of eddy properties in the Sulu Sea over the 1994–2015 period from BRAN. Figure S4: Seasonal cycles of eddy properties in the Sulawesi Sea over the 1994–2015 period from BRAN. Figure S5: Seasonal cycle of eddy properties in the Banda Sea over the 1994–2015 period from BRAN.

Author Contributions: Conceptualization, Z.H., Z.X. and M.F.; methodology, Z.H., M.F. and Q.L.; writing-original draft presentation, Z.H.; writing-review and editing, Z.X., M.F., B.Y. and Z.H. All authors have read and agreed to the published version of the manuscript.

Funding: This study was jointly supported by the Strategic Priority Research Program of Chinese Academy of Sciences (XDB42000000, XDA11010204), the National Natural Science Foundation of China (91858103, 92058202), CAS Key Research Program of Frontier Sciences (QYZDB-SSW-DQC024), the National Key Research and Development Program of China (2017YFA0604102, 2016YFC1402705), Key Deployment Project of Centre for Ocean Mega-Research of Science,Chinese Academy of Sciences (COMS2020Q07), the project jointly funded by the CAS and CSIRO (133244KYSB20190031).

Institutional Review Board Statement: Not applicable.

Informed Consent Statement: Not applicable.

Data Availability Statement: The datasets for this study are publicly available. The altimetry sea surface height, sea level anomalies and geostrophic velocities are provided by Archiving, Validation, and Interpolation of Satellite Oceanographic (AVISO) data and distributed by the Copernicus Marine Environment Monitoring Service (CMEMS) via https://marine.copernicus.eu/ (accessed on 4 March 2021). The Bluelink ReANalysis data is provided by National Computational Infrastructure thredds server of Australia via https://www.nci.org.au/ (accessed on 4 March 2021).

Acknowledgments: This research is partly supported by the Centre for Southern Hemisphere Oceans Research (CSHOR), which is a joint initiative between the Qingdao National Laboratory for Marine Science and Technology (QNLM), CSIRO, University of New South Wales and University of Tasmania. We would like to thank the four anonymous reviewers whose constructive comments have significantly improved the manuscript.

Conflicts of Interest: The authors declare no conflict of interest. The funders had no role in the design of the study; in the collection, analyses, or interpretation of data; in the writing of the manuscript, or in the decision to publish the results.

Appendix A

The seasonal cycle of mean circulation averaged above 300 m in the Indonesian Seas simulated by BRAN is shown in Figure A1. In the Sulu Sea, a cyclonic circulation prevails in boreal winter while an anticyclonic circulation exists in boreal summer, which resulted from seasonal local wind forcing and outflow via the Sibutu Passage [45]. MC intrudes in the Sulawesi Sea as a cyclonic loop structure through the whole year with a weaker intrusion in January (Figure A1a). There is an anticyclonic sub-basin scale circulation in the north of 4°N in the Sulawesi Sea. The other intrusion of MC concentrates in the northern Maluku Sea, while relatively weak northward flow exists in the central and southern Maluku Sea. In the Banda Sea, the currents in the upper circulation flows eastward during the northwesterly monsoon while flowing westward with relatively weak strength during the southeasterly monsoon, as also simulated by Zhu et al. [58].

Based on the complex upper-ocean current patterns, we selected four subregions with more frequent eddy genesis events. The detailed range of the four rectangular subregions, respectively, in the Sulu Sea, Sulawesi Sea, Maluku Sea and Banda Sea (Figure A1) are as follows:

1. Sulu Box: $119°E \sim 123°E$, $7°N \sim 10°N$;
2. Sulawesi Box: $119°E \sim 125°E$, $2°N \sim 5°N$;
3. Maluku Box: $125°E \sim 127°E$, $2°S \sim 1°N$;
4. Banda Box: $122°E \sim 131°E$, $4°S \sim 7.5°S$.

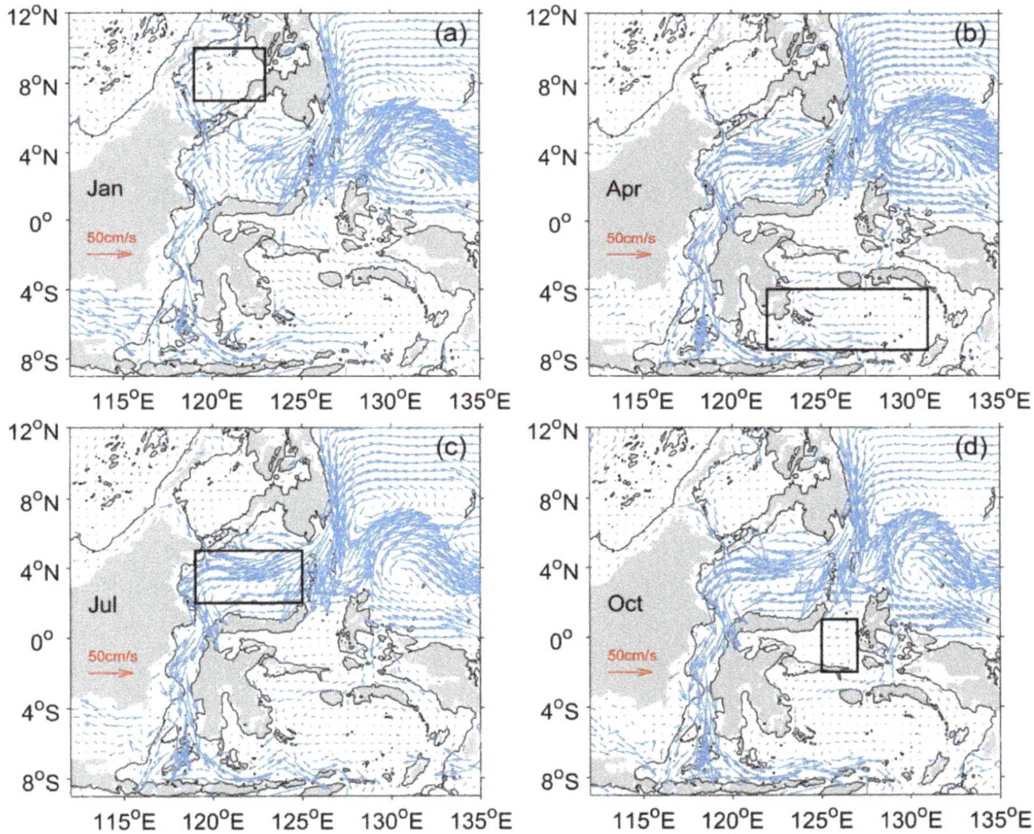

Figure A1. Mean currents (blue arrows) averaged above 300 m depth in (**a**) January, (**b**) April, (**c**) July and (**d**) October from the BRAN outputs. The solid line is the 200 m isobath. Location of the four subregions in the Sulu Sea, Sulawesi Sea, Maluku Sea and Banda Sea. The boxes are (**a**) Sulu Box, (**b**) Banda Box, (**c**) Sulawesi Box and (**d**) Maluku Box, respectively. For regions shallower than 300 m, the average is for the whole water column.

Appendix B

To evaluate whether BRAN reproduces realistic mesoscale eddies in the Sulu Sea, Sulawesi Sea, Maluku Sea and Banda Sea, we also detect eddies from the daily SLA snapshots of BRAN over the 1994–2015 period. The eddies with lifetime longer than 10 days and amplitude larger than 2 cm are considered for the analysis. The spatial distributions of eddy genesis and decay events from AVISO and BRAN over the 1994–2015 period are show in Figure A2. A total of 843 and 649 eddy tracks are detected from AVISO and BRAN over this period, respectively. The proportion of eddy tracks are 14% (22%), 53% (50%), 7% (1%) and 26% (27%) in the Sulu Sea, Sulawesi Sea, Maluku Sea and Banda Sea, respectively. The distributions of both eddy generation and eddy decay share a similar pattern between AVISO and BRAN in the Sulu Sea, Sulawesi Sea and Banda Sea. However, the numbers of eddy tracks in these four seas from BRAN are smaller than those from AVISO. This may result from the difference of data resolution, as is also found in the Subtropical Countercurrent regions from OFES outputs [59].

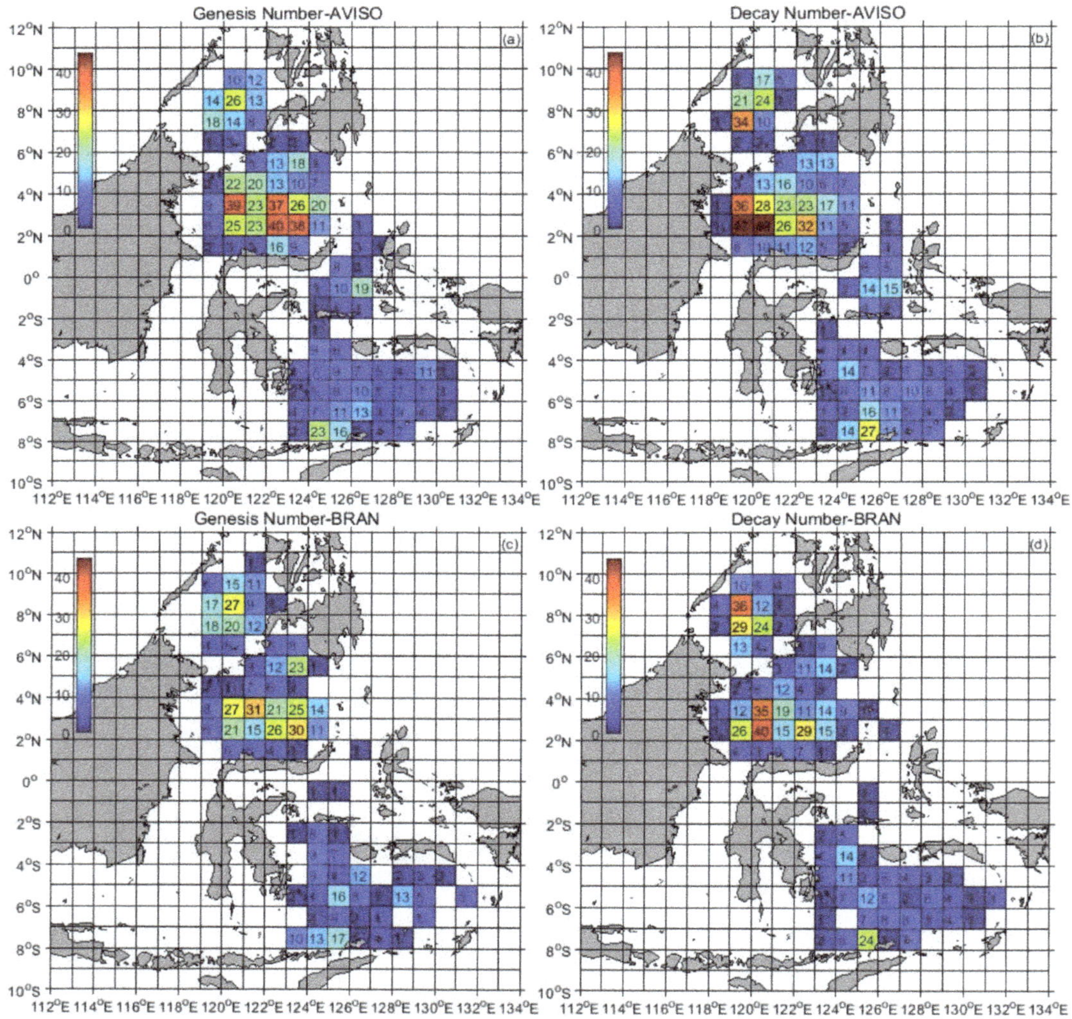

Figure A2. Spatial distribution of eddy genesis and decay events over the 1994–2015 period from AVISO (the first row) and BRAN (the second row): (**a**) genesis number from AVISO, (**b**) decay number from AVISO, (**c**) genesis number from BRAN and (**d**) decay number from BRAN. The unit is the number of events.

To further demonstrate the performance of BRAN in terms of mesoscale eddies, the comparison for the trajectories of eddies detected from AVISO and BRAN are performed. We find that BRAN can reproduce the realistic trajectories of mesoscale eddies in the regions we are interested, with the similar pathways and generation locations. Some examples in the Sulu Sea, Sulawesi Sea and Banda Sea are shown in Figure A3. Besides, BRAN can reasonably capture the spatial and temporal variations of eddy properties in the Sulu Sea, Sulawesi Sea and Banda Sea (see Supplementary Materials Figures S1–S5 for more details). Thus, we believe that BRAN is a credible reanalysis dataset to investigate the eddy generation mechanisms.

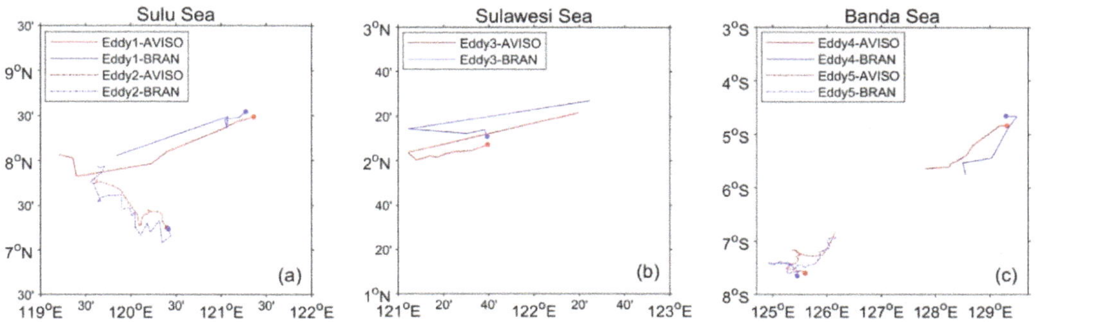

Figure A3. Trajectories of same eddies from AVISO and BRAN in the (**a**) Sulu Sea, (**b**) Sulawesi Sea and (**c**) Banda Sea. The red line represents the trajectory of an eddy observed in AVISO. The blue line represents the trajectory of the corresponding eddy simulated by BRAN. The dots represent the initial locations of each trajectory.

References

1. Fu, L.L.; Chelton, D.B.; Le Traon, P.Y.; Morrow, R. Eddy dynamics from satellite altimetry. *Oceanography* **2010**, *23*, 15–25. [CrossRef]
2. Chelton, D.B.; Schlax, M.G.; Samelson, R.M. Global observations of nonlinear mesoscale eddies. *Prog. Oceanogr.* **2011**, *91*, 167–216. [CrossRef]
3. Von Storch, J.S.; Eden, C.; Fast, I.; Haak, H.; Hernández-Deckers, D.; Maier-Reimer, E.; Marotzke, J.; Stammer, D. An estimate of the Lorenz energy cycle for the World Ocean based on the 1/10°STORM/NCEP simulation. *J. Phys. Oceanogr.* **2012**, *42*, 2185–2205. [CrossRef]
4. Kang, D.; Curchitser, E.N. Energetics of eddy-mean flow interactions in the gulf stream region. *J. Phys. Oceanogr.* **2015**, *45*, 1103–1120. [CrossRef]
5. Yan, X.; Kang, D.; Curchitser, E.N.; Pang, C. Energetics of eddy-mean flow interactions along the western boundary currents in the North Pacific. *J. Phys. Oceanogr.* **2019**, *49*, 789–810. [CrossRef]
6. Feng, M.; Wijffels, S.; Godfrey, S.; Meyers, G. Do eddies play a role in the momentum balance of the Leeuwin Current? *J. Phys. Oceanogr.* **2005**, *35*, 964–975. [CrossRef]
7. Zheng, Z.W.; Ho, C.R.; Kuo, N.J. Mechanism of weakening of west Luzon eddy during La Niña years. *Geophys. Res. Lett.* **2007**, *34*, 1–5. [CrossRef]
8. Cheng, Y.H.; Ho, C.R.; Zheng, Q.; Kuo, N.J. Statistical characteristics of mesoscale eddies in the north pacific derived from satellite altimetry. *Remote Sens.* **2014**, *6*, 5164–5183. [CrossRef]
9. Pegliasco, C.; Chaigneau, A.; Morrow, R. Main eddy vertical structures observed in the four major Eastern Boundary Upwelling Systems. *J. Geophys. Res. Oceans* **2015**, *120*, 6008–6033. [CrossRef]
10. Dong, D.; Brandt, P.; Chang, P.; Schütte, F.; Yang, X.; Yan, J.; Zeng, J. Mesoscale Eddies in the Northwestern Pacific Ocean: Three-Dimensional Eddy Structures and Heat/Salt Transports. *J. Geophys. Res. Oceans* **2017**, *122*, 9795–9813. [CrossRef]
11. Cheng, Y.H.; Ho, C.R.; Zheng, Q.; Qiu, B.; Hu, J.; Kuo, N.J. Statistical features of eddies approaching the Kuroshio east of Taiwan Island and Luzon Island. *J. Oceanogr.* **2017**, *73*, 427–438. [CrossRef]
12. Zhang, N.; Liu, G.; Liu, Q.; Zheng, S.; Perrie, W. Spatiotemporal Variations of Mesoscale Eddies in the Southeast Indian Ocean. *J. Geophys. Res. Oceans* **2020**, *125*, 1–18. [CrossRef]
13. Shen, D.; Li, X.; Wang, J.; Bao, S.; Pietrafesa, L.J. Dynamical ocean responses to Typhoon Malakas (2016) in the vicinity of Taiwan. *J. Geophys. Res. Oceans* **2020**. [CrossRef]
14. Stammer, D. On eddy characteristics, eddy transports, and mean flow properties. *J. Phys. Oceanogr.* **1998**, *28*, 727–739. [CrossRef]
15. Chen, G.; Han, G. Contrasting Short-Lived With Long-Lived Mesoscale Eddies in the Global Ocean. *J. Geophys. Res. Oceans* **2019**, *124*, 3149–3167. [CrossRef]
16. Laxenaire, R.; Speich, S.; Blanke, B.; Chaigneau, A.; Pegliasco, C.; Stegner, A. Anticyclonic Eddies Connecting the Western Boundaries of Indian and Atlantic Oceans. *J. Geophys. Res. Oceans* **2018**, *123*, 7651–7677. [CrossRef]
17. Faghmous, J.H.; Frenger, I.; Yao, Y.; Warmka, R.; Lindell, A.; Kumar, V. A daily global mesoscale ocean eddy dataset from satellite altimetry. *Sci. Data* **2015**, *2*, 1–16. [CrossRef]

18. Chaigneau, A.; Gizolme, A.; Grados, C. Mesoscale eddies off Peru in altimeter records: Identification algorithms and eddy spatio-temporal patterns. *Prog. Oceanogr.* **2008**, *79*, 106–119. [CrossRef]
19. Aguedjou, H.M.A.; Dadou, I.; Chaigneau, A.; Morel, Y.; Alory, G. Eddies in the Tropical Atlantic Ocean and Their Seasonal Variability. *Geophys. Res. Lett.* **2019**, *46*, 12156–12164. [CrossRef]
20. Feng, M.; Zhang, N.; Liu, Q.; Wijffels, S. The Indonesian throughflow, its variability and centennial change. *Geosci. Lett.* **2018**, *5*. [CrossRef]
21. Sprintall, J.; Gordon, A.L.; Wijffels, S.E.; Feng, M.; Hu, S.; Koch-Larrouy, A.; Phillips, H.; Nugroho, D.; Napitu, A.; Pujiana, K.; et al. Detecting change in the Indonesian seas. *Front. Mar. Sci.* **2019**, *6*. [CrossRef]
22. Lukas, R.; Firing, E.; Hacker, P.; Richardson, P.L.; Collins, C.A.; Fine, R.; Gammon, R. Observations of the Mindanao Current during the western equatorial Pacific Ocean circulation study. *J. Geophys. Res.* **1991**, *96*, 7089–7104. [CrossRef]
23. Yuan, D.; Li, X.; Wang, Z.; Li, Y.; Wang, J.; Yang, Y.; Hu, X.; Tan, S.; Zhou, H.; Wardana, A.K.; et al. Observed transport variations in the Maluku Channel of the Indonesian seas associated with western boundary current changes. *J. Phys. Oceanogr.* **2018**, *48*, 1803–1813. [CrossRef]
24. Gordon, A.L.; Napitu, A.; Huber, B.A.; Gruenburg, L.K.; Pujiana, K.; Agustiadi, T.; Kuswardani, A.; Mbay, N.; Setiawan, A. Makassar Strait Throughflow Seasonal and Interannual Variability: An Overview. *J. Geophys. Res. Oceans* **2019**, *124*, 3724–3736. [CrossRef]
25. Van Aken, H.M.; Brodjonegoro, I.S.; Jaya, I. The deep-water motion through the Lifamatola Passage and its contribution to the Indonesian throughflow. *Deep. Res. Part I Oceanogr. Res. Pap.* **2009**, *56*, 1203–1216. [CrossRef]
26. Fang, G.; Susanto, R.D.; Wirasantosa, S.; Qiao, F.; Supangat, A.; Fan, B.; Wei, Z.; Sulistiyo, B.; Li, S. Volume, heat, and freshwater transports from the South China Sea to Indonesian seas in the boreal winter of 2007-2008. *J. Geophys. Res. Oceans* **2010**, *115*, 1–11. [CrossRef]
27. Gordon, A.L.; Huber, B.A.; Metzger, E.J.; Susanto, R.D.; Hurlburt, H.E.; Adi, T.R. South China Sea throughflow impact on the Indonesian throughflow. *Geophys. Res. Lett.* **2012**, *39*, 1–7. [CrossRef]
28. Qiu, B.; Mao, M.; Kashino, Y. Intraseasonal variability in the Indo-Pacific throughflow and the regions surrounding the Indonesian Seas. *J. Phys. Oceanogr.* **1999**, *29*, 1599–1618. [CrossRef]
29. Masumoto, Y.; Kagimoto, T.; Yoshida, M.; Fukuda, M.; Hirose, N.; Yamagata, T. Intraseasonal eddies in the Sulawesi Sea simulated in an ocean general circulation model. *Geophys. Res. Lett.* **2001**, *28*, 1631–1634. [CrossRef]
30. Kashino, Y.; Watanabe, H.; Herunadi, B.; Aoyama, M.; Hartoyo, D. Current variability at the Pacific entrance of the Indonesian Throughflow. *J. Geophys. Res. Oceans* **1999**, *104*, 11021–11035. [CrossRef]
31. Chen, X.; Qiu, B.; Chen, S.; Cheng, X.; Qi, Y. Interannual Modulations of the 50-Day Oscillations in the Celebes Sea: Dynamics and Impact. *J. Geophys. Res. Oceans* **2018**, *123*, 4666–4679. [CrossRef]
32. Kartadikaria, A.R.; Miyazawa, Y.; Nadaoka, K.; Watanabe, A. Existence of eddies at crossroad of the Indonesian seas. *Ocean Dyn.* **2012**, *62*, 31–44. [CrossRef]
33. Yang, C.; Chen, X.; Cheng, X.; Qiu, B. Annual versus semi-annual eddy kinetic energy variability in the Celebes Sea. *J. Oceanogr.* **2020**, *76*, 401–418. [CrossRef]
34. He, Y.; Feng, M.; Xie, J.; Liu, J.; Chen, Z.; Xu, J.; Fang, W.; Cai, S. Spatiotemporal Variations of Mesoscale Eddies in the Sulu Sea. *J. Geophys. Res. Oceans* **2017**, *122*, 7867–7879. [CrossRef]
35. Liu, Y.; Chen, G.; Sun, M.; Liu, S.; Tian, F. A parallel SLA-based algorithm for global mesoscale eddy identification. *J. Atmos. Oceans Technol.* **2016**, *33*, 2743–2754. [CrossRef]
36. Gulakaram, V.S.; Vissa, N.K.; Bhaskaran, P.K. Characteristics and vertical structure of oceanic mesoscale eddies in the Bay of Bengal. *Dyn. Atmos. Oceans* **2020**, *89*, 101131. [CrossRef]
37. Lagerloef, G.S.E.; Mitchum, G.T.; Lukas, R.B.; Niiler, P.P. Tropical Pacific near-surface currents estimated from altimeter, wind, and drifter data. *J. Geophys. Res. Oceans* **1999**, *104*, 23313–23326. [CrossRef]
38. Arbic, B.K.; Scott, R.B.; Chelton, D.B.; Richman, J.G.; Shriver, J.F. Effects of stencil width on surface ocean geostrophic velocity and vorticity estimation from gridded satellite altimeter data. *J. Geophys. Res. Oceans* **2012**, *117*, 1–18. [CrossRef]
39. Yuan, D.; Han, W.; Hu, D. Anti-cyclonic eddies northwest of Luzon in summer-fall observed by satellite altimeters. *Geophys. Res. Lett.* **2007**, *34*, 1–6. [CrossRef]
40. Oke, P.R.; Sakov, P.; Cahill, M.L.; Dunn, J.R.; Fiedler, R.; Griffin, D.A.; Mansbridge, J.V.; Ridgway, K.R.; Schiller, A. Towards a dynamically balanced eddy-resolving ocean reanalysis: BRAN3. *Ocean Model.* **2013**, *67*, 52–70. [CrossRef]
41. Schiller, A.; Oke, P.R.; Brassington, G.; Entel, M.; Fiedler, R.; Griffin, D.A.; Mansbridge, J.V. Eddy-resolving ocean circulation in the Asian-Australian region inferred from an ocean reanalysis effort. *Prog. Oceanogr.* **2008**, *76*, 334–365. [CrossRef]
42. He, Z.; Feng, M.; Wang, D.; Slawinski, D. Contribution of the Karimata Strait transport to the Indonesian Throughflow as seen from a data assimilation model. *Cont. Shelf Res.* **2015**, *92*, 16–22. [CrossRef]
43. Feng, M.; Benthuysen, J.; Zhang, N.; Slawinski, D. Freshening anomalies in the Indonesian throughflow and impacts on the Leeuwin Current during 2010–2011. *Geophys. Res. Lett.* **2015**, *42*, 8555–8562. [CrossRef]
44. Chu, P.C.; Edmons, N.L.; Fan, C. Dynamical mechanisms for the South China Sea seasonal circulation and thermohaline variabilities. *J. Phys. Oceanogr.* **1999**, *29*, 2971–2989. [CrossRef]
45. Cai, S.; He, Y.; Wang, S.; Long, X. Seasonal upper circulation in the Sulu Sea from satellite altimetry data and a numerical model. *J. Geophys. Res. Oceans* **2009**, *114*, 1–14. [CrossRef]

46. Chelton, D.B.; Schlax, M.G.; Samelson, R.M.; de Szoeke, R.A. Global observations of large oceanic eddies. *Geophys. Res. Lett.* **2007**, *34*, 1–5. [CrossRef]
47. Zhan, P.; Subramanian, A.C.; Yao, F.; Hoteit, I. Eddies in the Red Sea: A statistical and dynamical study. *J. Geophys. Res. Oceans* **2014**, *119*, 3909–3925. [CrossRef]
48. Nencioli, F.; Dong, C.; Dickey, T.; Washburn, L.; Mcwilliams, J.C. A vector geometry-based eddy detection algorithm and its application to a high-resolution numerical model product and high-frequency radar surface velocities in the Southern California Bight. *J. Atmos. Oceans Technol.* **2010**, *27*, 564–579. [CrossRef]
49. Penven, P.; Echevin, V.; Pasapera, J.; Colas, F.; Tam, J. Average circulation, seasonal cycle, and mesoscale dynamics of the Peru Current System: A modeling approach. *J. Geophys. Res. C Oceans* **2005**, *110*, 1–21. [CrossRef]
50. Oey, L.Y. Loop current and deep eddies. *J. Phys. Oceanogr.* **2008**, *38*, 1426–1449. [CrossRef]
51. Dufau, C.; Orsztynowicz, M.; Dibarboure, G.; Morrow, R.; Le Traon, P.-Y. Mesoscale resolution capability of altimetry: Present and future. *J. Geophys. Res. Oceans* **2016**, *121*, 4910–4927. [CrossRef]
52. Pujol, M.I.; Faugère, Y.; Taburet, G.; Dupuy, S.; Pelloquin, C.; Ablain, M.; Picot, N. DUACS DT2014: The new multi-mission altimeter data set reprocessed over 20 years. *Ocean Sci.* **2016**, *12*, 1067–1090. [CrossRef]
53. Cai, S.; Su, J.; Gan, Z.; Liu, Q. The numerical study of the South China Sea upper circulation characteristics and its dynamic mechanism, in winter. *Cont. Shelf Res.* **2002**, *22*, 2247–2264. [CrossRef]
54. Bracco, A.; Pedlosky, J. Vortex generation by topography in locally unstable baroclinic flows. *J. Phys. Oceanogr.* **2003**, *33*, 207–219. [CrossRef]
55. Cushman-Roisin, B.; Beckers, J.-M. *Introduction to Geophysical Fluid Dynamics: Physical and Numerical Aspects*, 2nd ed.; Academic Press: Cambridge, MA, USA, 2011; ISBN 978-0-12-088759-0.
56. Castruccio, F.S.; Curchitser, E.N.; Kleypas, J.A. A model for quantifying oceanic transport and mesoscale variability in the Coral Triangle of the Indonesian/Philippines Archipelago. *J. Geophys. Res. Oceans* **2013**, *118*, 6123–6144. [CrossRef]
57. Liu, Q.; Feng, M.; Wang, D. ENSO-induced interannual variability in the southeastern South China Sea. *J. Oceanogr.* **2011**, *67*, 127–133. [CrossRef]
58. Zhu, Y.; Wang, L.; Wang, Y.; Xu, T.; Li, S.; Cao, G.; Wei, Z.; Qu, T. Stratified Circulation in the Banda Sea and Its Causal Mechanism. *J. Geophys. Res. Oceans* **2019**, *124*, 7030–7045. [CrossRef]
59. Xu, A.; Yu, F.; Nan, F. Study of subsurface eddy properties in northwestern Pacific Ocean based on an eddy-resolving OGCM. *Ocean Dyn.* **2019**, *69*, 463–474. [CrossRef]

remote sensing

MDPI

Article

Radiation Sensitivity Analysis of Ocean Wake Information Detection System Based on Visible Light Remote Sensing

Shipeng Ying [1,2,3], Hongsong Qu [1,3,*], Shuping Tao [1,3], Liangliang Zheng [1,3] and Xiaobin Wu [1,2,3]

1 Changchun Institute of Optics, Fine Mechanics and Physics, Chinese Academy of Sciences, Changchun 130033, China
2 University of Chinese Academy of Sciences, Beijing 100049, China
3 Key Laboratory of Space-Based Dynamic & Rapid Optical Imaging Technology, Chinese Academy of Sciences, Changchun 130033, China
* Correspondence: quhongsong@ciomp.ac.cn

Abstract: Various ships and submerged moving objects in the ocean are key targets of numerous remote sensors. Wake has developed into one of the key detection targets of ocean visible light remote sensing as the visible trail information left by moving objects on the ocean surface. In the situation of slow ship speed, deep draft, and the existence of air clouds and fog, the wake target signal is weak, and the signal-to-noise ratio is low due to the low reflectivity of the sea surface and the interference of the background waves on the sea surface. This paper analyzes the radiative sensitivity of visible light imaging systems for the most crucial wake detection indicator in order to address the aforementioned issues. The noise equivalent reflectance difference, which is widely used to describe radiative sensitivity in engineering, is derived and numerically simulated by establishing the imaging link model based on TDICCD. We calculated the noise equivalent reflectivity difference for eight bands commonly used in ocean remote sensing; results show that the index is generally on the order of 10^{-4}, and with the increase in the central wavelength, the value of noise equivalent reflectance difference also shows a downward trend and is stable within a certain value range. This research provides theoretical guidance for the engineering design of a visible spectrum imaging system for wake detection, aids in improving the imaging system's capacity to detect weak wake signals, and provides a basis for subsequent wake detection and enhancement processing, removal of false wakes, and retrieval of ship information.

Keywords: wake detection; radiation sensitivity; noise equivalent reflectance difference

Citation: Ying, S.; Qu, H.; Tao, S.; Zheng, L.; Wu, X. Radiation Sensitivity Analysis of Ocean Wake Information Detection System Based on Visible Light Remote Sensing. *Remote Sens.* **2022**, *14*, 4054. https://doi.org/10.3390/rs14164054

Academic Editors: Xiaofeng Li, Chung-Ru Ho and Antony K. Liu

Received: 4 July 2022
Accepted: 16 August 2022
Published: 19 August 2022

1. Introduction

In the context of the rapid development of remote sensing technology, along with the urgent need for marine monitoring technology, remote sensing technology has a good application prospect in ocean detection. When the marine dynamic target moves, different types of wakes will be formed. Common marine wakes can be classified as three categories based on their formation mechanisms, surface wave wake, turbulent wake, and internal wave wake [1]. According to statistics, wake has the characteristics of wide range and long existence time. The interaction of the ship with the sea water, as well as the rotation of the propeller, are the main factors that influence the formation of wake. From the ship's bow to a distance away from the stern, a clear and identifiable trace will be formed, and it has a strong scattering effect on electromagnetic waves. Therefore, wake detection can be carried out by a combination of various means. Compared with the common infrared imaging and SAR imaging methods, the image obtained by visible spectrum imaging has the advantages of high resolution and high contrast [2]. Therefore, we can obtain the wake information straight from the image and invert the dynamic target position information and its motion parameters, so as to better realize the strategic goal of wide-area and precise search on the sea surface, which has important application value.

Wake wave height is affected by sea surface wind speed and ship parameters. Due to the factors of sea surface background wave and low sea surface reflectivity, wake characteristics are relatively weak, which increases the difficulty of detection. Meanwhile, in the wake detection of visible spectrum, under the influence of environmental factors such as atmosphere and illumination and imaging system factors such as noise, the image quality after wake imaging will be further degraded, which is not conducive to the subsequent processing of the wake target. Therefore, for the visible light imaging system, how to improve imaging quality and achieve clear imaging of weak wake targets is a pressing issue that must be solved.

Recent research on wake targets is mainly based on two aspects. On the one hand, it is based on the characteristics of the wake, the establishment of different types of wake models, and the simulation of wake images [3–7]. On the other hand, it mainly focuses on image processing research on weak wake images, such as enhancing weak wakes through denoising and enhancement methods, so as to perform wake detection and extraction to obtain the speed, heading, and other parameters of marine moving targets [8–11]. However, if the wake target is too weak to be detected by the imaging system, that is, the wake information cannot be seen in the image, the subsequent image processing methods will not work. Therefore, in-depth research into the visible spectrum imaging system is required in order to discover the key to wake detection and ensure the existence of wake targets in image, which is one of the crucial issues that should be addressed in engineering practice.

In the design of most imaging systems in the visible spectrum, indicators such as spatial resolution and spectral resolution are often emphasized, but for wake detection, the most important indicator is radiative resolution (also known as radiative sensitivity), which refers to the minimum radiation difference that the remote sensor can distinguish when receiving spectral radiation signals, namely for the ability to distinguish between adjacent peaks and troughs in the wake. The smaller the indicator value, the stronger the ability of the imaging system to distinguish weak targets, and the higher the probability of detecting the wake target. In engineering practice, the indicator of noise equivalent reflectivity difference is often used to evaluate the imaging system in the visible spectrum [12], but this indicator has a wide range of applications; it is necessary to analyze its influencing factors in combination with the imaging mechanism of the wake target and to improve the radiative sensitivity through the parameter selection of the imaging system, so as to improve the imaging system's ability to clearly image weak wakes.

This paper concentrates on the issue of clear imaging of the wake in the visible spectrum, and studies the imaging system in the visible spectrum mentioned above. Firstly, the typical Kelvin wake wave height is simulated by the point source disturbance model combined with the Michell thin ship theory; then, the time delay and integration charge coupled devices' (TDICCD) imaging system's imaging link model is established. Based on this, the indicator of noise equivalent reflectivity difference commonly used to describe radiative sensitivity in engineering is deduced and numerically simulated. Finally, the main factors affecting this indicator and the methods to improve the radiative sensitivity of the system are discussed.

2. Models and Methods

2.1. Kelvin Wake Model

Kelvin wake is one of the most common detected wakes in visible spectrum remote sensing applied to ocean monitoring, so this paper takes the Kelvin wake as an example to study. In 1887, Lord Kelvin pointed out that the trace generated by the ship is a surface gravity wave containing two types of wave systems, transverse wave and divergent wave [13]. Transverse wave and divergent wave within the wake angle of about 16° and 19.5° interfere with each other to form Kelvin arms, which are the main features of Kelvin wakes in optical images. Figure 1 shows the structure diagram of the ship's Kelvin wake. It can be seen from the figure that the transverse wave propagates in the opposite direction of the ship's sailing direction, while the divergent wave propagates in the vertical direction of the ship's sailing direction.

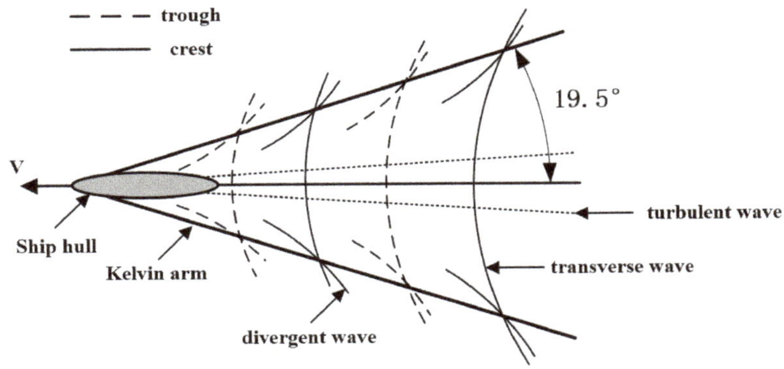

Figure 1. Kelvin wake diagram.

Assuming that the ship sails along the negative x-axis at velocity U, the corresponding Kelvin wake wave height can be described as a linear superposition of many free surface waves with different propagation directions, amplitudes, and frequencies [14].

$$\varsigma(x,y) = \mathrm{Re}\int_{-\frac{\pi}{2}}^{\frac{\pi}{2}} F(\theta)\exp[-ik_0\sec^2\theta(x\cos\theta + y\sin\theta)]d\theta \tag{1}$$

where Re is a symbol representing the real part, $k_0 = g/U^2$, g is the acceleration of gravity, and $F(\theta)$ represents the free spectrum of the ship. The free spectrum $F(\theta)$ of a ship can usually be described by the Michell thin ship theory [15], which holds that the intensity of water flow generated by the ship's transverse center point source has a certain proportional relationship with the local slope of the hull:

$$F(\theta) = \frac{2k_0}{\pi}\sec^3\theta \iint \frac{\partial y(x,z)}{\partial z}\exp\left(k_0 z\sec^2\theta + k_0 x\sec\theta\right)dxdz \tag{2}$$

where $y(x,z)$ is the characteristic equation describing the shape of the hull, which is a function of the offset position x and the draft z. For the convenience of calculation, the ship is simplified as an ellipsoid.

$$y(x,z) = \begin{cases} b(1 - x^2/l^2) & -d \le z \le 0, -l < x < l \\ 0 & z < d \end{cases} \tag{3}$$

where d is the draft depth, b is half the width of the ship, and l is half the length of the ship.

In fact, the front and rear parts of the ship have different effects on the water surface, and the stern of the ship has a certain drag effect on the water surface. Therefore, in order to be more realistic, the parameter of viscosity coefficient C is added. The integral term of Formula (1) is divided into two parts, which represent the influence of the bow and stern on the sea surface, respectively, and the part representing the stern is multiplied by the viscosity coefficient C = 0.6 to obtain the Kelvin wake wave height expression.

$$\begin{aligned} &\varsigma(x,y) = \tfrac{4b}{\pi k_0 l}\textstyle\int_{-\frac{\pi}{2}}^{\frac{\pi}{2}} [1 - \exp(-k_0 d\sec\theta)]\sin[k_0\sec^2\theta(x\cos\theta + y\sin\theta)]d\theta \\ &\varsigma(x,y) = \varsigma(x-l,y) + C\varsigma(x+l,y) \end{aligned} \tag{4}$$

Factors including sea surface wind speed, ship speed, ship size, and draft depth affect the Kelvin wake wave height. When the Kelvin wake appears on the sea surface, its fluctuation causes the roughness of the sea surface to change, modifying the facet's slope. The slope of an ideal reflecting surface is the result of adding the Kelvin wake slope and the wind wave slope, according to the paper by Liu et al. [16]. Therefore, the distribution of sea surface reflectivity can be obtained according to the Cox–Munk model [17]:

$$R = \frac{\rho P\left(S_x - S_{kx}, S_y - S_{ky}\right)}{4\cos\theta_r \cos^4\theta_n} \tag{5}$$

where ρ is the Fresnel reflectivity, $P\left(S_x - S_{kx}, S_y - S_{ky}\right)$ is the probability density function of the wave slope, θ_r is the reflection zenith angle, and θ_n is the inclination angle of the facet.

The change in Kelvin wake wave height leads to similar reflectivity changes in the crest and trough regions of the sea surface, and the changes in reflectivity at different positions on sea surface result in changes in reflected radiation on sea surface, thus affecting the information received by the detector. The presence or absence of reflection discrepancies between crests and troughs determines whether or not the wake characteristics may be seen in the designated wake imaging region. With an increase in the reflection difference between the crest and trough, wake features become more significant. According to the paper by Song et al. [18], in which this phenomenon was analyzed, the minimum reflectance difference between the crests and troughs of the imaging area are defined as the reflectance resolution R_r. In general, the reflectance resolution of wake targets is around 10^{-3}~10^{-4}, which may be lower if sea conditions are complex. Therefore, for visible imaging system, how to detect the wake target with such weak radiation difference is a big problem.

2.2. Imaging Theoretical Modeling

In applications for airborne and spaceborne remote sensing, with the increasing level of remote sensing information application requirements, the imaging performance requirements of cameras are also becoming higher and higher, but the improvement of spatial resolution or spectral resolution will always be accompanied by a decrease in radiative resolution. In order to solve this problem, more high-resolution cameras choose TDICCD as their detector. The sensor's own line-by-line energy accumulation capability can solve the issue of weak energy in a single integration, and it has obvious benefits in an ocean's visible spectrum remote sensing with weak incident energy. As a result, the imaging link modeling process is mainly analyzed for TDICCD in this section, including its imaging principle, noise model, and imaging link analysis, which provides the basis for the derivation of the radiative sensitivity model in the next section.

2.2.1. TDICCD Imaging Principle

The full name of TDICCD is time delay and integration charge coupled devices, which is an application form of area array CCD. The number of columns equals the number of pixels in a row, and the number of rows is the integral grade M of the TDICCD. Its working principle is shown in Figure 2. In the flight direction of the imaging system, each row of CCD photosensitive pixels images the same target. By imaging the same target multiple times at different times and accumulating charges, the photosensitive ability of the pixels is improved, thereby improving the ability to identify and acquire the target. TDICCD not only enhances the exposure of signals but also increases the noise. However, since signals increase linearly, whereas noise increases nonlinearly, the SNR of the system will also increase with the increase in integral grade, thus improving the sensitivity of the imaging system.

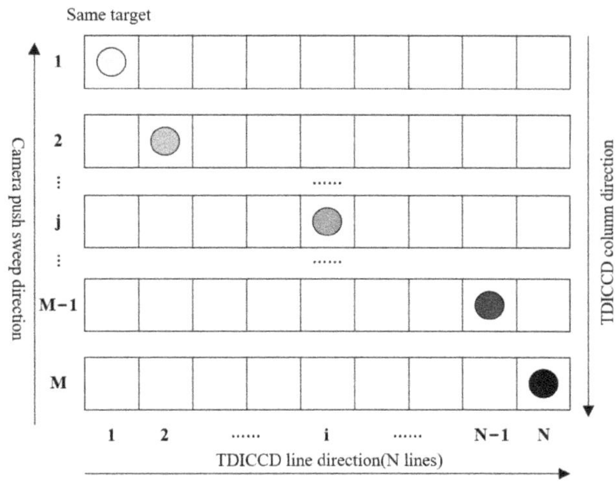

Figure 2. TDICCD imaging principle.

2.2.2. Noise Model

Noise can affect the image quality of the camera. The existence of noise will not only cause the degradation of image quality and reduce its SNR but also affect the subsequent image-based wake enhancement algorithm, resulting in inaccurate wake extraction and even algorithm failure when the noise is severe. Therefore, whether to improve image quality or to ensure the accuracy of subsequent algorithms, it is necessary to reduce or eliminate noise as much as possible.

The noise in the TDICCD imaging process mainly includes photon shot noise, dark current noise, readout noise, and fixed pattern noise. However, wake detection relies on the radiation difference between the crest and trough, and since the difference between them is very small, there may be only a few gray values reflected on the image, so the suppressed fixed pattern noise may also affect subsequent image processing, so the fixed pattern noise cannot be ignored in this paper.

Photon shot noise is the inherent noise of TDICCD devices and is caused by random fluctuations in the number of photons reaching the sensor. Its expression satisfies the discrete Poisson distribution function, and its equivalent electron is:

$$\sigma_{shot} = \sqrt{\overline{n}} = \sqrt{N_{signal}} \tag{6}$$

where $\overline{n}$ is the average charge number and N_{signal} is the signal charge number.

Dark current noise is a quantity closely related to detector temperature. When there is no signal input, a detector with a certain temperature produces a dark current offset due to the irregular thermal motion of electrons, forming dark current noise, and its equivalent electron is:

$$\sigma_{dark} = \sqrt{N_{dark}} \tag{7}$$

where N_{dark} is the number of electrons in the dark current.

Readout noise mainly includes reset noise and amplifier noise, and its equivalent noise electron is:

$$\sigma_{rms} = \frac{N_{rms}}{K_c} \tag{8}$$

where N_{rms} is the root mean square noise; K_c is the charge conversion scaling factor.

The noise that the detector pixel produces with different outputs for the same input due to the difference in the bias or response of each detector pixel is called fixed pattern noise, also called response non-uniformity, which is defined as [19]:

$$U = \frac{\sigma_R}{R} \tag{9}$$

where σ_R is the mean square error of the response rate of each pixel; R is the average response rate of each pixel.

Non-uniformity is manifested as fixed pattern noise, and the fixed pattern noise of each pixel is described by the number of electrons as:

$$\sigma_{FPN} = UN_{signal} \tag{10}$$

Therefore, the total noise electrons in the system can be obtained as:

$$N = \sqrt{\sigma_{shot}^2 + \sigma_{dark}^2 + \sigma_{rms}^2 + \sigma_{FPN}^2} \tag{11}$$

For the imaging system using a TDICCD sensor, when the integral grade is M, the quantity of noise electrons satisfies the following relation:

$$N_{noise} = \sqrt{M\sigma_{shot}^2 + M\sigma_{dark}^2 + M\sigma_{rms}^2 + M\sigma_{FPN}^2} = \sqrt{M}N \tag{12}$$

2.2.3. Wake Imaging Link Model

Lighting conditions in remote sensing imaging applications can be roughly divided into two categories. One is that the sun elevation angle is appropriate and the illumination is relatively sufficient, and ideal imaging effects can be achieved by setting a reasonable integral grade or integration time; another is the weak lighting condition when the sun elevation angle is low, and some objects are drowned out by noise in the image because the radiance is too low. In order to improve the imaging sensitivity and signal-to-noise ratio under low-light conditions, it is necessary to start with the imaging link and further explore high-sensitivity imaging measures.

The four main components of the sea surface radiance that the detector obtains are the sea surface radiation itself, the scattering of background radiation to the sky by the sea, the scattering of radiation from the sea to the sun, and the radiation of the atmospheric path above the sea surface, as shown in Figure 3. The formula for the total energy of radiation reaching the detector from the sea surface can be obtained:

$$L = \tau_a\left[\varepsilon L_{bb}(\lambda; T) + \rho\left(L_{sky} + L_{sun}\right)\right] + L_{path} \tag{13}$$

where τ_a is the atmospheric transmittance between the sea surface and detector; ε is the sea surface emissivity; ρ is the sea surface reflectivity; $L_{bb}(\lambda; T)$ is the blackbody radiant exitance with temperature T and wavelength λ; L_{sky} is the sky incident irradiance on the sea surface; L_{sun} is the sun incident irradiance on the sea surface; and L_{path} is the atmospheric path irradiance.

At room temperature, the radiant exitance of the blackbody in the wavelength range of visible light is approximately zero [20]. Therefore, radiation from the sea surface itself is ignored in the visible wavelength range. At the same time, since the magnitude of sunlight reflection is much higher than that of sky light reflection and atmospheric path radiation (the experimental data are shown in Figure A1 of Appendix A), we only consider the reflection of sunlight by the sea surface. So Equation (13) can be simplified as:

$$L = \tau_a \rho L_{sun} \tag{14}$$

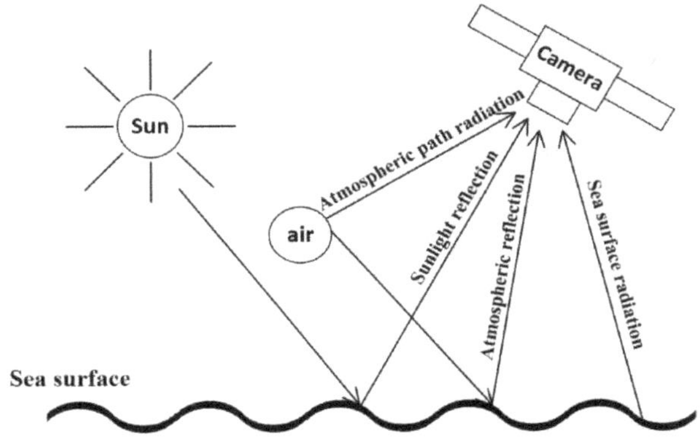

Figure 3. Ocean radiative transfer model.

When the light reaches the TDICCD sensor through the optical system, the illumination received by its photosensitive element is:

$$E_{TDI-CCD} = \frac{\pi \tau_o L}{4F^2} \tag{15}$$

where F represents the F number of the optical system, and τ_o is the total transmittance of the optical system.

For a multispectral camera, when the integral grade is M, the quantity of signal electrons generated by a detector in a certain spectral band is:

$$\begin{aligned} N_M &= \frac{E_{TDI-CCD} A_d M T_{\text{int}} \lambda_c \eta \Delta\lambda}{\pi h c} \\ &= \frac{\rho L_{sun} A_d \tau_a \tau_0 M T_{\text{int}} \lambda_c \eta \Delta\lambda}{4F^2 h c} \end{aligned} \tag{16}$$

where A_d is the area of the TDICCD detector; T_{int} is the integration time; λ_c is the center wavelength; η is the quantum efficiency; $\Delta\lambda$ is the spectral band width; h is Planck's constant; and c is the speed of light.

According to the noise model, the noise of M times is simulated, and the quantity of noise electrons integrated for M times is added to the quantity of electrons in the pixel signal. The whole quantity of electrons generated is:

$$N_{total} = N_{signal} + \sum_{1}^{M} N \tag{17}$$

The calculation formula of the digital image signal obtained by the signal charge after correlated double sampling, preamplification, and analog-to-digital conversion circuit is:

$$DN = \frac{N_{total} \cdot 2^n}{N_{fullwell}} \tag{18}$$

where n is the quantization bits and $N_{fullwell}$ is the full well capacity of the TDICCD.

The choice of quantization bits will affect the imaging quality. High quantization bits can reduce quantization noise and increase the grayscale range of the image, thereby increasing the grayscale difference between the crests and troughs of the wake, which is more conducive to distinguishing the wake target from the image. However, higher quantization bits will greatly increase the amount of data, so we should not blindly pursue high quantization bits. For the high reflectivity resolution required by wake detection, the quantization bit is generally 12~14 bit.

The imaging model can be stated as follows in light of the analysis above:

$$DN = f(\rho, F, M, n) \tag{19}$$

The wake imaging system's output digital signal value DN is influenced by the F number, quantization bits n, integral grade M, and reflectivity ρ. The imaging system must be capable of distinguishing the radiation difference between the crest and trough in order to produce a clear image of the wake. The imaging system's output signal should satisfy the following requirements when the reflectivity of the crest and trough is ρ_c and ρ_t:

$$\Delta DN = DN_c - DN_t \geq 1 \tag{20}$$

where ΔDN is the output digital signal's difference value. When ΔDN is not less than 1, the digital response of the crest and the trough (that is, the gray value of the corresponding image) is not exactly the same. In this instance, the wakes in the image may exhibit overall characteristics of the peaks and troughs, and further image processing can amplify this difference and make it easier to detect. ΔDN typically needs to be substantially higher than 1, as a big value of ΔDN in engineering applications represents a clear wake. After determining the reflectivity of the wake, the optical system's F number, quantization number, and integral grade can be carefully chosen to produce a clear image of the wake target.

2.3. Radiative Sensitivity Modeling

Radiometric resolution refers to the responsiveness of a remote sensor to distinguish subtle changes in the input radiation, that is, radiometric sensitivity. In visible light, near-infrared, and short-wave infrared bands, the noise equivalent reflectance difference $NE\Delta\rho$ is usually used to represent the radiative sensitivity of the remote sensor; in medium-wave and long-wave infrared, the noise equivalent temperature difference $NE\Delta T$ is usually used to represent the remote sensor radiative sensitivity [21]. Referring to the design of most visible ocean remote sensors, the requirement of noise equivalent reflectance difference is generally $NE\Delta\rho < 5 \times 10^{-4}$, but this requirement is mainly for ocean water color remote sensing, not for wake detection. Therefore, the noise equivalent reflectance difference of our visible spectral imaging system for wake detection is analyzed.

$NE\Delta\rho$ is the smallest detectable change in reflectivity of a ground target and is defined as the change in reflectivity of a ground target source required to produce a signal equivalent to system noise. It can be expressed as:

$$NE\Delta\rho = \frac{\rho}{(S/N)} \tag{21}$$

where ρ is the reflectivity change required to generate signal S; S is the signal; and N is noise.

In general, the condition defined by the SNR is a target with a reflectivity ρ, which is:

$$SNR_\rho = \frac{N_\rho}{N_{noise}} \tag{22}$$

If the signal difference between two targets with different reflectivity (that is, the crests and troughs of the wake) is less than or equal to the noise at this time, the two targets cannot be distinguished. Therefore, by calculating the SNR of the difference between two target signals with different reflectivity, it can be obtained from Equation (16):

$$\begin{aligned} SNR_{\Delta\rho} &= \frac{N_c - N_t}{N_{noise}} \\ &= \frac{L_{sun} A_d \tau_e \tau_0 M T_{\text{int}} \lambda_c \eta \Delta\lambda}{4F^2 hc}(\rho_c - \rho_t) \\ &= \frac{L_{sun} A_d \tau_e \tau_0 M T_{\text{int}} \lambda_c \eta \Delta\lambda}{4F^2 hc}\Delta\rho \end{aligned} \tag{23}$$

According to the above definition, we can obtain another important index, the noise equivalent reflectivity $NE\Delta\rho$. In the above equation, if $SNR_{\Delta\rho}$ is as small as 1, the two targets with different reflectivity cannot be distinguished. So when $SNR_{\Delta\rho}$ equals 1, the resulting $\Delta\rho$ is $NE\Delta\rho$. Therefore, $NE\Delta\rho$ of the multispectral camera is obtained:

$$NE\Delta\rho_M = \frac{4F^2 N_{noise}hc}{\tau_o \tau_a A_d L_{sun} T_{\text{int}} M \lambda_c \eta \Delta\lambda} \tag{24}$$

$NE\Delta\rho$ is a key parameter. $NE\Delta\rho$ is a crucial metric to measure the capability of the visible imaging system. When detecting the same wake target, the higher the $NE\Delta\rho$ (that is, the smaller the value), the bigger the difference between the distinguishable wake crests and troughs and the better the capacity to detect the wake.

3. Simulation

3.1. Kelvin Wake Model

The ship's Kelvin wake is simulated using the mathematical model of the Kelvin wake mentioned above, and the simulation results are displayed in Figure 4. In the simulation, it is assumed that the ship moves in the negative direction of the x-axis, the size of the ship is 80 m × 20 m, the draft is 3 m, and the simulation area is 800 m × 800 m.

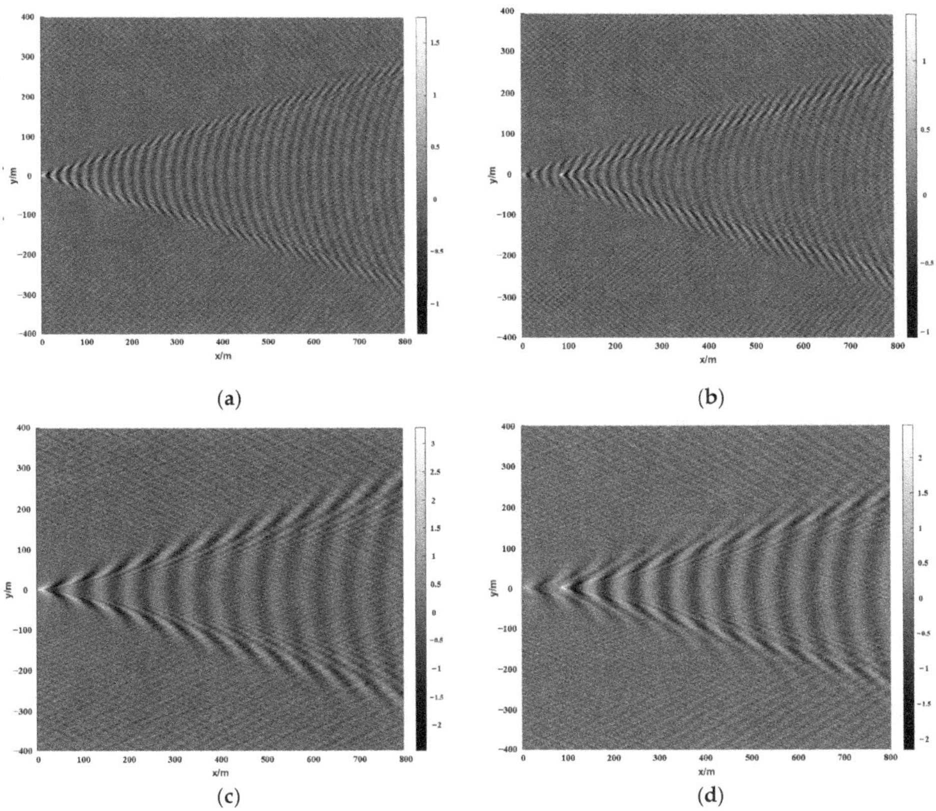

Figure 4. Simulation results of Kelvin wake under different conditions: (**a**) U = 6 m/s, viscous effects are not considered; (**b**) U = 6 m/s, considering the viscous effect; (**c**) U = 10 m/s, viscous effects are not considered; (**d**) U = 10 m/s, considering the viscous effect.

Figure 4a,c are the Kelvin wakes without considering the influence of viscosity at ship speeds of 6 m/s and 10 m/s. It is obvious from the simulation results that the Kelvin wake is mainly composed of two wave systems, the transverse wave propagating in the x-direction and diffuse wave propagating in the y-direction, and both the transverse wave and the diffuse wave show significant periodic characteristics. Additionally, the wave height and wavelength of the Kelvin wake increase to a certain extent with the increase in ship speed, and the difference between the wave crest and trough becomes more obvious, which is more conducive to the detection of the imaging system. Figure 4b,d show the Kelvin wake of the ship considering the viscous effect at 6 m/s and 10 m/s speed, respectively. The figure demonstrates that the Kelvin wake consists of point sources in the bow and stern, and the superposition of the two parts of the wake will also have a certain impact on the detection. Therefore, in order to facilitate calculation, the influence of viscosity is not taken into account by the Kelvin wake model in the follow-up study of this paper.

Different ship size, speed, draft, and other parameters will produce different Kelvin wakes. The most direct effect of the Kelvin wake is to adjust the roughness of the sea surface and affect the reflectivity of the sea surface. For the detector, how to design the key parameters to achieve clear imaging of the above Kelvin wakes is another content of this paper. So, we derive and simulate the radiative sensitivity of the system.

3.2. Radiative Sensitivity Model

Through the analysis of the radiative sensitivity model, we obtain the expression of the $NE\Delta\rho$. Then, we roughly calculate the $NE\Delta\rho$ of common spectral bands in marine remote sensing detection, which also provides a theoretical basis for the corresponding indicators of visible light imaging systems in engineering practice. The parameters required in the calculation process and the way to obtain it are shown in Figure 5.

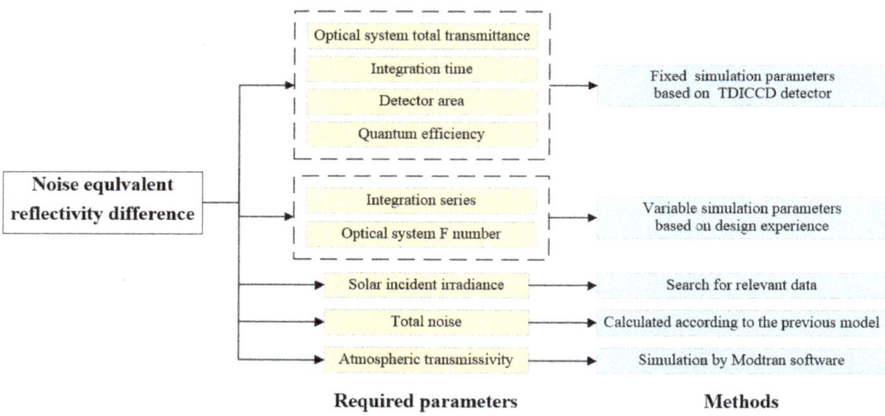

Figure 5. Calculation process of noise equivalent reflectivity difference.

Firstly, set the imaging parameters that need to be determined according to the parameters of the TDICCD detector, including optical system transmittance, integration time, detector area, and quantum efficiency. Table 1 shows the imaging simulation parameters; the solar irradiance outside the atmosphere in the common bands of ocean remote sensing is shown in Table 2.

MODTRAN software was used for the analysis of atmospheric transmittance. The weather model was sunny and cloudless in the experiment. The observation location was the South China Sea, and the aerosol model was the Navy aerosol model. At the observation height of 5km, the atmospheric transmittance of the visible light and near-infrared bands when the solar zenith angle is 0° and the detection system is vertically observed is shown in Figure 6.

Table 1. Fixed Simulation Parameters.

Parameters	Value
Optical system total transmittance	0.7
Detector pixel size (μm)	17.5 × 17.5
Pixel number	3072
Quantum efficiency	0.5
Integration time (ms)	4

Table 2. The solar irradiance outside the atmosphere in different bands.

Central Wavelength (nm)	Irradiance $\left(\mathbf{W} \cdot \mathbf{m}^{-2} \cdot \mu\mathbf{m}^{-1}\right)$
412	1760
443	1877
490	1950
520	1933
565	1705
670	1456
750	1235
865	958

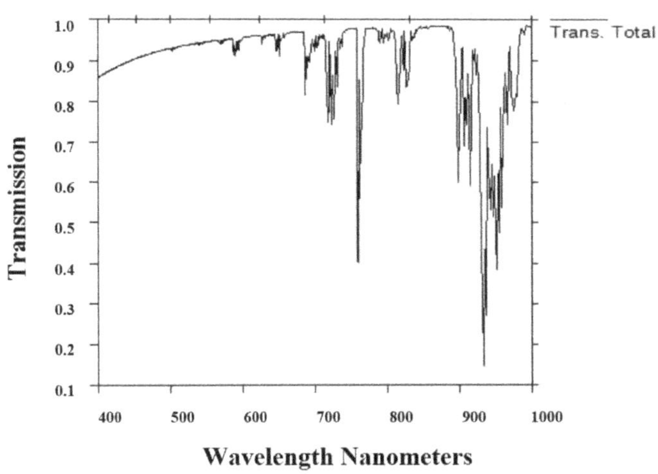

Figure 6. Simulation results of atmospheric transmittance.

According to the engineering practice experience, the F number of the imaging system is selected as 3, 6, and 10, the integral grade is selected as 16, and the noise electron number is simulated according to Equations (6)–(12). Formula (24) is used to calculate the noise equivalent reflectivity difference, and the calculation results are shown in Figure 7.

The figure demonstrates that the calculated noise equivalent reflectivity difference is generally in the order of 10^{-4}, and with the increase in the central wavelength, the value of noise equivalent reflectance difference also shows a downward trend and is stable within a certain value range. At the same time, with the decrease in the F number of the imaging system, the value of the noise equivalent reflectivity difference also decreases. Therefore, in engineering practice, under the premise of balancing the processing difficulty and cost, the F number of the imaging system should be reduced as much as possible to improve the radiative sensitivity of the system, thereby increasing the probability of detecting wake targets.

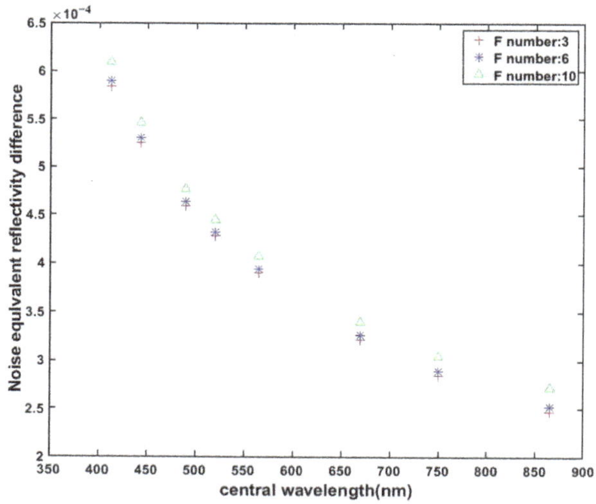

Figure 7. The noise equivalent reflectivity difference in different bands.

4. Discussion

According to the above calculation process, the factors affecting the radiative sensitivity of the imaging system are mainly two aspects: the optical system and the image sensor. The influencing parameters of the optical system include the system F number and the total transmittance of the optical system; the influencing parameters of the image sensor include integration time, integral grade, noise, and so on. Based on the above influencing factors, the methods to improve the radiative sensitivity of the system are discussed.

4.1. Improve the Optical System

In the previous section, the impact of the system F number on radiative sensitivity was analyzed. Reducing the system F number, that is, increasing the aperture of the optical system, can improve the energy converging on the image plane, thereby improving the radiative sensitivity of the system. However, the increase in aperture will increase the development cost and processing difficulty of the system, so in practical application, the system aperture is usually determined first, and the system is improved by improving other influencing factors. The total transmittance of an optical system is typically a function of the lens's transmittance, the transmittance of semi-reflecting and semi-transmitting, and the transmittance of the neutral variable filter in the center of the field of view, which can be improved by optimizing the lens material and coating. At the same time, it is also necessary to put forward requirements for the optical–mechanical structure design of the system to reduce the influence of stray light in the system.

4.2. Reduce Camera Noise

According to the analysis in Section 2.2.2, the size of the noise is closely related to the number of incident photons. Assuming that the inhomogeneity of the system is 1%, Figure 8 shows the change in the RMS electron number of noise when incident photons are different.

Figure 8 shows that when there are few incident photons, the system noise is mainly determined by readout and dark current noise. They have nothing to do with the number of incident photons and are mainly determined by the electronic parameters of the detector. When the number of incident photons is large, the noise is mainly determined by the photon shot noise, and other radiation sources besides target radiation should be reduced as far as possible. When the number of incident photons is large, the fixed graphic noise is the main influence, and it is particularly important to reduce the pixel non-uniformity.

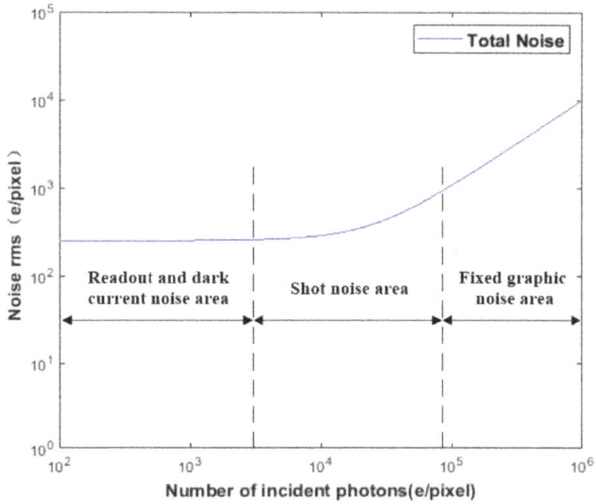

Figure 8. The total noise electrons in the system.

For the detection of marine targets, the overall irradiance of the sea surface is low, so it is necessary to adjust the parameters such as the integral grade and gain to make the irradiance obtained by the imaging system within a high range to meet the imaging needs. Therefore, the number of incident photons in the system will be in a high range, and the fixed pattern noise of the system cannot be ignored. It is necessary to reduce the non-uniformity of the imaging system as much as possible through the radiometric calibration and subsequent image processing and provide preprocessing operations for the recognition and enhancement of wake images.

4.3. Increase the Pixel Area

The sensitivity of the camera is also proportional to the quantum efficiency of the detector and the pixel area. The quantum efficiency is difficult to improve due to the influence of the semiconductor material and the wavelength of the incident light. Therefore, the pixel area can be indirectly increased by the binning method to improve the sensitivity of the detector. A binning operation on multiple pixels increases the total number of signals of merged pixels, which is more conducive to detecting wake targets under weak light conditions. Meanwhile, pixel binning also improves the image SNR. The conventional imaging mode and the 2×2 binning imaging mode for pixels are shown in Figure 9. When the binning operation is used for both directions at the same time, the proportion of the image remains unchanged, but the spatial resolution will decrease accordingly.

However, there are some limitations with pixel binning. Although pixel binning can improve the radiative resolution under some conditions, it also decreases the signal level. Pixel binning is used in high-resolution sensors, which can improve the radiative sensitivity and will not have a big impact on the image quality when the signal level decreases. In the case of moderate resolution sensors or even lower, pixel binning may have a big impact on the image quality when the signal level decreases, and it is difficult to obtain the wake information through subsequent image processing. Therefore, pixel binning is not suitable for all sensors and needs to be considered comprehensively.

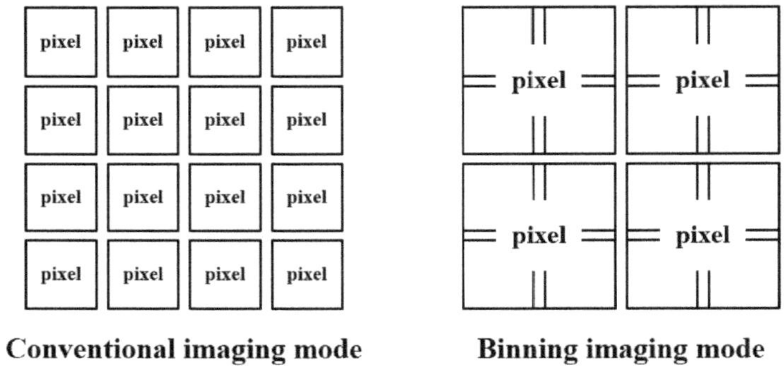

Figure 9. Conventional imaging and binning imaging schematic diagram.

In the above discussion section, some methods to improve the radiation sensitivity are proposed by analyzing the optical system and image sensor. However, the difficulty and cost should be considered comprehensively in the design of the imaging system. Therefore, we selected some methods in the above discussion which can be used in the imaging system design of wake detection. First, the F number of the optical system can be increased. Since the focal length of the system remains unchanged, increasing the F number of the system is equivalent to increasing the aperture of the system. Then, after the CCD sensor is selected, the non-uniformity of the imaging system can be corrected to reduce its total noise. Finally, the binning operation can also be performed on its pixels to achieve the purpose of improving the radiative sensitivity of the system.

In addition, there is a mutual constraint between radiative resolution and spatial resolution. Spatial resolution is the smallest pixel that is distinguishable on the ground. Generally speaking, when the spectral resolution remains the same, the minimum resolvable pixel increases with the increasing instantaneous field of view, while the spatial resolution decreases. At the same time, the greater the instantaneous radiation energy, the stronger the ability to detect the weak energy difference, namely, the higher the radiative resolution. In order to improve the radiative resolution, it is necessary to sacrifice part of the spatial resolution, so as to improve the wake detection capability of the system. Therefore, in the research and application of remote sensing technology, it is necessary to find a balance between high spatial resolution and high radiative resolution, in order to achieve the desired effect.

5. Conclusions

In this paper, an analysis of the radiative sensitivity of an imaging system in the visible spectrum is carried out. The Kelvin wake wave height is simulated by the point source disturbance model combined with the Michell thin ship theory. The TDICCD imaging system's imaging link model is established. Based on this, the noise equivalent reflectivity difference is deduced, and numerical simulation is carried out. Results show that this indicator is generally in the order of 10^{-4}, and with the increase in the central wavelength, the value of the indicator shows a downward trend and is stable within a certain value range. The main factors affecting this index and the methods to improve the radiative sensitivity of the system are discussed, which can be used to optimize and improve the system in the future. This research helps to improve the ability of the visible spectrum imaging system to detect weak wake signals and paves the way for subsequent wake detection and enhancement processing, removal of false wakes, and retrieval of ship information. It should be noted that this paper mainly discussed and analyzed the Kelvin wake as an example. The noise equivalent reflectance difference model is also applicable to turbulent wake, vortex, and internal wave wake, and different types of wakes can be calculated separately in the future.

Author Contributions: Conceptualization, S.Y. and H.Q.; methodology, S.Y. and H.Q.; software, S.Y.; writing—original draft preparation, S.Y. and H.Q.; writing—review and editing, S.Y., H.Q., S.T., L.Z. and X.W.; funding acquisition, H.Q. All authors have read and agreed to the published version of the manuscript.

Funding: This research was funded by the National Natural Science Foundation of China under Grant 62075219 and the Key Technological Research Projects of Jilin Province, China under Grant 20190303094SF.

Data Availability Statement: Not applicable.

Acknowledgments: The authors would like to thank the anonymous reviewers for their valuable comments.

Conflicts of Interest: The authors declare no conflict of interest.

Abbreviations

The following abbreviations are used in this manuscript:

TDICCD	Time delay and integration charge coupled devices
SNR	Signal-to-noise ratio
$NE\Delta\rho$	The noise equivalent reflectance difference
$NE\Delta T$	The noise equivalent temperature difference

Appendix A

We calculated the three kinds of radiation by MODTRAN software, and the calculation results are as follows:

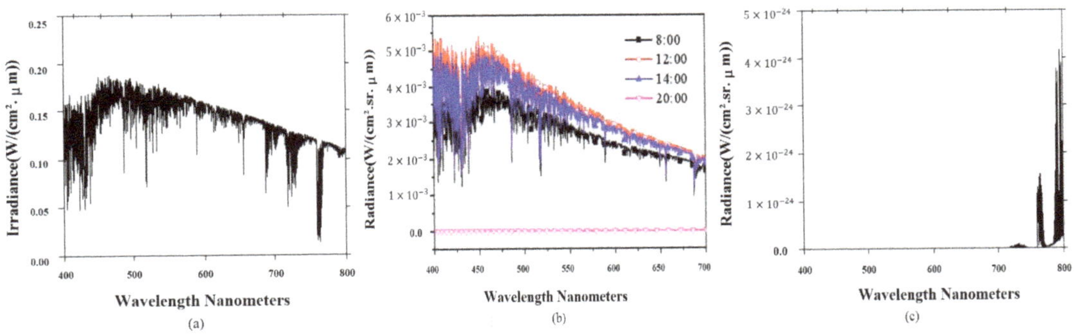

Figure A1. Sky and solar radiation: (**a**) solar irradiance; (**b**) the radiance of sky background radiation; (**c**) the radiance of atmospheric path.

References

1. Tunaley, J.; Buller, E.; Wu, K.; Rey, M. The simulation of the SAR image of a ship wake. *IEEE Trans. Geosci. Remote Sens.* **1991**, *29*, 149–156. [CrossRef]
2. Liu, Y.; Deng, R. Ship wakes in optical images. *J. Atmos. Ocean. Technol.* **2018**, *35*, 1633–1648. [CrossRef]
3. Xue, F.; Jin, W.; Qiu, S.; Yang, J. Wake Features of Moving Submerged Bodies and Motion State Inversion of Submarines. *IEEE Access* **2020**, *8*, 12713–12724. [CrossRef]
4. Xue, F.; Jin, W.; Qiu, S.; Yang, J. Airborne optical polarization imaging for observation of submarine Kelvin wakes on the sea surface: Imaging chain and simulation. *ISPRS J. Photogramm. Remote Sens.* **2021**, *178*, 136–154. [CrossRef]
5. Zilman, G.; Zapolski, A.; Marom, M. On detectability of a ship's Kelvin wake in simulated SAR images of rough sea surface. *IEEE Trans. Geosci. Remote Sens.* **2014**, *53*, 609–619. [CrossRef]
6. Sun, X.; Cai, M.; Wang, J.; Liu, C. Numerical Simulation of the Kelvin Wake Patterns. *Appl. Sci.* **2022**, *12*, 6265. [CrossRef]
7. Ahmadibebi, A.; Jones, B.; Shirkhodaie, A. Physics-based wake modeling for marine vehicles activity recognition based on simulated synthetic aperture radar. In Proceedings of the SPIE, Ocean Sensing and Monitoring XIV, Orlando, FL, USA, 3 April–13 June 2022; Volume 12118, pp. 107–119.
8. Zhao, H.; Ji, Z.; Zhang, Y.; Sun, X.; Song, P.; Li, Y. Mid-infrared imaging system based on polarizers for detecting marine targets covered in sun glint. *Opt. Express* **2016**, *24*, 16396–16409. [CrossRef] [PubMed]

9. Song, M.; Qu, H.; Zhang, G.; Tao, S.; Jin, G. variational model for sea image enhancement. *Remote Sens.* **2018**, *10*, 1313. [CrossRef]
10. Zhu, S.; Chen, Z.; Kang, L. Satellite-borne SAR low-contrast signals wave wake image enhancement based on a bilateral rapid filtering visualization algorithm. In Proceedings of the 2019 IEEE International Conference on Signal Processing, Communications and Computing (ICSPCC), Dalian, China, 20–22 September 2019; pp. 1–3.
11. Srivastava, R.; Christmas, J. Analysis of Sea Waves and Ship Wake Detection. In Proceedings of the OCEANS 2022-Chennai, Chennai, India, 21–24 February 2022; pp. 1–10.
12. Wang, J.; Wang, Y.; Li, C. Noise model of hyperspectral imaging system and influence on radiation sensitivity. *J. Remote Sens.* **2010**, *14*, 607–620.
13. Kelvin, L. On ship waves. *Proc. Inst. Mech. Engrs.* **1887**, *38*, 409–434.
14. Oumansour, K.; Wang, Y.; Saillard, J. Multifrequency SAR observation of a ship wake. *IEEE Proc. -Radar Sonar Navig.* **1996**, *143*, 275–280. [CrossRef]
15. Wang, H. Spectral comparisons of ocean waves and Kelvin ship waves. In Proceedings of the Seventh Offshore Mechanics and Arctic Engineering Symposium, New York, NY, USA; 1988; Volume 2, pp. 253–261.
16. Liu, Y.; Deng, R.; Zhao, J. Simulation of Kelvin wakes in optical images of rough sea surface. *Appl. Ocean. Res.* **2019**, *89*, 36–43. [CrossRef]
17. Cox, C.; Munk, W. Measurement of the roughness of the sea surface from photographs of the sun's glitter. *Josa* **1954**, *44*, 838–850. [CrossRef]
18. Song, M.; Wang, S.; Zhao, P.; Chen, Y.; Wang, J. Modeling Kelvin wake imaging mechanism of visible spectral remote sensing. *Appl. Ocean. Res.* **2021**, *113*, 102712. [CrossRef]
19. Wang, Y. Influence of Infrared Detector Response Nonuniformity on System Sensitivity. *Infrared Laser Eng.* **2006**, *3*, 258–261.
20. Wu, X.; Smith, W.L. Emissivity of rough sea surface for 8–13 μm: Modeling and verification. *Appl. Opt.* **1997**, *36*, 2609–2619. [CrossRef] [PubMed]
21. Han, X. Calculation of radiation sensitivity for spaceborne multispectral scanner. *Infrared Res. (Ser. A)* **1988**, *3*, 213–217.

remote sensing

MDPI

Article

Mapping the Coastal Upwelling East of Taiwan Using Geostationary Satellite Data

Zhi Huang [1], Jianyu Hu [1,2,*] and Weian Shi [1]

[1] State Key Laboratory of Marine Environmental Science, College of Ocean and Earth Sciences, Xiamen University, Xiamen 361102, China; zhihuang77@hotmail.com (Z.H.); 22320190154048@stu.xmu.edu.cn (W.S.)

[2] Southern Marine Science and Engineering Guangdong Laboratory (Zhuhai), Zhuhai 519000, China

* Correspondence: hujy@xmu.edu.cn

Abstract: Coastal upwelling is important for coastal ecosystems and the blue economy because of its large productivity and large potential for catching fish. However, coastal upwelling along the Taiwan east coast has received little attention from the research community. This study used five-year daily Himawari-8 geostationary satellite sea surface temperature data to map the coastal upwelling east of Taiwan during the summer monsoon season. We applied a semi-automatic image process technique based on the topographic position index for the quantitative upwelling mapping. The results show clear evidence of seasonal coastal upwelling along the entire Taiwan east coast, mainly under the influence of upwelling-favorable southwesterly/southerly winds. There are three broad upwelling centers along the Taiwan east coast: north, central, and south. The upwelling around the northern center has the longest upwelling season, lasting from May to September. The upwelling extents are larger between June and August during the height of the summer monsoon.

Keywords: coastal upwelling; Himawari-8; sea surface temperature; Taiwan; topographic position index; upwelling index; mapping

Citation: Huang, Z.; Hu, J.; Shi, W. Mapping the Coastal Upwelling East of Taiwan Using Geostationary Satellite Data. *Remote Sens.* **2021**, *13*, 170. https://doi.org/10.3390/rs13020170

Received: 2 December 2020
Accepted: 4 January 2021
Published: 6 January 2021

1. Introduction

Coastal upwelling is important for coastal ecosystems and the blue economy, because of its elevated productivity and large potential for fish catch resulted from the uplifting of nutrient-rich water towards the sea surface [1,2]. Major coastal upwelling systems are found around the world [1–6]. There are numerous upwelling hotspots in these major coastal upwelling systems, a number of which are known to occur in the China seas [3]. In their comprehensive review, ref. [3] identified 12 major upwelling regions in the China seas. Four of these upwelling regions are located in the Taiwan Strait (Southwestern Taiwan Strait, Northwestern Taiwan Strait, Taiwan Bank, and Penghu Islands), and one is located offshore Northeast of Taiwan.

Numerous studies have been focused on the offshore upwelling to the Northeast of Taiwan, which is associated with a cyclonic cold dome [7–10]. Little literature, however, has indicated the existence of coastal upwelling along the Taiwan east coast.

Taiwan Island is under the subtropical monsoon regime. Hence, the southwesterly/southerly (northeasterly/northerly) winds dominate in summer (winter). Although the southwesterly/southerly summer monsoon (June–August) is usually weaker than the northeasterly/northerly winter monsoon [11], the upwelling-favorable winds are still expected to induce upwelling along the Taiwan east coast, according to the classical Ekman transport theory. A similar mechanism has indeed been identified for the coastal upwelling in the northwestern and southwestern Taiwan Strait along the Fujian coast [3,12,13]. In addition, the Taiwan east coast is also influenced by the poleward flowing Kuroshio Current. However, the Kuroshio Current migrates offshore in summer [9,14]; and as a result, its potential influence on the summer upwelling along the Taiwan east coast is unclear and could be limited or secondary.

This study aimed to investigate the wind-driven coastal upwelling along the east coast of Taiwan using Himawari-8 sea surface temperature (SST) data. The objectives of this study include, (1) to identify significant upwelling-favorable wind events in the summers of recent several years; (2) to map the extent of the coastal upwelling using daily Himawari-8 SST data and examine its relationship with the wind events; and (3) to quantitatively investigate the characteristics of the coastal upwelling.

Remotely sensed SST data have often been used to detect and map upwelling that exhibits local SST anomaly on the sea surface [4,9,10,15,16]. A coastal upwelling event often develops rapidly after a distinct onset and lasts from several days to some weeks, depending on the local condition. The monthly-composited Moderate Resolution Imaging Spectroradiometer (MODIS) SST data used in [4] does not have an adequate temporal resolution to capture the development of an upwelling event. The daily MODIS and Advanced Very High Resolution Radiometer (AVHRR) SST data used in [15,16] are also not suitable for daily upwelling mapping because they often suffer incomplete spatial coverage due to clouds and much lower observation frequency (e.g., twice daily). The Himawari-8 SST data, on the other hand, offer unique advantages in the study of the daily development of upwelling events because of its reliability and high spatial (~2 km) and temporal resolutions (10 min) [17,18]. The merits of the Himawari-8 SST data have been recognized in the two recent upwelling studies [9,10], which used Himawari-8 SST data to investigate the upwelling off Northeastern Taiwan. In [9], the upwelling was only visually examined without attempting to explicitly map the upwelling extent; while in [10], a relatively complex gradient-based edge detection algorithm was used to map the upwelling. In this study, however, we used a scale-independent and relatively straightforward semi-automatic method developed and proved reliable in a recent upwelling study [4]. The near-real-time Himawari-8 SST data and the mapping technique developed in this study have the potential to be operationalized and thus provide a valuable monitoring tool for coastal upwelling.

2. Data and Methods

2.1. Study Area, Study Period, and Its Climate

The climate of Taiwan Island is predominantly affected by the East Asian monsoon. Hence, the weak southwesterly/southerly (strong northeasterly/northerly) winds usually prevail in summer (winter) [11]. The waters surrounding Taiwan Island are influenced by three major ocean currents [19]. Taiwan's east coast is mainly affected by the northward-flowing Kuroshio Current; while the Taiwan Strait is also affected by the intrusion of the Kuroshio Current, in addition to the southward flowing China Coastal Current in winter and the northward-flowing extension of the South China Sea Warm Current (or the Taiwan Strait Warm Current) throughout the year. In general, the SST is lower in the Taiwan Strait than that in the east of Taiwan, with an annual mean of >26 °C [20]. The salinity ranges 31–34 psu and has a relatively large spatial-temporal variability in the Taiwan Strait; while, it remains relatively stable east of Taiwan [20].

In this study, the study area was located to the east of Taiwan, with two sub-areas: one 20 km buffer and the other 20–70 km buffer from the east coast (Figure 1). The coastal upwelling is defined as the upwelling area mainly within the 20 km buffer, but may extend to the 20–70 km buffer. The study period is from May to September each year, between 2015 and 2019. Among these months, June to August is the main summer monsoon season; while, May and September are the transition period.

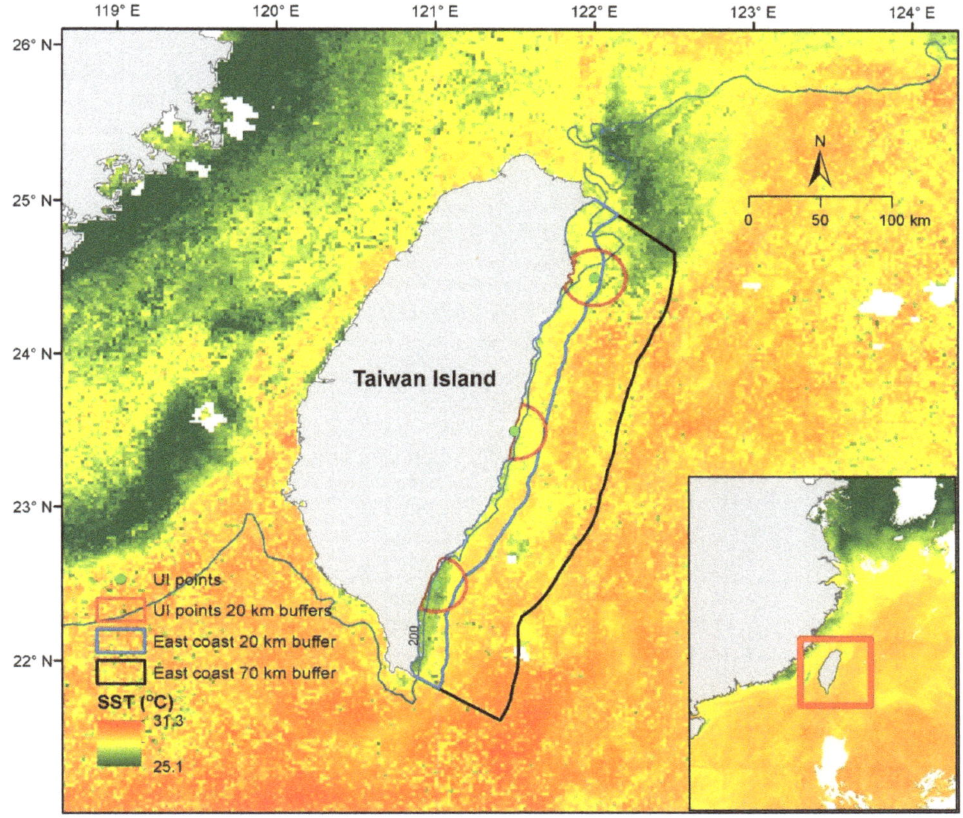

Figure 1. The study area is the coastal area of eastern Taiwan Island; the thick blue and black lines indicate the 20 km and 70 km buffers from the east coast of Taiwan; the green dots indicate the locations where the upwelling index (UI) was calculated using the CFS wind data; the red lines indicate the 20 km buffer around the UI points; the thin blue line is the 200 m isobath representing the approximate boundary of the continental shelf. The background image is the H-8 SST image on 19 July 2016.

2.2. Wind Data, Upwelling Index and Significant Upwelling-favorable Wind Events

We used wind data to calculate an upwelling index (UI). The Ekman-based upwelling index has been widely used for upwelling studies [2,4–6]. The wind data were obtained from CFSv2 (Climate Forecast System version 2) data of NCEP (National Center for Environmental Prediction) [21]. Based on ground observation, satellite data assimilation and reanalysis, the Climate Forecast System (CFS) is a global high-resolution coupling system of atmospheric-sea-land. The wind data have a spatial resolution of $0.5° \times 0.5°$ and a temporal resolution of 6 h.

The daily composited wind data were first calculated from the 6-hourly data. The daily UI (in $m^2\ s^{-1}$) was then calculated using the following equations:

$$UI = \frac{\tau}{f\,\rho_w} \cos(\alpha - \beta) \tag{1}$$

$$\tau = \rho_a\, C_d\, V^2 \tag{2}$$

$$f = 2\,\omega\,\sin(\varphi) \tag{3}$$

$$C_d = (0.8 + 0.065\,V) \times 0.001 \tag{4}$$

where τ is the wind stress, ρ_a is the air density (1.22 kg m^{-3}), ρ_w is the seawater density (1026 kg m^{-3}), C_d is the drag coefficient calculated using the empirical equation of [22], f is the Coriolis parameter (s^{-1}), α is the wind direction, β is the general shoreline orientation, ω is the rotation rate of the Earth (7.2921×10^{-5} rads^{-1}), φ is the latitude, and V is the wind speed (in m s^{-1}).

We calculated UI at three coastal locations (Figure 1), representing the northern, central and southern sections of the Taiwan east coast. The β values were set at 18°, 15°, and 20° from the northern aspect, for the northern, central, and southern locations, respectively. The significant upwelling-favorable wind events from May to September each year were then identified using the following criteria:

1. $\geq$5 days of positive UI, satisfying one of the following conditions:
    a. $\geq$5 days of consecutive positive UI;
    b. $\geq$3 days of consecutive positive UI, followed by 1 day of negative UI, then followed by $\geq$2 days of consecutive positive UI;
    c. $\geq$4 days of consecutive positive UI, followed by 2 days of negative UI, then followed by $\geq$3 days of consecutive positive UI;
    d. $\geq$5 days of consecutive positive UI, followed by 3 days of negative UI, then followed by $\geq$5 days of consecutive positive UI.
2. The cumulative UI for the first five days $\geq$2.0 m^2 s^{-1}

These criteria consider both the duration and intensity in the definition of a significant upwelling-favorable wind event. Note that a significant wind event is allowed to expand a few days into April or October to maintain the integrity of the event.

2.3. Himawari-8 SST Data and Upwelling Mapping

The Himawari-8 meteorological satellite was launched by the Japan Meteorological Agency (JMA) in October 2014 [17]. The Himawari-8 is a new generation geostationary satellite carrying an Advanced Himawari Imager (AHI), capable of providing geophysical data at a spatial resolution of ~2 km and a temporal resolution of 10-min full-disk frequency [17]. The Himawari-8 (H-8) SST data used in this study were processed by the Japan Aerospace Exploration Agency (JAXA) based on a quasi-physical SST algorithm and a Bayesian cloud screen method [18]. In [18], being evaluated against the buoy data which were measured at 20–30 cm, the ocean surface skin temperature of the H-8 SST data have a bias of −0.16 °C and a Root Mean Square Difference of ~0.59 °C. After considering the "cool skin effect" of the skin temperate as compared with the buoy temperature, the accuracy of the H-8 SST data is deemed reliable.

Specifically, we used cloud-free daily H-8 SST data processed to Level 3, with the highest quality-levels of 4 and 5, representing skin SST according to the GHRSST standard and format. Note that in 2015 the time-series starts from 13 July when the H-8 data became available.

Upwelling often exhibits a colder SST signature than the adjacent area [5,15,16,23–28]. This negative local SST anomaly is the foundation of mapping upwelling using SST data. Ref. [4] demonstrated that the topographic position index (*TPI*) [29] was an appropriate method to identify and map the negative local SST anomaly in the upwelling study. In this study, a similar semi-automatic technique based on the *TPI* method was developed for mapping the coastal upwelling off the Taiwan east coast from the daily H-8 SST data.

The *TPI* is a local-based image processing algorithm [29] that has been successfully used to map the ocean currents [30,31] and the coastal upwelling [4]. In this study, the *TPI* was calculated from the daily H-8 SST data using a nominated circular window with a radius of 50 cells (~100 km). This window size was deemed large enough to capture the coastal upwelling east of Taiwan.

To map upwelling areas, firstly we selected areas of negative *TPI* values using a pre-defined threshold, satisfying the following condition:

$$TPI \leq -S\_SD(TPI) \times \theta \quad (5)$$

where the *S_SD(TPI)* is the spatial standard deviation of the *TPI* image; θ is a threshold which was set at 0.75 in this study. Next, we further selected only those areas that have an areal extent greater than 100 km^2 as suitable upwelling candidates.

It was assumed that an upwelling area should have a reasonably large local SST anomaly (*SST_A*). The local SST anomaly of an upwelling candidate was calculated using the equation below:

$$SST\_A = SST_b - SST_c \quad (6)$$

where SST_c is the mean SST value of an upwelling candidate and SST_b is the mean SST value of the 20 km buffer around (but excluding) the upwelling candidate. An upwelling candidate was removed from the candidate set if its SST anomaly is less than a threshold, for example:

$$SST\_A < \delta \quad (7)$$

where δ was set at 0.5 °C in this study.

The threshold values of θ and δ were selected through a trial-and-error process and the experience gained from a previous upwelling study [4]. In the final step, only the upwelling candidates that are entirely or mainly within the 20 km buffer from the coast were regarded as coastal upwelling.

To examine the relationship between the upwelling-favorable winds and the coastal upwelling mapped from the H-8 SST data, a 20 km buffer was generated for each of the three wind locations (Figure 1). Within the 20 km buffer, one of three upwelling status was determined from the SST mapping results: (1) detected, when upwelling was detected within the buffer; (2) not-detected, when non-upwelling was detected within the buffer; (3) uncertain, when more than 50% of the buffer has no data in H-8 SST. Because the surface upwelling signature could lag behind a favorable wind event due to the travel time taken to uplift the deep water, the relationship was examined with three additional days extended to the end of each significant upwelling-favorable wind event.

2.4. Analyzing Upwelling Characteristics

After mapping the coastal upwelling east of Taiwan, we were able to quantitatively analyze several upwelling characteristics. As in [4], we calculated the area of influence (AoI), SST anomaly (*SST_A*), and chlorophyll-*a* ratio (Chla_R) of the coastal upwelling. These three upwelling properties represent different proxies of upwelling strength.

The AoI was calculated as the areal extent of the mapped upwelling areas. The *SST_A* was calculated using Equation (6) as the difference between the mean SST value of the upwelling areas and the mean SST value of the study area (i.e., the 70 km buffer from the coast). Similarly, Chla_R was calculated as the ratio between the mean chlorophyll-*a* concentration of the upwelling area and the mean chlorophyll-*a* concentration of the study area. The data used to calculate Chla_R were the daily composites of H-8 chlorophyll-*a* data processed by the JAXA to Level 3 using the algorithm developed in [32]. The spatial resolution of the chlorophyll-*a* data is ~5 km. Note that these three characteristics were not calculated for those days when upwelling could only be partially mapped due to the cloud coverage.

3. Results

As examples, the H-8 SST image on 19 July 2016, the *TPI* image calculated from the SST image, and the corresponding chlorophyll-*a* image are displayed in Figures 1 and 2. Two upwelling areas were identified along the central and southern sections of the Taiwan east coast, with lower SST values, negative *TPI* values, and higher chlorophyll-*a* concentrations than the offshore area.

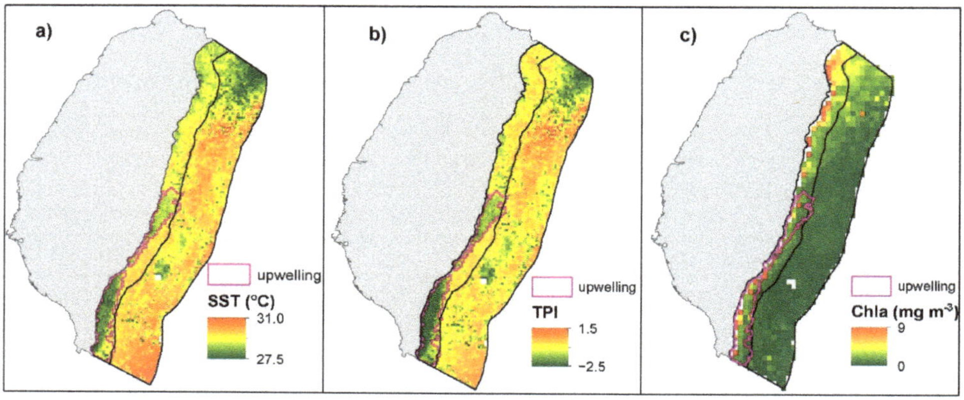

Figure 2. (**a**) The H-8 SST image on 19 July 2016, (**b**) the *TPI* image calculated from (**a**), (**c**) the H-8 chlorophyll-*a* image on 19 July 2016. The purple polygons indicate the upwelling areas mapped.

3.1. Coastal Upwelling with Significant Upwelling-Favorable Wind Events

Each year between May and September during 2015–2019, several significant upwelling-favorable wind events were identified with variable durations along the east coast of Taiwan (Figures 3–5; Table 1). The wind events with the longest durations occur between June and August, lasting for more than 40 days (Figures 3–5; Table 1). At the northern location, the significant wind events usually start in late April and last into early October; while, at the southern and central locations, the wind events usually start in May or June and last only to mid-September (Figures 3–5). During the entire study period, the northern location experienced more such events (21 events) and lasted more days (340 days) than the central (14 events and 245 days) and southern (15 events and 265 days) locations (Table 2). On average, the southern and northern locations were influenced by much stronger upwelling-favorable winds than the central location, with the mean UI values of 1.37, 1.21, and 0.56 $m^2\ s^{-1}$, respectively (Table 2).

The results show clear co-occurrence between the significant upwelling-favorable wind events and the detection of surface upwelling signature from the H-8 SST data, with an overall detection rate of 86.6% ± 17.1% (Figures 3–5; Table 1). Specifically, the upwelling detection rates are 92.9% ± 9.8%, 77.0% ± 22.9%, and 86.7% ± 15.7% for the north, the central, and the south locations, respectively (Table 1). Indeed, the surface upwelling signature often persisted several days after the end of the wind events (Figures 3–5). Uncertainties often occurred on the dates of strong winds (Figures 3–5), which were likely associated with overcast or stormy weather conditions preventing the H-8 sensor from acquiring cloud-free data.

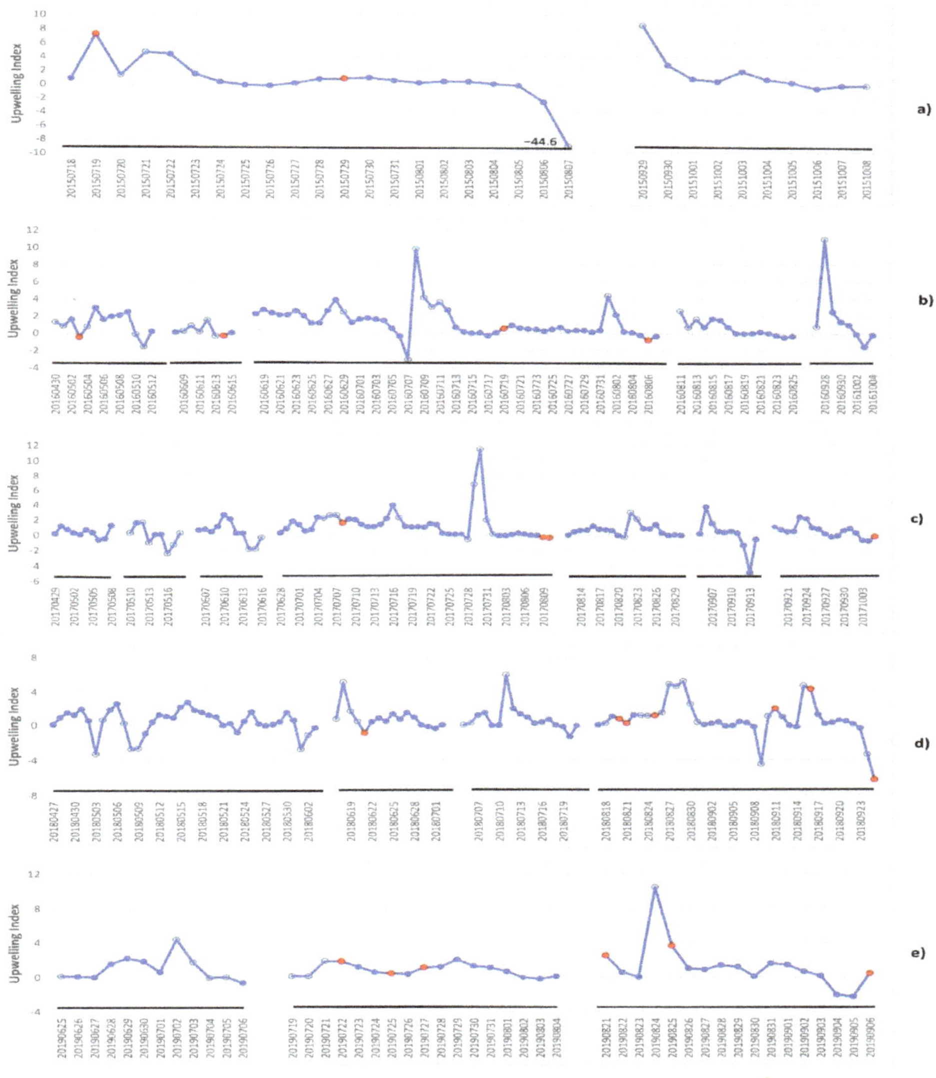

Figure 3. Time-series of the significant upwelling-favorable wind events at the northern location; (**a**) 2015, (**b**) 2016, (**c**) 2017, (**d**) 2018, (**e**) 2019. The solid blue (red) circle indicates that the upwelling (non-upwelling) was detected at the location; while, the un-filled circle indicates that the upwelling status was uncertain because more than 50% of the 20 km buffer has no data in H-8 SST. The labelled value(s) are actual UI value(s) that could not be properly plotted.

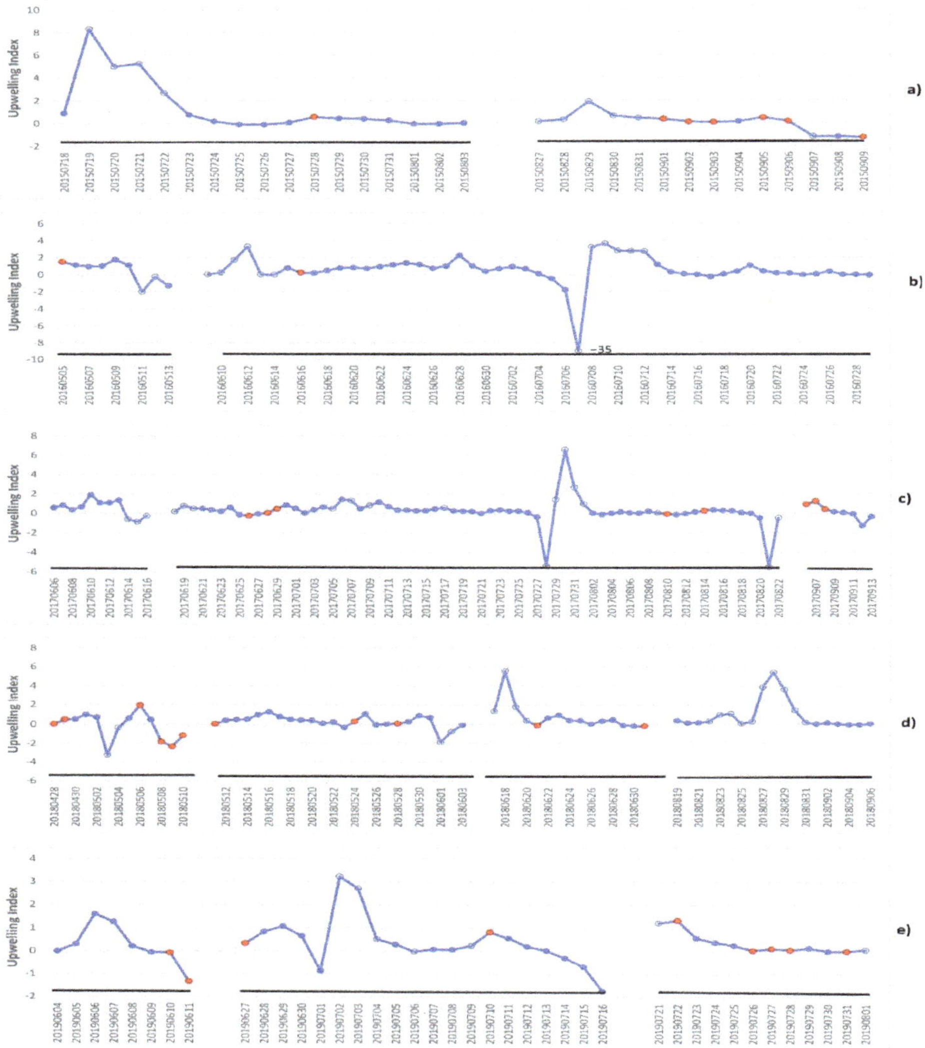

Figure 4. Time-series of the significant upwelling-favorable wind events at the central location; (**a**) 2015, (**b**) 2016, (**c**) 2017, (**d**) 2018, (**e**) 2019. Symbols as Figure 3. The labelled value(s) are actual UI value(s) that could not be properly plotted.

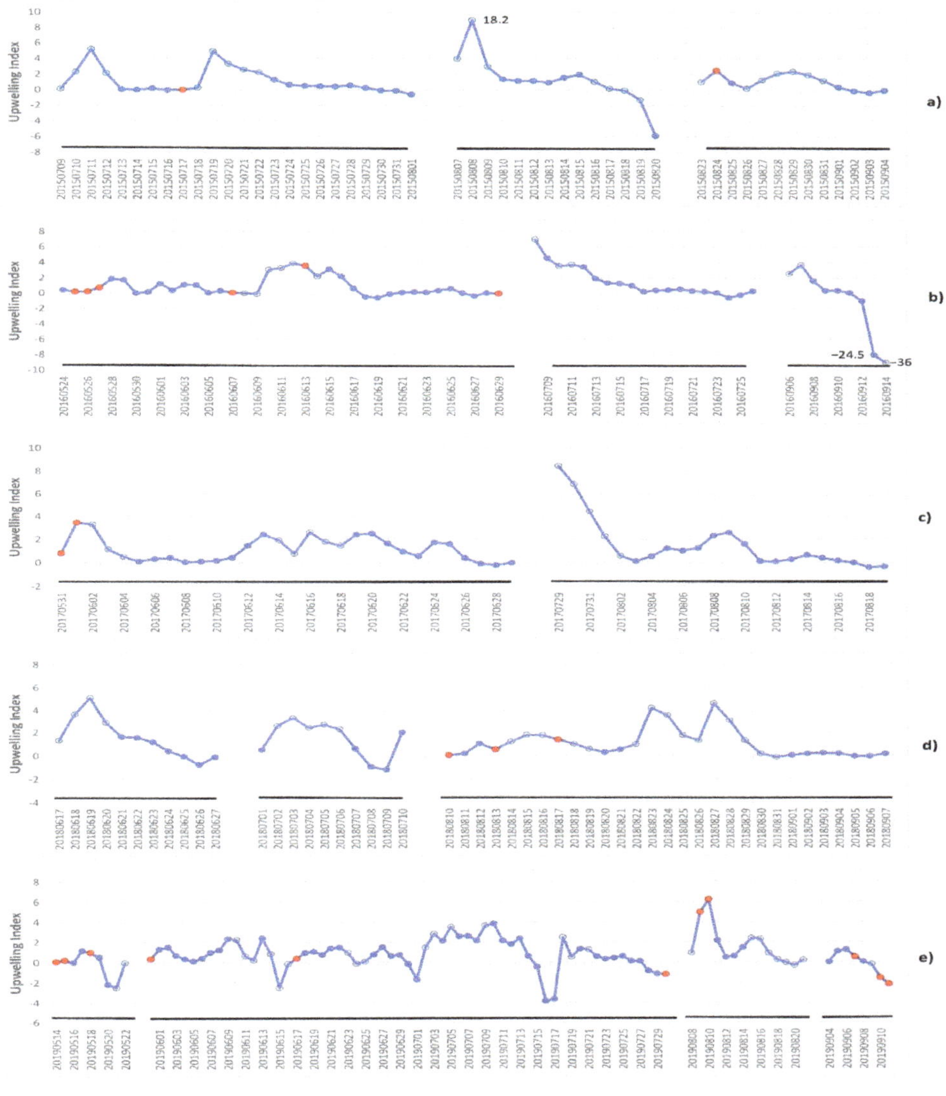

Figure 5. Time-series of the significant upwelling-favorable wind events at the southern location; (**a**) 2015, (**b**) 2016, (**c**) 2017, (**d**) 2018, (**e**) 2019. Symbols as Figure 3. The labelled value(s) are actual UI value(s) that could not be properly plotted.

Table 1. The statistics of individual significant upwelling-favorable wind events and upwelling mapped from the H-8 SST data.

Location	ID	Year	Number of Wind-Event Days	Upwelling Days	Non-Upwelling Days	UI5 [1] ($m^2\ s^{-1}$)	Mean UI ($m^2\ s^{-1}$)	Detection Rate [2]
North	1	2015	18	16	2	18.54	1.36	88.89
	2		7	8	0	14.42	2.19	100.00
	3	2016	10	7	1	4.08	1.53	87.50
	4		5	2	1	3.07	0.61	66.67
	5		48	42	2	11.82	1.48	95.45
	6		12	12	0	7.53	0.85	100.00
	7		5	6	0	16.68	3.34	100.00
	8	2017	7	10	0	2.52	0.51	100.00
	9		6	3	0	2.71	0.47	100.00
	10		8	8	0	5.80	1.06	100.00
	11		41	32	3	5.12	1.58	91.43
	12		16	16	0	3.06	0.85	100.00
	13		7	9	0	6.68	1.08	100.00
	14		14	14	1	5.62	0.86	93.33
	15	2018	35	31	0	5.76	0.60	100.00
	16		13	11	1	7.31	1.08	91.67
	17		14	14	0	3.27	1.10	100.00
	18		37	21	6	2.58	1.12	77.78
	19	2019	9	7	0	4.12	1.42	100.00
	20		14	11	3	5.26	1.01	78.57
	21		14	11	3	17.08	1.81	78.57
Centre	1	2015	14	12	1	22.16	1.76	92.31
	2		11	2	6	3.56	0.45	25.00
	3	2016	6	6	1	6.55	1.28	85.71
	4		48	38	1	5.35	0.09	97.44
	5	2017	8	8	0	4.38	0.98	100.00
	6		63	46	5	2.24	0.38	90.20
	7		5	5	3	2.89	0.58	62.50
	8	2018	10	5	6	2.83	0.23	45.45
	9		21	19	3	2.47	0.46	86.36
	10		12	9	2	9.10	1.04	81.82
	11		16	9	0	2.02	1.14	100.00
	12	2019	5	6	2	3.40	0.68	75.00
	13		17	13	2	2.08	0.64	86.67
	14		9	5	5	3.78	0.46	50.00
South	1	2015	21	14	1	10.02	1.34	93.33
	2		11	7	0	27.78	3.15	100.00
	3		10	5	1	5.98	1.41	83.33
	4	2016	34	25	6	3.53	0.94	80.65
	5		16	15	0	22.17	1.87	100.00
	6		6	6	0	8.58	1.44	100.00
	7	2017	27	20	2	9.26	1.30	90.91
	8		19	17	0	22.61	1.84	100.00
	9	2018	8	7	0	14.60	2.22	100.00
	10		7	4	0	11.51	2.06	100.00
	11		26	11	3	2.91	1.21	78.57
	12	2019	6	4	3	2.46	0.50	57.14
	13		58	42	3	4.23	0.92	93.33
	14		11	4	2	14.78	2.07	66.67
	15		5	4	3	2.88	0.58	57.14

Note: [1] The cumulative UI for the first five days of the upwelling event; [2] Defined as the ratio between the Upwelling-Days and the sum of the Upwelling-Days and Non-Upwelling Days.

Table 2. The statistics of the significant upwelling-favorable wind events and the mapping of coastal upwelling at the three locations of east Taiwan, during the five extended summer seasons (May to September) between 2015 and 2019.

Location	Number of Wind Events	Number of Wind-Event Days	Mean (Median) UI ($m^2 s^{-1}$)	Upwelling Days	Non-Upwelling Days	Uncertainty Days
North	21	340	1.21 (0.80)	291	23	89
Central	14	245	0.56 (0.40)	183	37	67
South	15	265	1.37 (0.99)	185	24	101

3.2. Upwelling Maps

To examine the overall spatial pattern of the coastal upwelling east of Taiwan, we generated an upwelling frequency map by combining the daily upwelling maps of those upwelling dates that have complete H-8 SST data (n = 291) during the entire study period (Figure 6a). The overall spatial pattern shows that during the significant wind events, upwelling can occur along the entire east coast of Taiwan (Figure 6a). In general, further away from the coast, less frequent coastal upwelling occurs. Although upwelling could extend more than 20 km offshore, most of the upwelling occurs within 20 km off the coast (Figure 6a). In particular, the northern and southernmost coastal areas experience upwelling more often, with a frequency greater than 70% (Figure 6a). In terms of spatial extent, the areas with high (70–90%), moderate (30–70%) and low (10–30%) upwelling frequency are 15% (~1250 km^2), 44% (~3820 km^2) and 41% (~3570 km^2), respectively (Figure 7a).

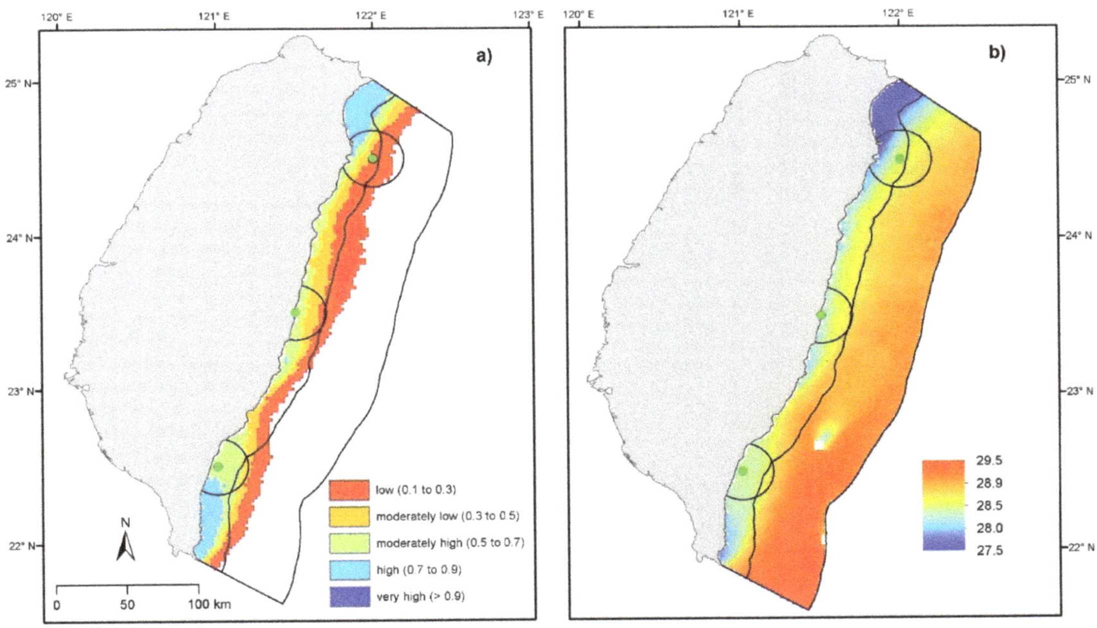

Figure 6. The overall upwelling frequency map (**a**) vs the overall temporally-averaged SST map (**b**); generated by combining the daily upwelling maps and SST maps of all upwelling dates (n = 291) together. The frequency is divided into five categories: low, moderately low, moderately high, high, and very high.

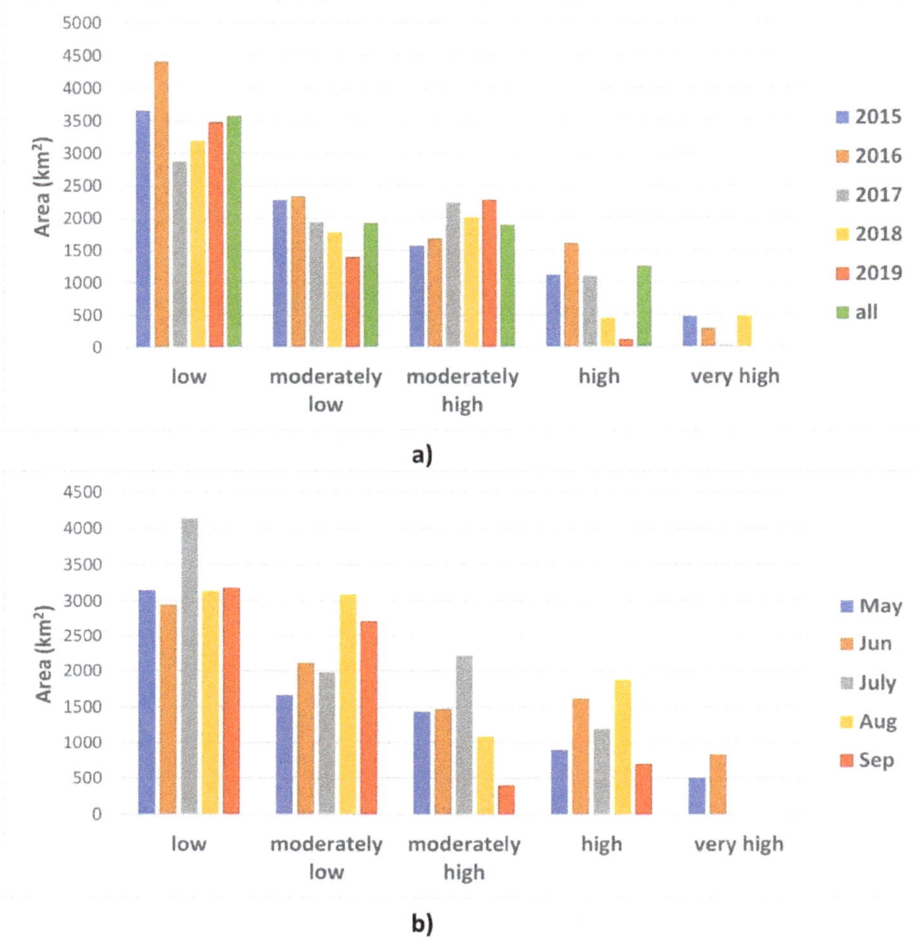

Figure 7. The areas of the upwelling frequency maps generated for (**a**) the inter-annual variation (Figure 9) and (**b**) the monthly variation (Figure 8).

There are substantial intra-seasonal variations in upwelling extent (Figure 8). The upwelling area increases substantially from May (~7648 km^2) to June (~8976 km^2) and July (~9516 km^2), especially at the northern section of the Taiwan east coast, then decreases slightly in August (~9176 km^2) and further reduces in September (~6968 km^2). Three upwelling centers (north, central, and south), with greater than 70% of upwelling frequency, are clearly visible from May to July (Figure 8a–c). The central upwelling center disappears in August; and by September, only the northern upwelling center still remains (Figure 8d,e). In May, the coastal upwelling almost always (>90% frequency) occurs at the northern center, as well as at a tiny area near the southern tip of the southern upwelling center (Figure 8a). In June, the extents of the three upwelling centers expand, with an increased and a reduced area experiencing a very-high frequency of upwelling at the southern center and the northern center, respectively (Figure 8b). In terms of spatial extent, the area experiencing very-high frequency upwelling occupies ~500 km^2 and ~830 km^2 in May and June, respectively (Figure 7b). The largest area with a high frequency of upwelling occurs in August (~1870 km^2), followed by June (~1620 km^2).

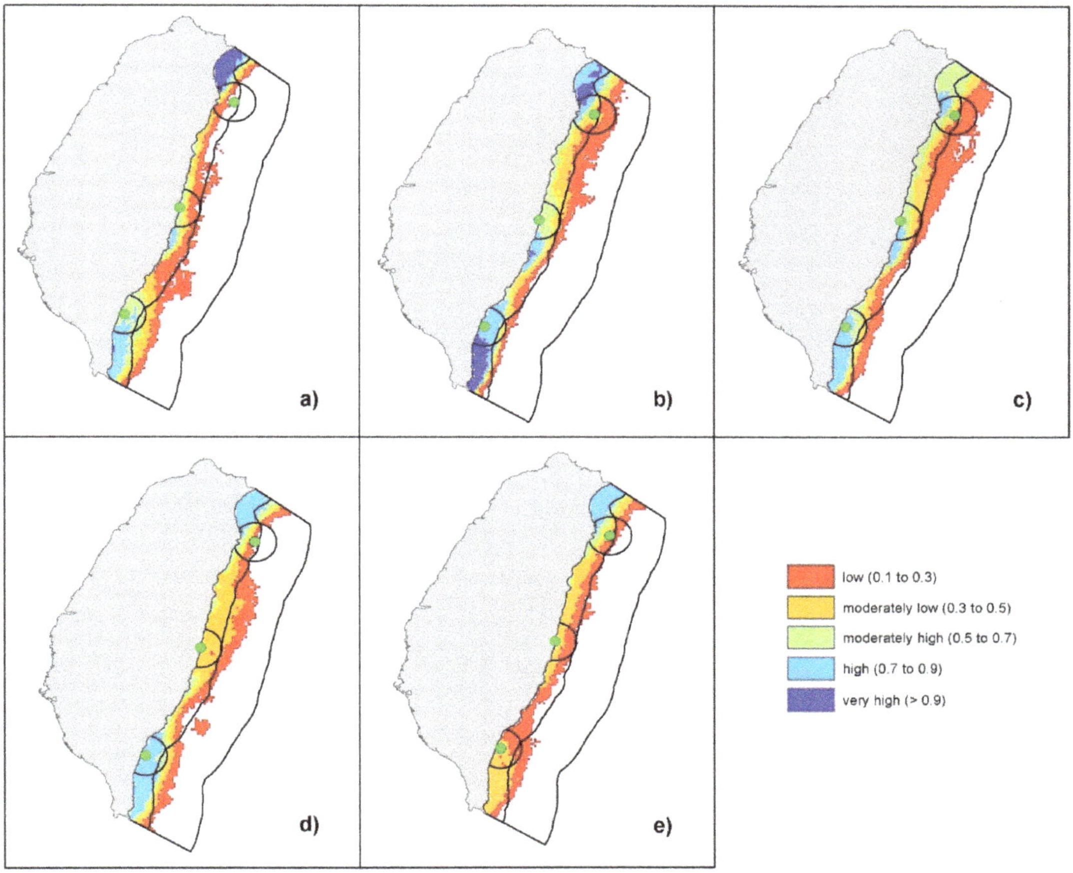

Figure 8. The monthly upwelling frequency map were generated by combining the daily upwelling maps of the corresponding months; (**a**) May, (**b**) June, (**c**) July, (**d**) August, (**e**) September. The frequency is divided into five categories: low, moderately low, moderately high, high, and very high.

There are also substantial interannual variations in upwelling extent (Figures 7a and 9). The total upwelling area was the largest in 2016 (~10330 km^2), followed by in 2015 (~9120 km^2), and the smallest in 2019 (~7260 km^2). The three upwelling centers were clearly visible in 2016, 2017, and 2018 (Figure 9b–d). The very-high frequency of upwelling occurred at part of the southern section in 2015 and 2016, and at part of the northern section in 2018 (Figure 9a,b,d). Overall, the coastal upwelling occurred less frequently in 2019 than that in other years. In total, the year 2016 had the largest upwelling area with high and very-high upwelling frequency (~1900 km^2), followed by the year 2015 (~1600 km^2) (Figure 7a).

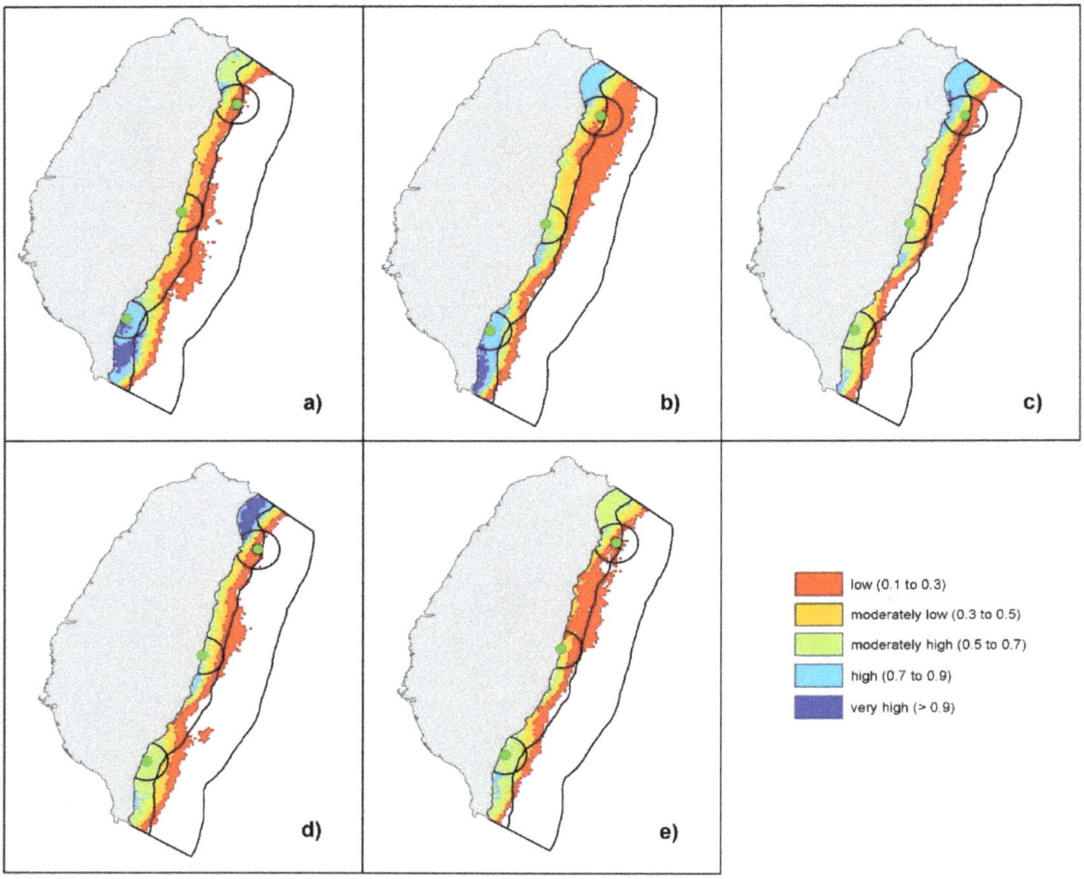

Figure 9. The inter-annual upwelling frequency maps were generated by combining the daily upwelling maps of the corresponding years; (**a**) 2015, (**b**) 2016, (**c**) 2017, (**d**) 2018, (**e**) 2019. The frequency is divided into five categories: low, moderately low, moderately high, high, and very high.

3.3. *Upwelling Characteristics*

The monthly mean upwelling area is highest in June (~5200 km^2) and lowest in September (~3100 km^2) (Figure 10a). Statistically, the monthly mean upwelling area is significantly higher in June than that in September or in May; while, it is significantly higher in July and August than that in September. The monthly mean SST anomaly between the upwelling area and the study area (the 70 km buffer) is largest (0.99 °C) in May but smallest (0.76 °C) in August (Figure 10b). Statistically, the monthly mean upwelling's *SST_A* is significantly higher in May and September than that in August. In terms of chlorophyll-*a* ratio between the upwelling area and the study area, the monthly mean Chla_R is the largest (3.1) in September, which is significantly higher than that (2.1–2.2) in all other months statistically (Figure 10c).

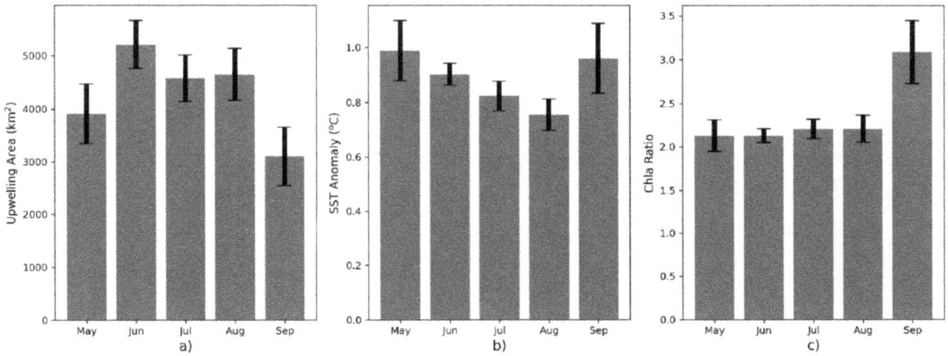

Figure 10. The monthly variation of upwelling's characteristics; (**a**) the upwelling area, (**b**) the upwelling's SST anomaly, (**c**) the upwelling's chlorophyll-*a* ratio. The error bars indicate 95% confidence intervals.

In terms of interannual variation, the annual mean upwelling area was highest in 2016 (~5125 km^2) and continued to reduce in the following three years (Figure 11a). Statistically, the annual mean upwelling area in 2016 was significantly higher than that in all of the following three years; while, the annual mean upwelling area (~4270 km^2) in 2017 was also significantly higher than that (~3250 km^2) in 2019. The annual mean SST anomaly was the largest (1.0 °C) in 2018 and the lowest (0.81 °C) in 2017 (Figure 11b). Their difference was statistically significant. The annual mean Chla_R was the highest (2.5) in 2018, and the lowest (2.2) in 2016 (Figure 11c). However, the differences between these four years were not statistically significant.

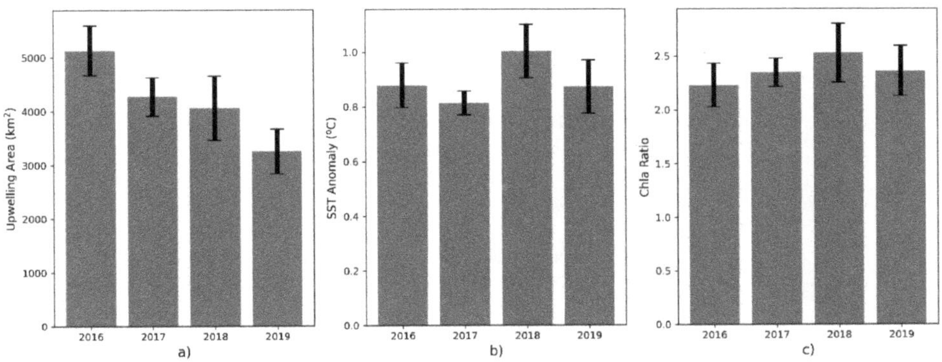

Figure 11. The interannual variations of the upwelling characteristics; (**a**) upwelling area, (**b**) SST anomaly, (**c**) Chla ratio. The year 2015 is not included as it is an incomplete year of data. The error bars indicate 95% confidence intervals.

4. Discussion

This study shows clear evidence of seasonal coastal upwelling along the Taiwan east coast (Figures 6a and 8). The coastal upwelling is mainly detected during the summer monsoon season when the southwesterly/southerly winds are upwelling-favorable (Figures 3–5). The coastal upwelling is hence believed to be wind-driven, with a similar mechanism as the coastal upwelling along the Fujian coast on the western side of the Taiwan Strait [3,12,13]. There are two pieces of additional evidence to support this finding. Firstly, a greater number of days with significant upwelling-favorable winds is associated with a greater number of upwelling days, both at the level of individual wind events (r = 0.96; n = 50; $p < 0.001$; Figure 12a) and at the level of the upwelling centers (r = 0.98; n = 3; $p < 0.01$; Table 2). Secondly, a stronger wind-driven upwelling index is also associated with a higher upwelling detection rate (r = 0.80; n = 3; $p < 0.1$; Table 2), although the positive correlation at the level of individual wind events is relatively weak (r = 0.40; n = 50; $p < 0.01$; Figure 12b).

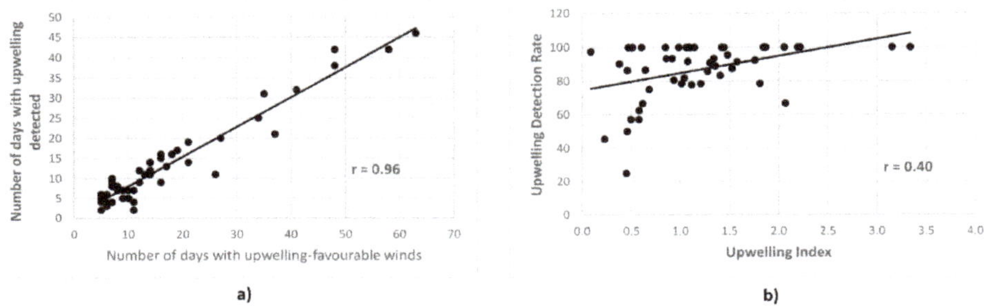

Figure 12. (**a**) Correlation between the number of days with upwelling-favorable winds and the number of days with upwelling detected from the H-8 SST data (n = 50); (**b**) correlation between the wind-driven upwelling index and the upwelling detection rate from the H-8 SST data. Each point represents one significant upwelling-favorable wind event (n = 50).

Although upwelling can be detected along the entire east coast of Taiwan, it generally occurs around three upwelling centers (Figures 6a, 8 and 9). However, there are clear spatial and temporal variabilities among the three upwelling centers. The northern and southern upwelling centers are persistent upwelling centers during the summer monsoon season; while, the central upwelling center, which can be absent in some months and years, is much weaker in intensity (Figures 8 and 9; Tables 1 and 2). Upwelling occurs more often at the northern and southern sections of the coast, most likely due to a larger number of days with significant upwelling-favorable winds and an overall stronger upwelling index (Tables 1 and 2). In particular, the northern upwelling center usually has the longest upwelling season, lasting through the entire extended summer season from May to September (Figure 8), which is most likely because of the prolonged summer monsoon season at this location (Figure 3). The upwelling's spatial patterns mapped in this study (Figures 6a, 8 and 9) generally agree with those of the temporally-averaged SST maps (Figures 6b, 13 and 14). In particular, the three upwelling centers, which have lower SST than the offshore area, are also clearly visible in the temporally-averaged SST maps. This confirms the reliability of the semi-automatic upwelling mapping technique used in this study.

The upwelling areas are larger in the main summer monsoon season between June and August than those in the transition period of May and September (Figures 8 and 10a). This is likely due to the higher upwelling index in these months (Figure 15). Larger upwelling areas in June–August coincide with lower upwelling SST anomaly (Figure 10a,b). This might be because upwelling in June–August could be sourced from relatively shallower (and warmer) water than that in May and September, due to a stronger summer thermocline barrier [33]. A much higher chlorophyll-*a* ratio in September than that in other months (Figure 10c) is due to higher chlorophyll-*a* concentrations (up to 0.7 mg m^{-3}) in the upwelling areas. This could be due to, at the beginning of autumn, the strong summer thermocline barrier becomes weaker, and surface cooling leads to deeper mixed layer depth [33], and as a result, nutrients from deeper waters can be uplifted to the surface during an upwelling event in early autumn. It should be noted that, however, this study could not separate chlorophyll-*a* of the terrestrial source from that of the upwelling source [9].

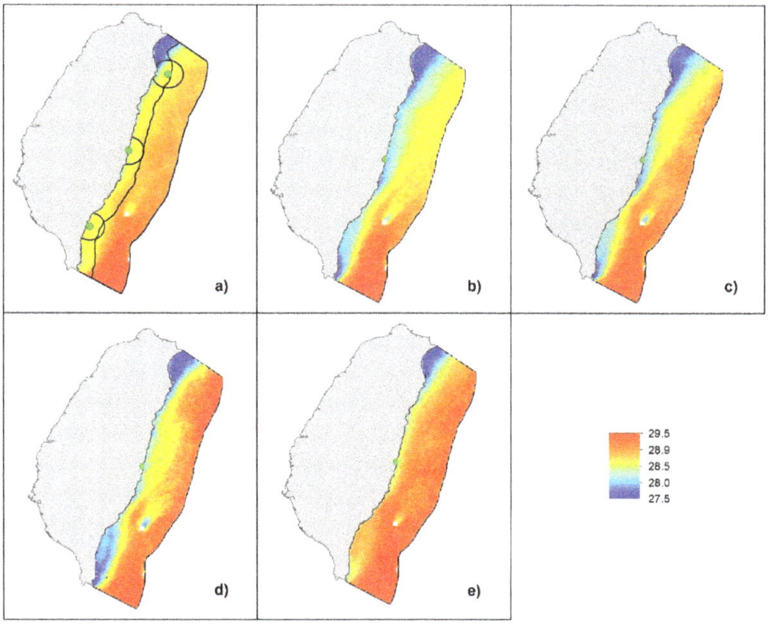

Figure 13. The monthly-averaged SST maps were generated by combining the daily SST maps of the corresponding months; (**a**) May, (**b**) June, (**c**) July, (**d**) August, (**e**) September.

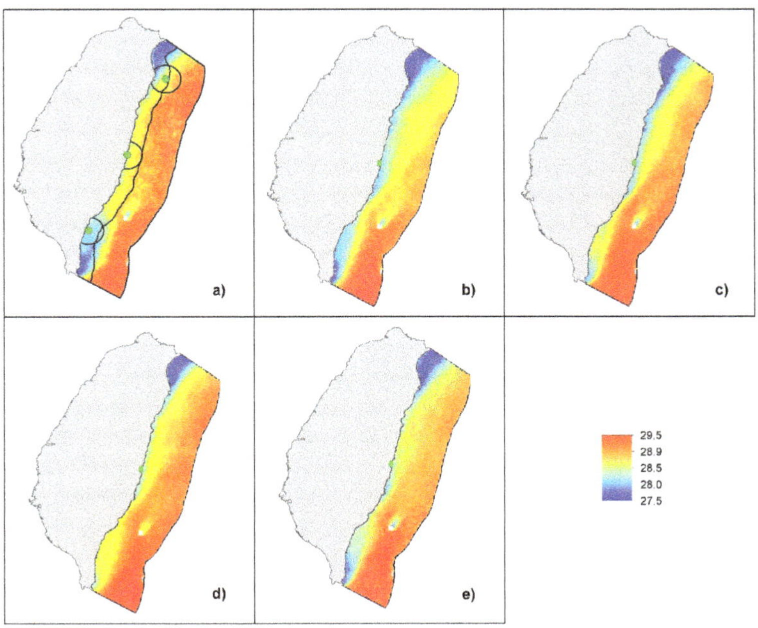

Figure 14. The annually-averaged SST maps were generated by combining the daily SST maps of the corresponding years; (**a**) 2015, (**b**) 2016, (**c**) 2017, (**d**) 2018, (**e**) 2019.

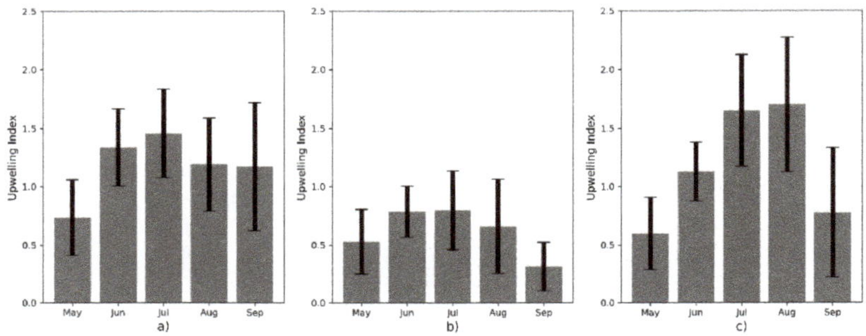

Figure 15. Monthly mean wind-driven upwelling index; (**a**) north, (**b**) central, (**c**) south. The error bars indicate 95% confidence intervals.

The time-series of this study is not sufficient to examine the interannual variability of the upwelling characteristics. According to the Ocean Niño Index, in the summer of 2016, the strong El Niño condition was quickly transitioned into the La Niña condition. This might have something to do with the strongest upwelling activities detected in the summer season of 2016 measured by the upwelling area (Figure 11a). In contrast, weak El Niño conditions during late spring and early summer of 2019 might have contributed to the weakest upwelling activities in the summer season of 2019 (Figure 11a). Similar ENSO influence on the upwelling of the Taiwan Strait has been documented in previous studies [3].

The role of the Kuroshio Current in the summer coastal upwelling east of Taiwan remains unclear. The incursion of the Kuroshio Current to the East China Sea plays an important role in the upwelling off Northeastern Taiwan [7–10]. A similar mechanism could help form and enhance the northern upwelling center of east Taiwan. In addition, the intrusion of the Kuroshio Current into the South China Sea through the Luzon Strait may form an anticyclonic loop current off Southern Taiwan [34–37]. The influence of the Kuroshio Current to the southern upwelling center of east Taiwan, however, is likely to be limited because this loop current more likely occurs in winter [36].

5. Conclusions

Coastal upwelling is important for coastal ecosystems and the blue economy. This study demonstrates the advantages of using Himawari-8 SST data to monitor the daily development of upwelling events because of its reliability and high spatial (~2 km) and temporal (10 min) resolutions. The semi-automatic, topographic position index based image processing method is simple, intuitive, and scale-independent. The method is able to map the spatial extent of the upwelling SST signature and, as a result, to quantitatively investigate the upwelling characteristics such as the area of influence, SST anomaly, and chlorophyll-*a* concentrations. Most importantly, this study offers the pioneering insight into the summer upwelling along the entire Taiwan east coast. The coastal upwelling east of Taiwan is believed to be driven by the southwesterly/southerly summer monsoon winds, with a similar mechanism as the coastal upwelling along the Fujian coast on the western side of the Taiwan Strait.

In summary, the key findings of this study are:

- Wind-driven upwelling occurs along the entire Taiwan east coast during the summer monsoon season;
- There are three board upwelling centers along the Taiwan east coast: north, central, and south;

- The upwelling around the northern center has the longest upwelling season, lasting from May to September;
- The upwelling extents are larger between June and August during the height of the summer monsoon.

Author Contributions: Conceptualization, Z.H. and J.H.; methodology, Z.H. and W.S.; writing—original draft preparation, Z.H.; writing—review and editing, J.H. and Z.H. All authors have read and agreed to the published version of the manuscript.

Funding: This study was jointly supported by the National Natural Science Foundation of China (91958203, 41776027 and 41730533). Z.H.'s visit to Xiamen University was supported by the MEL Visiting Fellowship (MELRS1767).

Data Availability Statement: The datasets for this research are publicly available. The wind data are provided by National Centers for Environmental Information, NOAA via https://www.ncdc.noaa.gov/data-access/model-data/model-datasets/climate-forecast-system-version2-cfsv2. The Himawari-8 SST and chlorophyll-*a* data are provided by GHRSST and JAXA/EORC via https://www.eorc.jaxa.jp/ptree/index.html. We appreciate the use of these publically available datasets.

Acknowledgments: We would like to thank the three anonymous reviewers whose constructive comments have significantly improved the manuscript.

Conflicts of Interest: The authors declare no conflict of interest. The funders had no role in the design of the study; in the collection, analyses, or interpretation of data; in the writing of the manuscript, or in the decision to publish the results.

References

1. McGregor, H.V.; Dima, M.; Fischer, H.W.; Mulitza, S. Rapid 20th-century increase in coastal upwelling off Northwest Africa. *Science* **2007**, *315*, 637–639. [CrossRef] [PubMed]
2. Wang, D.; Gouhier, T.C.; Menge, B.A.; Ganguly, A.R. Intensification and spatial homogenization of coastal upwelling under climate change. *Nature* **2015**, *518*, 390–394. [CrossRef] [PubMed]
3. Hu, J.; Wang, X.H. Progress on upwelling studies in the China seas. *Rev. Geophys.* **2016**, *54*, 653–673. [CrossRef]
4. Huang, Z.; Wang, X.H. Mapping the spatial and temporal variability of the upwelling systems of the Australian southeastern coast using 14-year of MODIS data. *Remote Sens. Environ.* **2019**, *227*, 90–109. [CrossRef]
5. Kampf, J.; Doubell, M.; Griffin, D.; Matthews, R.L.; Ward, T.M. Evidence of a large seasonal coastal upwelling system along the southern shelf of Australia. *Geophys. Res. Lett.* **2004**, *31*, L09310. [CrossRef]
6. Varela, R.; Alvarez, I.; Santos, F.; de Castro, M.; Gomez-Gesteira, M. Has upwelling strengthened along worldwide coasts over 1982–2010? *Sci. Rep.* **2015**, *5*, 10016. [CrossRef]
7. Chang, Y.L.; Wu, C.R.; Oey, L.Y. Bimodal behavior of the seasonal upwelling off the northeastern coast of Taiwan. *J. Geophys. Res. Ocean.* **2009**, *114*, C03027. [CrossRef]
8. Chang, Y.L.; Oey, L.Y.; Wu, C.R.; Lu, H.F. Why are there upwelling on the northern shelf of Taiwan under northeasterly winds? *J. Phys. Oceanogr.* **2010**, *40*, 1405–1417. [CrossRef]
9. Chung, H.W.; Liu, C.C. Spatiotemporal variation of cold eddies in the upwelling zone off Northeastern Taiwan revealed by the Geostationary satellite imagery of ocean color and sea surface temperature. *Sustainability* **2019**, *11*, 6979. [CrossRef]
10. Yin, W.; Huang, D. Short-term variations in the surface upwelling off Northeastern Taiwan observed via satellite data. *J. Geophys. Res. Ocean.* **2019**, *124*, 939–954. [CrossRef]
11. Lyu, K.W.; Hu, J.Y.; Yang, X.Y. Spatial patterns in seasonal variability of sea surface wind over the South China Sea and its adjacent ocean. *J. Trop. Oceanogr.* **2012**, *31*, 41–47. (In Chinese with English Abstract)
12. Hu, J.; Kawamura, H.; Hong, H.; Pan, W. A review of research on the upwelling in the Taiwan Strait. *Bull. Mar. Sci.* **2003**, *73*, 605–628.
13. Jiang, Y.; Chai, F.; Wan, Z.; Zhang, X.; Hong, H. Characteristics and mechanisms of the upwelling in the southern Taiwan Strait: A three-dimensional numerical model study. *J. Oceanogr.* **2011**, *67*, 699–708. [CrossRef]
14. Chao, S.Y. Circulation of the East China Sea, a numerical study. *J. Oceanogr.* **1991**, *42*, 273–295.
15. McClatchie, S.; Middleton, J.F.; Ward, T.M. Water mass analysis and alongshore variation in upwelling intensity in the eastern Great Australian Bight. *J. Geophys. Res. Ocean.* **2006**, *111*, C08007. [CrossRef]
16. Roughan, M.; Middleton, J.H. A comparison of observed upwelling mechanisms off the east coast of Australia. *Cont. Shelf Res.* **2002**, *22*, 2551–2572. [CrossRef]
17. Bessho, K.; Date, K.; Hayashi, M.; Ikeda, A.; Imai, T.; Inoue, H.; Kumagai, Y.; Miyakawa, T.; Murata, H.; Ohno, T.; et al. An introduction to Himawari-8/9—Japan's new-generation geostationary meteorological satellites. *J. Meteorol. Soc. Jpn. Ser. II* **2016**, *94*, 151–183.

18. Kurihara, Y.; Murakami, H.; Kachi, M. Sea surface temperature from the new Japanese geostationary meteorological Himawari-8 satellite. *Geophys. Res. Lett.* **2016**, *43*, 1234–1240. [CrossRef]
19. Hu, J.; Kawamura, H.; Li, C.; Hong, H.; Jiang, Y. Review on current and seawater volume transport through the Taiwan Strait. *J. Oceanogr.* **2010**, *66*, 591–610. [CrossRef]
20. Li, G.; Wan, R.; Li, B.; Wei, X.; Tang, J. Using remote sensing for study of temperature and salinity of surface sea water surrounding Taiwan. *Mar. Geol. Front.* **2017**, *33*, 41–46. (In Chinese with English Abstract)
21. Saha, S.; Moorthi, S.; Wu, X.; Wang, J.; Nadiga, S.; Tripp, P.; Behringer, D.; Hou, Y.T.; Chuang, H.Y.; Iredell, M.; et al. The NCEP Climate Forecast System Version 2. *J. Clim.* **2014**, *27*, 2185–2208. [CrossRef]
22. Toba, Y.; Iida, N.; Kawamura, H.; Ebuchi, N.; Jones, I.S.F. Wave dependence of sea surface wind stress. *J. Phys. Oceanogr.* **1990**, *20*, 705–721. [CrossRef]
23. Dabuleviciene, T.; Kozlov, I.; Vaiciute, D.; Dailidiene, I. Remote sensing of coastal upwelling in the south-eastern Baltic Sea: Statistical properties and implications for the coastal environment. *Remote Sens.* **2018**, *10*, 1752. [CrossRef]
24. Gill, P.C.; Morrice, M.G.; Page, B.; Pirzl, R.; Levings, A.H.; Coyne, M. Blue whale habitat selection and within-season distribution in a regional upwelling system off southern Australia. *Mar. Ecol. Prog. Ser.* **2011**, *421*, 243–263. [CrossRef]
25. Oke, P.R.; Griffin, D. The cold-core eddy and strong upwelling off the coast of New South Wales in early 2007. *Deep-Sea Res. II* **2011**, *58*, 574–591. [CrossRef]
26. Oke, P.R.; Middleton, J.H. Nutrient enrichment off Port Stephens: The role of the East Australian Current. *Cont. Shelf Res.* **2001**, *21*, 587–606. [CrossRef]
27. Roughan, M.; Oke, P.R.; Middleton, J.F. A modelling study of the climatological current field and the trajectories of upwelled particles in the east Australian current. *J. Phys. Oceanogr.* **2003**, *33*, 2551–2564. [CrossRef]
28. Willis, J.; Hobday, A.J. Influence of upwelling on movement of southern bluefin tuna (*Thunnus maccoyii*) in the Great Australian Bight. *Mar. Freshw. Res.* **2007**, *58*, 699–708. [CrossRef]
29. Weiss, A.D. Topographic Position and Landforms Analysis. In Proceedings of the ESRI International User Conference, San Diego, CA, USA, 9–13 July 2001.
30. Huang, Z.; Feng, M. Remotely sensed spatial and temporal variability of the Leeuwin Current using MODIS data. *Remote Sens. Environ.* **2015**, *166*, 214–232. [CrossRef]
31. Xie, S.; Huang, Z.; Wang, X. Quantitative mapping of the East Australian Current encroachment using time series Himawari-8 sea surface temperature data. *J. Geophys. Res. Ocean.* **2020**, *125*, e2019JC015647. [CrossRef]
32. Murakami, H. Ocean color estimation by Himawari-8/AHI. In Proceedings of the SPIE 9878, Remote Sensing of the Oceans and Inland Waters: Techniques, Applications, and Challenges, New Delhi, India, 4–7 April 2016; Curran Associates, Inc.: Red Hook, NY, USA, 2016; p. 987810.
33. Zhang, W.Z.; Wang, H.; Chai, F.; Qiu, G. Physical drivers of chlorophyll variability in the open South China Sea. *J. Geophys. Res. Ocean.* **2016**, *121*, 7123–7140. [CrossRef]
34. Caruso, M.J.; Gawarkiewicz, G.G.; Beardsley, R.C. Interannual variability of the Kuroshio intrusion in the South China Sea. *J. Oceanogr.* **2006**, *62*, 559–575. [CrossRef]
35. Hu, J.; Kawamura, H.; Hong, H.; Qi, Y. A review on the currents in the South China Sea: Seasonal circulation, South China Sea Warm Current and Kuroshio intrusion. *J. Oceanogr.* **2000**, *56*, 607–624. [CrossRef]
36. Nan, F.; Xue, H.; Yu, F. Kuroshio intrusion into the South China Sea: A review. *Prog. Oceanogr.* **2015**, *137*, 314–333. [CrossRef]
37. Xue, H.; Chai, F.; Pettigrew, N.; Xu, D.; Shi, M.; Xu, J. Kuroshio intrusion and the circulation in the South China Sea. *J. Geophys. Res. Ocean.* **2004**, *109*, C02017. [CrossRef]

remote sensing

MDPI

Letter

Springtime Upwelling and Its Formation Mechanism in Coastal Waters of Manaung Island, Myanmar

Yuhui Li [1], Yun Qiu [1,2,3,*], Jianyu Hu [2,4], Cherry Aung [5], Xinyu Lin [1] and Yue Dong [1]

1 Third Institute of Oceanography, Ministry of Natural Resources, Xiamen 361005, China; liyuhui@tio.org.cn (Y.L.); linxinyu@tio.org.cn (X.L.); dongyue@tio.org.cn (Y.D.)
2 Southern Marine Science and Engineering Guangdong Laboratory (Zhuhai), Zhuhai 519082, China; hujy@xmu.edu.cn
3 Laboratory for Regional Oceanography and Numerical Modeling, Qingdao National Laboratory for Marine Science and Technology, Qingdao 266237, China
4 State Key Laboratory of Marine Environmental Science, College of Ocean and Earth Sciences, Xiamen University, Xiamen 361102, China
5 Marine Science Department, Pathein University, Pathein 10014, Myanmar; missaungmarine@gmail.com
* Correspondence: qiuyun@tio.org.cn

Received: 30 September 2020; Accepted: 16 November 2020; Published: 18 November 2020

Abstract: Multisource satellite remote sensing data and the World Ocean Atlas 2018 (WOA18) temperature and salinity dataset have been used to analyze the spatial distribution, variability and possible forcing mechanisms of the upwelling off Manaung Island, Myanmar. Signals of upwelling exist off the coasts of Manaung Island, in western Myanmar during spring. It appears in February, reaches its peak in March and decays in May. Low-temperature (<28.3 °C) and high-salinity (>31.8 psu) water at the surface of this upwelling zone is caused by the upwelling of seawater from a depth below 100 m. The impact of the upwelling on temperature is more significant in the subsurface layer than that in the surface layer. In contrast, the impact of the upwelling on salinity in the surface layer is more significant. Further research reveals that the remote forcing from the equator predominantly induces the evolution of the upwelling, while the local wind forcing also contributes to strengthen the intensity of the upwelling during spring.

Keywords: coastal waters of Myanmar; upwelling; monsoon; remote equatorial forcing

1. Introduction

Upwelling usually refers to the upward movement of water, caused by the divergence of the flow in the surface layer of the ocean [1]. Since deeper water is usually enriched with nutrients, it tends to increase a supply of nutrients to upper oceanic layers and forms the basis for the high productivity of upwelling regions. Consequently, upwelling areas are among the most fertile regions of the global ocean [2]. The production and its variability over this coastal upwelling system are a key concern for the fishing community, since they may affect the day-to-day livelihood of the coastal population and are important for the Indian Ocean rim countries due to their developing country status [3]. Moreover, upwelling is also an important factor modulating regional and global climate. For example, the upwelling along the western coast of Java-Sumatra has changed sea surface temperature (SST) of the warm pool in the eastern Indian Ocean and caused anomalous atmospheric circulation, thereby affecting intraseasonal to decadal variabilities of the tropical climates [4–9]. Therefore, the understanding physical processes and their variabilities in the upwelling systems is important because it gives us crucial information regarding variability of a marine ecosystem and its regional climate [10,11].

The Bay of Bengal (BOB) is dominated by the South Asian monsoon. The southwest monsoon prevails in summer (June–August), while the northeast monsoon prevails in winter

(December–February), and the summer monsoon is stronger than the winter monsoon [12]. Upwelling in the BOB mostly occurs in the southwest monsoon period, indicating that it is only a seasonal phenomenon in the Bay [9]. Although the upwelling in the BOB or in the Arabian Sea is mainly driven by the monsoon, the former is much weaker than the latter [13]. This may be principally caused by two factors. Firstly, as a main forcing of the upwelling in either sea areas, the southwest monsoon is much stronger in the Arabian Sea than that in the BOB [14]. Secondly, strong salinity stratification is formed near the surface over the Bay due to abundant rainfall and a large amount of runoff input along its northern coasts [15], which greatly suppresses the intensity of upwelling in the upper layer [16].

In the Indian Ocean, there are four major upwelling systems, including the western Arabian Sea (WAS), the Java and Sumatra coasts (JC), the Seychelles-Chagos thermocline ridge (SCTR) and the Southeastern Bay of Bengal [3]. Many previous works have been done to understand the variabilities of upwelling in WAS, JC and SCTR with timescales ranging from intraseasonal to decadal (e.g., [9,17,18]). In contrast, the existing studies mainly focused on the seasonal variability of the upwelling over the Bay of Bengal due to the sparseness of observational data.

Upwelling in the BOB can be classified into two types. One is coastal upwelling, which mainly occurs along the southern coast of Sri Lanka and the eastern coast of the Indian Peninsula, with relatively fixed locations (Figure 1a). The other is open ocean upwelling generally associated with the activities of cold eddies. The coastal upwelling off southern Sri Lanka occurs during the southwest monsoon, with increased chlorophyll concentrations (>5 mg m^{-3}), and alongshore wind stress is its main cause. In addition, the southwest monsoon, blocked by the island of Sri Lanka, forms a strong positive wind stress curl on the southeastern coast of the island [19]. The upward Ekman pumping induced by the positive curl also makes an important contribution to the development of this upwelling [19–21]. The Sri Lanka cold eddy that forms east of Sri Lanka 5–10° N, 83–87° E during the southwest monsoon is induced by the local wind stress with a positive curl [22–24].

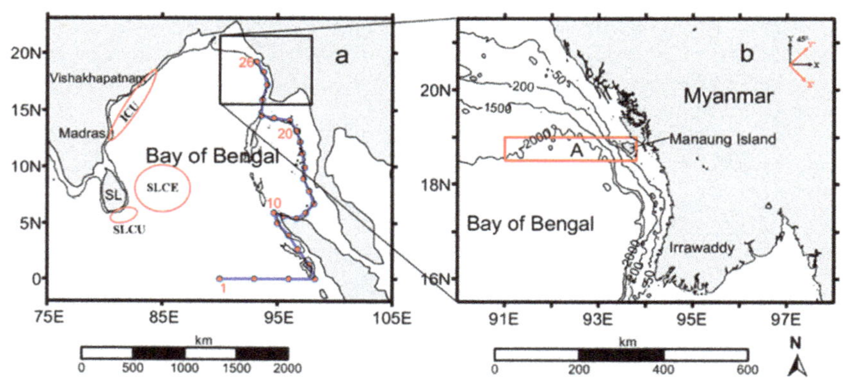

Figure 1. (**a**) Geographical location of the Bay of Bengal, and (**b**) topography of the northwestern coasts of Myanmar. The blue line in Figure 1a indicates the propagation path of the equatorial Kelvin wave and coastal Kelvin wave. The red dots in the blue line denote the selected stations of sea level anomaly (SLA) to track the propagation of Kevin wave along the equator and eastern boundary of the Bay. In Figure 1a, the India coastal upwelling, Sri Lanka coastal upwelling, Sri Lanka Cold Eddy and Sri Lanka are denoted by ICU, SLCU, SLCE and SL, respectively. In Figure 1b, the red box represents the study region A (91.0–93.5° E, 18.5–19.0° N), and the black contours are the isobaths of 50, 200, 1500 and 2000 m.

The strong upwelling around the Madras coast was first reported in 1964 from in-situ data [25], and the upwelled water was found to come from a 30 m layer induced by the strong winds along the coast of Madras. Based on hydrographic data collected during the summer monsoon of 1989, the upwelling zone off the eastern Indian Peninsula (Figure 1a) is found to extend in coastal waters from Madras to Visakhapatnam in summer [26,27], within a range of approximately 40 km offshore. Alongshore winds are an important dynamic mechanism for its formation [15]. Below this upwelling band, downwelling is often observed suggesting the presence of an undercurrent [15].

As mentioned above, previous studies mainly focused on the upwelling in the western boundary of the Bay [19–28]. In contrast, there are very few reports on the upwelling in its eastern boundary. Yesaki and Jantarapagdee [29] noted wind-induced upwelling over the continental shelf off the west coast of Thailand. The observations of La Fond [30] indicated that during April–May 1963, SST in the northeastern Bay exceeded 28 °C while it was relatively low (<27 °C) in the northern coasts off Myanmar, reflecting the possible impact of cold water upwelling from the subsurface. Follow-up studies also demonstrated the low-temperature zone located at the northwestern coasts of Myanmar 15–20° N during the winter monsoon, which was considered a sign of upwelling [12,31]. This conclusion was also supported by the satellite-derived Chlorophyll-*a* (Chl-*a*) data [32]. Akester [33] inferred that the high primary productivity in the continental shelf off Myanmar 16–20° N was caused by the upwelling there in April 2015, and the further evidence of upwelling, including low-temperature, high-salinity and low-dissolved oxygen, was shown in the coastal region.

In summary, although the existing studies have discovered signs of upwelling in the northwestern coasts of Myanmar in winter and spring, most studies were only based on scattered observations from a single cruise. Until now, little has been known about the spatial distribution and seasonal evolution of the upwelling, and there is a lack of research on its dynamic mechanisms. Therefore, in this paper, satellite observations and the World Ocean Atlas 2018 (WOA18) temperature and salinity data were used to analyze the characteristics and causes of the upwelling in the northwestern coast of Myanmar during the winter monsoon. The remainder of this paper is organized as follows: Section 2 introduces the data and method used in this study; Section 3 analyzes the spatial distribution and evolution process of the upwelling in the study area; Section 4 presents the dynamic mechanisms of the upwelling; discussions and, finally, conclusions are stated in Sections 5 and 6, respectively.

2. Data and Method

2.1. Data

The World Ocean Atlas 2018 (WOA18) has been newly released by the Ocean Climate Laboratory of the National Centers for Environmental Information (NCEI) and the National Oceanic and Atmospheric Administration (NOAA) of the U.S. The WOA18 collected temperature and salinity samples obtained by all ship-deployed Conductivity-Temperature-Depth (CTD) packages, profiling floats, moored and drifting buoys, gliders and undulating oceanographic recorder profiles. Using an objective analysis technique, the raw data are processed into gridded data with a horizontal resolution of 0.25° and 102 vertical levels above 5500 m (including 21 layers, at 5 m spacing, for the top 100 m). Data are presented for climatological composite periods (annual, seasonal, monthly, seasonal and monthly difference fields from the annual mean field, and the number of observations) at 102 standard depths. Standard error of the mean fields is binned into several ranges depending on the depth level. In our study region (90.5–97.5° E, 15.5–21.5° N), the error of temperature in upper 100 m layer ranges from 0.00 to 0.89 °C, and salinity error ranges between 0.00 and 0.33 psu [34,35]. In this study, climatological monthly temperature and salinity were used in the analysis of the upwelling distribution.

The Moderate Resolution Imaging Spectroradiometer (MODIS) daily Chl-*a* data were provided by the U.S. National Aeronautics and Space Administration (NASA). The chlorophyll concentration was derived using the OC3 algorithm [36]. There is a root mean squared error (RMSE) of 1.228 mg m^{-3} for MODIS Chl-*a* against in situ observations [37]. MODIS Chl-*a* with a horizontal resolution of 4 km [38] spanning between 1 January 2003 and 31 March 2020, were used to study seasonal evolution of the upwelling within the study area.

Sea surface wind data were from the cross-calibrated multi-platform (CCMP) product provided by the NASA's Physical Oceanography Distributed Active Archive Center (PODAAC). CCMP is a multisource fusion surface wind data derived from satellite observations, including measurements of Special Sensor Microwave Imager (SSM/I), Advanced Microwave Scanning Radiometer for Earth Observing System (AMSR-E) and Tropical Rainfall Measuring Mission Microwave Imager (TMI). To create the CCMP winds, an enhanced variational analysis method (VAM) performs quality control and combines all available Remote Sensing Systems (RSS) cross-calibrated wind data with available conventional ship and buoy data and European Centre for Medium-Range Weather Forecasts (ECMWF) analyses. Overall, the CCMP analysis has the best overall fit to the in-situ observations with an error speed difference ranging from 1.6 m s^{-1} versus ships to 0.6 m s^{-1} versus the higher-quality TAO buoys. This is also seen in the RMSE direction fit, which ranges from 11.5° to 7.0°. This dataset has a horizontal resolution of 0.25° × 0.25° and a temporal resolution of 6 h from the period 1993 to 2016 [39].

Daily sea level anomaly (SLA) data were provided by Copernicus Marine Environment Monitoring Service (CMEMS), with a horizontal resolution of 0.25° × 0.25°. This gridded data is derived from the multi-satellite measurements of TOPEX/Poseidon (T/P), European Remote Sensing Satellite-1 (ERS-1) and European Remote Sensing Satellite-2 (ERS-2). There is an error of ~0.02 m for the SLA data [40]. The SLA data we used are from between 1 January 1993 and 31 March 2020. These two data sets of CCMP wind and CMEMS SLA were averaged to obtain weekly and monthly values and then were used to analyze the dynamic mechanisms of the upwelling.

2.2. Method to Identify Area of the Upwelling and Its Intensity

Water that rises to the surface as a result of upwelling is characterized by low-temperature and high-salinity. Both SST and SSS have been extensively used as an indicator of upwelling in many previous studies (e.g., [41,42]). Correspondingly, upwelling areas were identified by a consideration of both SST and SSS from WOA18 in our study. As we shall see below that the region around Myanmar coasts with SST below 28.3 °C has a relatively high salinity. Thus, the regions around the coast of Sri Lanka with SST less than 28.3 °C are used to describe the regions affected by the coastal upwelling.

In addition, the upwelled water from subsurface is typically rich in nutrients. In the upwelling region, surface water has relatively high Chl-*a* concentration (e.g., [19,41]), besides its typical relatively cold and saline nature. Thus, Chl-*a* concentration is used as an indicator for the intensity of upwelling in previous works (e.g., [43]). In this study, climatological weekly Chl-*a* concentration averaged in the upwelling region during January–April is constructed from the daily MODIS Chl-*a* data and then it is used to explore the evolution processes of the intensity of the upwelling. A larger magnitude of Chl-*a* concentration denotes a stronger upwelling.

3. Spatial-Temporal Distribution of the Coastal Upwelling off Manaung Island Using WOA18 Data

The sample amounts of temperature and salinity in the surface layer for each 1° × 1° grid during spring (March to May) are shown in Figure 2a,b, respectively. Generally speaking, observations of temperature and salinity have a good coverage over the study region. In the coastal areas, the temperature and salinity samples around Manaung Island are 7–18 and 6–16, respectively, and both temperature and salinity samples in open ocean are far larger than that in the nearshore areas. Thus, WOA18 provides adequate data for this study to investigate the spatial structure of the upwelling in the northwestern coasts of Myanmar.

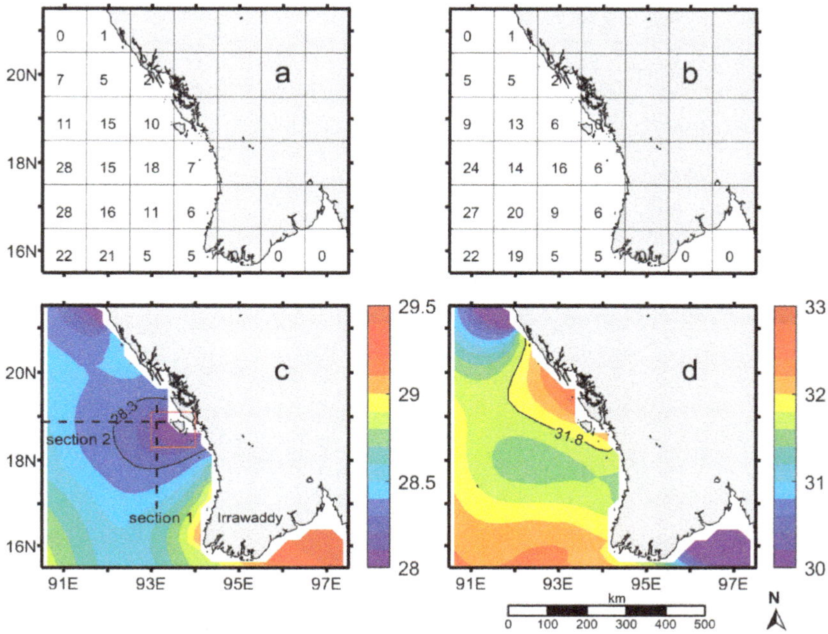

Figure 2. Sample amounts of surface temperature (**a**) and salinity (**b**) in spring (March-May) in a 1° × 1° grid; (**c**) and (**d**) are climatological surface temperature (unit: °C) and salinity (unit: psu) in the same period. In Figure 2c, the dotted lines indicate the locations of Section 1 (93.375° E, 16.875–19.125° N) and Section 2 (90.125–93.375° E, 18.875° N), the red box (93.0–94.0° E, 18.5–19.0° N) represents the core area of upwelling, and the solid line is the 28.3 °C isotherm. The solid line in Figure 2d is the 31.8 psu isohaline. Temperature and salinity data are from climatological monthly data set of World Ocean Atlas 2018 (WOA18).

Distributions of surface temperature and salinity in spring are shown in Figure 2c,d, respectively. SST in the study region is relatively high (>28 °C), owing to strong shortwave radiation during spring [44]. The highest SST (29.5 °C) appears in the coastal areas around the estuary of the Irrawaddy River. A low-temperature zone is found distributed along the northwestern coasts of Myanmar, with a cold center located off the coast of Manaung Island. Taken the 28.3 °C isotherm as the boundary of the upwelling, the upwelling zone appears as "tongue-shaped", extending across 17.7–19.3° N offshore of the Manaung Island, to the east of 92.3° E. Affected by a large amount of runoff and precipitation in the northern Bay, salinity of the study area is relatively low (<33.0 psu). The lowest salinity (~30 psu) occurs offshore of the northwestern Myanmar (Figure 2d). Similar to the low-temperature zone, a high-salinity zone is also formed in the waters around the Manaung Island, with salinity exceeding 31.8 psu, and the high-salinity center is located in the northwestern coast of Manaung Island. The high-salinity (>31.8 psu) and low-temperature (<28.3 °C) zones do not completely overlap, with the high-salinity zone occurring slightly to the north of the low-temperature zone. This low-temperature (<28.3 °C) and high-salinity (>31.8 psu) water at the surface may reflect the occurrence of upwelling near the Manaung Island in spring. This is broadly consistent with the location of upwelling signals revealed by Akester [33] using in-situ data obtained in April 2015.

Figure 3 shows the climatologically sectional distributions of temperature and salinity in Section 1 (along 93.375° N) and Section 2 (along 18.875° E) in spring (the locations of the two sections are shown in Figure 2c), illustrating the vertical structure of the upwelling. Both the isotherms and isohalines in the two sections have a trend of rising upwards in the coastal areas of the Manaung Island from about the 100 m layer of the open ocean, indicative of upwelling in the nearshore regions. The remarkable rise of cold subsurface water mainly occurs in coastal waters east of 92° E and north of 17.5° N. The amplitude of the isotherm uplift is significantly smaller in the near-surface layer (above 30 m layer) than that in the subsurface layer (30–70 m). This means that the impact of upwelling on temperature is more pronounced in the subsurface than that in the surface layer. This may be caused by two factors. Firstly, the subsurface layer lies within the seasonal thermocline and the vertical temperature gradient there is larger than that in the near-surface layer, and thus the impact of upwelling on temperature is more significant in the subsurface layer. Secondly, the intensity of upwelling may be partially suppressed by strong salinity stratification in the near-surface layer [16], which further weakens upwelling near the surface and thus results in weaker uplift of isotherm there. The sectional distribution of salinity also reveals upwelling of high-salinity water in the subsurface layer off the coast of Manaung Island (Figure 3b). However, in contrast to temperature, the upwelling of high-salinity water is more significant in the upper layer above 30 m than that in the subsurface layer. This is mainly because both the freshwater input from river discharge and the precipitation into the surface make the vertical salinity gradient larger in the upper layer than that in the subsurface layer [45]. Therefore, even under the condition of a weaker uplift of isotherms in the near-surface layer (Figure 3a), the uplift of isohalines is much more significant in the upper layer.

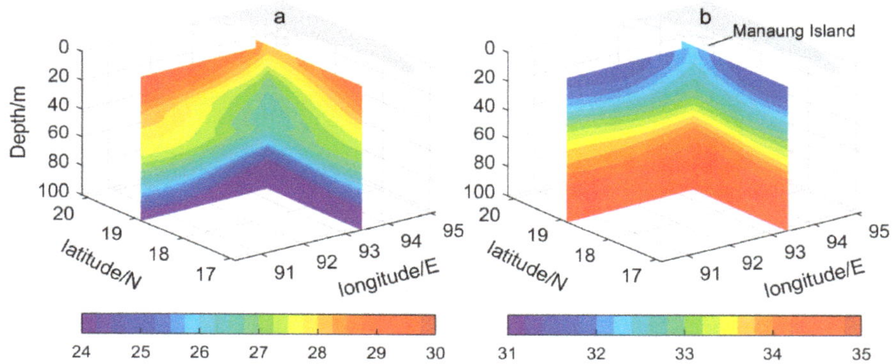

Figure 3. The climatologically sectional distributions of (**a**) temperature (unit: °C) and (**b**) salinity (unit: psu) of two sections in spring. The locations of the two sections are shown in Figure 2c.

Seasonal evolution for the coastal upwelling off Manaung Island is further explored using WOA18 climatologically monthly temperature and salinity data (Figure 4). In February, a high-salinity zone starts to form in the surface off the coasts of Manaung Island, while the low-temperature signal is not significant. This is mainly because the southward flowing low-temperature water from the northern Bay along the coasts during that period [46] covers the upwelled cold water from subsurface. In fact, both the low-temperature and high-salinity signals of the upwelling are clearly presented in the subsurface layers between 40 and 80 m around the coastal areas of Manaung Island in February (Figure 5). In March, the extent of low-temperature and high-salinity water is the largest, indicating that the coastal upwelling reaches its peak in this month. Subsequently, the extent of low-temperature and high-salinity water significantly shrinks in April and disappears in May. Thus, April and May are periods for the decay and dissipation of the coastal upwelling of Manaung Island.

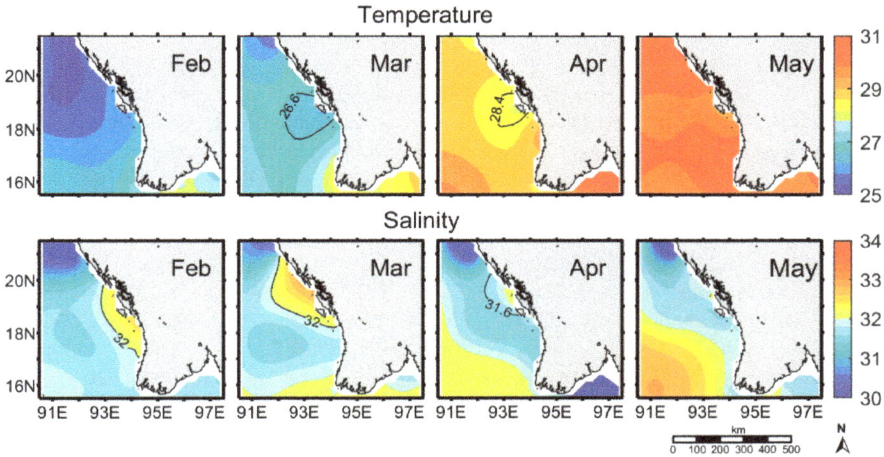

Figure 4. The climatological distributions of surface temperature (upper panels, unit: °C) and salinity (lower panels, unit: psu) in the waters of Myanmar from February to May. Temperature and salinity are from climatological monthly data of WOA18.

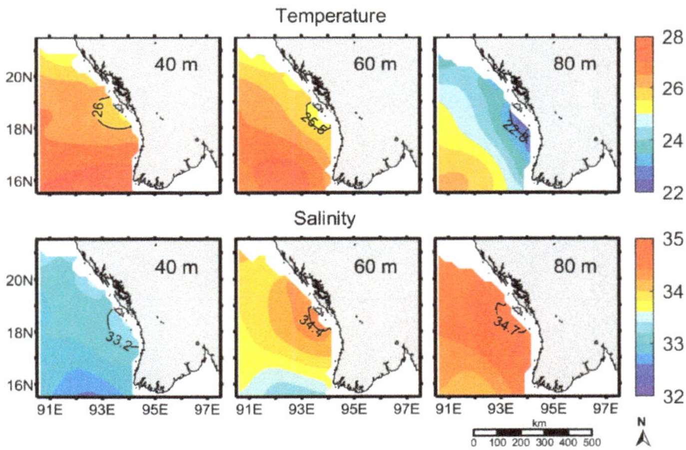

Figure 5. The climatological distributions of temperature (upper panels, unit: °C) and salinity (lower panels, unit: psu) in the 40, 60 and 80 m layers in the waters of Myanmar in February. Temperature and salinity are from climatological monthly data of WOA18.

4. Dynamic Mechanisms for Seasonal Evolution of the Coastal Upwelling of Manaung Island Using Satellite Measurements

As is well known, local surface wind is an important factor influencing the generation and evolution of coastal upwelling [28,47]. Therefore, the role of local wind on the formation and evolution of the coastal upwelling of Manaung Island is explored here.

Affected by the South Asian monsoon, the predominant wind in the northern Bay is northwesterly/northerly during winter and spring [12]. The direction of wind stress is roughly parallel to the NW-SE orientation of the coastline of northwestern Myanmar (Figure not shown). According to the classic Ekman transport theory, i.e., in the Northern Hemisphere, the wind blows in a direction parallel to the coast (on the left side of the wind direction), and it causes net movement of surface water at about 90 degrees to the right of the wind direction. Because the surface water flows away from the coast, the water is replaced with water from below [48]. Thus, the northwesterly wind stress will cause an offshore Ekman transport there, thereby triggering the upwelled subsurface water to the surface layer as a compensatory for the offshore transport, which induces the formation of coastal upwelling. To facilitate further analysis, the coordinate axes are rotated clockwise by 45°, and the wind stress is decomposed into directions parallel to and perpendicular to the coasts, with the southeast direction along the coastline being positive (as shown in Figure 1b). During the northeast monsoon (December to April of the following year), the coast of Manaung Island experiences the northwesterly wind, favorable to the formation of upwelling. The wind stress is generally weak (0.01–0.02 N m^{-2}) and reaches its maximum (approximately 0.02 N m^{-2}) in January (Figure 6a). At the same time, the wind stress curl reaches the maximum (approximately 2×10^{-7} N m^{-3}) (Figure 6b). Then, the wind direction reverses to a southeasterly one during the monsoon transition period in May, which is not favorable to the formation of upwelling. Thus, local wind stress may contribute to the formation of the upwelling in spring, although its intensity is relatively weak, and the timing of its maximum is two months out of phase with the peak of upwelling (in March). That is to say that although the local wind stress contributes to the development of upwelling, there may be other factors (i.e., remote forcing from the equator) that have important contributions to the development of the coastal upwelling of Manaung Island, as we shall see below.

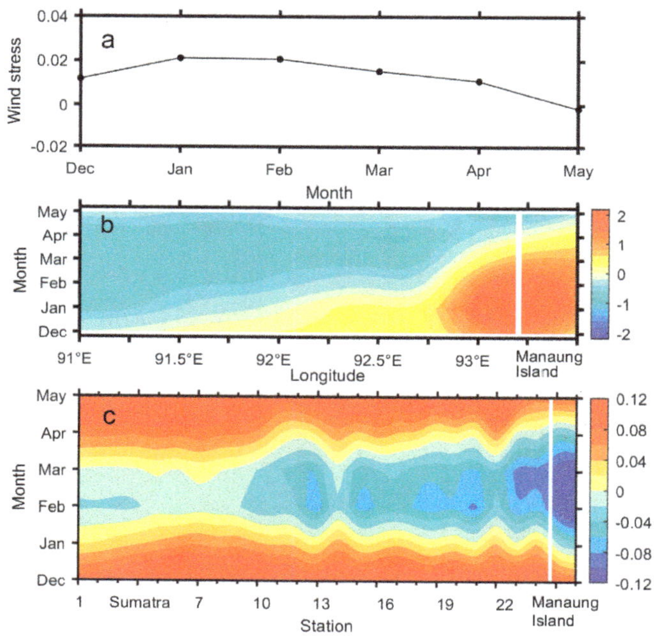

Figure 6. (**a**) Time series of climatological monthly alongshore wind stress of Manaung Island (the southeast direction along the coastline is positive, unit: N m^{-2}); (**b**) Time–Longitude diagram of climatological monthly wind stress curl averaged over area A (unit: 10^{-7} N m^{-3}); (**c**) Time–station diagram of climatological monthly SLA (unit: m). The stations and the range of area A are shown in Figure 1a,b, respectively. All climatological monthly mean data is averaged from the daily/6 h data for the period 1993 to 2016.

Besides local wind, remote forcing from the equator is also the main forcing that induces seasonal variabilities of the BOB. In particular, it plays a dominant role at the eastern and northern boundary of the Bay [49–51]. It affects the interior Bay via the planetary waves (including coastal Kelvin waves and the westward propagating Rossby waves triggered by the equatorial processes) [52]. Referring to Cheng et al. [51], the propagation path of the planetary waves from the equator to the interior Bay along the 200 m isobath (the blue line in Figure 1a) is selected, and the SLA evolutions at each station of December and May of the following year are illustrated (Figure 6c). The upwelling Kelvin wave originates from the equator, propagates into the eastern boundary of the BOB in early February and reaches the coastal region of Manaung Island in early March, resulting in a remarkable decrease in SLA there (Figure 6c). This further amplifies the upwelling induced by the local wind stress (Figure 6a,b), so that the upwelling peaks in March. Subsequently, the downwelling Kelvin wave, triggered by equatorial westerlies in early April, reaches the Manaung Island in early May, thereby making an important contribution to the decay of the upwelling.

Rao et al. [53] demonstrated that the aforementioned equatorial upwelling (downwelling) Kelvin wave is driven by equatorial easterlies (westerlies) wind. The lead-lag correlation analysis between SLA off the Manaung Island (Station 26 shown in Figure 1a) and that at the central equator (Station 1 shown in Figure 1a) shows that the correlation coefficient is positive and reaches maximum of 0.64 (significant at 99% confidence level) when variation of SLA at Manaung Island (SLA_n) lags behind SLA in the central equator (SLA_s) by 32 days (Figure 7). This means that it takes about 32 d for the signal of SLA variations propagating from the central equator to the Manaung Island. The mean propagation speed of the equatorial Kelvin wave or the coastal Kelvin wave is approximately 1.69 m s^{-1}, close to the typical speed of the second baroclinic mode (1.79 m s^{-1}) [50,54].

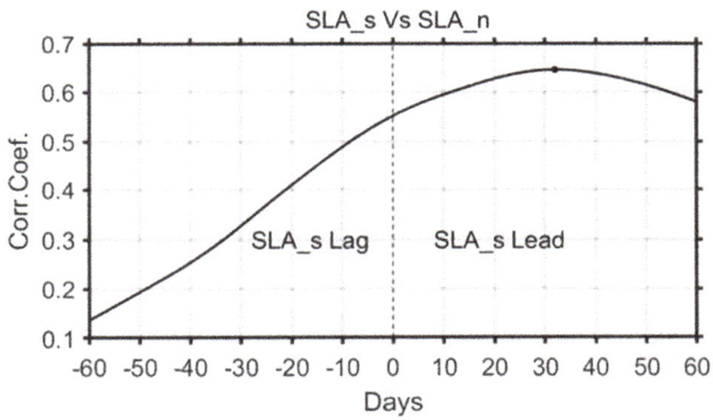

Figure 7. Lead/lag correlation between SLA at Station 26 (SLA_n) and that at Station 1 (SLA_s) using data from January 2003 to December 2016. The x-coordinate represents the number of days, a positive value of which indicates a leading SLA_s, and the y-coordinate represents the correlation coefficient.

The above analysis has demonstrated that both local wind and equatorial forcing have important impacts on the formation and evolution of the coastal upwelling of Manaung Island. The concentration of Chl-*a* is another important indicator of the upwelling intensity [43]. The MODIS images of Chl-*a* for the upwelling period between February and March in 2020 (Figure 8) demonstrate that the short-term evolution of Chl-*a* distribution in the upwelling zone off Manaung Island is consistent with that of SLA. This consistency underpins that both Chl-*a* and SLA can well characterize in the intensity of the coastal upwelling of Manaung Island.

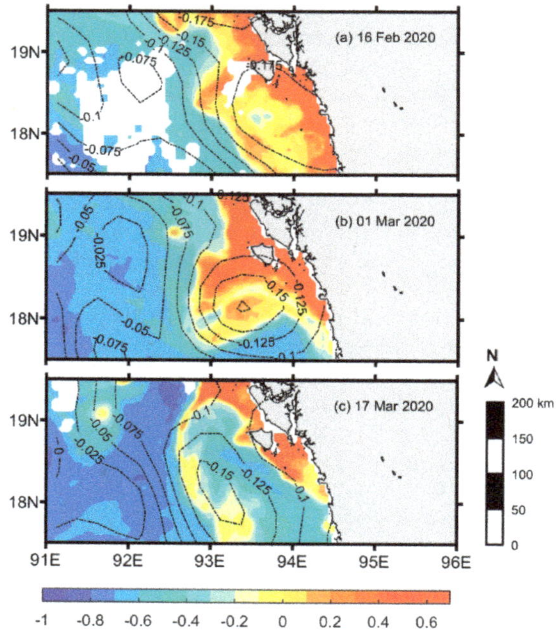

Figure 8. Logarithm (Log_{10}) of Chl-*a* concentrations (shaded, unit: mg m^{-3}) and SLA (contours, unit: m) in the waters of Manaung Island for (**a**) 16 February, (**b**) 1 March, and (**c**) 17 March in 2020. Blank areas indicate missing data.

The daily MODIS Chl-*a* data in the core area of upwelling (the red box in Figure 2c), from January to April, between 2003 and 2020, were composited and then bin-averaged to obtain the climatologically weekly averaged Chl-*a* time series from January to April. The SLA_n, SLA_s, alongshore wind stress, and wind stress curl were also analyzed using the same method. Then, all these time series were normalized by 1 standard deviation of each (Figure 9). The concentration of Chl-*a* in the upwelling zone (blue line) continues to increase from January, reaching its peak in early March, and then it decreases during late March and throughout April (Figure 9). This evolution pattern is consistent with variations in the upwelling intensity revealed by the WOA18 temperature and salinity data. As there is a significant correlation between Chl-*a* concentration and SLA_n with a correlation coefficient of −0.76, SLA_n serves as a good indicator for the intensity of the upwelling at Manaung Island. Thus, the relative contribution of each influencing factor to variations in the upwelling was evaluated by calculating its correlation with SLA_n. The results of correlation analysis (Table 1) demonstrate that there is a high positive correlation between weekly SLA_n and the weekly SLA_s with SLA_s leading SLA_n by 35 days, and their correlation coefficient is 0.60, significant at 95% confidence level. In contrast, SLA_n is negatively correlated with alongshore wind stress and local wind stress curl, with correlation coefficients of −0.35 and −0.33, respectively. These two correlations are not statistically significant ($p > 0.05$), which indicates that the local wind stress in the upwelling region of Manaung Island has a relatively small impact on the variation in SLA_n. Therefore, the correlation between SLA_n and the equatorial forcing is much stronger than the local wind (0.60 vs. −0.35/−0.33), underpinning that although both local wind and equatorial forcing have impacts on evolution of the upwelling at Manaung Island, the remote forcing from the equator plays a more important role than the local wind does.

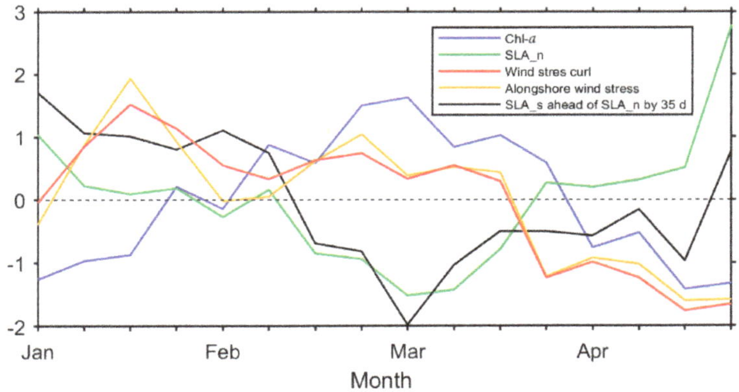

Figure 9. Time series of climatologically weekly Chl-*a* concentrations, wind stress curl, alongshore wind stress, SLA_n and SLA_s ahead of SLA_n by 35 d from January to April (all data are averaged during 2003–2016 and normalized by one standard deviation).

Table 1. The correlation coefficients between indicator of the upwelling (SLA_n and Chl-*a*) and forcing factors.

Indicator of the Upwelling	Indicator of the Upwelling/Forcing Factor	Correlation Coefficient
SLA_n	Chl-*a*	−0.76
SLA_n	Wind stress curl	−0.35
SLA_n	SLA_s ahead of SLA_n by 35 days	0.60
SLA_n	Alongshore wind stress	−0.33

5. Discussions

In the upwelling region, upwelled supplies of nitrate and the tremendous blooms of phytoplankton they support, render these regimes the "marine ranchings" [55]. This phytoplankton biomass then feeds into productive food chains which support a significant share of the biological resources that humans harvest from the ocean, and indeed, attract commercial fishers and fisheries [56]. Therefore, the coastal upwelling has a profound impact on coastal populations, and it is particularly important for developing countries (e.g., Myanmar). Despite of the importance of the upwelling, until now, little has been known about the coastal upwelling off Myanmar. This is mainly due to the sparseness of observational data and to the multi-scale aspects of upwelling that are difficult to measure and simulate in climate/ocean models.

Our analysis demonstrates that the center of the upwelling is located at the coastal region off Manaung Island during spring. This is generally consistent with the location of the upwelling reported in the previous studies [30,32,33]. However, seasonal evolution characteristics were not discussed in these previous works due to the scatted measurements from a single cruise used in these studies. In contrast, our study indicates that the upwelling generates in February, then peaks in March and finally decays in May. Based on the satellite observations, the influence of local winds and remote equatorial forcing on the evolution of the upwelling is further discussed by the correlation analysis for the first time. Nevertheless, the physical mechanisms of the upwelling are preliminary. Models will be helpful to isolate the dynamics and assess the relative contribution of various processes in the future study. Our results would contribute to the understanding of coastal processes (including the coastal upwelling) off Myanmar and regional climate variability and their prediction.

Note that the Indian Ocean Dipole (IOD) and El Niño and Southern Oscillation (ENSO) are major climate modes that affect the interannual variability in the northern Indian Ocean [57–59]. They can cause interannual variabilities in the regional upwelling in the Indian Ocean by inducing variations in the sea surface wind field [59,60]. For example, significantly interannual variability of the Java–Sumatra upwelling is primarily driven by local winds and remote equatorial Indian Ocean winds associated with IOD/ENSO events [9,18,61]. As our analysis has shown, the coastal upwelling off Myanmar is mainly caused by the remote equatorial forcing and to a lesser degree by the local winds. Both these two forcing indicate significant interannual variabilities caused by IOD and ENSO events [58]. However, how and to what extent the IOD and ENSO events affect the interannual variabilities of the upwelling off Myanmar are still unknown and await further study. A series of cruises are suggested to be carried out in the upwelling region off Myanmar coasts during normal years and IOD/ENSO years to address these important scientific issues.

6. Conclusions

In this paper, WOA18 temperature and salinity datasets, combined with the MODIS Chl-*a* data, SLA and the CCMP wind data, were used to analyze the spatial structure and seasonal evolution of the coastal upwelling around the Manaung Island. Possible dynamic mechanisms leading to the evolution process of the upwelling were also explored for the first time. The upwelling near the Manaung Island has been identified, which begins in February, reaches its peak in March and disappears in May. Low-temperature (<28.3 °C) and high-salinity (>31.8 psu) water at the surface of this upwelling zone mainly upwells from a depth of 100 m. The impact of upwelling on temperature is found to be more significant in the subsurface layer than that in the surface layer. In contrast, the impact of upwelling on salinity in the surface layer is more significant. This different effect of the upwelling on the temperature and salinity in upper layer is possibly caused by the strong salinity stratification there.

During winter and spring, the local wind near the Manaung Island is dominated by northwesterly/northerly wind. It tends to induce offshore transport in the upper water volume, and consequently is favorable to the formation of coastal upwelling. In addition, the upwelling Kelvin wave, driven by the equatorial easterly wind stress during January-March, propagates along the eastern boundary of the Bay of Bengal, which promotes uplifting of the thermocline and halocline in the coastal regions around the Manaung Island, further enhancing the upwelling. In April, the downwelling Kelvin wave, propagating into the eastern boundary of the Bay from the equator, deepens the thermocline in the waters around the Manaung Island, and simultaneously the local wind is weakening. Both these factors may play an important role in the dissipation of upwelling in this area. Correlation analysis further demonstrates that the remote equatorial forcing exerts a more significant effect on the upwelling variability than the local wind does. Although our study is qualitative, the physical processes are robust features of this region and their quantification will be a major challenge for future observational as well as modeling studies.

Despite the joint effects of remote equatorial forcing and local wind can well explain the generation and evolution of the coastal upwelling around the Manaung Island, we note that the center of the upwelling is located at the Manaung Island where the offshore bottom topography near the Manaung Island has the largest gradient towards the coast. This sharp change in topography may have a significant impact on the distribution of the upwelling center [62], which awaits further analysis in the future.

Author Contributions: Conceptualization, Y.Q.; methodology, Y.Q. and Y.L.; validation, Y.Q. and Y.L.; formal analysis, Y.L.; data curation, Y.Q. and Y.L.; writing—original draft preparation, Y.L. and Y.Q.; writing—review and editing, Y.Q., J.H., C.A., X.L., Y.D.; visualization, Y.L.; funding acquisition, Y.Q. All authors have read and agreed to the published version of the manuscript.

Funding: This study is supported by grants from the Scientific Research Foundation of Third Institute of Oceanography, Ministry of Natural Resources (Nos. 2018001, 2018030 and 2017012); the Ministry of Natural Resources Program on Global Change and Air-Sea interactions (Nos. GASI-IPOVAI-02 and GASI-IPOVAI-03); the National Natural Science Foundation of China (Nos. 41776027, 41906013 and 41276034).

Acknowledgments: We benefited from numerous datasets made freely available, including MODIS data, WOA18 dataset, CCMP wind and CMEMS products.

Conflicts of Interest: The authors declare no conflict of interest.

References

1. Nigam, T.; Pant, V.; Prakash, K.R. Impact of Indian ocean dipole on the coastal upwelling features off the southwest coast of India. *Ocean Dyn.* **2018**, *68*, 663–676. [CrossRef]
2. Ryther, J.H. Photosynthesis and fish production in the sea. *Science* **1969**, *166*, 72–76. [CrossRef]
3. Sreeush, M.G.; Valsala, V.; Pentakota, S.; Prasad, K.V.S.R.; Murtugudde, R. Biological production in the Indian Ocean upwelling zones—Part 1: Refined estimation via the use of a variable compensation depth in ocean carbon models. *Biogeosciences* **2018**, *15*, 1895–1918. [CrossRef]
4. Sardeshmukh, P.D.; Hoskins, B.J. The generation of global rotational flow by steady idealized tropical divergence. *J. Atmos. Sci.* **1988**, *45*, 1228–1251. [CrossRef]
5. Webster, P.J.; Lukas, R. TOGA COARE: The coupled ocean atmosphere response experiment. *Bull. Am. Meteorol. Soc.* **1992**, *73*, 1377–1416. [CrossRef]
6. Yu, J.-Y.; Mechoso, C.R.; McWilliams, J.C.; Arakawa, A. Impacts of the Indian Ocean on the ENSO cycle. *Geophys. Res. Lett.* **2002**, *29*, 1204. [CrossRef]
7. Izumo, T.; Vialard, J.; Lengaigne, M.; Montegut, C.D.B.; Behera, S.K.; Luo, J.-J.; Cravatte, S.; Masson, S.; Yamagata, T. Influence of the state of the Indian Ocean Dipole on the following year's El Niño. *Nat. Geosci.* **2010**, *3*, 168–172. [CrossRef]
8. Kim, S.T.; Yu, J.Y.; Lu, M.M. The distinct behaviors of Pacific and Indian Ocean warm pool properties on seasonal and interannual time scales. *J. Geophys. Res.* **2012**, *117*, D05128. [CrossRef]
9. Zhang, X.L.; Han, W.Q. Effects of climate modes on interannual variability of upwelling in the tropical Indian Ocean. *J. Clim.* **2020**, *33*, 1547–1573. [CrossRef]
10. Colwell, R.R. Global climate and infectious disease: The cholera paradigm. *Science* **1996**, *274*, 2025–2031. [CrossRef]
11. Harwell, C.; Kim, K.; Burkholder, J.; Colwell, R.; Epstein, P.R.; Grimes, D.; Hofmann, E.E.; Lipp, E.K.; Osterhaus, A.; Overshreet, R.M. Emerging marine diseases-climate links and anthropogenic factors. *Science* **1999**, *285*, 1505–1510. [CrossRef] [PubMed]
12. Varkey, M.J.; Murty, V.S.N.; Suryanarayana, A. Physical oceanography of the bay of Bengal and Andaman Sea. *Oceanogr. Mar. Biol. Annu. Rev.* **1996**, *34*, 1–70.
13. Liu, H.X.; Ke, Z.X.; Song, X.Y.; Tan, Y.H.; Huang, L.M.; Lin, Q. Primary production in the Bay of Bengal during spring intermonsoon period. *Acta Ecol. Sin.* **2011**, *31*, 7007–7012. [CrossRef]
14. Prasanna Kumar, S.; Muraleedharan, P.M.; Prasad, T.G.; Gauns, M.; Ramaiah, N.; de Souza, S.N.; Sardesai, S.; Madhupratap, M. Why is the Bay of Bengal less productive during summer monsoon compared to the Arabian Sea? *Geophys. Res. Lett.* **2002**, *29*. [CrossRef]
15. Shetye, S.R.; Shenoi, S.S.C.; Gouveia, A.D.; Michael, G.S.; Sundar, D.; Nampoothiri, G. Wind driven coastal upwelling along the western boundary of the Bay of Bengal during the southwest monsoon. *Cont. Shelf Res.* **1991**, *11*, 1397–1408. [CrossRef]
16. Rao, S.A.; Gopalakrishna, V.V.; Shetye, S.R.; Yamagata, T. Why were cool SST anomalies absent in the Bay of Bengal during the 1997 Indian Ocean Dipole Event? *Geophys. Res. Lett.* **2002**, *29*. [CrossRef]
17. Susanto, R.D.; Gordon, A.L.; Zheng, Q.N. Upwelling along the coasts of Java and Sumatra and its relation to ENSO. *Geophys. Res. Lett.* **2001**, *28*, 1599–1602. [CrossRef]
18. Chen, G.X.; Han, W.Q.; Li, Y.L.; Wang, D.X.; Shinoda, T. Intraseasonal variability of upwelling in the equatorial Eastern Indian Ocean. *J. Geophys. Res.* **2015**, *120*, 7598–7615. [CrossRef]
19. Li, Y.H.; Qiu, Y.; Hu, J.Y.; Aung, C.; Lin, X.Y.; Jing, C.S.; Zhang, J.P. The strong upwelling event in the southern coast of Sri Lanka in 2013 and its relationship with Indian Ocean Dipole events. *J. Clim.* **2020**. under review.
20. Vinayachandran, P.N.; Chauhan, P.; Mohan, M.; Nayak, S. Biological response of the sea around Sri Lanka to summer monsoon. *Geophys. Res. Lett.* **2004**, *31*, L01302. [CrossRef]
21. De Vos, A.; Pattiaratchi, C.; Wijeratne, E. Surface circulation and upwelling patterns around Sri Lanka. *Biogeosciences* **2014**, *11*, 5909. [CrossRef]
22. Vinayachandran, P.N.; Yamagata, T. Monsoon response of the sea around Sri Lanka: Generation of thermal domes and anticyclonic vortices. *J. Phys. Oceanogr.* **1998**, *28*, 1946–1960. [CrossRef]

23. Burns, J.M.; Subrahmanyam, B.; Murty, V.S.N. On the dynamics of the Sri Lanka Dome in the Bay of Bengal. *J. Geophys. Res. Oceans* **2017**, *122*, 7737–7750. [CrossRef]
24. Cullen, K.E.; Shroyer, E.L. Seasonality and interannual variability of the Sri Lanka dome. *Deep Sea Res. II* **2019**, *168*, 104642. [CrossRef]
25. Murty, C.S.; Varadachari, V.V.R. Upwelling along the east coast of India. *Bull. Natl. Inst. Sci. India* **1968**, *36*, 80–86.
26. Anand, S.P.; Murty, C.B.; Jayaraman, R.; Aggarwal, B.M. Distribution of temperature and oxygen in the Arabian Sea and Bay of Bengal during the monsoon season. *Bull. Natl. Inst. Sci. India* **1968**, *38*, 1–24.
27. De Souza, S.N.; Naqvi, S.W.A.; Reddy, C.V.G. Distribution of nutrients in the western Bay of Bengal. *Indian J. Mar. Sci.* **1981**, *10*, 327–331.
28. Yapa, K.K.A.S. Upwelling phenomena in the southern coastal waters of Sri Lanka during southwest monsoon period as seen from MODIS. *Sri. Lanka J. Phys.* **2009**, *10*, 7–15. [CrossRef]
29. Yesaki, M.; Jantarapagdee, P. Wind stress and sea temperature changes off the west coast of Thailand. In *Special Publication on the Occasion of the Tenth Anniversary*; Marine Biological Centre: Phuket, Thailand, 1981; pp. 27–42.
30. Lafond, E.C.; Lafond, K.G. Studies on oceanic circulation in the Bay of Bengal. *Bull. Natl. Inst. Sci. India* **1968**, *38*, 164–183.
31. Rao, D.P. A comparative study of some physical processes governing the potential productivity of the Bay of Bengal and Arabian Sea. Ph.D. Thesis, Andhra University, Waltair, India, 1977.
32. Suwannathatsa, S.; Wongwises, P.; Vongvisessomjai, S.; Wannawong, W.; Saetae, D. Phytoplankton tracking by oceanic model and satellite data in the Bay of Bengal and Andaman Sea. *APCBEE Procedia* **2012**, *2*, 183–189. [CrossRef]
33. Akester, M.J. Productivity and coastal fisheries biomass yields of the northeast coastal waters of the Bay of Bengal Large Marine Ecosystem. *Deep Sea Res. Part II Top. Stud. Oceanogr.* **2019**, *163*, 46–56. [CrossRef]
34. Zweng, M.M.; Reagan, J.R.; Seidov, D.; Boyer, T.P.; Locarnini, R.A.; Garcia, H.E.; Mishonov, A.V.; Baranova, O.K.; Weathers, K.W.; Paver, C.R.; et al. *World Ocean Atlas 2018, Volume 2: Salinity*; A. Mishonov Technical, Ed.; Silver Spring, Department of Commerce: Montgomery, MD, USA, 2018.
35. Locarnini, R.A.; Mishonov, A.V.; Baranova, O.K.; Boyer, T.P.; Zweng, M.M.; Garcia, H.E.; Reagan, J.R.; Seidov, D.; Weathers, K.W.; Paver, C.R.; et al. *World Ocean Atlas 2018. Volume 1: Temperature*; A. Mishonov Technical, Ed.; Silver Spring, Department of Commerce: Montgomery, MD, USA, 2018.
36. O'Reilly, J.E.; Maritorena, S.; Mitchell, B.; Siegel, D.; Carder, K.; Garver, S.; Kahru, M.; McClain, C. Ocean color chlorophyll algorithms for SeaWiFS. *J. Geophys. Res. Ocean* **1998**, *103*, 24937–24953. [CrossRef]
37. Cui, T.W.; Zhang, J.; Tang, J.W.; Sathyendranath, S.; Groom, S.; Ma, Y.; Zhao, W.; Song, Q.J. Assessment of satellite ocean color products of MERIS, MODIS and SeaWiFS along the East China Coast (in the Yellow Sea and East China Sea). *J. Photogramm. Remote Sens.* **2014**, *87*, 137–151. [CrossRef]
38. Esaias, W.E.; Abbott, M.R.; Barton, I.; Brown, O.B.; Campbell, J.W.; Carder, K.L.; Clark, D.K.; Evans, R.H.; Hoge, F.E.; Gordon, H.R.; et al. An overview of MODIS capabilities for ocean science observations. *IEEE Trans. Geosci. Remote* **1998**, *36*, 1250–1265. [CrossRef]
39. Atlas, R.; Hoffman, R.N.; Ardizzone, J.; Leidner, S.M.; Jusem, J.C.; Smith, D.K.; Gombos, D. A cross-calibrated, multiplatform ocean surface wind velocity product for meteorological and oceanographic applications. *Bull. Amer. Meteor. Soc.* **2011**, *92*, 157–174. [CrossRef]
40. Ducet, N.; Traon, P.Y.L.; Reverdin, G. Global high-resolution mapping of ocean circulation from TOPEX/Poseidon and ERS-1 and -2. *J. Geophys. Res.* **2000**, *105*, 19477–19498. [CrossRef]
41. Jing, Z.Y.; Qi, Y.Q.; Du, Y. Upwelling in the continental shelf of northern South China Sea associated with 1997–1998 El Niño. *J. Geophys. Res* **2011**, *116*, C02033. [CrossRef]
42. Akhir, M.F.; Daryabor, F.; Husain, M.L.; Tangang, F.; Qiao, F.L. Evidence of upwelling along peninsular malaysia during Southwest Monsoon. *Open J. Mar. Sci.* **2015**, *5*, 273–279. [CrossRef]
43. Xie, S.P.; Xie, Q.; Wang, D.X.; Liu, W.T. Summer upwelling in the South China Sea and its role in regional climate variations. *J. Geophys. Res.* **2003**, 108. [CrossRef]
44. Thangaprakash, V.P.; Girishkumar, M.S.; Suprit, K.; Suresh Kumar, N.; Chaudhuri, D.; Dinesh, K.; Kumar, A.; Shivaprasad, S.; Ravichandran, M.; Thomas Farrar, J.; et al. What controls seasonal evolution of sea surface temperature in the Bay of Bengal? *Oceanography* **2016**, *29*, 202–213. [CrossRef]

45. Li, K.P.; Wang, H.Y.; Yang, Y.; Yu, W.D.; Li, L.L. Observed characteristics and mechanisms of temperature inversion in the northern Bay of Bengal. *Acta Oceanol. Sin.* **2016**, *38*, 22–31. [CrossRef]
46. Qiu, Y.; Li, L. Annual variation of geostrophic circulation in the Bay of Bengal observed with TOPEX/Poseidon altimeter data. *Acta Oceanol. Sin.* **2007**, *29*, 39–46.
47. Hu, J.Y.; Kawamura, H.; Hong, H.S.; Pan, W.R. A review of research on the upwelling in the Taiwan strait. *Bull. Mar. Sci.* **2003**, *73*, 605–628.
48. Price, J.F.; Weller, R.A.; Schudlich, R.R. Wind-driven ocean currents and Ekman transport. *Science* **1987**, *238*, 1534–1538. [CrossRef]
49. Han, W.Q.; Webster, P.J. Forcing mechanisms of sea level interannual variability in the Bay of Bengal. *J. Phys. Oceanogr.* **2002**, *32*, 216–239. [CrossRef]
50. Cravatte, S.; Picaut, J.; Eldin, G. Second and first baroclinic Kelvin modes in the equatorial Pacific at intraseasonal timescales. *J. Geophys. Res. Oceans* **2003**, *108*, 3266. [CrossRef]
51. Cheng, X.H.; Xie, S.P.; Mccreary, J.P. Intraseasonal variability of sea surface height in the Bay of Bengal. *J. Geophys. Res. Oceans* **2013**, *188*, 816–830. [CrossRef]
52. Cheng, X.H.; Mccreary, J.P.; Qiu, B. Intraseasonal-to-semiannual variability of sea-surface height in the astern, equatorial Indian Ocean and Southern Bay of Bengal. *J. Geophys. Res. Oceans* **2017**, *122*, 4051–4067. [CrossRef]
53. Rao, R.R.; Girish Kumar, M.S.; Ravichandran, M.; Rao, A.R.; Gopalakrishna, V.V.; Thadathil, P. Interannual variability of Kelvin wave propagation in the wave guides of the equatorial Indian Ocean, the coastal Bay of Bengal and the southeastern Arabian Sea during 1993–2006. *Deep Sea Res. Part I Oceanogr. Res. Pap.* **2010**, *57*, 1–13. [CrossRef]
54. Gent, P.R.; O'Neill, K.; Cane, M.A. A model of the semiannual oscillation in the Equatorial Indian Ocean. *J. Phys. Oceanogr.* **1983**, *13*, 148. [CrossRef]
55. Huang, B.Q.; Xiang, W.G.; Zeng, X.B.; Chiang, K.P.; Tian, H.J.; Hu, J.; Lan, W.L.; Hong, H.S. Phytoplankton growth and microzooplankton grazing in a subtropical coastal upwelling system in the Taiwan Strait. *Cont. Shelf Res.* **2011**, *31*, S48–S56. [CrossRef]
56. Pauly, D.; Christensen, V. Primary production required to sustain global fisheries. *Nature* **1995**, *374*, 255–257. [CrossRef]
57. Saji, N.H.; Goswami, B.N.; Vinayachandran, P.N.; Yamagata, T. A dipole mode in the tropical Indian Ocean. *Nature* **1999**, *401*, 360–363. [CrossRef]
58. Yu, W.D.; Xiang, B.Q.; Liu, L.; Liu, N. Understanding the origins of interannual thermocline variations in the tropical Indian Ocean. *Geophys. Res. Lett.* **2005**, *32*, L24706. [CrossRef]
59. Saji, N.H.; Yamagata, T. Possible impacts of Indian Ocean Dipole mode events on global climate. *Clim. Res.* **2003**, *25*, 151–169. [CrossRef]
60. Anil, N.; Kumar, M.R.R.; Sajeev, R.; Saji, P.K. Role of distinct flavours of IOD events on Indian summer monsoon. *Nat. Hazards* **2016**, *82*, 1317–1326. [CrossRef]
61. Horii, T.; Ueki, I.; Ando, K. Coastal upwelling events along the southern coast of Java during the 2008 positive Indian Ocean Dipole. *J. Oceanogr.* **2008**, *74*, 499–508. [CrossRef]
62. Gan, J.P.; Cheung, A.; Guo, X.G.; Li, L. Intensified upwelling over a widened shelf in the northeastern South China Sea. *J. Geophys. Res.* **2009**, *114*, C09019. [CrossRef]

remote sensing

Communication

Remotely Sensed Seasonal Shoreward Intrusion of the East Australian Current: Implications for Coastal Ocean Dynamics

Senyang Xie [1,*], Zhi Huang [2] and Xiao Hua Wang [1]

1 The Sino-Australian Research Consortium for Coastal Management, School of Science, The University of New South Wales at the Australian Defence Force Academy, Canberra 2600, Australia; x.h.wang@unsw.edu.au
2 National Earth and Marine Observations Branch, Geoscience Australia, Canberra 2600, Australia; zhi.huang@ga.gov.au
* Correspondence: senyang.xie@student.adfa.edu.au

Abstract: For decades, the presence of a seasonal intrusion of the East Australian Current (EAC) has been disputed. In this study, with a Topographic Position Index (TPI)-based image processing technique, we use a 26-year satellite Sea Surface Temperature (SST) dataset to quantitatively map the EAC off northern New South Wales (NSW, Australia, 28–32°S and ~154°E). Our mapping products have enabled direct measurement ("distance" and "area") of the EAC's shoreward intrusion, and the results show that the EAC intrusion exhibits seasonal cycles, moving closer to the coast in austral summer than in winter. The maximum EAC-to-coast distance usually occurs during winter, ranging from 30 to 40 km. In contrast, the minimum distance usually occurs during summer, ranging from 15 to 25 km. Further spatial analyses indicate that the EAC undergoes a seasonal shift upstream of 29°40′S and seasonal widening downstream. This is the first time that the seasonality of the EAC intrusion has been confirmed by long-term remote-sensing observation. The findings provide new insights into seasonal upwelling and shelf circulation previously observed off the NSW coast.

Keywords: satellite remote sensing; quantitative mapping; spatial analysis; the East Australian Current; New South Wales; coastal upwelling; shelf circulation

Citation: Xie, S.; Huang, Z.; Wang, X.H. Remotely Sensed Seasonal Shoreward Intrusion of the East Australian Current: Implications for Coastal Ocean Dynamics. *Remote Sens.* **2021**, *13*, 854. https://doi.org/10.3390/rs13050854

Academic Editors: Chung-Ru Ho, Antony K. Liu and Xiaofeng Li

Received: 19 January 2021
Accepted: 22 February 2021
Published: 25 February 2021

1. Introduction

Originating from the equator, the East Australian Current (EAC) is a highly dynamic western boundary layer of the South Pacific Gyre, and it is characterized by warmer Sea Surface Temperature (SST) off the eastern coast of Australia [1]. Among Western Boundary Currents (WBCs), the EAC is unique, featuring very high spatiotemporal variability [1–6]. Along its main path off the south-east margin of Australia, the EAC frequently encroaches onto the continental shelf [1]. Such shoreward intrusion significantly changes continental shelf-slope biophysical dynamics in the region [7–10]. Usually, the EAC intrusion drives coastal bottom layer uplift or upwelling through Ekman pumping, which in turn changes the shelf-slope temperature and nutrient dynamics [7,8,11,12]. Additionally, the EAC intrusion intensifies surface alongshore flow, generates vertical (surface-bottom) current sheer and eventually induces circulations on the adjacent continental shelf [4]. The ecological ramifications of the EAC's shoreward intrusion are thus wide ranging and far reaching [10,13–17].

The EAC is a dynamic eddy-current system that features periodic intrinsic meander (every 20–45 days) and larger-amplitude fluctuation (every 60–100 days), which is associated with the EAC's eddy shedding [4,5,12,18–24]. Recently, submesoscale frontal eddies were observed being generated every ~7 days on the inshore edge of the EAC, which add additional complexity to this dynamic system [25,26]. A more recent study also demonstrated that the EAC is an eddy-dominant system with very high variability [6]. The latest quantitative mapping study by Xie et al. [1] confirmed the dynamic nature of the EAC system.

The EAC is such an energetic oceanographic feature that its low-frequency (e.g., seasonal) variability is usually overshadowed and hence rarely detected [1,4,18]. For example, Schaeffer et al. [27] inferred from mooring array current data that the EAC's shoreward intrusion occurs all year round without any clear seasonal cycles. Further, Schaeffer et al. [12] suggested that the EAC intrusion features a high-frequency cycle (every 90–100 days), as is inferred from current-driven bottom layer transport. However, both current velocity and bottom layer transport are indeed not a precise representation of the EAC's shoreward intrusion. In a new mapping study, Xie et al. [1] quantified the EAC's shoreward intrusion ("area" and "distance") using Himawari-8 SST data of high temporal resolution. Using the quantitative mapping results, their study confirmed the high-frequency EAC intrusion, as suggested by Schaeffer et al. [12,27]. Their study also indicated the seasonality of the EAC intrusion based on the two seasonal cycles between 2015 and 2017. In addition, indirect evidence such as the seasonality of coastal upwelling and shelf circulation off northern New South Wales (NSW) has hinted at the EAC's seasonal shoreward intrusion, as the regional wind patterns are unlikely the driver [8,28–31].

To date, no long-term and direct observation of the EAC's seasonal shoreward intrusion has been provided. The two-year dataset provided by Xie et al. is rather limited and lacks robustness and statistical power for the investigation of seasonal-scale variability [1]. This limits our ability to fully understand the EAC and the adjacent shelf-slope hydrodynamics. This study attempts to fill the research gap by conducting quantitative mapping of the EAC using a 26-year Advanced Very-High-Resolution Radiometer (AVHRR) SST dataset spanning between 1992 and 2018 (Section 2).

Remotely sensed SST images have enabled the quantitative and semiautomatic mapping of ocean currents such as the Leeuwin Current [32] and the EAC [1] as they have warmer SST signatures [33,34], which can be distinguished from background ocean using a TPI technique [35]. By analyzing the spatiotemporal patterns of the EAC mapping results from the long-term SST dataset, we aim to provide direct evidence and the underlying mechanism of the EAC's seasonal shoreward intrusion (Sections 3 and 4) and demonstrate its impacts on coastal ocean dynamics (Section 4).

2. Materials and Methods

Our study area covers the coastal ocean off northern NSW between 28 and 32.5°S, upstream of the typical EAC separation point at 32–33°S [36,37] (Figure 1a). In this area, the EAC is most significant and continuous [6,36], and its SST signature is therefore most recognizable [1] (Figure 1a). Along the south-east Australian margin, the continental shelf is narrow (20–50 km), and the slope is very steep, with water depth increases dramatically from ~200 m (at the shelf-break) to 2000–4000 m (Figure 1a).

We used monthly Advanced Very-High-Resolution Radiometer (AVHRR) SST images for the period between April 1992 and March 2018 to map the EAC. Available from the Integrated Marine Observing System (IMOS, Australia), this dataset contains Level 3 foundation SST (SSTfnd) products derived from AVHRR observations on all available National Oceanic and Atmospheric Administration (NOAA, USA) polar-orbiting satellites [38]. The SSTfnd is obtained by adding a constant 0.17 °C to the SSTskin measurements following the removal of measurements with low surface wind speeds (<6m/s by day and <2m/s at night). Validation against buoy SSTfnd observations for the central date indicate typical 2014 biases of <0.03 °C and standard deviations of 0.6 °C. The spatial resolution is 2 km, and each grid represents the monthly average of all the highest-quality SST observations [39,40].

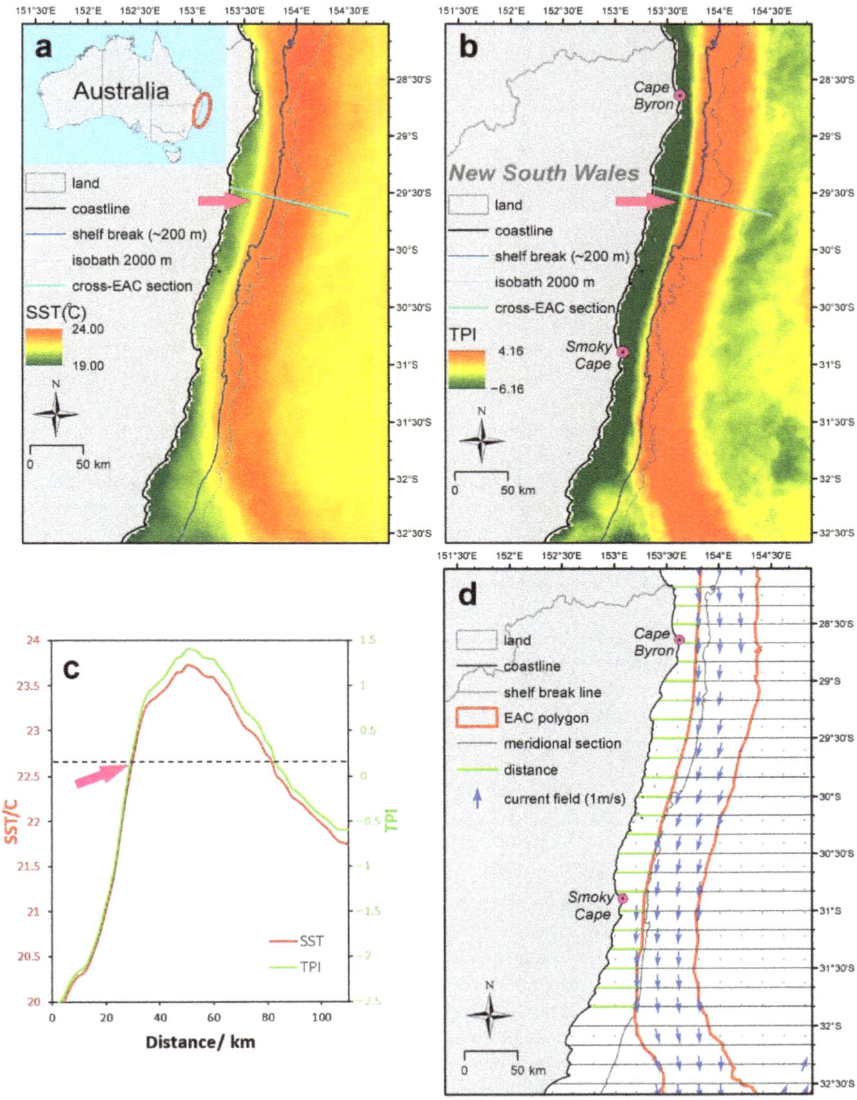

Figure 1. Study area and East Australian Current (EAC) mapping. (**a**) A monthly Advanced Very-High-Resolution Radiometer (AVHRR) Sea Surface Temperature (SST) image of September 2009. (**b**) Topographic Position Index (TPI) calculation of the SST image shown in (a). The arrow (in purple) indicates the boundary between the EAC and shelf waters at 29°30′S. (**c**) SST profile (red curve) and corresponding TPI value (green curve) along the cross-EAC section (green line) depicted in (a,b). The dashed line represents the TPI threshold (0.17). (**d**) Vectorization of the mapped EAC's SST signature (red polygon) and calculation of the "area" and "distance" of the on-shelf EAC waters. The current field for September 2009 is shown in the background (Integrated Marine Observing System (IMOS) Ocean Current data, with a spatial resolution of ~10 km). In (a,c), significantly lower water temperature (in green) in the coastal area indicates an upwelling event.

The Topographic Position Index (TPI) [35] is a local-based image processing algorithm that has been successfully applied to map large-scale ocean currents [1,32] and other oceanographic features such as coastal upwelling [28,41]. In this study, as in [1], we calculated the TPI (Equations (1)–(3)) of each SST image to capture the strong zonal temperature gradient between the EAC and its surrounding waters (Figure 1b,c). Then, a TPI threshold (mean + 0.5 std) was applied to extract the EAC from the background ocean (Figure 1c,d). The derived EAC was then vectorized for subsequent spatial analysis (Figure 1d). For detailed TPI calculation and mapping procedures for the EAC, one can refer to [1].

$$\mathrm{TPI(x,\ y)} = \mathrm{SST(x,\ y)} - \mathrm{M(x,\ y)} \tag{1}$$

$$\mathrm{M(x,\ y)} = \frac{\sum_{j=-n}^{n}\sum_{i=-n}^{n}\mathrm{SST(x-i,\ y-j)}}{N^2} \tag{2}$$

$$n = \frac{N-1}{2} \tag{3}$$

where (x, y) is the position of an image pixel, M is the average SST of a neighborhood centered at (x, y) and N is the pixel number of the neighborhood. We used a square window of 75 × 75 pixels to calculate TPI (N = 75), which is slightly larger than the typical width of the EAC's core flow.

Two indices, area and distance, were utilized to quantify the EAC's shoreward intrusion [1] (Figure 1d). First, the spatial extent of EAC components between the coastline and the shelf-break line was defined as the "area" index (hereafter referred to as "area"). Second, to measure the EAC's proximity to the coast, we drew meridional sections at 1/6 degree intervals. The lengths of the section components between the coastline and the inshore edge of the on-shelf EAC waters (green lines in Figure 1d) were then averaged to obtain the EAC's "distance" index (hereafter referred to as "distance"). After that, we conducted wavelet analyses [42,43] to decompose the time series of these two indices into time-frequency space so that we can determine the dominant frequencies in the time-series signal.

Further, we investigated the possible mechanisms associated with the EAC's shoreward intrusion (i.e., variability of the EAC's width and path). Firstly, as in [1], the EAC's width was calculated by averaging all the lengths of meridional sections bounded by the inner and outer boundaries of the EAC (Figure 1d). Secondly, we produced quantitative maps of the EAC by combining all the monthly EAC maps from 1992 to 2018, statistically showing the EAC's location, frequency, main path and centerline.

3. Results

3.1. The EAC's Shoreward Intrusion: Time Series, Wavelet Analysis and Statistics

Time series (1992–2018) of the "area" and "distance" of the EAC's shoreward intrusion are shown in Figure 2(a1,b1). In general, high-frequency fluctuations were observed throughout the "area" and "distance" time series, indicating that the EAC intrusion can occur all year round. On average, the area index is 3169 ± 1772 km^2, occupying 17.98±10.05% of the continental shelf (Figure 1a). In extreme EAC intrusion events, the EAC was observed occupying more than 40% of the shelf (e.g., 41.46% in January 1997, 43.55% in April 2005 and 46.55% in February 2016, as highlighted (red dots) in Figure 2(a1)). Overall, the distance index is 26.87 ± 4.29 km. During extreme events, the distance is typically reduced to less than 20 km. For example, in the above-mentioned extreme cases, the distances were 18.02, 17.48 and 18.80 km, respectively (Figure 2(b1)).

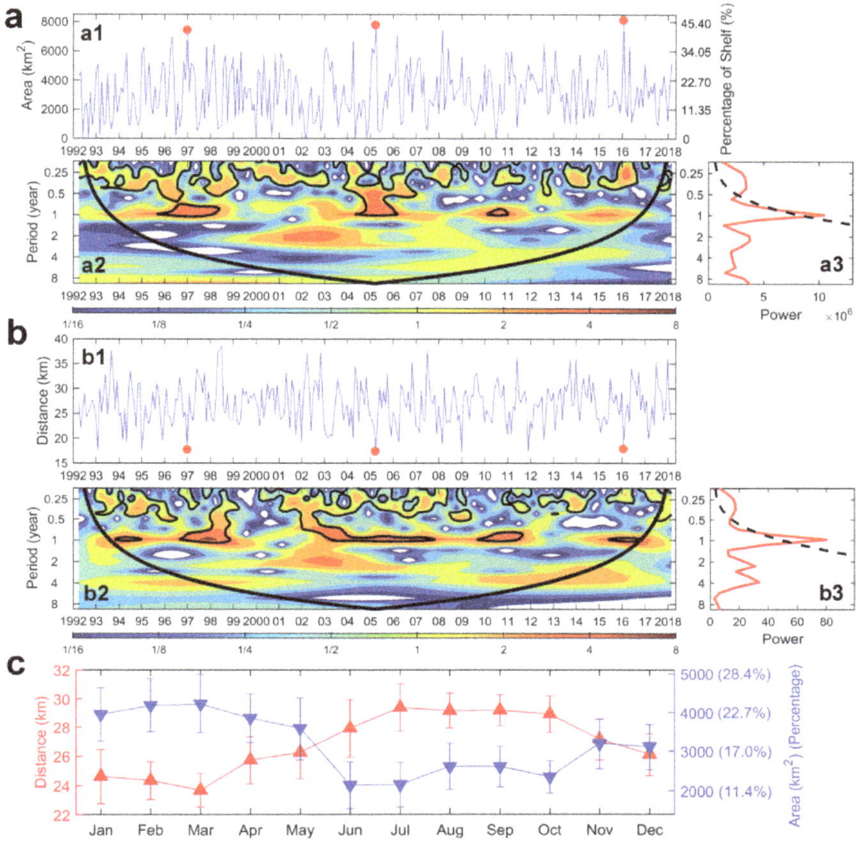

Figure 2. (**a1**) Time series of the EAC's intrusion area (left Y-axis) and the corresponding area percentage of the continental shelf (right Y-axis). Red dots represent three extreme EAC intrusion events mentioned in the text. (**a2**) Local wavelet power spectrum and (**a3**) global wavelet power spectrum (red line). In (**a2**), the color bar indicates wavelet power levels; the bold black curve indicates the "cone of influence" below which edge effects become important. The black contours in (**a2**) and the dashed black line in (**a3**) denote the 95% confidence level using a chi-square (χ^2) test. (**b1–b3**) Corresponding plots for the "distance" index. (**c**) Monthly average of the distance (red) and area (percentage) (blue) of on-shelf EAC waters. The error bars indicate 95% Confidence Intervals (CIs) (n = 26).

The results of the wavelet analyses are shown in Figure 2(a2,a3,b2,b3). Generally, high-frequency variability at the period of around 0.25 years (i.e., ~90 days) was observed throughout the entire time series of both "area" and "distance," as shown by the black contours in the local power spectra (Figure 2(a2,b2)) and the small peaks in the global power spectra (Figure 2(a3,b3)). This is not surprising because the EAC encroachment has been associated with its high-frequency intrinsic oscillation and eddy shedding in a mapping study using six-day composited Himawari-8 SST images [1]. However, in our study, the power of the high-frequency signal is considerably weaker, which is most likely due to the use of the monthly averaging SST data.

This study focuses on the lower-frequency variability of the EAC's shoreward intrusion. Significantly, a clear annual signal was detected from the wavelet spectra (Figure 2b,c). The local spectra (Figure 2(a2,b2)) identify a consistent higher-power band with a period of one year throughout the time series. This one-year periodicity is also clearly identified as the highest peak in the global power spectra (Figure 2(a3,b3)). From 1992 to 2018, this annual signal is continuous and statistically significant (based on chi-square test) except for just some short periods (e.g., 2000–2002 and 2012–2013; Figure 2(a2,b2)). This indeed demonstrates that, while the EAC intrudes shoreward at higher-frequencies (e.g., every 60–100 days) all year round (Figure 2) [1,4,12], the intrusion also exhibits seasonal cycles (detailed below).

The monthly variability of the EAC's shoreward intrusion is shown in Figure 2c. Generally, both "area" and "distance" of the EAC intrusion undergo a clear seasonal cycle, confirming that the EAC is closer to the coast in austral summer than in winter. Specifically, the area (percentage) reaches its maximum during January (3955 $\pm$ 692 km^2; 22.44 $\pm$ 3.93%), February (4186 $\pm$ 688 km^2; 23.75 $\pm$ 3.91%), and March (4224 $\pm$ 750 km^2; 23.96 $\pm$ 4.26%). Correspondingly, this is a period when the distance drops to its minimum, being 24.63 $\pm$ 1.85 km, 24.35 $\pm$ 1.28 km and 23.69 $\pm$ 1.15 km, respectively. In contrast, the area (percentage) is lowest in June (2149 $\pm$ 586 km^2; 12.19 $\pm$ 3.32%) and July (2161 $\pm$ 556 km^2; 12.26 $\pm$ 3.16%) when the maximum distance was observed (29.38 $\pm$ 1.65 km in July). The overall (26 years) monthly mean indicates that the EAC waters could be ~10 km closer to the coast in summer than in winter.

In Figure 3, we compare the maximum and minimum EAC-to-coast distances in winter and summer, respectively, from 1992 to 2017. In general, the maximum EAC-to-coast distance occurs during winter (blue line), ranging from 30 to 40 km. In contrast (red line), the minimum distance usually occurs during summer, ranging from 15 to 25 km. The difference (black line) between the maximum distance in winter and the minimum distance in summer is 10.86 km (mean), with a standard deviation of 3.41 km. As the continental shelf off southeast Australia is narrow (~25 km) [44], such seasonal intrusion of the EAC could exert significant influence on coastal hydrodynamics in this region (discussions in Sections 4.2 and 4.3).

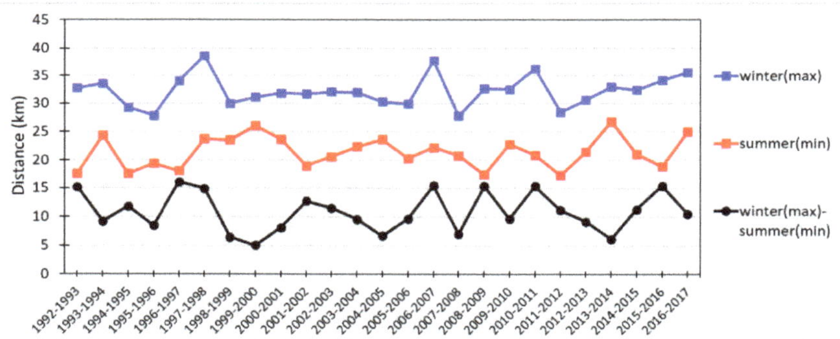

Figure 3. Comparisons of the maximum and minimum EAC-to-coast distances in winter and summer, respectively, from 1992 to 2017. The maximum (minimum) distance in winter (summer) was obtained from the monthly data (Figure 2(b1)) of June, July and August (December, January and February) in each year.

3.2. *Quantitative Maps of the EAC: Location, Frequency, Main Path and Centerline*

Spatially, the EAC's shoreward intrusion is directly associated with changes in the EAC's path and/or width. This is demonstrated in the long-term composite maps of the EAC (Figure 4). Along the north NSW coast (28–32°S) and onshore of the shelf-break, both areal extent and frequency of EAC intrusion are considerably larger in summer than in winter (Figure 4b,c). This coincides with the seasonal shift of the EAC's main path, represented by an area with an EAC frequency > 50%, and its centerline, with the seasonal widening of the EAC (Figure 4d).

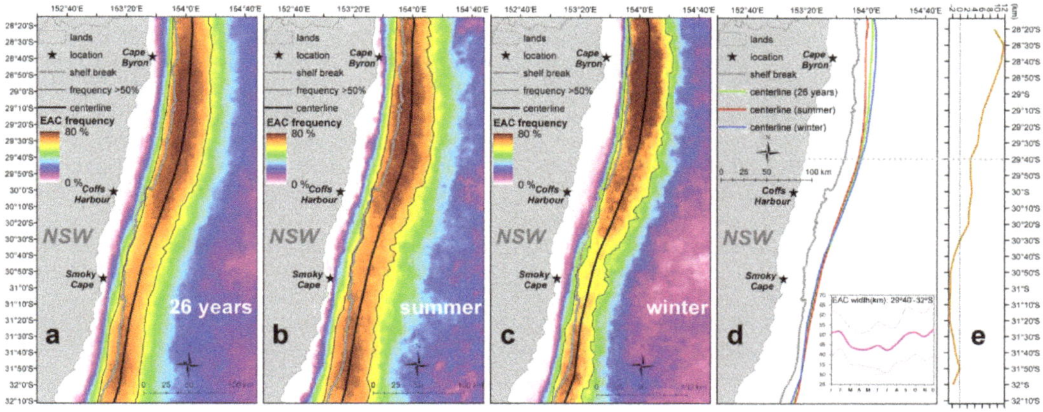

Figure 4. Quantitative maps of the EAC: location, frequency, main path and centerline. (**a**) Location and corresponding frequency of the EAC generated by combining all the monthly maps of EAC from 1992 to 2018. The dual black lines delineate the main path of the EAC with a frequency > 50%. The bold black line is a centerline of the EAC's main path. (**b**) The EAC's main path generated by combining all the monthly EAC maps in summers and (**c**) in winters from 1992 to 2018. (**d**) The centerlines of (a) (26 years), (b) (summer) and (c) (winter). The bottom right panel provides the intra-annual variation of the EAC's width downstream of 29°40′S (mean width (bold pink line) ± standard deviation (dotted pink line)), x-labels "J" to "D" represent January to December. (**e**) Meridional displacement (km) of the centerline from 28°20′S to 32°S (comparison between summer and winter, with the positive value denoting the centerline (summer) being closer to the coast and negative value further offshore). The vertical dotted line denotes zero displacement. The horizontal dotted gray line denotes the latitude of 29°40′S upstream of which the centerline exhibits notable seasonal shifts.

Specifically, downstream of 29°40′S, the three centerlines (summer, winter and 26-year) generally overlap, with insignificant meridional (cross-shelf) displacement of 2.03 ± 0.98 km between summer and winter (Figure 4d,e). However, in this area, the EAC's width exhibits significant seasonality, being broadest (52.85 ± 13.44 km) in December (austral summer) and narrowest (42.44 ± 11.50 km) in July (austral winter) (Figure 4d). This seasonal broadening of the EAC is also clearly shown in Figure 4b,c, where the EAC's main path, downstream of 29°40′S, is notably narrower (~10 km) in winter than in summer.

In contrast, upstream of 29°40′S, the EAC's path (centerline) undergoes considerable seasonal shift (Figure 4d,e). The centerline is centered at ~154°0′E and is on average ~8 km closer to the coast in summer than in winter. At 28°30′S, we observed the maximum shoreward displacement of 11.67 km (Figure 4e). In terms of the EAC's width, we noted that it is similar in summer (44.87 ± 12.80 km) and winter (46.05 ± 12.41 km) in this area.

4. Discussion

4.1. On the Mechanism of EAC's Seasonal Intrusion: Shift or Widening?

For decades, it has been disputed as to whether the EAC exhibits a seasonal shoreward intrusion. Although it is challenging to detect the low-frequency variability of this dynamic eddy-current system, previous research does provide some useful insights. For instance, from a broader synoptic structure of the EAC based on steric height data, Ridgway and Godfrey [3] hinted at the seasonal shift of the EAC's axis (being closer to the coast in summer). However, in a recent study using four years' HF radar observation, Archer et al. [4] demonstrated that the EAC axis does not exhibit any significant seasonal displacement. Instead, Archer et al. [4] suggested that the EAC's seasonal shoreward movement is due to the EAC's widening (5–15 km) during summer. Nevertheless, the finding only reflects the EAC component within a rather limited area between 30 and 31°S. In a recent mapping study, Xie et al. [1] also identified seasonality in the EAC's shoreward intrusion, which coincides with the EAC's seasonal broadening (10–15 km). However, their study period (2015–2017) is too short to statistically demonstrate the EAC's seasonality.

In this study, in the area downstream of ~29°40′S (Figure 4d,e), our results agree with Archer et al. [4], showing that seasonal shift of the EAC's path is insignificant with mean displacement between seasons of ~2 km. In addition, our results indicated that, in summer, the EAC exhibits a considerable widening of ~10 km (Figure 4b–d), a magnitude that is consistent with the above recent studies [1,4]. As our results were derived from the quantitative mapping using a 26-year-long dataset with a large spatial coverage, we are confident that, downstream of ~29°40′S, the seasonal shoreward intrusion of the EAC is due to its seasonal widening. Theoretically, the width of the western boundary layer will increase if the lateral viscosity increases [45]. This could occur as a consequence of an increase in eddy activity in a WBC system [46]. Downstream, the seasonal widening of the EAC is likely due to such seasonal eddy activity, in light of the mounting evidence of increased eddy kinetic energy in this area during summer [4,6,47–49].

Upstream of ~29°40′S (Figure 4d,e), however, our results reveal a different mechanism, indicating that seasonal shift of the EAC's path is the main driver of the EAC's seasonal shoreward intrusion, as previously alluded by Ridgway and Godfrey [3]. We observed an insignificant difference (~1 km) of the EAC width between summer and winter (Section 4.2). In contrast, we observed a considerable shift of the EAC's path (centerline) which is on average ~8 km closer to the coast in summer than in winter (~154°0′E, Figure 4d,e). These results demonstrated that, upstream of ~29°40′S, the EAC's seasonal shoreward intrusion is mainly due to the seasonal shift of the EAC's path. Such seasonal shift (i.e., onshore/offshore current transport across the ***f*/*H*** contours in summer/winter) reflects the seasonal advection of potential vorticity (APV [50]) near the continental shelf at ~154°0′E as demonstrated by Bhatt [51] using BRAN2.1 reanalysis data [52,53]. According to Bhatt [51], the Joint Effect of Baroclinity and Relief (JEBAR [54,55]) contributes significantly to the seasonal APV in this area (~154°0′E), where the role of wind stress curl is significantly weaker.

4.2. Implications for Coastal Upwelling

The EAC is a major driving force of coastal upwelling along the eastern margin of Australia, with the intensity of upwelling being proportional to the EAC's proximity and strength [11,12,29]. Figure 1a–c indicates an upwelling event occurred in the coastal area where both water temperature (color scale: green; 19–20 °C) and TPI value (color scale: green; negative) were significantly lower than that of the sea water further offshore. The EAC can be significantly accelerated (enhanced southward advection) through either cross-shelf encroachment into shallow waters or topographic acceleration where the flow path narrows. This in turn causes an increase in bottom stress and an extension of the bottom boundary layer (BBL) shut-down time. As a result, the prolonged Ekman pumping via the BBL forms coastal upwelling [8,11,29,56]. This mechanism of current-driven upwelling has been confirmed by recent observations from mooring arrays, ocean gliders and HF radars [12,27,55,57].

Due to the EAC's high-frequency shoreward intrusion [1,4,12], the EAC-driven coastal water uplift actually occurs all year round off south-east Australia [12,14,27,28,30]. However, this "all-year-round" upwelling could also exhibit a seasonal cycle, as hinted by previous studies (e.g., [28–30]). For example, Oke and Middleton found a greater occurrence of thermal fronts off the north NSW coast during spring and summer periods [29]. Rossi et al. also found that, between 25 and 32°S, current-driven upwelling typically maximizes in summer or spring [30]. However, they attributed such seasonal upwelling simply to the seasonal cycle of the EAC's southward transport, which is strongest in summer and weakest in winter [3]. More recently, Huang and Wang observed a similar seasonal pattern of upwelling along the north coast of NSW using 14 years of MODIS SST data [28]. They suggest that such seasonality is more likely current-forced than wind forced because of the lack of strong and persistent northerly winds. In fact, the above findings have hinted at the role of the EAC in the seasonal coastal upwelling. In this study, through direct measurement, we were able to provide robust evidence of the EAC's seasonal intrusion into the coastal water. Accordingly, we are confident to suggest that the intensified coastal upwelling during summer is mainly due to the combined influence of increased shoreward proximity and southward transport of the EAC.

4.3. Implications for Shelf Circulation

The EAC's shoreward intrusion also drives circulations on the continental shelf [4,8,27]. On the shelf, the surface current varies linearly with the EAC's shoreward proximity [4]. As there is mounting evidence showing that the EAC's shoreward intrusion features high frequency and a large amplitude [1,12], it is not surprising that the flow pattern on the shelf exhibits very high variability [4].

However, the seasonal shoreward intrusion of the EAC, as identified in this study, could also considerably change the shelf dynamics off northern NSW. In fact, Wood et al. observed significant seasonality in the vertical shear of alongshore flow velocity, with large vertical current shear occurring in summer but very small current sheer in winter [31]. They attributed such seasonal current sheer simply to seasonal changes in temperature gradient across the shelf (i.e., the thermal wind effect). They also excluded the role of winds as the local wind pattern indeed prevents the formation of a steady seasonal shelf circulation. Although the thermal wind theory has provided a convincing argument for the relatively weak current shear in winter, the strong current sheer in summer was only partially explained by the theory [31]. The EAC's seasonal shoreward intrusion, as demonstrated in our study, therefore provides a deeper understanding of the seasonal current sheer on the shelf. In summer, as the EAC moves further shoreward, surface along-stream flow is significantly intensified through the EAC's downstream advection [4]. The enhanced surface flow in turn contributes to a stronger vertical current sheer (circulation) on the shelf. Accordingly, we suggest the seasonal shelf circulation indeed reflects a joint effect of increased EAC shoreward intrusion and increased cross-shelf temperature gradient (thermal wind effect) in summer [31].

4.4. Impacts of Climate Processes on the EAC Encroachment?

Climate processes such as El Niño/Southern Oscillation (ENSO) could play a role in the interannual to decadal variability of the EAC's current transport and east coast sea level [58]. However, whether climate processes have an impact on the EAC's spatiotemporal variability and the extent of the impact remains unresolved. From the time-series EAC encroachment between 1992 and 2018 (Figure 2), we noted that the EAC's shoreward intrusion exhibits some interannual variability. However, we are not able to relate the variability to external climate signals such as El Niño and La Niña. In fact, the main pathway of the ENSO's influence occurs through the Indonesian seas and around a waveguide around the western and southern margins of Australia [59]. As such, a previous study has shown that the EAC transport undergoes variations on interannual timescales, but the ENSO's signal is very weak in the observations [33]. However, excluding the influences of the climate on the EAC intrusion requires more careful examinations. Indeed, detecting climate signals from the EAC's spatial variations is challenging, because the climate signals could be overshadowed by the EAC's high-frequency intrinsic fluctuations [1,4,18]. We propose that long-term, sustained and systematic in situ and remotely sensed observations with a high temporal resolution are crucial for future investigations over the influence of climate processes on the dynamics and spatiotemporal variability of the EAC system.

5. Conclusions

For decades, identifying the low-frequency variability of the EAC was hindered by its high-frequency intrinsic fluctuations. This study provides direct measurement ("area" and "distance") of the EAC's shoreward intrusion using EAC maps generated from monthly AVHRR SST images and with a robust TPI-based mapping technique. Subsequent spatial and temporal analyses on 26 years of intrusion measurements and EAC maps show that the EAC's shoreward intrusion undergoes a consistent and significant seasonal cycle, which is associated with the EAC's seasonal shift upstream and seasonal widening downstream. To our knowledge, this is the first time that the seasonality of the EAC's shoreward intrusion has been quantified and analyzed. Importantly, our results have provided new insights into the seasonal upwelling and shelf circulation previously observed in the study area. We suggest that the EAC is the main driver of the seasonal ocean dynamics off northern NSW.

In summary, the key findings of this study are as follows:

- The EAC undertakes a seasonal shoreward intrusion of ~8 km upstream of 29°40′S′;
- The EAC undertakes a seasonal widening of ~10 km downstream of 29°40′S;
- The minimum EAC-to-coast distance usually occurs during summer, ranging from 15 to 25 km; and
- The maximum EAC-to-coast distance occurs during winter, ranging from 30 to 40 km.

Author Contributions: Conceptualization, S.X. and Z.H.; methodology, S.X. and Z.H.; writing—original draft preparation, S.X.; writing—review and editing, S.X., Z.H. and X.H.W. All authors have read and agreed to the published version of the manuscript.

Funding: This research received no external funding.

Data Availability Statement: The SST dataset was downloaded from the Australian Ocean Data Network (AODN) (https://portal.aodn.org.au/ (accessed on 1 July 2019)), which is a joint effort from the Bureau of Meteorology and the Integrated Marine Observing System (IMOS) (http://imos.org.au/ (accessed on 1 July 2019)). IMOS is enabled by the National Collaborative Research Infrastructure Strategy (NCRIS). It is operated by a consortium of institutions as an unincorporated joint venture, with the University of Tasmania as Lead Agent. MATLAB scripts for wavelet analysis were provided by C. Torrence and G. Compo (http://atoc.colorado.edu/research/wavelets/ (accessed on 10 September 2019)). Produced and analyzed in this work, a 26-year-long dataset containing direct measurement of the EAC's shoreward intrusion is freely available in Mendeley Data (http://dx.doi.org/10.17632/j3yb23z3m9.1 (accessed on 25 September 2019)).

Acknowledgments: This is publication No. 90 of the Sino-Australian Research Consortium for Coastal Management (previously the Sino-Australian Research Centre for Coastal Management). We would like to thank Scott Nichol of Geoscience Australia for his valuable comments. The contribution from Zhi Huang is published with the permission of the CEO of Geoscience Australia. Senyang Xie is supported by a UNSW Tuition Fee Scholarship and the China Scholarship Council for his Ph.D. program. We would like to thank three anonymous reviewers for their constructive comments which have improved the manuscript.

Conflicts of Interest: The authors declare no conflict of interest.

References

1. Xie, S.; Huang, Z.; Wang, X.H.; Leplastrier, A. Quantitative Mapping of the East Australian Current Encroachment Using Time Series Himawari-8 Sea Surface Temperature Data. *J. Geophys. Res. Ocean.* **2020**, *125*, e2019JC015647. [CrossRef]
2. Ridgway, K.R.; Godfrey, J.S. Mass and heat budgets in the East Australian Current: A direct approach. *J. Geophys. Res.* **1994**, *99*, 3231–3248. [CrossRef]
3. Ridgway, K.R.; Godfrey, J.S. Seasonal cycle of the East Australian Current. *J. Geophys. Res. C Ocean.* **1997**, *102*, 22921–22936. [CrossRef]
4. Archer, M.R.; Roughan, M.; Keating, S.R.; Schaeffer, A. On the Variability of the East Australian Current: Jet Structure, Meandering, and Influence on Shelf Circulation. *J. Geophys. Res. Ocean.* **2017**, *122*, 8464–8481. [CrossRef]
5. Archer, M.R.; Keating, S.R.; Roughan, M.; Johns, W.E.; Lumpkin, R.; Beron-Vera, F.J.; Shay, L.K. The Kinematic Similarity of Two Western Boundary Currents Revealed by Sustained High-Resolution Observations. *Geophys Res Lett.* **2018**, *45*, 6176–6185. [CrossRef]
6. Oke, P.R.; Roughan, M.; Cetina-Heredia, P.; Pilo, G.S.; Ridgway, K.R.; Rykova, T.; Archer, M.R.; Coleman, R.C.; Kerry, C.G.; Rocha, C.; et al. Revisiting the circulation of the East Australian Current: Its path, separation, and eddy field. *Prog. Oceanogr.* **2019**, *176*, 102139. [CrossRef]
7. Oke, P.R.; Middleton, J.H. Nutrient enrichment off Port Stephens: The role of the East Australian Current. *Cont. Shelf Res.* **2001**, *21*, 587–606. [CrossRef]

8. Roughan, M.; Middleton, J.H. On the East Australian Current: Variability, encroachment, and upwelling. *J. Geophys. Res. C Ocean.* **2004**, *109*, C7. [CrossRef]
9. Everett, J.D.; Baird, M.E.; Roughan, M.; Suthers, I.M.; Doblin, M.A. Relative impact of seasonal and oceanographic drivers on surface chlorophyll a along a Western Boundary Current. *Prog. Oceanogr.* **2014**, *120*, 340–351. [CrossRef]
10. Brieva, D.; Ribbe, J.; Lemckert, C. Is the East Australian Current causing a marine ecological hot-spot and an important fisheries near Fraser Island, Australia? *Estuar. Coast. Shelf Sci.* **2015**, *153*, 121–134. [CrossRef]
11. Roughan, M.; Middleton, J.H. A comparison of observed upwelling mechanisms off the east coast of Australia. *Cont. Shelf Res.* **2002**, *22*, 2551–2572. [CrossRef]
12. Schaeffer, A.; Roughan, M.; Wood, J.E. Observed bottom boundary layer transport and uplift on the continental shelf adjacent to a western boundary current. *J. Geophys. Res. Ocean.* **2014**, *119*, 4922–4939. [CrossRef]
13. Condie, S.A.; Mansbridge, J.V.; Cahill, M.L. Contrasting local retention and cross-shore transports of the East Australian Current and the Leeuwin Current and their relative influences on the life histories of small pelagic fishes. *Deep. Res. Part II Top. Stud. Oceanogr.* **2011**, *58*, 606–615. [CrossRef]
14. Malcolm, H.A.; Davies, P.L.; Jordan, A.; Smith, S.D. Variation in sea temperature and the East Australian Current in the Solitary Islands region between 2001–2008. *Deep. Res. Part II Top. Stud. Oceanogr.* **2011**, *58*, 616–627. [CrossRef]
15. Roughan, M.; Macdonald, H.S.; Baird, M.E.; Glasby, T.M. Modelling coastal connectivity in a Western Boundary Current: Seasonal and inter-annual variability. *Deep. Res. Part II Top. Stud. Oceanogr.* **2011**, *58*, 628–644. [CrossRef]
16. Young, J.W.; Hobday, A.J.; Campbell, R.A.; Kloser, R.J.; Bonham, P.I.; Clementson, L.A.; Lansdell, M.J. The biological oceanography of the East Australian Current and surrounding waters in relation to tuna and billfish catches off eastern Australia. *Deep. Res. Part II Top. Stud. Oceanogr.* **2011**, *58*, 720–733. [CrossRef]
17. Schaeffer, A.; Roughan, M.; Jones, E.M.; White, D. Physical and biogeochemical spatial scales of variability in the East Australian Current separation from shelf glider measurements. *Biogeosciences* **2016**, *13*, 1967–1975. [CrossRef]
18. Bull, C.Y.S.; Kiss, A.E.; Jourdain, N.C.; England, M.H.; van Sebille, E. Wind Forced Variability in Eddy Formation, Eddy Shedding, and the Separation of the East Australian Current. *J. Geophys. Res. Ocean.* **2017**, *122*, 9980–9998. [CrossRef]
19. Bowen, M.M.; Wilkin, J.L.; Emery, W.J. Variability and forcing of the East Australian Current. *J. Geophys. Res. Ocean.* **2005**, *110*, C3. [CrossRef]
20. Mata, M.M.; Tomczak, M.; Wijffels, S.; Church, J.A. East Australian Current volume transports at 30°S: Estimates from the World Ocean Circulation Experiment hydrographic sections PR11/P6 and the PCM3 current meter array. *J. Geophys. Res. Ocean.* **2000**, *105*, 28509–28526. [CrossRef]
21. Mata, M.M.; Wijffels, S.E.; Church, J.A.; Tomczak, M. Eddy shedding and energy conversions in the East Australian Current. *J. Geophys. Res. Ocean.* **2006**, *111*, C9. [CrossRef]
22. Ridgway, K.R.; Coleman, R.C.; Bailey, R.J.; Sutton, P. Decadal variability of East Australian Current transport inferred from repeated high-density XBT transects, a CTD survey and satellite altimetry. *J. Geophys. Res. Ocean.* **2008**, *113*, C8. [CrossRef]
23. Schaeffer, A.; Archer, M.R.; Baumard, Q.; Roughan, M.; Kerry, C. An assessment of the East Australian Current as a renewable energy resource. *J. Mar. Syst.* **2020**, *204*, 103285. [CrossRef]
24. Wilkin, J.L.; Zhang, W.G. Modes of mesoscale sea surface height and temperature variability in the East Australian current. *J. Geophys. Res. Ocean.* **2007**, *112*, C1. [CrossRef]
25. Roughan, M.; Keating, S.R.; Schaeffer, A.; Cetina Heredia, P.; Rocha, C.; Griffin, D.; Robertson, R.; Suthers, I.M. A tale of two eddies: The biophysical characteristics of two contrasting cyclonic eddies in the East Australian Current System. *J. Geophys. Res. Ocean.* **2017**, *122*, 2494–2518. [CrossRef]
26. Schaeffer, A.; Gramoulle, A.; Roughan, M.; Mantovanelli, A. Characterizing frontal eddies along the East Australian Current from HF radar observations. *J. Geophys. Res. Ocean.* **2017**, *122*, 3964–3980. [CrossRef]
27. Schaeffer, A.; Roughan, M.; Morris, B.D. Cross-Shelf Dynamics in a Western Boundary Current Regime: Implications for Upwelling. *J. Phys. Oceanogr.* **2013**, *43*, 1042–1059. [CrossRef]
28. Huang, Z.; Wang, X.H. Mapping the spatial and temporal variability of the upwelling systems of the Australian south-eastern coast using 14-year of MODIS data. *Remote Sens. Environ.* **2019**, *227*, 90–109. [CrossRef]
29. Oke, P.R.; Middleton, J.H. Topographically induced upwelling off Eastern Australia. *J. Phys. Oceanogr.* **2000**, *30*, 512–531. [CrossRef]
30. Rossi, V.; Schaeffer, A.; Wood, J.; Galibert, G.; Morris, B.; Sudre, J.; Waite, A.M. Seasonality of sporadic physical processes driving temperature and nutrient high-frequency variability in the coastal ocean off southeast Australia. *J. Geophys. Res. Ocean.* **2014**, *119*, 445–460. [CrossRef]
31. Wood, J.E.; Schaeffer, A.; Roughan, M.; Tate, P.M. Seasonal variability in the continental shelf waters off southeastern Australia: Fact or fiction? *Cont. Shelf Res.* **2016**, *112*, 92–103. [CrossRef]
32. Huang, Z.; Feng, M. Remotely sensed spatial and temporal variability of the Leeuwin Current using MODIS data. *Remote Sens. Environ.* **2015**, *166*, 214–232. [CrossRef]
33. Ridgway, K.R. Long-term trend and decadal variability of the southward penetration of the East Australian Current. *Geophys. Res. Lett.* **2007**, *34*, 1–5. [CrossRef]
34. Ridgway, K.R. Seasonal circulation around Tasmania: An interface between eastern and western boundary dynamics. *J. Geophys. Res. Ocean.* **2007**, *112*, C10. [CrossRef]

35. Weiss, A. Topographic position and landforms analysis. In Proceedings of the ESRI User Conference, San Diego, CA, USA, 9–13 June 2001.
36. Ridgway, K.R.; Dunn, J.R. Mesoscale structure of the mean East Australian Current System and its relationship with topography. *Prog. Oceanogr.* **2003**, *56*, 189–222. [CrossRef]
37. Cetina-Heredia, P.; Roughan, M.; Van Sebille, E.; Coleman, M.A. Long-term trends in the East Australian Current separation latitude and eddy driven transport. *J. Geophys. Res. Ocean.* **2014**, *119*, 4351–4366. [CrossRef]
38. Walton, C.C.; Pichel, W.G.; Sapper, J.F.; May, D.A. The development and operational application of nonlinear algorithms for the measurement of sea surface temperatures with the NOAA polar-orbiting environmental satellites. *J. Geophys. Res. Ocean.* **1998**, *103*, 27999–28012. [CrossRef]
39. Beggs, H.; Majewski, L.; Griffin, C.; Verein, R.; Sakov, P.; Huang, X.; Garde, L.; Tingwell, C. Report to GHRSST13 from Australia—Bluelink and IMOS. In Proceedings of the GHRSST XIII Science Team Meeting, Tokyo, Japan, 8 June 2012.
40. Beggs, H.; Zhong, A.; Warren, G.; Alves, O.; Brassington, G.; Pugh, T. RAMSSA-An operational, high-resolution, Regional Australian Multi-Sensor Sea surface temperature Analysis over the Australian region. *Aust. Meteorol. Oceanogr. J.* **2011**, *61*, 1. [CrossRef]
41. Huang, Z.; Hu, J.; Shi, W. Mapping the coastal upwelling east of taiwan using geostationary satellite data. *Remote Sens.* **2021**, *13*, 170. [CrossRef]
42. Torrence, C.; Compo, G.P. A Practical Guide to Wavelet Analysis. *Bull. Am. Meteorol. Soc.* **1998**, *79*, 61–78. [CrossRef]
43. Liu, Y.; San Liang, X.; Weisberg, R.H. Rectification of the bias in the wavelet power spectrum. *J. Atmos. Ocean. Technol.* **2007**, *24*, 2093–2102. [CrossRef]
44. Harris, P.; Heap, A.D.; Passlow, V.; Sbaffi, L.; Fellows, M.; Porter-Smith, R.; Buchanan, C.; Daniell, J. *Geomorphic Features of the Continental Margin of Australia*; Geoscience Australia: Canberra, Australia, 2005.
45. Munk, W.H. On the wind-driven ocean circulation. *J. Meteorol.* **1950**, *7*, 80–93. [CrossRef]
46. Beal, L.M.; Elipot, S. Broadening not strengthening of the Agulhas Current since the early 1990s. *Nature* **2016**, *540*, 570–573. [CrossRef] [PubMed]
47. Qiu, B.; Chen, S. Seasonal modulations in the Eddy Field of the South Pacific Ocean. *J. Phys. Oceanogr.* **2004**, *34*, 1515–1527. [CrossRef]
48. Everett, J.D.; Baird, M.E.; Oke, P.R.; Suthers, I.M. An avenue of eddies: Quantifying the biophysical properties of mesoscale eddies in the Tasman Sea. *Geophys. Res. Lett.* **2012**, *39*, 1–5. [CrossRef]
49. Kerry, C.; Roughan, M. Downstream Evolution of the East Australian Current System: Mean Flow, Seasonal, and Intra-annual Variability. *J. Geophys. Res. Ocean.* **2020**, *125*, e2019JC015227. [CrossRef]
50. Guo, X.; Hukuda, H.; Miyazawa, Y.; Yamagata, T. A triply nested ocean model for simulating the Kuroshio-Roles of horizontal resolution on JEBAR. *J. Phys. Oceanogr.* **2003**, *33*, 146–169. [CrossRef]
51. Bhatt, V. Modelling Dynamics of the East Australian Current and the Subtropical Mode Water Off East Coast of Australia. Ph.D. Thesis, The University of New South Wales, Canberra, Australia, 2010.
52. Oke, P.R.; Schiller, A.; Griffin, D.A.; Brassington, G.B. Ensemble data assimilation for an eddy-resolving ocean model of the Australian region. *Q. J. R. Meteorol. Soc.* **2005**, *131*, 3301–3311. [CrossRef]
53. Oke, P.R.; Brassington, G.B.; Griffin, D.A.; Schiller, A. The Bluelink ocean data assimilation system (BODAS). *Ocean Model.* **2008**, *21*, 46–70. [CrossRef]
54. Sarkisyan, A.S.; Ivanov, V.F. Joint Effect of Baroclinicity and Bottom Relief As an Important Factor in the Dynamics of Sea Currents. *Izv. Akad. Nauk SSSR, Fiz. Atmos. Okeana* **1971**, *7*, 173–188.
55. Mertz, G.; Wright, D.G. Interpretations of the JEBAR Term. *J. Phys. Oceanogr.* **1992**, *22*, 301–305. [CrossRef]
56. Roughan, M.; Oke, P.R.; Middleton, J.H. A modeling study of the climatological current field and the trajectories of upwelled particles in the East Australian Current. *J. Phys. Oceanogr.* **2003**, *33*, 2551–2564. [CrossRef]
57. Schaeffer, A.; Roughan, M. Influence of a western boundary current on shelf dynamics and upwelling from repeat glider deployments. *Geophys. Res. Lett.* **2015**, *42*, 121–128. [CrossRef]
58. Holbrook, N.J.; Goodwin, I.D.; McGregor, S.; Molina, E.; Power, S.B. ENSO to multi-decadal time scale changes in East Australian Current transports and Fort Denison sea level: Oceanic Rossby waves as the connecting mechanism. *Deep. Res. Part II Top. Stud. Oceanogr.* **2011**, *58*, 547–558. [CrossRef]
59. Wijffels, S.; Meyers, G. An intersection of oceanic waveguides: Variability in the Indonesian throughflow region. *J. Phys. Oceanogr.* **2004**, *34*, 1232–1253. [CrossRef]

remote sensing

MDPI

Communication

Injection of High Chlorophyll-a Waters by a Branch of Kuroshio Current into the Nutrient-Poor North Pacific Subtropical Gyre

Chun-Hoe Chow [1,*], Yi-Chen Lin [1], Wee Cheah [2] and Jen-Hua Tai [3]

1 Department of Marine Environmental Informatics, National Taiwan Ocean University, Keelung 202301, Taiwan; 00781056@email.ntou.edu.tw
2 Institute of Ocean and Earth Sciences, Universiti Malaya, Kuala Lumpur 50603, Malaysia; wee.cheah@um.edu.my
3 Research Center for Environmental Changes, Academia Sinica, Taipei 115, Taiwan; jhtai@gate.sinica.edu.tw
* Correspondence: kilmerchow@email.ntou.edu.tw

Abstract: An unusual eastward flow was observed branching out from the Kuroshio Current near the island of Taiwan in the western North Pacific in during the period June–July 2010. The branch meandered eastward approximately 21°N, carrying high chlorophyll-a (Chla) waters for over 1000 km from 125°E into the nutrient-poor North Pacific subtropical gyre (NPSG). The branch was warmer and fresher than the surrounding waters, with temperature–salinity properties resembling those of Kuroshio Current. Thus, we called it the eastward cross-shore Kuroshio branch (ECKB). Injecting fresher waters far into the central NPSG, the ECKB flowed at a mean surface speed of 0.5 m per second, as shown in satellite altimeters, a Lagrangian drifter, and the Japan-Meteorological-Agency (JMA) 137°E-meridian cruise transect. The mechanism of the ECKB was linked to a surface cyclonic wind anomaly to the north at approximately 22–24°N. The cyclonic wind anomaly cooled the ocean surface beneath it via Ekman suction and then enhanced the subtropical front to its south at approximately 21°N near the Kuroshio Current. The strengthened subtropical front subsequently induced an eastward flow that bifurcated from the main stream of the northward-flowing Kuroshio Current.

Keywords: Kuroshio branch; salinity; chlorophyll-a; North Pacific subtropical gyre; satellite observation; in situ observation

Citation: Chow, C.-H.; Lin, Y.-C.; Cheah, W.; Tai, J.-H. Injection of High Chlorophyll-a Waters by a Branch of Kuroshio Current into the Nutrient-Poor North Pacific Subtropical Gyre. *Remote Sens.* **2022**, *14*, 1531. https://doi.org/10.3390/rs14071531

Academic Editor: Emmanuel Devred

Received: 18 February 2022
Accepted: 20 March 2022
Published: 22 March 2022

1. Introduction

In the subtropical North Pacific, surface ocean circulation is dominated by a wind-driven, clockwise-circulating North Pacific Subtropical Gyre (NPSG). At the western boundary of the NPSG, a fast-flowing western boundary current called the Kuroshio Current (KC) [1] connects the North Pacific Current in the north and the North Equatorial Current (NEC) in the south near the equator (Figure 1a). The KC, meaning "black stream" in Japanese, obtained its name due to the black color or precisely the deep blue color of its waters. The deep blue color of the KC is mostly due to a lack of phytoplankton as a result of nutrient depletion especially in waters off the east of Taiwan [2].

Against the westward-flowing NEC, a slow surface countercurrent flows eastward at speeds of approximately 2–10 cm s^{-1} [3–5]. This slow current is the North Pacific Subtropical Countercurrent (STCC) (Figure 1a), which was first reported by Uda and Hasunuma [6] from in situ observations. The STCC is strong in late winter to spring (March–June) with a peak in June, however, it is weak in fall [7]. The STCC is much slower than the KC, and its footprint cannot be clearly observed in the climatology geostrophic flow in July (Figure 1b). Between the main stream of the northward-flowing KC and the eastward-flowing STCC, the Kuroshio recirculation (KR; Figure 1a) generally occurs east of the Taiwan Island during the interaction between the KC and the mesoscale eddies that propagate from the east [4,8–12]. Based on computer simulations, Chu et al. [4] concluded that the STCC originates from the east of the Luzon Strait at approximately 122.5°E on the

23.5-sigma isopycnal surface. The STCC generally emerges from the southeastward turning of the KR, which centers at approximately 124°E and 24°N [4].

The mechanism for the changes of the STCC has been linked to changes in the surface [5,13] and subsurface (below the mixed layer) [7] subtropical fronts. Over long-term time scales (mean state and a decadal time scale from 1965 to 2008), the STCC was anchored and maintained by the subtropical mode water (STMW) [14–17], whereas for shorter timescales (seasonal and interannual), the STCC strength was suggested to be determined by the subtropical fronts, which were influenced by surface wind forcing rather than the surface thermal condition [5,7,18]. In addition, a recent study by Zhang and Xue [19] indicated that the width of the Luzon Strait might also play a key role in the formation of the STCC.

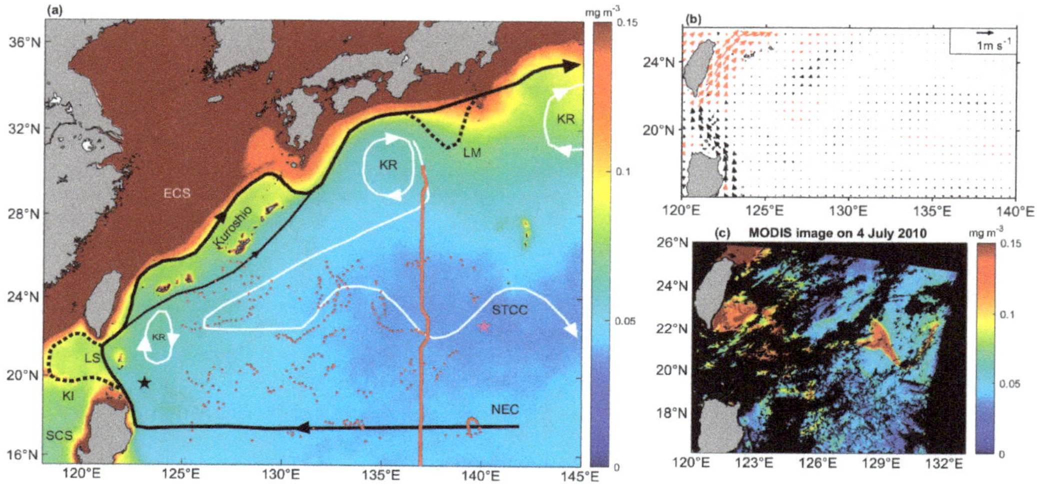

Figure 1. (**a**) Schematic of the Kuroshio Current (KC) circulation system [1,8,20] superimposed on the climatology 1/24°-gridded chlorophyll-a (Chl-a; color shaded) in July from 1998 to 2020. The black bold solid line with arrows shows the flow of the KC and its connection with the North Equatorial Current (NEC). The white circles show the clockwise Kuroshio recirculation (KR) near Japan [1] and Taiwan [8]. The white wave-pattern curve near 22°N indicates the eastward-flowing Subtropical Countercurrent (STCC) [6]. The black dashed curves show the large meandering (LM) of KC south of Japan [1] and the loop of Kuroshio intrusion (KI) via the Luzon Strait (LS) into the South China Sea (SCS) [21]. The meridional red bold line denotes the 137°E transect. The red dots show the locations of the Argo profiles used in this study. Black and magenta stars show the locations of background water masses near the KC and the STCC, respectively; (**b**) climatology of 1/4°-gridded geostrophic flow in July from 1998 to 2020 with red arrows indicating positive zonal velocities and black arrows indicating negative zonal velocities; (**c**) Aqua-MODIS Level 2 Chl-a image at 1 km spatial resolution taken on 4 July 2010.

Near the Luzon Strait, the KC typically bifurcates with one branch to the west (Figure 1a), intruding into the South China Sea (SCS) [21] during persistent northeasterlies in boreal winter. With the KC intruding into the SCS, the KR can be induced east of the Luzon Strait (Figure 1a) centering at approximately 123°E–124°E and 21°N–22°N [7], which is a region with many active anticyclonic eddies [22]. In contrast, without the Kuroshio intrusion (KI) in summer, the KR can be caused by the impinging of an anticyclonic eddy with the KC eastern flank [8].

In this study, we reported a filament of high chlorophyll-a (Chl-a) (>0.07 mg m^{-3}) which occurred in the active area of KR [8] and anticyclonic eddies [22] in July 2010, which can be observed via the Level 2 satellite image taken from the Moderate Resolution Imaging

Spectroradiometer (MODIS) sensor onboard the Aqua satellite (Figure 1b) and in daily images of the merged satellite ocean color product (Figure 2). Based on a suite of satellite and in situ observations, we found that the high Chl-a filament delineated the footprint of an unusual eastward flow (compared to the July climatology field in Figure 1b) branching out from the KC. This study designated the unusual eastward flow as the eastward cross-shore Kuroshio branch (ECKB) which could be observed in other years. In this study, we solely focused on the potential mechanisms that contribute to the 2010 ECKB and its impact on the oligotrophic NPSG.

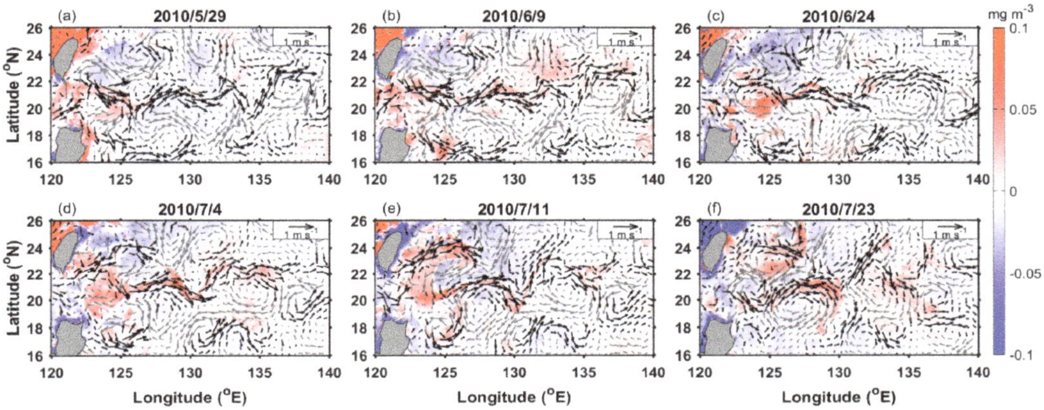

Figure 2. The filament evolution shown as daily 1/24°-gridded chlorophyll-a (Chl-a) anomalies (color shaded) for the selected days (**a**) 29 May, (**b**) 7 June, (**c**) 24 June, (**d**) 4 July, (**e**) 11 July and (**f**) 23 July 2010, in references to the 23 years (1998–2020) daily Chl-a climatological mean. The arrows show the geostrophic velocity anomalies (black for positive zonal velocities and gray for negative zonal velocities) based on the altimetry 1/4°-gridded product. The high Chl-a anomalies located along 21°N within 125°E and 135°E show the footprint of the eastward cross-shore Kuroshio branch (ECKB).

2. Data and Methods

As cloud-cover is a major issue in the study region, especially in summer (June and July), we used the daily Level 4 or "cloud-free" (interpolated) Copernicus-GlobColour Chl-a data, available since 1997, at 1/24° spatial resolution (https://doi.org/10.48670/moi-00100, accessed on 12 August 2020). The Chl-a data were produced by merging Chl-a from multiple ocean color sensors namely the Sea-Viewing Wide Field-of-View Sensor (SeaWiFS), MODIS, Medium Resolution Imaging Spectrometer (MERIS), Visible Infrared Imaging Radiometer Suite (VIIRS), and Ocean and Land Colour Instrument (OLCI). The "cloud-free" Chl-a data were calculated based on a combination of algorithms for different water types (oligotrophic, Chl-a dominated, and coastal turbid) and an optimal interpolation using the kriging method with regional anisotropic covariance models [23]. To ensure that the Level 4 Copernicus-GlobColour Chl-a was capturing real ocean color signals and not artifacts from interpolation, we cross-checked the Level 4 data with Level 2 Chl-a data (as shown in Figure 1c) obtained from NASA's Oceancolor Web (https://doi.org/10.5067/AQUA/MODIS/L2/OC/2018, accessed on 3 February 2022).

We used the altimeter ADT data at 1/4° spatial resolution from the Global Ocean Gridded Level 4 Sea Surface Heights and Derived Variables Reprocessed product (https://doi.org/10.48670/moi-00148, accessed on 9 July 2019), which provides the absolute geostrophic velocities and geostrophic velocity anomalies. The SST data were obtained from the OSTIA [24] global SST reprocessed product (https://doi.org/10.48670/moi-00168, accessed on 18 December 2021) that provides gap-free maps of foundation SST and ice concentration at 0.05° spatial resolution, based on satellite and in situ data. For winds, we used the monthly Cross-Calibrated Multi-Platform (CCMP) gridded 10 m winds available from July 1987 to April 2019, based on satellite, moored buoy, and model data, made available at 1/4° spatial resolution by the Remote Sensing Systems (RSS) at www.remss.com (accessed on 1 October 2020).

For in situ observations, we used data obtained from a Lagrangian surface drifter, Argo floats, and scientific cruises. These in situ observations were selected based on their locations that were along/near to the Chl-a filament from May to July 2010. The Argo ocean vertical profiles and the drifter-trajectory data were collected and made available by the Coriolis project and programs that contribute to the French operational oceanography program for in situ observations (http://www.coriolis.eu.org, accessed on 21 May 2020). Drifter and Argo floats were identified with their World Meteorological Organisation (WMO) numbers. The particular drifter and Argo floats were called with their WMO number hereafter. We only used those data of temperature and salinity with good-flagged quality from 361 Argo profiles (their locations are shown in Figure 1a). The surface current data of the drifter were used from May to June 2010 (only available until 24 June 2010), to compare with the satellite observations.

Additional in situ temperature and salinity profiles were obtained from the CTD casts measured by the Japan Meteorological Agency (JMA) along the 137°E meridional cruise transect (hereafter referred to as the 137°E transect), which was across the Chl-a filament during the period 14–16 July 2010. The JMA repeatedly carried out oceanographic and marine meteorological observations through research vessels in the western North Pacific. The data can be obtained from the web pages at https://www.data.jma.go.jp/gmd/kaiyou/db/vessel_obs/data-report/html/index_e.html (accessed on 25 November 2016) [25].

We used historical ocean profiles along and around the ECKB, obtained via Argo floats and the 137°E transect (Figure 1a), to observe the relative changes in water-mass properties. The observed ocean profiles were compared to the climatological field in the region of KC and NPSG (black and purple stars, respectively, in Figure 1a). In addition, observations from three Argo floats along the high Chl-a filament were used to detect fresher waters transported by the ECKB. From west to east along the ECKB, the WMO numbers of these three Argo floats were WMO 5901521, 5901512, and 2901588. The drifter WMO 22527 was carried along the ECKB from May to June 2010.

To study the wind forcing that drives the ECKB, we estimated the surface wind stress $(\overline{\tau_x}, \overline{\tau_y})$ using:

$$\overline{\tau}_x = \rho_a \times C_d \times spd_w \times u_w \tag{1}$$

$$\overline{\tau}_y = \rho_a \times C_d \times spd_w \times v_w \tag{2}$$

$$10^3 C_d = \frac{2.7}{spd_w} + 0.142 + 0.0764 \times spd_w \tag{3}$$

where the air density $\rho_a = 1.2$ kg m^{-3}, spd_w is the wind speed, u_w is the zonal wind velocity, v_w is the meridional wind velocity, and C_d is the drag coefficient that was computed by following Large and Pond [26]. Then, we followed the work by Qiu and Chen [5] to calculate the meridional Ekman velocity (v_{Ek}) averaged in the surface layer using $v_{Ek} = -\tau_x/(\rho_0 f H_0)$ where ρ_0 is the reference density, f is the Coriolis parameter, and H_0 (=150 m, by following [5]) is the thickness of the surface STCC layer, where the changes of STCC flow are largely confined [5]. Furthermore, we studied the ocean profiles obtained from the CTD casts by calculating the potential vorticity (PV) by ignoring the relative vorticity using:

$$\text{PV} = \frac{f}{\rho_0}\frac{\partial \rho}{\partial z} \tag{4}$$

where ρ is the ocean potential density, z is the depth (positive downward), ρ_0 is the mean ocean potential density between each depth, and f is the Coriolis parameter.

3. Results

3.1. Advection of Chl-a by Ocean Flow along 21°N

Figure 2 shows the selected daily anomalies of Chl-a referring to the daily climatology from 1998 to 2020. The Chl-a filament meandered at approximately 21°N and extended zonally by approximately 1100 km within 125°E and 135°E from 29 May to 23 July 2010. The daily anomalies of Chl-a concentrations are two times higher than the daily climatological mean along the Chl-a filament, displaying how large the anomalies of Chl-a concentration along the filament. To show Chl-a advection by ocean currents, Figure 2 also shows the geostrophic velocity anomalies during the same periods. The altimetry-obtained geostrophic flow matched with the high Chl-a filament, suggesting the Chl-a advection by the eastward-flowing ECKB, which can be seen along 21°N from every daily map shown in Figure 2.

On a monthly timescale, Figure 3 shows the monthly pattern of the Chl-a filament in June and July 2010. The meandering elongation of the Chl-a filament can still be clearly observed in the monthly averaged Chl-a distribution. In Figure 3a, the Chl-a filament starts from an anticyclonic eddy with a high Chl-a around the eddy edge [9] east of the Luzon Strait, matching the surface geostrophic flow obtained from the satellite and drifter observations in June. In Figure 3b, high Chl-a can be observed within the eastward-flowing ECKB, approximately within 125°E and 135°E in July. Note that these monthly patterns of high Chl-a filament and geostrophic flow are rarely observed east of the Luzon Strait in the July climatological field (Figure 1a,b).

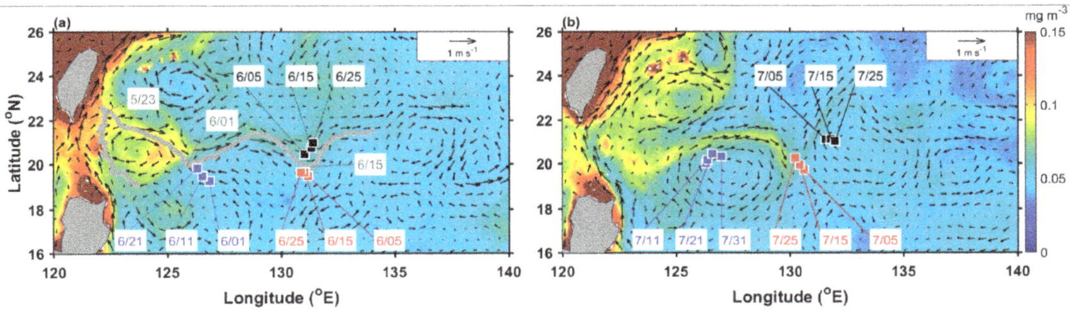

Figure 3. The monthly maps of 1/24°-gridded chlorophyll-a (Chl-a) concentrations (color shaded) and 1/4°-gridded absolute geostrophic velocities (arrows) in (**a**) June and (**b**) July 2010. The square symbols in blue, red, and black show the locations of Argo 5901521, 5901512, and 2901588, respectively, from west to east roughly along the high Chl-a filament. The gray curve in (**a**) shows the trajectory of drifter 22527 in May and June. The high Chl-a filament located along 21°N within 125°E and 135°E shows the footprint of the eastward cross-shore Kuroshio branch (ECKB). The dates (month/days) given in same colors point to the corresponding locations of the drifter and Argo floats.

In Figure 4, the eastward-flowing ECKB is represented by the trajectory of the drifter. The near-surface drifter flowed along the west–east band with large meridional gradients of ADT (Figure 4), where the eastward geostrophic flow was strong (Figure 3) at approximately 21°N. The consistency of altimetry and drifter observations show that the ocean currents along the high Chl-a filament were mainly determined by the geostrophic flow. The drifter moved northward between the KC and an anticyclonic eddy or KR in the east at the beginning of May (Figure 4a). Then, it turned southeastward when approaching 22°N, circling an anticyclonic eddy that was centered at approximately 123.4°E and 20.4°N until the end of May (Figure 4a). In June, the drifter moved further eastward, flowing along the curvature pattern of ADT contours (Figure 4b). Based on the drifter trajectory from May to June, the drifter was carried by the ECKB, which was the Kuroshio bifurcation to the east at approximately 22°N (Figure 4c).

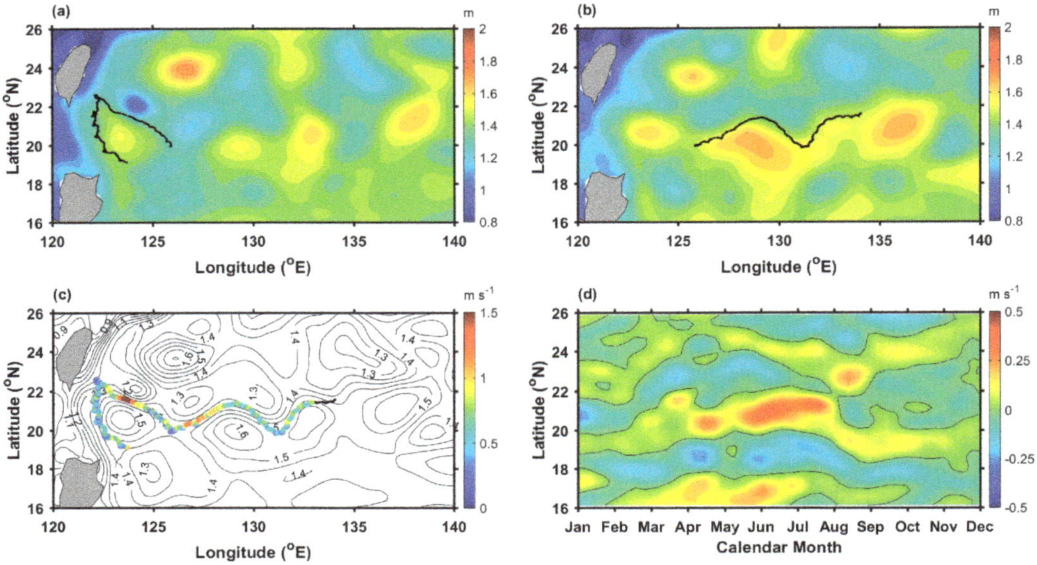

Figure 4. Spatial distribution of satellite-obtained absolute dynamic topography (ADT) averaged in (**a**) May, (**b**) June, and (**c**) from May to June (contours, units in m), in 2010, based on the altimetry 1/4°-gridded product. The black curve in (**a**,**b**) shows the trajectory of drifter 22527 in the relative month. The colored curve in (**c**) shows the drifter-measured current speed along the trajectory. (**d**) Hovmöller diagram of zonal geostrophic-current velocity averaged within 125° and 135°E, with the zero values of zonal geostrophic velocity shown by the thin contours.

The drifter speed can exceed 1 m s^{-1} along the ECKB (Figure 4c). Averaged within 125° and 135°E based on the satellite observation, the ECKB has a zonal speed reaching 0.5 m s^{-1}, which was close to the drifter speed at approximately 0.6 m s^{-1} averaged in the same region. According to the Hovmöller diagram of zonal geostrophic-current velocity shown in Figure 4d, the ECKB began in April, peaked in June and became weaker at the end of July, within 20°N and 22°N. Such a strong eastward flow can transport waters from the western boundary into the high-salinity NPSG where ocean currents are generally weak [27]. Moreover, the seasonal variability of the ECKB in 2010 was found to be similar to that of STCC, which is strong from late winter to spring (March–June), and weak in fall [7].

3.2. Injection of Fresher Waters by Ocean Flow along 21°N into the High-Salinity NPSG

A recent study by Yan et al. [28] showed that waters were getting fresher near the Luzon Strait in the NPSG. The horizontal advection of low-salinity anomalies associated with the KC system was found to be the main mechanism of the observed freshening.

From the time series of salinity and temperature profiles obtained from the three Argo floats along the ECKB from May to July (Figure 5a–c), fresher waters with low salinity below 34.6 ppt can be found above the mixed-layer depth (MLD) at approximately 50 m. Figure 5a–c show the minimum salinity near the surface detected in June and July via the Argo floats, implying fresher waters being transported further to the east of 130°E into the high-salinity region of NPSG, by the ECKB. Moreover, Figure 5a–c show that the MLDs were shallower than 50 m (reaching 25 m) and no MLD deepening could be found via the Argo floats along the Chl-a filament. Thus, this MLD analysis suggests that the Chl-a filament was mainly the result of advection by the ECKB, instead of local upwelling.

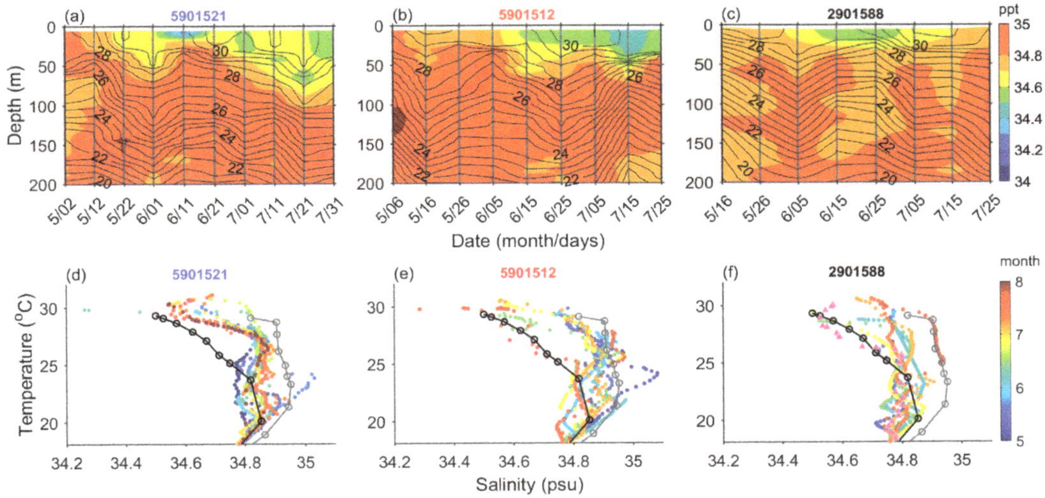

Figure 5. (**a–c**) Time series of salinity (color shaded) and temperature (contours) profiles corresponding to three selected Argo floats with the WMO numbers of 5901521, 5901512, and 2901588, from May to July 2010. The locations of these three Argo floats are marked in Figure 3. (**d–f**) Temperature–salinity (TS) diagrams corresponding to the profiles of three selected Argo floats shown in (**a–c**), respectively. Colors show the time in the calendar month. The magenta triangles in (**f**) show the TS properties obtained from the CTD casts near 137°E and 23°N in July 2010. Black and gray curves with circles show the background TS properties averaged in June and July, located at the black and magenta star, respectively, shown in Figure 1a. Note: Not all data points are shown in the figures for better clarity, but the scientific results are not affected by the unshown data points.

According to the temperature–salinity (TS) diagrams shown in Figure 5d–f, the Kuroshio water masses, which are fresher than those in the NPSG, can be detected along the ECKB. In Figure 5d, waters with salinity lower than 34.5 ppt could be found via the profiles of Argo 5901251 near 126°E (Figure 3a) during mid-June. In Figure 5e, low salinity waters could also be found to the east via the profiles of Argo 5901512 near 130°E (Figure 3b) during mid-July. The TS diagrams with lower salinity were similar to that of the KC background (see the black star in Figure 1a for the location) displayed by the black circled curves within 20–30 °C in Figure 5e,f, observed by Argo 5901512 and Argo 2901588 in July and June, respectively. Moreover, the fresher Kuroshio waters could be detected further to the east at 137°E (Figure 5f), as observed from the CTD measurements near 23°N along the 137°E transect. These TS properties show that the fresher Kuroshio waters were detected along the Chl-a filament, suggesting the injection of fresh and high-Chl-a-concentration waters by the ECKB into the high-salinity and nutrient-poor NPSG.

In Figure 6a,b, the ECKB was warmer and fresher above 50 m near 23°N than its surroundings, based on the CTD profiles along the 137°E transect. The temperature vertical structures concaved downward at the main axis of the ECKB, separating colder, denser waters north of 26°N and warmer, lighter waters south of 20°N (Figure 6a,c) above 200 m. In Figure 6c, the ECKB can be seen above the southern boundary of low PV waters, which might be the STMW [4] where its boundary is defined by the PV value at 2.5×10^{-10} s^{-2} [29]. The zonal geostrophic speed of the ECKB reached 0.5 m s^{-1} at the surface (Figure 6d), by assuming zero velocity at 1000 m. This zonal geostrophic speed estimated from the CTD casts was consistent with that obtained from the drifter and satellite observations west of 135°E.

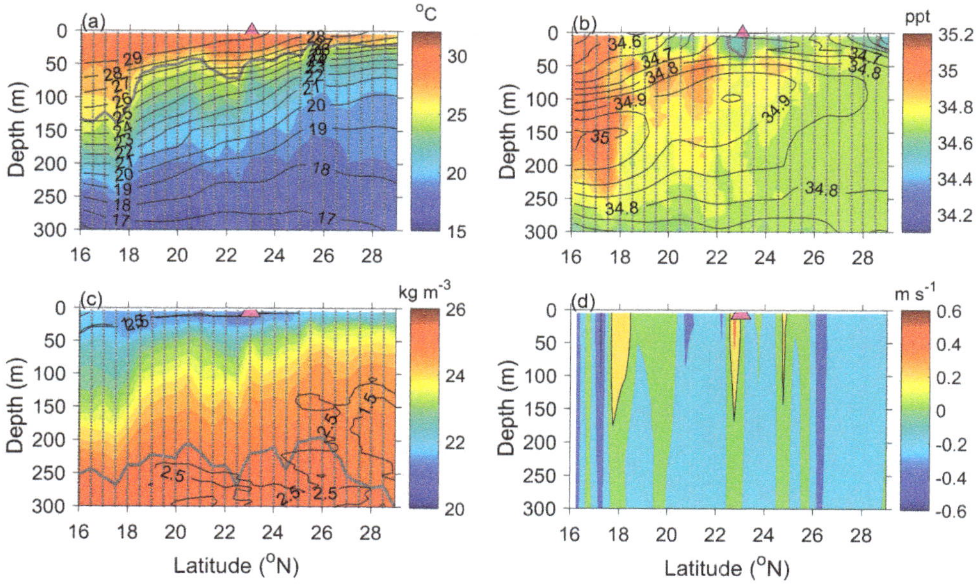

Figure 6. Profiles of (**a**) temperature, (**b**) salinity, (**c**) potential density, and (**d**) zonal geostrophic velocity, measured along the 137°E transect in July 2010. In (**a**–**c**), bold gray contours show the contour of temperature and salinity at 26 °C, 34.6 ppt, and 25 kg m^{-3}, respectively. Black thin contours show the July climatological field in (**a**,**b**), low potential vorticity at 1.5×10^{-10} s^{-2} and 2.5×10^{-10} s^{-2} in (**c**), and zonal geostrophic velocity at 0.2 m s^{-1} in (**d**). The purple triangles near the surface at approximately 23°N in all figures indicate the location of the CTD station, which corresponds to the purple triangles in the temperature-salinity diagram shown in Figure 5f. Gray dots in (**a**–**c**) show the location of CTD stations and depth resolution.

Warmer temperature (Figure 6a) and zonal geostrophic velocities (Figure 6d) near 23°N show the ECKB depth approaching 200 m. The main axis of the ECKB was approximately located at 23°N with a meridional width of approximately 50 km. The ECKB axis observed via the CTD measurements (Figure 6) matched the location of the high Chl-a filament and fast eastward geostrophic flow (Figures 2 and 3). The consistency of the ECKB information from different observations shows the existence of ECKB, which has not been previously reported. Figure 7 further shows the surface warming around the region of the high Chl-a filament in July 2010 (compared to the July climatology field from 1998 to 2020), and the daily coincidence of surface warming (>30 °C) with the high Chl-a filament on 12 July 2010. No signature of colder water upwelled from below could be observed in the upper mixed layer down to ~50 m (Figure 5a–c), suggesting that local upwelling in the ocean-front region is not the main mechanism causing the high Chl-a filament.

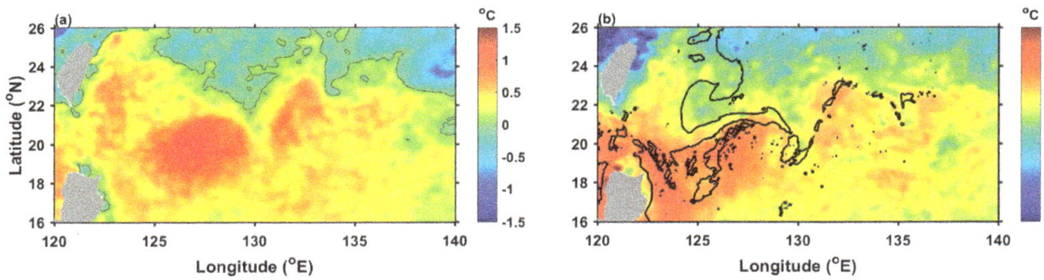

Figure 7. Spatial distribution of (**a**) monthly SST anomalies in July 2010 referring to the month climatology from 1998 to 2020, and (**b**) daily SST on 12 July 2010, based on the OISST product gridded at 0.05°. Thin curve in (**a**) show the zero contour of SST anomalies. Bold curve in (**b**) shows the Chl-a contour at 0.07 mg m^{-3}, representing the high Chl-a filament.

3.3. Mechanisms Driving the ECKB

In Figure 6a, the tilting of background isotherms can be observed at the subsurface below 50 m along the 137°E transect. The eastward flow would be induced under such conditions with isotherm tilting upward to the north, according to the thermal wind balance:

$$f\frac{\partial U_g}{\partial z} = -\alpha g\frac{\partial T}{\partial y} \quad (5)$$

where T is the ocean temperature, g is the gravity, and α is the thermal expansion coefficient. A negative value of $\partial T/\partial y$ indicates an enhanced isotherm tilting. Thus, the stronger the subtropical front (larger $\left|\frac{\partial T}{\partial y}\right|$), the faster the eastward flow near the surface is. The isotherm tilting is related to the subtropical front that was accompanied by colder waters in the north and warmer waters in the south.

Figure 8 shows the TS properties of different water masses separated by the subtropical front as observed via all available good-quality Argo floats in the research area from May to July 2010. Waters north of the front were fresher (Figure 8) than those in the south in May and June 2010. This comparison shows that the subtropical front was formed between two different water masses, which might be converged by the Ekman transport from the north and the south.

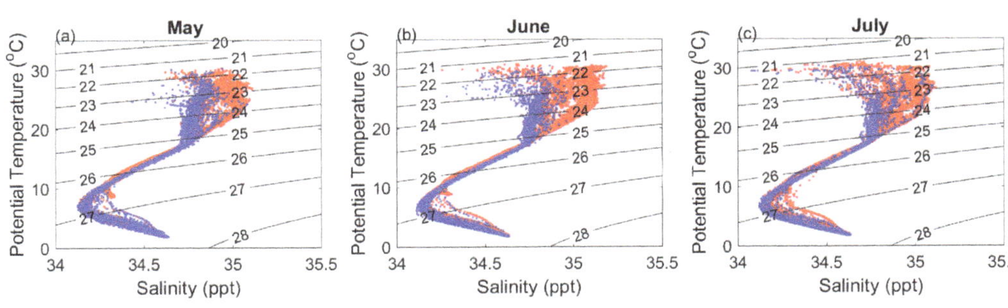

Figure 8. Temperature–salinity (TS) diagrams in (**a**) May, (**b**) June, and (**c**) July, obtained from the Argo floats with their locations shown in Figure 1b. Blue and red dots show the TS properties corresponding to the absolute dynamic topography (ADT) smaller and larger than 1.45 m (north and south of subtropical front), respectively. Black contours show the potential density.

Based on the satellite observations, the subtropical front variation could be observed in the spatial distribution of SST and meridional SST gradients (Figure 9) from May to July 2010. Figure 9a,b show large meridional SST gradients (negative signs) along the ECKB within 125°E and 135°E from May to June, representing the subtropical front. However, meridional SST gradients became smaller, and the SST distribution was more homogeneous in July than in the previous months (Figure 9c), showing the weakening of the subtropical front.

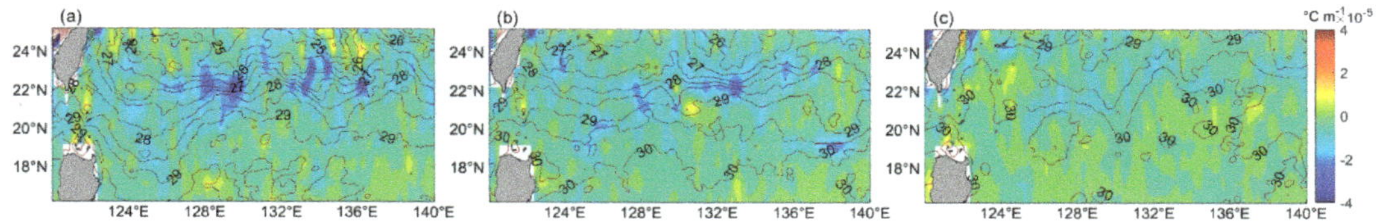

Figure 9. Spatial distribution of meridional SST gradient (color shaded) and SST (contours, units in °C) during (**a**) May, (**b**) June, and (**c**) July 2010, based on the OISST product gridded at 0.05°.

Overall, the subtropical front was represented by large meridional SST gradients that tilt slightly at a southwest–northeast orientation within 20°N and 25°N, consistent with the northernmost subtropical front reported by Kobashi et al. [15], who focused on the long-term mean of three subtropical fronts in the NPSG. Large meridional SST gradients could also be observed near the KC east of the Luzon Strait in May and June (Figure 9a,b), which have rarely been reported in previous studies. This implies that the elongation of the subtropical front can reach the western boundary in the North Pacific on time scales shorter than a season.

Qiu and Chen [5] found that meridional Ekman convergence ($\partial v_{Ek}/\partial y < 0$) mainly contributed to the strengthening of the subtropical front that enhanced the eastward-flowing STCC to produce more eddies via baroclinic instability on interannual time scales, according to the Ekman convergence, forcing:

$$\frac{\partial G}{\partial t} \approx -\frac{\partial}{\partial y}(v_{Ek}G) \tag{6}$$

where G is defined as meridional SST gradients ($-\partial T/\partial y$) and v_{Ek} is the meridional Ekman velocity. However, Kobashi and Xie [7] found that the Ekman suction enhanced by the strong cyclonic wind anomaly elevates the thermocline, decreases SST, and then induces the SST front, accelerating the STCC from late winter to spring (March–June) on interannual time scales.

To confirm the mechanisms enhancing the subtropical front near the KC, we estimated the $-\partial(v_{Ek}G)/\partial y$ term [5] and wind stress curl anomaly [7], averaged within 125–135°E and 20–23°N, from May to July 2010. Figure 10 shows the latitude–time changes of the relative forcing on the subtropical front. Positive wind stress curl and positive wind–stress–curl anomalies were formed north of 21°N from April to June 2010 (Figure 10a,b). The Ekman suction and divergence ($\partial v_{Ek}/\partial y > 0$) would occur beneath the cyclonic wind stress within 21–24°N according to the Ekman theory. Then, SST decreased and the SST anomalies reached −1 °C (Figure 10d) below the cyclonic wind stress anomaly, enhancing the subtropical front (Figure 10c) in the south and thus accelerating the ECKB via thermal wind balance [7]. On the other hand, the $-\partial(v_{Ek}G)/\partial y$ term has negative values ranging from -0.2×10^{-14} to -2×10^{-14} °C m^{-1} s^{-1} from May to June 2010, averaged within 125–135°E and 20–24°N (the area with negative meridional SST gradients, i.e., strong SST front). The negative $-\partial(v_{Ek}G)/\partial y$ term shows that the enhancing of the subtropical front was not attributed to the Ekman convergence forcing during the occurrence of the ECKB.

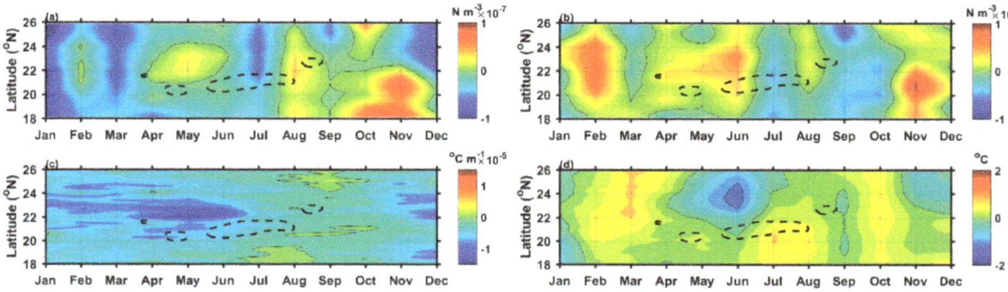

Figure 10. Hovmöller diagram of (**a**) surface wind stress curl, (**b**) surface wind stress curl anomalies, (**c**) meridional gradient of SST, and (**d**) SST anomalies, calculated from the monthly meridional profiles zonally averaged within 125° and 135°E, based on the CCMP wind and OISST data gridded at 1/4° and 0.05° resolution, respectively. Thin solid curves show zero contours. Dashed curves show the contour of zonal geostrophic-current velocity at 0.2 m s^{-1}. The monthly anomalies are herein referred to the monthly climatological fields from 1998 to 2018 due to the available CCMP wind data.

4. Discussion and Summary

This study reports an unusual eastward flow that injected fresher and high-Chl-a-concentration waters into the high-salinity, nutrient-poor NPSG during June and July 2010. The eastward flow was inferred to be the eastward cross-shore Kuroshio branch (ECKB) that sequentially (1) circled an anticyclonic eddy accompanied by high-Chl-a concentrations around the eddy edge; (2) meanderingly flowed eastward with a high Chl-a by following the thermal wind relation; and (3) injected the high-salinity NPSG with fresh waters in which TS properties were similar to those from the KC.

Observed via the satellite ocean-color images, the ECKB appeared as a filament of high-Chl-a concentrations, which were significantly higher than the climatological mean. A drifter trajectory showed that the KC bifurcated near 22°N to the east in May, and flowed eastward with a meandering pattern that matched with the spatial distribution of geostrophic flow and high-Chl-a concentrations in June and July 2010. Our analysis showed that the eastward flow transported the waters with the TS properties similar to those of the KC; thus, we called it the ECKB.

The ECKB carried warmer and fresher waters eastward into the NPSG. The mean geostrophic/Lagrangian speed of the ECKB was approximately 0.5 m s^{-1}, estimated via satellite altimeters, a drifter, and the JMA's 137°E transect. Based on the geostrophic speed at 0.2 m s^{-1}, the ECKB depth approached 200 m, just above the southern boundary of low PV waters (probably the low-PV STMW). The ECKB mechanism was found to be the same as that of the STCC, attributing to a cyclonic wind anomaly to the north at approximately 22–24°N. The cyclonic wind anomaly decreased SST beneath it via Ekman suction and then enhanced the SST front to its south at approximately 21°N. The enhanced SST front subsequently induced the ECKB that was bifurcated from the main stream of the northward-flowing KC.

The injection of warm, fresh, and high-Chl-a-concentration waters into oligotrophic gyres by the ECKB is important to the NPSG ecosystem but it is a largely unexplored issue. Thus, the findings from this study show that the nutrient-poor NPSG can benefit from the KC bifurcation in terms of injection of high-Chl-a-concentration waters, which could potentially influence the marine food webs in the NPSG. However, more in situ observations, are needed, especially on biogeochemical variables, if we are to better understand the impacts of high-Chl-a-concentration waters on higher trophic levels in the nutrient-poor NPSG. Additionally, this study tried to raise the attention of the oceanographic community that the ECKB could be found in other years, and that studying the ECKB interannual variability is important to further understand the connection between the northward-flowing KC and the eastward-flowing STCC.

Author Contributions: Conceptualization, C.-H.C., W.C. and J.-H.T.; methodology, C.-H.C.; software, C.-H.C. and Y.-C.L.; validation, C.-H.C. and Y.-C.L.; formal analysis, C.-H.C., W.C. and Y.-C.L.; investigation, C.-H.C., W.C. and Y.-C.L.; writing—original draft preparation, C.-H.C.; writing—review, revision, and editing, C.-H.C., W.C. and J.-H.T.; funding acquisition, C.-H.C. and W.C. All authors have read and agreed to the published version of the manuscript.

Funding: This research was funded by the Ministry of Science and Technology of Taiwan (R.O.C.), Grant MOST 109-2611-M-019-013 and MOST 110-2611-M-019-019, and by the Ministry of Higher Education of Malaysia FRSG grant (FRGS/1/2018/WAB09/UM/02/4) and the FIO-UM Joint Centre of Marine Science and Technology.

Institutional Review Board Statement: Not applicable.

Informed Consent Statement: Not applicable.

Data Availability Statement: The data presented in this study are available upon request from the corresponding author.

Acknowledgments: W.C. would like to thank the Ministry of Higher Education of Malaysia FRSG grant (FRGS/1/2018/WAB09/UM/02/4) and the FIO-UM Joint Centre of Marine Science and Technology. C.C.H. would like to acknowledge the financial support from the Ministry of Science and Technology of Taiwan (R.O.C.) grants (MOST 109-2611-M-019-013 and MOST 110-2611-M-019-019). This study was conducted using E.U. Copernicus Marine Service Information: <https://doi.org/10.48670/moi-00100, https://doi.org/10.48670/moi-00168, https://doi.org/10.48670/moi-00148, accessed on 31 January 2022>. We thank NASA OBPG for the MODIS-Aqua Level 2 data, Coriolis Operational Oceanography for providing the Argo and surface drifters data, and Remote Sensing Systems for the wind data. We are grateful to the captains and crews of the R/Vs Ryofu Maru and Keifu Maru, and the JMA for their long-term efforts in executing the 137°E repeated hydrographic transects in the North Pacific.

Conflicts of Interest: The authors declare no conflict of interest.

References

1. Nagai, T.; Saito, H.; Suzuki, K.; Takahashi, M. *Kuroshio Current: Physical, Biogeochemical, and Ecosystem Dynamic*; American Geophysical Union, Wiley: Washington, DC, USA, 2019; pp. 1–336. [CrossRef]
2. Chen, C.T.A.; Huang, T.H.; Wu, C.H.; Yang, H.; Guo, X. Variability of the nutrient stream near Kuroshio's origin. *Sci. Rep.* **2021**, *11*, 5080. [CrossRef] [PubMed]
3. Qiu, B. Seasonal eddy field modulation of the North Pacific subtropical countercurrent: TOPEX/Poseidon observations and theory. *J. Phys. Oceanogr.* **1999**, *29*, 2471–2486. [CrossRef]
4. Chu, P.C.; Li, R.; You, X. Northwest Pacific subtropical countercurrent on isopycnal surface in summer. *Geophys. Res. Lett.* **2002**, *29*, 1842. [CrossRef]
5. Qiu, B.; Chen, S. Interannual variability of the North Pacific Subtropical Countercurrent and its associated mesoscale eddy field. *J. Phys. Oceanogr.* **2010**, *40*, 213–225. [CrossRef]
6. Uda, M.; Hasunuma, K. The eastward subtropical countercurrent in the western North Pacific Ocean. *J. Oceanogr. Soc. Japan* **1969**, *25*, 201–210. [CrossRef]
7. Kobashi, F.; Xie, S.-P. Interannual variability of the North Pacific Subtropical Countercurrent: Role of local ocean–atmosphere interaction. *J. Oceanogr.* **2012**, *68*, 113. [CrossRef]
8. Yang, K.-C.; Wang, J.; Lee, C.M.; Ma, B.; Lien, R.-C.; Jan, S.; Yang, Y.J.; Chang, M.-H. Two mechanisms cause dual velocity maxima in the Kuroshio east of Taiwan. *Oceanography* **2015**, *28*, 64–73. [CrossRef]
9. Shih, Y.-Y.; Hung, C.-C.; Gong, G.-C.; Chung, W.C.; Wang, Y.-H.; Lee, I.-H.; Chen, K.-S.; Ho, C.-Y. Enhanced Particulate Organic Carbon Export at Eddy Edges in the Oligotrophic Western North Pacific Ocean. *PLoS ONE* **2015**, *10*, e0131538. [CrossRef]
10. Hsu, P.C.; Lin, C.C.; Huang, S.J.; Ho, C.R. Effects of cold eddy on Kuroshio meander and its surface properties, east of Taiwan. *IEEE J. Sel. Top. Appl. Earth Obs. Remote Sens.* **2016**, *9*, 5055–5063. [CrossRef]
11. Cheng, Y.-H.; Chang, M.-H.; Ko, D.S.; Jan, S.; Andres, M.; Kirincich, A.; Yang, Y.J.; Tai, J.-H. Submesoscale eddy and frontal instabilities in the Kuroshio interacting with a cape south of Taiwan. *J. Geophys. Res. Ocean.* **2020**, *124*, e2020JC016123. [CrossRef]
12. Tseng, Y.-H.; Lu, C.-Y.; Zheng, Q.; Ho, C.-R. Characteristic Analysis of Sea Surface Currents around Taiwan Island from CODAR Observations. *Remote Sens.* **2021**, *13*, 3025. [CrossRef]
13. Chang, Y.; Oey, L. Instability of the North Pacific Subtropical Countercurrent. *J. Phys. Oceanogr.* **2014**, *44*, 818–833. [CrossRef]
14. Yasuda, T.; Hanawa, K. Decadal changes in the mode waters in the midlatitude North Pacific. *J. Phys. Oceanogr.* **1997**, *27*, 858–870. [CrossRef]

15. Kobashi, F.; Mitsudera, H.; Xie, S.-P. Three subtropical fronts in the North Pacific: Observational evidence for mode water-induced subsurface frontogenesis. *J. Geophys. Res.* **2006**, *111*, C09033. [CrossRef]
16. Qiu, B.; Chen, S. Decadal variability in the formation of the North Pacific Subtropical Mode Water: Oceanic versus atmospheric control. *J. Phys. Oceanogr.* **2006**, *36*, 1365–1380. [CrossRef]
17. Sugimoto, S.; Hanawa, K. Impact of Aleutian Low activity on the STMW formation in the Kuroshio recirculation gyre region. Geophys. *Res. Lett.* **2010**, *37*, L03606. [CrossRef]
18. Takeuchi, K. Numerical study of the seasonal variations of the Subtropical Front and the Subtropical Countercurrent. *J. Phys. Oceanogr.* **1986**, *16*, 919–926. [CrossRef]
19. Zhang, Z.; Xue, H. The possible formation mechanism of the Subtropical Countercurrent in the Pacific Ocean. In Proceedings of the EGU General Assembly 2020, Online, 4–8 May 2020. EGU2020-6635. [CrossRef]
20. Talley, L.; Pickard, G.; Emery, W.; Swift, J. *Descriptive Physical Oceanography*, 6th ed.; Elsevier: San Diefo, CA, USA, 2011; pp. 308–311. [CrossRef]
21. Nan, F.; Xue, H.; Yu, F. Kuroshio intrusion into the South China Sea: A review. *Prog. Oceanogr.* **2015**, *137 Pt A*, 314–333. [CrossRef]
22. Trott, C.B.; Metzger, E.J.; Yu, Z. Investigating mesoscale eddy characteristics in the Luzon Strait region using altimetry. *Ocean Dyn.* **2021**, *71*, 679–698. [CrossRef]
23. Saulquin, B.; Gohin, F.; d'Andon, O.F. Interpolated fields of satellite-derived multi-algorithm chlorophyll-a estimates at global and European scales in the frame of the European Copernicus-Marine Environment Monitoring Service. *J. Oper. Oceanogr.* **2019**, *12*, 47–57. [CrossRef]
24. Good, S.; Fiedler, E.; Mao, C.; Martin, M.J.; Maycock, A.; Reid, R.; Roberts-Jones, J.; Searle, T.; Waters, J.; While, J.; et al. The Current Configuration of the OSTIA System for Operational Production of Foundation Sea Surface Temperature and Ice Concentration Analyses. *Remote Sens.* **2020**, *12*, 720. [CrossRef]
25. Oka, E.; Ishii, M.; Nakano, T.; Suga, T.; Kouketsu, S.; Miyamoto, M.; Nakano, H.; Qiu, B.; Sugimoto, S.; Takatani, Y. Fifty years of the 137°E repeat hydrographic section in the western North Pacific Ocean. *J. Oceanogr.* **2018**, *74*, 115–145. [CrossRef]
26. Large, W.G.; Pond, S. Sensible and latent heat flux measurements over the ocean. *J. Phys. Oceanogr.* **1982**, *12*, 464–482. [CrossRef]
27. Pedlosky, J. The dynamics of the oceanic subtropical gyres. *Science* **1990**, *248*, 316–322. [CrossRef] [PubMed]
28. Yan, Y.; Xu, D.; Qi, Y.; Gan, J. Observations of Freshening in the Northwest Pacific Subtropical Gyre near Luzon Strait. *Atmos.-Ocean* **2012**, *50* (Suppl. 1), 92–102. [CrossRef]
29. Xu, L.; Li, P.; Xie, S.-P.; Liu, Q.; Liu, C.; Gao, W. Observing mesoscale eddy effects on mode-water subduction and transport in the North Pacific. *Nat. Commun.* **2016**, *7*, 10505. [CrossRef]

remote sensing

MDPI

Article

A Strong Kuroshio Intrusion into the South China Sea and Its Accompanying Cold-Core Anticyclonic Eddy in Winter 2020–2021

Zhenyu Sun [1,2], Jianyu Hu [1,2,*], Zhaozhang Chen [1], Jia Zhu [1], Longqi Yang [1], Xirong Chen [1] and Xuewen Wu [1]

1 State Key Laboratory of Marine and Environmental Science, College of Ocean and Earth Sciences, Xiamen University, Xiamen 361102, China; sunzy@xmu.edu.cn (Z.S.); zzchen@xmu.edu.cn (Z.C.); zhujia@xmu.edu.cn (J.Z.); yanglq@xmu.edu.cn (L.Y.); cxr0819@xmu.edu.cn (X.C.); wxw790726@xmu.edu.cn (X.W.)

2 Southern Marine Science and Engineering Guangdong Laboratory (Zhuhai), Zhuhai 519000, China

* Correspondence: hujy@xmu.edu.cn

Abstract: Multiple remote sensing datasets, combined with in-situ drifter observations, were used to analyze the Kuroshio intrusion through the Luzon Strait (LS). The results showed that a strong Kuroshio Current Loop (KCL) and accompanying anticyclonic eddy (ACE) existed in winter 2020–2021. As quantitatively demonstrated by the Double Index (DI), the Kuroshio Warm Eddy Index (KWI) had low values during a long sustained period compared to those in all other years in the available historical records. Remarkable kinematic properties (i.e., amplitude, diameter, propagated distance, lifespan and propagating speed) of the accompanying ACE were extracted by automatic eddy detection algorithms, showing that the ACE had a maximum diameter of 381 km and a peak amplitude of 50 cm, which significantly exceeded the previous statistics in winter. The orographic negative wind stress curl southwest of Taiwan Island and the westward Ekman transport through the LS during the winter half year of 2020–2021 both had large values beyond their historical maxima. Hence wind forcing is regarded as the primary forcing mechanism during this event. Alternating cyclonic eddies (CEs) and ACEs approaching on the east of the LS were identified, indicating that the interaction between the Kuroshio and the impinging CEs at proper locations made extra contributions to enhancing the KCL. The accompanying ACE had a distinctive feature of a cold-core structure at the surface layer, so as to be categorized as a cold core ACE (CC-ACE), and the temperature difference between the cold core and outer warm ring was maintained for three months. The generation and long duration of the CC-ACE may be due to the sustaining entrainment supported by the warm water from Kuroshio intrusion and the Northwest Luzon Coastal Current (NWLCC) successively.

Keywords: Kuroshio intrusion; Kuroshio Current Loop; cold-core anticyclonic eddy

Citation: Sun, Z.; Hu, J.; Chen, Z.; Zhu, J.; Yang, L.; Chen, X.; Wu, X. A Strong Kuroshio Intrusion into the South China Sea and Its Accompanying Cold-Core Anticyclonic Eddy in Winter 2020–2021. *Remote Sens.* **2021**, *13*, 2645. https://doi.org/10.3390/rs13142645

Academic Editor: Yukiharu Hisaki

Received: 24 May 2021
Accepted: 2 July 2021
Published: 6 July 2021

1. Introduction

The Luzon Strait (LS; Figure 1) is a primary channel for the mass and energy exchange between the Western Pacific (WP) and the South China Sea (SCS). The Kuroshio intrusion from the WP, with warm and high-salinity water, has a significant influence on the circulation and stratification of the SCS [1,2]. The seasonal variability shows a stronger intrusion in winter than in summer [3], and the surface water can reach the interior SCS especially in winter [4].

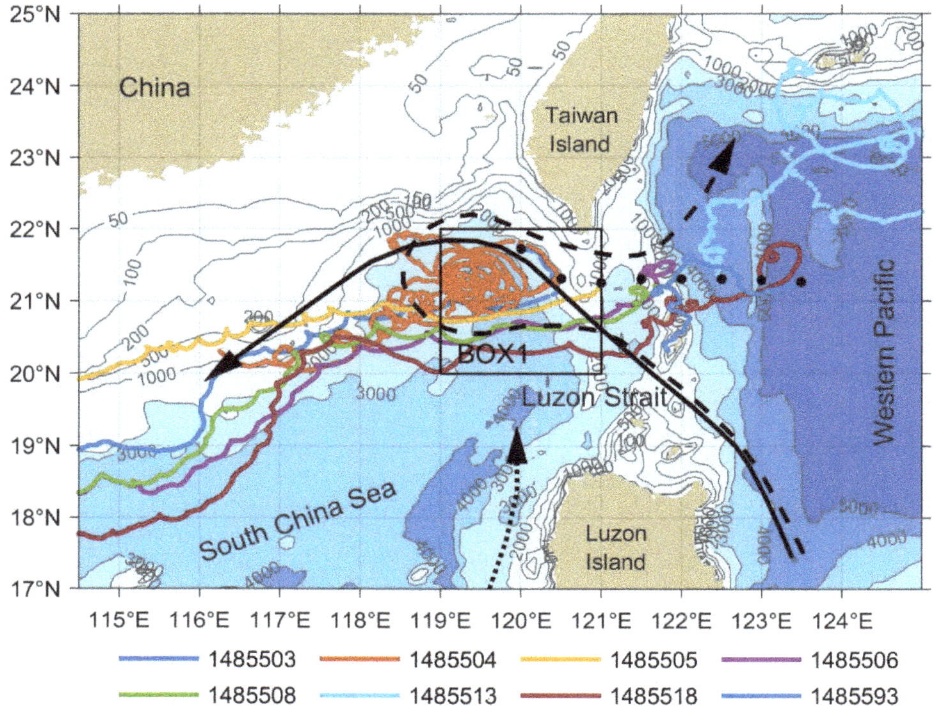

Figure 1. Topography (filled contours) in the LS and its surrounding area. The rectangle of BOX1 is the integral areas for the DI. Colored lines are the trajectories of the eight drifters deployed in December 2020. The SCSBK, KCL and NWLCC are denoted as black solid arrow, dashed arrow and dotted arrow, respectively.

The Kuroshio intrusion is reported to have different spatial patterns. Qiu et al. [5] illustrated the SCS Branch of Kuroshio (SCSBK) as a west-flowing current with relatively high speed and steady direction on the continental slope in the northern SCS. Li and Wu [6] suggested that the Kuroshio often intrudes into the SCS by forming an anticyclonic loop and raised the concept of the Kuroshio Current Loop (KCL). Caruso et al. [7] depicted five different types of Kuroshio intrusion paths into the SCS, including a small anticyclonic bend, the SCSBK, the KCL, a detached anticyclonic eddy (ACE), and a cyclonic loop.

To quantitatively identify different paths of the Kuroshio intrusion, the Kuroshio SCS Index (KSI) was developed by Nan et al. [8]. Their results showed that the Kuroshio bending and the net inflow through the LS decrease from the looping path to the leaking path, and then to the leaping path. More recently, the Double Index (DI), which consists of the Kuroshio Warm Eddy Index (KWI) and the Kuroshio Cold Eddy Index (KCI), was proposed by Huang et al. [9], to identify the spatial patterns of the Kuroshio Warm Eddy Path (KWEP) and the Kuroshio Cold Eddy Path (KCEP), respectively. This DI can overcome the cancelling problem between the positive and negative geostrophic vorticities, so as to reduce the missing and misjudged Kuroshio path events.

The ACE accompanied by or shed from the KCL has also been extensively studied. The eddy kinetic energy (EKE) by satellite altimeter indicated that the region southwest of Taiwan Island has a high probability of eddy occurrences [10,11]. In-situ hydrographic observations have revealed that the ACE could be originated from the Kuroshio [12,13]. Systematic censuses on the ACE [11,14–17] have been made by means of automatic eddy detection algorithms, to extract the ACE's features including location, diameter, amplitude, lifespan, propagation speed and distance. The statistics showed that more ACE shedding occurs in winter than in other seasons, and most of the shed ACEs propagate to the west with speeds similar to the first-mode baroclinic Rossby wave.

Mechanisms for the occurrence of the KCL and ACE and their properties (location, amplitude, sustaining period, etc.) still remain debatable. Metzger and Hurlburt [18] showed the nondeterministic nature of eddy shedding from the KCL, and no significant correlation could be found between the yearly Kuroshio intrusion and the oceanic and atmospheric environments. Yuan et al. [19] suggested that the anticyclonic intrusion of the Kuroshio is a transient phenomenon rather than a persistent circulation pattern. As a western boundary current flowing by the gap of the LS, multiple steady states of the Kuroshio and nonlinear hysteresis of the intrusion have also been demonstrated [20,21].

Wind stress (WS) and wind stress curl (WSC) are regarded as important mechanisms to generate meso-scale eddies in the SCS [22,23]. Wang et al. [24] suggested that the orographic wind jets during the wintertime monsoon and the gaps in the mountainous island chain along the eastern boundary of the SCS can spin up cyclonic eddies (CEs) and ACEs, and the region southwest of Taiwan is one of the regions with the coexistence of negative WSC and ACE occurrences. Jia et al. [15] demonstrated the linkage between ACE shedding and the wintertime monsoon, indicating the ACE shedding occurs within one month after the integrated Ekman transport through the LS exceeds the volume roughly corresponding to a mesoscale eddy. Wu et al. [25] concluded that the northwestward Ekman drift, due to northeasterly wind in winter, intensifies the Kuroshio intrusion into the SCS, and the WSC off southwest Taiwan is chiefly responsible for the Kuroshio intrusion.

Satellite observations have shown the abundance of generally westward-propagating eddies in the subtropical regions in the WP, and their effects on the Kuroshio and the SCS have received substantial attention [26–28]. Nonlinear Rossby eddies have been evidenced to penetrate through the LS into the SCS by cruise observation [29] and altimeter data [30], but the penetration has to take the form of coupling with the Kuroshio, instead of freely westward propagation. The approaching CEs and ACEs may reduce or enhance the Kuroshio transport [26,27], or lead to convergence and divergence upstream and downstream [31]. These effects further change the Kuroshio intrusion pathway in the LS. However, the correlation between the Kuroshio intrusion path and eddy activities in the WP is not statistically significant [32], and strong impinging eddies do not always have strong effects on the Kuroshio looping path [33].

ACE (CE) is usually associated with a warm (cold) core caused by eddy-induced convergence (divergence) motion. However, there are also some abnormal ACEs (CEs) with surface cold (warm) cores in the global ocean, hereby named cold-core ACEs (warm-core CEs). The surface cold (warm) core does not change the dynamic nature of the anticyclonic (cyclonic) spinning of the ACE (CE), because the spinning is mostly supported by the stratification at the subsurface layer. Such abnormal eddies are observed in different regions in the Pacific, and the Kuroshio Extension region is the most active area for cold-core ACEs (CC-ACEs) and warm-core CEs (WC-CEs) [34–36]. However, the ACEs shed from the KCL have rarely been reported to have a cold-core structure in previous studies. As summarized by Sun et al. [36], two possible mechanisms for generating the abnormal eddies are: (1) instability during the decay stage and (2) eddy-eddy interaction and horizontal entrainment. An alternative mechanism is the local generation by eddy–wind interaction as demonstrated by McGillicuddy [37].

The present paper provides a case study on a strong Kuroshio intrusion and its accompanying CC-ACE in winter 2020–2021. By a batch of surface drifters deployed in the LS, we captured notable surface intrusion of Kuroshio into the SCS, as well as a cold-core structure of the accompanying ACE. We used multiple remote sensing data to analyze the strong amplitude and other distinctive features of this event. The occurrence of strong KCL and ACE in winter 2020–2021 further demonstrated the energetic interannual variability of the Kuroshio intrusion, and we took this opportunity to verify the mechanisms responsible for these events, including the wind forcing and mesoscale eddy activities mentioned above.

2. Materials and Methods

Multiple remote sensing datasets were used in the present study, including sea surface height (SSH), sea surface temperature (SST) and sea surface wind (SSW), with the available time span from 1993 to 2021. Reprocessed datasets were preferably selected for better quality, while near real-time ones were used as alternative for the most recent months.

The level-4 product of daily global gridded SSH is processed by the DUACS multi-mission altimeter data processing system, incorporating data sources from all altimeter missions: Jason-3, Sentinel-3A, HY-2A, Saral/AltiKa, Cryosat-2, Jason-2, Jason-1, T/P, ENVISAT, GFO and ERS1/2. The SSH maps are presented in the forms of both sea level anomaly (SLA) and absolute dynamic topography (ADT), with a spatial resolution of $0.25^\circ \times 0.25^\circ$. The geostrophic currents derived from ADT are also provided.

The SST data used here were the Foundation SST by the Operational Sea Surface Temperature and Ice Analysis (OSTIA) system [38]. In-situ and satellite observations from both infrared and microwave radiometers were combined to produce the daily maps of the SST for the global ocean, with a high resolution of 0.05° (approximately 6 km).

For SSW data, the IFREMER CERSAT Global Blended Mean Wind Fields include wind components (meridional and zonal), wind module, WS, and WSC. The wind fields are six-hourly with a $0.25^\circ \times 0.25^\circ$ spatial resolution over global oceans.

During a cruise expedition through the LS en-route to the WP in December 2020, we deployed a batch of self-developed satellite-tracked surface drifters, named Surface Current Experiment (SUCE) drifters. These drifters were of identical design and configuration as the standard Global Drifter Program (GDP) drifters [39], with a surface floating sphere and a holey-sock drogue. Horizontal surface current velocities were derived based on the drifter trajectories, and SST samplings were acquired based on the integrated temperature sensor. In this study, we used observation data of eight drifters (Table 1) in the LS and surrounding area. The data records were reprocessed to a sampling interval of 1 h, in accordance with the data quality control procedures of GDP [40].

Table 1. SUCE drifters deployed in the LS in December 2020.

Drifter No.	Deployment Date	Deployment Longitude	Deployment Latitude	End Date [1]
1485503	24 December 2020	120.0°E	21.8°N	30 January 2021
1485504	24 December 2020	120.5°E	21.3°N	31 March 2021
1485505	26 December 2020	121.0°E	21.3°N	15 February 2021
1485506	26 December 2020	121.5°E	21.3°N	30 January 2021
1485508	26 December 2020	122.0°E	21.3°N	30 January 2021
1485513	26 December 2020	122.5°E	21.3°N	31 March 2021
1485518	26 December 2020	123.0°E	21.3°N	30 January 2021
1485593	26 December 2020	123.5°E	21.3°N	8 January 2021

[1] End date of data used in Figure 1.

3. Results

3.1. DI Performance

According to Huang et al. [9], the DI is calculated by Equations (1) and (2), as the integral of positive part and negative part of the surface geostrophic vorticity in the area of 119°E–121°E and 20°N–22°N (BOX1 in Figure 1), respectively:

$$KCI = \oiint sign\left(\frac{\partial v}{\partial x} - \frac{\partial u}{\partial y}\right)\left(\frac{\partial v}{\partial x} - \frac{\partial u}{\partial y}\right)dA \tag{1}$$

$$KWI = \oiint sign\left(-\left(\frac{\partial v}{\partial x} - \frac{\partial u}{\partial y}\right)\right)\left(\frac{\partial v}{\partial x} - \frac{\partial u}{\partial y}\right)dA \tag{2}$$

where u and v are the zonal and meridional components of the geostrophic currents derived from ADT, A is the integral area, and $sign(x)$ is the sign function defined as follows:

$$sign(x) = \begin{cases} 1, & x \geq 0 \\ 0, & x < 0 \end{cases} \tag{3}$$

The time-series of the daily KCI and KWI from 1993 to 2021 are shown in Figure 2. The winter 2020–2021 had a very strong KWEP event (KWI going lower than its standard deviation from its mean). During a typical KWEP, the main Kuroshio enters the SCS in the middle part and outflows in the northern part of the LS in an anticyclonic pattern with a warm eddy southwest of Taiwan Island [9]. From the whole 28-year time series, we choose nine strong KWEP events which met the criterion of having sustained periods longer than 40 consecutive days, and listed in Table 2 are their main features including sustained period, minimum KWI value and integral KWI value. Although the 2020–2021 event had a minimum KWI value of -4.38×10^5 m^2/s, slightly weaker than that of the 1996–1997 event, the sustained period (as long as 90 days) and integral value (-316.0×10^5 m^2/s· day) both significantly exceeded all other events. Such KWI performance indicated a remarkable KCL in the LS, as well as a prolonged ACE southwest of Taiwan Island. The detailed evolution of the DI in winter 2020–2021 will be discussed in the following sections.

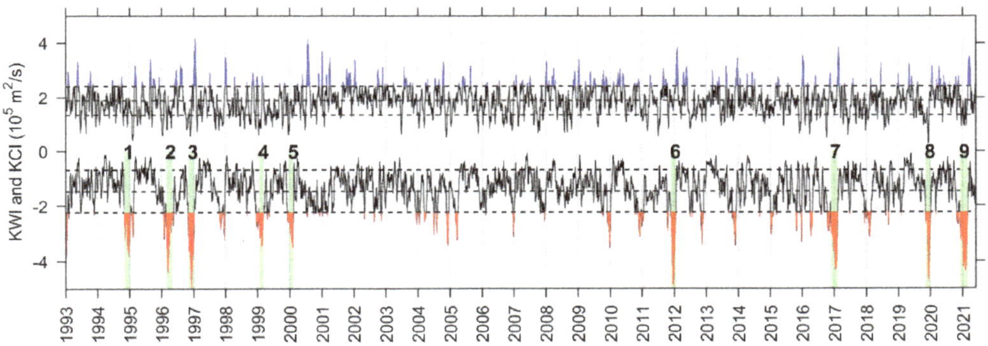

Figure 2. Time series of daily KCI and KWI from 1993 to 2021. The ticks on the time axis stand for the beginning of the year denoted by corresponding labels. Horizontal dash lines stand for the mean value (μ) and standard deviation span (μ ± σ) of the two indices. KCI values larger than μ + σ are shaded in blue, and KWI values less than μ − σ are shaded in red. Nine strong KWEP events are numbered from 1 to 9, and their time spans are shaded in green.

Table 2. Strong KWEP events from 1993 to 2021 and their characteristics.

Event No.	Start Date	End Date	Period (day)	Minimum KWI [1]	Integral KWI [2]
1	2 November 1994	10 January 1995	69	−3.82	−212.6
2	22 February 1996	4 May 1996	72	−4.43	−230.8
3	30 October 1996	8 January 1997	70	−4.96	−269.2
4	18 January 1999	12 March 1999	53	−3.49	−164.3
5	20 December 1999	14 February 2000	56	−3.55	−165.1
6	27 November 2011	24 January 2012	58	−4.89	−217.8
7	24 November 2016	10 February 2017	78	−4.35	−274.5
8	8 November 2019	5 January 2020	58	−4.76	−190.7
9	4 December 2020	4 March 2021	90	−4.38	−316.0

[1] minimum KWI in 10^5 m^2/s, [2] Integral of KWI during the event in 10^5 m^2/s ·day.

3.2. Detailed Evolution

To further investigate the evolution of the KCL and its accompanying ACE, we present the time series of monthly averaged ADT and the derivative surface geostrophic current from October 2020 to March 2021 in Figure 3. In October 2020, the Kuroshio had strong currents with a speed up to 1.0 m/s across the LS and along the east coast of Taiwan Island. The main axis of the current slightly bent towards the west in the northern LS, and a current loop pattern (i.e., KCL) started to form. From November 2020 to January 2021, the KCL was fully established. The intrusion current in the middle of the LS gradually turned from northwestward to westward, and reached a maximum speed of 1.0 m/s in December, making a strong inflow jet toward the SCS. Closed ADT contours were gradually developed to make an ACE (denoted as A1 in Figure 3) accompanying the KCL, and it kept intensifying with rising ADT at the eddy core.

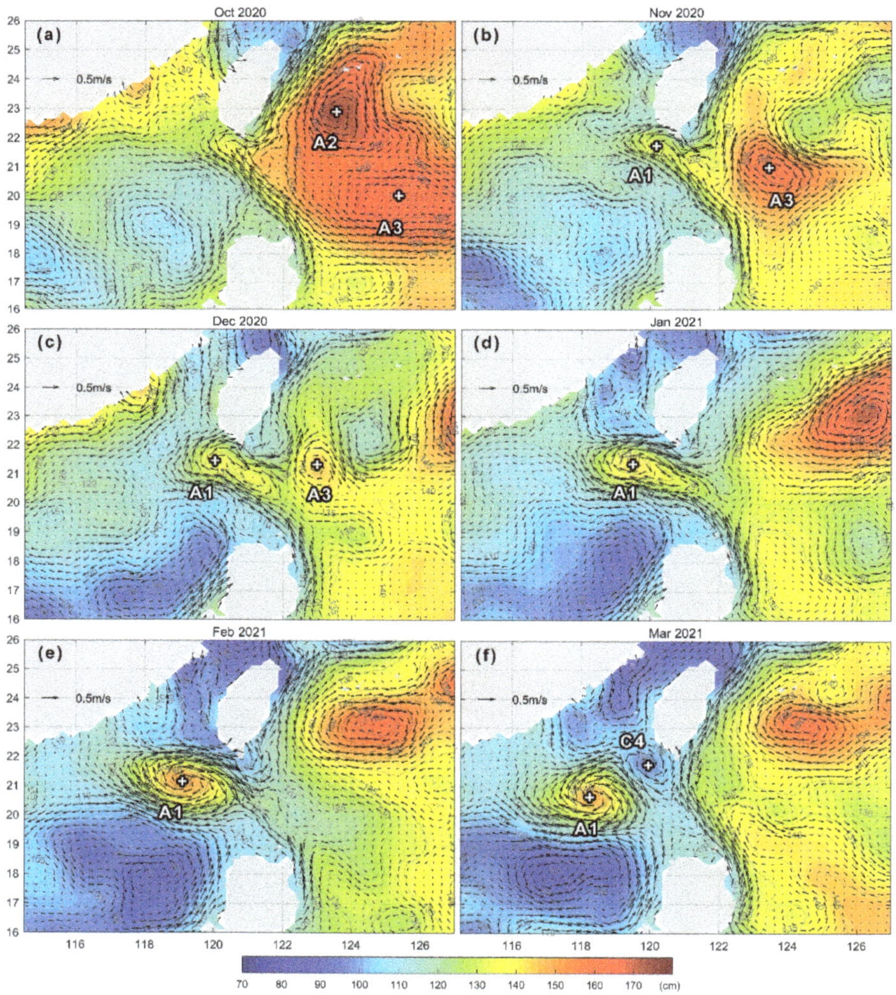

Figure 3. Time series of monthly averaged ADT (filled contours) and the derivative surface geostrophic current (black vectors) in the LS and the surrounding area. The panels of (**a**–**f**) are sequentially for the months from October 2020 to March 2021. A1~A3 are three ACEs, and C4 is a CE.

In February 2021, the ACE reached its peak state, with a maximum ADT of 150 cm at the center, and a maximum current speed of 1.2 m/s at the edge. As slowly propagating to the southwest, the eddy started to detach from the KCL, and the Kuroshio current became significantly weaker in the LS area. In March 2021, the ACE was completely detached from the KCL, and continued to propagate southwestward, maintaining its high ADT amplitude and current speed. A CE (C4 in Figure 3f) formed immediately on the east, which was once the original position of the ACE. The Kuroshio path finished converting from the KWEP to the KCEP, with a minimal bending of its main axis in the middle of the LS, which can also be verified by the detailed evolution of the KWI and KCI from February to March 2021 in Figure 4b.

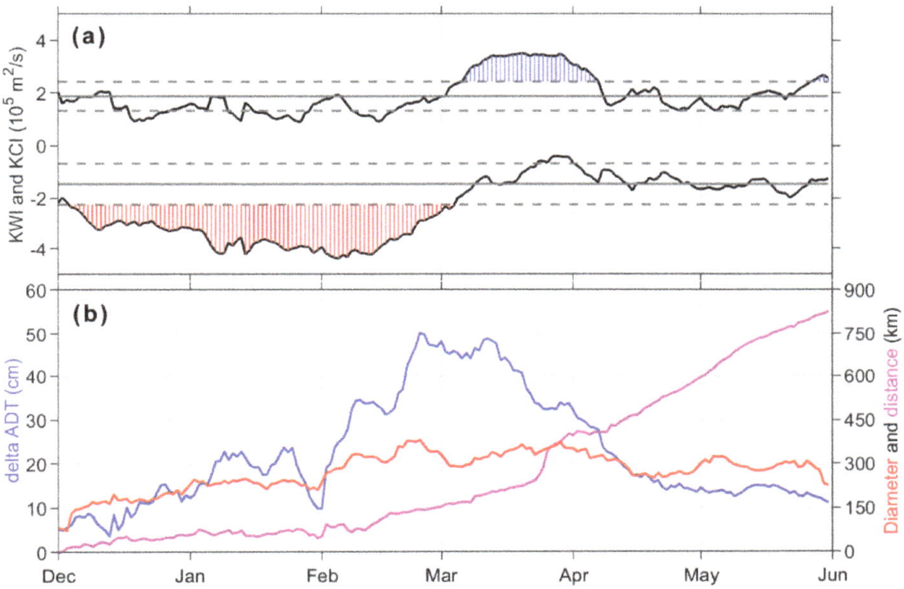

Figure 4. Time series of daily the KCI and KWI (**a**) and properties of the ACE (**b**) from December 2020 to May 2021. The ticks on the time axis stand for the beginning of the month denoted by corresponding labels. Horizontal solid lines in panel (**a**) stand for the mean value (μ), and horizontal dash lines stand for standard deviation span (μ ± σ) of the two indices. KCI values larger than μ + σ are shaded in blue, and KWI values less than μ − σ are shaded in red. In panel (**b**), ADT difference (delta ADT) between the eddy center and edge, diameter of the eddy and distance propagated from the start position are denoted as blue line, red line and magenta line, respectively.

Accordingly, the detailed evolution of DI is shown in Figure 4a. As indicated by the KWI and KCI, the Kuroshio maintained the KWEP for the whole three months from December 2020 to February 2021, and immediately converted to a KCEP in March 2021.

The automatic eddy detection algorithm by geometric criterion [41,42] was adopted to extract the ACE and its kinematic properties. The eddy identification process was performed in each daily ADT map, through two stages: (1) The identification of the local ADT maximum corresponding to the eddy center and (2) the selection of closed ADT contours associated with each eddy. The outermost contour embedding only one eddy center was considered as the eddy edge. The eddy amplitude is defined as the difference between the ADT at the center and that along the edge, while the eddy diameter D is defined as the diameter of a circle with the same area as the eddy area (A_E) enclosed by the eddy edge:

$$D = \sqrt{4A_E/\pi}, \tag{4}$$

The evolution of the above kinematic properties (i.e., amplitude, diameter and propagated distance from the starting position) of the ACE is shown in Figure 4b. We also summarized the statistics of the eddy properties (i.e., amplitude, diameter, propagated distance, lifespan and propagating speed) provided by five previous studies [11,14–17] about the wintertime ACE shed from the KCL in Table 3, together with the estimates in the present study to make a comparison in the following paragraphs.

Table 3. Kinematic properties of the ACEs shed from the KCL in winter, estimated by previous and the present studies.

	Wang et al. [11]	Guo et al. [14]	Jia et al. [15]	Nan et al. [16]	Wang et al. [17]	Present Study
Diameter [1]	A: 244 -	A: 225 M: 300	A: 160 M: 200	A: 128 M: 162	A: 166 M: 320	381
Amplitude [2]	A: 12 M: 32	- -	- -	A: 12 M: 20	A: 11 M: 47	50
Distance [3]	A: 195 -	- -	- -	A: 433 M: 1879	A: 218 M: 1020	>820
Lifespan [4]	A: 108 -	- M: 57	- -	A: 75 M: 273	A: 29 M: 135	>182
Speed [5]	A: 2.1 -	A: 4.5 M: 9.0	A: 10.0 M: 16.0	A: 6.4 M: 11.0	A: 8.3 M: 35.0	5.2

[1] maximum diameter in km; [2] maximum amplitude in cm; [3] maximum distance propagated in km; [4] lifespan in day; [5] average propagating speed in cm/s; A stands for average value, and M stands for maximum value.

The amplitude of the ACE in winter 2020–2021 exhibited a sustained but undulating increase, from as low as 5 cm at the beginning of December, all the way to the peak stage of 50 cm at the end of February, and started to decrease in March. Meanwhile, the eddy diameter had a similar increasing trend matched with the growth of amplitude, reaching a maximum value of 381 km in the peak stage, and basically maintained this scale afterwards. The peak amplitude and diameter in the present study are remarkably higher and larger than the average amplitudes ranging from 11 to 12 cm and the average diameters from 128 to 244 km by all listed previous studies. More importantly, they also exceeded the maximum amplitude of 47 cm and the maximum diameter of 320 km based on the most recent statistics by Wang et al. [17], becoming the strongest ACE shed from the KCL in winter in the available historical records.

By the end of May 2021, this ACE had propagated 820 km from its original position southwestward during a time span of 182 days, hence the average propagating speed being 5.2 cm/s, which is a moderate speed among the average speeds in the listed statistics. The current lifespan and propagated distance had already exceeded the average lifespan from 29 to 108 days and the average distance from 195 to 433 km in Table 3, but remained less than the maximum lifespan of 273 days and the maximum distance of 1879 km by Nan et al. [16]. The time series in the present study stops here, but with the remaining scale and amplitude, a considerable long lifespan and propagating distance in the future months can be expected.

3.3. *Surface Cold-Core Structure*

Another distinctive feature of the ACE in winter 2020–2021 is the cold-core structure at the surface layer, as revealed by the time series of monthly averaged SST shown in Figure 5. Hence, this eddy can be categorized as a CC-ACE, which is uncommon since it is apparently against the convergence and downwelling nature of ACEs.

In December 2020, the overall SST gradient from north to south in the study area had already been established, so the Kuroshio water and the SCS water at the same latitude could be directly distinguished by the SST distribution. The warm tongue injected by the Kuroshio current into the SCS had an SST higher than 26.0 °C and bent to the right on the southwest of Taiwan corresponding to the current loop pattern (i.e., KCL). However, the SST contours were not closed at the eddy center.

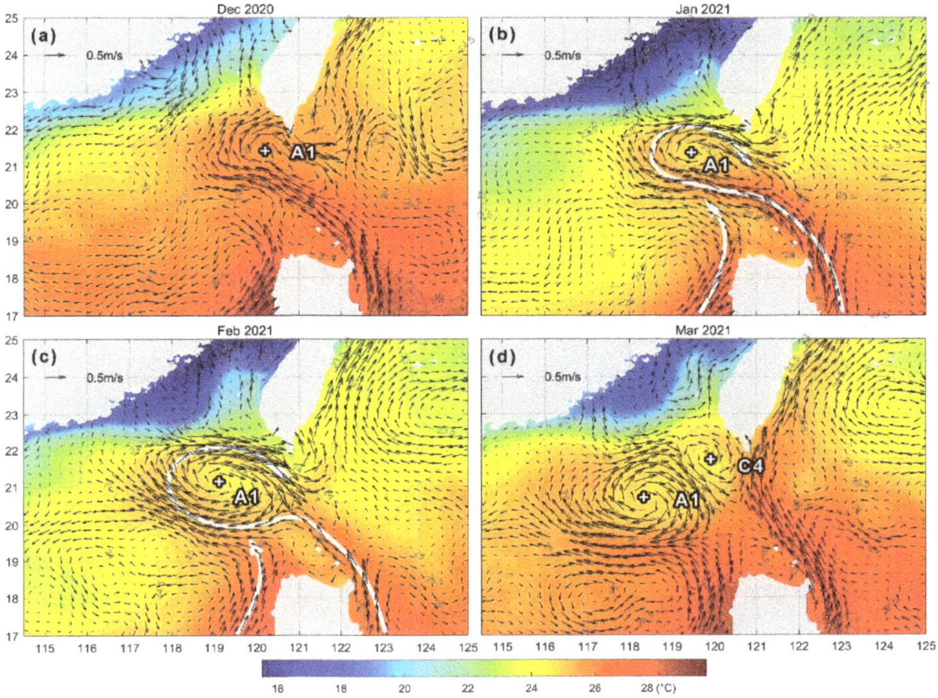

Figure 5. Time series of monthly averaged geostrophic current (vectors) superimposed on monthly averaged SST (filled contours) in the LS and the surrounding area. The panels of (**a–d**) are sequentially for the months from December 2020 to March 2021. The locations of ACE A1 and CE C4 are denoted. The KCL and NWLCC are marked as thick white arrows in (**b**,**c**).

In January 2021, the structure of a cold core enclosed by a warm ring was clearly formed. The cold core had an SST of 24.7 °C, significantly lower than that along the outer ring of approximately 25.5 °C. The warm water was mainly supplied by the westward jet of the Kuroshio (i.e., KCL) through the LS between 20°N and 21°N. The cold-core structure was maintained until February, with a center SST of 24.2 °C. However, the outer ring of warm water expanded to a larger scale. The Northwest Luzon Coastal Current (NWLCC) [43] had also joined with the Kuroshio intrusion to supply warm water.

In March 2021, the warm ring pattern started to collapse, leaving a faint cold core at the eddy center. Moreover, another core with an even lower SST induced by the CE C4 on the east emerged.

3.4. Verification by Drifters

We used the in-situ SST and surface current by drifter observations to verify the features of the Kuroshio path and the accompanying ACE. Figure 1 shows the overall trajectories of the eight drifters deployed in December 2020. Among them, seven drifters were entrained by the strong westward intrusion jet between 20°N and 21°N through the LS into the SCS, and only one drifter stayed in the WP. In particular, Drifter 1485504 (orange line in Figure 1) was tightly trapped by the ACE, and stayed at the eddy core with anticyclonic spinning for the rest of the eddy lifespan, indicating a strong surface convergence.

The drifter observation during the beginning days from 24 December 2020 to 7 January 2021 is shown in Figure 6. Strong westward currents along the trajectories were observed in the middle of the LS between 20°N and 21°N, with a maximum current speed of 1.6 m/s. The geostrophic current by ADT at the corresponding period showed the same pattern of intrusion flow in the LS, but the maximum speed was only 0.7 m/s. Note that the current observation by drifters is the total surface current, including the components of geostrophic current, Ekman current, tidal current, etc. Given that the tidal current in the middle of the LS is not strong (less than 0.2 m/s; e.g., [44]), the Ekman component in the strong intrusion took an important part.

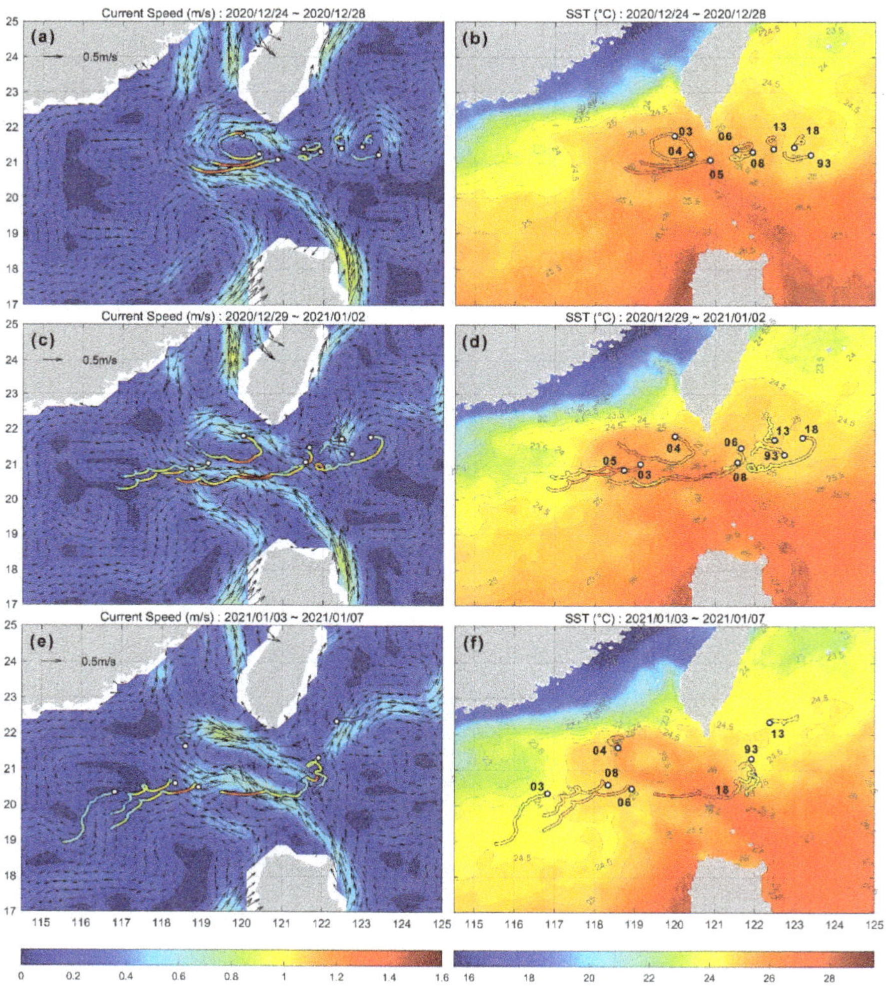

Figure 6. (**a**,**c**,**e**) Floating speed along the drifter trajectories from 24 December 2020 to 7 January 2021, being divided into three 5-day segments and superimposed on the corresponding 5-day averaged geostrophic current speed (colored shading) and geostrophic current vector derived from ADT; (**b**,**d**,**f**) In-situ SST by drifters along the trajectories during the same segments, superimposed on the corresponding 5-day averaged SST (colored shading) by remote sensing. Black circles are the starting points of drifters at each segment, and the last two digits of the drifter number (i.e., 04 is for Drifter 1485504) are used to label drifters at each segment.

The high SST of the intruding warm tongue and the outer warm ring of the ACE were also observed by drifters. The maximum in-situ SST in the middle of the LS was 27.4 °C, even higher than the counterpart by remote sensing, which is probably due to the smoothing effect in the gridded SST product.

Similar to the eddy center indicated by ADT, the daily position of the cold core was also identified at the point with minimum SST around the eddy area. The migration of the eddy center and the cold core center from December 2020 to March 2021 are compared in Figure 7a, together with the floating position of Drifter 1485504 trapped in the ACE. Along with the southwestward propagation of the ACE, the cold core center kept following the eddy center for the entire time span, within a maximum deviation of approximately 50 km. Around 20 days after the deployment of Drifter 1485504, it was tightly trapped around the eddy center and the cold core center, with a spinning radius of approximately 50 km. The floating trajectory was closer to the cold core center than to the ADT center. Due to the relatively coarse resolution of the ADT product and the wide interval between satellite tracks, the cold core by remote-sensing SST provided a better way for tracing the ACE's migration.

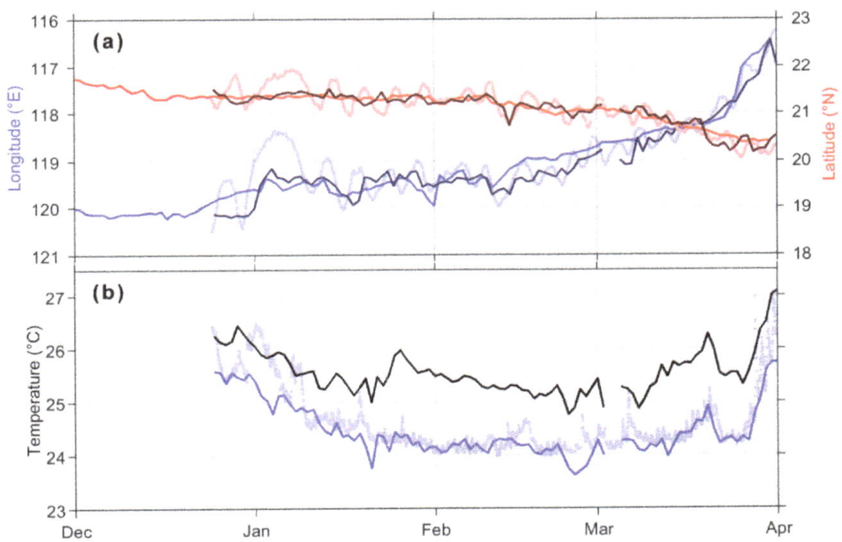

Figure 7. (**a**) Time series of the eddy center position indicated by ADT (blue and red lines), and cold core center by remote-sensing SST (dark blue and dark red lines), and the floating position of Drifter 1485504 (light blue and light red lines); (**b**) Time series of SST at the cold core center (blue line) and outer ring (black line) by remote sensing, and the in-situ SST by Drifter 1485504 (light blue line). The ticks on the time axis stand for the beginning of the month denoted by corresponding labels.

The SST at the cold core center and along the outer warm ring are compared in Figure 7b, together with the in-situ SST by Drifter 1485504. The temperature difference between the cold core and the outer warm ring was established and maintained since late December 2020, and reached a maximum value of 1.7 °C in late January 2021. The drifter SST closely followed that at the core center since mid-January 2021, as a result of the tight trapping by the ACE.

4. Discussion

The above results demonstrated the strong KCL intrusion and accompanying ACE in winter 2020–2021. We analyze the contributing factors for this event, in the aspects of local wind forcing and approaching eddies from the WP.

4.1. Wind Forcing

Figure 8 shows the monthly averaged SSW and WSC from September 2020 to February 2021. Under the control of the wintertime northeast monsoon, the WSC around the LS exhibited a strong dipole on the southwest of Taiwan Island, due to the orographic effect [24]. Intensive positive and negative WSCs were established and maintained from September 2020 to January 2021, and started to decrease in February 2021. The pattern and location of the dipole were basically the same as the monthly climatology, but the negative WSC reached a minimum value of -6×10^{-7} N/m^3 in December 2020, largely exceeding the climatology minimum value of -2×10^{-7} N/m^3. The location of the negative WSC basically covered the BOX1 area in Figure 1, and was also consistent with the locations of the KCL and ACE in the generation and rising phases. The spin-up effect by the negative WSC was in favor of the negative surface current vorticity in this region.

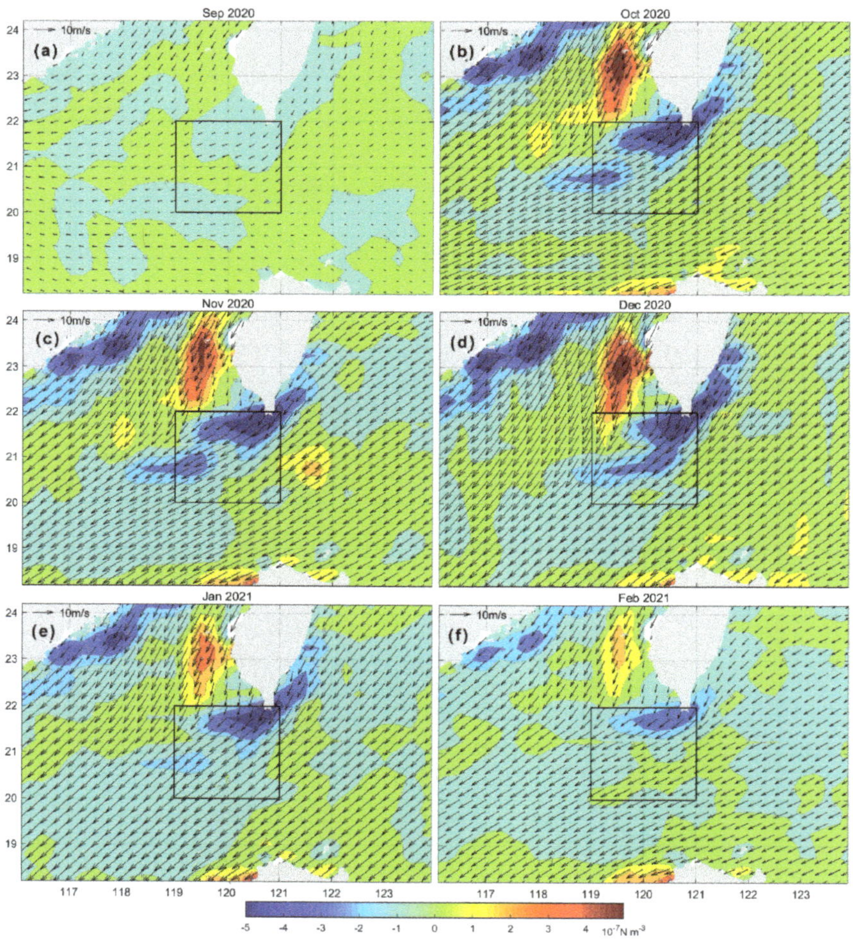

Figure 8. Time series of monthly averaged SSW (black vectors, m/s) and WSC (filled contours, 10^{-7} N/m^3) in the LS and surrounding area. The panels of (**a**–**f**) are sequentially for the months from September 2020 to February 2021. The black rectangle is the same as BOX1 in Figure 1.

In analogy to the KCI and KWI based on geostrophic current, we can also define the KCI_{WS} and KWI_{WS} as the integral of the positive and negative parts of WSC in the same area of 119°E–121°E and 20°N–22°N (BOX1 in Figures 1 and 8):

$$KCI_{WS} = \oiint sign(C)\ C\ dA \quad (5)$$

$$KWI_{WS} = \oiint sign(-C)\ C\ dA \quad (6)$$

where C is the sea surface WSC. The time-series of the daily KCI_{WS} and KWI_{WS} from 1993 to 2021 is shown in Figure 9a. Low KWI_{WS} values beyond the standard deviation only occurred in the winter half of the year from October to March in most of the years. Although winter 2020–2021 did not have the lowest daily KWI_{WS} value across the whole time series, its averaged value over the winter half of the year reached as low as -1.2×10^4 N/m, significantly exceeding those of all other years. Since wind forcing could be an important generation mechanism for the eddy activity in the SCS [22–24], it is evident that the strong and sustaining negative WSC southwest of Taiwan made unusual contributions to the generation and maintenance of the KCL and ACE in winter 2020–2021.

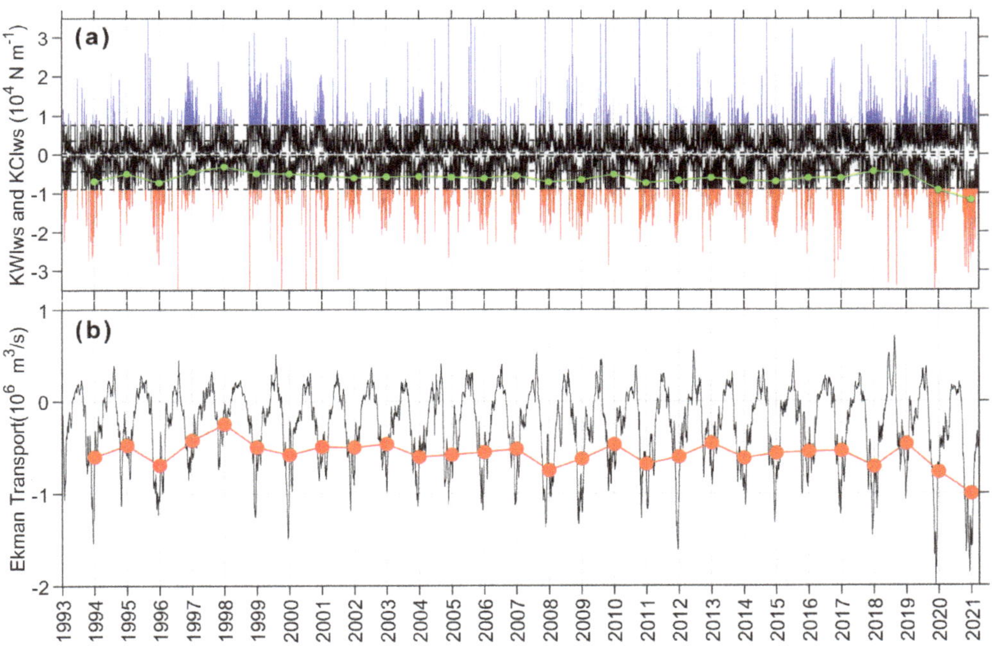

Figure 9. (**a**) Time series of daily KCI_{WS} and KWI_{WS} (black lines) from 1993 to 2021. Horizontal dash lines stand for the mean value (μ) and standard deviation span ($\mu \pm \sigma$) of the two indices. KCI_{WS} larger than $\mu + \sigma$ are shaded in blue, and KWI_{WS} lower than $\mu - \sigma$ are shaded in red. The green dot line is the averaged KWI_{WS} in the winter half of the year over October to March; (**b**) Time-series of daily Ekman transport (black line) through the LS from 1993 to 2021. Negative value means westward transport into the SCS. The red dot line is the averaged Ekman transport in the winter half of the year over October to March. The ticks on the time axis stand for the beginning of the year denoted by corresponding labels.

It has been suggested by Farris and Wimbush [45] that local winds have a direct effect on the Kuroshio intrusion, and the expansion of the KCL is largely determined by the time-integrated WS in the LS. Jia et al. [15] further demonstrated that the ACE shedding from the KCL occurs within one month after the integrated Ekman transport through the LS exceeds a volume of approximately 2 × 10^{12} m^3. The Ekman transport through the LS (across the meridional section at 121°E) was calculated based on the WS data and is presented in Figure 9b. Similar to the orographic WSC, the Ekman transport in 2020–2021 also had strong and sustaining inflow in the winter half of the year. The averaged Ekman transport over the winter half of the year reached −1.0 × 10^6 m^3/s, rated as the strongest Ekman transport inter-annually based on the available WS observation. Such strong transport will easily accumulate the volume beyond 2 × 10^{12} m^3 in less than one month, so as to expand the KCL and induce the ACE shedding in winter 2020–2021.

4.2. Impinging Mesoscale Eddies

Eddy activities in the WP east of the LS during autumn and winter 2020 were examined to discern their influence on the Kuroshio. In general, CEs may reduce the Kuroshio transport by affecting the zonal gradient of the SSH [26,27] or by the resulting upstream convergence and downstream divergence [31]. The reduced Kuroshio transport further provides a favorable condition for its intrusion into the SCS according to the nonlinear hysteresis theory [20,21], and thus induces a westward extension of the Kuroshio pathway into the SCS to form the KCL. Meanwhile, ACEs have the opposite effect, decreasing the looping path.

Since the KCL and ACE event in this study was sustained for approximately four months from November 2020 to February 2021, there were quite a few CEs and ACEs impinging into the Kuroshio at the segment across the LS. As shown in Figure 10, three CEs (C1~C3) and one ACE (A3) propagated westward and finally approached on the east of the LS in different months. Hence, C1~C3 tended to enhance the KCL at their respective time, while A3 tended to decrease the KCL. According to composite analysis by previous studies [33,46], a typical strong KCL corresponds to an impinging CE at a key interaction region around 20°N and 123°E, and the KCL reaches its maximum when the CE arrives at the interaction region after westward propagation. The C3 in February 2021 in our study was apparently such a case, and the looping path accompanied by ACE A1 at the corresponding time was also approximately at the maximum state, which agrees well with the composite pattern. However, C1 and C2 were not so typical, given their location and relatively small scale.

The alternating CEs and ACEs in this region had evolution periods of approximately one month, and the enhancing/decreasing effects of the eddies also faded away within one month according to the composite analysis. While the KCL event in our study exhibited a stable evolution and a rather long period of approximately four months. Thus, the contribution by the eddy activities should not be regarded as the dominant factor for the KCL and ACE event in winter 2020–2021.

Besides the mechanisms discussed above, there are some other factors influencing the performance of the KCL and ACE, such as the intensity of the upstream Kuroshio east of Luzon Island and the North Equatorial Current bifurcation latitude. However, no significant correlation could be found between the inter-annual Kuroshio intrusion and these oceanic and atmospheric environments [18]. It is also beyond this study's scope to provide a clear relationship to link the KCL and ACE events in the 28-year historical records with the aforementioned influencing factors.

As for the case in winter 2020–2021, because the forcing indicators (KWI_{WS} and Ekman transport) and the resultant index performance (integral KWI and sustained period) had values lower than their respective historical extremes in the same year, we suggest that the overwhelming wind forcing was the primary mechanism during this event, with or without the contribution of other factors. A similar case can be found for the KWEP event in winter 2019–2020 (Event 8 in Table 2), when the average KWI_{WS} was −0.92 × 10^4 N/m, and the Ekman transport was −0.76 × 10^6 m^3/s, both were the second lowest values in the available records. However, when wind forcing was not overwhelming enough in other years, such as the case in winter 2007–2008, there may have been no KCL and ACE occurrences due to the balancing between favorable and unfavorable mechanisms.

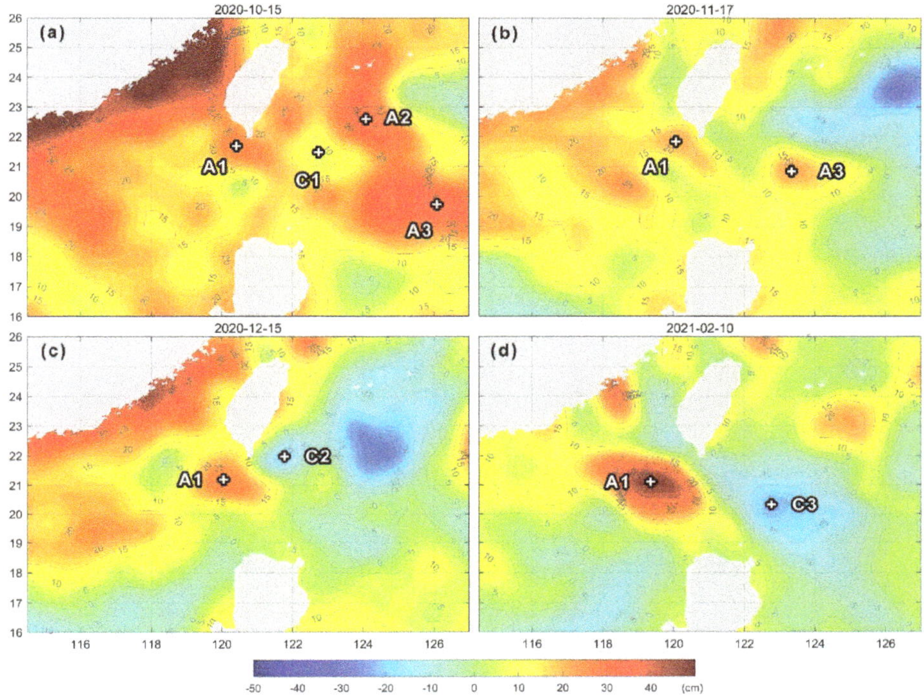

Figure 10. SLA distribution and eddy activities during winter 2020–2021. (**a–d**) are sequentially for 15 October, 17 November, 15 December 2020 and 10 February 2021. A1~A3 are ACEs, and C1~C3 are CEs.

4.3. *Mechanism for the Surface Cold-Core Structure*

The cold-core structure of the ACE shed from the KCL in this study was not common in previous events. As demonstrated by McGillicuddy [37], such a cold-core structure at the surface layer can be generated by local cyclonic wind forcing upon ACEs. Since the CC-ACE in winter 2020–2021 was mainly covered by the negative orographic WSC southeast of Taiwan, it was not a case of such eddy–wind interaction formation mechanism. Given the strong surface flow by drifter observation and Ekman transport in the LS, this CC-ACE event was attributed to the horizontal entrainment [36] of local cold water by the warm KCL water.

In general, most abnormal eddies (CC-ACEs and WC-CEs) cannot survive longer than two weeks [35]. Although it was under the unfavorable condition of strong anticyclonic wind forcing, the CC-ACE in winter 2020–2021 lasted for as long as three months. The long duration of entrainment was stably supported by the warm water from the Kuroshio intrusion and the NWLCC successively, making it a rare case. The continuous surface warm water supply demonstrated the intensity and stability of the Kuroshio intrusion from another aspect.

5. Conclusions

Multiple remote sensing datasets including SSH, SST and SSW, combined with in-situ drifter observations, were used to analyze the Kuroshio intrusion through the LS. The results showed a strong KCL and accompanying ACE in winter 2020–2021.

As quantitatively demonstrated by the DI, the KWI had an integral value of -316.0×10^5 m^2/s·day during the sustained period as long as 90 days, depressed lower than the values in all other years, making it the strongest and longest KCL event in the available historical records.

The KCL started to form in October 2020, and was fully established and maintained from November 2020 to January 2021, and finally converted to a KCEP from February to March 2021 along with the accompanying ACE's detachment.

Remarkable kinematic properties (i.e., amplitude, diameter, propagated distance, lifespan and propagating speed) of the shed ACE were extracted by the automatic eddy detection algorithm. It had a maximum diameter of 381 km and a peak amplitude of 50 cm, which significantly exceeded the maximum values of 320 km and 47 cm by previous statistics, rated as the strongest ACE shed from the KCL in winter.

The contributing mechanisms for the KCL and ACE event were analyzed. The orographic negative WSC southwest of Taiwan and the westward Ekman transport through the LS during the winter half of the year of 2020–2021 both had large values beyond their historical maxima. Hence wind forcing is regarded as the primary mechanism during this event. Alternating CEs and ACEs approaching on the east of the LS were identified, so the interaction between the Kuroshio and the impinging CEs at the proper locations made additional contributions to the enhancement of the KCL.

The accompanying ACE had a distinctive feature of a cold-core structure at the surface layer, so as to be categorized as a CC-ACE, which has rarely been reported in relation to the KCL previously. The temperature difference between the cold core and outer warm ring was maintained for three months and reached a maximum value of 1.7 °C in late January 2021. Given the unfavorable condition of strong anticyclonic wind forcing, the generation and long duration of the CC-ACE was due to the sustained horizontal entrainment supported by the warm water from both the Kuroshio intrusion and the NWLCC successively.

Author Contributions: Conceptualization, J.H.; methodology, Z.S.; software, J.Z.; validation, Z.C.; formal analysis, L.Y.; investigation, X.C. and X.W.; writing—original draft preparation, Z.S.; writing—review and editing, J.H.; visualization, Z.S. All authors have read and agreed to the published version of the manuscript.

Funding: This study was jointly supported by the National Natural Science Foundation of China (41776027, 91958203 and 92058205).

Data Availability Statement: The remote sensing datasets for this research are publicly available, and downloaded from ftp://my.cmems-du.eu and ftp://nrt.cmems-du.eu. The SSH data are at the subdirectory of/Core/SEALEVEL_GLO_PHY_L4_NRT_OBSERVATIONS_008_046; the SST data are at the subdirectory of/Core/SST_GLO_SST_L4_NRT_OBSERVATIONS_010_001; the SSW data are at the subdirectory of/Core/WIND_GLO_WIND_L4_NRT_OBSERVATIONS_012_004. We appreciate the provision of these publicly available datasets.

Acknowledgments: We thank three anonymous reviewers and the Editor for insightful comments that help to improve an earlier version of this manuscript.

Conflicts of Interest: The authors declare no conflict of interest. The funders had no role in the design of the study; in the collection, analyses, or interpretation of data; in the writing of the manuscript, or in the decision to publish the results.

References

1. Metzger, E.J.; Hurlburt, H.E. Coupled dynamics of the South China Sea, the Sulu Sea, and the Pacific Ocean. *J. Geophys. Res. Space Phys.* **1996**, *101*, 12331–12352. [CrossRef]
2. Qu, T. Upper-Layer circulation in the South China Sea. *J. Phys. Oceanogr.* **2000**, *30*, 1450–1460. [CrossRef]
3. Shaw, P.-T. The seasonal variation of the intrusion of the Philippine sea water into the South China Sea. *J. Geophys. Res. Space Phys.* **1991**, *96*, 821–827. [CrossRef]
4. Centurioni, L.R.; Niiler, P.P.; Lee, D.-K. Observations of Inflow of Philippine Sea Surface Water into the South China Sea through the Luzon Strait. *J. Phys. Oceanogr.* **2004**, *34*, 113–121. [CrossRef]
5. Qiu, D.; Yang, T.; Guo, Z. A west-flowing current in the northern part of the South China Sea in summer. *J. Trop. Oceanogr.* **1984**, *3*, 65–73.
6. Li, L.; Wu, B. A Kuroshio loop in South China Sea?—On circulations of the northeastern South China Sea. *J. Oceanogr. Taiwan Strait* **1989**, *8*, 89–95.

7. Caruso, M.J.; Gawarkiewicz, G.G.; Beardsley, R.C. Interannual variability of the Kuroshio intrusion in the South China Sea. *J. Oceanogr.* **2006**, *62*, 559–575. [CrossRef]
8. Nan, F.; Xue, H.; Chai, F.; Shi, L.; Shi, M.; Guo, P. Identification of different types of Kuroshio intrusion into the South China Sea. *Ocean Dyn.* **2011**, *61*, 1291–1304. [CrossRef]
9. Huang, Z.; Liu, H.; Hu, J.; Lin, P. A double-index method to classify Kuroshio intrusion paths in the Luzon Strait. *Adv. Atmos. Sci.* **2016**, *33*, 715–729. [CrossRef]
10. Hwang, C.; Chen, S.-A. Circulations and eddies over the South China Sea derived from TOPEX/Poseidon altimetry. *J. Geophys. Res. Space Phys.* **2000**, *105*, 23943–23965. [CrossRef]
11. Wang, G.; Su, J.; Chu, P. Mesoscale eddies in the South China Sea observed with altimeter data. *Geophys. Res. Lett.* **2003**, *30*, 2121. [CrossRef]
12. Li, L.; Nowlin, W.D.; Jilan, S. Anticyclonic rings from the Kuroshio in the South China Sea. *Deep Sea Res. Part I Oceanogr. Res. Pap.* **1998**, *45*, 1469–1482. [CrossRef]
13. Wang, D.; Xu, H.; Lin, J.; Hu, J. Anticyclonic eddies in the northeastern South China Sea during winter 2003/2004. *J. Oceanogr.* **2008**, *64*, 925–935. [CrossRef]
14. Guo, J.; Yuan, Y.; Xiong, X.; Guo, B. Statistics of the mesoscale eddies on both sides of the Luzon Strait. *Adv. Mar. Sci.* **2007**, *25*, 139–148.
15. Jia, Y.; Chassignet, E.P. Seasonal variation of eddy shedding from the Kuroshio intrusion in the Luzon Strait. *J. Oceanogr.* **2011**, *67*, 601–611. [CrossRef]
16. Nan, F.; Xue, H.; Xiu, P.; Chai, F.; Shi, M.; Guo, P. Oceanic eddy formation and propagation southwest of Taiwan. *J. Geophys. Res. Space Phys.* **2011**, *116*, 12045. [CrossRef]
17. Wang, D.; Fang, G.; Qiu, T. The characteristics of eddies shedding from Kuroshio in the Luzon Strait. *Oceanol. Limnol. Sin.* **2017**, *48*, 673–681.
18. Metzger, E.J.; Hurlburt, H.E. The Nondeterministic Nature of Kuroshio Penetration and Eddy Shedding in the South China Sea. *J. Phys. Oceanogr.* **2001**, *31*, 1712–1732. [CrossRef]
19. Yuan, D.; Han, W.; Hu, D. Surface Kuroshio path in the Luzon Strait area derived from satellite remote sensing data. *J. Geophys. Res. Space Phys.* **2006**, *111*. [CrossRef]
20. Sheremet, V.A. Hysteresis of a Western Boundary Current Leaping across a Gap. *J. Phys. Oceanogr.* **2001**, *31*, 1247–1259. [CrossRef]
21. Yuan, D.; Wang, Z. Hysteresis and Dynamics of a Western Boundary Current Flowing by a Gap Forced by Impingement of Mesoscale Eddies. *J. Phys. Oceanogr.* **2011**, *41*, 878–888. [CrossRef]
22. Gan, J.; Qu, T. Coastal jet separation and associated flow variability in the southwest South China Sea. *Deep Sea Res. Part I Oceanogr. Res. Pap.* **2008**, *55*, 1–19. [CrossRef]
23. Xiu, P.; Chai, F.; Shi, L.; Xue, H.; Chao, Y. A census of eddy activities in the South China Sea during 1993–2007. *J. Geophys. Res. Space Phys.* **2010**, *115*, 03012. [CrossRef]
24. Wang, G.; Chen, D.; Su, J. Winter Eddy Genesis in the Eastern South China Sea due to Orographic Wind Jets. *J. Phys. Oceanogr.* **2008**, *38*, 726–732. [CrossRef]
25. Wu, C.-R.; Hsin, Y.-C. The forcing mechanism leading to the Kuroshio intrusion into the South China Sea. *J. Geophys. Res. Space Phys.* **2012**, *117*. [CrossRef]
26. Jie, Z.; De-Hai, L. Response of the Kuroshio Current to Eddies in the Luzon Strait. *Atmos. Ocean. Sci. Lett.* **2010**, *3*, 160–164. [CrossRef]
27. Lien, R.-C.; Ma, B.; Cheng, Y.-H.; Ho, C.-R.; Qiu, B.; Lee, C.M.; Chang, M.-H. Modulation of Kuroshio transport by mesoscale eddies at the Luzon Strait entrance. *J. Geophys. Res. Oceans* **2014**, *119*, 2129–2142. [CrossRef]
28. Zhong, L.; Hua, L.; Luo, D. The Eddy–Mean Flow Interaction and the Intrusion of Western Boundary Current into the South China Sea–Type Basin in an Idealized Model. *J. Phys. Oceanogr.* **2016**, *46*, 2493–2527. [CrossRef]
29. Hu, J.; Zheng, Q.; Sun, Z.; Tai, C.-K. Penetration of nonlinear Rossby eddies into South China Sea evidenced by cruise data. *J. Geophys. Res. Space Phys.* **2012**, *117*. [CrossRef]
30. Zheng, Q.; Tai, C.-K.; Hu, J.; Lin, H.; Zhang, R.-H.; Su, F.-C.; Yang, X. Satellite altimeter observations of nonlinear Rossby eddy–Kuroshio interaction at the Luzon Strait. *J. Oceanogr.* **2011**, *67*, 365–376. [CrossRef]
31. Chang, Y.-L.; Miyazawa, Y.; Guo, X. Effects of the STCC eddies on the Kuroshio based on the 20-year JCOPE2 reanalysis results. *Prog. Oceanogr.* **2015**, *135*, 64–76. [CrossRef]
32. Nan, F.; Xue, H.; Chai, F.; Wang, D.; Yu, F.; Shi, M.; Guo, P.; Xiu, P. Weakening of the Kuroshio Intrusion into the South China Sea over the Past Two Decades. *J. Clim.* **2013**, *26*, 8097–8110. [CrossRef]
33. Yang, Q.; Liu, H.; Lin, P. The effect of oceanic mesoscale eddies on the looping path of the Kuroshio intrusion in the Luzon Strait. *Sci. Rep.* **2020**, *10*, 636. [CrossRef]
34. Yasuda, I.; Ito, S.-I.; Shimizu, Y.; Ichikawa, K.; Ueda, K.-I.; Honma, T.; Uchiyama, M.; Watanabe, K.; Sunou, N.; Tanaka, K.; et al. Cold-Core Anticyclonic Eddies South of the Bussol' Strait in the Northwestern Subarctic Pacific. *J. Phys. Oceanogr.* **2000**, *30*, 1137–1157. [CrossRef]
35. Ji, J.; Dong, C.; Zhang, B.; Liu, Y. An oceanic eddy statistical comparison using multiple observational data in the Kuroshio Extension region. *Acta Oceanol. Sin.* **2017**, *36*, 1–7. [CrossRef]

36. Sun, W.; Dong, C.; Tan, W.; He, Y. Statistical Characteristics of Cyclonic Warm-Core Eddies and Anticyclonic Cold-Core Eddies in the North Pacific Based on Remote Sensing Data. *Remote Sens.* **2019**, *11*, 208. [CrossRef]
37. McGillicuddy, D. Formation of Intrathermocline Lenses by Eddy–Wind Interaction. *J. Phys. Oceanogr.* **2015**, *45*, 606–612. [CrossRef]
38. Good, S.; Fiedler, E.; Mao, C.; Martin, M.J.; Maycock, A.; Reid, R.; Roberts-Jones, J.; Searle, T.; Waters, J.; While, J.; et al. The Current Configuration of the OSTIA System for Operational Production of Foundation Sea Surface Temperature and Ice Concentration Analyses. *Remote Sens.* **2020**, *12*, 720. [CrossRef]
39. Lumpkin, R.; Özgökmen, T.; Centurioni, L. Advances in the Application of Surface Drifters. *Annu. Rev. Mar. Sci.* **2017**, *9*, 59–81. [CrossRef]
40. Elipot, S.; Lumpkin, R.; Perez, R.C.; Lilly, J.M.; Early, J.J.; Sykulski, A. A global surface drifter data set at hourly resolution. *J. Geophys. Res. Oceans* **2016**, *121*, 2937–2966. [CrossRef]
41. Chaigneau, A.; Eldin, G.; Dewitte, B. Eddy activity in the four major upwelling systems from satellite altimetry (1992–2007). *Prog. Oceanogr.* **2009**, *83*, 117–123. [CrossRef]
42. Chelton, D.B.; Schlax, M.G.; Samelson, R.M. Global observations of nonlinear mesoscale eddies. *Prog. Oceanogr.* **2011**, *91*, 167–216. [CrossRef]
43. Hu, J.; Kawamura, H.; Hong, H.; Qi, Y. A Review on the Currents in the South China Sea: Seasonal Circulation, South China Sea Warm Current and Kuroshio Intrusion. *J. Oceanogr.* **2000**, *56*, 607–624. [CrossRef]
44. Zu, T.; Gan, J.; Erofeeva, S.Y. Numerical study of the tide and tidal dynamics in the South China Sea. *Deep Sea Res. Part I Oceanogr. Res. Pap.* **2008**, *55*, 137–154. [CrossRef]
45. Farris, A.; Wimbush, M. Wind-induced Kuroshio intrusion into the South China Sea. *J. Oceanogr.* **1996**, *52*, 771–784. [CrossRef]
46. Zhang, Z.; Zhao, W.; Qiu, B.; Tian, J. Anticyclonic Eddy Sheddings from Kuroshio Loop and the Accompanying Cyclonic Eddy in the Northeastern South China Sea. *J. Phys. Oceanogr.* **2017**, *47*, 1243–1259. [CrossRef]

remote sensing

Article

Variations in Flow Patterns in the Northern Taiwan Strait Observed by Satellite-Tracked Drifters

Ching-Yuan Lu [1], Po-Chun Hsu [2], Quanan Zheng [3] and Chung-Ru Ho [1,*]

[1] Department of Marine Environmental Informatics, National Taiwan Ocean University, 2 Pei-Ning Road, Keelung 202301, Taiwan; 20681002@mail.ntou.edu.tw
[2] Center for Space and Remote Sensing Research, National Central University, 300 Zhongda Road, Zhongli District, Taoyuan 320317, Taiwan; hpochun@csrsr.ncu.edu.tw
[3] Department of Atmospheric and Oceanic Science, University of Maryland, College Park, MD 20742, USA; qzheng2@umd.edu
* Correspondence: b0211@mail.ntou.edu.tw

Abstract: This study investigates the variations in flow patterns in the northern Taiwan Strait in summer using high-frequency (HF) radar measurements, satellite-tracked drifter trajectories and numerical models. There is an obvious interaction between intra-diurnal tides and ocean currents in northwestern Taiwan. When the tide changes between high tide and low tide, the change in direction of the nearshore flow occurs before the change in the offshore flow. Drifter trajectories show that there are three different drifting paths in the Taiwan Strait in summer. One path is along the west coast of Taiwan from the southwest coast to the northeast coast. Another path is the same as the first one but leads northward to the East China Sea instead of eastward to the northeast coast of Taiwan. The other path exists along the west coast of Taiwan, some distance out, after being deflected by the bottom ridge. The regional ocean modeling system model was used in this study to clarify the influencing factors that lead to these three paths. The results of multiple simulations and HF radar data indicate that the bifurcation of the first two drift paths in northwestern Taiwan is caused by ebb and flood tide transitions. The different routes of the latter two paths are due to the significant speed difference between the nearshore current and the offshore current approximately 45 km from the coast.

Keywords: Taiwan Strait; flow pattern; high-frequency radar; drifter; tide

Citation: Lu, C.-Y.; Hsu, P.-C.; Zheng, Q.; Ho, C.-R. Variations in Flow Patterns in the Northern Taiwan Strait Observed by Satellite-Tracked Drifters. *Remote Sens.* **2022**, *14*, 2154. https://doi.org/10.3390/rs14092154

Academic Editor: Yukiharu Hisaki

Received: 25 March 2022
Accepted: 29 April 2022
Published: 30 April 2022

1. Introduction

The Taiwan Strait (TS), located between China and Taiwan, is a narrow passage that connects to the South China Sea (SCS) in the south and the East China Sea (ECS) in the north. It has been an important waterway since ancient times. As shown in Figure 1, TS is a shallow strait about 60 m in depth on average. There is a deeper water channel in the TS, named Penghu Channel, located along the southwest coast of Taiwan; the bottom ridge of Changyun Rise (CYR) is north of it. The currents in the TS may be influenced by monsoon and long-term winds. Wind fields in the TS are dominated by the East Asian monsoon, which is southwesterly from May to August and northeasterly from September to April [1]. Previous studies mentioned that seasonal variations in volume transport in the TS are related to the reversal of the monsoon [2] and wind stress along the TS [3]. Additionally, the TS is strongly affected by other ocean currents from southern water [4]. These ocean currents include the remnants of the SCS warm current from the SCS and a branch of Kuroshio from the Luzon Strait [5,6]. Tidal currents also contribute to the flow in the Taiwan Strait [7]. In short, the currents in the TS are influenced by the complex topography, the monsoon wind, tides, and ocean currents from south of the TS.

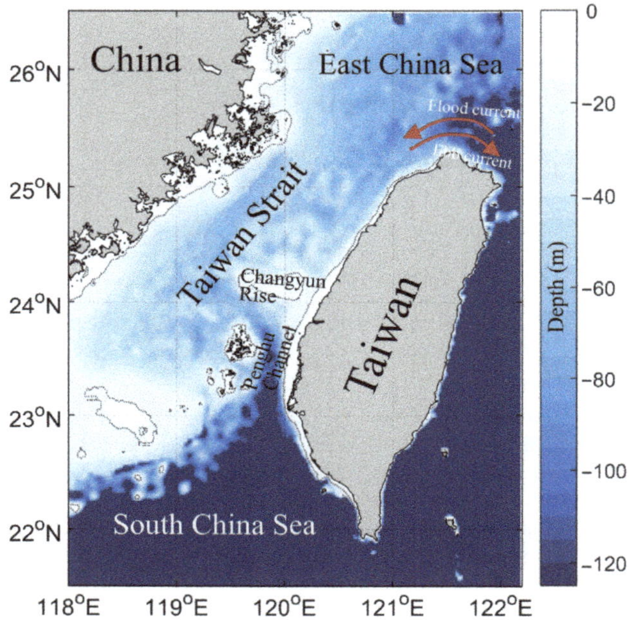

Figure 1. Topography around the TS.

Penghu Channel, which is the deepest passage in the TS, causes most of the water to converge and flow northward. Previous observations from shipboard ACDP show that the Penghu Channel is the major pathway for northward current in the TS, and the velocity is about 1 m/s after removing the tidal effect in the upper 50 m during summer [8]. Therefore, the Penghu Channel is an important entrance to the TS. In addition, the effect of the semidiurnal tide is quite significant in the TS. In the flood tide period, the water enters through the south and north entrances of the TS, and it leaves in the ebb tide period. The tidal range in the northern TS is larger than that in the southern TS [9]. Additionally, the tidal current is larger on the Taiwanese coast than on the mainland coast. The maximum amplitude is 0.80 m/s at the two entrances, and the minimum amplitude is 0.20 m/s in the middle of the TS [10]. Due to the strong tidal effect, the currents at the two entrances of the TS must be affected by the tidal currents. The sea surface temperature and chlorophyll could be moved westward and turned eastward with tidal current according to satellite images [11]. It is well known that the main current flows northward in the TS during summer. However, this phenomenon is based on long-term observations. The above results indicate that the tide, which changes between flood and ebb tides twice a day, might cause significant intra-diurnal variability in the flow pattern, especially at the north end of the TS.

Most previous studies observed flow patterns based on Eulerian descriptions. However, it is hard to depict the path of flow over time. On the contrary, satellite-tracked drifters can provide direct evidence to present the flow pattern in Lagrangian descriptions. A previous study divided the near-surface circulation into four kinds of patterns in winter based on the trajectories of drifters collected before 2007 [12]. In summer, most drifters travel northward through the TS to the ECS. It seems that the flow pattern in the TS was a steady northward flow [12]. However, after passing through the Penghu Channel at a high speed, the seawater immediately encounters the shallow area of CYR. With the influence of complex topography and the strong tidal effect, they may cause complicated changes in local flow fields. Therefore, if the variations in flow patterns in the northern TS can be clarified, it can assist in navigation safety, rescue, or tracking of marine debris. To understand the detailed effects of monsoon, tide, and current interactions, we used satellite-tracked drifters and HF radar to analyze variations in flow patterns and numerical models to find the causes of different current paths in the TS.

The remainder of this paper is organized as follows. Section 2 describes the data and methods. Section 3 presents the characteristics of surface currents with HF radar data and drifters. Section 4 presents the simulations of the nearshore current during summer under several conditions. Section 5 discusses the factors that cause the different current paths. Finally, Section 6 summarizes the main results.

2. Data and Methodology

2.1. High-Frequency Radar Data

The ocean surface current data used in this study are one-hour temporal resolution and 10 km spatial resolution data provided by the Taiwan Ocean Radar Observing System (TOROS) using the Coastal Ocean Dynamics Application Radar (CODAR) developed by the Taiwan Ocean Research Institute (TORI). There are 19 CODAR stations, including 13 sets of long-range 5 MHz systems and six sets of 13/24 MHz systems along the coast of Taiwan Island with the date period from January 2013 to December 2020. Unfortunately, some of the CODAR stations stopped supplying data after December 2020 because of the problems with devices and the lack of components. These high-frequency radars work on the principle of radio wave backscatter and Bragg scattering by analyzing the Doppler frequency shift from the first-order Doppler peak of the sea-echo reflected from the ocean surface. The Doppler frequency shift is due to the ocean current and gravity wave velocities [13]. The phase speed of the gravity wave in deep water is $\sqrt{(g\lambda/2\pi)}$, where g is the gravitational acceleration, and λ is the wavelength of gravity waves. Due to the Bragg scattering, the wavelength of the gravity wave measured by the CODAR is half the radar wavelength. The ocean velocity in the radial direction is calculated based on the difference between the measured velocities of radar and gravity waves. A single radar can only measure the radial velocity; the current vectors need to be determined by multiple CODAR sites.

2.2. Satellite-Tracked Drifter Trajectories

Satellite-tracked drifter trajectory data were downloaded from the National Oceanic and Atmospheric Administration (NOAA) Global Drifter Program database (GDP database) [14]. After the quality control and optimal interpolation procedures by the Drifter Data Assembly Center at Atlantic Oceanographic and Meteorological Laboratory (AOML), the data were interpolated for 6-h intervals. This dataset includes position (longitude, latitude, and time), sea surface temperature, and velocity. The drifter drogue is 15-m long and the bottom of the drogue is about 20-m deep [15]. A total of 30 trajectories were collected from March 1989 to December 2020. These started at the SCS, passed through the Penghu Channel, and entered the ECS north of Taiwan.

2.3. Numerical Model

This study employed the Regional Ocean Modeling System (ROMS) model to better understand the dynamic process and physical mechanism of the coastal flow field in the TS. ROMS is a three-dimensional, realistic bathymetry and free-surface ocean model used to simulate mesoscale and small-scale ocean phenomena around Taiwan [16,17]. The model domain was 118–124°E and 20–26.5°N, using a horizontal resolution of 0.05°. Vertically, there are ten sigma coordinate levels. The bathymetry of the model was extracted from the ETOPO1 database. The initial conditions of the model were set to zero and forced by climatological data from the Comprehensive Ocean-Atmosphere Data Set with open boundaries. The amplitudes and phases of the tidal constituents were derived from Oregon State University global models of ocean tides, TPXO7 [18], along with ten parameters (M2, S2, N2, K2, K1, O1, P1, Q1, Mf, and Mm). The model period was 30 days. The temporal resolution of the model output was one hour. The area in the Kuroshio region east of Taiwan that is in deeper water was not discussed in this study because the simulation focused on the continental shelf in shallow water.

3. Characteristics of Surface Currents in the North Side of TS

3.1. Tide-Current Interaction Observed by HF Radar Data

It is worth noting that the flow direction of the ebb tidal current at the north end of the TS was the same as the background current in summer, but the flood tidal current faced the opposite direction (see Figure 1). Therefore, there was complicated intra-diurnal variability in currents at the north end of the TS during the tidal period. The intra-diurnal variability in surface currents was hard to measure continuously by either satellite altimetry or ship-based sensors. However, coastal HF radar stations could provide high temporal and spatial resolution ocean surface current data around Taiwan. It is possible to observe the variations in nearshore current with the tidal effect; even the tide-current interaction in the TS. Figure 2 shows the average flow fields of the CODAR data in the summer (June to August) of 2017. Due to the southwest monsoon winds and narrow terrain, the background flow field of the TS was dominated by the northeastward current. After passing through the Penghu Channel, the ocean current was deflected by the bottom ridge of CYR and then continued to flow northward in the middle of the TS and entered the ECS. Interestingly, there was a current branch before the TS current entered the ECS. One of the tributaries flowed to the south of the ECS, while the other deflected eastward at a slower speed.

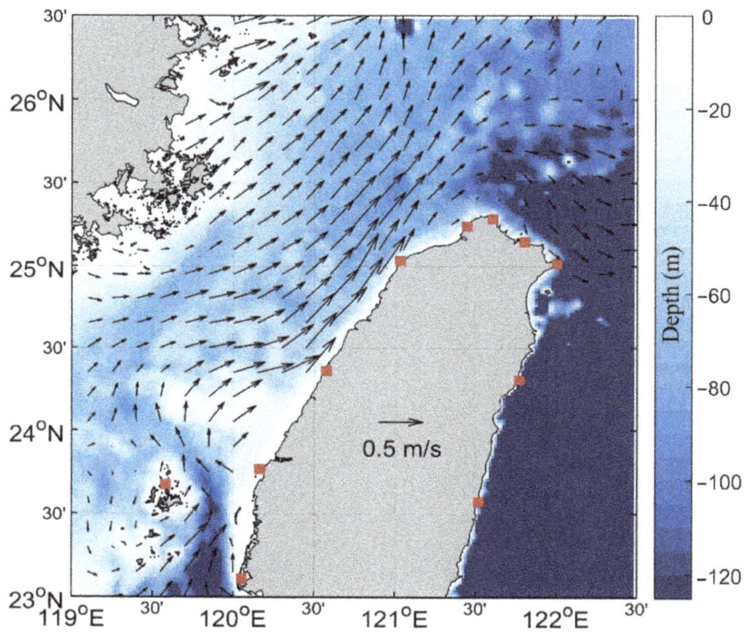

Figure 2. The average flow field in the summer of 2017 obtained from HF radar data. The red squares are the positions of CODAR stations around Taiwan.

In this study, we calculated six-hour average flow fields during different tidal periods based on the Linshanbi tidal station (green square in Figures 3 and 4), which was established by the Central Weather Bureau of Taiwan. Figure 3a shows the average flow field of the six-hour period after high tide in the summer of 2017. Figure 3b shows the average flow field for the six-hour period from one hour after high tide. Figure 3f shows the average current of the six-hour period starting from five hours after high tide in the summer of 2017. As in Figures 3 and 4 show the average flow fields of the six-hour period after low tide in the summer of 2017. According to short-time average flow fields during different tidal periods, there was complicated intra-diurnal variability. In particular, when the period of average current spanned low tide, there was obvious tide–current interaction during the flood tide, which flowed in the opposite direction to the average flow fields in the TS during summer (Figures 3e,f and 4a–d). We also divided the ocean above northern Taiwan into two areas to explore the changes in flow directions nearshore and offshore.

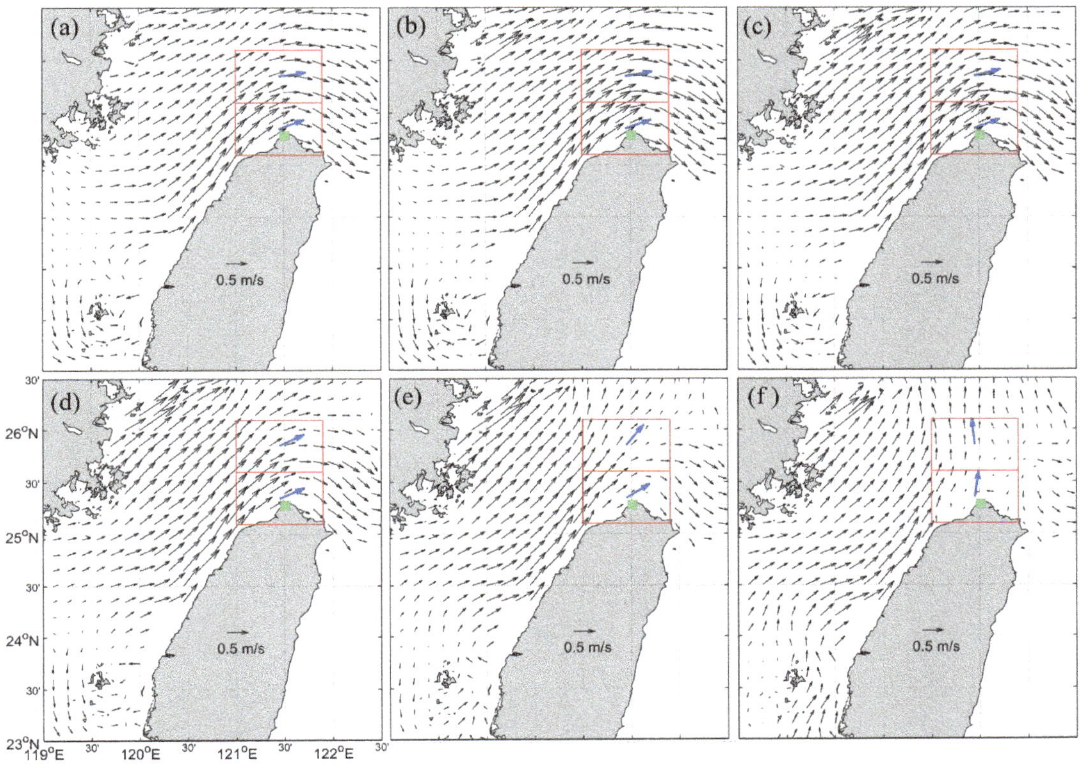

Figure 3. Average flow fields of six-hour periods (**a**) after high tide, (**b**) from one hour after high tide, (**c**) from two hours after high tide, (**d**) from three hours after high tide, (**e**) from four hours after high tide, and (**f**) from five hours after high tide in the summer of 2017. The green square is the tidal station (Linshanbi, located at 25.2839°N, 121.5103°E). The blue arrows represent the averages of flow directions north of Taiwan.

The flow speeds nearshore and offshore during the ebb period could reach 0.52 and 0.41 m/s, respectively (Figure 3b), and the flow directions were almost the same. On the other hand, because the direction of flow in the TS in summer was opposite to that of the flood current, the flow speeds nearshore and offshore were only 0.26 and 0.23 m/s, respectively (Figure 4b), and flowed in different directions. Furthermore, the average flow field began to change in the six-hour periods starting from four hours after high tide (Figure 3e) and three hours after low tide (Figure 4d). As shown in Figure 3e, the offshore current began to turn northward, while the nearshore current still flowed eastward. The flow speeds nearshore and offshore were only 0.24 and 0.20 m/s, respectively. In Figure 4d, the background flow in summer gradually dominated the flow field when the flood current weakened. At this time, the flow speeds nearshore and offshore were only 0.08 and 0.11 m/s, respectively. Table 1 shows the details of flow speeds north of Taiwan displayed in Figures 3 and 4.

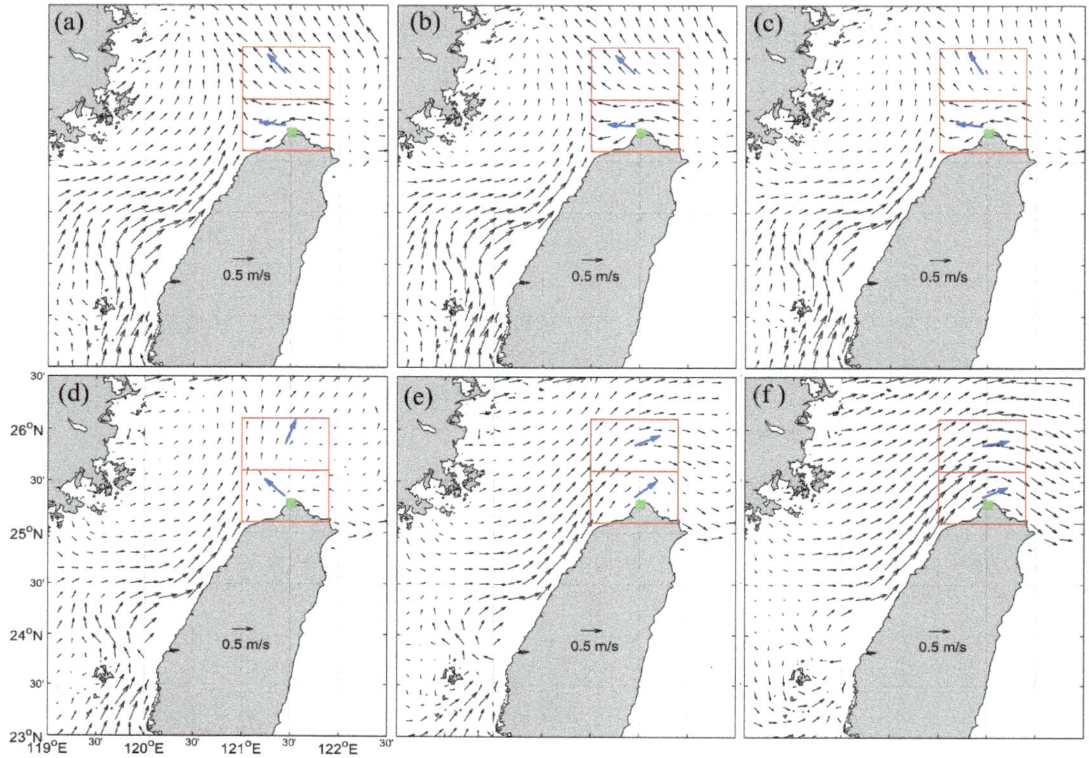

Figure 4. Average flow fields of six-hour periods (**a**) after low tide, (**b**) from one hour after low tide, (**c**) from two hours after low tide, (**d**) from three hours after low tide, (**e**) from four hours after low tide, and (**f**) from five hours after low tide in the summer of 2017. The green square is the tidal station (Linshanbi, located at 25.2839°N, 121.5103°E). The blue arrows represent the averages of flow directions north of Taiwan.

Table 1. The average speed of surface current within the red frame of Figures 3 and 4.

		Average Velocity of Red Frames in Figures 3 and 4 (m/s)					
		(a)	(b)	(c)	(d)	(e)	(f)
Figure 3	offshore	0.37	0.42	0.38	0.29	0.20	0.18
	nearshore	0.44	0.52	0.51	0.41	0.24	0.08
Figure 4	offshore	0.25	0.23	0.16	0.11	0.20	0.32
	nearshore	0.22	0.26	0.20	0.08	0.15	0.33

To sum up, in summer, the velocity of the nearshore current was faster than that of the offshore current during the flood tide period in northern Taiwan, but there was no obvious difference during the ebb tide period. Additionally, the flow speed during the ebb period was about twice as fast as during the flood period. In addition, the flow direction started to change when the tidal period crossed high tides or low tides. At the same time, the flow speed decreased significantly.

3.2. Flow Paths Observed by Drifter Trajectories

This study monitored 30 drifters that drifted northward through the Penghu Channel in the whole GDP database as of December 2020, and most of them drifted through the TS between May and August. There were 24 drifters whose drifting paths could be classified into three types (Figure 5) based on the trajectories; except for six drifters with strange drifting paths that were hard to classify, the classification conditions were as follows:

Path 1: (1) The drifters passed through the Penghu Channel and across the bottom ridge. (2) Then, they drifted close to the west coast of Taiwan. (3) Additionally, they drifted eastward to the northeast coast of Taiwan.

Path 2: (1) The drifters passed through the Penghu Channel and across the bottom ridge. (2) Then, they drifted close to the west coast of Taiwan. (3) They drifted to the south ECS instead of eastward to the northeast coast of Taiwan.

Path 3: (1) The drifters passed through the Penghu Channel and bypassed the bottom ridge. (2) They drifted along the west coast of Taiwan at a distance from the coast. (3) They drifted to the south ECS instead of eastward to the northeast coast of Taiwan.

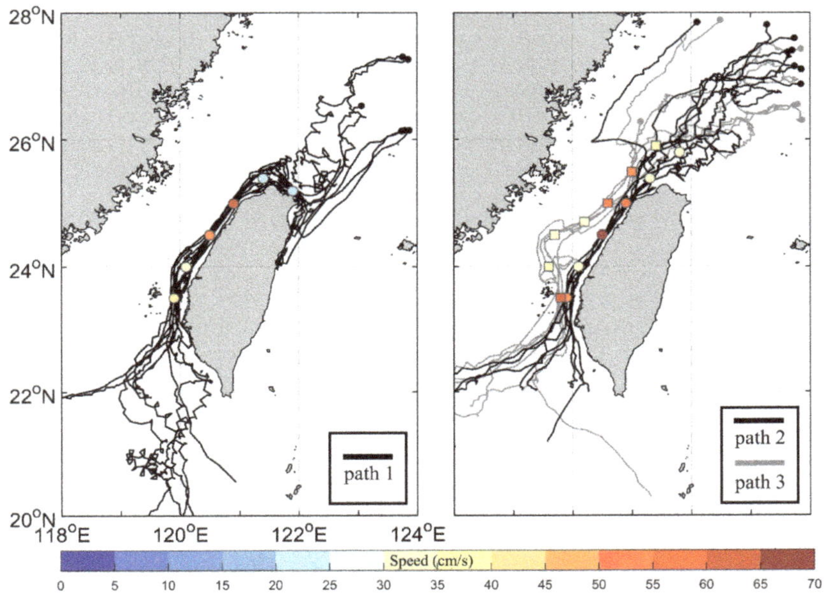

Figure 5. Trajectories of several satellite-tracked drifters that drifted clockwise around Taiwan. The black and gray circles represent the drifter destinations. The colored patches represent the average speeds of the drifting paths.

There were three different flow patterns driving drifters in the TS during summer. Path 1 shows these drifters drifted through the Penghu Channel with an average speed of 0.41 m/s. Then they drifted across the sea surface of the bottom ridge of CYR at 0.32 m/s and drifted northward along the western coast of Taiwan at 0.54–0.65 m/s. When these drifters arrived at the northwestern Taiwanese coast, they slowed down and deflected eastward to the northeast end of Taiwan at 0.20–0.22 m/s. Path 2 was similar to path 1. These drifters also drifted through the Penghu Channel and across the sea surface of the bottom ridge at 0.53 m/s and 0.57–0.68 m/s, respectively. However, they continued to drift northward to the ECS at 0.35 m/s instead of deflecting northward to the northeast of Taiwan. Path 3 was quite different from paths 1 and 2. These drifters drifted through the Penghu Channel with an average speed of 0.63 m/s. Notice that these trajectories were farther away from the west coast of Taiwan than paths 1 and 2. They did not cross the bottom ridge but bypassed it to the middle of the TS at 0.35–0.40 m/s. These drifters drifted to the ECS at 0.33–0.60 m/s through the middle of the TS, far from the coast of Taiwan.

In summary, we found that there were ten, eight, and six drifter trajectories that drifted along paths 1, 2, and 3, respectively, though a total of 30 drifters passed through the Penghu Channel in this study. This means that the probabilities of drifters using the three drifting paths were 33.3%, 26.7%, and 20.0%, respectively. The drifters that flowed along path 1 drifted along the coast of Taiwan until they arrived at the northeast end of Taiwan. Path 2 was similar to path 1 but extended to the ECS instead of deflecting eastward. The destination of path 3 was the same as that of path 2, but path 3 was farther from the coast than that of path 2.

4. Simulations of Nearshore Current in the TS during Summer

In the previous section, we found that there were three types of different flow patterns observed by drifters in the TS. The difference in destination between path 1 and path 2 may be caused by the tide–current interaction. As shown in Figure 3a–d, the flow direction of the ebb tidal current is the same as that of the background current in the northern TS, and this flow direction is similar to the direction of path 1. However, the flow direction of the offshore current started to flow northeastward during the next six-hour period (Figure 3e). In this period, the nearshore current was similar in direction to path 1 but the offshore current was more like path 2. These results show that the transition of ebb and flood tides could cause a different current path in the northern TS. Most of the drifters were drifting through the TS between May and August (summer) (Figure 5). To further explore the influences of tides on ocean surface currents in the TS, the ROMS ocean model was used to simulate the flow field in summer in the TS.

This study used the summer climatological wind to drive the ocean current with open boundaries in the model (Figure 6). Figure 7 presents the average surface flow fields of the ROMS simulation, and Figure 8 shows comparisons of flow speed and direction between ROMS outputs and CODAR observations from 23.5°N–26°N and 119.5°E–122°E. The average surface flow field was divided into five latitude bands at every 0.5° of latitude. The results show that the flow fields of ROMS were close to the average flow fields of CODAR in most latitude bands. To intuitively present the modeled flow fields, we used the particle tracking function in ROMS to simulate drifter trajectories. Two points in the Penghu Channel, the nearshore point (23.60°N, 120.05°E) and the offshore point (23.60°N, 119.85°E), were selected to release the floats (Figure 6). When the model was stable (after 72 h), the simulated floats were continuously released every hour for 360 h. Therefore, 360 simulated floats were released at each point. The depth of release for the simulated floats was 10 m because there was a 15-m-long drogue under the satellite-tracked drifters. Figure 9 shows the probability density distribution of the simulated floating trajectories released from the Penghu Channel. We found that the trajectories of the floats were affected by the distance from the starting position to the shore. The simulated floats that were released from the nearshore position (point N in Figure 6) could drift northward along the west coast of Taiwan. More than 40% of the floats drifted eastward to the north of Taiwan, and about 35% of the floats drifted northward to the ECS. The remaining floats were stopped by the coast. However, the simulated floats released from the offshore position (point O in Figure 6) kept their distance from the coast of Taiwan and drifted northward until arriving in the ECS. More than 50% of the floats drifted northward to the ECS, whereas less than 5% drifted to the north of Taiwan. The remaining floats were stopped by the coast.

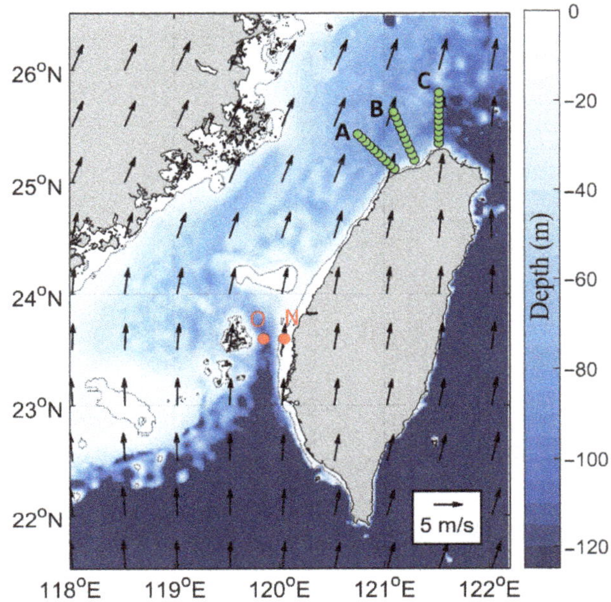

Figure 6. Wind field and seabed terrain set in ROMS model. The red points represent the starting points of the simulated floats. Point N means nearshore and point O means offshore. The green points represent the transects of the vertical profile of ocean flow.

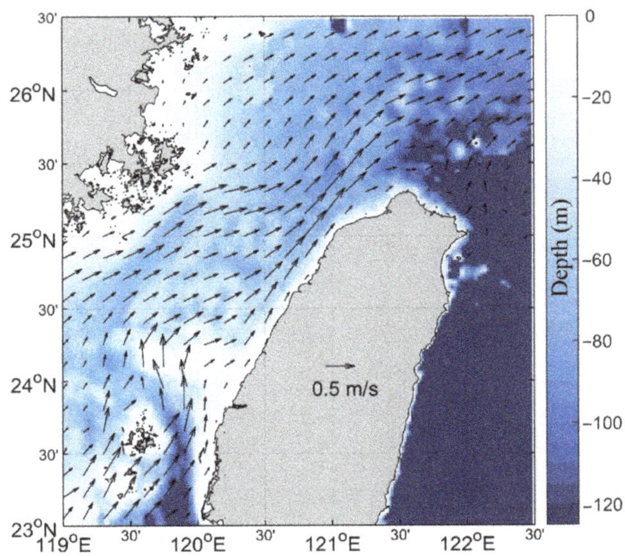

Figure 7. Average surface flow fields of the ROMS simulation in summer.

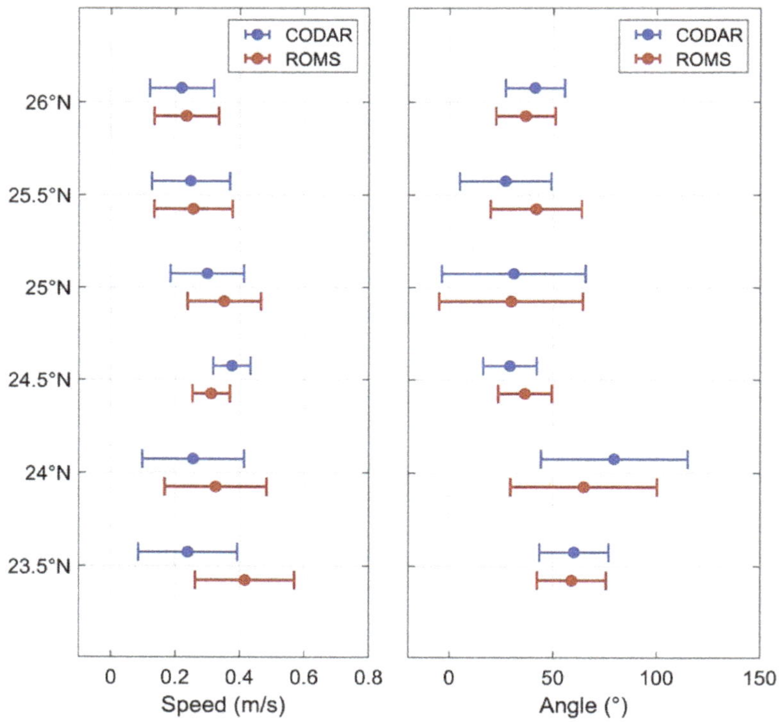

Figure 8. Comparisons of average flow fields between ROMS and CODAR data. The dots represent the mean and the bars represent the standard deviation of each latitude band.

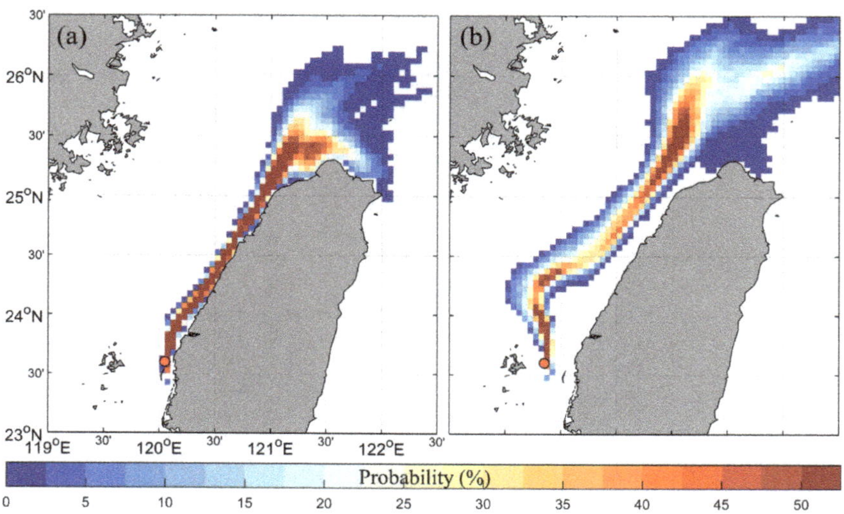

Figure 9. Probability density distribution of simulated floating trajectories released from (**a**) nearshore and (**b**) offshore points in the ROMS simulation with tidal forcing turned on.

To confirm the influence of tides on the ocean current in the TS, we performed another simulation with the same model settings as above but without the tidal forcing. The simulated floats were also deployed at nearshore and offshore points in the Penghu Channel, and the probability density distributions are shown in Figure 10. The results show that the trajectories of simulated floats were similar regardless of whether the tidal forcing was turned on or off before the floats flowed into the ECS. However, there was a significant difference after the floats passed through the TS. More than 30% of the floats released from the nearshore drifted eastward to the north end of Taiwan, but there were a few cases of floats drifting northward to the ECS. The simulated floats released from offshore drifted northeastward after passing northwestern Taiwan.

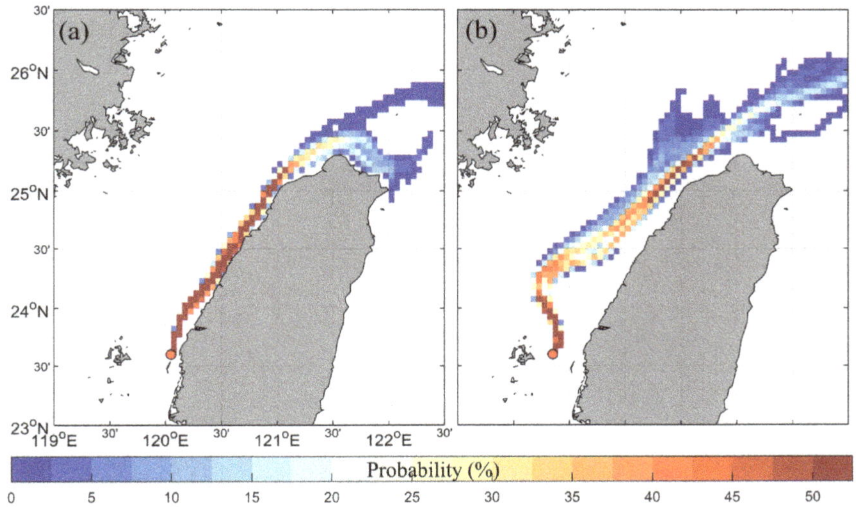

Figure 10. Probability density distributions of simulated trajectories of floats deployed from (**a**) nearshore and (**b**) offshore points without tidal forcing.

Furthermore, there were three transects designed to present the vertical profile of the model current (Figure 6), and the results are shown in Figure 11. The vertical stratification profile of the ocean current along with transect A indicates that the speeds of the nearshore currents were higher than those of the offshore currents, regardless of whether the tidal forcing was turned on or off (Figure 11a,d). The nearshore currents were more concentrated and closer to the coast with the depth until the water depth reached about 30 m. On the other hand, the flow profiles with tidal forcing were more dispersed than those without tidal forcing, whether in transect A or B. The flow profiles along transect C were generated with two high-velocity peaks when the tidal forcing was turned on in the model (Figure 11c). The locations of the two high-velocity peaks are the same as the branching sites of the probability density distribution in northwestern Taiwan (Figure 9a). Thus, it is possible to infer that these ocean flow branches are caused by the transition between ebb and flood tides.

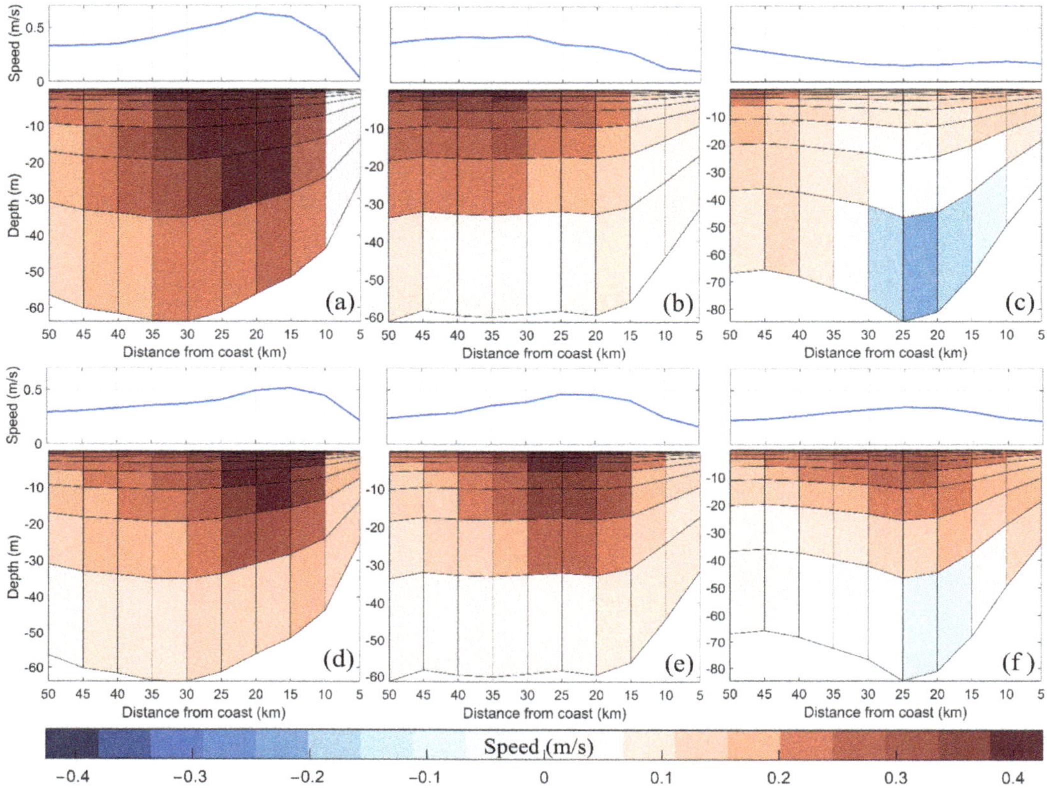

Figure 11. Vertical stratification profile of ocean current along transect A (**a**,**d**), along transect B (**b**,**e**), and along transect C (**c**,**f**) simulated using the ROMS model with (**a**–**c**) or without (**d**–**f**) tidal forcing. The solid blue line represents the surface speed of each vertical profile. Red (Blue) shading represents the direction toward the inside (outside) the plane of the paper (or screen).

5. Discussion

In summer, the main ocean current in the TS is a northward flow consisting of the SCS warm current and the Kuroshio branch current [5,6]. These currents flow northward to the southern ECS after passing by northwestern Taiwan. However, there are obvious transitions of ebb and flood tides from the northern TS to the north end of Taiwan twice a day. The flow direction of the ebb current in northern TS is the same as that of the main current in summer, but the flow direction of the flood current is completely opposite. Therefore, there is complicated intra-diurnal variability in currents in northwestern Taiwan. The variability is hard to observe from the long-term average flow field data. However, the average ocean surface current in a six-hour period from CODAR data suggested that there is diurnal tide–current interaction in the northern TS during summer (Figures 3 and 4). We found that the flow direction started to change when the tidal period crossed high and low tides. The flow direction of the nearshore current would change before the offshore current. Moreover, the speed of the nearshore current is higher than that of the offshore current during the flood tide in northern Taiwan, but there was no obvious difference during the ebb period.

In this study, we classified three different flow patterns according to the drifter trajectories that drifted northward through the Penghu Channel. Drifting path 1 is similar to path 2, as both of them are driven northward by the nearshore current, but they separate in northwestern Taiwan (Figure 5). There is an intra-diurnal tide–current interaction caused by the transitions of ebb and flood tides. Therefore, we infer that the difference between these two drifting paths was caused by a tidal effect. To confirm the influence of the tide, we simulated numerous float trajectories in ocean models. The probability density distributions show that drifting paths 1 and 2 occur with tidal forcing (Figure 9a). The distribution of simulated floats that drifted northward is almost the same as that of those that drifted eastward. However, the probability density distribution only shows drifting path 1 in the absence of tidal forcing (Figure 10a). Few floats could drift northward to the ECS when released at the nearshore position. Drifting path 1 appeared when there was no tidal influence in northwestern Taiwan. However, the ocean current flow eastward in northern Taiwan will be hindered by the flood current, forcing the ocean current to flow northward to the ECS (path 2). After the flood period, the ocean current can flow unhindered to northern Taiwan during the ebb period, as the trajectory of path 1. Thus, path 1 and path 2 will alternately appear in the northern TS as the ebb and flood tides' transition.

The drifting path 1 and path 2 can flow unimpeded by the bottom ridge of CYR after passing through the Penghu Channel, but path 3 is different from them. A previous study simulated the northward near-surface current (at 15 m) in the Penghu Channel in summer, and this current is relatively unimpeded by CYR; only the near-bottom current is deflected anti-cyclonically [19]. In another study, the results of a numerical model showed that the northward current appears to be relatively unimpeded by CYR and bifurcates slightly near the surface (at 20 m) [20]. The results of drifters and simulated floats in this study showed that whether the currents can be deflected by CYR is mainly affected by the distance from the shore. The drifters and floats that pass through the Penghu Channel near the shore will cross over the CYR, whereas those far from the shore are deflected by it. The discrepancy between our results and those of the previous study [19] was caused by the different speeds of the surface current. The surface current speed of that study was up to 1.5 m/s over CYR, so the surface current can be relatively unimpeded by the CYR [19]. However, the flow speeds of simulations in this study were 0.25–0.50 m/s, the same as that of the previous study [20]. Moreover, the drifting speeds of drifters were 0.32–0.68 m/s over CYR (Figure 5), which is also close to our simulation results.

The trajectory of path 3 north of CYR (at 24.5–25°N) is also different from the trajectories of the other two paths. It seems that there are two different current paths. One flows northward along the shore, and the other flows northward at a distance from the shore. Figure 12 shows the average speeds of the cross-shore distribution by the west coast of Taiwan from three datasets, including CODAR, ROMS, and drifters during summer. It should be noted that the average speeds of drifters in Figure 12 were calculated from all drifters passing through the TS in the summer. According to the average velocity of the cross-shore distribution, we found that all three datasets present a rapid alongshore current (Figure 12). It is worth noting that the speed magnitudes of CODAR and ROMS at about 45 km from the coast were reduced to half of those along the coast. Additionally, the velocity magnitude of drifters was even less than 0.10 m/s at a distance of more than 50 km. The results show that there is a significant difference in speed between nearshore and offshore currents. The drifters drifting along path 3 cannot easily blend with those drifting along path 1 and path 2. To sum up, after passing through the Penghu Channel, path 3 is deflected by CYR and flows northward at a distance from the west coast of Taiwan. This geographical factor makes path 3 different from the first two drifting paths, so it flows northward in the middle of the TS to the south of the ECS.

On the other hand, the trajectories of path 2 and path 3 drift northward to the south ECS after passing through the TS. They appear to exhibit a near-inertial oscillation in northeastern Taiwan (Figure 5). According to the inertial oscillation period $T_f = \pi/\Omega \sin\theta$, where Ω is the Earth's rotation rate and θ is latitude, the period is 27.3 h at 26°N. Figure 13 shows the rotary power spectrum of HF radar flow fields in northeastern Taiwan (26°N, 122°E), none of which are close to the inertial oscillation but show a strong semidiurnal tidal period. A previous study also observed that drifter trajectories in the south ECS were oscillated and trapped by strong tidal currents [21]. Therefore, it is speculated that these oscillations were caused by the semidiurnal tide.

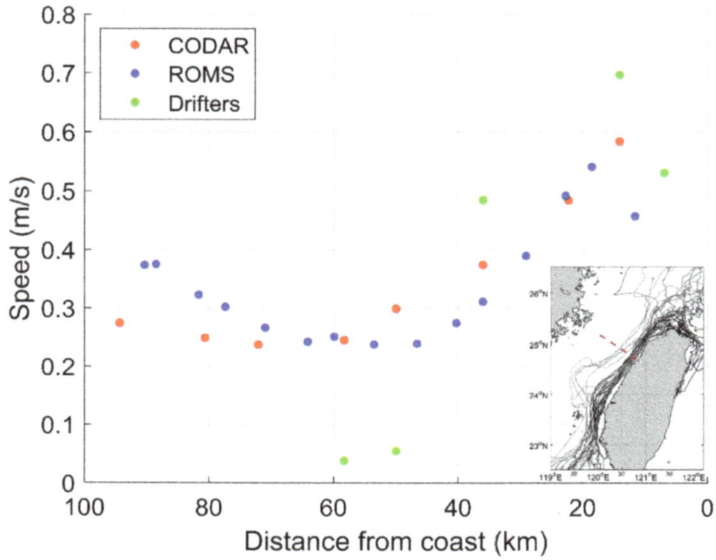

Figure 12. Average alongshore speeds of the cross-shore distributions from three datasets during summer. The red dotted line represents the transect for the cross-shore distribution.

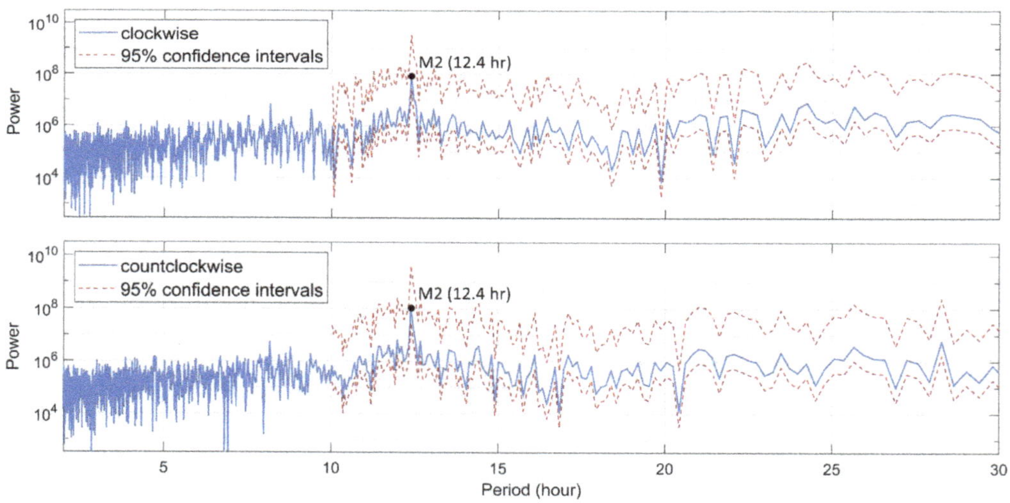

Figure 13. Rotary power spectrum of HF radar in northern Taiwan (26°N, 122°E).

6. Conclusions

In this study, we used the HF radar data to find the intra-diurnal tide-current interaction in the north entrance of the TS from average ocean surface currents of six-hour periods in summer. We found that the flow direction of the nearshore current changes before that of the offshore current after high tide and low tide. Moreover, the speed of the nearshore current is faster than that of the offshore current during the flood tide at the north end of Taiwan, but there is no obvious difference during the ebb period.

Due to the tidal effect and complex topography, there are several flow patterns in the TS. To clarify the variations in flow patterns in the TS, we collected 30 satellite tracking drifters that drifted northward through the Penghu Channel. There were 24 drifters among them that could be classified as having one of three types of drifting paths based on their trajectories. Path 1 presents a flow pattern along the west coast of Taiwan from the southwest coast to the northeast coast. Path 2 is the same as path 1 but flows northward to the ECS instead of eastward to the northeast coast of Taiwan. Path 3 presents a flow pattern that moves along the west coast of Taiwan at a distance from the coast after being deflected by the CYR. Each of the drifting paths has a unique trajectory and different drifting speeds.

The difference between paths 1 and 2 is the bifurcation by northwestern Taiwan, where there is a strong tidal effect. From the results of HF radar and several simulations, we confirmed that the factor that causes the difference is the transition between ebb and flood tides. The current flow eastward in northern Taiwan is hindered by flood currents, thereby forcing the current to flow northward to the ECS. The current can flow unhindered to the north side of Taiwan during the ebb tide period. On the other hand, paths 2 and 3 have the same destination with different trajectories in the TS. To find out the reason, we investigated the alongshore velocities of cross-shore distributions from three datasets, including HF radar, satellite drifter, and ocean model datasets. The results of the alongshore velocity show there is a clear difference in speed between nearshore and offshore currents at about 45 km distance from the coast—the speed is reduced to half of that along the coast. This geographical factor makes path 3 different from the first two: it flows northward from the middle side of the TS to the southern part of the ECS after being deflected by CYR.

In conclusion, there are intra-diurnal tide–current interactions in the north entrance of the TS, as observed by HF radar data. Additionally, there are three different flow patterns in the northern TS in summer. This study clarified the causes of these three flow patterns with HF radar data and ocean simulations. The findings will aid in the safety of ship navigation, search and rescue, and tracking of marine debris. However, we do not know the characteristics of the water masses in different currents or even the flow patterns in other seasons. There are still many deficiencies in this study, and more research is needed.

Author Contributions: Conceptualization, C.-Y.L., Q.Z. and C.-R.H.; methodology, C.-Y.L.; software, C.-Y.L. and P.-C.H.; validation, C.-Y.L., P.-C.H. and C.-R.H.; investigation, C.-Y.L.; resources, C.-R.H.; data curation, C.-Y.L. and P.-C.H.; writing—original draft preparation, C.-Y.L.; writing—review and editing, C.-Y.L., Q.Z. and C.-R.H.; visualization, C.-Y.L.; supervision, C.-R.H.; project administration, C.-R.H.; funding acquisition, C.-R.H. All authors have read and agreed to the published version of the manuscript.

Funding: This research and the APC were funded by the Ministry of Science and Technology of Taiwan under grant MOST 110-2611-M-019-001.

Data Availability Statement: The topography data are publicly available in the ETOPO1 Global Relief Model (https://www.ngdc.noaa.gov/mgg/global/ (accessed on 16 March 2020)). The surface current data are not publicly available. They can be obtained by request from TORI (https://www.tori.narl.org.tw/ETORI/eDefault.aspx (accessed on 3 January 2019)). The tracking drifter data are publicly available in the GDP database (https://www.aoml.noaa.gov/phod/gdp/index.php (accessed on 22 April 2021)). The wind data are provided by the Comprehensive Ocean–Atmosphere Data Set (COADS). The simulated ocean current and floats data are generated by the ROMS ocean model. This model is publicly available on the website of Coastal and Regional Ocean Community (https://www.croco-ocean.org/download/roms_agrif-project/ (accessed on 3 January 2020)). The version of the simulation used in this study is ROMS_AGRIF based on ROMS. The tidal station data were obtained from the Central Weather Bureau of Taiwan (CWB, https://www.cwb.gov.tw/eng/ (accessed on 25 January 2022)).

Acknowledgments: The authors thank NOAA, TORI, and CWB for providing the data sets, and the ROMS community for providing the ROMS model results. Three anonymous reviewers provided valuable comments and suggestions to help improve this paper.

Conflicts of Interest: The authors declare no conflict of interest.

References

1. Oey, L.Y.; Chang, Y.L.; Lin, Y.C.; Chang, M.C.; Varlamov, S.; Miyazawa, S. Cross Flows in the Taiwan Strait in winter. *J. Phys. Oceanogr.* **2014**, *44*, 801–817. [CrossRef]
2. Wu, C.R.; Hsin, Y.C. Volume transport through the Taiwan Strait: A numerical study. *Terr. Atomos. Oceanic Sci.* **2005**, *16*, 377–391. [CrossRef]
3. Chen, H.W.; Liu, C.T.; Matsuno, T.; Ichikawa, K.; Fukudome, K.; Yang, Y.; Doong, D.J.; Tsai, W.L. Temporal variations of volume transport through the Taiwan Strait, as identified by three-year measurements. *Cont. Shelf Res.* **2016**, *114*, 41–53. [CrossRef]
4. Hu, J.Y.; Kawamura, H.; Li, C.Y.; Hong, H.S.; Jiang, Y.W. Review on current and seawater volume transport through the Taiwan Strait. *J. Oceanogr.* **2010**, *66*, 591–610. [CrossRef]
5. Chen, D.; Lian, E.; Shu, Y.; Yang, S.; Li, Y.; Li, C.; Liu, P.; Su, N. Origin of the springtime South China Sea warm current in the southwestern Taiwan Strait: Evidence from seawater oxygen isotope. *Sci. China Earth Sci.* **2020**, *63*, 1564–1576. [CrossRef]
6. Guan, B.X.; Fang, G.H. Winter counter-wind currents off the southeastern China coast: A review. *J. Oceanogr.* **2006**, *62*, 1–24. [CrossRef]
7. Tseng, Y.H.; Lu, C.Y.; Zheng, Q.; Ho, C.R. Characteristic analysis of sea surface currents around Taiwan Island from CODAR observations. *Remote Sens.* **2021**, *13*, 3025. [CrossRef]
8. Jan, S.; Chao, S.Y. Seasonal variation of volume transport in the major inflow region of the Taiwan Strait: The Penghu Channel. *Deep Sea Res. II Top. Stud. Oceanogr.* **2003**, *50*, 1117–1126. [CrossRef]
9. Fang, G.; Yang, J.; Zhao, X. A numerical model for the tides and tidal currents in the Taiwan Strait. *Acta Oceanol. Sin.* **1985**, *4*, 189–200.
10. Wang, Y.H.; Jan, S.; Wang, D.P. Transport and tidal current estimates in the Taiwan Strait from shipboard ADCP observations (1999–2001). *Est. Coast. Shelf Sci.* **2003**, *57*, 193–199. [CrossRef]
11. Hsu, P.C.; Lu, C.Y.; Hsu, T.W.; Ho, C.R. Diurnal to Seasonal Variations in Ocean Chlorophyll and Ocean Currents in the North of Taiwan Observed by Geostationary Ocean Color Imager and Coastal Radar. *Remote Sens.* **2020**, *12*, 2853. [CrossRef]
12. Qiu, Y.; Li, L.; Chen, C.T.A.; Guo, X.; Jing, C. Currents in the Taiwan Strait as observed by surface drifters. *J. Oceanogr.* **2011**, *67*, 395–404. [CrossRef]
13. Paduan, J.D.; Rosenfeld, L.K. Remotely sensed surface currents in Monterey Bay from shore-based HF radar (Coastal Ocean Dynamics Application Radar). *J. Geophys. Res. Oceans.* **1996**, *101*, 20669–20686. [CrossRef]
14. Lumpkin, R.; Centurioni, L. *Global Drifter Program Quality-Controlled 6-Hour Interpolated Data from Ocean Surface Drifting Buoys*; NOAA National Centers for Environmental Information: Washington, DC, USA, 2019.
15. Lumpkin, R.; Maximenko, N.; Pazos, M. Evaluating where and why drifters die. *J. Atmos. Oceanic Technol.* **2012**, *29*, 300–308. [CrossRef]
16. Yang, D.; Yin, B.; Liu, Z.; Bai, T.; Qi, J.; Chen, H. Numerical study on the pattern and origins of Kuroshio branches in the bottom water of southern East China Sea in summer. *J. Geophys. Res. Oceans* **2012**, *117*. [CrossRef]
17. Liu, X.; Wang, D.P.; Su, J.; Chen, D.; Lian, T.; Dong, C.; Liu, T. On the vorticity balance over steep slopes: Kuroshio intrusions northeast of Taiwan. *J. Phys. Oceanogr.* **2020**, *50*, 2089–2104. [CrossRef]
18. Egbert, G.D.; Erofeeva, S.Y. Efficient inverse modeling of barotropic ocean tides. *J. Atmos. Ocean. Technol.* **2002**, *19*, 183–204. [CrossRef]
19. Jan, S.; Wang, J.; Chern, C.S.; Chao, S.Y. Seasonal variation of the circulation in the Taiwan Strait. *J. Mar. Syst.* **2002**, *35*, 249–268. [CrossRef]
20. Wu, C.R.; Chao, S.Y.; Hsu, C. Transient, seasonal and interannual variability of the Taiwan Strait current. *J. Oceanogr.* **2007**, *63*, 821–833. [CrossRef]
21. Hsu, P.C.; Centurioni, L.; Shao, H.J.; Zheng, Q.; Lu, C.Y.; Hsu, T.W.; Tseng, R.S. Surface Current Variations and Oceanic Fronts in the Southern East China Sea: Drifter Experiments, Coastal Radar Applications, and Satellite Observations. *J. Geophys. Res. Oceans* **2021**, *126*, e2021JC017373. [CrossRef]

remote sensing

MDPI

Article

Oil Spill Detection and Mapping: A 50-Year Bibliometric Analysis

Rodrigo N. Vasconcelos [1,*], André T. Cunha Lima [2,3], Carlos A. D. Lentini [2,3,4,5], Garcia V. Miranda [2], Luís F. Mendonça [4,6], Marcus A. Silva [7], Elaine C. B. Cambuí [8], José M. Lopes [2,4] and Milton J. Porsani [3,5]

1 Programa de Pós Graduação em Modelagem em Ciências da Terra e do Ambiente—PPGM, Universidade Estadual de Feira de Santana—UEFS, Feira de Santana 44036-900, Bahia, Brazil

2 Departamento de Física da Terra e do Meio Ambiente, Campus Ondina, Instituto de Física, Universidade Federal da Bahia—UFBA, Salvador 40170-115, Bahia, Brazil; at.cunhalima@ufba.br (A.T.C.L.); clentini@ufba.br (C.A.D.L.); vivasm@ufba.br (G.V.M.); joseml@ufba.br (J.M.L.)

3 Centro Interdisciplinar de Energia e Ambiente (CIEnAm), Universidade Federal da Bahia—UFBA, Salvador 40170-115, Bahia, Brazil; porsani@ufba.br

4 Programa de Pós-Graduação em Geoquímica: Petróleo e Meio Ambiente (Pospetro), Instituto de Geociências (IGEO/UFBA), Universidade Federal da Bahia—UFBA, Salvador 40170-115, Bahia, Brazil; lfelipem@msn.com

5 Programa de Pós-Graduação em Geofísica, Instituto de Geociências (IGEO/UFBA), Universidade Federal da Bahia—UFBA, Salvador 40170-115, Bahia, Brazil

6 Departamento de Oceanografia, Campus Ondina, Instituto de Geociências, Universidade Federal da Bahia—UFBA, Salvador 40170-115, Bahia, Brazil

7 Centro de tecnología e Geociências, LITPEG - Laboratório Integrado de Tecnologia em Petróleo, Gás e Biocombustíveis, Campus Recife, Universidade Federal de Pernambuco—UFPE Recife, Pernambuco 50670-901, Brazil; marcus@ufpe.br

8 Mestrado Profissional em Ecologia Aplicada, Instituto de Biología, Universidade Federal da Bahia—UFBA, Salvador 40170-115, Bahia, Brazil; elainecambui@gmail.com

* Correspondence: rnvuefsppgm@gmail.com; Tel.: +55-(075)-3161-8807

Received: 1 October 2020; Accepted: 1 November 2020; Published: 6 November 2020

Abstract: Oil spill detection and mapping (OSPM) is an extremely relevant issue from a scientific point of view due to the environmental impact on coastal and marine ecosystems. In this study, we present a new approach to assess scientific literature for the past 50 years. In this sense, our study aims to perform a bibliometric and network analysis using a literature review on the application of OSPM to assess researchers and trends in this field of science. In methodological terms we used the Scopus base to search for articles in the literature, then we used bibliometric tools to access information and reveal quantifying patterns in this field of literature. Our results suggest that the detection of oil in the sea has undergone a great evolution in the last decades and there is a strong relationship between the technological evolution aimed at detection with the improvement of remote sensing data acquisition methods. The most relevant contributions in this field of science involved countries such as China, the United States, and Canada. We revealed aspects of great importance and interest in OSPM literature using a bibliometric and network approach to give a clear overview of this field's research trends.

Keywords: bibliometric analysis; remote sensing; oil slicks; oil detection

1. Introduction

Nowadays ocean oil spill is among the most significant environmental impacts [1], bringing damage to ecosystems and biodiversity while causing loss of the ecosystem process [2–4]. In addition to the adverse impacts on the environment, other sectors of society, such as the economy and public health, are documented in the literature [5–8], a fact that reinforces its importance especially lately, where several environmental disasters with oil spills have been observed [3,4].

Since the beginning of the popularization of remote sensing as a science, an increasing number of studies with oil spill detection and mapping (OSPM) have been using remotely detected data for monitoring, surveillance, or risk assessment and management [9–19]. Albeit there are well-formed theoretical and conceptual frameworks according to the remote sensing literature dealing with the detection of oil spills and mapping from in situ observations [20–28], there is still a lack in the scientific community showing the trending techniques and algorithms that systematically summarize this kind of information. OSPM is not a new field of research, indeed although it has been carried out since the 1970s [1,2], it is crucial to conduct a comprehensive overview to investigate and understand the underlying developing patterns in this field of research [3,4]. Subsequently, scientists have used bibliometrics based on mathematical and statistical tools to analyze publications, citations, journals in many disciplines and fields of study [29–32].

Bibliometrics can be commonly defined as a qualitative and quantitative analysis of research that is often used to assess the impact of an individual researcher, research groups, institutions, countries, or journals [33,34]. Its method can analyze several publications to efficiently find influential researchers, authors, journals, organizations, and countries [33,34].

In fact, it can also analyze information intuitively by mapping networks, such as co-word, co-authorship, and co-citation networks [29–31]. Co-word is the co-occurrence of terms extracted from either the title or abstract fields of a dataset to find a specific research topic [30,31]. Co-authorship describes intellectual collaboration in scientific research [31]. Co-citation is defined as two or more publications that are cited by the same manuscript [31].

In recent years, bibliometric network visualization has been used to evaluate the research trends of remote sensing data and its applications in different scientific fields [35,36]. Several studies show that published literature analysis could provide critical information about the research production and scientific quality. Moreover, we believe that describing trends and characteristics of the articles for a specific scientific field can trigger authors' and institutions' collaborations. The same can be done to OSPM studies, which increased significantly in the past few years due to the Deepwater Horizon disaster [3].

A few questions may arise based on our dataset and methods related to oil spill detection and mapping. For instance, which countries have mostly contributed to OSPM research? Is there any trend in OSPM-related publications over the decades? If so, since when? Who are the top researchers and institutions that have focused their work on OSPM? What are the most influential journals in the OSPM field? What are the main differences in terms of the semantic network topology more evident over decades? Therefore, this paper aims to perform bibliometric and network analysis using a literature review regarding the application of OSPM to appraise the research, trends, and characteristics. Additionally, we report the amount of oil spill, the number of accidents, and the main tanker spills in the last 50 years. To the best of our knowledge, no study has specifically investigated OSPM using this conceptual framework. The manuscript is outlined as follows. Section 2 describes the Material and Methods, such as the search strategies and validation, as well as the semantic network analysis. Section 3 shows the publishing trend results of OSPM, co-occurrence, co-authoring, and top-cited authors and journals per decades, followed by Discussion (Section 4) and Concluding Remarks.

2. Materials and Methods

To answer the questions proposed in our study, we built a unified methodological approach that is able to unite classical bibliometric analysis using qualitative descriptors. Moreover, we added the calculation of metrics of the topological structure of the networks. This approach has the advantage of including in the universe of analysis associated with bibliometrics the quantification of topological patterns using quantitative descriptors with proper indexes. That is, the approach used here involves the use of qualitative and quantitative descriptors giving greater robustness to answer our work questions described above. Our approach and steps can be seen in Figure 1. In terms of methodological approach our study is divided into two main stages: Phase 1 and Phase 2. The first phase includes choosing the search database, identifying relevant terms to the subject, and including filters in the search engine. Afterwards, we performed a manual screening reading all the titles and abstracts of the articles returned in the search, identifying and excluding possible articles that did not fit our objective (Phase 1). After these two steps, we performed all analyses (Phase 2) for two data sets: for all articles and for the 25 most cited ones (Table 1). Table 1 describes the relationship between the working questions and the analysis proposed in our methodological approach and source data to make it. In order to make it clear, Table 1 contains three main fields: the questions of interest, the analysis used to answer it, and what kind of data is used.

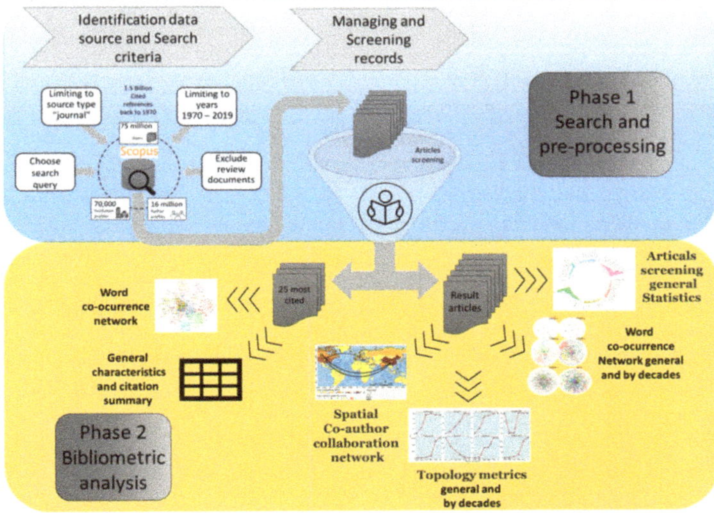

Figure 1. Logical scheme, methodological approach and data analysis for Phases 1 (blue color) and 2 (yellow color).

Table 1. Description of the relationship between working questions, analyses and data sources.

Questions	Analysis	Source Data
What is the publishing trends of OSPM?	General statistics/Word Co-ocurrence network/Co-author spatial network	All papers
Which countries have contributed to OSPM research?	General statistics/Co-author spatial network	All papers
What are the influential publications in the OSPM field?	General charactheristics and citation tables/Co-ocurrence network	25s Most cited
What are the influential journals in the OSPM field?	General statistics/General charactheristics and citation tables	All papers/25s Most cited
Who has contributed to OSPM research?	General statistics/General charactheristics and citation tables/Co-author spatial network	All papers/25s Most cited
What is the research focus in different periods?	Word Co-ocurrence network	All papers
What are the main differences in terms of the semantic network topology more evident over decades?	Topology metrics	All papers
What is the amount and the number of tanker spill in the last 50 years?	Plots of amount and number of tanker spill	Roser (2019)
What is the 20's main tanker spill in the last 50 years?	Table of 20s main tanker spill	Roser (2019)

2.1. Bibliographic Base

In this study, we choose the Scopus database. Scopus is the multidisciplinary largest database of abstracts and references in the literature, accounting for more than 24,000 peer-reviewed journals and 5000 international publishers and citations [37]. It has efficient analysis tools for retrieving and aggregating information and data exportation in multiple formats, which offers a comprehensive overview of the total volume of the world's research products in the areas of science, technology, medicine, social sciences, arts, and humanities. For these reasons, we employed Scopus as the database literature for this research.

This study limited our analysis to journal publications only, reducing bias caused by duplicate publications, minimizing false-positive results. Reviews, conference proceedings, book chapters, and books were not considered because they include works that might have been published more than once in different media sources (Figure 1).

2.2. Search Strategy and Screening Records

We considered documents with a year of publication between 1970 and 2019 (a 50-year study) and studies regarding the application of oil spill detection and mapping only. The search was implemented as follows: only article retrievals containing the words "Oil Spill detection" OR "Oil Spill mapping" in the title, abstract, or keywords were selected. Previous tests were performed with different words, terms and synonyms until we reached the used set of words and terms, which proved to be relevant to the scope of our study.

A pool of selected papers was followed by a manual screening of all retrieved documents to avoid word ambiguity. After that, a systematic review was performed in ten percent of the most cited articles. Within these articles, specific information was extracted, such as: sensor used, spatial and temporal resolutions, bands or feature space evaluated, number of analyzed images, country and region of interest, classification algorithms, mapped classes, or targets. Additionally to report the amount of oil spill, the number of accidents in the last 50 years in the world we use ITOPF data report [38] and the Max Roser data base [39].

2.3. Semantic and General Network Analysis

In this paper, the minimum number of occurrences of a keyword is once for titles, abstracts, and Keywords for all publications to build up a network. The network visualization was carried out using VOSviewer program [30–32,40]. This tool, specifically designed for bibliometric analysis, was employed to visualize the retrieved data of a specific search. VOSviewer can be used to construct networks of scientific publications, scientific journals, researchers, research organizations, countries, and keywords, for example. Items in these networks can be connected by co-authorship, co-occurrence, citation, bibliographic coupling, or co-citation links [31,32].

From a network point-of-view, items are represented by nodes and edges. In this study, nodes are objects as such word co-occurrence and countries. Between any pair of nodes there can be an edge. An edge is a connection or a relation between two nodes. The distance between two nodes in the visualization approximately indicates the relatedness of searching terms in terms of co-occurrence. The size of a label into a node is determined by the weight of an item in a network [31,32].

All articles from our search were analyzed according to their authorship, country, data sources, affiliation, and citations. Individual terms and words in titles, abstracts, and keywords were assessed by their frequency, and a ranking of publication sources by relevance was constructed. Authors' and countries' rankings were also related to productivity analyses. The total and the average number of citations were analyzed by country. Authorship analysis was carried out by computing the number of single-authored articles and multi-authored publications. We also calculated the number of articles per author.

In order to analyze word co-occurrence, all the following settings have the same weight and were flagged as one: number of keywords' occurrence, words and countries' co-authorship. For adequacy, a thesaurus file was built to avoid redundant terms. This text file is used to perform data cleaning when creating a map based on bibliographic data or text data. For more details, the reader should address [29] and [31,32] references.

To calculate all metrics related to semantic networks, we employed Gephi software [41,42]. It is an open-source platform for viewing and manipulating dynamic graphs and hierarchical trees, including all types of networks and complex systems. The user has the autonomy to modify the graphical outputs, as well as to interact with the structure of its network. Moreover, it is possible to add filters highlighting desired aspects and export the final results to SVG, PNG or PDF formats. Its principal function is to serve as a tool for data method analysis, building hypotheses, discovering of social and behavioral patterns, and isolation of essential structures within the hierarchical networks [41,42].

Graph theory uses mathematical structures that model relationships between objects of a specific group [43]. Graphs are defined as a set of vertices and edges [43]. In other nomenclatures, they can be addressed as networks, where the vertices are connected by the edges, determining a relationship of some kind between them, whether directed or undirected. A graph is represented by G = (V, E), where V is a set of vertices and E is a set of edges [43]. Different indexes of the networks' structures were also used to reveal the topological changes in semantic networks over the decades. A description of each index used can be seen in Table 2.

Table 2. Computed topological index and its description and characteristic.

Metrics	Description	Topological Network Metrics Characteristics
Edges Number	Act as the connections that link them to one another a series of connections (edges).	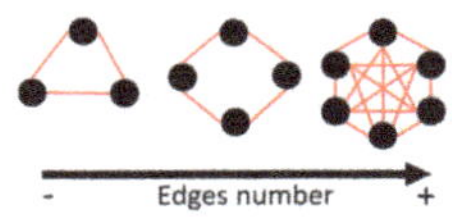
Nodes Number	Refers to the amount of information present in the network.	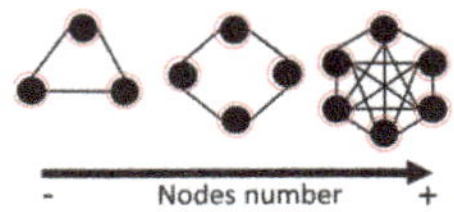
Average Clustering Coefficient	Measure the level at which the nodes are grouped together, as opposed to being equally or randomly connected across the network. Scores on this measure will have an inverse correlation with other statistics, including several of the centrality calculations, particularly when we are speaking at the global level (the entire graph).	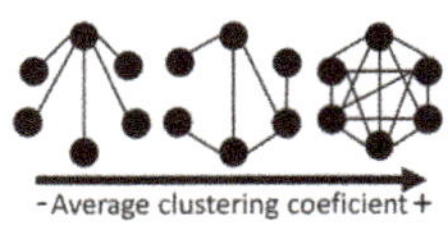
Average degree	Assess importance through the number of direct connections (degrees) one node has to other nodes. The assumption with degree centrality is that the number of connections is a key measure of importance or influence within the network. In an undirected network we do not have the luxury of determining whether one node exerts more or less influence in a relationship; we merely see that they are in fact connected and as such are weighted equally.	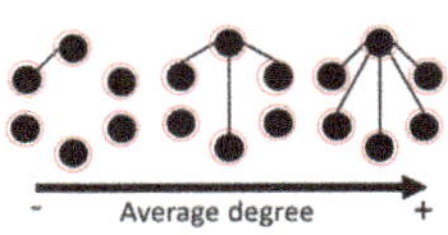

Table 2. *Cont.*

Average Path Length	The clustering coefficient for the word co-occurrence network refers to the probability or level at which the words are grouped together. Indicates how each word is connected to its neighborhood. Average clustering coefficient is the average value of the individual or local coefficients.	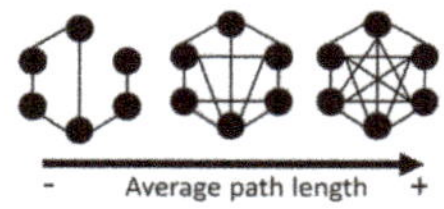
Connected Components	Number of distinct components within the network. When our network is fully connected, a value of 1 will be returned, so there is little need for this calculation. However, in very large networks it might be difficult to visually determine whether the network is fully connected, so we can use this function to ascertain the number of components.	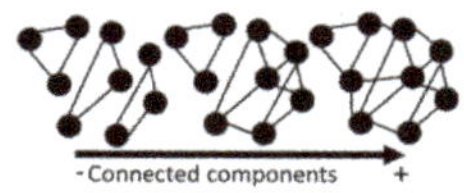
Graph Density	Measure of the level of connected edges within a network relative to the total possible value and is returned as a decimal value between zero and one. Graphs with values closer to one are typically considered to be dense graphs, while those closer to zero are termed as sparse graphs.	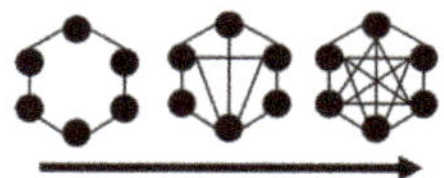
Network Diameter	Refers to the maximum number of connections required to traverse the graph. Another way to look at it is knowing how many steps it takes for the two most distant nodes in the network to reach one another	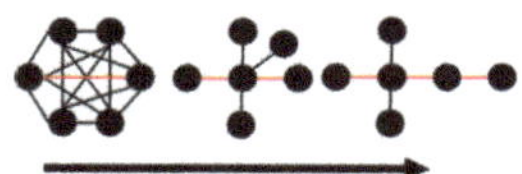

All plot charts were made on R version 3.6.2 [44], using IDE Rstudio, version 1.2.5033 [45], and the ggplot2 package, version 3.2.1 [46].

3. Results

3.1. Publishing Trend of OSPM

The total number of 235 documents was published between 1970–2019 (see Supplementary Materials Table S1 for details) with a mean and standard deviation of 4.7 ± 7.2 papers/year (Figure 2A,B). The number of cumulative publications in the later years increased significantly, being more than twenty-five by the year 2019 (Figure 2A). The first peak of productivity was evident from the mid-1990s, with a peak in 1996 with seven articles, which comprised 3% of the total productivity in the studied period. The highest peak of productivity was observed in 2016 and 2018. Both years accounted for 27 articles, which corresponds to 23% of the total productivity in that decade. This decadal period has the highest number of publications with 168, which corresponds to 71.5% of the total number of published documents in this 50-year time frame with a mean value of 16.8 ± 7.6 publications. This is followed by the 2000s with 17% (mean 4 ± 1.8 SD), 1990s, 8.1% (mean 1.9 ± 2.1 SD), 1980s, 2.1% (mean 0.5 ± 0.7 SD) and 1970s, 1.3% (mean 0.3 ± 0.5 SD). The majority of the retrieved documents were published in English (e.g., 93.2% = 231 manuscripts), followed by 15 documents in Chinese (~6.0%) and one document in Croatian (0.4%).

Figure 3 shows the top 10 manuscripts in terms of their distribution by subject area were Earth and Planetary Sciences (68 papers ~29.1%) Engineering (50 papers ~21.4%), Environmental Science (24 papers ~10%), Physics and Astronomy (22 papers ~9.3%), Computer Science (18 papers ~7.7%), Agricultural and Biological Sciences (13 papers ~5.6%), Chemistry (9 papers ~3.7%), Social Sciences (7 papers ~2.8%), Materials Science (5 papers ~2.3%), and Mathematics (5 papers ~2.1%) (Table S2 for details).

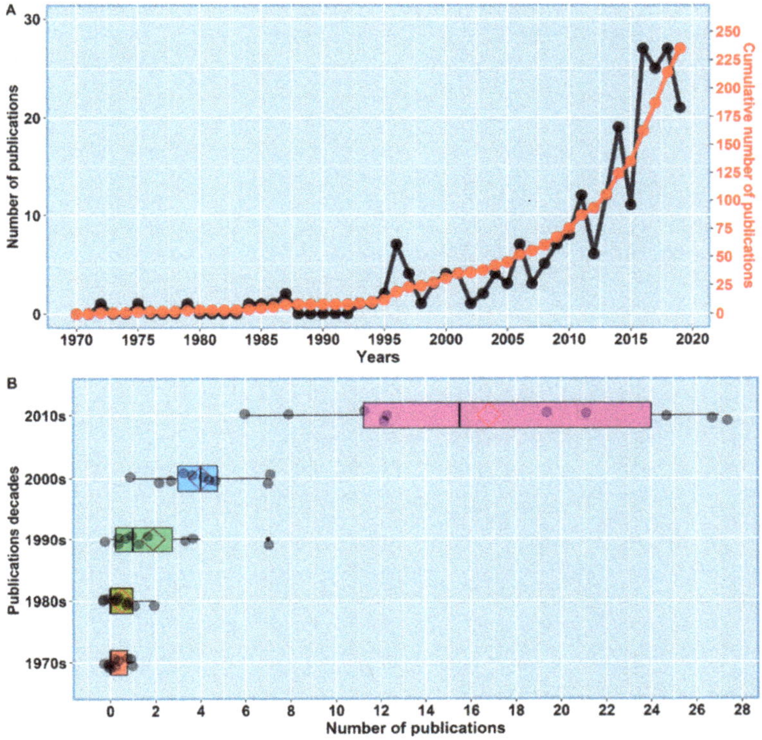

Figure 2. (**A**) Annual growth rate of OSPM publications (black curve, left y-axis) compared to the cumulative annual growth (red curve, right y-axis) rate of all the manuscripts indexed in the Scopus database per year (1970–2019). (**B**) Boxplot by decades. Red diamonds inside the polygons represent the mean.

3.2. Countries Contributed to OSPM Research

We constructed a bibliometric network to visualize collaboration nets and the distribution of publications across the world regarding the application of OSPM (Figure 4). Results show a total of forty-five researchers (45) from different countries with at least one connection with each other. China ranked first, with a total of 61 (i.e., 26% of the total) documents, followed by Italy (30 docs ~12.8% of the total), United States (26 docs ~11.1% of the total), India (16 docs ~6.8% of the total), Spain (14 docs ~6% of the total), Canada (13 docs ~5.5% of the total), Norway (12 docs ~5.1% of the total), Germany (10 docs ~4.3% of the total), Greece (9 docs ~3.8% of the total), United Kingdom (8 docs ~3.4% of the total), (Figures 3 and 4, Table S2).

Regarding the number of citations by countries, the United States ranked first, with a total of 1155 (27% of the total) documents, followed by Canada, Italy, Norway, China, Greece, Spain, United Kingdom, Germany and India, with 1030 (24% of the total), 976 (22,8% of the total), 519 (12.1% of the total), 436 (10.2% of the total), 269 (6.3% of the total), 161 (3.8% of the total), 119 (2.8% of the total) 105 (2.5% of the total) and 41 (1% of the total), respectively.

When only the number of total link strength is considered, described by the weight edges, the United States ranked first with a total of 18, followed by Italy, Canada, India, Greece, Germany, China, United Kingdom, Spain, Norway, with 14, 13, 10, 8, 6, 5, 5,2, 1 respectively. When we observe the relationship of scientific cooperation via publication between countries, the intensity of the links between the United States and China, the United States and Canada, Canada and China, Greece and Italy, the United Kingdom and Portugal, the United Kingdom and Italy is noteworthy (Figure 4). These connections are more evident between countries, indicating strong scientific cooperation on the subject.

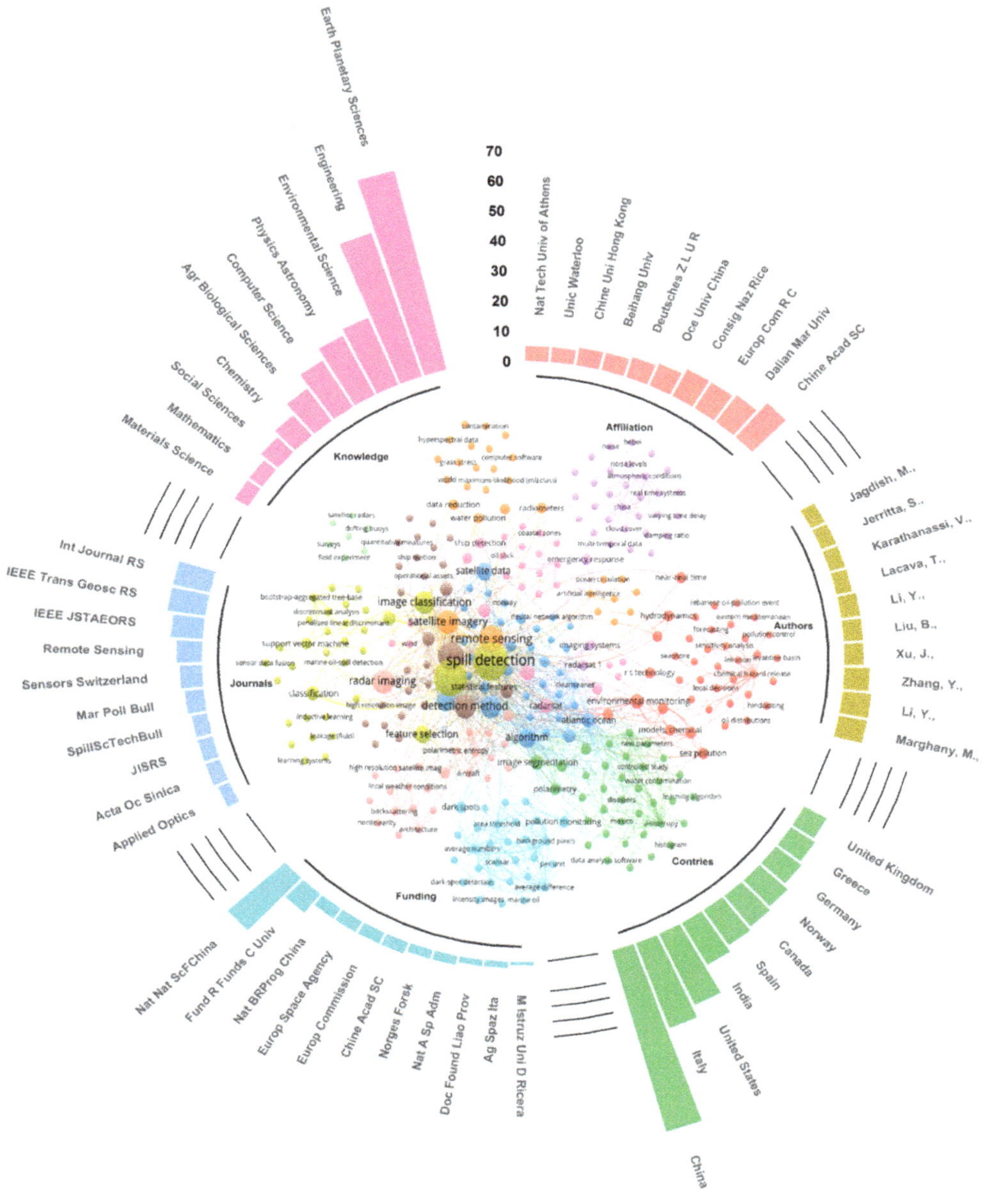

Figure 3. Word co-occurrence network built for the top 25 papers using words presented in titles, abstracts, keywords, and general feature information of the 50 years of documents published between 1970–2019.

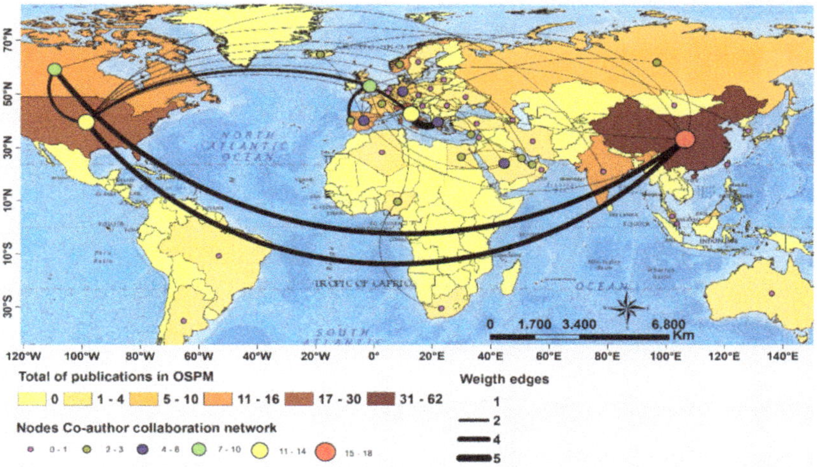

Figure 4. Total of publications on OSPM and co-authoring collaboration network by countries from documents published between 1970–2019.

3.3. Most Influential Publication Features in the OSPM Field

We evaluated the 25 most cited articles for the 50-year span, which accounted for ~10% of the total published papers between 1970–2019. The highest cited published papers can be seen in Tables S2 and S3. This subset represented 61.7% with 2.642 citations out of an overall total of 4.283 in the OSPM field (Tables S3 and S4). The top 10 most cited papers represented 45.1%, which corresponded to 1.923 citations (Table S4). The most cited article in the field of OSPM was from 1998 and had 760 quotations and represented 17.74% of all mentions. The decade that produced manuscripts with the highest number of citations was 1990 with 948 quotes (22.1%), followed by the 2000s and 2010s, with 919 and 18.1 775, respectively (Table S4).

The main terms and words most frequent in this top 25 subset were "spill detection"(24 occurrences), "SAR image" (21), "image analysis" (13), "remote sensing" (12), "detection method" (9), "radar imaging" (9), "satellite imagery" (9), "image classification" (8), "marine pollution" (8), and "algorithm" (6), respectively (Figure 3). The co-occurrence word networks presented 12 clusters totalizing 322 words. The largest cluster contained 40 words and in terms of conceptual approach it dealt with forecasting, pollution and prediction (Figure 3). Methodological approaches, evaluated targets, and sensor systems used in each of the top 25 articles can be seen in Table S3. Overall, all papers had the central objective of generating information regarding the detection of oil spills with different methodological approaches (Table S3). In all articles, oil spills were the main target to be identified, although in some studies other targets have been identified, for example, boats and soil contamination or even levels of pollution severity (Table S3). Except for one article, all the others used radar images to identify oil spills (Table S3). Only one of the manuscripts could be classified as an approach using hyperspectral images (Table S3). Moreover, in only three studies, it was possible to identify the use of a hybrid approach with the use of radar images and optical sensors to spot oil spills (Table S3). Only in one article we identified which sensor system was used (Table S2). In 12 studies, we observed the use of more than one type of radar image for oil spill detection. Of the total of sensor systems listed in the top 25 papers, the following stand out in many uses ranking ENVISAT ASAR with nine articles (17.3%), RADARSAT-1 with eight (15.4%), RADARSAT-2 seven (13.5%), ERS-2 SAR six (11.5), ERS-1 SAR five (9.6%), MODIS four (7.7%), MERIS three (5.8%), TerraSAR-X and UAVSAR with two, and 3.8% for each and Hyperspectral AISA, Landsat 7, Landsat 8, COSMO-SkyMed, ScanSAR, ALOS PALSAR with one paper (~2.0% for each one). The main methodological approaches used to detect oil spills are listed in Table S3. The methodological approaches that were most used were Artificial Neural Networks, Segmentation, Supervised Classification using different algorithms as such Support Vector Machine, Supervised Maximum Likelihood, Generalized Additive Model (Table S3).

There is a positive association of evolution and quality in detecting oil spills and the development and maturation of remote sensing as a science over the 25 most cited articles.

Through 50 years of peer-reviewed published material analyzed on this work using most cited 25 articles, we noticed a remarkable evolution in the digital processing image techniques, classification methods, and extracting relevant features useful to oil spill detection.

There was an association of evolution and quality improvement in detecting oil spills and the development and maturation of remote sensing as a science. Over the decades, the increasing number of published material was a clear sign of the gradual expansion in sensor systems' availability and the increment in the images' spatial and temporal resolution.

Another key point that prompted the search for refinement of oil spill detection methods was the occurrence of disasters. The search for more sophisticated techniques to determine the magnitude of the effects of these disasters served as a significant catalyst for improving detection methods. Besides that, it is important to remark that a few years ago satellite images could be costly but nowadays there are a great number of free images to download.

Over these 25 articles there was a consensus that SAR images provide a fundamental way for monitoring coastal and ocean waters, minimizing its potential adverse effects with a relatively extensive spatial coverage, short repeatability, day and night time, and in almost any climate condition. However, similar dark spots can arise from a range of unrelated meteorological (e.g., rain cells) and oceanographic phenomena (e.g., algae bloom), resulting in incorrect identification, known as look-alikes. Although there are limitations in the usage of optical images due to cloud cover, the auxiliary use of optical satellite images, such as multispectral and hyperspectral sensors, were pointed out as a plausible and suitable alternative in assisting SAR images in detecting oil spills from look-alike surface targets. Therefore, it is clear that there are still challenges associated with detecting oil spills with SAR-derived images in adverse weather conditions such as low wind and heavy rain, and biogenic films.

In specific situations, interpreter supervision was used as an alternative when SAR images were obtained in unfavorable conditions and in the absence of auxiliary data from optical sensor systems.

In the case of developing an automatic system for detecting oil spills, one should be cautious. Classifier balancing and training were the most common ones. Issues such as the low frequency of oil spills compared to the satellite-derived data remains a problem, even with the advance and abundance in the number of sensor systems and images available over the globe.

Machine learning and deep learning are the fundamental methods used over these 25 articles to identify real oil spills from the so-called look-alikes. On the other hand, processing in cloud computing environments was not evidenced over the 25 most cited articles, although it is a topic of great relevance.

According to the most influential publication features found in the neural network-related papers, the most used features in the top 25 articles are: area, perimeter, complexity, spreading, object standard deviation, background standard deviation, max contrast, difference, mean contrast, max gradient, mean gradient, gradient standard deviation, form factor, perimeter to area ratio, ratio between intensity and its standard deviation inside the dark area, granularity, shape, fractal dimension. They are widely used to classify a target as a real oil spill or just a look-alike.

3.4. *Influential Journals in the OSPM Field*

The highest number of published papers by journals can be seen in Figure 3. A total of 127 different journals, with an average of 2.54 publications per year. The top 10 journals in the context of OSPM are "IEEE Transactions on Geoscience and Remote Sensing" and "International Journal of Remote Sensing", with a total of 12 articles which corresponds to 5.1% of the total documents for each one, followed by "IEEE Journal of Selected Topics in Applied Earth Observations and Remote Sensing" (10 papers and

4.3% out of the total), "Remote Sensing" (8 papers and 3.4% out of the total), "Sensors Switzerland" (7 papers and 3% out of the total), "Marine Pollution Bulletin" (6 papers and 2.6% out of the total), both "Acta Oceanologica Sinica", "Journal of The Indian Society of Remote Sensing", and "Spill Science and Technology Bulletin" (5 papers and 3.0% out of the total for each one), and finally "Applied Optics" with four articles corresponding to 1.7% of the total (Figure 3, Table S2).

3.5. Authors Contributing to OSPM Research

Taking into account the number of publications by authors, we found 159 different authors with an average of 1.25 authors per paper. The top 10 articles per author in the context of OSPM can be seen in Figure 3. Li, Y., and Marghany, M., ranked first, with a total of 9 and 3.8% of the total documents for each, followed by Zhang, Y., with 8 (3.4% of the total), Li, Yy., Liu, B., and Xu, J., with a total of 6 and 2.6% of the total documents for each, Jagdish, M., Jerritta, S., Karathanassi, V., Lacava, T., with a total of 5 and 2.1% of the total documents for each, respectively (Figure 3, Table S2).

Considering the author's affiliation, 159 different institutions or organizations published at least one manuscript in the context of OSPM out of that 235 documents, with an average of 1.47 institutions or organizations per document. The top 10 affiliation institutions can be seen in Figure 3. It is possible to show that among this top 10 institutions, 50% are Chinese, 40% European, and 10% from Canada (Figure 3, Table S2). The Chinese Academy of Sciences ranked first, with a total of 15 documents (6.4% of the total), followed by Dalian Maritime University, Consiglio Nazionale delle Ricerche, European Commission Joint Research Centre, Ocean University of China, Deutsches Zentrum fur Luft- Und Raumfahrt, Chinese University of Hong Kong, Beihang University, National Technical University of Athens, University of Waterloo, with 11 (4.7% of the total), 10 (4.3% of the total), 10 (4.3% of the total), 8 (3.4% of the total), 7 (3.0% of the total), 6 (2.6% of the total), 6 (2.6% of the total), 5 (2.1% of the total) and 5 (2.1% of the total) documents, respectively (Figure 3, Table S2).

Similar to what was previously described, if one considered the number of institutions that funded scientific research related to OSPM, 59 names showed up, with an average of 4% institutions or organizations per publication. It is possible to show that among this top 10, 50% are Chinese, 20% European, 30% from the United States, and 10% each for Italy and Norway (Figure 3, Table S2). The National Natural Science Foundation of China ranked first, with a total of 24 documents (10.2% of the total), followed by Fundamental Research Funds for the Central Universities, Chinese Academy of Sciences, European Commission, European Space Agency, National Basic Research Program of China, National Aeronautics and Space Administration, Norges forskningsråd, Agenzia Spaziale Italiana, Doctoral Start-up Foundation of Liaoning Province, Ministero dell 'Istruzione Ministero dell' Università e della Ricerca, with 8 (3.4% of the total), 4 (1.7% of the total), 4 (1.7% of the total),4 (1.7% of the total),4 (1.7% of the total), 3 (1.3% of the total), 3 (1.3% of the total), 2 (0.9% of the total), 2 (0.9% of the total) and 1 (0.4% of the total) documents, respectively (Figure 3, Table S2).

3.6. Decadal Topology of Research Focus and Semantic Networks

Figure 5 shows a co-word occurrence network analysis of the keywords, which can be used to identify the state-of-the-art. The OSPM research topic has been categorized into five colored clusters, following composition over time: 1970s (4 clusters), 1980s (4 clusters), 1990s (10 clusters), 2000s (17 clusters), 2010s (50 clusters), and a comprehensive network for the 50 years with 58 groups (Figure 6).

The ten words with the highest number of co-occurrences between decadal networks and the worldwide network, in general, were spill detection SAR image, marine pollution, remote sensing, image analysis, detection method, satellite imagery, algorithm, neural network, feature selection. These words are a strong indication that they represent important conceptual and methodological constructs associated with OSPM research literature. The largest cluster of each of the co-word networks and the broad network, including all five decades, was associated with the methods used to detect and monitor oil spots, classification algorithms, types of images, and investigated targets (Figure 5).

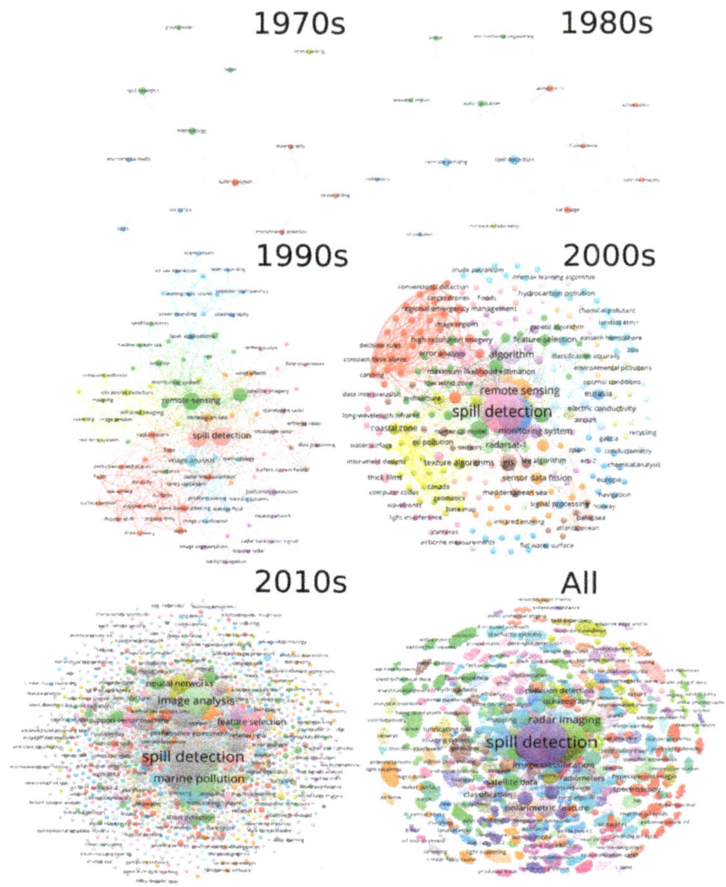

Figure 5. Word co-occurrence network presented in titles, abstracts and keywords for each decade (1970s, 1980s, 1990s, 2000s, and 2010s), as well as for all the scientific articles published in the 50-year time frame (1970–2019).

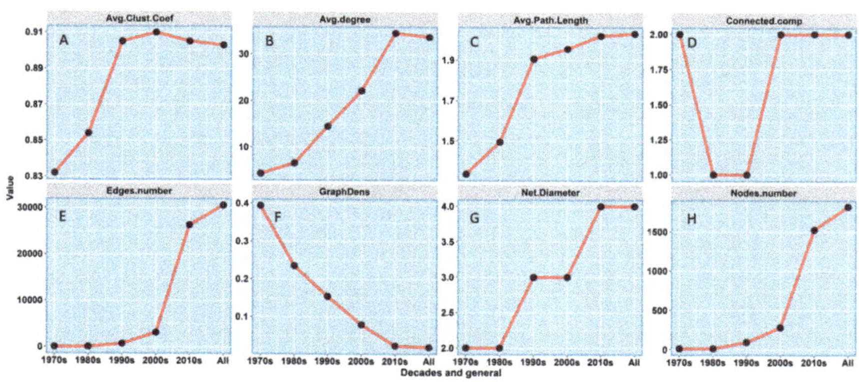

Figure 6. Similar to Figure 4, except for the topology metrics. Letters (**A–H**) indicate a topological metric described in Table 2.

In terms of topological structure, the co-word networks presented a gradual increase in the number of words and their links over the decades. However, there is a decrease in the network density and just a small change in the average clustering coefficient (i.e., ~0.07), especially in the last three decades. This topological behavior indicates that when a set of new terms is added to the field, they are more likely to be linked to the old terms than to each other. This suggests that the evolution of the concepts, method and techniques in the area is not abrupt; instead, they show a gradual increase. This behavior reveals an increase in the area's complexity, resulting in new approaches, themes, and interactions (Figures 5 and 6).

Two components divided the 1970s network. The larger one consists of 26 vertices (Figure 5), representing almost 95% of the network vertices, while the other one has just one vertex (~5%). However, the 1980s and 1990s have only one component, with 46 and 689 vertices (Figure 5). Regarding the 2000s and 2010s, and the overall network for all years, both decades presented two components, with the largest part comprising 46 and 689 vertices (Figure 6). However, these last two decades and the worldwide network of all years, show the largest components consisting of 3126, 26329, and 30529 vertices (Figure 6), representing approximately 98.9%, 99.7% and 99.7% of the network vertices respectively. The second-largest component of the co-word network for the 1970s, the 2000s, the 2010s and the global network was 0, 3, 4, and 4 vertices, representing approximately 0%, 0.10%, and 0.02% and 0.02% of the network vertices, respectively. To avoid confusion or a misleading result, one should keep in mind that the zero value means that the 1970s network has either only one node or no connection at all to another word. However, this same network gives an average centrality degree of 4.33 and a density equal to 0.4; that is, only 4% of the network's possible connections occur. Regarding the last four decades and the overall network, an average centrality degree of 6.57, 14.51, 22.19, 34.56, and 33.70 is observed, followed by a density value equal to 0.235, 0.154, 0.079, 0.023, and 0.019. These numbers correspond to 23.5, 15.4, 0.79, 0.23 and 0.19% of the possible connections in the network occur, respectively.

The 1970s network gives an average clustering coefficient of 0.8. The following four decades and the global network provide a coefficient of 0.85, 0.9, 0.9, 0.9, and 0.9, respectively.

The network diameter for the 1970s and the 1980s is the same and equals two, while the 1990s and the 2000s is a little higher (i.e., 3), and 2010s, as well as the worldwide network, amounts to four. The average path distance length between a pair of vertices for all the five decades increased over time: 1.3 in the (1970s), 1.5 (1980s), 1.9 (1990s), 1.95 (2000s), and 2.20 (2010s).

The most significant change in the network's topology occurred between the 1980s and 1990s, increasing the clustering coefficient and the network's average minimum path (Figure 6). This change occurred due to the inclusion of new techniques such as image analysis, perturbation techniques, and laser applications, mostly due to a new oceanographic approach shown in the dark blue module in the 1990s network.

3.7. Amount and the Number of Oil Spill in the Last 50 Years

This result is based on the ITOPF data report [38] and the Max Roser database [39]. A general negative trend in the number of oil spills from tankers indicates a marked decrease of accidents with time (Figure 7A). A similar trend can be also seen in the amount of oil spills with time (Figure 7B) until 2010, when the Deepwater Horizon resulted in the worst oil disaster in oil drilling history, and then in 2018 off the Shanghai coastline, China, when the oil tanker named SANCHI spilled over 113,000 tonnes of oil (Table 3).

According to Table 3, almost all the recorded oil spill incidents took place before the year 2000, except for the oil tanker named SANCHI (9th ranking position), which occurred in 2018 off the Shanghai coastline, China, and the Deepwater Horizon oil drilling platform (1st ranking position) in 2010 (Gulf of Mexico). Although the PRESTIGE (2002, off Galician waters—Spain) and the EXXON VALDEZ (1989, off Alaskan waters—USA) are some of the most worldwide well-known disasters, the total amount of oil spilled was less than the KATINA P oil tanker (Table 3). These two vessels were responsible for oil spill sizes of 63,000 and 37,000 tons, respectively.

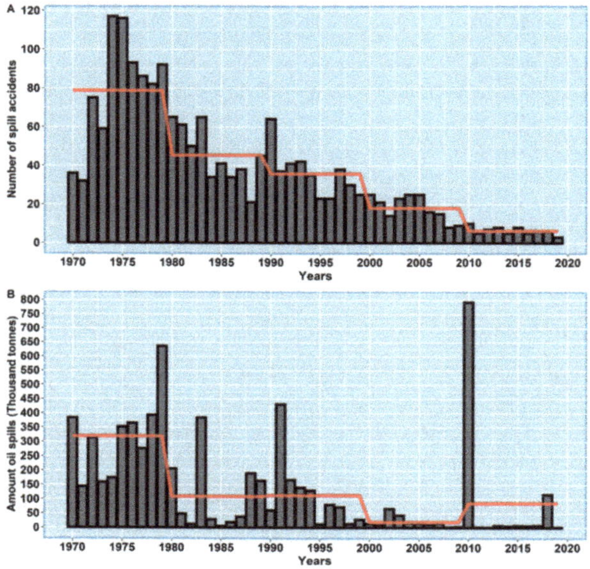

Figure 7. Number of spill events (**A**) and amount of oil spill (**B**, in thousand tonnes) over last 50 years (i.e., 1970–2019 record) based on ITOPF data report and the Max Roser database. Red line indicates the mean of each decade, respectively.

Table 3. Top 20 oil spill disasters around the world based on the amount of spill size for the 1970–2019 record based on the public media.

Shipname/Platform	Year	Location	Spill Size (Tonnes)
DEEPWATER HORIZON	2010	Macondo Prospect, Central Gulf of Mexico	780,000
ATLANTIC EMPRESS	1979	Off Tobago, West Indies	287,000
ABT SUMMER	1991	700 nautical miles off Angola	260,000
CASTILLO DE BELLVER	1983	Off Saldanha Bay, South Africa	252,000
AMOCO CADIZ	1978	Off Brittany, France	223,000
HAVEN	1991	Genoa, Italy	144,000
ODYSSEY	1988	700 nautical miles off Nova Scotia, Canada	132,000
SEA STAR	1972	Gulf of Oman	115,000
SANCHI*	2018	Off Shanghai, China	113,000
IRENES SERENADE	1980	Navarino Bay, Greece	100,000
URQUIOLA	1976	La Coruna, Spain	100,000
HAWAIIAN PATRIOT	1977	300 nautical miles off Honolulu	95,000
INDEPENDENTA	1979	Bosphorus, Turkey	95,000
JAKOB MAERSK	1975	Oporto, Portugal	88,000
BRAER	1993	Shetland Islands, UK	85,000
AEGEAN SEA	1992	La Coruna, Spain	74,000
SEA EMPRESS	1996	Milford Haven, UK	72,000
KHARK 5	1989	120 nautical miles off Atlantic coast of Morocco	70,000
NOVA	1985	Off Kharg Island, Gulf of Iran	70,000
KATINA P	1992	Off Maputo, Mozambique	67,000

4. Discussion

An increasing trend in the number of publications was observed over the years. However, this increase was more evident in the mid-1990s, with a peak in 1996, following an increasing trend until the current decade. Moreover, part of the previous years' production was higher than the general average of all years in some cases. This pattern can be explained by the increased frequency of the number of oil spills in the seas in the last decades [3,4], which has aroused the interest of scientists to create and improve new tools and useful technologies in the detection of oil spills in the seas. Another important point that influenced this increase can be seen by the rise in the number of sensor systems and the consequent increase in the availability of images to be used in the detection of spills [20–28].

From the perspective of subject categories of knowledge, the areas of Earth and Planetary Sciences, Engineering and Environmental Science were identified as the most influential in terms of the number of publications. This can be detected by the number of articles in each area and indicates that the field of OSPM is probably relevant in terms of the themes addressed by the frequency of publications [2–4].

The top four journals are all well-known journals in the field of remote sensing, IEEE Transactions on Geoscience and Remote Sensing, International Journal of Remote Sensing, IEEE Journal of Selected Topics in Applied Earth Observations, and Remote Sensing, all occupy a leading position concerning the number of articles published.

When considering the top four countries that stand out in terms of the number of publications, research support, and the number of authors, we found China, the United States, Canada, and Italy. This pattern can be evidenced in the co-authorship analysis of countries that determined that China, the United States, Canada, and Italy were at the center of international cooperation. Based on changing patterns and similarities in the presence of words and terms in the co-word networks over the decades, the general network of all years and the top 25 networks in terms of citations, it can be seen that oil spill detection, oil spill monitoring and remote sensing were the main foci in OSPM research over time.

An important observation regarding the use of sensor systems is that, in general, regardless of the decade or even in the case of the top 25 of the most cited articles, only one study did not use radar images to identify oil spills. Several other works corroborated this pattern and somehow drove the research in this field [2–4]. Another area that notoriously deserves more in-depth investigation is the use of hyperspectral images in the detection and monitoring of oil spills, although it has appeared in co-word networks in more recent decades. Optical sensor systems were less frequent; however, they were used and found in the word networks associated with the use of radar images, which determines the potential use of these images in conjunction with other sensor systems. In this sense, Modis and Landsat were the most frequent. When we consider classification methods and algorithms, undoubtedly, machine learning is the most frequently used technique. Algorithms such as artificial neural networks, Support Vector Machine, and Maximum Likelihood are the ones that stand out with higher frequency, regardless of the evaluated decade. It is clear that the OSPM field's evolution reflects the dynamics associated with the topology of all the networks due to the increase in the number of words, terms, and their interplay. In fact, it refers to the tendency of the increasing complexity of research in terms of objectives, the volume of published data, and interactions between countries. These topology changes indicate that the developed approaches are gradual without the appearance of new disruptive concepts with the previous themes, having their biggest change in the 1990s with the emergence of new techniques. All calculated indices indicate that this field of investigation is expanding rapidly and remains active over time.

5. Conclusions

In this article, we revisit the relevant literature on oil spill detection and mapping from the last fifty years (1970–2019). We revealed aspects of great importance and interest in this field of knowledge, using the qualitative and quantitative word association network approach to give a clear overview of this field's research trends for the past 50 years. We performed a bibliometric analysis with a systematic review. We found as an overall result that the main trends observed in this field of science, at this moment, are in frank expansion in terms of its conceptual, theoretical, and methodological approaches. The gradual increase in the number of new terms and words presented in the networks, as well as the intricate structure over the decades and their interactions, are also noticed.

Our results point to strong evidence that oil detection at sea has shown significant evolution in recent decades. This evolution is explained by the fact that there is a strong relationship between the technological evolution of detection and remote sensing data acquisition methods.

Among the countries that contributed most to this field of science, China, the United States and Canada stood out as the largest producers and disseminators of information in this research field.

Our study's approach, which involves bibliometric techniques and tools, and the use of metrics that describe the structure of the networks, depicted meaningful advances related to better highlighting the general and specific research trends in this field of science.

Supplementary Materials: The following are available online at http://www.mdpi.com/2072-4292/12/21/3647/s1, Table S1—The table containing the results of the search was carried out on the Scopus base between the years 1970–2019 without carrying out the screening. The table is structured with the following fields: Citation information (Author (s), Author (s) ID, Document title, Year, EID, Source title, volume, issue, pages, Citation count, Source and document type, Publication Stage, DOI, Access Type), Bibliographical information (Affiliations, Serial identifiers (e.g., ISSN), PubMed ID, Publisher, Editor (s), Language of original document, Correspondence address, Abbreviated source title), Abstract & keywords (Abstract, keywords, Index keywords), Funding details (Number, Acronym, Sponsor, Funding text), Table S2—Feature information statistics based on the 50 years of documents published between 1970–2019, Table S3—Top 25 most referenced publications by articles in the field of OSPM, Table S4—Citation number by decades and summary of the twenty-five most cited documents published for the 50-year period (1970–2019) in the field of OSPM.

Author Contributions: Conceptualization, R.N.V., A.T.C.L., C.A.D.L., G.V.M., J.M.L., L.F.M., M.J.P., M.A.S. and E.C.B.C.; methodology, R.N.V., A.T.C.L., C.A.D.L., G.V.M. and E.C.B.C.; software execution, R.N.V., A.T.C.L., C.A.D.L. and E.C.B.C.; writing—original draft preparation, R.N.V., C.A.D.L., A.T.C.L., and E.C.B.C.; writing—review and editing, R.N.V., C.A.D.L., A.T.C.L., G.V.M., J.M.L., L.F.M., M.J.P., M.A.S. and E.C.B.C.; supervision, R.N.V., C.A.D.L. and A.T.C.L.; funding acquisition, A.T.C.L., C.A.D.L. and M.J.P. All authors have read and agreed to the published version of the manuscript.

Funding: This work was funded by INCT-GP and MCTI/CNPQ/CAPES/FAPS N° 16/2014 process 465517/2014-5, PROGRAMA INCT and additive project entitled "Modeling, remote sensing, and preventive detection of oil/fuel accidents by MCTI/CNPQ/CAPES/FAPS 2019. During this work, the following authors were supported by research fellowship: RNV (CNPQ, process 103189/2020-3), JML (CNPQ, process 381139/2020-4), JGVM (CNPQ, process 307828/2018-2), ATCL (CNPQ, process 380653/2020-6), CADL (CNPQ, process 380671/2020-4), LFM (CNPQ, process 424495/2018-0 and 380652/2020-0).

Acknowledgments: We appreciate comments and suggestions from the anonymous reviewers that helped improve the quality and presentation of the manuscript.

Conflicts of Interest: The authors declare no conflict of interest.

References

1. Singh, H.; Bhardwaj, N.; Arya, S.K.; Khatri, M. Environmental impacts of oil spills and their remediation by magnetic nanomaterials. *Environ. Nanotechnol. Monit. Manag.* **2020**, *14*, 100305. [CrossRef]
2. Kingston, P.F. Long-term environmental impact of oil spills. *Spill Sci. Technol. Bull.* **2002**, *7*, 53–61. [CrossRef]
3. Beyer, J.; Trannum, H.C.; Bakke, T.; Hodson, P.V.; Collier, T.K. Environmental effects of the Deepwater Horizon oil spill: A review. *Mar. Pollut. Bull.* **2016**, *110*, 28–51. [CrossRef] [PubMed]
4. Li, P.; Cai, Q.; Lin, W.; Chen, B.; Zhang, B. Offshore oil spill response practices and emerging challenges. *Mar. Pollut. Bull.* **2016**, *110*, 6–27. [CrossRef] [PubMed]
5. Lawa, R.J.; Kelly, C. The impact of the "sea empress" oil spill. *Aquat. Living Resour.* **2004**, *17*, 389–394. [CrossRef]
6. Palinkas, L.; Downs, M.; Petterson, J.; Russell, J. Social, Cultural, and Psychological Impacts of the Exxon Valdez Oil Spill. *Hum. Organ.* **1993**, *52*, 1–13. [CrossRef]
7. Piatt, J.F.; Ford, G.R. How many seabirds were killed by the Exxon Valdez oil spill? *Am. Fish. Soc. Symp.* **1996**, *18*, 712–719.
8. Picou, J.S.; Gill, D.A.; Dyer, C.L.; Curry, E.W. Disruption and stress in an Alaskan fishing community: Initial and continuing impacts of the Exxon Valdez oil spill. *Organ. Environ.* **1992**, *6*, 235–257. [CrossRef]
9. Millard, J.P.; Arvesen, J.C. Airborne Optical Detection of Oil on Water. *Appl. Opt.* **1972**, *11*, 102–107. [CrossRef]
10. Gruner, K.; Bartsch, N.; Keydel, W.; Witte, F. Contributions to Oil-Spill Detection and Analysis with Radar and Microwave Radiometry: Results of the Archimedes II Campaign. *IEEE Trans. Geosci. Remote Sens.* **1987**, *6*, 677–690. [CrossRef]

11. Krestenitis, M.; Orfanidis, G.; Ioannidis, K.; Avgerinakis, K.; Vrochidis, S.; Kompatsiaris, I. Oil spill identification from satellite images using deep neural networks. *Remote Sens.* **2019**, *11*, 1762. [CrossRef]
12. Espedal, H.A.; Wahl, T. Satellite sar oil spill detection using wind history information. *Int. J. Remote Sens.* **1999**, *20*, 49–65. [CrossRef]
13. Del Frate, F.; Petrocchi, A.; Lichtenegger, J.; Calabresi, G. Neural networks for oil spill detection using ERS-SAR data. *IEEE Trans. Geosci. Remote Sens.* **2000**, *38*, 2282–2287. [CrossRef]
14. Shirvany, R.; Member, S.; Chabert, M. Ship and Oil-Spill Detection Using the Degree of Polarization in Linear and Hybrid/Compact Dual-Pol SAR. *IEEE J. Sel. Top. Appl. Earth Obs. Remote Sens.* **2012**, *5*, 885–892. [CrossRef]
15. Minchew, B.; Jones, C.E.; Holt, B. Polarimetric analysis of backscatter from the deepwater horizon oil spill using l-band synthetic aperture radar. *IEEE Trans. Geosci. Remote Sens.* **2012**, *50*, 3812–3830. [CrossRef]
16. Liu, P.; Li, Y.; Liu, B.; Chen, P.; Xu, J. Semi-automatic oil spill detection on X-band marine radar images using texture analysis, machine learning, and adaptive thresholding. *Remote Sens.* **2019**, *11*, 756. [CrossRef]
17. Tong, S.; Liu, X.; Chen, Q.; Zhang, Z.; Xie, G. Multi-feature based ocean oil spill detection for polarimetric SAR data using random forest and the self-similarity parameter. *Remote Sens.* **2019**, *11*, 451. [CrossRef]
18. Pisano, A.; Bignami, F.; Santoleri, R. Oil spill detection in glint-contaminated near-infrared MODIS imagery. *Remote Sens.* **2015**, *7*, 1112–1134. [CrossRef]
19. Lacava, T.; Ciancia, E.; Coviello, I.; Di Polito, C.; Grimaldi, C.S.L.; Pergola, N.; Satriano, V.; Temimi, M.; Zhao, J.; Tramutoli, V. A MODIS-based robust satellite technique (RST) for timely detection of oil spilled areas. *Remote Sens.* **2017**, *9*, 128. [CrossRef]
20. Brekke, C.; Solberg, A.H.S. Oil spill detection by satellite remote sensing. *Remote Sens. Environ.* **2005**, *95*, 1–13. [CrossRef]
21. Jha, M.N.; Levy, J.; Gao, Y. Advances in remote sensing for oil spill disaster management: State-of-the-art sensors technology for oil spill surveillance. *Sensors* **2008**, *8*, 236–255. [CrossRef] [PubMed]
22. Gens, R. Oceanographic applications of SAR remote sensing. *GISci. Remote Sens.* **2008**, *45*, 275–305. [CrossRef]
23. Topouzelis, K.N. Oil spill detection by SAR images: Dark formation detection, feature extraction and classification algorithms. *Sensors* **2008**, *8*, 6642–6659. [CrossRef]
24. Liu, Y.; MacFadyen, A.; Ji, Z.G.; Weisberg, R.H. Introduction to Monitoring and Modeling the Deepwater Horizon Oil Spill. *Monit. Model. Deep. Horiz. Oil Spill A Rec. Break. Enterp.* **2013**, *195*, 1–7. [CrossRef]
25. Ambituuni, A.; Amezaga, J.; Emeseh, E. Analysis of safety and environmental regulations for downstream petroleum industry operations in Nigeria: Problems and prospects. *Environ. Dev.* **2014**, *9*, 43–60. [CrossRef]
26. Catapano, I.; Affinito, A.; Bertolla, L.; Porsani, J.L.; Soldovieri, F. Oil spill monitoring via microwave tomography enhanced GPR surveys. *J. Appl. Geophys.* **2014**, *108*, 95–103. [CrossRef]
27. Fingas, M.; Brown, C.E. A review of oil spill remote sensing. *Sensors* **2018**, *18*, 91. [CrossRef]
28. Li, Y.; Chen, J.; Zhang, Y. Progress in Research on Marine Oil Spills Detection Using Synthetic Aperture Radar. *Dianzi Yu Xinxi Xuebao/J. Electron. Inf. Technol.* **2019**, *41*, 751–762. [CrossRef]
29. Van Eck, N.J.; Waltman, L.; Van Den Berg, J.; Kaymak, U. Visualizing the computational intelligence field. *IEEE Comput. Intell. Mag.* **2006**, *1*, 6–10. [CrossRef]
30. Van Eck, N.J.; Waltman, L. Bibliometric mapping of the computational intelligence field. *Int. J. Uncertain. Fuzziness Knowl. Based Syst.* **2007**, *15*, 625–645. [CrossRef]
31. van Eck, N.J.; Waltman, L. Software survey: VOSviewer, a computer program for bibliometric mapping. *Scientometrics* **2010**, *84*, 523–538. [CrossRef]
32. van Eck, N.J.; Waltman, L.; Noyons, E.C.M.; Buter, R.K. Automatic term identification for bibliometric mapping. *Scientometrics* **2010**, *82*, 581–596. [CrossRef]
33. Andres, A. *Measuring Academic Research*; Chandos Publishing: Oxford, UK, 2009. [CrossRef]
34. De Bellis, N. *Bibliometrics and Citation Analysis from the Science Citation Index to Cybermetrics*; De Bellis, N., Ed.; Scarecrow Press, Inc.: Lanham, MD, USA, 2009; ISBN 9780128019115.
35. Viana, J.; Santos, J.V.; Neiva, R.M.; Souza, J.; Duarte, L.; Teodoro, A.C.; Freitas, A. Remote sensing in human health: A 10-year bibliometric analysis. *Remote Sens.* **2017**, *9*, 1225. [CrossRef]
36. Hu, K.; Qi, K.; Guan, Q.; Wu, C.; Yu, J.; Qing, Y.; Zheng, J.; Wu, H.; Li, X. A scientometric visualization analysis for night-time light remote sensing research from 1991 to 2016. *Remote Sens.* **2017**, *9*, 802. [CrossRef]
37. Elsevier Content—How Scopus Works—Scopus—Elsevier Solutions. Available online: https://www.elsevier.com/solutions/scopus/how-scopus-works/content (accessed on 27 September 2020).

38. ITOPF. *Oil Tanker Spill Statistics*; Itopf Limited: London, UK, 2019.
39. Roser, M. Oil Spills. Available online: https://ourworldindata.org/oil-spills (accessed on 27 September 2020).
40. Van Eck, N.J.; Waltman, L. Text mining and visualization using VOSviewer. *arXiv* **2011**, arXiv:1109.2058.
41. Bastian, M.; Heymann, S.; Jacomy, M. Gephi: An open source software for exploring and manipulating networks. *Int. AAAI Conf. Weblogs Soc. Media* **2009**, *8*, 361–362.
42. Cherven, K. *Mastering Gephi Network Visualization. Produce Advanced Network Graphs in Gephi and Gain Valuable Insights into Your Network Datasets*; Packt Publishing: Birmingham, UK, 2015; ISBN 9781783987344.
43. Pachayappan, M.; Venkatesakumar, R. A Graph Theory Based Systematic Literature Network Analysis. *Theor. Econ. Lett.* **2018**, *8*, 960–980. [CrossRef]
44. R Core Team. *The R Project for Statistical Computing*; R Foundation for Statistical Computing: Vienna, Austria, 2019; pp. 1–12. Available online: http://www.R-Project.Org/ (accessed on 1 January 2019).
45. RStudio RStudio. Open Source & Professional Software for Data Science Teams—RStudio. Available online: https://rstudio.com/ (accessed on 27 September 2020).
46. Wickham, H. *Ggplot2: Elegant Graphics for Data Analysis*; Version 3.6.1.; Springer: New York, NY, USA, 2016; ISBN 978-3-319-24277-4.

MDPI
St. Alban-Anlage 66
4052 Basel
Switzerland
Tel. +41 61 683 77 34
Fax +41 61 302 89 18
www.mdpi.com

Remote Sensing Editorial Office
E-mail: remotesensing@mdpi.com
www.mdpi.com/journal/remotesensing

www.ingramcontent.com/pod-product-compliance
Lightning Source LLC
Chambersburg PA
CBHW041932260926
51850CB00018B/69
* 9 7 8 3 0 3 6 5 6 4 3 8 8 *